To the Student

As you begin, you may feel anxious about the number of theorems, definitions, procedures, and equations. You may wonder if you can learn it all in time. Don't worry—your concerns are normal. This textbook was written with you in mind. If you attend class, work hard, and read and study this text, you will build the knowledge and skills you need to be successful. Here's how you can use the text to your benefit.

Read Carefully

When you get busy, it's easy to skip reading and go right to the problems. Don't . . . the text has a large number of examples and clear explanations to help you break down the mathematics into easy-to-understand steps. Reading will provide you with a clearer understanding, beyond simple memorization. Read before class (not after) so you can ask questions about anything you didn't understand. You'll be amazed at how much more you'll get out of class if you do this.

Use the Features

I use many different methods in the classroom to communicate. Those methods, when incorporated into the text, are called "features." The features serve many purposes, from providing timely review of material you learned before (just when you need it) to providing organized review sessions to help you prepare for quizzes and tests. Take advantage of the features and you will master the material.

To make this easier, we've provided a brief guide to getting the most from this text. Refer to "Prepare for Class," "Practice," and "Review" on the following three pages. Spend fifteen minutes reviewing the guide and familiarizing yourself with the features by flipping to the page numbers provided. Then, as you read, use them. This is the best way to make the most of your text.

Please do not hesitate to contact me, through Pearson Education, with any questions, comments, or suggestions for improving this text. I look forward to hearing from you, and good luck with all of your studies.

Best Wishes!

Michael Sullivan

Prepare for Class "Read the Book"

Feature	Description	Benefit	Page
Every Chapter Opener begins with . . .			
Chapter-Opening Topic & Project	Each chapter begins with a discussion of a topic of current interest and ends with a related project.	The Project lets you apply what you learned to solve a problem related to the topic.	402
Internet-Based Projects	The projects allow for the integration of spreadsheet technology that you will need to be a productive member of the workforce.	The projects give you an opportunity to collaborate and use mathematics to deal with issues of current interest.	503
Every Section begins with . . .			
Learning Objectives 2	Each section begins with a list of objectives. Objectives also appear in the text where the objective is covered.	These focus your studying by emphasizing what's most important and where to find it.	423
Sections contain . . .			
PREPARING FOR THIS SECTION	Most sections begin with a list of key concepts to review with page numbers.	Ever forget what you've learned? This feature highlights previously learned material to be used in this section. Review it, and you'll always be prepared to move forward.	423
Now Work the 'Are You Prepared?' Problems	Problems that assess whether you have the prerequisite knowledge for the upcoming section.	Not sure you need the Preparing for This Section review? Work the 'Are You Prepared?' problems. If you get one wrong, you'll know exactly what you need to review and where to review it!	423, 434
Now Work PROBLEMS	These follow most examples and direct you to a related exercise.	We learn best by doing. You'll solidify your understanding of examples if you try a similar problem right away, to be sure you understand what you've just read.	430, 435
WARNING	Warnings are provided in the text.	These point out common mistakes and help you to avoid them.	456
Exploration and Seeing the Concept	These graphing utility activities foreshadow a concept or solidify a concept just presented.	You will obtain a deeper and more intuitive understanding of theorems and definitions.	418, 443
In Words	These provide alternative descriptions of select definitions and theorems.	Does math ever look foreign to you? This feature translates math into plain English.	440
Calculus	These appear next to information essential for the study of calculus.	Pay attention–if you spend extra time now, you'll do better later!	205, 407, 431
SHOWCASE EXAMPLES	These examples provide "how-to" instruction by offering a guided, step-by-step approach to solving a problem.	With each step presented on the left and the mathematics displayed on the right, you can immediately see how each step is employed.	334
Model It! Examples and Problems	These examples and problems require you to build a mathematical model from either a verbal description or data. The homework Model It! problems are marked by purple headings.	It is rare for a problem to come in the form *"Solve the following equation."* Rather, the equation must be developed based on an explanation of the problem. These problems require you to develop models that will allow you to describe the problem mathematically and suggest a solution to the problem.	447, 475

Practice "Work the Problems"

Feature	Description	Benefit	Page
'Are You Prepared?' Problems	These assess your retention of the prerequisite material you'll need. Answers are given at the end of the section exercises. This feature is related to the Preparing for This Section feature.	Do you always remember what you've learned? Working these problems is the best way to find out. If you get one wrong, you'll know exactly what you need to review and where to review it!	434, 440
Concepts and Vocabulary	These short-answer questions, mainly Fill-in-the-Blank, Multiple-Choice and True/False items, assess your understanding of key definitions and concepts in the current section.	It is difficult to learn math without knowing the language of mathematics. These problems test your understanding of the formulas and vocabulary.	434
Skill Building	Correlated with section examples, these problems provide straightforward practice.	It's important to dig in and develop your skills. These problems provide you with ample opportunity to do so.	434–436
Mixed Practice	These problems offer comprehensive assessment of the skills learned in the section by asking problems that relate to more than one concept or objective. These problems may also require you to utilize skills learned in previous sections.	Learning mathematics is a building process. Many concepts are interrelated. These problems help you see how mathematics builds on itself and also see how the concepts tie together.	436–437
Applications and Extensions	These problems allow you to apply your skills to real-world problems. They also allow you to extend concepts learned in the section.	You will see that the material learned within the section has many uses in everyday life.	437–439
Explaining Concepts: Discussion and Writing	"Discussion and Writing" problems are colored red. They support class discussion, verbalization of mathematical ideas, and writing and research projects.	To verbalize an idea, or to describe it clearly in writing, shows real understanding. These problems nurture that understanding. Many are challenging, but you'll get out what you put in.	439
NEW! Retain Your Knowledge	These problems allow you to practice content learned earlier in the course.	Remembering how to solve all the different kinds of problems that you encounter throughout the course is difficult. This practice helps you remember.	439
Now Work PROBLEMS	Many examples refer you to a related homework problem. These related problems are marked by a pencil and orange numbers.	If you get stuck while working problems, look for the closest Now Work problem, and refer to the related example to see if it helps.	432, 435, 436
Review Exercises	Every chapter concludes with a comprehensive list of exercises to pratice. Use the list of objectives to determine the objective and examples that correspond to the problems.	Work these problems to ensure that you understand all the skills and concepts of the chapter. Think of it as a comprehensive review of the chapter.	499–501

Review "Study for Quizzes and Tests"

Feature	Description	Benefit	Page
The Chapter Review at the end of each chapter contains . . .			
Things to Know	A detailed list of important theorems, formulas, and definitions from the chapter.	Review these and you'll know the most important material in the chapter!	497–498
You Should Be Able to . . .	Contains a complete list of objectives by section, examples that illustrate the objective, and practice exercises that test your understanding of the objective.	Do the recommended exercises and you'll have mastered the key material. If you get something wrong, go back and work through the example listed and try again.	498–499
Review Exercises	These provide comprehensive review and practice of key skills, matched to the Learning Objectives for each section.	Practice makes perfect. These problems combine exercises from all sections, giving you a comprehensive review in one place.	499–501
Chapter Test	About 15–20 problems that can be taken as a Chapter Test. Be sure to take the Chapter Test under test conditions—no notes!	Be prepared. Take the sample practice test under test conditions. This will get you ready for your instructor's test. If you get a problem wrong, you can watch the Chapter Test Prep Video.	502
Cumulative Review	These problem sets appear at the end of each chapter, beginning with Chapter 2. They combine problems from previous chapters, providing an ongoing cumulative review. When you use them in conjunction with the Retain Your Knowledge problems, you will be ready for the final exam.	These problem sets are really important. Completing them will ensure that you are not forgetting anything as you go. This will go a long way toward keeping you primed for the final exam.	502–503
Chapter Projects	The Chapter Projects apply to what you've learned in the chapter. Additional projects are available on the Instructor's Resource Center (IRC).	The Chapter Projects give you an opportunity to apply what you've learned in the chapter to the opening topic. If your instructor allows, these make excellent opportunities to work in a group, which is often the best way of learning math.	503–504
Internet-Based Projects	In selected chapters, a Web-based project is given.	These projects give you an opportunity to collaborate and use mathematics to deal with issues of current interest by using the Internet to research and collect data.	503

Achieve Your Potential

The author, Michael Sullivan, has developed specific content in MyMathLab® to ensure you have many resources to help you achieve success in mathematics - and beyond! The MyMathLab features described here will help you:

- **Review** math skills and concepts you may have forgotten
- **Retain** new concepts as you move through your math course
- **Develop** skills that will help with your transition to college

Adaptive Study Plan

The Study Plan will help you study more efficiently and effectively.

Your performance and activity are assessed continually in real time, providing a personalized experience based on your individual needs.

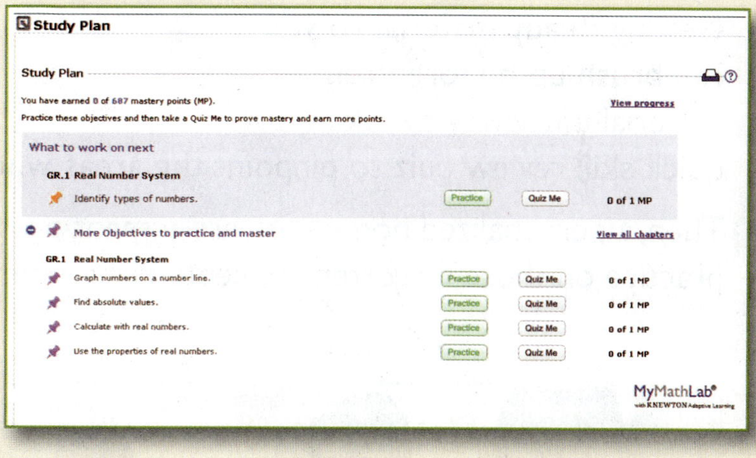

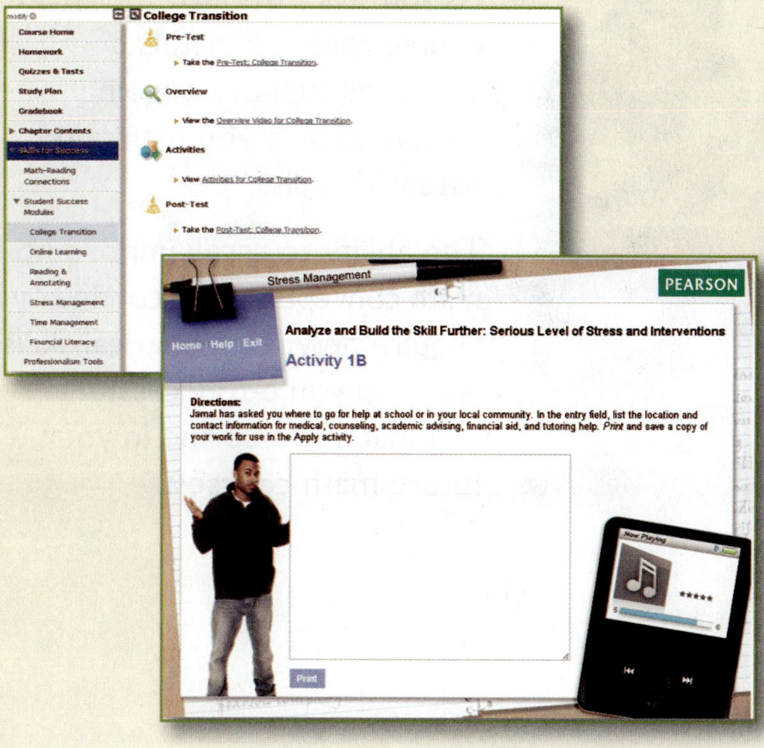

Skills for Success

The Skills for Success Modules support your continued success in college. These modules provide tutorials and guidance on a variety of topics, including transitioning to college, online learning, time management, and more.

Additional content is provided to help with the development of professional skills such as resume writing and interview preparation.

Getting Ready

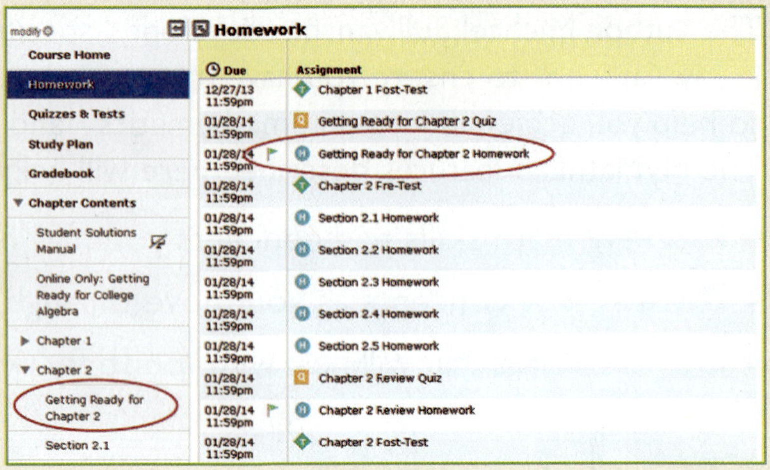

Are you frustrated when you know you learned a math concept in the past, but you can't quite remember the skill when it's time to use it? Don't worry!

The author has included Getting Ready material so you can brush up on forgotten material efficiently by taking a quick skill review quiz to pinpoint the areas where you need help.

Then, a personalized homework assignment provides additional practice on those forgotten concepts, right when you need it.

Retain Your Knowledge

As you work through your math course, these MyMathLab® exercises support ongoing review to help you maintain essential skills.

The ability to recall important math concepts as you continually acquire new mathematical skills will help you be successful in this math course and in your future math courses.

PEARSON ALWAYS LEARNING

Michael Sullivan

College Algebra

Fourth Custom Edition for University of New Mexico

Taken from:
College Algebra, Tenth Edition
by Michael Sullivan

Cover Art: Courtesy of PhotoDisc/Getty Images.

Taken from:

College Algebra, Tenth Edition
by Michael Sullivan

New York, New York 10013

This special edition published in cooperation with Pearson Learning Solutions.

Pearson Learning Solutions, 330 Hudson Street, New York, New York 10013
A Pearson Education Company
www.pearsoned.com

Printed in the United States of America

1 2 3 4 5 6 7 8 9 10 V003 18 17 16 15

000200010271970290

AJ

PEARSON

ISBN 10: 1-323-11616-8
ISBN 13: 978-1-323-11616-6

Contents

Appendix Graphing Utilities A1

Dedicated to the Memory of Mary

Three Distinct Series

Students have different goals, learning styles, and levels of preparation. Instructors have different teaching philosophies, styles, and techniques. Rather than write one series to fit all, the Sullivans have written three distinct series. All share the same goal—to develop a high level of mathematical understanding and an appreciation for the way mathematics can describe the world around us. The manner of reaching that goal, however, differs from series to series.

Contemporary Series, Tenth Edition

The Contemporary Series is the most traditional in approach yet modern in its treatment of precalculus mathematics. Graphing utility coverage is optional and can be included or excluded at the discretion of the instructor: *College Algebra, Algebra & Trigonometry, Trigonometry: A Unit Circle Approach, Precalculus.*

Enhanced with Graphing Utilities Series, Sixth Edition

This series provides a thorough integration of graphing utilities into topics, allowing students to explore mathematical concepts and encounter ideas usually studied in later courses. Using technology, the approach to solving certain problems differs from the Contemporary Series, while the emphasis on understanding concepts and building strong skills does not: *College Algebra, Algebra & Trigonometry, Precalculus*.

Concepts through Functions Series, Third Edition

This series differs from the others, utilizing a functions approach that serves as the organizing principle tying concepts together. Functions are introduced early in various formats. This approach supports the Rule of Four, which states that functions are represented symbolically, numerically, graphically, and verbally. Each chapter introduces a new type of function and then develops all concepts pertaining to that particular function. The solutions of equations and inequalities, instead of being developed as stand-alone topics, are developed in the context of the underlying functions. Graphing utility coverage is optional and can be included or excluded at the discretion of the instructor: *College Algebra; Precalculus, with a Unit Circle Approach to Trigonometry; Precalculus, with a Right Triangle Approach to Trigonometry*.

The Contemporary Series

College Algebra, Tenth Edition

This text provides a contemporary approach to college algebra, with three chapters of review material preceding the chapters on functions. Graphing calculator usage is provided, but is optional. After completing this book, a student will be adequately prepared for trigonometry, finite mathematics, and business calculus.

Algebra & Trigonometry, Tenth Edition

This text contains all the material in *College Algebra*, but also develops the trigonometric functions using a right triangle approach and showing how it relates to the unit circle approach. Graphing techniques are emphasized, including a thorough discussion of polar coordinates, parametric equations, and conics using polar coordinates. Graphing calculator usage is provided, but is optional. After completing this book, a student will be adequately prepared for finite mathematics, business calculus, and engineering calculus.

Precalculus, Tenth Edition

This text contains one review chapter before covering the traditional precalculus topic of functions and their graphs, polynomial and rational functions, and exponential and logarithmic functions. The trigonometric functions are introduced using a unit circle approach and showing how it relates to the right triangle approach. Graphing techniques are emphasized, including a thorough discussion of polar coordinates, parametric equations, and conics using polar coordinates. Graphing calculator usage is provided, but is optional. The final chapter provides an introduction to calculus, with a discussion of the limit, the derivative, and the integral of a function. After completing this book, a student will be adequately prepared for finite mathematics, business calculus, and engineering calculus.

Trigonometry: a Unit Circle Approach, Tenth Edition

This text, designed for stand-alone courses in trigonometry, develops the trigonometric functions using a unit circle approach and showing how it relates to the right triangle approach. Graphing techniques are emphasized, including a thorough discussion of polar coordinates, parametric equations, and conics using polar coordinates. Graphing calculator usage is provided, but is optional. After completing this book, a student will be adequately prepared for finite mathematics, business calculus, and engineering calculus.

Preface to the Instructor

As a professor of mathematics at an urban public university for 35 years, I understand the varied needs of College Algebra students. Students range from being underprepared, with little mathematical background and a fear of mathematics, to being highly prepared and motivated. For some, this is their final course in mathematics. For others, it is preparation for future mathematics courses. I have written this text with both groups in mind.

A tremendous benefit of authoring a successful series is the broad-based feedback I receive from teachers and students who have used previous editions. I am sincerely grateful for their support. Virtually every change to this edition is the result of their thoughtful comments and suggestions. I hope that I have been able to take their ideas and, building upon a successful foundation of the ninth edition, make this series an even better learning and teaching tool for students and teachers.

Features in the Tenth Edition

A descriptive list of the many special features of *College Algebra* can be found on the endpapers in the front of this text.

This list places the features in their proper context, as building blocks of an overall learning system that has been carefully crafted over the years to help students get the most out of the time they put into studying. Please take the time to review this and to discuss it with your students at the beginning of your course. My experience has been that when students utilize these features, they are more successful in the course.

New to the Tenth Edition

- **Retain Your Knowledge** This new category of problems in the exercise set are based on the article "To Retain New Learning, Do the Math" published in the *Edurati Review*. In this article, Kevin Washburn suggests that "the more students are required to recall new content or skills, the better their memory will be." It is frustrating when students cannot recall skills learned earlier in the course. To alleviate this recall problem, we have created "Retain Your Knowledge" problems. These are problems considered to be "final exam material" that students can use to maintain their skills. All the answers to these problems appear in the back of the text, and all are programmed in MyMathLab.
- **Guided Lecture Notes** Ideal for online, emporium/redesign courses, inverted classrooms, or traditional lecture classrooms. These lecture notes help students take thorough, organized, and understandable notes as they watch the Author in Action videos. They ask students to complete definitions, procedures, and examples based on the content of the videos and text. In addition, experience suggests that students learn by doing and understanding the why/how of the concept or property. Therefore, many sections will have an exploration activity to motivate student learning. These explorations introduce the topic and/or connect it to either a real-world application or a previous section. For example, when the vertical-line test is discussed in Section 3.2, after the theorem statement, the notes ask the students to explain why the vertical-line test works by using the definition of a function. This challenge helps students process the information at a higher level of understanding.
- **Illustrations** Many of the figures now have captions to help connect the illustrations to the explanations in the body of the text.
- **TI Screen Shots** In this edition all the screen shots from the ninth edition have been replaced with screen shots using a TI-84Plus C. These updated screen shots help students visualize concepts clearly and help make stronger connections between equations, data, and graphs in full color.
- **Chapter Projects,** which apply the concepts of each chapter to a real-world situation, have been enhanced to give students an up-to-the-minute experience. Many projects are new and Internet-based, requiring the student to research information online in order to solve problems.
- **Exercise Sets** All the exercises in the text have been reviewed and analyzed for this edition, some have been removed, and new ones have been added. All time-sensitive problems have been updated to the most recent information available. The problem sets remain classified according to purpose.

  The ***'Are You Prepared?'*** problems have been improved to better serve their purpose as a just-in-time review of concepts that the student will need to apply in the upcoming section.

  The ***Concepts and Vocabulary*** problems have been expanded and now include multiple-choice exercises. Together with the fill-in-the-blank and True/False problems, these exercises have been written to serve as reading quizzes.

  Skill Building problems develop the student's computational skills with a large selection of exercises that are directly related to the objectives of the section. ***Mixed Practice*** problems offer a comprehensive assessment of skills that relate to more than one objective. Often these require skills learned earlier in the course.

  Applications and Extensions problems have been updated. Further, many new application-type exercises have been added, especially ones involving information and data drawn from sources the student will recognize, to improve relevance and timeliness.

  The ***Explaining Concepts: Discussion and Writing*** exercises have been improved and expanded to provide more opportunity for classroom discussion and group projects.

New to this edition, ***Retain Your Knowledge*** exercises consist of a collection of four problems in each exercise set that are based on material learned earlier in the course. They serve to keep information that has already been learned "fresh" in the mind of the student. Answers to all these problems appear in the Student Edition.

The ***Review Exercises*** in the Chapter Review have been streamlined, but they remain tied to the clearly expressed objectives of the chapter. Answers to all these problems appear in the Student Edition.

- **Annotated Instructor's Edition** As a guide, the author's suggestions for homework assignments are indicated by a blue underscore below the problem number. These problems are assignable in the MyMathLab as part of a "Ready-to-Go" course.

Content Changes in the Tenth Edition

- **Section 3.1** The objective Find the Difference Quotient of a Function has been added.
- **Section 5.1** The subsection Behavior of the Graph of a Polynomial Function Near a Zero has been removed.
- **Section 5.3** A subsection has been added that discusses the role of multiplicity of the zeros of the denominator of a rational function as it relates to the graph near a vertical asymptote.
- **Section 5.5** The objective Use Descartes' Rule of Signs has been included.
- **Section 5.5** The theorem Bounds on the Zeros of a Polynomial Function is now based on the traditional method of using synthetic division.

Using the Tenth Edition Effectively with Your Syllabus

To meet the varied needs of diverse syllabi, this text contains more content than is likely to be covered in a *College Algebra* course. As the chart illustrates, this text has been organized with flexibility of use in mind. Within a given chapter, certain sections are optional (see the details that follow the figure below) and can be omitted without loss of continuity.

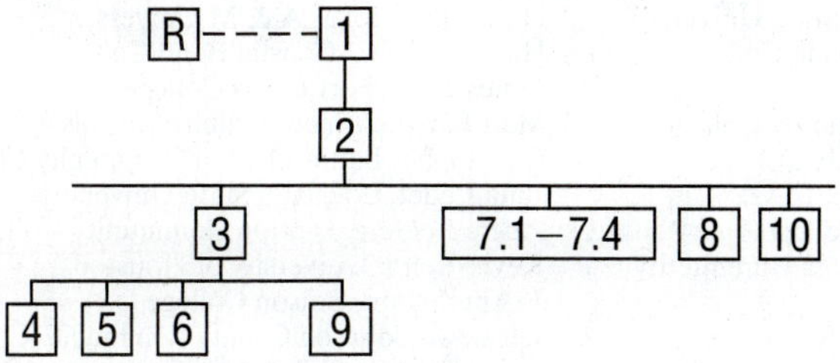

Chapter R Review
This chapter consists of review material. It may be used as the first part of the course or later as a just-in-time review when the content is required. Specific references to this chapter occur throughout the text to assist in the review process.

Chapter 1 Equations and Inequalities
Primarily a review of Intermediate Algebra topics, this material is a prerequisite for later topics. The coverage of complex numbers and quadratic equations with a negative discriminant is optional and may be postponed or skipped entirely without loss of continuity.

Chapter 2 Graphs
This chapter lays the foundation for functions. Section 2.5 is optional.

Chapter 3 Functions and Their Graphs
Perhaps the most important chapter. Section 3.6 is optional.

Chapter 4 Linear and Quadratic Functions
Topic selection depends on your syllabus. Sections 4.2 and 4.4 may be omitted without loss of continuity.

Chapter 5 Polynomial and Rational Functions
Topic selection depends on your syllabus.

Chapter 6 Exponential and Logarithmic Functions
Sections 6.1–6.6 follow in sequence. Sections 6.7, 6.8, and 6.9 are optional.

Chapter 7 Analytic Geometry
Sections 7.1–7.4 follow in sequence.

Chapter 8 Systems of Equations and Inequalities
Sections 8.2–8.7 may be covered in any order, but each requires Section 8.1. Section 8.8 requires Section 8.7.

Chapter 9 Sequences; Induction; The Binomial Theorem
There are three independent parts: Sections 9.1–9.3; Section 9.4; and Section 9.5.

Chapter 10 Counting and Probability
The sections follow in sequence.

Acknowledgments

Textbooks are written by authors, but evolve from an idea to final form through the efforts of many people. It was Don Dellen who first suggested this text and series to me. Don is remembered for his extensive contributions to publishing and mathematics.

Thanks are due to the following people for their assistance and encouragement to the preparation of this edition:

- From Pearson Education: Anne Kelly for her substantial contributions, ideas, and enthusiasm; Dawn Murrin, for her unmatched talent at getting the details right; Joseph Colella for always getting the reviews and pages to me on time; Peggy McMahon for directing the always difficult production process; Rose Kernan for handling liaison between the compositor and author; Peggy Lucas for her genuine interest in marketing this text;

Chris Hoag for her continued support and genuine interest; Greg Tobin for his leadership and commitment to excellence; and the Pearson Math and Science Sales team, for their continued confidence and personal support of my texts.

- Accuracy checkers: C. Brad Davis, who read the entire manuscript and accuracy checked answers. His attention to detail is amazing; Timothy Britt, for creating the Solutions Manuals and accuracy checking answers.

Finally, I offer my grateful thanks to the dedicated users and reviewers of my texts, whose collective insights form the backbone of each textbook revision.

James Africh, College of DuPage
Steve Agronsky, Cal Poly State University
Gererdo Aladro, Florida International University
Grant Alexander, Joliet Junior College
Dave Anderson, South Suburban College
Richard Andrews, Florida A&M University
Joby Milo Anthony, University of Central Florida
James E. Arnold, University of Wisconsin-Milwaukee
Adel Arshaghi, Center for Educational Merit
Carolyn Autray, University of West Georgia
Agnes Azzolino, Middlesex County College
Wilson P. Banks, Illinois State University
Sudeshna Basu, Howard University
Dale R. Bedgood, East Texas State University
Beth Beno, South Suburban College
Carolyn Bernath, Tallahassee Community College
Rebecca Berthiaume, Edison State College
William H. Beyer, University of Akron
Annette Blackwelder, Florida State University
Richelle Blair, Lakeland Community College
Kevin Bodden, Lewis and Clark College
Jeffrey Boerner, University of Wisconsin-Stout
Barry Booten, Florida Atlantic University
Larry Bouldin, Roane State Community College
Bob Bradshaw, Ohlone College
Trudy Bratten, Grossmont College
Tim Bremer, Broome Community College
Tim Britt, Jackson State Community College
Michael Brook, University of Delaware
Joanne Brunner, Joliet Junior College
Warren Burch, Brevard Community College
Mary Butler, Lincoln Public Schools
Melanie Butler, West Virginia University
Jim Butterbach, Joliet Junior College
William J. Cable, University of Wisconsin-Stevens Point
Lois Calamia, Brookdale Community College
Jim Campbell, Lincoln Public Schools
Roger Carlsen, Moraine Valley Community College
Elena Catoiu, Joliet Junior College
Mathews Chakkanakuzhi, Palomar College
Tim Chappell, Penn Valley Community College
John Collado, South Suburban College
Alicia Collins, Mesa Community College
Nelson Collins, Joliet Junior College
Rebecca Connell, Troy University
Jim Cooper, Joliet Junior College
Denise Corbett, East Carolina University
Carlos C. Corona, San Antonio College
Theodore C. Coskey, South Seattle Community College
Rebecca Connell, Troy University
Donna Costello, Plano Senior High School
Paul Crittenden, University of Nebraska at Lincoln
John Davenport, East Texas State University
Faye Dang, Joliet Junior College
Antonio David, Del Mar College
Stephanie Deacon, Liberty University
Duane E. Deal, Ball State University
Jerry DeGroot, Purdue North Central
Timothy Deis, University of Wisconsin-Platteville
Joanna DelMonaco, Middlesex Community College
Vivian Dennis, Eastfield College
Deborah Dillon, R. L. Turner High School
Guesna Dohrman, Tallahassee Community College
Cheryl Doolittle, Iowa State University
Karen R. Dougan, University of Florida
Jerrett Dumouchel, Florida Community College at Jacksonville
Louise Dyson, Clark College
Paul D. East, Lexington Community College
Don Edmondson, University of Texas-Austin
Erica Egizio, Joliet Junior College
Jason Eltrevoog, Joliet Junior College
Christopher Ennis, University of Minnesota
Kathy Eppler, Salt Lake Community College
Ralph Esparza, Jr., Richland College
Garret J. Etgen, University of Houston
Scott Fallstrom, Shoreline Community College
Pete Falzone, Pensacola Junior College
Arash Farahmand, Skyline College
W.A. Ferguson, University of Illinois-Urbana/Champaign
Iris B. Fetta, Clemson University
Mason Flake, student at Edison Community College
Timothy W. Flood, Pittsburg State University
Robert Frank, Westmoreland County Community College
Merle Friel, Humboldt State University
Richard A. Fritz, Moraine Valley Community College
Dewey Furness, Ricks College
Mary Jule Gabiou, North Idaho College
Randy Gallaher, Lewis and Clark College
Tina Garn, University of Arizona
Dawit Getachew, Chicago State University
Wayne Gibson, Rancho Santiago College
Loran W. Gierhart, University of Texas at San Antonio and Palo Alto College
Robert Gill, University of Minnesota Duluth
Nina Girard, University of Pittsburgh at Johnstown
Sudhir Kumar Goel, Valdosta State University
Adrienne Goldstein, Miami Dade College, Kendall Campus
Joan Goliday, Sante Fe Community College
Lourdes Gonzalez, Miami Dade College, Kendall Campus
Frederic Gooding, Goucher College
Donald Goral, Northern Virginia Community College
Sue Graupner, Lincoln Public Schools
Mary Beth Grayson, Liberty University
Jennifer L. Grimsley, University of Charleston
Ken Gurganus, University of North Carolina
James E. Hall, University of Wisconsin-Madison
Judy Hall, West Virginia University
Edward R. Hancock, DeVry Institute of Technology
Julia Hassett, DeVry Institute, Dupage
Christopher Hay-Jahans, University of South Dakota
Michah Heibel, Lincoln Public Schools
LaRae Helliwell, San Jose City College
Celeste Hernandez, Richland College
Gloria P. Hernandez, Louisiana State University at Eunice
Brother Herron, Brother Rice High School
Robert Hoburg, Western Connecticut State University
Lynda Hollingsworth, Northwest Missouri State University
Deltrye Holt, Augusta State University
Charla Holzbog, Denison High School
Lee Hruby, Naperville North High School
Miles Hubbard, St. Cloud State University
Kim Hughes, California State College-San Bernardino
Stanislav, Jabuka, University of Nevada, Reno
Ron Jamison, Brigham Young University
Richard A. Jensen, Manatee Community College
Glenn Johnson, Middlesex Community College
Sandra G. Johnson, St. Cloud State University
Tuesday Johnson, New Mexico State University
Susitha Karunaratne, Purdue University North Central
Moana H. Karsteter, Tallahassee Community College
Donna Katula, Joliet Junior College
Arthur Kaufman, College of Staten Island
Thomas Kearns, North Kentucky University
Jack Keating, Massasoit Community College
Shelia Kellenbarger, Lincoln Public Schools
Rachael Kenney, North Carolina State University
John B. Klassen, North Idaho College
Debra Kopcso, Louisiana State University
Lynne Kowski, Raritan Valley Community College
Yelena Kravchuk, University of Alabama at Birmingham
Ray S. Kuan, Skyline College
Keith Kuchar, Manatee Community College
Tor Kwembe, Chicago State University
Linda J. Kyle, Tarrant Country Jr. College
H.E. Lacey, Texas A & M University
Harriet Lamm, Coastal Bend College
James Lapp, Fort Lewis College
Matt Larson, Lincoln Public Schools
Christopher Lattin, Oakton Community College
Julia Ledet, Lousiana State University
Adele LeGere, Oakton Community College
Kevin Leith, University of Houston
JoAnn Lewin, Edison College
Jeff Lewis, Johnson County Community College
Janice C. Lyon, Tallahassee Community College
Jean McArthur, Joliet Junior College
Virginia McCarthy, Iowa State University
Karla McCavit, Albion College
Michael McClendon, University of Central Oklahoma
Tom McCollow, DeVry Institute of Technology

Marilyn McCollum, North Carolina State University
Jill McGowan, Howard University
Will McGowant, Howard University
Angela McNulty, Joliet Junior College
Laurence Maher, North Texas State University
Jay A. Malmstrom, Oklahoma City Community College
Rebecca Mann, Apollo High School
Lynn Marecek, Santa Ana College
Sherry Martina, Naperville North High School
Alec Matheson, Lamar University
Nancy Matthews, University of Oklahoma
James Maxwell, Oklahoma State University-Stillwater
Marsha May, Midwestern State University
James McLaughlin, West Chester University
Judy Meckley, Joliet Junior College
David Meel, Bowling Green State University
Carolyn Meitler, Concordia University
Samia Metwali, Erie Community College
Rich Meyers, Joliet Junior College
Eldon Miller, University of Mississippi
James Miller, West Virginia University
Michael Miller, Iowa State University
Kathleen Miranda, SUNY at Old Westbury
Chris Mirbaha, The Community College of Baltimore County
Val Mohanakumar, Hillsborough Community College
Thomas Monaghan, Naperville North High School
Miguel Montanez, Miami Dade College, Wolfson Campus
Maria Montoya, Our Lady of the Lake University
Susan Moosai, Florida Atlantic University
Craig Morse, Naperville North High School
Samad Mortabit, Metropolitan State University
Pat Mower, Washburn University
Tammy Muhs, University of Central Florida
A. Muhundan, Manatee Community College
Jane Murphy, Middlesex Community College
Richard Nadel, Florida International University
Gabriel Nagy, Kansas State University
Bill Naegele, South Suburban College
Karla Neal, Lousiana State University
Lawrence E. Newman, Holyoke Community College
Dwight Newsome, Pasco-Hernando Community College
Denise Nunley, Maricopa Community Colleges
James Nymann, University of Texas-El Paso
Mark Omodt, Anoka-Ramsey Community College
Seth F. Oppenheimer, Mississippi State University
Leticia Oropesa, University of Miami
Linda Padilla, Joliet Junior College
Sanja Pantic, University of Illinois at Chicago
E. James Peake, Iowa State University
Kelly Pearson, Murray State University
Dashamir Petrela, Florida Atlantic University
Philip Pina, Florida Atlantic University
Charlotte Pisors, Baylor University
Michael Prophet, University of Northern Iowa
Laura Pyzdrowski, West Virginia University
Carrie Quesnell, Weber State University
Neal C. Raber, University of Akron
Thomas Radin, San Joaquin Delta College
Aibeng Serene Radulovic, Florida Atlantic University
Ken A. Rager, Metropolitan State College
Kenneth D. Reeves, San Antonio College
Elsi Reinhardt, Truckee Meadows Community College
Jose Remesar, Miami Dade College, Wolfson Campus
Jane Ringwald, Iowa State University
Douglas F. Robertson, University of Minnesota, MPLS
Stephen Rodi, Austin Community College
William Rogge, Lincoln Northeast High School
Howard L. Rolf, Baylor University
Mike Rosenthal, Florida International University
Phoebe Rouse, Lousiana State University
Edward Rozema, University of Tennessee at Chattanooga
Dennis C. Runde, Manatee Community College
Alan Saleski, Loyola University of Chicago
Susan Sandmeyer, Jamestown Community College
Brenda Santistevan, Salt Lake Community College
Linda Schmidt, Greenville Technical College
Ingrid Scott, Montgomery College
A.K. Shamma, University of West Florida
Zachery Sharon, University of Texas at San Antonio
Martin Sherry, Lower Columbia College
Carmen Shershin, Florida International University
Tatrana Shubin, San Jose State University
Anita Sikes, Delgado Community College
Timothy Sipka, Alma College
Charlotte Smedberg, University of Tampa
Lori Smellegar, Manatee Community College
Gayle Smith, Loyola Blakefield
Cindy Soderstrom, Salt Lake Community College
Leslie Soltis, Mercyhurst College
John Spellman, Southwest Texas State University
Karen Spike, University of North Carolina
Rajalakshmi Sriram, Okaloosa-Walton Community College
Katrina Staley, North Carolina Agricultural and Technical State University
Becky Stamper, Western Kentucky University
Judy Staver, Florida Community College-South
Robin Steinberg, Pima Community College
Neil Stephens, Hinsdale South High School
Sonya Stephens, Florida A&M Univeristy
Patrick Stevens, Joliet Junior College
John Sumner, University of Tampa
Matthew TenHuisen, University of North Carolina, Wilmington
Christopher Terry, Augusta State University
Diane Tesar, South Suburban College
Tommy Thompson, Brookhaven College
Martha K. Tietze, Shawnee Mission Northwest High School
Richard J. Tondra, Iowa State University
Florentina Tone, University of West Florida
Suzanne Topp, Salt Lake Community College
Marilyn Toscano, University of Wisconsin, Superior
Marvel Townsend, University of Florida
Jim Trudnowski, Carroll College
Robert Tuskey, Joliet Junior College
Mihaela Vajiac, Chapman University-Orange
Julia Varbalow, Thomas Nelson Community College-Leesville
Richard G. Vinson, University of South Alabama
Jorge Viola-Prioli, Florida Atlantic University
Mary Voxman, University of Idaho
Jennifer Walsh, Daytona Beach Community College
Donna Wandke, Naperville North High School
Timothy L. Warkentin, Cloud County Community College
Melissa J. Watts, Virginia State University
Hayat Weiss, Middlesex Community College
Kathryn Wetzel, Amarillo College
Darlene Whitkenack, Northern Illinois University
Suzanne Williams, Central Piedmont Community College
Larissa Williamson, University of Florida
Christine Wilson, West Virginia University
Brad Wind, Florida International University
Anna Wiodarczyk, Florida International University
Mary Wolyniak, Broome Community College
Canton Woods, Auburn University
Tamara S. Worner, Wayne State College
Terri Wright, New Hampshire Community Technical College, Manchester
Aletheia Zambesi, University of West Florida
George Zazi, Chicago State University
Steve Zuro, Joliet Junior College

Michael Sullivan
Chicago State University

Resources for Success

MyMathLab® Online Course (access code required)

MyMathLab delivers **proven results** in helping individual students succeed. It provides **engaging experiences** that personalize, stimulate, and measure learning for each student. And it comes from an **experienced partner** with educational expertise and an eye on the future. MyMathLab helps prepare students and gets them thinking more conceptually and visually through the following features:

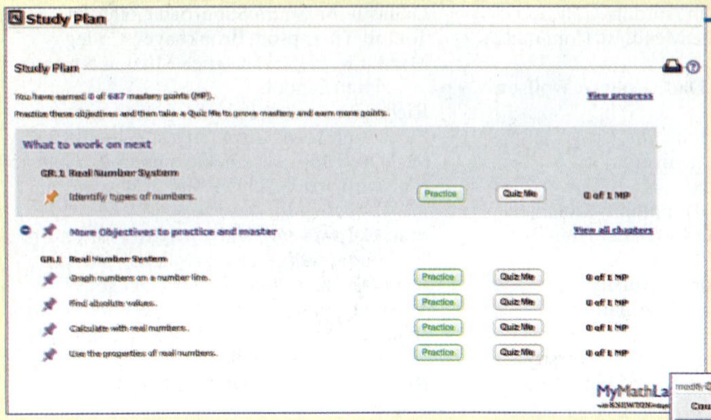

Adaptive Study Plan

The Study Plan makes studying more efficient and effective for every student. Performance and activity are assessed continually in real time. The data and analytics are used to provide personalized content–reinforcing concepts that target each student's strengths and weaknesses.

Getting Ready

Students refresh prerequisite topics through assignable skill review quizzes and personalized homework integrated in MyMathLab.

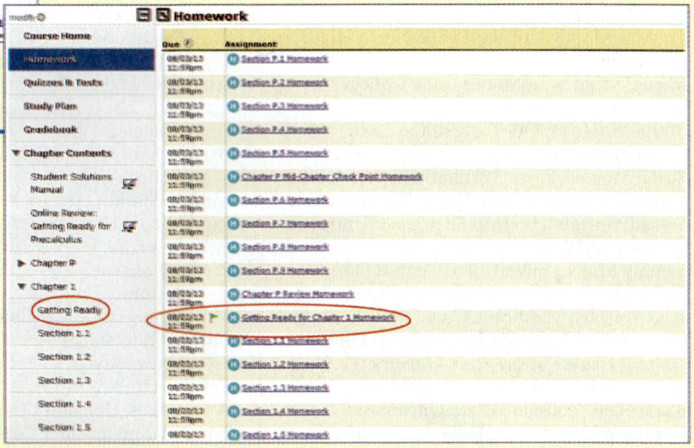

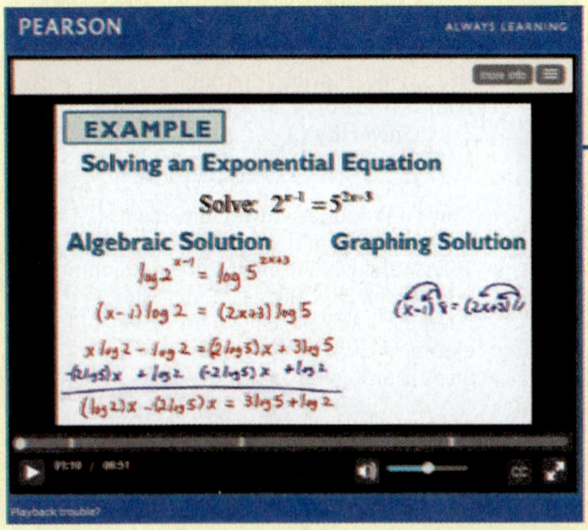

Video Assessment

Video assessment is tied to key Author in Action videos to check students' conceptual understanding of important math concepts.

Enhanced Graphing Functionality

New functionality within the graphing utility allows graphing of 3-point quadratic functions, 4-point cubic graphs, and transformations in exercises.

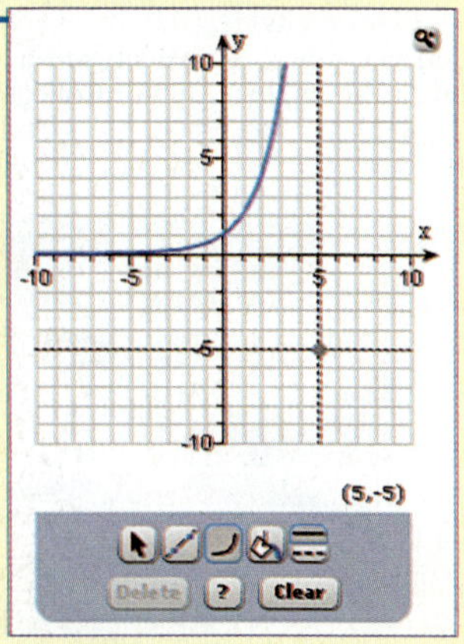

Skills for Success Modules are integrated within the MyMathLab course to help students succeed in collegiate courses and prepare for future professions.

Retain Your Knowledge These new exercises support ongoing review at the course level and help students maintain essential skills.

Instructor Resources

Additional resources can be downloaded from **www.pearsonhighered.com** or hardcopy resources can be ordered from your sales representative.

Ready to Go MyMathLab® Course

Now it is even easier to get started with MyMathLab. The Ready to Go MyMathLab course option includes author-chosen preassigned homework, integrated review, and more.

TestGen®

TestGen® (www.pearsoned.com/testgen) enables instructors to build, edit, print, and administer tests using a computerized bank of questions developed to cover all the objectives of the text.

PowerPoint® Lecture Slides

Fully editable slides correlated with the text.

Annotated Instructor's Edition

Shorter answers are on the page beside the exercises. Longer answers are in the back of the text.

Instructor Solutions Manual

Includes fully worked solutions to all exercises in the text.

Mini Lecture Notes

Includes additional examples and helpful teaching tips, by section.

Online Chapter Projects

Additional projects that give students an opportunity to apply what they learned in the chapter.

Student Resources

Additional resources to enhance student success:

Lecture Video

Author in Action videos are actual classroom lectures with fully worked out examples presented by Michael Sullivan, III. All video is assignable within MyMathlab.

Chapter Test Prep Videos

Students can watch instructors work through step-by-step solutions to all chapter test exercises from the text. These are available in MyMathlab and on YouTube.

Student Solutions Manual

Provides detailed worked-out solutions to odd-numbered exercises.

Guided Lecture Notes

These lecture notes assist students in taking thorough, organized, and understandable notes while watching Author in Action videos. Students actively participate in learning the how/why of important concepts through explorations and activities. The Guided Lecture Notes are available as PDF's and customizable Word files in MyMathLab. They can also be packaged with the text and the MyMathLab access code.

Algebra Review

Four chapters of Intermediate Algebra review. Perfect for a slower-paced course or for individual review.

Applications Index

Review

R

A Look Ahead •••

Chapter R, as the title states, contains review material. Your instructor may choose to cover all or part of it as a regular chapter at the beginning of your course or later as a just-in-time review when the content is required. Regardless, when information in this chapter is needed, a specific reference to this chapter will be made so you can review.

Outline

R.1 Real Numbers

PREPARING FOR THIS TEXT *Before getting started, read "To the Student" at the front of this text.*

OBJECTIVES
1 Work with Sets (p. 2)
2 Classify Numbers (p. 4)
3 Evaluate Numerical Expressions (p. 8)
4 Work with Properties of Real Numbers (p. 9)

1 Work with Sets

A **set** is a well-defined collection of distinct objects. The objects of a set are called its **elements**. By **well-defined**, we mean that there is a rule that enables us to determine whether a given object is an element of the set. If a set has no elements, it is called the **empty set**, or **null set**, and is denoted by the symbol $\varnothing$.

For example, the set of **digits** consists of the collection of numbers 0, 1, 2, 3, 4, 5, 6, 7, 8, and 9. If we use the symbol D to denote the set of digits, then we can write

$$D = \{0, 1, 2, 3, 4, 5, 6, 7, 8, 9\}$$

In this notation, the braces { } are used to enclose the objects, or **elements**, in the set. This method of denoting a set is called the **roster method**. A second way to denote a set is to use **set-builder notation**, where the set D of digits is written as

$$D = \{ \quad x \quad | \ x \text{ is a digit}\}$$

Read as "D is the set of all x such that x is a digit."

EXAMPLE 1 **Using Set-builder Notation and the Roster Method**

(a) $E = \{x | x \text{ is an even digit}\} = \{0, 2, 4, 6, 8\}$
(b) $O = \{x | x \text{ is an odd digit}\} = \{1, 3, 5, 7, 9\}$

Because the elements of a set are distinct, we never repeat elements. For example, we would never write $\{1, 2, 3, 2\}$; the correct listing is $\{1, 2, 3\}$. Because a set is a collection, the order in which the elements are listed is immaterial. $\{1, 2, 3\}$, $\{1, 3, 2\}$, $\{2, 1, 3\}$, and so on, all represent the same set.

If every element of a set A is also an element of a set B, then A is a **subset** of B, which is denoted $A \subseteq B$. If two sets A and B have the same elements, then A **equals** B, which is denoted $A = B$.

For example, $\{1, 2, 3\} \subseteq \{1, 2, 3, 4, 5\}$ and $\{1, 2, 3\} = \{2, 3, 1\}$.

DEFINITION

If A and B are sets, the **intersection** of A with B, denoted $A \cap B$, is the set consisting of elements that belong to both A and B. The **union** of A with B, denoted $A \cup B$, is the set consisting of elements that belong to either A or B, or both.

EXAMPLE 2 **Finding the Intersection and Union of Sets**

Let $A = \{1, 3, 5, 8\}$, $B = \{3, 5, 7\}$, and $C = \{2, 4, 6, 8\}$. Find:

(a) $A \cap B$ (b) $A \cup B$ (c) $B \cap (A \cup C)$

Solution

(a) $A \cap B = \{1,3,5,8\} \cap \{3,5,7\} = \{3,5\}$

(b) $A \cup B = \{1,3,5,8\} \cup \{3,5,7\} = \{1,3,5,7,8\}$

(c) $B \cap (A \cup C) = \{3,5,7\} \cap [\{1,3,5,8\} \cup \{2,4,6,8\}]$
$= \{3,5,7\} \cap \{1,2,3,4,5,6,8\} = \{3,5\}$ ●

Now Work PROBLEM 15

Usually, in working with sets, we designate a **universal set** U, the set consisting of all the elements that we wish to consider. Once a universal set has been designated, we can consider elements of the universal set not found in a given set.

DEFINITION

If A is a set, the **complement** of A, denoted $\overline{A}$, is the set consisting of all the elements in the universal set that are not in A.*

EXAMPLE 3 Finding the Complement of a Set

If the universal set is $U = \{1,2,3,4,5,6,7,8,9\}$ and if $A = \{1,3,5,7,9\}$, then $\overline{A} = \{2,4,6,8\}$. ●

It follows from the definition of complement that $A \cup \overline{A} = U$ and $A \cap \overline{A} = \varnothing$. Do you see why?

Now Work PROBLEM 19

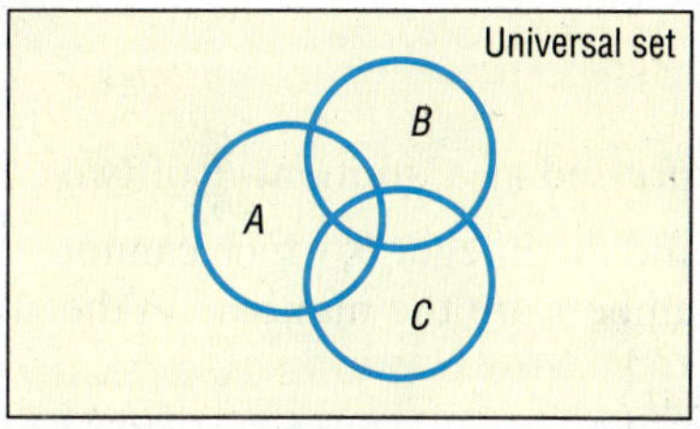

Figure 1 Venn diagram

It is often helpful to draw pictures of sets. Such pictures, called **Venn diagrams**, represent sets as circles enclosed in a rectangle, which represents the universal set. Such diagrams often help us to visualize various relationships among sets. See Figure 1.

If we know that $A \subseteq B$, we might use the Venn diagram in Figure 2(a). If we know that A and B have no elements in common—that is, if $A \cap B = \varnothing$—we might use the Venn diagram in Figure 2(b). The sets A and B in Figure 2(b) are said to be **disjoint**.

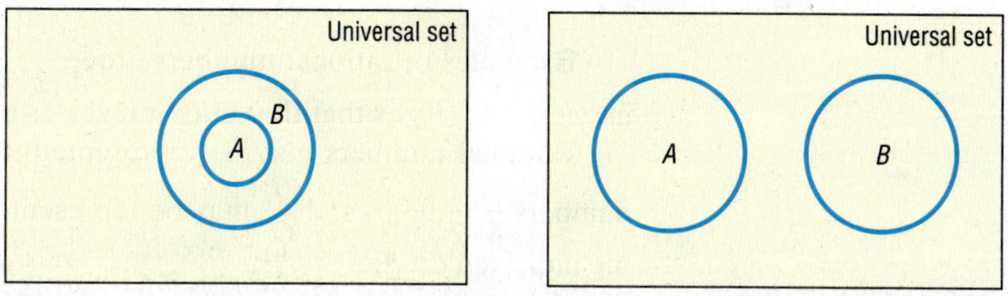

Figure 2 (a) $A \subseteq B$ subset (b) $A \cap B = \varnothing$ disjoint sets

Figures 3(a), 3(b), and 3(c) use Venn diagrams to illustrate the definitions of intersection, union, and complement, respectively.

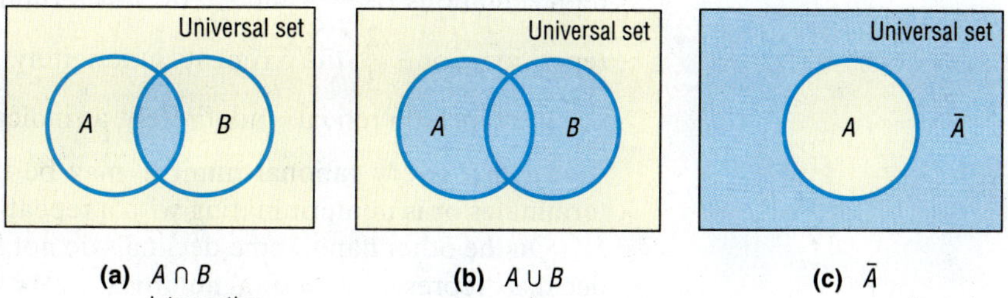

Figure 3 (a) $A \cap B$ intersection (b) $A \cup B$ union (c) $\bar{A}$ complement

*Some books use the notation A' for the complement of A.

2 Classify Numbers

It is helpful to classify the various kinds of numbers that we deal with as sets. The **counting numbers**, or **natural numbers**, are the numbers in the set $\{1, 2, 3, 4, \ldots\}$. (The three dots, called an **ellipsis**, indicate that the pattern continues indefinitely.) As their name implies, these numbers are often used to count things. For example, there are 26 letters in our alphabet; there are 100 cents in a dollar. The **whole numbers** are the numbers in the set $\{0, 1, 2, 3, \ldots\}$ —that is, the counting numbers together with 0. The set of counting numbers is a subset of the set of whole numbers.

DEFINITION

The **integers** are the set of numbers $\{\ldots, -3, -2, -1, 0, 1, 2, 3, \ldots\}$.

These numbers are useful in many situations. For example, if your checking account has \$10 in it and you write a check for \$15, you can represent the current balance as −\$5.

Each time we expand a number system, such as from the whole numbers to the integers, we do so in order to be able to handle new, and usually more complicated, problems. The integers enable us to solve problems requiring both positive and negative counting numbers, such as profit/loss, height above/below sea level, temperature above/below 0°F, and so on.

But integers alone are not sufficient for *all* problems. For example, they do not answer the question "What part of a dollar is 38 cents?" To answer such a question, we enlarge our number system to include *rational numbers*. For example, $\frac{38}{100}$ answers the question "What part of a dollar is 38 cents?"

DEFINITION

A **rational number** is a number that can be expressed as a quotient $\frac{a}{b}$ of two integers. The integer a is called the **numerator**, and the integer b, which cannot be 0, is called the **denominator**. The rational numbers are the numbers in the set $\left\{x \,\middle|\, x = \frac{a}{b}, \text{ where } a, b \text{ are integers and } b \neq 0\right\}$.

Examples of rational numbers are $\frac{3}{4}, \frac{5}{2}, \frac{0}{4}, -\frac{2}{3}$, and $\frac{100}{3}$. Since $\frac{a}{1} = a$ for any integer a, it follows that the set of integers is a subset of the set of rational numbers.

Rational numbers may be represented as **decimals**. For example, the rational numbers $\frac{3}{4}, \frac{5}{2}, -\frac{2}{3}$, and $\frac{7}{66}$ may be represented as decimals by merely carrying out the indicated division:

$$\frac{3}{4} = 0.75 \qquad \frac{5}{2} = 2.5 \qquad -\frac{2}{3} = -0.666\ldots = -0.\overline{6} \qquad \frac{7}{66} = 0.1060606\ldots = 0.1\overline{06}$$

Notice that the decimal representations of $\frac{3}{4}$ and $\frac{5}{2}$ terminate, or end. The decimal representations of $-\frac{2}{3}$ and $\frac{7}{66}$ do not terminate, but they do exhibit a pattern of repetition. For $-\frac{2}{3}$, the 6 repeats indefinitely, as indicated by the bar over the 6; for $\frac{7}{66}$, the block 06 repeats indefinitely, as indicated by the bar over the 06. It can be shown that every rational number may be represented by a decimal that either terminates or is nonterminating with a repeating block of digits, and vice versa.

On the other hand, some decimals do not fit into either of these categories. Such decimals represent **irrational numbers**. Every irrational number may be represented by a decimal that neither repeats nor terminates. In other words, irrational numbers cannot be written in the form $\frac{a}{b}$, where a, b are integers and $b \neq 0$.

Irrational numbers occur naturally. For example, consider the isosceles right triangle whose legs are each of length 1. See Figure 4. The length of the hypotenuse is $\sqrt{2}$, an irrational number.

Also, the number that equals the ratio of the circumference C to the diameter d of any circle, denoted by the symbol π (the Greek letter pi), is an irrational number. See Figure 5.

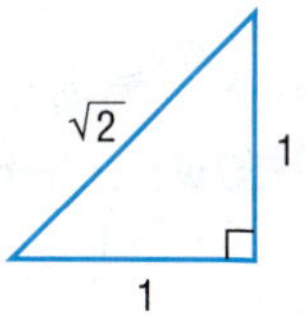

Figure 4

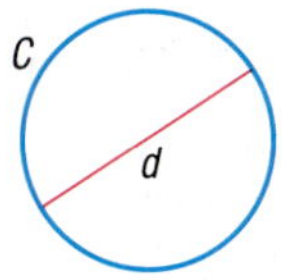

Figure 5 $\pi = \dfrac{C}{d}$

DEFINITION

The set of **real numbers** is the union of the set of rational numbers with the set of irrational numbers.

Figure 6 shows the relationship of various types of numbers.*

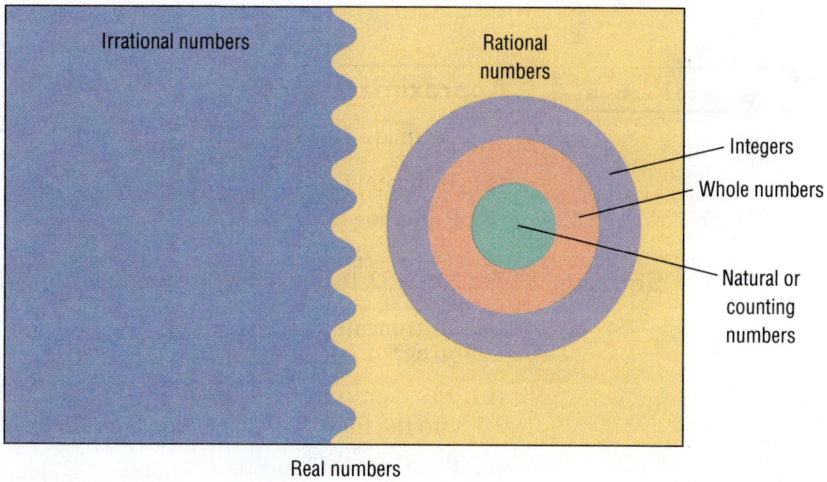

Figure 6

EXAMPLE 4 Classifying the Numbers in a Set

List the numbers in the set

$$\left\{-3, \frac{4}{3}, 0.12, \sqrt{2}, \pi, 10, 2.151515\ldots \text{(where the block 15 repeats)}\right\}$$

that are

(a) Natural numbers (b) Integers (c) Rational numbers
(d) Irrational numbers (e) Real numbers

Solution

(a) 10 is the only natural number.
(b) -3 and 10 are integers.
(c) $-3, 10, \frac{4}{3}, 0.12$, and $2.151515\ldots$ are rational numbers.
(d) $\sqrt{2}$ and π are irrational numbers.
(e) All the numbers listed are real numbers. ●

Now Work PROBLEM 25

*The set of real numbers is a subset of the set of complex numbers. We discuss complex numbers in Chapter 1, Section 1.3.

Approximations

Every decimal may be represented by a real number (either rational or irrational), and every real number may be represented by a decimal.

In practice, the decimal representation of an irrational number is given as an approximation. For example, using the symbol ≈ (read as "approximately equal to"), we can write

$$\sqrt{2} \approx 1.4142 \qquad \pi \approx 3.1416$$

In approximating decimals, we either *round off* or *truncate* to a given number of decimal places.* The number of places establishes the location of the *final digit* in the decimal approximation.

Truncation: Drop all of the digits that follow the specified final digit in the decimal.

Rounding: Identify the specified final digit in the decimal. If the next digit is 5 or more, add 1 to the final digit; if the next digit is 4 or less, leave the final digit as it is. Then truncate following the final digit.

EXAMPLE 5

Approximating a Decimal to Two Places

Approximate 20.98752 to two decimal places by

(a) Truncating

(b) Rounding

Solution For 20.98752, the final digit is 8, since it is two decimal places from the decimal point.

(a) To truncate, we remove all digits following the final digit 8. The truncation of 20.98752 to two decimal places is 20.98.

(b) The digit following the final digit 8 is the digit 7. Since 7 is 5 or more, we add 1 to the final digit 8 and truncate. The rounded form of 20.98752 to two decimal places is 20.99.

EXAMPLE 6

Approximating a Decimal to Two and Four Places

Number	Rounded to Two Decimal Places	Rounded to Four Decimal Places	Truncated to Two Decimal Places	Truncated to Four Decimal Places
(a) 3.14159	3.14	3.1416	3.14	3.1415
(b) 0.056128	0.06	0.0561	0.05	0.0561
(c) 893.46125	893.46	893.4613	893.46	893.4612

Now Work PROBLEM 29

Calculators

Calculators are incapable of displaying decimals that contain a large number of digits. For example, some calculators are capable of displaying only eight digits. When a number requires more than eight digits, the calculator either truncates or rounds.

* Sometimes we say "correct to a given number of decimal places" instead of "truncate."

To see how your calculator handles decimals, divide 2 by 3. How many digits do you see? Is the last digit a 6 or a 7? If it is a 6, your calculator truncates; if it is a 7, your calculator rounds.

There are different kinds of calculators. An **arithmetic** calculator can only add, subtract, multiply, and divide numbers; therefore, this type is not adequate for this course. **Scientific** calculators have all the capabilities of arithmetic calculators and also contain **function keys** labeled ln, log, sin, cos, tan, x^y, inv, and so on. As you proceed through this text, you will discover how to use many of the function keys. **Graphing** calculators have all the capabilities of scientific calculators and contain a screen on which graphs can be displayed.

For those who have access to a graphing calculator, we have included comments, examples, and exercises marked with a 🖩, indicating that a graphing calculator is required. We have also included an appendix that explains some of the capabilities of a graphing calculator. The 🖩 comments, examples, and exercises may be omitted without loss of continuity, if so desired.

Operations

In algebra, we use letters such as x, y, a, b, and c to represent numbers. The symbols used in algebra for the operations of addition, subtraction, multiplication, and division are $+, -, \cdot$, and $/$. The words used to describe the results of these operations are **sum**, **difference, product**, and **quotient**. Table 1 summarizes these ideas.

Table 1

Operation	Symbol	Words
Addition	$a + b$	Sum: *a* plus *b*
Subtraction	$a - b$	Difference: *a* minus *b*
Multiplication	$a \cdot b, (a) \cdot b, a \cdot (b), (a) \cdot (b)$, $ab, (a)b, a(b), (a)(b)$	Product: *a* times *b*
Division	a/b or $\frac{a}{b}$	Quotient: *a* divided by *b*

In algebra, we generally avoid using the multiplication sign $\times$ and the division sign $\div$ so familiar in arithmetic. Notice also that when two expressions are placed next to each other without an operation symbol, as in ab, or in parentheses, as in $(a)(b)$, it is understood that the expressions, called **factors**, are to be multiplied.

We also prefer not to use mixed numbers in algebra. When mixed numbers are used, addition is understood; for example, $2\frac{3}{4}$ means $2 + \frac{3}{4}$. In algebra, use of a mixed number may be confusing because the absence of an operation symbol between two terms is generally taken to mean multiplication. The expression $2\frac{3}{4}$ is therefore written instead as 2.75 or as $\frac{11}{4}$.

The symbol $=$, called an **equal sign** and read as "equals" or "is," is used to express the idea that the number or expression on the left of the equal sign is equivalent to the number or expression on the right.

EXAMPLE 7 **Writing Statements Using Symbols**

(a) The sum of 2 and 7 equals 9. In symbols, this statement is written as $2 + 7 = 9$.

(b) The product of 3 and 5 is 15. In symbols, this statement is written as $3 \cdot 5 = 15$.

Now Work PROBLEM 41

3 Evaluate Numerical Expressions

Consider the expression $2 + 3 \cdot 6$. It is not clear whether we should add 2 and 3 to get 5, and then multiply by 6 to get 30; or first multiply 3 and 6 to get 18, and then add 2 to get 20. To avoid this ambiguity, we have the following agreement.

In Words
Multiply first, then add.

> We agree that whenever the two operations of addition and multiplication separate three numbers, the multiplication operation will always be performed first, followed by the addition operation.

For $2 + 3 \cdot 6$, then, we have

$$2 + 3 \cdot 6 = 2 + 18 = 20$$

EXAMPLE 8 Finding the Value of an Expression

Evaluate each expression.

(a) $3 + 4 \cdot 5$ (b) $8 \cdot 2 + 1$ (c) $2 + 2 \cdot 2$

Solution (a) $3 + 4 \cdot 5 = 3 + 20 = 23$ ↑ Multiply first

(b) $8 \cdot 2 + 1 = 16 + 1 = 17$ ↑ Multiply first

(c) $2 + 2 \cdot 2 = 2 + 4 = 6$ ●

Now Work PROBLEM 53

When we want to indicate adding 3 and 4 and then multiplying the result by 5, we use parentheses and write $(3 + 4) \cdot 5$. Whenever parentheses appear in an expression, it means "perform the operations within the parentheses first!"

EXAMPLE 9 Finding the Value of an Expression

(a) $(5 + 3) \cdot 4 = 8 \cdot 4 = 32$

(b) $(4 + 5) \cdot (8 - 2) = 9 \cdot 6 = 54$ ●

When we divide two expressions, as in

$$\frac{2 + 3}{4 + 8}$$

it is understood that the division bar acts like parentheses; that is,

$$\frac{2 + 3}{4 + 8} = \frac{(2 + 3)}{(4 + 8)}$$

Rules for the Order of Operations

1. Begin with the innermost parentheses and work outward. Remember that in dividing two expressions, we treat the numerator and denominator as if they were enclosed in parentheses.
2. Perform multiplications and divisions, working from left to right.
3. Perform additions and subtractions, working from left to right.

EXAMPLE 10 **Finding the Value of an Expression**

Evaluate each expression.

(a) $8 \cdot 2 + 3$ (b) $5 \cdot (3 + 4) + 2$

(c) $\dfrac{2 + 5}{2 + 4 \cdot 7}$ (d) $2 + [4 + 2 \cdot (10 + 6)]$

Solution

(a) $8 \cdot 2 + 3 = 16 + 3 = 19$
↑ Multiply first

(b) $5 \cdot (3 + 4) + 2 = 5 \cdot 7 + 2 = 35 + 2 = 37$
↑ Parentheses first ↑ Multiply before adding

(c) $\dfrac{2 + 5}{2 + 4 \cdot 7} = \dfrac{2 + 5}{2 + 28} = \dfrac{7}{30}$

(d) $2 + [4 + 2 \cdot (10 + 6)] = 2 + [4 + 2 \cdot (16)]$
$= 2 + [4 + 32] = 2 + [36] = 38$

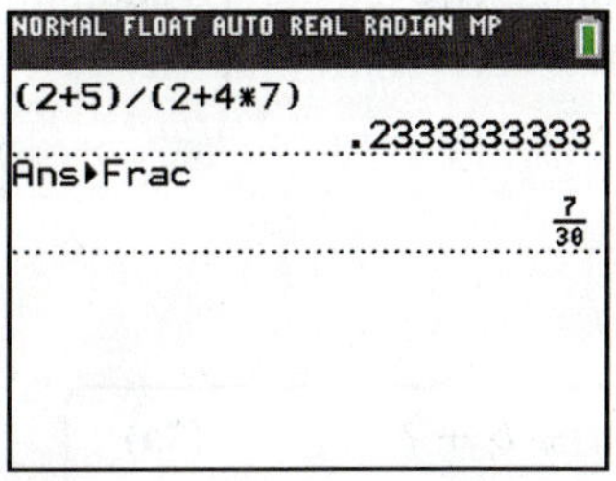

Figure 7

Be careful if you use a calculator. For Example 10(c), you need to use parentheses. See Figure 7.* If you don't, the calculator will compute the expression

$$2 + \frac{5}{2} + 4 \cdot 7 = 2 + 2.5 + 28 = 32.5$$

giving a wrong answer.

Now Work PROBLEMS 59 AND 67

4 Work with Properties of Real Numbers

The equal sign is used to mean that one expression is equivalent to another. Four important properties of equality are listed next. In this list, *a*, *b*, and *c* represent real numbers.

1. The **reflexive property** states that a number equals itself; that is, $a = a$.
2. The **symmetric property** states that if $a = b$, then $b = a$.
3. The **transitive property** states that if $a = b$ and $b = c$, then $a = c$.
4. The **principle of substitution** states that if $a = b$, then we may substitute b for a in any expression containing a.

Now, let's consider some other properties of real numbers.

EXAMPLE 11 **Commutative Properties**

(a) $3 + 5 = 8$
$5 + 3 = 8$
$3 + 5 = 5 + 3$

(b) $2 \cdot 3 = 6$
$3 \cdot 2 = 6$
$2 \cdot 3 = 3 \cdot 2$

This example illustrates the **commutative property** of real numbers, which states that the order in which addition or multiplication takes place does not affect the final result.

* Notice that we converted the decimal to its fraction form. Another option, when using a TI-84 Plus C, is to use the fraction template under the MATH button to enter the expression as it appears in Example 10(c). Consult your manual to see how to enter such expressions on your calculator.

Commutative Properties

$$a + b = b + a \tag{1a}$$

$$a \cdot b = b \cdot a \tag{1b}$$

Here, and in the properties listed next and on pages 11–13, a, b, and c represent real numbers.

EXAMPLE 12 Associative Properties

(a) $2 + (3 + 4) = 2 + 7 = 9$
$(2 + 3) + 4 = 5 + 4 = 9$
$2 + (3 + 4) = (2 + 3) + 4$

(b) $2 \cdot (3 \cdot 4) = 2 \cdot 12 = 24$
$(2 \cdot 3) \cdot 4 = 6 \cdot 4 = 24$
$2 \cdot (3 \cdot 4) = (2 \cdot 3) \cdot 4$

The way we add or multiply three real numbers does not affect the final result. Expressions such as $2 + 3 + 4$ and $3 \cdot 4 \cdot 5$ present no ambiguity, even though addition and multiplication are performed on one pair of numbers at a time. This property is called the **associative property**.

Associative Properties

$$a + (b + c) = (a + b) + c = a + b + c \tag{2a}$$

$$a \cdot (b \cdot c) = (a \cdot b) \cdot c = a \cdot b \cdot c \tag{2b}$$

Distributive Property

$$a \cdot (b + c) = a \cdot b + a \cdot c \tag{3a}$$

$$(a + b) \cdot c = a \cdot c + b \cdot c \tag{3b}$$

The **distributive property** may be used in two different ways.

EXAMPLE 13 Distributive Property

(a) $2 \cdot (x + 3) = 2 \cdot x + 2 \cdot 3 = 2x + 6$ *Use to remove parentheses.*

(b) $3x + 5x = (3 + 5)x = 8x$ *Use to combine two expressions.*

(c) $(x + 2)(x + 3) = x(x + 3) + 2(x + 3) = (x^2 + 3x) + (2x + 6)$
$= x^2 + (3x + 2x) + 6 = x^2 + 5x + 6$

Now Work PROBLEM 89

The real numbers 0 and 1 have unique properties called the *identity properties.*

EXAMPLE 14 Identity Properties

(a) $4 + 0 = 0 + 4 = 4$

(b) $3 \cdot 1 = 1 \cdot 3 = 3$

Identity Properties

$$0 + a = a + 0 = a \qquad \textbf{(4a)}$$

$$a \cdot 1 = 1 \cdot a = a \qquad \textbf{(4b)}$$

We call 0 the **additive identity** and 1 the **multiplicative identity**.

For each real number a, there is a real number $-a$, called the **additive inverse** of a, having the following property:

Additive Inverse Property

$$a + (-a) = -a + a = 0 \qquad \textbf{(5a)}$$

EXAMPLE 15

Finding an Additive Inverse

(a) The additive inverse of 6 is -6, because $6 + (-6) = 0$.

(b) The additive inverse of -8 is $-(-8) = 8$, because $-8 + 8 = 0$.

The additive inverse of a, that is, $-a$, is often called the *negative* of a or the *opposite* of a. The use of such terms can be dangerous, because they suggest that the additive inverse is a negative number, which may not be the case. For example, the additive inverse of -3, or $-(-3)$, equals 3, a positive number.

For each *nonzero* real number a, there is a real number $\frac{1}{a}$, called the **multiplicative inverse** of a, having the following property:

Multiplicative Inverse Property

$$a \cdot \frac{1}{a} = \frac{1}{a} \cdot a = 1 \quad \text{if } a \neq 0 \qquad \textbf{(5b)}$$

The multiplicative inverse $\frac{1}{a}$ of a nonzero real number a is also referred to as the **reciprocal** of a.

EXAMPLE 16

Finding a Reciprocal

(a) The reciprocal of 6 is $\frac{1}{6}$, because $6 \cdot \frac{1}{6} = 1$.

(b) The reciprocal of -3 is $\frac{1}{-3}$, because $-3 \cdot \frac{1}{-3} = 1$.

(c) The reciprocal of $\frac{2}{3}$ is $\frac{3}{2}$, because $\frac{2}{3} \cdot \frac{3}{2} = 1$.

With these properties for adding and multiplying real numbers, we can define the operations of subtraction and division as follows:

DEFINITION

The **difference** $a - b$, also read "a less b" or "a minus b," is defined as

$$a - b = a + (-b) \qquad \textbf{(6)}$$

To subtract b from a, add the opposite of b to a.

DEFINITION

If b is a nonzero real number, the **quotient** $\frac{a}{b}$, also read as "a divided by b" or "the ratio of a to b," is defined as

$$\frac{a}{b} = a \cdot \frac{1}{b} \quad \text{if } b \neq 0 \tag{7}$$

EXAMPLE 17

Working with Differences and Quotients

(a) $8 - 5 = 8 + (-5) = 3$

(b) $4 - 9 = 4 + (-9) = -5$

(c) $\frac{5}{8} = 5 \cdot \frac{1}{8}$

For any number a, the product of a times 0 is always 0; that is,

In Words
The result of multiplying by zero is zero.

Multiplication by Zero

$$a \cdot 0 = 0 \tag{8}$$

For a nonzero number a,

Division Properties

$$\frac{0}{a} = 0 \qquad \frac{a}{a} = 1 \quad \text{if } a \neq 0 \tag{9}$$

NOTE Division by 0 is not defined. One reason is to avoid the following difficulty: $\frac{2}{0} = x$ means to find x such that $0 \cdot x = 2$. But $0 \cdot x$ equals 0 for all x, so there is no unique number x such that $\frac{2}{0} = x$.

Rules of Signs

$$a(-b) = -(ab) \qquad (-a)b = -(ab) \qquad (-a)(-b) = ab$$
$$-(-a) = a \qquad \frac{a}{-b} = \frac{-a}{b} = -\frac{a}{b} \qquad \frac{-a}{-b} = \frac{a}{b} \tag{10}$$

EXAMPLE 18

Applying the Rules of Signs

(a) $2(-3) = -(2 \cdot 3) = -6$

(b) $(-3)(-5) = 3 \cdot 5 = 15$

(c) $\frac{3}{-2} = \frac{-3}{2} = -\frac{3}{2}$

(d) $\frac{-4}{-9} = \frac{4}{9}$

(e) $\frac{x}{-2} = \frac{1}{-2} \cdot x = -\frac{1}{2}x$

Cancellation Properties

$$ac = bc \quad \text{implies} \quad a = b \quad \text{if } c \neq 0$$
$$\frac{ac}{bc} = \frac{a}{b} \quad \text{if } b \neq 0, c \neq 0 \qquad \textbf{(11)}$$

EXAMPLE 19

Using the Cancellation Properties

(a) If $2x = 6$, then

$$\begin{aligned} 2x &= 6 \\ 2x &= 2 \cdot 3 && \text{Factor 6.} \\ x &= 3 && \text{Cancel the 2's.} \end{aligned}$$

(b) $\frac{18}{12} = \frac{3 \cdot \not{6}}{2 \cdot \not{6}} = \frac{3}{2}$

↑ Cancel the 6's.

NOTE We follow the common practice of using slash marks to indicate cancellations. ■

In Words

If a product equals 0, then one or both of the factors is 0.

Zero-Product Property

If $ab = 0$, then $a = 0$, or $b = 0$, or both. **(12)**

EXAMPLE 20

Using the Zero-Product Property

If $2x = 0$, then either $2 = 0$ or $x = 0$. Since $2 \neq 0$, it follows that $x = 0$.

Arithmetic of Quotients

$$\frac{a}{b} + \frac{c}{d} = \frac{ad}{bd} + \frac{bc}{bd} = \frac{ad + bc}{bd} \quad \text{if } b \neq 0, d \neq 0 \qquad \textbf{(13)}$$

$$\frac{a}{b} \cdot \frac{c}{d} = \frac{ac}{bd} \quad \text{if } b \neq 0, d \neq 0 \qquad \textbf{(14)}$$

$$\frac{\frac{a}{b}}{\frac{c}{d}} = \frac{a}{b} \cdot \frac{d}{c} = \frac{ad}{bc} \quad \text{if } b \neq 0, c \neq 0, d \neq 0 \qquad \textbf{(15)}$$

EXAMPLE 21

Adding, Subtracting, Multiplying, and Dividing Quotients

(a) $\frac{2}{3} + \frac{5}{2} = \frac{2 \cdot 2}{3 \cdot 2} + \frac{3 \cdot 5}{3 \cdot 2} = \frac{2 \cdot 2 + 3 \cdot 5}{3 \cdot 2} = \frac{4 + 15}{6} = \frac{19}{6}$

↑ By equation (13)

(b) $\frac{3}{5} - \frac{2}{3} = \frac{3}{5} + \left(-\frac{2}{3}\right) = \frac{3}{5} + \frac{-2}{3}$

↑ By equation (6) ↑ By equation (10)

$$= \frac{3 \cdot 3 + 5 \cdot (-2)}{5 \cdot 3} = \frac{9 + (-10)}{15} = \frac{-1}{15} = -\frac{1}{15}$$

↑ By equation (13)

NOTE Slanting the cancellation marks in different directions for different factors, as shown here, is a good practice to follow, since it will help in checking for errors. ■

(c) $\frac{8}{3} \cdot \frac{15}{4} = \frac{8 \cdot 15}{3 \cdot 4} = \frac{2 \cdot \cancel{4} \cdot \cancel{3} \cdot 5}{\cancel{3} \cdot \cancel{4} \cdot 1} = \frac{2 \cdot 5}{1} = 10$

By equation (14) By equation (11)

(d) $\frac{\frac{3}{5}}{\frac{7}{9}} = \frac{3}{5} \cdot \frac{9}{7} = \frac{3 \cdot 9}{5 \cdot 7} = \frac{27}{35}$

By equation (15) By equation (14) ●

NOTE In writing quotients, we shall follow the usual convention and write the quotient in lowest terms. That is, we write it so that any common factors of the numerator and the denominator have been removed using the cancellation properties, equation (11). As examples,

$$\frac{90}{24} = \frac{15 \cdot \cancel{6}}{4 \cdot \cancel{6}} = \frac{15}{4}$$

$$\frac{24x^2}{18x} = \frac{4 \cdot \cancel{6} \cdot x \cdot \cancel{x}}{3 \cdot \cancel{6} \cdot \cancel{x}} = \frac{4x}{3} \qquad x \neq 0$$ ■

Now Work PROBLEMS 69, 73, AND 83

Sometimes it is easier to add two fractions using *least common multiples* (LCM). The LCM of two numbers is the smallest number that each has as a common multiple.

EXAMPLE 22 **Finding the Least Common Multiple of Two Numbers**

Find the least common multiple of 15 and 12.

Solution To find the LCM of 15 and 12, we look at multiples of 15 and 12.

15, 30, 45, 60, 75, 90, 105, 120,. . .

12, 24, 36, 48, 60, 72, 84, 96, 108, 120,. . .

The *common* multiples are in blue. The *least* common multiple is 60. ●

EXAMPLE 23 **Using the Least Common Multiple to Add Two Fractions**

Find: $\frac{8}{15} + \frac{5}{12}$

Solution We use the LCM of the denominators of the fractions and rewrite each fraction using the LCM as a common denominator. The LCM of the denominators (12 and 15) is 60. Rewrite each fraction using 60 as the denominator.

$$\begin{aligned}\frac{8}{15} + \frac{5}{12} &= \frac{8}{15} \cdot \frac{4}{4} + \frac{5}{12} \cdot \frac{5}{5} \\ &= \frac{32}{60} + \frac{25}{60} \\ &= \frac{32 + 25}{60} \\ &= \frac{57}{60} \\ &= \frac{19}{20}\end{aligned}$$ ●

Now Work PROBLEM 77

Historical Feature

The real number system has a history that stretches back at least to the ancient Babylonians (1800 BC). It is remarkable how much the ancient Babylonian attitudes resemble our own. As we stated in the text, the fundamental difficulty with irrational numbers is that they cannot be written as quotients of integers or, equivalently, as repeating or terminating decimals. The Babylonians wrote their numbers in a system based on 60 in the same way that we write ours based on 10. They would carry as many places for π as the accuracy of the problem demanded, just as we now use

$$\pi \approx 3\frac{1}{7} \quad \text{or} \quad \pi \approx 3.1416 \quad \text{or} \quad \pi \approx 3.14159$$
$$\text{or} \quad \pi \approx 3.14159265358979$$

depending on how accurate we need to be.

Things were very different for the Greeks, whose number system allowed only rational numbers. When it was discovered that $\sqrt{2}$ was not a rational number, this was regarded as a fundamental flaw in the number concept. So serious was the matter that the Pythagorean Brotherhood (an early mathematical society) is said to have drowned one of its members for revealing this terrible secret. Greek mathematicians then turned away from the number concept, expressing facts about whole numbers in terms of line segments.

In astronomy, however, Babylonian methods, including the Babylonian number system, continued to be used. Simon Stevin (1548–1620), probably using the Babylonian system as a model, invented the decimal system, complete with rules of calculation, in 1585. [Others, for example, al-Kashi of Samarkand (d. 1429), had made some progress in the same direction.] The decimal system so effectively conceals the difficulties that the need for more logical precision began to be felt only in the early 1800s. Around 1880, Georg Cantor (1845–1918) and Richard Dedekind (1831–1916) gave precise definitions of real numbers. Cantor's definition, although more abstract and precise, has its roots in the decimal (and hence Babylonian) numerical system.

Sets and set theory were a spin-off of the research that went into clarifying the foundations of the real number system. Set theory has developed into a large discipline of its own, and many mathematicians regard it as the foundation upon which modern mathematics is built. Cantor's discoveries that infinite sets can also be counted and that there are different sizes of infinite sets are among the most astounding results of modern mathematics.

R.1 Assess Your Understanding

Concepts and Vocabulary

1. The numbers in the set $\left\{x \middle| x = \frac{a}{b}, \text{where } a, b \text{ are integers and } b \neq 0\right\}$ are called ________ numbers.
2. The value of the expression $4 + 5 \cdot 6 - 3$ is ___.
3. The fact that $2x + 3x = (2 + 3)x$ is a consequence of the __________ Property.
4. "The product of 5 and $x + 3$ equals 6" may be written as ____________.
5. The intersection of sets A and B is denoted by which of the following?
   (a) $A \cap B$ (b) $A \cup B$ (c) $A \subseteq B$ (d) $A \varnothing B$
6. Choose the correct name for the set of numbers $\{0, 1, 2, 3, \ldots\}$.
   (a) Counting numbers (b) Whole numbers
   (c) Integers (d) Irrational numbers
7. ***True or False*** Rational numbers have decimals that either terminate or are nonterminating with a repeating block of digits.
8. ***True or False*** The Zero-Product Property states that the product of any number and zero equals zero.
9. ***True or False*** The least common multiple of 12 and 18 is 6.
10. ***True or False*** No real number is both rational and irrational.

Skill Building

In Problems 11–22, use U = universal set = $\{0, 1, 2, 3, 4, 5, 6, 7, 8, 9\}$, A = $\{1, 3, 4, 5, 9\}$, B = $\{2, 4, 6, 7, 8\}$, and C = $\{1, 3, 4, 6\}$ to find each set.

11. $A \cup B$
12. $A \cup C$
13. $A \cap B$
14. $A \cap C$
15. $(A \cup B) \cap C$
16. $(A \cap B) \cup C$
17. $\overline{A}$
18. $\overline{C}$
19. $\overline{A \cap B}$
20. $\overline{B \cup C}$
21. $\overline{A} \cup \overline{B}$
22. $\overline{B} \cap \overline{C}$

In Problems 23–28, list the numbers in each set that are (a) Natural numbers, (b) Integers, (c) Rational numbers, (d) Irrational numbers, (e) Real numbers.

23. $A = \left\{-6, \frac{1}{2}, -1.333\ldots \text{(the 3's repeat)}, \pi, 2, 5\right\}$
24. $B = \left\{-\frac{5}{3}, 2.060606\ldots \text{(the block 06 repeats)}, 1.25, 0, 1, \sqrt{5}\right\}$
25. $C = \left\{0, 1, \frac{1}{2}, \frac{1}{3}, \frac{1}{4}\right\}$
26. $D = \{-1, -1.1, -1.2, -1.3\}$
27. $E = \left\{\sqrt{2}, \pi, \sqrt{2} + 1, \pi + \frac{1}{2}\right\}$
28. $F = \left\{-\sqrt{2}, \pi + \sqrt{2}, \frac{1}{2} + 10.3\right\}$

In Problems 29–40, approximate each number (a) rounded and (b) truncated to three decimal places.

29. 18.9526 **30.** 25.86134 **31.** 28.65319 **32.** 99.05249 **33.** 0.06291 **34.** 0.05388

35. 9.9985 **36.** 1.0006 **37.** $\frac{3}{7}$ **38.** $\frac{5}{9}$ **39.** $\frac{521}{15}$ **40.** $\frac{81}{5}$

In Problems 41–50, write each statement using symbols.

41. The sum of 3 and 2 equals 5.
42. The product of 5 and 2 equals 10.
43. The sum of x and 2 is the product of 3 and 4.
44. The sum of 3 and y is the sum of 2 and 2.
45. The product of 3 and y is the sum of 1 and 2.
46. The product of 2 and x is the product of 4 and 6.
47. The difference x less 2 equals 6.
48. The difference 2 less y equals 6.
49. The quotient x divided by 2 is 6.
50. The quotient 2 divided by x is 6.

In Problems 51–88, evaluate each expression.

51. $9 - 4 + 2$ **52.** $6 - 4 + 3$ **53.** $-6 + 4 \cdot 3$ **54.** $8 - 4 \cdot 2$

55. $4 + 5 - 8$ **56.** $8 - 3 - 4$ **57.** $4 + \frac{1}{3}$ **58.** $2 - \frac{1}{2}$

59. $6 - [3 \cdot 5 + 2 \cdot (3 - 2)]$ **60.** $2 \cdot [8 - 3(4 + 2)] - 3$ **61.** $2 \cdot (3 - 5) + 8 \cdot 2 - 1$ **62.** $1 - (4 \cdot 3 - 2 + 2)$

63. $10 - [6 - 2 \cdot 2 + (8 - 3)] \cdot 2$ **64.** $2 - 5 \cdot 4 - [6 \cdot (3 - 4)]$

65. $(5 - 3)\frac{1}{2}$ **66.** $(5 + 4)\frac{1}{3}$ **67.** $\frac{4 + 8}{5 - 3}$ **68.** $\frac{2 - 4}{5 - 3}$

69. $\frac{3}{5} \cdot \frac{10}{21}$ **70.** $\frac{5}{9} \cdot \frac{3}{10}$ **71.** $\frac{6}{25} \cdot \frac{10}{27}$ **72.** $\frac{21}{25} \cdot \frac{100}{3}$

73. $\frac{3}{4} + \frac{2}{5}$ **74.** $\frac{4}{3} + \frac{1}{2}$ **75.** $\frac{5}{6} + \frac{9}{5}$ **76.** $\frac{8}{9} + \frac{15}{2}$

77. $\frac{5}{18} + \frac{1}{12}$ **78.** $\frac{2}{15} + \frac{8}{9}$ **79.** $\frac{1}{30} - \frac{7}{18}$ **80.** $\frac{3}{14} - \frac{2}{21}$

81. $\frac{3}{20} - \frac{2}{15}$ **82.** $\frac{6}{35} - \frac{3}{14}$ **83.** $\dfrac{\frac{5}{18}}{\frac{11}{27}}$ **84.** $\dfrac{\frac{5}{21}}{\frac{2}{35}}$

85. $\frac{1}{2} \cdot \frac{3}{5} + \frac{7}{10}$ **86.** $\frac{2}{3} + \frac{4}{5} \cdot \frac{1}{6}$ **87.** $2 \cdot \frac{3}{4} + \frac{3}{8}$ **88.** $3 \cdot \frac{5}{6} - \frac{1}{2}$

In Problems 89–100, use the Distributive Property to remove the parentheses.

89. $6(x + 4)$ **90.** $4(2x - 1)$ **91.** $x(x - 4)$ **92.** $4x(x + 3)$

93. $2\left(\frac{3}{4}x - \frac{1}{2}\right)$ **94.** $3\left(\frac{2}{3}x + \frac{1}{6}\right)$ **95.** $(x + 2)(x + 4)$ **96.** $(x + 5)(x + 1)$

97. $(x - 2)(x + 1)$ **98.** $(x - 4)(x + 1)$ **99.** $(x - 8)(x - 2)$ **100.** $(x - 4)(x - 2)$

Explaining Concepts: Discussion and Writing

101. Explain to a friend how the Distributive Property is used to justify the fact that $2x + 3x = 5x$.

102. Explain to a friend why $2 + 3 \cdot 4 = 14$, whereas $(2 + 3) \cdot 4 = 20$.

103. Explain why $2(3 \cdot 4)$ is not equal to $(2 \cdot 3) \cdot (2 \cdot 4)$.

104. Explain why $\frac{4 + 3}{2 + 5}$ is not equal to $\frac{4}{2} + \frac{3}{5}$.

105. Is subtraction commutative? Support your conclusion with an example.

106. Is subtraction associative? Support your conclusion with an example.

107. Is division commutative? Support your conclusion with an example.

108. Is division associative? Support your conclusion with an example.

109. If $2 = x$, why does $x = 2$?

110. If $x = 5$, why does $x^2 + x = 30$?

111. Are there any real numbers that are both rational and irrational? Are there any real numbers that are neither? Explain your reasoning.

112. Explain why the sum of a rational number and an irrational number must be irrational.

113. A rational number is defined as the quotient of two integers. When written as a decimal, the decimal will either repeat or terminate. By looking at the denominator of the rational number, there is a way to tell in advance whether its decimal representation will repeat or terminate. Make a list of rational numbers and their decimals. See if you can discover the pattern. Confirm your conclusion by consulting books on number theory at the library. Write a brief essay on your findings.

114. The current time is 12 noon CST. What time (CST) will it be 12,997 hours from now?

115. Both $\frac{a}{0}$ $(a \neq 0)$ and $\frac{0}{0}$ are undefined, but for different reasons. Write a paragraph or two explaining the different reasons.

R.2 Algebra Essentials

OBJECTIVES

1. Graph Inequalities (p. 18)
2. Find Distance on the Real Number Line (p. 19)
3. Evaluate Algebraic Expressions (p. 20)
4. Determine the Domain of a Variable (p. 21)
5. Use the Laws of Exponents (p. 21)
6. Evaluate Square Roots (p. 23)
7. Use a Calculator to Evaluate Exponents (p. 24)
8. Use Scientific Notation (p. 24)

The Real Number Line

Real numbers can be represented by points on a line called the **real number line**. There is a one-to-one correspondence between real numbers and points on a line. That is, every real number corresponds to a point on the line, and each point on the line has a unique real number associated with it.

Pick a point on a line somewhere in the center, and label it O. This point, called the **origin**, corresponds to the real number 0. See Figure 8. The point 1 unit to the right of O corresponds to the number 1. The distance between 0 and 1 determines the **scale** of the number line. For example, the point associated with the number 2 is twice as far from O as 1. Notice that an arrowhead on the right end of the line indicates the direction in which the numbers increase. Points to the left of the origin correspond to the real numbers -1, -2, and so on. Figure 8 also shows the points associated with the rational numbers $-\frac{1}{2}$ and $\frac{1}{2}$ and with the irrational numbers $\sqrt{2}$ and π.

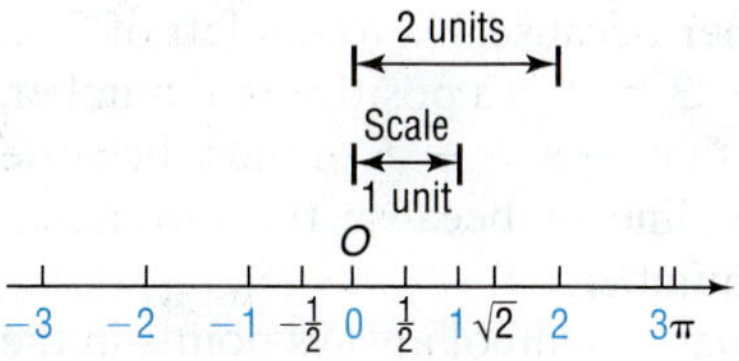

Figure 8 Real number line

DEFINITION

The real number associated with a point P is called the **coordinate** of P, and the line whose points have been assigned coordinates is called the **real number line**.

Now Work PROBLEM 13

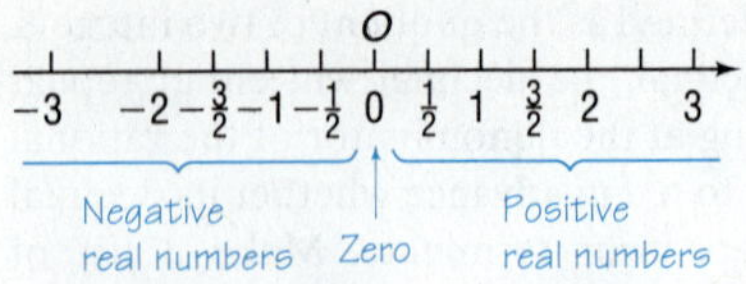

Figure 9

The real number line consists of three classes of real numbers, as shown in Figure 9.

1. The **negative real numbers** are the coordinates of points to the left of the origin O.
2. The real number **zero** is the coordinate of the origin O.
3. The **positive real numbers** are the coordinates of points to the right of the origin O.

Multiplication Properties of Positive and Negative Numbers

1. The product of two positive numbers is a positive number.
2. The product of two negative numbers is a positive number.
3. The product of a positive number and a negative number is a negative number.

1 Graph Inequalities

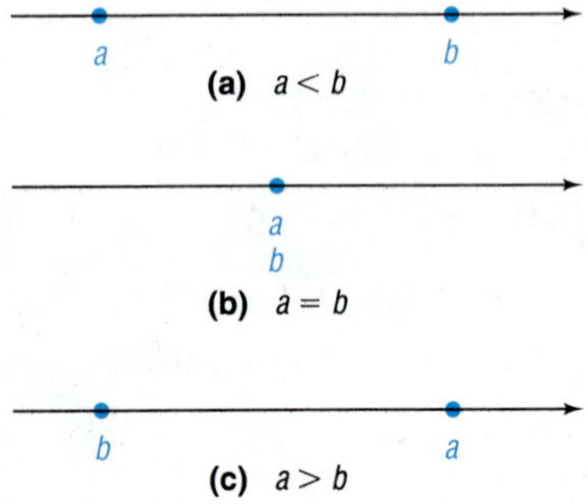

Figure 10

An important property of the real number line follows from the fact that, given two numbers (points) a and b, either a is to the left of b, or a is at the same location as b, or a is to the right of b. See Figure 10.

If a is to the left of b, then "a is less than b," which is written $a < b$. If a is to the right of b, then "a is greater than b," which is written $a > b$. If a is at the same location as b, then $a = b$. If a is either less than or equal to b, then $a \leq b$. Similarly, $a \geq b$ means that a is either greater than or equal to b. Collectively, the symbols $<, >, \leq$, and $\geq$ are called **inequality symbols**.

Note that $a < b$ and $b > a$ mean the same thing. It does not matter whether we write $2 < 3$ or $3 > 2$.

Furthermore, if $a < b$ or if $b > a$, then the difference $b - a$ is positive. Do you see why?

EXAMPLE 1 **Using Inequality Symbols**

(a) $3 < 7$ (b) $-8 > -16$ (c) $-6 < 0$
(d) $-8 < -4$ (e) $4 > -1$ (f) $8 > 0$

In Example 1(a), we conclude that $3 < 7$ either because 3 is to the left of 7 on the real number line or because the difference, $7 - 3 = 4$, is a positive real number.

Similarly, we conclude in Example 1(b) that $-8 > -16$ either because -8 lies to the right of -16 on the real number line or because the difference, $-8 - (-16) = -8 + 16 = 8$, is a positive real number.

Look again at Example 1. Note that the inequality symbol always points in the direction of the smaller number.

An **inequality** is a statement in which two expressions are related by an inequality symbol. The expressions are referred to as the **sides** of the inequality. Inequalities of the form $a < b$ or $b > a$ are called **strict inequalities**, whereas inequalities of the form $a \leq b$ or $b \geq a$ are called **nonstrict inequalities**.

Based on the discussion so far, we conclude that

$a > 0$ is equivalent to a is positive

$a < 0$ is equivalent to a is negative

We sometimes read $a > 0$ by saying that "a is positive." If $a \geq 0$, then either $a > 0$ or $a = 0$, and we may read this as "a is nonnegative."

Now Work PROBLEMS 17 AND 27

EXAMPLE 2

Graphing Inequalities

(a) On the real number line, graph all numbers x for which $x > 4$.
(b) On the real number line, graph all numbers x for which $x \leq 5$.

Solution

(a) See Figure 11. Notice that we use a left parenthesis to indicate that the number 4 is *not* part of the graph.

(b) See Figure 12. Notice that we use a right bracket to indicate that the number 5 *is* part of the graph.

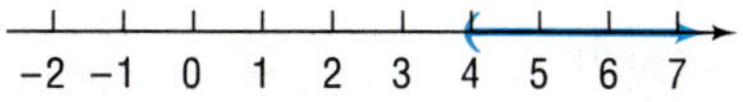

Figure 11 $x > 4$

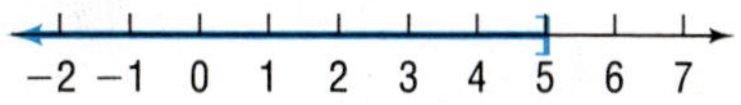

Figure 12 $x \leq 5$

Now Work PROBLEM 33

2 Find Distance on the Real Number Line

The *absolute value* of a number a is the distance from 0 to a on the number line. For example, -4 is 4 units from 0, and 3 is 3 units from 0. See Figure 13. That is, the absolute value of -4 is 4, and the absolute value of 3 is 3.

A more formal definition of absolute value is given next.

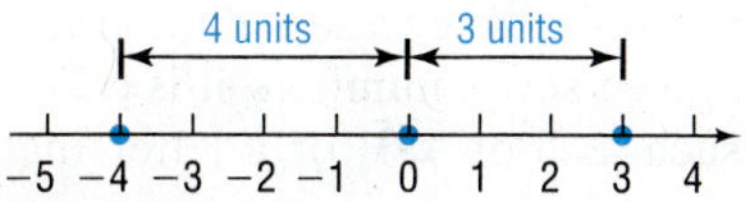

Figure 13

DEFINITION

The **absolute value** of a real number a, denoted by the symbol $|a|$, is defined by the rules

$$|a| = a \quad \text{if } a \geq 0 \qquad \text{and} \qquad |a| = -a \quad \text{if } a < 0$$

For example, because $-4 < 0$, the second rule must be used to get $|-4| = -(-4) = 4$.

EXAMPLE 3

Computing Absolute Value

(a) $|8| = 8$ (b) $|0| = 0$ (c) $|-15| = -(-15) = 15$

Look again at Figure 13. The distance from -4 to 3 is 7 units. This distance is the difference $3 - (-4)$, obtained by subtracting the smaller coordinate from the larger. However, since $|3 - (-4)| = |7| = 7$ and $|-4 - 3| = |-7| = 7$, we can use absolute value to calculate the distance between two points without being concerned about which is smaller.

DEFINITION

If P and Q are two points on a real number line with coordinates a and b, respectively, the **distance between P and Q**, denoted by $d(P, Q)$, is

$$d(P, Q) = |b - a|$$

Since $|b - a| = |a - b|$, it follows that $d(P, Q) = d(Q, P)$.

EXAMPLE 4 **Finding Distance on a Number Line**

Let P, Q, and R be points on a real number line with coordinates -5, 7, and -3, respectively. Find the distance

(a) between P and Q (b) between Q and R

Solution See Figure 14.

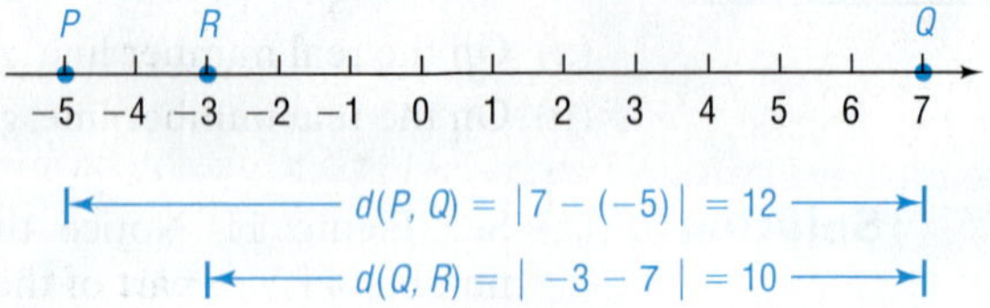

Figure 14

(a) $d(P, Q) = |7 - (-5)| = |12| = 12$
(b) $d(Q, R) = |-3 - 7| = |-10| = 10$

Now Work PROBLEM 39

3 Evaluate Algebraic Expressions

Remember, in algebra we use letters such as x, y, a, b, and c to represent numbers. If the letter used is to represent *any* number from a given set of numbers, it is called a **variable.** A **constant** is either a fixed number, such as 5 or $\sqrt{3}$, or a letter that represents a fixed (possibly unspecified) number.

Constants and variables are combined using the operations of addition, subtraction, multiplication, and division to form *algebraic expressions.* Examples of algebraic expressions include

$$x + 3 \qquad \frac{3}{1 - t} \qquad 7x - 2y$$

To evaluate an algebraic expression, substitute a numerical value for each variable.

EXAMPLE 5 **Evaluating an Algebraic Expression**

Evaluate each expression if $x = 3$ and $y = -1$.

(a) $x + 3y$ (b) $5xy$ (c) $\dfrac{3y}{2 - 2x}$ (d) $|-4x + y|$

Solution (a) Substitute 3 for x and -1 for y in the expression $x + 3y$.

$$x + 3y = 3 + 3(-1) = 3 + (-3) = 0$$

↑ $x = 3, y = -1$

(b) If $x = 3$ and $y = -1$, then

$$5xy = 5(3)(-1) = -15$$

(c) If $x = 3$ and $y = -1$, then

$$\frac{3y}{2 - 2x} = \frac{3(-1)}{2 - 2(3)} = \frac{-3}{2 - 6} = \frac{-3}{-4} = \frac{3}{4}$$

(d) If $x = 3$ and $y = -1$, then

$$|-4x + y| = |-4(3) + (-1)| = |-12 + (-1)| = |-13| = 13$$

Now Work PROBLEMS 41 AND 49

4 Determine the Domain of a Variable

In working with expressions or formulas involving variables, the variables may be allowed to take on values from only a certain set of numbers. For example, in the formula for the area A of a circle of radius r, $A = \pi r^2$, the variable r is necessarily restricted to the positive real numbers. In the expression $\frac{1}{x}$, the variable x cannot take on the value 0, since division by 0 is not defined.

DEFINITION

The set of values that a variable may assume is called the **domain of the variable**.

EXAMPLE 6 **Finding the Domain of a Variable**

The domain of the variable x in the expression

$$\frac{5}{x - 2}$$

is $\{x|x \neq 2\}$ since, if $x = 2$, the denominator becomes 0, which is not defined. •

EXAMPLE 7 **Circumference of a Circle**

In the formula for the circumference C of a circle of radius r,

$$C = 2\pi r$$

the domain of the variable r, representing the radius of the circle, is the set of positive real numbers, $\{r|r > 0\}$. The domain of the variable C, representing the circumference of the circle, is also the set of positive real numbers, $\{C|C > 0\}$. •

In describing the domain of a variable, we may use either set notation or words, whichever is more convenient.

Now Work PROBLEM 59

5 Use the Laws of Exponents

Integer exponents provide a shorthand notation for representing repeated multiplications of a real number. For example,

$$3^4 = 3 \cdot 3 \cdot 3 \cdot 3 = 81$$

Additionally, many formulas have exponents. For example,

- The formula for the horsepower rating H of an engine is

$$H = \frac{D^2 N}{2.5}$$

where D is the diameter of a cylinder and N is the number of cylinders.

- A formula for the resistance R of blood flowing in a blood vessel is

$$R = C\frac{L}{r^4}$$

where L is the length of the blood vessel, r is the radius, and C is a positive constant.

DEFINITION

If a is a real number and n is a positive integer, then the symbol a^n represents the product of n factors of a. That is,

$$a^n = \underbrace{a \cdot a \cdot \ldots \cdot a}_{n \text{ factors}} \qquad (1)$$

WARNING Be careful with minus signs and exponents.

$$-2^4 = -1 \cdot 2^4 = -16$$

whereas

$$(-2)^4 = (-2)(-2)(-2)(-2) = 16$$

In the definition it is understood that $a^1 = a$. Furthermore, $a^2 = a \cdot a$, $a^3 = a \cdot a \cdot a$, and so on. In the expression a^n, a is called the **base** and n is called the **exponent**, or **power**. We read a^n as "a raised to the power n" or as "a to the nth power." We usually read a^2 as "a squared" and a^3 as "a cubed."

In working with exponents, the operation of *raising to a power* is performed before any other operation. As examples,

$$4 \cdot 3^2 = 4 \cdot 9 = 36 \qquad 2^2 + 3^2 = 4 + 9 = 13$$
$$-2^4 = -16 \qquad 5 \cdot 3^2 + 2 \cdot 4 = 5 \cdot 9 + 2 \cdot 4 = 45 + 8 = 53$$

Parentheses are used to indicate operations to be performed first. For example,

$$(-2)^4 = (-2)(-2)(-2)(-2) = 16 \qquad (2 + 3)^2 = 5^2 = 25$$

DEFINITION

If $a \neq 0$, then

$$a^0 = 1$$

DEFINITION

If $a \neq 0$ and if n is a positive integer, then

$$a^{-n} = \frac{1}{a^n}$$

Whenever you encounter a negative exponent, think "reciprocal."

EXAMPLE 8 Evaluating Expressions Containing Negative Exponents

(a) $2^{-3} = \frac{1}{2^3} = \frac{1}{8}$ (b) $x^{-4} = \frac{1}{x^4}$ (c) $\left(\frac{1}{5}\right)^{-2} = \frac{1}{\left(\frac{1}{5}\right)^2} = \frac{1}{\frac{1}{25}} = 25$

Now Work PROBLEMS 77 AND 97

The following properties, called the **Laws of Exponents**, can be proved using the preceding definitions. In the list, a and b are real numbers, and m and n are integers.

THEOREM

Laws of Exponents

$$a^m a^n = a^{m+n} \qquad (a^m)^n = a^{mn} \qquad (ab)^n = a^n b^n$$

$$\frac{a^m}{a^n} = a^{m-n} = \frac{1}{a^{n-m}} \text{ if } a \neq 0 \qquad \left(\frac{a}{b}\right)^n = \frac{a^n}{b^n} \text{ if } b \neq 0$$

EXAMPLE 9 Using the Laws of Exponents

(a) $x^{-3} \cdot x^5 = x^{-3+5} = x^2 \quad x \neq 0$

(b) $(x^{-3})^2 = x^{-3 \cdot 2} = x^{-6} = \dfrac{1}{x^6} \quad x \neq 0$

(c) $(2x)^3 = 2^3 \cdot x^3 = 8x^3$

(d) $\left(\dfrac{2}{3}\right)^4 = \dfrac{2^4}{3^4} = \dfrac{16}{81}$

(e) $\dfrac{x^{-2}}{x^{-5}} = x^{-2-(-5)} = x^3 \quad x \neq 0$

Now Work PROBLEM 79

EXAMPLE 10 Using the Laws of Exponents

Write each expression so that all exponents are positive.

(a) $\dfrac{x^5y^{-2}}{x^3y} \quad x \neq 0, \quad y \neq 0$

(b) $\left(\dfrac{x^{-3}}{3y^{-1}}\right)^{-2} \quad x \neq 0, \quad y \neq 0$

Solution

(a) $\dfrac{x^5y^{-2}}{x^3y} = \dfrac{x^5}{x^3} \cdot \dfrac{y^{-2}}{y} = x^{5-3} \cdot y^{-2-1} = x^2y^{-3} = x^2 \cdot \dfrac{1}{y^3} = \dfrac{x^2}{y^3}$

(b) $\left(\dfrac{x^{-3}}{3y^{-1}}\right)^{-2} = \dfrac{(x^{-3})^{-2}}{(3y^{-1})^{-2}} = \dfrac{x^6}{3^{-2}(y^{-1})^{-2}} = \dfrac{x^6}{\frac{1}{9}y^2} = \dfrac{9x^6}{y^2}$

Now Work PROBLEM 89

6 Evaluate Square Roots

> **In Words**
> $\sqrt{36}$ means "give me the nonnegative number whose square is 36."

A real number is squared when it is raised to the power 2. The inverse of squaring is finding a **square root**. For example, since $6^2 = 36$ and $(-6)^2 = 36$, the numbers 6 and -6 are square roots of 36.

The symbol $\sqrt{\ }$, called a **radical sign**, is used to denote the **principal**, or nonnegative, square root. For example, $\sqrt{36} = 6$.

DEFINITION

If a is a nonnegative real number, the nonnegative number b such that $b^2 = a$ is the **principal square root** of a, and is denoted by $b = \sqrt{a}$.

The following comments are noteworthy:

1. Negative numbers do not have square roots (in the real number system), because the square of any real number is *nonnegative*. For example, $\sqrt{-4}$ is not a real number, because there is no real number whose square is -4.
2. The principal square root of 0 is 0, since $0^2 = 0$. That is, $\sqrt{0} = 0$.
3. The principal square root of a positive number is positive.
4. If $c \geq 0$, then $(\sqrt{c})^2 = c$. For example, $(\sqrt{2})^2 = 2$ and $(\sqrt{3})^2 = 3$.

EXAMPLE 11 Evaluating Square Roots

(a) $\sqrt{64} = 8$

(b) $\sqrt{\dfrac{1}{16}} = \dfrac{1}{4}$

(c) $(\sqrt{1.4})^2 = 1.4$

Examples 11(a) and (b) are examples of square roots of perfect squares, since $64 = 8^2$ and $\dfrac{1}{16} = \left(\dfrac{1}{4}\right)^2$.

Consider the expression $\sqrt{a^2}$. Since $a^2 \ge 0$, the principal square root of a^2 is defined whether $a > 0$ or $a < 0$. However, since the principal square root is nonnegative, we need an absolute value to ensure the nonnegative result. That is,

$$\sqrt{a^2} = |a| \qquad a \text{ any real number} \tag{2}$$

EXAMPLE 12

Using Equation (2)

(a) $\sqrt{(2.3)^2} = |2.3| = 2.3$ (b) $\sqrt{(-2.3)^2} = |-2.3| = 2.3$

(c) $\sqrt{x^2} = |x|$ ●

Now Work PROBLEM 85

7 Use a Calculator to Evaluate Exponents

Your calculator has either a caret key, $\boxed{\wedge}$, or an $\boxed{x^y}$ key, that is used for computations involving exponents.

EXAMPLE 13

Exponents on a Graphing Calculator

Evaluate: $(2.3)^5$

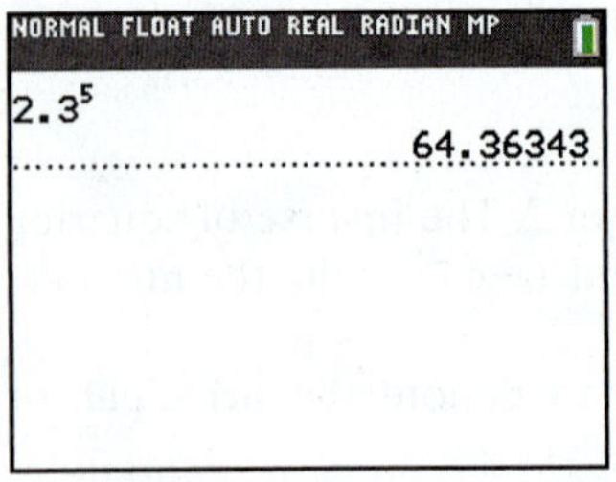

Figure 15

Solution Figure 15 shows the result using a TI-84 Plus C graphing calculator. ●

Now Work PROBLEM 115

8 Use Scientific Notation

Measurements of physical quantities can range from very small to very large. For example, the mass of a proton is approximately 0.00000000000000000000000000167 kilogram and the mass of Earth is about 5,980,000,000,000,000,000,000,000 kilograms. These numbers obviously are tedious to write down and difficult to read, so we use exponents to rewrite them.

DEFINITION

When a number has been written as the product of a number x, where $1 \le x < 10$, times a power of 10, it is said to be written in **scientific notation.**

In scientific notation,

$$\text{Mass of a proton} = 1.67 \times 10^{-27} \text{ kilogram}$$

$$\text{Mass of Earth} = 5.98 \times 10^{24} \text{ kilograms}$$

Converting a Decimal to Scientific Notation

To change a positive number into scientific notation:

1. Count the number N of places that the decimal point must be moved to arrive at a number x, where $1 \le x < 10$.
2. If the original number is greater than or equal to 1, the scientific notation is $x \times 10^N$. If the original number is between 0 and 1, the scientific notation is $x \times 10^{-N}$.

EXAMPLE 14 Using Scientific Notation

Write each number in scientific notation.

(a) 9582 (b) 1.245 (c) 0.285 (d) 0.000561

Solution (a) The decimal point in 9582 follows the 2. Count left from the decimal point

9 5 8 2 .
3 2 1

stopping after three moves, because 9.582 is a number between 1 and 10. Since 9582 is greater than 1, we write

$$9582 = 9.582 \times 10^3$$

(b) The decimal point in 1.245 is between the 1 and 2. Since the number is already between 1 and 10, the scientific notation for it is $1.245 \times 10^0 = 1.245$.

(c) The decimal point in 0.285 is between the 0 and the 2. We count

0 . 2 8 5
1

stopping after one move, because 2.85 is a number between 1 and 10. Since 0.285 is between 0 and 1, we write

$$0.285 = 2.85 \times 10^{-1}$$

(d) The decimal point in 0.000561 is moved as follows:

0 . 0 0 0 5 6 1
1 2 3 4

As a result,

$$0.000561 = 5.61 \times 10^{-4}$$

Now Work PROBLEM 121

EXAMPLE 15 Changing from Scientific Notation to Decimals

Write each number as a decimal.

(a) 2.1×10^4 (b) 3.26×10^{-5} (c) 1×10^{-2}

Solution

(a) $2.1 \times 10^4 = 2\ .\ 1\ 0\ 0\ 0 \times 10^4 = 21{,}000$
1 2 3 4

(b) $3.26 \times 10^{-5} = 0\ 0\ 0\ 0\ 0\ 3\ .\ 2\ 6 \times 10^{-5} = 0.0000326$
5 4 3 2 1

(c) $1 \times 10^{-2} = 0\ 0\ 1\ .\ 0 \times 10^{-2} = 0.01$
2 1

Now Work PROBLEM 129

EXAMPLE 16 **Using Scientific Notation**

(a) The diameter of the smallest living cell is only about 0.00001 centimeter (cm).* Express this number in scientific notation.

(b) The surface area of Earth is about 1.97×10^8 square miles.† Express the surface area as a whole number.

Solution (a) 0.00001 cm $= 1 \times 10^{-5}$ cm because the decimal point is moved five places and the number is less than 1.

(b) 1.97×10^8 square miles $= 197{,}000{,}000$ square miles. ●

Now Work PROBLEM 155

COMMENT On a calculator, a number such as 3.615×10^{12} is usually displayed as 3.615E12. ■

* *Powers of Ten,* Philip and Phylis Morrison.

† *2011 Information Please Almanac.*

Historical Feature

The word *algebra* is derived from the Arabic word *al-jabr*. This word is a part of the title of a ninth-century work, "Hisâb al-jabr w'al-muqâbalah," written by Mohammed ibn Músâ al-Khwârizmî. The word *al-jabr* means "a restoration," a reference to the fact that if a number is added to one side of an equation, then it must also be added to the other side in order to "restore" the equality. The title of the work, freely translated, is "The Science of Reduction and Cancellation." Of course, today, algebra has come to mean a great deal more.

R.2 Assess Your Understanding

Concepts and Vocabulary

1. A(n) ________ is a letter used in algebra to represent any number from a given set of numbers.
2. On the real number line, the real number zero is the coordinate of the ______.
3. An inequality of the form $a > b$ is called a(n) ______ inequality.
4. In the expression 2^4, the number 2 is called the ____ and 4 is called the ________________.
5. In scientific notation, $1234.5678 =$ ______________.
6. If a is a nonnegative real number, then which inequality statement best describes a?
   (a) $a < 0$ (b) $a > 0$ (c) $a \le 0$ (d) $a \ge 0$
7. Let a and b be non-zero real numbers and m and n be integers. Which of the following is not a law of exponents?
   (a) $\left(\dfrac{a}{b}\right)^n = \dfrac{a^n}{b^n}$ (b) $(a^m)^n = a^{m+n}$
   (c) $\dfrac{a^m}{a^n} = a^{m-n}$ (d) $(ab)^n = a^n b^n$
8. ***True or False*** The product of two negative real numbers is always greater than zero.
9. ***True or False*** The distance between two distinct points on the real number line is always greater than zero.
10. ***True or False*** The absolute value of a real number is always greater than zero.
11. ***True or False*** When a number is expressed in scientific notation, it is expressed as the product of a number x, $0 \le x < 1$, and a power of 10.
12. ***True or False*** To multiply two expressions having the same base, retain the base and multiply the exponents.

Skill Building

13. On the real number line, label the points with coordinates $0, 1, -1, \dfrac{5}{2}, -2.5, \dfrac{3}{4}$, and 0.25.

14. Repeat Problem 13 for the coordinates $0, -2, 2, -1.5, \dfrac{3}{2}, \dfrac{1}{3}$, and $\dfrac{2}{3}$.

In Problems 15–24, replace the question mark by $<$, $>$, or $=$, whichever is correct.

15. $\dfrac{1}{2}\ ?\ 0$
16. $5\ ?\ 6$
17. $-1\ ?\ -2$
18. $-3\ ?\ -\dfrac{5}{2}$
19. $\pi\ ?\ 3.14$
20. $\sqrt{2}\ ?\ 1.41$
21. $\dfrac{1}{2}\ ?\ 0.5$
22. $\dfrac{1}{3}\ ?\ 0.33$
23. $\dfrac{2}{3}\ ?\ 0.67$
24. $\dfrac{1}{4}\ ?\ 0.25$

In Problems 25–30, write each statement as an inequality.

25. x is positive **26.** z is negative **27.** x is less than 2

28. y is greater than -5 **29.** x is less than or equal to 1 **30.** x is greater than or equal to 2

In Problems 31–34, graph the numbers x on the real number line.

31. $x \geq -2$ **32.** $x < 4$ **33.** $x > -1$ **34.** $x \leq 7$

In Problems 35–40, use the given real number line to compute each distance.

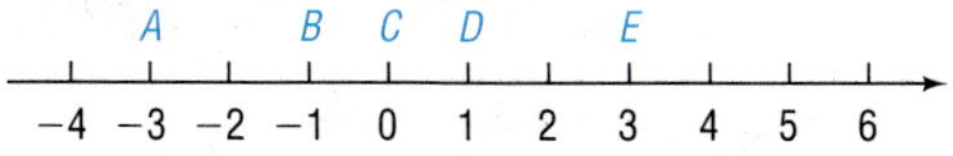

35. $d(C, D)$ **36.** $d(C, A)$ **37.** $d(D, E)$ **38.** $d(C, E)$ **39.** $d(A, E)$ **40.** $d(D, B)$

In Problems 41–48, evaluate each expression if $x = -2$ and $y = 3$.

41. $x + 2y$ **42.** $3x + y$ **43.** $5xy + 2$ **44.** $-2x + xy$

45. $\dfrac{2x}{x - y}$ **46.** $\dfrac{x + y}{x - y}$ **47.** $\dfrac{3x + 2y}{2 + y}$ **48.** $\dfrac{2x - 3}{y}$

In Problems 49–58, find the value of each expression if $x = 3$ and $y = -2$.

49. $|x + y|$ **50.** $|x - y|$ **51.** $|x| + |y|$ **52.** $|x| - |y|$ **53.** $\dfrac{|x|}{x}$

54. $\dfrac{|y|}{y}$ **55.** $|4x - 5y|$ **56.** $|3x + 2y|$ **57.** $||4x| - |5y||$ **58.** $3|x| + 2|y|$

In Problems 59–66, determine which of the values (a) through (d), if any, must be excluded from the domain of the variable in each expression.

(a) $x = 3$ (b) $x = 1$ (c) $x = 0$ (d) $x = -1$

59. $\dfrac{x^2 - 1}{x}$ **60.** $\dfrac{x^2 + 1}{x}$ **61.** $\dfrac{x}{x^2 - 9}$ **62.** $\dfrac{x}{x^2 + 9}$

63. $\dfrac{x^2}{x^2 + 1}$ **64.** $\dfrac{x^3}{x^2 - 1}$ **65.** $\dfrac{x^2 + 5x - 10}{x^3 - x}$ **66.** $\dfrac{-9x^2 - x + 1}{x^3 + x}$

In Problems 67–70, determine the domain of the variable x in each expression.

67. $\dfrac{4}{x - 5}$ **68.** $\dfrac{-6}{x + 4}$ **69.** $\dfrac{x}{x + 4}$ **70.** $\dfrac{x - 2}{x - 6}$

In Problems 71–74, use the formula $C = \frac{5}{9}(F - 32)$ for converting degrees Fahrenheit into degrees Celsius to find the Celsius measure of each Fahrenheit temperature.

71. $F = 32°$ **72.** $F = 212°$ **73.** $F = 77°$ **74.** $F = -4°$

In Problems 75–86, simplify each expression.

75. $(-4)^2$ **76.** -4^2 **77.** 4^{-2} **78.** -4^{-2} **79.** $3^{-6} \cdot 3^4$ **80.** $4^{-2} \cdot 4^3$

81. $(3^{-2})^{-1}$ **82.** $(2^{-1})^{-3}$ **83.** $\sqrt{25}$ **84.** $\sqrt{36}$ **85.** $\sqrt{(-4)^2}$ **86.** $\sqrt{(-3)^2}$

In Problems 87–96, simplify each expression. Express the answer so that all exponents are positive. Whenever an exponent is 0 or negative, we assume that the base is not 0.

87. $(8x^3)^2$ **88.** $(-4x^2)^{-1}$ **89.** $(x^2y^{-1})^2$ **90.** $(x^{-1}y)^3$ **91.** $\dfrac{x^2y^3}{xy^4}$

92. $\dfrac{x^{-2}y}{xy^2}$ **93.** $\dfrac{(-2)^3x^4(yz)^2}{3^2xy^3z}$ **94.** $\dfrac{4x^{-2}(yz)^{-1}}{2^3x^4y}$ **95.** $\left(\dfrac{3x^{-1}}{4y^{-1}}\right)^{-2}$ **96.** $\left(\dfrac{5x^{-2}}{6y^{-2}}\right)^{-3}$

In Problems 97–108, find the value of each expression if $x = 2$ and $y = -1$.

97. $2xy^{-1}$ 98. $-3x^{-1}y$ 99. $x^2 + y^2$ 100. x^2y^2

101. $(xy)^2$ 102. $(x + y)^2$ 103. $\sqrt{x^2}$ 104. $(\sqrt{x})^2$

105. $\sqrt{x^2 + y^2}$ 106. $\sqrt{x^2} + \sqrt{y^2}$ 107. x^y 108. y^x

109. Find the value of the expression $2x^3 - 3x^2 + 5x - 4$ if $x = 2$. What is the value if $x = 1$?

110. Find the value of the expression $4x^3 + 3x^2 - x + 2$ if $x = 1$. What is the value if $x = 2$?

111. What is the value of $\dfrac{(666)^4}{(222)^4}$?

112. What is the value of $(0.1)^3(20)^3$?

In Problems 113–120, use a calculator to evaluate each expression. Round your answer to three decimal places.

113. $(8.2)^6$ 114. $(3.7)^5$ 115. $(6.1)^{-3}$ 116. $(2.2)^{-5}$

117. $(-2.8)^6$ 118. $-(2.8)^6$ 119. $(-8.11)^{-4}$ 120. $-(8.11)^{-4}$

In Problems 121–128, write each number in scientific notation.

121. 454.2 122. 32.14 123. 0.013 124. 0.00421

125. 32,155 126. 21,210 127. 0.000423 128. 0.0514

In Problems 129–136, write each number as a decimal.

129. 6.15×10^4 130. 9.7×10^3 131. 1.214×10^{-3} 132. 9.88×10^{-4}

133. 1.1×10^8 134. 4.112×10^2 135. 8.1×10^{-2} 136. 6.453×10^{-1}

Applications and Extensions

In Problems 137–146, express each statement as an equation involving the indicated variables.

137. **Area of a Rectangle** The area A of a rectangle is the product of its length l and its width w.

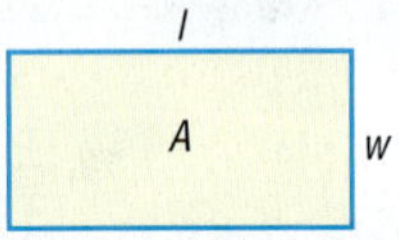

138. **Perimeter of a Rectangle** The perimeter P of a rectangle is twice the sum of its length l and its width w.

139. **Circumference of a Circle** The circumference C of a circle is the product of π and its diameter d.

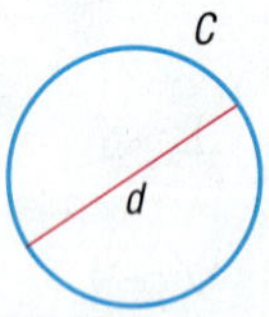

140. **Area of a Triangle** The area A of a triangle is one-half the product of its base b and its height h.

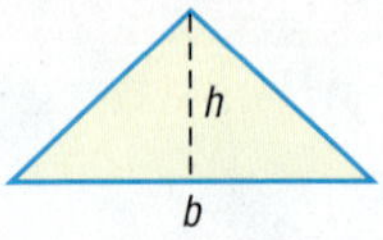

141. **Area of an Equilateral Triangle** The area A of an equilateral triangle is $\dfrac{\sqrt{3}}{4}$ times the square of the length x of one side.

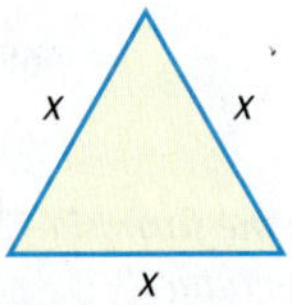

142. **Perimeter of an Equilateral Triangle** The perimeter P of an equilateral triangle is 3 times the length x of one side.

143. **Volume of a Sphere** The volume V of a sphere is $\dfrac{4}{3}$ times π times the cube of the radius r.

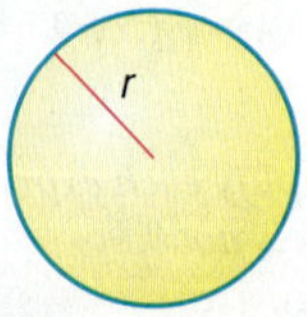

144. **Surface Area of a Sphere** The surface area S of a sphere is 4 times π times the square of the radius r.

145. Volume of a Cube The volume V of a cube is the cube of the length x of a side.

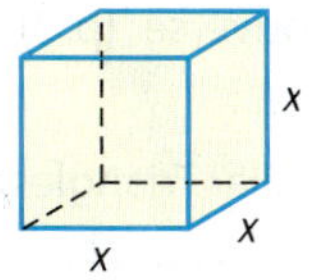

146. Surface Area of a Cube The surface area S of a cube is 6 times the square of the length x of a side.

147. Manufacturing Cost The weekly production cost C of manufacturing x watches is given by the formula $C = 4000 + 2x$, where the variable C is in dollars.
(a) What is the cost of producing 1000 watches?
(b) What is the cost of producing 2000 watches?

148. Balancing a Checkbook At the beginning of the month, Mike had a balance of \$210 in his checking account. During the next month, he deposited \$80, wrote a check for \$120, made another deposit of \$25, and wrote two checks: one for \$60 and the other for \$32. He was also assessed a monthly service charge of \$5. What was his balance at the end of the month?

In Problems 149 and 150, write an inequality using an absolute value to describe each statement.

149. x is at least 6 units from 4.

150. x is more than 5 units from 2.

151. U.S. Voltage In the United States, normal household voltage is 110 volts. It is acceptable for the actual voltage x to differ from normal by at most 5 volts. A formula that describes this is

$$|x - 110| \le 5$$

(a) Show that a voltage of 108 volts is acceptable.
(b) Show that a voltage of 104 volts is not acceptable.

152. Foreign Voltage In other countries, normal household voltage is 220 volts. It is acceptable for the actual voltage x to differ from normal by at most 8 volts. A formula that describes this is

$$|x - 220| \le 8$$

(a) Show that a voltage of 214 volts is acceptable.
(b) Show that a voltage of 209 volts is not acceptable.

153. Making Precision Ball Bearings The FireBall Company manufactures ball bearings for precision equipment. One of its products is a ball bearing with a stated radius of 3 centimeters (cm). Only ball bearings with a radius within 0.01 cm of this stated radius are acceptable. If x is the radius of a ball bearing, a formula describing this situation is

$$|x - 3| \le 0.01$$

(a) Is a ball bearing of radius $x = 2.999$ acceptable?
(b) Is a ball bearing of radius $x = 2.89$ acceptable?

154. Body Temperature Normal human body temperature is 98.6°F. A temperature x that differs from normal by at least 1.5°F is considered unhealthy. A formula that describes this is

$$|x - 98.6| \ge 1.5$$

(a) Show that a temperature of 97°F is unhealthy.
(b) Show that a temperature of 100°F is not unhealthy.

155. Distance from Earth to Its Moon The distance from Earth to the Moon is about 4×10^8 meters.* Express this distance as a whole number.

156. Height of Mt. Everest The height of Mt. Everest is 8848 meters.* Express this height in scientific notation.

157. Wavelength of Visible Light The wavelength of visible light is about 5×10^{-7} meter.* Express this wavelength as a decimal.

158. Diameter of an Atom The diameter of an atom is about 1×10^{-10} meter.* Express this diameter as a decimal.

159. Diameter of Copper Wire The smallest commercial copper wire is about 0.0005 inch in diameter.† Express this diameter using scientific notation.

160. Smallest Motor The smallest motor ever made is less than 0.05 centimeter wide.† Express this width using scientific notation.

161. Astronomy One light-year is defined by astronomers to be the distance that a beam of light will travel in 1 year (365 days). If the speed of light is 186,000 miles per second, how many miles are in a light-year? Express your answer in scientific notation.

162. Astronomy How long does it take a beam of light to reach Earth from the Sun when the Sun is 93,000,000 miles from Earth? Express your answer in seconds, using scientific notation.

163. Does $\frac{1}{3}$ equal 0.333? If not, which is larger? By how much?

164. Does $\frac{2}{3}$ equal 0.666? If not, which is larger? By how much?

Explaining Concepts: Discussion and Writing

165. Is there a positive real number "closest" to 0?

166. Number game I'm thinking of a number! It lies between 1 and 10; its square is rational and lies between 1 and 10. The number is larger than π. Correct to two decimal places (that is, truncated to two decimal places), name the number. Now think of your own number, describe it, and challenge a fellow student to name it.

167. Write a brief paragraph that illustrates the similarities and differences between "less than" ($<$) and "less than or equal to" ($\le$).

168. Give a reason why the statement $5 < 8$ is true.

* *Powers of Ten*, Philip and Phylis Morrison.
† *2011 Information Please Almanac.*

R.3 Geometry Essentials

OBJECTIVES
1 Use the Pythagorean Theorem and Its Converse (p. 30)
2 Know Geometry Formulas (p. 31)
3 Understand Congruent Triangles and Similar Triangles (p. 32)

1 Use the Pythagorean Theorem and Its Converse

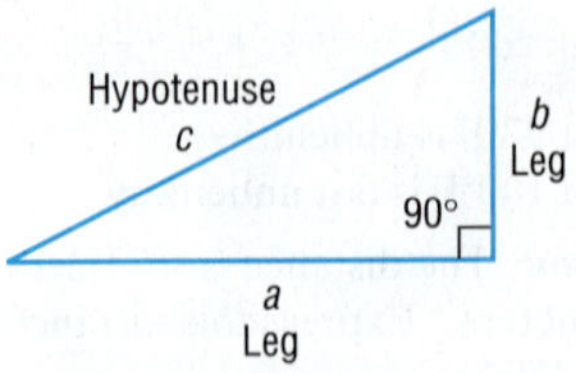

Figure 16 A right triangle

The *Pythagorean Theorem* is a statement about *right triangles*. A **right triangle** is one that contains a **right angle**—that is, an angle of 90°. The side of the triangle opposite the 90° angle is called the **hypotenuse**; the remaining two sides are called **legs**. In Figure 16 we have used c to represent the length of the hypotenuse and a and b to represent the lengths of the legs. Notice the use of the symbol ⊾ to show the 90° angle. We now state the Pythagorean Theorem.

PYTHAGOREAN THEOREM

In a right triangle, the square of the length of the hypotenuse is equal to the sum of the squares of the lengths of the legs. That is, in the right triangle shown in Figure 16,

$$c^2 = a^2 + b^2 \tag{1}$$

A proof of the Pythagorean Theorem is given at the end of this section.

EXAMPLE 1 Finding the Hypotenuse of a Right Triangle

In a right triangle, one leg has length 4 and the other has length 3. What is the length of the hypotenuse?

Solution Since the triangle is a right triangle, we use the Pythagorean Theorem with $a = 4$ and $b = 3$ to find the length c of the hypotenuse. From equation (1),

$$c^2 = a^2 + b^2$$
$$c^2 = 4^2 + 3^2 = 16 + 9 = 25$$
$$c = \sqrt{25} = 5$$

Now Work PROBLEM 15

The converse of the Pythagorean Theorem is also true.

CONVERSE OF THE PYTHAGOREAN THEOREM

In a triangle, if the square of the length of one side equals the sum of the squares of the lengths of the other two sides, the triangle is a right triangle. The 90° angle is opposite the longest side.

A proof is given at the end of this section.

EXAMPLE 2 Verifying That a Triangle Is a Right Triangle

Show that a triangle whose sides are of lengths 5, 12, and 13 is a right triangle. Identify the hypotenuse.

Solution Square the lengths of the sides.

$$5^2 = 25 \qquad 12^2 = 144 \qquad 13^2 = 169$$

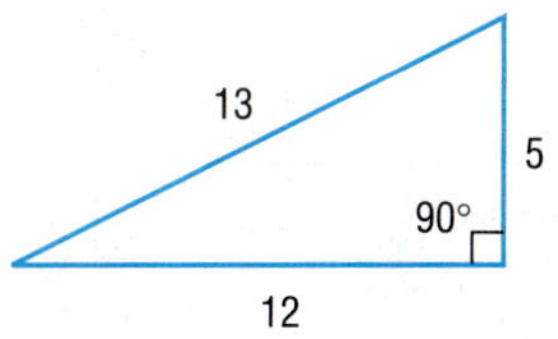

Figure 17

Notice that the sum of the first two squares (25 and 144) equals the third square (169). That is, because $5^2 + 12^2 = 13^2$, the triangle is a right triangle. The longest side, 13, is the hypotenuse. See Figure 17.

Now Work PROBLEM 23

EXAMPLE 3

Applying the Pythagorean Theorem

The tallest building in the world is Burj Khalifa in Dubai, United Arab Emirates, at 2717 feet and 163 floors. The observation deck is 1483 feet above ground level. How far can a person standing on the observation deck see (with the aid of a telescope)? Use 3960 miles for the radius of Earth.

Source: Council on Tall Buildings and Urban Habitat

Solution From the center of Earth, draw two radii: one through Burj Khalifa and the other to the farthest point a person can see from the observation deck. See Figure 18. Apply the Pythagorean Theorem to the right triangle.

Since 1 mile = 5280 feet, 1483 feet $= \frac{1483}{5280}$ mile. Then

$$d^2 + (3960)^2 = \left(3960 + \frac{1483}{5280}\right)^2$$

$$d^2 = \left(3960 + \frac{1483}{5280}\right)^2 - (3960)^2 \approx 2224.58$$

$$d \approx 47.17$$

A person can see more than 47 miles from the observation deck.

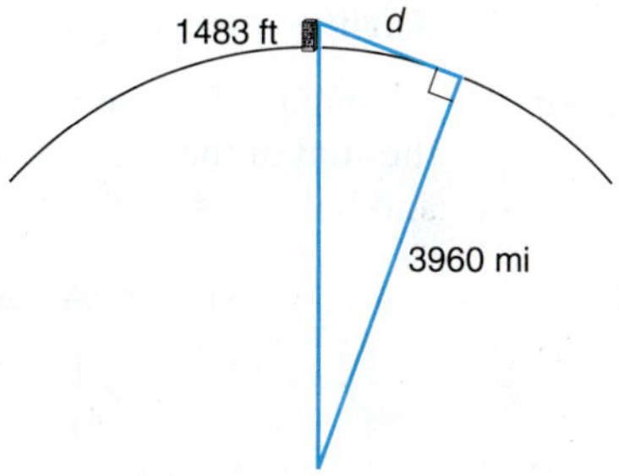

Figure 18

Now Work PROBLEM 55

2 Know Geometry Formulas

Certain formulas from geometry are useful in solving algebra problems.

For a rectangle of length l and width w,

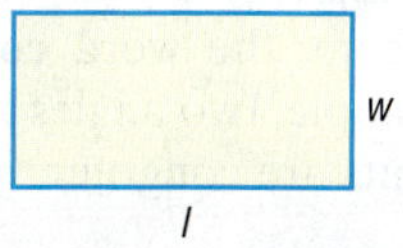

$$\text{Area} = lw \qquad \text{Perimeter} = 2l + 2w$$

For a triangle with base b and altitude h,

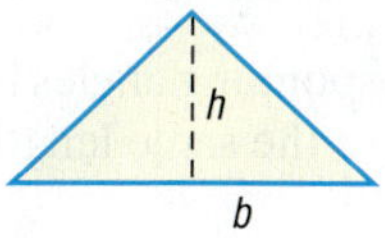

$$\text{Area} = \frac{1}{2}bh$$

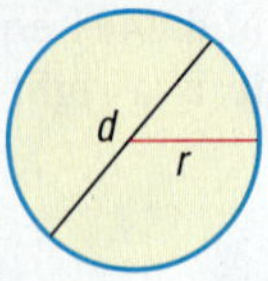

For a circle of radius r (diameter $d = 2r$),

$$\text{Area} = \pi r^2 \qquad \text{Circumference} = 2\pi r = \pi d$$

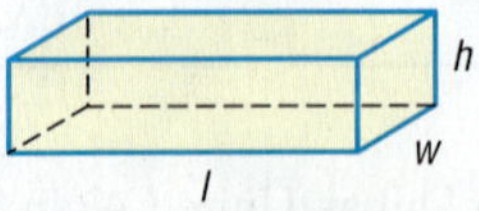

For a closed rectangular box of length l, width w, and height h,

$$\text{Volume} = lwh \qquad \text{Surface area} = 2lh + 2wh + 2lw$$

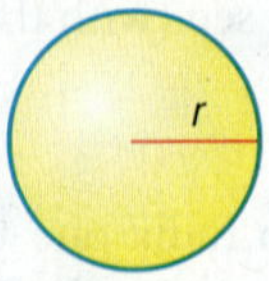

For a sphere of radius r,

$$\text{Volume} = \frac{4}{3}\pi r^3 \qquad \text{Surface area} = 4\pi r^2$$

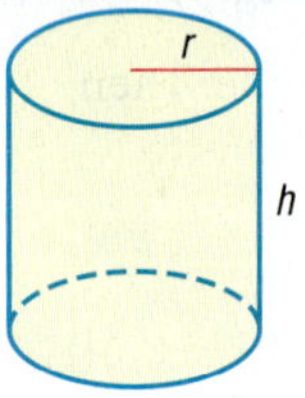

For a closed right circular cylinder of height h and radius r,

$$\text{Volume} = \pi r^2 h \qquad \text{Surface area} = 2\pi r^2 + 2\pi rh$$

Now Work PROBLEM 31

EXAMPLE 4

Using Geometry Formulas

A Christmas tree ornament is in the shape of a semicircle on top of a triangle. How many square centimeters (cm^2) of copper is required to make the ornament if the height of the triangle is 6 cm and the base is 4 cm?

Solution See Figure 19. The amount of copper required equals the shaded area. This area is the sum of the areas of the triangle and the semicircle. The triangle has height $h = 6$ and base $b = 4$. The semicircle has diameter $d = 4$, so its radius is $r = 2$.

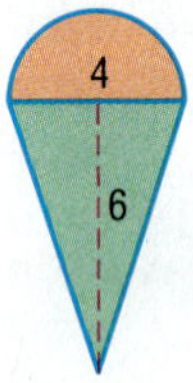

Figure 19

$$\begin{aligned}\text{Area} &= \text{Area of triangle} + \text{Area of semicircle}\\ &= \frac{1}{2}bh + \frac{1}{2}\pi r^2 = \frac{1}{2}(4)(6) + \frac{1}{2}\pi \cdot 2^2 \qquad b = 4;\ h = 6;\ r = 2\\ &= 12 + 2\pi \approx 18.28\ \text{cm}^2\end{aligned}$$

About 18.28 cm^2 of copper is required.

Now Work PROBLEM 49

3 Understand Congruent Triangles and Similar Triangles

In Words
Two triangles are congruent if they have the same size and shape.

Throughout the text we will make reference to triangles. We begin with a discussion of *congruent* triangles. According to dictionary.com, the word **congruent** means "coinciding exactly when superimposed." For example, two angles are congruent if they have the same measure, and two line segments are congruent if they have the same length.

DEFINITION Two triangles are **congruent** if each pair of corresponding angles have the same measure and each pair of corresponding sides are the same length.

In Figure 20, corresponding angles are equal and the corresponding sides are equal in length: $a = d$, $b = e$, and $c = f$. As a result, these triangles are congruent.

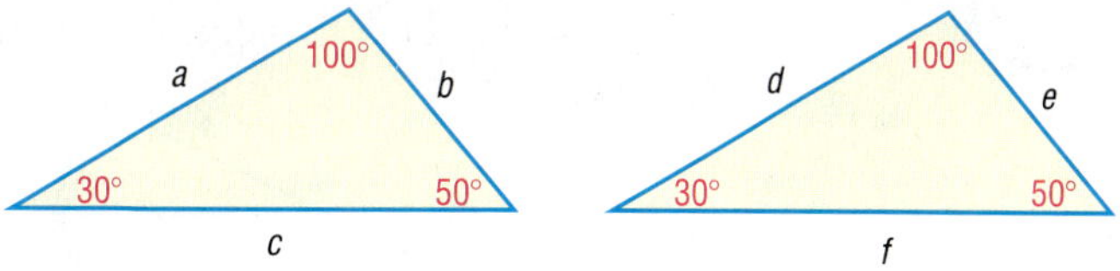

Figure 20 Congruent triangles

Actually, it is not necessary to verify that all three angles and all three sides are the same measure to determine whether two triangles are congruent.

Determining Congruent Triangles

1. **Angle–Side–Angle Case** Two triangles are congruent if two of the angles are equal and the lengths of the corresponding sides between the two angles are equal.

   For example, in Figure 21(a), the two triangles are congruent because two angles and the included side are equal.

2. **Side–Side–Side Case** Two triangles are congruent if the lengths of the corresponding sides of the triangles are equal.

   For example, in Figure 21(b), the two triangles are congruent because the three corresponding sides are all equal.

3. **Side–Angle–Side Case** Two triangles are congruent if the lengths of two corresponding sides are equal and the angles between the two sides are the same.

   For example, in Figure 21(c), the two triangles are congruent because two sides and the included angle are equal.

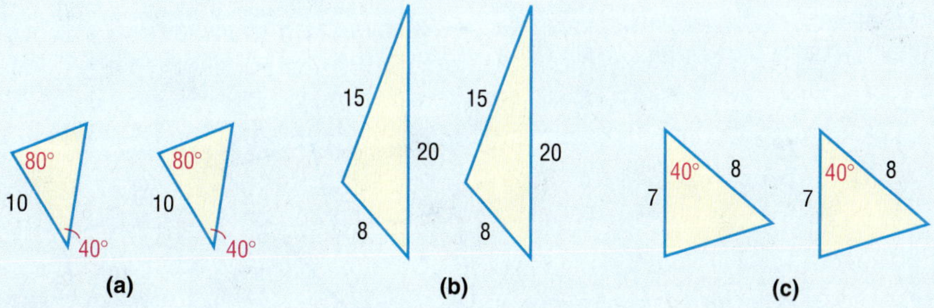

Figure 21

We contrast congruent triangles with *similar* triangles.

DEFINITION

Two triangles are **similar** if the corresponding angles are equal and the lengths of the corresponding sides are proportional.

In Words
Two triangles are similar if they have the same shape, but (possibly) different sizes.

For example, the triangles in Figure 22 (on the next page) are similar because the corresponding angles are equal. In addition, the lengths of the corresponding sides are proportional because each side in the triangle on the right is twice as long as each corresponding side in the triangle on the left. That is, the ratio of the corresponding sides is a constant: $\frac{d}{a} = \frac{e}{b} = \frac{f}{c} = 2$.

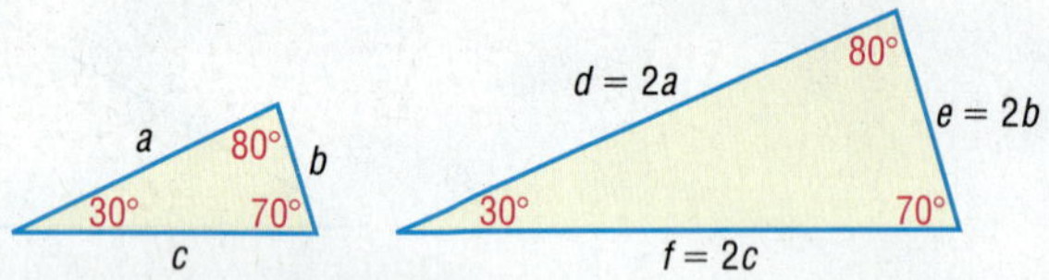

Figure 22 Similar triangles

It is not necessary to verify that all three angles are equal and all three sides are proportional to determine whether two triangles are congruent.

Determining Similar Triangles

1. **Angle–Angle Case** Two triangles are similar if two of the corresponding angles are equal.

   For example, in Figure 23(a), the two triangles are similar because two angles are equal.

2. **Side–Side–Side Case** Two triangles are similar if the lengths of all three sides of each triangle are proportional.

   For example, in Figure 23(b), the two triangles are similar because

$$\frac{10}{30} = \frac{5}{15} = \frac{6}{18} = \frac{1}{3}$$

3. **Side–Angle–Side Case** Two triangles are similar if two corresponding sides are proportional and the angles between the two sides are equal.

   For example, in Figure 23(c), the two triangles are similar because $\frac{4}{6} = \frac{12}{18} = \frac{2}{3}$ and the angles between the sides are equal.

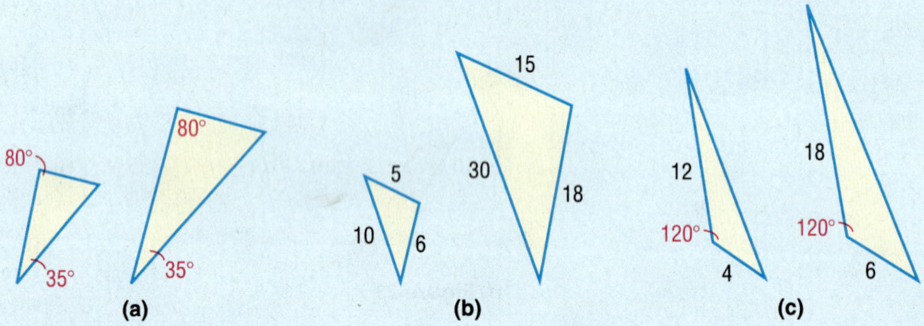

Figure 23

EXAMPLE 5

Using Similar Triangles

Given that the triangles in Figure 24 are similar, find the missing length x and the angles A, B, and C.

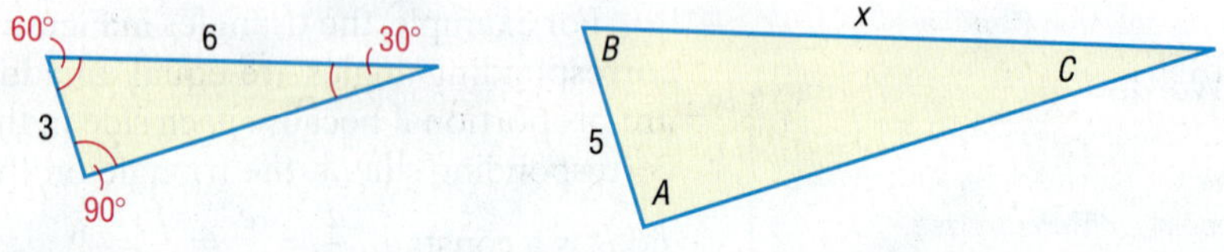

Figure 24

Solution Because the triangles are similar, corresponding angles are equal. So $A = 90°$, $B = 60°$, and $C = 30°$. Also, the corresponding sides are proportional. That is, $\frac{3}{5} = \frac{6}{x}$. We solve this equation for x.

$$\frac{3}{5} = \frac{6}{x}$$

$$5x \cdot \frac{3}{5} = 5x \cdot \frac{6}{x} \quad \text{Multiply both sides by } 5x.$$

$$3x = 30 \quad \text{Simplify.}$$

$$x = 10 \quad \text{Divide both sides by 3.}$$

The missing length is 10 units.

Now Work PROBLEM 43

Proof of the Pythagorean Theorem Begin with a square, each side of length $a + b$. In this square, form four right triangles, each having legs equal in length to a and b. See Figure 25. All these triangles are congruent (two sides and their included angle are equal). As a result, the hypotenuse of each is the same, say c, and the pink shading in Figure 25 indicates a square with an area equal to c^2.

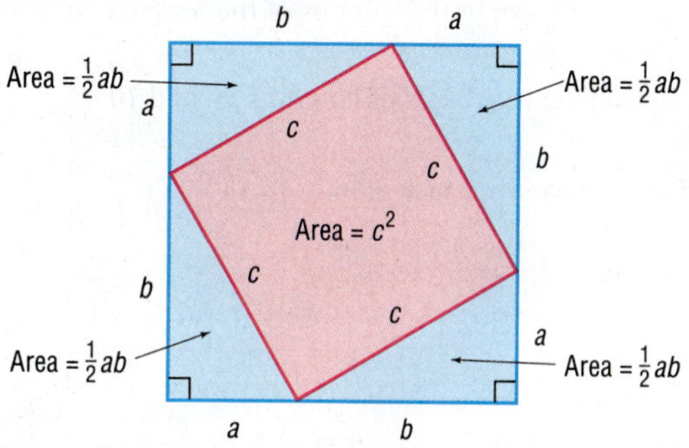

Figure 25

The area of the original square with sides $a + b$ equals the sum of the areas of the four triangles (each of area $\frac{1}{2}ab$) plus the area of the square with side c. That is,

$$(a + b)^2 = \frac{1}{2}ab + \frac{1}{2}ab + \frac{1}{2}ab + \frac{1}{2}ab + c^2$$

$$a^2 + 2ab + b^2 = 2ab + c^2$$

$$a^2 + b^2 = c^2$$

The proof is complete. ■

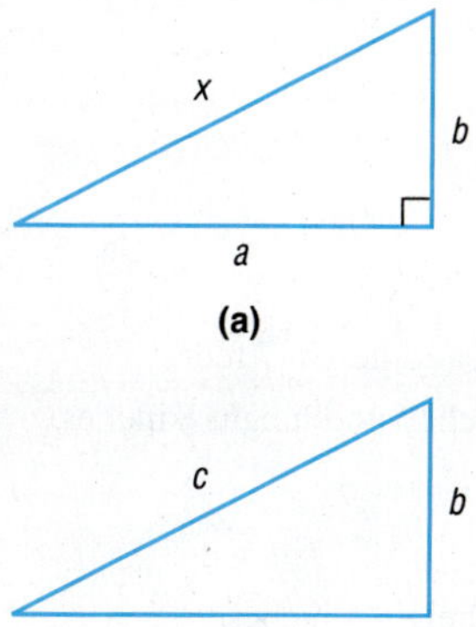

(a)

(b) $c^2 = a^2 + b^2$

Figure 26

Proof of the Converse of the Pythagorean Theorem Begin with two triangles: one a right triangle with legs a and b and the other a triangle with sides a, b, and c for which $c^2 = a^2 + b^2$. See Figure 26. By the Pythagorean Theorem, the length x of the third side of the first triangle is

$$x^2 = a^2 + b^2$$

But $c^2 = a^2 + b^2$. Then,

$$x^2 = c^2$$

$$x = c$$

The two triangles have the same sides and are therefore congruent. This means corresponding angles are equal, so the angle opposite side c of the second triangle equals 90°.

The proof is complete. ■

R.3 Assess Your Understanding

Concepts and Vocabulary

1. A(n) ____ triangle is one that contains an angle of 90 degrees. The longest side is called the __________.
2. For a triangle with base b and altitude h, a formula for the area A is __________.
3. The formula for the circumference C of a circle of radius r is ________.
4. Two triangles are _______ if corresponding angles are equal and the lengths of the corresponding sides are proportional.
5. Which of the following is not a case for determining congruent triangles?
   (a) Angle–Side–Angle (b) Side–Angle–Side
   (c) Angle–Angle–Angle (d) Side-Side-Side
6. Choose the formula for the volume of a sphere of radius r.
   (a) $\frac{4}{3}\pi r^2$ (b) $\frac{4}{3}\pi r^3$ (c) $4\pi r^3$ (d) $4\pi r^2$
7. ***True or False*** In a right triangle, the square of the length of the longest side equals the sum of the squares of the lengths of the other two sides.
8. ***True or False*** The triangle with sides of lengths 6, 8, and 10 is a right triangle.
9. ***True or False*** The surface area of a sphere of radius r is $\frac{4}{3}\pi r^2$.
10. ***True or False*** The triangles shown are congruent.

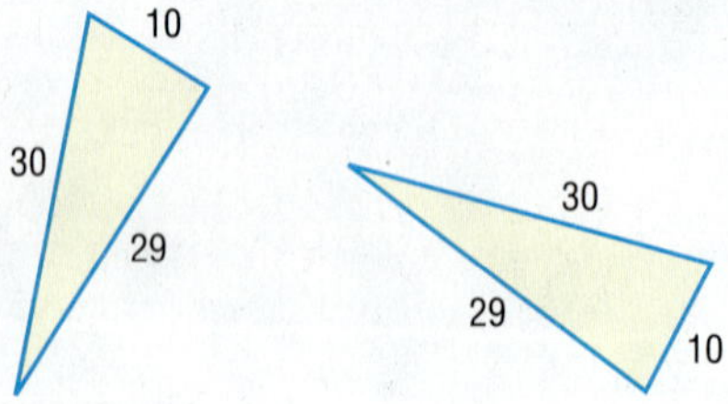

11. ***True or False*** The triangles shown are similar.

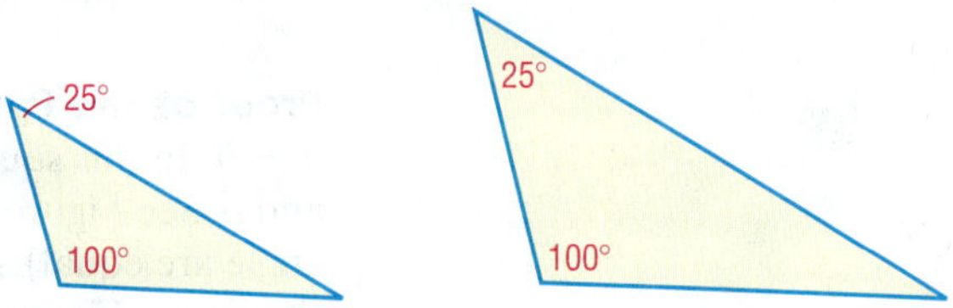

12. ***True or False*** The triangles shown are similar.

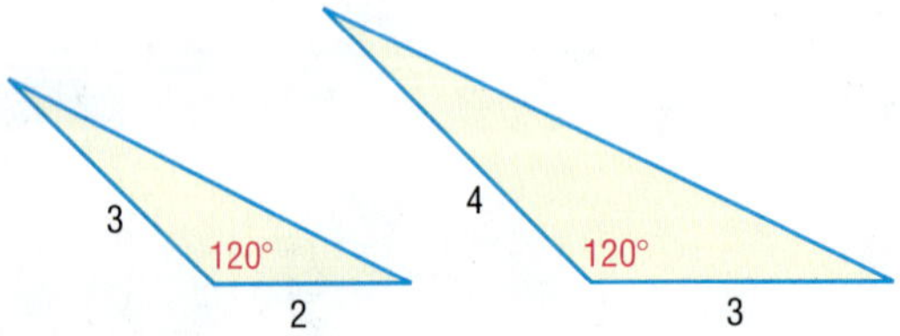

Skill Building

In Problems 13–18, the lengths of the legs of a right triangle are given. Find the hypotenuse.

13. $a = 5$, $b = 12$
14. $a = 6$, $b = 8$
15. $a = 10$, $b = 24$
16. $a = 4$, $b = 3$
17. $a = 7$, $b = 24$
18. $a = 14$, $b = 48$

In Problems 19–26, the lengths of the sides of a triangle are given. Determine which are right triangles. For those that are, identify the hypotenuse.

19. 3, 4, 5
20. 6, 8, 10
21. 4, 5, 6
22. 2, 2, 3
23. 7, 24, 25
24. 10, 24, 26
25. 6, 4, 3
26. 5, 4, 7
27. Find the area A of a rectangle with length 4 inches and width 2 inches.
28. Find the area A of a rectangle with length 9 centimeters and width 4 centimeters.
29. Find the area A of a triangle with height 4 inches and base 2 inches.
30. Find the area A of a triangle with height 9 centimeters and base 4 centimeters.
31. Find the area A and circumference C of a circle of radius 5 meters.
32. Find the area A and circumference C of a circle of radius 2 feet.
33. Find the volume V and surface area S of a closed rectangular box with length 8 feet, width 4 feet, and height 7 feet.
34. Find the volume V and surface area S of a closed rectangular box with length 9 inches, width 4 inches, and height 8 inches.
35. Find the volume V and surface area S of a sphere of radius 4 centimeters.
36. Find the volume V and surface area S of a sphere of radius 3 feet.
37. Find the volume V and surface area S of a closed right circular cylinder with radius 9 inches and height 8 inches.
38. Find the volume V and surface area S of a closed right circular cylinder with radius 8 inches and height 9 inches.

In Problems 39–42, find the area of the shaded region.

39.

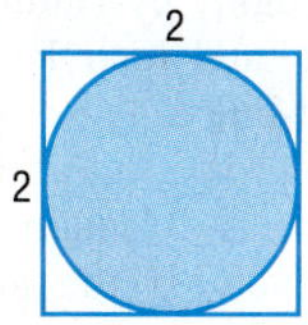

40.

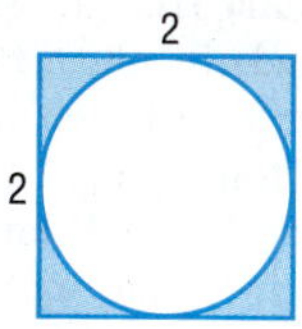

41.

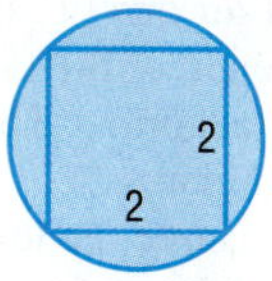

42.

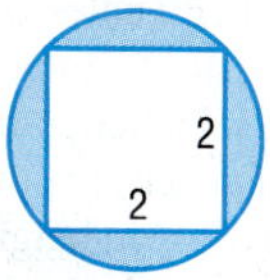

In Problems 43–46, the triangles in each pair are similar. Find the missing length x and the missing angles A, B, and C.

43.

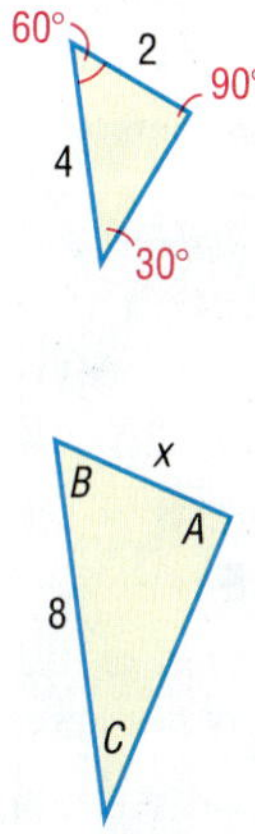

44.

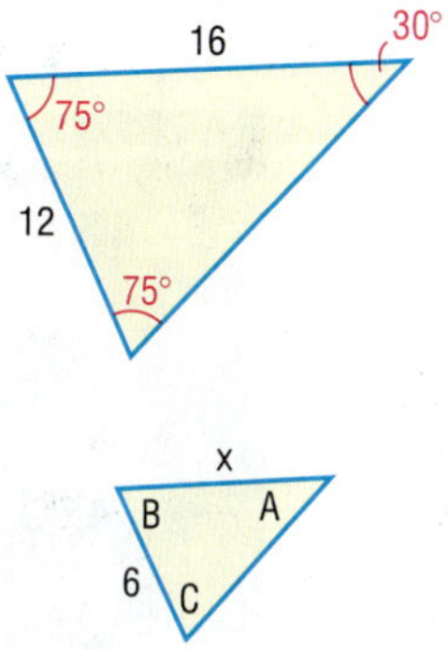

45.

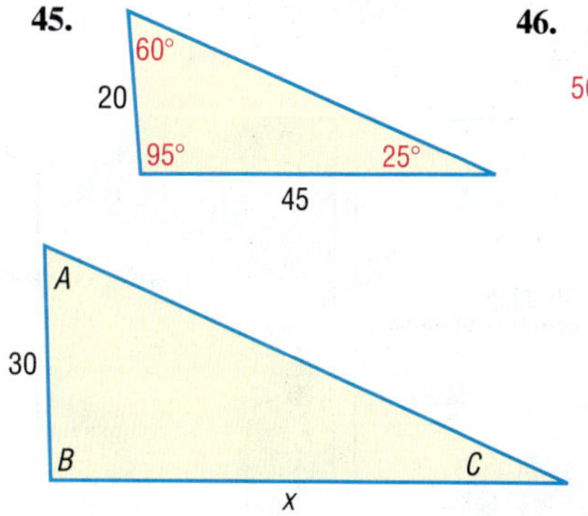

46.

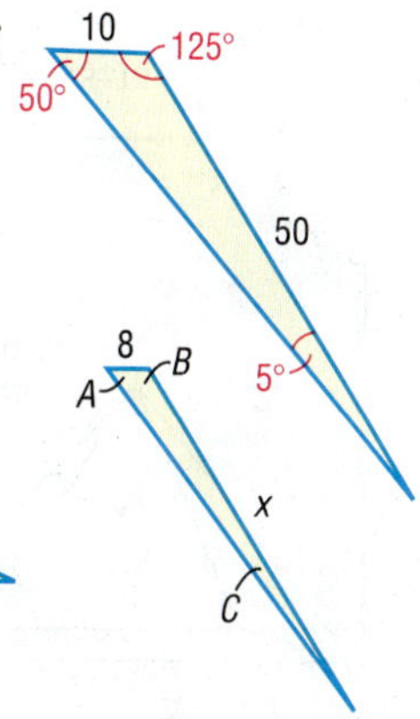

Applications and Extensions

47. How many feet has a wheel with a diameter of 16 inches traveled after four revolutions?

48. How many revolutions will a circular disk with a diameter of 4 feet have completed after it has rolled 20 feet?

49. In the figure shown, *ABCD* is a square, with each side of length 6 feet. The width of the border (shaded portion) between the outer square *EFGH* and *ABCD* is 2 feet. Find the area of the border.

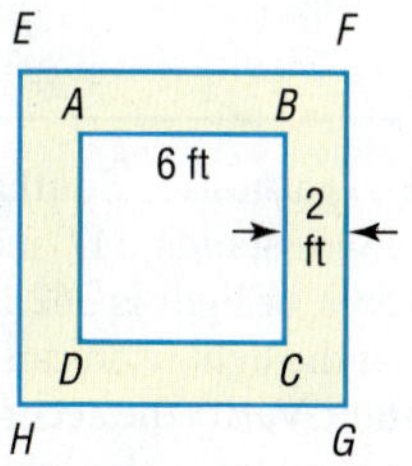

50. Refer to the figure. Square *ABCD* has an area of 100 square feet; square *BEFG* has an area of 16 square feet. What is the area of the triangle *CGF*?

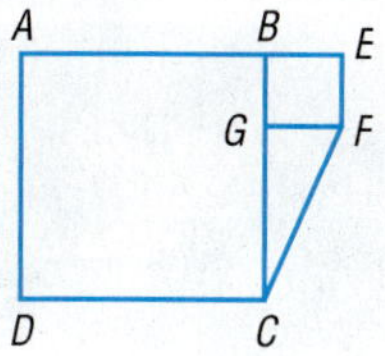

51. Architecture A **Norman window** consists of a rectangle surmounted by a semicircle. Find the area of the Norman window shown in the illustration. How much wood frame is needed to enclose the window?

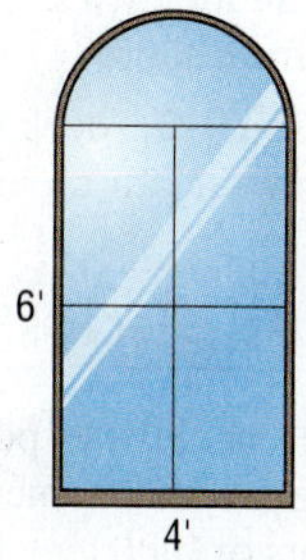

52. Construction A circular swimming pool that is 20 feet in diameter is enclosed by a wooden deck that is 3 feet wide. What is the area of the deck? How much fence is required to enclose the deck?

53. How Tall Is the Great Pyramid? The ancient Greek philosopher Thales of Miletus is reported on one occasion to have visited Egypt and calculated the height of the Great Pyramid of Cheops by means of shadow reckoning. Thales knew that each side of the base of the pyramid was 252 paces and that his own height was 2 paces. He measured the length of the pyramid's shadow to be 114 paces and determined the length of his shadow to be 3 paces. See the illustration. Using similar triangles, determine the height of the Great Pyramid in terms of the number of paces.

Source: Diggins, Julie E, *String Straightedge and Shadow: The Story of Geometry*, 2003, Whole Spirit Press, http://wholespiritpress.com.

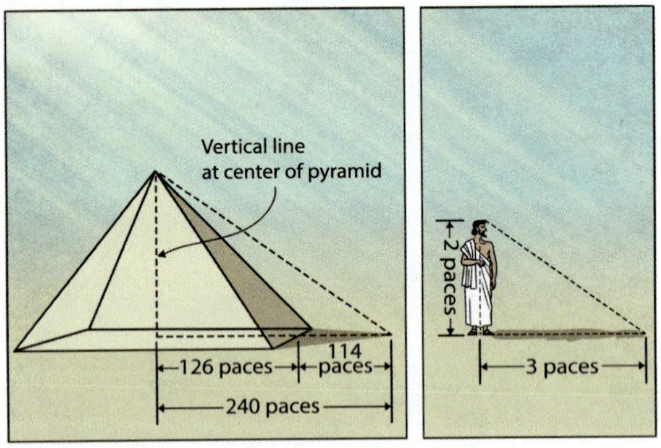

54. The Bermuda Triangle Karen is doing research on the Bermuda Triangle which she defines roughly by Hamilton, Bermuda; San Juan, Puerto Rico; and Fort Lauderdale, Florida. On her atlas Karen measures the straight-line distances from Hamilton to Fort Lauderdale, Fort Lauderdale to San Juan, and San Juan to Hamilton to be approximately 57 millimeters (mm), 58 mm, and 53.5 mm respectively. If the actual distance from Fort Lauderdale to San Juan is 1046 miles, approximate the actual distances from San Juan to Hamilton and from Hamilton to Fort Lauderdale.

In Problems 55–57, use the facts that the radius of Earth is 3960 miles and 1 mile = 5280 feet.

55. How Far Can You See? The conning tower of the U.S.S. *Silversides,* a World War II submarine now permanently stationed in Muskegon, Michigan, is approximately 20 feet above sea level. How far can you see from the conning tower?

56. How Far Can You See? A person who is 6 feet tall is standing on the beach in Fort Lauderdale, Florida, and looks out onto the Atlantic Ocean. Suddenly, a ship appears on the horizon. How far is the ship from shore?

57. How Far Can You See? The deck of a destroyer is 100 feet above sea level. How far can a person see from the deck? How far can a person see from the bridge, which is 150 feet above sea level?

58. Suppose that m and n are positive integers with $m > n$. If $a = m^2 - n^2$, $b = 2mn$, and $c = m^2 + n^2$, show that a, b, and c are the lengths of the sides of a right triangle. (This formula can be used to find the sides of a right triangle that are integers, such as 3, 4, 5; 5, 12, 13; and so on. Such triplets of integers are called **Pythagorean triples**.)

Explaining Concepts: Discussion and Writing

59. You have 1000 feet of flexible pool siding and intend to construct a swimming pool. Experiment with rectangular-shaped pools with perimeters of 1000 feet. How do their areas vary? What is the shape of the rectangle with the largest area? Now compute the area enclosed by a circular pool with a perimeter (circumference) of 1000 feet. What would be your choice of shape for the pool? If rectangular, what is your preference for dimensions? Justify your choice. If your only consideration is to have a pool that encloses the most area, what shape should you use?

60. The Gibb's Hill Lighthouse, Southampton, Bermuda, in operation since 1846, stands 117 feet high on a hill 245 feet high, so its beam of light is 362 feet above sea level. A brochure states that the light itself can be seen on the horizon about 26 miles distant. Verify the accuracy of this information. The brochure further states that ships 40 miles away can see the light and that planes flying at 10,000 feet can see it 120 miles away. Verify the accuracy of these statements. What assumption did the brochure make about the height of the ship?

R.4 Polynomials

We have described algebra as a generalization of arithmetic in which letters are used to represent real numbers. From now on, we shall use the letters at the end of the alphabet, such as x, y, and z, to represent variables and use the letters at the beginning of the alphabet, such as a, b, and c, to represent constants. In the expressions $3x + 5$ and $ax + b$, it is understood that x is a variable and that a and b are constants, even though the constants a and b are unspecified. As you will find out, the context usually makes the intended meaning clear.

1 Recognize Monomials

DEFINITION

NOTE The nonnegative integers are the integers 0, 1, 2, 3, ….

A **monomial** in one variable is the product of a constant and a variable raised to a nonnegative integer power. A monomial is of the form

$$ax^k$$

where a is a constant, x is a variable, and $k \geq 0$ is an integer. The constant a is called the **coefficient** of the monomial. If $a \neq 0$, then k is called the **degree** of the monomial.

EXAMPLE 1 **Examples of Monomials**

Monomial	Coefficient	Degree	
(a) $6x^2$	6	2	
(b) $-\sqrt{2}x^3$	$-\sqrt{2}$	3	
(c) 3	3	0	Since $3 = 3 \cdot 1 = 3x^0, x \neq 0$
(d) $-5x$	-5	1	Since $-5x = -5x^1$
(e) x^4	1	4	Since $x^4 = 1 \cdot x^4$

EXAMPLE 2 **Examples of Nonmonomial Expressions**

(a) $3x^{1/2}$ is not a monomial, since the exponent of the variable x is $\frac{1}{2}$, and $\frac{1}{2}$ is not a nonnegative integer.

(b) $4x^{-3}$ is not a monomial, since the exponent of the variable x is -3, and -3 is not a nonnegative integer.

Now Work PROBLEM 9

2 Recognize Polynomials

Two monomials with the same variable raised to the same power are called **like terms**. For example, $2x^4$ and $-5x^4$ are like terms. In contrast, the monomials $2x^3$ and $2x^5$ are not like terms.

We can add or subtract like terms using the Distributive Property. For example,

$$2x^2 + 5x^2 = (2 + 5)x^2 = 7x^2 \quad \text{and} \quad 8x^3 - 5x^3 = (8 - 5)x^3 = 3x^3$$

The sum or difference of two monomials having different degrees is called a **binomial**. The sum or difference of three monomials with three different degrees is called a **trinomial**. For example,

$x^2 - 2$ is a binomial.

$x^3 - 3x + 5$ is a trinomial.

$2x^2 + 5x^2 + 2 = 7x^2 + 2$ is a binomial.

DEFINITION

A **polynomial** in one variable is an algebraic expression of the form

$$a_nx^n + a_{n-1}x^{n-1} + \cdots + a_1x + a_0 \tag{1}$$

where $a_n, a_{n-1}, \ldots, a_1, a_0$ are constants,* called the **coefficients** of the polynomial, $n \geq 0$ is an integer, and x is a variable. If $a_n \neq 0$, it is called the **leading coefficient**, a_nx^n is called the **leading term**, and n is the **degree** of the polynomial.

In Words
A polynomial is a sum of monomials.

The monomials that make up a polynomial are called its **terms**. If all of the coefficients are 0, the polynomial is called the **zero polynomial**, which has no degree.

Polynomials are usually written in **standard form**, beginning with the nonzero term of highest degree and continuing with terms in descending order according to degree. If a power of x is missing, it is because its coefficient is zero.

EXAMPLE 3 **Examples of Polynomials**

Polynomial	Coefficients	Degree
$-8x^3 + 4x^2 - 6x + 2$	$-8, 4, -6, 2$	3
$3x^2 - 5 = 3x^2 + 0 \cdot x + (-5)$	$3, 0, -5$	2
$8 - 2x + x^2 = 1 \cdot x^2 + (-2)x + 8$	$1, -2, 8$	2
$5x + \sqrt{2} = 5x^1 + \sqrt{2}$	$5, \sqrt{2}$	1
$3 = 3 \cdot 1 = 3 \cdot x^0$	3	0
0	0	No degree

Although we have been using x to represent the variable, letters such as y and z are also commonly used.

$3x^4 - x^2 + 2$ is a polynomial (in x) of degree 4.

$9y^3 - 2y^2 + y - 3$ is a polynomial (in y) of degree 3.

$z^5 + \pi$ is a polynomial (in z) of degree 5.

Algebraic expressions such as

$$\frac{1}{x} \quad \text{and} \quad \frac{x^2 + 1}{x + 5}$$

*The notation a_n is read as "a sub n." The number n is called a **subscript** and should not be confused with an exponent. We use subscripts to distinguish one constant from another when a large or undetermined number of constants are required.

are not polynomials. The first is not a polynomial because $\frac{1}{x} = x^{-1}$ has an exponent that is not a nonnegative integer. Although the second expression is the quotient of two polynomials, the polynomial in the denominator has degree greater than 0, so the expression cannot be a polynomial.

Now Work PROBLEM 19

3 Add and Subtract Polynomials

Polynomials are added and subtracted by combining like terms.

EXAMPLE 4 **Adding Polynomials**

Find the sum of the polynomials:

$$8x^3 - 2x^2 + 6x - 2 \quad \text{and} \quad 3x^4 - 2x^3 + x^2 + x$$

Solution We shall find the sum in two ways.

Horizontal Addition: The idea here is to group the like terms and then combine them.

$$\begin{aligned}
&(8x^3 - 2x^2 + 6x - 2) + (3x^4 - 2x^3 + x^2 + x) \\
&\quad = 3x^4 + (8x^3 - 2x^3) + (-2x^2 + x^2) + (6x + x) - 2 \\
&\quad = 3x^4 + 6x^3 - x^2 + 7x - 2
\end{aligned}$$

Vertical Addition: The idea here is to vertically line up the like terms in each polynomial and then add the coefficients.

$$\begin{array}{rrrrr}
x^4 & x^3 & x^2 & x^1 & x^0 \\
 & 8x^3 & -\ 2x^2 & +\ 6x & -\ 2 \\
+\ 3x^4 & -\ 2x^3 & +\ x^2 & +\ x & \\
\hline
3x^4 & +\ 6x^3 & -\ x^2 & +\ 7x & -\ 2
\end{array}$$

●

We can subtract two polynomials horizontally or vertically as well.

EXAMPLE 5 **Subtracting Polynomials**

Find the difference: $(3x^4 - 4x^3 + 6x^2 - 1) - (2x^4 - 8x^2 - 6x + 5)$

Solution ***Horizontal Subtraction:***

$$\begin{aligned}
&(3x^4 - 4x^3 + 6x^2 - 1) - (2x^4 - 8x^2 - 6x + 5) \\
&\quad = 3x^4 - 4x^3 + 6x^2 - 1 + \underbrace{(-2x^4 + 8x^2 + 6x - 5)}_{\text{Be sure to change the sign of each term in the second polynomial.}} \\
&\quad \underset{\text{Group like terms.}}{=} (3x^4 - 2x^4) + (-4x^3) + (6x^2 + 8x^2) + 6x + (-1 - 5) \\
&\quad = x^4 - 4x^3 + 14x^2 + 6x - 6
\end{aligned}$$

COMMENT Vertical subtraction will be used when we divide polynomials. ■

Vertical Subtraction: We line up like terms, change the sign of each coefficient of the second polynomial, and add.

$$\begin{array}{rrrrrr} & x^4 & x^3 & x^2 & x^1 & x^0 \\ & 3x^4 & -4x^3 & +6x^2 & & -1 \\ - & [2x^4 & & -8x^2 & -6x & +5] \\ \hline \end{array} = \begin{array}{rrrrr} x^4 & x^3 & x^2 & x^1 & x^0 \\ 3x^4 & -4x^3 & +6x^2 & & -1 \\ -2x^4 & & +8x^2 & +6x & -5 \\ \hline x^4 & -4x^3 & +14x^2 & +6x & -6 \end{array}$$

●

Which of these methods to use for adding and subtracting polynomials is up to you. To save space, we shall most often use the horizontal format.

Now Work PROBLEM 31

4 Multiply Polynomials

Two monomials may be multiplied using the Laws of Exponents and the Commutative and Associative Properties. For example,

$$(2x^3) \cdot (5x^4) = (2 \cdot 5) \cdot (x^3 \cdot x^4) = 10x^{3+4} = 10x^7$$

Products of polynomials are found by repeated use of the Distributive Property and the Laws of Exponents. Again, you have a choice of horizontal or vertical format.

EXAMPLE 6 Multiplying Polynomials

Find the product: $(2x + 5)(x^2 - x + 2)$

Solution ***Horizontal Multiplication:***

$$(2x + 5)(x^2 - x + 2) = 2x(x^2 - x + 2) + 5(x^2 - x + 2)$$

↑ Distributive Property

$$= (2x \cdot x^2 - 2x \cdot x + 2x \cdot 2) + (5 \cdot x^2 - 5 \cdot x + 5 \cdot 2)$$

↑ Distributive Property

$$= (2x^3 - 2x^2 + 4x) + (5x^2 - 5x + 10)$$

↑ Law of Exponents

$$= 2x^3 + 3x^2 - x + 10$$

↑ Combine like terms.

Vertical Multiplication: The idea here is very much like multiplying a two-digit number by a three-digit number.

$$\begin{array}{rrrrl} & & x^2 & -\ x & +\ 2 \\ & & & 2x & +\ 5 \\ \hline & 2x^3 & -2x^2 & +4x & \\ (+) & & 5x^2 & -5x & +10 \\ \hline & 2x^3 & +3x^2 & -\ x & +10 \end{array}$$

This line is $2x(x^2 - x + 2)$.
This line is $5(x^2 - x + 2)$.
Sum of the above two lines ●

Now Work PROBLEM 47

5 Know Formulas for Special Products

Certain products, which we call **special products**, occur frequently in algebra. We can calculate them easily using the **FOIL** (*F*irst, *O*uter, *I*nner, *L*ast) method of multiplying two binomials.

$$\begin{aligned}(ax + b)(cx + d) &= ax(cx + d) + b(cx + d)\\ &= \underbrace{ax \cdot cx}_{\text{First}} + \underbrace{ax \cdot d}_{\text{Outer}} + \underbrace{b \cdot cx}_{\text{Inner}} + \underbrace{b \cdot d}_{\text{Last}}\\ &= acx^2 + adx + bcx + bd\\ &= acx^2 + (ad + bc)x + bd\end{aligned}$$

EXAMPLE 7

Using FOIL

(a) $(x - 3)(x + 3) = \underset{F}{x^2} + \underset{O}{3x} - \underset{I}{3x} - \underset{L}{9} = x^2 - 9$

(b) $(x + 2)^2 = (x + 2)(x + 2) = x^2 + 2x + 2x + 4 = x^2 + 4x + 4$

(c) $(x - 3)^2 = (x - 3)(x - 3) = x^2 - 3x - 3x + 9 = x^2 - 6x + 9$

(d) $(x + 3)(x + 1) = x^2 + x + 3x + 3 = x^2 + 4x + 3$

(e) $(2x + 1)(3x + 4) = 6x^2 + 8x + 3x + 4 = 6x^2 + 11x + 4$

Now Work PROBLEMS 49 AND 57

Some products have been given special names because of their form. The following special products are based on Examples 7(a), (b), and (c).

Difference of Two Squares

$$(x - a)(x + a) = x^2 - a^2 \qquad \textbf{(2)}$$

Squares of Binomials, or Perfect Squares

$$(x + a)^2 = x^2 + 2ax + a^2 \qquad \textbf{(3a)}$$

$$(x - a)^2 = x^2 - 2ax + a^2 \qquad \textbf{(3b)}$$

EXAMPLE 8

Using Special Product Formulas

(a) $(x - 5)(x + 5) = x^2 - 5^2 = x^2 - 25$ — Difference of two squares

(b) $(x + 7)^2 = x^2 + 2 \cdot 7 \cdot x + 7^2 = x^2 + 14x + 49$ — Square of a binomial

(c) $(2x + 1)^2 = (2x)^2 + 2 \cdot 1 \cdot 2x + 1^2 = 4x^2 + 4x + 1$ — Notice that we used $2x$ in place of x in formula (3a).

(d) $(3x - 4)^2 = (3x)^2 - 2 \cdot 4 \cdot 3x + 4^2 = 9x^2 - 24x + 16$ — Replace x by $3x$ in formula (3b).

Now Work PROBLEMS 67, 69, AND 71

Let's look at some more examples that lead to general formulas.

EXAMPLE 9 **Cubing a Binomial**

(a) $(x+2)^3 = (x+2)(x+2)^2 = (x+2)(x^2+4x+4)$ Formula (3a)

$= (x^3+4x^2+4x) + (2x^2+8x+8)$

$= x^3+6x^2+12x+8$

(b) $(x-1)^3 = (x-1)(x-1)^2 = (x-1)(x^2-2x+1)$ Formula (3b)

$= (x^3-2x^2+x) - (x^2-2x+1)$

$= x^3-3x^2+3x-1$

Cubes of Binomials, or Perfect Cubes

$$(x+a)^3 = x^3+3ax^2+3a^2x+a^3 \quad \textbf{(4a)}$$

$$(x-a)^3 = x^3-3ax^2+3a^2x-a^3 \quad \textbf{(4b)}$$

Now Work PROBLEM 87

EXAMPLE 10 **Forming the Difference of Two Cubes**

$$(x-1)(x^2+x+1) = x(x^2+x+1) - 1(x^2+x+1)$$

$$= x^3+x^2+x-x^2-x-1$$

$$= x^3-1$$

EXAMPLE 11 **Forming the Sum of Two Cubes**

$$(x+2)(x^2-2x+4) = x(x^2-2x+4) + 2(x^2-2x+4)$$

$$= x^3-2x^2+4x+2x^2-4x+8$$

$$= x^3+8$$

Examples 10 and 11 lead to two more special products.

Difference of Two Cubes

$$(x-a)(x^2+ax+a^2) = x^3-a^3 \quad \textbf{(5)}$$

Sum of Two Cubes

$$(x+a)(x^2-ax+a^2) = x^3+a^3 \quad \textbf{(6)}$$

6 Divide Polynomials Using Long Division

The procedure for dividing two polynomials is similar to the procedure for dividing two integers.

EXAMPLE 12 Dividing Two Integers

Divide 842 by 15.

Solution

$$\begin{array}{r l} & 56 \quad \leftarrow \text{Quotient} \\ \text{Divisor} \rightarrow 15 & \overline{)842} \quad \leftarrow \text{Dividend} \\ & \underline{75} \quad \leftarrow 5 \cdot 15 \text{ (subtract)} \\ & 92 \\ & \underline{90} \quad \leftarrow 6 \cdot 15 \text{ (subtract)} \\ & 2 \quad \leftarrow \text{Remainder} \end{array}$$

So, $\dfrac{842}{15} = 56 + \dfrac{2}{15}$.

In the long-division process detailed in Example 12, the number 15 is called the **divisor**, the number 842 is called the **dividend**, the number 56 is called the **quotient**, and the number 2 is called the **remainder**.

To check the answer obtained in a division problem, multiply the quotient by the divisor and add the remainder. The answer should be the dividend.

(Quotient) (Divisor) + Remainder = Dividend

For example, we can check the results obtained in Example 12 as follows:

$$(56)(15) + 2 = 840 + 2 = 842$$

To divide two polynomials, we first must write each polynomial in standard form. The process then follows a pattern similar to that of Example 12. The next example illustrates the procedure.

EXAMPLE 13 Dividing Two Polynomials

Find the quotient and the remainder when

$$3x^3 + 4x^2 + x + 7 \quad \text{is divided by} \quad x^2 + 1$$

Solution Each polynomial is in standard form. The dividend is $3x^3 + 4x^2 + x + 7$, and the divisor is $x^2 + 1$.

NOTE Remember, a polynomial is in standard form when its terms are written in descending powers of x. ■

STEP 1: Divide the leading term of the dividend, $3x^3$, by the leading term of the divisor, x^2. Enter the result, $3x$, over the term $3x^3$, as follows:

$$\begin{array}{r l} & 3x \\ x^2 + 1 & \overline{)3x^3 + 4x^2 + x + 7} \end{array}$$

STEP 2: Multiply $3x$ by $x^2 + 1$, and enter the result below the dividend.

$$\begin{array}{r l} & 3x \\ x^2 + 1 & \overline{)3x^3 + 4x^2 + x + 7} \\ & \underline{3x^3 + 3x} \quad \leftarrow 3x \cdot (x^2 + 1) = 3x^3 + 3x \end{array}$$

↑ Align the $3x$ term under the x to make the next step easier.

STEP 3: Subtract and bring down the remaining terms.

$$\begin{array}{r l} & 3x \\ x^2 + 1 & \overline{)3x^3 + 4x^2 + x + 7} \\ & \underline{3x^3 + 3x} \quad \leftarrow \text{Subtract (change the signs and add).} \\ & 4x^2 - 2x + 7 \quad \leftarrow \text{Bring down the } 4x^2 \text{ and the } 7. \end{array}$$

Step 4: Repeat Steps 1–3 using $4x^2 - 2x + 7$ as the dividend.

$$
\begin{array}{rrrrl}
 & 3x & +4 & & \\
x^2+1 \,\big) & 3x^3 + 4x^2 & + x & + 7 & \\
 & 3x^3 & + 3x & & \\
\hline
 & 4x^2 & - 2x & + 7 & \leftarrow \text{Divide } 4x^2 \text{ by } x^2 \text{ to get } 4. \\
 & 4x^2 & & + 4 & \leftarrow \text{Multiply } x^2+1 \text{ by } 4\text{; subtract.} \\
\hline
 & & -2x & + 3 &
\end{array}
$$

COMMENT If the degree of the divisor is greater than the degree of the dividend, then the process ends. ■

Since x^2 does not divide $-2x$ evenly (that is, the result is not a monomial), the process ends. The quotient is $3x + 4$, and the remainder is $-2x + 3$.

✓**Check:** (Quotient) (Divisor) + Remainder

$$= (3x + 4)(x^2 + 1) + (-2x + 3)$$
$$= 3x^3 + 3x + 4x^2 + 4 + (-2x + 3)$$
$$= 3x^3 + 4x^2 + x + 7 = \text{Dividend}$$

Then

$$\frac{3x^3 + 4x^2 + x + 7}{x^2 + 1} = 3x + 4 + \frac{-2x + 3}{x^2 + 1}$$ ●

The next example combines the steps involved in long division.

EXAMPLE 14 Dividing Two Polynomials

Find the quotient and the remainder when

$$x^4 - 3x^3 + 2x - 5 \quad \text{is divided by} \quad x^2 - x + 1$$

Solution In setting up this division problem, it is necessary to leave a space for the missing x^2 term in the dividend.

$$
\begin{array}{lrl}
 & x^2 - 2x - 3 & \leftarrow \text{Quotient} \\
\text{Divisor} \rightarrow \; x^2 - x + 1 \,\big) & x^4 - 3x^3 \qquad + 2x - 5 & \leftarrow \text{Dividend} \\
\text{Subtract} \rightarrow & x^4 - x^3 + x^2 \qquad\qquad & \\
\hline
 & -2x^3 - x^2 + 2x - 5 & \\
\text{Subtract} \rightarrow & -2x^3 + 2x^2 - 2x \qquad & \\
\hline
 & -3x^2 + 4x - 5 & \\
\text{Subtract} \rightarrow & -3x^2 + 3x - 3 & \\
\hline
 & x - 2 & \leftarrow \text{Remainder}
\end{array}
$$

✓**Check:** (Quotient) (Divisor) + Remainder

$$= (x^2 - 2x - 3)(x^2 - x + 1) + x - 2$$
$$= x^4 - x^3 + x^2 - 2x^3 + 2x^2 - 2x - 3x^2 + 3x - 3 + x - 2$$
$$= x^4 - 3x^3 + 2x - 5 = \text{Dividend}$$

As a result,

$$\frac{x^4 - 3x^3 + 2x - 5}{x^2 - x + 1} = x^2 - 2x - 3 + \frac{x - 2}{x^2 - x + 1}$$ ●

The process of dividing two polynomials leads to the following result:

THEOREM

Let Q be a polynomial of positive degree, and let P be a polynomial whose degree is greater than or equal to the degree of Q. The remainder after dividing P by Q is either the zero polynomial or a polynomial whose degree is less than the degree of the divisor Q.

Now Work PROBLEM 95

7 Work with Polynomials in Two Variables

A **monomial in two variables** x and y has the form $ax^n y^m$, where a is a constant, x and y are variables, and n and m are nonnegative integers. The **degree** of a monomial is the sum of the powers of the variables.

For example,

$$2xy^3, \quad x^2y^2, \quad \text{and} \quad x^3y$$

are all monomials that have degree 4.

A **polynomial in two variables** x and y is the sum of one or more monomials in two variables. The **degree of a polynomial** in two variables is the highest degree of all the monomials with nonzero coefficients.

EXAMPLE 15 **Examples of Polynomials in Two Variables**

$3x^2 + 2x^3y + 5$	$\pi x^3 - y^2$	$x^4 + 4x^3y - xy^3 + y^4$
Two variables, degree is 4.	Two variables, degree is 3.	Two variables, degree is 4.

Multiplying polynomials in two variables is handled in the same way as multiplying polynomials in one variable.

EXAMPLE 16 **Using a Special Product Formula**

To multiply $(2x - y)^2$, use the Squares of Binomials formula (3b) with $2x$ instead of x and with y instead of a.

$$(2x - y)^2 = (2x)^2 - 2 \cdot y \cdot 2x + y^2$$
$$= 4x^2 - 4xy + y^2$$

Now Work PROBLEM 81

R.4 Assess Your Understanding

Concepts and Vocabulary

1. The polynomial $3x^4 - 2x^3 + 13x^2 - 5$ is of degree __. The leading coefficient is __.
2. $(x^2 - 4)(x^2 + 4) =$ _______.
3. $(x - 2)(x^2 + 2x + 4) =$ _______.
4. The monomials that make up a polynomial are called which of the following?
   (a) terms (b) variables (c) factors (d) coefficients
5. Choose the degree of the monomial $3x^4y^2$.
   (a) 3 (b) 8 (c) 6 (d) 2
6. ***True or False*** $4x^{-2}$ is a monomial of degree -2.
7. ***True or False*** The degree of the product of two nonzero polynomials equals the sum of their degrees.
8. ***True or False*** $(x + a)(x^2 + ax + a) = x^3 + a^3$.

Skill Building

In Problems 9–18, tell whether the expression is a monomial. If it is, name the variable(s) and the coefficient, and give the degree of the monomial. If it is not a monomial, state why not.

9. $2x^3$ **10.** $-4x^2$ **11.** $\frac{8}{x}$ **12.** $-2x^{-3}$ **13.** $-2xy^2$

14. $5x^2y^3$ **15.** $\frac{8x}{y}$ **16.** $-\frac{2x^2}{y^3}$ **17.** $x^2 + y^2$ **18.** $3x^2 + 4$

In Problems 19–28, tell whether the expression is a polynomial. If it is, give its degree. If it is not, state why not.

19. $3x^2 - 5$ **20.** $1 - 4x$ **21.** 5 **22.** $-\pi$ **23.** $3x^2 - \frac{5}{x}$

24. $\frac{3}{x} + 2$ **25.** $2y^3 - \sqrt{2}$ **26.** $10z^2 + z$ **27.** $\frac{x^2 + 5}{x^3 - 1}$ **28.** $\frac{3x^3 + 2x - 1}{x^2 + x + 1}$

In Problems 29–48, add, subtract, or multiply, as indicated. Express your answer as a single polynomial in standard form.

29. $(x^2 + 4x + 5) + (3x - 3)$ **30.** $(x^3 + 3x^2 + 2) + (x^2 - 4x + 4)$

31. $(x^3 - 2x^2 + 5x + 10) - (2x^2 - 4x + 3)$ **32.** $(x^2 - 3x - 4) - (x^3 - 3x^2 + x + 5)$

33. $(6x^5 + x^3 + x) + (5x^4 - x^3 + 3x^2)$ **34.** $(10x^5 - 8x^2) + (3x^3 - 2x^2 + 6)$

35. $(x^2 - 3x + 1) + 2(3x^2 + x - 4)$ **36.** $-2(x^2 + x + 1) + (-5x^2 - x + 2)$

37. $6(x^3 + x^2 - 3) - 4(2x^3 - 3x^2)$ **38.** $8(4x^3 - 3x^2 - 1) - 6(4x^3 + 8x - 2)$

39. $(x^2 - x + 2) + (2x^2 - 3x + 5) - (x^2 + 1)$ **40.** $(x^2 + 1) - (4x^2 + 5) + (x^2 + x - 2)$

41. $9(y^2 - 3y + 4) - 6(1 - y^2)$ **42.** $8(1 - y^3) + 4(1 + y + y^2 + y^3)$

43. $x(x^2 + x - 4)$ **44.** $4x^2(x^3 - x + 2)$

45. $-2x^2(4x^3 + 5)$ **46.** $5x^3(3x - 4)$

47. $(x + 1)(x^2 + 2x - 4)$ **48.** $(2x - 3)(x^2 + x + 1)$

In Problems 49–66, multiply the polynomials using the FOIL method. Express your answer as a single polynomial in standard form.

49. $(x + 2)(x + 4)$ **50.** $(x + 3)(x + 5)$ **51.** $(2x + 5)(x + 2)$

52. $(3x + 1)(2x + 1)$ **53.** $(x - 4)(x + 2)$ **54.** $(x + 4)(x - 2)$

55. $(x - 3)(x - 2)$ **56.** $(x - 5)(x - 1)$ **57.** $(2x + 3)(x - 2)$

58. $(2x - 4)(3x + 1)$ **59.** $(-2x + 3)(x - 4)$ **60.** $(-3x - 1)(x + 1)$

61. $(-x - 2)(-2x - 4)$ **62.** $(-2x - 3)(3 - x)$ **63.** $(x - 2y)(x + y)$

64. $(2x + 3y)(x - y)$ **65.** $(-2x - 3y)(3x + 2y)$ **66.** $(x - 3y)(-2x + y)$

In Problems 67–90, multiply the polynomials using the special product formulas. Express your answer as a single polynomial in standard form.

67. $(x - 7)(x + 7)$ **68.** $(x - 1)(x + 1)$ **69.** $(2x + 3)(2x - 3)$ **70.** $(3x + 2)(3x - 2)$

71. $(x + 4)^2$ **72.** $(x + 5)^2$ **73.** $(x - 4)^2$ **74.** $(x - 5)^2$

75. $(3x + 4)(3x - 4)$ **76.** $(5x - 3)(5x + 3)$ **77.** $(2x - 3)^2$ **78.** $(3x - 4)^2$

79. $(x + y)(x - y)$
80. $(x + 3y)(x - 3y)$
81. $(3x + y)(3x - y)$
82. $(3x + 4y)(3x - 4y)$
83. $(x + y)^2$
84. $(x - y)^2$
85. $(x - 2y)^2$
86. $(2x + 3y)^2$
87. $(x - 2)^3$
88. $(x + 1)^3$
89. $(2x + 1)^3$
90. $(3x - 2)^3$

In Problems 91–106, find the quotient and the remainder. Check your work by verifying that

$$(\textit{Quotient})(\textit{Divisor}) + \textit{Remainder} = \textit{Dividend}$$

91. $4x^3 - 3x^2 + x + 1$ divided by $x + 2$
92. $3x^3 - x^2 + x - 2$ divided by $x + 2$
93. $4x^3 - 3x^2 + x + 1$ divided by x^2
94. $3x^3 - x^2 + x - 2$ divided by x^2
95. $5x^4 - 3x^2 + x + 1$ divided by $x^2 + 2$
96. $5x^4 - x^2 + x - 2$ divided by $x^2 + 2$
97. $4x^5 - 3x^2 + x + 1$ divided by $2x^3 - 1$
98. $3x^5 - x^2 + x - 2$ divided by $3x^3 - 1$
99. $2x^4 - 3x^3 + x + 1$ divided by $2x^2 + x + 1$
100. $3x^4 - x^3 + x - 2$ divided by $3x^2 + x + 1$
101. $-4x^3 + x^2 - 4$ divided by $x - 1$
102. $-3x^4 - 2x - 1$ divided by $x - 1$
103. $1 - x^2 + x^4$ divided by $x^2 + x + 1$
104. $1 - x^2 + x^4$ divided by $x^2 - x + 1$
105. $x^3 - a^3$ divided by $x - a$
106. $x^5 - a^5$ divided by $x - a$

Explaining Concepts: Discussion and Writing

107. Explain why the degree of the product of two nonzero polynomials equals the sum of their degrees.
108. Explain why the degree of the sum of two polynomials of different degrees equals the larger of their degrees.
109. Give a careful statement about the degree of the sum of two polynomials of the same degree.
110. Do you prefer adding two polynomials using the horizontal method or the vertical method? Write a brief position paper defending your choice.
111. Do you prefer to memorize the rule for the square of a binomial $(x + a)^2$ or to use FOIL to obtain the product? Write a brief position paper defending your choice.

R.5 Factoring Polynomials

OBJECTIVES
1 Factor the Difference of Two Squares and the Sum and Difference of Two Cubes (p. 50)
2 Factor Perfect Squares (p. 51)
3 Factor a Second-Degree Polynomial: $x^2 + Bx + C$ (p. 52)
4 Factor by Grouping (p. 53)
5 Factor a Second-Degree Polynomial: $Ax^2 + Bx + C,\ A \neq 1$ (p. 54)
6 Complete the Square (p. 56)

Consider the following product:

$$(2x + 3)(x - 4) = 2x^2 - 5x - 12$$

The two polynomials on the left side are called **factors** of the polynomial on the right side. Expressing a given polynomial as a product of other polynomials—that is, finding the factors of a polynomial—is called **factoring**.

We shall restrict our discussion here to factoring polynomials in one variable into products of polynomials in one variable, where all coefficients are integers. We call this **factoring over the integers**.

Any polynomial can be written as the product of 1 times itself or as -1 times its additive inverse. If a polynomial cannot be written as the product of two other polynomials (excluding 1 and -1), then the polynomial is **prime**. When a polynomial has been written as a product consisting only of prime factors, it is **factored completely**. Examples of prime polynomials (over the integers) are

$$2, \quad 3, \quad 5, \quad x, \quad x + 1, \quad x - 1, \quad 3x + 4, \quad x^2 + 4$$

COMMENT Over the real numbers, $3x + 4$ factors into $3\left(x + \frac{4}{3}\right)$. It is the noninteger $\frac{4}{3}$ that causes $3x + 4$ to be prime over the integers. In most instances, we will be factoring over the integers. ■

The first factor to look for in a factoring problem is a common monomial factor present in each term of the polynomial. If one is present, use the Distributive Property to factor it out. Continue factoring out monomial factors until none are left.

EXAMPLE 1 **Identifying Common Monomial Factors**

Polynomial	Common Monomial Factor	Remaining Factor	Factored Form
$2x + 4$	2	$x + 2$	$2x + 4 = 2(x + 2)$
$3x - 6$	3	$x - 2$	$3x - 6 = 3(x - 2)$
$2x^2 - 4x + 8$	2	$x^2 - 2x + 4$	$2x^2 - 4x + 8 = 2(x^2 - 2x + 4)$
$8x - 12$	4	$2x - 3$	$8x - 12 = 4(2x - 3)$
$x^2 + x$	x	$x + 1$	$x^2 + x = x(x + 1)$
$x^3 - 3x^2$	x^2	$x - 3$	$x^3 - 3x^2 = x^2(x - 3)$
$6x^2 + 9x$	$3x$	$2x + 3$	$6x^2 + 9x = 3x(2x + 3)$ ●

Notice that once all common monomial factors have been removed from a polynomial, the remaining factor is either a prime polynomial of degree 1 or a polynomial of degree 2 or higher. (Do you see why?)

Now Work PROBLEM 9

1 Factor the Difference of Two Squares and the Sum and Difference of Two Cubes

When you factor a polynomial, first check for common monomial factors. Then see whether you can use one of the special formulas discussed in the previous section.

Difference of Two Squares	$x^2 - a^2 = (x - a)(x + a)$
Perfect Squares	$x^2 + 2ax + a^2 = (x + a)^2$
	$x^2 - 2ax + a^2 = (x - a)^2$
Sum of Two Cubes	$x^3 + a^3 = (x + a)(x^2 - ax + a^2)$
Difference of Two Cubes	$x^3 - a^3 = (x - a)(x^2 + ax + a^2)$

EXAMPLE 2 **Factoring the Difference of Two Squares**

Factor completely: $x^2 - 4$

Solution Note that $x^2 - 4$ is the difference of two squares, x^2 and 2^2.

$$x^2 - 4 = (x - 2)(x + 2)$$ ●

EXAMPLE 3 **Factoring the Difference of Two Cubes**

Factor completely: $x^3 - 1$

Solution Because $x^3 - 1$ is the difference of two cubes, x^3 and 1^3,

$$x^3 - 1 = (x - 1)(x^2 + x + 1)$$

●

EXAMPLE 4 **Factoring the Sum of Two Cubes**

Factor completely: $x^3 + 8$

Solution Because $x^3 + 8$ is the sum of two cubes, x^3 and 2^3,

$$x^3 + 8 = (x + 2)(x^2 - 2x + 4)$$

●

EXAMPLE 5 **Factoring the Difference of Two Squares**

Factor completely: $x^4 - 16$

Solution Because $x^4 - 16$ is the difference of two squares, $x^4 = (x^2)^2$ and $16 = 4^2$,

$$x^4 - 16 = (x^2 - 4)(x^2 + 4)$$

But $x^2 - 4$ is also the difference of two squares. Then,

$$x^4 - 16 = (x^2 - 4)(x^2 + 4) = (x - 2)(x + 2)(x^2 + 4)$$

●

Now Work PROBLEMS 19 AND 37

2 Factor Perfect Squares

When the first term and third term of a trinomial are both positive and are perfect squares, such as x^2, $9x^2$, 1, and 4, check to see whether the trinomial is a perfect square.

EXAMPLE 6 **Factoring a Perfect Square**

Factor completely: $x^2 + 6x + 9$

Solution The first term, x^2, and the third term, $9 = 3^2$, are perfect squares. Because the middle term, $6x$, is twice the product of x and 3, we have a perfect square.

$$x^2 + 6x + 9 = (x + 3)^2$$

●

EXAMPLE 7 **Factoring a Perfect Square**

Factor completely: $9x^2 - 6x + 1$

Solution The first term, $9x^2 = (3x)^2$, and the third term, $1 = 1^2$, are perfect squares. Because the middle term, $-6x$, is -2 times the product of $3x$ and 1, we have a perfect square.

$$9x^2 - 6x + 1 = (3x - 1)^2$$

●

EXAMPLE 8 **Factoring a Perfect Square**

Factor completely: $25x^2 + 30x + 9$

Solution The first term, $25x^2 = (5x)^2$, and the third term, $9 = 3^2$, are perfect squares. Because the middle term, $30x$, is twice the product of $5x$ and 3, we have a perfect square.

$$25x^2 + 30x + 9 = (5x + 3)^2$$

●

Now Work PROBLEMS 29 AND 103

If a trinomial is not a perfect square, it may be possible to factor it using the technique discussed next.

3 Factor a Second-Degree Polynomial: $x^2 + Bx + C$

The idea behind factoring a second-degree polynomial like $x^2 + Bx + C$ is to see whether it can be made equal to the product of two (possibly equal) first-degree polynomials.

For example, consider

$$(x + 3)(x + 4) = x^2 + 7x + 12$$

The factors of $x^2 + 7x + 12$ are $x + 3$ and $x + 4$. Notice the following:

$$x^2 + 7x + 12 = (x + 3)(x + 4)$$

12 is the product of 3 and 4

7 is the sum of 3 and 4

In general, if $x^2 + Bx + C = (x + a)(x + b) = x^2 + (a + b)x + ab$, then $ab = C$ and $a + b = B$.

To factor a second-degree polynomial $x^2 + Bx + C$, find integers whose product is C and whose sum is B. That is, if there are numbers a, b, where $ab = C$ and $a + b = B$, then

$$x^2 + Bx + C = (x + a)(x + b)$$

EXAMPLE 9 **Factoring a Trinomial**

Factor completely: $x^2 + 7x + 10$

Solution First determine all pairs of integers whose product is 10, and then compute their sums.

Integers whose product is 10	1, 10	−1, −10	2, 5	−2, −5
Sum	11	−11	7	−7

The integers 2 and 5 have a product of 10 and add up to 7, the coefficient of the middle term. As a result,

$$x^2 + 7x + 10 = (x + 2)(x + 5)$$

EXAMPLE 10 **Factoring a Trinomial**

Factor completely: $x^2 - 6x + 8$

Solution First determine all pairs of integers whose product is 8, and then compute each sum.

Integers whose product is 8	1, 8	−1, −8	2, 4	−2, −4
Sum	9	−9	6	−6

Since −6 is the coefficient of the middle term,

$$x^2 - 6x + 8 = (x - 2)(x - 4)$$

EXAMPLE 11 **Factoring a Trinomial**

Factor completely: $x^2 - x - 12$

Solution First determine all pairs of integers whose product is -12, and then compute each sum.

Integers whose product is -12	1, −12	−1, 12	2, −6	−2, 6	3, −4	−3, 4
Sum	−11	11	−4	4	−1	1

Since -1 is the coefficient of the middle term,

$$x^2 - x - 12 = (x + 3)(x - 4)$$

EXAMPLE 12 **Factoring a Trinomial**

Factor completely: $x^2 + 4x - 12$

Solution The integers -2 and 6 have a product of -12 and have the sum 4. So,

$$x^2 + 4x - 12 = (x - 2)(x + 6)$$

To avoid errors in factoring, always check your answer by multiplying it out to see whether the result equals the original expression.

When none of the possibilities works, the polynomial is prime.

EXAMPLE 13 **Identifying a Prime Polynomial**

Show that $x^2 + 9$ is prime.

Solution First list the pairs of integers whose product is 9, and then compute their sums.

Integers whose product is 9	1, 9	−1, −9	3, 3	−3, −3
Sum	10	−10	6	−6

Since the coefficient of the middle term in $x^2 + 9 = x^2 + 0x + 9$ is 0 and none of the sums equals 0, we conclude that $x^2 + 9$ is prime.

Example 13 demonstrates a more general result:

THEOREM

Any polynomial of the form $x^2 + a^2$, a real, is prime.

Now Work PROBLEMS 43 AND 87

4 Factor by Grouping

Sometimes a common factor does not occur in every term of the polynomial but does occur in each of several groups of terms that together make up the polynomial. When this happens, the common factor can be factored out of each group by means of the Distributive Property. This technique is called **factoring by grouping**.

EXAMPLE 14 **Factoring by Grouping**

Factor completely by grouping: $(x^2 + 2)x + (x^2 + 2) \cdot 3$

Solution Notice the common factor $x^2 + 2$. Applying the Distributive Property yields

$$(x^2 + 2)x + (x^2 + 2) \cdot 3 = (x^2 + 2)(x + 3)$$

Since $x^2 + 2$ and $x + 3$ are prime, the factorization is complete.

The next example shows a factoring problem that occurs in calculus.

EXAMPLE 15 **Factoring by Grouping**

Factor completely by grouping: $3(x-1)^2(x+2)^4 + 4(x-1)^3(x+2)^3$

Solution Here, $(x-1)^2(x+2)^3$ is a common factor of both $3(x-1)^2(x+2)^4$ and $4(x-1)^3(x+2)^3$. As a result,

$$\begin{aligned} 3(x-1)^2(x+2)^4 + 4(x-1)^3(x+2)^3 &= (x-1)^2(x+2)^3[3(x+2)+4(x-1)] \\ &= (x-1)^2(x+2)^3[3x+6+4x-4] \\ &= (x-1)^2(x+2)^3(7x+2) \end{aligned}$$

EXAMPLE 16 **Factoring by Grouping**

Factor completely by grouping: $x^3 - 4x^2 + 2x - 8$

Solution To see whether factoring by grouping will work, group the first two terms and the last two terms. Then look for a common factor in each group. In this example, factor x^2 from $x^3 - 4x^2$ and 2 from $2x - 8$. The remaining factor in each case is the same, $x - 4$. This means that factoring by grouping will work, as follows:

$$\begin{aligned} x^3 - 4x^2 + 2x - 8 &= (x^3 - 4x^2) + (2x - 8) \\ &= x^2(x-4) + 2(x-4) \\ &= (x-4)(x^2+2) \end{aligned}$$

Since $x^2 + 2$ and $x - 4$ are prime, the factorization is complete.

Now Work PROBLEMS 55 AND 131

5 Factor a Second-Degree Polynomial: $Ax^2 + Bx + C$, $A \neq 1$

To factor a second-degree polynomial $Ax^2 + Bx + C$, when $A \neq 1$ and A, B, and C have no common factors, follow these steps:

Steps for Factoring $Ax^2 + Bx + C$, When $A \neq 1$ and A, B, and C Have No Common Factors

STEP 1: Find the value of AC.

STEP 2: Find a pair of integers whose product is AC and that add up to B. That is, find a and b such that $ab = AC$ and $a + b = B$.

STEP 3: Write $Ax^2 + Bx + C = Ax^2 + ax + bx + C$.

STEP 4: Factor this last expression by grouping.

EXAMPLE 17 **Factoring a Trinomial**

Factor completely: $2x^2 + 5x + 3$

Solution Comparing $2x^2 + 5x + 3$ to $Ax^2 + Bx + C$, we find that $A = 2$, $B = 5$, and $C = 3$.

STEP 1: The value of AC is $2 \cdot 3 = 6$.

STEP 2: Determine the pairs of integers whose product is $AC = 6$ and compute their sums.

Integers whose product is 6	1, 6	−1, −6	2, 3	−2, −3
Sum	7	−7	5	−5

STEP 3: The integers whose product is 6 that add up to $B = 5$ are 2 and 3.

$$2x^2 + 5x + 3 = 2x^2 + 2x + 3x + 3$$

STEP 4: Factor by grouping.

$$\begin{aligned} 2x^2 + 2x + 3x + 3 &= (2x^2 + 2x) + (3x + 3) \\ &= 2x(x + 1) + 3(x + 1) \\ &= (x + 1)(2x + 3) \end{aligned}$$

As a result,

$$2x^2 + 5x + 3 = (x + 1)(2x + 3)$$

EXAMPLE 18 Factoring a Trinomial

Factor completely: $2x^2 - x - 6$

Solution Comparing $2x^2 - x - 6$ to $Ax^2 + Bx + C$, we find that $A = 2, B = -1$, and $C = -6$.

STEP 1: The value of AC is $2 \cdot (-6) = -12$.
STEP 2: Determine the pairs of integers whose product is $AC = -12$ and compute their sums.

Integers whose product is − 12	1, −12	−1, 12	2, −6	−2, 6	3, −4	−3, 4
Sum	−11	11	−4	4	−1	1

STEP 3: The integers whose product is -12 that add up to $B = -1$ are -4 and 3.

$$2x^2 - x - 6 = 2x^2 - 4x + 3x - 6$$

STEP 4: Factor by grouping.

$$\begin{aligned} 2x^2 - 4x + 3x - 6 &= (2x^2 - 4x) + (3x - 6) \\ &= 2x(x - 2) + 3(x - 2) \\ &= (x - 2)(2x + 3) \end{aligned}$$

As a result,

$$2x^2 - x - 6 = (x - 2)(2x + 3)$$

Now Work PROBLEM 61

SUMMARY

Type of Polynomial	Method	Example
Any polynomial	Look for common monomial factors. (Always do this first!)	$6x^2 + 9x = 3x(2x + 3)$
Binomials of degree 2 or higher	Check for a special product:	
	Difference of two squares, $x^2 - a^2$	$x^2 - 16 = (x - 4)(x + 4)$
	Difference of two cubes, $x^3 - a^3$	$x^3 - 64 = (x - 4)(x^2 + 4x + 16)$
	Sum of two cubes, $x^3 + a^3$	$x^3 + 27 = (x + 3)(x^2 - 3x + 9)$
Trinomials of degree 2	Check for a perfect square, $(x \pm a)^2$	$x^2 + 8x + 16 = (x + 4)^2$ $x^2 - 10x + 25 = (x - 5)^2$
	Factoring $x^2 + Bx + C$ (p. 52)	$x^2 - x - 2 = (x - 2)(x + 1)$
	Factoring $Ax^2 + Bx + C$ (p. 54)	$6x^2 + x - 1 = (2x + 1)(3x - 1)$
Four or more terms	Grouping	$2x^3 - 3x^2 + 4x - 6 = (2x - 3)(x^2 + 2)$

6 Complete the Square

The idea behind completing the square in one variable is to "adjust" an expression of the form $x^2 + bx$ to make it a perfect square. Perfect squares are trinomials of the form

$$x^2 + 2ax + a^2 = (x + a)^2 \text{ or } x^2 - 2ax + a^2 = (x - a)^2$$

For example, $x^2 + 6x + 9$ is a perfect square because $x^2 + 6x + 9 = (x + 3)^2$. And $p^2 - 12p + 36$ is a perfect square because $p^2 - 12p + 36 = (p - 6)^2$.

So how do we "adjust" $x^2 + bx$ to make it a perfect square? We do it by adding a number. For example, to make $x^2 + 6x$ a perfect square, add 9. But how do we know to add 9? If we divide the coefficient of the first-degree term, 6, by 2, and then square the result, we obtain 9. This approach works in general.

WARNING To use $\left(\frac{1}{2}b\right)^2$ to complete the square, the coefficient of the x^2 term must be 1.

Completing the Square of $x^2 + bx$

Identify the coefficient of the first-degree term. Multiply this coefficient by $\frac{1}{2}$ and then square the result. That is, determine the value of b in $x^2 + bx$ and compute $\left(\frac{1}{2}b\right)^2$.

EXAMPLE 19 **Completing the Square**

Determine the number that must be added to each expression to complete the square. Then factor the expression.

Start	Add	Result	Factored Form
$y^2 + 8y$	$\left(\frac{1}{2} \cdot 8\right)^2 = 16$	$y^2 + 8y + 16$	$(y + 4)^2$
$x^2 + 12x$	$\left(\frac{1}{2} \cdot 12\right)^2 = 36$	$x^2 + 12x + 36$	$(x + 6)^2$
$a^2 - 20a$	$\left(\frac{1}{2} \cdot (-20)\right)^2 = 100$	$a^2 - 20a + 100$	$(a - 10)^2$
$p^2 - 5p$	$\left(\frac{1}{2} \cdot (-5)\right)^2 = \frac{25}{4}$	$p^2 - 5p + \frac{25}{4}$	$\left(p - \frac{5}{2}\right)^2$

Notice that the factored form of a perfect square is either

$$x^2 + bx + \left(\frac{b}{2}\right)^2 = \left(x + \frac{b}{2}\right)^2 \quad \text{or} \quad x^2 - bx + \left(\frac{b}{2}\right)^2 = \left(x - \frac{b}{2}\right)^2$$

Now Work PROBLEM 73

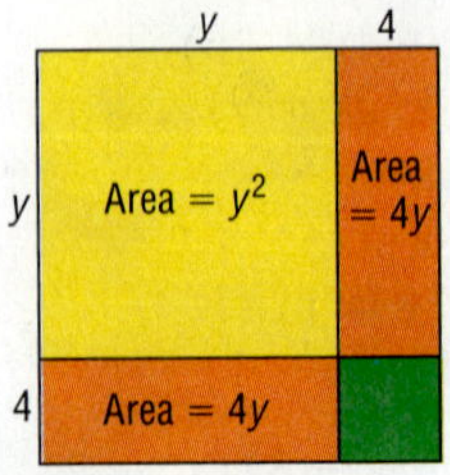

Figure 27

Are you wondering why we refer to making an expression a perfect square as "completing the square"? Look at the square in Figure 27. Its area is $(y + 4)^2$. The yellow area is y^2 and each orange area is $4y$ (for a total area of $8y$). The sum of these areas is $y^2 + 8y$. To complete the square, we need to add the area of the green region: $4 \cdot 4 = 16$. As a result, $y^2 + 8y + 16 = (y + 4)^2$.

R.5 Assess Your Understanding

Concepts and Vocabulary

1. If factored completely, $3x^3 - 12x =$ ______________.
2. If a polynomial cannot be written as the product of two other polynomials (excluding 1 and -1), then the polynomial is said to be ______.
3. For $x^2 + Bx + C = (x + a)(x + b)$, which of the following must be true?
   (a) $ab = B$ and $a + b = C$
   (b) $a + b = C$ and $a - b = B$
   (c) $ab = C$ and $a + b = B$
   (d) $ab = B$ and $a - b = C$
4. Choose the best description of $x^2 - 64$.
   (a) Prime (b) Difference of two squares
   (c) Difference of two cubes (d) Perfect Square
5. Choose the complete factorization of $4x^2 - 8x - 60$.
   (a) $2(x + 3)(x - 5)$ (b) $4(x^2 - 2x - 15)$
   (c) $(2x + 6)(2x - 10)$ (d) $4(x + 3)(x - 5)$
6. To complete the square of $x^2 + bx$, use which of the following?
   (a) $(2b)^2$ (b) $2b^2$ (c) $\left(\frac{1}{2}b\right)^2$ (d) $\frac{1}{2}b^2$
7. ***True or False*** The polynomial $x^2 + 4$ is prime.
8. ***True or False*** $3x^3 - 2x^2 - 6x + 4 = (3x - 2)(x^2 + 2)$.

Skill Building

In Problems 9–18, factor each polynomial by removing the common monomial factor.

9. $3x + 6$ 10. $7x - 14$ 11. $ax^2 + a$ 12. $ax - a$ 13. $x^3 + x^2 + x$
14. $x^3 - x^2 + x$ 15. $2x^2 - 2x$ 16. $3x^2 - 3x$ 17. $3x^2y - 6xy^2 + 12xy$ 18. $60x^2y - 48xy^2 + 72x^3y$

In Problems 19–26, factor the difference of two squares.

19. $x^2 - 1$ 20. $x^2 - 4$ 21. $4x^2 - 1$ 22. $9x^2 - 1$
23. $x^2 - 16$ 24. $x^2 - 25$ 25. $25x^2 - 4$ 26. $36x^2 - 9$

In Problems 27–36, factor the perfect squares.

27. $x^2 + 2x + 1$ 28. $x^2 - 4x + 4$ 29. $x^2 + 4x + 4$ 30. $x^2 - 2x + 1$
31. $x^2 - 10x + 25$ 32. $x^2 + 10x + 25$ 33. $4x^2 + 4x + 1$ 34. $9x^2 + 6x + 1$
35. $16x^2 + 8x + 1$ 36. $25x^2 + 10x + 1$

In Problems 37–42, factor the sum or difference of two cubes.

37. $x^3 - 27$ 38. $x^3 + 125$ 39. $x^3 + 27$ 40. $27 - 8x^3$ 41. $8x^3 + 27$ 42. $64 - 27x^3$

In Problems 43–54, factor each polynomial.

43. $x^2 + 5x + 6$ 44. $x^2 + 6x + 8$ 45. $x^2 + 7x + 6$ 46. $x^2 + 9x + 8$
47. $x^2 + 7x + 10$ 48. $x^2 + 11x + 10$ 49. $x^2 - 10x + 16$ 50. $x^2 - 17x + 16$
51. $x^2 - 7x - 8$ 52. $x^2 - 2x - 8$ 53. $x^2 + 7x - 8$ 54. $x^2 + 2x - 8$

In Problems 55–60, factor by grouping.

55. $2x^2 + 4x + 3x + 6$ 56. $3x^2 - 3x + 2x - 2$ 57. $2x^2 - 4x + x - 2$
58. $3x^2 + 6x - x - 2$ 59. $6x^2 + 9x + 4x + 6$ 60. $9x^2 - 6x + 3x - 2$

In Problems 61–72, factor each polynomial.

61. $3x^2 + 4x + 1$ 62. $2x^2 + 3x + 1$ 63. $2z^2 + 5z + 3$ 64. $6z^2 + 5z + 1$
65. $3x^2 + 2x - 8$ 66. $3x^2 + 10x + 8$ 67. $3x^2 - 2x - 8$ 68. $3x^2 - 10x + 8$
69. $3x^2 + 14x + 8$ 70. $3x^2 - 14x + 8$ 71. $3x^2 + 10x - 8$ 72. $3x^2 - 10x - 8$

In Problems 73–78, determine what number should be added to complete the square of each expression. Then factor each expression.

73. $x^2 + 10x$ 74. $p^2 + 14p$ 75. $y^2 - 6y$
76. $x^2 - 4x$ 77. $x^2 - \frac{1}{2}x$ 78. $x^2 + \frac{1}{3}x$

Mixed Practice

In Problems 79–126, factor each polynomial completely. If the polynomial cannot be factored, say it is prime.

79. $x^2 - 36$
80. $x^2 - 9$
81. $2 - 8x^2$
82. $3 - 27x^2$
83. $x^2 + 11x + 10$
84. $x^2 + 5x + 4$
85. $x^2 - 10x + 21$
86. $x^2 - 6x + 8$
87. $4x^2 - 8x + 32$
88. $3x^2 - 12x + 15$
89. $x^2 + 4x + 16$
90. $x^2 + 12x + 36$
91. $15 + 2x - x^2$
92. $14 + 6x - x^2$
93. $3x^2 - 12x - 36$
94. $x^3 + 8x^2 - 20x$
95. $y^4 + 11y^3 + 30y^2$
96. $3y^3 - 18y^2 - 48y$
97. $4x^2 + 12x + 9$
98. $9x^2 - 12x + 4$
99. $6x^2 + 8x + 2$
100. $8x^2 + 6x - 2$
101. $x^4 - 81$
102. $x^4 - 1$
103. $x^6 - 2x^3 + 1$
104. $x^6 + 2x^3 + 1$
105. $x^7 - x^5$
106. $x^8 - x^5$
107. $16x^2 + 24x + 9$
108. $9x^2 - 24x + 16$
109. $5 + 16x - 16x^2$
110. $5 + 11x - 16x^2$
111. $4y^2 - 16y + 15$
112. $9y^2 + 9y - 4$
113. $1 - 8x^2 - 9x^4$
114. $4 - 14x^2 - 8x^4$
115. $x(x + 3) - 6(x + 3)$
116. $5(3x - 7) + x(3x - 7)$
117. $(x + 2)^2 - 5(x + 2)$
118. $(x - 1)^2 - 2(x - 1)$
119. $(3x - 2)^3 - 27$
120. $(5x + 1)^3 - 1$
121. $3(x^2 + 10x + 25) - 4(x + 5)$
122. $7(x^2 - 6x + 9) + 5(x - 3)$
123. $x^3 + 2x^2 - x - 2$
124. $x^3 - 3x^2 - x + 3$
125. $x^4 - x^3 + x - 1$
126. $x^4 + x^3 + x + 1$

Applications and Extensions

In Problems 127–136, expressions that occur in calculus are given. Factor each expression completely.

127. $2(3x + 4)^2 + (2x + 3) \cdot 2(3x + 4) \cdot 3$
128. $5(2x + 1)^2 + (5x - 6) \cdot 2(2x + 1) \cdot 2$
129. $2x(2x + 5) + x^2 \cdot 2$
130. $3x^2(8x - 3) + x^3 \cdot 8$
131. $2(x + 3)(x - 2)^3 + (x + 3)^2 \cdot 3(x - 2)^2$
132. $4(x + 5)^3(x - 1)^2 + (x + 5)^4 \cdot 2(x - 1)$
133. $(4x - 3)^2 + x \cdot 2(4x - 3) \cdot 4$
134. $3x^2(3x + 4)^2 + x^3 \cdot 2(3x + 4) \cdot 3$
135. $2(3x - 5) \cdot 3(2x + 1)^3 + (3x - 5)^2 \cdot 3(2x + 1)^2 \cdot 2$
136. $3(4x + 5)^2 \cdot 4(5x + 1)^2 + (4x + 5)^3 \cdot 2(5x + 1) \cdot 5$
137. Show that $x^2 + 4$ is prime.
138. Show that $x^2 + x + 1$ is prime.

Explaining Concepts: Discussion and Writing

139. Make up a polynomial that factors into a perfect square.

140. Explain to a fellow student what you look for first when presented with a factoring problem. What do you do next?

R.6 Synthetic Division

OBJECTIVE 1 Divide Polynomials Using Synthetic Division (p. 58)

1 Divide Polynomials Using Synthetic Division

To find the quotient as well as the remainder when a polynomial of degree 1 or higher is divided by $x - c$, a shortened version of long division, called **synthetic division**, makes the task simpler.

To see how synthetic division works, first consider long division for dividing the polynomial $2x^3 - x^2 + 3$ by $x - 3$.

$$\begin{array}{r l}
2x^2 + 5x \quad + 15 & \leftarrow \text{Quotient} \\
x - 3\overline{)2x^3 - x^2 \qquad\quad + 3} & \\
2x^3 - 6x^2 \qquad\qquad\quad & \\
\hline
5x^2 \qquad\qquad\quad & \\
5x^2 - 15x \qquad\quad & \\
\hline
15x + 3 & \\
15x - 45 & \\
\hline
48 & \leftarrow \text{Remainder}
\end{array}$$

Check: (Divisor) · (Quotient) + Remainder

$$\begin{aligned}
&= (x - 3)(2x^2 + 5x + 15) + 48 \\
&= 2x^3 + 5x^2 + 15x - 6x^2 - 15x - 45 + 48 \\
&= 2x^3 - x^2 + 3
\end{aligned}$$

The process of synthetic division arises from rewriting the long division in a more compact form, using simpler notation. For example, in the long division above, the terms in blue are not really necessary because they are identical to the terms directly above them. With these terms removed, we have

$$\begin{array}{r}
2x^2 + 5x \quad + 15 \\
x - 3\overline{)2x^3 - x^2 \qquad\quad + 3} \\
-6x^2 \qquad\qquad\quad \\
\hline
5x^2 \qquad\qquad\quad \\
-15x \qquad\quad \\
\hline
15x \qquad\quad \\
-45 \\
\hline
48
\end{array}$$

Most of the x's that appear in this process can also be removed, provided that we are careful about positioning each coefficient. In this regard, we will need to use 0 as the coefficient of x in the dividend, because that power of x is missing. Now we have

$$\begin{array}{r}
2x^2 + 5x + 15 \\
x - 3\overline{)2 \quad -1 \quad 0 \quad 3} \\
-6 \qquad\qquad \\
\hline
5 \qquad\qquad \\
-15 \qquad \\
\hline
15 \qquad \\
-45 \\
\hline
48
\end{array}$$

We can make this display more compact by moving the lines up until the numbers in blue align horizontally.

$$\begin{array}{r r r r l}
 & 2x^2 & + 5x & + 15 & \text{Row 1} \\
x - 3\overline{)2} & -1 & 0 & 3 & \text{Row 2} \\
 & -6 & -15 & -45 & \text{Row 3} \\
\hline
\bigcirc & 5 & 15 & 48 & \text{Row 4}
\end{array}$$

Because the leading coefficient of the divisor is always 1, the leading coefficient of the dividend will also be the leading coefficient of the quotient. So we place the leading coefficient of the quotient, 2, in the circled position. Now, the first three numbers in row 4 are precisely the coefficients of the quotient, and the last number in row 4 is the remainder. Since row 1 is not really needed, we can compress the process to three rows, where the bottom row contains both the coefficients of the quotient and the remainder.

$$\begin{array}{r|rrrrl} x-3 & 2 & -1 & 0 & 3 & \text{Row 1} \\ & & -6 & -15 & -45 & \text{Row 2 (subtract)} \\ \hline & 2 & 5 & 15 & 48 & \text{Row 3} \end{array}$$

Recall that the entries in row 3 are obtained by subtracting the entries in row 2 from those in row 1. Rather than subtracting the entries in row 2, we can change the sign of each entry and add. With this modification, our display will look like this:

$$\begin{array}{r|rrrrl} x-3 & 2 & -1 & 0 & 3 & \text{Row 1} \\ & & 6 & 15 & 45 & \text{Row 2 (add)} \\ \hline & 2 & 5 & 15 & 48 & \text{Row 3} \end{array}$$

Notice that the entries in row 2 are three times the prior entries in row 3. Our last modification to the display replaces the $x - 3$ by 3. The entries in row 3 give the quotient and the remainder, as shown next.

$$\begin{array}{r|rrrrl} 3 & 2 & -1 & 0 & 3 & \text{Row 1} \\ & & 6 & 15 & 45 & \text{Row 2 (add)} \\ \hline & 2 & 5 & 15 & 48 & \text{Row 3} \end{array}$$

Quotient: $2x^2 + 5x + 15$ Remainder: 48

Let's go through an example step by step.

EXAMPLE 1

Using Synthetic Division to Find the Quotient and Remainder

Use synthetic division to find the quotient and remainder when

$$x^3 - 4x^2 - 5 \quad \text{is divided by} \quad x - 3$$

Solution

STEP 1: Write the dividend in descending powers of x. Then copy the coefficients, remembering to insert a 0 for any missing powers of x.

$$\begin{array}{rrrrl} 1 & -4 & 0 & -5 & \text{Row 1} \end{array}$$

STEP 2: Insert the usual division symbol. In synthetic division, the divisor is of the form $x - c$, and c is the number placed to the left of the division symbol. Here, since the divisor is $x - 3$, insert 3 to the left of the division symbol.

$$\begin{array}{r|rrrrl} 3 & 1 & -4 & 0 & -5 & \text{Row 1} \end{array}$$

STEP 3: Bring the 1 down two rows, and enter it in row 3.

$$\begin{array}{r|rrrrl} 3 & 1 & -4 & 0 & -5 & \text{Row 1} \\ & \downarrow & & & & \text{Row 2} \\ \hline & 1 & & & & \text{Row 3} \end{array}$$

STEP 4: Multiply the latest entry in row 3 by 3, and place the result in row 2, one column over to the right.

$$\begin{array}{r|rrrrl} 3 & 1 & -4 & 0 & -5 & \text{Row 1} \\ & & 3 & & & \text{Row 2} \\ \hline & 1 & & & & \text{Row 3} \end{array}$$

STEP 5: Add the entry in row 2 to the entry above it in row 1, and enter the sum in row 3.

$$\begin{array}{r|rrrrl} 3 & 1 & -4 & 0 & -5 & \text{Row 1} \\ & & 3 & & & \text{Row 2} \\ \hline & 1 & -1 & & & \text{Row 3} \end{array}$$

Step 6: Repeat Steps 4 and 5 until no more entries are available in row 1.

$$\begin{array}{r|rrrrl} 3 & 1 & -4 & 0 & -5 & \text{Row 1} \\ & & 3 & -3 & -9 & \text{Row 2} \\ \hline & 1 & -1 & -3 & -14 & \text{Row 3} \end{array}$$

Step 7: The final entry in row 3, the -14, is the remainder; the other entries in row 3, the 1, -1, and -3, are the coefficients (in descending order) of a polynomial whose degree is 1 less than that of the dividend. This is the quotient. That is,

$$\text{Quotient} = x^2 - x - 3 \qquad \text{Remainder} = -14$$

✓**Check:** (Divisor) (Quotient) + Remainder

$$= (x - 3)(x^2 - x - 3) + (-14)$$
$$= (x^3 - x^2 - 3x - 3x^2 + 3x + 9) + (-14)$$
$$= x^3 - 4x^2 - 5 = \text{Dividend}$$

Let's do an example in which all seven steps are combined.

EXAMPLE 2 Using Synthetic Division to Verify a Factor

Use synthetic division to show that $x + 3$ is a factor of

$$2x^5 + 5x^4 - 2x^3 + 2x^2 - 2x + 3$$

Solution The divisor is $x + 3 = x - (-3)$, so place -3 to the left of the division symbol. Then the row 3 entries will be multiplied by -3, entered in row 2, and added to row 1.

$$\begin{array}{r|rrrrrrl} -3 & 2 & 5 & -2 & 2 & -2 & 3 & \text{Row 1} \\ & & -6 & 3 & -3 & 3 & -3 & \text{Row 2} \\ \hline & 2 & -1 & 1 & -1 & 1 & 0 & \text{Row 3} \end{array}$$

Because the remainder is 0, we have

(Divisor) (Quotient) + Remainder

$$= (x + 3)(2x^4 - x^3 + x^2 - x + 1) = 2x^5 + 5x^4 - 2x^3 + 2x^2 - 2x + 3$$

As we see, $x + 3$ is a factor of $2x^5 + 5x^4 - 2x^3 + 2x^2 - 2x + 3$.

As Example 2 illustrates, the remainder after division gives information about whether the divisor is, or is not, a factor. We shall have more to say about this in Chapter 5.

Now Work PROBLEMS 9 AND 19

R.6 Assess Your Understanding

Concepts and Vocabulary

1. To check division, the expression that is being divided, the dividend, should equal the product of the _______ and the _______ plus the _______.

2. To divide $2x^3 - 5x + 1$ by $x + 3$ using synthetic division, the first step is to write ___) __________.

3. Choose the division problem that cannot be done using synthetic division.
   (a) $2x^3 - 4x^2 + 6x - 8$ is divided by $x - 8$
   (b) $x^4 - 3$ is divided by $x + 1$
   (c) $x^5 + 3x^2 - 9x + 2$ is divided by $x + 10$
   (d) $x^4 - 5x^3 + 3x^2 - 9x + 13$ is divided by $x^2 + 5$

4. Choose the correct conclusion based on the following synthetic division:
$$\begin{array}{r|rrrr} -5 & 2 & 3 & -38 & -15 \\ & & -10 & 35 & 15 \\ \hline & 2 & -7 & -3 & 0 \end{array}$$
   (a) $x + 5$ is a factor of $2x^3 + 3x^2 - 38x - 15$
   (b) $x - 5$ is a factor of $2x^3 + 3x^2 - 38x - 15$
   (c) $x + 5$ is not a factor of $2x^3 + 3x^2 - 38x - 15$
   (d) $x - 5$ is not a factor of $2x^3 + 3x^2 - 38x - 15$

5. ***True or False*** In using synthetic division, the divisor is always a polynomial of degree 1, whose leading coefficient is 1.

6. ***True or False***
$$\begin{array}{r|rrrr} -2 & 5 & 3 & 2 & 1 \\ & & -10 & 14 & -32 \\ \hline & 5 & -7 & 16 & -31 \end{array}$$
means $\dfrac{5x^3 + 3x^2 + 2x + 1}{x + 2} = 5x^2 - 7x + 16 + \dfrac{-31}{x + 2}$.

Skill Building

In Problems 7–18, use synthetic division to find the quotient and remainder when:

7. $x^3 - x^2 + 2x + 4$ is divided by $x - 2$
8. $x^3 + 2x^2 - 3x + 1$ is divided by $x + 1$
9. $3x^3 + 2x^2 - x + 3$ is divided by $x - 3$
10. $-4x^3 + 2x^2 - x + 1$ is divided by $x + 2$
11. $x^5 - 4x^3 + x$ is divided by $x + 3$
12. $x^4 + x^2 + 2$ is divided by $x - 2$
13. $4x^6 - 3x^4 + x^2 + 5$ is divided by $x - 1$
14. $x^5 + 5x^3 - 10$ is divided by $x + 1$
15. $0.1x^3 + 0.2x$ is divided by $x + 1.1$
16. $0.1x^2 - 0.2$ is divided by $x + 2.1$
17. $x^5 - 1$ is divided by $x - 1$
18. $x^5 + 1$ is divided by $x + 1$

In Problems 19–28, use synthetic division to determine whether $x - c$ is a factor of the given polynomial.

19. $4x^3 - 3x^2 - 8x + 4$; $x - 2$
20. $-4x^3 + 5x^2 + 8$; $x + 3$
21. $3x^4 - 6x^3 - 5x + 10$; $x - 2$
22. $4x^4 - 15x^2 - 4$; $x - 2$
23. $3x^6 + 82x^3 + 27$; $x + 3$
24. $2x^6 - 18x^4 + x^2 - 9$; $x + 3$
25. $4x^6 - 64x^4 + x^2 - 15$; $x + 4$
26. $x^6 - 16x^4 + x^2 - 16$; $x + 4$
27. $2x^4 - x^3 + 2x - 1$; $x - \frac{1}{2}$
28. $3x^4 + x^3 - 3x + 1$; $x + \frac{1}{3}$

Applications and Extensions

29. Find the sum of a, b, c, and d if

$$\frac{x^3 - 2x^2 + 3x + 5}{x + 2} = ax^2 + bx + c + \frac{d}{x + 2}$$

Explaining Concepts: Discussion and Writing

30. When dividing a polynomial by $x - c$, do you prefer to use long division or synthetic division? Does the value of c make a difference to you in choosing? Give reasons.

R.7 Rational Expressions

OBJECTIVES
1. Reduce a Rational Expression to Lowest Terms (p. 62)
2. Multiply and Divide Rational Expressions (p. 63)
3. Add and Subtract Rational Expressions (p. 64)
4. Use the Least Common Multiple Method (p. 66)
5. Simplify Complex Rational Expressions (p. 68)

1 Reduce a Rational Expression to Lowest Terms

If we form the quotient of two polynomials, the result is called a **rational expression**. Some examples of rational expressions are

(a) $\frac{x^3 + 1}{x}$ (b) $\frac{3x^2 + x - 2}{x^2 + 5}$ (c) $\frac{x}{x^2 - 1}$ (d) $\frac{xy^2}{(x - y)^2}$

Expressions (a), (b), and (c) are rational expressions in one variable, x, whereas (d) is a rational expression in two variables, x and y.

Rational expressions are described in the same manner as rational numbers. In expression (a), the polynomial $x^3 + 1$ is the **numerator**, and x is the **denominator**. When the numerator and denominator of a rational expression contain no common factors (except 1 and -1), we say that the rational expression is **reduced to lowest terms, or simplified**.

The polynomial in the denominator of a rational expression cannot be equal to 0 because division by 0 is not defined. For example, for the expression $\frac{x^3 + 1}{x}$, x cannot take on the value 0. The domain of the variable x is $\{x \mid x \neq 0\}$.

A rational expression is reduced to lowest terms by factoring the numerator and the denominator completely and canceling any common factors using the Cancellation Property:

$$\frac{a\not{c}}{b\not{c}} = \frac{a}{b} \qquad \text{if } b \neq 0, c \neq 0 \qquad \textbf{(1)}$$

EXAMPLE 1 **Reducing a Rational Expression to Lowest Terms**

Reduce to lowest terms: $\dfrac{x^2 + 4x + 4}{x^2 + 3x + 2}$

Solution Begin by factoring the numerator and the denominator.

$$x^2 + 4x + 4 = (x + 2)(x + 2)$$
$$x^2 + 3x + 2 = (x + 2)(x + 1)$$

WARNING Apply the Cancellation Property only to rational expressions written in factored form. Be sure to cancel only common factors, not common terms! ■

Since a common factor, $x + 2$, appears, the original expression is not in lowest terms. To reduce it to lowest terms, use the Cancellation Property:

$$\frac{x^2 + 4x + 4}{x^2 + 3x + 2} = \frac{\cancel{(x + 2)}(x + 2)}{\cancel{(x + 2)}(x + 1)} = \frac{x + 2}{x + 1} \qquad x \neq -2, -1$$

●

EXAMPLE 2 **Reducing Rational Expressions to Lowest Terms**

Reduce each rational expression to lowest terms.

(a) $\dfrac{x^3 - 8}{x^3 - 2x^2}$ (b) $\dfrac{8 - 2x}{x^2 - x - 12}$

Solution

(a) $$\frac{x^3 - 8}{x^3 - 2x^2} = \frac{\cancel{(x - 2)}(x^2 + 2x + 4)}{x^2\cancel{(x - 2)}} = \frac{x^2 + 2x + 4}{x^2} \qquad x \neq 0, 2$$

(b) $$\frac{8 - 2x}{x^2 - x - 12} = \frac{2(4 - x)}{(x - 4)(x + 3)} = \frac{2(-1)\cancel{(x - 4)}}{\cancel{(x - 4)}(x + 3)} = \frac{-2}{x + 3} \qquad x \neq -3, 4$$

●

Now Work PROBLEM 7

2 Multiply and Divide Rational Expressions

The rules for multiplying and dividing rational expressions are the same as the rules for multiplying and dividing rational numbers. If $\frac{a}{b}$ and $\frac{c}{d}$, $b \neq 0$, $d \neq 0$, are two rational expressions, then

$$\frac{a}{b} \cdot \frac{c}{d} = \frac{ac}{bd} \qquad \text{if } b \neq 0, d \neq 0 \qquad \textbf{(2)}$$

$$\frac{\frac{a}{b}}{\frac{c}{d}} = \frac{a}{b} \cdot \frac{d}{c} = \frac{ad}{bc} \qquad \text{if } b \neq 0, c \neq 0, d \neq 0 \qquad \textbf{(3)}$$

In using equations (2) and (3) with rational expressions, be sure first to factor each polynomial completely so that common factors can be canceled. Leave your answer in factored form.

EXAMPLE 3 **Multiplying and Dividing Rational Expressions**

Perform the indicated operation and simplify the result. Leave your answer in factored form.

(a) $\dfrac{x^2-2x+1}{x^3+x}\cdot\dfrac{4x^2+4}{x^2+x-2}$ (b) $\dfrac{\dfrac{x+3}{x^2-4}}{\dfrac{x^2-x-12}{x^3-8}}$

Solution (a)

$$\frac{x^2-2x+1}{x^3+x}\cdot\frac{4x^2+4}{x^2+x-2}=\frac{(x-1)^2}{x(x^2+1)}\cdot\frac{4(x^2+1)}{(x+2)(x-1)}$$
$$=\frac{(x-1)^2(4)\cancel{(x^2+1)}}{x\cancel{(x^2+1)}(x+2)\cancel{(x-1)}}$$
$$=\frac{4(x-1)}{x(x+2)}\qquad x\neq -2,0,1$$

(b)

$$\frac{\dfrac{x+3}{x^2-4}}{\dfrac{x^2-x-12}{x^3-8}}=\frac{x+3}{x^2-4}\cdot\frac{x^3-8}{x^2-x-12}$$
$$=\frac{x+3}{(x-2)(x+2)}\cdot\frac{(x-2)(x^2+2x+4)}{(x-4)(x+3)}$$
$$=\frac{\cancel{(x+3)}\cancel{(x-2)}(x^2+2x+4)}{\cancel{(x-2)}(x+2)(x-4)\cancel{(x+3)}}$$
$$=\frac{x^2+2x+4}{(x+2)(x-4)}\qquad x\neq -3,-2,2,4$$

Now Work PROBLEMS 19 AND 27

3 Add and Subtract Rational Expressions

In Words
To add (or subtract) two rational expressions with the same denominator, keep the common denominator and add (or subtract) the numerators.

The rules for adding and subtracting rational expressions are the same as the rules for adding and subtracting rational numbers. If the denominators of two rational expressions to be added (or subtracted) are equal, then add (or subtract) the numerators and keep the common denominator.

If $\dfrac{a}{b}$ and $\dfrac{c}{b}$ are two rational expressions, then

$$\frac{a}{b}+\frac{c}{b}=\frac{a+c}{b}\qquad \frac{a}{b}-\frac{c}{b}=\frac{a-c}{b}\qquad \text{if } b\neq 0 \tag{4}$$

EXAMPLE 4 **Adding and Subtracting Rational Expressions with Equal Denominators**

Perform the indicated operation and simplify the result. Leave your answer in factored form.

(a) $\dfrac{2x^2-4}{2x+5}+\dfrac{x+3}{2x+5}\quad x\neq-\dfrac{5}{2}$ (b) $\dfrac{x}{x-3}-\dfrac{3x+2}{x-3}\quad x\neq 3$

Solution (a)

$$\frac{2x^2-4}{2x+5}+\frac{x+3}{2x+5}=\frac{(2x^2-4)+(x+3)}{2x+5}$$
$$=\frac{2x^2+x-1}{2x+5}=\frac{(2x-1)(x+1)}{2x+5}$$

(b) $\dfrac{x}{x-3} - \dfrac{3x+2}{x-3} = \dfrac{x-(3x+2)}{x-3} = \dfrac{x-3x-2}{x-3}$

$= \dfrac{-2x-2}{x-3} = \dfrac{-2(x+1)}{x-3}$ ●

EXAMPLE 5

Adding Rational Expressions Whose Denominators Are Additive Inverses of Each Other

Perform the indicated operation and simplify the result. Leave your answer in factored form.

$$\frac{2x}{x-3} + \frac{5}{3-x} \qquad x \neq 3$$

Solution Notice that the denominators of the two rational expressions are different. However, the denominator of the second expression is the additive inverse of the denominator of the first. That is,

$$3 - x = -x + 3 = -1 \cdot (x-3) = -(x-3)$$

Then

$$\frac{2x}{x-3} + \frac{5}{3-x} \underset{3-x=-(x-3)}{=} \frac{2x}{x-3} + \frac{5}{-(x-3)} \underset{\frac{a}{-b}=\frac{-a}{b}}{=} \frac{2x}{x-3} + \frac{-5}{x-3}$$

$$= \frac{2x+(-5)}{x-3} = \frac{2x-5}{x-3}$$ ●

Now Work PROBLEMS 39 AND 45

If the denominators of two rational expressions to be added or subtracted are not equal, we can use the general formulas for adding and subtracting rational expressions.

$$\frac{a}{b} + \frac{c}{d} = \frac{a \cdot d}{b \cdot d} + \frac{b \cdot c}{b \cdot d} = \frac{ad+bc}{bd} \qquad \text{if } b \neq 0, d \neq 0 \qquad \textbf{(5a)}$$

$$\frac{a}{b} - \frac{c}{d} = \frac{a \cdot d}{b \cdot d} - \frac{b \cdot c}{b \cdot d} = \frac{ad-bc}{bd} \qquad \text{if } b \neq 0, d \neq 0 \qquad \textbf{(5b)}$$

EXAMPLE 6

Adding and Subtracting Rational Expressions with Unequal Denominators

Perform the indicated operation and simplify the result. Leave your answer in factored form.

(a) $\dfrac{x-3}{x+4} + \dfrac{x}{x-2} \quad x \neq -4, 2$ (b) $\dfrac{x^2}{x^2-4} - \dfrac{1}{x} \quad x \neq -2, 0, 2$

Solution (a) $\dfrac{x-3}{x+4} + \dfrac{x}{x-2} \underset{(5a)}{=} \dfrac{x-3}{x+4} \cdot \dfrac{x-2}{x-2} + \dfrac{x+4}{x+4} \cdot \dfrac{x}{x-2}$

$$= \frac{(x-3)(x-2) + (x+4)(x)}{(x+4)(x-2)}$$

$$= \frac{x^2 - 5x + 6 + x^2 + 4x}{(x+4)(x-2)} = \frac{2x^2 - x + 6}{(x+4)(x-2)}$$

(b) $$\frac{x^2}{x^2-4} - \frac{1}{x} = \frac{x^2}{x^2-4}\cdot\frac{x}{x} - \frac{x^2-4}{x^2-4}\cdot\frac{1}{x} = \frac{x^2(x) - (x^2-4)(1)}{(x^2-4)(x)}$$

(5b)

$$= \frac{x^3 - x^2 + 4}{(x-2)(x+2)(x)}$$

Now Work PROBLEM 49

4 Use the Least Common Multiple Method

If the denominators of two rational expressions to be added (or subtracted) have common factors, we usually do not use the general rules given by equations (5a) and (5b). Just as with fractions, we apply the **least common multiple (LCM) method**. The LCM method uses the polynomial of least degree that has each denominator polynomial as a factor.

The LCM Method for Adding or Subtracting Rational Expressions

The Least Common Multiple (LCM) Method requires four steps:

STEP 1: Factor completely the polynomial in the denominator of each rational expression.

STEP 2: The LCM of the denominators is the product of each of these factors raised to a power equal to the greatest number of times that the factor occurs in the polynomials.

STEP 3: Write each rational expression using the LCM as the common denominator.

STEP 4: Add or subtract the rational expressions using equation (4).

We begin with an example that requires only Steps 1 and 2.

EXAMPLE 7 Finding the Least Common Multiple

Find the least common multiple of the following pair of polynomials:

$$x(x-1)^2(x+1) \quad \text{and} \quad 4(x-1)(x+1)^3$$

Solution **STEP 1:** The polynomials are already factored completely as

$$x(x-1)^2(x+1) \quad \text{and} \quad 4(x-1)(x+1)^3$$

STEP 2: Start by writing the factors of the left-hand polynomial. (Or you could start with the one on the right.)

$$x(x-1)^2(x+1)$$

Now look at the right-hand polynomial. Its first factor, 4, does not appear in our list, so we insert it.

$$4x(x-1)^2(x+1)$$

The next factor, $x - 1$, is already in our list, so no change is necessary. The final factor is $(x+1)^3$. Since our list has $x + 1$ to the first power only, we replace $x + 1$ in the list by $(x+1)^3$. The LCM is

$$4x(x-1)^2(x+1)^3$$

Notice that the LCM is, in fact, the polynomial of least degree that contains $x(x-1)^2(x+1)$ and $4(x-1)(x+1)^3$ as factors.

Now Work PROBLEM 55

EXAMPLE 8 Using the Least Common Multiple to Add Rational Expressions

Perform the indicated operation and simplify the result. Leave your answer in factored form.

$$\frac{x}{x^2+3x+2} + \frac{2x-3}{x^2-1} \qquad x \neq -2, -1, 1$$

Solution

STEP 1: Factor completely the polynomials in the denominators.

$$x^2+3x+2 = (x+2)(x+1)$$
$$x^2-1 = (x-1)(x+1)$$

STEP 2: The LCM is $(x+2)(x+1)(x-1)$. Do you see why?

STEP 3: Write each rational expression using the LCM as the denominator.

$$\frac{x}{x^2+3x+2} = \frac{x}{(x+2)(x+1)} = \frac{x}{(x+2)(x+1)} \cdot \frac{x-1}{x-1} = \frac{x(x-1)}{(x+2)(x+1)(x-1)}$$

↑ Multiply numerator and denominator by $x - 1$ to get the LCM in the denominator.

$$\frac{2x-3}{x^2-1} = \frac{2x-3}{(x-1)(x+1)} = \frac{2x-3}{(x-1)(x+1)} \cdot \frac{x+2}{x+2} = \frac{(2x-3)(x+2)}{(x-1)(x+1)(x+2)}$$

↑ Multiply numerator and denominator by $x + 2$ to get the LCM in the denominator.

STEP 4: Now add by using equation (4).

$$\frac{x}{x^2+3x+2} + \frac{2x-3}{x^2-1} = \frac{x(x-1)}{(x+2)(x+1)(x-1)} + \frac{(2x-3)(x+2)}{(x+2)(x+1)(x-1)}$$
$$= \frac{(x^2-x)+(2x^2+x-6)}{(x+2)(x+1)(x-1)}$$
$$= \frac{3x^2-6}{(x+2)(x+1)(x-1)} = \frac{3(x^2-2)}{(x+2)(x+1)(x-1)}$$

EXAMPLE 9 Using the Least Common Multiple to Subtract Rational Expressions

Perform the indicated operation and simplify the result. Leave your answer in factored form.

$$\frac{3}{x^2+x} - \frac{x+4}{x^2+2x+1} \qquad x \neq -1, 0$$

Solution

STEP 1: Factor completely the polynomials in the denominators.

$$x^2+x = x(x+1)$$
$$x^2+2x+1 = (x+1)^2$$

STEP 2: The LCM is $x(x+1)^2$.

STEP 3: Write each rational expression using the LCM as the denominator.

$$\frac{3}{x^2+x} = \frac{3}{x(x+1)} = \frac{3}{x(x+1)}\cdot\frac{x+1}{x+1} = \frac{3(x+1)}{x(x+1)^2}$$

$$\frac{x+4}{x^2+2x+1} = \frac{x+4}{(x+1)^2} = \frac{x+4}{(x+1)^2}\cdot\frac{x}{x} = \frac{x(x+4)}{x(x+1)^2}$$

STEP 4: Subtract, using equation (4).

$$\frac{3}{x^2+x} - \frac{x+4}{x^2+2x+1} = \frac{3(x+1)}{x(x+1)^2} - \frac{x(x+4)}{x(x+1)^2}$$

$$= \frac{3(x+1) - x(x+4)}{x(x+1)^2}$$

$$= \frac{3x+3-x^2-4x}{x(x+1)^2}$$

$$= \frac{-x^2-x+3}{x(x+1)^2}$$

Now Work PROBLEM 65

5 Simplify Complex Rational Expressions

When sums and/or differences of rational expressions appear as the numerator and/or denominator of a quotient, the quotient is called a **complex rational expression**.* For example,

$$\frac{1+\dfrac{1}{x}}{1-\dfrac{1}{x}} \quad \text{and} \quad \frac{\dfrac{x^2}{x^2-4}-3}{\dfrac{x-3}{x+2}-1}$$

are complex rational expressions. To **simplify** a complex rational expression means to write it as a rational expression reduced to lowest terms. This can be accomplished in either of two ways.

Simplifying a Complex Rational Expression

OPTION 1: Treat the numerator and denominator of the complex rational expression separately, performing whatever operations are indicated and simplifying the results. Follow this by simplifying the resulting rational expression.

OPTION 2: Find the LCM of the denominators of all rational expressions that appear in the complex rational expression. Multiply the numerator and denominator of the complex rational expression by the LCM and simplify the result.

We use both options in the next example. By carefully studying each option, you can discover situations in which one may be easier to use than the other.

EXAMPLE 10 **Simplifying a Complex Rational Expression**

Simplify: $$\frac{\dfrac{1}{2}+\dfrac{3}{x}}{\dfrac{x+3}{4}} \qquad x \neq -3, 0$$

* Some texts use the term **complex fraction.**

Solution *Option 1:* First, we perform the indicated operation in the numerator, and then we divide.

$$\frac{\frac{1}{2}+\frac{3}{x}}{\frac{x+3}{4}} = \frac{\frac{1\cdot x+2\cdot 3}{2\cdot x}}{\frac{x+3}{4}} = \frac{\frac{x+6}{2x}}{\frac{x+3}{4}} = \frac{x+6}{2x}\cdot\frac{4}{x+3}$$

Rule for adding quotients; Rule for dividing quotients

$$= \frac{(x+6)\cdot 4}{2\cdot x\cdot(x+3)} = \frac{\cancel{2}\cdot 2\cdot(x+6)}{\cancel{2}\cdot x\cdot(x+3)} = \frac{2(x+6)}{x(x+3)}$$

Rule for multiplying quotients

Option 2: The rational expressions that appear in the complex rational expression are

$$\frac{1}{2},\ \frac{3}{x},\ \frac{x+3}{4}$$

The LCM of their denominators is $4x$. We multiply the numerator and denominator of the complex rational expression by $4x$ and then simplify.

$$\frac{\frac{1}{2}+\frac{3}{x}}{\frac{x+3}{4}} = \frac{4x\cdot\left(\frac{1}{2}+\frac{3}{x}\right)}{4x\cdot\left(\frac{x+3}{4}\right)} = \frac{4x\cdot\frac{1}{2}+4x\cdot\frac{3}{x}}{\frac{4x\cdot(x+3)}{4}}$$

Multiply the numerator and denominator by $4x$. Use the Distributive Property in the numerator.

$$= \frac{\cancel{2}\cdot 2x\cdot\frac{1}{\cancel{2}}+4\cancel{x}\cdot\frac{3}{\cancel{x}}}{\frac{\cancel{4}x\cdot(x+3)}{\cancel{4}}} = \frac{2x+12}{x(x+3)} = \frac{2(x+6)}{x(x+3)}$$

Simplify; Factor

●

EXAMPLE 11 Simplifying a Complex Rational Expression

Simplify: $\dfrac{\frac{x^2}{x-4}+2}{\frac{2x-2}{x}-1} \qquad x \neq 0, 2, 4$

Solution We will use Option 1.

$$\frac{\frac{x^2}{x-4}+2}{\frac{2x-2}{x}-1} = \frac{\frac{x^2}{x-4}+\frac{2(x-4)}{x-4}}{\frac{2x-2}{x}-\frac{x}{x}} = \frac{\frac{x^2+2x-8}{x-4}}{\frac{2x-2-x}{x}}$$

$$= \frac{\frac{(x+4)(x-2)}{x-4}}{\frac{x-2}{x}} = \frac{(x+4)\cancel{(x-2)}}{x-4}\cdot\frac{x}{\cancel{x-2}}$$

$$= \frac{(x+4)\cdot x}{x-4}$$

●

Now Work PROBLEM 75

Application

EXAMPLE 12 **Solving an Application in Electricity**

An electrical circuit contains two resistors connected in parallel, as shown in Figure 28. If these two resistors provide resistance of R_1 and R_2 ohms, respectively, their combined resistance R is given by the formula

$$R = \frac{1}{\frac{1}{R_1} + \frac{1}{R_2}}$$

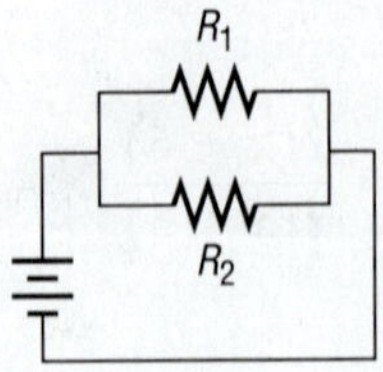

Figure 28

Express R as a rational expression; that is, simplify the right-hand side of this formula. Evaluate the rational expression if $R_1 = 6$ ohms and $R_2 = 10$ ohms.

Solution We will use Option 2. If we consider 1 as the fraction $\frac{1}{1}$, the rational expressions in the complex rational expression are

$$\frac{1}{1}, \quad \frac{1}{R_1}, \quad \frac{1}{R_2}$$

The LCM of the denominators is R_1R_2. We multiply the numerator and denominator of the complex rational expression by R_1R_2 and simplify.

$$\frac{1}{\frac{1}{R_1} + \frac{1}{R_2}} = \frac{1 \cdot R_1R_2}{\left(\frac{1}{R_1} + \frac{1}{R_2}\right) \cdot R_1R_2} = \frac{R_1R_2}{\frac{1}{R_1} \cdot R_1R_2 + \frac{1}{R_2} \cdot R_1R_2} = \frac{R_1R_2}{R_2 + R_1}$$

So,

$$R = \frac{R_1R_2}{R_2 + R_1}$$

If $R_1 = 6$ and $R_2 = 10$, then

$$R = \frac{6 \cdot 10}{10 + 6} = \frac{60}{16} = \frac{15}{4} \text{ ohms}$$

R.7 Assess Your Understanding

Concepts and Vocabulary

1. When the numerator and denominator of a rational expression contain no common factors (except 1 and -1), the rational expression is in ______ ______.

2. LCM is an abbreviation for ______ ______ ______.

3. Choose the statement that is not true. Assume $b \neq 0, c \neq 0$, and $d \neq 0$ as necessary.
(a) $\frac{ac}{bc} = \frac{a}{b}$ (b) $\frac{a}{b} + \frac{c}{b} = \frac{a+c}{b}$
(c) $\frac{a}{b} - \frac{c}{d} = \frac{ad - bc}{bd}$ (d) $\frac{\frac{a}{b}}{\frac{c}{d}} = \frac{ac}{bd}$

4. Choose the rational expression that simplifies to -1.
(a) $\frac{a-b}{b-a}$ (b) $\frac{a-b}{a-b}$
(c) $\frac{a+b}{a-b}$ (d) $\frac{b-a}{b+a}$

5. ***True or False*** The rational expression $\frac{2x^3 - 4x}{x - 2}$ is reduced to lowest terms.

6. ***True or False*** The LCM of $2x^3 + 6x^2$ and $6x^4 + 4x^3$ is $4x^3(x + 1)$.

Skill Building

In Problems 7–18, reduce each rational expression to lowest terms.

7. $\frac{3x + 9}{x^2 - 9}$
8. $\frac{4x^2 + 8x}{12x + 24}$
9. $\frac{x^2 - 2x}{3x - 6}$
10. $\frac{15x^2 + 24x}{3x^2}$
11. $\frac{24x^2}{12x^2 - 6x}$
12. $\frac{x^2 + 4x + 4}{x^2 - 4}$
13. $\frac{y^2 - 25}{2y^2 - 8y - 10}$
14. $\frac{3y^2 - y - 2}{3y^2 + 5y + 2}$
15. $\frac{x^2 + 4x - 5}{x^2 - 2x + 1}$
16. $\frac{x - x^2}{x^2 + x - 2}$
17. $\frac{x^2 + 5x - 14}{2 - x}$
18. $\frac{2x^2 + 5x - 3}{1 - 2x}$

In Problems 19–36, perform the indicated operation and simplify the result. Leave your answer in factored form.

19. $\dfrac{3x+6}{5x^2}\cdot\dfrac{x}{x^2-4}$ **20.** $\dfrac{3}{2x}\cdot\dfrac{x^2}{6x+10}$ **21.** $\dfrac{4x^2}{x^2-16}\cdot\dfrac{x^3-64}{2x}$ **22.** $\dfrac{12}{x^2+x}\cdot\dfrac{x^3+1}{4x-2}$

23. $\dfrac{4x-8}{-3x}\cdot\dfrac{12}{12-6x}$ **24.** $\dfrac{6x-27}{5x}\cdot\dfrac{2}{4x-18}$ **25.** $\dfrac{x^2-3x-10}{x^2+2x-35}\cdot\dfrac{x^2+4x-21}{x^2+9x+14}$

26. $\dfrac{x^2+x-6}{x^2+4x-5}\cdot\dfrac{x^2-25}{x^2+2x-15}$ **27.** $\dfrac{\dfrac{6x}{x^2-4}}{\dfrac{3x-9}{2x+4}}$ **28.** $\dfrac{\dfrac{12x}{5x+20}}{\dfrac{4x^2}{x^2-16}}$

29. $\dfrac{\dfrac{8x}{x^2-1}}{\dfrac{10x}{x+1}}$ **30.** $\dfrac{\dfrac{x-2}{4x}}{\dfrac{x^2-4x+4}{12x}}$ **31.** $\dfrac{\dfrac{4-x}{4+x}}{\dfrac{4x}{x^2-16}}$ **32.** $\dfrac{\dfrac{3+x}{3-x}}{\dfrac{x^2-9}{9x^3}}$

33. $\dfrac{\dfrac{x^2+7x+12}{x^2-7x+12}}{\dfrac{x^2+x-12}{x^2-x-12}}$ **34.** $\dfrac{\dfrac{x^2+7x+6}{x^2+x-6}}{\dfrac{x^2+5x-6}{x^2+5x+6}}$ **35.** $\dfrac{\dfrac{2x^2-x-28}{3x^2-x-2}}{\dfrac{4x^2+16x+7}{3x^2+11x+6}}$ **36.** $\dfrac{\dfrac{9x^2+3x-2}{12x^2+5x-2}}{\dfrac{9x^2-6x+1}{8x^2-10x-3}}$

In Problems 37–54, perform the indicated operation and simplify the result. Leave your answer in factored form.

37. $\dfrac{x}{2}+\dfrac{5}{2}$ **38.** $\dfrac{3}{x}-\dfrac{6}{x}$ **39.** $\dfrac{x^2}{2x-3}-\dfrac{4}{2x-3}$

40. $\dfrac{3x^2}{2x-1}-\dfrac{9}{2x-1}$ **41.** $\dfrac{x+1}{x-3}+\dfrac{2x-3}{x-3}$ **42.** $\dfrac{2x-5}{3x+2}+\dfrac{x+4}{3x+2}$

43. $\dfrac{3x+5}{2x-1}-\dfrac{2x-4}{2x-1}$ **44.** $\dfrac{5x-4}{3x+4}-\dfrac{x+1}{3x+4}$ **45.** $\dfrac{4}{x-2}+\dfrac{x}{2-x}$

46. $\dfrac{6}{x-1}-\dfrac{x}{1-x}$ **47.** $\dfrac{4}{x-1}-\dfrac{2}{x+2}$ **48.** $\dfrac{2}{x+5}-\dfrac{5}{x-5}$

49. $\dfrac{x}{x+1}+\dfrac{2x-3}{x-1}$ **50.** $\dfrac{3x}{x-4}+\dfrac{2x}{x+3}$ **51.** $\dfrac{x-3}{x+2}-\dfrac{x+4}{x-2}$

52. $\dfrac{2x-3}{x-1}-\dfrac{2x+1}{x+1}$ **53.** $\dfrac{x}{x^2-4}+\dfrac{1}{x}$ **54.** $\dfrac{x-1}{x^3}+\dfrac{x}{x^2+1}$

In Problems 55–62, find the LCM of the given polynomials.

55. $x^2-4,\ x^2-x-2$ **56.** $x^2-x-12,\ x^2-8x+16$ **57.** $x^3-x,\ x^2-x$

58. $3x^2-27,\ 2x^2-x-15$ **59.** $4x^3-4x^2+x,\ 2x^3-x^2,\ x^3$ **60.** $x-3,\ x^2+3x,\ x^3-9x$

61. $x^3-x,\ x^3-2x^2+x,\ x^3-1$ **62.** $x^2+4x+4,\ x^3+2x^2,\ (x+2)^3$

In Problems 63–74, perform the indicated operations and simplify the result. Leave your answer in factored form.

63. $\dfrac{x}{x^2-7x+6}-\dfrac{x}{x^2-2x-24}$ **64.** $\dfrac{x}{x-3}-\dfrac{x+1}{x^2+5x-24}$ **65.** $\dfrac{4x}{x^2-4}-\dfrac{2}{x^2+x-6}$

66. $\dfrac{3x}{x-1}-\dfrac{x-4}{x^2-2x+1}$ **67.** $\dfrac{3}{(x-1)^2(x+1)}+\dfrac{2}{(x-1)(x+1)^2}$ **68.** $\dfrac{2}{(x+2)^2(x-1)}-\dfrac{6}{(x+2)(x-1)^2}$

69. $\dfrac{x+4}{x^2-x-2}-\dfrac{2x+3}{x^2+2x-8}$ **70.** $\dfrac{2x-3}{x^2+8x+7}-\dfrac{x-2}{(x+1)^2}$ **71.** $\dfrac{1}{x}-\dfrac{2}{x^2+x}+\dfrac{3}{x^3-x^2}$

72. $\dfrac{x}{(x-1)^2}+\dfrac{2}{x}-\dfrac{x+1}{x^3-x^2}$ **73.** $\dfrac{1}{h}\left(\dfrac{1}{x+h}-\dfrac{1}{x}\right)$ **74.** $\dfrac{1}{h}\left[\dfrac{1}{(x+h)^2}-\dfrac{1}{x^2}\right]$

In Problems 75–86, perform the indicated operations and simplify the result. Leave your answer in factored form.

75. $\dfrac{1+\dfrac{1}{x}}{1-\dfrac{1}{x}}$

76. $\dfrac{4+\dfrac{1}{x^2}}{3-\dfrac{1}{x^2}}$

77. $\dfrac{2-\dfrac{x+1}{x}}{3+\dfrac{x-1}{x+1}}$

78. $\dfrac{1-\dfrac{x}{x+1}}{2-\dfrac{x-1}{x}}$

79. $\dfrac{\dfrac{x+4}{x-2}-\dfrac{x-3}{x+1}}{x+1}$

80. $\dfrac{\dfrac{x-2}{x+1}-\dfrac{x}{x-2}}{x+3}$

81. $\dfrac{\dfrac{x-2}{x+2}+\dfrac{x-1}{x+1}}{\dfrac{x}{x+1}-\dfrac{2x-3}{x}}$

82. $\dfrac{\dfrac{2x+5}{x}-\dfrac{x}{x-3}}{\dfrac{x^2}{x-3}-\dfrac{(x+1)^2}{x+3}}$

83. $1-\dfrac{1}{1-\dfrac{1}{x}}$

84. $1-\dfrac{1}{1-\dfrac{1}{1-x}}$

85. $\dfrac{2(x-1)^{-1}+3}{3(x-1)^{-1}+2}$

86. $\dfrac{4(x+2)^{-1}-3}{3(x+2)^{-1}-1}$

In Problems 87–94, expressions that occur in calculus are given. Reduce each expression to lowest terms.

87. $\dfrac{(2x+3)\cdot 3-(3x-5)\cdot 2}{(3x-5)^2}$

88. $\dfrac{(4x+1)\cdot 5-(5x-2)\cdot 4}{(5x-2)^2}$

89. $\dfrac{x\cdot 2x-(x^2+1)\cdot 1}{(x^2+1)^2}$

90. $\dfrac{x\cdot 2x-(x^2-4)\cdot 1}{(x^2-4)^2}$

91. $\dfrac{(3x+1)\cdot 2x-x^2\cdot 3}{(3x+1)^2}$

92. $\dfrac{(2x-5)\cdot 3x^2-x^3\cdot 2}{(2x-5)^2}$

93. $\dfrac{(x^2+1)\cdot 3-(3x+4)\cdot 2x}{(x^2+1)^2}$

94. $\dfrac{(x^2+9)\cdot 2-(2x-5)\cdot 2x}{(x^2+9)^2}$

Applications and Extensions

95. **The Lensmaker's Equation** The focal length f of a lens with index of refraction n is

$$\frac{1}{f}=(n-1)\left[\frac{1}{R_1}+\frac{1}{R_2}\right]$$

where R_1 and R_2 are the radii of curvature of the front and back surfaces of the lens. Express f as a rational expression. Evaluate the rational expression for $n = 1.5$, $R_1 = 0.1$ meter, and $R_2 = 0.2$ meter.

96. **Electrical Circuits** An electrical circuit contains three resistors connected in parallel. If these three resistors provide resistance of R_1, R_2, and R_3 ohms, respectively, their combined resistance R is given by the formula

$$\frac{1}{R}=\frac{1}{R_1}+\frac{1}{R_2}+\frac{1}{R_3}$$

Express R as a rational expression. Evaluate R for $R_1 = 5$ ohms, $R_2 = 4$ ohms, and $R_3 = 10$ ohms.

Explaining Concepts: Discussion and Writing

97. The following expressions are called **continued fractions**:

$$1+\frac{1}{x},\quad 1+\frac{1}{1+\dfrac{1}{x}},\quad 1+\frac{1}{1+\dfrac{1}{1+\dfrac{1}{x}}},\quad 1+\frac{1}{1+\dfrac{1}{1+\dfrac{1}{1+\dfrac{1}{x}}}},\ldots$$

Each simplifies to an expression of the form

$$\frac{ax+b}{bx+c}$$

Trace the successive values of a, b, and c as you "continue" the fraction. Can you discover the patterns that these values follow? Go to the library and research Fibonacci numbers. Write a report on your findings.

98. Explain to a fellow student when you would use the LCM method to add two rational expressions. Give two examples of adding two rational expressions: one in which you use the LCM and the other in which you do not.

99. Which of the two options given in the text for simplifying complex rational expressions do you prefer? Write a brief paragraph stating the reasons for your choice.

R.8 *n*th Roots; Rational Exponents

PREPARING FOR THIS SECTION *Before getting started, review the following:*

- Exponents, Square Roots (Section R.2, pp. 21–24)

Now Work the **'Are You Prepared?'** problems on page 78.

OBJECTIVES

1 Work with *n*th Roots

DEFINITION

The **principal *n*th root of a real number *a***, $n \geq 2$ an integer, symbolized by $\sqrt[n]{a}$, is defined as follows:

$$\sqrt[n]{a} = b \quad \text{means} \quad a = b^n$$

where $a \geq 0$ and $b \geq 0$ if n is even and a, b are any real numbers if n is odd.

In Words
The symbol $\sqrt[n]{a}$ means "give me the number that, when raised to the power n, equals a."

Notice that if a is negative and n is even, then $\sqrt[n]{a}$ is not defined. When it is defined, the principal nth root of a number is unique.

The symbol $\sqrt[n]{a}$ for the principal nth root of a is called a **radical**; the integer n is called the **index**, and a is called the **radicand**. If the index of a radical is 2, we call $\sqrt[2]{a}$ the **square root** of a and omit the index 2 by simply writing $\sqrt{a}$. If the index is 3, we call $\sqrt[3]{a}$ the **cube root** of a.

EXAMPLE 1 **Simplifying Principal *n*th Roots**

(a) $\sqrt[3]{8} = \sqrt[3]{2^3} = 2$

(b) $\sqrt[3]{-64} = \sqrt[3]{(-4)^3} = -4$

(c) $\sqrt[4]{\dfrac{1}{16}} = \sqrt[4]{\left(\dfrac{1}{2}\right)^4} = \dfrac{1}{2}$

(d) $\sqrt[6]{(-2)^6} = |-2| = 2$

These are examples of **perfect roots**, since each simplifies to a rational number. Notice the absolute value in Example 1(d). If n is even, then the principal nth root must be nonnegative.

In general, if $n \geq 2$ is an integer and a is a real number, we have

$$\sqrt[n]{a^n} = a \quad \text{if } n \geq 3 \text{ is odd} \qquad \textbf{(1a)}$$

$$\sqrt[n]{a^n} = |a| \quad \text{if } n \geq 2 \text{ is even} \qquad \textbf{(1b)}$$

Now Work PROBLEM 11

Radicals provide a way of representing many irrational real numbers. For example, it can be shown that there is no rational number whose square is 2. Using radicals, we can say that $\sqrt{2}$ is the positive number whose square is 2.

EXAMPLE 2

Using a Calculator to Approximate Roots

Use a calculator to approximate $\sqrt[5]{16}$.

Solution Figure 29 shows the result using a TI-84 Plus C graphing calculator. ●

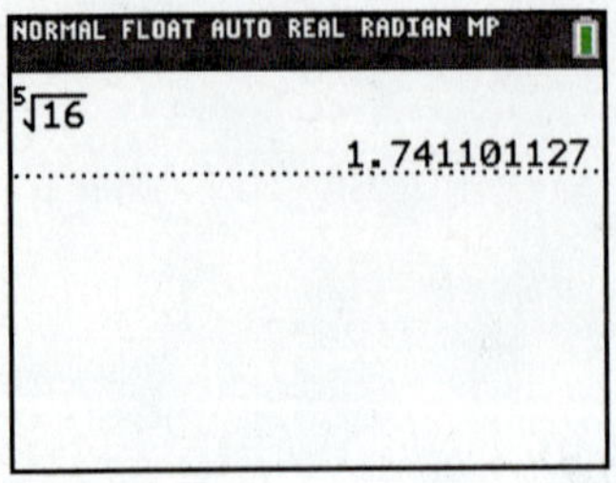

Figure 29

Now Work PROBLEM 113

2 Simplify Radicals

Let $n \geq 2$ and $m \geq 2$ denote integers, and let a and b represent real numbers. Assuming that all radicals are defined, we have the following properties:

Properties of Radicals

$$\sqrt[n]{ab} = \sqrt[n]{a}\,\sqrt[n]{b} \qquad \textbf{(2a)}$$

$$\sqrt[n]{\frac{a}{b}} = \frac{\sqrt[n]{a}}{\sqrt[n]{b}} \qquad b \neq 0 \qquad \textbf{(2b)}$$

$$\sqrt[n]{a^m} = (\sqrt[n]{a})^m \qquad \textbf{(2c)}$$

When used in reference to radicals, the direction to "simplify" will mean to remove from the radicals any perfect roots that occur as factors.

EXAMPLE 3

Simplifying Radicals

(a) $\sqrt{32} = \sqrt{16 \cdot 2} = \sqrt{16} \cdot \sqrt{2} = 4\sqrt{2}$

↑ Factor out 16, a perfect square. ↑ (2a)

(b) $\sqrt[3]{16} = \sqrt[3]{8 \cdot 2} = \sqrt[3]{8} \cdot \sqrt[3]{2} = \sqrt[3]{2^3} \cdot \sqrt[3]{2} = 2\sqrt[3]{2}$

↑ Factor out 8, a perfect cube. ↑ (2a)

(c) $\sqrt[3]{-16x^4} = \sqrt[3]{-8 \cdot 2 \cdot x^3 \cdot x} = \sqrt[3]{(-8x^3)(2x)}$

↑ Factor perfect cubes inside radical. ↑ Group perfect cubes.

$$= \sqrt[3]{(-2x)^3 \cdot 2x} = \sqrt[3]{(-2x)^3} \cdot \sqrt[3]{2x} = -2x\sqrt[3]{2x}$$

↑ (2a)

(d) $\sqrt[4]{\dfrac{16x^5}{81}} = \sqrt[4]{\dfrac{2^4x^4x}{3^4}} = \sqrt[4]{\left(\dfrac{2x}{3}\right)^4 \cdot x} = \sqrt[4]{\left(\dfrac{2x}{3}\right)^4} \cdot \sqrt[4]{x} = \left|\dfrac{2x}{3}\right| \sqrt[4]{x}$ ●

Now Work PROBLEMS 15 AND 21

Two or more radicals can be combined, provided that they have the same index and the same radicand. Such radicals are called **like radicals**.

EXAMPLE 4

Combining Like Radicals

(a) $-8\sqrt{12} + \sqrt{3} = -8\sqrt{4 \cdot 3} + \sqrt{3}$

$$= -8 \cdot \sqrt{4}\,\sqrt{3} + \sqrt{3}$$

$$= -16\sqrt{3} + \sqrt{3} = -15\sqrt{3}$$

(b) $\sqrt[3]{8x^4} + \sqrt[3]{-x} + 4\sqrt[3]{27x} = \sqrt[3]{2^3x^3x} + \sqrt[3]{-1 \cdot x} + 4\sqrt[3]{3^3x}$

$= \sqrt[3]{(2x)^3} \cdot \sqrt[3]{x} + \sqrt[3]{-1} \cdot \sqrt[3]{x} + 4\sqrt[3]{3^3} \cdot \sqrt[3]{x}$

$= 2x\sqrt[3]{x} - 1 \cdot \sqrt[3]{x} + 12\sqrt[3]{x}$

$= (2x + 11)\sqrt[3]{x}$

Now Work PROBLEM 39

3 Rationalize Denominators

When radicals occur in quotients, it is customary to rewrite the quotient so that the new denominator contains no radicals. This process is referred to as **rationalizing the denominator**.

The idea is to multiply by an appropriate expression so that the new denominator contains no radicals. For example:

If a Denominator Contains the Factor	Multiply by	To Obtain a Denominator Free of Radicals
$\sqrt{3}$	$\sqrt{3}$	$(\sqrt{3})^2 = 3$
$\sqrt{3} + 1$	$\sqrt{3} - 1$	$(\sqrt{3})^2 - 1^2 = 3 - 1 = 2$
$\sqrt{2} - 3$	$\sqrt{2} + 3$	$(\sqrt{2})^2 - 3^2 = 2 - 9 = -7$
$\sqrt{5} - \sqrt{3}$	$\sqrt{5} + \sqrt{3}$	$(\sqrt{5})^2 - (\sqrt{3})^2 = 5 - 3 = 2$
$\sqrt[3]{4}$	$\sqrt[3]{2}$	$\sqrt[3]{4} \cdot \sqrt[3]{2} = \sqrt[3]{8} = 2$

In rationalizing the denominator of a quotient, be sure to multiply both the numerator and the denominator by the expression.

EXAMPLE 5 **Rationalizing Denominators**

Rationalize the denominator of each expression:

(a) $\dfrac{1}{\sqrt{3}}$ (b) $\dfrac{5}{4\sqrt{2}}$ (c) $\dfrac{\sqrt{2}}{\sqrt{3} - 3\sqrt{2}}$

Solution (a) The denominator contains the factor $\sqrt{3}$, so we multiply the numerator and denominator by $\sqrt{3}$ to obtain

$$\frac{1}{\sqrt{3}} = \frac{1}{\sqrt{3}} \cdot \frac{\sqrt{3}}{\sqrt{3}} = \frac{\sqrt{3}}{(\sqrt{3})^2} = \frac{\sqrt{3}}{3}$$

(b) The denominator contains the factor $\sqrt{2}$, so we multiply the numerator and denominator by $\sqrt{2}$ to obtain

$$\frac{5}{4\sqrt{2}} = \frac{5}{4\sqrt{2}} \cdot \frac{\sqrt{2}}{\sqrt{2}} = \frac{5\sqrt{2}}{4(\sqrt{2})^2} = \frac{5\sqrt{2}}{4 \cdot 2} = \frac{5\sqrt{2}}{8}$$

(c) The denominator contains the factor $\sqrt{3} - 3\sqrt{2}$, so we multiply the numerator and denominator by $\sqrt{3} + 3\sqrt{2}$ to obtain

$$\frac{\sqrt{2}}{\sqrt{3} - 3\sqrt{2}} = \frac{\sqrt{2}}{\sqrt{3} - 3\sqrt{2}} \cdot \frac{\sqrt{3} + 3\sqrt{2}}{\sqrt{3} + 3\sqrt{2}} = \frac{\sqrt{2}(\sqrt{3} + 3\sqrt{2})}{(\sqrt{3})^2 - (3\sqrt{2})^2}$$

$$= \frac{\sqrt{2}\sqrt{3} + 3(\sqrt{2})^2}{3 - 18} = \frac{\sqrt{6} + 6}{-15} = -\frac{6 + \sqrt{6}}{15}$$

Now Work PROBLEM 53

4 Simplify Expressions with Rational Exponents

Radicals are used to define rational exponents.

DEFINITION

If a is a real number and $n \geq 2$ is an integer, then

$$a^{1/n} = \sqrt[n]{a} \qquad (3)$$

provided that $\sqrt[n]{a}$ exists.

Note that if n is even and $a < 0$, then $\sqrt[n]{a}$ and $a^{1/n}$ do not exist.

EXAMPLE 6 **Writing Expressions Containing Fractional Exponents as Radicals**

(a) $4^{1/2} = \sqrt{4} = 2$

(b) $8^{1/2} = \sqrt{8} = 2\sqrt{2}$

(c) $(-27)^{1/3} = \sqrt[3]{-27} = -3$

(d) $16^{1/3} = \sqrt[3]{16} = 2\sqrt[3]{2}$

DEFINITION

If a is a real number and m and n are integers containing no common factors, with $n \geq 2$, then

$$a^{m/n} = \sqrt[n]{a^m} = (\sqrt[n]{a})^m \qquad (4)$$

provided that $\sqrt[n]{a}$ exists.

We have two comments about equation (4):

1. The exponent $\frac{m}{n}$ must be in lowest terms, and $n \geq 2$ must be positive.
2. In simplifying the rational expression $a^{m/n}$, either $\sqrt[n]{a^m}$ or $(\sqrt[n]{a})^m$ may be used, the choice depending on which is easier to simplify. Generally, taking the root first, as in $(\sqrt[n]{a})^m$, is easier.

EXAMPLE 7 **Using Equation (4)**

(a) $4^{3/2} = (\sqrt{4})^3 = 2^3 = 8$

(b) $(-8)^{4/3} = (\sqrt[3]{-8})^4 = (-2)^4 = 16$

(c) $(32)^{-2/5} = (\sqrt[5]{32})^{-2} = 2^{-2} = \frac{1}{4}$

(d) $25^{6/4} = 25^{3/2} = (\sqrt{25})^3 = 5^3 = 125$

Now Work PROBLEM 63

It can be shown that the Laws of Exponents hold for rational exponents. The next example illustrates using the Laws of Exponents to simplify.

EXAMPLE 8 **Simplifying Expressions Containing Rational Exponents**

Simplify each expression. Express your answer so that only positive exponents occur. Assume that the variables are positive.

(a) $(x^{2/3}y)(x^{-2}y)^{1/2}$

(b) $\left(\frac{2x^{1/3}}{y^{2/3}}\right)^{-3}$

(c) $\left(\frac{9x^2y^{1/3}}{x^{1/3}y}\right)^{1/2}$

Solution (a) $(x^{2/3}y)(x^{-2}y)^{1/2} = (x^{2/3}y)[(x^{-2})^{1/2}y^{1/2}]$

$$= x^{2/3}yx^{-1}y^{1/2}$$
$$= (x^{2/3}\cdot x^{-1})(y\cdot y^{1/2})$$
$$= x^{-1/3}y^{3/2}$$
$$= \frac{y^{3/2}}{x^{1/3}}$$

(b) $\left(\dfrac{2x^{1/3}}{y^{2/3}}\right)^{-3} = \left(\dfrac{y^{2/3}}{2x^{1/3}}\right)^{3} = \dfrac{(y^{2/3})^3}{(2x^{1/3})^3} = \dfrac{y^2}{2^3(x^{1/3})^3} = \dfrac{y^2}{8x}$

(c) $\left(\dfrac{9x^2y^{1/3}}{x^{1/3}y}\right)^{1/2} = \left(\dfrac{9x^{2-(1/3)}}{y^{1-(1/3)}}\right)^{1/2} = \left(\dfrac{9x^{5/3}}{y^{2/3}}\right)^{1/2} = \dfrac{9^{1/2}(x^{5/3})^{1/2}}{(y^{2/3})^{1/2}} = \dfrac{3x^{5/6}}{y^{1/3}}$ ●

Now Work PROBLEM 83

The next two examples illustrate some algebra that you will need to know for certain calculus problems.

EXAMPLE 9 Writing an Expression as a Single Quotient

Write the following expression as a single quotient in which only positive exponents appear.

$$(x^2+1)^{1/2} + x\cdot\frac{1}{2}(x^2+1)^{-1/2}\cdot 2x$$

Solution

$$(x^2+1)^{1/2} + x\cdot\frac{1}{2}(x^2+1)^{-1/2}\cdot 2x = (x^2+1)^{1/2} + \frac{x^2}{(x^2+1)^{1/2}}$$
$$= \frac{(x^2+1)^{1/2}(x^2+1)^{1/2} + x^2}{(x^2+1)^{1/2}}$$
$$= \frac{(x^2+1) + x^2}{(x^2+1)^{1/2}}$$
$$= \frac{2x^2+1}{(x^2+1)^{1/2}}$$ ●

Now Work PROBLEM 89

EXAMPLE 10 Factoring an Expression Containing Rational Exponents

Factor and simplify: $\frac{4}{3}x^{1/3}(2x+1) + 2x^{4/3}$

Solution Begin by writing $2x^{4/3}$ as a fraction with 3 as the denominator.

$$\frac{4}{3}x^{1/3}(2x+1) + 2x^{4/3} = \frac{4x^{1/3}(2x+1)}{3} + \frac{6x^{4/3}}{3} \underset{\text{Add the two fractions}}{=} \frac{4x^{1/3}(2x+1) + 6x^{4/3}}{3}$$

$$\underset{\text{2 and } x^{1/3} \text{ are common factors}}{=} \frac{2x^{1/3}[2(2x+1)+3x]}{3} \underset{\text{Simplify}}{=} \frac{2x^{1/3}(7x+2)}{3}$$ ●

Now Work PROBLEM 101

Historical Note

The radical sign, $\sqrt{\ }$, was first used in print by Christoff Rudolff in 1525. It is thought to be the manuscript form of the letter *r* (for the Latin word *radix* = *root*), although this has not been quite conclusively confirmed. It took a long time for $\sqrt{\ }$ to become the standard symbol for a square root and much longer to standardize $\sqrt[3]{\ }$, $\sqrt[4]{\ }$, $\sqrt[5]{\ }$, and so on. The indexes of the root were placed in every conceivable position, with

$$\sqrt{8}^{\,3}, \quad \sqrt{③8}, \quad \text{and} \quad \underset{3}{\sqrt{8}}$$

all being variants for $\sqrt[3]{8}$. The notation $\sqrt{\ }\sqrt{16}$ was popular for $\sqrt[4]{16}$. By the 1700s, the index had settled where we now put it.

The bar on top of the present radical symbol, as follows,

$$\sqrt{a^2 + 2ab + b^2}$$

is the last survivor of the **vinculum**, a bar placed atop an expression to indicate what we would now indicate with parentheses. For example,

$$a\overline{b + c} = a(b + c)$$

R.8 Assess Your Understanding

'Are You Prepared?' *Answers are given at the end of these exercises. If you get a wrong answer, read the pages in red.*

1. $(-3)^2 = \_\_;\ -3^2 = \_\_\_$ (pp. 21–24)

2. $\sqrt{16} = \_\_;\ \sqrt{(-4)^2} = \_\_$ (pp. 21–24)

Concepts and Vocabulary

3. In the symbol $\sqrt[n]{a}$, the integer n is called the ________.

4. We call $\sqrt[3]{a}$ the _______ _______ of a.

5. Let $n \ge 2$ and $m \ge 2$ be integers, and let a and b be real numbers. Which of the following is not a property of radicals? Assume all radicals are defined.

(a) $\sqrt[n]{\dfrac{a}{b}} = \dfrac{\sqrt[n]{a}}{\sqrt[n]{b}}$ (b) $\sqrt[n]{a+b} = \sqrt[n]{a} + \sqrt[n]{b}$

(c) $\sqrt[n]{ab} = \sqrt[n]{a}\sqrt[n]{b}$ (d) $\sqrt[n]{a^m} = (\sqrt[n]{a})^m$

6. If a is a real number and $n \ge 2$ is an integer, then which of the following expressions is equivalent to $\sqrt[n]{a}$, provided that it exists?

(a) a^{-n} (b) a^n (c) $\dfrac{1}{a^n}$ (d) $a^{1/n}$

7. Which of the following phrases best defines like radicals?
(a) Radical expressions that have the same index
(b) Radical expressions that have the same radicand
(c) Radical expressions that have the same index and the same radicand
(d) Radical expressions that have the same variable

8. To rationalize the denominator of the expression $\dfrac{\sqrt{2}}{1-\sqrt{3}}$, multiply both the numerator and the denominator by which of the following?

(a) $\sqrt{3}$ (b) $\sqrt{2}$ (c) $1 + \sqrt{3}$ (d) $1 - \sqrt{3}$

9. ***True or False*** $\sqrt[5]{-32} = -2$

10. ***True or False*** $\sqrt[4]{(-3)^4} = -3$

Skill Building

In Problems 11–48, simplify each expression. Assume that all variables are positive when they appear.

11. $\sqrt[3]{27}$ **12.** $\sqrt[4]{16}$ **13.** $\sqrt[3]{-8}$ **14.** $\sqrt[3]{-1}$

15. $\sqrt{8}$ **16.** $\sqrt[3]{54}$ **17.** $\sqrt[3]{-8x^4}$ **18.** $\sqrt[4]{48x^5}$

19. $\sqrt[4]{x^{12}y^8}$ **20.** $\sqrt[5]{x^{10}y^5}$ **21.** $\sqrt[4]{\dfrac{x^9y^7}{xy^3}}$ **22.** $\sqrt[3]{\dfrac{3xy^2}{81x^4y^2}}$

23. $\sqrt{36x}$ **24.** $\sqrt{9x^5}$ **25.** $\sqrt[4]{162x^9y^{12}}$ **26.** $\sqrt[3]{-40x^{14}y^{10}}$

27. $\sqrt{3x^2}\sqrt{12x}$ **28.** $\sqrt{5x}\sqrt{20x^3}$ **29.** $(\sqrt{5}\sqrt[3]{9})^2$ **30.** $(\sqrt[3]{3}\sqrt{10})^4$

31. $(3\sqrt{6})(2\sqrt{2})$ **32.** $(5\sqrt{8})(-3\sqrt{3})$ **33.** $3\sqrt{2} + 4\sqrt{2}$ **34.** $6\sqrt{5} - 4\sqrt{5}$

35. $-\sqrt{18} + 2\sqrt{8}$ **36.** $2\sqrt{12} - 3\sqrt{27}$ **37.** $(\sqrt{3}+3)(\sqrt{3}-1)$ **38.** $(\sqrt{5}-2)(\sqrt{5}+3)$

39. $5\sqrt[3]{2} - 2\sqrt[3]{54}$ **40.** $9\sqrt[3]{24} - \sqrt[3]{81}$ **41.** $(\sqrt{x}-1)^2$ **42.** $(\sqrt{x}+\sqrt{5})^2$

43. $\sqrt[3]{16x^4} - \sqrt[3]{2x}$ **44.** $\sqrt[4]{32x} + \sqrt[4]{2x^5}$ **45.** $\sqrt{8x^3} - 3\sqrt{50x}$ **46.** $3x\sqrt{9y} + 4\sqrt{25y}$

47. $\sqrt[3]{16x^4y} - 3x\sqrt[3]{2xy} + 5\sqrt[3]{-2xy^4}$ **48.** $8xy - \sqrt{25x^2y^2} + \sqrt[3]{8x^3y^3}$

In Problems 49–62, rationalize the denominator of each expression. Assume that all variables are positive when they appear.

49. $\dfrac{1}{\sqrt{2}}$ **50.** $\dfrac{2}{\sqrt{3}}$ **51.** $\dfrac{-\sqrt{3}}{\sqrt{5}}$ **52.** $\dfrac{-\sqrt{3}}{\sqrt{8}}$

53. $\dfrac{\sqrt{3}}{5-\sqrt{2}}$ **54.** $\dfrac{\sqrt{2}}{\sqrt{7}+2}$ **55.** $\dfrac{2-\sqrt{5}}{2+3\sqrt{5}}$ **56.** $\dfrac{\sqrt{3}-1}{2\sqrt{3}+3}$

57. $\dfrac{5}{\sqrt{2}-1}$ **58.** $\dfrac{-3}{\sqrt{5}+4}$ **59.** $\dfrac{5}{\sqrt[3]{2}}$ **60.** $\dfrac{-2}{\sqrt[3]{9}}$

61. $\dfrac{\sqrt{x+h}-\sqrt{x}}{\sqrt{x+h}+\sqrt{x}}$ **62.** $\dfrac{\sqrt{x+h}+\sqrt{x-h}}{\sqrt{x+h}-\sqrt{x-h}}$

In Problems 63–78, simplify each expression.

63. $8^{2/3}$ **64.** $4^{3/2}$ **65.** $(-27)^{1/3}$ **66.** $16^{3/4}$

67. $16^{3/2}$ **68.** $25^{3/2}$ **69.** $9^{-3/2}$ **70.** $16^{-3/2}$

71. $\left(\dfrac{9}{8}\right)^{3/2}$ **72.** $\left(\dfrac{27}{8}\right)^{2/3}$ **73.** $\left(\dfrac{8}{9}\right)^{-3/2}$ **74.** $\left(\dfrac{8}{27}\right)^{-2/3}$

75. $(-1000)^{-1/3}$ **76.** $-25^{-1/2}$ **77.** $\left(-\dfrac{64}{125}\right)^{-2/3}$ **78.** $-81^{-3/4}$

In Problems 79–86, simplify each expression. Express your answer so that only positive exponents occur. Assume that the variables are positive.

79. $x^{3/4}x^{1/3}x^{-1/2}$ **80.** $x^{2/3}x^{1/2}x^{-1/4}$ **81.** $(x^3y^6)^{1/3}$ **82.** $(x^4y^8)^{3/4}$

83. $\dfrac{(x^2y)^{1/3}(xy^2)^{2/3}}{x^{2/3}y^{2/3}}$ **84.** $\dfrac{(xy)^{1/4}(x^2y^2)^{1/2}}{(x^2y)^{3/4}}$ **85.** $\dfrac{(16x^2y^{-1/3})^{3/4}}{(xy^2)^{1/4}}$ **86.** $\dfrac{(4x^{-1}y^{1/3})^{3/2}}{(xy)^{3/2}}$

Applications and Extensions

In Problems 87–100, expressions that occur in calculus are given. Write each expression as a single quotient in which only positive exponents and/or radicals appear.

87. $\dfrac{x}{(1+x)^{1/2}}+2(1+x)^{1/2} \quad x>-1$

88. $\dfrac{1+x}{2x^{1/2}}+x^{1/2} \quad x>0$

89. $2x(x^2+1)^{1/2}+x^2\cdot\dfrac{1}{2}(x^2+1)^{-1/2}\cdot 2x$

90. $(x+1)^{1/3}+x\cdot\dfrac{1}{3}(x+1)^{-2/3} \quad x\neq-1$

91. $\sqrt{4x+3}\cdot\dfrac{1}{2\sqrt{x-5}}+\sqrt{x-5}\cdot\dfrac{1}{5\sqrt{4x+3}} \quad x>5$

92. $\dfrac{\sqrt[3]{8x+1}}{3\sqrt[3]{(x-2)^2}}+\dfrac{\sqrt[3]{x-2}}{24\sqrt[3]{(8x+1)^2}} \quad x\neq 2, x\neq-\dfrac{1}{8}$

93. $\dfrac{\sqrt{1+x}-x\cdot\dfrac{1}{2\sqrt{1+x}}}{1+x} \quad x>-1$

94. $\dfrac{\sqrt{x^2+1}-x\cdot\dfrac{2x}{2\sqrt{x^2+1}}}{x^2+1}$

95. $\dfrac{(x+4)^{1/2}-2x(x+4)^{-1/2}}{x+4} \quad x>-4$

96. $\dfrac{(9-x^2)^{1/2}+x^2(9-x^2)^{-1/2}}{9-x^2} \quad -3<x<3$

97. $\dfrac{\dfrac{x^2}{(x^2-1)^{1/2}}-(x^2-1)^{1/2}}{x^2} \quad x<-1 \text{ or } x>1$

98. $\dfrac{(x^2+4)^{1/2}-x^2(x^2+4)^{-1/2}}{x^2+4}$

99. $\dfrac{\dfrac{1+x^2}{2\sqrt{x}}-2x\sqrt{x}}{(1+x^2)^2} \quad x>0$

100. $\dfrac{2x(1-x^2)^{1/3}+\dfrac{2}{3}x^3(1-x^2)^{-2/3}}{(1-x^2)^{2/3}} \quad x\neq-1, x\neq 1$

In Problems 101–110, expressions that occur in calculus are given. Factor each expression. Express your answer so that only positive exponents occur.

101. $(x+1)^{3/2} + x \cdot \frac{3}{2}(x+1)^{1/2} \quad x \geq -1$

102. $(x^2+4)^{4/3} + x \cdot \frac{4}{3}(x^2+4)^{1/3} \cdot 2x$

103. $6x^{1/2}(x^2+x) - 8x^{3/2} - 8x^{1/2} \quad x \geq 0$

104. $6x^{1/2}(2x+3) + x^{3/2} \cdot 8 \quad x \geq 0$

105. $3(x^2+4)^{4/3} + x \cdot 4(x^2+4)^{1/3} \cdot 2x$

106. $2x(3x+4)^{4/3} + x^2 \cdot 4(3x+4)^{1/3}$

107. $4(3x+5)^{1/3}(2x+3)^{3/2} + 3(3x+5)^{4/3}(2x+3)^{1/2} \quad x \geq -\frac{3}{2}$

108. $6(6x+1)^{1/3}(4x-3)^{3/2} + 6(6x+1)^{4/3}(4x-3)^{1/2} \quad x \geq \frac{3}{4}$

109. $3x^{-1/2} + \frac{3}{2}x^{1/2} \quad x > 0$

110. $8x^{1/3} - 4x^{-2/3} \quad x \neq 0$

In Problems 111–118, use a calculator to approximate each radical. Round your answer to two decimal places.

111. $\sqrt{2}$

112. $\sqrt{7}$

113. $\sqrt[3]{4}$

114. $\sqrt[3]{-5}$

115. $\dfrac{2+\sqrt{3}}{3-\sqrt{5}}$

116. $\dfrac{\sqrt{5}-2}{\sqrt{2}+4}$

117. $\dfrac{3\sqrt[3]{5}-\sqrt{2}}{\sqrt{3}}$

118. $\dfrac{2\sqrt{3}-\sqrt[3]{4}}{\sqrt{2}}$

119. Calculating the Amount of Gasoline in a Tank A Shell station stores its gasoline in underground tanks that are right circular cylinders lying on their sides. See the illustration. The volume V of gasoline in the tank (in gallons) is given by the formula

$$V = 40h^2\sqrt{\frac{96}{h} - 0.608}$$

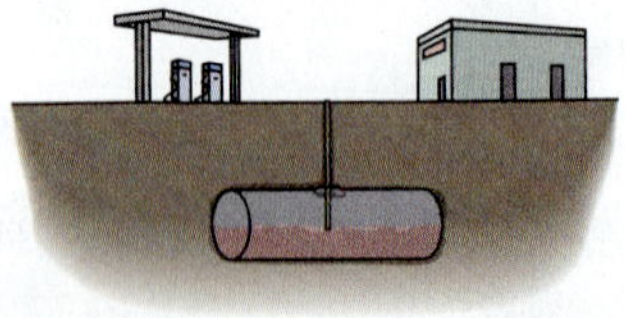

where h is the height of the gasoline (in inches) as measured on a depth stick.

(a) If $h = 12$ inches, how many gallons of gasoline are in the tank?

(b) If $h = 1$ inch, how many gallons of gasoline are in the tank?

120. Inclined Planes The final velocity v of an object in feet per second (ft/sec) after it slides down a frictionless inclined plane of height h feet is

$$v = \sqrt{64h + v_0^2}$$

where v_0 is the initial velocity (in ft/sec) of the object.

(a) What is the final velocity v of an object that slides down a frictionless inclined plane of height 4 feet? Assume that the initial velocity is 0.

(b) What is the final velocity v of an object that slides down a frictionless inclined plane of height 16 feet? Assume that the initial velocity is 0.

(c) What is the final velocity v of an object that slides down a frictionless inclined plane of height 2 feet with an initial velocity of 4 ft/sec?

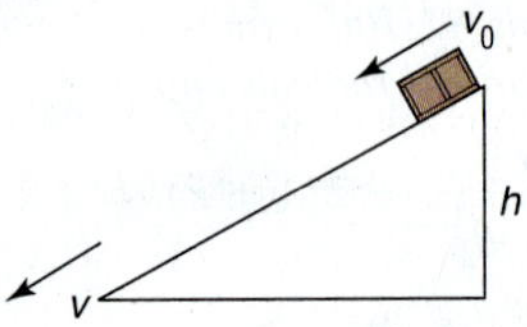

Problems 121 and 122 require the following information.

Period of a Pendulum *The period T, in seconds, of a pendulum of length l, in feet, may be approximated using the formula*

$$T = 2\pi\sqrt{\frac{l}{32}}$$

In Problems 121 and 122, express your answer both as a square root and as a decimal.

121. Find the period T of a pendulum whose length is 64 feet.

122. Find the period T of a pendulum whose length is 16 feet.

Explaining Concepts: Discussion and Writing

123. Give an example to show that $\sqrt{a^2}$ is not equal to a. Use it to explain why $\sqrt{a^2} = |a|$.

'Are You Prepared?' Answers

1. 9; −9 **2.** 4; 4

Equations and Inequalities

1

Financing a Purchase

Whenever we make a major purchase, such as an automobile or a house, we often need to finance the purchase by borrowing money from a lending institution, such as a bank. Have you ever wondered how the bank determines the monthly payment? How much total interest will be paid over the course of the loan? What roles do the rate of interest and the length of the loan play?

—See the Internet-based Chapter Project I—

A Look Ahead •••

Chapter 1, Equations and Inequalities, reviews many topics covered in Intermediate Algebra. If your instructor decides to exclude complex numbers from the course, don't be alarmed. The book has been designed so that the topic of complex numbers can be included or excluded without any confusion later on.

Outline

1.1 Linear Equations

PREPARING FOR THIS SECTION *Before getting started, review the following:*

- Properties of Real Numbers (Section R.1, pp. 9–13)
- Domain of a Variable (Section R.2, p. 21)

Now Work the **'Are You Prepared?'** problems on page 90.

OBJECTIVES
1 Solve a Linear Equation (p. 84)
2 Solve Equations That Lead to Linear Equations (p. 86)
3 Solve Problems That Can Be Modeled by Linear Equations (p. 87)

An **equation in one variable** is a statement in which two expressions, at least one containing the variable, are set equal. The expressions are called the **sides** of the equation. Since an equation is a statement, it may be true or false, depending on the value of the variable. Unless otherwise restricted, the admissible values of the variable are those in the domain of the variable. These admissible values of the variable, if any, that result in a true statement are called **solutions**, or **roots**, of the equation. To **solve an equation** means to find all the solutions of the equation.

For example, the following are all equations in one variable, x:

$$x + 5 = 9 \qquad x^2 + 5x = 2x - 2 \qquad \frac{x^2 - 4}{x + 1} = 0 \qquad \sqrt{x^2 + 9} = 5$$

The first of these statements, $x + 5 = 9$, is true when $x = 4$ and false for any other choice of x. That is, 4 is a solution of the equation $x + 5 = 9$. We also say that 4 **satisfies** the equation $x + 5 = 9$, because, when 4 is substituted for x, a true statement results.

Sometimes an equation will have more than one solution. For example, the equation

$$\frac{x^2 - 4}{x + 1} = 0$$

has $x = -2$ and $x = 2$ as solutions.

Usually, we will write the solution of an equation in set notation. This set is called the **solution set** of the equation. For example, the solution set of the equation $x^2 - 9 = 0$ is $\{-3, 3\}$.

Some equations have no real solution. For example, $x^2 + 9 = 5$ has no real solution, because there is no real number whose square, when added to 9, equals 5.

An equation that is satisfied for every value of the variable for which both sides are defined is called an **identity**. For example, the equation

$$3x + 5 = x + 3 + 2x + 2$$

is an identity, because this statement is true for any real number x.

One method for solving an equation is to replace the original equation by a succession of **equivalent equations**, equations having the same solution set, until an equation with an obvious solution is obtained.

For example, consider the following succession of equivalent equations:

$$2x + 3 = 13$$
$$2x = 10$$
$$x = 5$$

We conclude that the solution set of the original equation is $\{5\}$.

How do we obtain equivalent equations? In general, there are five ways.

Procedures That Result in Equivalent Equations

1. Interchange the two sides of the equation:

   Replace $3 = x$ by $x = 3$

2. Simplify the sides of the equation by combining like terms, eliminating parentheses, and so on:

   Replace $(x + 2) + 6 = 2x + (x + 1)$
   by $x + 8 = 3x + 1$

3. Add or subtract the same expression on both sides of the equation:

   Replace $3x - 5 = 4$
   by $(3x - 5) + 5 = 4 + 5$

4. Multiply or divide both sides of the equation by the same nonzero expression:

   Replace $\frac{3x}{x - 1} = \frac{6}{x - 1} \quad x \neq 1$
   by $\frac{3x}{x - 1} \cdot (x - 1) = \frac{6}{x - 1} \cdot (x - 1)$

5. If one side of the equation is 0 and the other side can be factored, then we may use the Zero-Product Property* and set each factor equal to 0:

   Replace $x(x - 3) = 0$
   by $x = 0$ or $x - 3 = 0$

WARNING Squaring both sides of an equation does not necessarily lead to an equivalent equation. For example, $x = 3$ has one solution, but $x^2 = 9$ has two solutions, $x = -3$ and $x = 3$.

Whenever it is possible to solve an equation in your head, do so. For example,

The solution of $2x = 8$ is $x = 4$.
The solution of $3x - 15 = 0$ is $x = 5$.

Now Work PROBLEM 11

EXAMPLE 1

Solving an Equation

Solve the equation: $3x - 5 = 4$

Solution Replace the original equation by a succession of equivalent equations.

$$3x - 5 = 4$$
$$(3x - 5) + 5 = 4 + 5 \quad \text{Add 5 to both sides.}$$
$$3x = 9 \quad \text{Simplify.}$$
$$\frac{3x}{3} = \frac{9}{3} \quad \text{Divide both sides by 3.}$$
$$x = 3 \quad \text{Simplify.}$$

The last equation, $x = 3$, has the single solution 3. All these equations are equivalent, so 3 is the only solution of the original equation, $3x - 5 = 4$.

*The Zero-Product Property says that if $ab = 0$, then $a = 0$ or $b = 0$ or both equal 0.

Check: Check the solution by substituting 3 for x in the original equation.

$$\begin{aligned} 3x - 5 &= 4 \\ 3(3) - 5 &\stackrel{?}{=} 4 \\ 9 - 5 &\stackrel{?}{=} 4 \\ 4 &= 4 \end{aligned}$$

The solution checks. The solution set is $\{3\}$.

Now Work PROBLEM 25

Steps for Solving Equations

STEP 1: List any restrictions on the domain of the variable.

STEP 2: Simplify the equation by replacing the original equation by a succession of equivalent equations using the procedures listed earlier.

STEP 3: If the result of Step 2 is a product of factors equal to 0, use the Zero-Product Property and set each factor equal to 0 (procedure 5).

STEP 4: Check your solution(s).

1 Solve a Linear Equation

Linear equations are equations such as

$$3x + 12 = 0 \qquad -2x + 5 = 0 \qquad \frac{1}{2}x - \sqrt{3} = 0$$

DEFINITION

A **linear equation in one variable** is an equation equivalent in form to

$$ax + b = 0$$

where a and b are real numbers and $a \neq 0$.

Sometimes a linear equation is called a **first-degree equation**, because the left side is a polynomial in x of degree 1.

It is relatively easy to solve a linear equation. The idea is to *isolate* the variable:

$$\begin{aligned} ax + b &= 0 && a \neq 0 \\ ax &= -b && \text{Subtract } b \text{ from both sides.} \\ x &= \frac{-b}{a} && \text{Divide both sides by } a, a \neq 0. \end{aligned}$$

The linear equation $ax + b = 0, a \neq 0$, has the single solution given by the formula $x = -\frac{b}{a}$.

EXAMPLE 2 **Solving a Linear Equation**

Solve the equation: $\frac{1}{2}(x + 5) - 4 = \frac{1}{3}(2x - 1)$

Solution To clear the equation of fractions, multiply both sides by 6, the least common multiple (LCM) of the denominators of the fractions $\frac{1}{2}$ and $\frac{1}{3}$.

$$\frac{1}{2}(x+5) - 4 = \frac{1}{3}(2x-1)$$

$$6\left[\frac{1}{2}(x+5) - 4\right] = 6\left[\frac{1}{3}(2x-1)\right] \quad \text{Multiply both sides by 6, the LCM of 2 and 3.}$$

$$3(x+5) - 6\cdot 4 = 2(2x-1) \quad \text{Use the Distributive Property on the left and the Associative Property on the right.}$$

$$3x + 15 - 24 = 4x - 2 \quad \text{Use the Distributive Property.}$$

$$3x - 9 = 4x - 2 \quad \text{Combine like terms.}$$

$$3x - 9 + 9 = 4x - 2 + 9 \quad \text{Add 9 to each side.}$$

$$3x = 4x + 7 \quad \text{Simplify.}$$

$$3x - 4x = 4x + 7 - 4x \quad \text{Subtract } 4x \text{ from each side.}$$

$$-x = 7 \quad \text{Simplify.}$$

$$x = -7 \quad \text{Multiply both sides by } -1.$$

✓**Check:** Substitute -7 for x in the expressions on the left and right sides of the original equation, and simplify. If the two expressions are equal, the solution checks.

$$\frac{1}{2}(x+5) - 4 = \frac{1}{2}(-7+5) - 4 = \frac{1}{2}(-2) - 4 = -1 - 4 = -5$$

$$\frac{1}{3}(2x-1) = \frac{1}{3}[2(-7) - 1] = \frac{1}{3}(-14-1) = \frac{1}{3}(-15) = -5$$

Since the two expressions are equal, the solution checks. The solution set is $\{-7\}$. ●

Now Work PROBLEM 35

EXAMPLE 3 Solving a Linear Equation Using a Calculator

Solve the equation: $2.78x + \dfrac{2}{17.931} = 54.06$

Round the answer to two decimal places.

Solution To avoid rounding errors, solve for x before using a calculator.

$$2.78x + \frac{2}{17.931} = 54.06$$

$$2.78x = 54.06 - \frac{2}{17.931} \quad \text{Subtract } \frac{2}{17.931} \text{ from each side.}$$

$$x = \frac{54.06 - \frac{2}{17.931}}{2.78} \quad \text{Divide each side by 2.78.}$$

Now use your calculator. The solution, rounded to two decimal places, is 19.41.

✓**Check:** Store the calculator solution 19.40592134 in memory, and proceed to evaluate $2.78x + \dfrac{2}{17.931}$.

$$(2.78)(19.40592134) + \frac{2}{17.931} = 54.06$$

●

Now Work PROBLEM 67

2 Solve Equations That Lead to Linear Equations

EXAMPLE 4 **Solving an Equation That Leads to a Linear Equation**

Solve the equation: $(2y + 1)(y - 1) = (y + 5)(2y - 5)$

Solution

$$\begin{aligned}(2y + 1)(y - 1) &= (y + 5)(2y - 5) \\ 2y^2 - y - 1 &= 2y^2 + 5y - 25 && \text{Multiply and combine like terms.} \\ -y - 1 &= 5y - 25 && \text{Subtract } 2y^2 \text{ from each side.} \\ -y &= 5y - 24 && \text{Add 1 to each side.} \\ -6y &= -24 && \text{Subtract } 5y \text{ from each side.} \\ y &= 4 && \text{Divide both sides by } -6.\end{aligned}$$

Check: $(2y + 1)(y - 1) = [2(4) + 1](4 - 1) = (8 + 1)(3) = (9)(3) = 27$

$(y + 5)(2y - 5) = (4 + 5)[2(4) - 5] = (9)(8 - 5) = (9)(3) = 27$

The two expressions are equal, so the solution checks. The solution set is $\{4\}$. ●

EXAMPLE 5 **Solving an Equation That Leads to a Linear Equation**

Solve the equation: $\dfrac{3}{x - 2} = \dfrac{1}{x - 1} + \dfrac{7}{(x - 1)(x - 2)}$

Solution First, notice that the domain of the variable is $\{x \mid x \neq 1, x \neq 2\}$. Clear the equation of fractions by multiplying both sides by the least common multiple of the denominators of the three fractions, $(x - 1)(x - 2)$.

$$\frac{3}{x - 2} = \frac{1}{x - 1} + \frac{7}{(x - 1)(x - 2)}$$

$$(x - 1)\cancel{(x - 2)}\frac{3}{\cancel{x - 2}} = (x - 1)(x - 2)\left[\frac{1}{x - 1} + \frac{7}{(x - 1)(x - 2)}\right] \quad \text{Multiply both sides by } (x - 1)(x - 2)\text{; cancel on the left.}$$

$$3x - 3 = \cancel{(x - 1)}(x - 2)\frac{1}{\cancel{x - 1}} + \cancel{(x - 1)(x - 2)}\frac{7}{\cancel{(x - 1)(x - 2)}} \quad \text{Use the Distributive Property on each side; cancel on the right.}$$

$$\begin{aligned}3x - 3 &= (x - 2) + 7 && \text{Simplify.} \\ 3x - 3 &= x + 5 && \text{Combine like terms.} \\ 2x &= 8 && \text{Add 3 to each side; subtract } x \text{ from each side.} \\ x &= 4 && \text{Divide by 2.}\end{aligned}$$

Check: $\dfrac{3}{x - 2} = \dfrac{3}{4 - 2} = \dfrac{3}{2}$

$$\frac{1}{x - 1} + \frac{7}{(x - 1)(x - 2)} = \frac{1}{4 - 1} + \frac{7}{(4 - 1)(4 - 2)} = \frac{1}{3} + \frac{7}{3 \cdot 2} = \frac{2}{6} + \frac{7}{6} = \frac{9}{6} = \frac{3}{2}$$

The solution checks. The solution set is $\{4\}$. ●

Now Work PROBLEM 61

EXAMPLE 6 **An Equation with No Solution**

Solve the equation: $\dfrac{3x}{x - 1} + 2 = \dfrac{3}{x - 1}$

Solution First, note that the domain of the variable is $\{x \mid x \neq 1\}$. Since the two quotients in the equation have the same denominator, $x - 1$, simplify by multiplying both sides by $x - 1$. The resulting equation is equivalent to the original equation, since we are multiplying by $x - 1$, which is not 0. (Remember, $x \neq 1$.)

$$\frac{3x}{x-1} + 2 = \frac{3}{x-1}$$

$$\left(\frac{3x}{x-1} + 2\right) \cdot (x-1) = \frac{3}{\cancel{x-1}} \cdot \cancel{(x-1)} \quad \text{Multiply both sides by } x - 1\text{; cancel on the right.}$$

$$\frac{3x}{\cancel{x-1}} \cdot \cancel{(x-1)} + 2 \cdot (x-1) = 3 \quad \text{Use the Distributive Property on the left side; cancel on the left.}$$

$$3x + 2x - 2 = 3 \quad \text{Simplify.}$$

$$5x - 2 = 3 \quad \text{Combine like terms.}$$

$$5x = 5 \quad \text{Add 2 to each side.}$$

$$x = 1 \quad \text{Divide both sides by 5.}$$

The solution appears to be 1. But recall that $x = 1$ is not in the domain of the variable, so this value must be discarded. The equation has no solution. The solution set is $\varnothing$. ●

NOTE Example 6 illustrates the appearance of an *extraneous solution*. This will be discussed more in Section 1.4. ■

Now Work PROBLEM 51

EXAMPLE 7 **Converting to Fahrenheit from Celsius**

In the United States we measure temperature in both degrees Fahrenheit (°F) and degrees Celsius (°C), which are related by the formula $C = \frac{5}{9}(F - 32)$. What are the Fahrenheit temperatures corresponding to Celsius temperatures of 0°, 10°, 20°, and 30°C?

Solution We could solve four equations for F by replacing C each time by 0, 10, 20, and 30. Instead, it is much easier and faster first to solve the equation $C = \frac{5}{9}(F - 32)$ for F and then to substitute in the values of C.

$$C = \frac{5}{9}(F - 32)$$

$$9C = 5(F - 32) \quad \text{Multiply both sides by 9.}$$

$$9C = 5F - 160 \quad \text{Use the Distributive Property.}$$

$$5F - 160 = 9C \quad \text{Interchange sides.}$$

$$5F = 9C + 160 \quad \text{Add 160 to each side.}$$

$$F = \frac{9}{5}C + 32 \quad \text{Divide both sides by 5.}$$

Now do the required arithmetic.

$$0°\text{C}: \quad F = \frac{9}{5}(0) + 32 = 32°\text{F}$$

$$10°\text{C}: \quad F = \frac{9}{5}(10) + 32 = 50°\text{F}$$

$$20°\text{C}: \quad F = \frac{9}{5}(20) + 32 = 68°\text{F}$$

$$30°\text{C}: \quad F = \frac{9}{5}(30) + 32 = 86°\text{F}$$

●

NOTE The icon is a Model It! icon. It indicates that the discussion or problem involves modeling. ■

3 Solve Problems That Can Be Modeled by Linear Equations

Although each situation has its unique features, we can provide an outline of the steps to follow when solving applied problems.

Steps for Solving Applied Problems

STEP 1: Read the problem carefully, perhaps two or three times. Pay particular attention to the question being asked in order to identify what you are looking for. Identify any relevant formulas you may need ($d = rt$, $A = \pi r^2$, etc.). If you can, determine realistic possibilities for the answer.

STEP 2: Assign a letter (variable) to represent what you are looking for, and, if necessary, express any remaining unknown quantities in terms of this variable.

STEP 3: Make a list of all the known facts, and translate them into mathematical expressions. These may take the form of an equation (or, later, an inequality) involving the variable. The equation (or inequality) is called the **model**. If possible, draw an appropriately labeled diagram to assist you. Sometimes a table or chart helps.

STEP 4: Solve the equation for the variable, and then answer the question, usually using a complete sentence.

STEP 5: Check the answer with the facts in the problem. If it agrees, congratulations! If it does not agree, try again.

NOTE It is a good practice to choose a variable that reminds you of the unknown. For example, use t for time. ■

EXAMPLE 8 Investments

A total of $18,000 is invested, some in stocks and some in bonds. If the amount invested in bonds is half that invested in stocks, how much is invested in each category?

Step-by-Step Solution

Step 1: Determine what you are looking for.

We are being asked to find the amount of two investments. These amounts must total $18,000. (Do you see why?)

Step 2: Assign a variable to represent what you are looking for. If necessary, express any remaining unknown quantities in terms of this variable.

If x equals the amount invested in stocks, then the rest of the money, $18{,}000 - x$, is the amount invested in bonds.

Step 3: Translate the English into mathematical statements. It may be helpful to draw a figure that represents the situation. Sometimes a table can be used to organize the information. Use the information to build your model.

Set up a table:

Amount in Stocks	Amount in Bonds	Reason
x	$18{,}000 - x$	Total invested is $18,000.

We also know that:

Total amount invested in bonds is one-half that in stocks

$$18{,}000 - x = \frac{1}{2}(x)$$

Step 4: Solve the equation and answer the original question.

$$18{,}000 - x = \frac{1}{2}x$$

$$18{,}000 = x + \frac{1}{2}x \quad \text{Add } x \text{ to both sides.}$$

$$18{,}000 = \frac{3}{2}x \quad \text{Simplify.}$$

$$\left(\frac{2}{3}\right)18{,}000 = \left(\frac{2}{3}\right)\left(\frac{3}{2}x\right) \quad \text{Multiply both sides by } \frac{2}{3}.$$

$$12{,}000 = x \quad \text{Simplify.}$$

So $12,000 is invested in stocks, and $18,000 − $12,000 = $6000 is invested in bonds.

Step 5: Check your answer with the facts presented in the problem.

The total invested is $12,000 + $6000 = $18,000, and the amount in bonds, $6000, is half that in stocks, $12,000. ●

Now Work PROBLEM 83

EXAMPLE 9 Determining an Hourly Wage

Shannon grossed \$725 one week by working 52 hours. Her employer pays time-and-a-half for all hours worked in excess of 40 hours. With this information, can you determine Shannon's regular hourly wage?

Solution

STEP 1: We are looking for an hourly wage. Our answer will be expressed in dollars per hour.

STEP 2: Let x represent the regular hourly wage, measured in dollars per hour. Then $1.5x$ is the overtime hourly wage.

STEP 3: Set up a table:

	Hours Worked	Hourly Wage	Salary
Regular	40	x	$40x$
Overtime	12	$1.5x$	$12(1.5x) = 18x$

The sum of regular salary plus overtime salary will equal \$725. From the table, $40x + 18x = 725$.

STEP 4:

$$\begin{aligned} 40x + 18x &= 725 \\ 58x &= 725 \\ x &= 12.50 \end{aligned}$$

Shannon's regular hourly wage is \$12.50 per hour.

STEP 5: Forty hours yields a salary of $40(12.50) = \$500$, and 12 hours of overtime yields a salary of $12(1.5)(12.50) = \$225$, for a total of \$725.

Now Work PROBLEM 85

SUMMARY

Steps for Solving a Linear Equation

To solve a linear equation, follow these steps:

STEP 1: List any restrictions on the variable.

STEP 2: If necessary, clear the equation of fractions by multiplying both sides by the least common multiple (LCM) of the denominators of all the fractions.

STEP 3: Remove all parentheses and simplify.

STEP 4: Collect all terms containing the variable on one side and all remaining terms on the other side.

STEP 5: Simplify and solve.

STEP 6: Check your solution(s).

Historical Feature

Solving equations is among the oldest of mathematical activities, and efforts to systematize this activity determined much of the shape of modern mathematics.

Consider the following problem and its solution using only words: Solve the problem of how many apples Jim has, given that

"Bob's five apples and Jim's apples together make twelve apples" by thinking,

"Jim's apples are all twelve apples less Bob's five apples" and then concluding,

"Jim has seven apples."

The mental steps translated into algebra are

$$\begin{aligned} 5 + x &= 12 \\ x &= 12 - 5 \\ &= 7 \end{aligned}$$

The solution of this problem using only words is the earliest form of algebra. Such problems were solved exactly this way in Babylonia in 1800 BC. We know almost nothing of mathematical work before this date, although most authorities believe the sophistication of the earliest known texts indicates that there must have been a long period of previous development. The method of writing out equations in words persisted for thousands of years, and although it now seems extremely cumbersome, it was used very effectively by many generations of mathematicians. The Arabs developed a good deal of the theory of cubic equations while writing out all the equations in words. About AD 1500, the tendency to abbreviate words in the written equations began to lead in the direction of modern notation; for example, the Latin word *et* (meaning "and") developed into the plus sign, +. Although the occasional use of letters to represent variables dates back to AD 1200, the practice did not become common until about AD 1600. Development thereafter was rapid, and by 1635 algebraic notation did not differ essentially from what we use now.

1.1 Assess Your Understanding

'Are You Prepared?' *Answers are given at the end of these exercises. If you get a wrong answer, read the pages listed in red.*

1. The fact that $2(x + 3) = 2x + 6$ is attributable to the ________ Property. (pp. 9–13)
2. The fact that $3x = 0$ implies that $x = 0$ is a result of the ________ Property. (pp. 9–13)
3. The domain of the variable in the expression $\frac{x}{x-4}$ is ________. (p. 21)

Concepts and Vocabulary

4. ***True or False*** Multiplying both sides of an equation by any number results in an equivalent equation.
5. An equation that is satisfied for every value of the variable for which both sides are defined is called a(n) ________.
6. An equation of the form $ax + b = 0$ is called a(n) ____ equation or a(n) ________ equation.
7. ***True or False*** The solution of the equation $3x - 8 = 0$ is $\frac{3}{8}$.
8. ***True or False*** Some equations have no solution.
9. An admissible value for the variable that makes the equation a true statement is called a(n) ______ of the equation.
   (a) identity (b) solution (c) degree (d) model
10. A chemist mixes 10 liters of a 20% solution with x liters of a 35% solution. Which of the following expressions represents the total number of liters in the mixture?
   (a) x (b) $20 - x$ (c) $\frac{35}{x}$ (d) $10 + x$

Skill Building

In Problems 11–18, mentally solve each equation.

11. $7x = 21$
12. $6x = -24$
13. $3x + 15 = 0$
14. $6x + 18 = 0$
15. $2x - 3 = 0$
16. $3x + 4 = 0$
17. $\frac{1}{3}x = \frac{5}{12}$
18. $\frac{2}{3}x = \frac{9}{2}$

In Problems 19–66, solve each equation.

19. $3x + 4 = x$
20. $2x + 9 = 5x$
21. $2t - 6 = 3 - t$
22. $5y + 6 = -18 - y$
23. $6 - x = 2x + 9$
24. $3 - 2x = 2 - x$
25. $3 + 2n = 4n + 7$
26. $6 - 2m = 3m + 1$
27. $2(3 + 2x) = 3(x - 4)$
28. $3(2 - x) = 2x - 1$
29. $8x - (3x + 2) = 3x - 10$
30. $7 - (2x - 1) = 10$
31. $\frac{3}{2}x + 2 = \frac{1}{2} - \frac{1}{2}x$
32. $\frac{1}{3}x = 2 - \frac{2}{3}x$
33. $\frac{1}{2}x - 5 = \frac{3}{4}x$
34. $1 - \frac{1}{2}x = 6$
35. $\frac{2}{3}p = \frac{1}{2}p + \frac{1}{3}$
36. $\frac{1}{2} - \frac{1}{3}p = \frac{4}{3}$
37. $0.9t = 0.4 + 0.1t$
38. $0.9t = 1 + t$
39. $\frac{x+1}{3} + \frac{x+2}{7} = 2$
40. $\frac{2x+1}{3} + 16 = 3x$
41. $\frac{2}{y} + \frac{4}{y} = 3$
42. $\frac{4}{y} - 5 = \frac{5}{2y}$
43. $\frac{1}{2} + \frac{2}{x} = \frac{3}{4}$
44. $\frac{3}{x} - \frac{1}{3} = \frac{1}{6}$
45. $(x + 7)(x - 1) = (x + 1)^2$
46. $(x + 2)(x - 3) = (x + 3)^2$
47. $x(2x - 3) = (2x + 1)(x - 4)$
48. $x(1 + 2x) = (2x - 1)(x - 2)$
49. $z(z^2 + 1) = 3 + z^3$
50. $w(4 - w^2) = 8 - w^3$
51. $\frac{x}{x-2} + 3 = \frac{2}{x-2}$
52. $\frac{2x}{x+3} = \frac{-6}{x+3} - 2$
53. $\frac{2x}{x^2-4} = \frac{4}{x^2-4} - \frac{3}{x+2}$
54. $\frac{x}{x^2-9} + \frac{4}{x+3} = \frac{3}{x^2-9}$
55. $\frac{x}{x+2} = \frac{3}{2}$
56. $\frac{3x}{x-1} = 2$
57. $\frac{5}{2x-3} = \frac{3}{x+5}$
58. $\frac{-4}{x+4} = \frac{-3}{x+6}$
59. $\frac{6t+7}{4t-1} = \frac{3t+8}{2t-4}$
60. $\frac{8w+5}{10w-7} = \frac{4w-3}{5w+7}$

61. $\dfrac{4}{x-2} = \dfrac{-3}{x+5} + \dfrac{7}{(x+5)(x-2)}$

62. $\dfrac{-4}{2x+3} + \dfrac{1}{x-1} = \dfrac{1}{(2x+3)(x-1)}$

63. $\dfrac{2}{y+3} + \dfrac{3}{y-4} = \dfrac{5}{y+6}$

64. $\dfrac{5}{5z-11} + \dfrac{4}{2z-3} = \dfrac{-3}{5-z}$

65. $\dfrac{x}{x^2-1} - \dfrac{x+3}{x^2-x} = \dfrac{-3}{x^2+x}$

66. $\dfrac{x+1}{x^2+2x} - \dfrac{x+4}{x^2+x} = \dfrac{-3}{x^2+3x+2}$

In Problems 67–70, use a calculator to solve each equation. Round the solution to two decimal places.

67. $3.2x + \dfrac{21.3}{65.871} = 19.23$

68. $6.2x - \dfrac{19.1}{83.72} = 0.195$

69. $14.72 - 21.58x = \dfrac{18}{2.11}x + 2.4$

70. $18.63x - \dfrac{21.2}{2.6} = \dfrac{14}{2.32}x - 20$

Applications and Extensions

In Problems 71–74, solve each equation. The letters a, b, and c are constants.

71. $ax - b = c, \quad a \neq 0$

72. $1 - ax = b, \quad a \neq 0$

73. $\dfrac{x}{a} + \dfrac{x}{b} = c, \quad a \neq 0, b \neq 0, a \neq -b$

74. $\dfrac{a}{x} + \dfrac{b}{x} = c, \quad c \neq 0$

75. Find the number a for which $x = 4$ is a solution of the equation

$$x + 2a = 16 + ax - 6a$$

76. Find the number b for which $x = 2$ is a solution of the equation

$$x + 2b = x - 4 + 2bx$$

Problems 77–82 list some formulas that occur in applications. Solve each formula for the indicated variable.

77. **Electricity** $\dfrac{1}{R} = \dfrac{1}{R_1} + \dfrac{1}{R_2}$ for R

78. **Finance** $A = P(1 + rt)$ for r

79. **Mechanics** $F = \dfrac{mv^2}{R}$ for R

80. **Chemistry** $PV = nRT$ for T

81. **Mathematics** $S = \dfrac{a}{1-r}$ for r

82. **Mechanics** $v = -gt + v_0$ for t

83. **Finance** A total of $20,000 is to be invested, some in bonds and some in certificates of deposit (CDs). If the amount invested in bonds is to exceed that in CDs by $3000, how much will be invested in each type of investment?

84. **Finance** A total of $10,000 is to be divided between Sean and George, with George to receive $3000 less than Sean. How much will each receive?

85. **Computing Hourly Wages** Sandra, who is paid time-and-a-half for hours worked in excess of 40 hours, had gross weekly wages of $546 for 48 hours worked. What is her regular hourly rate?

86. **Computing Hourly Wages** Leigh is paid time-and-a-half for hours worked in excess of 40 hours and double-time for hours worked on Sunday. If Leigh had gross weekly wages of $627 for working 50 hours, 4 of which were on Sunday, what is her regular hourly rate?

87. **Computing Grades** Going into the final exam, which will count as two tests, Brooke has test scores of 80, 83, 71, 61, and 95. What score does Brooke need on the final in order to have an average score of 80?

88. **Computing Grades** Going into the final exam, which will count as two-thirds of the final grade, Mike has test scores of 86, 80, 84, and 90. What minimum score does Mike need on the final in order to earn a B, which requires an average score of 80? What does he need to earn an A, which requires an average of 90?

89. **Business: Discount Pricing** A builder of tract homes reduced the price of a model by 15%. If the new price is $425,000, what was its original price? How much can be saved by purchasing the model?

90. **Business: Discount Pricing** A car dealer, at a year-end clearance, reduces the list price of last year's models by 15%. If a certain four-door model has a discounted price of $8000, what was its list price? How much can be saved by purchasing last year's model?

91. **Personal Finance: Concession Markup** A movie theater marks up the candy it sells by 275%. If a box of candy sells for $3.00 at the theater, how much did the theater pay for the box?

92. Personal Finance: Cost of a Car The suggested list price of a new car is \$18,000. The dealer's cost is 85% of list. How much will you pay if the dealer is willing to accept \$100 over cost for the car?

93. Business: Theater Attendance The manager of the Coral Theater wants to know whether the majority of its patrons are adults or children. One day in July, 5200 tickets were sold and the receipts totaled \$29,961. The adult admission is \$7.50, and the children's admission is \$4.50. How many adult patrons were there?

94. Business: Discount Pricing A wool suit, discounted by 30% for a clearance sale, has a price tag of \$399. What was the suit's original price?

95. Geometry The perimeter of a rectangle is 60 feet. Find its length and width if the length is 8 feet longer than the width.

96. Geometry The perimeter of a rectangle is 42 meters. Find its length and width if the length is twice the width.

97. Counting Calories Herschel uses an app on his smartphone to keep track of his daily calories from meals. One day his calories from breakfast were 125 more than his calories from lunch, and his calories from dinner were 300 less than twice his calories from lunch. If his total caloric intake from meals was 2025, determine his calories for each meal.

98. Counting Calories Tyshira tracks her net calories (calories taken in minus calories burned) as part of her fitness program. For one particular day, her net intake was 1480 calories. Her lunch calories were half her breakfast calories, and her dinner calories were 200 more than her breakfast calories. She ate 120 less calories in snacks than for breakfast, and she burned 700 calories by exercising on her elliptical. How many calories did she take in from snacks?

99. Sharing the Cost of a Pizza Judy and Tom agree to share the cost of an \$18 pizza based on how much each ate. If Tom ate $\frac{2}{3}$ the amount that Judy ate, how much should each pay? [**Hint:** Some pizza may be left.]

Tom's portion

Judy's portion

Explaining Concepts: Discussion and Writing

100. What Is Wrong? One step in the following list contains an error. Identify it and explain what is wrong.

$$x = 2 \quad (1)$$
$$3x - 2x = 2 \quad (2)$$
$$3x = 2x + 2 \quad (3)$$
$$x^2 + 3x = x^2 + 2x + 2 \quad (4)$$
$$x^2 + 3x - 10 = x^2 + 2x - 8 \quad (5)$$
$$(x - 2)(x + 5) = (x - 2)(x + 4) \quad (6)$$
$$x + 5 = x + 4 \quad (7)$$
$$1 = 0 \quad (8)$$

101. The equation

$$\frac{5}{x+3} + 3 = \frac{8+x}{x+3}$$

has no solution, yet when we go through the process of solving it, we obtain $x = -3$. Write a brief paragraph to explain what causes this to happen.

102. Make up an equation that has no solution and give it to a fellow student to solve. Ask the fellow student to write a critique of your equation.

'Are You Prepared?' Answers

1. Distributive **2.** Zero-Product **3.** $\{x \mid x \neq 4\}$

1.2 Quadratic Equations

PREPARING FOR THIS SECTION *Before getting started, review the following:*

- Factoring Polynomials (Section R.5, pp. 49–55)
- Zero-Product Property (Section R.1, p. 13)
- Square Roots (Section R.2, pp. 23–24)
- Complete the Square (Section R.5, p. 56)

Now Work the **'Are You Prepared?'** problems on page 101.

OBJECTIVES
1. Solve a Quadratic Equation by Factoring (p. 93)
2. Solve a Quadratic Equation by Completing the Square (p. 95)
3. Solve a Quadratic Equation Using the Quadratic Formula (p. 96)
4. Solve Problems That Can Be Modeled by Quadratic Equations (p. 99)

Quadratic equations are equations such as

$$2x^2 + x + 8 = 0$$
$$3x^2 - 5x + 6 = 0$$
$$x^2 - 9 = 0$$

DEFINITION

A **quadratic equation** is an equation equivalent to one of the form

$$ax^2 + bx + c = 0 \quad (1)$$

where a, b, and c are real numbers and $a \neq 0$.

A quadratic equation written in the form $ax^2 + bx + c = 0$ is said to be in **standard form**.

Sometimes, a quadratic equation is called a **second-degree equation** because, when it is in standard form, the left side is a polynomial of degree 2. We shall discuss three ways of solving quadratic equations: by factoring, by completing the square, and by using the quadratic formula.

1 Solve a Quadratic Equation by Factoring

When a quadratic equation is written in standard form, it may be possible to factor the expression on the left side into the product of two first-degree polynomials. Then, by using the Zero-Product Property and setting each factor equal to 0, the resulting linear equations can be solved to obtain the solutions of the quadratic equation.

EXAMPLE 1 **Solving a Quadratic Equation by Factoring**

Solve the equations:

(a) $x^2 + 6x = 0$ (b) $2x^2 = x + 3$

Solution (a) The equation is in the standard form specified in equation (1). The left side may be factored as

$$x^2 + 6x = 0$$
$$x(x + 6) = 0 \quad \text{Factor.}$$

Using the Zero-Product Property, set each factor equal to 0 and then solve the resulting first-degree equations.

$$x = 0 \quad \text{or} \quad x + 6 = 0 \quad \text{Zero-Product Property}$$
$$x = 0 \quad \text{or} \quad x = -6 \quad \text{Solve.}$$

The solution set is $\{0, -6\}$.

(b) Place the equation $2x^2 = x + 3$ in standard form by subtracting x and 3 from both sides.

$$2x^2 = x + 3$$
$$2x^2 - x - 3 = 0 \quad \text{Subtract } x \text{ and 3 from both sides.}$$

The left side may now be factored as

$$(2x - 3)(x + 1) = 0 \quad \text{Factor.}$$

so that

$$2x - 3 = 0 \quad \text{or} \quad x + 1 = 0 \quad \text{Zero-Product Property}$$
$$x = \frac{3}{2} \qquad x = -1 \quad \text{Solve.}$$

The solution set is $\left\{-1, \frac{3}{2}\right\}$.

When the left side factors into two linear equations with the same solution, the quadratic equation is said to have a **repeated solution**. This solution is also called a **root of multiplicity 2**, or a **double root**.

EXAMPLE 2 **Solving a Quadratic Equation by Factoring**

Solve the equation: $9x^2 - 6x + 1 = 0$

Solution This equation is already in standard form, and the left side can be factored.

$$9x^2 - 6x + 1 = 0$$
$$(3x - 1)(3x - 1) = 0$$

so

$$x = \frac{1}{3} \quad \text{or} \quad x = \frac{1}{3}$$

This equation has only the repeated solution $\frac{1}{3}$. The solution set is $\left\{\frac{1}{3}\right\}$.

Now Work PROBLEMS 13 AND 23

The Square Root Method

Suppose that we wish to solve the quadratic equation

$$x^2 = p \tag{2}$$

where $p \geq 0$ is a nonnegative number. Proceeding as in the earlier examples,

$$x^2 - p = 0 \quad \text{Put in standard form.}$$
$$(x - \sqrt{p})(x + \sqrt{p}) = 0 \quad \text{Factor (over the real numbers).}$$
$$x = \sqrt{p} \quad \text{or} \quad x = -\sqrt{p} \quad \text{Solve.}$$

We have the following result:

If $x^2 = p$ and $p \geq 0$, then $x = \sqrt{p}$ or $x = -\sqrt{p}$. **(3)**

When statement (3) is used, it is called the **Square Root Method**. In statement (3), note that if $p > 0$ the equation $x^2 = p$ has two solutions, $x = \sqrt{p}$ and $x = -\sqrt{p}$. We usually abbreviate these solutions as $x = \pm\sqrt{p}$, which is read as "x equals plus or minus the square root of p."

For example, the two solutions of the equation

$$x^2 = 4$$

are

$$x = \pm\sqrt{4} \quad \text{Use the Square Root Method.}$$

and, since $\sqrt{4} = 2$, we have

$$x = \pm 2$$

The solution set is $\{-2, 2\}$.

EXAMPLE 3 **Solving a Quadratic Equation Using the Square Root Method**

Solve each equation.

(a) $x^2 = 5$ (b) $(x - 2)^2 = 16$

Solution (a) Use the Square Root Method to get

$$x^2 = 5$$
$$x = \pm\sqrt{5} \quad \text{Use the Square Root Method.}$$
$$x = \sqrt{5} \quad \text{or} \quad x = -\sqrt{5}$$

The solution set is $\{-\sqrt{5}, \sqrt{5}\}$.

(b) Use the Square Root Method to get

$$(x-2)^2 = 16$$

$$x-2 = \pm\sqrt{16} \quad \text{Use the Square Root Method.}$$

$$x-2 = \pm 4 \quad \sqrt{16}=4$$

$$x-2=4 \quad \text{or} \quad x-2=-4$$

$$x=6 \quad \text{or} \quad x=-2$$

The solution set is $\{-2, 6\}$.

Now Work PROBLEM 33

2 Solve a Quadratic Equation by Completing the Square

EXAMPLE 4 Solving a Quadratic Equation by Completing the Square

Solve by completing the square: $x^2 + 5x + 4 = 0$

Solution Always begin this procedure by rearranging the equation so that the constant is on the right side.

$$x^2 + 5x + 4 = 0$$

$$x^2 + 5x = -4$$

NOTE If the coefficient of the square term is not 1, divide both sides by the coefficient of the square term before attempting to complete the square. ■

Since the coefficient of x^2 is 1, complete the square on the left side by adding $\left(\frac{1}{2}\cdot 5\right)^2 = \frac{25}{4}$. Of course, in an equation, whatever is added to the left side must also be added to the right side. So add $\frac{25}{4}$ to *both* sides.

$$x^2 + 5x + \frac{25}{4} = -4 + \frac{25}{4} \quad \text{Add } \frac{25}{4} \text{ to both sides.}$$

$$\left(x+\frac{5}{2}\right)^2 = \frac{9}{4} \quad \text{Factor the left side.}$$

$$x + \frac{5}{2} = \pm\sqrt{\frac{9}{4}} \quad \text{Use the Square Root Method.}$$

$$x + \frac{5}{2} = \pm\frac{3}{2} \quad \sqrt{\frac{9}{4}} = \frac{\sqrt{9}}{\sqrt{4}} = \frac{3}{2}$$

$$x = -\frac{5}{2} \pm \frac{3}{2}$$

$$x = -\frac{5}{2} + \frac{3}{2} = -1 \quad \text{or} \quad x = -\frac{5}{2} - \frac{3}{2} = -4$$

The solution set is $\{-4, -1\}$.

THE SOLUTION OF THE EQUATION IN EXAMPLE 4 CAN ALSO BE OBTAINED BY FACTORING. REWORK EXAMPLE 4 USING FACTORING.

The next example illustrates an equation that cannot be solved by factoring.

EXAMPLE 5 Solving a Quadratic Equation by Completing the Square

Solve by completing the square: $2x^2 - 8x - 5 = 0$

Solution First, rewrite the equation so that the constant is on the right side.

$$2x^2 - 8x - 5 = 0$$

$$2x^2 - 8x = 5 \quad \text{Add 5 to both sides.}$$

Next, divide both sides by 2 so that the coefficient of x^2 is 1. (This enables us to complete the square at the next step.)

$$x^2 - 4x = \frac{5}{2}$$

Finally, complete the square by adding 4 to both sides.

$$x^2 - 4x + 4 = \frac{5}{2} + 4 \quad \text{Add 4 to both sides.}$$

$$(x-2)^2 = \frac{13}{2} \quad \text{Factor on the left; simplify on the right.}$$

$$x - 2 = \pm\sqrt{\frac{13}{2}} \quad \text{Use the Square Root Method.}$$

$$x - 2 = \pm\frac{\sqrt{26}}{2} \quad \sqrt{\frac{13}{2}} = \frac{\sqrt{13}}{\sqrt{2}} = \frac{\sqrt{13}}{\sqrt{2}} \cdot \frac{\sqrt{2}}{\sqrt{2}} = \frac{\sqrt{26}}{2}$$

$$x = 2 \pm \frac{\sqrt{26}}{2}$$

The solution set is $\left\{2 - \frac{\sqrt{26}}{2}, 2 + \frac{\sqrt{26}}{2}\right\}$.

NOTE If we wanted an approximation, say rounded to two decimal places, of these solutions, we would use a calculator to get $\{-0.55, 4.55\}$.

Now Work PROBLEM 37

3 Solve a Quadratic Equation Using the Quadratic Formula

We can use the method of completing the square to obtain a general formula for solving any quadratic equation

$$ax^2 + bx + c = 0 \qquad a \neq 0$$

NOTE There is no loss in generality to assume that $a > 0$, since if $a < 0$ we can multiply by -1 to obtain an equivalent equation with a positive leading coefficient.

As in Examples 4 and 5, rearrange the terms as

$$ax^2 + bx = -c \qquad a > 0$$

Since $a > 0$, divide both sides by a to get

$$x^2 + \frac{b}{a}x = -\frac{c}{a}$$

Now the coefficient of x^2 is 1. To complete the square on the left side, add the square of $\frac{1}{2}$ of the coefficient of x; that is, add

$$\left(\frac{1}{2} \cdot \frac{b}{a}\right)^2 = \frac{b^2}{4a^2}$$

to both sides. Then

$$x^2 + \frac{b}{a}x + \frac{b^2}{4a^2} = \frac{b^2}{4a^2} - \frac{c}{a}$$

$$\left(x + \frac{b}{2a}\right)^2 = \frac{b^2 - 4ac}{4a^2} \qquad \frac{b^2}{4a^2} - \frac{c}{a} = \frac{b^2}{4a^2} - \frac{4ac}{4a^2} = \frac{b^2 - 4ac}{4a^2} \tag{4}$$

Provided that $b^2 - 4ac \geq 0$, we can now use the Square Root Method to get

$$x + \frac{b}{2a} = \pm\sqrt{\frac{b^2 - 4ac}{4a^2}}$$

$$x + \frac{b}{2a} = \frac{\pm\sqrt{b^2 - 4ac}}{2a}$$

The square root of a quotient equals the quotient of the square roots. Also, $\sqrt{4a^2} = 2a$ since $a > 0$.

$$x = -\frac{b}{2a} \pm \frac{\sqrt{b^2 - 4ac}}{2a} \quad \text{Add } -\frac{b}{2a} \text{ to both sides.}$$

$$= \frac{-b \pm \sqrt{b^2 - 4ac}}{2a} \quad \text{Combine the quotients on the right.}$$

What if $b^2 - 4ac$ is negative? Then equation (4) states that the left expression (a real number squared) equals the right expression (a negative number). Since this is impossible for real numbers, we conclude that if $b^2 - 4ac < 0$, the quadratic equation has no *real* solution. (We discuss quadratic equations for which the quantity $b^2 - 4ac < 0$ in detail in the next section.)

THEOREM

Quadratic Formula

Consider the quadratic equation

$$ax^2 + bx + c = 0 \qquad a \neq 0$$

If $b^2 - 4ac < 0$, this equation has no real solution.
If $b^2 - 4ac \geq 0$, the real solution(s) of this equation is (are) given by the **quadratic formula**:

$$x = \frac{-b \pm \sqrt{b^2 - 4ac}}{2a} \tag{5}$$

The quantity $\mathbf{b^2 - 4ac}$ is called the **discriminant** of the quadratic equation, because its value tells us whether the equation has real solutions. In fact, it also tells us how many solutions to expect.

Discriminant of a Quadratic Equation

For a quadratic equation $ax^2 + bx + c = 0, a \neq 0$:

1. If $b^2 - 4ac > 0$, there are two unequal real solutions.
2. If $b^2 - 4ac = 0$, there is a repeated solution, a double root.
3. If $b^2 - 4ac < 0$, there is no real solution.

When asked to find the real solutions, of a quadratic equation, always evaluate the discriminant first to see if there are any real solutions.

EXAMPLE 6 Solving a Quadratic Equation Using the Quadratic Formula

Use the quadratic formula to find the real solutions, if any, of the equation

$$3x^2 - 5x + 1 = 0$$

Solution The equation is in standard form, so compare it to $ax^2 + bx + c = 0$ to find a, b, and c.

$$3x^2 - 5x + 1 = 0$$
$$ax^2 + bx + c = 0 \quad a = 3, b = -5, c = 1$$

With $a = 3$, $b = -5$, and $c = 1$, evaluate the discriminant $b^2 - 4ac$.

$$b^2 - 4ac = (-5)^2 - 4(3)(1) = 25 - 12 = 13$$

Since $b^2 - 4ac > 0$, there are two real solutions, which can be found using the quadratic formula.

$$x = \frac{-b \pm \sqrt{b^2 - 4ac}}{2a} = \frac{-(-5) \pm \sqrt{13}}{2(3)} = \frac{5 \pm \sqrt{13}}{6}$$

The solution set is $\left\{\frac{5 - \sqrt{13}}{6}, \frac{5 + \sqrt{13}}{6}\right\}$.

EXAMPLE 7 Solving a Quadratic Equation Using the Quadratic Formula

Use the quadratic formula to find the real solutions, if any, of the equation

$$\frac{25}{2}x^2 - 30x + 18 = 0$$

Solution The equation is given in standard form. However, to simplify the arithmetic, clear the fraction.

$$\frac{25}{2}x^2 - 30x + 18 = 0$$

$$25x^2 - 60x + 36 = 0 \quad \text{Clear fraction; multiply by 2.}$$

$$ax^2 + bx + c = 0 \quad \text{Compare to standard form.}$$

With $a = 25$, $b = -60$, and $c = 36$, evaluate the discriminant.

$$b^2 - 4ac = (-60)^2 - 4(25)(36) = 3600 - 3600 = 0$$

The equation has a repeated solution, which is found by using the quadratic formula.

$$x = \frac{-b \pm \sqrt{b^2 - 4ac}}{2a} = \frac{60 \pm \sqrt{0}}{50} = \frac{60}{50} = \frac{6}{5}$$

The solution set is $\left\{\frac{6}{5}\right\}$.

EXAMPLE 8 Solving a Quadratic Equation Using the Quadratic Formula

Use the quadratic formula to find the real solutions, if any, of the equation

$$3x^2 + 2 = 4x$$

Solution The equation, as given, is not in standard form.

$$3x^2 + 2 = 4x$$

$$3x^2 - 4x + 2 = 0 \quad \text{Put in standard form.}$$

$$ax^2 + bx + c = 0 \quad \text{Compare to standard form.}$$

With $a = 3$, $b = -4$, and $c = 2$, the discriminant is

$$b^2 - 4ac = (-4)^2 - 4(3)(2) = 16 - 24 = -8$$

Since $b^2 - 4ac < 0$, the equation has no real solution.

Now Work PROBLEMS 47 AND 57

EXAMPLE 9 **Solving a Quadratic Equation Using the Quadratic Formula**

Find the real solutions, if any, of the equation: $9 + \frac{3}{x} - \frac{2}{x^2} = 0,\ x \neq 0$

Solution In its present form, the equation

$$9 + \frac{3}{x} - \frac{2}{x^2} = 0$$

is not a quadratic equation. However, it can be transformed into one by multiplying each side by x^2. The result is

$$9x^2 + 3x - 2 = 0$$

Although we multiplied each side by x^2, we know that $x^2 \neq 0$ (do you see why?), so this quadratic equation is equivalent to the original equation.

Using $a = 9$, $b = 3$, and $c = -2$, the discriminant is

$$b^2 - 4ac = 3^2 - 4(9)(-2) = 9 + 72 = 81$$

Since $b^2 - 4ac > 0$, the new equation has two real solutions.

$$x = \frac{-b \pm \sqrt{b^2 - 4ac}}{2a} = \frac{-3 \pm \sqrt{81}}{2(9)} = \frac{-3 \pm 9}{18}$$

$$x = \frac{-3 + 9}{18} = \frac{6}{18} = \frac{1}{3} \quad \text{or} \quad x = \frac{-3 - 9}{18} = \frac{-12}{18} = -\frac{2}{3}$$

The solution set is $\left\{-\frac{2}{3}, \frac{1}{3}\right\}$.

SUMMARY

Procedure for Solving a Quadratic Equation

To solve a quadratic equation, first put it in standard form:

$$ax^2 + bx + c = 0$$

Then:

STEP 1: Identify a, b, and c.

STEP 2: Evaluate the discriminant, $b^2 - 4ac$.

STEP 3: (a) If the discriminant is negative, the equation has no real solution.
(b) If the discriminant is zero, the equation has one real solution, a double root.
(c) If the discriminant is positive, the equation has two distinct real solutions.

If you can easily spot factors, use the factoring method to solve the equation. Otherwise, use the quadratic formula or the method of completing the square.

4 Solve Problems That Can Be Modeled by Quadratic Equations

Many applied problems require the solution of a quadratic equation. Let's look at one that you will probably see again in a slightly different form if you study calculus.

EXAMPLE 10 **Constructing a Box**

From each corner of a square piece of sheet metal, remove a square of side 9 centimeters. Turn up the edges to form an open box. If the box is to hold 144 cubic centimeters (cm^3), what should be the dimensions of the piece of sheet metal?

Solution Use Figure 1 as a guide. We have labeled by x the length of a side of the square piece of sheet metal. The box will be of height 9 centimeters, and its square base will measure $x - 18$ on each side. The volume V (Length × Width × Height) of the box is therefore

$$V = (x - 18)(x - 18) \cdot 9 = 9(x - 18)^2$$

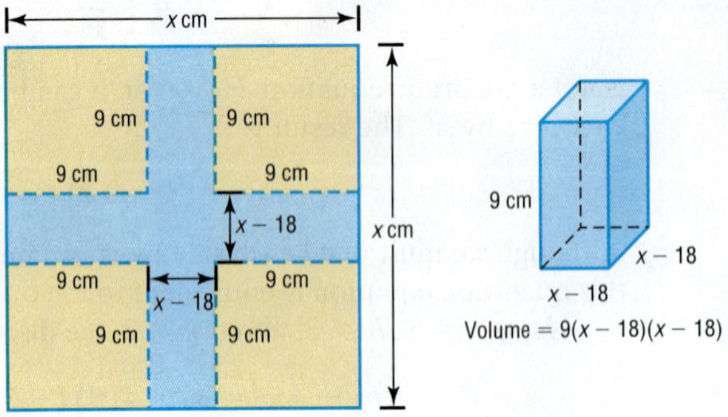

Figure 1

Since the volume of the box is to be 144 cm³, we have

$$\begin{aligned} 9(x - 18)^2 &= 144 && V = 144 \\ (x - 18)^2 &= 16 && \text{Divide each side by 9.} \\ x - 18 &= \pm 4 && \text{Use the Square Root Method.} \\ x &= 18 \pm 4 \\ x = 22 \quad &\text{or} \quad x = 14 \end{aligned}$$

Discard the solution $x = 14$ (do you see why?) and conclude that the sheet metal should be 22 centimeters by 22 centimeters. ●

✓**Check:** If we take a piece of sheet metal 22 centimeters by 22 centimeters, cut out a 9-centimeter square from each corner, and fold up the edges, we get a box whose dimensions are 9 by 4 by 4, with volume $9 \times 4 \times 4 = 144$ cm³, as required.

Now Work PROBLEM 99

Historical Feature

Problems using quadratic equations are found in the oldest known mathematical literature. Babylonians and Egyptians were solving such problems before 1800 BC. Euclid solved quadratic equations geometrically in his *Data* (300 BC), and the Hindus and Arabs gave rules for solving any quadratic equation with real roots. Because negative numbers were not freely used before AD 1500, there were several different types of quadratic equations, each with its own rule. Thomas Harriot (1560–1621) introduced the method of factoring to obtain solutions, and François Viète (1540–1603) introduced a method that is essentially completing the square.

Until modern times it was usual to neglect the negative roots (if there were any), and equations involving square roots of negative quantities were regarded as unsolvable until the 1500s.

Historical Problems

1. *One of al-Khwărízmí solutions* Solve $x^2 + 12x = 85$ by drawing the square shown. The area of the four white rectangles and the yellow square is $x^2 + 12x$. We then set this expression equal to 85 to get the equation $x^2 + 12x = 85$. If we add the four blue squares, we will have a larger square of known area. Complete the solution.

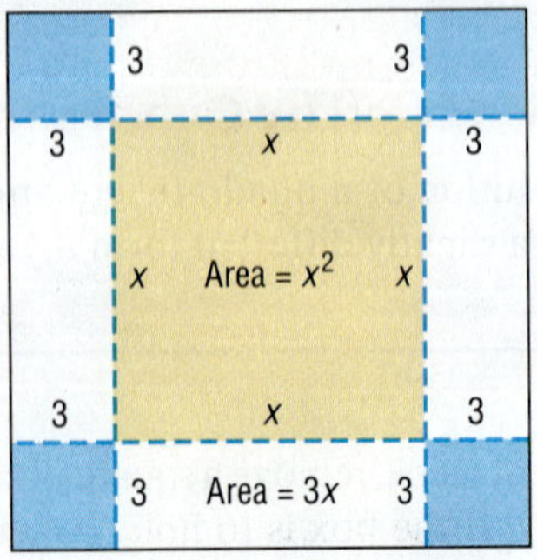

2. *Viète's method* Solve $x^2 + 12x - 85 = 0$ by letting $x = u + z$. Then

$$(u + z)^2 + 12(u + z) - 85 = 0$$
$$u^2 + (2z + 12)u + (z^2 + 12z - 85) = 0$$

Now select z so that $2z + 12 = 0$ and finish the solution.

3. *Another method to get the quadratic formula* Look at equation (4) on page 96. Rewrite the right side as $\left(\dfrac{\sqrt{b^2 - 4ac}}{2a}\right)^2$ and then subtract it from each side. The right side is now 0 and the left side is a difference of two squares. If you factor this difference of two squares, you will easily be able to get the quadratic formula, and moreover, the quadratic expression is factored, which is sometimes useful.

1.2 Assess Your Understanding

'Are You Prepared?' *Answers are given at the end of these exercises. If you get a wrong answer, read the pages listed in red.*

1. Factor: $x^2 - 5x - 6$ (pp. 49–55)
2. Factor: $2x^2 - x - 3$ (pp. 49–55)
3. The solution set of the equation $(x - 3)(3x + 5) = 0$ is ______. (p. 13)
4. ***True or False*** $\sqrt{x^2} = |x|$. (pp. 23–24)
5. Complete the square of $x^2 + 5x$. Factor the new expression. (p. 56)

Concepts and Vocabulary

6. The quantity $b^2 - 4ac$ is called the ______ of a quadratic equation. If it is ______, the equation has no real solution.
7. ***True or False*** Quadratic equations always have two real solutions.
8. ***True or False*** If the discriminant of a quadratic equation is positive, then the equation has two solutions that are negatives of one another.
9. A quadratic equation is sometimes called a ______ equation.
   (a) first-degree (b) second-degree
   (c) third-degree (d) fourth-degree
10. Which of the following quadratic equations is in standard form?
   (a) $x^2 - 7x = 5$ (b) $9 = x^2$
   (c) $(x + 5)(x - 4) = 0$ (d) $0 = 5x^2 - 6x - 1$

Skill Building

In Problems 11–30, solve each equation by factoring.

11. $x^2 - 9x = 0$
12. $x^2 + 4x = 0$
13. $x^2 - 25 = 0$
14. $x^2 - 9 = 0$
15. $z^2 + z - 6 = 0$
16. $v^2 + 7v + 6 = 0$
17. $2x^2 - 5x - 3 = 0$
18. $3x^2 + 5x + 2 = 0$
19. $3t^2 - 48 = 0$
20. $2y^2 - 50 = 0$
21. $x(x - 8) + 12 = 0$
22. $x(x + 4) = 12$
23. $4x^2 + 9 = 12x$
24. $25x^2 + 16 = 40x$
25. $6(p^2 - 1) = 5p$
26. $2(2u^2 - 4u) + 3 = 0$
27. $6x - 5 = \frac{6}{x}$
28. $x + \frac{12}{x} = 7$
29. $\frac{4(x - 2)}{x - 3} + \frac{3}{x} = \frac{-3}{x(x - 3)}$
30. $\frac{5}{x + 4} = 4 + \frac{3}{x - 2}$

In Problems 31–36, solve each equation by the Square Root Method.

31. $x^2 = 25$
32. $x^2 = 36$
33. $(x - 1)^2 = 4$
34. $(x + 2)^2 = 1$
35. $(2y + 3)^2 = 9$
36. $(3z - 2)^2 = 4$

In Problems 37–42, solve each equation by completing the square.

37. $x^2 + 4x = 21$
38. $x^2 - 6x = 13$
39. $x^2 - \frac{1}{2}x - \frac{3}{16} = 0$
40. $x^2 + \frac{2}{3}x - \frac{1}{3} = 0$
41. $3x^2 + x - \frac{1}{2} = 0$
42. $2x^2 - 3x - 1 = 0$

In Problems 43–66, find the real solutions, if any, of each equation. Use the quadratic formula.

43. $x^2 - 4x + 2 = 0$
44. $x^2 + 4x + 2 = 0$
45. $x^2 - 4x - 1 = 0$
46. $x^2 + 6x + 1 = 0$
47. $2x^2 - 5x + 3 = 0$
48. $2x^2 + 5x + 3 = 0$
49. $4y^2 - y + 2 = 0$
50. $4t^2 + t + 1 = 0$
51. $4x^2 = 1 - 2x$
52. $2x^2 = 1 - 2x$
53. $4x^2 = 9x$
54. $5x = 4x^2$
55. $9t^2 - 6t + 1 = 0$
56. $4u^2 - 6u + 9 = 0$
57. $\frac{3}{4}x^2 - \frac{1}{4}x - \frac{1}{2} = 0$

58. $\frac{2}{3}x^2 - x - 3 = 0$

59. $\frac{5}{3}x^2 - x = \frac{1}{3}$

60. $\frac{3}{5}x^2 - x = \frac{1}{5}$

61. $2x(x + 2) = 3$

62. $3x(x + 2) = 1$

63. $4 - \frac{1}{x} - \frac{2}{x^2} = 0$

64. $4 + \frac{1}{x} - \frac{1}{x^2} = 0$

65. $\frac{3x}{x - 2} + \frac{1}{x} = 4$

66. $\frac{2x}{x - 3} + \frac{1}{x} = 4$

In Problems 67–72, find the real solutions, if any, of each equation. Use the quadratic formula and a calculator. Express any solutions rounded to two decimal places.

67. $x^2 - 4.1x + 2.2 = 0$

68. $x^2 + 3.9x + 1.8 = 0$

69. $x^2 + \sqrt{3}x - 3 = 0$

70. $x^2 + \sqrt{2}x - 2 = 0$

71. $\pi x^2 - x - \pi = 0$

72. $\pi x^2 + \pi x - 2 = 0$

In Problems 73–78, use the discriminant to determine whether each quadratic equation has two unequal real solutions, a repeated real solution (a double root), or no real solution, without solving the equation.

73. $2x^2 - 6x + 7 = 0$

74. $x^2 + 4x + 7 = 0$

75. $9x^2 - 30x + 25 = 0$

76. $25x^2 - 20x + 4 = 0$

77. $3x^2 + 5x - 8 = 0$

78. $2x^2 - 3x - 7 = 0$

Mixed Practice

In Problems 79–92, find the real solutions, if any, of each equation. Use any method.

79. $x^2 - 5 = 0$

80. $x^2 - 6 = 0$

81. $16x^2 - 8x + 1 = 0$

82. $9x^2 - 12x + 4 = 0$

83. $10x^2 - 19x - 15 = 0$

84. $6x^2 + 7x - 20 = 0$

85. $2 + z = 6z^2$

86. $2 = y + 6y^2$

87. $x^2 + \sqrt{2}x = \frac{1}{2}$

88. $\frac{1}{2}x^2 = \sqrt{2}x + 1$

89. $x^2 + x = 4$

90. $x^2 + x = 1$

91. $\frac{x}{x - 2} + \frac{2}{x + 1} = \frac{7x + 1}{x^2 - x - 2}$

92. $\frac{3x}{x + 2} + \frac{1}{x - 1} = \frac{4 - 7x}{x^2 + x - 2}$

Applications and Extensions

93. Pythagorean Theorem How many right triangles have a hypotenuse that measures $2x + 3$ meters and legs that measure $2x - 5$ meters and $x + 7$ meters? What are their dimensions?

94. Pythagorean Theorem How many right triangles have a hypotenuse that measures $4x + 5$ inches and legs that measure $3x + 13$ inches and x inches? What are the dimensions of the triangle(s)?

95. Dimensions of a Window The area of the opening of a rectangular window is to be 143 square feet. If the length is to be 2 feet more than the width, what are the dimensions?

96. Dimensions of a Window The area of a rectangular window is to be 306 square centimeters. If the length exceeds the width by 1 centimeter, what are the dimensions?

97. Geometry Find the dimensions of a rectangle whose perimeter is 26 meters and whose area is 40 square meters.

98. Watering a Field An adjustable water sprinkler that sprays water in a circular pattern is placed at the center of a square field whose area is 1250 square feet (see the figure). What is the shortest radius setting that can be used if the field is to be completely enclosed within the circle?

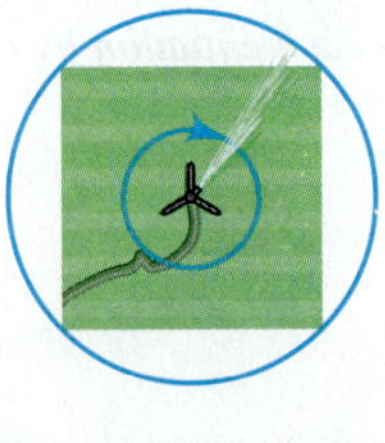

99. Constructing a Box An open box is to be constructed from a square piece of sheet metal by removing a square of side 1 foot from each corner and turning up the edges. If the box is to hold 4 cubic feet, what should be the dimensions of the sheet metal?

100. Constructing a Box Rework Problem 99 if the piece of sheet metal is a rectangle whose length is twice its width.

101. Physics A ball is thrown vertically upward from the top of a building 96 feet tall with an initial velocity of

80 feet per second. The distance s (in feet) of the ball from the ground after t seconds is $s = 96 + 80t - 16t^2$.

(a) After how many seconds does the ball strike the ground?

(b) After how many seconds will the ball pass the top of the building on its way down?

102. Physics An object is propelled vertically upward with an initial velocity of 20 meters per second. The distance s (in meters) of the object from the ground after t seconds is $s = -4.9t^2 + 20t$.

(a) When will the object be 15 meters above the ground?

(b) When will it strike the ground?

(c) Will the object reach a height of 100 meters?

103. Reducing the Size of a Candy Bar A jumbo chocolate bar with a rectangular shape measures 12 centimeters in length, 7 centimeters in width, and 3 centimeters in thickness. Due to escalating costs of cocoa, management decides to reduce the volume of the bar by 10%. To accomplish this reduction, management decides that the new bar should have the same 3-centimeter thickness, but the length and width of each should be reduced an equal number of centimeters. What should be the dimensions of the new candy bar?

104. Reducing the Size of a Candy Bar Rework Problem 103 if the reduction is to be 20%.

105. Constructing a Border around a Pool A circular pool measures 10 feet across. One cubic yard of concrete is to be used to create a circular border of uniform width around the pool. If the border is to have a depth of 3 inches, how wide will the border be? (1 cubic yard = 27 cubic feet) See the illustration.

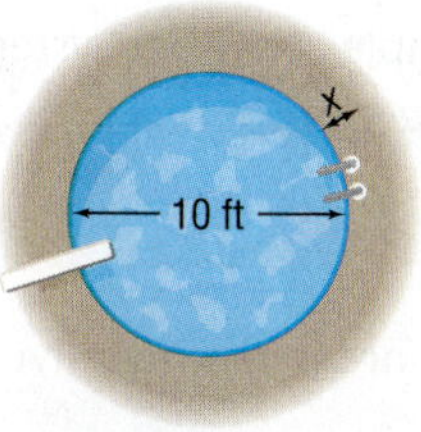

106. Constructing a Border around a Pool Rework Problem 105 if the depth of the border is 4 inches.

107. Constructing a Border around a Garden A landscaper, who just completed a rectangular flower garden measuring 6 feet by 10 feet, orders 1 cubic yard of premixed cement, all of which is to be used to create a border of uniform width around the garden. If the border is to have a depth of 3 inches, how wide will the border be? (1 cubic yard = 27 cubic feet)

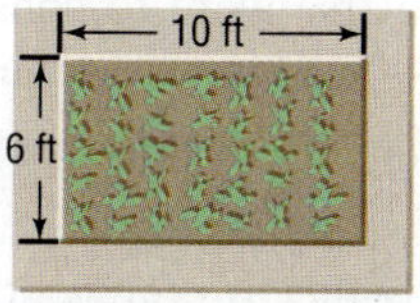

108. Dimensions of a Patio A contractor orders 8 cubic yards of premixed cement, all of which is to be used to pour a patio that will be 4 inches thick. If the length of the patio is specified to be twice the width, what will be the patio dimensions? (1 cubic yard = 27 cubic feet)

109. Comparing Tablets The screen size of a tablet is determined by the length of the diagonal of the rectangular screen. The 9.7-inch iPad Air™ comes in a 4:3 format, which means that the ratio of the length to the width of the rectangular screen is 4:3. What is the area of the iPad's screen? What is the area of a 10-inch Google Nexus™ if its screen is in a 16:10 format? Which screen is larger? **(Hint:** If x is the length of a 4:3 format screen, then $\frac{3}{4}x$ is the width.)

iPad mini 4:3

Google Nexus 16:10

110. Comparing Tablets Refer to Problem 109. Find the screen area of a 7.9-inch iPad mini with Retina™ in a 4:3 format, and compare it with an 8-inch Dell Venue Pro™ if its screen is in a 16:9 format. Which screen is larger?

111. Field Design A football field is sloped from the center toward the sides for drainage. The height h, in feet, of the field, x feet from the side, is given by $h = -0.00025x^2 + 0.04x$. Find the height of the field a distance of 35 feet from the side. Round to the nearest tenth of a foot.

112. College Value The difference, d, in median earnings, in \$1000s, between high school graduates and college graduates can be approximated by $d = -0.002x^2 + 0.319x + 7.512$, where x is the number of years after 1965. Based on this model, estimate to the nearest year when the difference in median earnings was \$15,000. (*Source: Current Population Survey*)

113. Student Working A study found that a student's GPA, g, is related to the number of hours worked each week, h, by the equation $g = -0.0006h^2 + 0.015h + 3.04$. Estimate the number of hours worked each week for a student with a GPA of 2.97. Round to the nearest whole hour.

114. Fraternity Purchase A fraternity wants to buy a new LED Smart TV that costs \$1470. If 7 members of the fraternity are not able to contribute, the share for the remaining members increases by \$5. How many members are in the fraternity?

115. The sum of the consecutive integers $1, 2, 3, \dots, n$ is given by the formula $\frac{1}{2}n(n + 1)$. How many consecutive integers, starting with 1, must be added to get a sum of 703?

116. Geometry If a polygon of n sides has $\frac{1}{2}n(n - 3)$ diagonals, how many sides will a polygon with 65 diagonals have? Is there a polygon with 80 diagonals?

117. Show that the sum of the roots of a quadratic equation is $-\frac{b}{a}$.

118. Show that the product of the roots of a quadratic equation is $\frac{c}{a}$.

119. Find k such that the equation $kx^2 + x + k = 0$ has a repeated real solution.

120. Find k such that the equation $x^2 - kx + 4 = 0$ has a repeated real solution.

121. Show that the real solutions of the equation $ax^2 + bx + c = 0$ are the negatives of the real solutions of the equation $ax^2 - bx + c = 0$. Assume that $b^2 - 4ac \geq 0$.

122. Show that the real solutions of the equation $ax^2 + bx + c = 0$ are the reciprocals of the real solutions of the equation $cx^2 + bx + a = 0$. Assume that $b^2 - 4ac \geq 0$.

Explaining Concepts: Discussion and Writing

123. Which of the following pairs of equations are equivalent? Explain.
(a) $x^2 = 9$; $x = 3$ (b) $x = \sqrt{9}$; $x = 3$
(c) $(x - 1)(x - 2) = (x - 1)^2$; $x - 2 = x - 1$

124. Describe three ways that you might solve a quadratic equation. State your preferred method; explain why you chose it.

125. Explain the benefits of evaluating the discriminant of a quadratic equation before attempting to solve it.

126. Create three quadratic equations: one having two distinct real solutions, one having no real solution, and one having exactly one real solution.

127. The word *quadratic* seems to imply four (*quad*), yet a quadratic equation is an equation that involves a polynomial of degree 2. Investigate the origin of the term *quadratic* as it is used in the expression *quadratic equation*. Write a brief essay on your findings.

'Are You Prepared?' Answers

1. $(x - 6)(x + 1)$ 2. $(2x - 3)(x + 1)$ 3. $\left\{-\frac{5}{3}, 3\right\}$ 4. True 5. $x^2 + 5x + \frac{25}{4} = \left(x + \frac{5}{2}\right)^2$

1.3 Complex Numbers; Quadratic Equations in the Complex Number System*

PREPARING FOR THIS SECTION *Before getting started, review the following:*

- Classification of Numbers (Section R.1, pp. 4–5)
- Rationalizing Denominators (Section R.8, p. 75)

Now Work the **'Are You Prepared?'** problems on page 111.

OBJECTIVES 1 Add, Subtract, Multiply, and Divide Complex Numbers (p. 105)
2 Solve Quadratic Equations in the Complex Number System (p. 109)

Complex Numbers

One property of a real number is that its square is nonnegative. For example, there is no real number x for which

$$x^2 = -1$$

To remedy this situation, we introduce a new number called the *imaginary unit.*

DEFINITION The **imaginary unit**, which we denote by ***i***, is the number whose square is -1. That is,

$$i^2 = -1$$

This should not surprise you. If our universe were to consist only of integers, there would be no number x for which $2x = 1$. This was remedied by introducing numbers such as $\frac{1}{2}$ and $\frac{2}{3}$, the *rational numbers*. If our universe were to consist only of rational numbers, there would be no x whose square equals 2. That is, there would be no number x for which $x^2 = 2$. To remedy this, we introduced numbers such as $\sqrt{2}$ and $\sqrt[3]{5}$, the *irrational numbers*. Recall that the *real numbers* consist of the rational numbers and the irrational numbers. Now, if our universe were to consist only of real numbers, then there would be no number x whose square is -1. To remedy this, we introduce the number i, whose square is -1.

*This section may be omitted without any loss of continuity.

In the progression outlined, each time we encountered a situation that was unsuitable, a new number system was introduced to remedy the situation. The number system that results from introducing the number i is called the **complex number system.**

DEFINITION

Complex numbers are numbers of the form $\boldsymbol{a + bi}$, where a and b are real numbers. The real number a is called the **real part** of the number $a + bi$; the real number b is called the **imaginary part** of $a + bi$; and i is the imaginary unit, so $i^2 = -1$.

For example, the complex number $-5 + 6i$ has the real part -5 and the imaginary part 6.

When a complex number is written in the form $a + bi$, where a and b are real numbers, it is in **standard form**. However, if the imaginary part of a complex number is negative, such as in the complex number $3 + (-2)i$, we agree to write it instead in the form $3 - 2i$.

Also, the complex number $a + 0i$ is usually written merely as a. This serves to remind us that the real numbers are a subset of the complex numbers. Similarly, the complex number $0 + bi$ is usually written as bi. Sometimes the complex number bi is called a **pure imaginary number**.

1 Add, Subtract, Multiply, and Divide Complex Numbers

Equality, addition, subtraction, and multiplication of complex numbers are defined so as to preserve the familiar rules of algebra for real numbers. Two complex numbers are equal if and only if their real parts are equal and their imaginary parts are equal.

Equality of Complex Numbers

$$a + bi = c + di \quad \text{if and only if} \quad a = c \text{ and } b = d \tag{1}$$

Two complex numbers are added by forming the complex number whose real part is the sum of the real parts and whose imaginary part is the sum of the imaginary parts.

Sum of Complex Numbers

$$(a + bi) + (c + di) = (a + c) + (b + d)i \tag{2}$$

To subtract two complex numbers, use this rule:

Difference of Complex Numbers

$$(a + bi) - (c + di) = (a - c) + (b - d)i \tag{3}$$

EXAMPLE 1 **Adding and Subtracting Complex Numbers**

(a) $(3 + 5i) + (-2 + 3i) = [3 + (-2)] + (5 + 3)i = 1 + 8i$

(b) $(6 + 4i) - (3 + 6i) = (6 - 3) + (4 - 6)i = 3 + (-2)i = 3 - 2i$ •

Now Work PROBLEM 15

Products of complex numbers are calculated as illustrated in Example 2.

EXAMPLE 2 **Multiplying Complex Numbers**

$$(5 + 3i)\cdot(2 + 7i) = 5\cdot(2 + 7i) + 3i(2 + 7i) = 10 + 35i + 6i + 21i^2$$

(Distributive Property; Distributive Property)

$$= 10 + 41i + 21(-1)$$

($i^2 = -1$)

$$= -11 + 41i$$

Based on the procedure of Example 2, the **product** of two complex numbers is defined as follows:

Product of Complex Numbers

$$(a + bi)\cdot(c + di) = (ac - bd) + (ad + bc)i \qquad (4)$$

Do not bother to memorize formula (4). Instead, whenever it is necessary to multiply two complex numbers, follow the usual rules for multiplying two binomials, as in Example 2, remembering that $i^2 = -1$. For example,

$$(2i)(2i) = 4i^2 = 4(-1) = -4$$

$$(2 + i)(1 - i) = 2 - 2i + i - i^2 = 3 - i$$

Now Work PROBLEM 21

Algebraic properties for addition and multiplication, such as the commutative, associative, and distributive properties, hold for complex numbers. The property that every nonzero complex number has a multiplicative inverse, or reciprocal, requires a closer look.

DEFINITION If $z = a + bi$ is a complex number, then its **conjugate**, denoted by $\bar{z}$, is defined as

$$\bar{z} = \overline{a + bi} = a - bi$$

NOTE The conjugate of a complex number can be found by changing the sign of the imaginary part. ■

For example, $\overline{2 + 3i} = 2 - 3i$ and $\overline{-6 - 2i} = -6 + 2i$.

EXAMPLE 3 **Multiplying a Complex Number by Its Conjugate**

Find the product of the complex number $z = 3 + 4i$ and its conjugate $\bar{z}$.

Solution Since $\bar{z} = 3 - 4i$, we have

$$z\bar{z} = (3 + 4i)(3 - 4i) = 9 - 12i + 12i - 16i^2 = 9 + 16 = 25$$

The result obtained in Example 3 has an important generalization.

THEOREM The product of a complex number and its conjugate is a nonnegative real number. That is, if $z = a + bi$, then

$$z\bar{z} = a^2 + b^2 \qquad (5)$$

Proof If $z = a + bi$, then

$$z\bar{z} = (a + bi)(a - bi) = a^2 - (bi)^2 = a^2 - b^2i^2 = a^2 + b^2 \qquad \blacksquare$$

To express the reciprocal of a nonzero complex number z in standard form, multiply the numerator and denominator of $\frac{1}{z}$ by $\bar{z}$. That is, if $z = a + bi$ is a nonzero complex number, then

$$\frac{1}{a + bi} = \frac{1}{z} = \frac{1}{z} \cdot \frac{\bar{z}}{\bar{z}} = \frac{\bar{z}}{z\bar{z}} = \frac{a - bi}{a^2 + b^2}$$

↑ Use (5).

$$= \frac{a}{a^2 + b^2} - \frac{b}{a^2 + b^2}i$$

EXAMPLE 4 **Writing the Reciprocal of a Complex Number in Standard Form**

Write $\frac{1}{3 + 4i}$ in standard form $a + bi$; that is, find the reciprocal of $3 + 4i$.

Solution The idea is to multiply the numerator and denominator by the conjugate of $3 + 4i$, that is, by the complex number $3 - 4i$. The result is

$$\frac{1}{3 + 4i} = \frac{1}{3 + 4i} \cdot \frac{3 - 4i}{3 - 4i} = \frac{3 - 4i}{9 + 16} = \frac{3}{25} - \frac{4}{25}i \qquad \bullet$$

To express the quotient of two complex numbers in standard form, multiply the numerator and denominator of the quotient by the conjugate of the denominator.

EXAMPLE 5 **Writing the Quotient of Two Complex Numbers in Standard Form**

Write each of the following in standard form.

(a) $\frac{1 + 4i}{5 - 12i}$ (b) $\frac{2 - 3i}{4 - 3i}$

Solution (a) $$\frac{1 + 4i}{5 - 12i} = \frac{1 + 4i}{5 - 12i} \cdot \frac{5 + 12i}{5 + 12i} = \frac{5 + 12i + 20i + 48i^2}{25 + 144}$$

$$= \frac{-43 + 32i}{169} = -\frac{43}{169} + \frac{32}{169}i$$

(b) $$\frac{2 - 3i}{4 - 3i} = \frac{2 - 3i}{4 - 3i} \cdot \frac{4 + 3i}{4 + 3i} = \frac{8 + 6i - 12i - 9i^2}{16 + 9} = \frac{17 - 6i}{25} = \frac{17}{25} - \frac{6}{25}i \qquad \bullet$$

Now Work PROBLEM 29

EXAMPLE 6 **Writing Other Expressions in Standard Form**

If $z = 2 - 3i$ and $w = 5 + 2i$, write each of the following expressions in standard form.

(a) $\frac{z}{w}$ (b) $\overline{z + w}$ (c) $z + \bar{z}$

Solution (a) $\dfrac{z}{w} = \dfrac{z \cdot \overline{w}}{w \cdot \overline{w}} = \dfrac{(2-3i)(5-2i)}{(5+2i)(5-2i)} = \dfrac{10 - 4i - 15i + 6i^2}{25+4}$

$= \dfrac{4-19i}{29} = \dfrac{4}{29} - \dfrac{19}{29}i$

(b) $\overline{z+w} = \overline{(2-3i)+(5+2i)} = \overline{7-i} = 7+i$

(c) $z + \overline{z} = (2-3i) + (2+3i) = 4$

The conjugate of a complex number has certain general properties that will be useful later.

For a real number $a = a + 0i$, the conjugate is $\overline{a} = \overline{a + 0i} = a - 0i = a$.

THEOREM

The conjugate of a real number is the real number itself.

Other properties that are direct consequences of the definition of the conjugate are given next. In each statement, z and w represent complex numbers.

THEOREM

The conjugate of the conjugate of a complex number is the complex number itself.

$$\overline{(\overline{z})} = z \qquad \textbf{(6)}$$

The conjugate of the sum of two complex numbers equals the sum of their conjugates.

$$\overline{z + w} = \overline{z} + \overline{w} \qquad \textbf{(7)}$$

The conjugate of the product of two complex numbers equals the product of their conjugates.

$$\overline{z \cdot w} = \overline{z} \cdot \overline{w} \qquad \textbf{(8)}$$

The proofs of equations (6), (7), and (8) are left as exercises.

Powers of *i*

The **powers of *i*** follow a pattern that is useful to know.

$i^1 = i$	$i^5 = i^4 \cdot i = 1 \cdot i = i$
$i^2 = -1$	$i^6 = i^4 \cdot i^2 = -1$
$i^3 = i^2 \cdot i = -1 \cdot i = -i$	$i^7 = i^4 \cdot i^3 = -i$
$i^4 = i^2 \cdot i^2 = (-1)(-1) = 1$	$i^8 = i^4 \cdot i^4 = 1$

And so on. The powers of i repeat with every fourth power.

EXAMPLE 7 **Evaluating Powers of *i***

(a) $i^{27} = i^{24} \cdot i^3 = (i^4)^6 \cdot i^3 = 1^6 \cdot i^3 = -i$

(b) $i^{101} = i^{100} \cdot i^1 = (i^4)^{25} \cdot i = 1^{25} \cdot i = i$

EXAMPLE 8

Writing the Power of a Complex Number in Standard Form

Write $(2 + i)^3$ in standard form.

Solution Use the special product formula for $(x + a)^3$.

$$(x + a)^3 = x^3 + 3ax^2 + 3a^2x + a^3$$

NOTE Another way to find $(2 + i)^3$ is to multiply out $(2 + i)^2(2 + i)$. ■

Using this special product formula,

$$\begin{aligned}(2 + i)^3 &= 2^3 + 3\cdot i\cdot 2^2 + 3\cdot i^2\cdot 2 + i^3\\ &= 8 + 12i + 6(-1) + (-i)\\ &= 2 + 11i\end{aligned}$$

●

Now Work PROBLEM 43

2 Solve Quadratic Equations in the Complex Number System

Quadratic equations with a negative discriminant have no real number solution. However, if we extend our number system to allow complex numbers, quadratic equations will always have a solution. Since the solution to a quadratic equation involves the square root of the discriminant, we begin with a discussion of square roots of negative numbers.

DEFINITION

If N is a positive real number, we define the **principal square root of $-N$**, denoted by $\sqrt{-N}$, as

$$\sqrt{-N} = \sqrt{N}i$$

where i is the imaginary unit and $i^2 = -1$.

WARNING In writing $\sqrt{-N} = \sqrt{N}i$, be sure to place i outside the $\sqrt{\ }$ symbol. ■

EXAMPLE 9

Evaluating the Square Root of a Negative Number

(a) $\sqrt{-1} = \sqrt{1}i = i$

(b) $\sqrt{-4} = \sqrt{4}i = 2i$

(c) $\sqrt{-8} = \sqrt{8}i = 2\sqrt{2}i$ ●

EXAMPLE 10

Solving Equations

Solve each equation in the complex number system.

(a) $x^2 = 4$ (b) $x^2 = -9$

Solution

(a) $x^2 = 4$

$x = \pm\sqrt{4} = \pm 2$

The equation has two solutions, -2 and 2. The solution set is $\{-2, 2\}$.

(b) $x^2 = -9$

$x = \pm\sqrt{-9} = \pm\sqrt{9}i = \pm 3i$

The equation has two solutions, $-3i$ and $3i$. The solution set is $\{-3i, 3i\}$. ●

Now Work PROBLEMS 51 AND 55

WARNING When working with square roots of negative numbers, do not set the square root of a product equal to the product of the square roots (which can be done with positive real numbers). To see why, look at this calculation: We know that $\sqrt{100} = 10$. However, it is also true that $100 = (-25)(-4)$, so

$$10 = \sqrt{100} = \sqrt{(-25)(-4)} = \sqrt{-25}\sqrt{-4} = (\sqrt{25}i)(\sqrt{4}i) = (5i)(2i) = 10i^2 = -10$$

↑ Here is the error.

Because we have defined the square root of a negative number, we can now restate the quadratic formula without restriction.

THEOREM

Quadratic Formula

In the complex number system, the solutions of the quadratic equation $ax^2 + bx + c = 0$, where a, b, and c are real numbers and $a \neq 0$, are given by the formula

$$x = \frac{-b \pm \sqrt{b^2 - 4ac}}{2a} \qquad (9)$$

EXAMPLE 11 **Solving a Quadratic Equation in the Complex Number System**

Solve the equation $x^2 - 4x + 8 = 0$ in the complex number system.

Solution Here $a = 1, b = -4, c = 8$, and $b^2 - 4ac = (-4)^2 - 4(1)(8) = -16$. Using equation (9), we find that

$$x = \frac{-(-4) \pm \sqrt{-16}}{2(1)} = \frac{4 \pm \sqrt{16}i}{2} = \frac{4 \pm 4i}{2} = \frac{2(2 \pm 2i)}{2} = 2 \pm 2i$$

The equation has two solutions: $2 - 2i$ and $2 + 2i$.

The solution set is $\{2 - 2i, 2 + 2i\}$.

✓**Check:** $2 + 2i$: $(2 + 2i)^2 - 4(2 + 2i) + 8 = 4 + 8i + 4i^2 - 8 - 8i + 8$
$= 4 + 4i^2$
$= 4 - 4 = 0$

$2 - 2i$: $(2 - 2i)^2 - 4(2 - 2i) + 8 = 4 - 8i + 4i^2 - 8 + 8i + 8$
$= 4 - 4 = 0$

Now Work PROBLEM 61

The discriminant $b^2 - 4ac$ of a quadratic equation still serves as a way to determine the character of the solutions.

Character of the Solutions of a Quadratic Equation

In the complex number system, consider a quadratic equation $ax^2 + bx + c = 0$ with real coefficients.

1. If $b^2 - 4ac > 0$, the equation has two unequal real solutions.
2. If $b^2 - 4ac = 0$, the equation has a repeated real solution, a double root.
3. If $b^2 - 4ac < 0$, the equation has two complex solutions that are not real. The solutions are conjugates of each other.

The third conclusion in the display is a consequence of the fact that if $b^2 - 4ac = -N < 0$, then by the quadratic formula, the solutions are

$$x = \frac{-b + \sqrt{b^2 - 4ac}}{2a} = \frac{-b + \sqrt{-N}}{2a} = \frac{-b + \sqrt{N}i}{2a} = \frac{-b}{2a} + \frac{\sqrt{N}}{2a}i$$

and

$$x = \frac{-b - \sqrt{b^2 - 4ac}}{2a} = \frac{-b - \sqrt{-N}}{2a} = \frac{-b - \sqrt{N}i}{2a} = \frac{-b}{2a} - \frac{\sqrt{N}}{2a}i$$

which are conjugates of each other.

EXAMPLE 12 **Determining the Character of the Solutions of a Quadratic Equation**

Without solving, determine the character of the solutions of each equation.

(a) $3x^2 + 4x + 5 = 0$ (b) $2x^2 + 4x + 1 = 0$ (c) $9x^2 - 6x + 1 = 0$

Solution (a) Here $a = 3, b = 4$, and $c = 5$, so $b^2 - 4ac = 16 - 4(3)(5) = -44$. The solutions are two complex numbers that are not real and are conjugates of each other.

(b) Here $a = 2, b = 4$, and $c = 1$, so $b^2 - 4ac = 16 - 8 = 8$. The solutions are two unequal real numbers.

(c) Here $a = 9, b = -6$, and $c = 1$, so $b^2 - 4ac = 36 - 4(9)(1) = 0$. The solution is a repeated real number—that is, a double root. ●

Now Work PROBLEM 75

1.3 Assess Your Understanding

'Are You Prepared?' *Answers are given at the end of these exercises. If you get a wrong answer, read the pages listed in red.*

1. Name the integers and the rational numbers in the set $\left\{-3, 0, \sqrt{2}, \frac{6}{5}, \pi\right\}$. (pp. 4–5)
2. ***True or False*** Rational numbers and irrational numbers are in the set of real numbers. (pp. 4–5)
3. Rationalize the denominator of $\frac{3}{2 + \sqrt{3}}$. (p. 75)

Concepts and Vocabulary

4. In the complex number $5 + 2i$, the number 5 is called the ______ part; the number 2 is called the ______ part; the number i is called the ______ ______.
5. ***True or False*** The conjugate of $2 + 5i$ is $-2 - 5i$.
6. ***True or False*** All real numbers are complex numbers.
7. ***True or False*** If $2 - 3i$ is a solution of a quadratic equation with real coefficients, then $-2 + 3i$ is also a solution.
8. Which of the following is the principal square root of -4?
(a) $-2i$ (b) $2i$ (c) -2 (d) 2
9. Which operation involving complex numbers requires the use of a conjugate?
(a) division (b) multiplication
(c) subtraction (d) addition
10. Powers of i repeat every ______ power.
(a) second (b) third (c) fourth (d) fifth

Skill Building

In Problems 11–48, perform the indicated operation, and write each expression in the standard form $a + bi$.

11. $(2 - 3i) + (6 + 8i)$
12. $(4 + 5i) + (-8 + 2i)$
13. $(-3 + 2i) - (4 - 4i)$
14. $(3 - 4i) - (-3 - 4i)$
15. $(2 - 5i) - (8 + 6i)$
16. $(-8 + 4i) - (2 - 2i)$
17. $3(2 - 6i)$
18. $-4(2 + 8i)$
19. $2i(2 - 3i)$
20. $3i(-3 + 4i)$
21. $(3 - 4i)(2 + i)$
22. $(5 + 3i)(2 - i)$
23. $(-6 + i)(-6 - i)$
24. $(-3 + i)(3 + i)$
25. $\frac{10}{3 - 4i}$
26. $\frac{13}{5 - 12i}$
27. $\frac{2 + i}{i}$
28. $\frac{2 - i}{-2i}$
29. $\frac{6 - i}{1 + i}$
30. $\frac{2 + 3i}{1 - i}$
31. $\left(\frac{1}{2} + \frac{\sqrt{3}}{2}i\right)^2$
32. $\left(\frac{\sqrt{3}}{2} - \frac{1}{2}i\right)^2$
33. $(1 + i)^2$
34. $(1 - i)^2$

35. i^{23} 36. i^{14} 37. i^{-15} 38. i^{-23} 39. $i^6 - 5$

40. $4 + i^3$ 41. $6i^3 - 4i^5$ 42. $4i^3 - 2i^2 + 1$ 43. $(1 + i)^3$ 44. $(3i)^4 + 1$

45. $i^7(1 + i^2)$ 46. $2i^4(1 + i^2)$ 47. $i^6 + i^4 + i^2 + 1$ 48. $i^7 + i^5 + i^3 + i$

In Problems 49–54, perform the indicated operations, and express your answer in the form a + bi.

49. $\sqrt{-4}$ 50. $\sqrt{-9}$ 51. $\sqrt{-25}$

52. $\sqrt{-64}$ 53. $\sqrt{(3 + 4i)(4i - 3)}$ 54. $\sqrt{(4 + 3i)(3i - 4)}$

In Problems 55–74, solve each equation in the complex number system.

55. $x^2 + 4 = 0$ 56. $x^2 - 4 = 0$ 57. $x^2 - 16 = 0$ 58. $x^2 + 25 = 0$

59. $x^2 - 6x + 13 = 0$ 60. $x^2 + 4x + 8 = 0$ 61. $x^2 - 6x + 10 = 0$ 62. $x^2 - 2x + 5 = 0$

63. $8x^2 - 4x + 1 = 0$ 64. $10x^2 + 6x + 1 = 0$ 65. $5x^2 + 1 = 2x$ 66. $13x^2 + 1 = 6x$

67. $x^2 + x + 1 = 0$ 68. $x^2 - x + 1 = 0$ 69. $x^3 - 8 = 0$ 70. $x^3 + 27 = 0$

71. $x^4 = 16$ 72. $x^4 = 1$ 73. $x^4 + 13x^2 + 36 = 0$ 74. $x^4 + 3x^2 - 4 = 0$

In Problems 75–80, without solving, determine the character of the solutions of each equation in the complex number system.

75. $3x^2 - 3x + 4 = 0$ 76. $2x^2 - 4x + 1 = 0$ 77. $2x^2 + 3x = 4$

78. $x^2 + 6 = 2x$ 79. $9x^2 - 12x + 4 = 0$ 80. $4x^2 + 12x + 9 = 0$

81. $2 + 3i$ is a solution of a quadratic equation with real coefficients. Find the other solution.

82. $4 - i$ is a solution of a quadratic equation with real coefficients. Find the other solution.

In Problems 83–86, $z = 3 - 4i$ and $w = 8 + 3i$. Write each expression in the standard form a + bi.

83. $z + \bar{z}$ 84. $w - \bar{w}$ 85. $z\bar{z}$ 86. $\overline{z - w}$

Applications and Extensions

87. **Electrical Circuits** The impedance Z, in ohms, of a circuit element is defined as the ratio of the phasor voltage V, in volts, across the element to the phasor current I, in amperes, through the elements. That is, $Z = \frac{V}{I}$. If the voltage across a circuit element is $18 + i$ volts and the current through the element is $3 - 4i$ amperes, determine the impedance.

88. **Parallel Circuits** In an ac circuit with two parallel pathways, the total impedance Z, in ohms, satisfies the formula $\frac{1}{Z} = \frac{1}{Z_1} + \frac{1}{Z_2}$, where Z_1 is the impedance of the first pathway and Z_2 is the impedance of the second pathway. Determine the total impedance if the impedances of the two pathways are $Z_1 = 2 + i$ ohms and $Z_2 = 4 - 3i$ ohms.

89. Use $z = a + bi$ to show that $z + \bar{z} = 2a$ and $z - \bar{z} = 2bi$.

90. Use $z = a + bi$ to show that $\bar{\bar{z}} = z$.

91. Use $z = a + bi$ and $w = c + di$ to show that $\overline{z + w} = \bar{z} + \bar{w}$.

92. Use $z = a + bi$ and $w = c + di$ to show that $\overline{z \cdot w} = \bar{z} \cdot \bar{w}$.

Explaining Concepts: Discussion and Writing

93. Explain to a friend how you would add two complex numbers and how you would multiply two complex numbers. Explain any differences between the two explanations.

94. Write a brief paragraph that compares the method used to rationalize the denominator of a radical expression and the method used to write the quotient of two complex numbers in standard form.

95. Use an Internet search engine to investigate the origins of complex numbers. Write a paragraph describing what you find, and present it to the class.

96. Explain how the method of multiplying two complex numbers is related to multiplying two binomials.

97. **What Went Wrong?** A student multiplied $\sqrt{-9}$ and $\sqrt{-9}$ as follows:

$$\begin{aligned}\sqrt{-9} \cdot \sqrt{-9} &= \sqrt{(-9)(-9)}\\ &= \sqrt{81}\\ &= 9\end{aligned}$$

The instructor marked the problem incorrect. Why?

'Are You Prepared?' Answers

1. Integers: $\{-3, 0\}$; rational numbers: $\left\{-3, 0, \frac{6}{5}\right\}$ 2. True 3. $3(2 - \sqrt{3})$

1.4 Radical Equations; Equations Quadratic in Form; Factorable Equations

PREPARING FOR THIS SECTION *Before getting started, review the following:*

- Square Roots (Section R.2, pp. 23–24)
- Factoring Polynomials (Section R.5, pp. 49–55)
- *n*th Roots; Rational Exponents (Section R.8, pp. 73–77)

Now Work the **'Are You Prepared?'** problems on page 117.

OBJECTIVES
1. Solve Radical Equations (p. 113)
2. Solve Equations Quadratic in Form (p. 114)
3. Solve Equations by Factoring (p. 116)

1 Solve Radical Equations

When the variable in an equation occurs in a square root, cube root, and so on—that is, when it occurs in a radical—the equation is called a **radical equation**. Sometimes a suitable operation will change a radical equation to one that is linear or quadratic. A commonly used procedure is to isolate the most complicated radical on one side of the equation and then eliminate it by raising each side to a power equal to the index of the radical. Care must be taken, however, because apparent solutions that are not, in fact, solutions of the original equation may result. These are called **extraneous solutions**. Therefore, we need to check all answers when working with radical equations, and we check them in the *original* equation.

EXAMPLE 1 **Solving a Radical Equation**

Find the real solutions of the equation: $\sqrt[3]{2x-4} - 2 = 0$

Solution The equation contains a radical whose index is 3. Isolate it on the left side.

$$\sqrt[3]{2x-4} - 2 = 0$$

$$\sqrt[3]{2x-4} = 2 \quad \text{Add 2 to both sides.}$$

Now raise each side to the third power (the index of the radical is 3) and solve.

$$\left(\sqrt[3]{2x-4}\right)^3 = 2^3 \quad \text{Raise each side to the power 3.}$$

$$2x - 4 = 8 \quad \text{Simplify.}$$

$$2x = 12 \quad \text{Add 4 to both sides.}$$

$$x = 6 \quad \text{Divide both sides by 2.}$$

Check: $\sqrt[3]{2(6)-4} - 2 = \sqrt[3]{12-4} - 2 = \sqrt[3]{8} - 2 = 2 - 2 = 0$

The solution set is $\{6\}$. •

Now Work PROBLEM 9

EXAMPLE 2 **Solving a Radical Equation**

Find the real solutions of the equation: $\sqrt{x-1} = x - 7$

Solution Square both sides since the index of a square root is 2.

$$\sqrt{x-1} = x - 7$$

$$\left(\sqrt{x-1}\right)^2 = (x-7)^2 \quad \text{Square both sides.}$$

$$x - 1 = x^2 - 14x + 49 \quad \text{Remove parentheses.}$$

$$x^2 - 15x + 50 = 0 \quad \text{Put in standard form.}$$

$$(x - 10)(x - 5) = 0 \qquad \text{Factor.}$$
$$x = 10 \quad \text{or} \quad x = 5 \qquad \text{Apply the Zero-Product Property and solve.}$$

Check:

$x = 10$: $\sqrt{x - 1} = \sqrt{10 - 1} = \sqrt{9} = 3$ and $x - 7 = 10 - 7 = 3$

$x = 5$: $\sqrt{x - 1} = \sqrt{5 - 1} = \sqrt{4} = 2$ and $x - 7 = 5 - 7 = -2$

The solution $x = 5$ does not check, so it is extraneous; the only solution of the equation is $x = 10$. The solution set is $\{10\}$. ●

Now Work PROBLEM 19

Sometimes it is necessary to raise each side to a power more than once in order to solve a radical equation.

EXAMPLE 3 Solving a Radical Equation

Find the real solutions of the equation: $\sqrt{2x + 3} - \sqrt{x + 2} = 2$

Solution First, isolate the more complicated radical expression (in this case, $\sqrt{2x + 3}$) on the left side.

$$\sqrt{2x + 3} = \sqrt{x + 2} + 2$$

Now square both sides (the index of the radical on the left is 2).

$$\left(\sqrt{2x + 3}\right)^2 = \left(\sqrt{x + 2} + 2\right)^2 \qquad \text{Square both sides.}$$
$$2x + 3 = \left(\sqrt{x + 2}\right)^2 + 4\sqrt{x + 2} + 4 \qquad \text{Multiply out.}$$
$$2x + 3 = x + 2 + 4\sqrt{x + 2} + 4 \qquad \text{Simplify.}$$
$$2x + 3 = x + 6 + 4\sqrt{x + 2} \qquad \text{Combine like terms.}$$

Because the equation still contains a radical, isolate the remaining radical on the right side and again square both sides.

$$x - 3 = 4\sqrt{x + 2} \qquad \text{Isolate the radical on the right side.}$$
$$(x - 3)^2 = 16(x + 2) \qquad \text{Square both sides.}$$
$$x^2 - 6x + 9 = 16x + 32 \qquad \text{Multiply out.}$$
$$x^2 - 22x - 23 = 0 \qquad \text{Put in standard form.}$$
$$(x - 23)(x + 1) = 0 \qquad \text{Factor.}$$
$$x = 23 \quad \text{or} \quad x = -1 \qquad \text{Apply the Zero-Product Property and solve.}$$

The original equation appears to have the solution set $\{-1, 23\}$. However, we have not yet checked.

Check:

$x = 23$: $\sqrt{2x + 3} - \sqrt{x + 2} = \sqrt{2(23) + 3} - \sqrt{23 + 2} = \sqrt{49} - \sqrt{25} = 7 - 5 = 2$

$x = -1$: $\sqrt{2x + 3} - \sqrt{x + 2} = \sqrt{2(-1) + 3} - \sqrt{-1 + 2} = \sqrt{1} - \sqrt{1} = 1 - 1 = 0$

The equation has only one solution, 23; the solution -1 is extraneous. The solution set is $\{23\}$. ●

Now Work PROBLEM 31

2 Solve Equations Quadratic in Form

The equation $x^4 + x^2 - 12 = 0$ is not quadratic in x, but it is quadratic in x^2. That is, if we let $u = x^2$, we get $u^2 + u - 12 = 0$, a quadratic equation. This equation can be solved for u, and in turn, by using $u = x^2$, we can find the solutions x of the original equation.

In general, if an appropriate substitution u transforms an equation into one of the form

$$au^2 + bu + c = 0 \qquad a \neq 0$$

then the original equation is called an **equation of the quadratic type** or an **equation quadratic in form**.

The difficulty of solving such an equation lies in the determination that the equation is, in fact, quadratic in form. After you are told an equation is quadratic in form, it is easy enough to see it, but some practice is needed to enable you to recognize such equations on your own.

EXAMPLE 4

Solving an Equation Quadratic in Form

Find the real solutions of the equation: $(x + 2)^2 + 11(x + 2) - 12 = 0$

Solution For this equation, let $u = x + 2$. Then $u^2 = (x + 2)^2$, and the original equation,

$$(x + 2)^2 + 11(x + 2) - 12 = 0$$

becomes

$$\begin{aligned} u^2 + 11u - 12 &= 0 && \text{Let } u = x + 2. \text{ Then } u^2 = (x+2)^2. \\ (u + 12)(u - 1) &= 0 && \text{Factor.} \\ u = -12 \quad \text{or} \quad u &= 1 && \text{Solve.} \end{aligned}$$

But we want to solve for x. Because $u = x + 2$, we have

$$\begin{aligned} x + 2 = -12 \quad &\text{or} \quad x + 2 = 1 \\ x = -14 \quad & \qquad\quad x = -1 \end{aligned}$$

Check: $x = -14$: $(-14 + 2)^2 + 11(-14 + 2) - 12$
$= (-12)^2 + 11(-12) - 12 = 144 - 132 - 12 = 0$

$x = -1$: $(-1 + 2)^2 + 11(-1 + 2) - 12 = 1 + 11 - 12 = 0$

The original equation has the solution set $\{-14, -1\}$.

EXAMPLE 5

Solving an Equation Quadratic in Form

Find the real solutions of the equation: $(x^2 - 1)^2 + (x^2 - 1) - 12 = 0$

Solution For the equation $(x^2 - 1)^2 + (x^2 - 1) - 12 = 0$, let $u = x^2 - 1$ so that $u^2 = (x^2 - 1)^2$. Then the original equation,

$$(x^2 - 1)^2 + (x^2 - 1) - 12 = 0$$

becomes

$$\begin{aligned} u^2 + u - 12 &= 0 && \text{Let } u = x^2 - 1. \text{ Then } u^2 = (x^2 - 1)^2. \\ (u + 4)(u - 3) &= 0 && \text{Factor.} \\ u = -4 \quad \text{or} \quad u &= 3 && \text{Solve.} \end{aligned}$$

But remember that we want to solve for x. Because $u = x^2 - 1$, we have

$$\begin{aligned} x^2 - 1 = -4 \quad &\text{or} \quad x^2 - 1 = 3 \\ x^2 = -3 \quad & \qquad\quad x^2 = 4 \end{aligned}$$

The first of these has no real solution; the second has the solution set $\{-2, 2\}$.

Check: $x = -2$: $((-2)^2 - 1)^2 + ((-2)^2 - 1) - 12 = 9 + 3 - 12 = 0$

$x = 2$: $((2)^2 - 1)^2 + ((2)^2 - 1) - 12 = 9 + 3 - 12 = 0$

The original equation has the solution set $\{-2, 2\}$.

EXAMPLE 6 **Solving an Equation Quadratic in Form**

Find the real solutions of the equation: $x + 2\sqrt{x} - 3 = 0$

Solution For the equation $x + 2\sqrt{x} - 3 = 0$, let $u = \sqrt{x}$. Then $u^2 = x$, and the original equation,

$$x + 2\sqrt{x} - 3 = 0$$

becomes

$$u^2 + 2u - 3 = 0 \quad \text{Let } u = \sqrt{x}. \text{ Then } u^2 = x.$$

$$(u + 3)(u - 1) = 0 \quad \text{Factor.}$$

$$u = -3 \quad \text{or} \quad u = 1 \quad \text{Solve.}$$

Since $u = \sqrt{x}$, we have $\sqrt{x} = -3$ or $\sqrt{x} = 1$. The first of these, $\sqrt{x} = -3$, has no real solution, since the principal square root of a real number is never negative. The second, $\sqrt{x} = 1$, has the solution $x = 1$.

✓**Check:** $1 + 2\sqrt{1} - 3 = 1 + 2 - 3 = 0$

The original equation has the solution set $\{1\}$. ●

ANOTHER METHOD FOR SOLVING EXAMPLE 6 WOULD BE TO TREAT IT AS A RADICAL EQUATION. SOLVE IT THIS WAY FOR PRACTICE.

The idea should now be clear. If an equation contains an expression and that same expression squared, make a substitution for the expression. You may get a quadratic equation.

Now Work PROBLEM 53

3 Solve Equations by Factoring

We have already solved certain quadratic equations using factoring. Let's look at examples of other kinds of equations that can be solved by factoring.

EXAMPLE 7 **Solving an Equation by Factoring**

Solve the equation: $x^4 = 4x^2$

Solution Begin by collecting all terms on one side. This results in 0 on one side and an expression to be factored on the other.

$$x^4 = 4x^2$$

$$x^4 - 4x^2 = 0$$

$$x^2(x^2 - 4) = 0 \quad \text{Factor.}$$

$$x^2 = 0 \quad \text{or} \quad x^2 - 4 = 0 \quad \text{Apply the Zero-Product Property.}$$

$$x = 0 \quad \text{or} \quad x^2 = 4$$

$$x = 0 \quad \text{or} \quad x = -2 \quad \text{or} \quad x = 2$$

✓**Check:** $x = -2$: $(-2)^4 = 16$ and $4(-2)^2 = 16$ — -2 is a solution.

$x = 0$: $0^4 = 0$ and $4 \cdot 0^2 = 0$ — 0 is a solution.

$x = 2$: $2^4 = 16$ and $4 \cdot 2^2 = 16$ — 2 is a solution.

The solution set is $\{-2, 0, 2\}$. ●

EXAMPLE 8 **Solving an Equation by Factoring**

Solve the equation: $x^3 - x^2 - 4x + 4 = 0$

Solution Do you recall the method of factoring by grouping? (If not, review pp. 53–54.) Group the terms of $x^3 - x^2 - 4x + 4 = 0$ as follows:

$$(x^3 - x^2) - (4x - 4) = 0$$

Factor out x^2 from the first grouping and 4 from the second.

$$x^2(x - 1) - 4(x - 1) = 0$$

This reveals the common factor $(x - 1)$, so we have

$$(x^2 - 4)(x - 1) = 0$$

$$(x - 2)(x + 2)(x - 1) = 0 \quad \text{Factor again.}$$

$$x - 2 = 0 \text{ or } x + 2 = 0 \text{ or } x - 1 = 0 \quad \text{Set each factor equal to 0.}$$

$$x = 2 \qquad x = -2 \qquad x = 1 \quad \text{Solve.}$$

Check:

$x = -2$: $(-2)^3 - (-2)^2 - 4(-2) + 4 = -8 - 4 + 8 + 4 = 0$ — −2 is a solution.

$x = 1$: $1^3 - 1^2 - 4(1) + 4 = 1 - 1 - 4 + 4 = 0$ — 1 is a solution.

$x = 2$: $2^3 - 2^2 - 4(2) + 4 = 8 - 4 - 8 + 4 = 0$ — 2 is a solution.

The solution set is $\{-2, 1, 2\}$.

Now Work PROBLEM 81

1.4 Assess Your Understanding

'Are You Prepared?' *Answers are given at the end of these exercises. If you get a wrong answer, read the pages listed in red.*

1. ***True or False*** The principal square root of any nonnegative real number is always nonnegative. (pp. 23–24)
2. $\sqrt[3]{-8} =$ ___ (pp. 73–77)
3. Factor $6x^3 - 2x^2$ (pp. 49–55)

Concepts and Vocabulary

4. ***True or False*** Factoring can only be used to solve quadratic equations or equations that are quadratic in form.
5. If u is an expression that involves x, then the equation $au^2 + bu + c = 0, a \neq 0$, is called an equation ________ ________.
6. ***True or False*** Radical equations sometimes have extraneous solutions.
7. An apparent solution that does not satisfy the original equation is called a(n) ________ solution.
   (a) extraneous (b) imaginary
   (c) radical (d) conditional
8. Which equation is likely to require squaring each side more than once?
   (a) $\sqrt{x+2} = \sqrt{3x-5}$ (b) $x^4 - 3x^2 = 10$
   (c) $\sqrt{x+1} + \sqrt{x-4} = 8$ (d) $\sqrt{3x+1} = 5$

Skill Building

In Problems 9–42, find the real solutions of each equation.

9. $\sqrt{2t-1} = 1$
10. $\sqrt{3t+4} = 2$
11. $\sqrt{3t+4} = -6$
12. $\sqrt{5t+3} = -2$
13. $\sqrt[3]{1-2x} - 3 = 0$
14. $\sqrt[3]{1-2x} - 1 = 0$
15. $\sqrt[5]{x^2+2x} = -1$
16. $\sqrt[4]{x^2+16} = \sqrt{5}$
17. $x = 8\sqrt{x}$
18. $x = 3\sqrt{x}$
19. $\sqrt{15-2x} = x$
20. $\sqrt{12-x} = x$
21. $x = 2\sqrt{x-1}$
22. $x = 2\sqrt{-x-1}$
23. $\sqrt{x^2-x-4} = x+2$
24. $\sqrt{3-x+x^2} = x-2$
25. $3 + \sqrt{3x+1} = x$
26. $2 + \sqrt{12-2x} = x$
27. $\sqrt{3(x+10)} - 4 = x$
28. $\sqrt{1-x} - 3 = x+2$
29. $\sqrt{2x+3} - \sqrt{x+1} = 1$
30. $\sqrt{3x+7} + \sqrt{x+2} = 1$
31. $\sqrt{3x+1} - \sqrt{x-1} = 2$
32. $\sqrt{3x-5} - \sqrt{x+7} = 2$

33. $\sqrt{3 - 2\sqrt{x}} = \sqrt{x}$ **34.** $\sqrt{10 + 3\sqrt{x}} = \sqrt{x}$ **35.** $(3x + 1)^{1/2} = 4$

36. $(3x - 5)^{1/2} = 2$ **37.** $(5x - 2)^{1/3} = 2$ **38.** $(2x + 1)^{1/3} = -1$

39. $(x^2 + 9)^{1/2} = 5$ **40.** $(x^2 - 16)^{1/2} = 9$ **41.** $x^{3/2} - 3x^{1/2} = 0$ **42.** $x^{3/4} - 9x^{1/4} = 0$

In Problems 43–74, find the real solutions of each equation.

43. $x^4 - 5x^2 + 4 = 0$ **44.** $x^4 - 10x^2 + 25 = 0$ **45.** $3x^4 - 2x^2 - 1 = 0$

46. $2x^4 - 5x^2 - 12 = 0$ **47.** $x^6 + 7x^3 - 8 = 0$ **48.** $x^6 - 7x^3 - 8 = 0$

49. $(x+2)^2 + 7(x+2) + 12 = 0$ **50.** $(2x+5)^2 - (2x+5) - 6 = 0$ **51.** $(3x+4)^2 - 6(3x+4) + 9 = 0$

52. $(2-x)^2 + (2-x) - 20 = 0$ **53.** $2(s+1)^2 - 5(s+1) = 3$ **54.** $3(1-y)^2 + 5(1-y) + 2 = 0$

55. $x - 4x\sqrt{x} = 0$ **56.** $x + 8\sqrt{x} = 0$ **57.** $x + \sqrt{x} = 20$

58. $x + \sqrt{x} = 6$ **59.** $t^{1/2} - 2t^{1/4} + 1 = 0$ **60.** $z^{1/2} - 4z^{1/4} + 4 = 0$

61. $4x^{1/2} - 9x^{1/4} + 4 = 0$ **62.** $x^{1/2} - 3x^{1/4} + 2 = 0$ **63.** $\sqrt[4]{5x^2 - 6} = x$

64. $\sqrt[4]{4 - 5x^2} = x$ **65.** $x^2 + 3x + \sqrt{x^2 + 3x} = 6$ **66.** $x^2 - 3x - \sqrt{x^2 - 3x} = 2$

67. $\dfrac{1}{(x+1)^2} = \dfrac{1}{x+1} + 2$ **68.** $\dfrac{1}{(x-1)^2} + \dfrac{1}{x-1} = 12$ **69.** $3x^{-2} - 7x^{-1} - 6 = 0$

70. $2x^{-2} - 3x^{-1} - 4 = 0$ **71.** $2x^{2/3} - 5x^{1/3} - 3 = 0$ **72.** $3x^{4/3} + 5x^{2/3} - 2 = 0$

73. $\left(\dfrac{v}{v+1}\right)^2 + \dfrac{2v}{v+1} = 8$ **74.** $\left(\dfrac{y}{y-1}\right)^2 = 6\left(\dfrac{y}{y-1}\right) + 7$

In Problems 75–90, find the real solutions of each equation by factoring.

75. $x^3 - 9x = 0$ **76.** $x^4 - x^2 = 0$ **77.** $4x^3 = 3x^2$ **78.** $x^5 = 4x^3$

79. $x^3 + x^2 - 20x = 0$ **80.** $x^3 + 6x^2 - 7x = 0$ **81.** $x^3 + x^2 - x - 1 = 0$

82. $x^3 + 4x^2 - x - 4 = 0$ **83.** $x^3 - 3x^2 - 4x + 12 = 0$ **84.** $x^3 - 3x^2 - x + 3 = 0$

85. $2x^3 + 4 = x^2 + 8x$ **86.** $3x^3 + 4x^2 = 27x + 36$ **87.** $5x^3 + 45x = 2x^2 + 18$

88. $3x^3 + 12x = 5x^2 + 20$ **89.** $x(x^2 - 3x)^{1/3} + 2(x^2 - 3x)^{4/3} = 0$ **90.** $3x(x^2 + 2x)^{1/2} - 2(x^2 + 2x)^{3/2} = 0$

In Problems 91–96, find the real solutions of each equation. Use a calculator to express any solutions rounded to two decimal places.

91. $x - 4x^{1/2} + 2 = 0$ **92.** $x^{2/3} + 4x^{1/3} + 2 = 0$ **93.** $x^4 + \sqrt{3}x^2 - 3 = 0$

94. $x^4 + \sqrt{2}x^2 - 2 = 0$ **95.** $\pi(1 + t)^2 = \pi + 1 + t$ **96.** $\pi(1 + r)^2 = 2 + \pi(1 + r)$

Mixed Practice

97. If $k = \dfrac{x+3}{x-3}$ and $k^2 - k = 12$, find x. **98.** If $k = \dfrac{x+3}{x-4}$ and $k^2 - 3k = 28$, find x.

Applications

99. Physics: Using Sound to Measure Distance The distance to the surface of the water in a well can sometimes be found by dropping an object into the well and measuring the time elapsed until a sound is heard. If t_1 is the time (measured in seconds) that it takes for the object to strike the water, then t_1 will obey the equation $s = 16t_1^2$, where s is the distance (measured in feet). It follows that $t_1 = \dfrac{\sqrt{s}}{4}$. Suppose that t_2 is the time that it takes for the sound of the impact to reach your ears. Because sound waves are known to travel at a speed of approximately 1100 feet per second, the time t_2 to travel the distance s will be $t_2 = \dfrac{s}{1100}$. See the illustration.

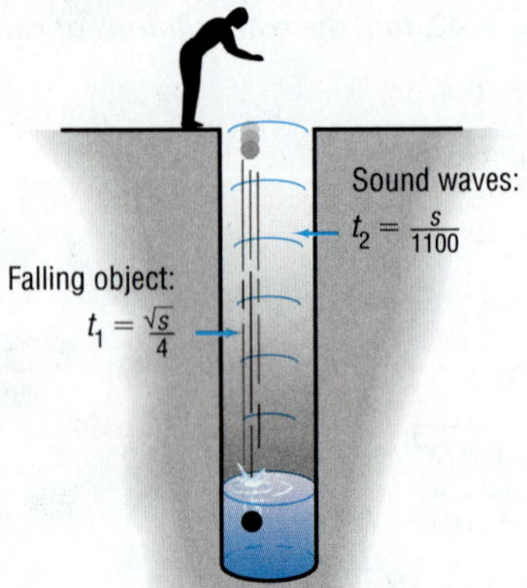

Now $t_1 + t_2$ is the total time that elapses from the moment that the object is dropped to the moment that a sound is heard. We have the equation

$$\text{Total time elapsed} = \frac{\sqrt{s}}{4} + \frac{s}{1100}$$

Find the distance to the water's surface if the total time elapsed from dropping a rock to hearing it hit water is 4 seconds.

100. Crushing Load A civil engineer relates the thickness T, in inches, and height H, in feet, of a square wooden pillar to its crushing load L, in tons, using the model $T = \sqrt[4]{\dfrac{LH^2}{25}}$. If a square wooden pillar is 4 inches thick and 10 feet high, what is its crushing load?

101. Foucault's Pendulum The period of a pendulum is the time it takes the pendulum to make one full swing back and forth. The period T, in seconds, is given by the formula $T = 2\pi\sqrt{\dfrac{l}{32}}$, where l is the length, in feet, of the pendulum. In 1851, Jean Bernard Leon Foucault demonstrated the axial rotation of Earth using a large pendulum that he hung in the Panthéon in Paris. The period of Foucault's pendulum was approximately 16.5 seconds. What was its length?

Explaining Concepts: Discussion and Writing

102. Make up a radical equation that has no solution.

103. Make up a radical equation that has an extraneous solution.

104. Discuss the step in the solving process for radical equations that leads to the possibility of extraneous solutions. Why is there no such possibility for linear and quadratic equations?

105. What Went Wrong? On an exam, Jane solved the equation $\sqrt{2x + 3} - x = 0$ and wrote that the solution set was $\{-1, 3\}$. Jane received 3 out of 5 points for the problem. Jane asks you why she received 3 out of 5 points. Provide an explanation.

'Are You Prepared?' Answers

1. True **2.** -2 **3.** $2x^2(3x - 1)$

1.5 Solving Inequalities

PREPARING FOR THIS SECTION *Before getting started, review the following:*

- Algebra Essentials (Section R.2, pp. 17–26)

Now Work the **'Are You Prepared?'** problems on page 127.

OBJECTIVES
1 Use Interval Notation (p. 120)
2 Use Properties of Inequalities (p. 121)
3 Solve Inequalities (p. 123)
4 Solve Combined Inequalities (p. 124)

Suppose that a and b are two real numbers and $a < b$. The notation $a < x < b$ means that x is a number *between* a and b. The expression $a < x < b$ is equivalent to the two inequalities $a < x$ and $x < b$. Similarly, the expression $a \le x \le b$ is equivalent to the two inequalities $a \le x$ and $x \le b$. The remaining two possibilities, $a \le x < b$ and $a < x \le b$, are defined similarly.

Although it is acceptable to write $3 \ge x \ge 2$, it is preferable to reverse the inequality symbols and write instead $2 \le x \le 3$ so that the values go from smaller to larger, reading from left to right.

A statement such as $2 \le x \le 1$ is false because there is no number x for which $2 \le x$ and $x \le 1$. Finally, never mix inequality symbols, as in $2 \le x \ge 3$.

1 Use Interval Notation

Let a and b represent two real numbers with $a < b$.

DEFINITION

An **open interval**, denoted by **(a, b)**, consists of all real numbers x for which $a < x < b$.

A **closed interval**, denoted by **$[a, b]$**, consists of all real numbers x for which $a \leq x \leq b$.

The **half-open**, or **half-closed**, **intervals** are **$(a, b]$**, consisting of all real numbers x for which $a < x \leq b$, and **$[a, b)$**, consisting of all real numbers x for which $a \leq x < b$.

In Words

The notation $[a, b]$ represents all real numbers between a and b, inclusive. The notation (a, b) represents all real numbers between a and b, not including either a or b.

In each of these definitions, a is called the **left endpoint** and b the **right endpoint** of the interval.

The symbol ∞ (read as "infinity") is not a real number, but notation used to indicate unboundedness in the positive direction. The symbol $-\infty$ (read as "negative infinity") also is not a real number, but notation used to indicate unboundedness in the negative direction. The symbols ∞ and $-\infty$ are used to define five other kinds of intervals:

$[a, \infty)$	Consists of all real numbers x for which $x \geq a$
(a, ∞)	Consists of all real numbers x for which $x > a$
$(-\infty, a]$	Consists of all real numbers x for which $x \leq a$
$(-\infty, a)$	Consists of all real numbers x for which $x < a$
$(-\infty, \infty)$	Consists of all real numbers

Note that ∞ and $-\infty$ are never included as endpoints, since neither is a real number.

Table 1 summarizes interval notation, corresponding inequality notation, and their graphs.

Table 1

Interval	Inequality	Graph
The open interval (a, b)	$a < x < b$	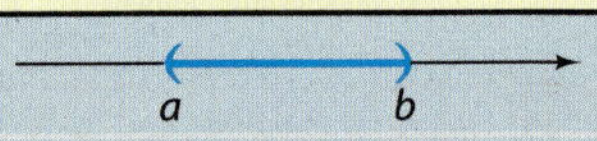
The closed interval $[a, b]$	$a \leq x \leq b$	
The half-open interval $[a, b)$	$a \leq x < b$	
The half-open interval $(a, b]$	$a < x \leq b$	
The interval $[a, \infty)$	$x \geq a$	
The interval (a, ∞)	$x > a$	
The interval $(-\infty, a]$	$x \leq a$	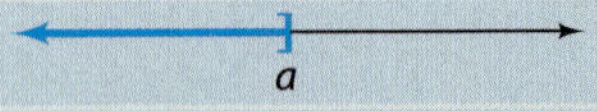
The interval $(-\infty, a)$	$x < a$	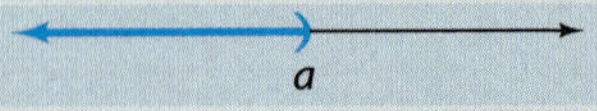
The interval $(-\infty, \infty)$	All real numbers	

EXAMPLE 1 **Writing Inequalities Using Interval Notation**

Write each inequality using interval notation.

(a) $1 \leq x \leq 3$ (b) $-4 < x < 0$ (c) $x > 5$ (d) $x \leq 1$

Solution

(a) $1 \leq x \leq 3$ describes all real numbers x between 1 and 3, inclusive. In interval notation, we write $[1, 3]$.

(b) In interval notation, $-4 < x < 0$ is written $(-4, 0)$.

(c) In interval notation, $x > 5$ is written $(5, \infty)$.
(d) In interval notation, $x \leq 1$ is written $(-\infty, 1]$.

EXAMPLE 2 **Writing Intervals Using Inequality Notation**

Write each interval as an inequality involving x.

(a) $[1, 4)$ (b) $(2, \infty)$ (c) $[2, 3]$ (d) $(-\infty, -3]$

Solution
(a) $[1, 4)$ consists of all real numbers x for which $1 \leq x < 4$.
(b) $(2, \infty)$ consists of all real numbers x for which $x > 2$.
(c) $[2, 3]$ consists of all real numbers x for which $2 \leq x \leq 3$.
(d) $(-\infty, -3]$ consists of all real numbers x for which $x \leq -3$.

Now Work PROBLEMS 13, 25, AND 33

2 Use Properties of Inequalities

The product of two positive real numbers is positive, the product of two negative real numbers is positive, and the product of 0 and 0 is 0. For any real number a, the value of a^2 is 0 or positive; that is, a^2 is nonnegative. This is called the **nonnegative property**.

In Words
The square of a real number is never negative.

Nonnegative Property
For any real number a,

$$a^2 \geq 0 \qquad (1)$$

When the same number is added to both sides of an inequality, an equivalent inequality is obtained. For example, since $3 < 5$, then $3 + 4 < 5 + 4$ or $7 < 9$. This is called the **addition property** of inequalities.

In Words
The addition property states that the sense, or direction, of an inequality remains unchanged if the same number is added to each side.

Addition Property of Inequalities
For real numbers a, b, and c,

$$\text{If } a < b, \text{ then } a + c < b + c. \qquad (2a)$$
$$\text{If } a > b, \text{ then } a + c > b + c. \qquad (2b)$$

Figure 2 illustrates the addition property (2a). In Figure 2(a), we see that a lies to the left of b. If c is positive, then $a + c$ and $b + c$ lie c units to the right of a and c units to the right of b, respectively. Consequently, $a + c$ must lie to the left of $b + c$; that is, $a + c < b + c$. Figure 2(b) illustrates the situation if c is negative.

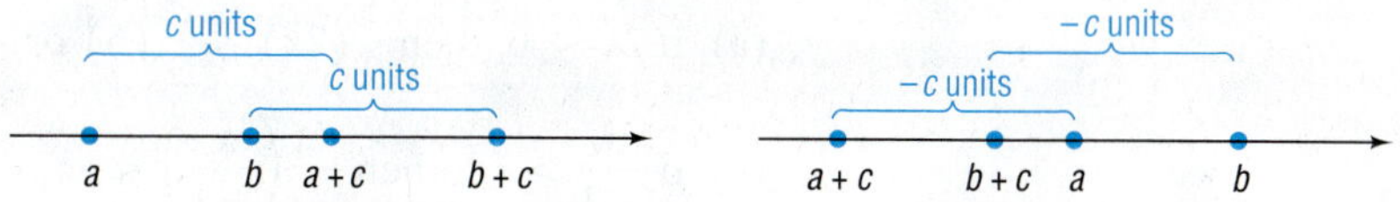

(a) If $a<b$ and $c>0$, then $a+c<b+c$.
(b) If $a<b$ and $c<0$, then $a+c<b+c$.

Figure 2 Addition property of inequalities

DRAW AN ILLUSTRATION SIMILAR TO FIGURE 2 THAT ILLUSTRATES THE ADDITION PROPERTY (2b).

EXAMPLE 3 **Addition Property of Inequalities**

(a) If $x < -5$, then $x + 5 < -5 + 5$ or $x + 5 < 0$.

(b) If $x > 2$, then $x + (-2) > 2 + (-2)$ or $x - 2 > 0$.

Now Work PROBLEM 41

EXAMPLE 4 **Multiplying an Inequality by a Positive Number**

Express as an inequality the result of multiplying each side of the inequality $3 < 7$ by 2.

Solution Begin with

$$3 < 7$$

Multiplying each side by 2 yields the numbers 6 and 14, so we have

$$6 < 14$$

EXAMPLE 5 **Multiplying an Inequality by a Negative Number**

Express as an inequality the result of multiplying each side of the inequality $9 > 2$ by -4.

Solution Begin with

$$9 > 2$$

Multiplying each side by -4 yields the numbers -36 and -8, so we have

$$-36 < -8$$

In Words
Multiplying by a negative number reverses the inequality.

Note that the effect of multiplying both sides of $9 > 2$ by the negative number -4 is that the direction of the inequality symbol is reversed.

Examples 4 and 5 illustrate the following general **multiplication properties** for inequalities:

In Words
The multiplication properties state that the sense, or direction, of an inequality *remains the same* if each side is multiplied by a *positive* real number, whereas the direction *is reversed* if each side is multiplied by a *negative* real number.

Multiplication Properties for Inequalities

For real numbers a, b, and c,

If $a < b$ and if $c > 0$, then $ac < bc$.
If $a < b$ and if $c < 0$, then $ac > bc$. **(3a)**

If $a > b$ and if $c > 0$, then $ac > bc$.
If $a > b$ and if $c < 0$, then $ac < bc$. **(3b)**

EXAMPLE 6 **Multiplication Property of Inequalities**

(a) If $2x < 6$, then $\frac{1}{2}(2x) < \frac{1}{2}(6)$ or $x < 3$.

(b) If $\frac{x}{-3} > 12$, then $-3\left(\frac{x}{-3}\right) < -3(12)$ or $x < -36$.

(c) If $-4x < -8$, then $\frac{-4x}{-4} > \frac{-8}{-4}$ or $x > 2$.

(d) If $-x > 8$, then $(-1)(-x) < (-1)(8)$ or $x < -8$.

Now Work PROBLEM 47

In Words
The reciprocal property states that the reciprocal of a positive real number is positive and that the reciprocal of a negative real number is negative.

Reciprocal Property for Inequalities

If $a > 0$, then $\frac{1}{a} > 0$. If $\frac{1}{a} > 0$, then $a > 0$. **(4a)**

If $a < 0$, then $\frac{1}{a} < 0$. If $\frac{1}{a} < 0$, then $a < 0$. **(4b)**

3 Solve Inequalities

An **inequality in one variable** is a statement involving two expressions, at least one containing the variable, separated by one of the inequality symbols: $<, \leq, >$, or $\geq$. To **solve an inequality** means to find all values of the variable for which the statement is true. These values are called **solutions** of the inequality.

For example, the following are all inequalities involving one variable x:

$$x + 5 < 8 \qquad 2x - 3 \geq 4 \qquad x^2 - 1 \leq 3 \qquad \frac{x+1}{x-2} > 0$$

As with equations, one method for solving an inequality is to replace it by a series of equivalent inequalities until an inequality with an obvious solution, such as $x < 3$, is obtained. Equivalent inequalities are obtained by applying some of the same properties that are used to find equivalent equations. The addition property and the multiplication properties for inequalities form the basis for the following procedures.

Procedures That Leave the Inequality Symbol Unchanged

1. Simplify both sides of the inequality by combining like terms and eliminating parentheses:

$$\text{Replace} \quad x + 2 + 6 > 2x + 5(x + 1)$$
$$\text{by} \quad x + 8 > 7x + 5$$

2. Add or subtract the same expression on both sides of the inequality:

$$\text{Replace} \quad 3x - 5 < 4$$
$$\text{by} \quad (3x - 5) + 5 < 4 + 5$$

3. Multiply or divide both sides of the inequality by the same *positive* expression:

$$\text{Replace} \quad 4x > 16 \quad \text{by} \quad \frac{4x}{4} > \frac{16}{4}$$

Procedures That Reverse the Sense or Direction of the Inequality Symbol

1. Interchange the two sides of the inequality:

$$\text{Replace} \quad 3 < x \quad \text{by} \quad x > 3$$

2. Multiply or divide both sides of the inequality by the same *negative* expression:

$$\text{Replace} \quad -2x > 6 \quad \text{by} \quad \frac{-2x}{-2} < \frac{6}{-2}$$

As the examples that follow illustrate, we solve inequalities using many of the same steps that we would use to solve equations. In writing the solution of an inequality, either set notation or interval notation may be used, whichever is more convenient.

EXAMPLE 7 **Solving an Inequality**

Solve the inequality $3 - 2x < 5$, and graph the solution set.

Solution

$$3 - 2x < 5$$

$$3 - 2x - 3 < 5 - 3 \quad \text{Subtract 3 from both sides.}$$

$$-2x < 2 \quad \text{Simplify.}$$

$$\frac{-2x}{-2} > \frac{2}{-2} \quad \text{Divide both sides by } -2. \text{ (The sense of the inequality symbol is reversed.)}$$

$$x > -1 \quad \text{Simplify.}$$

Figure 3 $x > -1$

The solution set is $\{x \mid x > -1\}$ or, using interval notation, all numbers in the interval $(-1, \infty)$. See Figure 3 for the graph. ●

EXAMPLE 8 **Solving an Inequality**

Solve the inequality $4x + 7 \geq 2x - 3$, and graph the solution set.

Solution

$$4x + 7 \geq 2x - 3$$

$$4x + 7 - 7 \geq 2x - 3 - 7 \quad \text{Subtract 7 from both sides.}$$

$$4x \geq 2x - 10 \quad \text{Simplify.}$$

$$4x - 2x \geq 2x - 10 - 2x \quad \text{Subtract } 2x \text{ from both sides.}$$

$$2x \geq -10 \quad \text{Simplify.}$$

$$\frac{2x}{2} \geq \frac{-10}{2} \quad \text{Divide both sides by 2. (The direction of the inequality symbol is unchanged.)}$$

$$x \geq -5 \quad \text{Simplify.}$$

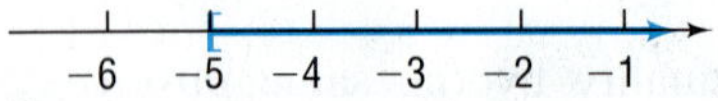

Figure 4 $x \geq -5$

The solution set is $\{x \mid x \geq -5\}$ or, using interval notation, all numbers in the interval $[-5, \infty)$. See Figure 4 for the graph. ●

Now Work PROBLEM 55

4 Solve Combined Inequalities

EXAMPLE 9 **Solving a Combined Inequality**

Solve the inequality $-5 < 3x - 2 < 1$, and graph the solution set.

Solution Recall that the inequality

$$-5 < 3x - 2 < 1$$

is equivalent to the two inequalities

$$-5 < 3x - 2 \quad \text{and} \quad 3x - 2 < 1$$

Solve each of these inequalities separately.

$$\begin{array}{rll} -5 < 3x - 2 & & 3x - 2 < 1 \\ -5 + 2 < 3x - 2 + 2 & \text{Add 2 to both sides.} & 3x - 2 + 2 < 1 + 2 \\ -3 < 3x & \text{Simplify.} & 3x < 3 \\ \dfrac{-3}{3} < \dfrac{3x}{3} & \text{Divide both sides by 3.} & \dfrac{3x}{3} < \dfrac{3}{3} \\ -1 < x & \text{Simplify.} & x < 1 \end{array}$$

The solution set of the original pair of inequalities consists of all x for which

$$-1 < x \quad \text{and} \quad x < 1$$

Figure 5 $-1 < x < 1$

This may be written more compactly as $\{x \mid -1 < x < 1\}$. In interval notation, the solution is $(-1, 1)$. See Figure 5 for the graph. ●

Observe in the preceding process that solving each of the two inequalities required exactly the same steps. A shortcut to solving the original inequality algebraically is to deal with the two inequalities at the same time, as follows:

$$\begin{array}{rcll} -5 < & 3x - 2 & < 1 & \\ -5 + 2 < & 3x - 2 + 2 & < 1 + 2 & \text{Add 2 to each part.} \\ -3 < & 3x & < 3 & \text{Simplify.} \\ \dfrac{-3}{3} < & \dfrac{3x}{3} & < \dfrac{3}{3} & \text{Divide each part by 3.} \\ -1 < & x & < 1 & \text{Simplify.} \end{array}$$

EXAMPLE 10 Solving a Combined Inequality

Solve the inequality $-1 \le \dfrac{3 - 5x}{2} \le 9$, and graph the solution set.

Solution

$$\begin{array}{rcll} -1 \le & \dfrac{3 - 5x}{2} & \le 9 & \\ 2(-1) \le & 2\left(\dfrac{3 - 5x}{2}\right) & \le 2(9) & \text{Multiply each part by 2 to remove the denominator.} \\ -2 \le & 3 - 5x & \le 18 & \text{Simplify.} \\ -2 - 3 \le & 3 - 5x - 3 & \le 18 - 3 & \text{Subtract 3 from each part to isolate the term containing } x. \\ -5 \le & -5x & \le 15 & \text{Simplify.} \\ \dfrac{-5}{-5} \ge & \dfrac{-5x}{-5} & \ge \dfrac{15}{-5} & \text{Divide each part by } -5 \text{ (reverse the direction of each inequality symbol).} \\ 1 \ge & x & \ge -3 & \text{Simplify.} \\ -3 \le & x & \le 1 & \text{Reverse the order so that the numbers get larger as you read from left to right.} \end{array}$$

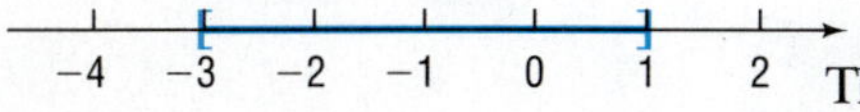

Figure 6 $-3 \le x \le 1$

The solution set is $\{x \mid -3 \le x \le 1\}$, that is, all x in the interval $[-3, 1]$. Figure 6 illustrates the graph. ●

Now Work PROBLEM 75

EXAMPLE 11

Using the Reciprocal Property to Solve an Inequality

Solve the inequality $(4x - 1)^{-1} > 0$, and graph the solution set.

Solution Recall that $(4x - 1)^{-1} = \dfrac{1}{4x - 1}$. The Reciprocal Property states that when $\dfrac{1}{a} > 0$, then $a > 0$.

$$(4x - 1)^{-1} > 0$$

$$\frac{1}{4x - 1} > 0$$

$$4x - 1 > 0 \quad \text{Reciprocal Property}$$

$$4x > 1 \quad \text{Add 1 to both sides.}$$

$$x > \frac{1}{4} \quad \text{Divide both sides by 4.}$$

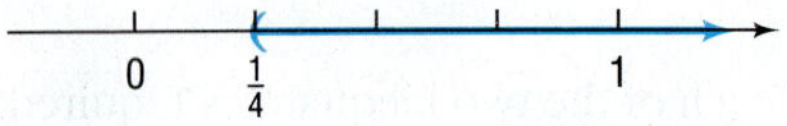

Figure 7 $x > \frac{1}{4}$

The solution set is $\left\{x \middle| x > \frac{1}{4}\right\}$, that is, all x in the interval $\left(\frac{1}{4}, \infty\right)$. Figure 7 illustrates the graph.

Now Work PROBLEM 85

EXAMPLE 12

Creating Equivalent Inequalities

If $-1 < x < 4$, find a and b so that $a < 2x + 1 < b$.

Solution The idea here is to change the middle part of the combined inequality from x to $2x + 1$, using properties of inequalities.

$$-1 < x < 4$$

$$-2 < 2x < 8 \quad \text{Multiply each part by 2.}$$

$$-1 < 2x + 1 < 9 \quad \text{Add 1 to each part.}$$

Now we see that $a = -1$ and $b = 9$.

Now Work PROBLEM 95

Application

EXAMPLE 13

Physics: Ohm's Law

In electricity, Ohm's law states that $E = IR$, where E is the voltage (in volts), I is the current (in amperes), and R is the resistance (in ohms). An air-conditioning unit is rated at a resistance of 10 ohms. If the voltage varies from 110 to 120 volts, inclusive, what corresponding range of current will the air conditioner draw?

Solution The voltage lies between 110 and 120, inclusive, so

$$110 \le E \le 120$$

$$110 \le IR \le 120 \quad \text{Ohm's law, } E = IR$$

$$110 \le I(10) \le 120 \quad R = 10$$

$$\frac{110}{10} \le \frac{I(10)}{10} \le \frac{120}{10} \quad \text{Divide each part by 10.}$$

$$11 \le I \le 12 \quad \text{Simplify.}$$

The air conditioner will draw between 11 and 12 amperes of current, inclusive.

1.5 Assess Your Understanding

'Are You Prepared?' *Answers are given at the end of these exercises. If you get a wrong answer, read the pages listed in red.*

1. Graph the inequality: $x \geq -2$. (pp. 17–26)

2. ***True or False*** $-5 > -3$ (pp. 17–26)

Concepts and Vocabulary

3. A(n) ________ __________, denoted $[a, b]$, consists of all real numbers x for which $a \leq x \leq b$.

4. The ______________ ___________ state that the sense, or direction, of an inequality remains the same if each side is multiplied by a positive number, while the direction is reversed if each side is multiplied by a negative number.

In Problems 5–8, assume that $a < b$ and $c < 0$.

5. ***True or False*** $a + c < b + c$

6. ***True or False*** $a - c < b - c$

7. ***True or False*** $ac > bc$

8. ***True or False*** $\frac{a}{c} < \frac{b}{c}$

9. ***True or False*** The square of any real number is always nonnegative.

10. ***True or False*** A half-closed interval must have an endpoint of either $-\infty$ or ∞.

11. Which of the following will change the direction, or sense, of an inequality?
(a) Dividing each side by a positive number
(b) Interchanging sides
(c) Adding a negative number to each side
(d) Subtracting a positive number from each side

12. Which pair of inequalities is equivalent to $0 < x \leq 3$?
(a) $x > 0$ and $x \geq 3$ (b) $x < 0$ and $x \geq 3$
(c) $x > 0$ and $x \leq 3$ (d) $x < 0$ and $x \leq 3$

Skill Building

In Problems 13–18, express the graph shown in blue using interval notation. Also express each as an inequality involving x.

13.
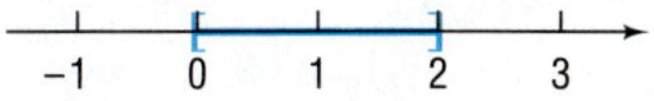

14.
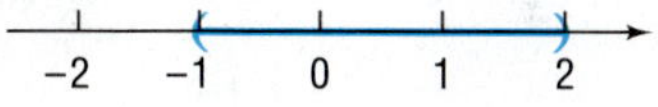

15.
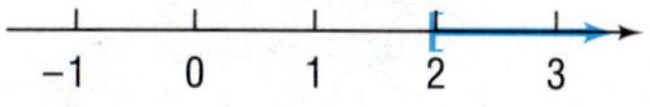

16.

17.
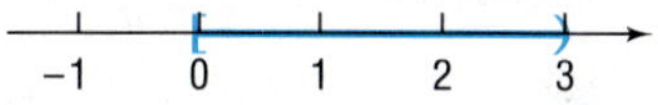

18.
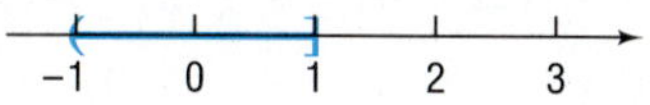

In Problems 19–24, an inequality is given. Write the inequality obtained by:
(a) Adding 3 to each side of the given inequality.
(b) Subtracting 5 from each side of the given inequality.
(c) Multiplying each side of the given inequality by 3.
(d) Multiplying each side of the given inequality by −2.

19. $3 < 5$ **20.** $2 > 1$ **21.** $4 > -3$ **22.** $-3 > -5$ **23.** $2x + 1 < 2$ **24.** $1 - 2x > 5$

In Problems 25–32, write each inequality using interval notation, and graph each inequality on the real number line.

25. $0 \leq x \leq 4$ **26.** $-1 < x < 5$ **27.** $4 \leq x < 6$ **28.** $-2 < x < 0$

29. $x \geq 4$ **30.** $x \leq 5$ **31.** $x < -4$ **32.** $x > 1$

In Problems 33–40, write each interval as an inequality involving x, and graph each inequality on the real number line.

33. $[2, 5]$ **34.** $(1, 2)$ **35.** $(-3, -2)$ **36.** $[0, 1)$

37. $[4, \infty)$ **38.** $(-\infty, 2]$ **39.** $(-\infty, -3)$ **40.** $(-8, \infty)$

In Problems 41–54, fill in the blank with the correct inequality symbol.

41. If $x < 5$, then $x - 5$ ___ 0.

42. If $x < -4$, then $x + 4$ ___ 0.

43. If $x > -4$, then $x + 4$ ___ 0.

44. If $x > 6$, then $x - 6$ ___ 0.

45. If $x \geq -4$, then $3x$ ___ -12.

46. If $x \leq 3$, then $2x$ ___ 6.

47. If $x > 6$, then $-2x$ ___ -12.

48. If $x > -2$, then $-4x$ ___ 8.

49. If $x \geq 5$, then $-4x$ ___ -20.

50. If $x \leq -4$, then $-3x$ ___ 12.

51. If $2x > 6$, then x ___ 3.

52. If $3x \leq 12$, then x ___ 4.

53. If $\frac{1}{2}x \leq 3$, then x ___ -6.

54. If $-\frac{1}{4}x > 1$, then x ___ -4.

In Problems 55–92, solve each inequality. Express your answer using set notation or interval notation. Graph the solution set.

55. $x + 1 < 5$

56. $x - 6 < 1$

57. $1 - 2x \le 3$

58. $2 - 3x \le 5$

59. $3x - 7 > 2$

60. $2x + 5 > 1$

61. $3x - 1 \ge 3 + x$

62. $2x - 2 \ge 3 + x$

63. $-2(x + 3) < 8$

64. $-3(1 - x) < 12$

65. $4 - 3(1 - x) \le 3$

66. $8 - 4(2 - x) \le -2x$

67. $\frac{1}{2}(x - 4) > x + 8$

68. $3x + 4 > \frac{1}{3}(x - 2)$

69. $\frac{x}{2} \ge 1 - \frac{x}{4}$

70. $\frac{x}{3} \ge 2 + \frac{x}{6}$

71. $0 \le 2x - 6 \le 4$

72. $4 \le 2x + 2 \le 10$

73. $-5 \le 4 - 3x \le 2$

74. $-3 \le 3 - 2x \le 9$

75. $-3 < \frac{2x - 1}{4} < 0$

76. $0 < \frac{3x + 2}{2} < 4$

77. $1 < 1 - \frac{1}{2}x < 4$

78. $0 < 1 - \frac{1}{3}x < 1$

79. $(x + 2)(x - 3) > (x - 1)(x + 1)$

80. $(x - 1)(x + 1) > (x - 3)(x + 4)$

81. $x(4x + 3) \le (2x + 1)^2$

82. $x(9x - 5) \le (3x - 1)^2$

83. $\frac{1}{2} \le \frac{x + 1}{3} < \frac{3}{4}$

84. $\frac{1}{3} < \frac{x + 1}{2} \le \frac{2}{3}$

85. $(4x + 2)^{-1} < 0$

86. $(2x - 1)^{-1} > 0$

87. $(2 - 7x)^{-1} \ge 5$

88. $2(3x + 5)^{-1} \le -3$

89. $0 < \frac{2}{x} < \frac{3}{5}$

90. $0 < \frac{4}{x} < \frac{2}{3}$

91. $0 < (2x - 4)^{-1} < \frac{1}{2}$

92. $0 < (3x + 6)^{-1} < \frac{1}{3}$

Applications and Extensions

In Problems 93–102, find a and b.

93. If $-1 < x < 1$, then $a < x + 4 < b$.

94. If $-3 < x < 2$, then $a < x - 6 < b$.

95. If $2 < x < 3$, then $a < -4x < b$.

96. If $-4 < x < 0$, then $a < \frac{1}{2}x < b$.

97. If $0 < x < 4$, then $a < 2x + 3 < b$.

98. If $-3 < x < 3$, then $a < 1 - 2x < b$.

99. If $-3 < x < 0$, then $a < \frac{1}{x + 4} < b$.

100. If $2 < x < 4$, then $a < \frac{1}{x - 6} < b$.

101. If $6 < 3x < 12$, then $a < x^2 < b$.

102. If $0 < 2x < 6$, then $a < x^2 < b$.

103. What is the domain of the variable in the expression $\sqrt{3x + 6}$?

104. What is the domain of the variable in the expression $\sqrt{8 + 2x}$?

105. A young adult may be defined as someone older than 21 but less than 30 years of age. Express this statement using inequalities.

106. Middle-aged may be defined as being 40 or more and less than 60. Express this statement using inequalities.

107. Life Expectancy The Social Security Administration determined that an average 30-year-old male in 2014 could expect to live at least 51.9 more years and that an average 30-year-old female in 2014 could expect to live at least 55.6 more years.

(a) To what age could an average 30-year-old male expect to live? Express your answer as an inequality.

(b) To what age could an average 30-year-old female expect to live? Express your answer as an inequality.

(c) Who can expect to live longer, a male or a female? By how many years?

Source: Social Security Administration, 2014

108. General Chemistry For a certain ideal gas, the volume V (in cubic centimeters) equals 20 times the temperature T (in degrees Celsius). If the temperature varies from 80° to 120°C, inclusive, what is the corresponding range of the volume of the gas?

109. Real Estate A real estate agent agrees to sell an apartment complex according to the following commission schedule: \$45,000 plus 25% of the selling price in excess of \$900,000. Assuming that the complex will sell at some price between

\$900,000 and \$1,100,000, inclusive, over what range does the agent's commission vary? How does the commission vary as a percent of selling price?

110. Sales Commission A used car salesperson is paid a commission of \$25 plus 40% of the selling price in excess of owner's cost. The owner claims that used cars typically sell for at least owner's cost plus \$200 and at most owner's cost plus \$3000. For each sale made, over what range can the salesperson expect the commission to vary?

111. Federal Tax Withholding The percentage method of withholding for federal income tax (2014) states that a single person whose weekly wages, after subtracting withholding allowances, are over \$753, but not over \$1762, shall have \$97.75 plus 25% of the excess over \$753 withheld. Over what range does the amount withheld vary if the weekly wages vary from \$900 to \$1100, inclusive?

Source: Employer's Tax Guide. Internal Revenue Service, 2014.

112. Exercising Sue wants to lose weight. For healthy weight loss, the American College of Sports Medicine (ACSM) recommends 200 to 300 minutes of exercise per week. For the first six days of the week, Sue exercised 40, 45, 0, 50, 25, and 35 minutes. How long should Sue exercise on the seventh day in order to stay within the ACSM guidelines?

113. Electricity Rates Commonwealth Edison Company's charge for electricity in January 2014 was 8.21¢ per kilowatt-hour. In addition, each monthly bill contains a customer charge of \$15.37. If last year's bills ranged from a low of \$72.84 to a high of \$237.04, over what range did usage vary (in kilowatt-hours)?

Source: Commonwealth Edison Co., 2014.

114. Water Bills The Village of Oak Lawn charges homeowners \$57.07 per quarter-year plus \$5.81 per 1000 gallons for water usage in excess of 10,000 gallons. In 2014 one homeowner's quarterly bill ranged from a high of \$150.03 to a low of \$97.74. Over what range did water usage vary?

Source: Village of Oak Lawn, Illinois, January 2014.

115. Markup of a New Car The markup over dealer's cost of a new car ranges from 12% to 18%. If the sticker price is \$18,000, over what range will the dealer's cost vary?

116. IQ Tests A standard intelligence test has an average score of 100. According to statistical theory, of the people who take the test, the 2.5% with the highest scores will have scores of more than 1.96σ above the average, where σ (sigma, a number called the **standard deviation**) depends on the nature of the test. If $\sigma = 12$ for this test and there is (in principle) no upper limit to the score possible on the test, write the interval of possible test scores of the people in the top 2.5%.

117. Computing Grades In your Economics 101 class, you have scores of 68, 82, 87, and 89 on the first four of five tests. To get a grade of B, the average of the first five test scores must be greater than or equal to 80 and less than 90.

(a) Solve an inequality to find the least score you can get on the last test and still earn a B.

(b) What score do you need if the fifth test counts double?

118. "Light" Foods For food products to be labeled "light," the U.S. Food and Drug Administration requires that the altered product must either contain at least one-third fewer calories than the regular product or contain at least one-half less fat than the regular product. If a serving of Miracle Whip® Light contains 20 calories and 1.5 grams of fat, then what must be true about either the number of calories or the grams of fat in a serving of regular Miracle Whip®?

119. Arithmetic Mean If $a < b$, show that $a < \dfrac{a+b}{2} < b$. The number $\dfrac{a+b}{2}$ is called the **arithmetic mean** of a and b.

120. Refer to Problem 119. Show that the arithmetic mean of a and b is equidistant from a and b.

121. Geometric Mean If $0 < a < b$, show that $a < \sqrt{ab} < b$. The number $\sqrt{ab}$ is called the **geometric mean** of a and b.

122. Refer to Problems 119 and 121. Show that the geometric mean of a and b is less than the arithmetic mean of a and b.

123. Harmonic Mean For $0 < a < b$, let h be defined by

$$\frac{1}{h} = \frac{1}{2}\left(\frac{1}{a} + \frac{1}{b}\right)$$

Show that $a < h < b$. The number h is called the **harmonic mean** of a and b.

124. Refer to Problems 119, 121, and 123. Show that the harmonic mean of a and b equals the geometric mean squared, divided by the arithmetic mean.

125. Another Reciprocal Property Prove that if $0 < a < b$, then $0 < \dfrac{1}{b} < \dfrac{1}{a}$.

Explaining Concepts: Discussion and Writing

126. Make up an inequality that has no solution. Make up an inequality that has exactly one solution.

127. The inequality $x^2 + 1 < -5$ has no real solution. Explain why.

128. Do you prefer to use inequality notation or interval notation to express the solution to an inequality? Give your reasons. Are there particular circumstances when you prefer one to the other? Cite examples.

129. How would you explain to a fellow student the underlying reason for the multiplication properties for inequalities (page 122)? That is, the sense or direction of an inequality remains the same if each side is multiplied by a positive real number, whereas the direction is reversed if each side is multiplied by a negative real number.

'Are You Prepared?' Answers

1.

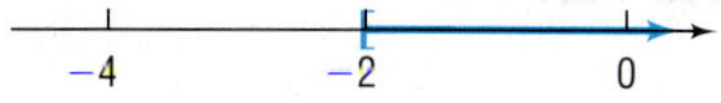

2. False

1.6 Equations and Inequalities Involving Absolute Value

PREPARING FOR THIS SECTION *Before getting started, review the following:*

- Algebra Essentials (Chapter R, Section R.2, pp. 17–26)

Now Work the 'Are You Prepared?' problems on page 132.

OBJECTIVES 1 Solve Equations Involving Absolute Value (p. 130)
2 Solve Inequalities Involving Absolute Value (p. 130)

1 Solve Equations Involving Absolute Value

Recall that on the real number line, the absolute value of a equals the distance from the origin to the point whose coordinate is a. For example, there are two points whose distance from the origin is 5 units, -5 and 5. So the equation $|x| = 5$ will have the solution set $\{-5, 5\}$. This leads to the following result:

THEOREM

If a is a positive real number and if u is any algebraic expression, then

$$|u| = a \quad \text{is equivalent to} \quad u = a \quad \text{or} \quad u = -a \qquad \textbf{(1)}$$

Equation (1) requires that a be a positive number. If $a = 0$, equation (1) becomes $|u| = 0$ which is equivalent to $u = 0$. If a is less than zero, the equation has no real solution. To see why, consider the equation $|x| = -2$. Because there is no real number whose distance from 0 on the real number line is negative, this equation has no real solution.

EXAMPLE 1 **Solving an Equation Involving Absolute Value**

Solve the equations:

(a) $|x + 4| = 13$ (b) $|2x - 3| + 2 = 7$

Solution (a) This follows the form of equation (1), where $u = x + 4$. There are two possibilities:

$$x + 4 = 13 \quad \text{or} \quad x + 4 = -13$$
$$x = 9 \quad \text{or} \quad x = -17$$

The solution set is $\{-17, 9\}$.

(b) The equation $|2x - 3| + 2 = 7$ is not in the form of equation (1). Proceed as follows:

$$|2x - 3| + 2 = 7$$
$$|2x - 3| = 5 \qquad \text{Subtract 2 from each side.}$$
$$2x - 3 = 5 \quad \text{or} \quad 2x - 3 = -5 \qquad \text{Apply (1).}$$
$$2x = 8 \quad \text{or} \quad 2x = -2 \qquad \text{Add 3 to both sides.}$$
$$x = 4 \quad \text{or} \quad x = -1 \qquad \text{Divide both sides by 2.}$$

The solution set is $\{-1, 4\}$. •

Now Work PROBLEM 11

2 Solve Inequalities Involving Absolute Value

EXAMPLE 2 **Solving an Inequality Involving Absolute Value**

Solve the inequality: $|x| < 4$

Solution We are looking for all points whose coordinate x is a distance less than 4 units from the origin. See Figure 8 for an illustration. Because any number x between -4 and 4 satisfies the condition $|x| < 4$, the solution set consists of all numbers x for which $-4 < x < 4$, that is, all x in the interval $(-4, 4)$. ●

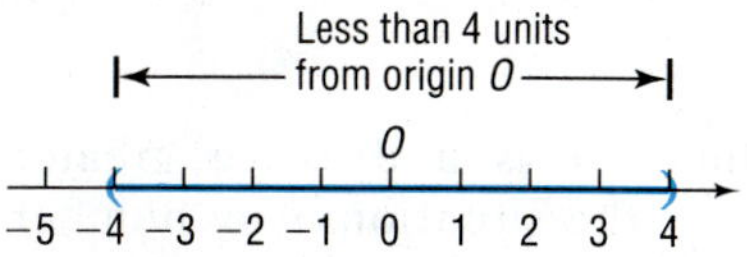

Figure 8 $|x| < 4$

THEOREM If a is a positive number and if u is an algebraic expression, then

$|u| < a$ is equivalent to $-a < u < a$ **(2)**

$|u| \le a$ is equivalent to $-a \le u \le a$ **(3)**

In other words, $|u| < a$ is equivalent to $-a < u$ and $u < a$.

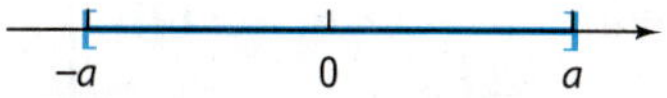

Figure 9 $|u| \le a, a > 0$

See Figure 9 for an illustration of statement (3).

EXAMPLE 3

Solving an Inequality Involving Absolute Value

Solve the inequality $|2x + 4| \le 3$, and graph the solution set.

Solution

$$|2x + 4| \le 3$$ This follows the form of statement (3); the expression $u = 2x + 4$ is inside the absolute value bars.

$$-3 \le 2x + 4 \le 3$$ Apply statement (3).

$$-3 - 4 \le 2x + 4 - 4 \le 3 - 4$$ Subtract 4 from each part.

$$-7 \le 2x \le -1$$ Simplify.

$$\frac{-7}{2} \le \frac{2x}{2} \le \frac{-1}{2}$$ Divide each part by 2.

$$-\frac{7}{2} \le x \le -\frac{1}{2}$$ Simplify.

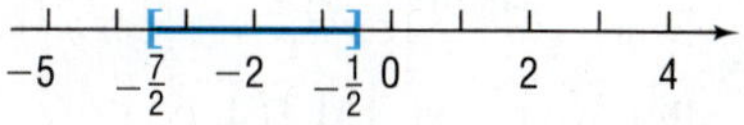

Figure 10 $|2x + 4| \le 3$

The solution set is $\left\{x \middle| -\frac{7}{2} \le x \le -\frac{1}{2}\right\}$, that is, all x in the interval $\left[-\frac{7}{2}, -\frac{1}{2}\right]$. See Figure 10 for the graph of the solution set. ●

EXAMPLE 4

Solving an Inequality Involving Absolute Value

Solve the inequality $|1 - 4x| < 5$, and graph the solution set.

Solution

$$|1 - 4x| < 5$$ This expression follows the form of statement (2); the expression $u = 1 - 4x$ is inside the absolute value bars.

$$-5 < 1 - 4x < 5$$ Apply statement (2).

$$-5 - 1 < 1 - 4x - 1 < 5 - 1$$ Subtract 1 from each part.

$$-6 < -4x < 4$$ Simplify.

$$\frac{-6}{-4} > \frac{-4x}{-4} > \frac{4}{-4}$$ Divide each part by -4, which reverses the direction of the inequality symbols.

$$\frac{3}{2} > x > -1$$ Simplify.

$$-1 < x < \frac{3}{2}$$ Rearrange the ordering.

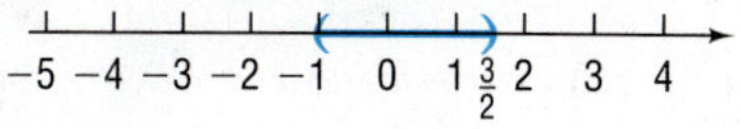

Figure 11 $|1 - 4x| < 5$

The solution set is $\left\{x \middle| -1 < x < \frac{3}{2}\right\}$, that is, all x in the interval $\left(-1, \frac{3}{2}\right)$. See Figure 11 for the graph of the solution set. ●

Now Work PROBLEM 41

EXAMPLE 5 Solving an Inequality Involving Absolute Value

Solve the inequality $|x| > 3$, and graph the solution set.

Solution We are looking for all points whose coordinate x is a distance greater than 3 units from the origin. Figure 12 illustrates the situation. Any number x less than -3 or greater than 3 satisfies the condition $|x| > 3$. The solution set consists of all numbers x for which $x < -3$ or $x > 3$, that is, all x in $(-\infty, -3) \cup (3, \infty)$.*

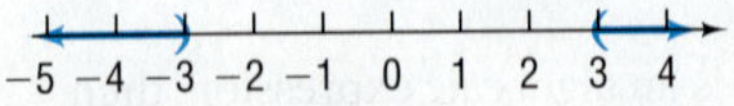

Figure 12 $|x| > 3$

THEOREM

If a is a positive number and u is an algebraic expression, then

$$|u| > a \quad \text{is equivalent to} \quad u < -a \quad \text{or} \quad u > a \qquad (4)$$

$$|u| \geq a \quad \text{is equivalent to} \quad u \leq -a \quad \text{or} \quad u \geq a \qquad (5)$$

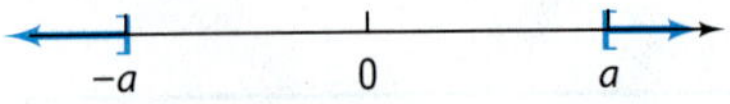

Figure 13 $|u| \geq a, a > 0$

See Figure 13 for an illustration of statement (5).

EXAMPLE 6 Solving an Inequality Involving Absolute Value

Solve the inequality $|2x - 5| > 3$, and graph the solution set.

Solution

$|2x - 5| > 3$ — This follows the form of statement (4); the expression $u = 2x - 5$ is inside the absolute value bars.

$2x - 5 < -3$ or $2x - 5 > 3$ — Apply statement (4).

$2x - 5 + 5 < -3 + 5$ or $2x - 5 + 5 > 3 + 5$ — Add 5 to each part.

$2x < 2$ or $2x > 8$ — Simplify.

$\dfrac{2x}{2} < \dfrac{2}{2}$ or $\dfrac{2x}{2} > \dfrac{8}{2}$ — Divide each part by 2.

$x < 1$ or $x > 4$ — Simplify.

The solution set is $\{x \mid x < 1 \text{ or } x > 4\}$, that is, all x in $(-\infty, 1) \cup (4, \infty)$. See Figure 14 for the graph of the solution set.

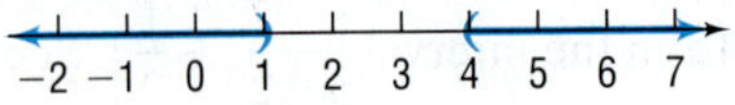

Figure 14 $|2x - 5| > 3$

WARNING A common error to be avoided is to attempt to write the solution $x < 1$ or $x > 4$ as the combined inequality $1 > x > 4$, which is incorrect, since there are no numbers x for which $1 > x$ and $x > 4$. Another common error is to "mix" the symbols and write $1 < x > 4$, which makes no sense.

Now Work PROBLEM 45

*Recall that the symbol $\cup$ stands for the union of two sets. Refer to page 2 if necessary.

1.6 Assess Your Understanding

'Are You Prepared?' *Answers are given at the end of these exercises. If you get a wrong answer, read the pages listed in red.*

1. $|-2| =$ ____ (p. 19)

2. ***True or False*** $|x| \geq 0$ for any real number x. (p. 19)

Concepts and Vocabulary

3. The solution set of the equation $|x| = 5$ is { ________ }.
4. The solution set of the inequality $|x| < 5$ is $\{x \mid$ ________ $\}$.
5. ***True or False*** The equation $|x| = -2$ has no solution.
6. ***True or False*** The inequality $|x| \geq -2$ has the set of real numbers as its solution set.
7. Which of the following pairs of inequalities is equivalent to $|x| > 4$?
   (a) $x > -4$ and $x < 4$ (b) $x < -4$ and $x < 4$
   (c) $x > -4$ or $x > 4$ (d) $x < -4$ or $x > 4$
8. Which of the following has no solution?
   (a) $|x| < -5$ (b) $|x| \leq 0$ (c) $|x| > 0$ (d) $|x| \geq 0$

Skill Building

In Problems 9–36, find the real solutions, if any, of each equation.

9. $|2x| = 6$ **10.** $|3x| = 12$ **11.** $|2x + 3| = 5$ **12.** $|3x - 1| = 2$

13. $|1 - 4t| + 8 = 13$ **14.** $|1 - 2z| + 6 = 9$ **15.** $|-2x| = |8|$ **16.** $|-x| = |1|$

17. $|-2|x = 4$ **18.** $|3|x = 9$ **19.** $\frac{2}{3}|x| = 9$ **20.** $\frac{3}{4}|x| = 9$

21. $\left|\frac{x}{3} + \frac{2}{5}\right| = 2$ **22.** $\left|\frac{x}{2} - \frac{1}{3}\right| = 1$ **23.** $|u - 2| = -\frac{1}{2}$ **24.** $|2 - v| = -1$

25. $4 - |2x| = 3$ **26.** $5 - \left|\frac{1}{2}x\right| = 3$ **27.** $|x^2 - 9| = 0$ **28.** $|x^2 - 16| = 0$

29. $|x^2 - 2x| = 3$ **30.** $|x^2 + x| = 12$ **31.** $|x^2 + x - 1| = 1$ **32.** $|x^2 + 3x - 2| = 2$

33. $\left|\frac{3x - 2}{2x - 3}\right| = 2$ **34.** $\left|\frac{2x + 1}{3x + 4}\right| = 1$ **35.** $|x^2 + 3x| = |x^2 - 2x|$ **36.** $|x^2 - 2x| = |x^2 + 6x|$

In Problems 37–64, solve each inequality. Express your answer using set notation or interval notation. Graph the solution set.

37. $|2x| < 8$ **38.** $|3x| < 15$ **39.** $|3x| > 12$ **40.** $|2x| > 6$

41. $|x - 2| + 2 < 3$ **42.** $|x + 4| + 3 < 5$ **43.** $|3t - 2| \le 4$ **44.** $|2u + 5| \le 7$

45. $|2x - 3| \ge 2$ **46.** $|3x + 4| \ge 2$ **47.** $|1 - 4x| - 7 < -2$ **48.** $|1 - 2x| - 4 < -1$

49. $|1 - 2x| > 3$ **50.** $|2 - 3x| > 1$ **51.** $|-4x| + |-5| \le 1$ **52.** $|-x| - |4| \le 2$

53. $|-2x| > |-3|$ **54.** $|-x - 2| \ge 1$ **55.** $-|2x - 1| \ge -3$ **56.** $-|1 - 2x| \ge -3$

57. $|2x| < -1$ **58.** $|3x| \ge 0$ **59.** $|5x| \ge -1$ **60.** $|6x| < -2$

61. $\left|\frac{2x + 3}{3} - \frac{1}{2}\right| < 1$ **62.** $3 - |x + 1| < \frac{1}{2}$ **63.** $5 + |x - 1| > \frac{1}{2}$ **64.** $\left|\frac{2x - 3}{2} + \frac{1}{3}\right| > 1$

Applications and Extensions

65. Body Temperature "Normal" human body temperature is 98.6°F. If a temperature x that differs from normal by at least 1.5° is considered unhealthy, write the condition for an unhealthy temperature x as an inequality involving an absolute value, and solve for x.

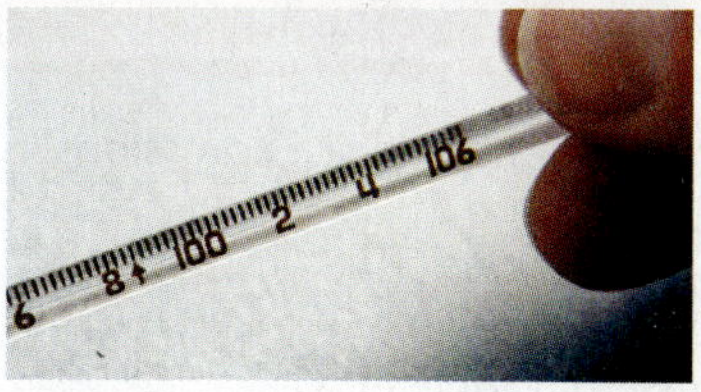

66. Household Voltage In the United States, normal household voltage is 110 volts. However, it is not uncommon for actual voltage to differ from normal voltage by at most 5 volts. Express this situation as an inequality involving an absolute value. Use x as the actual voltage and solve for x.

67. Reading Books A HuffPost/YouGov poll conducted September 27–28, 2013, found that Americans read an average of 13.6 books per year. Suppose HuffPost/YouGov is 99% confident that the result from this poll is off by fewer than 1.8 books from the actual average x. Express this situation as an inequality involving absolute value, and solve the inequality for x to determine the interval in which the actual average is likely to fall. [**Note:** In statistics, this interval is called a 99% **confidence interval.**]

68. Speed of Sound According to data from the Hill Aerospace Museum (Hill Air Force Base, Utah), the speed of sound varies depending on altitude, barometric pressure, and temperature. For example, at 20,000 feet, 13.75 inches of mercury, and −12.3°F, the speed of sound is about 707 miles per hour, but the speed can vary from this result by as much as 55 miles per hour as conditions change.

(a) Express this situation as an inequality involving an absolute value.

(b) Using x for the speed of sound, solve for x to find an interval for the speed of sound.

69. Express the fact that x differs from 3 by less than $\frac{1}{2}$ as an inequality involving an absolute value. Solve for x.

70. Express the fact that x differs from -4 by less than 1 as an inequality involving an absolute value. Solve for x.

71. Express the fact that x differs from -3 by more than 2 as an inequality involving an absolute value. Solve for x.

72. Express the fact that x differs from 2 by more than 3 as an inequality involving an absolute value. Solve for x.

In Problems 73–78, find a and b.

73. If $|x - 1| < 3$, then $a < x + 4 < b$.

74. If $|x + 2| < 5$, then $a < x - 2 < b$.

75. If $|x + 4| \le 2$, then $a \le 2x - 3 \le b$.

76. If $|x - 3| \le 1$, then $a \le 3x + 1 \le b$.

77. If $|x - 2| \le 7$, then $a \le \dfrac{1}{x - 10} \le b$.

78. If $|x + 1| \le 3$, then $a \le \dfrac{1}{x + 5} \le b$.

79. Show that: if $a > 0$, $b > 0$, and $\sqrt{a} < \sqrt{b}$, then $a < b$.
[**Hint:** $b - a = (\sqrt{b} - \sqrt{a})(\sqrt{b} + \sqrt{a})$.]

80. Show that $a \le |a|$.

81. Prove the *triangle inequality* $|a + b| \le |a| + |b|$.
[**Hint:** Expand $|a + b|^2 = (a + b)^2$, and use the result of Problem 80.]

82. Prove that $|a - b| \ge |a| - |b|$.
[**Hint:** Apply the triangle inequality from Problem 81 to $|a| = |(a - b) + b|$.]

83. If $a > 0$, show that the solution set of the inequality
$$x^2 < a$$
consists of all numbers x for which
$$-\sqrt{a} < x < \sqrt{a}$$

84. If $a > 0$, show that the solution set of the inequality
$$x^2 > a$$
consists of all numbers x for which
$$x < -\sqrt{a} \quad \text{or} \quad x > \sqrt{a}$$

In Problems 85–92, use the results found in Problems 83 and 84 to solve each inequality.

85. $x^2 < 1$

86. $x^2 < 4$

87. $x^2 \ge 9$

88. $x^2 \ge 1$

89. $x^2 \le 16$

90. $x^2 \le 9$

91. $x^2 > 4$

92. $x^2 > 16$

93. Solve $|3x - |2x + 1|| = 4$.

94. Solve $|x + |3x - 2|| = 2$.

Explaining Concepts: Discussion and Writing

95. The equation $|x| = -2$ has no solution. Explain why.

96. The inequality $|x| > -0.5$ has all real numbers as solutions. Explain why.

97. The inequality $|x| > 0$ has as solution set $\{x | x \ne 0\}$. Explain why.

'Are You Prepared?' Answers

1. 2

2. True

1.7 Problem Solving: Interest, Mixture, Uniform Motion, Constant Rate Job Applications

OBJECTIVES

1 Translate Verbal Descriptions into Mathematical Expressions (p. 135)
2 Solve Interest Problems (p. 136)
3 Solve Mixture Problems (p. 137)
4 Solve Uniform Motion Problems (p. 138)
5 Solve Constant Rate Job Problems (p. 140)

Applied (word) problems do not come in the form "Solve the equation . . . . " Instead, they supply information using words, a verbal description of the real problem. So, to solve applied problems, we must be able to translate the verbal description into the language of mathematics. This can be done by using variables to represent unknown quantities and then finding relationships (such as equations) that involve these variables. The process of doing all this is called **mathematical modeling**. An equation or inequality that describes a relationship among the variables is called a **model**.

Any solution to the mathematical problem must be checked against the mathematical problem, the verbal description, and the real problem. See Figure 15 for an illustration of the **modeling process**.

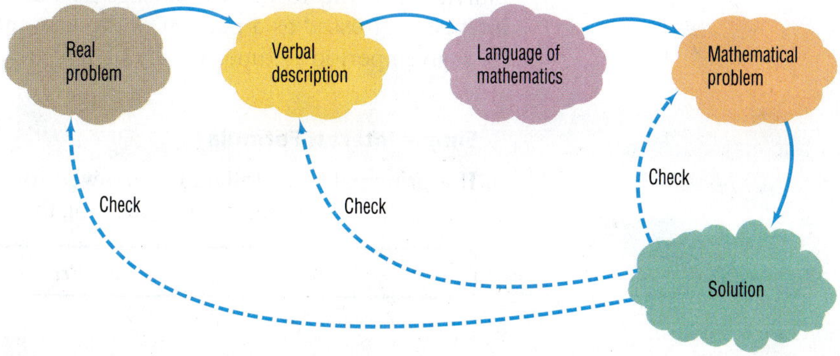

Figure 15

1 Translate Verbal Descriptions into Mathematical Expressions

EXAMPLE 1 **Translating Verbal Descriptions into Mathematical Expressions**

(a) For uniform motion, the average speed of an object equals the distance traveled divided by the time required.

Translation: If r is the speed, d the distance, and t the time, then $r = \frac{d}{t}$.

(b) Let x denote a number.
The number 5 times as large as x is $5x$.
The number 3 less than x is $x - 3$.
The number that exceeds x by 4 is $x + 4$.
The number that, when added to x, gives 5 is $5 - x$.

Now Work PROBLEM 9

Always check the units used to measure the variables of an applied problem. In Example (1a), if r is measured in miles per hour, then the distance d must be expressed in miles, and the time t must be expressed in hours. It is a good practice to check units to be sure that they are consistent and make sense.

The steps to follow for solving applied problems, given earlier, are repeated next.

Steps for Solving Applied Problems

STEP 1: Read the problem carefully, perhaps two or three times. Pay particular attention to the question being asked in order to identify what you are looking for. Identify any relevant formulas you may need ($d = rt$, $A = \pi r^2$, etc.). If you can, determine realistic possibilities for the answer.

STEP 2: Assign a letter (variable) to represent what you are looking for, and if necessary, express any remaining unknown quantities in terms of this variable.

STEP 3: Make a list of all the known facts, and translate them into mathematical expressions. These may take the form of an equation or an inequality involving the variable. If possible, draw an appropriately labeled diagram to assist you. Sometimes, creating a table or chart helps.

STEP 4: Solve for the variable, and then answer the question.

STEP 5: Check the answer with the facts in the problem. If it agrees, congratulations! If it does not agree, try again.

2 Solve Interest Problems

Interest is money paid for the use of money. The total amount borrowed (whether by an individual from a bank in the form of a loan, or by a bank from an individual in the form of a savings account) is called the **principal**. The **rate of interest**, expressed as a percent, is the amount charged for the use of the principal for a given period of time, usually on a yearly (that is, per annum) basis.

Simple Interest Formula

If a principal of P dollars is borrowed for a period of t years at a per annum interest rate r, expressed as a decimal, the interest I charged is

$$I = Prt \tag{1}$$

Interest charged according to formula (1) is called **simple interest**. When using formula (1), be sure to express r as a decimal. For example, if the rate of interest is 4%, then $r = 0.04$.

EXAMPLE 2 **Finance: Computing Interest on a Loan**

Suppose that Juanita borrows \$500 for 6 months at the simple interest rate of 9% per annum. What is the interest that Juanita will be charged on the loan? How much does Juanita owe after 6 months?

Solution The rate of interest is given per annum, so the actual time that the money is borrowed must be expressed in years. The interest charged would be the principal, \$500, times the rate of interest ($9\% = 0.09$), times the time in years, $\frac{1}{2}$:

$$\text{Interest charged} = I = Prt = (500)(0.09)\left(\frac{1}{2}\right) = \$22.50$$

After 6 months, Juanita will owe what she borrowed plus the interest:

$$\$500 + \$22.50 = \$522.50$$

EXAMPLE 3 **Financial Planning**

Candy has \$70,000 to invest and wants an annual return of \$2800, which requires an overall rate of return of 4%. She can invest in a safe, government-insured certificate of deposit, but it pays only 2%. To obtain 4%, she agrees to invest some of her money in noninsured corporate bonds paying 7%. How much should be placed in each investment to achieve her goal?

Solution **STEP 1:** The question is asking for two dollar amounts: the principal to invest in the corporate bonds and the principal to invest in the certificate of deposit.

STEP 2: Let b represent the amount (in dollars) to be invested in the bonds. Then $70{,}000 - b$ is the amount that will be invested in the certificate. (Do you see why?)

STEP 3: We set up a table:

	Principal (\$)	Rate	Time (yr)	Interest (\$)
Bonds	b	$7\% = 0.07$	1	$0.07b$
Certificate	$70{,}000 - b$	$2\% = 0.02$	1	$0.02(70{,}000 - b)$
Total	70,000	$4\% = 0.04$	1	$0.04(70{,}000) = 2800$

Since the combined interest from the investments is equal to the total interest, we have

$$\text{Bond interest} + \text{Certificate interest} = \text{Total interest}$$
$$0.07b + 0.02(70{,}000 - b) = 2800$$

(Note that the units are consistent: the unit is dollars on each side.)

STEP 4: $0.07b + 1400 - 0.02b = 2800$
$$0.05b = 1400$$
$$b = 28{,}000$$

Candy should place \$28,000 in the bonds and \$70,000 − \$28,000 = \$42,000 in the certificate.

✓ **STEP 5:** The interest on the bonds after 1 year is 0.07(\$28,000) = \$1960; the interest on the certificate after 1 year is 0.02(\$42,000) = \$840. The total annual interest is \$2800, the required amount. ●

Now Work PROBLEM 19

3 Solve Mixture Problems

Oil refineries sometimes produce gasoline that is a blend of two or more types of fuel; bakeries occasionally blend two or more types of flour for their bread. These problems are referred to as **mixture problems** because they combine two or more quantities to form a mixture.

EXAMPLE 4

Blending Coffees

The manager of a Starbucks store decides to experiment with a new blend of coffee. She will mix some B grade Colombian coffee that sells for \$5 per pound with some A grade Arabica coffee that sells for \$10 per pound to get 100 pounds of the new blend. The selling price of the new blend is to be \$7 per pound, and there is to be no difference in revenue between selling the new blend and selling the other types. How many pounds of the B grade Colombian coffee and how many pounds of the A grade Arabica coffees are required?

Solution Let c represent the number of pounds of the B grade Colombian coffee. Then $100 - c$ equals the number of pounds of the A grade Arabica coffee. See Figure 16.

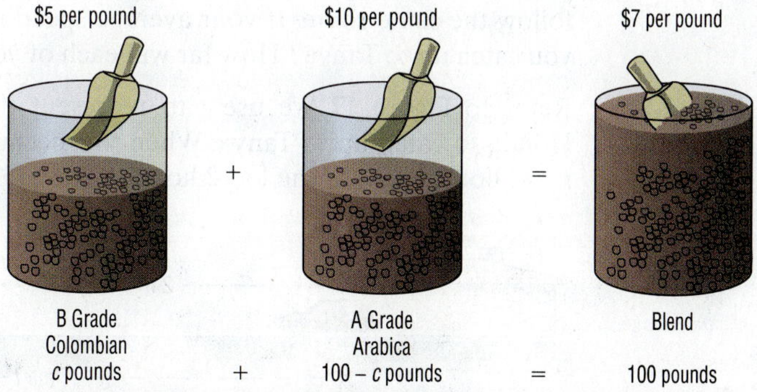

Figure 16

Since there is to be no difference in revenue between selling the A and B grades separately and selling the blend, we have

Revenue from B grade	+	Revenue from A grade	=	Revenue from blend
{Price per pound of B grade} {Pounds of B grade}	+	{Price per pound of A grade} {Pounds of A grade}	=	{Price per pound of blend} {Pounds of blend}
\$5 · c	+	\$10 · $(100 - c)$	=	\$7 · 100

Now solve the equation:

$$
\begin{aligned}
5c + 10(100 - c) &= 700 \\
5c + 1000 - 10c &= 700 \\
-5c &= -300 \\
c &= 60
\end{aligned}
$$

The manager should blend 60 pounds of B grade Colombian coffee with $100 - 60 = 40$ pounds of A grade Arabica coffee to get the desired blend.

✓**Check:** The 60 pounds of B grade coffee would sell for $(\$5)(60) = \$300$, and the 40 pounds of A grade coffee would sell for $(\$10)(40) = \$400$; the total revenue, \$700, equals the revenue obtained from selling the blend, as desired. ●

Now Work PROBLEM 23

4 Solve Uniform Motion Problems

Objects that move at a constant speed are said to be in **uniform motion**. When the average speed of an object is known, it can be interpreted as that object's constant speed. For example, a bicyclist traveling at an average speed of 25 miles per hour can be considered in uniform motion with a constant speed of 25 miles per hour.

Uniform Motion Formula

If an object moves at an average speed (rate) r, the distance d covered in time t is given by the formula

$$d = rt \qquad (2)$$

That is, Distance = Rate · Time.

EXAMPLE 5 **Physics: Uniform Motion**

Tanya, who is a long-distance runner, runs at an average speed of 8 miles per hour (mi/h). Two hours after Tanya leaves your house, you leave in your Honda and follow the same route. If your average speed is 40 mi/h, how long will it be before you catch up to Tanya? How far will each of you be from your home?

Solution Refer to Figure 17. We use t to represent the time (in hours) that it takes the Honda to catch up to Tanya. When this occurs, the total time elapsed for Tanya is $t + 2$ hours because she left 2 hours earlier.

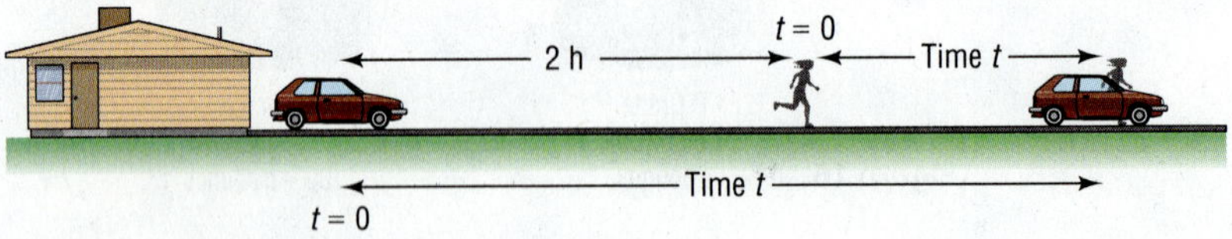

Figure 17

Set up the following table:

	Rate mi/h	Time h	Distance mi
Tanya	8	$t + 2$	$8(t + 2)$
Honda	40	t	$40t$

The distance traveled is the same for both, which leads to the equation

$$\begin{aligned} 8(t+2) &= 40t \\ 8t + 16 &= 40t \\ 32t &= 16 \\ t &= \frac{1}{2} \text{ hour} \end{aligned}$$

It will take the Honda $\frac{1}{2}$ hour to catch up to Tanya. Each will have gone 20 miles.

Check: In 2.5 hours, Tanya travels a distance of $(2.5)(8) = 20$ miles. In $\frac{1}{2}$ hour, the Honda travels a distance of $\left(\frac{1}{2}\right)(40) = 20$ miles.

EXAMPLE 6 Physics: Uniform Motion

A motorboat heads upstream a distance of 24 miles on a river whose current is running at 3 miles per hour (mi/h). The trip up and back takes 6 hours. Assuming that the motorboat maintained a constant speed relative to the water, what was its speed?

Solution See Figure 18. Use r to represent the constant speed of the motorboat relative to the water. Then the true speed going upstream is $r - 3$ mi/h, and the true speed going downstream is $r + 3$ mi/h. Since Distance = Rate · Time, then Time $= \dfrac{\text{Distance}}{\text{Rate}}$. Set up a table.

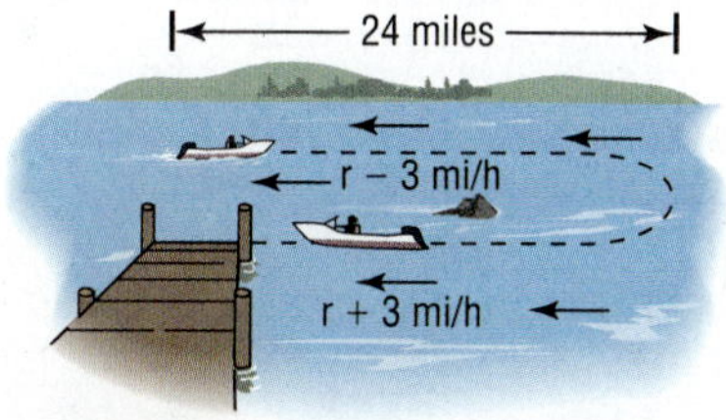

Figure 18

	Rate mi/h	Distance mi	Time h = Distance / Rate
Upstream	$r - 3$	24	$\frac{24}{r-3}$
Downstream	$r + 3$	24	$\frac{24}{r+3}$

The total time up and back is 6 hours, which gives the equation

$$\frac{24}{r-3} + \frac{24}{r+3} = 6$$

$$\frac{24(r+3) + 24(r-3)}{(r-3)(r+3)} = 6 \qquad \text{Add the quotients on the left.}$$

$$\frac{48r}{r^2 - 9} = 6 \qquad \text{Simplify.}$$

$$48r = 6(r^2 - 9) \qquad \text{Multiply both sides by } r^2 - 9.$$

$$6r^2 - 48r - 54 = 0 \qquad \text{Place in standard form.}$$

$$r^2 - 8r - 9 = 0 \qquad \text{Divide by 6.}$$

$$(r-9)(r+1) = 0 \qquad \text{Factor.}$$

$$r = 9 \text{ or } r = -1 \qquad \text{Apply the Zero-Product Property and solve.}$$

Discard the solution $r = -1$ mi/h and conclude that the speed of the motorboat relative to the water is 9 mi/h.

Now Work PROBLEM 29

5 Solve Constant Rate Job Problems

Here we look at jobs that are performed at a **constant rate**. The assumption is that if a job can be done in t units of time, then $\frac{1}{t}$ of the job is done in 1 unit of time. In other words, if a job takes 4 hours, then $\frac{1}{4}$ of the job is done in 1 hour.

EXAMPLE 7 **Working Together to Do a Job**

At 10 AM Danny is asked by his father to weed the garden. From past experience, Danny knows that this will take him 4 hours, working alone. His older brother Mike, when it is his turn to do this job, requires 6 hours. Since Mike wants to go golfing with Danny and has a reservation for 1 PM, he agrees to help Danny. Assuming no gain or loss of efficiency, when will they finish if they work together? Can they make the golf date?

Solution Set up Table 2. In 1 hour, Danny does $\frac{1}{4}$ of the job, and in 1 hour, Mike does $\frac{1}{6}$ of the job. Let t be the time (in hours) that it takes them to do the job together. In 1 hour, then, $\frac{1}{t}$ of the job is completed. Reason as follows:

Table 2

	Hours to Do Job	Part of Job Done in 1 Hour
Danny	4	$\frac{1}{4}$
Mike	6	$\frac{1}{6}$
Together	t	$\frac{1}{t}$

$$\left(\begin{array}{c}\text{Part done by Danny}\\ \text{in 1 hour}\end{array}\right)+\left(\begin{array}{c}\text{Part done by Mike}\\ \text{in 1 hour}\end{array}\right)=\left(\begin{array}{c}\text{Part done together}\\ \text{in 1 hour}\end{array}\right)$$

From Table 2,

$$\frac{1}{4}+\frac{1}{6}=\frac{1}{t} \quad \text{The model.}$$

$$\frac{3}{12}+\frac{2}{12}=\frac{1}{t} \quad \text{LCD} = 12 \text{ on the left.}$$

$$\frac{5}{12}=\frac{1}{t} \quad \text{Simplify.}$$

$$5t = 12 \quad \text{Multiply both sides by } 12t.$$

$$t=\frac{12}{5} \quad \text{Divide each side by 5.}$$

Working together, Mike and Danny can do the job in $\frac{12}{5}$ hours, or 2 hours, 24 minutes. They should make the golf date, since they will finish at 12:24 PM.

Now Work PROBLEM 35

1.7 Assess Your Understanding

Concepts and Vocabulary

1. The process of using variables to represent unknown quantities and then finding relationships that involve these variables is referred to as ________ ________.
2. The money paid for the use of money is ________.
3. Objects that move at a constant speed are said to be in ________ ________.
4. ***True or False*** The amount charged for the use of principal for a given period of time is called the rate of interest.
5. ***True or False*** If an object moves at an average speed r, the distance d covered in time t is given by the formula $d = rt$.
6. Suppose that you want to mix two coffees in order to obtain 100 pounds of a blend. If x represents the number of pounds of coffee A, write an algebraic expression that represents the number of pounds of coffee B.

   (a) $100 - x$ (b) $x - 100$ (c) $100x$ (d) $100 + x$
7. Which of the following is the simple interest formula?

   (a) $I = \frac{rt}{P}$ (b) $I = Prt$ (c) $I = \frac{P}{rt}$ (d) $I = P + rt$
8. If it takes 5 hours to complete a job, what fraction of the job is done in 1 hour?

   (a) $\frac{4}{5}$ (b) $\frac{5}{4}$ (c) $\frac{1}{5}$ (d) $\frac{1}{4}$

Applications and Extensions

In Problems 9–18, translate each sentence into a mathematical equation. Be sure to identify the meaning of all symbols.

9. **Geometry** The area of a circle is the product of the number π and the square of the radius.

10. **Geometry** The circumference of a circle is the product of the number π and twice the radius.

11. **Geometry** The area of a square is the square of the length of a side.

12. **Geometry** The perimeter of a square is four times the length of a side.

13. **Physics** Force equals the product of mass and acceleration.

14. **Physics** Pressure is force per unit area.

15. **Physics** Work equals force times distance.

16. **Physics** Kinetic energy is one-half the product of the mass and the square of the velocity.

17. **Business** The total variable cost of manufacturing x dishwashers is \$150 per dishwasher times the number of dishwashers manufactured.

18. **Business** The total revenue derived from selling x dishwashers is \$250 per dishwasher times the number of dishwashers sold.

19. **Financial Planning** Betsy, a recent retiree, requires \$6000 per year in extra income. She has \$50,000 to invest and can invest in B-rated bonds paying 15% per year or in a certificate of deposit (CD) paying 7% per year. How much money should Betsy invest in each to realize exactly \$6000 in interest per year.

20. **Financial Planning** After 2 years, Betsy (see Problem 19) finds that she will now require \$7000 per year. Assuming that the remaining information is the same, how should the money be reinvested?

21. **Banking** A bank loaned out \$12,000, part of it at the rate of 8% per year and the rest at the rate of 18% per year. If the interest received totaled \$1000, how much was loaned at 8%?

22. **Banking** Wendy, a loan officer at a bank, has \$1,000,000 to lend and is required to obtain an average return of 18% per year. If she can lend at the rate of 19% or at the rate of 16%, how much can she lend at the 16% rate and still meet her requirement?

23. **Blending Teas** The manager of a store that specializes in selling tea decides to experiment with a new blend. She will mix some Earl Grey tea that sells for \$5 per pound with some Orange Pekoe tea that sells for \$3 per pound to get 100 pounds of the new blend. The selling price of the new blend is to be \$4.50 per pound, and there is to be no difference in revenue between selling the new blend and selling the other types. How many pounds of the Earl Grey tea and of the Orange Pekoe tea are required?

24. **Business: Blending Coffee** A coffee manufacturer wants to market a new blend of coffee that sells for \$3.90 per pound by mixing two coffees that sell for \$2.75 and \$5 per pound, respectively. What amounts of each coffee should be blended to obtain the desired mixture?

[**Hint:** Assume that the total weight of the desired blend is 100 pounds.]

25. **Business: Mixing Nuts** A nut store normally sells cashews for \$9.00 per pound and almonds for \$3.50 per pound. But at the end of the month the almonds had not sold well, so, in order to sell 60 pounds of almonds, the manager decided to mix the 60 pounds of almonds with some cashews and sell the mixture for \$7.50 per pound. How many pounds of cashews should be mixed with the almonds to ensure no change in the revenue?

26. **Business: Mixing Candy** A candy store sells boxes of candy containing caramels and cremes. Each box sells for \$12.50 and holds 30 pieces of candy (all pieces are the same size). If the caramels cost \$0.25 to produce and the cremes cost \$0.45 to produce, how many of each should be in a box to yield a profit of \$3?

27. **Physics: Uniform Motion** A motorboat can maintain a constant speed of 16 miles per hour relative to the water. The boat travels upstream to a certain point in 20 minutes; the return trip takes 15 minutes. What is the speed of the current? See the figure.

28. **Physics: Uniform Motion** A motorboat heads upstream on a river that has a current of 3 miles per hour. The trip upstream takes 5 hours, and the return trip takes 2.5 hours. What is the speed of the motorboat? (Assume that the boat maintains a constant speed relative to the water.)

29. **Physics: Uniform Motion** A motorboat maintained a constant speed of 15 miles per hour relative to the water in going 10 miles upstream and then returning. The total time for the trip was 1.5 hours. Use this information to find the speed of the current.

30. **Physics: Uniform Motion** Two cars enter the Florida Turnpike at Commercial Boulevard at 8:00 AM, each heading for Wildwood. One car's average speed is 10 miles per hour more than the other's. The faster car arrives at Wildwood at 11:00 AM, $\frac{1}{2}$ hour before the other car. What was the average speed of each car? How far did each travel?

31. **Moving Walkways** The speed of a moving walkway is typically about 2.5 feet per second. Walking on such a moving walkway, it takes Karen a total of 40 seconds to travel 50 feet with the movement of the walkway and then back again against the movement of the walkway. What is Karen's normal walking speed?

Source: Answers.com

32. High-Speed Walkways Toronto's Pearson International Airport has a high-speed version of a moving walkway. If Liam walks while riding this moving walkway, he can travel 280 meters in 60 seconds less time than if he stands still on the moving walkway. If Liam walks at a normal rate of 1.5 meters per second, what is the speed of the walkway?

Source: *Answers.com*

33. Tennis A regulation doubles tennis court has an area of 2808 square feet. If it is 6 feet longer than twice its width, determine the dimensions of the court.

Source: *United States Tennis Association*

34. Laser Printers It takes an HP LaserJet M451dw laser printer 16 minutes longer to complete an 840-page print job by itself than it takes an HP LaserJet CP4025dn to complete the same job by itself. Together the two printers can complete the job in 15 minutes. How long does it take each printer to complete the print job alone? What is the speed of each printer?

Source: Hewlett-Packard

35. Working Together on a Job Trent can deliver his newspapers in 30 minutes. It takes Lois 20 minutes to do the same route. How long would it take them to deliver the newspapers if they worked together?

36. Working Together on a Job Patrice, by himself, can paint four rooms in 10 hours. If he hires April to help, they can do the same job together in 6 hours. If he lets April work alone, how long will it take her to paint four rooms?

37. Enclosing a Garden A gardener has 46 feet of fencing to be used to enclose a rectangular garden that has a border 2 feet wide surrounding it. See the figure.

(a) If the length of the garden is to be twice its width, what will be the dimensions of the garden?
(b) What is the area of the garden?
(c) If the length and width of the garden are to be the same, what will be the dimensions of the garden?
(d) What will be the area of the square garden?

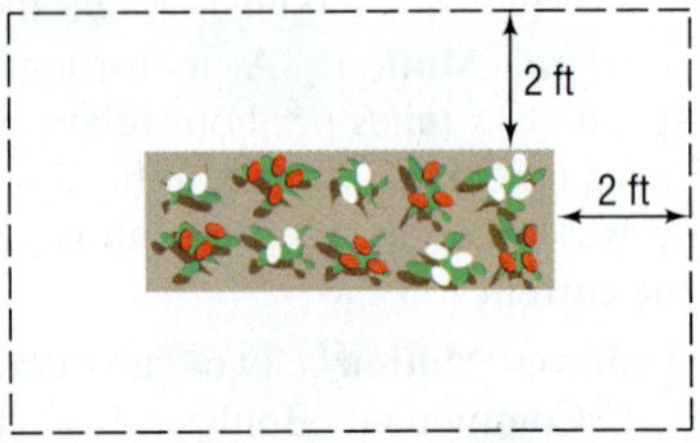

38. Construction A pond is enclosed by a wooden deck that is 3 feet wide. The fence surrounding the deck is 100 feet long.

(a) If the pond is square, what are its dimensions?
(b) If the pond is rectangular and the length of the pond is to be three times its width, what are its dimensions?
(c) If the pond is circular, what is its diameter?
(d) Which pond has the larger area?

39. Football A tight end can run the 100-yard dash in 12 seconds. A defensive back can do it in 10 seconds. The tight end catches a pass at his own 20-yard line with the defensive back at the 15-yard line. (See the figure-top, right.) If no other players are nearby, at what yard line will the defensive back catch up to the tight end?

[**Hint:** At time $t = 0$, the defensive back is 5 yards behind the tight end.]

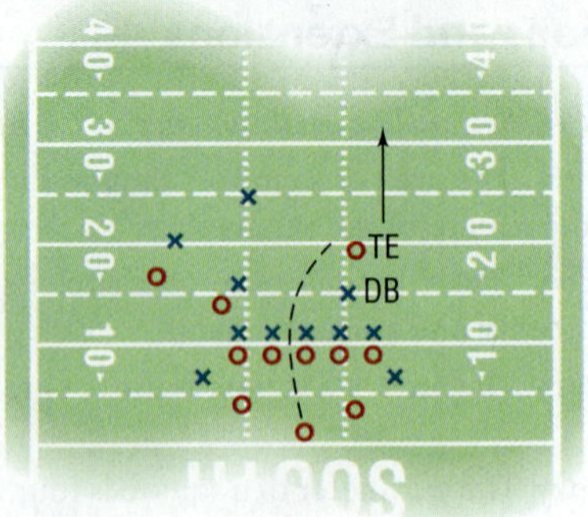

40. Computing Business Expense Therese, an outside salesperson, uses her car for both business and pleasure. Last year, she traveled 30,000 miles, using 900 gallons of gasoline. Her car gets 40 miles per gallon on the highway and 25 in the city. She can deduct all highway travel, but no city travel, on her taxes. How many miles should Therese deduct as a business expense?

41. Mixing Water and Antifreeze How much water should be added to 1 gallon of pure antifreeze to obtain a solution that is 60% antifreeze?

42. Mixing Water and Antifreeze The cooling system of a certain foreign-made car has a capacity of 15 liters. If the system is filled with a mixture that is 40% antifreeze, how much of this mixture should be drained and replaced by pure antifreeze so that the system is filled with a solution that is 60% antifreeze?

43. Chemistry: Salt Solutions How much water must be evaporated from 32 ounces of a 4% salt solution to make a 6% salt solution?

44. Chemistry: Salt Solutions How much water must be evaporated from 240 gallons of a 3% salt solution to produce a 5% salt solution?

45. Purity of Gold The purity of gold is measured in karats, with pure gold being 24 karats. Other purities of gold are expressed as proportional parts of pure gold. Thus, 18-karat gold is $\frac{18}{24}$, or 75% pure gold; 12-karat gold is $\frac{12}{24}$, or 50% pure gold; and so on. How much 12-karat gold should be mixed with pure gold to obtain 60 grams of 16-karat gold?

46. Chemistry: Sugar Molecules A sugar molecule has twice as many atoms of hydrogen as it does oxygen and one more atom of carbon than of oxygen. If a sugar molecule has a total of 45 atoms, how many are oxygen? How many are hydrogen?

47. Running a Race Mike can run the mile in 6 minutes, and Dan can run the mile in 9 minutes. If Mike gives Dan a head start of 1 minute, how far from the start will Mike pass Dan? How long does it take? See the figure.

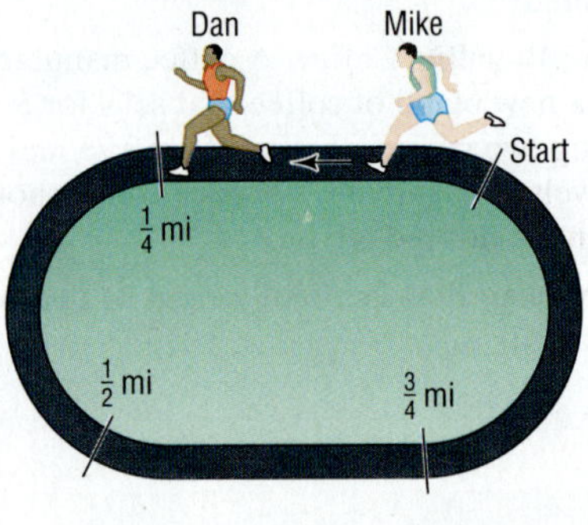

48. Range of an Airplane An air rescue plane averages 300 miles per hour in still air. It carries enough fuel for 5 hours of flying time. If, upon takeoff, it encounters a head wind of 30 mi/h, how far can it fly and return safely? (Assume that the wind remains constant.)

49. Emptying Oil Tankers An oil tanker can be emptied by the main pump in 4 hours. An auxiliary pump can empty the tanker in 9 hours. If the main pump is started at 9 AM, when should the auxiliary pump be started so that the tanker is emptied by noon?

50. Cement Mix A 20-pound bag of Economy brand cement mix contains 25% cement and 75% sand. How much pure cement must be added to produce a cement mix that is 40% cement?

51. Emptying a Tub A bathroom tub will fill in 15 minutes with both faucets open and the stopper in place. With both faucets closed and the stopper removed, the tub will empty in 20 minutes. How long will it take for the tub to fill if both faucets are open and the stopper is removed?

52. Using Two Pumps A 5-horsepower (hp) pump can empty a pool in 5 hours. A smaller, 2-hp pump empties the same pool in 8 hours. The pumps are used together to begin emptying this pool. After two hours, the 2-hp pump breaks down. How long will it take the larger pump to finish emptying the pool?

53. A Biathlon Suppose that you have entered an 87-mile biathlon that consists of a run and a bicycle race. During your run, your average speed is 6 miles per hour, and during your bicycle race, your average speed is 25 miles per hour. You finish the race in 5 hours. What is the distance of the run? What is the distance of the bicycle race?

54. Cyclists Two cyclists leave a city at the same time, one going east and the other going west. The westbound cyclist bikes 5 mph faster than the eastbound cyclist. After 6 hours they are 246 miles apart. How fast is each cyclist riding?

55. Comparing Olympic Heroes In the 2012 Olympics, Usain Bolt of Jamaica won the gold medal in the 100-meter race with a time of 9.69 seconds. In the 1896 Olympics, Thomas Burke of the United States won the gold medal in the 100-meter race in 12.0 seconds. If they ran in the same race, repeating their respective times, by how many meters would Bolt beat Burke?

56. Constructing a Coffee Can A 39-ounce can of Hills Bros.® coffee requires 188.5 square inches of aluminum. If its height is 7 inches, what is its radius? **[Hint:** The surface area S of a right cylinder is $S = 2\pi r^2 + 2\pi rh$, where r is the radius and h is the height.]

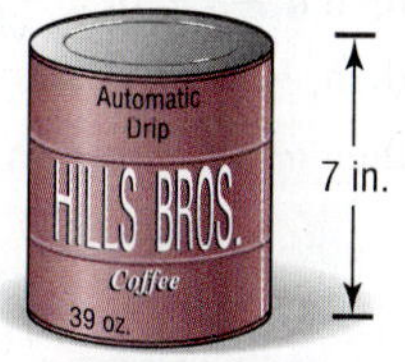

Explaining Concepts: Discussion and Writing

57. Critical Thinking You are the manager of a clothing store and have just purchased 100 dress shirts for $20.00 each. After 1 month of selling the shirts at the regular price, you plan to have a sale giving 40% off the original selling price. However, you still want to make a profit of $4 on each shirt at the sale price. What should you price the shirts at initially to ensure this? If, instead of 40% off at the sale, you give 50% off, by how much is your profit reduced?

58. Critical Thinking Make up a word problem that requires solving a linear equation as part of its solution. Exchange problems with a friend. Write a critique of your friend's problem.

59. Critical Thinking Without solving, explain what is wrong with the following mixture problem: How many liters of 25% ethanol should be added to 20 liters of 48% ethanol to obtain a solution of 58% ethanol? Now go through an algebraic solution. What happens?

60. Computing Average Speed In going from Chicago to Atlanta, a car averages 45 miles per hour, and in going from Atlanta to Miami, it averages 55 miles per hour. If Atlanta is halfway between Chicago and Miami, what is the average speed from Chicago to Miami? Discuss an intuitive solution. Write a paragraph defending your intuitive solution. Then solve the problem algebraically. Is your intuitive solution the same as the algebraic one? If not, find the flaw.

61. Speed of a Plane On a recent flight from Phoenix to Kansas City, a distance of 919 nautical miles, the plane arrived 20 minutes early. On leaving the aircraft, I asked the captain, "What was our tail wind?" He replied, "I don't know, but our ground speed was 550 knots." Has enough information been provided for you to find the tail wind? If possible, find the tail wind. (1 knot = 1 nautical mile per hour)

Chapter Review

Things to Know

Quadratic formula (pp. 97 and 110)

If $ax^2 + bx + c = 0, a \neq 0$, then $x = \dfrac{-b \pm \sqrt{b^2 - 4ac}}{2a}$.

If $b^2 - 4ac < 0$, there are no real solutions.

Discriminant (pp. 97 and 110)

If $b^2 - 4ac > 0$, there are two unequal real solutions.

If $b^2 - 4ac = 0$, there is one repeated real solution, a double root.

If $b^2 - 4ac < 0$, there are no real solutions, but there are two distinct complex solutions that are not real; the complex solutions are conjugates of each other.

Interval notation (p. 120)

$[a, b]$	$\{x \mid a \le x \le b\}$	$[a, \infty)$	$\{x \mid x \ge a\}$
$[a, b)$	$\{x \mid a \le x < b\}$	(a, ∞)	$\{x \mid x > a\}$
$(a, b]$	$\{x \mid a < x \le b\}$	$(-\infty, a]$	$\{x \mid x \le a\}$
(a, b)	$\{x \mid a < x < b\}$	$(-\infty, a)$	$\{x \mid x < a\}$
		$(-\infty, \infty)$	All real numbers

Properties of inequalities

Addition property (p. 121)
If $a < b$, then $a + c < b + c$.
If $a > b$, then $a + c > b + c$.

Multiplication properties (p. 122)
(a) If $a < b$ and if $c > 0$, then $ac < bc$.
If $a < b$ and if $c < 0$, then $ac > bc$.
(b) If $a > b$ and if $c > 0$, then $ac > bc$.
If $a > b$ and if $c < 0$, then $ac < bc$.

Reciprocal properties (p. 123)
If $a > 0$, then $\frac{1}{a} > 0$.
If $a < 0$, then $\frac{1}{a} < 0$.
If $\frac{1}{a} > 0$, then $a > 0$.
If $\frac{1}{a} < 0$, then $a < 0$.

Absolute value

If $|u| = a$, $a > 0$, then $u = -a$ or $u = a$. (p. 130)

If $|u| \le a$, $a > 0$, then $-a \le u \le a$. (p. 131)

If $|u| \ge a$, $a > 0$, then $u \le -a$ or $u \ge a$. (p. 132)

Objectives

Section		You should be able to . . .	Examples	Review Exercises
1.1	1	Solve a linear equation (p. 84)	1–3	1–3, 7, 8
	2	Solve equations that lead to linear equations (p. 86)	4–6	4
	3	Solve problems that can be modeled by linear equations (p. 87)	8, 9	45, 57
1.2	1	Solve a quadratic equation by factoring (p. 93)	1, 2	6, 9, 21, 22
	2	Solve a quadratic equation by completing the square (p. 95)	4, 5	5, 6, 9, 10, 13, 21, 22
	3	Solve a quadratic equation using the quadratic formula (p. 96)	6–9	5, 6, 9, 10, 13, 21, 22
	4	Solve problems that can be modeled by quadratic equations (p. 99)	10	50, 54, 56
1.3	1	Add, subtract, multiply, and divide complex numbers (p. 105)	1–8	35–39
	2	Solve quadratic equations in the complex number system (p. 109)	9–12	40–43
1.4	1	Solve radical equations (p. 113)	1–3	11, 12, 15–19, 23
	2	Solve equations quadratic in form (p. 114)	4–6	14, 20
	3	Solve equations by factoring (p. 116)	7, 8	26, 27
1.5	1	Use interval notation (p. 120)	1, 2	28–34
	2	Use properties of inequalities (p. 121)	3–6	28–34, 47
	3	Solve inequalities (p. 123)	7, 8	28, 47
	4	Solve combined inequalities (p. 124)	9, 10	29, 30
1.6	1	Solve equations involving absolute value (p. 130)	1	24, 25
	2	Solve inequalities involving absolute value (p. 130)	2–6	31–34
1.7	1	Translate verbal descriptions into mathematical expressions (p. 135)	1	44
	2	Solve interest problems (p. 136)	2, 3	45
	3	Solve mixture problems (p. 137)	4	53, 55
	4	Solve uniform motion problems (p. 138)	5, 6	46, 48, 49, 59, 60
	5	Solve constant rate job problems (p. 140)	7	51, 52, 58

Review Exercises

In Problems 1–27, find the real solutions, if any, of each equation. (Where they appear, a, b, m, and n are positive constants.)

1. $2 - \frac{x}{3} = 8$

2. $-2(5 - 3x) + 8 = 4 + 5x$

3. $\frac{3x}{4} - \frac{x}{3} = \frac{1}{12}$

4. $\frac{x}{x-1} = \frac{6}{5} \quad x \neq 1$

5. $x(1 - x) = 6$

6. $x(1 + x) = 6$

7. $\frac{1}{2}\left(x - \frac{1}{3}\right) = \frac{3}{4} - \frac{x}{6}$

8. $\frac{1 - 3x}{4} = \frac{x + 6}{3} + \frac{1}{2}$

9. $(x - 1)(2x + 3) = 3$

10. $2x + 3 = 4x^2$

11. $\sqrt[3]{x^2 - 1} = 2$

12. $\sqrt{1 + x^3} = 3$

13. $x(x + 1) + 2 = 0$

14. $x^4 - 5x^2 + 4 = 0$

15. $\sqrt{2x - 3} + x = 3$

16. $\sqrt[4]{2x + 3} = 2$

17. $\sqrt{x + 1} + \sqrt{x - 1} = \sqrt{2x + 1}$

18. $\sqrt{2x - 1} - \sqrt{x - 5} = 3$

19. $2x^{1/2} - 3 = 0$

20. $x^{-6} - 7x^{-3} - 8 = 0$

21. $x^2 + m^2 = 2mx + (nx)^2 \quad n \neq 1$

22. $10a^2x^2 - 2abx - 36b^2 = 0$

23. $\sqrt{x^2 + 3x + 7} - \sqrt{x^2 - 3x + 9} + 2 = 0$

24. $|2x + 3| = 7$

25. $|2 - 3x| + 2 = 9$

26. $2x^3 = 3x^2$

27. $2x^3 + 5x^2 - 8x - 20 = 0$

In Problems 28–34, solve each inequality. Express your answer using set notation or interval notation. Graph the solution set.

28. $\frac{2x - 3}{5} + 2 \leq \frac{x}{2}$

29. $-9 \leq \frac{2x + 3}{-4} \leq 7$

30. $2 < \frac{3 - 3x}{12} < 6$

31. $|3x + 4| < \frac{1}{2}$

32. $|2x - 5| \geq 9$

33. $2 + |2 - 3x| \leq 4$

34. $1 - |2 - 3x| < -4$

In Problems 35–39, use the complex number system and write each expression in the standard form a + bi.

35. $(6 + 3i) - (2 - 4i)$

36. $4(3 - i) + 3(-5 + 2i)$

37. $\frac{3}{3 + i}$

38. i^{50}

39. $(2 + 3i)^3$

In Problems 40–43, solve each equation in the complex number system.

40. $x^2 + x + 1 = 0$

41. $2x^2 + x - 2 = 0$

42. $x^2 + 3 = x$

43. $x(1 - x) = 6$

---

44. Translate the following statement into a mathematical expression: The total cost C of manufacturing x bicycles in one day is \$50,000 plus \$95 times the number of bicycles manufactured.

45. Financial Planning Steve, a recent retiree, requires \$5000 per year in extra income. He has \$70,000 to invest and can invest in A-rated bonds paying 8% per year or in a certificate of deposit (CD) paying 5% per year. How much money should be invested in each to realize exactly \$5000 in interest per year?

46. Lightning and Thunder A flash of lightning is seen, and the resulting thunderclap is heard 3 seconds later. If the speed of sound averages 1100 feet per second, how far away is the storm?

47. **Physics: Intensity of Light** The intensity I (in candlepower) of a certain light source obeys the equation $I = \frac{900}{x^2}$, where x is the distance (in meters) from the light. Over what range of distances can an object be placed from this light source so that the range of intensity of light is from 1600 to 3600 candlepower, inclusive?

48. **Extent of Search and Rescue** A search plane has a cruising speed of 250 miles per hour and carries enough fuel for at most 5 hours of flying. If there is a wind that averages 30 miles per hour and the direction of the search is with the wind one way and against it the other, how far can the search plane travel before it has to turn back?

49. **Rescue at Sea** A life raft, set adrift from a sinking ship 150 miles offshore, travels directly toward a Coast Guard station at the rate of 5 miles per hour. At the time that the raft is set adrift, a rescue helicopter is dispatched from the Coast Guard station. If the helicopter's average speed is 90 miles per hour, how long will it take the helicopter to reach the life raft?

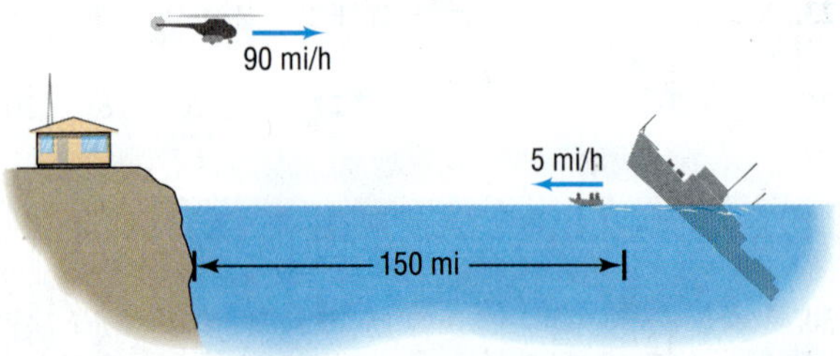

50. **Physics** An object is thrown down from the top of a building 1280 feet tall with an initial velocity of 32 feet per second. The distance s (in feet) of the object from the ground after t seconds is $s = 1280 - 32t - 16t^2$.
    (a) When will the object strike the ground?
    (b) What is the height of the object after 4 seconds?

51. **Working Together to Get a Job Done** Clarissa and Shawna, working together, can paint the exterior of a house in 6 days. Clarissa by herself can complete this job in 5 days less than Shawna. How long will it take Clarissa to complete the job by herself?

52. **Emptying a Tank** Two pumps of different sizes, working together, can empty a fuel tank in 5 hours. The larger pump can empty this tank in 4 hours less than the smaller one. If the larger pump is out of order, how long will it take the smaller one to do the job alone?

53. **Chemistry: Salt Solutions** How much water should be added to 64 ounces of a 10% salt solution to make a 2% salt solution?

54. **Geometry** The diagonal of a rectangle measures 10 inches. If the length is 2 inches more than the width, find the dimensions of the rectangle.

55. **Chemistry: Mixing Acids** A laboratory has 60 cubic centimeters (cm^3) of a solution that is 40% HCl acid. How many cubic centimeters of a 15% solution of HCl acid should be mixed with the 60 cm^3 of 40% acid to obtain a solution of 25% HCl? How much of the 25% solution is there?

56. **Framing a Painting** An artist has 50 inches of oak trim to frame a painting. The frame is to have a border 3 inches wide surrounding the painting.
    (a) If the painting is square, what are its dimensions? What are the dimensions of the frame?
    (b) If the painting is rectangular with a length twice its width, what are the dimensions of the painting? What are the dimensions of the frame?

57. **Finance** An inheritance of $900,000 is to be divided among Scott, Alice, and Tricia in the following manner: Alice is to receive $\frac{3}{4}$ of what Scott gets, while Tricia gets $\frac{1}{2}$ of what Scott gets. How much does each receive?

58. **Utilizing Copying Machines** A new copying machine can do a certain job in 1 hour less than an older copier. Together they can do this job in 72 minutes. How long would it take the older copier by itself to do the job?

59. **Evening Up a Race** In a 100-meter race, Todd crosses the finish line 5 meters ahead of Scott. To even things up, Todd suggests to Scott that they race again, this time with Todd lining up 5 meters behind the start.
    (a) Assuming that Todd and Scott run at the same pace as before, does the second race end in a tie?
    (b) If not, who wins?
    (c) By how many meters does he win?
    (d) How far back should Todd start so that the race ends in a tie?

    After running the race a second time, Scott, to even things up, suggests to Todd that he (Scott) line up 5 meters in front of the start.
    (e) Assuming again that they run at the same pace as in the first race, does the third race result in a tie?
    (f) If not, who wins?
    (g) By how many meters?
    (h) How far ahead should Scott start so that the race ends in a tie?

60. **Physics: Uniform Motion** A man is walking at an average speed of 4 miles per hour alongside a railroad track. A freight train, going in the same direction at an average speed of 30 miles per hour, requires 5 seconds to pass the man. How long is the freight train? Give your answer in feet.

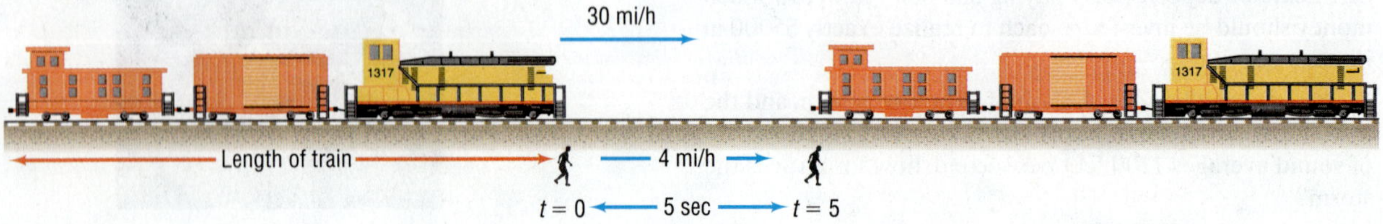

Chapter Test

CHAPTER **Test Prep** VIDEOS The Chapter Test Prep Videos are step-by-step solutions available in MyMathLab, or on this text's YouTube Channel. Flip back to the Resources for Success page for a link to this text's YouTube channel.

In Problems 1–7, solve each equation.

1. $\frac{2x}{3} - \frac{x}{2} = \frac{5}{12}$ **2.** $x(x-1) = 6$ **3.** $x^4 - 3x^2 - 4 = 0$ **4.** $\sqrt{2x-5} + 2 = 4$

5. $|2x-3| + 7 = 10$ **6.** $3x^3 + 2x^2 - 12x - 8 = 0$ **7.** $3x^2 - x + 1 = 0$

In Problems 8–10, solve each inequality. Express your answer using interval notation. Graph the solution set.

8. $-3 \le \frac{3x-4}{2} \le 6$ **9.** $|3x+4| < 8$ **10.** $2 + |2x-5| \ge 9$

11. Write $\frac{-2}{3-i}$ in the standard form $a + bi$.

12. Solve the equation $4x^2 - 4x + 5 = 0$ in the complex number system.

13. Blending Coffee A coffee house has 20 pounds of a coffee that sells for $4 per pound. How many pounds of a coffee that sells for $8 per pound should be mixed with the 20 pounds of $4-per-pound coffee to obtain a blend that will sell for $5 per pound? How much of the $5-per-pound coffee is there to sell?

Chapter Projects

Internet-based Project

I. Financing a Purchase At some point in your life, you are likely to need to borrow money to finance a purchase. For example, most of us will finance the purchase of a car or a home. What is the mathematics behind financing a purchase? When you borrow money from a bank, the bank uses a rather complex equation (or formula) to determine how much you need to pay each month to repay the loan. There are a number of variables that determine the monthly payment. These variables include the amount borrowed, the interest rate, and the length of the loan. The interest rate is based on current economic conditions, the length of the loan, the type of item being purchased, and your credit history.

The following formula gives the monthly payment P required to pay off a loan amount L at an annual interest rate r, expressed as a decimal, but usually given as a percent. The time t, measured in months, is the length of the loan. For example, a 30-year loan requires $12 \times 30 = 360$ monthly payments.

$$P = L\left[\frac{\frac{r}{12}}{1-\left(1+\frac{r}{12}\right)^{-t}}\right]$$

P = monthly payment
L = loan amount
r = annual rate of interest expressed as a decimal
t = length of loan, in months

1. Interest rates change daily. Many websites post current interest rates on loans. Go to *www.bankrate.com* (or some other website that posts lenders' interest rates) and find the current best interest rate on a 60-month new-car purchase loan. Use this rate to determine the monthly payment on a $30,000 automobile loan.
2. Determine the total amount paid for the loan by multiplying the loan payment by the term of the loan.
3. Determine the total amount of interest paid by subtracting the loan amount from the total amount paid from question 2.
4. More often than not, we decide how much of a payment we can afford and use that information to determine the loan amount. Suppose you can afford a monthly payment of $500. Use the interest rate from question 1 to determine the maximum amount you can borrow. If you have $5000 to put down on the car, what is the maximum value of a car you can purchase?
5. Repeat questions 1 through 4 using a 72-month new-car purchase loan, a 60-month used-car purchase loan, and a 72-month used-car purchase loan.

6. We can use the power of a spreadsheet, such as Excel, to create a loan amortization schedule. A loan amortization schedule is a list of the monthly payments, a breakdown of interest and principal, along with a current loan balance. Create a loan amortization schedule for each of the four loan scenarios discussed on the previous page, using the following as a guide. You may want to use an Internet search engine to research specific keystrokes for creating an amortization schedule in a spreadsheet. We supply a sample spreadsheet with formulas included as a guide. Use the spreadsheet to verify your results from questions 1 through 5.

Loan Information		Payment Number	Payment Amount	Interest	Principal	Balance	Total Interest Paid
Loan Amount	$30,000.00	1	=PMT($B$3/12,$B$5,–$B$2,0)	=B2*$B$3/12	=E2–F2	=B2–G2	=B2*$B$3/12
Annual Interest Rate	0.045	2	=PMT($B$3/12,$B$5,–$B$2,0)	=H2*$B$3/12	=E3–F3	=H2–G3	=I2+F3
Length of Loan (years)	5	3	=PMT($B$3/12,$B$5,–$B$2,0)	=H3*$B$3/12	=E4–F4	=H3–G4	=I3+F4
Number of Payments	=B4*12	⋮	⋮	⋮	⋮	⋮	⋮

7. Go to an online automobile website such as www.cars.com, www.edmunds.com, or www.autobytel.com. Research the types of vehicles you can afford for a monthly payment of $500. Decide on a vehicle you would purchase based on your analysis in questions 1–6. Be sure to justify your decision, and include the impact the term of the loan has on your decision. You might consider other factors in your decision, such as expected maintenance costs and insurance costs.

The following project is also available on the Instructor's Resource Center (IRC):

II. Project at Motorola **How Many Cellular Phones Can I Make?** An industrial engineer uses a model involving equations to be sure production levels meet customer demand.

Graphs

2

How to Value a House

Two things to consider in valuing a home are, first, how does it compare to similar homes that have sold recently? Is the asking price fair? And second, what value do you place on the advertised features and amenities? Yes, other people might value them highly, but do you?

Zestimate home valuation, RealEstateABC.com, and Reply.com are among the many algorithmic (generated by a computer model) starting points in figuring out the value of a home. They show you how the home is priced relative to other homes in the area, but you need to add in all the things that only someone who has seen the house knows. You can do that using My Estimator, and then you create your own estimate and see how it stacks up against the asking price.

Looking at "Comps"

Knowing whether an asking price is fair will be important when you're ready to make an offer on a house. It will be even more important when your mortgage lender hires an appraiser to determine whether the house is worth the loan you want.

Check with your agent, Zillow.com, propertyshark.com, or other websites to see recent sales of homes in the area that are similar, or comparable, to what you're looking for. Print them out and keep these "comps" in a three-ring binder; you'll be referring to them quite a bit.

Note that "recent sales" usually means within the last six months. A sales price from a year ago may bear little or no relation to what is going on in your area right now. In fact, some lenders will not accept comps older than three months.

Market activity also determines how easy or difficult it is to find accurate comps. In a "hot" or busy market, with sales happening all the time, you're likely to have lots of comps to choose from. In a less active market, finding reasonable comps becomes harder. And if the home you're looking at has special design features, finding a comparable property is harder still. It's also necessary to know what's going on in a given sub-segment. Maybe large, high-end homes are selling like hotcakes, but owners of smaller houses are staying put, or vice versa.

Source: *http://allmyhome.blogspot.com/2008/07/how-to-value-house.html*

—See the Internet-based Chapter Project—

••• A Look Back

Chapter R and Chapter 1 review skills from intermediate algebra.

A Look Ahead •••

Here we connect algebra and geometry using the rectangular coordinate system. In the 1600s, algebra had developed to the point that René Descartes (1596–1650) and Pierre de Fermat (1601–1665) were able to use rectangular coordinates to translate geometry problems into algebra problems, and vice versa. This enabled both geometers and algebraists to gain new insights into their subjects, which had been thought to be separate but now were seen as connected.

Outline

2.1 The Distance and Midpoint Formulas

PREPARING FOR THIS SECTION *Before getting started, review the following:*

- Algebra Essentials (Chapter R, Section R.2, pp. 17–26)
- Geometry Essentials (Chapter R, Section R.3, pp. 30–35)

Now Work the **'Are You Prepared?'** problems on page 154.

OBJECTIVES 1 Use the Distance Formula (p. 151)
2 Use the Midpoint Formula (p. 153)

Rectangular Coordinates

We locate a point on the real number line by assigning it a single real number, called the *coordinate of the point.* For work in a two-dimensional plane, we locate points by using two numbers.

Begin with two real number lines located in the same plane: one horizontal and the other vertical. The horizontal line is called the ***x*-axis**, the vertical line the ***y*-axis**, and the point of intersection the **origin *O***. See Figure 1. Assign coordinates to every point on these number lines using a convenient scale. In mathematics, we usually use the same scale on each axis, but in applications, different scales appropriate to the application may be used.

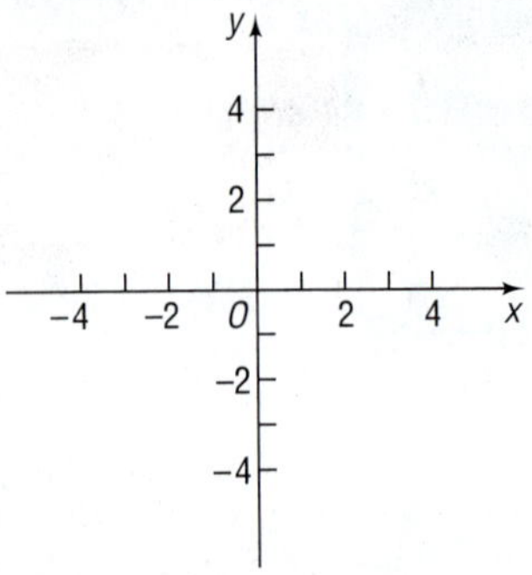

Figure 1 *xy*-Plane

The origin O has a value of 0 on both the x-axis and the y-axis. Points on the x-axis to the right of O are associated with positive real numbers, and those to the left of O are associated with negative real numbers. Points on the y-axis above O are associated with positive real numbers, and those below O are associated with negative real numbers. In Figure 1, the x-axis and y-axis are labeled as x and y, respectively, and an arrow at the end of each axis is used to denote the positive direction.

The coordinate system described here is called a **rectangular** or **Cartesian*** **coordinate system**. The plane formed by the x-axis and y-axis is sometimes called the ***xy*-plane**, and the x-axis and y-axis are referred to as the **coordinate axes**.

Any point P in the xy-plane can be located by using an **ordered pair** (x, y) of real numbers. Let x denote the signed distance of P from the y-axis (*signed* means that if P is to the right of the y-axis, then $x > 0$, and if P is to the left of the y-axis, then $x < 0$); and let y denote the signed distance of P from the x-axis. The ordered pair (x, y), also called the **coordinates** of P, gives us enough information to locate the point P in the plane.

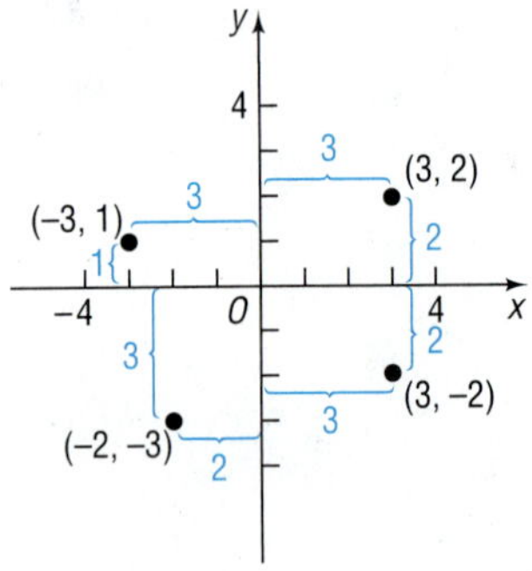

Figure 2

For example, to locate the point whose coordinates are $(-3, 1)$, go 3 units along the x-axis to the left of O and then go straight up 1 unit. We **plot** this point by placing a dot at this location. See Figure 2, in which the points with coordinates $(-3, 1)$, $(-2, -3)$, $(3, -2)$, and $(3, 2)$ are plotted.

The origin has coordinates $(0, 0)$. Any point on the x-axis has coordinates of the form $(x, 0)$, and any point on the y-axis has coordinates of the form $(0, y)$.

If (x, y) are the coordinates of a point P, then x is called the ***x*-coordinate**, or **abscissa**, of P, and y is the ***y*-coordinate**, or **ordinate**, of P. We identify the point P by its coordinates (x, y) by writing $P = (x, y)$. Usually, we will simply say "the point (x, y)" rather than "the point whose coordinates are (x, y)."

The coordinate axes divide the xy-plane into four sections called **quadrants**, as shown in Figure 3. In quadrant I, both the x-coordinate and the y-coordinate of all points are positive; in quadrant II, x is negative and y is positive; in quadrant III, both x and y are negative; and in quadrant IV, x is positive and y is negative. Points on the coordinate axes belong to no quadrant.

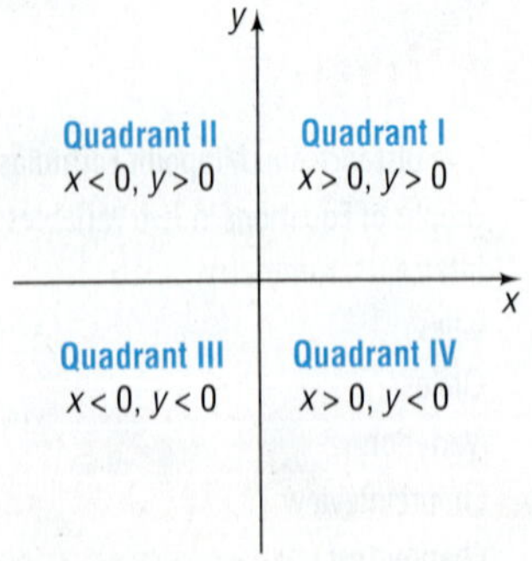

Figure 3

Now Work PROBLEM 15

* Named after René Descartes (1596–1650), a French mathematician, philosopher, and theologian.

COMMENT On a graphing calculator, you can set the scale on each axis. Once this has been done, you obtain the **viewing rectangle.** See Figure 4 for a typical viewing rectangle. You should now read Section 1, *The Viewing Rectangle*, in the Appendix.

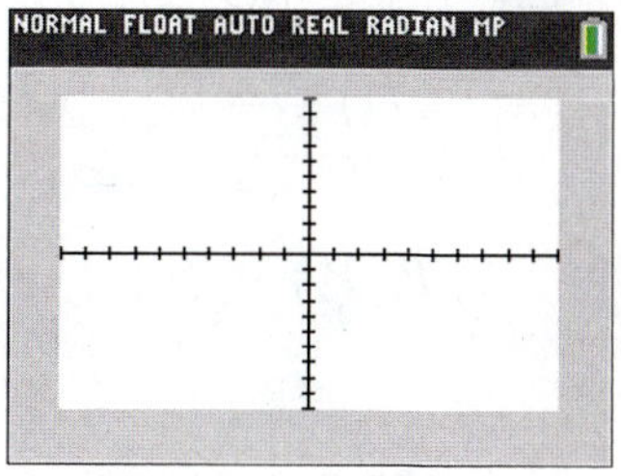

Figure 4 TI-84 Plus C Standard Viewing Rectangle

1 Use the Distance Formula

If the same units of measurement (such as inches, centimeters, and so on) are used for both the x-axis and y-axis, then all distances in the xy-plane can be measured using this unit of measurement.

EXAMPLE 1 **Finding the Distance between Two Points**

Find the distance d between the points $(1, 3)$ and $(5, 6)$.

Solution First plot the points $(1, 3)$ and $(5, 6)$ and connect them with a straight line. See Figure 5(a). To find the length d, begin by drawing a horizontal line from $(1, 3)$ to $(5, 3)$ and a vertical line from $(5, 3)$ to $(5, 6)$, forming a right triangle, as shown in Figure 5(b). One leg of the triangle is of length 4 (since $|5 - 1| = 4$), and the other is of length 3 (since $|6 - 3| = 3$). By the Pythagorean Theorem, the square of the distance d that we seek is

$$d^2 = 4^2 + 3^2 = 16 + 9 = 25$$
$$d = \sqrt{25} = 5$$

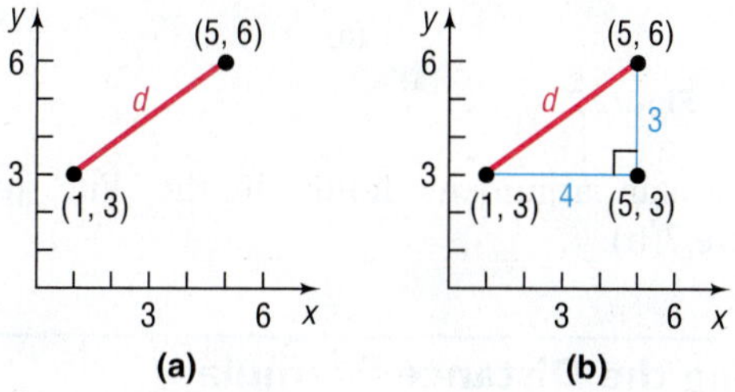

Figure 5

The **distance formula** provides a straightforward method for computing the distance between two points.

THEOREM

Distance Formula

The distance between two points $P_1 = (x_1, y_1)$ and $P_2 = (x_2, y_2)$, denoted by $d(P_1, P_2)$, is

$$d(P_1, P_2) = \sqrt{(x_2 - x_1)^2 + (y_2 - y_1)^2} \quad \textbf{(1)}$$

In Words

To compute the distance between two points, find the difference of the x-coordinates, square it, and add this to the square of the difference of the y-coordinates. The square root of this sum is the distance.

Proof of the Distance Formula Let (x_1, y_1) denote the coordinates of point P_1 and let (x_2, y_2) denote the coordinates of point P_2. Assume that the line joining P_1 and P_2 is neither horizontal nor vertical. Refer to Figure 6(a) on page 152. The coordinates of P_3 are (x_2, y_1). The horizontal distance from P_1 to P_3 is the absolute value of the difference of the x-coordinates, $|x_2 - x_1|$. The vertical distance from P_3 to P_2 is the absolute value of the difference of the y-coordinates, $|y_2 - y_1|$. See Figure 6(b). The distance $d(P_1, P_2)$ is the length of the hypotenuse of the right triangle, so, by the Pythagorean Theorem, it follows that

$$\begin{aligned}[d(P_1, P_2)]^2 &= |x_2 - x_1|^2 + |y_2 - y_1|^2 \\ &= (x_2 - x_1)^2 + (y_2 - y_1)^2 \\ d(P_1, P_2) &= \sqrt{(x_2 - x_1)^2 + (y_2 - y_1)^2}\end{aligned}$$

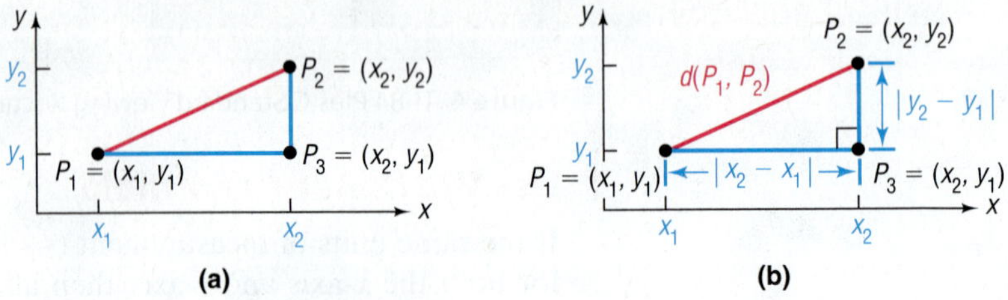

Figure 6

Now, if the line joining P_1 and P_2 is horizontal, then the y-coordinate of P_1 equals the y-coordinate of P_2; that is, $y_1 = y_2$. Refer to Figure 7(a). In this case, the distance formula (1) still works, because for $y_1 = y_2$, it reduces to

$$d(P_1, P_2) = \sqrt{(x_2 - x_1)^2 + 0^2} = \sqrt{(x_2 - x_1)^2} = |x_2 - x_1|$$

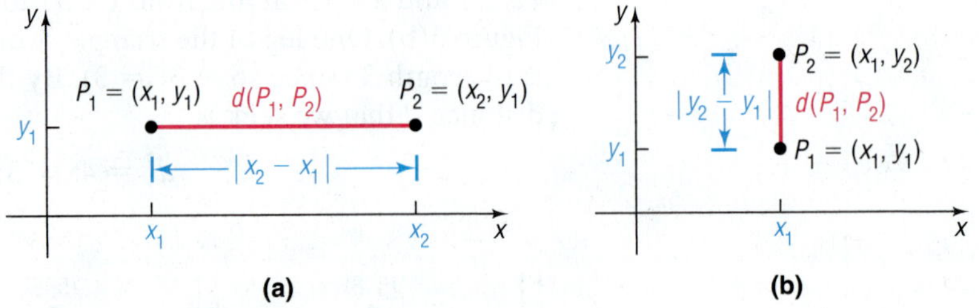

Figure 7

A similar argument holds if the line joining P_1 and P_2 is vertical. See Figure 7(b). ■

EXAMPLE 2 **Using the Distance Formula**

Find the distance d between the points $(-4, 5)$ and $(3, 2)$.

Solution Using the distance formula, equation (1), reveals that the distance d is

$$\begin{aligned}d = \sqrt{[3 - (-4)]^2 + (2 - 5)^2} &= \sqrt{7^2 + (-3)^2} \\ &= \sqrt{49 + 9} = \sqrt{58} \approx 7.62\end{aligned}$$ ●

Now Work PROBLEMS 19 AND 23

The distance between two points $P_1 = (x_1, y_1)$ and $P_2 = (x_2, y_2)$ is never a negative number. Also, the distance between two points is 0 only when the points are identical—that is, when $x_1 = x_2$ and $y_1 = y_2$. And, because $(x_2 - x_1)^2 = (x_1 - x_2)^2$ and $(y_2 - y_1)^2 = (y_1 - y_2)^2$, it makes no difference whether the distance is computed from P_1 to P_2 or from P_2 to P_1; that is, $d(P_1, P_2) = d(P_2, P_1)$.

The introduction to this chapter mentioned that rectangular coordinates enable us to translate geometry problems into algebra problems, and vice versa. The next example shows how algebra (the distance formula) can be used to solve geometry problems.

EXAMPLE 3 **Using Algebra to Solve Geometry Problems**

Consider the three points $A = (-2, 1)$, $B = (2, 3)$, and $C = (3, 1)$.

(a) Plot each point and form the triangle ABC.
(b) Find the length of each side of the triangle.
(c) Show that the triangle is a right triangle.
(d) Find the area of the triangle.

Solution

(a) Figure 8 shows the points A, B, C and the triangle ABC.

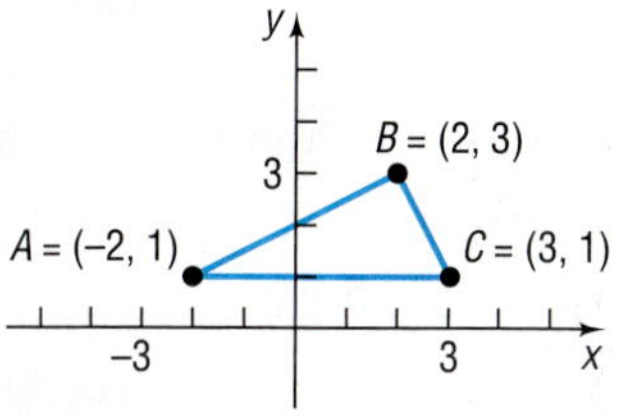

Figure 8

(b) To find the length of each side of the triangle, use the distance formula, equation (1).

$$d(A, B) = \sqrt{[2 - (-2)]^2 + (3 - 1)^2} = \sqrt{16 + 4} = \sqrt{20} = 2\sqrt{5}$$
$$d(B, C) = \sqrt{(3 - 2)^2 + (1 - 3)^2} = \sqrt{1 + 4} = \sqrt{5}$$
$$d(A, C) = \sqrt{[3 - (-2)]^2 + (1 - 1)^2} = \sqrt{25 + 0} = 5$$

(c) If the sum of the squares of the lengths of two of the sides equals the square of the length of the third side, then the triangle is a right triangle. Looking at Figure 8, it seems reasonable to conjecture that the angle at vertex B might be a right angle. We shall check to see whether

$$[d(A, B)]^2 + [d(B, C)]^2 = [d(A, C)]^2$$

Using the results in part (b) yields

$$[d(A, B)]^2 + [d(B, C)]^2 = (2\sqrt{5})^2 + (\sqrt{5})^2$$
$$= 20 + 5 = 25 = [d(A, C)]^2$$

It follows from the converse of the Pythagorean Theorem that triangle ABC is a right triangle.

(d) Because the right angle is at vertex B, the sides AB and BC form the base and height of the triangle. Its area is

$$\text{Area} = \frac{1}{2}(\text{Base})(\text{Height}) = \frac{1}{2}(2\sqrt{5})(\sqrt{5}) = 5 \text{ square units}$$

Now Work PROBLEM 31

2 Use the Midpoint Formula

We now derive a formula for the coordinates of the **midpoint of a line segment**. Let $P_1 = (x_1, y_1)$ and $P_2 = (x_2, y_2)$ be the endpoints of a line segment, and let $M = (x, y)$ be the point on the line segment that is the same distance from P_1 as it is from P_2. See Figure 9. The triangles P_1AM and MBP_2 are congruent. [Do you see why? $d(P_1, M) = d(M, P_2)$ is given; also, $\angle AP_1M = \angle BMP_2$* and $\angle P_1MA = \angle MP_2B$. Thus, we have angle–side–angle.] Because triangles P_1AM and MBP_2 are congruent, corresponding sides are equal in length. That is,

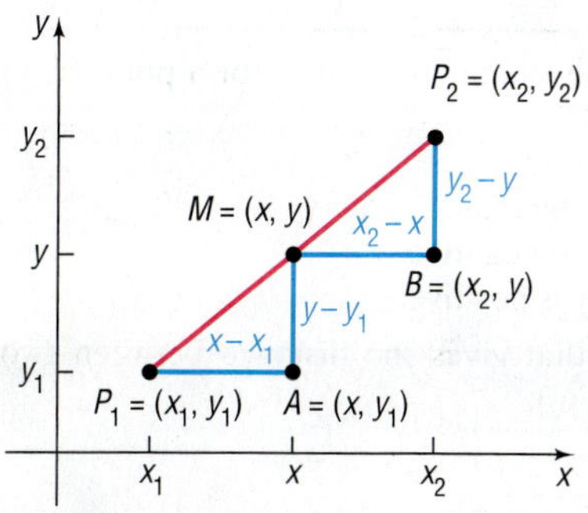

Figure 9

$$x - x_1 = x_2 - x \quad \text{and} \quad y - y_1 = y_2 - y$$
$$2x = x_1 + x_2 \qquad 2y = y_1 + y_2$$
$$x = \frac{x_1 + x_2}{2} \qquad y = \frac{y_1 + y_2}{2}$$

*A postulate from geometry states that the transversal $\overline{P_1P_2}$ forms congruent corresponding angles with the parallel line segments $\overline{P_1A}$ and $\overline{MB}$.

THEOREM

Midpoint Formula

The midpoint $M = (x, y)$ of the line segment from $P_1 = (x_1, y_1)$ to $P_2 = (x_2, y_2)$ is

$$M = (x, y) = \left(\frac{x_1 + x_2}{2}, \frac{y_1 + y_2}{2}\right) \tag{2}$$

In Words

To find the midpoint of a line segment, average the x-coordinates of the endpoints, and average the y-coordinates of the endpoints.

EXAMPLE 4 **Finding the Midpoint of a Line Segment**

Find the midpoint of the line segment from $P_1 = (-5, 5)$ to $P_2 = (3, 1)$. Plot the points P_1 and P_2 and their midpoint.

Solution Apply the midpoint formula (2) using $x_1 = -5$, $y_1 = 5$, $x_2 = 3$, and $y_2 = 1$. Then the coordinates (x, y) of the midpoint M are

$$x = \frac{x_1 + x_2}{2} = \frac{-5 + 3}{2} = -1 \quad \text{and} \quad y = \frac{y_1 + y_2}{2} = \frac{5 + 1}{2} = 3$$

That is, $M = (-1, 3)$. See Figure 10. ●

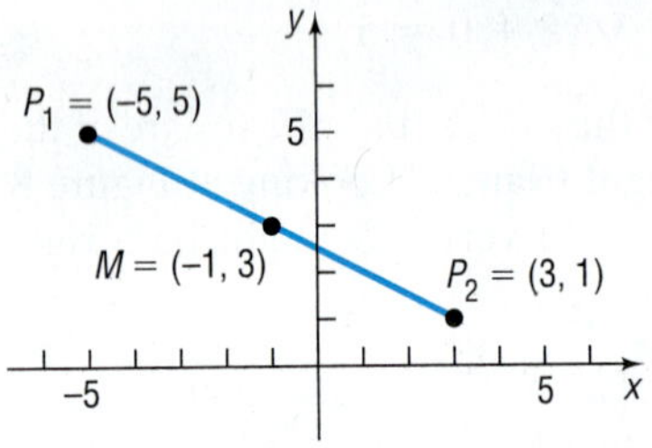

Figure 10

Now Work PROBLEM 37

2.1 Assess Your Understanding

'Are You Prepared?' *Answers are given at the end of these exercises. If you get a wrong answer, read the pages listed in red.*

1. On the real number line, the origin is assigned the number ________. (p. 17)
2. If −3 and 5 are the coordinates of two points on the real number line, the distance between these points is ________. (pp. 19–20)
3. If 3 and 4 are the legs of a right triangle, the hypotenuse is ________. (p. 30)
4. Use the converse of the Pythagorean Theorem to show that a triangle whose sides are of lengths 11, 60, and 61 is a right triangle. (pp. 30–31)
5. The area A of a triangle whose base is b and whose altitude is h is $A =$ ______. (p. 31)
6. ***True or False*** Two triangles are congruent if two angles and the included side of one equals two angles and the included side of the other. (pp. 32–33).

Concepts and Vocabulary

7. If (x, y) are the coordinates of a point P in the xy-plane, then x is called the ____________________ of P, and y is the ____________________ of P.
8. The coordinate axes divide the xy-plane into four sections called ____________.
9. If three distinct points P, Q, and R all lie on a line, and if $d(P, Q) = d(Q, R)$, then Q is called the ____________ of the line segment from P to R.
10. ***True or False*** The distance between two points is sometimes a negative number.
11. ***True or False*** The point $(-1, 4)$ lies in quadrant IV of the Cartesian plane.
12. ***True or False*** The midpoint of a line segment is found by averaging the x-coordinates and averaging the y-coordinates of the endpoints.
13. Which of the following statements is true for a point (x, y) that lies in quadrant III?
    (a) Both x and y are positive.
    (b) Both x and y are negative.
    (c) x is positive, and y is negative.
    (d) x is negative, and y is positive.
14. Choose the formula that gives the distance between two points (x_1, y_1) and (x_2, y_2).
    (a) $\sqrt{(x_2 - x_1)^2 + (y_2 - y_1)^2}$
    (b) $\sqrt{(x_2 + x_1)^2 - (y_2 + y_1)^2}$
    (c) $\sqrt{(x_2 - x_1)^2 - (y_2 - y_1)^2}$
    (d) $\sqrt{(x_2 + x_1)^2 + (y_2 + y_1)^2}$

Skill Building

In Problems 15 and 16, plot each point in the xy-plane. Tell in which quadrant or on what coordinate axis each point lies.

15. (a) $A = (-3, 2)$ (b) $B = (6, 0)$ (c) $C = (-2, -2)$ (d) $D = (6, 5)$ (e) $E = (0, -3)$ (f) $F = (6, -3)$

16. (a) $A = (1, 4)$ (b) $B = (-3, -4)$ (c) $C = (-3, 4)$ (d) $D = (4, 1)$ (e) $E = (0, 1)$ (f) $F = (-3, 0)$

17. Plot the points $(2, 0)$, $(2, -3)$, $(2, 4)$, $(2, 1)$, and $(2, -1)$. Describe the set of all points of the form $(2, y)$, where y is a real number.

18. Plot the points $(0, 3)$, $(1, 3)$, $(-2, 3)$, $(5, 3)$, and $(-4, 3)$. Describe the set of all points of the form $(x, 3)$, where x is a real number.

In Problems 19–30, find the distance $d(P_1, P_2)$ between the points P_1 and P_2.

19.

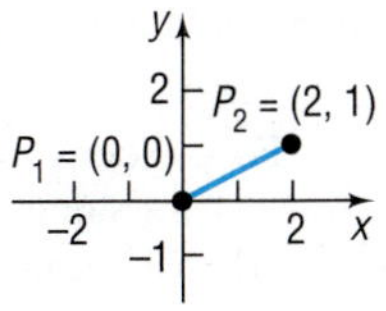

20.

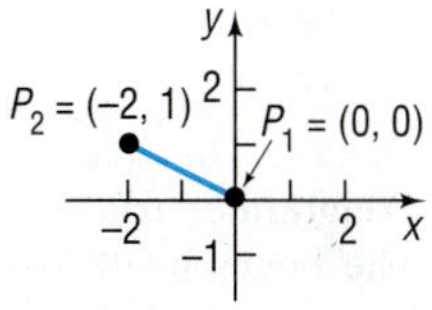

21.

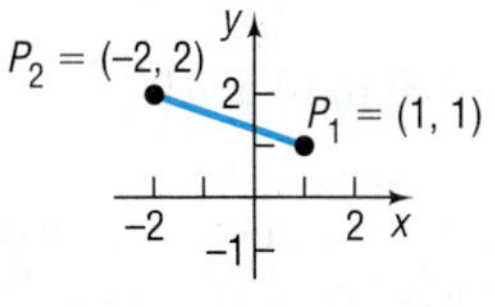

22.

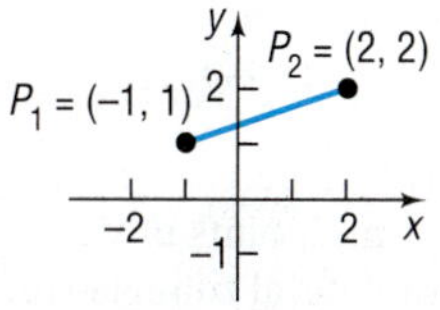

23. $P_1 = (3, -4)$; $P_2 = (5, 4)$

24. $P_1 = (-1, 0)$; $P_2 = (2, 4)$

25. $P_1 = (-3, 2)$; $P_2 = (6, 0)$

26. $P_1 = (2, -3)$; $P_2 = (4, 2)$

27. $P_1 = (4, -3)$; $P_2 = (6, 4)$

28. $P_1 = (-4, -3)$; $P_2 = (6, 2)$

29. $P_1 = (a, b)$; $P_2 = (0, 0)$

30. $P_1 = (a, a)$; $P_2 = (0, 0)$

In Problems 31–36, plot each point and form the triangle ABC. Show that the triangle is a right triangle. Find its area.

31. $A = (-2, 5)$; $B = (1, 3)$; $C = (-1, 0)$

32. $A = (-2, 5)$; $B = (12, 3)$; $C = (10, -11)$

33. $A = (-5, 3)$; $B = (6, 0)$; $C = (5, 5)$

34. $A = (-6, 3)$; $B = (3, -5)$; $C = (-1, 5)$

35. $A = (4, -3)$; $B = (0, -3)$; $C = (4, 2)$

36. $A = (4, -3)$; $B = (4, 1)$; $C = (2, 1)$

In Problems 37–44, find the midpoint of the line segment joining the points P_1 and P_2.

37. $P_1 = (3, -4)$; $P_2 = (5, 4)$

38. $P_1 = (-2, 0)$; $P_2 = (2, 4)$

39. $P_1 = (-3, 2)$; $P_2 = (6, 0)$

40. $P_1 = (2, -3)$; $P_2 = (4, 2)$

41. $P_1 = (4, -3)$; $P_2 = (6, 1)$

42. $P_1 = (-4, -3)$; $P_2 = (2, 2)$

43. $P_1 = (a, b)$; $P_2 = (0, 0)$

44. $P_1 = (a, a)$; $P_2 = (0, 0)$

Applications and Extensions

45. If the point $(2, 5)$ is shifted 3 units to the right and 2 units down, what are its new coordinates?

46. If the point $(-1, 6)$ is shifted 2 units to the left and 4 units up, what are its new coordinates?

47. Find all points having an x-coordinate of 3 whose distance from the point $(-2, -1)$ is 13.
(a) By using the Pythagorean Theorem.
(b) By using the distance formula.

48. Find all points having a y-coordinate of -6 whose distance from the point $(1, 2)$ is 17.
(a) By using the Pythagorean Theorem.
(b) By using the distance formula.

49. Find all points on the x-axis that are 6 units from the point $(4, -3)$.

50. Find all points on the y-axis that are 6 units from the point $(4, -3)$.

51. Suppose that $A = (2, 5)$ are the coordinates of a point in the xy-plane.
(a) Find the coordinates of the point if A is shifted 3 units to the left and 4 units down.
(b) Find the coordinates of the point if A is shifted 2 units to the left and 8 units up.

52. Plot the points $A = (-1, 8)$ and $M = (2, 3)$ in the xy-plane. If M is the midpoint of a line segment AB, find the coordinates of B.

53. The midpoint of the line segment from P_1 to P_2 is $(-1, 4)$. If $P_1 = (-3, 6)$, what is P_2?

54. The midpoint of the line segment from P_1 to P_2 is $(5, -4)$. If $P_2 = (7, -2)$, what is P_1?

55. Geometry The **medians** of a triangle are the line segments from each vertex to the midpoint of the opposite side (see the figure). Find the lengths of the medians of the triangle with vertices at $A = (0, 0)$, $B = (6, 0)$, and $C = (4, 4)$.

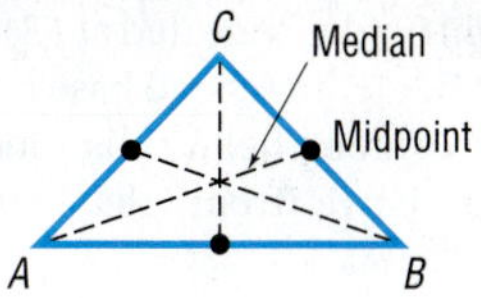

56. Geometry An **equilateral triangle** is one in which all three sides are of equal length. If two vertices of an equilateral triangle are $(0, 4)$ and $(0, 0)$, find the third vertex. How many of these triangles are possible?

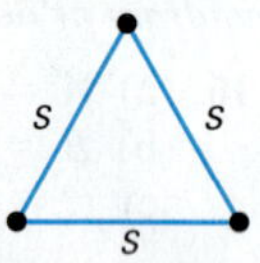

57. Geometry Find the midpoint of each diagonal of a square with side of length s. Draw the conclusion that the diagonals of a square intersect at their midpoints.

[**Hint:** Use $(0, 0)$, $(0, s)$, $(s, 0)$, and (s, s) as the vertices of the square.]

58. Geometry Verify that the points $(0,0)$, $(a,0)$, and $\left(\frac{a}{2}, \frac{\sqrt{3}a}{2}\right)$ are the vertices of an equilateral triangle. Then show that the midpoints of the three sides are the vertices of a second equilateral triangle (refer to Problem 56).

*In Problems 59–62, find the length of each side of the triangle determined by the three points P_1, P_2, and P_3. State whether the triangle is an isosceles triangle, a right triangle, neither of these, or both. (An **isosceles triangle** is one in which at least two of the sides are of equal length.)*

59. $P_1 = (2, 1);\ P_2 = (-4, 1);\ P_3 = (-4, -3)$

60. $P_1 = (-1, 4);\ P_2 = (6, 2);\ P_3 = (4, -5)$

61. $P_1 = (-2, -1);\ P_2 = (0, 7);\ P_3 = (3, 2)$

62. $P_1 = (7, 2);\ P_2 = (-4, 0);\ P_3 = (4, 6)$

63. Baseball A major league baseball "diamond" is actually a square 90 feet on a side (see the figure). What is the distance directly from home plate to second base (the diagonal of the square)?

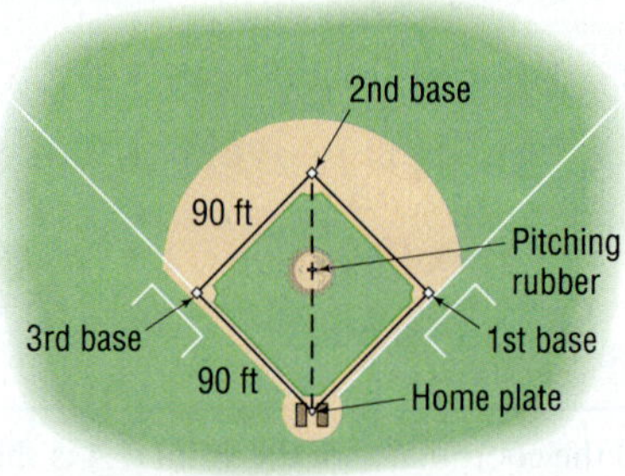

64. Little League Baseball The layout of a Little League playing field is a square 60 feet on a side. How far is it directly from home plate to second base (the diagonal of the square)?

Source: Little League Baseball, Official Regulations and Playing Rules, 2014.

65. Baseball Refer to Problem 63. Overlay a rectangular coordinate system on a major league baseball diamond so that the origin is at home plate, the positive x-axis lies in the direction from home plate to first base, and the positive y-axis lies in the direction from home plate to third base.

(a) What are the coordinates of first base, second base, and third base? Use feet as the unit of measurement.

(b) If the right fielder is located at $(310, 15)$, how far is it from the right fielder to second base?

(c) If the center fielder is located at $(300, 300)$, how far is it from the center fielder to third base?

66. Little League Baseball Refer to Problem 64. Overlay a rectangular coordinate system on a Little League baseball diamond so that the origin is at home plate, the positive x-axis lies in the direction from home plate to first base, and the positive y-axis lies in the direction from home plate to third base.

(a) What are the coordinates of first base, second base, and third base? Use feet as the unit of measurement.

(b) If the right fielder is located at $(180, 20)$, how far is it from the right fielder to second base?

(c) If the center fielder is located at $(220, 220)$, how far is it from the center fielder to third base?

67. Distance between Moving Objects A Ford Focus and a Freightliner truck leave an intersection at the same time. The Focus heads east at an average speed of 30 miles per hour, while the truck heads south at an average speed of 40 miles per hour. Find an expression for their distance apart d (in miles) at the end of t hours.

68. Distance of a Moving Object from a Fixed Point A hot-air balloon, headed due east at an average speed of 15 miles per hour and at a constant altitude of 100 feet, passes over an intersection (see the figure). Find an expression for the distance d (measured in feet) from the balloon to the intersection t seconds later.

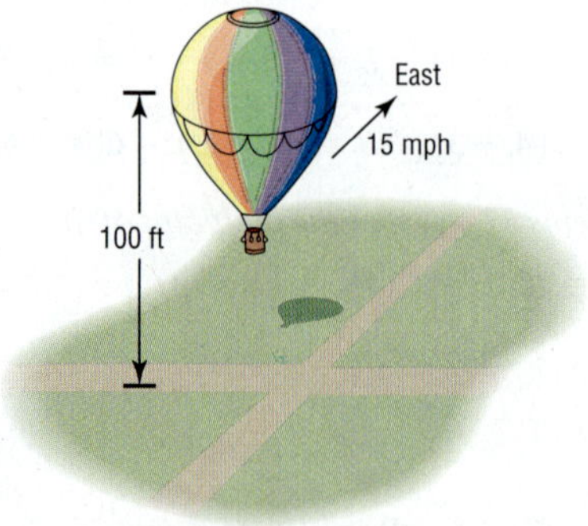

69. Drafting Error When a draftsman draws three lines that are to intersect at one point, the lines may not intersect as intended and subsequently will form an **error triangle**. If this error triangle is long and thin, one estimate for the location of the desired point is the midpoint of the shortest side. The figure shows one such error triangle.

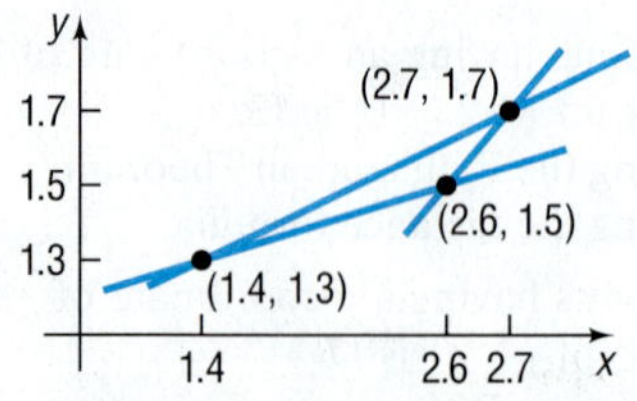

(a) Find an estimate for the desired intersection point.

(b) Find the length of the median for the midpoint found in part (a). See Problem 55.

70. Net Sales The figure on page 157 illustrates how net sales of Wal-Mart Stores, Inc., grew from 2007 through 2013. Use the midpoint formula to estimate the net sales of Wal-Mart Stores, Inc., in 2010. How does your result compare to the reported value of \$405 billion?

Source: Wal-Mart Stores, Inc., 2013 Annual Report

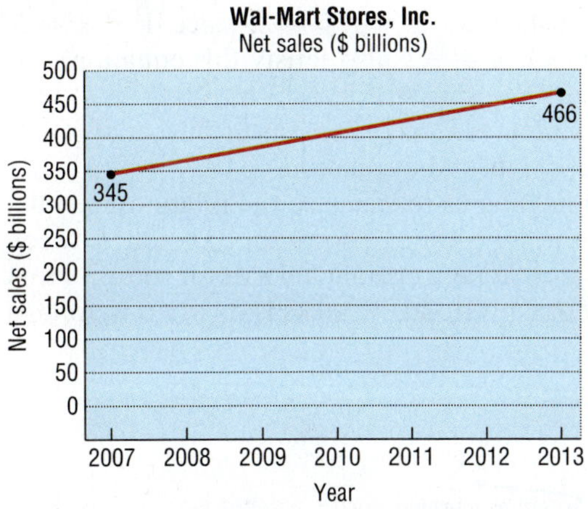

71. Poverty Threshold Poverty thresholds are determined by the U.S. Census Bureau. A poverty threshold represents the minimum annual household income for a family not to be considered poor. In 2003, the poverty threshold for a family of four with two children under the age of 18 years was \$18,660. In 2013, the poverty threshold for a family of four with two children under the age of 18 years was \$23,624. Assuming that poverty thresholds increase in a straight-line fashion, use the midpoint formula to estimate the poverty threshold for a family of four with two children under the age of 18 in 2008. How does your result compare to the actual poverty threshold in 2008 of \$21,834?

Source: *U.S. Census Bureau*

Explaining Concepts: Discussion and Writing

72. Write a paragraph that describes a Cartesian plane. Then write a second paragraph that describes how to plot points in the Cartesian plane. Your paragraphs should include the terms "coordinate axes," "ordered pair," "coordinates," "plot," "x-coordinate," and "y-coordinate."

Retain Your Knowledge

Problems 73–76 are based on material learned earlier in the course. The purpose of these problems is to keep the material fresh in your mind so that you are better prepared for the final exam.

73. Determine the domain of the variable x in the expression $\dfrac{3x + 1}{2x - 5}$.

74. Find the real solution(s), if any, of the equation $3x^2 - 7x - 20 = 0$.

75. Multiply $(7 + 3i)(1 - 2i)$. Write the answer in the form $a + bi$.

76. Solve the inequality $5(x - 3) + 2x \geq 6(2x - 3) - 7$. Express the solution using interval notation. Graph the solution set.

'Are You Prepared?' Answers

1. 0 **2.** 8 **3.** 5 **4.** $11^2 + 60^2 = 121 + 3600 = 3721 = 61^2$ **5.** $A = \frac{1}{2}bh$ **6.** True

2.2 Graphs of Equations in Two Variables; Intercepts; Symmetry

PREPARING FOR THIS SECTION *Before getting started, review the following:*

- Solving Linear Equations (Section 1.1, pp. 82–85)
- Solve a Quadratic Equation by Factoring (Section 1.2, pp. 93–94)

Now Work the **'Are You Prepared?'** problems on page 164.

OBJECTIVES
1 Graph Equations by Plotting Points (p. 157)
2 Find Intercepts from a Graph (p. 159)
3 Find Intercepts from an Equation (p. 160)
4 Test an Equation for Symmetry with Respect to the x-Axis, the y-Axis, and the Origin (p. 160)
5 Know How to Graph Key Equations (p. 163)

1 Graph Equations by Plotting Points

An **equation in two variables**, say x and y, is a statement in which two expressions involving x and y are equal. The expressions are called the **sides** of the equation. Since an equation is a statement, it may be true or false, depending on the value of the variables. Any values of x and y that result in a true statement are said to **satisfy** the equation.

For example, the following are all equations in two variables x and y:

$$x^2 + y^2 = 5 \qquad 2x - y = 6 \qquad y = 2x + 5 \qquad x^2 = y$$

The first of these, $x^2 + y^2 = 5$, is satisfied for $x = 1, y = 2$, since $1^2 + 2^2 = 5$. Other choices of x and y, such as $x = -1, y = -2$, also satisfy this equation. It is not satisfied for $x = 2$ and $y = 3$, since $2^2 + 3^2 = 4 + 9 = 13 \neq 5$.

The **graph of an equation in two variables** x and y consists of the set of points in the xy-plane whose coordinates (x, y) satisfy the equation.

Graphs play an important role in helping us to visualize the relationships that exist between two variables or quantities. Figure 11 shows the relation between the level of risk in a stock portfolio and the average annual rate of return. From the graph, we can see that when 30% of a portfolio of stocks is invested in foreign companies, risk is minimized.

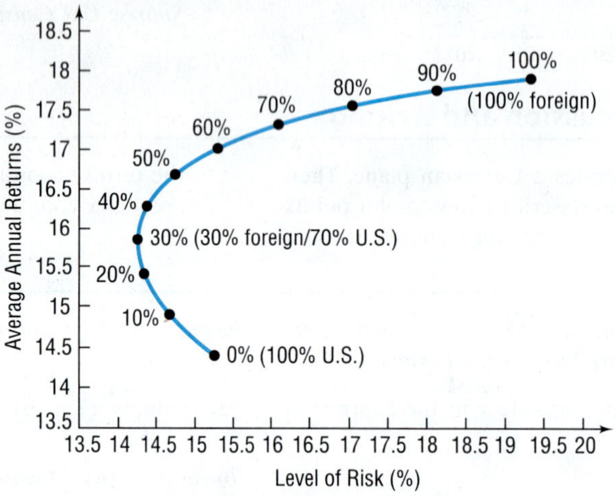

Figure 11
Source: T. Rowe Price

EXAMPLE 1 Determining Whether a Point Is on the Graph of an Equation

Determine if the following points are on the graph of the equation $2x - y = 6$.

(a) $(2, 3)$ (b) $(2, -2)$

Solution

(a) For the point $(2, 3)$, check to see whether $x = 2, y = 3$ satisfies the equation $2x - y = 6$.

$$2x - y = 2(2) - 3 = 4 - 3 = 1 \neq 6$$

The equation is not satisfied, so the point $(2, 3)$ is not on the graph of $2x - y = 6$.

(b) For the point $(2, -2)$,

$$2x - y = 2(2) - (-2) = 4 + 2 = 6$$

The equation is satisfied, so the point $(2, -2)$ is on the graph of $2x - y = 6$.

Now Work PROBLEM 13

EXAMPLE 2 Graphing an Equation by Plotting Points

Graph the equation: $y = 2x + 5$

Solution

The graph consists of all points (x, y) that satisfy the equation. To locate some of these points (and get an idea of the pattern of the graph), assign some numbers to x, and find corresponding values for y.

If	Then	Point on Graph
$x = 0$	$y = 2(0) + 5 = 5$	$(0, 5)$
$x = 1$	$y = 2(1) + 5 = 7$	$(1, 7)$
$x = -5$	$y = 2(-5) + 5 = -5$	$(-5, -5)$
$x = 10$	$y = 2(10) + 5 = 25$	$(10, 25)$

By plotting these points and then connecting them, we obtain the graph of the equation (a *line*), as shown in Figure 12.

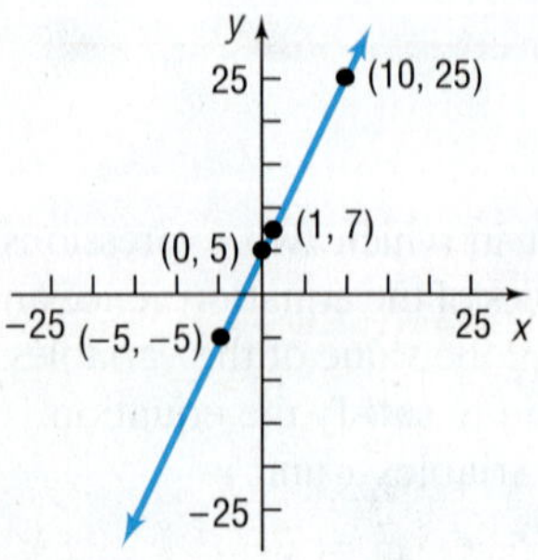

Figure 12 $y = 2x + 5$

EXAMPLE 3 **Graphing an Equation by Plotting Points**

Graph the equation: $y = x^2$

Solution Table 1 provides several points on the graph. Plotting these points and connecting them with a smooth curve gives the graph (a *parabola*) shown in Figure 13.

Table 1

x	$y = x^2$	(x, y)
−4	16	(−4, 16)
−3	9	(−3, 9)
−2	4	(−2, 4)
−1	1	(−1, 1)
0	0	(0, 0)
1	1	(1, 1)
2	4	(2, 4)
3	9	(3, 9)
4	16	(4, 16)

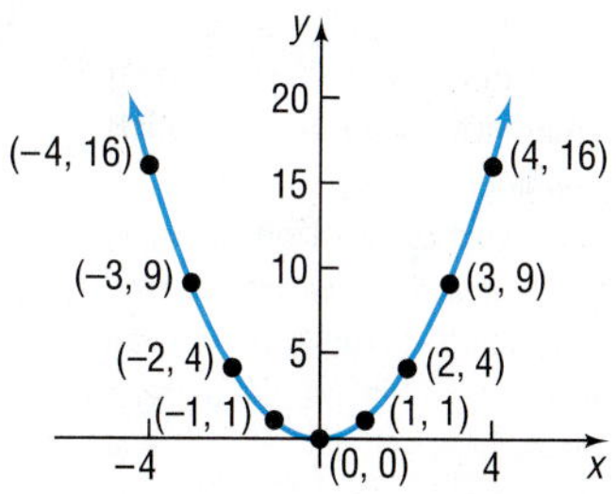

Figure 13 $y = x^2$

The graphs of the equations shown in Figures 12 and 13 do not show all points. For example, in Figure 12, the point $(20, 45)$ is a part of the graph of $y = 2x + 5$, but it is not shown. Since the graph of $y = 2x + 5$ can be extended out indefinitely, we use arrows to indicate that the pattern shown continues. It is important when illustrating a graph to present enough of the graph so that any viewer of the illustration will "see" the rest of it as an obvious continuation of what is actually there. This is referred to as a **complete graph**.

One way to obtain the complete graph of an equation is to plot enough points on the graph for a pattern to become evident. Then these points are connected with a smooth curve following the suggested pattern. But how many points are sufficient? Sometimes knowledge about the equation tells us. For example, we will learn in the next section that if an equation is of the form $y = mx + b$, then its graph is a line. In this case, only two points are needed to obtain the graph.

COMMENT Another way to obtain the graph of an equation is to use a graphing utility. Read Section 2, *Using a Graphing Utility to Graph Equations*, in the Appendix.

One purpose of this book is to investigate the properties of equations in order to decide whether a graph is complete. Sometimes we shall graph equations by plotting points. Shortly, we shall investigate various techniques that will enable us to graph an equation without plotting so many points.

Two techniques that sometimes reduce the number of points required to graph an equation involve finding *intercepts* and checking for *symmetry*.

2 Find Intercepts from a Graph

The points, if any, at which a graph crosses or touches the coordinate axes are called the **intercepts**. See Figure 14. The x-coordinate of a point at which the graph crosses or touches the x-axis is an **x-intercept**, and the y-coordinate of a point at which the graph crosses or touches the y-axis is a **y-intercept**.

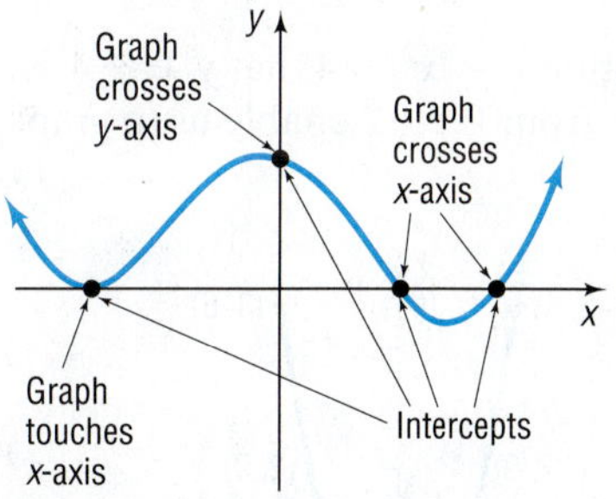

Figure 14

EXAMPLE 4 **Finding Intercepts from a Graph**

Find the intercepts of the graph in Figure 15. What are its x-intercepts? What are its y-intercepts?

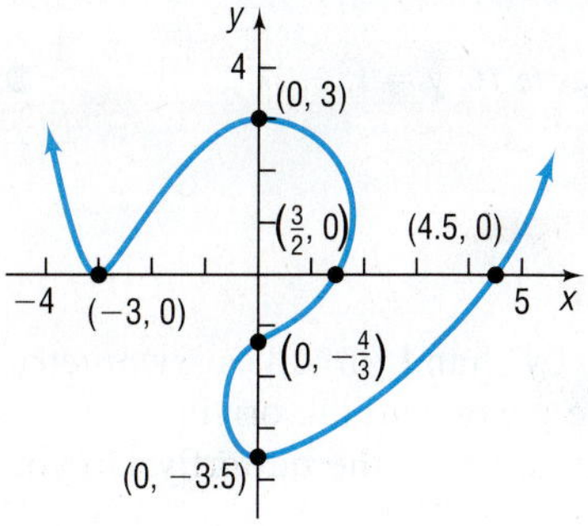

Figure 15

Solution The intercepts of the graph are the points

$$(-3, 0),\quad (0, 3),\quad \left(\frac{3}{2}, 0\right),\quad \left(0, -\frac{4}{3}\right),\quad (0, -3.5),\quad (4.5, 0)$$

The x-intercepts are $-3, \frac{3}{2}$, and 4.5; the y-intercepts are $-3.5, -\frac{4}{3}$, and 3.

In Example 4, note the following usage: If the type of intercept (x- versus y-) is not specified, then report the intercept as an ordered pair. However, if the type of intercept is specified, then report the coordinate of the specified intercept. For

x-intercepts, report the x-coordinate of the intercept; for y-intercepts, report the y-coordinate of the intercept.

Now Work PROBLEM 41(a)

3 Find Intercepts from an Equation

The intercepts of a graph can be found from its equation by using the fact that points on the x-axis have y-coordinates equal to 0, and points on the y-axis have x-coordinates equal to 0.

COMMENT For many equations, finding intercepts may not be so easy. In such cases, a graphing utility can be used. Read the first part of Section 3, *Using a Graphing Utility to Locate Intercepts and Check for Symmetry*, in the Appendix, to find out how to locate intercepts using a graphing utility. ■

Procedure for Finding Intercepts

1. To find the x-intercept(s), if any, of the graph of an equation, let $y = 0$ in the equation and solve for x, where x is a real number.
2. To find the y-intercept(s), if any, of the graph of an equation, let $x = 0$ in the equation and solve for y, where y is a real number.

EXAMPLE 5 **Finding Intercepts from an Equation**

Find the x-intercept(s) and the y-intercept(s) of the graph of $y = x^2 - 4$. Then graph $y = x^2 - 4$ by plotting points.

Solution To find the x-intercept(s), let $y = 0$ and obtain the equation

$$x^2 - 4 = 0 \quad y = x^2 - 4 \text{ with } y = 0$$

$$(x + 2)(x - 2) = 0 \quad \text{Factor.}$$

$$x + 2 = 0 \quad \text{or} \quad x - 2 = 0 \quad \text{Zero-Product Property}$$

$$x = -2 \quad \text{or} \quad x = 2 \quad \text{Solve.}$$

The equation has two solutions, -2 and 2. The x-intercepts are -2 and 2.

To find the y-intercept(s), let $x = 0$ in the equation.

$$y = x^2 - 4$$
$$= 0^2 - 4 = -4$$

The y-intercept is -4.

Since $x^2 \geq 0$ for all x, we deduce from the equation $y = x^2 - 4$ that $y \geq -4$ for all x. This information, the intercepts, and the points from Table 2 enable us to graph $y = x^2 - 4$. See Figure 16.

Table 2

x	$y = x^2 - 4$	(x, y)
−3	5	(−3, 5)
−1	−3	(−1, −3)
1	−3	(1, −3)
3	5	(3, 5)

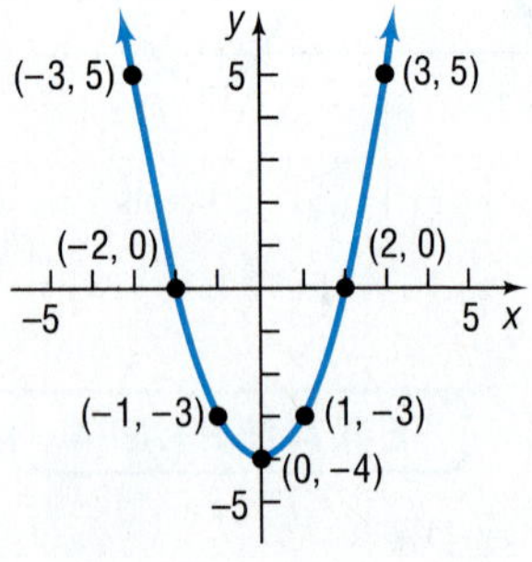

Figure 16 $y = x^2 - 4$ ●

Now Work PROBLEM 23

4 Test an Equation for Symmetry with Respect to the x-Axis, the y-Axis, and the Origin

Another helpful tool for graphing equations by hand involves *symmetry*, particularly symmetry with respect to the x-axis, the y-axis, and the origin.

Symmetry often occurs in nature. Consider the picture of the butterfly. Do you see the symmetry?

DEFINITION

A graph is said to be **symmetric with respect to the x-axis** if, for every point (x, y) on the graph, the point $(x, -y)$ is also on the graph.

Figure 17 illustrates the definition. Note that when a graph is symmetric with respect to the x-axis, the part of the graph above the x-axis is a reflection (or mirror image) of the part below it, and vice versa.

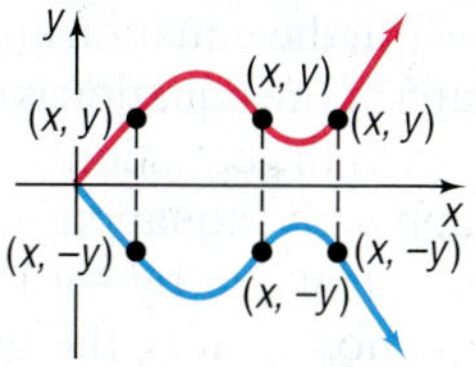

Figure 17
Symmetry with respect to the x-axis

EXAMPLE 6

Points Symmetric with Respect to the *x*-Axis

If a graph is symmetric with respect to the x-axis, and the point $(3, 2)$ is on the graph, then the point $(3, -2)$ is also on the graph.

●

DEFINITION

A graph is said to be **symmetric with respect to the y-axis** if, for every point (x, y) on the graph, the point $(-x, y)$ is also on the graph.

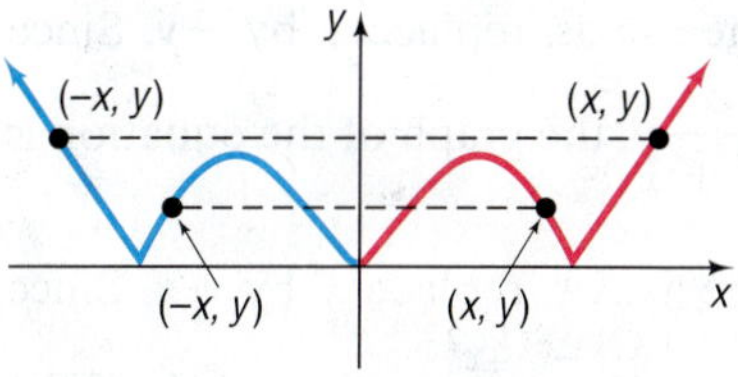

Figure 18
Symmetry with respect to the y-axis

Figure 18 illustrates the definition. When a graph is symmetric with respect to the y-axis, the part of the graph to the right of the y-axis is a reflection of the part to the left of it, and vice versa.

EXAMPLE 7

Points Symmetric with Respect to the *y*-Axis

If a graph is symmetric with respect to the y-axis and the point $(5, 8)$ is on the graph, then the point $(-5, 8)$ is also on the graph.

●

DEFINITION

A graph is said to be **symmetric with respect to the origin** if, for every point (x, y) on the graph, the point $(-x, -y)$ is also on the graph.

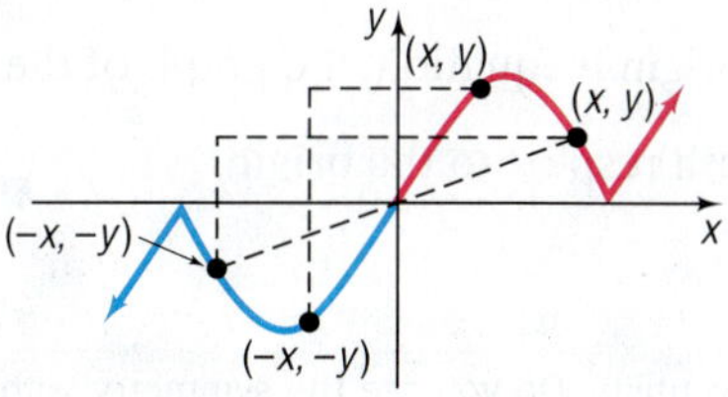

Figure 19
Symmetry with respect to the origin

Figure 19 illustrates the definition. Symmetry with respect to the origin may be viewed in three ways:

1. As a reflection about the y-axis, followed by a reflection about the x-axis
2. As a projection along a line through the origin so that the distances from the origin are equal
3. As half of a complete revolution about the origin

EXAMPLE 8

Points Symmetric with Respect to the Origin

If a graph is symmetric with respect to the origin, and the point $(4, 2)$ is on the graph, then the point $(-4, -2)$ is also on the graph.

●

Now Work PROBLEMS 31 AND 41(b)

When the graph of an equation is symmetric with respect to a coordinate axis or the origin, the number of points that you need to plot in order to see the pattern is reduced. For example, if the graph of an equation is symmetric with respect to the

y-axis, then once points to the right of the y-axis are plotted, an equal number of points on the graph can be obtained by reflecting them about the y-axis. Because of this, before we graph an equation, we should first determine whether it has any symmetry. The following tests are used for this purpose.

Tests for Symmetry

To test the graph of an equation for symmetry with respect to the

x-Axis Replace y by $-y$ in the equation and simplify. If an equivalent equation results, the graph of the equation is symmetric with respect to the x-axis.

y-Axis Replace x by $-x$ in the equation and simplify. If an equivalent equation results, the graph of the equation is symmetric with respect to the y-axis.

Origin Replace x by $-x$ and y by $-y$ in the equation and simplify. If an equivalent equation results, the graph of the equation is symmetric with respect to the origin.

EXAMPLE 9 **Testing an Equation for Symmetry**

Test $y = \dfrac{4x^2}{x^2+1}$ for symmetry.

Solution *x-Axis:* To test for symmetry with respect to the x-axis, replace y by $-y$. Since $-y = \dfrac{4x^2}{x^2+1}$ is not equivalent to $y = \dfrac{4x^2}{x^2+1}$, the graph of the equation is not symmetric with respect to the x-axis.

y-Axis: To test for symmetry with respect to the y-axis, replace x by $-x$. Since $y = \dfrac{4(-x)^2}{(-x)^2+1} = \dfrac{4x^2}{x^2+1}$ is equivalent to $y = \dfrac{4x^2}{x^2+1}$, the graph of the equation is symmetric with respect to the y-axis.

Origin: To test for symmetry with respect to the origin, replace x by $-x$ and y by $-y$.

$$-y = \frac{4(-x)^2}{(-x)^2+1} \quad \text{Replace } x \text{ by } -x \text{ and } y \text{ by } -y.$$

$$-y = \frac{4x^2}{x^2+1} \quad \text{Simplify.}$$

$$y = -\frac{4x^2}{x^2+1} \quad \text{Multiply both sides by } -1.$$

Since the result is not equivalent to the original equation, the graph of the equation $y = \dfrac{4x^2}{x^2+1}$ is not symmetric with respect to the origin. ●

Seeing the Concept

Figure 20 shows the graph of $y = \dfrac{4x^2}{x^2+1}$ using a graphing utility. Do you see the symmetry with respect to the y-axis?

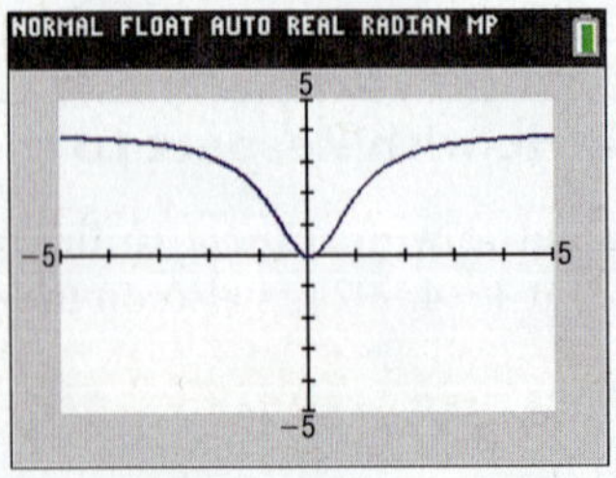

Figure 20 $y = \dfrac{4x^2}{x^2+1}$

Now Work PROBLEM 61

5 Know How to Graph Key Equations

The next three examples use intercepts, symmetry, and point plotting to obtain the graphs of key equations. It is important to know the graphs of these key equations because we use them later. The first of these is $y = x^3$.

EXAMPLE 10 Graphing the Equation $y = x^3$ by Finding Intercepts, Checking for Symmetry, and Plotting Points

Graph the equation $y = x^3$ by plotting points. Find any intercepts and check for symmetry first.

Solution First, find the intercepts. When $x = 0$, then $y = 0$; and when $y = 0$, then $x = 0$. The origin $(0, 0)$ is the only intercept. Now test for symmetry.

x-Axis: Replace y by $-y$. Since $-y = x^3$ is not equivalent to $y = x^3$, the graph is not symmetric with respect to the x-axis.

y-Axis: Replace x by $-x$. Since $y = (-x)^3 = -x^3$ is not equivalent to $y = x^3$, the graph is not symmetric with respect to the y-axis.

Origin: Replace x by $-x$ and y by $-y$. Since $-y = (-x)^3 = -x^3$ is equivalent to $y = x^3$ (multiply both sides by -1), the graph is symmetric with respect to the origin.

To graph $y = x^3$, use the equation to obtain several points on the graph. Because of the symmetry, we need to locate only points on the graph for which $x \geq 0$. See Table 3. Since $(1, 1)$ is on the graph, and the graph is symmetric with respect to the origin, the point $(-1, -1)$ is also on the graph. Plot the points from Table 3 and use the symmetry. Figure 21 shows the graph.

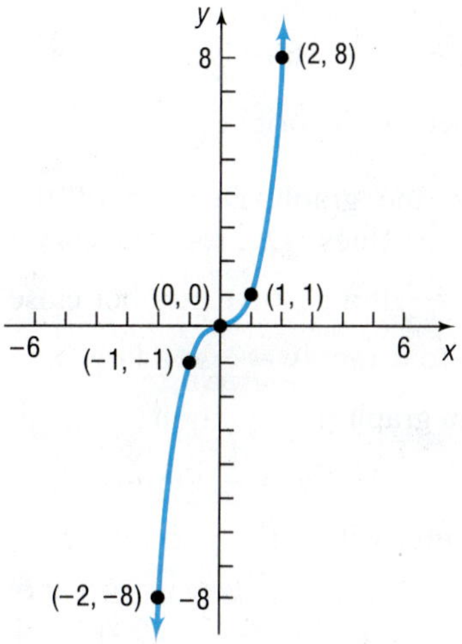

Figure 21 $y = x^3$

Table 3

x	$y = x^3$	(x, y)
0	0	(0, 0)
1	1	(1, 1)
2	8	(2, 8)
3	27	(3, 27)

EXAMPLE 11 Graphing the Equation $x = y^2$

(a) Graph the equation $x = y^2$. Find any intercepts and check for symmetry first.

(b) Graph $x = y^2$, $y \geq 0$.

Solution (a) The lone intercept is $(0, 0)$. The graph is symmetric with respect to the x-axis. (Do you see why? Replace y by $-y$.) Figure 22 shows the graph.

(b) If we restrict y so that $y \geq 0$, the equation $x = y^2$, $y \geq 0$, may be written equivalently as $y = \sqrt{x}$. The portion of the graph of $x = y^2$ in quadrant I is therefore the graph of $y = \sqrt{x}$. See Figure 23.

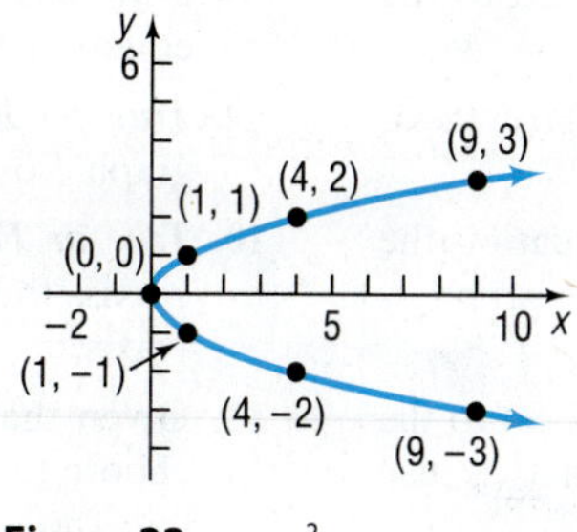

Figure 22 $x = y^2$

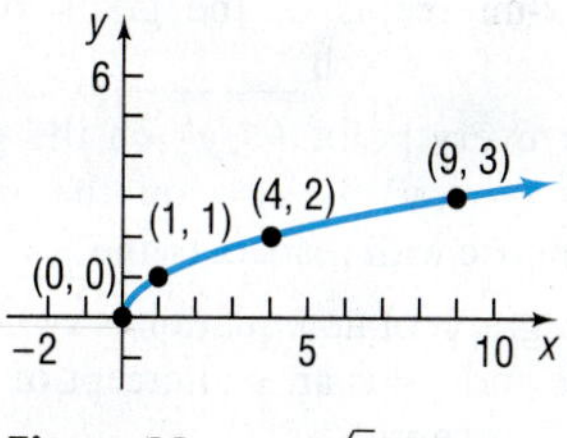

Figure 23 $y = \sqrt{x}$

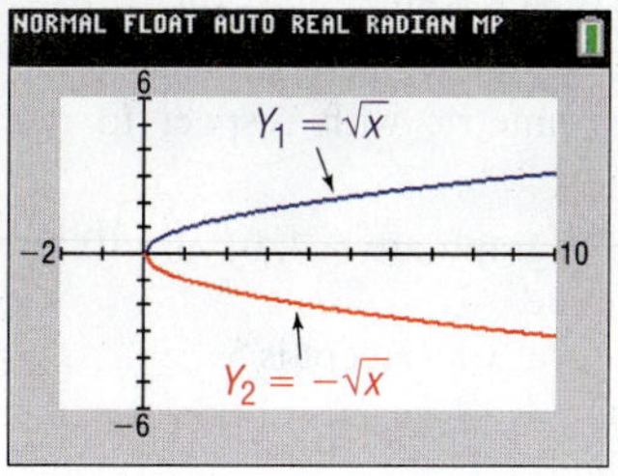

Figure 24

COMMENT To see the graph of the equation $x = y^2$ on a graphing calculator, you will need to graph two equations: $Y_1 = \sqrt{x}$ and $Y_2 = -\sqrt{x}$. We discuss why in Chapter 3. See Figure 24.

EXAMPLE 12

Graphing the Equation $y = \frac{1}{x}$

Graph the equation $y = \frac{1}{x}$. First, find any intercepts and check for symmetry.

Solution Check for intercepts first. If we let $x = 0$, we obtain 0 in the denominator, which makes y undefined. We conclude that there is no y-intercept. If we let $y = 0$, we get the equation $\frac{1}{x} = 0$, which has no solution. We conclude that there is no x-intercept. The graph of $y = \frac{1}{x}$ does not cross or touch the coordinate axes.

Next check for symmetry:

x-Axis: Replacing y by $-y$ yields $-y = \frac{1}{x}$, which is not equivalent to $y = \frac{1}{x}$.

y-Axis: Replacing x by $-x$ yields $y = \frac{1}{-x} = -\frac{1}{x}$, which is not equivalent to $y = \frac{1}{x}$.

Origin: Replacing x by $-x$ and y by $-y$ yields $-y = -\frac{1}{x}$, which is equivalent to $y = \frac{1}{x}$. The graph is symmetric with respect to the origin.

Table 4

x	$y = \frac{1}{x}$	(x, y)
$\frac{1}{10}$	10	$\left(\frac{1}{10}, 10\right)$
$\frac{1}{3}$	3	$\left(\frac{1}{3}, 3\right)$
$\frac{1}{2}$	2	$\left(\frac{1}{2}, 2\right)$
1	1	(1, 1)
2	$\frac{1}{2}$	$\left(2, \frac{1}{2}\right)$
3	$\frac{1}{3}$	$\left(3, \frac{1}{3}\right)$
10	$\frac{1}{10}$	$\left(10, \frac{1}{10}\right)$

Now set up Table 4, listing several points on the graph. Because of the symmetry with respect to the origin, we use only positive values of x. From Table 4 we infer that if x is a large and positive number, then $y = \frac{1}{x}$ is a positive number close to 0. We also infer that if x is a positive number close to 0, then $y = \frac{1}{x}$ is a large and positive number. Armed with this information, we can graph the equation.

Figure 25 illustrates some of these points and the graph of $y = \frac{1}{x}$. Observe how the absence of intercepts and the existence of symmetry with respect to the origin were utilized. ●

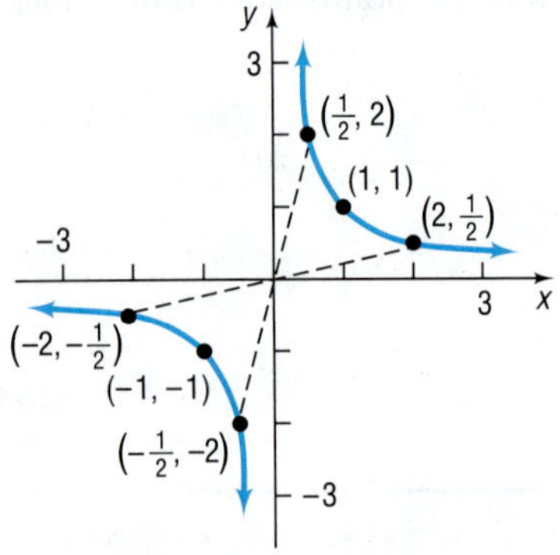

Figure 25 $y = \frac{1}{x}$

COMMENT Refer to Example 2 in the Appendix, Section 3, for the graph of $y = \frac{1}{x}$ found using a graphing utility. ■

2.2 Assess Your Understanding

'Are You Prepared?' *Answers are given at the end of these exercises. If you get a wrong answer, read the pages listed in red.*

1. Solve the equation $2(x + 3) - 1 = -7$. (pp. 82–85)
2. Solve the equation $x^2 - 9 = 0$. (pp. 93–94)

Concepts and Vocabulary

3. The points, if any, at which a graph crosses or touches the coordinate axes are called ________.
4. The x-intercepts of the graph of an equation are those x-values for which ________.
5. If for every point (x, y) on the graph of an equation the point $(-x, y)$ is also on the graph, then the graph is symmetric with respect to the ________.
6. If the graph of an equation is symmetric with respect to the y-axis and -4 is an x-intercept of this graph, then __ is also an x-intercept.
7. If the graph of an equation is symmetric with respect to the origin and $(3, -4)$ is a point on the graph, then ________ is also a point on the graph.
8. ***True or False*** To find the y-intercepts of the graph of an equation, let $x = 0$ and solve for y.
9. ***True or False*** The y-coordinate of a point at which the graph crosses or touches the x-axis is an x-intercept.
10. ***True or False*** If a graph is symmetric with respect to the x-axis, then it cannot be symmetric with respect to the y-axis.
11. Given that the intercepts of a graph are (–4, 0) and (0, 5), choose the statement that is true.
    (a) The y-intercept is –4, and the x-intercept is 5.
    (b) The y-intercepts are –4 and 5.
    (c) The x-intercepts are –4 and 5.
    (d) The x-intercept is –4, and the y-intercept is 5.

12. To test whether the graph of an equation is symmetric with respect to the origin, replace _______ in the equation and simplify. If an equivalent equation results, then the graph is symmetric with respect to the origin.

(a) x by $-x$
(b) y by $-y$
(c) x by $-x$ and y by $-y$
(d) x by $-y$ and y by $-x$

Skill Building

In Problems 13–18, determine which of the given points are on the graph of the equation.

13. Equation: $y = x^4 - \sqrt{x}$
Points: $(0, 0)$; $(1, 1)$; $(2, 4)$

14. Equation: $y = x^3 - 2\sqrt{x}$
Points: $(0, 0)$; $(1, 1)$; $(1, -1)$

15. Equation: $y^2 = x^2 + 9$
Points: $(0, 3)$; $(3, 0)$; $(-3, 0)$

16. Equation: $y^3 = x + 1$
Points: $(1, 2)$; $(0, 1)$; $(-1, 0)$

17. Equation: $x^2 + y^2 = 4$
Points: $(0, 2)$; $(-2, 2)$; $(\sqrt{2}, \sqrt{2})$

18. Equation: $x^2 + 4y^2 = 4$
Points: $(0, 1)$; $(2, 0)$; $\left(2, \frac{1}{2}\right)$

In Problems 19–30, find the intercepts and graph each equation by plotting points. Be sure to label the intercepts.

19. $y = x + 2$
20. $y = x - 6$
21. $y = 2x + 8$
22. $y = 3x - 9$
23. $y = x^2 - 1$
24. $y = x^2 - 9$
25. $y = -x^2 + 4$
26. $y = -x^2 + 1$
27. $2x + 3y = 6$
28. $5x + 2y = 10$
29. $9x^2 + 4y = 36$
30. $4x^2 + y = 4$

In Problems 31–40, plot each point. Then plot the point that is symmetric to it with respect to (a) the x-axis; (b) the y-axis; (c) the origin.

31. $(3, 4)$
32. $(5, 3)$
33. $(-2, 1)$
34. $(4, -2)$
35. $(5, -2)$
36. $(-1, -1)$
37. $(-3, -4)$
38. $(4, 0)$
39. $(0, -3)$
40. $(-3, 0)$

In Problems 41–52, the graph of an equation is given. (a) Find the intercepts. (b) Indicate whether the graph is symmetric with respect to the x-axis, the y-axis, or the origin.

41.
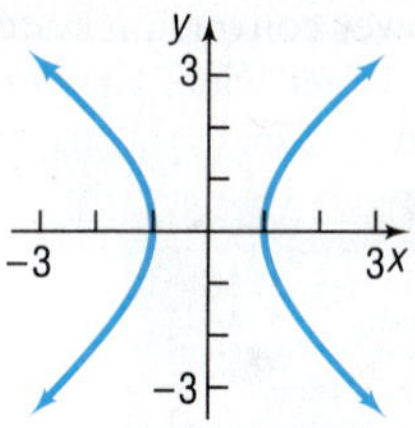

42.
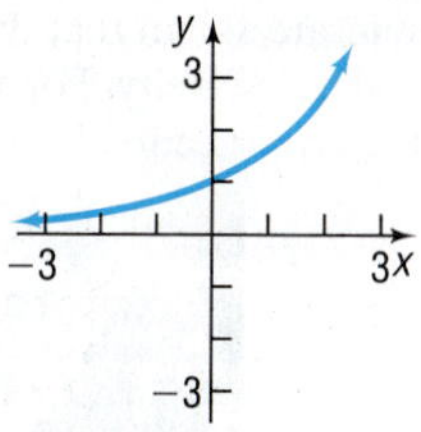

43.
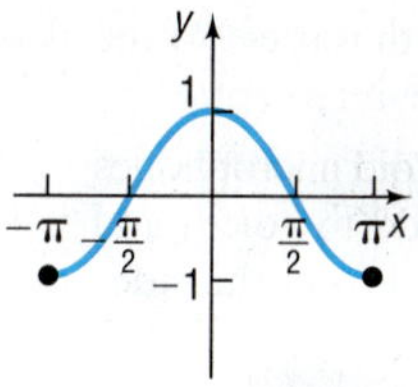

44.
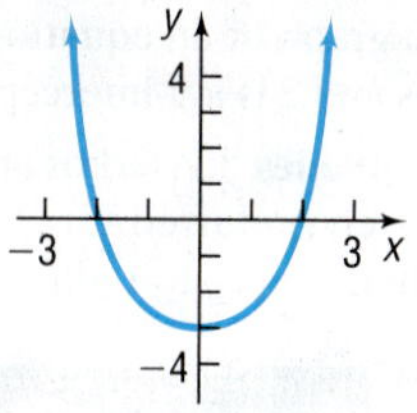

45.
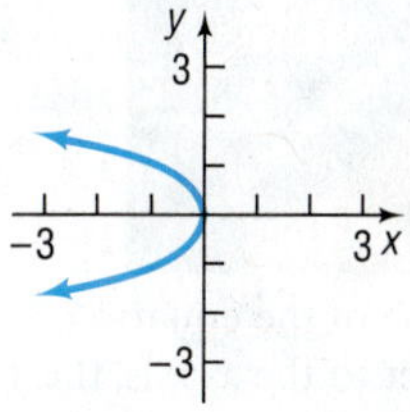

46.
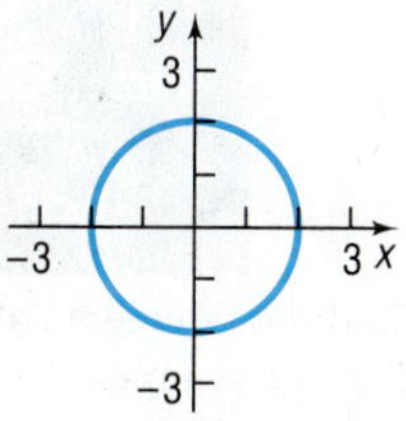

47.
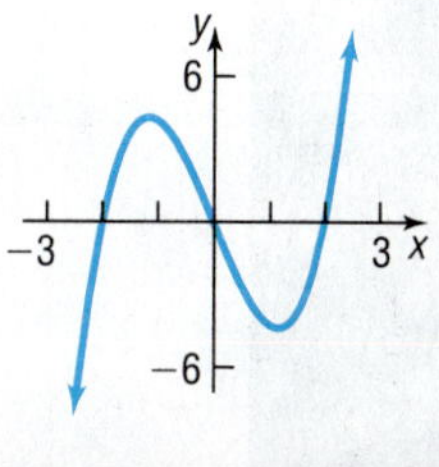

48.
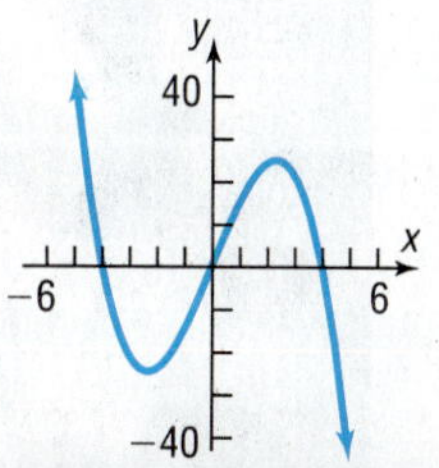

49.
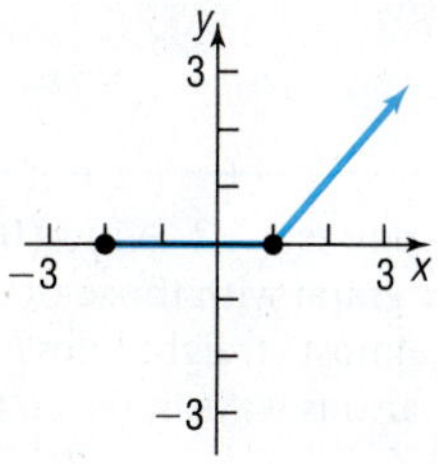

50.
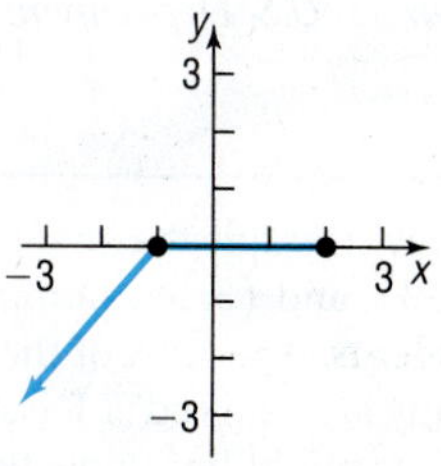

51.
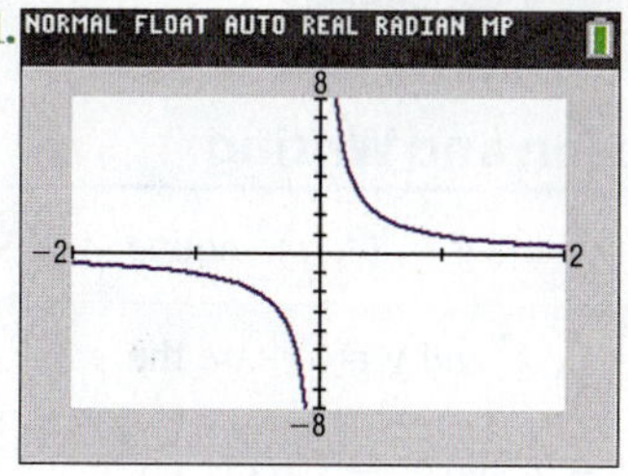

52.
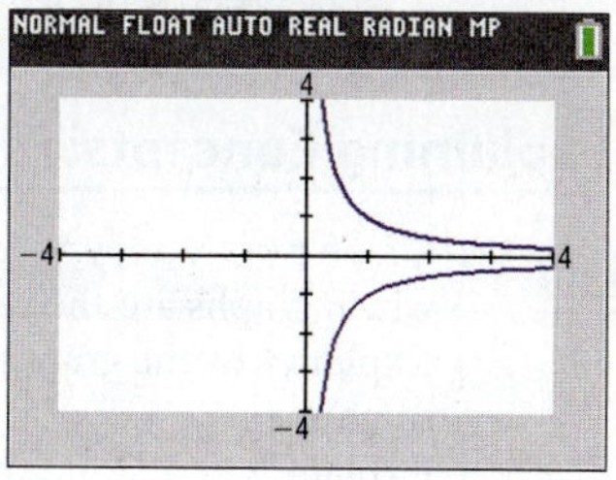

In Problems 53–56, draw a complete graph so that it has the type of symmetry indicated.

53. y-axis
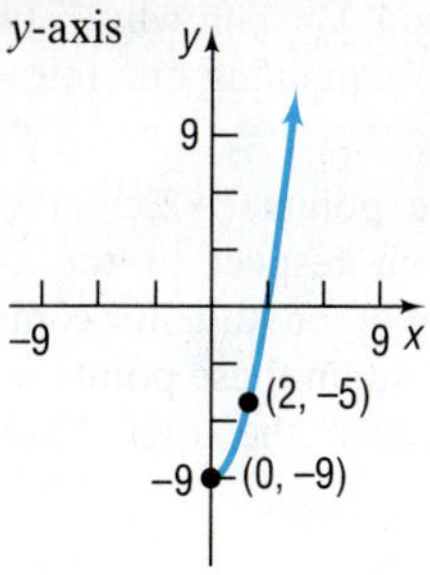

54. x-axis
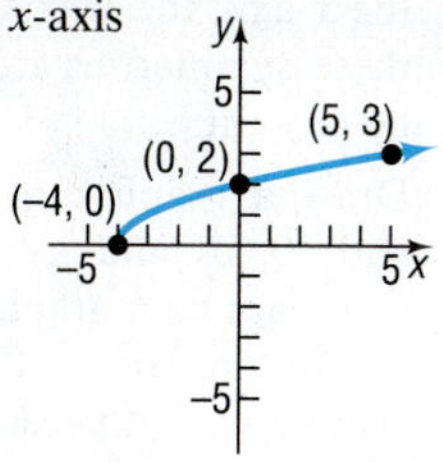

55. Origin
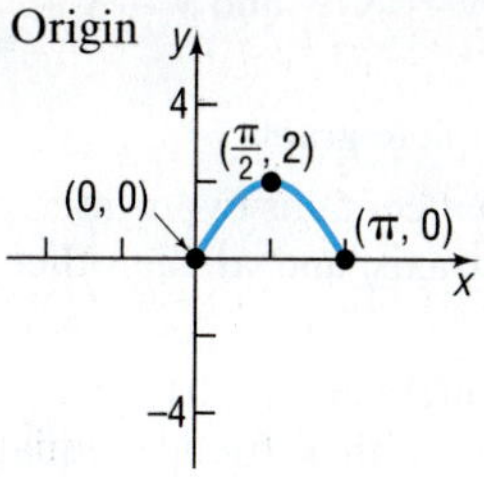

56. y-axis
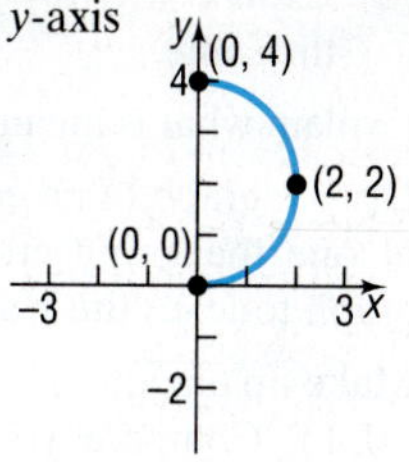

In Problems 57–72, list the intercepts and test for symmetry.

57. $y^2 = x + 4$

58. $y^2 = x + 9$

59. $y = \sqrt[3]{x}$

60. $y = \sqrt[5]{x}$

61. $x^2 + y - 9 = 0$

62. $x^2 - y - 4 = 0$

63. $9x^2 + 4y^2 = 36$

64. $4x^2 + y^2 = 4$

65. $y = x^3 - 27$

66. $y = x^4 - 1$

67. $y = x^2 - 3x - 4$

68. $y = x^2 + 4$

69. $y = \dfrac{3x}{x^2 + 9}$

70. $y = \dfrac{x^2 - 4}{2x}$

71. $y = \dfrac{-x^3}{x^2 - 9}$

72. $y = \dfrac{x^4 + 1}{2x^5}$

In Problems 73–76, draw a quick sketch of the graph of each equation.

73. $y = x^3$

74. $x = y^2$

75. $y = \sqrt{x}$

76. $y = \dfrac{1}{x}$

77. If $(a, 4)$ is a point on the graph of $y = x^2 + 3x$, what is a?

78. If $(a, -5)$ is a point on the graph of $y = x^2 + 6x$, what is a?

Applications and Extensions

79. Given that the point (1, 2) is on the graph of an equation that is symmetric with respect to the origin, what other point is on the graph?

80. If the graph of an equation is symmetric with respect to the y-axis and 6 is an x-intercept of this graph, name another x-intercept.

81. If the graph of an equation is symmetric with respect to the origin and -4 is an x-intercept of this graph, name another x-intercept.

82. If the graph of an equation is symmetric with respect to the x-axis and 2 is a y-intercept, name another y-intercept.

83. Microphones In studios and on stages, cardioid microphones are often preferred for the richness they add to voices and for their ability to reduce the level of sound from the sides and rear of the microphone. Suppose one such cardioid pattern is given by the equation $(x^2 + y^2 - x)^2 = x^2 + y^2$.

(a) Find the intercepts of the graph of the equation.

(b) Test for symmetry with respect to the x-axis, the y-axis, and the origin.

Source: *www.notaviva.com*

84. Solar Energy The solar electric generating systems at Kramer Junction, California, use parabolic troughs to heat a heat-transfer fluid to a high temperature. This fluid is used to generate steam that drives a power conversion system to produce electricity. For troughs 7.5 feet wide, an equation for the cross section is $16y^2 = 120x - 225$.

(a) Find the intercepts of the graph of the equation.

(b) Test for symmetry with respect to the x-axis, the y-axis, and the origin.

Source: *U.S. Department of Energy*

Explaining Concepts: Discussion and Writing

85. (a) Graph $y = \sqrt{x^2}$, $y = x$, $y = |x|$, and $y = (\sqrt{x})^2$, noting which graphs are the same.

(b) Explain why the graphs of $y = \sqrt{x^2}$ and $y = |x|$ are the same.

(c) Explain why the graphs of $y = x$ and $y = (\sqrt{x})^2$ are not the same.

(d) Explain why the graphs of $y = \sqrt{x^2}$ and $y = x$ are not the same.

86. Explain what is meant by a complete graph.

87. Draw a graph of an equation that contains two x-intercepts; at one the graph crosses the x-axis, and at the other the graph touches the x-axis.

88. Make up an equation with the intercepts $(2, 0)$, $(4, 0)$, and $(0, 1)$. Compare your equation with a friend's equation. Comment on any similarities.

89. Draw a graph that contains the points $(-2, -1)$, $(0, 1)$, $(1, 3)$, and $(3, 5)$. Compare your graph with those of other students. Are most of the graphs almost straight lines? How many are "curved"? Discuss the various ways in which these points might be connected.

90. An equation is being tested for symmetry with respect to the x-axis, the y-axis, and the origin. Explain why, if two of these symmetries are present, the remaining one must also be present.

91. Draw a graph that contains the points $(-2, 5)$, $(-1, 3)$, and $(0, 2)$ and is symmetric with respect to the y-axis. Compare your graph with those of other students; comment on any similarities. Can a graph contain these points and be symmetric with respect to the x-axis? the origin? Why or why not?

Retain Your Knowledge

Problems 92–95 are based on material learned earlier in the course. The purpose of these problems is to keep the material fresh in your mind so that you are better prepared for the final exam.

92. Find the value of $\frac{x+y}{x-y}$ if $x = 6$ and $y = -2$.

93. Factor $3x^2 - 30x + 75$ completely.

94. Simplify: $\sqrt{-196}$

95. Solve $x^2 - 8x + 4 = 0$ by completing the square.

'Are You Prepared?' Answers

1. $\{-6\}$ **2.** $\{-3, 3\}$

2.3 Lines

In this section we study a certain type of equation that contains two variables, called a *linear equation,* and its graph, a *line*.

1 Calculate and Interpret the Slope of a Line

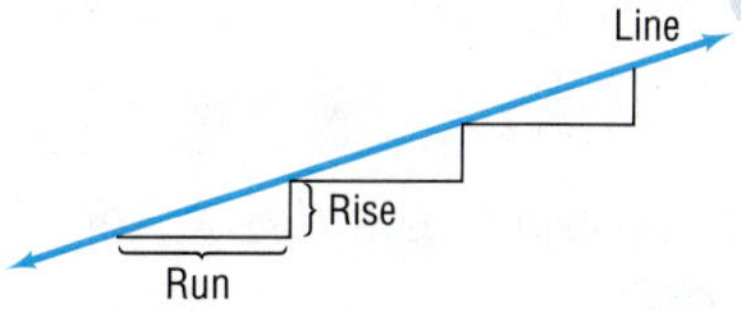

Figure 26

Consider the staircase illustrated in Figure 26. Each step contains exactly the same horizontal **run** and the same vertical **rise**. The ratio of the rise to the run, called the *slope*, is a numerical measure of the steepness of the staircase. For example, if the run is increased and the rise remains the same, the staircase becomes less steep. If the run is kept the same but the rise is increased, the staircase becomes more steep. This important characteristic of a line is best defined using rectangular coordinates.

DEFINITION

Let $P = (x_1, y_1)$ and $Q = (x_2, y_2)$ be two distinct points. If $x_1 \neq x_2$, the **slope *m*** of the nonvertical line L containing P and Q is defined by the formula

$$m = \frac{y_2 - y_1}{x_2 - x_1} \qquad x_1 \neq x_2 \qquad \textbf{(1)}$$

If $x_1 = x_2$, then L is a **vertical line** and the slope m of L is **undefined** (since this results in division by 0).

Figure 27(a) on page 168 provides an illustration of the slope of a nonvertical line; Figure 27(b) illustrates a vertical line.

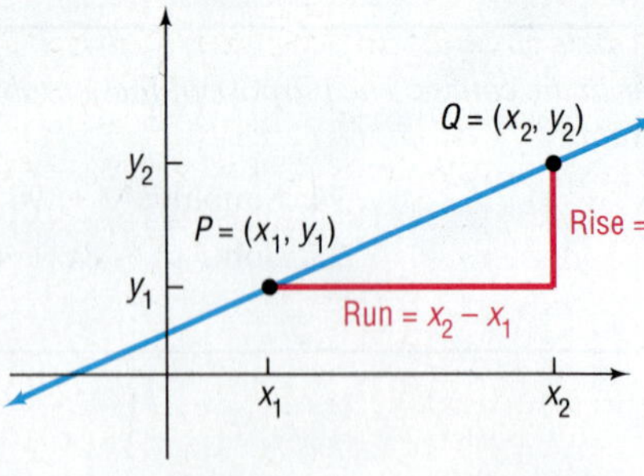

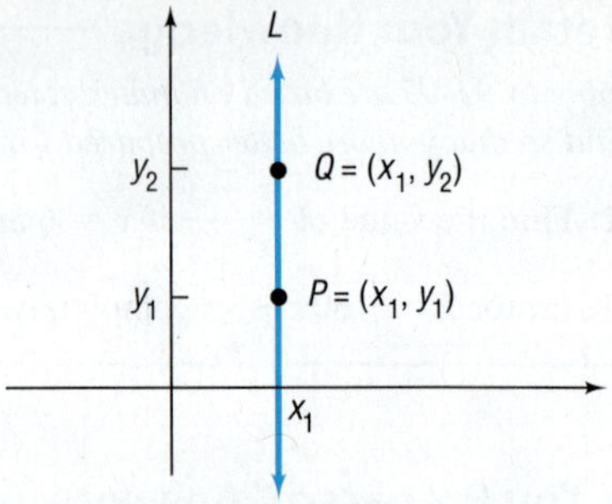

Figure 27 **(a)** Slope of L is $m = \dfrac{y_2 - y_1}{x_2 - x_1}$, $x_1 \neq x_2$ **(b)** Slope is undefined; L is vertical

> **In Words**
> The symbol Δ is the Greek uppercase letter delta. In mathematics, Δ is read "change in," so $\dfrac{\Delta y}{\Delta x}$ is read "change in y divided by change in x."

As Figure 27(a) illustrates, the slope m of a nonvertical line may be viewed as

$$m = \frac{y_2 - y_1}{x_2 - x_1} = \frac{\text{Rise}}{\text{Run}} \quad \text{or as } m = \frac{y_2 - y_1}{x_2 - x_1} = \frac{\text{Change in } y}{\text{Change in } x} = \frac{\Delta y}{\Delta x}$$

That is, the slope m of a nonvertical line measures the amount y changes when x changes from x_1 to x_2. The expression $\dfrac{\Delta y}{\Delta x}$ is called the **average rate of change** of y with respect to x.

Two comments about computing the slope of a nonvertical line may prove helpful:

1. Any two distinct points on the line can be used to compute the slope of the line. (See Figure 28 for justification.) Since any two distinct points can be used to compute the slope of a line, the average rate of change of a line is always the same number.

Figure 28
Triangles ABC and PQR are similar (equal angles), so ratios of corresponding sides are proportional. Then

Slope using P and Q $= \dfrac{y_2 - y_1}{x_2 - x_1} = \dfrac{d(B, C)}{d(A, C)}$ = Slope using A and B

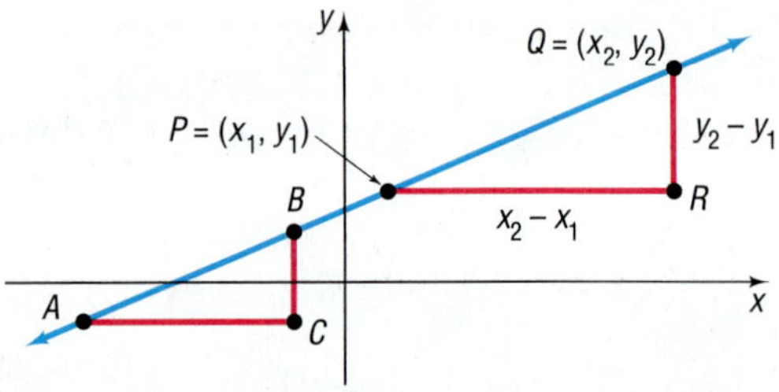

2. The slope of a line may be computed from $P = (x_1, y_1)$ to $Q = (x_2, y_2)$ or from Q to P because

$$\frac{y_2 - y_1}{x_2 - x_1} = \frac{y_1 - y_2}{x_1 - x_2}$$

EXAMPLE 1

Finding and Interpreting the Slope of a Line Given Two Points

The slope m of the line containing the points $(1, 2)$ and $(5, -3)$ may be computed as

$$m = \frac{-3 - 2}{5 - 1} = \frac{-5}{4} = -\frac{5}{4} \quad \text{or as} \quad m = \frac{2 - (-3)}{1 - 5} = \frac{5}{-4} = -\frac{5}{4}$$

For every 4-unit change in x, y will change by -5 units. That is, if x increases by 4 units, then y will decrease by 5 units. The average rate of change of y with respect to x is $-\dfrac{5}{4}$.

Now Work PROBLEMS 13 AND 19

EXAMPLE 2 **Finding the Slopes of Various Lines Containing the Same Point (2, 3)**

Compute the slopes of the lines L_1, L_2, L_3, and L_4 containing the following pairs of points. Graph all four lines on the same set of coordinate axes.

$$\begin{aligned} L_1&: P = (2,3) \quad Q_1 = (-1,-2)\\ L_2&: P = (2,3) \quad Q_2 = (3,-1)\\ L_3&: P = (2,3) \quad Q_3 = (5,3)\\ L_4&: P = (2,3) \quad Q_4 = (2,5)\end{aligned}$$

Solution Let m_1, m_2, m_3, and m_4 denote the slopes of the lines L_1, L_2, L_3, and L_4, respectively. Then

$$m_1 = \frac{-2-3}{-1-2} = \frac{-5}{-3} = \frac{5}{3}$$ A rise of 5 divided by a run of 3

$$m_2 = \frac{-1-3}{3-2} = \frac{-4}{1} = -4$$

$$m_3 = \frac{3-3}{5-2} = \frac{0}{3} = 0$$

m_4 is undefined because $x_1 = x_2 = 2$

The graphs of these lines are given in Figure 29.

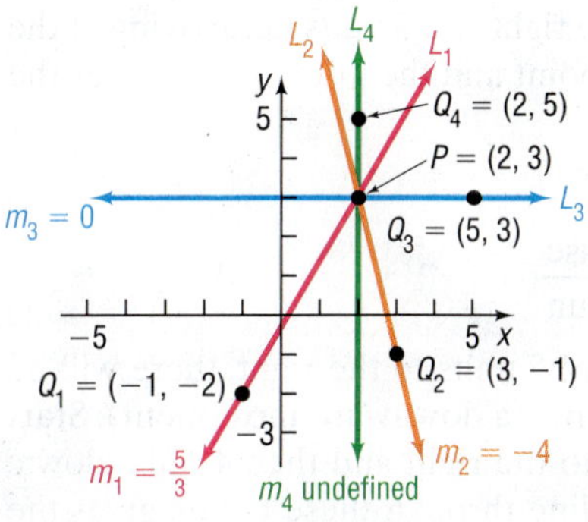

Figure 29

Figure 29 illustrates the following facts:

1. When the slope of a line is positive, the line slants upward from left to right (L_1).
2. When the slope of a line is negative, the line slants downward from left to right (L_2).
3. When the slope is 0, the line is horizontal (L_3).
4. When the slope is undefined, the line is vertical (L_4).

Seeing the Concept

On the same screen, graph the following equations:

$Y_1 = 0$ Slope of line is 0.

$Y_2 = \frac{1}{4}x$ Slope of line is $\frac{1}{4}$.

$Y_3 = \frac{1}{2}x$ Slope of line is $\frac{1}{2}$.

$Y_4 = x$ Slope of line is 1.

$Y_5 = 2x$ Slope of line is 2.

$Y_6 = 6x$ Slope of line is 6.

See Figure 30.

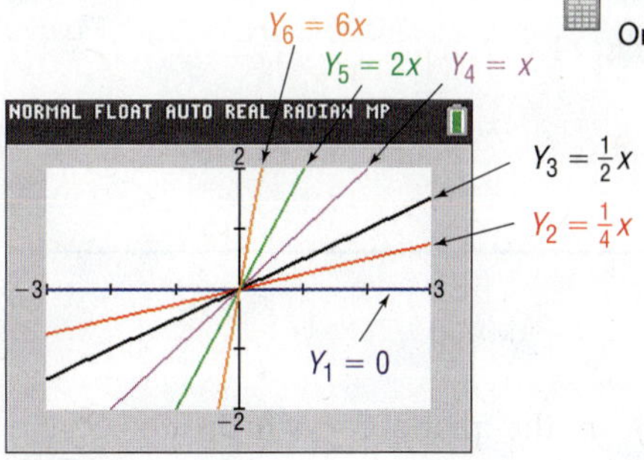

Figure 30

Seeing the Concept

On the same screen, graph the following equations:

$Y_1 = 0$ Slope of line is 0.

$Y_2 = -\frac{1}{4}x$ Slope of line is $-\frac{1}{4}$.

$Y_3 = -\frac{1}{2}x$ Slope of line is $-\frac{1}{2}$.

$Y_4 = -x$ Slope of line is -1.

$Y_5 = -2x$ Slope of line is -2.

$Y_6 = -6x$ Slope of line is -6.

See Figure 31.

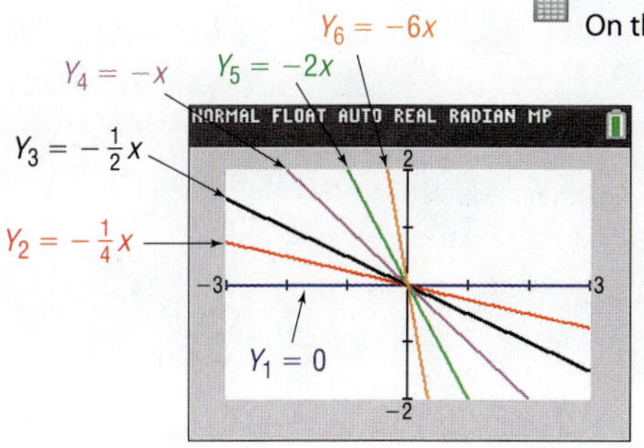

Figure 31

Figures 30 and 31 on page 169 illustrate that the closer the line is to the vertical position, the greater the magnitude of the slope.

2 Graph Lines Given a Point and the Slope

EXAMPLE 3 **Graphing a Line Given a Point and a Slope**

Draw a graph of the line that contains the point $(3, 2)$ and has a slope of:

(a) $\frac{3}{4}$ (b) $-\frac{4}{5}$

Solution (a) Slope $= \frac{\text{Rise}}{\text{Run}}$. The slope $\frac{3}{4}$ means that for every horizontal movement (run) of 4 units to the right, there will be a vertical movement (rise) of 3 units. Start at the given point $(3, 2)$ and move 4 units to the right and 3 units up, arriving at the point $(7, 5)$. Drawing the line through this point and the point $(3, 2)$ gives the graph. See Figure 32.

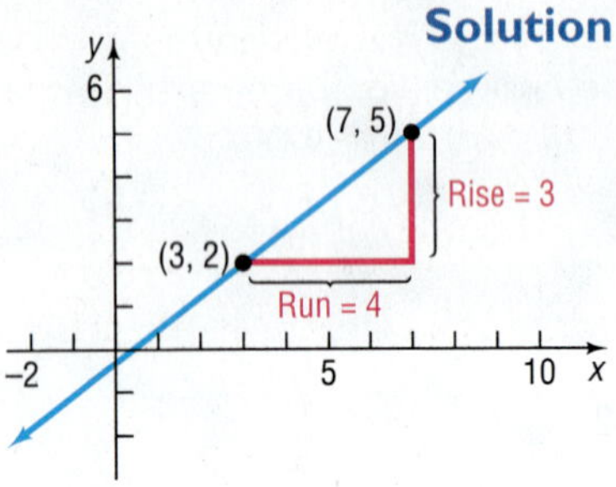

Figure 32

(b) The fact that the slope is

$$-\frac{4}{5} = \frac{-4}{5} = \frac{\text{Rise}}{\text{Run}}$$

means that for every horizontal movement of 5 units to the right, there will be a corresponding vertical movement of -4 units (a downward movement). Start at the given point $(3, 2)$ and move 5 units to the right and then 4 units down, arriving at the point $(8, -2)$. Drawing the line through these points gives the graph. See Figure 33.

Alternatively, consider that

$$-\frac{4}{5} = \frac{4}{-5} = \frac{\text{Rise}}{\text{Run}}$$

so for every horizontal movement of -5 units (a movement to the left), there will be a corresponding vertical movement of 4 units (upward). This approach leads to the point $(-2, 6)$, which is also on the graph of the line in Figure 33.

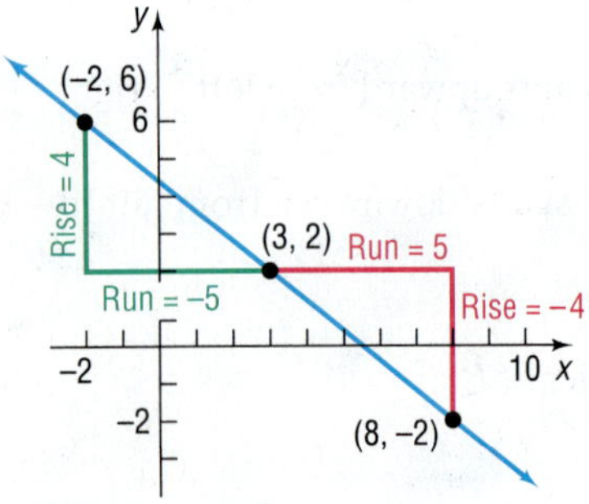

Figure 33

Now Work PROBLEM 25 (graph the line)

3 Find the Equation of a Vertical Line

EXAMPLE 4 **Graphing a Line**

Graph the equation: $x = 3$

Solution To graph $x = 3$, we find all points (x, y) in the plane for which $x = 3$. No matter what y-coordinate is used, the corresponding x-coordinate always equals 3. Consequently, the graph of the equation $x = 3$ is a vertical line with x-intercept 3 and undefined slope. See Figure 34.

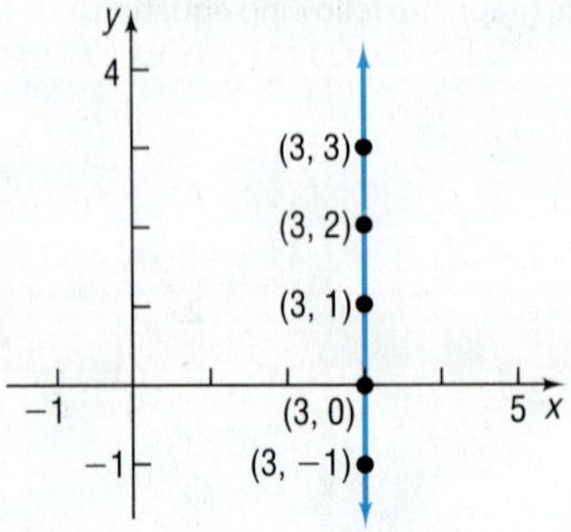

Figure 34 $x = 3$

Example 4 suggests the following result:

THEOREM

Equation of a Vertical Line

A vertical line is given by an equation of the form

$$x = a$$

where a is the x-intercept.

COMMENT To graph an equation using a graphing utility, we need to express the equation in the form $y = \{expression \text{ in } x\}$. But $x = 3$ cannot be put in this form. To overcome this, most graphing utilities have special commands for drawing vertical lines. DRAW, LINE, PLOT, and VERT are among the more common ones. Consult your manual to determine the correct methodology for your graphing utility. ■

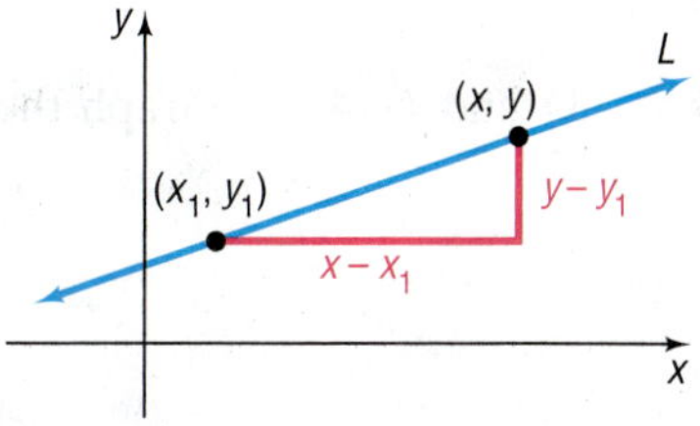

Figure 35

4 Use the Point–Slope Form of a Line; Identify Horizontal Lines

Let L be a nonvertical line with slope m that contains the point (x_1, y_1). See Figure 35. For any other point (x, y) on L, we have

$$m = \frac{y - y_1}{x - x_1} \quad \text{or} \quad y - y_1 = m(x - x_1)$$

THEOREM

Point–Slope Form of an Equation of a Line

An equation of a nonvertical line with slope m that contains the point (x_1, y_1) is

$$y - y_1 = m(x - x_1) \qquad \textbf{(2)}$$

EXAMPLE 5 **Using the Point–Slope Form of a Line**

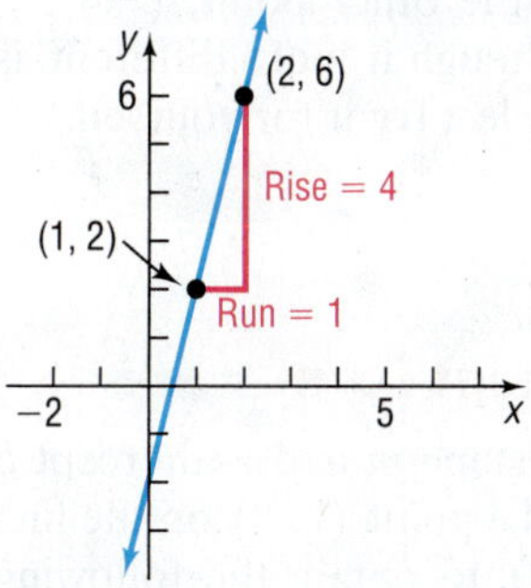

Figure 36 $y = 4x - 2$

An equation of the line with slope 4 that contains the point $(1, 2)$ can be found by using the point–slope form with $m = 4$, $x_1 = 1$, and $y_1 = 2$.

$$y - y_1 = m(x - x_1)$$
$$y - 2 = 4(x - 1) \qquad m = 4, x_1 = 1, y_1 = 2$$
$$y = 4x - 2 \qquad \text{Solve for } y.$$

See Figure 36 for the graph. ●

Now Work PROBLEM 25 (find the point-slope form)

EXAMPLE 6 **Finding the Equation of a Horizontal Line**

Find an equation of the horizontal line containing the point $(3, 2)$.

Solution Because all the y-values are equal on a horizontal line, the slope of a horizontal line is 0. To get an equation, use the point–slope form with $m = 0$, $x_1 = 3$, and $y_1 = 2$.

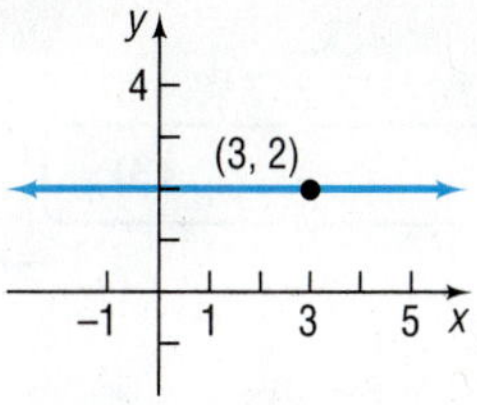

Figure 37 $y = 2$

$$y - y_1 = m(x - x_1)$$
$$y - 2 = 0 \cdot (x - 3) \qquad m = 0, x_1 = 3, \text{ and } y_1 = 2$$
$$y - 2 = 0$$
$$y = 2$$

See Figure 37 for the graph. ●

Example 6 suggests the following result:

THEOREM **Equation of a Horizontal Line**

A horizontal line is given by an equation of the form

$$y = b$$

where b is the y-intercept.

5 Find the Equation of a Line Given Two Points

EXAMPLE 7 **Finding an Equation of a Line Given Two Points**

Find an equation of the line containing the points $(2, 3)$ and $(-4, 5)$. Graph the line.

Solution First compute the slope of the line.

$$m = \frac{5-3}{-4-2} = \frac{2}{-6} = -\frac{1}{3} \qquad m = \frac{y_2 - y_1}{x_2 - x_1}$$

Use the point $(2, 3)$ and the slope $m = -\frac{1}{3}$ to get the point–slope form of the equation of the line.

$$y - 3 = -\frac{1}{3}(x - 2) \qquad y - y_1 = m(x - x_1)$$

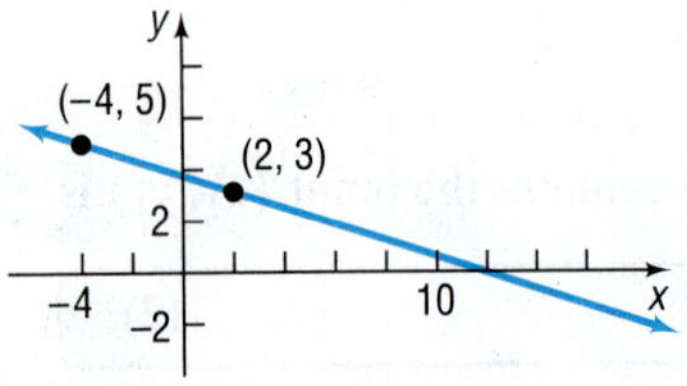

Figure 38 $y - 3 = -\frac{1}{3}(x - 2)$

See Figure 38 for the graph.

In the solution to Example 7, we could have used the other point, $(-4, 5)$, instead of the point $(2, 3)$. The equation that results, although it looks different, is equivalent to the equation that we obtained in the example. (Try it for yourself.)

Now Work PROBLEM 39

6 Write the Equation of a Line in Slope–Intercept Form

Another useful equation of a line is obtained when the slope m and y-intercept b are known. In this event, both the slope m of the line and a point $(0, b)$ on the line are known; then use the point–slope form, equation (2), to obtain the following equation:

$$y - b = m(x - 0) \quad \text{or} \quad y = mx + b$$

THEOREM **Slope–Intercept Form of an Equation of a Line**

An equation of a line with slope m and y-intercept b is

$$y = mx + b \tag{3}$$

Now Work PROBLEMS 47 AND 53 (express answer in slope–intercept form)

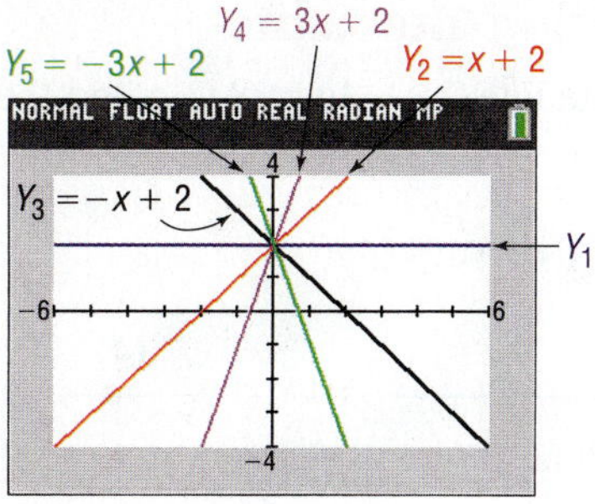

Figure 39 $y = mx + 2$

Seeing the Concept

To see the role that the slope m plays, graph the following lines on the same screen.

$$Y_1 = 2 \quad m = 0$$
$$Y_2 = x + 2 \quad m = 1$$
$$Y_3 = -x + 2 \quad m = -1$$
$$Y_4 = 3x + 2 \quad m = 3$$
$$Y_5 = -3x + 2 \quad m = -3$$

See Figure 39. What do you conclude about the lines $y = mx + 2$?

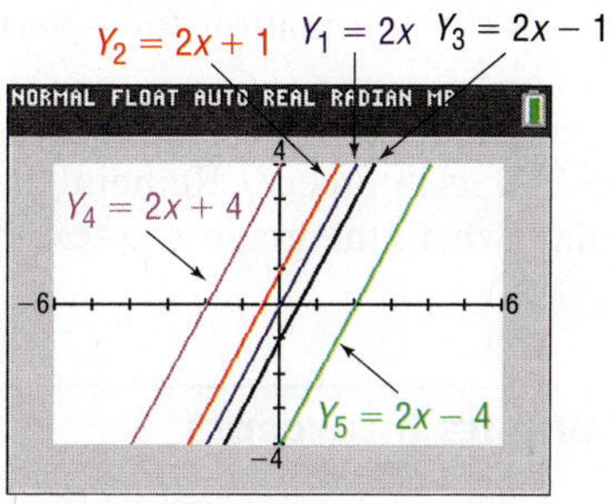

Figure 40 $y = 2x + b$

Seeing the Concept

To see the role of the y-intercept b, graph the following lines on the same screen.

$$Y_1 = 2x \quad b = 0$$
$$Y_2 = 2x + 1 \quad b = 1$$
$$Y_3 = 2x - 1 \quad b = -1$$
$$Y_4 = 2x + 4 \quad b = 4$$
$$Y_5 = 2x - 4 \quad b = -4$$

See Figure 40. What do you conclude about the lines $y = 2x + b$?

7 Identify the Slope and y-Intercept of a Line from Its Equation

When the equation of a line is written in slope–intercept form, it is easy to find the slope m and y-intercept b of the line. For example, suppose that the equation of a line is

$$y = -2x + 7$$

Compare this equation to $y = mx + b$.

$$\begin{array}{ccc} y = & -2x & + \; 7 \\ & \uparrow & \uparrow \\ y = & mx & + \; b \end{array}$$

The slope of this line is -2 and its y-intercept is 7.

Now Work PROBLEM 73

EXAMPLE 8 Finding the Slope and y-Intercept

Find the slope m and y-intercept b of the equation $2x + 4y = 8$. Graph the equation.

Solution To obtain the slope and y-intercept, write the equation in slope–intercept form by solving for y.

$$2x + 4y = 8$$
$$4y = -2x + 8$$
$$y = -\frac{1}{2}x + 2 \quad y = mx + b$$

The coefficient of x, $-\frac{1}{2}$, is the slope, and the constant, 2, is the y-intercept. Graph the line with y-intercept 2 and with slope $-\frac{1}{2}$. Starting at the point $(0, 2)$, go to the right 2 units and then down 1 unit to the point $(2, 1)$. Draw the line through these points. See Figure 41.

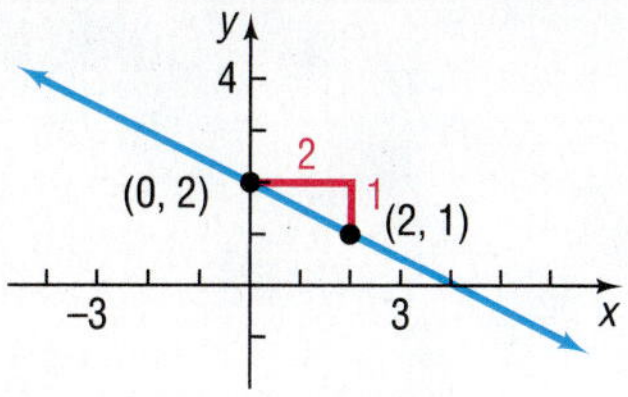

Figure 41 $y = -\frac{1}{2}x + 2$

Now Work PROBLEM 79

8 Graph Lines Written in General Form Using Intercepts

Refer to Example 8. The form of the equation of the line $2x + 4y = 8$ is called the *general form*.

DEFINITION

The equation of a line is in **general form*** when it is written as

$$Ax + By = C \tag{4}$$

where A, B, and C are real numbers and A and B are not both 0.

If $B = 0$ in equation (4), then $A \neq 0$ and the graph of the equation is a vertical line: $x = \frac{C}{A}$. If $B \neq 0$ in equation (4), then we can solve the equation for y and write the equation in slope–intercept form as we did in Example 8.

Another approach to graphing equation (4) is to find its intercepts. Remember, the intercepts of the graph of an equation are the points where the graph crosses or touches a coordinate axis.

EXAMPLE 9 Graphing an Equation in General Form Using Its Intercepts

Graph the equation $2x + 4y = 8$ by finding its intercepts.

Solution To obtain the x-intercept, let $y = 0$ in the equation and solve for x.

$$\begin{aligned} 2x + 4y &= 8 \\ 2x + 4(0) &= 8 \quad \text{Let } y = 0. \\ 2x &= 8 \\ x &= 4 \quad \text{Divide both sides by 2.} \end{aligned}$$

The x-intercept is 4, and the point $(4, 0)$ is on the graph of the equation.

To obtain the y-intercept, let $x = 0$ in the equation and solve for y.

$$\begin{aligned} 2x + 4y &= 8 \\ 2(0) + 4y &= 8 \quad \text{Let } x = 0. \\ 4y &= 8 \\ y &= 2 \quad \text{Divide both sides by 4.} \end{aligned}$$

The y-intercept is 2, and the point $(0, 2)$ is on the graph of the equation.

Plot the points $(4, 0)$ and $(0, 2)$ and draw the line through the points. See Figure 42.

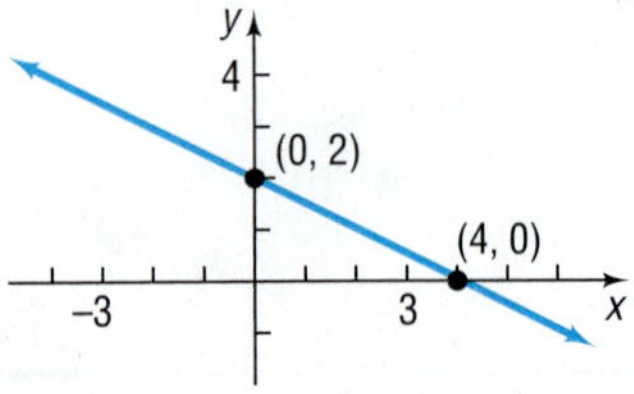

Figure 42 $2x + 4y = 8$

Now Work PROBLEM 93

Every line has an equation that is equivalent to an equation written in general form. For example, a vertical line whose equation is

$$x = a$$

can be written in the general form

$$1 \cdot x + 0 \cdot y = a \quad A = 1, B = 0, C = a$$

A horizontal line whose equation is

$$y = b$$

can be written in the general form

$$0 \cdot x + 1 \cdot y = b \quad A = 0, B = 1, C = b$$

*Some texts use the term **standard form**.

Lines that are neither vertical nor horizontal have general equations of the form

$$Ax + By = C \quad A \neq 0 \text{ and } B \neq 0$$

Because the equation of every line can be written in general form, any equation equivalent to equation (4) is called a **linear equation**.

9 Find Equations of Parallel Lines

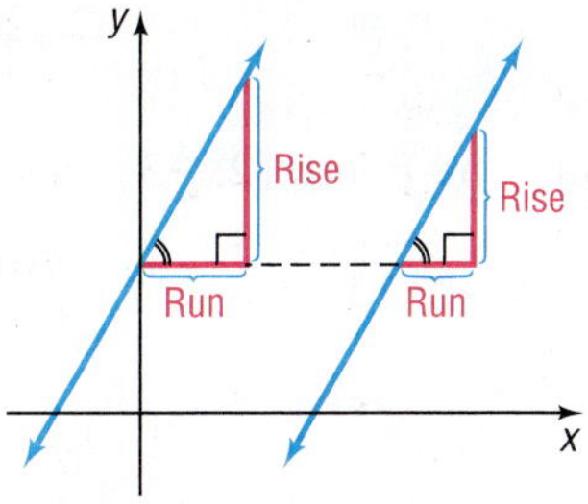

Figure 43 Parallel lines

When two lines (in the plane) do not intersect (that is, they have no points in common), they are **parallel**. Look at Figure 43. There we have drawn two parallel lines and have constructed two right triangles by drawing sides parallel to the coordinate axes. The right triangles are similar. (Do you see why? Two angles are equal.) Because the triangles are similar, the ratios of corresponding sides are equal.

THEOREM Criteria for Parallel Lines

Two nonvertical lines are parallel if and only if their slopes are equal and they have different y-intercepts.

The use of the phrase "if and only if" in the preceding theorem means that actually two statements are being made, one the converse of the other.

If two nonvertical lines are parallel, then their slopes are equal and they have different y-intercepts.

If two nonvertical lines have equal slopes and they have different y-intercepts, then they are parallel.

EXAMPLE 10 **Showing That Two Lines Are Parallel**

Show that the lines given by the following equations are parallel.

$$L_1: \ 2x + 3y = 6 \qquad L_2: \ 4x + 6y = 0$$

Solution To determine whether these lines have equal slopes and different y-intercepts, write each equation in slope–intercept form.

$$L_1: \ 2x + 3y = 6 \qquad\qquad L_2: \ 4x + 6y = 0$$

$$3y = -2x + 6 \qquad\qquad 6y = -4x$$

$$y = -\frac{2}{3}x + 2 \qquad\qquad y = -\frac{2}{3}x$$

$$\text{Slope} = -\frac{2}{3}; \quad y\text{-intercept} = 2 \qquad\qquad \text{Slope} = -\frac{2}{3}; \quad y\text{-intercept} = 0$$

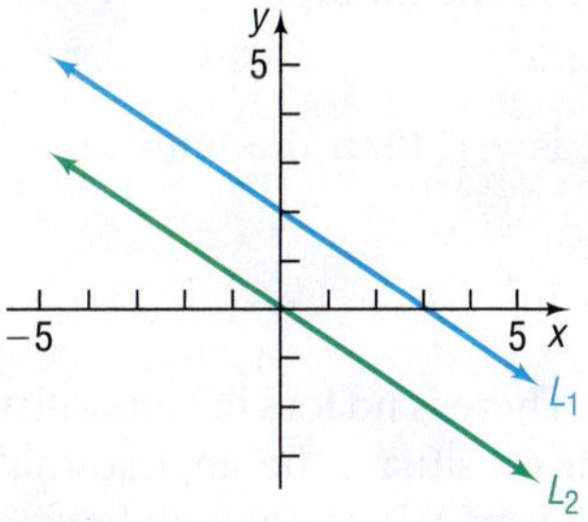

Figure 44

Because these lines have the same slope, $-\frac{2}{3}$, but different y-intercepts, the lines are parallel. See Figure 44.

EXAMPLE 11 **Finding a Line That Is Parallel to a Given Line**

Find an equation for the line that contains the point $(2, -3)$ and is parallel to the line $2x + y = 6$.

Solution Since the two lines are to be parallel, the slope of the line being sought equals the slope of the line $2x + y = 6$. Begin by writing the equation of the line $2x + y = 6$ in slope–intercept form.

$$2x + y = 6$$

$$y = -2x + 6$$

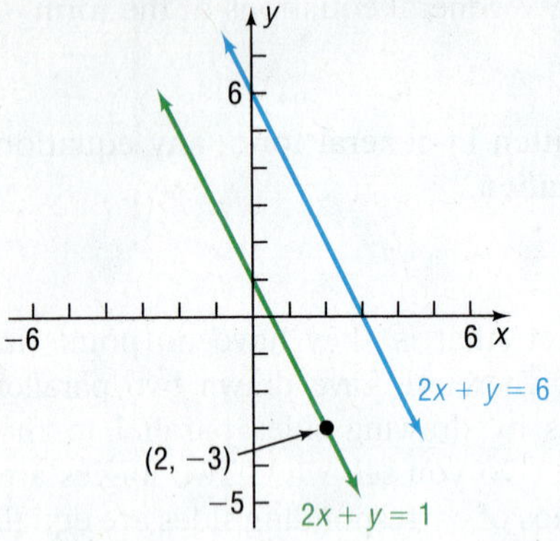

Figure 45

The slope is -2. Since the line being sought also has slope -2 and contains the point $(2, -3)$, use the point–slope form to obtain its equation.

$$y - y_1 = m(x - x_1) \quad \text{Point–slope form}$$
$$y - (-3) = -2(x - 2) \quad m = -2, x_1 = 2, y_1 = -3$$
$$y + 3 = -2x + 4 \quad \text{Simplify.}$$
$$y = -2x + 1 \quad \text{Slope–intercept form}$$
$$2x + y = 1 \quad \text{General form}$$

This line is parallel to the line $2x + y = 6$ and contains the point $(2, -3)$. See Figure 45.

Now Work PROBLEM 61

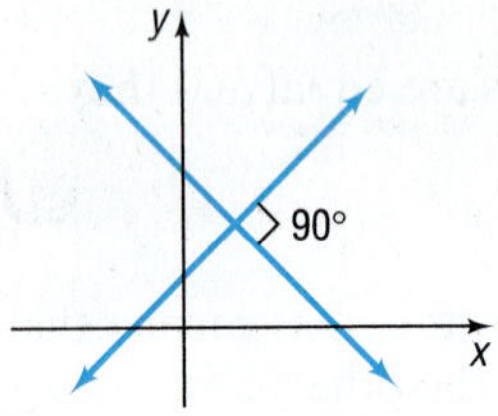

Figure 46 Perpendicular lines

10 Find Equations of Perpendicular Lines

When two lines intersect at a right angle (90°), they are **perpendicular**. See Figure 46.

The following result gives a condition, in terms of their slopes, for two lines to be perpendicular.

THEOREM

Criterion for Perpendicular Lines

Two nonvertical lines are perpendicular if and only if the product of their slopes is -1.

Here we shall prove the "only if" part of the statement:

If two nonvertical lines are perpendicular, then the product of their slopes is -1.

In Problem 130 you are asked to prove the "if" part of the theorem:

If two nonvertical lines have slopes whose product is -1, then the lines are perpendicular.

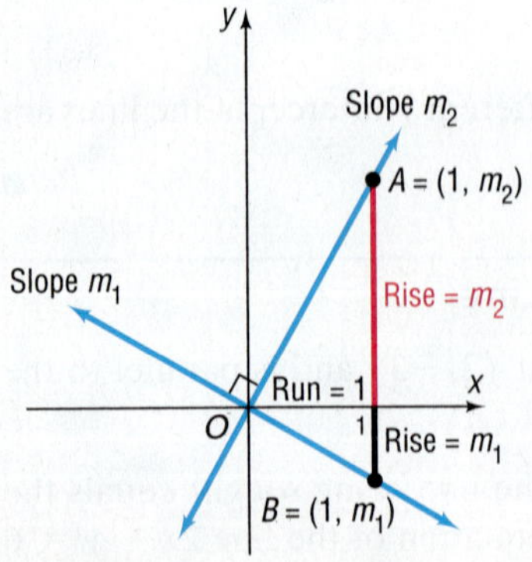

Figure 47

Proof Let m_1 and m_2 denote the slopes of the two lines. There is no loss in generality (that is, neither the angle nor the slopes are affected) if we situate the lines so that they meet at the origin. See Figure 47. The point $A = (1, m_2)$ is on the line having slope m_2, and the point $B = (1, m_1)$ is on the line having slope m_1. (Do you see why this must be true?)

Suppose that the lines are perpendicular. Then triangle OAB is a right triangle. As a result of the Pythagorean Theorem, it follows that

$$[d(O, A)]^2 + [d(O, B)]^2 = [d(A, B)]^2 \quad \textbf{(5)}$$

Using the distance formula, the squares of these distances are

$$[d(O, A)]^2 = (1 - 0)^2 + (m_2 - 0)^2 = 1 + m_2^2$$
$$[d(O, B)]^2 = (1 - 0)^2 + (m_1 - 0)^2 = 1 + m_1^2$$
$$[d(A, B)]^2 = (1 - 1)^2 + (m_2 - m_1)^2 = m_2^2 - 2m_1m_2 + m_1^2$$

Using these facts in equation (5), we get

$$(1 + m_2^2) + (1 + m_1^2) = m_2^2 - 2m_1m_2 + m_1^2$$

which, upon simplification, can be written as

$$m_1m_2 = -1$$

If the lines are perpendicular, the product of their slopes is -1. ■

You may find it easier to remember the condition for two nonvertical lines to be perpendicular by observing that the equality $m_1m_2 = -1$ means that m_1 and m_2 are negative reciprocals of each other; that is, $m_1 = -\frac{1}{m_2}$ and $m_2 = -\frac{1}{m_1}$.

EXAMPLE 12 **Finding the Slope of a Line Perpendicular to Another Line**

If a line has slope $\frac{3}{2}$, any line having slope $-\frac{2}{3}$ is perpendicular to it. ●

EXAMPLE 13 **Finding the Equation of a Line Perpendicular to a Given Line**

Find an equation of the line that contains the point $(1, -2)$ and is perpendicular to the line $x + 3y = 6$. Graph the two lines.

Solution First write the equation of the given line in slope–intercept form to find its slope.

$$x + 3y = 6$$

$$3y = -x + 6 \qquad \text{Proceed to solve for } y.$$

$$y = -\frac{1}{3}x + 2 \qquad \text{Place in the form } y = mx + b.$$

The given line has slope $-\frac{1}{3}$. Any line perpendicular to this line will have slope 3.

Because the point $(1, -2)$ is on this line with slope 3, use the point–slope form of the equation of a line.

$$y - y_1 = m(x - x_1) \qquad \text{Point–slope form}$$

$$y - (-2) = 3(x - 1) \qquad m = 3, x_1 = 1, y_1 = -2$$

$$y + 2 = 3(x - 1)$$

To obtain other forms of the equation, proceed as follows:

$$y + 2 = 3x - 3 \qquad \text{Simplify.}$$

$$y = 3x - 5 \qquad \text{Slope–intercept form}$$

$$3x - y = 5 \qquad \text{General form}$$

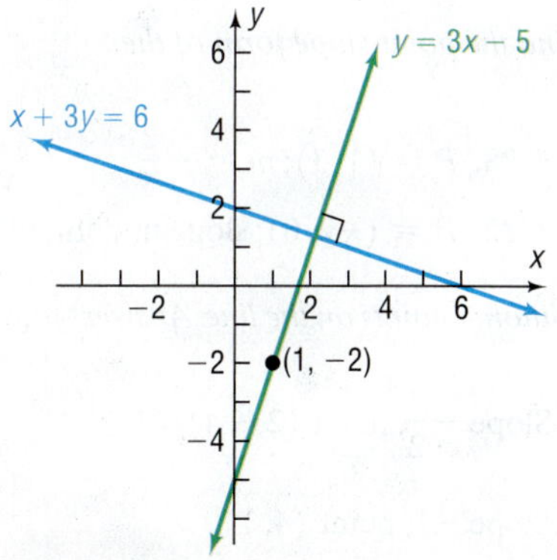

Figure 48

Figure 48 shows the graphs. ●

Now Work PROBLEM 67

WARNING Be sure to use a square screen when you use a graphing calculator to graph perpendicular lines. Otherwise, the angle between the two lines will appear distorted. A discussion of square screens is given in Section 5 of the Appendix. ■

2.3 Assess Your Understanding

Concepts and Vocabulary

1. The slope of a vertical line is ________; the slope of a horizontal line is ____.
2. For the line $2x + 3y = 6$, the x-intercept is ____ and the y-intercept is ____.
3. ***True or False*** The equation $3x + 4y = 6$ is written in general form.
4. ***True or False*** The slope of the line $2y = 3x + 5$ is 3.
5. ***True or False*** The point $(1, 2)$ is on the line $2x + y = 4$.
6. Two nonvertical lines have slopes m_1 and m_2, respectively. The lines are parallel if ________ and the ________ are unequal; the lines are perpendicular if ________.
7. The lines $y = 2x + 3$ and $y = ax + 5$ are parallel if $a =$ ____.
8. The lines $y = 2x - 1$ and $y = ax + 2$ are perpendicular if $a =$ ____.
9. ***True or False*** Perpendicular lines have slopes that are reciprocals of one another.
10. Choose the formula for finding the slope m of a nonvertical line that contains the two distinct points (x_1, y_1) and (x_2, y_2).
    (a) $m = \dfrac{y_2 - x_2}{y_1 - x_1}$ $\quad x_1 \neq y_1$
    (b) $m = \dfrac{y_2 - x_1}{x_2 - y_1}$ $\quad y_1 \neq x_2$
    (c) $m = \dfrac{x_2 - x_1}{y_2 - y_1}$ $\quad y_1 \neq y_2$
    (d) $m = \dfrac{y_2 - y_1}{x_2 - x_1}$ $\quad x_1 \neq x_2$
11. If a line slants downward from left to right, then which of the following describes its slope?
    (a) positive (b) zero
    (c) negative (d) undefined
12. Choose the correct statement about the graph of the line $y = -3$.
    (a) The graph is vertical with x-intercept -3.
    (b) The graph is horizontal with y-intercept -3.
    (c) The graph is vertical with y-intercept -3.
    (d) The graph is horizontal with x-intercept -3.

Skill Building

In Problems 13–16, (a) find the slope of the line and (b) interpret the slope.

13.
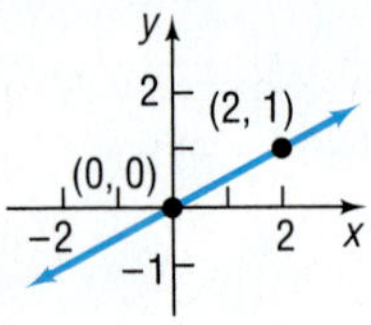

14.
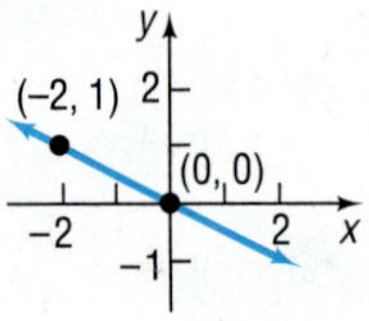

15.
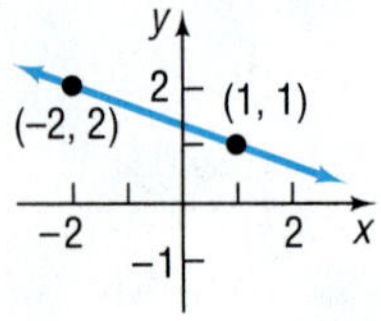

16.
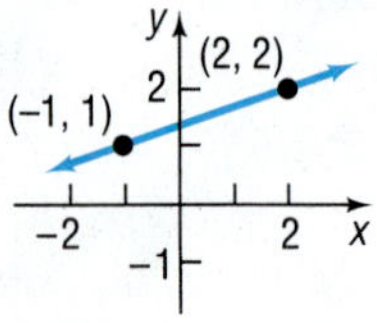

In Problems 17–24, plot each pair of points and determine the slope of the line containing them. Graph the line.

17. $(2, 3)$; $(4, 0)$
18. $(4, 2)$; $(3, 4)$
19. $(-2, 3)$; $(2, 1)$
20. $(-1, 1)$; $(2, 3)$
21. $(-3, -1)$; $(2, -1)$
22. $(4, 2)$; $(-5, 2)$
23. $(-1, 2)$; $(-1, -2)$
24. $(2, 0)$; $(2, 2)$

In Problems 25–32, graph the line that contains the point P and has slope m. In Problems 25–30, find the point-slope form of the equation of the line. In Problems 31 and 32, find an equation of the line.

25. $P = (1, 2)$; $m = 3$
26. $P = (2, 1)$; $m = 4$
27. $P = (2, 4)$; $m = -\dfrac{3}{4}$
28. $P = (1, 3)$; $m = -\dfrac{2}{5}$
29. $P = (-1, 3)$; $m = 0$
30. $P = (2, -4)$; $m = 0$
31. $P = (0, 3)$; slope undefined
32. $P = (-2, 0)$; slope undefined

In Problems 33–38, the slope and a point on a line are given. Use this information to locate three additional points on the line. Answers may vary. [**Hint:** *It is not necessary to find the equation of the line. See Example 3.*]

33. Slope 4; point $(1, 2)$
34. Slope 2; point $(-2, 3)$
35. Slope $-\dfrac{3}{2}$; point $(2, -4)$
36. Slope $\dfrac{4}{3}$; point $(-3, 2)$
37. Slope -2; point $(-2, -3)$
38. Slope -1; point $(4, 1)$

In Problems 39–46, find an equation of the line L.

39.
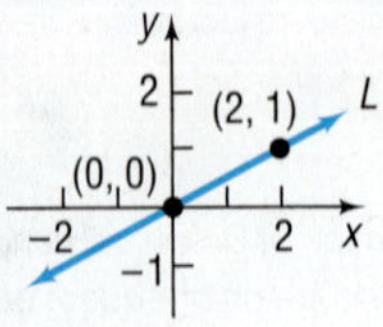

40.
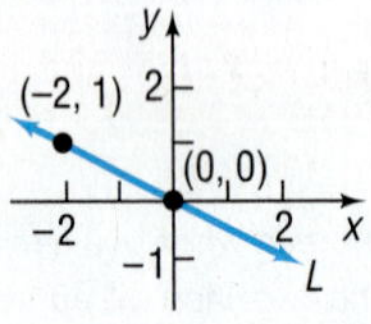

41.
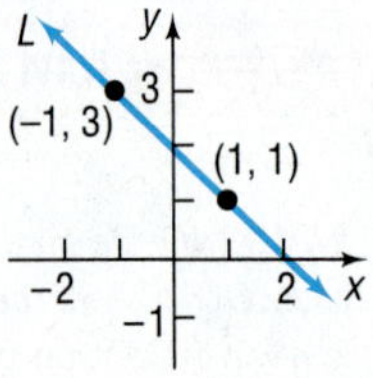

42.
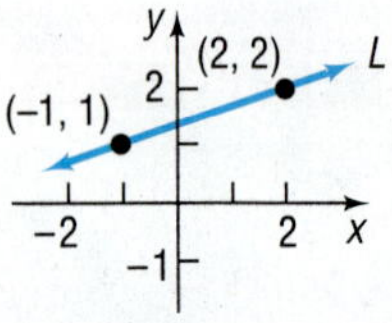

43.

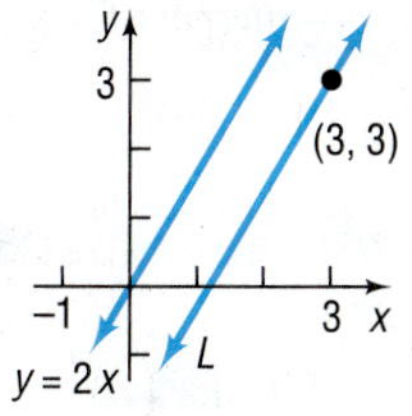

L is parallel to $y = 2x$

44.

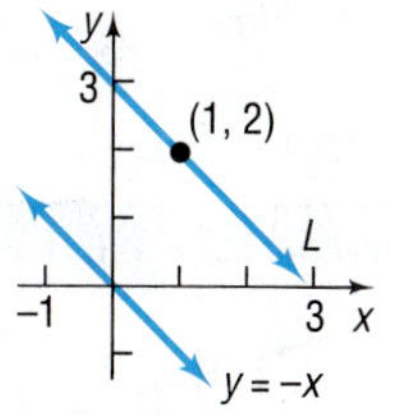

L is parallel to $y = -x$

45.

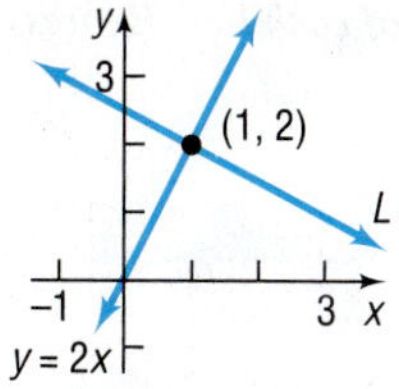

L is perpendicular to $y = 2x$

46.

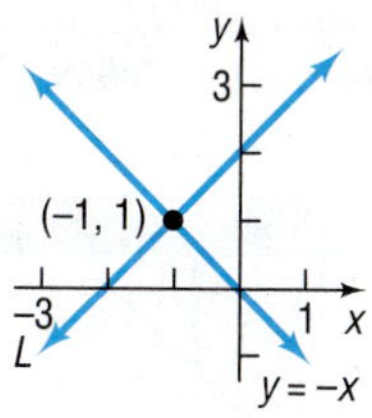

L is perpendicular to $y = -x$

In Problems 47–72, find an equation for the line with the given properties. Express your answer using either the general form or the slope–intercept form of the equation of a line, whichever you prefer.

47. Slope $= 3$; containing the point $(-2, 3)$

48. Slope $= 2$; containing the point $(4, -3)$

49. Slope $= -\frac{2}{3}$; containing the point $(1, -1)$

50. Slope $= \frac{1}{2}$; containing the point $(3, 1)$

51. Containing the points $(1, 3)$ and $(-1, 2)$

52. Containing the points $(-3, 4)$ and $(2, 5)$

53. Slope $= -3$; y-intercept $= 3$

54. Slope $= -2$; y-intercept $= -2$

55. x-intercept $= 2$; y-intercept $= -1$

56. x-intercept $= -4$; y-intercept $= 4$

57. Slope undefined; containing the point $(2, 4)$

58. Slope undefined; containing the point $(3, 8)$

59. Horizontal; containing the point $(-3, 2)$

60. Vertical; containing the point $(4, -5)$

61. Parallel to the line $y = 2x$; containing the point $(-1, 2)$

62. Parallel to the line $y = -3x$; containing the point $(-1, 2)$

63. Parallel to the line $2x - y = -2$; containing the point $(0, 0)$

64. Parallel to the line $x - 2y = -5$; containing the point $(0, 0)$

65. Parallel to the line $x = 5$; containing the point $(4, 2)$

66. Parallel to the line $y = 5$; containing the point $(4, 2)$

67. Perpendicular to the line $y = \frac{1}{2}x + 4$; containing the point $(1, -2)$

68. Perpendicular to the line $y = 2x - 3$; containing the point $(1, -2)$

69. Perpendicular to the line $2x + y = 2$; containing the point $(-3, 0)$

70. Perpendicular to the line $x - 2y = -5$; containing the point $(0, 4)$

71. Perpendicular to the line $x = 8$; containing the point $(3, 4)$

72. Perpendicular to the line $y = 8$; containing the point $(3, 4)$

In Problems 73–92, find the slope and y-intercept of each line. Graph the line.

73. $y = 2x + 3$
74. $y = -3x + 4$
75. $\frac{1}{2}y = x - 1$
76. $\frac{1}{3}x + y = 2$
77. $y = \frac{1}{2}x + 2$
78. $y = 2x + \frac{1}{2}$
79. $x + 2y = 4$
80. $-x + 3y = 6$
81. $2x - 3y = 6$
82. $3x + 2y = 6$
83. $x + y = 1$
84. $x - y = 2$
85. $x = -4$
86. $y = -1$
87. $y = 5$
88. $x = 2$
89. $y - x = 0$
90. $x + y = 0$
91. $2y - 3x = 0$
92. $3x + 2y = 0$

In Problems 93–102, (a) find the intercepts of the graph of each equation and (b) graph the equation.

93. $2x + 3y = 6$
94. $3x - 2y = 6$
95. $-4x + 5y = 40$
96. $6x - 4y = 24$
97. $7x + 2y = 21$
98. $5x + 3y = 18$
99. $\frac{1}{2}x + \frac{1}{3}y = 1$
100. $x - \frac{2}{3}y = 4$
101. $0.2x - 0.5y = 1$
102. $-0.3x + 0.4y = 1.2$

103. Find an equation of the x-axis.

104. Find an equation of the y-axis.

In Problems 105–108, the equations of two lines are given. Determine whether the lines are parallel, perpendicular, or neither.

105. $y = 2x - 3$
$y = 2x + 4$

106. $y = \frac{1}{2}x - 3$
$y = -2x + 4$

107. $y = 4x + 5$
$y = -4x + 2$

108. $y = -2x + 3$
$y = -\frac{1}{2}x + 2$

In Problems 109–112, write an equation of each line. Express your answer using either the general form or the slope–intercept form of the equation of a line, whichever you prefer.

109.

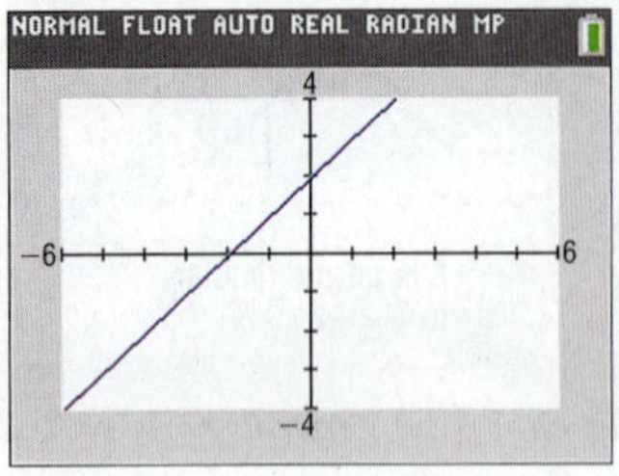

110.

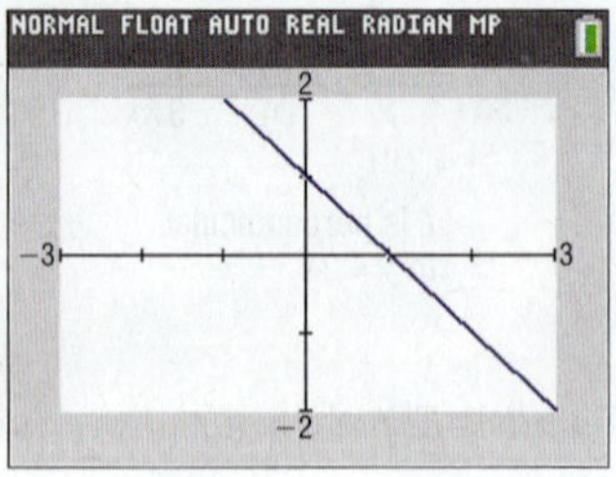

111.

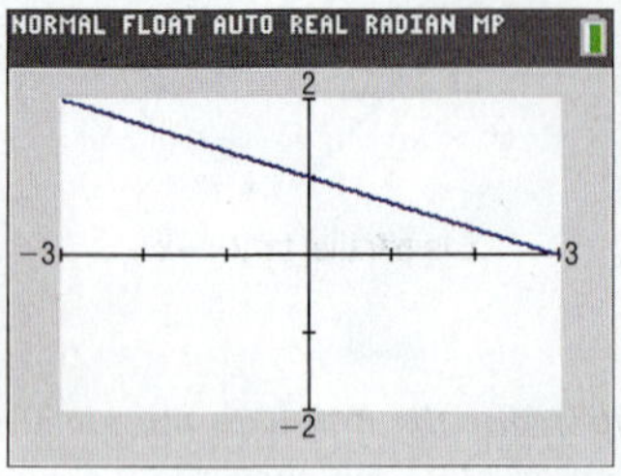

112.

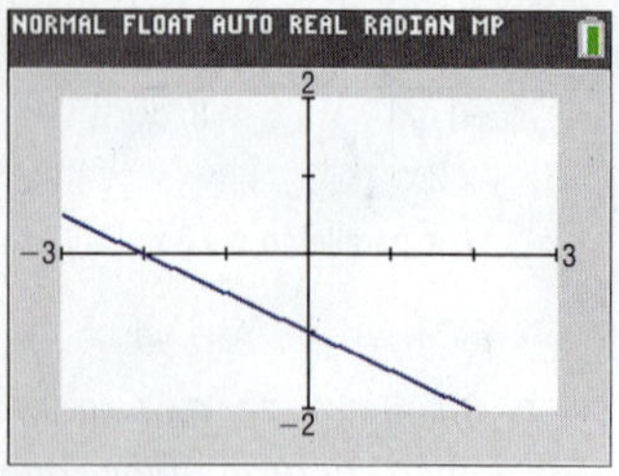

Applications and Extensions

113. Geometry Use slopes to show that the triangle whose vertices are $(-2, 5)$, $(1, 3)$, and $(-1, 0)$ is a right triangle.

114. Geometry Use slopes to show that the quadrilateral whose vertices are $(1, -1)$, $(4, 1)$, $(2, 2)$, and $(5, 4)$ is a parallelogram.

115. Geometry Use slopes to show that the quadrilateral whose vertices are $(-1, 0)$, $(2, 3)$, $(1, -2)$, and $(4, 1)$ is a rectangle.

116. Geometry Use slopes and the distance formula to show that the quadrilateral whose vertices are $(0, 0)$, $(1, 3)$, $(4, 2)$, and $(3, -1)$ is a square.

117. Truck Rentals A truck rental company rents a moving truck for one day by charging \$39 plus \$0.60 per mile. Write a linear equation that relates the cost C, in dollars, of renting the truck to the number x of miles driven. What is the cost of renting the truck if the truck is driven 110 miles? 230 miles?

118. Cost Equation The **fixed costs** of operating a business are the costs incurred regardless of the level of production. Fixed costs include rent, fixed salaries, and costs of leasing machinery. The **variable costs** of operating a business are the costs that change with the level of output. Variable costs include raw materials, hourly wages, and electricity. Suppose that a manufacturer of jeans has fixed daily costs of \$500 and variable costs of \$8 for each pair of jeans manufactured. Write a linear equation that relates the daily cost C, in dollars, of manufacturing the jeans to the number x of jeans manufactured. What is the cost of manufacturing 400 pairs of jeans? 740 pairs?

119. Cost of Driving a Car The annual fixed costs of owning a small sedan are \$4462, assuming the car is completely paid for. The cost to drive the car is approximately \$0.17 per mile. Write a linear equation that relates the cost C and the number x of miles driven annually.

Source: *AAA*, April 2013

120. Wages of a Car Salesperson Dan receives \$375 per week for selling new and used cars at a car dealership in Oak Lawn, Illinois. In addition, he receives 5% of the profit on any sales that he generates. Write a linear equation that represents Dan's weekly salary S when he has sales that generate a profit of x dollars.

121. Electricity Rates in Illinois Commonwealth Edison Company supplies electricity to residential customers for a monthly customer charge of \$15.37 plus 8.21 cents per kilowatt-hour for up to 800 kilowatt-hours (kW-hr).

(a) Write a linear equation that relates the monthly charge C, in dollars, to the number x of kilowatt-hours used in a month, $0 \le x \le 800$.
(b) Graph this equation.
(c) What is the monthly charge for using 200 kilowatt-hours?
(d) What is the monthly charge for using 500 kilowatt-hours?
(e) Interpret the slope of the line.

Source: *Commonwealth Edison Company*, January 2014.

122. Electricity Rates in Florida Florida Power & Light Company supplies electricity to residential customers for a monthly customer charge of \$7.24 plus 9.07 cents per kilowatt-hour for up to 1000 kilowatt-hours.

(a) Write a linear equation that relates the monthly charge C, in dollars, to the number x of kilowatt-hours used in a month, $0 \le x \le 1000$.
(b) Graph this equation.
(c) What is the monthly charge for using 200 kilowatt-hours?
(d) What is the monthly charge for using 500 kilowatt-hours?
(e) Interpret the slope of the line.

Source: *Florida Power & Light Company*, March 2014.

123. Measuring Temperature The relationship between Celsius (°C) and Fahrenheit (°F) degrees of measuring temperature is linear. Find a linear equation relating °C and °F if 0°C corresponds to 32°F and 100°C corresponds to 212°F. Use the equation to find the Celsius measure of 70°F.

124. Measuring Temperature The Kelvin (K) scale for measuring temperature is obtained by adding 273 to the Celsius temperature.

(a) Write a linear equation relating K and °C.
(b) Write a linear equation relating K and °F (see Problem 123).

125. Access Ramp A wooden access ramp is being built to reach a platform that sits 30 inches above the floor. The ramp drops 2 inches for every 25-inch run.

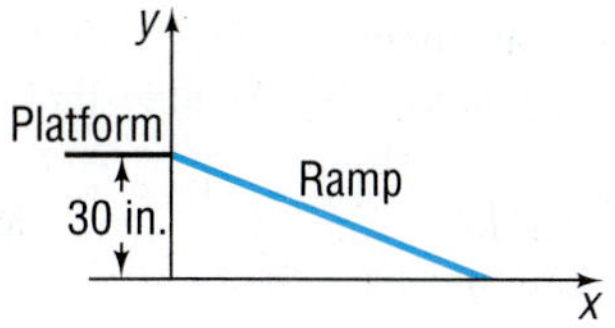

(a) Write a linear equation that relates the height y of the ramp above the floor to the horizontal distance x from the platform.
(b) Find and interpret the x-intercept of the graph of your equation.
(c) Design requirements stipulate that the maximum run be 30 feet and that the maximum slope be a drop of 1 inch for each 12 inches of run. Will this ramp meet the requirements? Explain.
(d) What slopes could be used to obtain the 30-inch rise and still meet design requirements?

Source: www.adaptiveaccess.com/wood_ramps.php

126. Cigarette Use A report in the Child Trends DataBase indicated that in 2000, 20.6% of twelfth grade students reported daily use of cigarettes. In 2012, 9.3% of twelfth grade students reported daily use of cigarettes.
(a) Write a linear equation that relates the percent y of twelfth grade students who smoke cigarettes daily to the number x of years after 2000.
(b) Find the intercepts of the graph of your equation.
(c) Do these intercepts have meaningful interpretation?
(d) Use your equation to predict the percent for the year 2025. Is this result reasonable?

Source: www.childtrendsdatabank.org

127. Product Promotion A cereal company finds that the number of people who will buy one of its products in the first month that the product is introduced is linearly related to the amount of money it spends on advertising. If it spends \$40,000 on advertising, then 100,000 boxes of cereal will be sold, and if it spends \$60,000, then 200,000 boxes will be sold.
(a) Write a linear equation that relates the amount A spent on advertising to the number x of boxes the company aims to sell.
(b) How much expenditure on advertising is needed to sell 300,000 boxes of cereal?
(c) Interpret the slope.

128. Show that the line containing the points (a, b) and (b, a), $a \neq b$, is perpendicular to the line $y = x$. Also show that the midpoint of (a, b) and (b, a) lies on the line $y = x$.

129. The equation $2x - y = C$ defines a **family of lines**, one line for each value of C. On one set of coordinate axes, graph the members of the family when $C = -4$, $C = 0$, and $C = 2$. Can you draw a conclusion from the graph about each member of the family?

130. Prove that if two nonvertical lines have slopes whose product is -1, then the lines are perpendicular. [**Hint:** Refer to Figure 47 and use the converse of the Pythagorean Theorem.]

Explaining Concepts: Discussion and Writing

131. Which of the following equations might have the graph shown? (More than one answer is possible.)
(a) $2x + 3y = 6$
(b) $-2x + 3y = 6$
(c) $3x - 4y = -12$
(d) $x - y = 1$
(e) $x - y = -1$
(f) $y = 3x - 5$
(g) $y = 2x + 3$
(h) $y = -3x + 3$

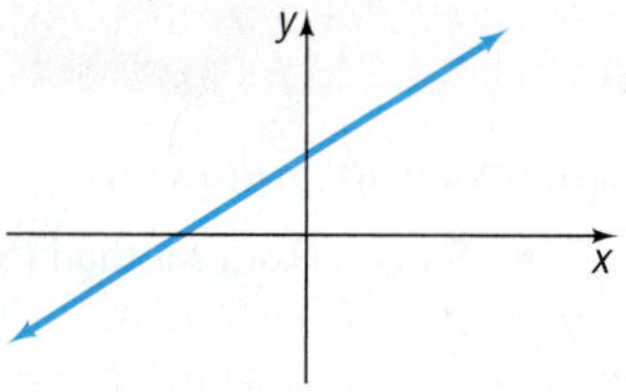

132. Which of the following equations might have the graph shown? (More than one answer is possible.)
(a) $2x + 3y = 6$
(b) $2x - 3y = 6$
(c) $3x + 4y = 12$
(d) $x - y = 1$
(e) $x - y = -1$
(f) $y = -2x - 1$
(g) $y = -\frac{1}{2}x + 10$
(h) $y = x + 4$

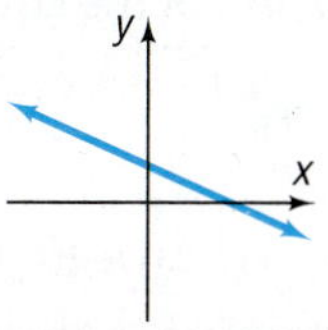

133. The figure shows the graph of two parallel lines. Which of the following pairs of equations might have such a graph?
(a) $x - 2y = 3$
$x + 2y = 7$
(b) $x + y = 2$
$x + y = -1$
(c) $x - y = -2$
$x - y = 1$
(d) $x - y = -2$
$2x - 2y = -4$
(e) $x + 2y = 2$
$x + 2y = -1$

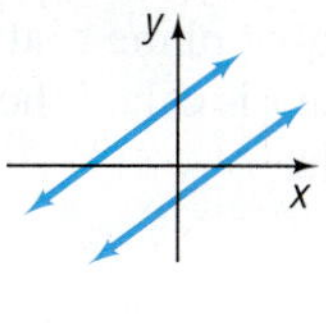

134. The figure shows the graph of two perpendicular lines. Which of the following pairs of equations might have such a graph?
(a) $y - 2x = 2$
$y + 2x = -1$
(b) $y - 2x = 0$
$2y + x = 0$
(c) $2y - x = 2$
$2y + x = -2$
(d) $y - 2x = 2$
$x + 2y = -1$
(e) $2x + y = -2$
$2y + x = -2$

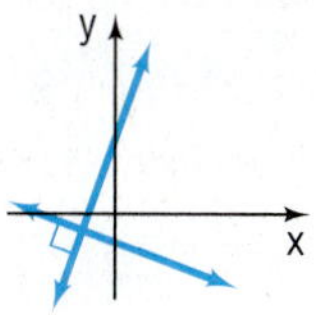

135. *m* is for Slope The accepted symbol used to denote the slope of a line is the letter m. Investigate the origin of this practice. Begin by consulting a French dictionary and looking up the French word *monter*. Write a brief essay on your findings.

136. Grade of a Road The term *grade* is used to describe the inclination of a road. How is this term related to the notion of slope of a line? Is a 4% grade very steep? Investigate the grades of some mountainous roads and determine their slopes. Write a brief essay on your findings.

137. Carpentry Carpenters use the term *pitch* to describe the steepness of staircases and roofs. How is pitch related to slope? Investigate typical pitches used for stairs and for roofs. Write a brief essay on your findings.

138. Can the equation of every line be written in slope–intercept form? Why?

139. Does every line have exactly one x-intercept and one y-intercept? Are there any lines that have no intercepts?

140. What can you say about two lines that have equal slopes and equal y-intercepts?

141. What can you say about two lines with the same x-intercept and the same y-intercept? Assume that the x-intercept is not 0.

142. If two distinct lines have the same slope but different x-intercepts, can they have the same y-intercept?

143. If two distinct lines have the same y-intercept but different slopes, can they have the same x-intercept?

144. Which form of the equation of a line do you prefer to use? Justify your position with an example that shows that your choice is better than another. Have reasons.

145. What Went Wrong? A student is asked to find the slope of the line joining $(-3, 2)$ and $(1, -4)$. He states that the slope is $\frac{3}{2}$. Is he correct? If not, what went wrong?

Retain Your Knowledge

Problems 146–149 are based on material learned earlier in the course. The purpose of these problems is to keep the material fresh in your mind so that you are better prepared for the final exam.

146. Simplify $\left(\dfrac{x^2y^{-3}}{x^4y^5}\right)^{-2}$. Assume $x \neq 0$ and $y \neq 0$. Express the answer so that all exponents are positive.

147. The lengths of the legs of a right triangle are $a = 8$ and $b = 15$. Find the hypotenuse.

148. Solve the equation: $(x - 3)^2 + 25 = 49$

149. Solve $|2x - 5| + 7 < 10$. Express the answer using set notation or interval notation. Graph the solution set.

2.4 Circles

PREPARING FOR THIS SECTION *Before getting started, review the following:*

- Completing the Square (Chapter R, Section R.5, p. 56)
- Square Root Method (Section 1.2, pp. 94–95)

Now Work the **'Are You Prepared?'** problems on page 186.

OBJECTIVES
1. Write the Standard Form of the Equation of a Circle (p. 182)
2. Graph a Circle (p. 183)
3. Work with the General Form of the Equation of a Circle (p. 184)

1 Write the Standard Form of the Equation of a Circle

One advantage of a coordinate system is that it enables us to translate a geometric statement into an algebraic statement, and vice versa. Consider, for example, the following geometric statement that defines a circle.

DEFINITION

A **circle** is a set of points in the xy-plane that are a fixed distance r from a fixed point (h, k). The fixed distance r is called the **radius**, and the fixed point (h, k) is called the **center** of the circle.

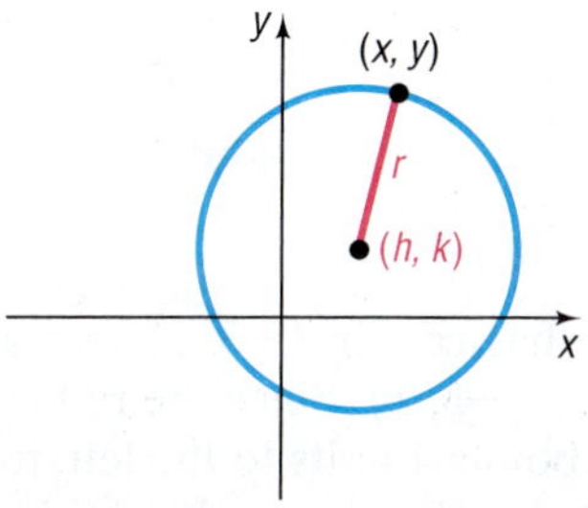

Figure 49 $(x-h)^2 + (y-k)^2 = r^2$

Figure 49 shows the graph of a circle. To find the equation, let (x, y) represent the coordinates of any point on a circle with radius r and center (h, k). Then the distance between the points (x, y) and (h, k) must always equal r. That is, by the distance formula,

$$\sqrt{(x-h)^2 + (y-k)^2} = r$$

or, equivalently,

$$(x-h)^2 + (y-k)^2 = r^2$$

DEFINITION

The **standard form of an equation of a circle** with radius r and center (h, k) is

$$(x-h)^2 + (y-k)^2 = r^2 \qquad \textbf{(1)}$$

THEOREM

The standard form of an equation of a circle of radius r with center at the origin $(0, 0)$ is

$$x^2 + y^2 = r^2$$

DEFINITION

If the radius $r = 1$, the circle whose center is at the origin is called the **unit circle** and has the equation

$$x^2 + y^2 = 1$$

See Figure 50. Notice that the graph of the unit circle is symmetric with respect to the x-axis, the y-axis, and the origin.

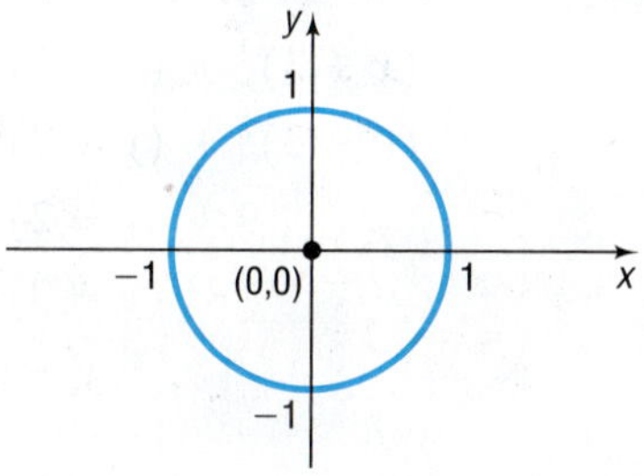

Figure 50
Unit circle $x^2 + y^2 = 1$

EXAMPLE 1 **Writing the Standard Form of the Equation of a Circle**

Write the standard form of the equation of the circle with radius 5 and center $(-3, 6)$.

Solution Substitute the values $r = 5$, $h = -3$, and $k = 6$ into equation (1).

$$(x-h)^2 + (y-k)^2 = r^2$$
$$(x+3)^2 + (y-6)^2 = 25$$

●

Now Work PROBLEM 9

2 Graph a Circle

EXAMPLE 2 **Graphing a Circle**

Graph the equation: $(x+3)^2 + (y-2)^2 = 16$

Solution Since the equation is in the form of equation (1), its graph is a circle. To graph the equation, compare the given equation to the standard form of the equation of a circle. The comparison yields information about the circle.

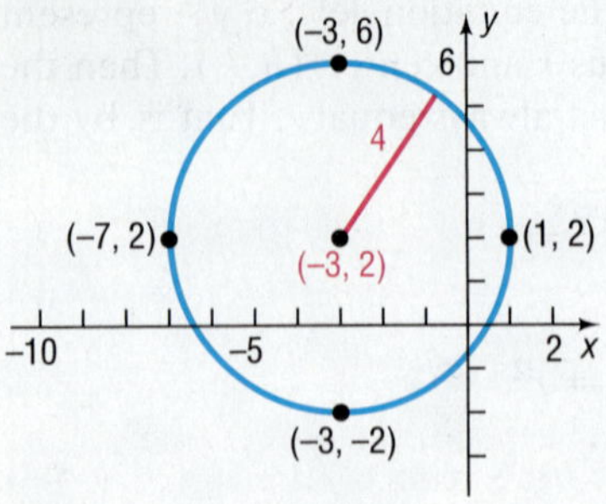

Figure 51 $(x+3)^2+(y-2)^2=16$

$$(x+3)^2+(y-2)^2=16$$
$$(x-(-3))^2+(y-2)^2=4^2$$
$$(x-h)^2+(y-k)^2=r^2$$

We see that $h=-3, k=2$, and $r=4$. The circle has center $(-3,2)$ and a radius of 4 units. To graph this circle, first plot the center $(-3,2)$. Since the radius is 4, we can locate four points on the circle by plotting points 4 units to the left, to the right, up, and down from the center. These four points can then be used as guides to obtain the graph. See Figure 51. ●

Now Work PROBLEMS 25(a) AND (b)

EXAMPLE 3

Finding the Intercepts of a Circle

For the circle $(x+3)^2+(y-2)^2=16$, find the intercepts, if any, of its graph.

Solution This is the equation discussed and graphed in Example 2. To find the x-intercepts, if any, let $y=0$. Then

> **In Words**
> The symbol $\pm$ is read "plus or minus." It means to add and subtract the quantity following the $\pm$ symbol. For example, 5 ± 2 means "$5-2=3$ or $5+2=7$."

$$(x+3)^2+(y-2)^2=16$$
$$(x+3)^2+(0-2)^2=16 \quad y=0$$
$$(x+3)^2+4=16 \quad \text{Simplify.}$$
$$(x+3)^2=12 \quad \text{Simplify.}$$
$$x+3=\pm\sqrt{12} \quad \text{Apply the Square Root Method.}$$
$$x=-3\pm 2\sqrt{3} \quad \text{Solve for } x.$$

The x-intercepts are $-3-2\sqrt{3}\approx -6.46$ and $-3+2\sqrt{3}\approx 0.46$.

To find the y-intercepts, if any, let $x=0$. Then

$$(x+3)^2+(y-2)^2=16$$
$$(0+3)^2+(y-2)^2=16$$
$$9+(y-2)^2=16$$
$$(y-2)^2=7$$
$$y-2=\pm\sqrt{7}$$
$$y=2\pm\sqrt{7}$$

The y-intercepts are $2-\sqrt{7}\approx -0.65$ and $2+\sqrt{7}\approx 4.65$.

Look back at Figure 51 to verify the approximate locations of the intercepts. ●

Now Work PROBLEM 25(c)

3 Work with the General Form of the Equation of a Circle

If we eliminate the parentheses from the standard form of the equation of the circle given in Example 2, we get

$$(x+3)^2+(y-2)^2=16$$
$$x^2+6x+9+y^2-4y+4=16$$

which simplifies to

$$x^2+y^2+6x-4y-3=0 \qquad \textbf{(2)}$$

It can be shown that any equation of the form

$$x^2+y^2+ax+by+c=0$$

has a graph that is a circle, is a point, or has no graph at all. For example, the graph of the equation $x^2+y^2=0$ is the single point $(0,0)$. The equation $x^2+y^2+5=0$, or $x^2+y^2=-5$, has no graph, because sums of squares of real numbers are never negative.

DEFINITION

When its graph is a circle, the equation

$$x^2 + y^2 + ax + by + c = 0$$

is the **general form of the equation of a circle.**

Now Work PROBLEM 15

If an equation of a circle is in general form, we use the method of completing the square to put the equation in standard form so that we can identify its center and radius.

EXAMPLE 4 Graphing a Circle Whose Equation Is in General Form

Graph the equation: $x^2 + y^2 + 4x - 6y + 12 = 0$

Solution Group the terms involving x, group the terms involving y, and put the constant on the right side of the equation. The result is

$$(x^2 + 4x) + (y^2 - 6y) = -12$$

Next, complete the square of each expression in parentheses. Remember that any number added on the left side of the equation must also be added on the right.

$$(x^2 + 4x + 4) + (y^2 - 6y + 9) = -12 + 4 + 9$$

$$\left(\frac{4}{2}\right)^2 = 4 \qquad \left(\frac{-6}{2}\right)^2 = 9$$

$$(x + 2)^2 + (y - 3)^2 = 1 \quad \text{Factor.}$$

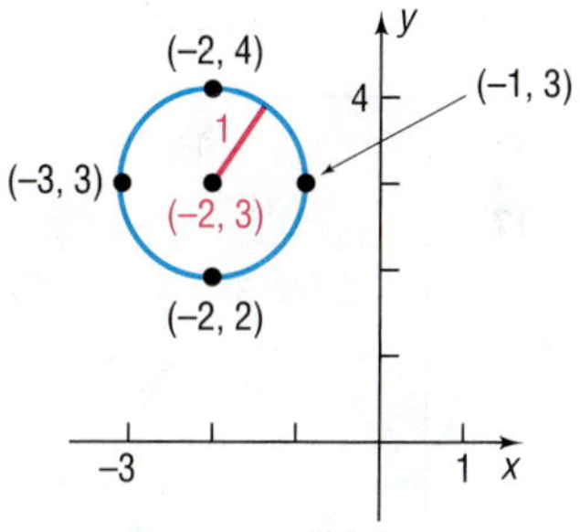

Figure 52 $(x + 2)^2 + (y - 3)^2 = 1$

This equation is the standard form of the equation of a circle with radius 1 and center $(-2, 3)$. To graph the equation, use the center $(-2, 3)$ and the radius 1. See Figure 52.

Now Work PROBLEM 29

EXAMPLE 5 Using a Graphing Utility to Graph a Circle

Graph the equation: $x^2 + y^2 = 4$

Solution This is the equation of a circle with center at the origin and radius 2. To graph this equation, solve for y.

$$x^2 + y^2 = 4$$

$$y^2 = 4 - x^2 \quad \text{Subtract } x^2 \text{ from each side.}$$

$$y = \pm\sqrt{4 - x^2} \quad \text{Apply the Square Root Method to solve for } y.$$

There are two equations to graph: first graph $Y_1 = \sqrt{4 - x^2}$ and then graph $Y_2 = -\sqrt{4 - x^2}$ on the same square screen. (Your circle will appear oval if you do not use a square screen.*) See Figure 53.

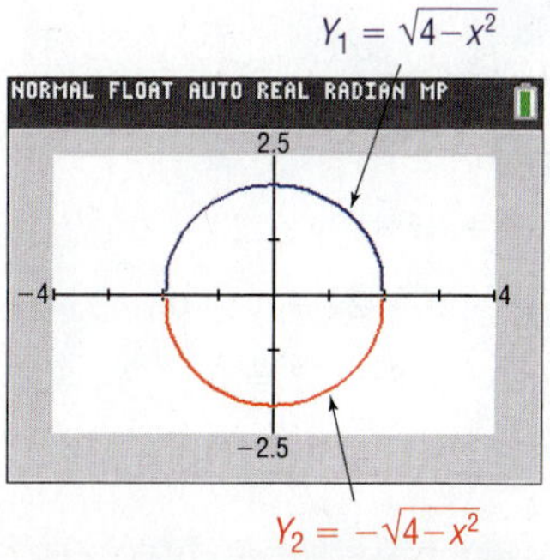

Figure 53 $x^2 + y^2 = 4$

Overview

The discussion in Sections 2.3 and 2.4 about lines and circles dealt with two main types of problems that can be generalized as follows:

1. Given an equation, classify it and graph it.
2. Given a graph, or information about a graph, find its equation.

This text deals with both types of problems. We shall study various equations, classify them, and graph them. The second type of problem is usually more difficult to solve than the first.

*The square screen ratio for the TI-84 Plus C calculator is 8:5.

2.4 Assess Your Understanding

Are You Prepared? *Answers are given at the end of these exercises. If you get a wrong answer, read the pages listed in red.*

1. To complete the square of $x^2 + 10x$, you would ______ (*add/subtract*) the number ______. (p. 56)

2. Use the Square Root Method to solve the equation $(x - 2)^2 = 9$. (pp. 94–95)

Concepts and Vocabulary

3. ***True or False*** Every equation of the form

$$x^2 + y^2 + ax + by + c = 0$$

has a circle as its graph.

4. For a circle, the ______ is the distance from the center to any point on the circle.

5. ***True or False*** The radius of the circle $x^2 + y^2 = 9$ is 3.

6. ***True or False*** The center of the circle

$$(x + 3)^2 + (y - 2)^2 = 13$$

is $(3, -2)$.

7. Choose the equation of a circle with radius 6 and center $(3, -5)$.
(a) $(x - 3)^2 + (y + 5)^2 = 6$
(b) $(x + 3)^2 + (y - 5)^2 = 36$
(c) $(x + 3)^2 + (y - 5)^2 = 6$
(d) $(x - 3)^2 + (y + 5)^2 = 36$

8. The equation of a circle can be changed from general form to standard from by doing which of the following?
(a) completing the squares
(b) solving for x
(c) solving for y
(d) squaring both sides

Skill Building

In Problems 9–12, find the center and radius of each circle. Write the standard form of the equation.

9.
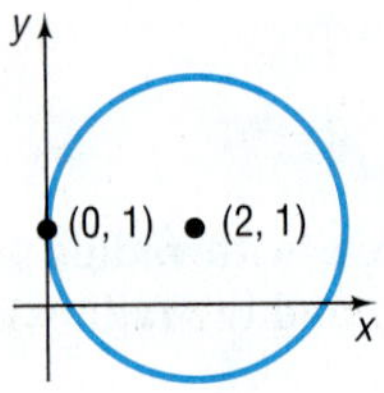

10.
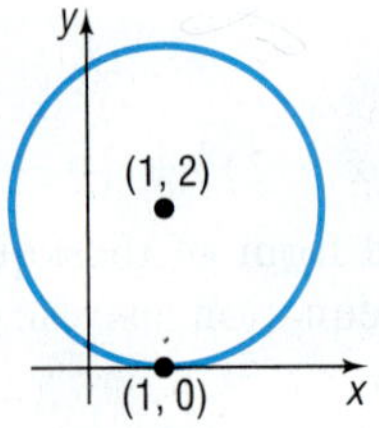

11.
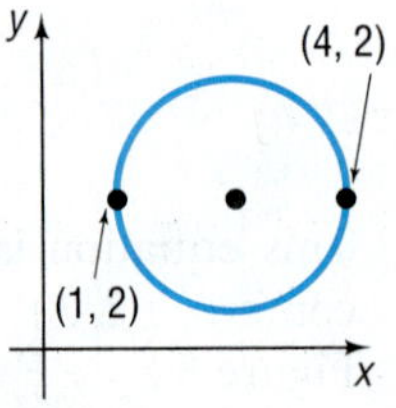

12.
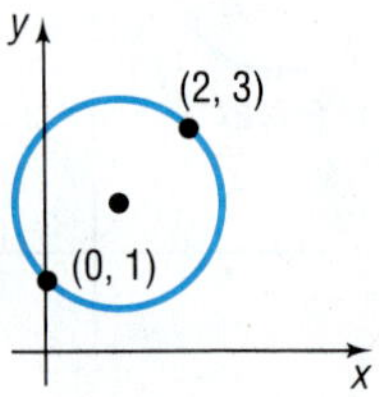

In Problems 13–22, write the standard form of the equation and the general form of the equation of each circle of radius r and center (h, k). Graph each circle.

13. $r = 2;\quad (h, k) = (0, 0)$
14. $r = 3;\quad (h, k) = (0, 0)$
15. $r = 2;\quad (h, k) = (0, 2)$
16. $r = 3;\quad (h, k) = (1, 0)$
17. $r = 5;\quad (h, k) = (4, -3)$
18. $r = 4;\quad (h, k) = (2, -3)$
19. $r = 4;\quad (h, k) = (-2, 1)$
20. $r = 7;\quad (h, k) = (-5, -2)$
21. $r = \frac{1}{2};\quad (h, k) = \left(\frac{1}{2}, 0\right)$
22. $r = \frac{1}{2};\quad (h, k) = \left(0, -\frac{1}{2}\right)$

In Problems 23–36, (a) find the center (h, k) and radius r of each circle; (b) graph each circle; (c) find the intercepts, if any.

23. $x^2 + y^2 = 4$
24. $x^2 + (y - 1)^2 = 1$
25. $2(x - 3)^2 + 2y^2 = 8$
26. $3(x + 1)^2 + 3(y - 1)^2 = 6$
27. $x^2 + y^2 - 2x - 4y - 4 = 0$
28. $x^2 + y^2 + 4x + 2y - 20 = 0$
29. $x^2 + y^2 + 4x - 4y - 1 = 0$
30. $x^2 + y^2 - 6x + 2y + 9 = 0$
31. $x^2 + y^2 - x + 2y + 1 = 0$
32. $x^2 + y^2 + x + y - \frac{1}{2} = 0$
33. $2x^2 + 2y^2 - 12x + 8y - 24 = 0$
34. $2x^2 + 2y^2 + 8x + 7 = 0$
35. $2x^2 + 8x + 2y^2 = 0$
36. $3x^2 + 3y^2 - 12y = 0$

In Problems 37–44, find the standard form of the equation of each circle.

37. Center at the origin and containing the point $(-2, 3)$
38. Center $(1, 0)$ and containing the point $(-3, 2)$
39. Center $(2, 3)$ and tangent to the x-axis
40. Center $(-3, 1)$ and tangent to the y-axis
41. With endpoints of a diameter at $(1, 4)$ and $(-3, 2)$
42. With endpoints of a diameter at $(4, 3)$ and $(0, 1)$
43. Center $(-1, 3)$ and tangent to the line $y = 2$
44. Center $(4, -2)$ and tangent to the line $x = 1$

In Problems 45–48, match each graph with the correct equation.

(a) $(x-3)^2+(y+3)^2=9$ (b) $(x+1)^2+(y-2)^2=4$ (c) $(x-1)^2+(y+2)^2=4$ (d) $(x+3)^2+(y-3)^2=9$

45.

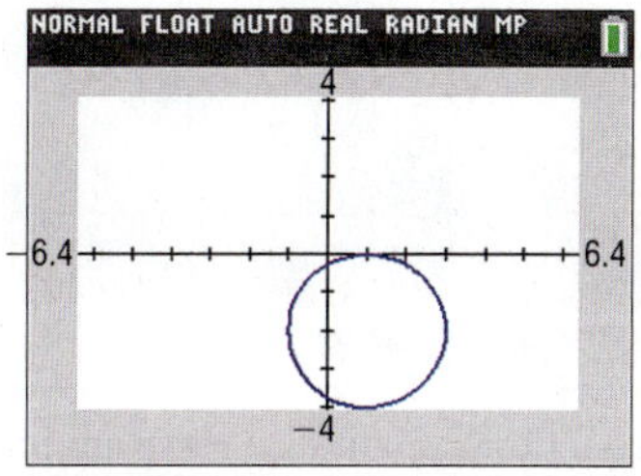

46.

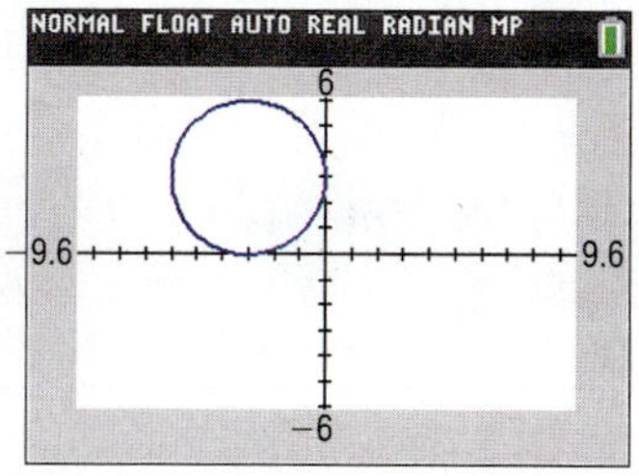

47.

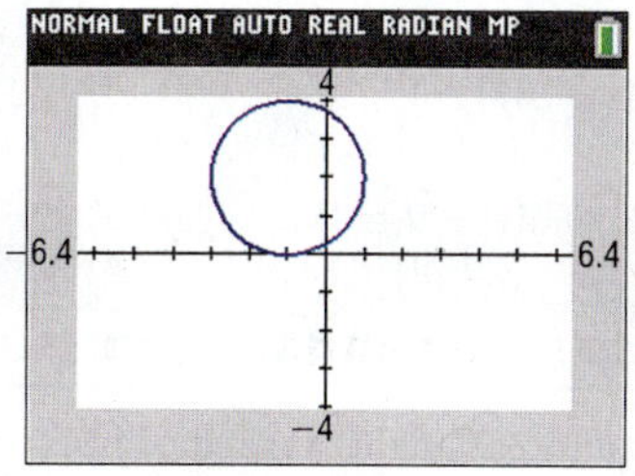

48.

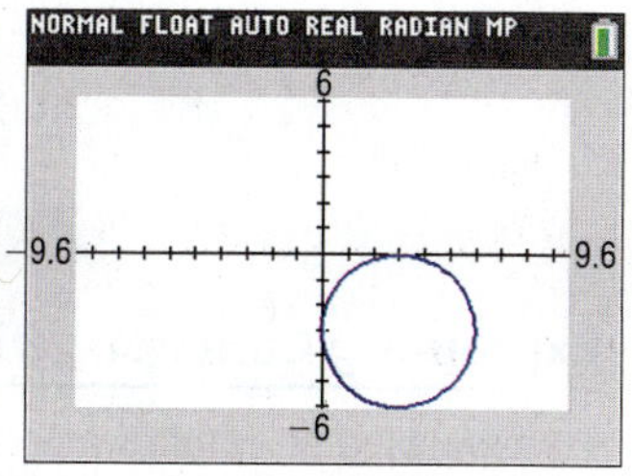

Applications and Extensions

49. Find the area of the square in the figure.

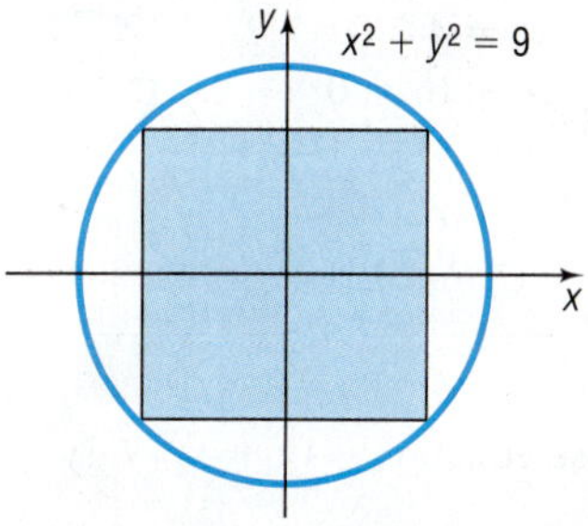

50. Find the area of the blue shaded region in the figure, assuming the quadrilateral inside the circle is a square.

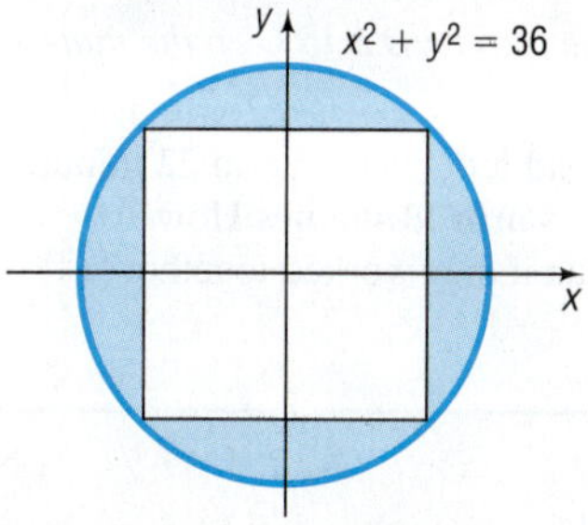

51. Ferris Wheel The original Ferris wheel was built in 1893 by Pittsburgh, Pennsylvania bridge builder George W. Ferris. The Ferris wheel was originally built for the 1893 World's Fair in Chicago, but it was also later reconstructed for the 1904 World's Fair in St. Louis. It had a maximum height of 264 feet and a wheel diameter of 250 feet. Find an equation for the wheel if the center of the wheel is on the y-axis.

Source: *guinnessworldrecords.com*

52. Ferris Wheel Opening in 2014 in Las Vegas, The High Roller observation wheel has a maximum height of 550 feet and a diameter of 520 feet, with one full rotation taking approximately 30 minutes. Find an equation for the wheel if the center of the wheel is on the y-axis.

Source: *Las Vegas Review Journal*

53. Weather Satellites Earth is represented on a map of a portion of the solar system so that its surface is the circle with equation $x^2+y^2+2x+4y-4091=0$. A weather satellite circles 0.6 unit above Earth with the center of its circular orbit at the center of Earth. Find the equation for the orbit of the satellite on this map.

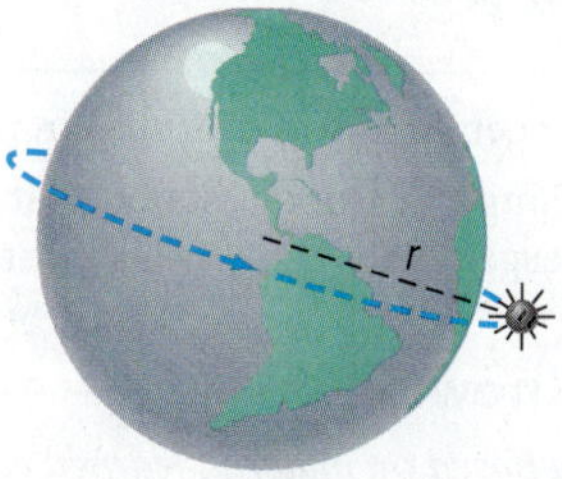

54. The **tangent line** to a circle may be defined as the line that intersects the circle in a single point, called the **point of tangency**. See the figure.

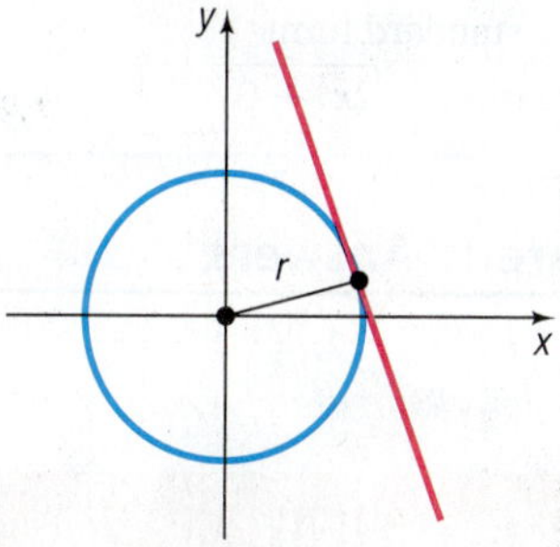

If the equation of the circle is $x^2+y^2=r^2$ and the equation of the tangent line is $y=mx+b$, show that:

(a) $r^2(1+m^2)=b^2$

[**Hint:** The quadratic equation $x^2+(mx+b)^2=r^2$ has exactly one solution.]

(b) The point of tangency is $\left(\dfrac{-r^2m}{b}, \dfrac{r^2}{b}\right)$.

(c) The tangent line is perpendicular to the line containing the center of the circle and the point of tangency.

55. The Greek Method The Greek method for finding the equation of the tangent line to a circle uses the fact that at any point on a circle, the lines containing the center and the tangent line are perpendicular (see Problem 54). Use this method to find an equation of the tangent line to the circle $x^2+y^2=9$ at the point $(1, 2\sqrt{2})$.

56. Use the Greek method described in Problem 55 to find an equation of the tangent line to the circle $x^2+y^2-4x+6y+4=0$ at the point $(3, 2\sqrt{2}-3)$.

57. Refer to Problem 54. The line $x - 2y + 4 = 0$ is tangent to a circle at $(0, 2)$. The line $y = 2x - 7$ is tangent to the same circle at $(3, -1)$. Find the center of the circle.

58. Find an equation of the line containing the centers of the two circles

$$x^2 + y^2 - 4x + 6y + 4 = 0$$

and

$$x^2 + y^2 + 6x + 4y + 9 = 0$$

59. If a circle of radius 2 is made to roll along the x-axis, what is an equation for the path of the center of the circle?

60. If the circumference of a circle is 6π, what is its radius?

Explaining Concepts: Discussion and Writing

61. Which of the following equations might have the graph shown? (More than one answer is possible.)

(a) $(x - 2)^2 + (y + 3)^2 = 13$
(b) $(x - 2)^2 + (y - 2)^2 = 8$
(c) $(x - 2)^2 + (y - 3)^2 = 13$
(d) $(x + 2)^2 + (y - 2)^2 = 8$
(e) $x^2 + y^2 - 4x - 9y = 0$
(f) $x^2 + y^2 + 4x - 2y = 0$
(g) $x^2 + y^2 - 9x - 4y = 0$
(h) $x^2 + y^2 - 4x - 4y = 4$

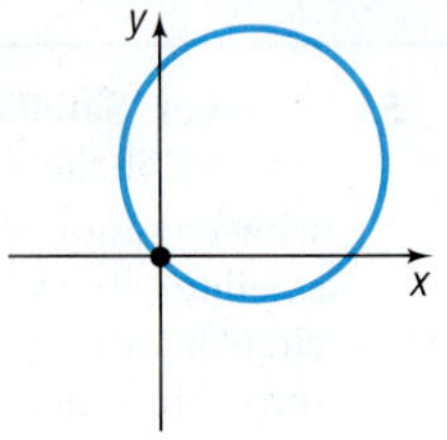

62. Which of the following equations might have the graph shown? (More than one answer is possible.)

(a) $(x - 2)^2 + y^2 = 3$
(b) $(x + 2)^2 + y^2 = 3$
(c) $x^2 + (y - 2)^2 = 3$
(d) $(x + 2)^2 + y^2 = 4$
(e) $x^2 + y^2 + 10x + 16 = 0$
(f) $x^2 + y^2 + 10x - 2y = 1$
(g) $x^2 + y^2 + 9x + 10 = 0$
(h) $x^2 + y^2 - 9x - 10 = 0$

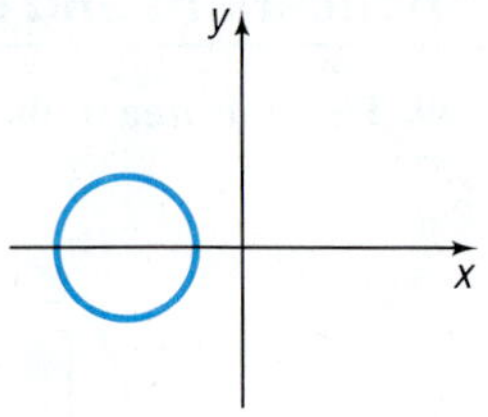

63. Explain how the center and radius of a circle can be used to graph the circle.

64. **What Went Wrong?** A student stated that the center and radius of the graph whose equation is $(x + 3)^2 + (y - 2)^2 = 16$ are $(3, -2)$ and 4, respectively. Why is this incorrect?

Retain Your Knowledge

Problems 65–68 are based on material learned earlier in the course. The purpose of these problems is to keep the material fresh in your mind so that you are better prepared for the final exam.

65. Find the area and circumference of a circle of radius 13 cm.

66. Multiply $(3x - 2)(x^2 - 2x + 3)$. Express the answer as a polynomial in standard form.

67. Solve the equation: $\sqrt{2x^2 + 3x - 1} = x + 1$

68. Aaron can load a delivery van in 22 minutes. Elizabeth can load the same van in 28 minutes. How long would it take them to load the van if they worked together?

'Are You Prepared?' Answers

1. add; 25 **2.** $\{-1, 5\}$

2.5 Variation

OBJECTIVES
1 Construct a Model Using Direct Variation (p. 189)
2 Construct a Model Using Inverse Variation (p. 189)
3 Construct a Model Using Joint Variation or Combined Variation (p. 190)

When a mathematical model is developed for a real-world problem, it often involves relationships between quantities that are expressed in terms of proportionality:

Force is proportional to acceleration.

When an ideal gas is held at a constant temperature, pressure and volume are inversely proportional.

The force of attraction between two heavenly bodies is inversely proportional to the square of the distance between them.

Revenue is directly proportional to sales.

Each of the preceding statements illustrates the idea of **variation**, or how one quantity varies in relation to another quantity. Quantities may vary *directly*, *inversely*, or *jointly*.

1 Construct a Model Using Direct Variation

DEFINITION

Let x and y denote two quantities. Then y **varies directly** with x, or y is **directly proportional to** x, if there is a nonzero number k such that

$$y = kx$$

The number k is called the **constant of proportionality**.

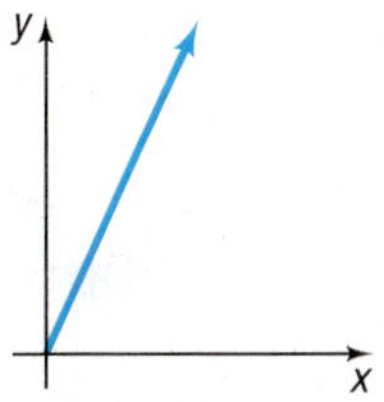

Figure 54 $y = kx; k > 0, x \geq 0$

The graph in Figure 54 illustrates the relationship between y and x if y varies directly with x and $k > 0$, $x \geq 0$. Note that the constant of proportionality is, in fact, the slope of the line.

If two quantities vary directly, then knowing the value of each quantity in one instance enables us to write a formula that is true in all cases.

EXAMPLE 1 Mortgage Payments

The monthly payment p on a mortgage varies directly with the amount borrowed B. If the monthly payment on a 30-year mortgage is \$6.65 for every \$1000 borrowed, find a formula that relates the monthly payment p to the amount borrowed B for a mortgage with these terms. Then find the monthly payment p when the amount borrowed B is \$120,000.

Solution Because p varies directly with B, we know that

$$p = kB$$

for some constant k. Because $p = 6.65$ when $B = 1000$, it follows that

$$6.65 = k(1000)$$

$$k = 0.00665 \quad \text{Solve for } k.$$

Since $p = kB$,

$$p = 0.00665B$$

In particular, when $B = \$120{,}000$,

$$p = 0.00665(\$120{,}000) = \$798$$

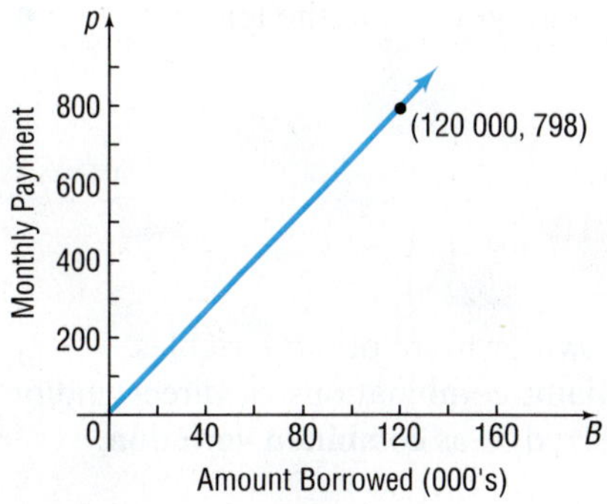

Figure 55

Figure 55 illustrates the relationship between the monthly payment p and the amount borrowed B.

Now Work PROBLEMS 5 AND 23

2 Construct a Model Using Inverse Variation

DEFINITION

Let x and y denote two quantities. Then y **varies inversely** with x, or y is **inversely proportional to** x, if there is a nonzero constant k such that

$$y = \frac{k}{x}$$

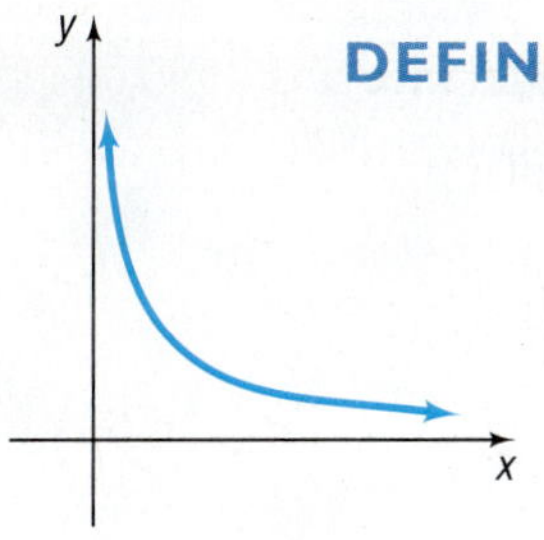

Figure 56 $y = \frac{k}{x}; k > 0, x > 0$

The graph in Figure 56 illustrates the relationship between y and x if y varies inversely with x and $k > 0$, $x > 0$.

EXAMPLE 2 **Maximum Weight That Can Be Supported by a Piece of Pine**

See Figure 57. The maximum weight W that can be safely supported by a 2-inch by 4-inch piece of pine varies inversely with its length l. Experiments indicate that the maximum weight that a 10-foot-long 2-by-4 piece of pine can support is 500 pounds. Write a general formula relating the maximum weight W (in pounds) to length l (in feet). Find the maximum weight W that can be safely supported by a length of 25 feet.

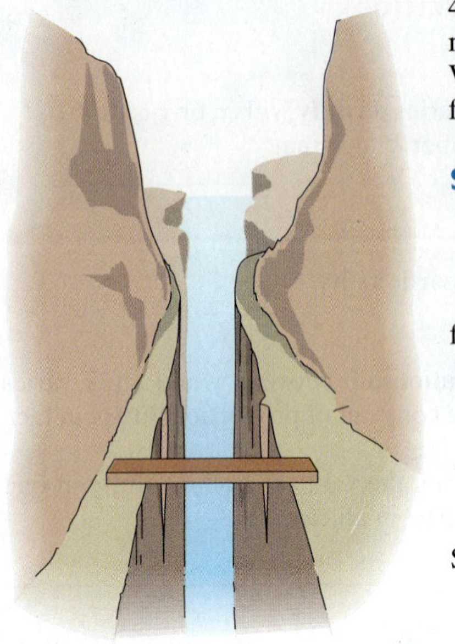

Figure 57

Solution Because W varies inversely with l, we know that

$$W = \frac{k}{l}$$

for some constant k. Because $W = 500$ when $l = 10$, we have

$$500 = \frac{k}{10}$$

$$k = 5000$$

Since $W = \frac{k}{l}$,

$$W = \frac{5000}{l}$$

In particular, the maximum weight W that can be safely supported by a piece of pine 25 feet in length is

$$W = \frac{5000}{25} = 200 \text{ pounds}$$

Figure 58 illustrates the relationship between the weight W and the length l.

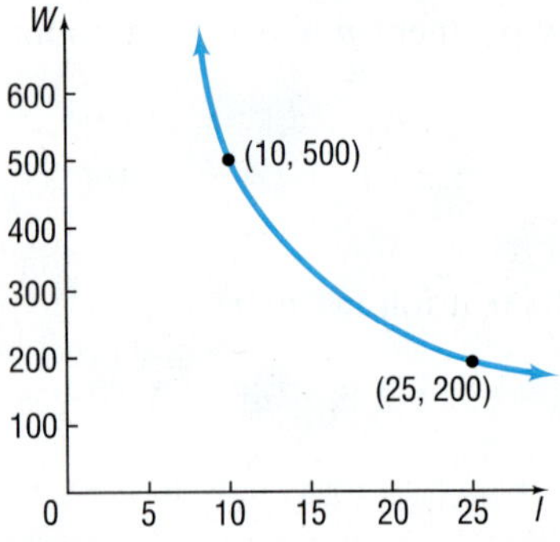

Figure 58 $W = \frac{5000}{l}$

Now Work PROBLEM 33

3 Construct a Model Using Joint Variation or Combined Variation

When a variable quantity Q is proportional to two or more other variables, we say that Q **varies jointly** with these quantities. Finally, combinations of direct and/or inverse variation may occur. This is usually referred to as **combined variation.**

EXAMPLE 3 **Loss of Heat through a Wall**

The loss of heat through a wall varies jointly with the area of the wall and the difference between the inside and outside temperatures and varies inversely with the thickness of the wall. Write an equation that relates these quantities.

Solution Begin by assigning symbols to represent the quantities:

L = Heat loss $\quad$ T = Temperature difference

A = Area of wall $\quad$ d = Thickness of wall

Then

$$L = k\frac{AT}{d}$$

where k is the constant of proportionality.

In direct or inverse variation, the quantities that vary may be raised to powers. For example, in the early seventeenth century, Johannes Kepler (1571–1630) discovered that the square of the period of revolution T of a planet around the Sun varies directly with the cube of its mean distance a from the Sun. That is, $T^2 = ka^3$, where k is the constant of proportionality.

EXAMPLE 4 **Force of the Wind on a Window**

The force F of the wind on a flat surface positioned at a right angle to the direction of the wind varies jointly with the area A of the surface and the square of the speed v of the wind. A wind of 30 miles per hour blowing on a window measuring 4 feet by 5 feet has a force of 150 pounds. See Figure 59. What force does a wind of 50 miles per hour exert on a window measuring 3 feet by 4 feet?

Figure 59

Solution Since F varies jointly with A and v^2, we have

$$F = kAv^2$$

where k is the constant of proportionality. We are told that $F = 150$ when $A = 4 \cdot 5 = 20$ and $v = 30$. Then

$$150 = k(20)(900) \quad F = kAv^2, F = 150, A = 20, v = 30$$

$$k = \frac{1}{120}$$

Since $F = kAv^2$,

$$F = \frac{1}{120}Av^2$$

For a wind of 50 miles per hour blowing on a window whose area is $A = 3 \cdot 4 = 12$ square feet, the force F is

$$F = \frac{1}{120}(12)(2500) = 250 \text{ pounds}$$

Now Work PROBLEM 41

2.5 Assess Your Understanding

Concepts and Vocabulary

1. If x and y are two quantities, then y is directly proportional to x if there is a nonzero number k such that ________.

2. ***True or False*** If y varies directly with x, then $y = \frac{k}{x}$, where k is a constant.

3. Which equation represents a joint variation model?
   (a) $y = 5x$ (b) $y = 5xzw$
   (c) $y = \frac{5}{x}$ (d) $y = \frac{5xz}{w}$

4. Choose the best description for the model $y = \frac{kx}{z}$, if k is a nonzero constant.
   (a) y varies jointly with x and z.
   (b) y is inversely proportional to x and z.
   (c) y varies directly with x and inversely with z.
   (d) y is directly proportion to z and inversely proportional to x.

Skill Building

In Problems 5–16, write a general formula to describe each variation.

5. y varies directly with x; $y = 2$ when $x = 10$

6. v varies directly with t; $v = 16$ when $t = 2$

7. A varies directly with x^2; $A = 4\pi$ when $x = 2$

8. V varies directly with x^3; $V = 36\pi$ when $x = 3$

9. F varies inversely with d^2; $F = 10$ when $d = 5$

10. y varies inversely with $\sqrt{x}$; $y = 4$ when $x = 9$

11. z varies directly with the sum of the squares of x and y; $z = 5$ when $x = 3$ and $y = 4$

12. T varies jointly with the cube root of x and the square of d; $T = 18$ when $x = 8$ and $d = 3$

13. M varies directly with the square of d and inversely with the square root of x; $M = 24$ when $x = 9$ and $d = 4$

14. z varies directly with the sum of the cube of x and the square of y; $z = 1$ when $x = 2$ and $y = 3$

15. The square of T varies directly with the cube of a and inversely with the square of d; $T = 2$ when $a = 2$ and $d = 4$

16. The cube of z varies directly with the sum of the squares of x and y; $z = 2$ when $x = 9$ and $y = 4$

Applications and Extensions

In Problems 17–22, write an equation that relates the quantities.

17. Geometry The volume V of a sphere varies directly with the cube of its radius r. The constant of proportionality is $\frac{4\pi}{3}$.

18. Geometry The square of the length of the hypotenuse c of a right triangle varies jointly with the sum of the squares of the lengths of its legs a and b. The constant of proportionality is 1.

19. Geometry The area A of a triangle varies jointly with the product of the lengths of the base b and the height h. The constant of proportionality is $\frac{1}{2}$.

20. Geometry The perimeter p of a rectangle varies jointly with the sum of the lengths of its sides l and w. The constant of proportionality is 2.

21. Physics: Newton's Law The force F (in newtons) of attraction between two bodies varies jointly with their masses m and M (in kilograms) and inversely with the square of the distance d (in meters) between them. The constant of proportionality is $G = 6.67 \times 10^{-11}$.

22. Physics: Simple Pendulum The **period** of a pendulum is the time required for one oscillation; the pendulum is usually referred to as **simple** when the angle made to the vertical is less than 5°. The period T of a simple pendulum (in seconds) varies directly with the square root of its length l (in feet). The constant of proportionality is $\frac{2\pi}{\sqrt{32}}$.

23. Mortgage Payments The monthly payment p on a mortgage varies directly with the amount borrowed B. If the monthly payment on a 30-year mortgage is \$6.49 for every \$1000 borrowed, find a linear equation that relates the monthly payment p to the amount borrowed B for a mortgage with these terms. Then find the monthly payment p when the amount borrowed B is \$145,000.

24. Mortgage Payments The monthly payment p on a mortgage varies directly with the amount borrowed B. If the monthly payment on a 15-year mortgage is \$8.99 for every \$1000 borrowed, find a linear equation that relates the monthly payment p to the amount borrowed B for a mortgage with these terms. Then find the monthly payment p when the amount borrowed B is \$175,000.

25. Physics: Falling Objects The distance s that an object falls is directly proportional to the square of the time t of the fall. If an object falls 16 feet in 1 second, how far will it fall in 3 seconds? How long will it take an object to fall 64 feet?

26. Physics: Falling Objects The velocity v of a falling object is directly proportional to the time t of the fall. If, after 2 seconds, the velocity of the object is 64 feet per second, what will its velocity be after 3 seconds?

27. Physics: Stretching a Spring The elongation E of a spring balance varies directly with the applied weight W (see the figure). If $E = 3$ when $W = 20$, find E when $W = 15$.

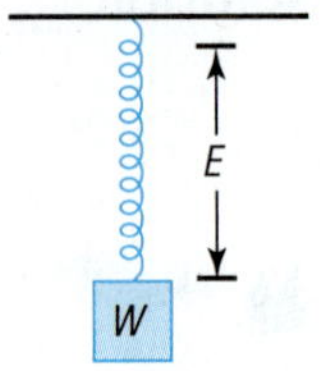

28. Physics: Vibrating String The rate of vibration of a string under constant tension varies inversely with the length of the string. If a string is 48 inches long and vibrates 256 times per second, what is the length of a string that vibrates 576 times per second?

29. Revenue Equation At the corner Shell station, the revenue R varies directly with the number g of gallons of gasoline sold. If the revenue is \$47.40 when the number of gallons sold is 12, find a linear equation that relates revenue R to the number g of gallons of gasoline sold. Then find the revenue R when the number of gallons of gasoline sold is 10.5.

30. Cost Equation The cost C of roasted almonds varies directly with the number A of pounds of almonds purchased. If the cost is \$23.75 when the number of pounds of roasted almonds purchased is 5, find a linear equation that relates the cost C to the number A of pounds of almonds purchased. Then find the cost C when the number of pounds of almonds purchased is 3.5.

31. Demand Suppose that the demand D for candy at the movie theater is inversely related to the price p.

(a) When the price of candy is \$2.75 per bag, the theater sells 156 bags of candy. Express the demand for candy in terms of its price.

(b) Determine the number of bags of candy that will be sold if the price is raised to \$3 a bag.

32. Driving to School The time t that it takes to get to school varies inversely with your average speed s.

(a) Suppose that it takes you 40 minutes to get to school when your average speed is 30 miles per hour. Express the driving time to school in terms of average speed.

(b) Suppose that your average speed to school is 40 miles per hour. How long will it take you to get to school?

33. Pressure The volume of a gas V held at a constant temperature in a closed container varies inversely with its pressure P. If the volume of a gas is 600 cubic centimeters (cm^3) when the pressure is 150 millimeters of mercury (mm Hg), find the volume when the pressure is 200 mm Hg.

34. Resistance The current i in a circuit is inversely proportional to its resistance Z measured in ohms. Suppose that when the current in a circuit is 30 amperes, the resistance is 8 ohms. Find the current in the same circuit when the resistance is 10 ohms.

35. Weight The weight of an object above the surface of Earth varies inversely with the square of the distance from the center of Earth. If Maria weighs 125 pounds when she is on the surface of Earth (3960 miles from the center), determine Maria's weight when she is at the top of Mount McKinley (3.8 miles from the surface of Earth).

36. Weight of a Body The weight of a body above the surface of Earth varies inversely with the square of the distance from the center of Earth. If a certain body weighs 55 pounds when it is 3960 miles from the center of Earth, how much will it weigh when it is 3965 miles from the center?

37. Geometry The volume V of a right circular cylinder varies jointly with the square of its radius r and its height h. The constant of proportionality is π. See the figure. Write an equation for V.

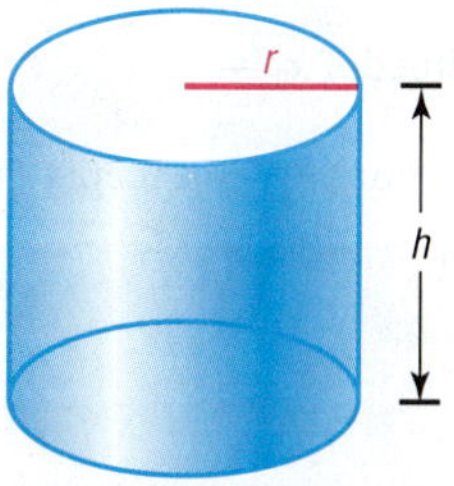

38. Geometry The volume V of a right circular cone varies jointly with the square of its radius r and its height h. The constant of proportionality is $\frac{\pi}{3}$. See the figure. Write an equation for V.

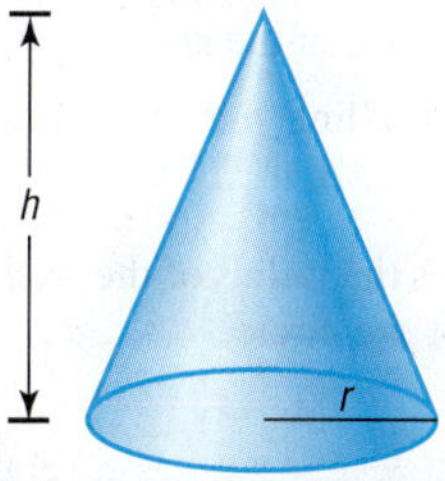

39. Intensity of Light The intensity I of light (measured in foot-candles) varies inversely with the square of the distance from the bulb. Suppose that the intensity of a 100-watt light bulb at a distance of 2 meters is 0.075 foot-candle. Determine the intensity of the bulb at a distance of 5 meters.

40. Force of the Wind on a Window The force exerted by the wind on a plane surface varies jointly with the area of the surface and the square of the velocity of the wind. If the force on an area of 20 square feet is 11 pounds when the wind velocity is 22 miles per hour, find the force on a surface area of 47.125 square feet when the wind velocity is 36.5 miles per hour.

41. Horsepower The horsepower (hp) that a shaft can safely transmit varies jointly with its speed (in revolutions per minute, rpm) and the cube of its diameter. If a shaft of a certain material 2 inches in diameter can transmit 36 hp at 75 rpm, what diameter must the shaft have in order to transmit 45 hp at 125 rpm?

42. Chemistry: Gas Laws The volume V of an ideal gas varies directly with the temperature T and inversely with the pressure P. Write an equation relating V, T, and P using k as the constant of proportionality. If a cylinder contains oxygen at a temperature of 300 K and a pressure of 15 atmospheres in a volume of 100 liters, what is the constant of proportionality k? If a piston is lowered into the cylinder, decreasing the volume occupied by the gas to 80 liters and raising the temperature to 310 K, what is the gas pressure?

43. Physics: Kinetic Energy The kinetic energy K of a moving object varies jointly with its mass m and the square of its velocity v. If an object weighing 25 kilograms and moving with a velocity of 10 meters per second has a kinetic energy of 1250 joules, find its kinetic energy when the velocity is 15 meters per second.

44. Electrical Resistance of a Wire The electrical resistance of a wire varies directly with the length of the wire and inversely with the square of the diameter of the wire. If a wire 432 feet long and 4 millimeters in diameter has a resistance of 1.24 ohms, find the length of a wire of the same material whose resistance is 1.44 ohms and whose diameter is 3 millimeters.

45. Measuring the Stress of Materials The stress in the material of a pipe subject to internal pressure varies jointly with the internal pressure and the internal diameter of the pipe and inversely with the thickness of the pipe. The stress is 100 pounds per square inch when the diameter is 5 inches, the thickness is 0.75 inch, and the internal pressure is 25 pounds per square inch. Find the stress when the internal pressure is 40 pounds per square inch if the diameter is 8 inches and the thickness is 0.50 inch.

46. Safe Load for a Beam The maximum safe load for a horizontal rectangular beam varies jointly with the width of the beam and the square of the thickness of the beam and inversely with its length. If an 8-foot beam will support up to 750 pounds when the beam is 4 inches wide and 2 inches thick, what is the maximum safe load in a similar beam 10 feet long, 6 inches wide, and 2 inches thick?

Explaining Concepts: Discussion and Writing

47. In the early seventeenth century, Johannes Kepler discovered that the square of the period T of the revolution of a planet around the Sun varies directly with the cube of its mean distance a from the Sun. Go to the library and research this law and Kepler's other two laws. Write a brief paper about these laws and Kepler's place in history.

48. Using a situation that has not been discussed in the text, write a real-world problem that you think involves two variables that vary directly. Exchange your problem with another student's to solve and critique.

49. Using a situation that has not been discussed in the text, write a real-world problem that you think involves two variables that vary inversely. Exchange your problem with another student's to solve and critique.

50. Using a situation that has not been discussed in the text, write a real-world problem that you think involves three variables that vary jointly. Exchange your problem with another student's to solve and critique.

Retain Your Knowledge

Problems 51–54 are based on material learned earlier in the course. The purpose of these problems is to keep the material fresh in your mind so that you are better prepared for the final exam.

51. Factor $3x^3 + 25x^2 - 12x - 100$ completely.

52. Add $\frac{5}{x+3} + \frac{x-2}{x^2+7x+12}$ and simplify the result.

53. Simplify: $\left(\frac{4}{25}\right)^{3/2}$

54. Rationalize the denominator of $\frac{3}{\sqrt{7}-2}$.

Chapter Review

Things to Know

Formulas

Distance formula (p. 151)	$d = \sqrt{(x_2 - x_1)^2 + (y_2 - y_1)^2}$
Midpoint formula (p. 154)	$(x, y) = \left(\frac{x_1 + x_2}{2}, \frac{y_1 + y_2}{2}\right)$
Slope (p. 167)	$m = \frac{y_2 - y_1}{x_2 - x_1}$ if $x_1 \neq x_2$; undefined if $x_1 = x_2$
Parallel lines (p. 175)	Equal slopes $(m_1 = m_2)$ and different y-intercepts $(b_1 \neq b_2)$
Perpendicular lines (p. 176)	Product of slopes is -1 $(m_1 \cdot m_2 = -1)$
Direct variation (p. 189)	$y = kx$
Inverse variation (p. 189)	$y = \frac{k}{x}$

Equations of Lines and Circles

Vertical line (p. 171)	$x = a$; a is the x-intercept
Horizontal line (p. 172)	$y = b$; b is the y-intercept
Point–slope form of the equation of a line (p. 171)	$y - y_1 = m(x - x_1)$; m is the slope of the line, (x_1, y_1) is a point on the line
Slope–intercept form of the equation of a line (p. 172)	$y = mx + b$; m is the slope of the line, b is the y-intercept
General form of the equation of a line (p. 174)	$Ax + By = C$; A, B not both 0
Standard form of the equation of a circle (p. 183)	$(x - h)^2 + (y - k)^2 = r^2$; r is the radius of the circle, (h, k) is the center of the circle
Equation of the unit circle (p. 183)	$x^2 + y^2 = 1$
General form of the equation of a circle (p. 185)	$x^2 + y^2 + ax + by + c = 0$, with restrictions on a, b, and c

Objectives

Section		You should be able to ...	Examples	Review Exercises
2.1	1	Use the distance formula (p. 151)	1–3	1(a)–3(a), 29, 30(a), 31
	2	Use the midpoint formula (p. 153)	4	1(b)–3(b), 31
2.2	1	Graph equations by plotting points (p. 157)	1–3	4
	2	Find intercepts from a graph (p. 159)	4	5
	3	Find intercepts from an equation (p. 160)	5	6–10
	4	Test an equation for symmetry with respect to the x-axis, the y-axis, and the origin (p. 160)	6–9	6–10
	5	Know how to graph key equations (p. 163)	10–12	26, 27
2.3	1	Calculate and interpret the slope of a line (p. 167)	1, 2	1(c)–3(c), 1(d)–3(d), 32
	2	Graph lines given a point and the slope (p. 170)	3	28
	3	Find the equation of a vertical line (p. 170)	4	17

Section		You should be able to ...	Examples	Review Exercises
	4	Use the point–slope form of a line; identify horizontal lines (p. 171)	5, 6	16
	5	Find the equation of a line given two points (p. 172)	7	18, 19
	6	Write the equation of a line in slope–intercept form (p. 172)	8	16, 18–21
	7	Identify the slope and y-intercept of a line from its equation (p. 173)	8	22, 23
	8	Graph lines written in general form using intercepts (p. 174)	9	24, 25
	9	Find equations of parallel lines (p. 175)	10, 11	20
	10	Find equations of perpendicular lines (p. 176)	12, 13	21, 30(b)
2.4	1	Write the standard form of the equation of a circle (p. 182)	1	11, 12, 31
	2	Graph a circle (p. 183)	2, 3, 5	13–15
	3	Work with the general form of the equation of a circle (p. 184)	4	14, 15
2.5	1	Construct a model using direct variation (p. 189)	1	33
	2	Construct a model using inverse variation (p. 189)	2	34
	3	Construct a model using joint or combined variation (p. 190)	3, 4	35

Review Exercises

In Problems 1–3, find the following for each pair of points:
(a) The distance between the points
(b) The midpoint of the line segment connecting the points
(c) The slope of the line containing the points
(d) Interpret the slope found in part (c)

1. $(0, 0); (4, 2)$
2. $(1, -1); (-2, 3)$
3. $(4, -4); (4, 8)$
4. Graph $y = x^2 + 4$ by plotting points.
5. List the intercepts of the graph below.

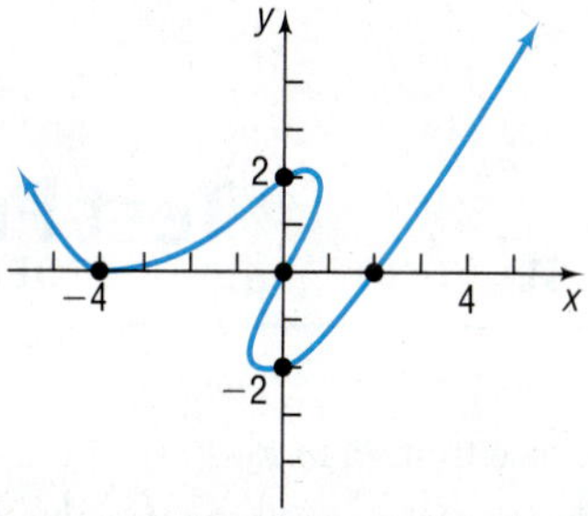

In Problems 6–10, list the intercepts and test for symmetry with respect to the x-axis, the y-axis, and the origin.

6. $2x = 3y^2$
7. $x^2 + 4y^2 = 16$
8. $y = x^4 + 2x^2 + 1$
9. $y = x^3 - x$
10. $x^2 + x + y^2 + 2y = 0$

In Problems 11 and 12, find the standard form of the equation of the circle whose center and radius are given.

11. $(h, k) = (-2, 3); r = 4$
12. $(h, k) = (-1, -2); r = 1$

In Problems 13–15, find the center and radius of each circle. Graph each circle. Find the intercepts, if any, of each circle.

13. $x^2 + (y - 1)^2 = 4$
14. $x^2 + y^2 - 2x + 4y - 4 = 0$
15. $3x^2 + 3y^2 - 6x + 12y = 0$

In Problems 16–21, find an equation of the line having the given characteristics. Express your answer using either the general form or the slope–intercept form of the equation of a line, whichever you prefer.

16. Slope $= -2$; containing the point $(3, -1)$
17. Vertical; containing the point $(-3, 4)$
18. y-intercept $= -2$; containing the point $(5, -3)$
19. Containing the points $(3, -4)$ and $(2, 1)$
20. Parallel to the line $2x - 3y = -4$; containing the point $(-5, 3)$
21. Perpendicular to the line $x + y = 2$; containing the point $(4, -3)$

In Problems 22 and 23, find the slope and y-intercept of each line. Graph the line, labeling any intercepts.

22. $4x - 5y = -20$
23. $\frac{1}{2}x - \frac{1}{3}y = -\frac{1}{6}$

In Problems 24 and 25, find the intercepts and graph each line.

24. $2x - 3y = 12$
25. $\frac{1}{2}x + \frac{1}{3}y = 2$

26. Sketch a graph of $y = x^3$.

27. Sketch a graph of $y = \sqrt{x}$.

28. Graph the line with slope $\frac{2}{3}$ containing the point $(1, 2)$.

29. Show that the points $A = (3, 4), B = (1, 1)$, and $C = (-2, 3)$ are the vertices of an isosceles triangle.

30. Show that the points $A = (-2, 0), B = (-4, 4)$, and $C = (8, 5)$ are the vertices of a right triangle in two ways:
(a) By using the converse of the Pythagorean Theorem
(b) By using the slopes of the lines joining the vertices

31. The endpoints of the diameter of a circle are $(-3, 2)$ and $(5, -6)$. Find the center and radius of the circle. Write the standard equation of this circle.

32. Show that the points $A = (2, 5), B = (6, 1)$, and $C = (8, -1)$ lie on a line by using slopes.

33. **Mortgage Payments** The monthly payment p on a mortgage varies directly with the amount borrowed B. If the monthly payment on a 30-year mortgage is \$854.00 when \$130,000 is borrowed, find an equation that relates the monthly payment p to the amount borrowed B for a mortgage with these terms. Then find the monthly payment p when the amount borrowed B is \$165,000.

34. **Weight of a Body** The weight of a body varies inversely with the square of its distance from the center of Earth. Assuming that the radius of Earth is 3960 miles, how much would a man weigh at an altitude of 1 mile above Earth's surface if he weighs 200 pounds on Earth's surface?

35. **Heat Loss** The amount of heat transferred per hour through a glass window varies jointly with the surface area of the window and the difference in temperature between the areas separated by the glass. A window with a surface area of 7.5 square feet loses 135 Btu per hour when the temperature difference is 40°F. How much heat is lost per hour for a similar window with a surface are of 12 square feet when the temperature difference is 35°F?

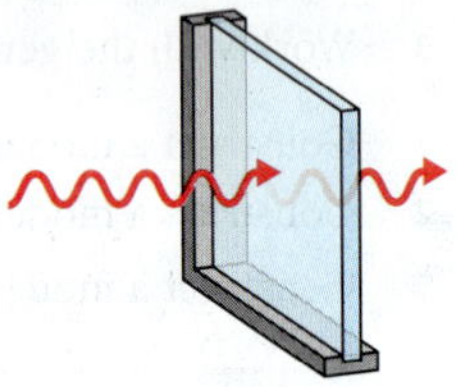

Chapter Test

CHAPTER **Test Prep** VIDEOS

The Chapter Test Prep Videos are step-by-step solutions available in MyMathLab, or on this text's YouTube Channel. Flip back to the Resources for Success page for a link to this text's YouTube channel.

In Problems 1–3, use $P_1 = (-1, 3)$ and $P_2 = (5, -1)$.

1. Find the distance from P_1 to P_2.
2. Find the midpoint of the line segment joining P_1 and P_2.
3. (a) Find the slope of the line containing P_1 and P_2.
   (b) Interpret this slope.
4. Graph $y = x^2 - 9$ by plotting points.
5. Sketch the graph of $y^2 = x$.
6. List the intercepts and test for symmetry: $x^2 + y = 9$
7. Write the slope–intercept form of the line with slope -2 containing the point $(3, -4)$. Graph the line.
8. Write the general form of the circle with center $(4, -3)$ and radius 5.
9. Find the center and radius of the circle $x^2 + y^2 + 4x - 2y - 4 = 0$. Graph this circle.
10. For the line $2x + 3y = 6$, find a line parallel to it containing the point $(1, -1)$. Also find a line perpendicular to it containing the point $(0, 3)$.
11. **Resistance Due to a Conductor** The resistance (in ohms) of a circular conductor varies directly with the length of the conductor and inversely with the square of the radius of the conductor. If 50 feet of wire with a radius of 6×10^{-3} inch has a resistance of 10 ohms, what would be the resistance of 100 feet of the same wire if the radius were increased to 7×10^{-3} inch?

Cumulative Review

In Problems 1–8, find the real solution(s), if any, of each equation.

1. $3x - 5 = 0$
2. $x^2 - x - 12 = 0$
3. $2x^2 - 5x - 3 = 0$
4. $x^2 - 2x - 2 = 0$
5. $x^2 + 2x + 5 = 0$
6. $\sqrt{2x + 1} = 3$
7. $|x - 2| = 1$
8. $\sqrt{x^2 + 4x} = 2$

In Problems 9 and 10, solve each equation in the complex number system.

9. $x^2 = -9$
10. $x^2 - 2x + 5 = 0$

In Problems 11–14, solve each inequality. Graph the solution set.

11. $2x - 3 \le 7$
12. $-1 < x + 4 < 5$
13. $|x - 2| \le 1$
14. $|2 + x| > 3$

15. Find the distance between the points $P = (-1, 3)$ and $Q = (4, -2)$. Find the midpoint of the line segment from P to Q.

16. Which of the following points are on the graph of $y = x^3 - 3x + 1$?
(a) $(-2, -1)$ (b) $(2, 3)$ (c) $(3, 1)$

17. Sketch the graph of $y = x^3$.

18. Find the equation of the line containing the points $(-1, 4)$ and $(2, -2)$. Express your answer in slope–intercept form.

19. Find the equation of the line perpendicular to the line $y = 2x + 1$ and containing the point $(3, 5)$. Express your answer in slope–intercept form and graph the line.

20. Graph the equation $x^2 + y^2 - 4x + 8y - 5 = 0$.

Chapter Project

Internet-based Project

Determining the Selling Price of a Home Determining how much to pay for a home is one of the more difficult decisions that must be made when purchasing a home. There are many factors that play a role in a home's value. Location, size, number of bedrooms, number of bathrooms, lot size, and building materials are just a few. Fortunately, the website Zillow.com has developed its own formula for predicting the selling price of a home. This information is a great tool for predicting the actual sale price. For example, the data below show the "zestimate"—the selling price of a home as predicted by the folks at Zillow—and the actual selling price of the home, for homes in Oak Park, Illinois.

Zestimate ($ thousands)	Sale Price ($ thousands)
291.5	268
320	305
371.5	375
303.5	283
351.5	350
314	275
332.5	356
295	300
313	285
368	385

The graph below, called a scatter diagram, shows the points (291.5, 268), (320, 305), . . . , (368, 385) in a Cartesian plane. From the graph, it appears that the data follow a linear relation.

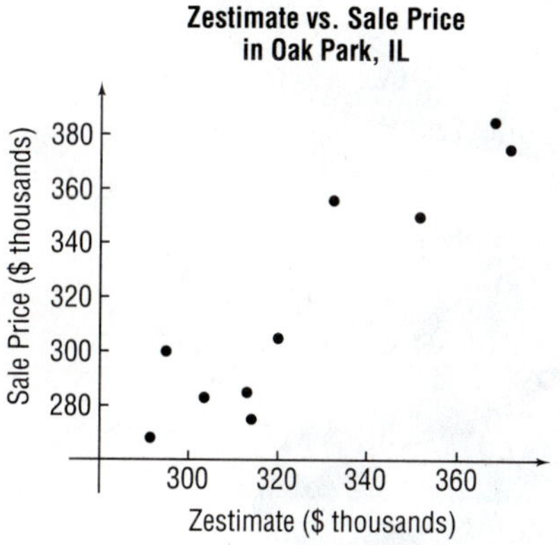

1. Imagine drawing a line through the data that appears to fit the data well. Do you believe the slope of the line would be positive, negative, or close to zero? Why?

2. Pick two points from the scatter diagram. Treat the zestimate as the value of x, and treat the sale price as the corresponding value of y. Find the equation of the line through the two points you selected.

3. Interpret the slope of the line.

4. Use your equation to predict the selling price of a home whose zestimate is $335,000.

5. Do you believe it would be a good idea to use the equation you found in part 2 if the zestimate were $950,000? Why or why not?

6. Choose a location in which you would like to live. Go to www.zillow.com and randomly select at least ten homes that have recently sold.
   (a) Draw a scatter diagram of your data.
   (b) Select two points from the scatter diagram and find the equation of the line through the points.
   (c) Interpret the slope.
   (d) Find a home from the Zillow website that interests you under the "Make Me Move" option for which a zestimate is available. Use your equation to predict the sale price based on the zestimate.

3 Functions and Their Graphs

Choosing a Wireless Data Plan

Most consumers choose a cellular provider first and then select an appropriate data plan from that provider. The choice as to the type of plan selected depends on your use of the device. For example, is online gaming important? Do you want to stream audio or video? The mathematics learned in this chapter can help you decide what plan is best suited to your particular needs.

—See the Internet-based Chapter Project—

Outline

••• A Look Back

So far, our discussion has focused on techniques for graphing equations containing two variables.

A Look Ahead •••

In this chapter, we look at a special type of equation involving two variables called a *function*. This chapter deals with what a function is, how to graph functions, properties of functions, and how functions are used in applications. The word *function* apparently was introduced by René Descartes in 1637. For him, a function was simply any positive integral power of a variable *x*. Gottfried Wilhelm Leibniz (1646–1716), who always emphasized the geometric side of mathematics, used the word *function* to denote any quantity associated with a curve, such as the coordinates of a point on the curve. Leonhard Euler (1707–1783) employed the word to mean any equation or formula involving variables and constants. His idea of a function is similar to the one most often seen in courses that precede calculus. Later, the use of functions in investigating heat flow equations led to a very broad definition that originated with Lejeune Dirichlet (1805–1859), which describes a function as a correspondence between two sets. That is the definition used in this text.

3.1 Functions

PREPARING FOR THIS SECTION *Before getting started, review the following:*

- Intervals (Section 1.5, pp. 120–121)
- Solving Inequalities (Section 1.5, pp. 123–126)
- Evaluating Algebraic Expressions, Domain of a Variable (Chapter R, Section R.2, pp. 20–23)
- Rationalizing Denominators (Chapter R, Section R.8, p. 75)

Now Work the 'Are You Prepared?' problems on page 210.

OBJECTIVES
1 Determine Whether a Relation Represents a Function (p. 199)
2 Find the Value of a Function (p. 202)
3 Find the Difference Quotient of a Function (p. 205)
4 Find the Domain of a Function Defined by an Equation (p. 206)
5 Form the Sum, Difference, Product, and Quotient of Two Functions (p. 208)

1 Determine Whether a Relation Represents a Function

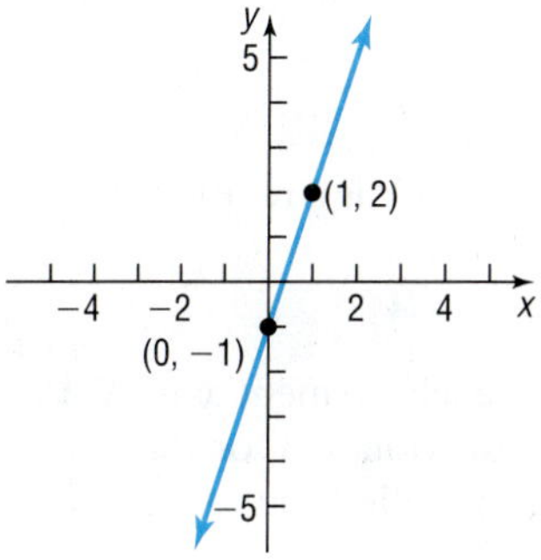

Figure 1 $y = 3x - 1$

Often there are situations where the value of one variable is somehow linked to the value of another variable. For example, an individual's level of education is linked to annual income. Engine size is linked to gas mileage. When the value of one variable is related to the value of a second variable, we have a *relation*. A **relation** is a correspondence between two sets. If x and y are two elements, one from each of these sets, and if a relation exists between x and y, then we say that x **corresponds** to y or that y **depends on** x, and we write $x \rightarrow y$.

There are a number of ways to express relations between two sets. For example, the equation $y = 3x - 1$ shows a relation between x and y. It says that if we take some number x, multiply it by 3, and then subtract 1, we obtain the corresponding value of y. In this sense, x serves as the **input** to the relation, and y is the **output** of the relation. This relation, expressed as a graph, is shown in Figure 1.

The set of all inputs for a relation is called the **domain** of the relation, and the set of all outputs is called the **range**.

In addition to being expressed as equations and graphs, relations can be expressed through a technique called *mapping*. A **map** illustrates a relation as a set of inputs with an arrow drawn from each element in the set of inputs to the corresponding element in the set of outputs. **Ordered pairs** can be used to represent $x \rightarrow y$ as (x, y).

EXAMPLE 1 **Maps and Ordered Pairs as Relations**

Figure 2 shows a relation between states and the number of representatives each state has in the House of Representatives. (*Source:* www.house.gov). The relation might be named "number of representatives."

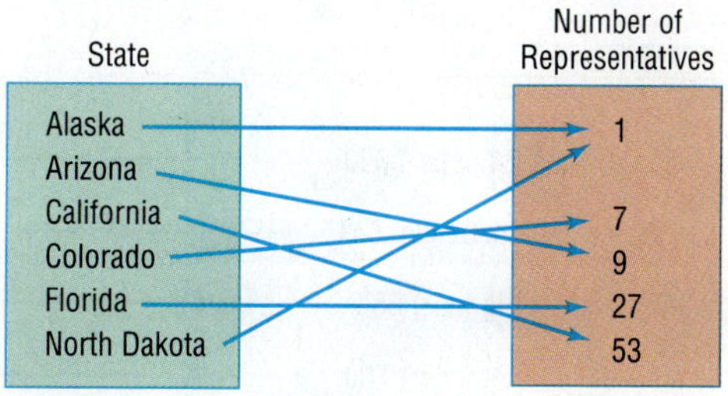

Figure 2 Number of representatives

In this relation, Alaska corresponds to 1, Arizona corresponds to 9, and so on. Using ordered pairs, this relation would be expressed as

$$\{(\text{Alaska}, 1), (\text{Arizona}, 9), (\text{California}, 53), (\text{Colorado}, 7), (\text{Florida}, 27), (\text{North Dakota}, 1)\}$$

The domain of the relation is {Alaska, Arizona, California, Colorado, Florida, North Dakota}, and the range is {1, 7, 9, 27, 53}. Note that the output "1" is listed only once in the range. ●

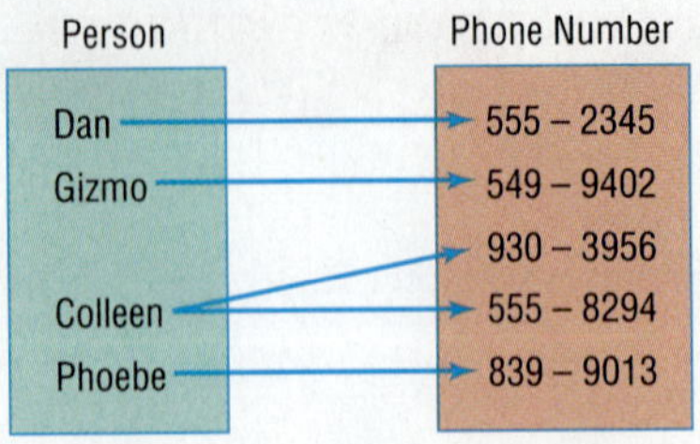

Figure 3 Phone numbers

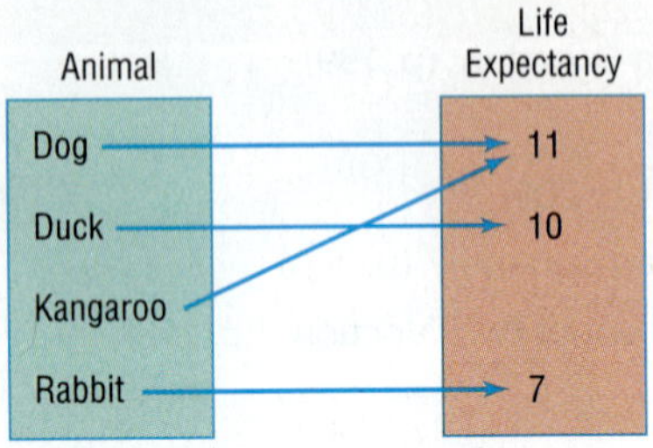

Figure 4 Animal life expectancy

One of the most important concepts in algebra is the *function*. A function is a special type of relation. To understand the idea behind a function, let's revisit the relation presented in Example 1. If we were to ask, "How many representatives does Alaska have?" you would respond "1." In fact, each input *state* corresponds to a single output *number of representatives*.

Let's consider a second relation, one that involves a correspondence between four people and their phone numbers. See Figure 3. Notice that Colleen has two telephone numbers. There is no single answer to the question "What is Colleen's phone number?"

Let's look at one more relation. Figure 4 is a relation that shows a correspondence between type of *animal* and *life expectancy*. If asked to determine the life expectancy of a dog, we would all respond, "11 years." If asked to determine the life expectancy of a rabbit, we would all respond, "7 years."

Notice that the relations presented in Figures 2 and 4 have something in common. What is it? In both of these relations, each input corresponds to exactly one output. This leads to the definition of a *function*.

DEFINITION

Let X and Y be two nonempty sets.* A **function** from X into Y is a relation that associates with each element of X exactly one element of Y.

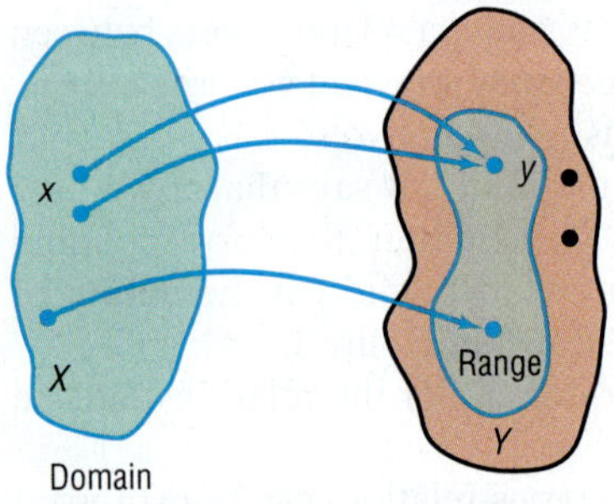

Figure 5

The set X is called the domain of the function. For each element x in X, the corresponding element y in Y is called the **value** of the function at x, or the **image** of x. The set of all images of the elements in the domain is called the range of the function. See Figure 5.

Since there may be some elements in Y that are not the image of some x in X, it follows that the range of a function may be a subset of Y, as shown in Figure 5.

Not all relations between two sets are functions. The next example shows how to determine whether a relation is a function.

EXAMPLE 2 **Determining Whether a Relation Is a Function**

For each relation in Figures 6, 7, and 8, state the domain and range. Then determine whether the relation is a function.

(a) See Figure 6. For this relation, the input is the number of calories in a fast-food sandwich, and the output is the fat content (in grams).

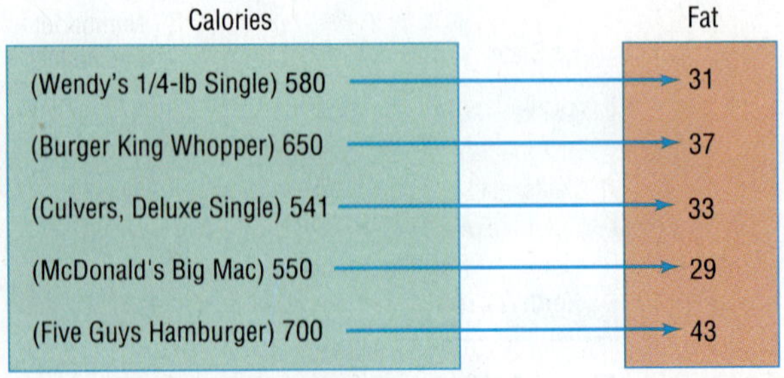

Figure 6 Fat content
Source: Each company's Web site

*The sets X and Y will usually be sets of real numbers, in which case a (real) function results. The two sets can also be sets of complex numbers, and then we have defined a complex function. In the broad definition (proposed by Lejeune Dirichlet), X and Y can be any two sets.

(b) See Figure 7. For this relation, the inputs are gasoline stations in Harris County, Texas, and the outputs are the price per gallon of unleaded regular in March 2014.

(c) See Figure 8. For this relation, the inputs are the weight (in carats) of pear-cut diamonds and the outputs are the price (in dollars).

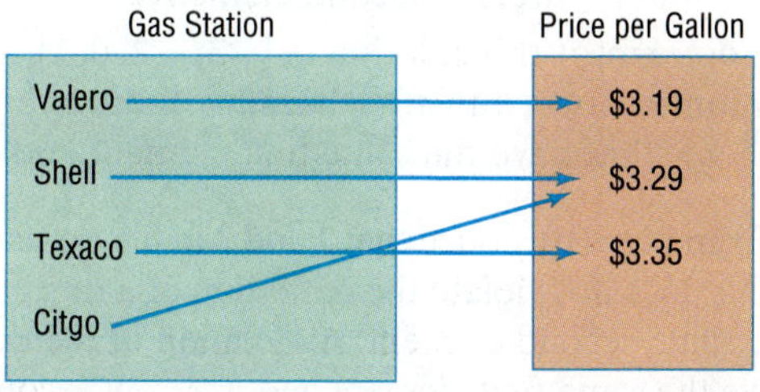

Figure 7 Unleaded price per gallon

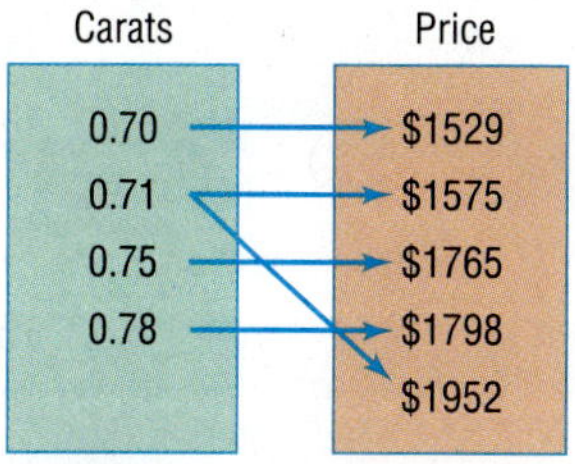

Figure 8 Diamond price
Source: Used with permission of Diamonds.com

Solution (a) The domain of the relation is {541, 550, 580, 650, 700}, and the range of the relation is {29, 31, 33, 37, 43}. The relation in Figure 6 is a function because each element in the domain corresponds to exactly one element in the range.

(b) The domain of the relation is {Citgo, Shell, Texaco, Valero}. The range of the relation is {\$3.19, \$3.29, \$3.35}. The relation in Figure 7 is a function because each element in the domain corresponds to exactly one element in the range. Notice that it is okay for more than one element in the domain to correspond to the same element in the range (Shell and Citgo both sell gas for \$3.29 a gallon).

(c) The domain of the relation is {0.70, 0.71, 0.75, 0.78} and the range is {\$1529, \$1575, \$1765, \$1798, \$1952}. The relation in Figure 8 is not a function because not every element in the domain corresponds to exactly one element in the range. If a 0.71-carat diamond is chosen from the domain, a single price cannot be assigned to it. ●

Now Work PROBLEM 19

The idea behind a function is its predictability. If the input is known, we can use the function to determine the output. With "nonfunctions," we don't have this predictability. Look back at Figure 6. If asked, "How many grams of fat are in a 580-calorie sandwich?" we could use the correspondence to answer, "31." Now consider Figure 8. If asked, "What is the price of a 0.71-carat diamond?" we could not give a single response because two outputs result from the single input "0.71." For this reason, the relation in Figure 8 is not a function.

> **In Words**
> For a function, no input has more than one output. The domain of a function is the set of all inputs; the range is the set of all outputs.

We may also think of a function as a set of ordered pairs (x, y) in which no ordered pairs have the same first element and different second elements. The set of all first elements x is the domain of the function, and the set of all second elements y is its range. Each element x in the domain corresponds to exactly one element y in the range.

EXAMPLE 3 Determining Whether a Relation Is a Function

For each relation, state the domain and range. Then determine whether the relation is a function.

(a) $\{(1,4), (2,5), (3,6), (4,7)\}$

(b) $\{(1,4), (2,4), (3,5), (6,10)\}$

(c) $\{(-3,9), (-2,4), (0,0), (1,1), (-3,8)\}$

Solution (a) The domain of this relation is {1, 2, 3, 4}, and its range is {4, 5, 6, 7}. This relation is a function because there are no ordered pairs with the same first element and different second elements.

(b) The domain of this relation is {1, 2, 3, 6}, and its range is {4, 5, 10}. This relation is a function because there are no ordered pairs with the same first element and different second elements.

(c) The domain of this relation is $\{-3, -2, 0, 1\}$, and its range is {0, 1, 4, 8, 9}. This relation is not a function because there are two ordered pairs, $(-3, 9)$ and $(-3, 8)$, that have the same first element and different second elements. ●

In Example 3(b), notice that 1 and 2 in the domain both have the same image in the range. This does not violate the definition of a function; two different first elements can have the same second element. A violation of the definition occurs when two ordered pairs have the same first element and different second elements, as in Example 3(c).

Now Work PROBLEM 23

Up to now we have shown how to identify when a relation is a function for relations defined by mappings (Example 2) and ordered pairs (Example 3). But relations can also be expressed as equations. The circumstances under which equations are functions are discussed next.

To determine whether an equation, where y depends on x, is a function, it is often easiest to solve the equation for y. If any value of x in the domain corresponds to more than one y, the equation does not define a function; otherwise, it does define a function.

EXAMPLE 4 **Determining Whether an Equation Is a Function**

Determine whether the equation $y = 2x - 5$ defines y as a function of x.

Solution The equation tells us to take an input x, multiply it by 2, and then subtract 5. For any input x, these operations yield only one output y, so the equation is a function. For example, if $x = 1$, then $y = 2(1) - 5 = -3$. If $x = 3$, then $y = 2(3) - 5 = 1$. The graph of the equation $y = 2x - 5$ is a line with slope 2 and y-intercept -5. The function is called a *linear function*.

EXAMPLE 5 **Determining Whether an Equation Is a Function**

Determine whether the equation $x^2 + y^2 = 1$ defines y as a function of x.

Solution To determine whether the equation $x^2 + y^2 = 1$, which defines the unit circle, is a function, solve the equation for y.

$$x^2 + y^2 = 1$$

$$y^2 = 1 - x^2$$

$$y = \pm\sqrt{1 - x^2}$$

For values of x for which $-1 < x < 1$, two values of y result. For example, if $x = 0$, then $y = \pm 1$, so two different outputs result from the same input. This means that the equation $x^2 + y^2 = 1$ does not define a function. ●

Now Work PROBLEM 37

2 Find the Value of a Function

Functions are often denoted by letters such as f, F, g, G, and others. If f is a function, then for each number x in its domain, the corresponding image in the range is designated by the symbol $f(x)$, read as "f of x" or as "f at x." We refer to $f(x)$ as the **value of f at the number x**; $f(x)$ is the number that results when x is given and the function f is applied; $f(x)$ is the output corresponding to x or $f(x)$ is the image of x; $f(x)$

does *not* mean "f times x." For example, the function given in Example 4 may be written as $y = f(x) = 2x - 5$. Then $f(1) = -3$ and $f(3) = 1$.

Figure 9 illustrates some other functions. Notice that in every function, for each x in the domain, there is one value in the range.

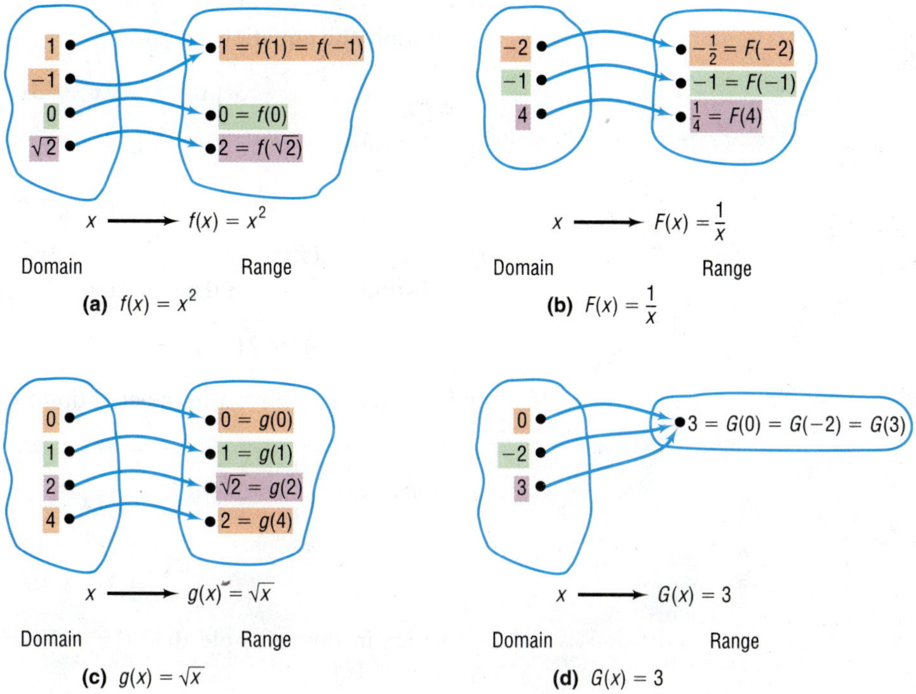

Figure 9

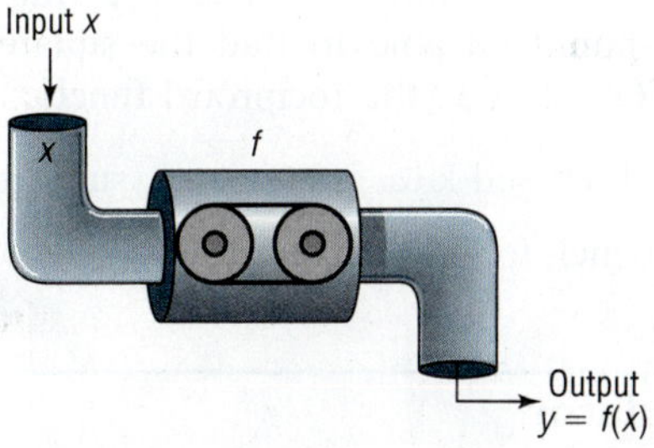

Figure 10 Input/output machine

Sometimes it is helpful to think of a function f as a machine that receives as input a number from the domain, manipulates it, and outputs a value. See Figure 10. The restrictions on this input/output machine are as follows:

1. It accepts only numbers from the domain of the function.
2. For each input, there is exactly one output (which may be repeated for different inputs).

For a function $y = f(x)$, the variable x is called the **independent variable**, because it can be assigned any of the permissible numbers from the domain. The variable y is called the **dependent variable**, because its value depends on x.

Any symbols can be used to represent the independent and dependent variables. For example, if f is the *cube function,* then f can be given by $f(x) = x^3$ or $f(t) = t^3$ or $f(z) = z^3$. All three functions are the same. Each says to cube the independent variable to get the output. In practice, the symbols used for the independent and dependent variables are based on common usage, such as using C for cost in business.

The independent variable is also called the **argument** of the function. Thinking of the independent variable as an argument can sometimes make it easier to find the value of a function. For example, if f is the function defined by $f(x) = x^3$, then f tells us to cube the argument. Thus $f(2)$ means to cube 2, $f(a)$ means to cube the number a, and $f(x + h)$ means to cube the quantity $x + h$.

EXAMPLE 6

Finding Values of a Function

For the function f defined by $f(x) = 2x^2 - 3x$, evaluate

(a) $f(3)$ (b) $f(x) + f(3)$ (c) $3f(x)$ (d) $f(-x)$
(e) $-f(x)$ (f) $f(3x)$ (g) $f(x + 3)$

Solution (a) Substitute 3 for x in the equation for f, $f(x) = 2x^2 - 3x$, to get

$$f(3) = 2(3)^2 - 3(3) = 18 - 9 = 9$$

The image of 3 is 9.

(b) $f(x) + f(3) = (2x^2 - 3x) + (9) = 2x^2 - 3x + 9$

(c) Multiply the equation for f by 3.

$$3f(x) = 3(2x^2 - 3x) = 6x^2 - 9x$$

(d) Substitute $-x$ for x in the equation for f and simplify.

$$f(-x) = 2(-x)^2 - 3(-x) = 2x^2 + 3x$$ Notice the use of parentheses here.

(e) $-f(x) = -(2x^2 - 3x) = -2x^2 + 3x$

(f) Substitute $3x$ for x in the equation for f and simplify.

$$f(3x) = 2(3x)^2 - 3(3x) = 2(9x^2) - 9x = 18x^2 - 9x$$

(g) Substitute $x + 3$ for x in the equation for f and simplify.

$$\begin{aligned} f(x+3) &= 2(x+3)^2 - 3(x+3) \\ &= 2(x^2 + 6x + 9) - 3x - 9 \\ &= 2x^2 + 12x + 18 - 3x - 9 \\ &= 2x^2 + 9x + 9 \end{aligned}$$

Notice in this example that $f(x+3) \neq f(x) + f(3)$, $f(-x) \neq -f(x)$, and $3f(x) \neq f(3x)$.

Now Work PROBLEM 43

Most calculators have special keys that allow you to find the value of certain commonly used functions. For example, you should be able to find the square function $f(x) = x^2$, the square root function $f(x) = \sqrt{x}$, the reciprocal function $f(x) = \frac{1}{x} = x^{-1}$, and many others that will be discussed later in this text (such as $\ln x$ and $\log x$). Verify the results of Example 7, which follows, on your calculator.

EXAMPLE 7 Finding Values of a Function on a Calculator

(a) $f(x) = x^2$ $\quad f(1.234) = 1.234^2 = 1.522756$

(b) $F(x) = \frac{1}{x}$ $\quad F(1.234) = \frac{1}{1.234} \approx 0.8103727715$

(c) $g(x) = \sqrt{x}$ $\quad g(1.234) = \sqrt{1.234} \approx 1.110855526$

COMMENT Graphing calculators can be used to evaluate any function. Figure 11 shows the result obtained in Example 6(a) on a TI-84 Plus C graphing calculator with the function to be evaluated, $f(x) = 2x^2 - 3x$, in Y_1.

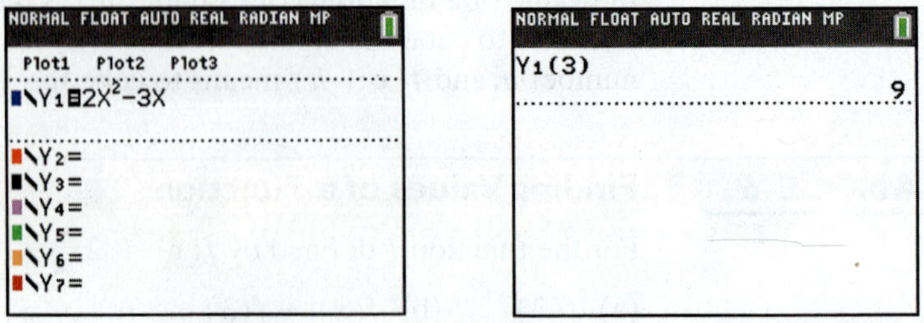

Figure 11 Evaluating $f(x) = 2x^2 - 3x$ for $x = 3$

Implicit Form of a Function

COMMENT The explicit form of a function is the form required by a graphing calculator. ■

In general, when a function f is defined by an equation in x and y, we say that the function f is given **implicitly**. If it is possible to solve the equation for y in terms of x, then we write $y = f(x)$ and say that the function is given **explicitly**. For example,

Implicit Form	Explicit Form
$3x + y = 5$	$y = f(x) = -3x + 5$
$x^2 - y = 6$	$y = f(x) = x^2 - 6$
$xy = 4$	$y = f(x) = \frac{4}{x}$

SUMMARY

Important Facts about Functions

(a) For each x in the domain of a function f, there is exactly one image $f(x)$ in the range; however, an element in the range can result from more than one x in the domain.

(b) f is the symbol that we use to denote the function. It is symbolic of the equation (rule) that we use to get from an x in the domain to $f(x)$ in the range.

(c) If $y = f(x)$, then x is called the independent variable or argument of f, and y is called the dependent variable or the value of f at x.

3 Find the Difference Quotient of a Function

An important concept in calculus involves looking at a certain quotient. For a given function $y = f(x)$, the inputs x and $x + h$, $h \neq 0$, result in the images $f(x)$ and $f(x + h)$. The quotient of their differences

$$\frac{f(x+h) - f(x)}{(x+h) - x} = \frac{f(x+h) - f(x)}{h}$$

with $h \neq 0$, is called the *difference quotient of f at x*.

DEFINITION

The **difference quotient** of a function f at x is given by

$$\frac{f(x+h) - f(x)}{h} \qquad h \neq 0 \qquad \textbf{(1)}$$

The difference quotient is used in calculus to define the derivative, which leads to applications such as the velocity of an object and optimization of resources.

When finding a difference quotient, it is necessary to simplify the expression in order to cancel the h in the denominator, as illustrated in the following example.

EXAMPLE 8 **Finding the Difference Quotient of a Function**

Find the difference quotient of each function.

(a) $f(x) = 2x^2 - 3x$

(b) $f(x) = \frac{4}{x}$

(c) $f(x) = \sqrt{x}$

Solution (a)

$$\frac{f(x+h)-f(x)}{h} = \frac{[2(x+h)^2 - 3(x+h)] - [2x^2 - 3x]}{h}$$

$\uparrow$ $f(x+h) = 2(x+h)^2 - 3(x+h)$

$$= \frac{2(x^2 + 2xh + h^2) - 3x - 3h - 2x^2 + 3x}{h} \quad \text{Simplify.}$$

$$= \frac{2x^2 + 4xh + 2h^2 - 3h - 2x^2}{h} \quad \text{Distribute and combine like terms.}$$

$$= \frac{4xh + 2h^2 - 3h}{h} \quad \text{Combine like terms.}$$

$$= \frac{h(4x + 2h - 3)}{h} \quad \text{Factor out } h.$$

$$= 4x + 2h - 3 \quad \text{Divide out the } h\text{'s.}$$

(b)

$$\frac{f(x+h)-f(x)}{h} = \frac{\frac{4}{x+h} - \frac{4}{x}}{h} \quad f(x+h) = \frac{4}{x+h}$$

$$= \frac{\frac{4x - 4(x+h)}{x(x+h)}}{h} \quad \text{Subtract.}$$

$$= \frac{4x - 4x - 4h}{x(x+h)h} \quad \text{Divide and distribute.}$$

$$= \frac{-4h}{x(x+h)h} \quad \text{Simplify.}$$

$$= -\frac{4}{x(x+h)} \quad \text{Divide out the factor } h.$$

(c)

$$\frac{f(x+h)-f(x)}{h} = \frac{\sqrt{x+h} - \sqrt{x}}{h} \quad f(x+h) = \sqrt{x+h}$$

$$= \frac{\sqrt{x+h} - \sqrt{x}}{h} \cdot \frac{\sqrt{x+h} + \sqrt{x}}{\sqrt{x+h} + \sqrt{x}} \quad \text{Rationalize the numerator.}$$

$$= \frac{(\sqrt{x+h})^2 - (\sqrt{x})^2}{h(\sqrt{x+h} + \sqrt{x})} \quad (A - B)(A + B) = A^2 - B^2$$

$$= \frac{h}{h(\sqrt{x+h} + \sqrt{x})} \quad (\sqrt{x+h})^2 - (\sqrt{x})^2 = x + h - x = h$$

$$= \frac{1}{\sqrt{x+h} + \sqrt{x}} \quad \text{Divide out the factor } h.$$

Now Work PROBLEM 79

4 Find the Domain of a Function Defined by an Equation

Often the domain of a function f is not specified; instead, only the equation defining the function is given. In such cases, we agree that the **domain of f** is the largest set of real numbers for which the value $f(x)$ is a real number. The domain of a function f is the same as the domain of the variable x in the expression $f(x)$.

EXAMPLE 9 **Finding the Domain of a Function**

Find the domain of each of the following functions.

(a) $f(x) = x^2 + 5x$ (b) $g(x) = \dfrac{3x}{x^2 - 4}$

(c) $h(t) = \sqrt{4 - 3t}$ (d) $F(x) = \dfrac{\sqrt{3x + 12}}{x - 5}$

Solution

(a) The function says to square a number and then add five times the number. Since these operations can be performed on any real number, the domain of f is the set of all real numbers.

(b) The function g says to divide $3x$ by $x^2 - 4$. Since division by 0 is not defined, the denominator $x^2 - 4$ can never be 0, so x can never equal -2 or 2. The domain of the function g is $\{x | x \neq -2, x \neq 2\}$.

In Words

The domain of g found in Example 9(b) is $\{x | x \neq -2, x \neq 2\}$. This notation is read, "The domain of the function g is the set of all real numbers x such that x does not equal -2 and x does not equal 2."

(c) The function h says to take the square root of $4 - 3t$. But only nonnegative numbers have real square roots, so the expression under the square root (the radicand) must be nonnegative (greater than or equal to zero). This requires that

$$\begin{aligned} 4 - 3t &\geq 0 \\ -3t &\geq -4 \\ t &\leq \frac{4}{3} \end{aligned}$$

The domain of h is $\left\{t \middle| t \leq \frac{4}{3}\right\}$, or the interval $\left(-\infty, \frac{4}{3}\right]$.

(d) The function F says to take the square root of $3x + 12$ and divide this result by $x - 5$. This requires that $3x + 12 \geq 0$, so $x \geq -4$, and also that $x - 5 \neq 0$, so $x \neq 5$. Combining these two restrictions, the domain of F is

$$\{x | x \geq -4, \quad x \neq 5\}.$$

The following steps may prove helpful for finding the domain of a function that is defined by an equation and whose domain is a subset of the real numbers.

Finding the Domain of a Function Defined by an Equation

1. Start with the domain as the set of real numbers.
2. If the equation has a denominator, exclude any numbers that give a zero denominator.
3. If the equation has a radical of even index, exclude any numbers that cause the expression inside the radical (the radicand) to be negative.

Now Work PROBLEM 55

If x is in the domain of a function f, we shall say that **f is defined at x**, or **$f(x)$ exists**. If x is not in the domain of f, we say that **f is not defined at x**, or **$f(x)$ does not exist**. For example, if $f(x) = \frac{x}{x^2 - 1}$, then $f(0)$ exists, but $f(1)$ and $f(-1)$ do not exist. (Do you see why?)

We have not said much about finding the range of a function. We will say more about finding the range when we look at the graph of a function in the next section. When a function is defined by an equation, it can be difficult to find the range. Therefore, we shall usually be content to find just the domain of a function when the function is defined by an equation. We shall express the domain of a function using inequalities, interval notation, set notation, or words, whichever is most convenient.

When we use functions in applications, the domain may be restricted by physical or geometric considerations. For example, the domain of the function f defined by $f(x) = x^2$ is the set of all real numbers. However, if f is used to obtain the area of a square when the length x of a side is known, then we must restrict the domain of f to the positive real numbers, since the length of a side can never be 0 or negative.

EXAMPLE 10 **Finding the Domain in an Application**

Express the area of a circle as a function of its radius. Find the domain.

Solution See Figure 12. The formula for the area A of a circle of radius r is $A = \pi r^2$. Using r to represent the independent variable and A to represent the dependent variable, the function expressing this relationship is

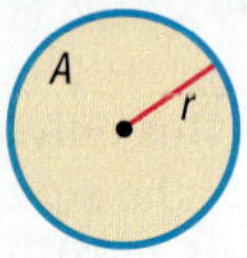

Figure 12 Circle of radius r

$$A(r) = \pi r^2$$

In this setting, the domain is $\{r | r > 0\}$. (Do you see why?) ●

Observe, in the solution to Example 10, that the symbol A is used in two ways: It is used to name the function, and it is used to symbolize the dependent variable. This double use is common in applications and should not cause any difficulty.

Now Work PROBLEM 97

5 Form the Sum, Difference, Product, and Quotient of Two Functions

Next we introduce some operations on functions. Functions, like numbers, can be added, subtracted, multiplied, and divided. For example, if $f(x) = x^2 + 9$ and $g(x) = 3x + 5$, then

$$f(x) + g(x) = (x^2 + 9) + (3x + 5) = x^2 + 3x + 14$$

The new function $y = x^2 + 3x + 14$ is called the *sum function* $f + g$. Similarly,

$$f(x) \cdot g(x) = (x^2 + 9)(3x + 5) = 3x^3 + 5x^2 + 27x + 45$$

The new function $y = 3x^3 + 5x^2 + 27x + 45$ is called the *product function* $f \cdot g$.

The general definitions are given next.

DEFINITION

If f and g are functions:
The **sum** $f + g$ is the function defined by

$$(f + g)(x) = f(x) + g(x)$$

In Words
Remember, the symbol ∩ stands for intersection. It means you should find the elements that are common to two sets.

The domain of $f + g$ consists of the numbers x that are in the domains of both f and g. That is, domain of $f + g =$ domain of $f \cap$ domain of g.

DEFINITION

The **difference** $f - g$ is the function defined by

$$(f - g)(x) = f(x) - g(x)$$

The domain of $f - g$ consists of the numbers x that are in the domains of both f and g. That is, domain of $f - g =$ domain of $f \cap$ domain of g.

DEFINITION

The **product** $f \cdot g$ is the function defined by

$$(f \cdot g)(x) = f(x) \cdot g(x)$$

The domain of $f \cdot g$ consists of the numbers x that are in the domains of both f and g. That is, domain of $f \cdot g =$ domain of $f \cap$ domain of g.

DEFINITION

The **quotient** $\frac{f}{g}$ is the function defined by

$$\left(\frac{f}{g}\right)(x) = \frac{f(x)}{g(x)} \qquad g(x) \neq 0$$

The domain of $\frac{f}{g}$ consists of the numbers x for which $g(x) \neq 0$ and that are in the domains of both f and g. That is,

$$\text{domain of } \frac{f}{g} = \{x | g(x) \neq 0\} \cap \text{domain of } f \cap \text{domain of } g$$

EXAMPLE 11 **Operations on Functions**

Let f and g be two functions defined as

$$f(x) = \frac{1}{x+2} \quad \text{and} \quad g(x) = \frac{x}{x-1}$$

Find the following functions, and determine the domain in each case.

(a) $(f+g)(x)$ (b) $(f-g)(x)$ (c) $(f \cdot g)(x)$ (d) $\left(\frac{f}{g}\right)(x)$

Solution The domain of f is $\{x | x \neq -2\}$ and the domain of g is $\{x | x \neq 1\}$.

(a) $$(f+g)(x) = f(x) + g(x) = \frac{1}{x+2} + \frac{x}{x-1}$$
$$= \frac{x-1}{(x+2)(x-1)} + \frac{x(x+2)}{(x+2)(x-1)} = \frac{x^2+3x-1}{(x+2)(x-1)}$$

The domain of $f+g$ consists of those numbers x that are in the domains of both f and g. Therefore, the domain of $f+g$ is $\{x | x \neq -2, x \neq 1\}$.

(b) $$(f-g)(x) = f(x) - g(x) = \frac{1}{x+2} - \frac{x}{x-1}$$
$$= \frac{x-1}{(x+2)(x-1)} - \frac{x(x+2)}{(x+2)(x-1)} = \frac{-(x^2+x+1)}{(x+2)(x-1)}$$

The domain of $f-g$ consists of those numbers x that are in the domains of both f and g. Therefore, the domain of $f-g$ is $\{x | x \neq -2, x \neq 1\}$.

(c) $$(f \cdot g)(x) = f(x) \cdot g(x) = \frac{1}{x+2} \cdot \frac{x}{x-1} = \frac{x}{(x+2)(x-1)}$$

The domain of $f \cdot g$ consists of those numbers x that are in the domains of both f and g. Therefore, the domain of $f \cdot g$ is $\{x | x \neq -2, x \neq 1\}$.

(d) $$\left(\frac{f}{g}\right)(x) = \frac{f(x)}{g(x)} = \frac{\frac{1}{x+2}}{\frac{x}{x-1}} = \frac{1}{x+2} \cdot \frac{x-1}{x} = \frac{x-1}{x(x+2)}$$

The domain of $\frac{f}{g}$ consists of the numbers x for which $g(x) \neq 0$ and that are in the domains of both f and g. Since $g(x) = 0$ when $x = 0$, we exclude 0 as well as -2 and 1 from the domain. The domain of $\frac{f}{g}$ is $\{x | x \neq -2, x \neq 0, x \neq 1\}$.

Now Work PROBLEM 67

In calculus, it is sometimes helpful to view a complicated function as the sum, difference, product, or quotient of simpler functions. For example,

$$F(x) = x^2 + \sqrt{x} \text{ is the sum of } f(x) = x^2 \text{ and } g(x) = \sqrt{x}.$$

$$H(x) = \frac{x^2 - 1}{x^2 + 1} \text{ is the quotient of } f(x) = x^2 - 1 \text{ and } g(x) = x^2 + 1.$$

SUMMARY

Function	A relation between two sets of real numbers so that each number x in the first set, the domain, has corresponding to it exactly one number y in the second set. A set of ordered pairs (x, y) or $(x, f(x))$ in which no first element is paired with two different second elements. The range is the set of y-values of the function that are the images of the x-values in the domain. A function f may be defined implicitly by an equation involving x and y or explicitly by writing $y = f(x)$.
Unspecified domain	If a function f is defined by an equation and no domain is specified, then the domain will be taken to be the largest set of real numbers for which the equation defines a real number.
Function notation	$y = f(x)$ f is a symbol for the function. x is the independent variable, or argument. y is the dependent variable. $f(x)$ is the value of the function at x, or the image of x.

3.1 Assess Your Understanding

'Are You Prepared?' *Answers are given at the end of these exercises. If you get a wrong answer, read the pages listed in red.*

1. The inequality $-1 < x < 3$ can be written in interval notation as ________. (pp. 120–121)
2. If $x = -2$, the value of the expression $3x^2 - 5x + \frac{1}{x}$ is ______. (pp. 20–23)
3. The domain of the variable in the expression $\frac{x-3}{x+4}$ is ______________. (pp. 20–23)
4. Solve the inequality: $3 - 2x > 5$. Graph the solution set. (pp. 123–126)
5. To rationalize the denominator of $\frac{3}{\sqrt{5} - 2}$, multiply the numerator and denominator by ________ (p. 75)
6. A quotient is considered rationalized if its denominator contains no ________. (p. 75)

Concepts and Vocabulary

7. If f is a function defined by the equation $y = f(x)$, then x is called the ___________ variable, and y is the _________ variable.
8. If the domain of f is all real numbers in the interval $[0, 7]$, and the domain of g is all real numbers in the interval $[-2, 5]$, then the domain of $f + g$ is all real numbers in the interval _______.
9. The domain of $\frac{f}{g}$ consists of numbers x for which $g(x)$ ___ 0 that are in the domains of both ___ and ___.
10. If $f(x) = x + 1$ and $g(x) = x^3$, then _________ $= x^3 - (x + 1)$.
11. ***True or False*** Every relation is a function.
12. ***True or False*** The domain of $(f \cdot g)(x)$ consists of the numbers x that are in the domains of both f and g.
13. ***True or False*** If no domain is specified for a function f, then the domain of f is taken to be the set of real numbers.
14. ***True or False*** The domain of the function $f(x) = \frac{x^2 - 4}{x}$ is $\{x \mid x \neq \pm 2\}$.
15. The set of all images of the elements in the domain of a function is called the ______.
(a) range (b) domain (c) solution set (d) function
16. The independent variable is sometimes referred to as the ______ of the function.
(a) range (b) value (c) argument (d) definition
17. The expression $\frac{f(x+h) - f(x)}{h}$ is called the ______ of f.
(a) radicand (b) image
(c) correspondence (d) difference quotient
18. When written as $y = f(x)$, a function is said to be defined ______.
(a) explicitly (b) consistently
(c) implicitly (d) rationally

Skill Building

In Problems 19–30, state the domain and range for each relation. Then determine whether each relation represents a function.

19.

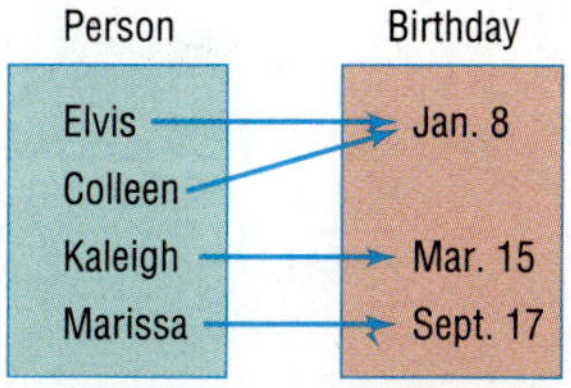

20.

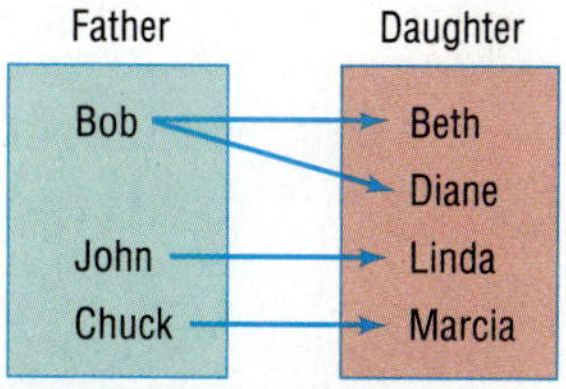

21.

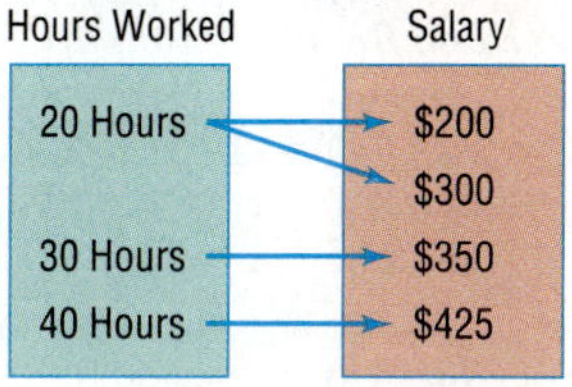

22.

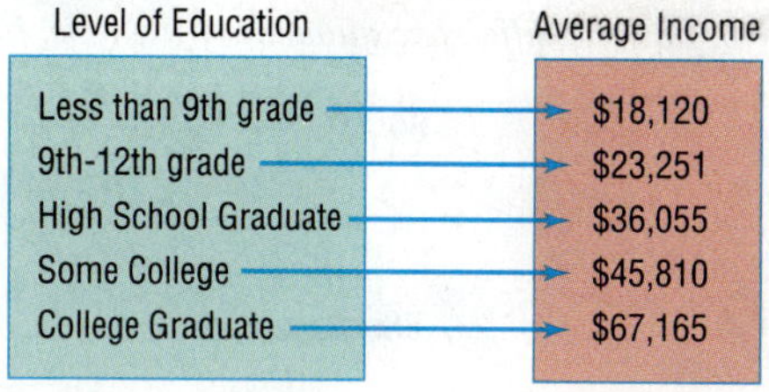

23. $\{(2,6),(-3,6),(4,9),(2,10)\}$ **24.** $\{(-2,5),(-1,3),(3,7),(4,12)\}$ **25.** $\{(1,3),(2,3),(3,3),(4,3)\}$

26. $\{(0,-2),(1,3),(2,3),(3,7)\}$ **27.** $\{(-2,4),(-2,6),(0,3),(3,7)\}$ **28.** $\{(-4,4),(-3,3),(-2,2),(-1,1),(-4,0)\}$

29. $\{(-2,4),(-1,1),(0,0),(1,1)\}$ **30.** $\{(-2,16),(-1,4),(0,3),(1,4)\}$

In Problems 31–42, determine whether the equation defines y as a function of x.

31. $y = 2x^2 - 3x + 4$ **32.** $y = x^3$ **33.** $y = \dfrac{1}{x}$ **34.** $y = |x|$

35. $y^2 = 4 - x^2$ **36.** $y = \pm\sqrt{1 - 2x}$ **37.** $x = y^2$ **38.** $x + y^2 = 1$

39. $y = \sqrt[3]{x}$ **40.** $y = \dfrac{3x - 1}{x + 2}$ **41.** $2x^2 + 3y^2 = 1$ **42.** $x^2 - 4y^2 = 1$

In Problems 43–50, find the following for each function:
(a) $f(0)$ *(b)* $f(1)$ *(c)* $f(-1)$ *(d)* $f(-x)$ *(e)* $-f(x)$ *(f)* $f(x+1)$ *(g)* $f(2x)$ *(h)* $f(x+h)$

43. $f(x) = 3x^2 + 2x - 4$ **44.** $f(x) = -2x^2 + x - 1$ **45.** $f(x) = \dfrac{x}{x^2 + 1}$ **46.** $f(x) = \dfrac{x^2 - 1}{x + 4}$

47. $f(x) = |x| + 4$ **48.** $f(x) = \sqrt{x^2 + x}$ **49.** $f(x) = \dfrac{2x + 1}{3x - 5}$ **50.** $f(x) = 1 - \dfrac{1}{(x + 2)^2}$

In Problems 51–66, find the domain of each function.

51. $f(x) = -5x + 4$ **52.** $f(x) = x^2 + 2$ **53.** $f(x) = \dfrac{x}{x^2 + 1}$ **54.** $f(x) = \dfrac{x^2}{x^2 + 1}$

55. $g(x) = \dfrac{x}{x^2 - 16}$ **56.** $h(x) = \dfrac{2x}{x^2 - 4}$ **57.** $F(x) = \dfrac{x - 2}{x^3 + x}$ **58.** $G(x) = \dfrac{x + 4}{x^3 - 4x}$

59. $h(x) = \sqrt{3x - 12}$ **60.** $G(x) = \sqrt{1 - x}$ **61.** $p(x) = \sqrt{\dfrac{2}{x - 1}}$ **62.** $f(x) = \dfrac{4}{\sqrt{x - 9}}$

63. $f(x) = \dfrac{x}{\sqrt{x - 4}}$ **64.** $f(x) = \dfrac{-x}{\sqrt{-x - 2}}$ **65.** $P(t) = \dfrac{\sqrt{t - 4}}{3t - 21}$ **66.** $h(z) = \dfrac{\sqrt{z + 3}}{z - 2}$

In Problems 67–76, for the given functions f and g, find the following. For parts (a)–(d), also find the domain.

(a) $(f + g)(x)$ *(b)* $(f - g)(x)$ *(c)* $(f \cdot g)(x)$ *(d)* $\left(\dfrac{f}{g}\right)(x)$

(e) $(f + g)(3)$ *(f)* $(f - g)(4)$ *(g)* $(f \cdot g)(2)$ *(h)* $\left(\dfrac{f}{g}\right)(1)$

67. $f(x) = 3x + 4;\ g(x) = 2x - 3$ **68.** $f(x) = 2x + 1;\ g(x) = 3x - 2$

69. $f(x) = x - 1;\ g(x) = 2x^2$ **70.** $f(x) = 2x^2 + 3;\ g(x) = 4x^3 + 1$

71. $f(x) = \sqrt{x};\ g(x) = 3x - 5$

72. $f(x) = |x|;\ g(x) = x$

73. $f(x) = 1 + \frac{1}{x};\ g(x) = \frac{1}{x}$

74. $f(x) = \sqrt{x-1};\ g(x) = \sqrt{4-x}$

75. $f(x) = \frac{2x+3}{3x-2};\ g(x) = \frac{4x}{3x-2}$

76. $f(x) = \sqrt{x+1};\ g(x) = \frac{2}{x}$

77. Given $f(x) = 3x + 1$ and $(f+g)(x) = 6 - \frac{1}{2}x$, find the function g.

78. Given $f(x) = \frac{1}{x}$ and $\left(\frac{f}{g}\right)(x) = \frac{x+1}{x^2 - x}$, find the function g.

In Problems 79–90, find the difference quotient of f; that is, find $\frac{f(x+h) - f(x)}{h}$, $h \neq 0$, *for each function. Be sure to simplify.*

79. $f(x) = 4x + 3$

80. $f(x) = -3x + 1$

81. $f(x) = x^2 - 4$

82. $f(x) = 3x^2 + 2$

83. $f(x) = x^2 - x + 4$

84. $f(x) = 3x^2 - 2x + 6$

85. $f(x) = \frac{1}{x^2}$

86. $f(x) = \frac{1}{x+3}$

87. $f(x) = \frac{2x}{x+3}$

88. $f(x) = \frac{5x}{x-4}$

89. $f(x) = \sqrt{x-2}$ [**Hint:** Rationalize the numerator.]

90. $f(x) = \sqrt{x+1}$

Applications and Extensions

91. Given $f(x) = x^2 - 2x + 3$, find the value(s) for x such that $f(x) = 11$.

92. Given $f(x) = \frac{5}{6}x - \frac{3}{4}$, find the value(s) for x such that $f(x) = -\frac{7}{16}$.

93. If $f(x) = 2x^3 + Ax^2 + 4x - 5$ and $f(2) = 5$, what is the value of A?

94. If $f(x) = 3x^2 - Bx + 4$ and $f(-1) = 12$, what is the value of B?

95. If $f(x) = \frac{3x+8}{2x-A}$ and $f(0) = 2$, what is the value of A?

96. If $f(x) = \frac{2x-B}{3x+4}$ and $f(2) = \frac{1}{2}$, what is the value of B?

97. Geometry Express the area A of a rectangle as a function of the length x if the length of the rectangle is twice its width.

98. Geometry Express the area A of an isosceles right triangle as a function of the length x of one of the two equal sides.

99. Constructing Functions Express the gross salary G of a person who earns \$14 per hour as a function of the number x of hours worked.

100. Constructing Functions Tiffany, a commissioned salesperson, earns \$100 base pay plus \$10 per item sold. Express her gross salary G as a function of the number x of items sold.

101. Population as a Function of Age The function

$$P(a) = 0.014a^2 - 5.073a + 327.287$$

represents the population P (in millions) of Americans that are a years of age or older in 2012.

(a) Identify the dependent and independent variables.
(b) Evaluate $P(20)$. Provide a verbal explanation of the meaning of $P(20)$.
(c) Evaluate $P(0)$. Provide a verbal explanation of the meaning of $P(0)$.

Source: U.S. Census Bureau

102. Number of Rooms The function

$$N(r) = -1.35r^2 + 15.45r - 20.71$$

represents the number N of housing units (in millions) in 2012 that had r rooms, where r is an integer and $2 \le r \le 9$.

(a) Identify the dependent and independent variables.
(b) Evaluate $N(3)$. Provide a verbal explanation of the meaning of $N(3)$.

103. Effect of Gravity on Earth If a rock falls from a height of 20 meters on Earth, the height H (in meters) after x seconds is approximately

$$H(x) = 20 - 4.9x^2$$

(a) What is the height of the rock when $x = 1$ second? $x = 1.1$ seconds? $x = 1.2$ seconds? $x = 1.3$ seconds?
(b) When is the height of the rock 15 meters? When is it 10 meters? When is it 5 meters?
(c) When does the rock strike the ground?

104. Effect of Gravity on Jupiter If a rock falls from a height of 20 meters on the planet Jupiter, its height H (in meters) after x seconds is approximately

$$H(x) = 20 - 13x^2$$

(a) What is the height of the rock when $x = 1$ second? $x = 1.1$ seconds? $x = 1.2$ seconds?
(b) When is the height of the rock 15 meters? When is it 10 meters? When is it 5 meters?
(c) When does the rock strike the ground?

105. Cost of Trans-Atlantic Travel A Boeing 747 crosses the Atlantic Ocean (3000 miles) with an airspeed of 500 miles per hour. The cost C (in dollars) per passenger is given by

$$C(x) = 100 + \frac{x}{10} + \frac{36{,}000}{x}$$

where x is the ground speed (airspeed $\pm$ wind).
(a) What is the cost per passenger for quiescent (no wind) conditions?
(b) What is the cost per passenger with a head wind of 50 miles per hour?
(c) What is the cost per passenger with a tail wind of 100 miles per hour?
(d) What is the cost per passenger with a head wind of 100 miles per hour?

106. Cross-sectional Area The cross-sectional area of a beam cut from a log with radius 1 foot is given by the function $A(x) = 4x\sqrt{1 - x^2}$, where x represents the length, in feet, of half the base of the beam. See the figure. Determine the cross-sectional area of the beam if the length of half the base of the beam is as follows:
(a) One-third of a foot
(b) One-half of a foot
(c) Two-thirds of a foot

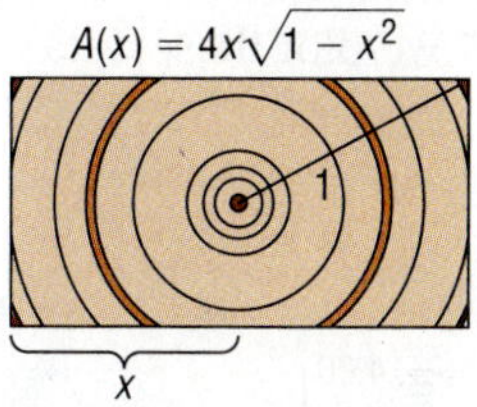

107. Economics The **participation rate** is the number of people in the labor force divided by the civilian population (excludes military). Let $L(x)$ represent the size of the labor force in year x, and $P(x)$ represent the civilian population in year x. Determine a function that represents the participation rate R as a function of x.

108. Crimes Suppose that $V(x)$ represents the number of violent crimes committed in year x and $P(x)$ represents the number of property crimes committed in year x. Determine a function T that represents the combined total of violent crimes and property crimes in year x.

109. Health Care Suppose that $P(x)$ represents the percentage of income spent on health care in year x and $I(x)$ represents income in year x. Determine a function H that represents total health care expenditures in year x.

110. Income Tax Suppose that $I(x)$ represents the income of an individual in year x before taxes and $T(x)$ represents the individual's tax bill in year x. Determine a function N that represents the individual's net income (income after taxes) in year x.

111. Profit Function Suppose that the revenue R, in dollars, from selling x cell phones, in hundreds, is $R(x) = -1.2x^2 + 220x$. The cost C, in dollars, of selling x cell phones, in hundreds, is $C(x) = 0.05x^3 - 2x^2 + 65x + 500$.
(a) Find the profit function, $P(x) = R(x) - C(x)$.
(b) Find the profit if $x = 15$ hundred cell phones are sold.
(c) Interpret $P(15)$.

112. Profit Function Suppose that the revenue R, in dollars, from selling x clocks is $R(x) = 30x$. The cost C, in dollars, of selling x clocks is $C(x) = 0.1x^2 + 7x + 400$.
(a) Find the profit function, $P(x) = R(x) - C(x)$.
(b) Find the profit if $x = 30$ clocks are sold.
(c) Interpret $P(30)$.

113. Stopping Distance When the driver of a vehicle observes an impediment, the total stopping distance involves both the reaction distance (the distance the vehicle travels while the driver moves his or her foot to the brake pedal) and the braking distance (the distance the vehicle travels once the brakes are applied). For a car traveling at a speed of v miles per hour, the reaction distance R, in feet, can be estimated by $R(v) = 2.2v$. Suppose that the braking distance B, in feet, for a car is given by $B(v) = 0.05v^2 + 0.4v - 15$.
(a) Find the stopping distance function

$$D(v) = R(v) + B(v).$$

(b) Find the stopping distance if the car is traveling at a speed of 60 mph.
(c) Interpret $D(60)$.

114. Some functions f have the property that $f(a + b) = f(a) + f(b)$ for all real numbers a and b. Which of the following functions have this property?
(a) $h(x) = 2x$ (b) $g(x) = x^2$
(c) $F(x) = 5x - 2$ (d) $G(x) = \dfrac{1}{x}$

Explaining Concepts: Discussion and Writing

115. Are the functions $f(x) = x - 1$ and $g(x) = \dfrac{x^2 - 1}{x + 1}$ the same? Explain.

116. Investigate when, historically, the use of the function notation $y = f(x)$ first appeared.

117. Find a function H that multiplies a number x by 3 and then subtracts the cube of x and divides the result by your age.

Retain Your Knowledge

Problems 118–121 are based on material learned earlier in the course. The purpose of these problems is to keep the material fresh in your mind so that you are better prepared for the final exam.

118. List the intercepts and test for symmetry: $(x + 12)^2 + y^2 = 16$

119. Determine which of the given points are on the graph of the equation. $y = 3x^2 - 8\sqrt{x}$
Points: $(-1, -5)$, $(4, 32)$, $(9, 171)$

120. How many pounds of lean hamburger that is 7% fat must be mixed with 12 pounds of ground chuck that is 20% fat to have a hamburger mixture that is 15% fat?

121. Solve $x^3 - 9x = 2x^2 - 18$.

'Are You Prepared?' Answers

1. $(-1, 3)$ **2.** 21.5 **3.** $\{x \mid x \neq -4\}$ **4.** $\{x \mid x < -1\}$

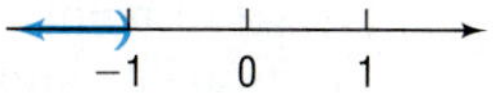

5. $\sqrt{5} + 2$ **6.** radicals

3.2 The Graph of a Function

PREPARING FOR THIS SECTION *Before getting started, review the following:*

- Graphs of Equations (Section 2.2, pp. 157–159)
- Intercepts (Section 2.2, pp. 159–160)

Now Work the **'Are You Prepared?'** problems on page 218.

OBJECTIVES 1 Identify the Graph of a Function (p. 214)
2 Obtain Information from or about the Graph of a Function (p. 215)

In applications, a graph often demonstrates more clearly the relationship between two variables than, say, an equation or table. For example, Table 1 shows the average price of gasoline in the United States for the years 1985–2014 (adjusted for inflation, based on 2014 dollars). If we plot these data and then connect the points, we obtain Figure 13.

Table 1

Year	Price	Year	Price	Year	Price
1985	2.55	1995	1.72	2005	2.74
1986	1.90	1996	1.80	2006	3.01
1987	1.89	1997	1.76	2007	3.19
1988	1.81	1998	1.49	2008	3.56
1989	1.87	1999	1.61	2009	2.58
1990	2.03	2000	2.03	2010	3.00
1991	1.91	2001	1.90	2011	3.69
1992	1.82	2002	1.76	2012	3.72
1993	1.74	2003	1.99	2013	3.54
1994	1.71	2004	2.31	2014	3.43

Source: *U.S. Energy Information Administration*

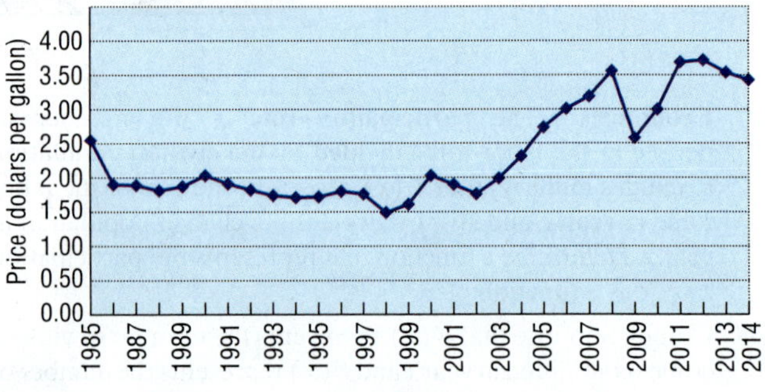

Figure 13 Average retail price of gasoline (2014 dollars)
Source: *U.S. Energy Information Administration*

We can see from the graph that the price of gasoline (adjusted for inflation) stayed roughly the same from 1986 to 1991 and rose rapidly from 2002 to 2008. The graph also shows that the lowest price occurred in 1998. To learn information such as this from an equation requires that some calculations be made.

Look again at Figure 13. The graph shows that for each date on the horizontal axis, there is only one price on the vertical axis. The graph represents a function, although the exact rule for getting from date to price is not given.

When a function is defined by an equation in x and y, the **graph of the function** is the graph of the equation; that is, it is the set of points (x, y) in the xy-plane that satisfy the equation.

1 Identify the Graph of a Function

> **In Words**
> If any vertical line intersects a graph at more than one point, the graph is not the graph of a function.

Not every collection of points in the xy-plane represents the graph of a function. Remember, for a function, each number x in the domain has exactly one image y in the range. This means that the graph of a function cannot contain two points with the same x-coordinate and different y-coordinates. Therefore, the graph of a function must satisfy the following **vertical-line test**.

THEOREM **Vertical-Line Test**

A set of points in the xy-plane is the graph of a function if and only if every vertical line intersects the graph in at most one point.

EXAMPLE 1

Identifying the Graph of a Function

Which of the graphs in Figure 14 are graphs of functions?

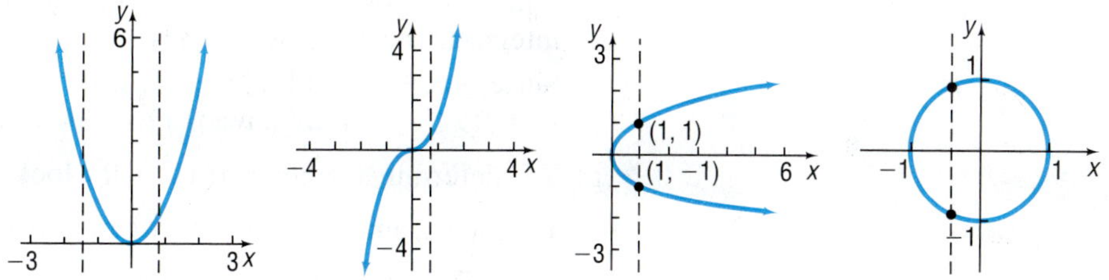

Figure 14 (a) $y = x^2$ (b) $y = x^3$ (c) $x = y^2$ (d) $x^2 + y^2 = 1$

Solution The graphs in Figures 14(a) and 14(b) are graphs of functions, because every vertical line intersects each graph in at most one point. The graphs in Figures 14(c) and 14(d) are not graphs of functions, because there is a vertical line that intersects each graph in more than one point. Notice in Figure 14(c) that the input 1 corresponds to two outputs, -1 and 1. This is why the graph does not represent a function. ●

Now Work PROBLEM 17

2 Obtain Information from or about the Graph of a Function

If (x, y) is a point on the graph of a function f, then y is the value of f at x; that is, $y = f(x)$. Also if $y = f(x)$, then (x, y) is a point on the graph of f. For example, if $(-2, 7)$ is on the graph of f, then $f(-2) = 7$, and if $f(5) = 8$, then the point $(5, 8)$ is on the graph of $y = f(x)$. The next example illustrates how to obtain information about a function if its graph is given.

EXAMPLE 2

Obtaining Information from the Graph of a Function

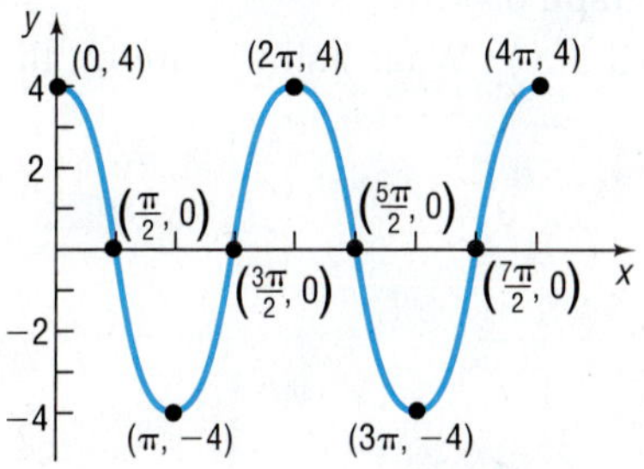

Figure 15

Let f be the function whose graph is given in Figure 15. (The graph of f might represent the distance y that the bob of a pendulum is from its *at-rest* position at time x. Negative values of y mean that the pendulum is to the left of the at-rest position, and positive values of y mean that the pendulum is to the right of the at-rest position.)

(a) What are $f(0)$, $f\left(\frac{3\pi}{2}\right)$, and $f(3\pi)$?
(b) What is the domain of f?
(c) What is the range of f?
(d) List the intercepts. (Recall that these are the points, if any, where the graph crosses or touches the coordinate axes.)
(e) How many times does the line $y = 2$ intersect the graph?
(f) For what values of x does $f(x) = -4$?
(g) For what values of x is $f(x) > 0$?

Solution

(a) Since $(0, 4)$ is on the graph of f, the y-coordinate 4 is the value of f at the x-coordinate 0; that is, $f(0) = 4$. In a similar way, when $x = \frac{3\pi}{2}$, then $y = 0$, so $f\left(\frac{3\pi}{2}\right) = 0$. When $x = 3\pi$, then $y = -4$, so $f(3\pi) = -4$.

(b) To determine the domain of f, notice that the points on the graph of f have x-coordinates between 0 and 4π, inclusive; and for each number x between 0 and 4π, there is a point $(x, f(x))$ on the graph. The domain of f is $\{x \mid 0 \leq x \leq 4\pi\}$ or the interval $[0, 4\pi]$.

(c) The points on the graph all have y-coordinates between -4 and 4, inclusive; and for each such number y, there is at least one number x in the domain. The range of f is $\{y \mid -4 \leq y \leq 4\}$ or the interval $[-4, 4]$.

(d) The intercepts are the points

$$(0,4), \left(\frac{\pi}{2}, 0\right), \left(\frac{3\pi}{2}, 0\right), \left(\frac{5\pi}{2}, 0\right), \text{ and } \left(\frac{7\pi}{2}, 0\right)$$

(e) Draw the horizontal line $y = 2$ on the graph in Figure 15. Notice that the line intersects the graph four times.

(f) Since $(\pi, -4)$ and $(3\pi, -4)$ are the only points on the graph for which $y = f(x) = -4$, we have $f(x) = -4$ when $x = \pi$ and $x = 3\pi$.

(g) To determine where $f(x) > 0$, look at Figure 15 and determine the x-values from 0 to 4π for which the y-coordinate is positive. This occurs on $\left[0, \frac{\pi}{2}\right) \cup \left(\frac{3\pi}{2}, \frac{5\pi}{2}\right) \cup \left(\frac{7\pi}{2}, 4\pi\right]$. Using inequality notation, $f(x) > 0$ for $0 \le x < \frac{\pi}{2}$ or $\frac{3\pi}{2} < x < \frac{5\pi}{2}$ or $\frac{7\pi}{2} < x \le 4\pi$. ●

When the graph of a function is given, its domain may be viewed as the shadow created by the graph on the x-axis by vertical beams of light. Its range can be viewed as the shadow created by the graph on the y-axis by horizontal beams of light. Try this technique with the graph given in Figure 15.

Now Work PROBLEMS 11 AND 15

EXAMPLE 3 **Obtaining Information about the Graph of a Function**

Consider the function: $f(x) = \dfrac{x+1}{x+2}$

(a) Find the domain of f.

(b) Is the point $\left(1, \frac{1}{2}\right)$ on the graph of f?

(c) If $x = 2$, what is $f(x)$? What point is on the graph of f?

(d) If $f(x) = 2$, what is x? What point is on the graph of f?

(e) What are the x-intercepts of the graph of f (if any)? What point(s) are on the graph of f?

Solution

(a) The domain of f is $\{x \mid x \neq -2\}$.

(b) When $x = 1$, then

$$f(1) = \frac{1+1}{1+2} = \frac{2}{3} \qquad f(x) = \frac{x+1}{x+2}$$

The point $\left(1, \frac{2}{3}\right)$ is on the graph of f; the point $\left(1, \frac{1}{2}\right)$ is not.

(c) If $x = 2$, then

$$f(2) = \frac{2+1}{2+2} = \frac{3}{4}$$

The point $\left(2, \frac{3}{4}\right)$ is on the graph of f.

(d) If $f(x) = 2$, then

$$\frac{x+1}{x+2} = 2 \qquad f(x) = 2$$

$$x + 1 = 2(x+2) \qquad \text{Multiply both sides by } x + 2.$$

$$x + 1 = 2x + 4 \qquad \text{Distribute.}$$

$$x = -3 \qquad \text{Solve for } x.$$

If $f(x) = 2$, then $x = -3$. The point $(-3, 2)$ is on the graph of f.

(e) The x-intercepts of the graph of f are the real solutions of the equation $f(x) = 0$ that are in the domain of f.

$$\frac{x+1}{x+2} = 0$$
$$x + 1 = 0 \quad \text{Multiply both sides by } x + 2.$$
$$x = -1 \quad \text{Subtract 1 from both sides.}$$

The only real solution of the equation $f(x) = \frac{x+1}{x+2} = 0$ is $x = -1$, so -1 is the only x-intercept. Since $f(-1) = 0$, the point $(-1, 0)$ is on the graph of f. ●

Now Work PROBLEM 27

EXAMPLE 4 Average Cost Function

The average cost $\overline{C}$ per computer of manufacturing x computers per day is given by the function

$$\overline{C}(x) = 0.56x^2 - 34.39x + 1212.57 + \frac{20{,}000}{x}$$

Determine the average cost of manufacturing:

(a) 30 computers in a day
(b) 40 computers in a day
(c) 50 computers in a day
(d) Graph the function $\overline{C} = \overline{C}(x)$, $0 < x \le 80$.
(e) Create a TABLE with TblStart = 1 and ΔTbl = 1. Which value of x minimizes the average cost?

Solution

(a) The average cost per computer of manufacturing $x = 30$ computers is

$$\overline{C}(30) = 0.56(30)^2 - 34.39(30) + 1212.57 + \frac{20{,}000}{30} = \$1351.54$$

(b) The average cost per computer of manufacturing $x = 40$ computers is

$$\overline{C}(40) = 0.56(40)^2 - 34.39(40) + 1212.57 + \frac{20{,}000}{40} = \$1232.97$$

(c) The average cost per computer of manufacturing $x = 50$ computers is

$$\overline{C}(50) = 0.56(50)^2 - 34.39(50) + 1212.57 + \frac{20{,}000}{50} = \$1293.07$$

(d) See Figure 16 for the graph of $\overline{C} = \overline{C}(x)$.

(e) With the function $\overline{C} = \overline{C}(x)$ in Y_1, we create Table 2. We scroll down until we find a value of x for which Y_1 is smallest. Table 3 shows that manufacturing $x = 41$ computers minimizes the average cost at \$1231.74 per computer.

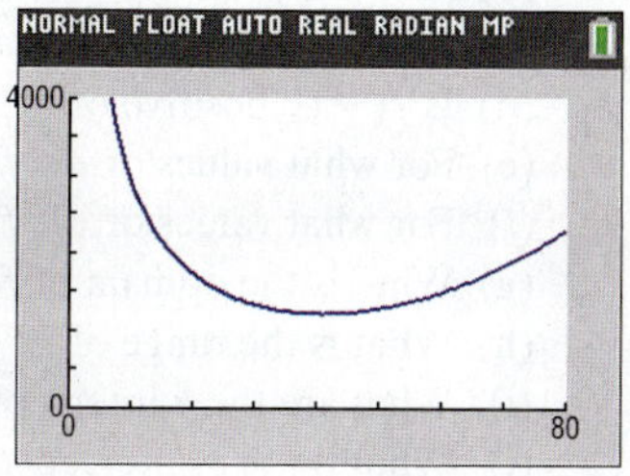

Figure 16 $\overline{C}(x) = 0.56x^2 - 34.39x + 1212.57 + \frac{20{,}000}{x}$

NORMAL FLOAT AUTO REAL RADIAN MP
PRESS ENTER TO EDIT

X	Y_1
1	21179
2	11146
3	7781.1
4	6084
5	5054.6
6	4359.7
7	3856.4
8	3473.3
9	3170.6
10	2924.7
11	2720.2

Y_1=0.56X²−34.39X+1212.57+2

Table 2

NORMAL FLOAT AUTO REAL RADIAN MP
PRESS ◆ TO EDIT FUNCTION

X	Y_1
38	1240.7
39	1235.9
40	1233
41	1231.7
42	1232.2
43	1234.4
44	1238.1
45	1243.5
46	1250.4
47	1258.8
48	1268.8

Y_1=1231.74487805

Table 3 ●

Now Work PROBLEM 35

SUMMARY

Graph of a Function The collection of points (x, y) that satisfies the equation $y = f(x)$.

Vertical-Line Test A collection of points is the graph of a function if and only if every vertical line intersects the graph in at most one point.

3.2 Assess Your Understanding

'Are You Prepared?' *Answers are given at the end of these exercises. If you get a wrong answer, read the pages listed in red.*

1. The intercepts of the equation $x^2 + 4y^2 = 16$ are ______. (pp. 159–160)

2. ***True or False*** The point $(-2, -6)$ is on the graph of the equation $x = 2y - 2$. (pp. 157–159)

Concepts and Vocabulary

3. A set of points in the xy-plane is the graph of a function if and only if every ______ line intersects the graph in at most one point.

4. If the point $(5, -3)$ is a point on the graph of f, then $f(\_\_) = \_\_\_$.

5. Find a so that the point $(-1, 2)$ is on the graph of $f(x) = ax^2 + 4$.

6. ***True or False*** Every graph represents a function.

7. ***True or False*** The graph of a function $y = f(x)$ always crosses the y-axis.

8. ***True or False*** The y-intercept of the graph of the function $y = f(x)$, whose domain is all real numbers, is $f(0)$.

9. If a function is defined by an equation in x and y, then the set of points (x, y) in the xy-plane that satisfy the equation is called ______.
(a) the domain of the function (b) the range of the function
(c) the graph of the function (d) the relation of the function

10. The graph of a function $y = f(x)$ can have more than one of which type of intercept?
(a) x-intercept (b) y-intercept (c) both (d) neither

Skill Building

11. Use the given graph of the function f to answer parts (a)–(n).

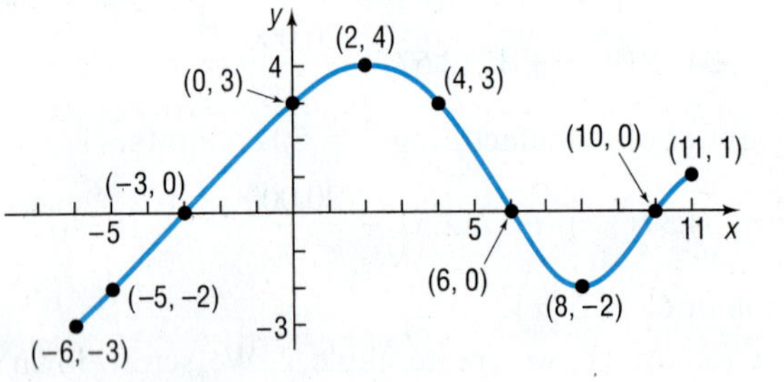

(a) Find $f(0)$ and $f(-6)$.
(b) Find $f(6)$ and $f(11)$.
(c) Is $f(3)$ positive or negative?
(d) Is $f(-4)$ positive or negative?
(e) For what values of x is $f(x) = 0$?
(f) For what values of x is $f(x) > 0$?
(g) What is the domain of f?
(h) What is the range of f?
(i) What are the x-intercepts?
(j) What is the y-intercept?
(k) How often does the line $y = \frac{1}{2}$ intersect the graph?
(l) How often does the line $x = 5$ intersect the graph?
(m) For what values of x does $f(x) = 3$?
(n) For what values of x does $f(x) = -2$?

12. Use the given graph of the function f to answer parts (a)–(n).

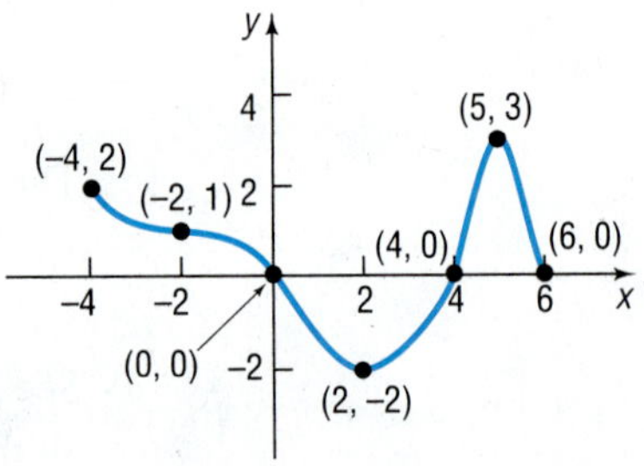

(a) Find $f(0)$ and $f(6)$.
(b) Find $f(2)$ and $f(-2)$.
(c) Is $f(3)$ positive or negative?
(d) Is $f(-1)$ positive or negative?
(e) For what values of x is $f(x) = 0$?
(f) For what values of x is $f(x) < 0$?
(g) What is the domain of f?
(h) What is the range of f?
(i) What are the x-intercepts?
(j) What is the y-intercept?
(k) How often does the line $y = -1$ intersect the graph?
(l) How often does the line $x = 1$ intersect the graph?
(m) For what value of x does $f(x) = 3$?
(n) For what value of x does $f(x) = -2$?

In Problems 13–24, determine whether the graph is that of a function by using the vertical-line test. If it is, use the graph to find:
(a) The domain and range *(b) The intercepts, if any* *(c) Any symmetry with respect to the x-axis, the y-axis, or the origin*

13.
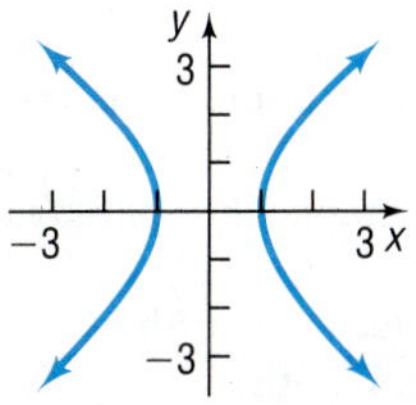

14.
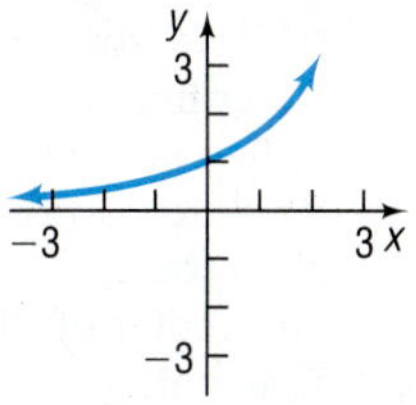

15.
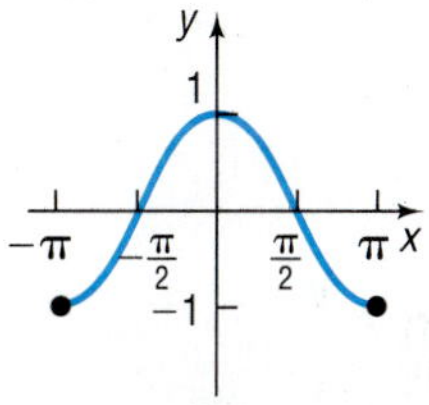

16.
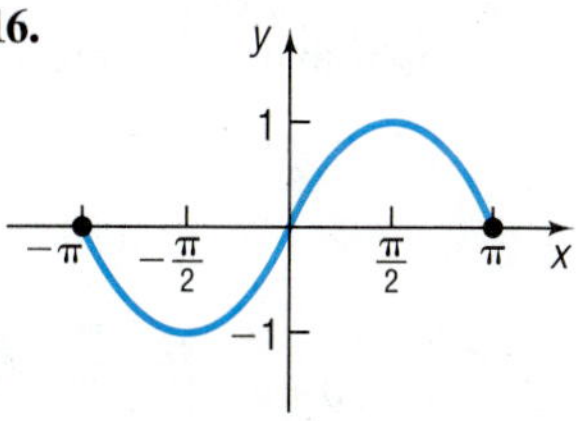

17.
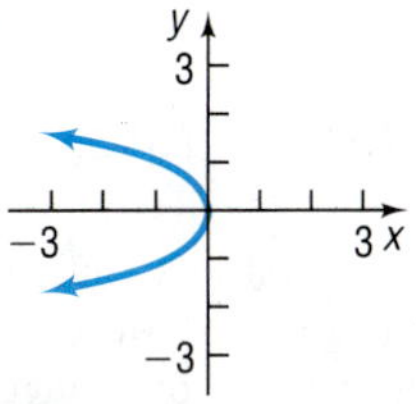

18.
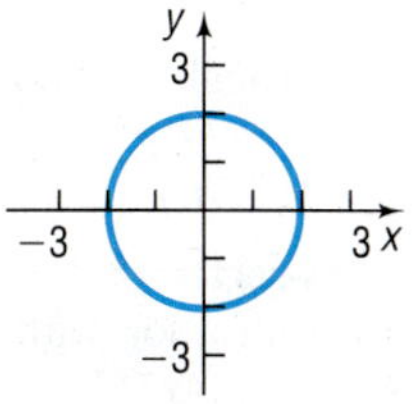

19.
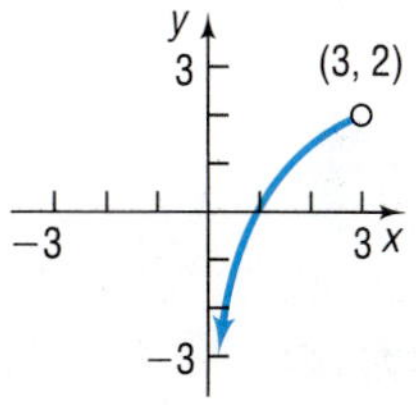

20.
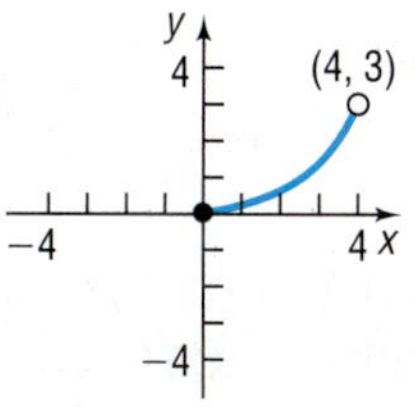

21.
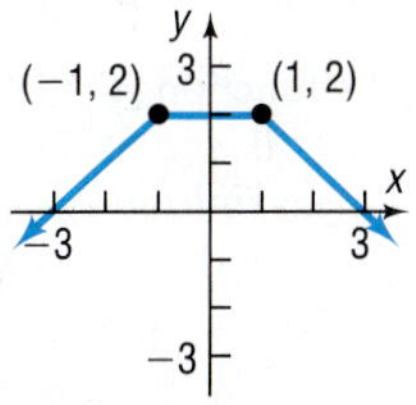

22.
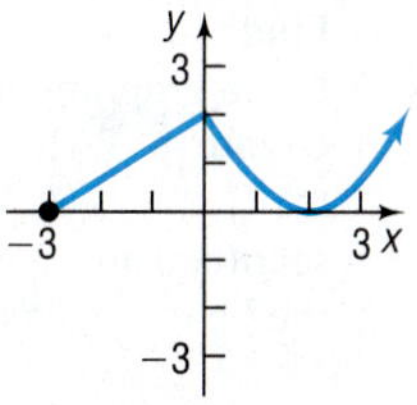

23.
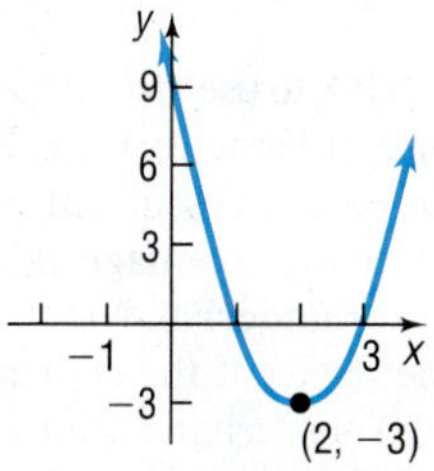

24.
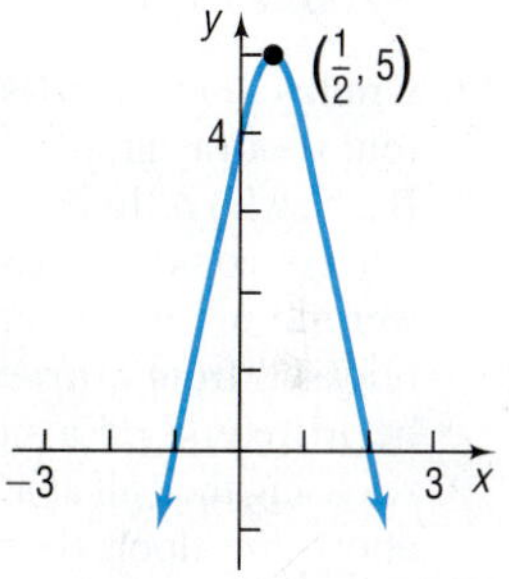

In Problems 25–30, answer the questions about the given function.

25. $f(x) = 2x^2 - x - 1$
(a) Is the point $(-1, 2)$ on the graph of f?
(b) If $x = -2$, what is $f(x)$? What point is on the graph of f?
(c) If $f(x) = -1$, what is x? What point(s) are on the graph of f?
(d) What is the domain of f?
(e) List the x-intercepts, if any, of the graph of f.
(f) List the y-intercept, if there is one, of the graph of f.

26. $f(x) = -3x^2 + 5x$
(a) Is the point $(-1, 2)$ on the graph of f?
(b) If $x = -2$, what is $f(x)$? What point is on the graph of f?
(c) If $f(x) = -2$, what is x? What point(s) are on the graph of f?
(d) What is the domain of f?
(e) List the x-intercepts, if any, of the graph of f.
(f) List the y-intercept, if there is one, of the graph of f.

27. $f(x) = \dfrac{x + 2}{x - 6}$
(a) Is the point $(3, 14)$ on the graph of f?
(b) If $x = 4$, what is $f(x)$? What point is on the graph of f?
(c) If $f(x) = 2$, what is x? What point(s) are on the graph of f?
(d) What is the domain of f?
(e) List the x-intercepts, if any, of the graph of f.
(f) List the y-intercept, if there is one, of the graph of f.

28. $f(x) = \dfrac{x^2 + 2}{x + 4}$
(a) Is the point $\left(1, \dfrac{3}{5}\right)$ on the graph of f?
(b) If $x = 0$, what is $f(x)$? What point is on the graph of f?
(c) If $f(x) = \dfrac{1}{2}$, what is x? What point(s) are on the graph of f?
(d) What is the domain of f?
(e) List the x-intercepts, if any, of the graph of f.
(f) List the y-intercept, if there is one, of the graph of f.

29. $f(x) = \dfrac{2x^2}{x^4 + 1}$
(a) Is the point $(-1, 1)$ on the graph of f?
(b) If $x = 2$, what is $f(x)$? What point is on the graph of f?
(c) If $f(x) = 1$, what is x? What point(s) are on the graph of f?
(d) What is the domain of f?
(e) List the x-intercepts, if any, of the graph of f.
(f) List the y-intercept, if there is one, of the graph of f.

30. $f(x) = \dfrac{2x}{x - 2}$
(a) Is the point $\left(\dfrac{1}{2}, -\dfrac{2}{3}\right)$ on the graph of f?
(b) If $x = 4$, what is $f(x)$? What point is on the graph of f?
(c) If $f(x) = 1$, what is x? What point(s) are on the graph of f?
(d) What is the domain of f?
(e) List the x-intercepts, if any, of the graph of f.
(f) List the y-intercept, if there is one, of the graph of f.

Applications and Extensions

31. The graphs of two functions, f and g, are illustrated. Use the graphs to answer parts (a)–(f).

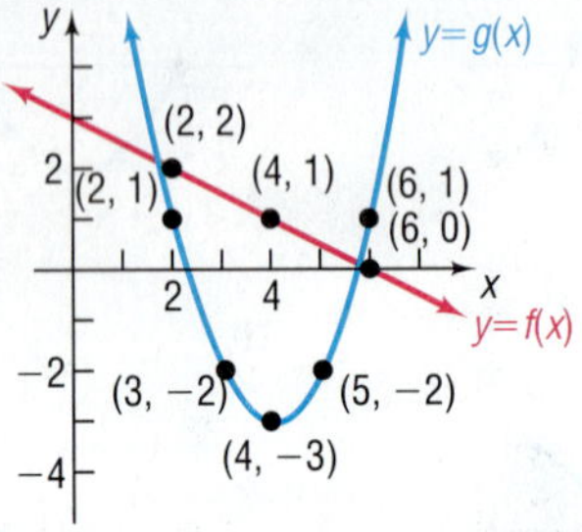

(a) $(f+g)(2)$ (b) $(f+g)(4)$
(c) $(f-g)(6)$ (d) $(g-f)(6)$
(e) $(f \cdot g)(2)$ (f) $\left(\dfrac{f}{g}\right)(4)$

32. Granny Shots The last player in the NBA to use an underhand foul shot (a "granny" shot) was Hall of Fame forward Rick Barry, who retired in 1980. Barry believes that current NBA players could increase their free-throw percentage if they were to use an underhand shot. Since underhand shots are released from a lower position, the angle of the shot must be increased. If a player shoots an underhand foul shot, releasing the ball at a 70-degree angle from a position 3.5 feet above the floor, then the path of the ball can be modeled by the function $h(x) = -\dfrac{136x^2}{v^2} + 2.7x + 3.5$, where h is the height of the ball above the floor, x is the forward distance of the ball in front of the foul line, and v is the initial velocity with which the ball is shot in feet per second.

(a) The center of the hoop is 10 feet above the floor and 15 feet in front of the foul line. Determine the initial velocity with which the ball must be shot in order for the ball to go through the hoop.
(b) Write the function for the path of the ball using the velocity found in part (a).
(c) Determine the height of the ball after it has traveled 9 feet in front of the foul line.
(d) Find additional points and graph the path of the basketball.

Source: The Physics of Foul Shots, Discover, *Vol. 21, No. 10, October 2000*

33. Free-throw Shots According to physicist Peter Brancazio, the key to a successful foul shot in basketball lies in the arc of the shot. Brancazio determined the optimal angle of the arc from the free-throw line to be 45 degrees. The arc also depends on the velocity with which the ball is shot. If a player shoots a foul shot, releasing the ball at a 45-degree angle from a position 6 feet above the floor, then the path of the ball can be modeled by the function

$$h(x) = -\frac{44x^2}{v^2} + x + 6$$

where h is the height of the ball above the floor, x is the forward distance of the ball in front of the foul line, and v is the initial velocity with which the ball is shot in feet per second. Suppose a player shoots a ball with an initial velocity of 28 feet per second.

(a) Determine the height of the ball after it has traveled 8 feet in front of the foul line.
(b) Determine the height of the ball after it has traveled 12 feet in front of the foul line.
(c) Find additional points and graph the path of the basketball.
(d) The center of the hoop is 10 feet above the floor and 15 feet in front of the foul line. Will the ball go through the hoop? Why or why not? If not, with what initial velocity must the ball be shot in order for the ball to go through the hoop?

Source: The Physics of Foul Shots, Discover, *Vol. 21, No. 10, October 2000*

34. Cross-sectional Area The cross-sectional area of a beam cut from a log with radius 1 foot is given by the function $A(x) = 4x\sqrt{1-x^2}$, where x represents the length, in feet, of half the base of the beam. See the figure.

(a) Find the domain of A.
(b) Use a graphing utility to graph the function $A = A(x)$.
(c) Create a TABLE with TblStart $= 0$ and ΔTbl $= 0.1$ for $0 \le x \le 1$. Which value of x maximizes the cross-sectional area? What should be the length of the base of the beam to maximize the cross-sectional area?

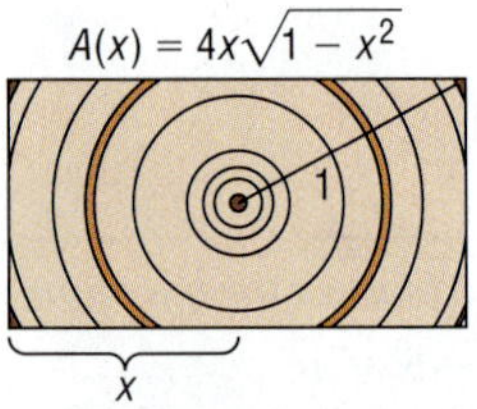

35. Motion of a Golf Ball A golf ball is hit with an initial velocity of 130 feet per second at an inclination of 45° to the horizontal. In physics, it is established that the height h of the golf ball is given by the function

$$h(x) = \frac{-32x^2}{130^2} + x$$

where x is the horizontal distance that the golf ball has traveled.

(a) Determine the height of the golf ball after it has traveled 100 feet.
(b) What is the height after it has traveled 300 feet?
(c) What is the height after it has traveled 500 feet?
(d) How far was the golf ball hit?
(e) Use a graphing utility to graph the function $h = h(x)$.
(f) Use a graphing utility to determine the distance that the ball has traveled when the height of the ball is 90 feet.
(g) Create a TABLE with TblStart = 0 and ΔTbl = 25. To the nearest 25 feet, how far does the ball travel before it reaches a maximum height? What is the maximum height?
(h) Adjust the value of ΔTbl until you determine the distance, to within 1 foot, that the ball travels before it reaches its maximum height.

36. Effect of Elevation on Weight If an object weighs m pounds at sea level, then its weight W (in pounds) at a height of h miles above sea level is given approximately by

$$W(h) = m\left(\frac{4000}{4000 + h}\right)^2$$

(a) If Amy weighs 120 pounds at sea level, how much will she weigh on Pikes Peak, which is 14,110 feet above sea level?
(b) Use a graphing utility to graph the function $W = W(h)$. Use $m = 120$ pounds.
(c) Create a Table with TblStart = 0 and ΔTbl = 0.5 to see how the weight W varies as h changes from 0 to 5 miles.
(d) At what height will Amy weigh 119.95 pounds?
(e) Does your answer to part (d) seem reasonable? Explain.

37. Cost of Trans-Atlantic Travel A Boeing 747 crosses the Atlantic Ocean (3000 miles) with an airspeed of 500 miles per hour. The cost C (in dollars) per passenger is given by

$$C(x) = 100 + \frac{x}{10} + \frac{36{,}000}{x}$$

where x is the groundspeed (airspeed $\pm$ wind).
(a) What is the cost when the groundspeed is 480 miles per hour? 600 miles per hour?
(b) Find the domain of C.
(c) Use a graphing utility to graph the function $C = C(x)$.
(d) Create a TABLE with TblStart = 0 and ΔTbl = 50.
(e) To the nearest 50 miles per hour, what groundspeed minimizes the cost per passenger?

38. Reading and Interpreting Graphs Let C be the function whose graph is given in the next column. This graph represents the cost C of manufacturing q computers in a day.
(a) Determine $C(0)$. Interpret this value.
(b) Determine $C(10)$. Interpret this value.
(c) Determine $C(50)$. Interpret this value.
(d) What is the domain of C? What does this domain imply in terms of daily production?
(e) Describe the shape of the graph.
(f) The point (30, 32 000) is called an *inflection point*. Describe the behavior of the graph around the inflection point.

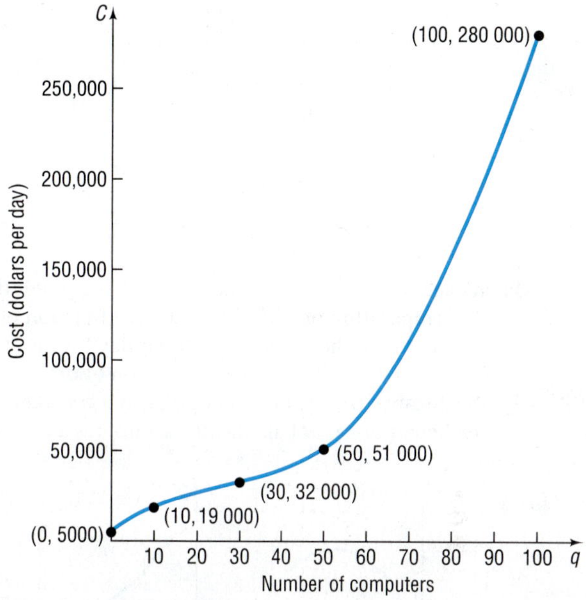

39. Reading and Interpreting Graphs Let C be the function whose graph is given below. This graph represents the cost C of using g gigabytes of data in a month for a shared-data family plan.
(a) Determine $C(0)$. Interpret this value.
(b) Determine $C(5)$. Interpret this value.
(c) Determine $C(15)$. Interpret this value.
(d) What is the domain of C? What does this domain imply in terms of the number of gigabytes?
(e) Describe the shape of the graph.

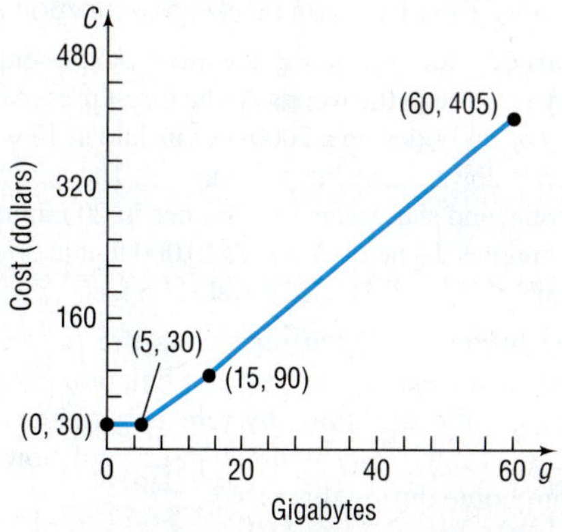

Explaining Concepts: Discussion and Writing

40. Describe how you would find the domain and range of a function if you were given its graph. How would your strategy change if you were given the equation defining the function instead of its graph?

41. How many x-intercepts can the graph of a function have? How many y-intercepts can the graph of a function have?

42. Is a graph that consists of a single point the graph of a function? Can you write the equation of such a function?

43. Match each of the following functions with the graph that best describes the situation.
(a) The cost of building a house as a function of its square footage
(b) The height of an egg dropped from a 300-foot building as a function of time
(c) The height of a human as a function of time
(d) The demand for Big Macs as a function of price
(e) The height of a child on a swing as a function of time

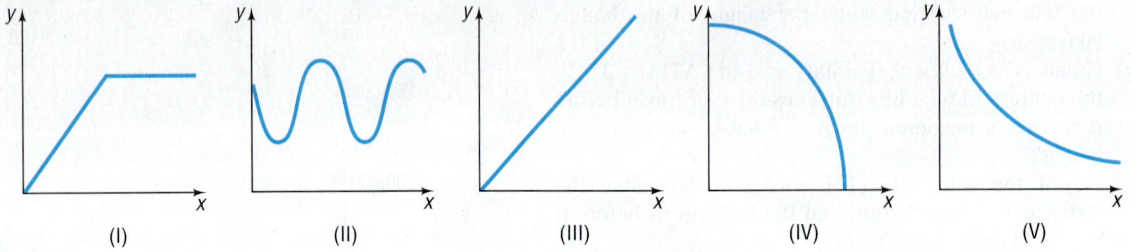

44. Match each of the following functions with the graph that best describes the situation.
(a) The temperature of a bowl of soup as a function of time
(b) The number of hours of daylight per day over a 2-year period
(c) The population of Florida as a function of time
(d) The distance traveled by a car going at a constant velocity as a function of time
(e) The height of a golf ball hit with a 7-iron as a function of time

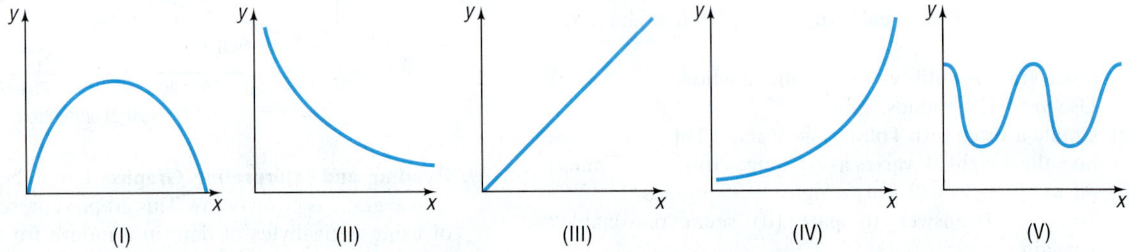

45. Consider the following scenario: Barbara decides to take a walk. She leaves home, walks 2 blocks in 5 minutes at a constant speed, and realizes that she forgot to lock the door. So Barbara runs home in 1 minute. While at her doorstep, it takes her 1 minute to find her keys and lock the door. Barbara walks 5 blocks in 15 minutes and then decides to jog home. It takes her 7 minutes to get home. Draw a graph of Barbara's distance from home (in blocks) as a function of time.

46. Consider the following scenario: Jayne enjoys riding her bicycle through the woods. At the forest preserve, she gets on her bicycle and rides up a 2000-foot incline in 10 minutes. She then travels down the incline in 3 minutes. The next 5000 feet is level terrain, and she covers the distance in 20 minutes. She rests for 15 minutes. Jayne then travels 10,000 feet in 30 minutes. Draw a graph of Jayne's distance traveled (in feet) as a function of time.

47. The following sketch represents the distance d (in miles) that Kevin was from home as a function of time t (in hours). Answer the questions by referring to the graph. In parts (a)–(g), how many hours elapsed and how far was Kevin from home during this time?

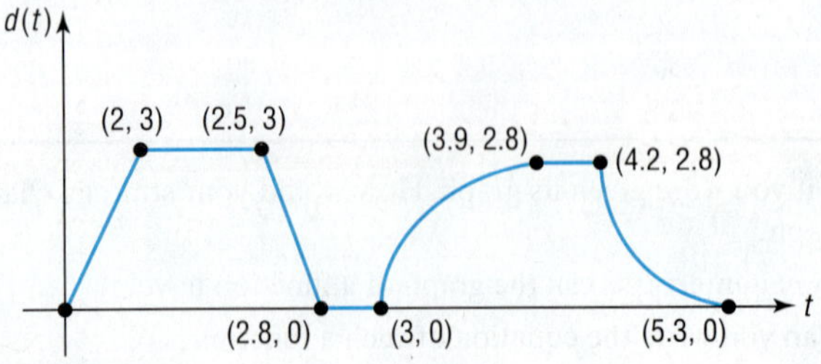

(a) From $t = 0$ to $t = 2$
(b) From $t = 2$ to $t = 2.5$
(c) From $t = 2.5$ to $t = 2.8$
(d) From $t = 2.8$ to $t = 3$
(e) From $t = 3$ to $t = 3.9$
(f) From $t = 3.9$ to $t = 4.2$
(g) From $t = 4.2$ to $t = 5.3$
(h) What is the farthest distance that Kevin was from home?
(i) How many times did Kevin return home?

48. The following sketch represents the speed v (in miles per hour) of Michael's car as a function of time t (in minutes).

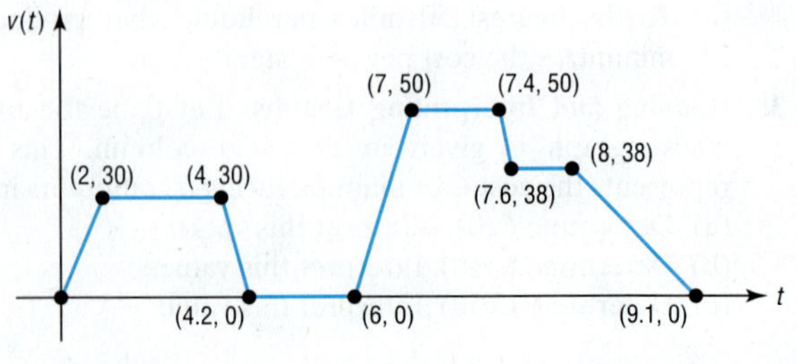

(a) Over what interval of time was Michael traveling fastest?
(b) Over what interval(s) of time was Michael's speed zero?
(c) What was Michael's speed between 0 and 2 minutes?
(d) What was Michael's speed between 4.2 and 6 minutes?
(e) What was Michael's speed between 7 and 7.4 minutes?
(f) When was Michael's speed constant?

49. Draw the graph of a function whose domain is $\{x \mid -3 \le x \le 8,\ x \ne 5\}$ and whose range is $\{y \mid -1 \le y \le 2,\ y \ne 0\}$. What point(s) in the rectangle $-3 \le x \le 8, -1 \le y \le 2$ cannot be on the graph? Compare your graph with those of other students. What differences do you see?

50. Is there a function whose graph is symmetric with respect to the x-axis? Explain.

51. Explain why the vertical-line test works.

Retain Your Knowledge

Problems 52–55 are based on material learned earlier in the course. The purpose of these problems is to keep the material fresh in your mind so that you are better prepared for the final exam.

52. Factor completely: $8x^2 + 24x + 18$

53. Find the distance between the points $(3, -6)$ and $(1, 0)$.

54. Write the equation of the line with slope $\frac{2}{3}$ that passes through the point $(-6, 4)$.

55. Subtract: $(4x^3 - 5x^2 + 2) - (3x^2 + 5x - 2)$

'Are You Prepared?' Answers

1. $(-4, 0), (4, 0), (0, -2), (0, 2)$

2. False

3.3 Properties of Functions

PREPARING FOR THIS SECTION *Before getting started, review the following:*

- Intervals (Section 1.5, pp. 120–121)
- Intercepts (Section 2.2, pp. 159–160)
- Slope of a Line (Section 2.3, pp. 167–169)
- Point–Slope Form of a Line (Section 2.3, p. 171)
- Symmetry (Section 2.2, pp. 160–162)

Now Work the **'Are You Prepared?'** problems on page 232.

OBJECTIVES

1. Determine Even and Odd Functions from a Graph (p. 223)
2. Identify Even and Odd Functions from an Equation (p. 225)
3. Use a Graph to Determine Where a Function Is Increasing, Decreasing, or Constant (p. 225)
4. Use a Graph to Locate Local Maxima and Local Minima (p. 226)
5. Use a Graph to Locate the Absolute Maximum and the Absolute Minimum (p. 227)
6. Use a Graphing Utility to Approximate Local Maxima and Local Minima and to Determine Where a Function Is Increasing or Decreasing (p. 229)
7. Find the Average Rate of Change of a Function (p. 230)

To obtain the graph of a function $y = f(x)$, it is often helpful to know certain properties that the function has and the impact of these properties on the way the graph will look.

1 Determine Even and Odd Functions from a Graph

The words *even* and *odd,* when applied to a function *f*, describe the symmetry that exists for the graph of the function.

A function f is even if and only if, whenever the point (x, y) is on the graph of f, the point $(-x, y)$ is also on the graph. Using function notation, we define an even function as follows:

DEFINITION

A function f is **even** if, for every number x in its domain, the number $-x$ is also in the domain and

$$f(-x) = f(x)$$

A function f is odd if and only if, whenever the point (x, y) is on the graph of f, the point $(-x, -y)$ is also on the graph. Using function notation, we define an odd function as follows:

DEFINITION

A function f is **odd** if, for every number x in its domain, the number $-x$ is also in the domain and

$$f(-x) = -f(x)$$

Refer to page 162, where the tests for symmetry are listed. The following results are then evident.

THEOREM

A function is even if and only if its graph is symmetric with respect to the y-axis. A function is odd if and only if its graph is symmetric with respect to the origin.

EXAMPLE 1 Determining Even and Odd Functions from the Graph

Determine whether each graph given in Figure 17 is the graph of an even function, an odd function, or a function that is neither even nor odd.

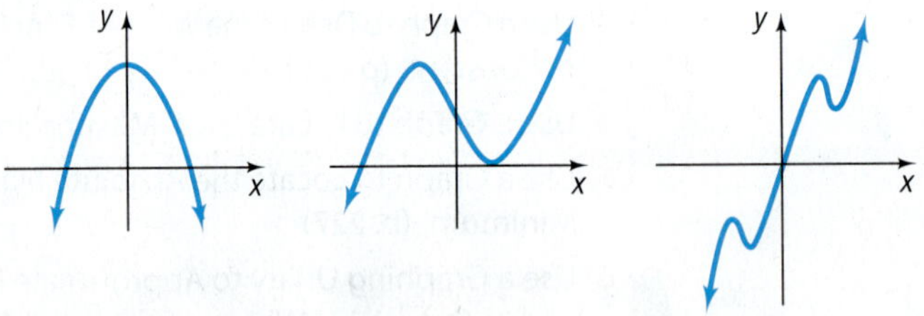

Figure 17 (a) (b) (c)

Solution (a) The graph in Figure 17(a) is that of an even function, because the graph is symmetric with respect to the y-axis.

(b) The function whose graph is given in Figure 17(b) is neither even nor odd, because the graph is neither symmetric with respect to the y-axis nor symmetric with respect to the origin.

(c) The function whose graph is given in Figure 17(c) is odd, because its graph is symmetric with respect to the origin. ●

Now Work PROBLEMS 25(a), (b), AND (d)

2 Identify Even and Odd Functions from an Equation

EXAMPLE 2 **Identifying Even and Odd Functions Algebraically**

Determine whether each of the following functions is even, odd, or neither. Then determine whether the graph is symmetric with respect to the y-axis, with respect to the origin, or neither.

(a) $f(x) = x^2 - 5$ (b) $g(x) = x^3 - 1$

(c) $h(x) = 5x^3 - x$ (d) $F(x) = |x|$

Solution (a) To determine whether f is even, odd, or neither, replace x by $-x$ in $f(x) = x^2 - 5$.

$$f(-x) = (-x)^2 - 5 = x^2 - 5 = f(x)$$

Since $f(-x) = f(x)$, the function is even, and the graph of f is symmetric with respect to the y-axis.

(b) Replace x by $-x$ in $g(x) = x^3 - 1$.

$$g(-x) = (-x)^3 - 1 = -x^3 - 1$$

Since $g(-x) \neq g(x)$ and $g(-x) \neq -g(x) = -(x^3 - 1) = -x^3 + 1$, the function is neither even nor odd. The graph of g is not symmetric with respect to the y-axis, nor is it symmetric with respect to the origin.

(c) Replace x by $-x$ in $h(x) = 5x^3 - x$.

$$h(-x) = 5(-x)^3 - (-x) = -5x^3 + x = -(5x^3 - x) = -h(x)$$

Since $h(-x) = -h(x)$, h is an odd function, and the graph of h is symmetric with respect to the origin.

(d) Replace x by $-x$ in $F(x) = |x|$.

$$F(-x) = |-x| = |-1| \cdot |x| = |x| = F(x)$$

Since $F(-x) = F(x)$, F is an even function, and the graph of F is symmetric with respect to the y-axis. ●

Now Work PROBLEM 37

3 Use a Graph to Determine Where a Function Is Increasing, Decreasing, or Constant

Consider the graph given in Figure 18. If you look from left to right along the graph of the function, you will notice that parts of the graph are going up, parts are going down, and parts are horizontal. In such cases, the function is described as *increasing*, *decreasing*, or *constant*, respectively.

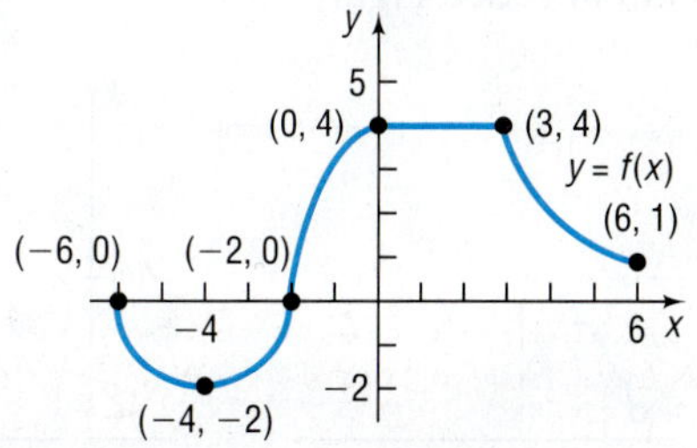

Figure 18

EXAMPLE 3 **Determining Where a Function Is Increasing, Decreasing, or Constant from Its Graph**

Determine the values of x for which the function in Figure 18 is increasing. Where is it decreasing? Where is it constant?

Solution When determining where a function is increasing, where it is decreasing, and where it is constant, we use strict inequalities involving the independent variable x, or we use open intervals* of x-coordinates. The function whose graph is given in Figure 18 is increasing on the open interval $(-4, 0)$, or for $-4 < x < 0$. The function is decreasing on the open intervals $(-6, -4)$ and $(3, 6)$, or for $-6 < x < -4$ and $3 < x < 6$. The function is constant on the open interval $(0, 3)$, or for $0 < x < 3$.

WARNING Describe the behavior of a graph in terms of its x-values. Do not say the graph in Figure 18 is increasing from the point $(-4, -2)$ to the point $(0, 4)$. Rather, say it is increasing on the interval $(-4, 0)$.

More precise definitions follow:

DEFINITIONS A function f is **increasing** on an open interval I if, for any choice of x_1 and x_2 in I, with $x_1 < x_2$, we have $f(x_1) < f(x_2)$.

A function f is **decreasing** on an open interval I if, for any choice of x_1 and x_2 in I, with $x_1 < x_2$, we have $f(x_1) > f(x_2)$.

A function f is **constant** on an open interval I if, for all choices of x in I, the values $f(x)$ are equal.

In Words

If a function is decreasing, then as the values of x get bigger, the values of the function get smaller. If a function is increasing, then as the values of x get bigger, the values of the function also get bigger. If a function is constant, then as the values of x get bigger, the values of the function remain unchanged.

Figure 19 illustrates the definitions. The graph of an increasing function goes up from left to right, the graph of a decreasing function goes down from left to right, and the graph of a constant function remains at a fixed height.

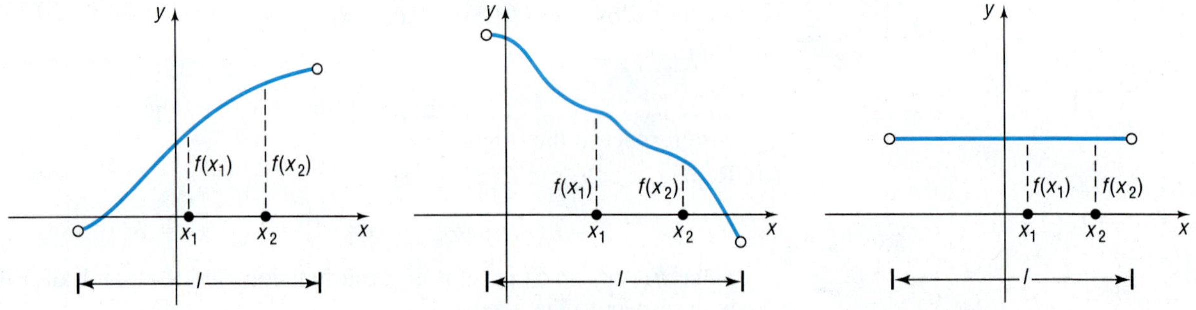

Figure 19

(a) For $x_1 < x_2$ in I, $f(x_1) < f(x_2)$; f is increasing on I.

(b) For $x_1 < x_2$ in I, $f(x_1) > f(x_2)$; f is decreasing on I.

(c) For all x in I, the values of f are equal; f is constant on I.

Now Work PROBLEMS 13, 15, 17, AND 25(c)

4 Use a Graph to Locate Local Maxima and Local Minima

Suppose f is a function defined on an open interval I containing c. If the value of f at c is greater than or equal to the values of f on I, then f has a *local maximum* at c.[†] See Figure 20(a).

If the value of f at c is less than or equal to the values of f on I, then f has a *local minimum* at c. See Figure 20(b).

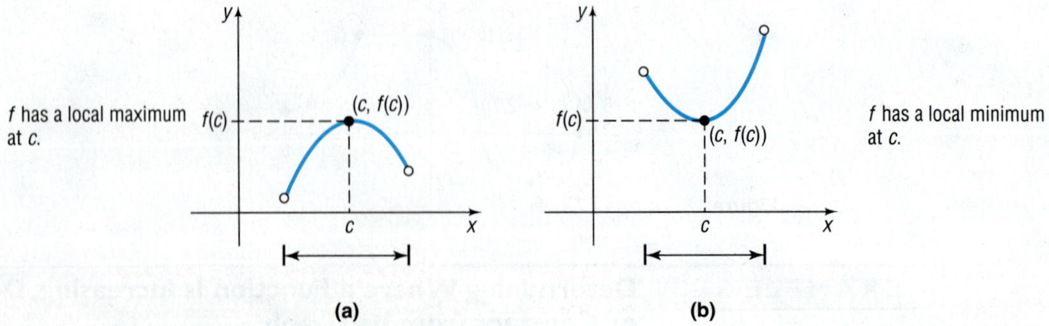

Figure 20 Local maximum and local minimum

*The open interval (a, b) consists of all real numbers x for which $a < x < b$.
[†]Some texts use the term *relative* instead of *local*.

DEFINITIONS

Let f be a function defined on some interval I.

A function f has a **local maximum** at c if there is an open interval in I containing c so that, for all x in this open interval, we have $f(x) \le f(c)$. We call $f(c)$ a **local maximum value of f**.

A function f has a **local minimum** at c if there is an open interval in I containing c so that, for all x in this open interval, we have $f(x) \ge f(c)$. We call $f(c)$ a **local minimum value of f**.

If f has a local maximum at c, then the value of f at c is greater than or equal to the values of f near c. If f has a local minimum at c, then the value of f at c is less than or equal to the values of f near c. The word *local* is used to suggest that it is only near c, not necessarily over the entire domain, that the value $f(c)$ has these properties.

EXAMPLE 4 Finding Local Maxima and Local Minima from the Graph of a Function and Determining Where the Function Is Increasing, Decreasing, or Constant

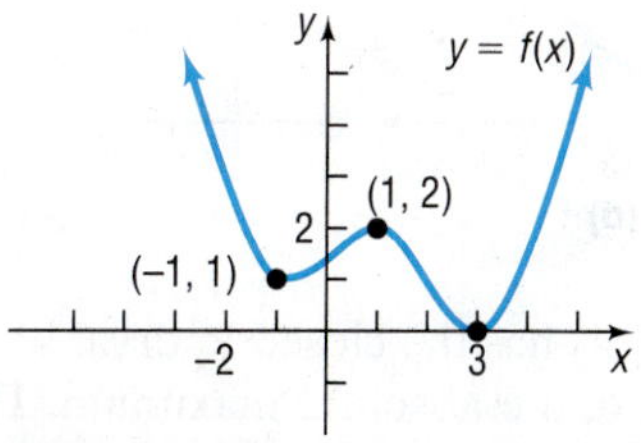

Figure 21

Figure 21 shows the graph of a function f.

(a) At what value(s) of x, if any, does f have a local maximum? List the local maximum values.

(b) At what value(s) of x, if any, does f have a local minimum? List the local minimum values.

(c) Find the intervals on which f is increasing. Find the intervals on which f is decreasing.

Solution The domain of f is the set of real numbers.

(a) f has a local maximum at 1, since for all x close to 1, we have $f(x) \le f(1)$. The local maximum value is $f(1) = 2$.

(b) f has local minima at -1 and at 3. The local minimum values are $f(-1) = 1$ and $f(3) = 0$.

(c) The function whose graph is given in Figure 21 is increasing for all values of x between -1 and 1 and for all values of x greater than 3. That is, the function is increasing on the intervals $(-1, 1)$ and $(3, \infty)$, or for $-1 < x < 1$ and $x > 3$. The function is decreasing for all values of x less than -1 and for all values of x between 1 and 3. That is, the function is decreasing on the intervals $(-\infty, -1)$ and $(1, 3)$, or for $x < -1$ and $1 < x < 3$.

WARNING The y-value is the local maximum value or local minimum value, and it occurs at some x-value. For example, in Figure 21, we say f has a local maximum at 1 and the local maximum value is 2.

Now Work PROBLEMS 19 AND 21

5 Use a Graph to Locate the Absolute Maximum and the Absolute Minimum

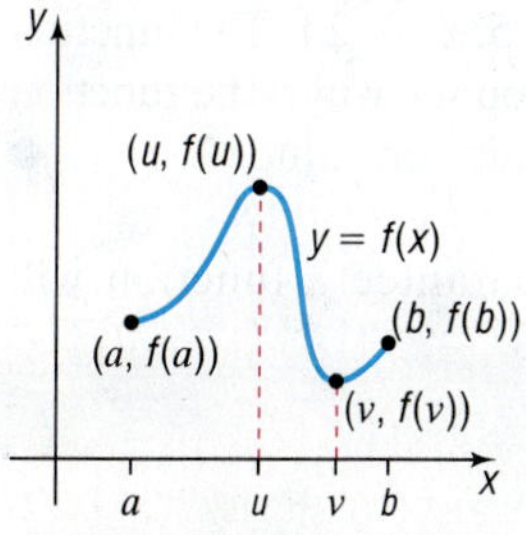

domain: $[a, b]$
for all x in $[a, b]$, $f(x) \le f(u)$
for all x in $[a, b]$, $f(x) \ge f(v)$
absolute maximum: $f(u)$
absolute minimum: $f(v)$

Figure 22

Look at the graph of the function f given in Figure 22. The domain of f is the closed interval $[a, b]$. Also, the largest value of f is $f(u)$ and the smallest value of f is $f(v)$. These are called, respectively, the *absolute maximum* and the *absolute minimum* of f on $[a, b]$.

DEFINITION Let f be a function defined on some interval I. If there is a number u in I for which $f(x) \le f(u)$ for all x in I, then f has an **absolute maximum at u**, and the number $f(u)$ is the **absolute maximum of f on I**.

If there is a number v in I for which $f(x) \ge f(v)$ for all x in I, then f has an **absolute minimum at v**, and the number $f(v)$ is the **absolute minimum of f on I**.

The absolute maximum and absolute minimum of a function f are sometimes called the **extreme values** of f on I.

The absolute maximum or absolute minimum of a function f may not exist. Let's look at some examples.

EXAMPLE 5 Finding the Absolute Maximum and the Absolute Minimum from the Graph of a Function

For each graph of a function $y = f(x)$ in Figure 23, find the absolute maximum and the absolute minimum, if they exist. Also, find any local maxima or local minima.

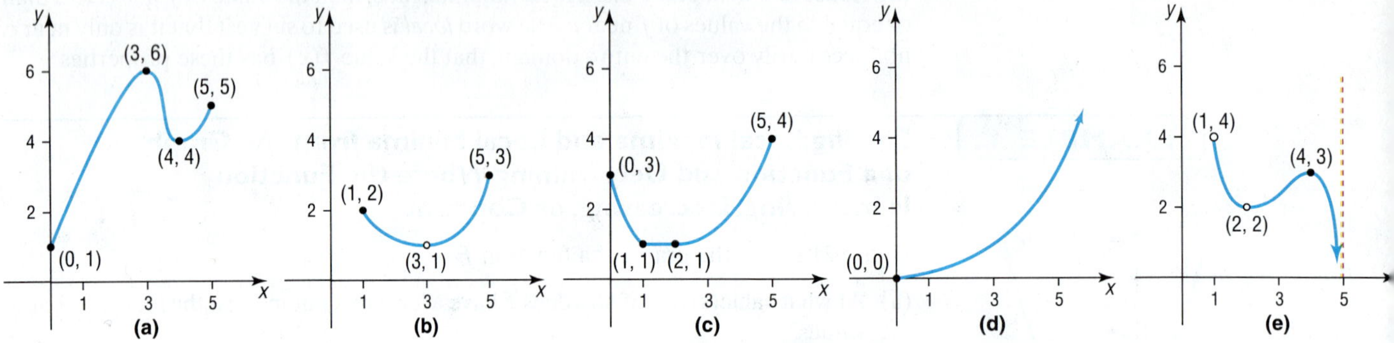

Figure 23

Solution

(a) The function f whose graph is given in Figure 23(a) has the closed interval $[0, 5]$ as its domain. The largest value of f is $f(3) = 6$, the absolute maximum. The smallest value of f is $f(0) = 1$, the absolute minimum. The function has a local maximum of 6 at $x = 3$ and a local minimum of 4 at $x = 4$.

(b) The function f whose graph is given in Figure 23(b) has the domain $\{x|1 \le x \le 5, x \ne 3\}$. Note that we exclude 3 from the domain because of the "hole" at (3, 1). The largest value of f on its domain is $f(5) = 3$, the absolute maximum. There is no absolute minimum. Do you see why? As you trace the graph, getting closer to the point (3, 1), there is no single smallest value. [As soon as you claim a smallest value, we can trace closer to (3, 1) and get a smaller value!] The function has no local maxima or minima.

(c) The function f whose graph is given in Figure 23(c) has the interval $[0, 5]$ as its domain. The absolute maximum of f is $f(5) = 4$. The absolute minimum is 1. Notice that the absolute minimum 1 occurs at any number in the interval $[1, 2]$. The function has a local minimum value of 1 at every x in the interval $[1, 2]$, but it has no local maximum value.

(d) The function f given in Figure 23(d) has the interval $[0, \infty)$ as its domain. The function has no absolute maximum; the absolute minimum is $f(0) = 0$. The function has no local maximum or local minimum.

(e) The function f in Figure 23(e) has the domain $\{x|1 < x < 5, x \ne 2\}$. The function has no absolute maximum and no absolute minimum. Do you see why? The function has a local maximum value of 3 at $x = 4$, but no local minimum value. ●

WARNING A function may have an absolute maximum or an absolute minimum at an endpoint but not a local maximum or a local minimum. Why? Local maxima and local minima are found over some open interval I, and this interval cannot be created around an endpoint. ■

In calculus, there is a theorem with conditions that guarantee a function will have an absolute maximum and an absolute minimum.

THEOREM

Extreme Value Theorem

If f is a continuous function* whose domain is a closed interval $[a, b]$, then f has an absolute maximum and an absolute minimum on $[a, b]$.

*Although a precise definition requires calculus, we'll agree for now that a continuous function is one whose graph has no gaps or holes and can be traced without lifting the pencil from the paper.

The absolute maximum (minimum) can be found by selecting the largest (smallest) value of f from the following list:

1. The values of f at any local maxima or local minima of f in $[a, b]$.
2. The value of f at each endpoint of $[a, b]$—that is, $f(a)$ and $f(b)$.

For example, the graph of the function f given in Figure 23(a) is continuous on the closed interval $[0, 5]$. The Extreme Value Theorem guarantees that f has extreme values on $[0, 5]$. To find them, we list

1. The value of f at the local extrema: $f(3) = 6, f(4) = 4$
2. The value of f at the endpoints: $f(0) = 1, f(5) = 5$

The largest of these, 6, is the absolute maximum; the smallest of these, 1, is the absolute minimum.

Now Work PROBLEM 49

6 Use a Graphing Utility to Approximate Local Maxima and Local Minima and to Determine Where a Function Is Increasing or Decreasing

To locate the exact value at which a function f has a local maximum or a local minimum usually requires calculus. However, a graphing utility may be used to approximate these values using the MAXIMUM and MINIMUM features.

EXAMPLE 6 Using a Graphing Utility to Approximate Local Maxima and Minima and to Determine Where a Function Is Increasing or Decreasing

(a) Use a graphing utility to graph $f(x) = 6x^3 - 12x + 5$ for $-2 < x < 2$. Approximate where f has a local maximum and where f has a local minimum.

(b) Determine where f is increasing and where it is decreasing.

Solution (a) Graphing utilities have a feature that finds the maximum or minimum point of a graph within a given interval. Graph the function f for $-2 < x < 2$. The MAXIMUM and MINIMUM commands require us to first determine the open interval I. The graphing utility will then approximate the maximum or minimum value in the interval. Using MAXIMUM, we find that the local maximum value is 11.53 and that it occurs at $x = -0.82$, rounded to two decimal places. See Figure 24(a). Using MINIMUM, we find that the local minimum value is -1.53 and that it occurs at $x = 0.82$, rounded to two decimal places. See Figure 24(b).

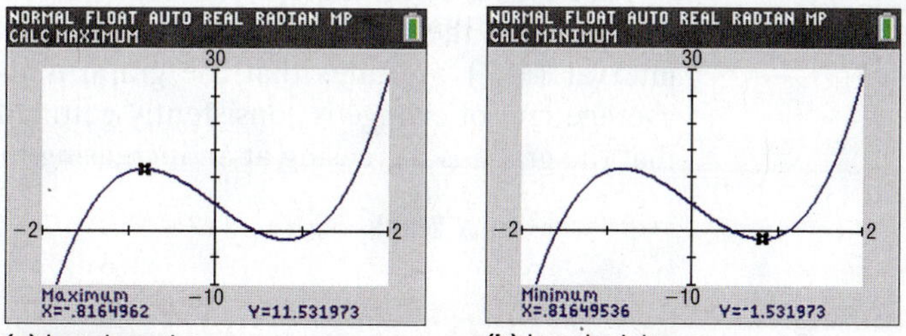

Figure 24 **(a)** Local maximum **(b)** Local minimum

(b) Looking at Figures 24(a) and (b), we see that the graph of f is increasing from $x = -2$ to $x = -0.82$ and from $x = 0.82$ to $x = 2$, so f is increasing on the intervals $(-2, -0.82)$ and $(0.82, 2)$, or for $-2 < x < -0.82$ and $0.82 < x < 2$. The graph is decreasing from $x = -0.82$ to $x = 0.82$, so f is decreasing on the interval $(-0.82, 0.82)$, or for $-0.82 < x < 0.82$. ●

Now Work PROBLEM 57

7 Find the Average Rate of Change of a Function

In Section 2.3, we said that the slope of a line can be interpreted as the average rate of change. To find the average rate of change of a function between any two points on its graph, calculate the slope of the line containing the two points.

DEFINITION

If a and b, $a \neq b$, are in the domain of a function $y = f(x)$, the **average rate of change of** f from a to b is defined as

$$\text{Average rate of change} = \frac{\Delta y}{\Delta x} = \frac{f(b) - f(a)}{b - a} \qquad a \neq b \qquad \textbf{(1)}$$

In Words
The symbol Δ is the Greek capital letter delta and is read "change in."

The symbol Δy in equation (1) is the "change in y," and Δx is the "change in x." The average rate of change of f is the change in y divided by the change in x.

EXAMPLE 7 **Finding the Average Rate of Change**

Find the average rate of change of $f(x) = 3x^2$:

(a) From 1 to 3 (b) From 1 to 5 (c) From 1 to 7

Solution (a) The average rate of change of $f(x) = 3x^2$ from 1 to 3 is

$$\frac{\Delta y}{\Delta x} = \frac{f(3) - f(1)}{3 - 1} = \frac{27 - 3}{3 - 1} = \frac{24}{2} = 12$$

(b) The average rate of change of $f(x) = 3x^2$ from 1 to 5 is

$$\frac{\Delta y}{\Delta x} = \frac{f(5) - f(1)}{5 - 1} = \frac{75 - 3}{5 - 1} = \frac{72}{4} = 18$$

(c) The average rate of change of $f(x) = 3x^2$ from 1 to 7 is

$$\frac{\Delta y}{\Delta x} = \frac{f(7) - f(1)}{7 - 1} = \frac{147 - 3}{7 - 1} = \frac{144}{6} = 24$$

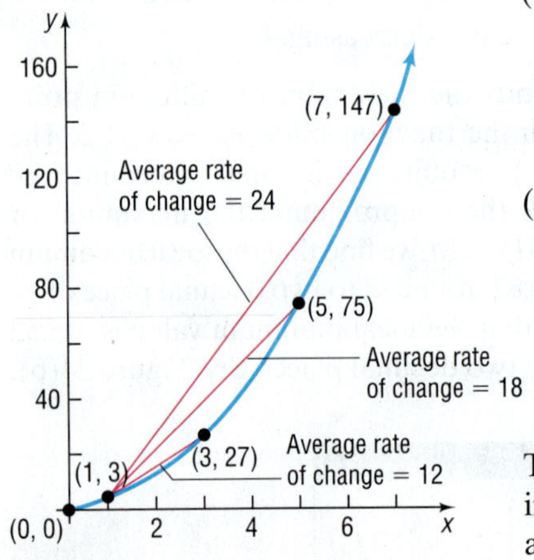

Figure 25 $f(x) = 3x^2$

See Figure 25 for a graph of $f(x) = 3x^2$. The function f is increasing for $x > 0$. The fact that the average rate of change is positive for any $x_1, x_2, x_1 \neq x_2$, in the interval $(1, 7)$ indicates that the graph is increasing on $1 < x < 7$. Further, the average rate of change is consistently getting larger for $1 < x < 7$, which indicates that the graph is increasing at an increasing rate.

Now Work PROBLEM 65

The Secant Line

The average rate of change of a function has an important geometric interpretation. Look at the graph of $y = f(x)$ in Figure 26. Two points are labeled on the graph: $(a, f(a))$ and $(b, f(b))$. The line containing these two points is called the **secant line**; its slope is

$$m_{\text{sec}} = \frac{f(b) - f(a)}{b - a}$$

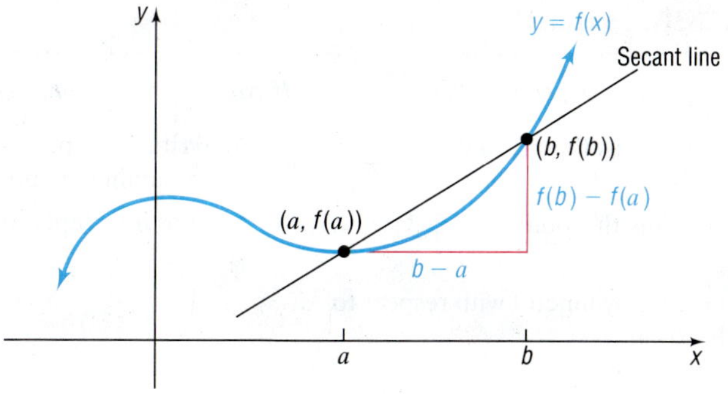

Figure 26 Secant line

THEOREM

Slope of the Secant Line

The average rate of change of a function from a to b equals the slope of the secant line containing the two points $(a, f(a))$ and $(b, f(b))$ on its graph.

EXAMPLE 8

Finding the Equation of a Secant Line

Suppose that $g(x) = 3x^2 - 2x + 3$.

(a) Find the average rate of change of g from -2 to 1.

(b) Find an equation of the secant line containing $(-2, g(-2))$ and $(1, g(1))$.

(c) Using a graphing utility, draw the graph of g and the secant line obtained in part (b) on the same screen.

Solution (a) The average rate of change of $g(x) = 3x^2 - 2x + 3$ from -2 to 1 is

$$\text{Average rate of change} = \frac{g(1) - g(-2)}{1 - (-2)}$$

$$= \frac{4 - 19}{3} \qquad g(1) = 3(1)^2 - 2(1) + 3 = 4,\ g(-2) = 3(-2)^2 - 2(-2) + 3 = 19$$

$$= -\frac{15}{3} = -5$$

(b) The slope of the secant line containing $(-2, g(-2)) = (-2, 19)$ and $(1, g(1)) = (1, 4)$ is $m_{sec} = -5$. Use the point–slope form to find an equation of the secant line.

$$y - y_1 = m_{sec}(x - x_1) \qquad \text{Point–slope form of the secant line}$$

$$y - 19 = -5(x - (-2)) \qquad x_1 = -2,\ y_1 = g(-2) = 19,\ m_{sec} = -5$$

$$y - 19 = -5x - 10 \qquad \text{Distribute.}$$

$$y = -5x + 9 \qquad \text{Slope-intercept form of the secant line}$$

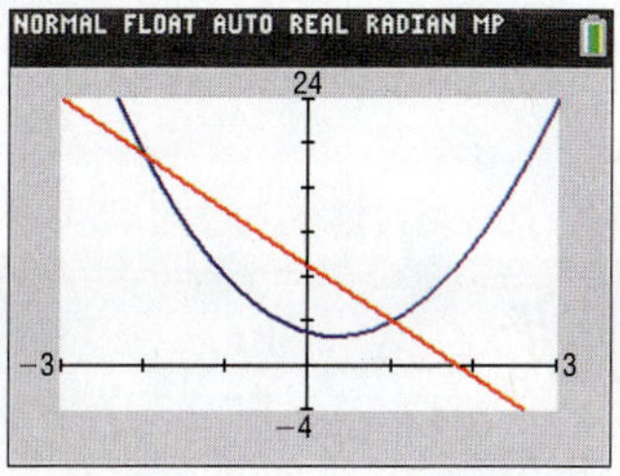

Figure 27 Graph of g and the secant line

(c) Figure 27 shows the graph of g along with the secant line $y = -5x + 9$. ●

Now Work PROBLEM 71

3.3 Assess Your Understanding

'Are You Prepared?' *Answers are given at the end of these exercises. If you get a wrong answer, read the pages listed in red.*

1. The interval $(2, 5)$ can be written as the inequality _____. (pp. 120–121)

2. The slope of the line containing the points $(-2, 3)$ and $(3, 8)$ is ___. (pp. 167–169)

3. Test the equation $y = 5x^2 - 1$ for symmetry with respect to the x-axis, the y-axis, and the origin. (pp. 160–162)

4. Write the point–slope form of the line with slope 5 containing the point $(3, -2)$. (p. 171)

5. The intercepts of the equation $y = x^2 - 9$ are _____. (pp. 159–160)

Concepts and Vocabulary

6. A function f is ________ on an open interval I if, for any choice of x_1 and x_2 in I, with $x_1 < x_2$, we have $f(x_1) < f(x_2)$.

7. A(n) ____ function f is one for which $f(-x) = f(x)$ for every x in the domain of f; a(n) ____ function f is one for which $f(-x) = -f(x)$ for every x in the domain of f.

8. ***True or False*** A function f is decreasing on an open interval I if, for any choice of x_1 and x_2 in I, with $x_1 < x_2$, we have $f(x_1) > f(x_2)$.

9. ***True or False*** A function f has a local maximum at c if there is an open interval I containing c such that for all x in I, $f(x) \le f(c)$.

10. ***True or False*** Even functions have graphs that are symmetric with respect to the origin.

11. An odd function is symmetric with respect to _____.
(a) the x-axis (b) the y-axis
(c) the origin (d) the line $y = x$

12. Which of the following intervals is required to guarantee a continuous function will have both an absolute maximum and an absolute minimum?
(a) (a, b) (b) $(a, b]$
(c) $[a, b)$ (d) $[a, b]$

Skill Building

In Problems 13–24, use the graph of the function f given.

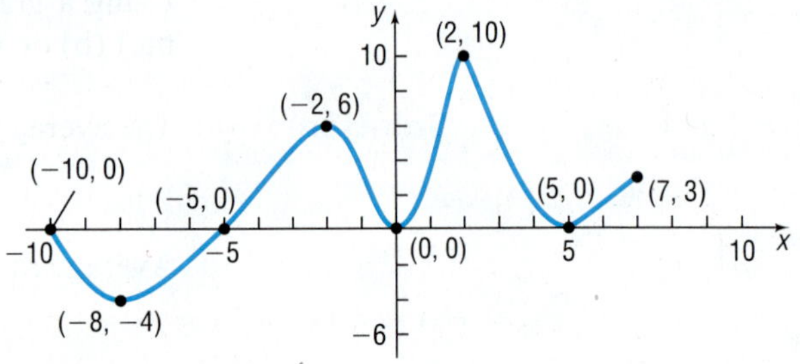

13. Is f increasing on the interval $(-8, -2)$?

14. Is f decreasing on the interval $(-8, -4)$?

15. Is f increasing on the interval $(-2, 6)$?

16. Is f decreasing on the interval $(2, 5)$?

17. List the interval(s) on which f is increasing.

18. List the interval(s) on which f is decreasing.

19. Is there a local maximum at 2? If yes, what is it?

20. Is there a local maximum at 5? If yes, what is it?

21. List the number(s) at which f has a local maximum. What are the local maximum values?

22. List the number(s) at which f has a local minimum. What are the local minimum values?

23. Find the absolute minimum of f on $[-10, 7]$.

24. Find the absolute maximum of f on $[-10, 7]$.

In Problems 25–32, the graph of a function is given. Use the graph to find:
(a) The intercepts, if any
(b) The domain and range
(c) The intervals on which the function is increasing, decreasing, or constant
(d) Whether the function is even, odd, or neither

25.

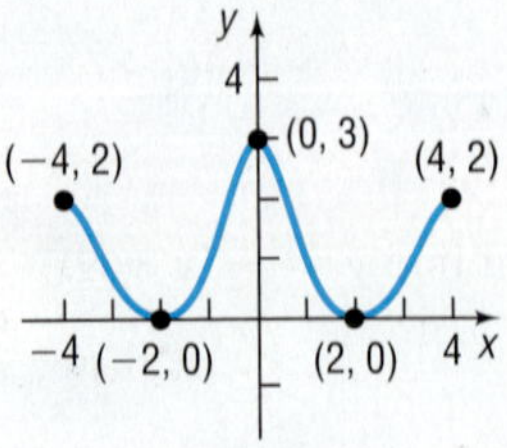

26.

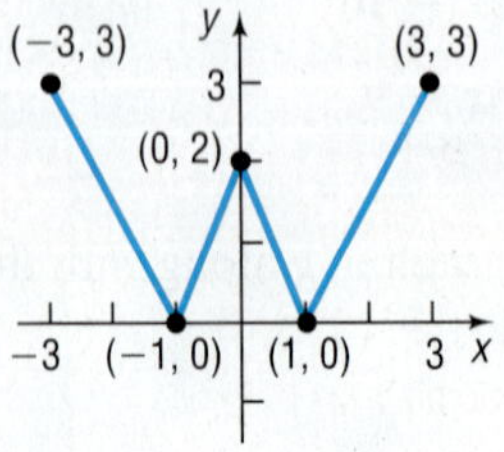

27.

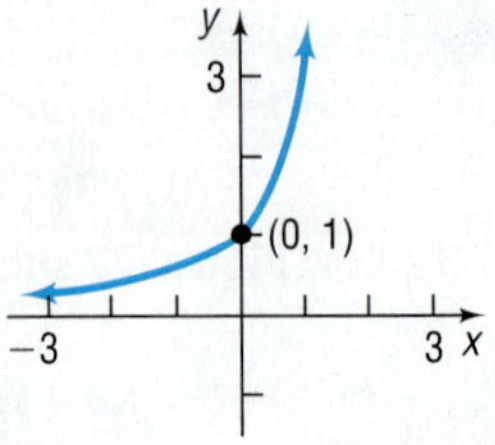

28.

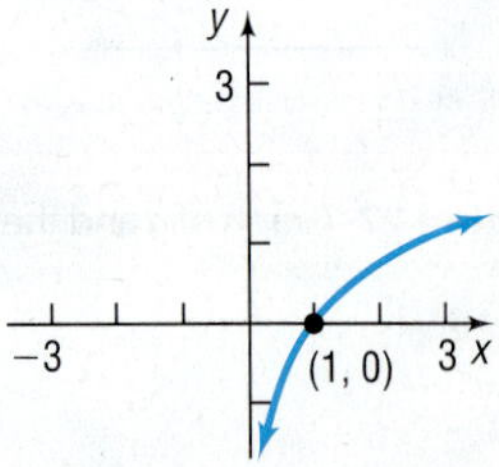

29.

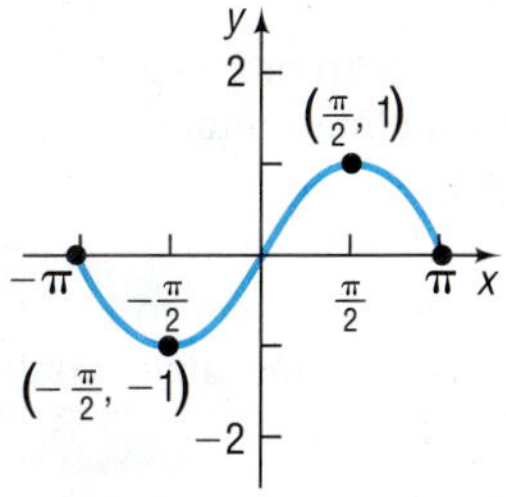

30.

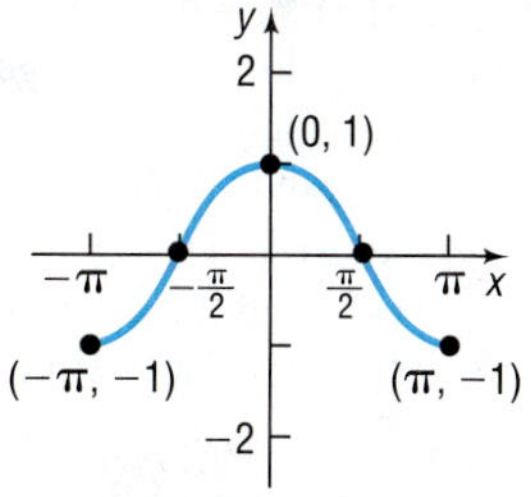

31.

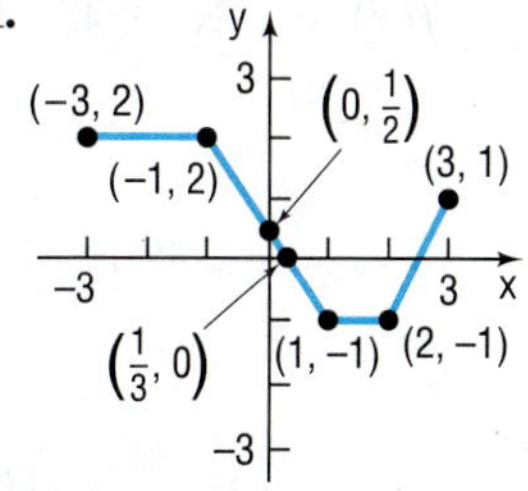

32.

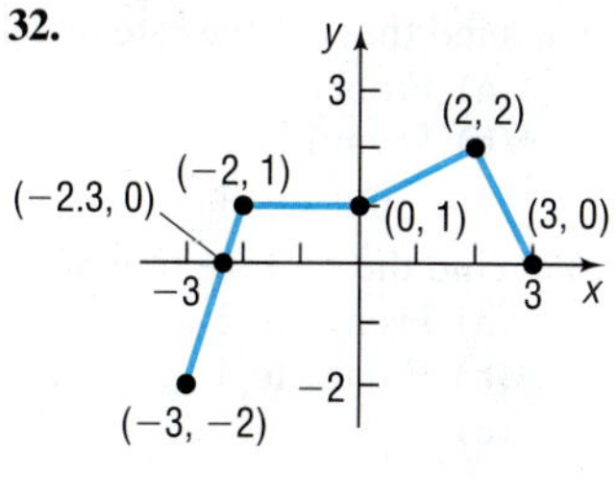

In Problems 33–36, the graph of a function f is given. Use the graph to find:
(a) The numbers, if any, at which f has a local maximum. What are the local maximum values?
(b) The numbers, if any, at which f has a local minimum. What are the local minimum values?

33.

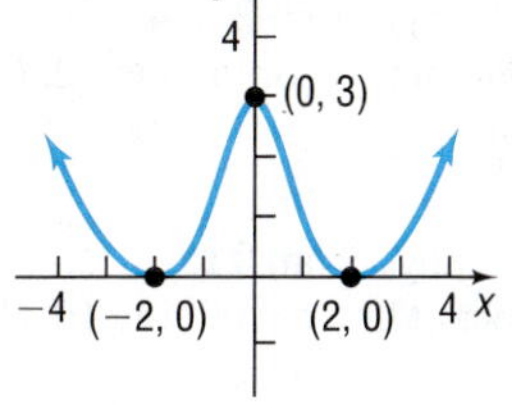

34.

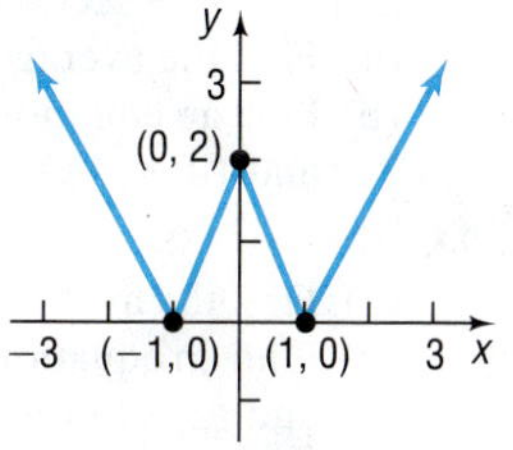

35.

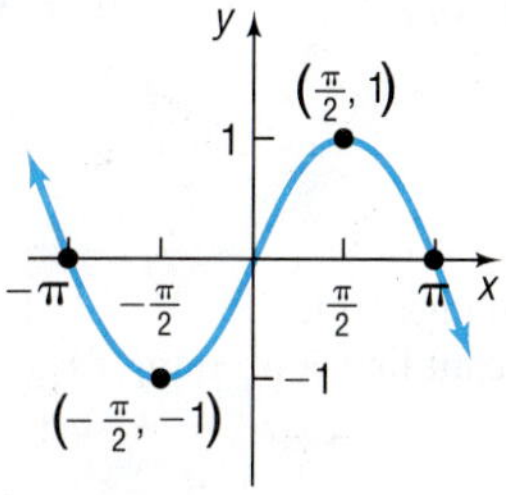

36.

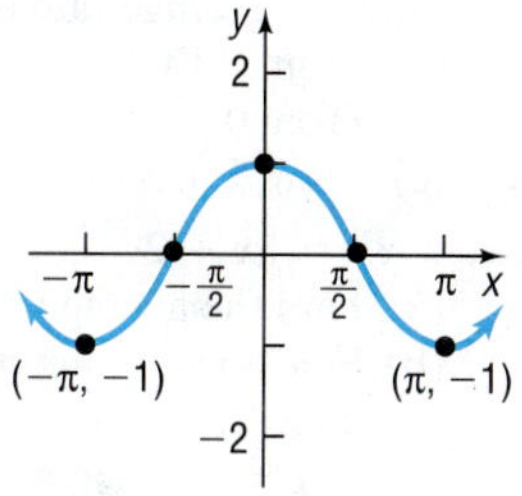

In Problems 37–48, determine algebraically whether each function is even, odd, or neither.

37. $f(x) = 4x^3$

38. $f(x) = 2x^4 - x^2$

39. $g(x) = -3x^2 - 5$

40. $h(x) = 3x^3 + 5$

41. $F(x) = \sqrt[3]{x}$

42. $G(x) = \sqrt{x}$

43. $f(x) = x + |x|$

44. $f(x) = \sqrt[3]{2x^2 + 1}$

45. $g(x) = \dfrac{1}{x^2}$

46. $h(x) = \dfrac{x}{x^2 - 1}$

47. $h(x) = \dfrac{-x^3}{3x^2 - 9}$

48. $F(x) = \dfrac{2x}{|x|}$

In Problems 49–56, for each graph of a function $y = f(x)$, find the absolute maximum and the absolute minimum, if they exist. Identify any local maximum values or local minimum values.

49.

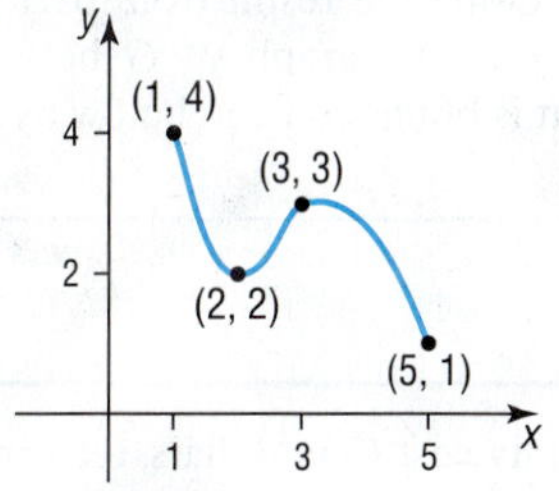

50.

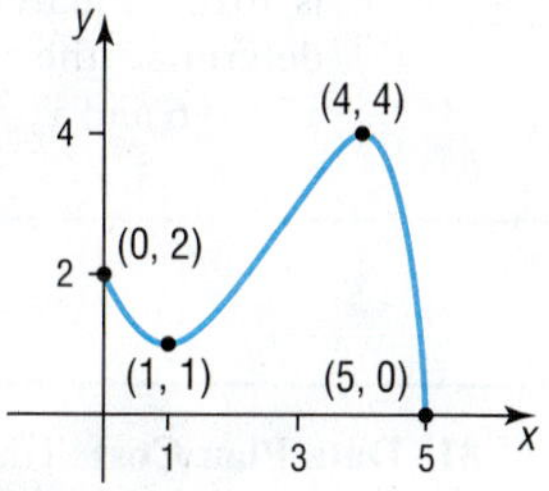

51.

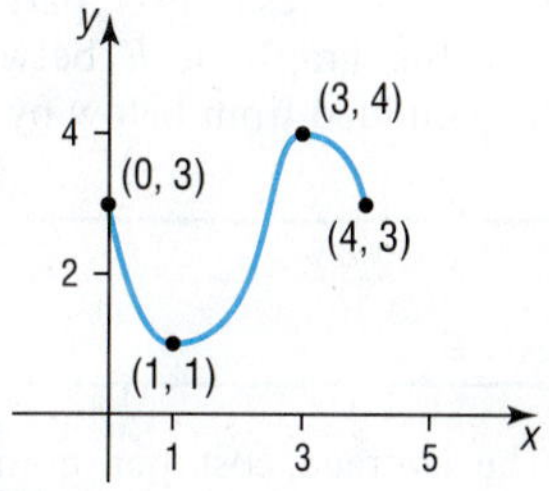

52.

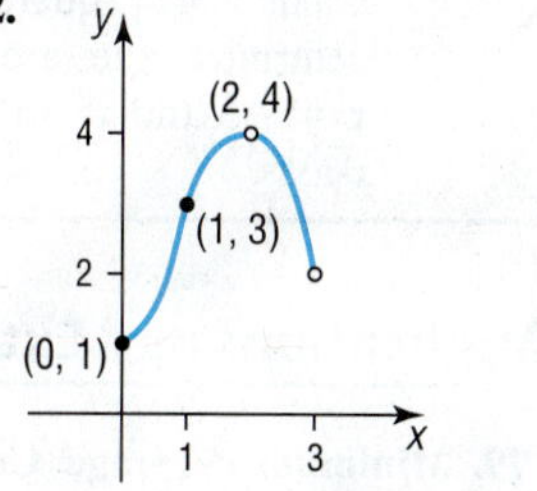

53.

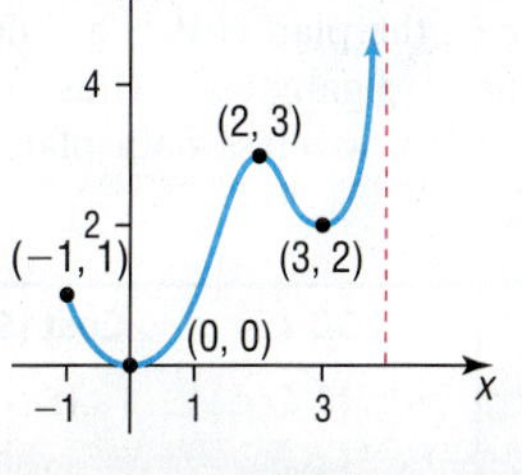

54.

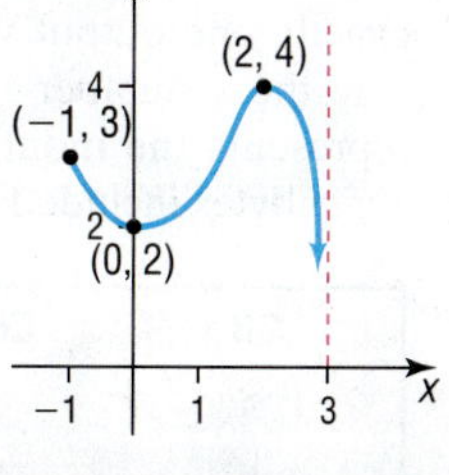

55.

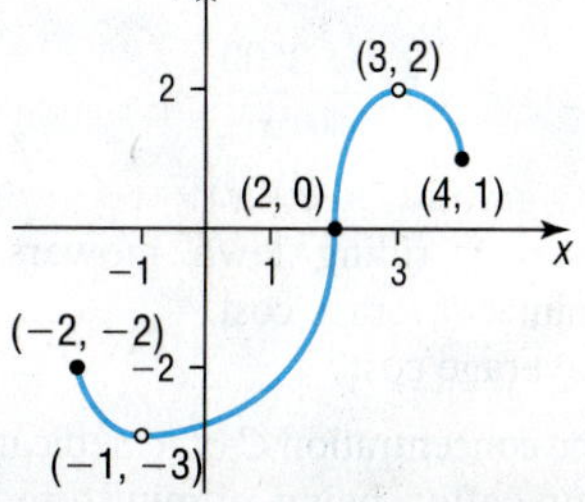

56.

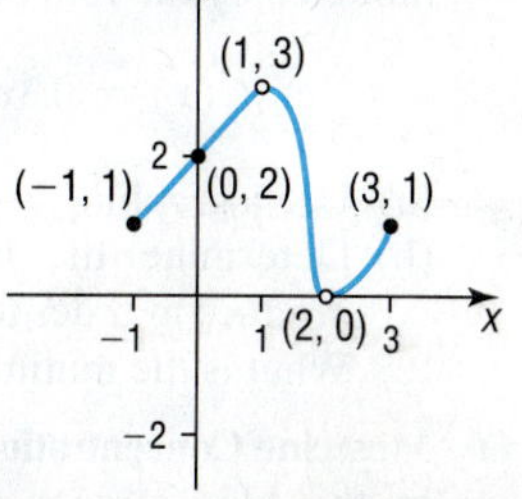

In Problems 57–64, use a graphing utility to graph each function over the indicated interval and approximate any local maximum values and local minimum values. Determine where the function is increasing and where it is decreasing. Round answers to two decimal places.

57. $f(x) = x^3 - 3x + 2 \quad (-2, 2)$

58. $f(x) = x^3 - 3x^2 + 5 \quad (-1, 3)$

59. $f(x) = x^5 - x^3 \quad (-2, 2)$

60. $f(x) = x^4 - x^2 \quad (-2, 2)$

61. $f(x) = -0.2x^3 - 0.6x^2 + 4x - 6 \quad (-6, 4)$

62. $f(x) = -0.4x^3 + 0.6x^2 + 3x - 2 \quad (-4, 5)$

63. $f(x) = 0.25x^4 + 0.3x^3 - 0.9x^2 + 3 \quad (-3, 2)$

64. $f(x) = -0.4x^4 - 0.5x^3 + 0.8x^2 - 2 \quad (-3, 2)$

65. Find the average rate of change of $f(x) = -2x^2 + 4$:
(a) From 0 to 2
(b) From 1 to 3
(c) From 1 to 4

66. Find the average rate of change of $f(x) = -x^3 + 1$:
(a) From 0 to 2
(b) From 1 to 3
(c) From -1 to 1

67. Find the average rate of change of $g(x) = x^3 - 2x + 1$:
(a) From -3 to -2
(b) From -1 to 1
(c) From 1 to 3

68. Find the average rate of change of $h(x) = x^2 - 2x + 3$:
(a) From -1 to 1
(b) From 0 to 2
(c) From 2 to 5

69. $f(x) = 5x - 2$
(a) Find the average rate of change from 1 to 3.
(b) Find an equation of the secant line containing $(1, f(1))$ and $(3, f(3))$.

70. $f(x) = -4x + 1$
(a) Find the average rate of change from 2 to 5.
(b) Find an equation of the secant line containing $(2, f(2))$ and $(5, f(5))$.

71. $g(x) = x^2 - 2$
(a) Find the average rate of change from -2 to 1.
(b) Find an equation of the secant line containing $(-2, g(-2))$ and $(1, g(1))$.

72. $g(x) = x^2 + 1$
(a) Find the average rate of change from -1 to 2.
(b) Find an equation of the secant line containing $(-1, g(-1))$ and $(2, g(2))$.

73. $h(x) = x^2 - 2x$
(a) Find the average rate of change from 2 to 4.
(b) Find an equation of the secant line containing $(2, h(2))$ and $(4, h(4))$.

74. $h(x) = -2x^2 + x$
(a) Find the average rate of change from 0 to 3.
(b) Find an equation of the secant line containing $(0, h(0))$ and $(3, h(3))$.

Mixed Practice

75. $g(x) = x^3 - 27x$
(a) Determine whether g is even, odd, or neither.
(b) There is a local minimum value of -54 at 3. Determine the local maximum value.

76. $f(x) = -x^3 + 12x$
(a) Determine whether f is even, odd, or neither.
(b) There is a local maximum value of 16 at 2. Determine the local minimum value.

77. $F(x) = -x^4 + 8x^2 + 8$
(a) Determine whether F is even, odd, neither.
(b) There is a local maximum value of 24 at $x = 2$. Determine a second local maximum value.
(c) Suppose the area under the graph of F between $x = 0$ and $x = 3$ that is bounded from below by the x-axis is 47.4 square units. Using the result from part (a), determine the area under the graph of F between $x = -3$ and $x = 0$ that is bounded from below by the x-axis.

78. $G(x) = -x^4 + 32x^2 + 144$
(a) Determine whether G is even, odd, or neither.
(b) There is a local maximum value of 400 at $x = 4$. Determine a second local maximum value.
(c) Suppose the area under the graph of G between $x = 0$ and $x = 6$ that is bounded from below by the x-axis is 1612.8 square units. Using the result from part (a), determine the area under the graph of G between $x = -6$ and $x = 0$ that is bounded from below by the x-axis.

Applications and Extensions

79. Minimum Average Cost The average cost per hour in dollars, $\overline{C}$, of producing x riding lawn mowers can be modeled by the function

$$\overline{C}(x) = 0.3x^2 + 21x - 251 + \frac{2500}{x}$$

(a) Use a graphing utility to graph $\overline{C} = \overline{C}(x)$.
(b) Determine the number of riding lawn mowers to produce in order to minimize average cost.
(c) What is the minimum average cost?

80. Medicine Concentration The concentration C of a medication in the bloodstream t hours after being administered is modeled by the function

$$C(t) = -0.002t^4 + 0.039t^3 - 0.285t^2 + 0.766t + 0.085$$

(a) After how many hours will the concentration be highest?
(b) A woman nursing a child must wait until the concentration is below 0.5 before she can feed him. After taking the medication, how long must she wait before feeding her child?

81. Data Plan Cost The monthly cost C, in dollars, for wireless data plans with x gigabytes of data included is shown in the table below. Since each input value for x corresponds to exactly one output value for C, the plan cost is a function of the number of data gigabytes. Thus $C(x)$ represents the monthly cost for a wireless data plan with x gigabytes included.

GB	Cost ($)	GB	Cost ($)
4	70	20	150
6	80	30	225
10	100	40	300
15	130	50	375

(a) Plot the points (4, 70), (6, 80), (10, 100), and so on in a Cartesian plane.
(b) Draw a line segment from the point (10, 100) to (30, 225). What does the slope of this line segment represent?

(c) Find the average rate of change of the monthly cost from 4 to 10 gigabytes.
(d) Find the average rate of change of the monthly cost from 10 to 30 gigabytes.
(e) Find the average rate of change of the monthly cost from 30 to 50 gigabytes.
(f) What is happening to the average rate of change as the gigabytes of data increase?

82. National Debt The size of the total debt owed by the United States federal government continues to grow. In fact, according to the Department of the Treasury, the debt per person living in the United States is approximately \$53,000 (or over \$140,000 per U.S. household). The following data represent the U.S. debt for the years 2001–2013. Since the debt D depends on the year y, and each input corresponds to exactly one output, the debt is a function of the year. So $D(y)$ represents the debt for each year y.

Year	Debt (billions of dollars)	Year	Debt (billions of dollars)
2001	5807	2008	10,025
2002	6228	2009	11,910
2003	6783	2010	13,562
2004	7379	2011	14,790
2005	7933	2012	16,066
2006	8507	2013	16,738
2007	9008		

Source: *www.treasurydirect.gov*

(a) Plot the points (2001, 5807), (2002, 6228), and so on in a Cartesian plane.
(b) Draw a line segment from the point (2001, 5807) to (2006, 8507). What does the slope of this line segment represent?
(c) Find the average rate of change of the debt from 2002 to 2004.
(d) Find the average rate of change of the debt from 2006 to 2008.
(e) Find the average rate of change of the debt from 2010 to 2012.
(f) What appears to be happening to the average rate of change as time passes?

83. *E. coli* Growth A strain of *E. coli* Beu 397-recA441 is placed into a nutrient broth at 30° Celsius and allowed to grow. The data shown in the table are collected. The population is measured in grams and the time in hours. Since population P depends on time t, and each input corresponds to exactly one output, we can say that population is a function of time. Thus $P(t)$ represents the population at time t.

Time (hours), t	Population (grams), P
0	0.09
2.5	0.18
3.5	0.26
4.5	0.35
6	0.50

(a) Find the average rate of change of the population from 0 to 2.5 hours.
(b) Find the average rate of change of the population from 4.5 to 6 hours.
(c) What is happening to the average rate of change as time passes?

84. e-Filing Tax Returns The Internal Revenue Service Restructuring and Reform Act (RRA) was signed into law by President Bill Clinton in 1998. A major objective of the RRA was to promote electronic filing of tax returns. The data in the table that follows show the percentage of individual income tax returns filed electronically for filing years 2004–2012. Since the percentage P of returns filed electronically depends on the filing year y, and each input corresponds to exactly one output, the percentage of returns filed electronically is a function of the filing year; so $P(y)$ represents the percentage of returns filed electronically for filing year y.

(a) Find the average rate of change of the percentage of e-filed returns from 2004 to 2006.
(b) Find the average rate of change of the percentage of e-filed returns from 2007 to 2009.
(c) Find the average rate of change of the percentage of e-filed returns from 2010 to 2012.
(d) What is happening to the average rate of change as time passes?

Year	Percentage of returns e-filed
2004	46.5
2005	51.1
2006	53.8
2007	57.1
2008	58.5
2009	67.2
2010	69.8
2011	77.2
2012	82.7

Source: Internal Revenue Service

85. For the function $f(x) = x^2$, compute the average rate of change:
(a) From 0 to 1
(b) From 0 to 0.5
(c) From 0 to 0.1
(d) From 0 to 0.01
(e) From 0 to 0.001
(f) Use a graphing utility to graph each of the secant lines along with f.
(g) What do you think is happening to the secant lines?
(h) What is happening to the slopes of the secant lines? Is there some number that they are getting closer to? What is that number?

86. For the function $f(x) = x^2$, compute the average rate of change:
(a) From 1 to 2
(b) From 1 to 1.5
(c) From 1 to 1.1
(d) From 1 to 1.01
(e) From 1 to 1.001
(f) Use a graphing utility to graph each of the secant lines along with f.
(g) What do you think is happening to the secant lines?
(h) What is happening to the slopes of the secant lines? Is there some number that they are getting closer to? What is that number?

Problems 87–94 require the following discussion of a secant line. The slope of the secant line containing the two points $(x, f(x))$ and $(x + h, f(x + h))$ on the graph of a function $y = f(x)$ may be given as

$$m_{\text{sec}} = \frac{f(x+h) - f(x)}{(x+h) - x} = \frac{f(x+h) - f(x)}{h}, \qquad h \neq 0$$

In calculus, this expression is called the **difference quotient of** ***f***.

(a) *Express the slope of the secant line of each function in terms of x and h. Be sure to simplify your answer.*
(b) *Find m_{sec} for $h = 0.5, 0.1$, and 0.01 at $x = 1$. What value does m_{sec} approach as h approaches 0?*
(c) *Find an equation for the secant line at $x = 1$ with $h = 0.01$.*
(d) *Use a graphing utility to graph f and the secant line found in part (c) in the same viewing window.*

87. $f(x) = 2x + 5$
88. $f(x) = -3x + 2$
89. $f(x) = x^2 + 2x$
90. $f(x) = 2x^2 + x$
91. $f(x) = 2x^2 - 3x + 1$
92. $f(x) = -x^2 + 3x - 2$
93 $f(x) = \frac{1}{x}$
94. $f(x) = \frac{1}{x^2}$

Explaining Concepts: Discussion and Writing

95. Draw the graph of a function that has the following properties: domain: all real numbers; range: all real numbers; intercepts: $(0, -3)$ and $(3, 0)$; a local maximum value of -2 is at -1; a local minimum value of -6 is at 2. Compare your graph with those of others. Comment on any differences.

96. Redo Problem 95 with the following additional information: increasing on $(-\infty, -1), (2, \infty)$; decreasing on $(-1, 2)$. Again compare your graph with others and comment on any differences.

97. How many x-intercepts can a function defined on an interval have if it is increasing on that interval? Explain.

98. Suppose that a friend of yours does not understand the idea of increasing and decreasing functions. Provide an explanation, complete with graphs, that clarifies the idea.

99. Can a function be both even and odd? Explain.

100. Using a graphing utility, graph $y = 5$ on the interval $(-3, 3)$. Use MAXIMUM to find the local maximum values on $(-3, 3)$. Comment on the result provided by the calculator.

101. A function f has a positive average rate of change on the interval $[2, 5]$. Is f increasing on $[2, 5]$? Explain.

102. Show that a constant function $f(x) = b$ has an average rate of change of 0. Compute the average rate of change of $y = \sqrt{4 - x^2}$ on the interval $[-2, 2]$. Explain how this can happen.

Retain Your Knowledge

Problems 103–106 are based on material learned earlier in the course. The purpose of these problems is to keep the material fresh in your mind so that you are better prepared for the final exam.

103. Write each number in scientific notation.
(a) 0.00000701
(b) 2,305,000,000

104. Simplify: $\sqrt{540}$

105. Solve: $14 < 5 - 3x \leq 29$

106. The shelf life of a perishable commodity varies inversely with the storage temperature. If the shelf life at 10°C is 33 days, what is the shelf life at 40°C?

'Are You Prepared?' Answers

1. $2 < x < 5$ **2.** 1 **3.** symmetric with respect to the y-axis **4.** $y + 2 = 5(x - 3)$ **5.** $(-3, 0), (3, 0), (0, -9)$

3.4 Library of Functions; Piecewise-defined Functions

PREPARING FOR THIS SECTION *Before getting started, review the following:*

- Intercepts (Section 2.2, pp. 159–160)
- Graphs of Key Equations (Section 2.2: Example 3, p. 159; Example 10, p. 163; Example 11, p. 163; Example 12, p. 164)

Now Work the **'Are You Prepared?'** problems on page 244.

OBJECTIVES 1 Graph the Functions Listed in the Library of Functions (p. 237)
2 Graph Piecewise-defined Functions (p. 242)

1 Graph the Functions Listed in the Library of Functions

First we introduce a few more functions, beginning with the *square root function.*

On page 163, we graphed the equation $y = \sqrt{x}$. Figure 28 shows a graph of the function $f(x) = \sqrt{x}$. Based on the graph, we have the following properties:

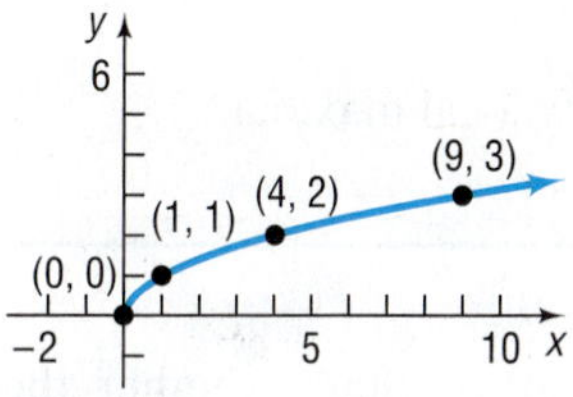

Figure 28 Square root function

Properties of $f(x) = \sqrt{x}$

1. The domain and the range are the set of nonnegative real numbers.
2. The x-intercept of the graph of $f(x) = \sqrt{x}$ is 0. The y-intercept of the graph of $f(x) = \sqrt{x}$ is also 0.
3. The function is neither even nor odd.
4. The function is increasing on the interval $(0, \infty)$.
5. The function has an absolute minimum of 0 at $x = 0$.

EXAMPLE 1 **Graphing the Cube Root Function**

(a) Determine whether $f(x) = \sqrt[3]{x}$ is even, odd, or neither. State whether the graph of f is symmetric with respect to the y-axis or symmetric with respect to the origin.

(b) Determine the intercepts, if any, of the graph of $f(x) = \sqrt[3]{x}$.

(c) Graph $f(x) = \sqrt[3]{x}$.

Solution (a) Because

$$f(-x) = \sqrt[3]{-x} = -\sqrt[3]{x} = -f(x)$$

the function is odd. The graph of f is symmetric with respect to the origin.

(b) The y-intercept is $f(0) = \sqrt[3]{0} = 0$. The x-intercept is found by solving the equation $f(x) = 0$.

$$f(x) = 0$$
$$\sqrt[3]{x} = 0 \quad f(x) = \sqrt[3]{x}$$
$$x = 0 \quad \text{Cube both sides of the equation.}$$

The x-intercept is also 0.

(c) Use the function to form Table 4 (on page 238) and obtain some points on the graph. Because of the symmetry with respect to the origin, we find only points (x, y) for which $x \geq 0$. Figure 29 shows the graph of $f(x) = \sqrt[3]{x}$.

Table 4

x	$y = f(x) = \sqrt[3]{x}$	(x, y)
0	0	(0, 0)
$\frac{1}{8}$	$\frac{1}{2}$	$\left(\frac{1}{8}, \frac{1}{2}\right)$
1	1	(1, 1)
2	$\sqrt[3]{2} \approx 1.26$	$(2, \sqrt[3]{2})$
8	2	(8, 2)

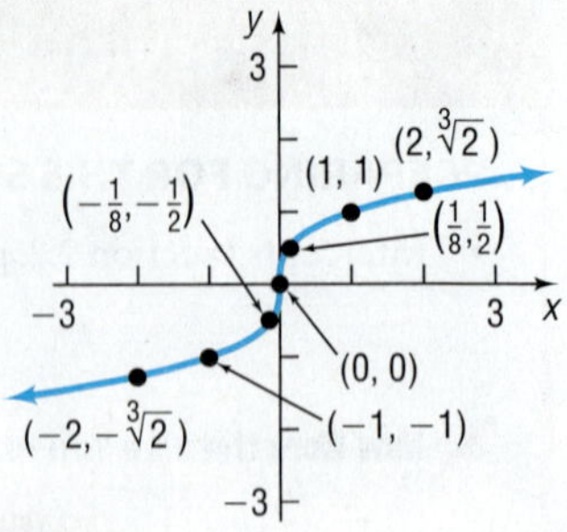

Figure 29 Cube Root Function

From the results of Example 1 and Figure 29, we have the following properties of the cube root function.

Properties of $f(x) = \sqrt[3]{x}$

1. The domain and the range are the set of all real numbers.
2. The x-intercept of the graph of $f(x) = \sqrt[3]{x}$ is 0. The y-intercept of the graph of $f(x) = \sqrt[3]{x}$ is also 0.
3. The function is odd. The graph is symmetric with respect to the origin.
4. The function is increasing on the interval $(-\infty, \infty)$.
5. The function does not have any local minima or any local maxima.

EXAMPLE 2 **Graphing the Absolute Value Function**

(a) Determine whether $f(x) = |x|$ is even, odd, or neither. State whether the graph of f is symmetric with respect to the y-axis, symmetric with respect to the origin, or neither.

(b) Determine the intercepts, if any, of the graph of $f(x) = |x|$.

(c) Graph $f(x) = |x|$.

Solution (a) Because

$$\begin{aligned} f(-x) &= |-x| \\ &= |x| = f(x) \end{aligned}$$

the function is even. The graph of f is symmetric with respect to the y-axis.

(b) The y-intercept is $f(0) = |0| = 0$. The x-intercept is found by solving the equation $f(x) = 0$, or $|x| = 0$. The x-intercept is 0.

(c) Use the function to form Table 5 and obtain some points on the graph. Because of the symmetry with respect to the y-axis, we only need to find points (x, y) for which $x \geq 0$. Figure 30 shows the graph of $f(x) = |x|$.

Table 5

x	$y = f(x) = \lvert x \rvert$	(x, y)
0	0	(0, 0)
1	1	(1, 1)
2	2	(2, 2)
3	3	(3, 3)

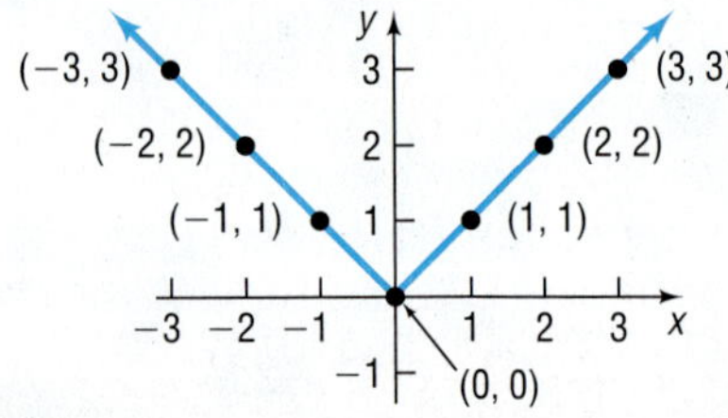

Figure 30 Absolute Value Function

From the results of Example 2 and Figure 30, we have the following properties of the absolute value function.

Properties of $f(x) = |x|$

1. The domain is the set of all real numbers. The range of f is $\{y \mid y \geq 0\}$.
2. The x-intercept of the graph of $f(x) = |x|$ is 0. The y-intercept of the graph of $f(x) = |x|$ is also 0.
3. The function is even. The graph is symmetric with respect to the y-axis.
4. The function is decreasing on the interval $(-\infty, 0)$. It is increasing on the interval $(0, \infty)$.
5. The function has an absolute minimum of 0 at $x = 0$.

Seeing the Concept

Graph $y = |x|$ on a square screen and compare what you see with Figure 30. Note that some graphing calculators use abs(x) for absolute value.

Below is a list of the key functions that we have discussed. In going through this list, pay special attention to the properties of each function, particularly to the shape of each graph. Knowing these graphs, along with key points on each graph, will lay the foundation for further graphing techniques.

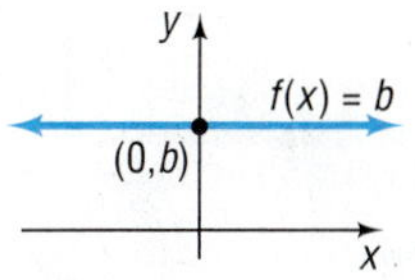

Figure 31 Constant Function

Constant Function

$$f(x) = b \qquad b \text{ is a real number}$$

See Figure 31.

The domain of a **constant function** is the set of all real numbers; its range is the set consisting of a single number b. Its graph is a horizontal line whose y-intercept is b. The constant function is an even function.

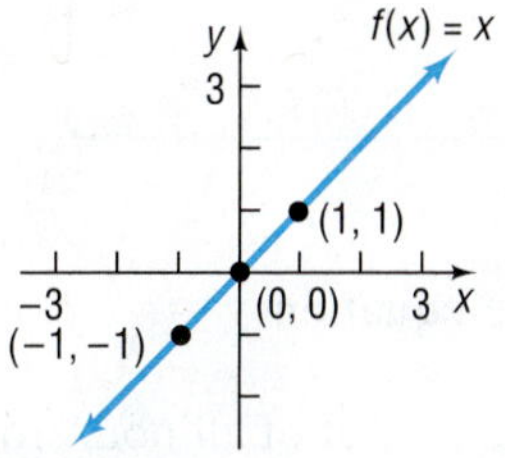

Figure 32 Identity Function

Identity Function

$$f(x) = x$$

See Figure 32.

The domain and the range of the **identity function** are the set of all real numbers. Its graph is a line whose slope is 1 and whose y-intercept is 0. The line consists of all points for which the x-coordinate equals the y-coordinate. The identity function is an odd function that is increasing over its domain. Note that the graph bisects quadrants I and III.

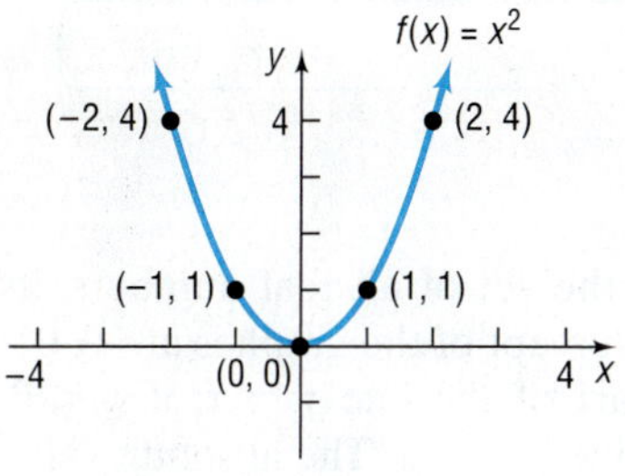

Figure 33 Square Function

Square Function

$$f(x) = x^2$$

See Figure 33.

The domain of the **square function** is the set of all real numbers; its range is the set of nonnegative real numbers. The graph of this function is a parabola whose intercept is at $(0, 0)$. The square function is an even function that is decreasing on the interval $(-\infty, 0)$ and increasing on the interval $(0, \infty)$.

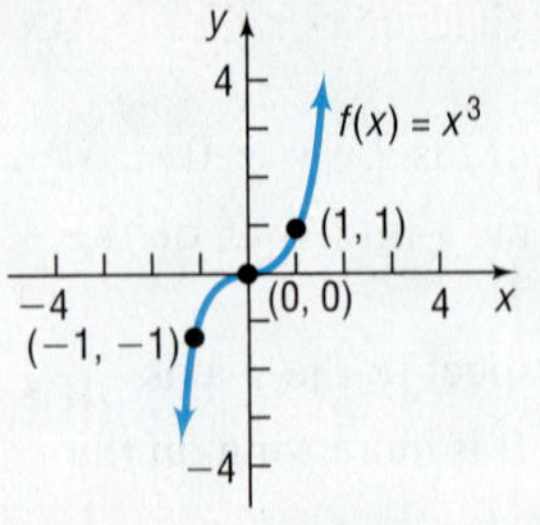

Figure 34 Cube Function

Cube Function

$$f(x) = x^3$$

See Figure 34.

The domain and the range of the **cube function** are the set of all real numbers. The intercept of the graph is at $(0, 0)$. The cube function is odd and is increasing on the interval $(-\infty, \infty)$.

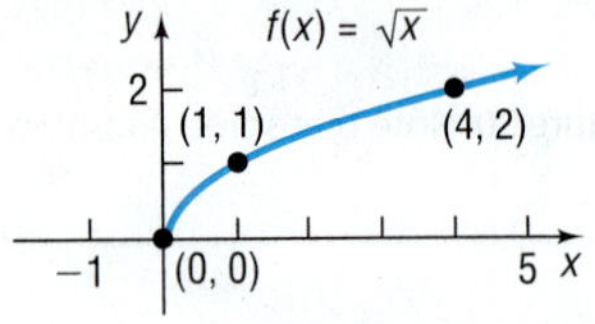

Figure 35 Square Root Function

Square Root Function

$$f(x) = \sqrt{x}$$

See Figure 35.

The domain and the range of the **square root function** are the set of nonnegative real numbers. The intercept of the graph is at $(0, 0)$. The square root function is neither even nor odd and is increasing on the interval $(0, \infty)$.

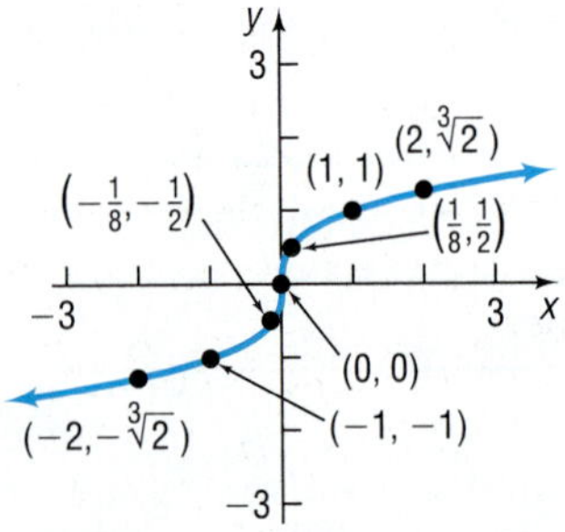

Figure 36 Cube Root Function

Cube Root Function

$$f(x) = \sqrt[3]{x}$$

See Figure 36.

The domain and the range of the **cube root function** are the set of all real numbers. The intercept of the graph is at $(0, 0)$. The cube root function is an odd function that is increasing on the interval $(-\infty, \infty)$.

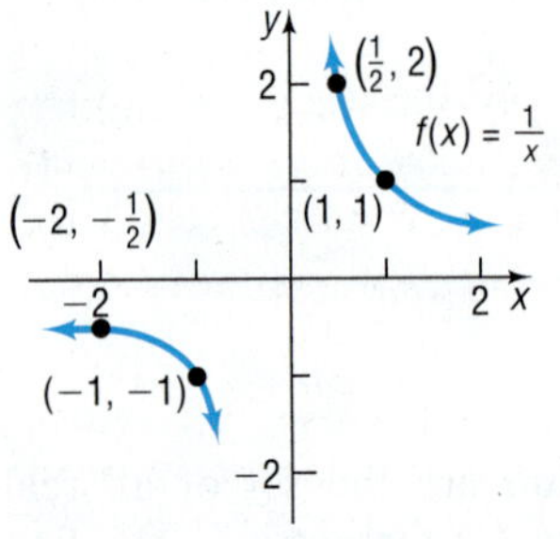

Figure 37 Reciprocal Function

Reciprocal Function

$$f(x) = \frac{1}{x}$$

Refer to Example 12, page 164, for a discussion of the equation $y = \frac{1}{x}$. See Figure 37.

The domain and the range of the **reciprocal function** are the set of all nonzero real numbers. The graph has no intercepts. The reciprocal function is decreasing on the intervals $(-\infty, 0)$ and $(0, \infty)$ and is an odd function.

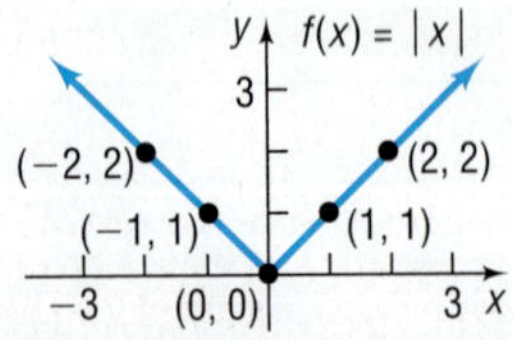

Figure 38 Absolute Value Function

Absolute Value Function

$$f(x) = |x|$$

See Figure 38.

The domain of the **absolute value function** is the set of all real numbers; its range is the set of nonnegative real numbers. The intercept of the graph is at $(0, 0)$. If $x \geq 0$, then $f(x) = x$, and the graph of f is part of the line $y = x$; if $x < 0$, then $f(x) = -x$, and the graph of f is part of the line $y = -x$. The absolute value function is an even function; it is decreasing on the interval $(-\infty, 0)$ and increasing on the interval $(0, \infty)$.

The notation $\text{int}(x)$ stands for the largest integer less than or equal to x. For example,

$$\text{int}(1) = 1, \quad \text{int}(2.5) = 2, \quad \text{int}\left(\frac{1}{2}\right) = 0, \quad \text{int}\left(-\frac{3}{4}\right) = -1, \quad \text{int}(\pi) = 3$$

This type of correspondence occurs frequently enough in mathematics that we give it a name.

DEFINITION Greatest Integer Function

$$f(x) = \text{int}(x)^* = \text{greatest integer less than or equal to } x$$

Table 6

x	$y = f(x) = \text{int}(x)$	(x, y)
-1	-1	$(-1, -1)$
$-\frac{1}{2}$	-1	$\left(-\frac{1}{2}, -1\right)$
$-\frac{1}{4}$	-1	$\left(-\frac{1}{4}, -1\right)$
0	0	$(0, 0)$
$\frac{1}{4}$	0	$\left(\frac{1}{4}, 0\right)$
$\frac{1}{2}$	0	$\left(\frac{1}{2}, 0\right)$
$\frac{3}{4}$	0	$\left(\frac{3}{4}, 0\right)$

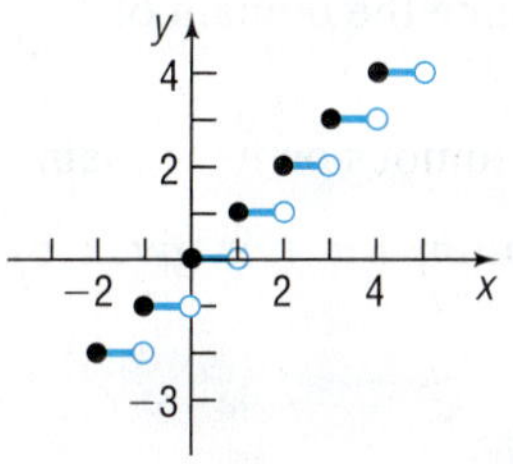

Figure 39 Greatest Integer Function

We obtain the graph of $f(x) = \text{int}(x)$ by plotting several points. See Table 6. For values of x, $-1 \le x < 0$, the value of $f(x) = \text{int}(x)$ is -1; for values of x, $0 \le x < 1$, the value of f is 0. See Figure 39 for the graph.

The domain of the **greatest integer function** is the set of all real numbers; its range is the set of integers. The y-intercept of the graph is 0. The x-intercepts lie in the interval $[0, 1)$. The greatest integer function is neither even nor odd. It is constant on every interval of the form $[k, k + 1)$, for k an integer. In Figure 39, a solid dot is used to indicate, for example, that at $x = 1$ the value of f is $f(1) = 1$; an open circle is used to illustrate that the function does not assume the value of 0 at $x = 1$.

Although a precise definition requires the idea of a limit (discussed in calculus), in a rough sense, a function is said to be **continuous** if its graph has no gaps or holes and can be drawn without lifting a pencil from the paper on which the graph is drawn. We contrast this with a *discontinuous* function. A function is **discontinuous** if its graph has gaps or holes and so cannot be drawn without lifting a pencil from the paper.

From the graph of the greatest integer function, we can see why it is also called a **step function**. At $x = 0$, $x = \pm 1$, $x = \pm 2$, and so on, this function is discontinuous because, at integer values, the graph suddenly "steps" from one value to another without taking on any of the intermediate values. For example, to the immediate left of $x = 3$, the y-coordinates of the points on the graph are 2, and at $x = 3$ and to the immediate right of $x = 3$, the y-coordinates of the points on the graph are 3. Consequently, the graph has gaps in it.

COMMENT When graphing a function using a graphing utility, typically you can choose either **connected mode**, in which points plotted on the screen are connected, making the graph appear without any breaks, or **dot mode**, in which only the points plotted appear. When graphing the greatest integer function with a graphing utility, it may be necessary to be in **dot mode**. This is to prevent the utility from "connecting the dots" when $f(x)$ changes from one integer value to the next. However, some utilities will display the gaps even when in "connected" mode. See Figure 40. ■

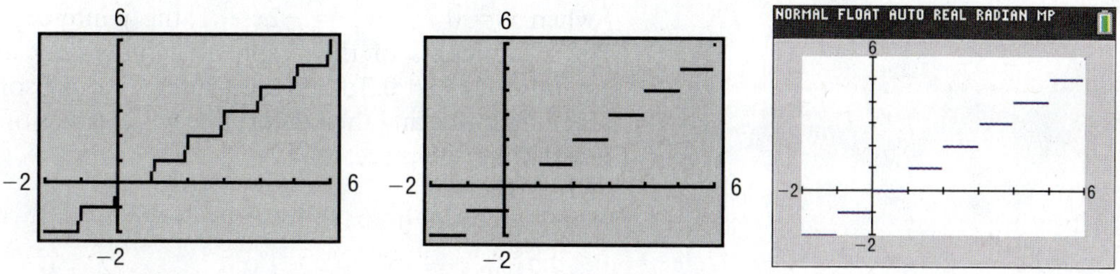

Figure 40 $f(x) = \text{int}(x)$ **(a)** TI-83 Plus, connected mode **(b)** TI-83 Plus, dot mode **(c)** TI-84 Plus C

*Some texts use the notation $f(x) = [x]$ instead of $\text{int}(x)$.

The functions discussed so far are basic. Whenever you encounter one of them, you should see a mental picture of its graph. For example, if you encounter the function $f(x) = x^2$, you should see in your mind's eye a picture like Figure 33.

Now Work PROBLEMS 11 THROUGH 18

2 Graph Piecewise-defined Functions

Sometimes a function is defined using different equations on different parts of its domain. For example, the absolute value function $f(x) = |x|$ is actually defined by two equations: $f(x) = x$ if $x \geq 0$ and $f(x) = -x$ if $x < 0$. For convenience, these equations are generally combined into one expression as

$$f(x) = |x| = \begin{cases} x & \text{if } x \geq 0 \\ -x & \text{if } x < 0 \end{cases}$$

When a function is defined by different equations on different parts of its domain, it is called a **piecewise-defined** function.

EXAMPLE 3 Analyzing a Piecewise-defined Function

The function f is defined as

$$f(x) = \begin{cases} -2x + 1 & \text{if } -3 \leq x < 1 \\ 2 & \text{if } x = 1 \\ x^2 & \text{if } x > 1 \end{cases}$$

(a) Find $f(-2), f(1)$, and $f(2)$.
(b) Determine the domain of f.
(c) Locate any intercepts.
(d) Graph f.
(e) Use the graph to find the range of f.
(f) Is f continuous on its domain?

Solution (a) To find $f(-2)$, observe that when $x = -2$, the equation for f is given by $f(x) = -2x + 1$. So

$$f(-2) = -2(-2) + 1 = 5$$

When $x = 1$, the equation for f is $f(x) = 2$. So,

$$f(1) = 2$$

When $x = 2$, the equation for f is $f(x) = x^2$. So

$$f(2) = 2^2 = 4$$

(b) To find the domain of f, look at its definition. Since f is defined for all x greater than or equal to -3, the domain of f is $\{x | x \geq -3\}$, or the interval $[-3, \infty)$.

(c) The y-intercept of the graph of the function is $f(0)$. Because the equation for f when $x = 0$ is $f(x) = -2x + 1$, the y-intercept is $f(0) = -2(0) + 1 = 1$. The x-intercepts of the graph of a function f are the real solutions to the equation $f(x) = 0$. To find the x-intercepts of f, solve $f(x) = 0$ for each "piece" of the function, and then determine what values of x, if any, satisfy the condition that defines the piece.

$$\begin{aligned} f(x) &= 0 \\ -2x + 1 &= 0 \quad -3 \leq x < 1 \\ -2x &= -1 \\ x &= \frac{1}{2} \end{aligned} \qquad \begin{aligned} f(x) &= 0 \\ 2 &= 0 \quad x = 1 \\ &\text{No solution} \end{aligned} \qquad \begin{aligned} f(x) &= 0 \\ x^2 &= 0 \quad x > 1 \\ x &= 0 \end{aligned}$$

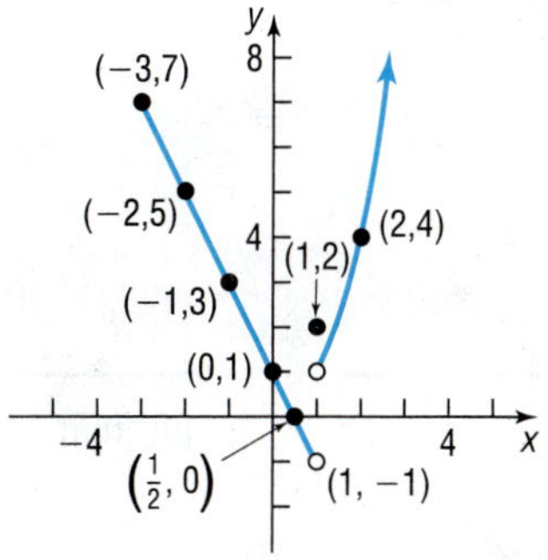

Figure 41

The first potential x-intercept, $x = \frac{1}{2}$, satisfies the condition $-3 \le x < 1$, so $x = \frac{1}{2}$ is an x-intercept. The second potential x-intercept, $x = 0$, does not satisfy the condition $x > 1$, so $x = 0$ is not an x-intercept. The only x-intercept is $\frac{1}{2}$. The intercepts are $(0, 1)$ and $\left(\frac{1}{2}, 0\right)$.

(d) To graph f, graph each "piece." First graph the line $y = -2x + 1$ and keep only the part for which $-3 \le x < 1$. Then plot the point $(1, 2)$ because, when $x = 1$, $f(x) = 2$. Finally, graph the parabola $y = x^2$ and keep only the part for which $x > 1$. See Figure 41.

(e) From the graph, we conclude that the range of f is $\{y \mid y > -1\}$, or the interval $(-1, \infty)$.

(f) The function f is not continuous because there is a "jump" in the graph at $x = 1$.

Now Work PROBLEM 31

EXAMPLE 4 **Cost of Electricity**

In the spring of 2014, Duke Energy Progress supplied electricity to residences in South Carolina for a monthly customer charge of \$6.50 plus 9.971¢ per kilowatt-hour (kWh) for the first 800 kWh supplied in the month and 8.971¢ per kWh for all usage over 800 kWh in the month.

(a) What is the charge for using 300 kWh in a month?

(b) What is the charge for using 1500 kWh in a month?

(c) If C is the monthly charge for x kWh, develop a model relating the monthly charge and kilowatt-hours used. That is, express C as a function of x.

Source: Duke Energy Progress, 2014

Solution (a) For 300 kWh, the charge is \$6.50 plus (9.971¢ = \$0.09971) per kWh. That is,

$$\text{Charge} = \$6.50 + \$0.09971(300) = \$36.41$$

(b) For 1500 kWh, the charge is \$6.50 plus 9.971¢ per kWh for the first 800 kWh plus 8.971¢ per kWh for the 700 in excess of 800. That is,

$$\text{Charge} = \$6.50 + \$0.09971(800) + \$0.08971\,(700) = \$\,149.07$$

(c) Let x represent the number of kilowatt-hours used. If $0 \le x \le 800$, then the monthly charge C (in dollars) can be found by multiplying x times \$0.09971 and adding the monthly customer charge of \$6.50. So if $0 \le x \le 800$, then

$$C(x) = 0.09971x + 6.50$$

For $x > 800$, the charge is $0.09971(800) + 6.50 + 0.08971(x - 800)$, since $(x - 800)$ equals the usage in excess of 800 kWh, which costs \$0.08971 per kWh. That is, if $x > 800$, then

$$\begin{aligned} C(x) &= 0.09971(800) + 6.50 + 0.08971(x - 800) \\ &= 79.768 + 6.50 + 0.08971x - 71.768 \\ &= 0.08971x + 14.50 \end{aligned}$$

The rule for computing C follows two equations:

$$C(x) = \begin{cases} 0.09971x + 6.50 & \text{if } 0 \le x \le 800 \\ 0.08971x + 14.50 & \text{if } x > 800 \end{cases}$$ The Model

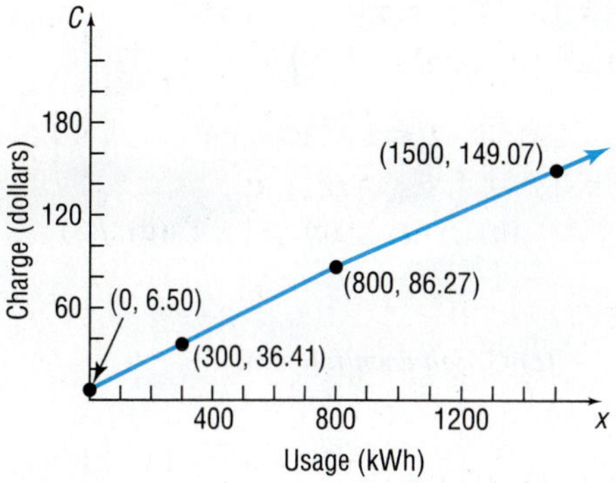

Figure 42

See Figure 42 for the graph. Note that the two "pieces" are linear, but they have different slopes (rates), and meet at the point (800, 86.27).

3.4 Assess Your Understanding

'Are You Prepared?' *Answers are given at the end of these exercises. If you get a wrong answer, read the pages listed in red.*

1. Sketch the graph of $y = \sqrt{x}$. (p. 163)

2. Sketch the graph of $y = \frac{1}{x}$. (p. 164)

3. List the intercepts of the equation $y = x^3 - 8$. (pp. 159–160)

Concepts and Vocabulary

4. The function $f(x) = x^2$ is decreasing on the interval ________.

5. When functions are defined by more than one equation, they are called ______________ functions.

6. ***True or False*** The cube function is odd and is increasing on the interval $(-\infty, \infty)$.

7. ***True or False*** The cube root function is odd and is decreasing on the interval $(-\infty, \infty)$.

8. ***True or False*** The domain and the range of the reciprocal function are the set of all real numbers.

9. Which of the following functions has a graph that is symmetric about the y-axis?
(a) $y = \sqrt{x}$ (b) $y = |x|$ (c) $y = x^3$ (d) $y = \frac{1}{x}$

10. Consider the following function.

$$f(x) = \begin{cases} 3x - 2 & \text{if} \quad x < 2 \\ x^2 + 5 & \text{if} \quad 2 \le x < 10 \\ 3 & \text{if} \quad x \ge 10 \end{cases}$$

Which "piece(s)" should be used to find the y-intercept?
(a) $3x - 2$ (b) $x^2 + 5$ (c) 3 (d) all three

Skill Building

In Problems 11–18, match each graph to its function.

A. Constant function | *B. Identity function* | *C. Square function* | *D. Cube function*
E. Square root function | *F. Reciprocal function* | *G. Absolute value function* | *H. Cube root function*

11. 

12.

13. 

14. 

15. 

16. 

17. 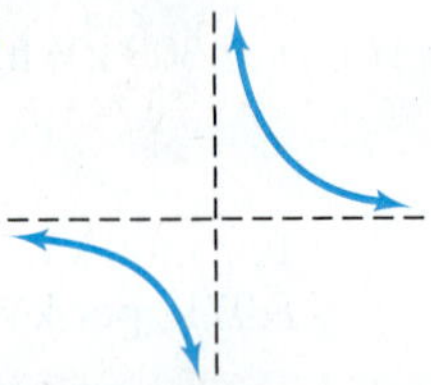

18. 

In Problems 19–26, sketch the graph of each function. Be sure to label three points on the graph.

19. $f(x) = x$ **20.** $f(x) = x^2$ **21.** $f(x) = x^3$ **22.** $f(x) = \sqrt{x}$

23. $f(x) = \frac{1}{x}$ **24.** $f(x) = |x|$ **25.** $f(x) = \sqrt[3]{x}$ **26.** $f(x) = 3$

27. If $f(x) = \begin{cases} x^2 & \text{if } x < 0 \\ 2 & \text{if } x = 0 \\ 2x + 1 & \text{if } x > 0 \end{cases}$

find: (a) $f(-2)$ (b) $f(0)$ (c) $f(2)$

28. If $f(x) = \begin{cases} -3x & \text{if } x < -1 \\ 0 & \text{if } x = -1 \\ 2x^2 + 1 & \text{if } x > -1 \end{cases}$

find: (a) $f(-2)$ (b) $f(-1)$ (c) $f(0)$

29. If $f(x) = \begin{cases} 2x - 4 & \text{if } -1 \le x \le 2 \\ x^3 - 2 & \text{if } 2 < x \le 3 \end{cases}$

find: (a) $f(0)$ (b) $f(1)$ (c) $f(2)$ (d) $f(3)$

30. If $f(x) = \begin{cases} x^3 & \text{if } -2 \le x < 1 \\ 3x + 2 & \text{if } 1 \le x \le 4 \end{cases}$

find: (a) $f(-1)$ (b) $f(0)$ (c) $f(1)$ (d) $f(3)$

In Problems 31–42:

(a) Find the domain of each function. *(b) Locate any intercepts.* *(c) Graph each function.*
(d) Based on the graph, find the range. *(e) Is f continuous on its domain?*

31. $f(x) = \begin{cases} 2x & \text{if } x \ne 0 \\ 1 & \text{if } x = 0 \end{cases}$

32. $f(x) = \begin{cases} 3x & \text{if } x \ne 0 \\ 4 & \text{if } x = 0 \end{cases}$

33. $f(x) = \begin{cases} -2x + 3 & \text{if } x < 1 \\ 3x - 2 & \text{if } x \ge 1 \end{cases}$

34. $f(x) = \begin{cases} x + 3 & \text{if } x < -2 \\ -2x - 3 & \text{if } x \geq -2 \end{cases}$

35. $f(x) = \begin{cases} x + 3 & \text{if } -2 \leq x < 1 \\ 5 & \text{if } x = 1 \\ -x + 2 & \text{if } x > 1 \end{cases}$

36. $f(x) = \begin{cases} 2x + 5 & \text{if } -3 \leq x < 0 \\ -3 & \text{if } x = 0 \\ -5x & \text{if } x > 0 \end{cases}$

37. $f(x) = \begin{cases} 1 + x & \text{if } x < 0 \\ x^2 & \text{if } x \geq 0 \end{cases}$

38. $f(x) = \begin{cases} \dfrac{1}{x} & \text{if } x < 0 \\ \sqrt[3]{x} & \text{if } x \geq 0 \end{cases}$

39. $f(x) = \begin{cases} |x| & \text{if } -2 \leq x < 0 \\ x^3 & \text{if } x > 0 \end{cases}$

40. $f(x) = \begin{cases} 2 - x & \text{if } -3 \leq x < 1 \\ \sqrt{x} & \text{if } x > 1 \end{cases}$

41. $f(x) = 2\,\text{int}(x)$

42. $f(x) = \text{int}(2x)$

In Problems 43–46, the graph of a piecewise-defined function is given. Write a definition for each function.

43.

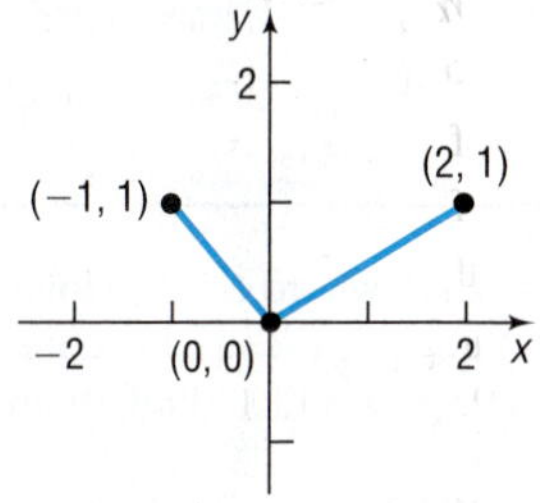

44.

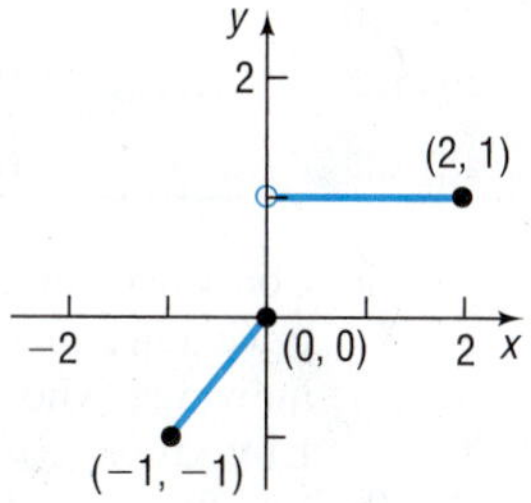

45.

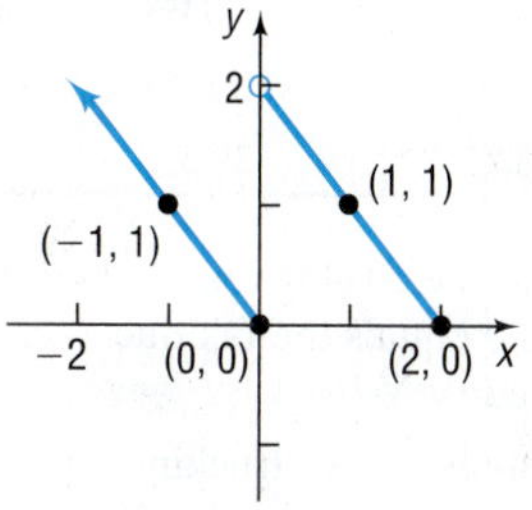

46.

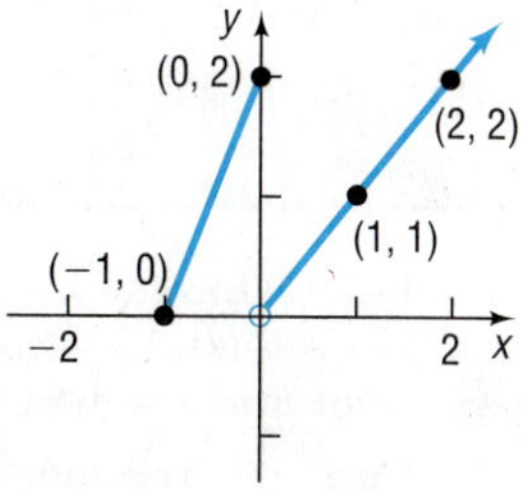

47. If $f(x) = \text{int}(2x)$, find

(a) $f(1.2)$ (b) $f(1.6)$ (c) $f(-1.8)$

48. If $f(x) = \text{int}\left(\dfrac{x}{2}\right)$, find

(a) $f(1.2)$ (b) $f(1.6)$ (c) $f(-1.8)$

Applications and Extensions

49. Tablet Service Sprint offers a monthly tablet plan for \$34.99. It includes 3 gigabytes of data and charges \$15 per gigabyte for additional gigabytes. The following function is used to compute the monthly cost for a subscriber.

$$C(x) = \begin{cases} 34.99 & \text{if } 0 \leq x \leq 3 \\ 15x - 10.01 & \text{if } x > 3 \end{cases}$$

Compute the monthly cost for each of the following gigabytes of use.

(a) 2 (b) 5 (c) 13

Source: Sprint, April 2014

50. Parking at O'Hare International Airport The short-term (no more than 24 hours) parking fee F (in dollars) for parking x hours on a weekday at O'Hare International Airport's main parking garage can be modeled by the function

$$F(x) = \begin{cases} 3 & \text{if } 0 < x \leq 3 \\ 5\,\text{int}(x + 1) + 1 & \text{if } 3 < x < 9 \\ 50 & \text{if } 9 \leq x \leq 24 \end{cases}$$

Determine the fee for parking in the short-term parking garage for

(a) 2 hours (b) 7 hours

(c) 15 hours (d) 8 hours and 24 minutes

Source: O'Hare International Airport

51. Cost of Natural Gas In March 2014, Laclede Gas had the rate schedule (on, right) for natural gas usage in single-family residences.

(a) What is the charge for using 20 therms in a month?

(b) What is the charge for using 150 therms in a month?

(c) Develop a function that models the monthly charge C for x therms of gas.

(d) Graph the function found in part (c).

Monthly service charge	\$19.50
Delivery charge	
First 30 therms	\$0.91686/therm
Over 30 therms	\$0
Natural gas cost	
First 30 therms	\$0.3313/therm
Over 30 therms	\$0.5757/therm

Source: Laclede Gas

52. Cost of Natural Gas In April 2014, Nicor Gas had the following rate schedule for natural gas usage in small businesses.

Monthly customer charge	\$72.60
Distribution charge	
1st 150 therms	\$0.1201/therm
Next 4850 therms	\$0.0549/therm
Over 5000 therms	\$0.0482/therm
Gas supply charge	\$0.68/therm

(a) What is the charge for using 1000 therms in a month?

(b) What is the charge for using 6000 therms in a month?

(c) Develop a function that models the monthly charge C for x therms of gas.

(d) Graph the function found in part (c).

Source: Nicor Gas, 2014

53. Federal Income Tax Two 2014 Tax Rate Schedules are given in the accompanying table. If x equals taxable income and y equals the tax due, construct a function $y = f(x)$ for Schedule X.

2014 Tax Rate Schedules											
Schedule X—Single						Schedule Y-1—Married Filing Jointly or Qualified Widow(er)					
If Taxable Income is Over	But Not Over	The Tax is This Amount		Plus This %	Of the Excess Over	If Taxable Income is Over	But Not Over	The Tax is This Amount		Plus This %	Of the Excess Over
\$0	\$9,075	\$0	+	10%	\$0	\$0	\$18,150	\$0	+	10%	\$0
9,075	36,900	907.50	+	15%	9,075	18,150	73,800	1,815	+	15%	18,150
36,900	89,350	5,081.25	+	25%	36,900	73,800	148,850	10,162.50	+	25%	73,800
89,350	186,350	18,193.75	+	28%	89,350	148,850	226,850	28,925.00	+	28%	148,850
186,350	405,100	45,353.75	+	33%	186,350	226,850	405,100	50,765.00	+	33%	226,850
405,100	406,750	117,541.25	+	35%	405,100	405,100	457,600	109,587.50	+	35%	405,100
406,750	—	118,188.75	+	39.6%	406,750	457,600	—	127,962.50	+	39.6%	457,600

54. Federal Income Tax Refer to the 2014 tax rate schedules. If x equals taxable income and y equals the tax due, construct a function $y = f(x)$ for Schedule Y-1.

55. Cost of Transporting Goods A trucking company transports goods between Chicago and New York, a distance of 960 miles. The company's policy is to charge, for each pound, \$0.50 per mile for the first 100 miles, \$0.40 per mile for the next 300 miles, \$0.25 per mile for the next 400 miles, and no charge for the remaining 160 miles.

(a) Graph the relationship between the cost of transportation in dollars and mileage over the entire 960-mile route.

(b) Find the cost as a function of mileage for hauls between 100 and 400 miles from Chicago.

(c) Find the cost as a function of mileage for hauls between 400 and 800 miles from Chicago.

56. Car Rental Costs An economy car rented in Florida from Enterprise® on a weekly basis costs \$185 per week. Extra days cost \$37 per day until the day rate exceeds the weekly rate, in which case the weekly rate applies. Also, any part of a day used counts as a full day. Find the cost C of renting an economy car as a function of the number x of days used, where $7 \le x \le 14$. Graph this function.

57. Mortgage Fees Fannie Mae charges a loan-level price adjustment (LLPA) on all mortgages, which represents a fee homebuyers seeking a loan must pay. The rate paid depends on the credit score of the borrower, the amount borrowed, and the loan-to-value (LTV) ratio. The LTV ratio is the ratio of amount borrowed to appraised value of the home. For example, a homebuyer who wishes to borrow \$250,000 with a credit score of 730 and an LTV ratio of 80% will pay 0.5% (0.005) of \$250,000, or \$1250. The table shows the LLPA for various credit scores and an LTV ratio of 80%.

Credit Score	Loan-Level Price Adjustment Rate
≤659	3.00%
660–679	2.50%
680–699	1.75%
700–719	1%
720–739	0.5%
≥740	0.25%

Source: Fannie Mae.

(a) Construct a function $C = C(s)$, where C is the loan-level price adjustment (LLPA) and s is the credit score of an individual who wishes to borrow \$300,000 with an 80% LTV ratio.

(b) What is the LLPA on a \$300,000 loan with an 80% LTV ratio for a borrower whose credit score is 725?

(c) What is the LLPA on a \$300,000 loan with an 80% LTV ratio for a borrower whose credit score is 670?

58. Minimum Payments for Credit Cards Holders of credit cards issued by banks, department stores, oil companies, and so on receive bills each month that state minimum amounts that must be paid by a certain due date. The minimum due depends on the total amount owed. One such credit card company uses the following rules: For a bill of less than \$10, the entire amount is due. For a bill of at least \$10 but less than \$500, the minimum due is \$10. A minimum of \$30 is due on a bill of at least \$500 but less than \$1000, a minimum of \$50 is due on a bill of at least \$1000 but less than \$1500, and a minimum of \$70 is due on bills of \$1500 or more. Find the function f that describes the minimum payment due on a bill of x dollars. Graph f.

59. Wind Chill The wind chill factor represents the air temperature at a standard wind speed that would produce the same heat loss as the given temperature and wind speed. One formula for computing the equivalent temperature is

$$W = \begin{cases} t & 0 \le v < 1.79 \\ 33 - \dfrac{(10.45 + 10\sqrt{v} - v)(33 - t)}{22.04} & 1.79 \le v \le 20 \\ 33 - 1.5958(33 - t) & v > 20 \end{cases}$$

where v represents the wind speed (in meters per second) and t represents the air temperature (°C). Compute the wind chill for the following:

(a) An air temperature of 10°C and a wind speed of 1 meter per second (m/sec)

(b) An air temperature of 10°C and a wind speed of 5 m/sec

(c) An air temperature of 10°C and a wind speed of 15 m/sec

(d) An air temperature of 10°C and a wind speed of 25 m/sec

(e) Explain the physical meaning of the equation corresponding to $0 \le v < 1.79$.

(f) Explain the physical meaning of the equation corresponding to $v > 20$.

60. Wind Chill Redo Problem 59(a)–(d) for an air temperature of $-10°C$.

61. First-class Mail In 2014 the U.S. Postal Service charged \$0.98 postage for first-class mail retail flats (such as an 8.5" by 11" envelope) weighing up to 1 ounce, plus \$0.21 for each additional ounce up to 13 ounces. First-class rates do not apply to flats weighing more than 13 ounces. Develop a model that relates C, the first-class postage charged, for a flat weighing x ounces. Graph the function.

Source: United States Postal Service

Explaining Concepts: Discussion and Writing

In Problems 62–69, use a graphing utility.

62. Exploration Graph $y = x^2$. Then on the same screen graph $y = x^2 + 2$, followed by $y = x^2 + 4$, followed by $y = x^2 - 2$. What pattern do you observe? Can you predict the graph of $y = x^2 - 4$? Of $y = x^2 + 5$?

63. Exploration Graph $y = x^2$. Then on the same screen graph $y = (x - 2)^2$, followed by $y = (x - 4)^2$, followed by $y = (x + 2)^2$. What pattern do you observe? Can you predict the graph of $y = (x + 4)^2$? Of $y = (x - 5)^2$?

64. Exploration Graph $y = |x|$. Then on the same screen graph $y = 2|x|$, followed by $y = 4|x|$, followed by $y = \frac{1}{2}|x|$. What pattern do you observe? Can you predict the graph of $y = \frac{1}{4}|x|$? Of $y = 5|x|$?

65. Exploration Graph $y = x^2$. Then on the same screen graph $y = -x^2$. Now try $y = |x|$ and $y = -|x|$. What do you conclude?

66. Exploration Graph $y = \sqrt{x}$. Then on the same screen graph $y = \sqrt{-x}$. Now try $y = 2x + 1$ and $y = 2(-x) + 1$. What do you conclude?

67. Exploration Graph $y = x^3$. Then on the same screen graph $y = (x - 1)^3 + 2$. Could you have predicted the result?

68. Exploration Graph $y = x^2$, $y = x^4$, and $y = x^6$ on the same screen. What do you notice is the same about each graph? What do you notice is different?

69. Exploration Graph $y = x^3$, $y = x^5$, and $y = x^7$ on the same screen. What do you notice is the same about each graph? What do you notice is different?

70. Consider the equation

$$y = \begin{cases} 1 & \text{if } x \text{ is rational} \\ 0 & \text{if } x \text{ is irrational} \end{cases}$$

Is this a function? What is its domain? What is its range? What is its y-intercept, if any? What are its x-intercepts, if any? Is it even, odd, or neither? How would you describe its graph?

71. Define some functions that pass through $(0, 0)$ and $(1, 1)$ and are increasing for $x \geq 0$. Begin your list with $y = \sqrt{x}$, $y = x$, and $y = x^2$. Can you propose a general result about such functions?

Retain Your Knowledge

Problems 72–75 are based on material learned earlier in the course. The purpose of these problems is to keep the material fresh in your mind so that you are better prepared for the final exam.

72. Simplify: $(3 + 2i)(4 - 5i)$

73. Find the center and radius of the circle $x^2 + y^2 = 6y + 16$.

74. Solve: $4x - 5(2x - 1) = 4 - 7(x + 1)$

75. Ethan has \$60,000 to invest. He puts part of the money in a CD that earns 3% simple interest per year and the rest in a mutual fund that earns 8% simple interest per year. How much did he invest in each if his earned interest the first year was \$3700?

'Are You Prepared?' Answers

1.

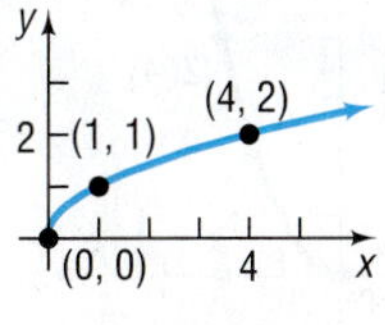

2.

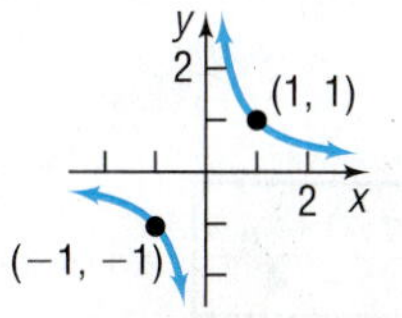

3. $(0, -8), (2, 0)$

3.5 Graphing Techniques: Transformations

OBJECTIVES
1 Graph Functions Using Vertical and Horizontal Shifts (p. 248)
2 Graph Functions Using Compressions and Stretches (p. 251)
3 Graph Functions Using Reflections about the x-Axis and the y-Axis (p. 253)

At this stage, if you were asked to graph any of the functions defined by $y = x$, $y = x^2$, $y = x^3$, $y = \sqrt{x}$, $y = \sqrt[3]{x}$, $y = \frac{1}{x}$, or $y = |x|$, your response should be, "Yes, I recognize these functions and know the general shapes of their graphs." (If this is not your answer, review the previous section, Figures 32 through 38.)

Sometimes we are asked to graph a function that is "almost" like one that we already know how to graph. In this section, we develop techniques for graphing such functions. Collectively, these techniques are referred to as **transformations**.

1 Graph Functions Using Vertical and Horizontal Shifts

EXAMPLE 1 Vertical Shift Up

Use the graph of $f(x) = x^2$ to obtain the graph of $g(x) = x^2 + 3$. Find the domain and range of g.

Solution Begin by obtaining some points on the graphs of f and g. For example, when $x = 0$, then $y = f(0) = 0$ and $y = g(0) = 3$. When $x = 1$, then $y = f(1) = 1$ and $y = g(1) = 4$. Table 7 lists these and a few other points on each graph. Notice that each y-coordinate of a point on the graph of g is 3 units larger than the y-coordinate of the corresponding point on the graph of f. We conclude that the graph of g is identical to that of f, except that it is shifted vertically up 3 units. See Figure 43.

Table 7

x	$y = f(x) = x^2$	$y = g(x) = x^2 + 3$
−2	4	7
−1	1	4
0	0	3
1	1	4
2	4	7

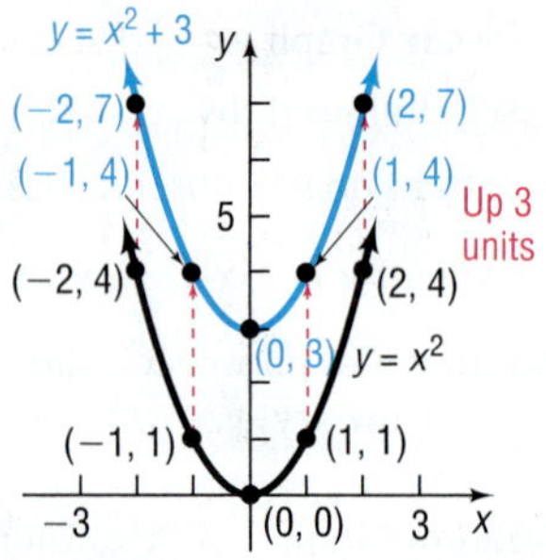

Figure 43

The domain of g is all real numbers, or $(-\infty, \infty)$. The range of g is $[3, \infty)$.

EXAMPLE 2 Vertical Shift Down

Use the graph of $f(x) = x^2$ to obtain the graph of $g(x) = x^2 - 4$. Find the domain and range of g.

Solution Table 8 lists some points on the graphs of f and g. Notice that each y-coordinate of g is 4 units less than the corresponding y-coordinate of f.

To obtain the graph of g from the graph of f, subtract 4 from each y-coordinate on the graph of f. The graph of g is identical to that of f, except that it is shifted down 4 units. See Figure 44.

Table 8

x	$y = f(x) = x^2$	$y = g(x) = x^2 - 4$
−2	4	0
−1	1	−3
0	0	−4
1	1	−3
2	4	0

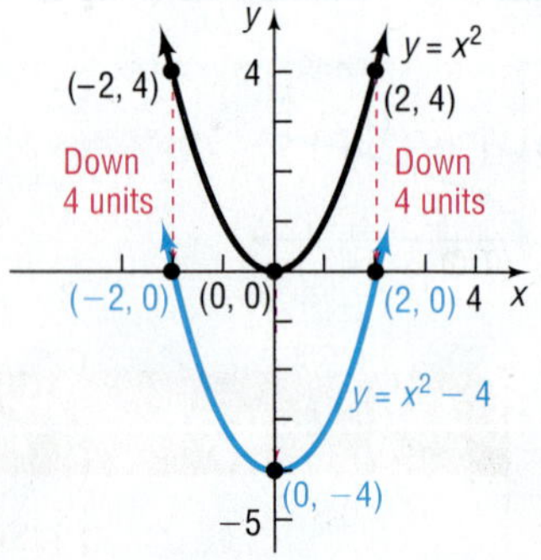

Figure 44

The domain of g is all real numbers, or $(-\infty, \infty)$. The range of g is $[-4, \infty)$.

Note that a vertical shift affects only the range of a function, not the domain. For example, the range of $f(x) = x^2$ is $[0, \infty)$. In Example 1 the range of g is $[3, \infty)$, whereas in Example 2 the range of g is $[-4, \infty)$. The domain for all three functions is all real numbers.

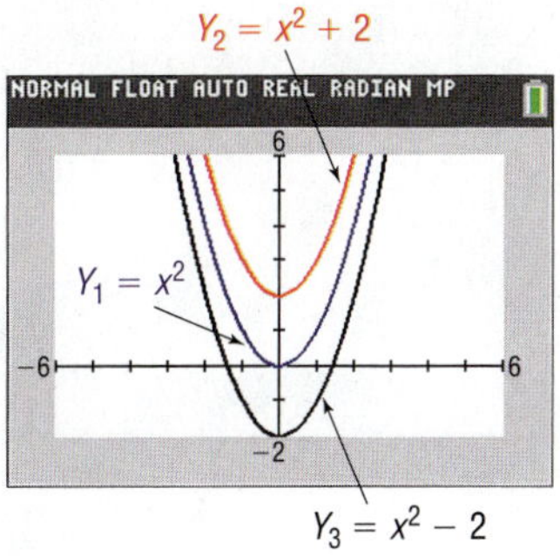

Figure 45

Exploration On the same screen, graph each of the following functions:

$$Y_1 = x^2$$
$$Y_2 = x^2 + 2$$
$$Y_3 = x^2 - 2$$

Figure 45 illustrates the graphs. You should have observed a general pattern. With $Y_1 = x^2$ on the screen, the graph of $Y_2 = x^2 + 2$ is identical to that of $Y_1 = x^2$, except that it is shifted vertically up 2 units. The graph of $Y_3 = x^2 - 2$ is identical to that of $Y_1 = x^2$, except that it is shifted vertically down 2 units.

We are led to the following conclusions:

> **In Words**
> For $y = f(x) + k, k > 0$, add k to each y-coordinate on the graph of $y = f(x)$ to shift the graph up k units.
> For $y = f(x) - k, k > 0$, subtract k from each y-coordinate to shift the graph down k units.

If a positive real number k is added to the output of a function $y = f(x)$, the graph of the new function $y = f(x) + k$ is the graph of f **shifted vertically up** k units.

If a positive real number k is subtracted from the output of a function $y = f(x)$, the graph of the new function $y = f(x) - k$ is the graph of f **shifted vertically down** k units.

Now Work PROBLEM 39

EXAMPLE 3

Horizontal Shift to the Right

Use the graph of $f(x) = \sqrt{x}$ to obtain the graph of $g(x) = \sqrt{x - 2}$. Find the domain and range of g.

Solution The function $g(x) = \sqrt{x - 2}$ is basically a square root function. Table 9 lists some points on the graphs of f and g. Note that when $f(x) = 0$, then $x = 0$, and when $g(x) = 0$, then $x = 2$. Also, when $f(x) = 2$, then $x = 4$, and when $g(x) = 2$, then $x = 6$. Notice that the x-coordinates on the graph of g are 2 units larger than the corresponding x-coordinates on the graph of f for any given y-coordinate. We conclude that the graph of g is identical to that of f, except that it is shifted horizontally 2 units to the right. See Figure 46.

Table 9

x	$y = f(x) = \sqrt{x}$	x	$y = g(x) = \sqrt{x - 2}$
0	0	2	0
1	1	3	1
4	2	6	2
9	3	11	3

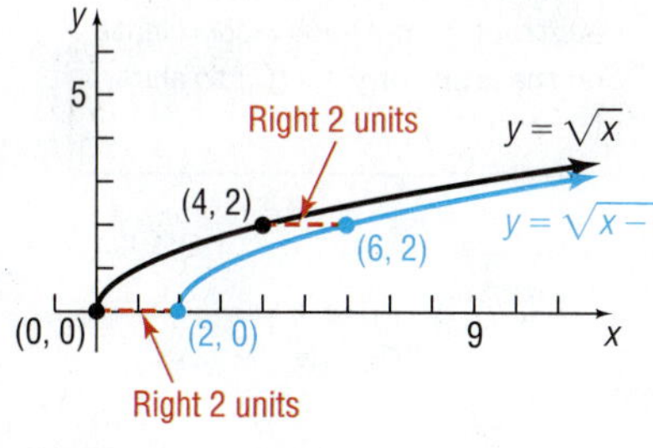

Figure 46

The domain of g is $[2, \infty)$ and the range is $[0, \infty)$. ●

EXAMPLE 4

Horizontal Shift to the Left

Use the graph of $f(x) = \sqrt{x}$ to obtain the graph of $g(x) = \sqrt{x + 4}$. Find the domain and range of g.

Solution The function $g(x) = \sqrt{x + 4}$ is basically a square root function. Its graph is the same as that of f, except that it is shifted horizontally 4 units to the left. See Figure 47 on page 250.

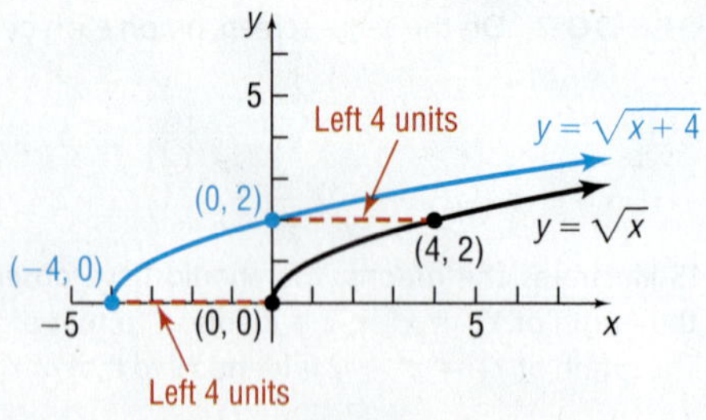

Figure 47

The domain of g is $[-4, \infty)$ and the range is $[0, \infty)$.

Now Work PROBLEM 43

Note that a horizontal shift affects only the domain of a function, not the range. For example, the domain of $f(x) = \sqrt{x}$ is $[0, \infty)$. In Example 3 the domain of g is $[2, \infty)$, whereas in Example 4 the domain of g is $[-4, \infty)$. The range for all three functions is $[0, \infty)$.

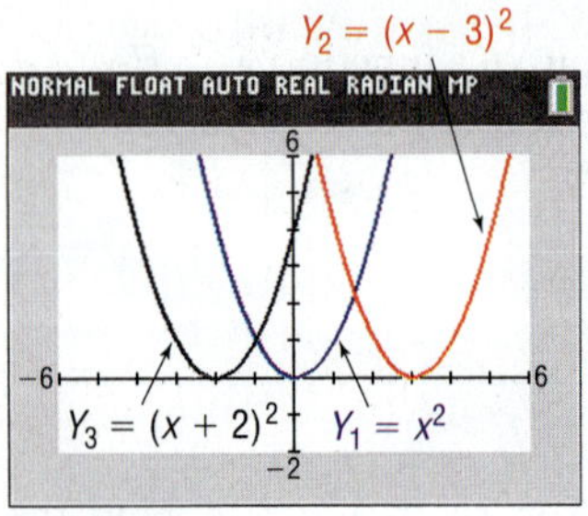

Figure 48

Exploration On the same screen, graph each of the following functions:

$$Y_1 = x^2$$
$$Y_2 = (x - 3)^2$$
$$Y_3 = (x + 2)^2$$

Figure 48 illustrates the graphs.

You should have observed the following pattern. With the graph of $Y_1 = x^2$ on the screen, the graph of $Y_2 = (x - 3)^2$ is identical to that of $Y_1 = x^2$, except that it is shifted horizontally to the right 3 units. The graph of $Y_3 = (x + 2)^2$ is identical to that of $Y_1 = x^2$, except that it is shifted horizontally to the left 2 units.

We are led to the following conclusions:

In Words

For $y = f(x - h)$, $h > 0$, add h to each x-coordinate on the graph of $y = f(x)$ to shift the graph right h units. For $y = f(x + h)$, $h > 0$, subtract h from each x-coordinate on the graph of $y = f(x)$ to shift the graph left h units.

If the argument x of a function f is replaced by $x - h$, $h > 0$, the graph of the new function $y = f(x - h)$ is the graph of f **shifted horizontally right** h units.

If the argument x of a function f is replaced by $x + h$, $h > 0$, the graph of the new function $y = f(x + h)$ is the graph of f **shifted horizontally left** h units.

Observe the distinction between vertical and horizontal shifts. The graph of $f(x) = x^3 + 2$ is obtained by shifting the graph of $y = x^3$ *up* 2 units, because we evaluate the cube function first and then add 2. The graph of $g(x) = (x + 2)^3$ is obtained by shifting the graph of $y = x^3$ *left* 2 units, because we add 2 to x before we evaluate the cube function.

Vertical and horizontal shifts are sometimes combined.

EXAMPLE 5 **Combining Vertical and Horizontal Shifts**

Graph the function $f(x) = |x + 3| - 5$. Find the domain and range of f.

Solution We graph f in steps. First, note that the rule for f is basically an absolute value function, so begin with the graph of $y = |x|$ as shown in Figure 49(a). Next, to get the graph of $y = |x + 3|$, shift the graph of $y = |x|$ horizontally 3 units to the left. See Figure 49(b). Finally, to get the graph of $y = |x + 3| - 5$, shift the graph of $y = |x + 3|$ vertically down 5 units. See Figure 49(c). Note the points plotted on each graph. Using key points can be helpful in keeping track of the transformation that has taken place.

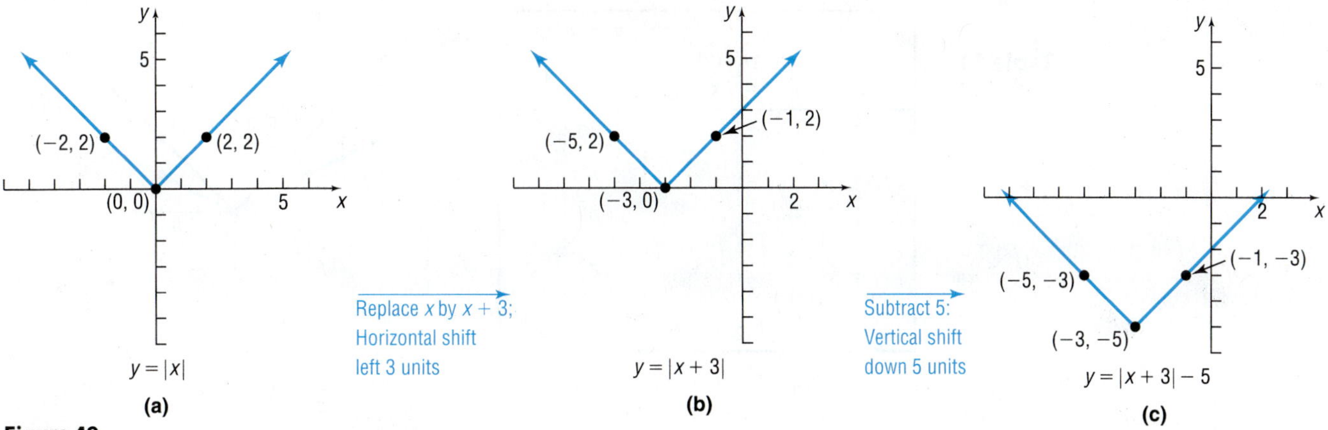

Figure 49

The domain of f is all real numbers, or $(-\infty, \infty)$. The range of f is $[-5, \infty)$. ●

Check: Graph $Y_1 = f(x) = |x + 3| - 5$ and compare the graph to Figure 49(c).

In Example 5, if the vertical shift had been done first, followed by the horizontal shift, the final graph would have been the same. Try it for yourself.

Now Work PROBLEMS 45 AND 69

2 Graph Functions Using Compressions and Stretches

EXAMPLE 6 **Vertical Stretch**

Use the graph of $f(x) = \sqrt{x}$ to obtain the graph of $g(x) = 2\sqrt{x}$.

Solution To see the relationship between the graphs of f and g, we form Table 10, listing points on each graph. For each x, the y-coordinate of a point on the graph of g is 2 times as large as the corresponding y-coordinate on the graph of f. The graph of $f(x) = \sqrt{x}$ is vertically stretched by a factor of 2 to obtain the graph of $g(x) = 2\sqrt{x}$. For example, $(1, 1)$ is on the graph of f, but $(1, 2)$ is on the graph of g. See Figure 50.

Table 10

x	$y = f(x) = \sqrt{x}$	$y = g(x) = 2\sqrt{x}$
0	0	0
1	1	2
4	2	4
9	3	6

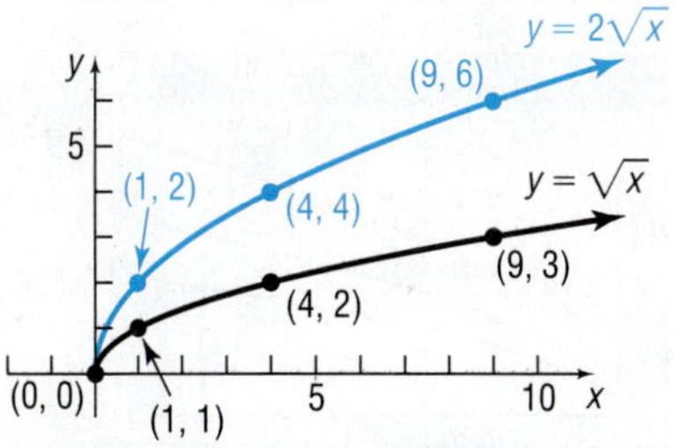

Figure 50 ●

EXAMPLE 7 **Vertical Compression**

Use the graph of $f(x) = |x|$ to obtain the graph of $g(x) = \frac{1}{2}|x|$.

Solution For each x, the y-coordinate of a point on the graph of g is $\frac{1}{2}$ as large as the corresponding y-coordinate on the graph of f. The graph of $f(x) = |x|$ is vertically compressed by a factor of $\frac{1}{2}$ to obtain the graph of $g(x) = \frac{1}{2}|x|$. For example, $(2, 2)$ is on the graph of f, but $(2, 1)$ is on the graph of g. See Table 11 and Figure 51 on page 252.

Table 11

x	$y = f(x) = \lvert x \rvert$	$y = g(x) = \frac{1}{2}\lvert x \rvert$
−2	2	1
−1	1	$\frac{1}{2}$
0	0	0
1	1	$\frac{1}{2}$
2	2	1

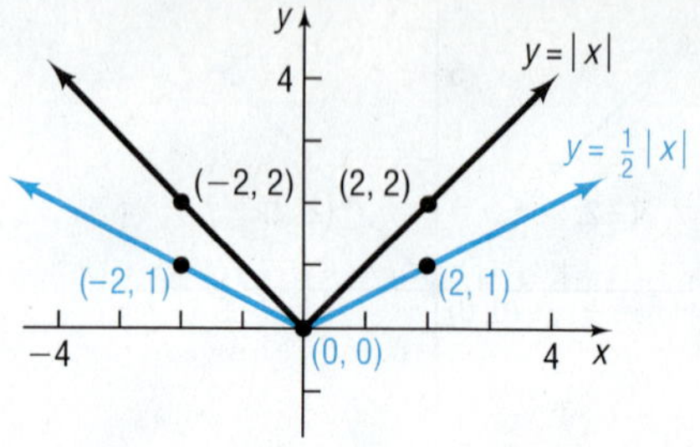

Figure 51

In Words

For $y = af(x)$, $a > 0$, the factor a is "outside" the function, so it affects the y-coordinates. Multiply each y-coordinate on the graph of $y = f(x)$ by a.

When the right side of a function $y = f(x)$ is multiplied by a positive number a, the graph of the new function $y = af(x)$ is obtained by multiplying each y-coordinate on the graph of $y = f(x)$ by a. The new graph is a **vertically compressed** (if $0 < a < 1$) or a **vertically stretched** (if $a > 1$) version of the graph of $y = f(x)$.

Now Work PROBLEM 47

What happens if the argument x of a function $y = f(x)$ is multiplied by a positive number a, creating a new function $y = f(ax)$? To find the answer, look at the following Exploration.

Exploration

On the same screen, graph each of the following functions:

$$Y_1 = f(x) = \sqrt{x} \qquad Y_2 = f(2x) = \sqrt{2x} \qquad Y_3 = f\left(\frac{1}{2}x\right) = \sqrt{\frac{1}{2}x} = \sqrt{\frac{x}{2}}$$

Create a table of values to explore the relation between the x- and y-coordinates of each function.

Result You should have obtained the graphs in Figure 52. Look at Table 12(a). Note that (1, 1), (4, 2), and (9, 3) are points on the graph of $Y_1 = \sqrt{x}$. Also, (0.5, 1), (2, 2), and (4.5, 3) are points on the graph of $Y_2 = \sqrt{2x}$. For a given y-coordinate, the x-coordinate on the graph of Y_2 is $\frac{1}{2}$ of the x-coordinate on Y_1.

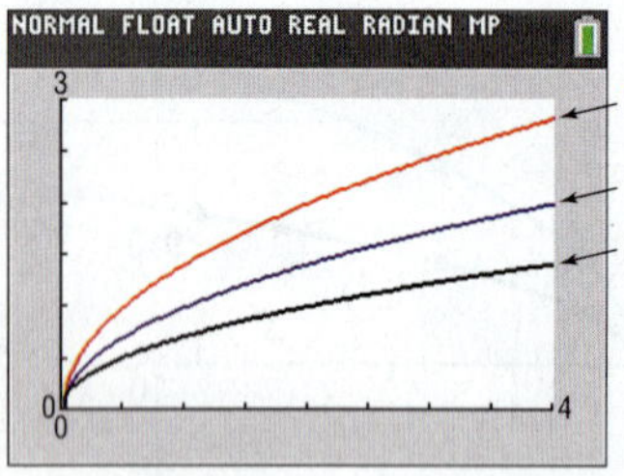

Figure 52

Table 12

NORMAL FLOAT AUTO REAL RADIAN MP
PRESS ENTER TO EDIT

X	Y1	Y2
0	0	0
.5	.70711	1
1	1	1.4142
2	1.4142	2
4	2	2.8284
4.5	2.1213	3
8	2.8284	4
9	3	4.2426
16	4	5.6569
12.5	3.5355	5
25	5	7.0711

Y2=√(2X)

(a)

NORMAL FLOAT AUTO REAL RADIAN MP
PRESS ENTER TO EDIT

X	Y1	Y3
0	0	0
1	1	.70711
2	1.4142	1
4	2	1.4142
8	2.8284	2
9	3	2.1213
16	4	2.8284
18	4.2426	3
25	5	3.5355
32	5.6569	4
50	7.0711	5

Y3=√(X/2)

(b)

We conclude that the graph of $Y_2 = \sqrt{2x}$ is obtained by multiplying the x-coordinate of each point on the graph of $Y_1 = \sqrt{x}$ by $\frac{1}{2}$. The graph of $Y_2 = \sqrt{2x}$ is the graph of $Y_1 = \sqrt{x}$ *compressed* horizontally.

Look at Table 12(b). Notice that (1, 1), (4, 2), and (9, 3) are points on the graph of $Y_1 = \sqrt{x}$. Also notice that (2, 1), (8, 2), and (18, 3) are points on the graph of $Y_3 = \sqrt{\frac{x}{2}}$. For a given y-coordinate, the x-coordinate on the graph of Y_3 is 2 times the x-coordinate on Y_1. We conclude that the graph of $Y_3 = \sqrt{\frac{x}{2}}$ is obtained by multiplying the x-coordinate of each point on the graph of $Y_1 = \sqrt{x}$ by 2. The graph of $Y_3 = \sqrt{\frac{x}{2}}$ is the graph of $Y_1 = \sqrt{x}$ *stretched* horizontally.

> **In Words**
> For $y = f(ax)$, $a > 0$, the factor a is "inside" the function, so it affects the x-coordinates. Multiply each x-coordinate on the graph of $y = f(x)$ by $\frac{1}{a}$.

Based on the Exploration, we have the following result:

If the argument x of a function $y = f(x)$ is multiplied by a positive number a, then the graph of the new function $y = f(ax)$ is obtained by multiplying each x-coordinate of $y = f(x)$ by $\frac{1}{a}$. A **horizontal compression** results if $a > 1$, and a **horizontal stretch** results if $0 < a < 1$.

Let's look at an example.

EXAMPLE 8

Graphing Using Stretches and Compressions

The graph of $y = f(x)$ is given in Figure 53. Use this graph to find the graphs of

(a) $y = 2f(x)$ (b) $y = f(3x)$

Solution

(a) The graph of $y = 2f(x)$ is obtained by multiplying each y-coordinate of $y = f(x)$ by 2. See Figure 54.

(b) The graph of $y = f(3x)$ is obtained from the graph of $y = f(x)$ by multiplying each x-coordinate of $y = f(x)$ by $\frac{1}{3}$. See Figure 55.

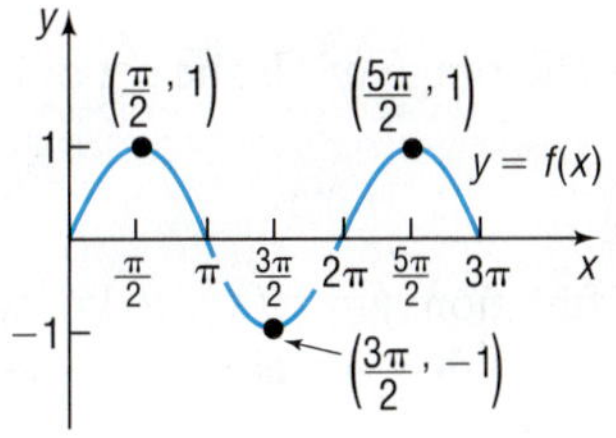

Figure 53 $y = f(x)$

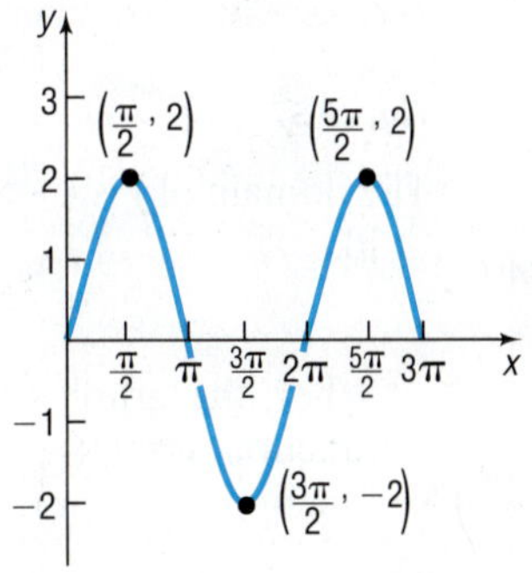

Figure 54 $y = 2f(x)$

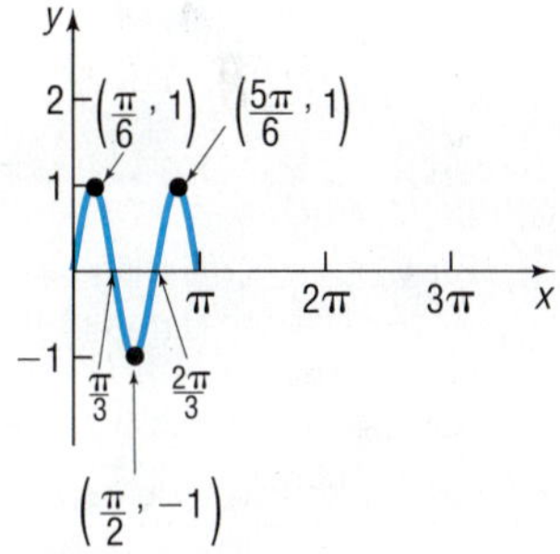

Figure 55 $y = f(3x)$

Now Work PROBLEMS 63(e) AND (g)

3 Graph Functions Using Reflections about the x-Axis and the y-Axis

EXAMPLE 9

Reflection about the x-Axis

Graph the function $f(x) = -x^2$. Find the domain and range of f.

Solution Begin with the graph of $y = x^2$, as shown in black in Figure 56. For each point (x, y) on the graph of $y = x^2$, the point $(x, -y)$ is on the graph of $y = -x^2$, as indicated in Table 13. Draw the graph of $y = -x^2$ by reflecting the graph of $y = x^2$ about the x-axis. See Figure 56.

Table 13

x	$y = x^2$	$y = -x^2$
−2	4	−4
−1	1	−1
0	0	0
1	1	−1
2	4	−4

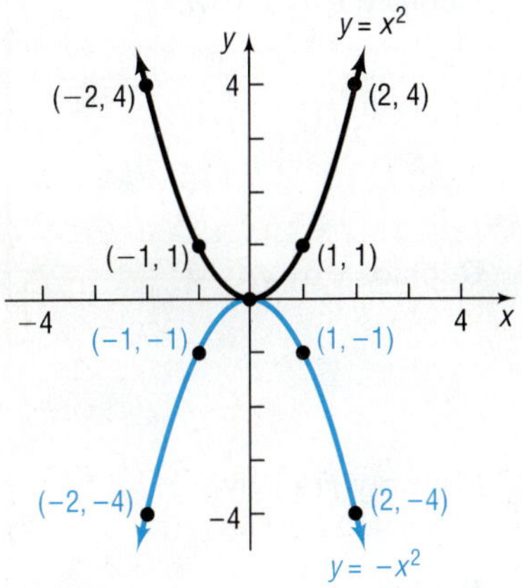

Figure 56

The domain of f is all real numbers, or $(-\infty, \infty)$. The range of f is $(-\infty, 0]$.

When the right side of the function $y = f(x)$ is multiplied by -1, the graph of the new function $y = -f(x)$ is the **reflection about the x-axis** of the graph of the function $y = f(x)$.

Now Work PROBLEM 49

EXAMPLE 10

Reflection about the y-Axis

Graph the function $f(x) = \sqrt{-x}$. Find the domain and range of f.

Solution To get the graph of $f(x) = \sqrt{-x}$, begin with the graph of $y = \sqrt{x}$, as shown in Figure 57. For each point (x, y) on the graph of $y = \sqrt{x}$, the point $(-x, y)$ is on the graph of $y = \sqrt{-x}$. Obtain the graph of $y = \sqrt{-x}$ by reflecting the graph of $y = \sqrt{x}$ about the y-axis. See Figure 57.

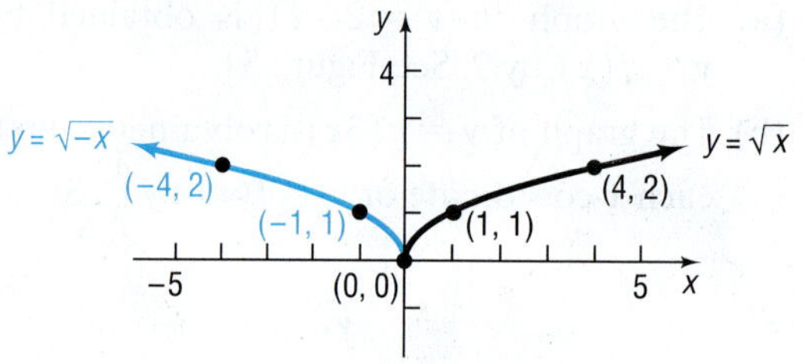

Figure 57

The domain of f is $(-\infty, 0]$. The range of f is the set of all nonnegative real numbers, or $[0, \infty)$.

In Words
For $y = -f(x)$, multiply each y-coordinate on the graph of $y = f(x)$ by -1.
For $y = f(-x)$, multiply each x-coordinate by -1.

When the graph of the function $y = f(x)$ is known, the graph of the new function $y = f(-x)$ is the **reflection about the y-axis** of the graph of the function $y = f(x)$.

SUMMARY OF GRAPHING TECHNIQUES

To Graph:	Draw the Graph of *f* and:	Functional Change to *f*(*x*)
Vertical shifts		
$y = f(x) + k,\ k > 0$	Raise the graph of f by k units.	Add k to $f(x)$.
$y = f(x) - k,\ k > 0$	Lower the graph of f by k units.	Subtract k from $f(x)$.
Horizontal shifts		
$y = f(x + h),\ h > 0$	Shift the graph of f to the left h units.	Replace x by $x + h$.
$y = f(x - h),\ h > 0$	Shift the graph of f to the right h units.	Replace x by $x - h$.
Compressing or stretching		
$y = af(x),\ a > 0$	Multiply each y-coordinate of $y = f(x)$ by a. Stretch the graph of f vertically if $a > 1$. Compress the graph of f vertically if $0 < a < 1$.	Multiply $f(x)$ by a.
$y = f(ax),\ a > 0$	Multiply each x-coordinate of $y = f(x)$ by $\frac{1}{a}$. Stretch the graph of f horizontally if $0 < a < 1$. Compress the graph of f horizontally if $a > 1$.	Replace x by ax.
Reflection about the x-axis		
$y = -f(x)$	Reflect the graph of f about the x-axis.	Multiply $f(x)$ by -1.
Reflection about the y-axis		
$y = f(-x)$	Reflect the graph of f about the y-axis.	Replace x by $-x$.

EXAMPLE 11 **Determining the Function Obtained from a Series of Transformations**

Find the function that is finally graphed after the following three transformations are applied to the graph of $y = |x|$.

1. Shift left 2 units
2. Shift up 3 units
3. Reflect about the y-axis

Solution

1. Shift left 2 units: Replace x by $x + 2$. $y = |x + 2|$
2. Shift up 3 units: Add 3. $y = |x + 2| + 3$
3. Reflect about the y-axis: Replace x by $-x$. $y = |-x + 2| + 3$ ●

Now Work PROBLEM 27

EXAMPLE 12 **Combining Graphing Procedures**

Graph the function $f(x) = \dfrac{3}{x - 2} + 1$. Find the domain and range of f.

Solution It is helpful to write f as $f(x) = 3\left(\dfrac{1}{x - 2}\right) + 1$. Now use the following steps to obtain the graph of f.

STEP 1: $y = \dfrac{1}{x}$ Reciprocal function

STEP 2: $y = 3 \cdot \left(\dfrac{1}{x}\right) = \dfrac{3}{x}$ Multiply by 3; vertical stretch by a factor of 3.

STEP 3: $y = \dfrac{3}{x - 2}$ Replace x by $x - 2$; horizontal shift to the right 2 units.

STEP 4: $y = \dfrac{3}{x - 2} + 1$ Add 1; vertical shift up 1 unit.

See Figure 58.

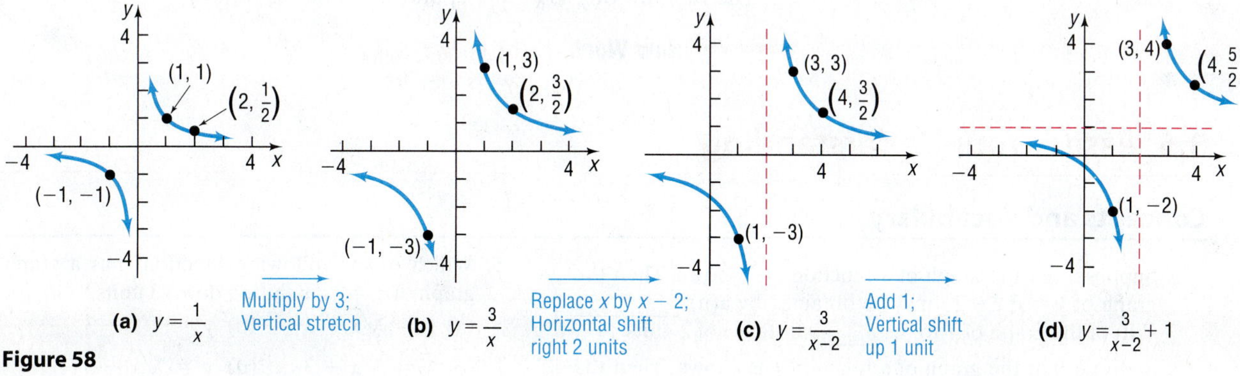

Figure 58

The domain of $y = \dfrac{1}{x}$ is $\{x | x \neq 0\}$ and its range is $\{y | y \neq 0\}$. Because we shifted right 2 units and up 1 unit to obtain f, the domain of f is $\{x | x \neq 2\}$ and its range is $\{y | y \neq 1\}$. ●

HINT: Although the order in which transformations are performed can be altered, consider using the following order for consistency:

1. Reflections
2. Compressions and stretches
3. Shifts

Other orderings of the steps shown in Example 12 would also result in the graph of f. For example, try this one:

Step 1: $y = \dfrac{1}{x}$ — Reciprocal function

Step 2: $y = \dfrac{1}{x-2}$ — Replace x by $x - 2$; horizontal shift to the right 2 units.

Step 3: $y = \dfrac{3}{x-2}$ — Multiply by 3; vertical stretch by a factor of 3.

Step 4: $y = \dfrac{3}{x-2} + 1$ — Add 1; vertical shift up 1 unit.

EXAMPLE 13 **Combining Graphing Procedures**

Graph the function $f(x) = \sqrt{1-x} + 2$. Find the domain and range of f.

Solution Because horizontal shifts require the form $x - h$, begin by rewriting $f(x)$ as $f(x) = \sqrt{1-x} + 2 = \sqrt{-(x-1)} + 2$. Now use the following steps.

Step 1: $y = \sqrt{x}$ — Square root function

Step 2: $y = \sqrt{-x}$ — Replace x by $-x$; reflect about the y-axis.

Step 3: $y = \sqrt{-(x-1)} = \sqrt{1-x}$ — Replace x by $x - 1$; horizontal shift to the right 1 unit.

Step 4: $y = \sqrt{1-x} + 2$ — Add 2; vertical shift up 2 units.

See Figure 59.

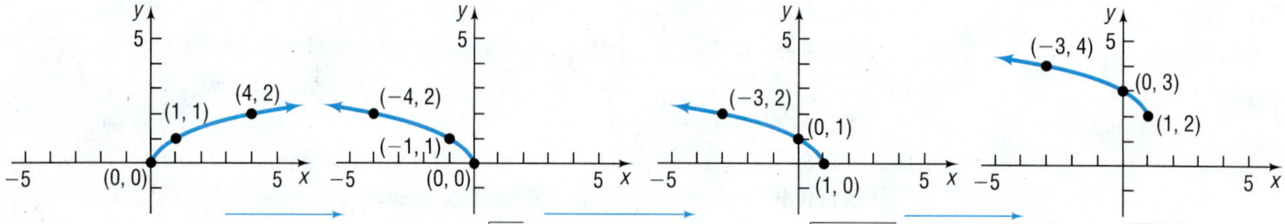

(a) $y = \sqrt{x}$ → Replace x by $-x$; Reflect about y-axis → (b) $y = \sqrt{-x}$ → Replace x by $x - 1$; Horizontal shift right 1 unit → (c) $y = \sqrt{-(x-1)} = \sqrt{-x+1} = \sqrt{1-x}$ → Add 2; Vertical shift up 2 units → (d) $y = \sqrt{1-x} + 2$

Figure 59

The domain of f is $(-\infty, 1]$ and the range is $[2, \infty)$. ●

Now Work PROBLEM 55

3.5 Assess Your Understanding

Concepts and Vocabulary

1. Suppose that the graph of a function f is known. Then the graph of $y = f(x-2)$ may be obtained by a(n) ________ shift of the graph of f to the ____ a distance of 2 units.
2. Suppose that the graph of a function f is known. Then the graph of $y = f(-x)$ may be obtained by a reflection about the ____-axis of the graph of the function $y = f(x)$.
3. ***True or False*** The graph of $y = \frac{1}{3}g(x)$ is the graph of $y = g(x)$ stretched by a factor of 3.
4. ***True or False*** The graph of $y = -f(x)$ is the reflection about the x-axis of the graph of $y = f(x)$.
5. Which of the following functions has a graph that is the graph of $y = \sqrt{x}$ shifted down 3 units?
   (a) $y = \sqrt{x+3}$ (b) $y = \sqrt{x} - 3$
   (c) $y = \sqrt{x} + 3$ (d) $y = \sqrt{x-3}$
6. Which of the following functions has a graph that is the graph of $y = f(x)$ compressed horizontally by a factor of 4?
   (a) $y = f(4x)$ (b) $y = f\left(\frac{1}{4}x\right)$
   (c) $y = 4f(x)$ (d) $y = \frac{1}{4}f(x)$

Skill Building

In Problems 7–18, match each graph to one of the following functions:

A. $y = x^2 + 2$ B. $y = -x^2 + 2$ C. $y = |x| + 2$ D. $y = -|x| + 2$

E. $y = (x - 2)^2$ F. $y = -(x + 2)^2$ G. $y = |x - 2|$ H. $y = -|x + 2|$

I. $y = 2x^2$ J. $y = -2x^2$ K. $y = 2|x|$ L. $y = -2|x|$

7.

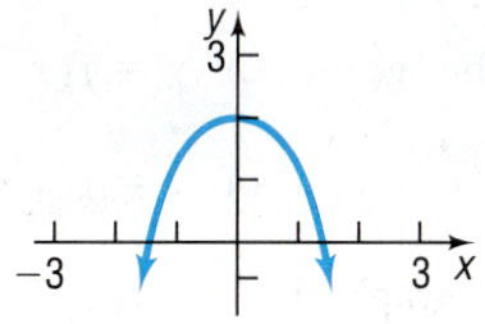

8.

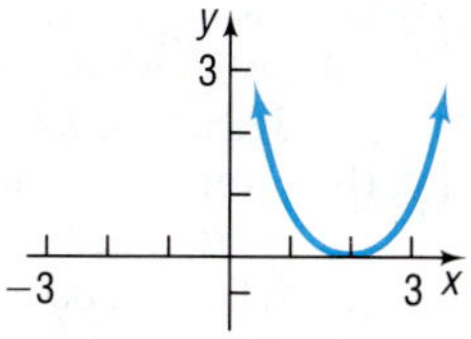

9.

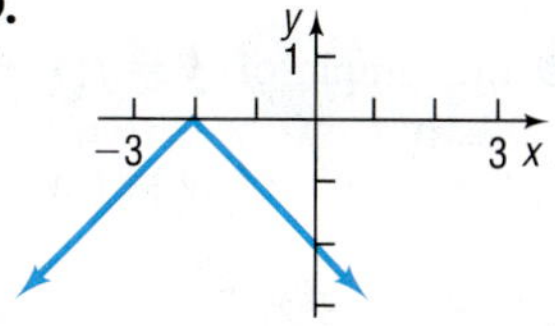

10.

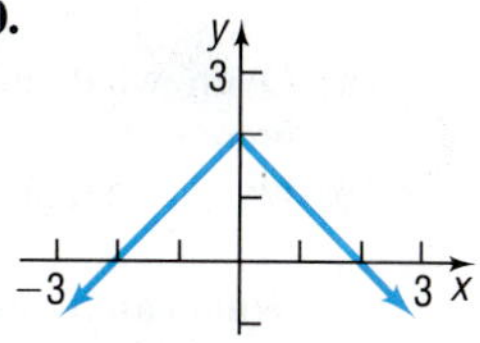

11.

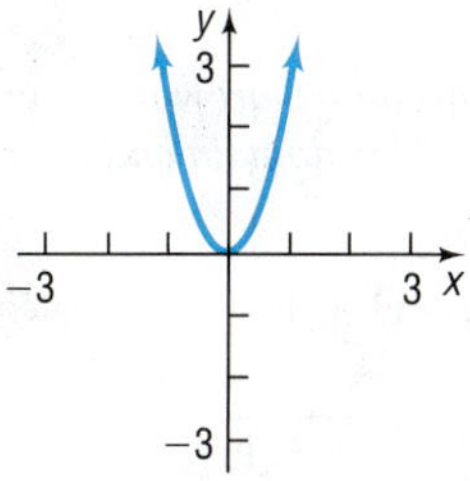

12.

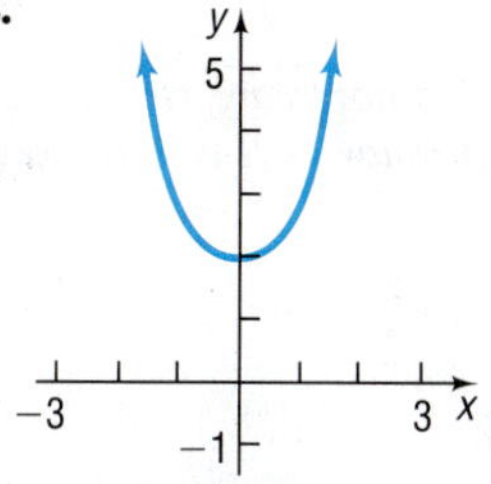

13.

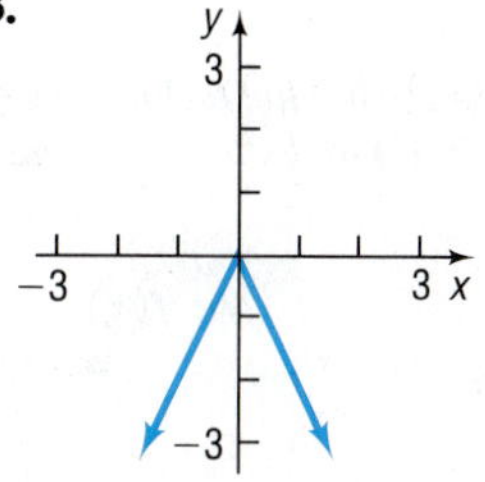

14.

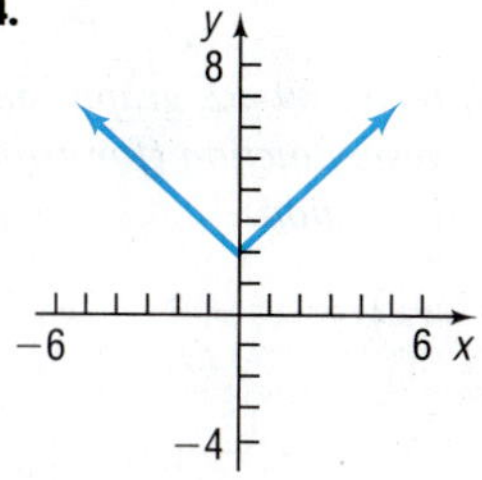

15.

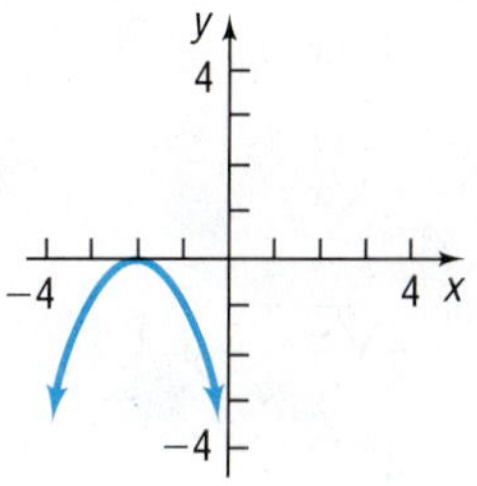

16.

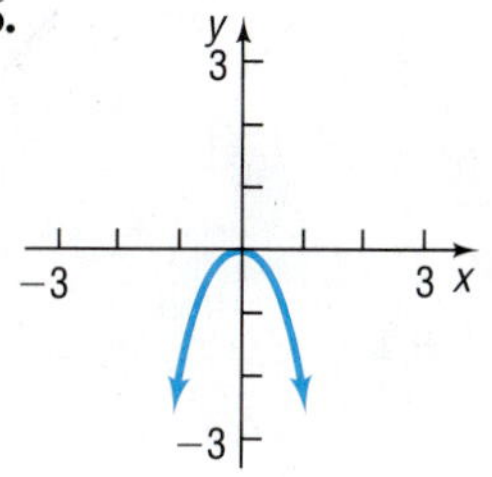

17.

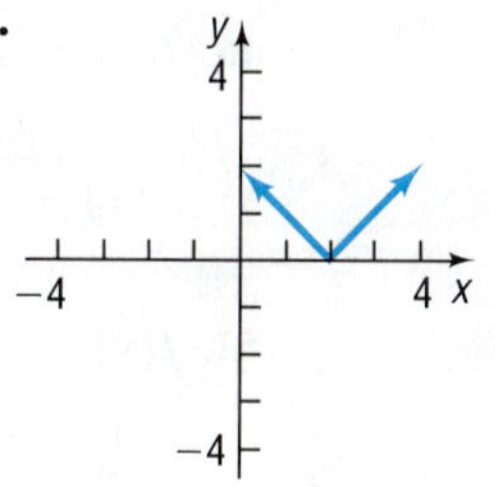

18.

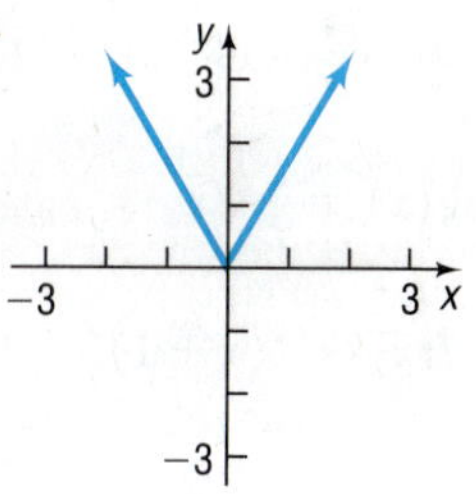

In Problems 19–26, write the function whose graph is the graph of $y = x^3$, but is:

19. Shifted to the right 4 units

20. Shifted to the left 4 units

21. Shifted up 4 units

22. Shifted down 4 units

23. Reflected about the y-axis

24. Reflected about the x-axis

25. Vertically stretched by a factor of 4

26. Horizontally stretched by a factor of 4

In Problems 27–30, find the function that is finally graphed after each of the following transformations is applied to the graph of $y = \sqrt{x}$ in the order stated.

27. (1) Shift up 2 units
(2) Reflect about the x-axis
(3) Reflect about the y-axis

28. (1) Reflect about the x-axis
(2) Shift right 3 units
(3) Shift down 2 units

29. (1) Reflect about the x-axis
(2) Shift up 2 units
(3) Shift left 3 units

30. (1) Shift up 2 units
(2) Reflect about the y-axis
(3) Shift left 3 units

31. If $(3, 6)$ is a point on the graph of $y = f(x)$, which of the following points must be on the graph of $y = -f(x)$?
(a) $(6, 3)$ (b) $(6, -3)$
(c) $(3, -6)$ (d) $(-3, 6)$

32. If $(3, 6)$ is a point on the graph of $y = f(x)$, which of the following points must be on the graph of $y = f(-x)$?
(a) $(6, 3)$ (b) $(6, -3)$
(c) $(3, -6)$ (d) $(-3, 6)$

33. If $(1, 3)$ is a point on the graph of $y = f(x)$, which of the following points must be on the graph of $y = 2f(x)$?
(a) $\left(1, \frac{3}{2}\right)$ (b) $(2, 3)$
(c) $(1, 6)$ (d) $\left(\frac{1}{2}, 3\right)$

34. If $(4, 2)$ is a point on the graph of $y = f(x)$, which of the following points must be on the graph of $y = f(2x)$?
(a) $(4, 1)$ (b) $(8, 2)$
(c) $(2, 2)$ (d) $(4, 4)$

35. Suppose that the x-intercepts of the graph of $y = f(x)$ are -5 and 3.
(a) What are the x-intercepts of the graph of $y = f(x + 2)$?
(b) What are the x-intercepts of the graph of $y = f(x - 2)$?
(c) What are the x-intercepts of the graph of $y = 4f(x)$?
(d) What are the x-intercepts of the graph of $y = f(-x)$?

36. Suppose that the x-intercepts of the graph of $y = f(x)$ are -8 and 1.
(a) What are the x-intercepts of the graph of $y = f(x + 4)$?
(b) What are the x-intercepts of the graph of $y = f(x - 3)$?
(c) What are the x-intercepts of the graph of $y = 2f(x)$?
(d) What are the x-intercepts of the graph of $y = f(-x)$?

37. Suppose that the function $y = f(x)$ is increasing on the interval $(-1, 5)$.
(a) Over what interval is the graph of $y = f(x + 2)$ increasing?
(b) Over what interval is the graph of $y = f(x - 5)$ increasing?
(c) What can be said about the graph of $y = -f(x)$?
(d) What can be said about the graph of $y = f(-x)$?

38. Suppose that the function $y = f(x)$ is decreasing on the interval $(-2, 7)$.
(a) Over what interval is the graph of $y = f(x + 2)$ decreasing?
(b) Over what interval is the graph of $y = f(x - 5)$ decreasing?
(c) What can be said about the graph of $y = -f(x)$?
(d) What can be said about the graph of $y = f(-x)$?

In Problems 39–62, graph each function using the techniques of shifting, compressing, stretching, and/or reflecting. Start with the graph of the basic function (for example, $y = x^2$) and show all stages. Be sure to show at least three key points. Find the domain and the range of each function.

39. $f(x) = x^2 - 1$

40. $f(x) = x^2 + 4$

41. $g(x) = x^3 + 1$

42. $g(x) = x^3 - 1$

43. $h(x) = \sqrt{x + 2}$

44. $h(x) = \sqrt{x + 1}$

45. $f(x) = (x - 1)^3 + 2$

46. $f(x) = (x + 2)^3 - 3$

47. $g(x) = 4\sqrt{x}$

48. $g(x) = \frac{1}{2}\sqrt{x}$

49. $f(x) = -\sqrt[3]{x}$

50. $f(x) = -\sqrt{x}$

51. $f(x) = 2(x + 1)^2 - 3$

52. $f(x) = 3(x - 2)^2 + 1$

53. $g(x) = 2\sqrt{x - 2} + 1$

54. $g(x) = 3|x + 1| - 3$

55. $h(x) = \sqrt{-x} - 2$

56. $h(x) = \frac{4}{x} + 2$

57. $f(x) = -(x + 1)^3 - 1$

58. $f(x) = -4\sqrt{x - 1}$

59. $g(x) = 2|1 - x|$

60. $g(x) = 4\sqrt{2 - x}$

61. $h(x) = \frac{1}{2x}$

62. $h(x) = \sqrt[3]{x - 1} + 3$

In Problems 63–66, the graph of a function f is illustrated. Use the graph of f as the first step toward graphing each of the following functions:

(a) $F(x) = f(x) + 3$
(b) $G(x) = f(x + 2)$
(c) $P(x) = -f(x)$
(d) $H(x) = f(x + 1) - 2$
(e) $Q(x) = \frac{1}{2}f(x)$
(f) $g(x) = f(-x)$
(g) $h(x) = f(2x)$

63.
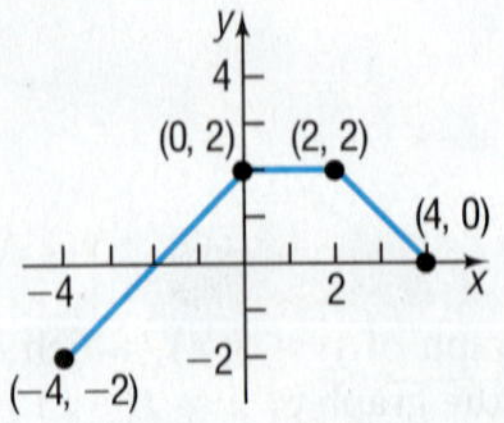

64.
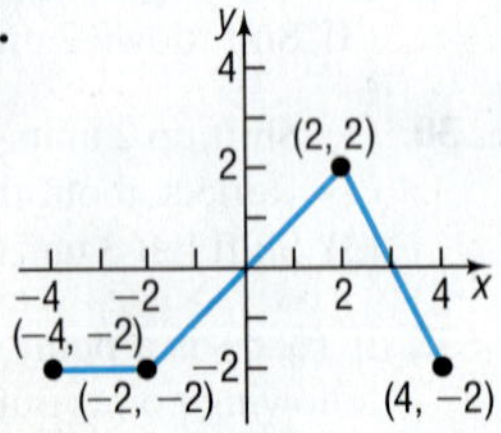

65.
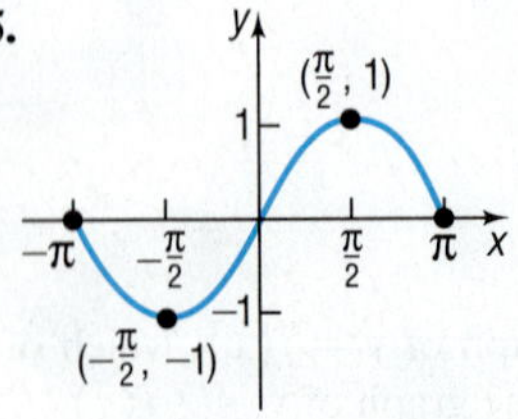

66.
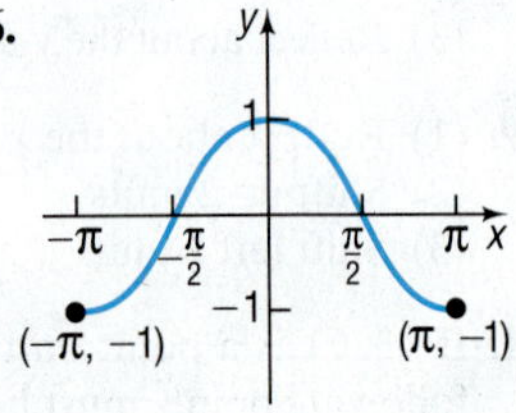

Mixed Practice

In Problems 67–74, complete the square of each quadratic expression. Then graph each function using the technique of shifting. (If necessary, refer to Chapter R, Section R.5 to review completing the square.)

67. $f(x) = x^2 + 2x$

68. $f(x) = x^2 - 6x$

69. $f(x) = x^2 - 8x + 1$

70. $f(x) = x^2 + 4x + 2$

71. $f(x) = 2x^2 - 12x + 19$

72. $f(x) = 3x^2 + 6x + 1$

73. $f(x) = -3x^2 - 12x - 17$

74. $f(x) = -2x^2 - 12x - 13$

Applications and Extensions

75. The graph of a function f is illustrated in the figure.
(a) Draw the graph of $y = |f(x)|$.
(b) Draw the graph of $y = f(|x|)$.

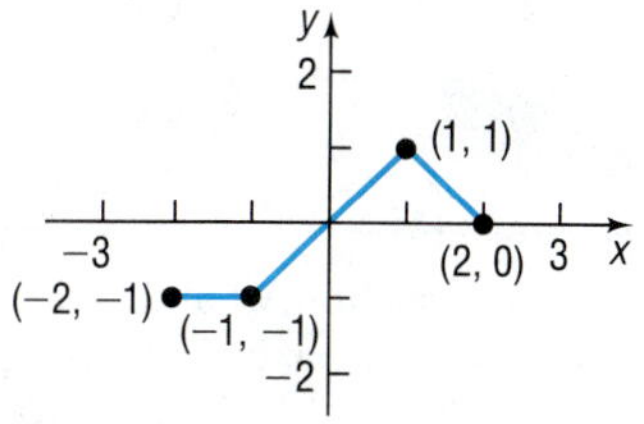

76. The graph of a function f is illustrated in the figure.
(a) Draw the graph of $y = |f(x)|$.
(b) Draw the graph of $y = f(|x|)$.

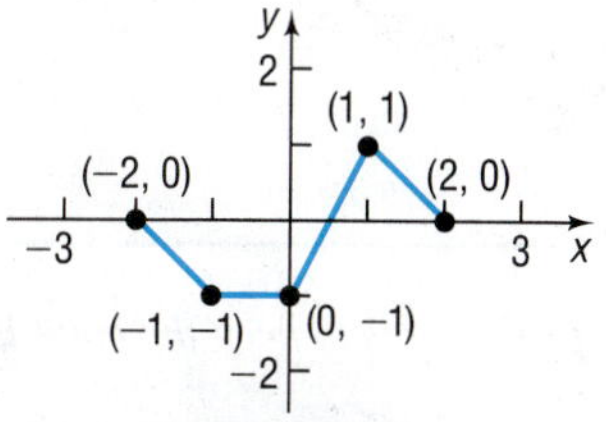

77. Suppose $(1, 3)$ is a point on the graph of $y = f(x)$.
(a) What point is on the graph of $y = f(x + 3) - 5$?
(b) What point is on the graph of $y = -2f(x - 2) + 1$?
(c) What point is on the graph of $y = f(2x + 3)$?

78. Suppose $(-3, 5)$ is a point on the graph of $y = g(x)$.
(a) What point is on the graph of $y = g(x + 1) - 3$?
(b) What point is on the graph of $y = -3g(x - 4) + 3$?
(c) What point is on the graph of $y = g(3x + 9)$?

79. Graph the following functions using transformations.
(a) $f(x) = \text{int}(-x)$ (b) $g(x) = -\text{int}(x)$

80. Graph the following functions using transformations
(a) $f(x) = \text{int}(x - 1)$ (b) $g(x) = \text{int}(1 - x)$

81. (a) Graph $f(x) = |x - 3| - 3$ using transformations.
(b) Find the area of the region that is bounded by f and the x-axis and lies below the x-axis.

82. (a) Graph $f(x) = -2|x - 4| + 4$ using transformations.
(b) Find the area of the region that is bounded by f and the x-axis and lies above the x-axis.

83. Thermostat Control Energy conservation experts estimate that homeowners can save 5% to 10% on winter heating bills by programming their thermostats 5 to 10 degrees lower while sleeping. In the graph (top, right), the temperature T (in degrees Fahrenheit) of a home is given as a function of time t (in hours after midnight) over a 24-hour period.
(a) At what temperature is the thermostat set during daytime hours? At what temperature is the thermostat set overnight?
(b) The homeowner reprograms the thermostat to $y = T(t) - 2$. Explain how this affects the temperature in the house. Graph this new function.
(c) The homeowner reprograms the thermostat to $y = T(t + 1)$. Explain how this affects the temperature in the house. Graph this new function.

Source: Roger Albright, *547 Ways to Be Fuel Smart*, 2000

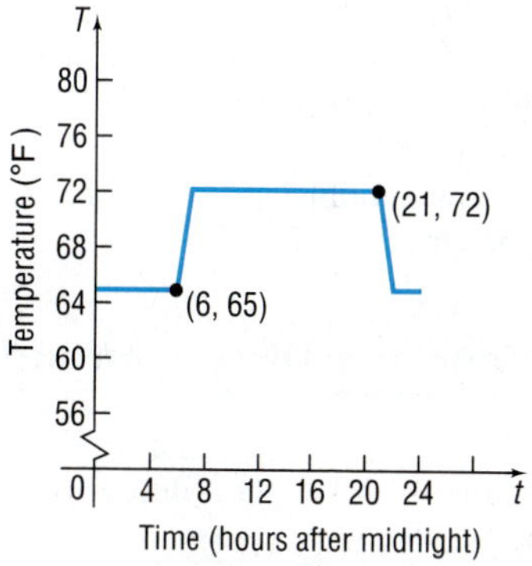

84. Digital Music Revenues The total projected worldwide digital music revenues R, in millions of dollars, for the years 2012 through 2017 can be estimated by the function

$$R(x) = 28.6x^2 + 300x + 4843$$

where x is the number of years after 2012.
(a) Find $R(0)$, $R(3)$, and $R(5)$ and explain what each value represents.
(b) Find $r(x) = R(x - 2)$.
(c) Find $r(2)$, $r(5)$, and $r(7)$ and explain what each value represents.
(d) In the model $r = r(x)$, what does x represent?
(e) Would there be an advantage in using the model r when estimating the projected revenues for a given year instead of the model R?

Source: IFPI Digital Music Report 2013

85. Temperature Measurements The relationship between the Celsius (°C) and Fahrenheit (°F) scales for measuring temperature is given by the equation

$$F = \frac{9}{5}C + 32$$

The relationship between the Celsius (°C) and Kelvin (K) scales is $K = C + 273$. Graph the equation $F = \frac{9}{5}C + 32$ using degrees Fahrenheit on the y-axis and degrees Celsius on the x-axis. Use the techniques introduced in this section to obtain the graph showing the relationship between Kelvin and Fahrenheit temperatures.

86. Period of a Pendulum The period T (in seconds) of a simple pendulum is a function of its length l (in feet) defined by the equation

$$T = 2\pi\sqrt{\frac{l}{g}}$$

where $g \approx 32.2$ feet per second per second is the acceleration due to gravity. (*Continued on page 260.*)

(a) Use a graphing utility to graph the function $T = T(l)$.
(b) Now graph the functions $T = T(l + 1)$, $T = T(l + 2)$, and $T = T(l + 3)$.
(c) Discuss how adding to the length l changes the period T.
(d) Now graph the functions $T = T(2l)$, $T = T(3l)$, and $T = T(4l)$.
(e) Discuss how multiplying the length l by factors of 2, 3, and 4 changes the period T.

87. The equation $y = (x - c)^2$ defines a *family of parabolas*, one parabola for each value of c. On one set of coordinate axes, graph the members of the family for $c = 0$, $c = 3$, and $c = -2$.

88. Repeat Problem 87 for the family of parabolas $y = x^2 + c$.

Explaining Concepts: Discussion and Writing

89. Suppose that the graph of a function f is known. Explain how the graph of $y = 4f(x)$ differs from the graph of $y = f(4x)$.

90. Suppose that the graph of a function f is known. Explain how the graph of $y = f(x) - 2$ differs from the graph of $y = f(x - 2)$.

91. The area under the curve $y = \sqrt{x}$ bounded from below by the x-axis and on the right by $x = 4$ is $\frac{16}{3}$ square units. Using the ideas presented in this section, what do you think is the area under the curve of $y = \sqrt{-x}$ bounded from below by the x-axis and on the left by $x = -4$? Justify your answer.

92. Explain how the range of the function $f(x) = x^2$ compares to the range of $g(x) = f(x) + k$.

93. Explain how the domain of $g(x) = \sqrt{x}$ compares to the domain of $g(x - k)$, where $k \geq 0$.

Retain Your Knowledge

Problems 94–97 are based on material learned earlier in the course. The purpose of these problems is to keep the material fresh in your mind so that you are better prepared for the final exam.

94. Determine the slope and y-intercept of the graph of $3x - 5y = 30$.

95. Simplify $\dfrac{(x^{-2}y^3)^4}{(x^2y^{-5})^{-2}}$

96. The amount of water used when taking a shower varies directly with the number of minutes the shower is run. If a 4-minute shower uses 7 gallons of water, how much water is used in a 9-minute shower?

97. List the intercepts and test for symmetry: $y^2 = x + 4$

3.6 Mathematical Models: Building Functions

OBJECTIVE 1 Build and Analyze Functions (p. 260)

1 Build and Analyze Functions

Real-world problems often result in mathematical models that involve functions. These functions need to be constructed or built based on the information given. In building functions, we must be able to translate the verbal description into the language of mathematics. This is done by assigning symbols to represent the independent and dependent variables and then by finding the function or rule that relates these variables.

EXAMPLE 1 Finding the Distance from the Origin to a Point on a Graph

Let $P = (x, y)$ be a point on the graph of $y = x^2 - 1$.

(a) Express the distance d from P to the origin O as a function of x.
(b) What is d if $x = 0$?
(c) What is d if $x = 1$?
(d) What is d if $x = \frac{\sqrt{2}}{2}$?

(e) Use a graphing utility to graph the function $d = d(x), x \geq 0$. Rounding to two decimal places, find the value(s) of x at which d has a local minimum. [This gives the point(s) on the graph of $y = x^2 - 1$ closest to the origin.]

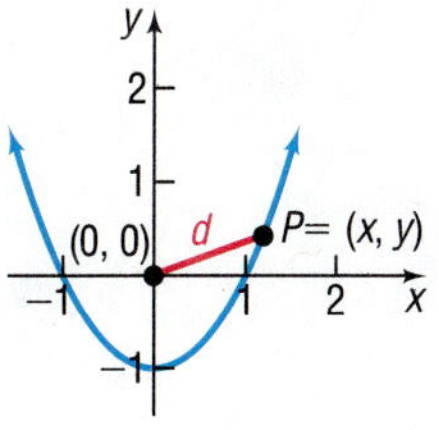

Figure 60 $y = x^2 - 1$

Solution (a) Figure 60 illustrates the graph of $y = x^2 - 1$. The distance d from P to O is

$$d = \sqrt{(x-0)^2 + (y-0)^2} = \sqrt{x^2 + y^2}$$

Since P is a point on the graph of $y = x^2 - 1$, substitute $x^2 - 1$ for y. Then

$$d(x) = \sqrt{x^2 + (x^2-1)^2} = \sqrt{x^4 - x^2 + 1}$$

The distance d is expressed as a function of x.

(b) If $x = 0$, the distance d is

$$d(0) = \sqrt{0^4 - 0^2 + 1} = \sqrt{1} = 1$$

(c) If $x = 1$, the distance d is

$$d(1) = \sqrt{1^4 - 1^2 + 1} = 1$$

(d) If $x = \frac{\sqrt{2}}{2}$, the distance d is

$$d\left(\frac{\sqrt{2}}{2}\right) = \sqrt{\left(\frac{\sqrt{2}}{2}\right)^4 - \left(\frac{\sqrt{2}}{2}\right)^2 + 1} = \sqrt{\frac{1}{4} - \frac{1}{2} + 1} = \frac{\sqrt{3}}{2}$$

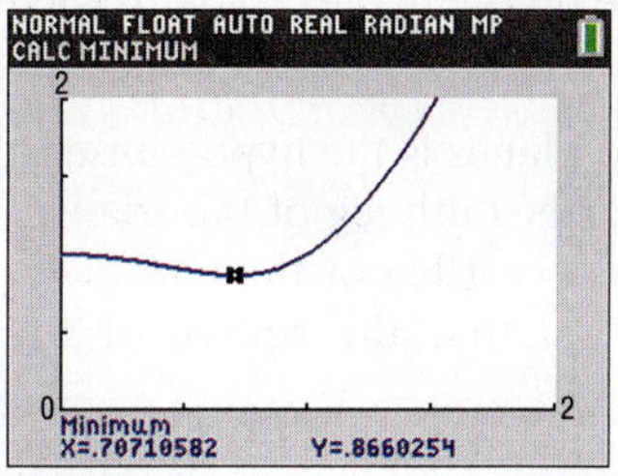

Figure 61 $d(x) = \sqrt{x^4 - x^2 + 1}$

(e) Figure 61 shows the graph of $Y_1 = \sqrt{x^4 - x^2 + 1}$. Using the MINIMUM feature on a graphing utility, we find that when $x \approx 0.71$ the value of d is smallest. The local minimum is $d \approx 0.87$ rounded to two decimal places. Since $d(x)$ is even, it follows by symmetry that when $x \approx -0.71$, the value of d is also a local minimum. Since $(\pm 0.71)^2 - 1 \approx -0.50$, the points $(-0.71, -0.50)$ and $(0.71, -0.50)$ on the graph of $y = x^2 - 1$ are closest to the origin. ●

Now Work PROBLEM 1

EXAMPLE 2 Area of a Rectangle

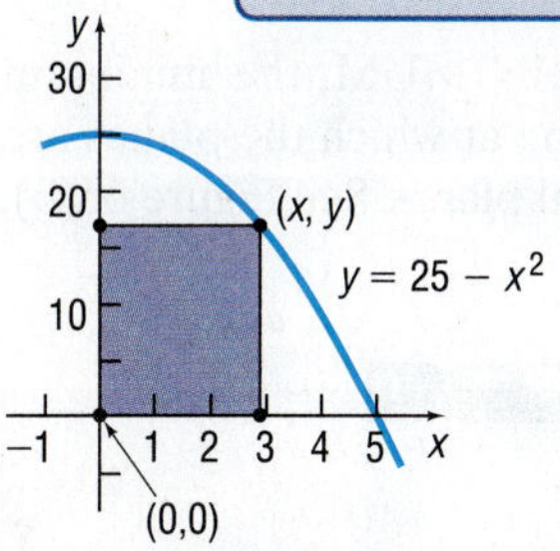

Figure 62

A rectangle has one corner in quadrant I on the graph of $y = 25 - x^2$, another at the origin, a third on the positive y-axis, and the fourth on the positive x-axis. See Figure 62.

(a) Express the area A of the rectangle as a function of x.
(b) What is the domain of A?
(c) Graph $A = A(x)$.
(d) For what value of x is the area largest?

Solution (a) The area A of the rectangle is $A = xy$, where $y = 25 - x^2$. Substituting this expression for y, we obtain $A(x) = x(25 - x^2) = 25x - x^3$.

(b) Since (x, y) is in quadrant I, we have $x > 0$. Also, $y = 25 - x^2 > 0$, which implies that $x^2 < 25$, so $-5 < x < 5$. Combining these restrictions, we have the domain of A as $\{x \mid 0 < x < 5\}$, or $(0, 5)$ using interval notation.

(c) See Figure 63 on page 262 for the graph of $A = A(x)$.

(d) Using MAXIMUM, we find that the maximum area is 48.11 square units at $x = 2.89$ units, each rounded to two decimal places. See Figure 64.

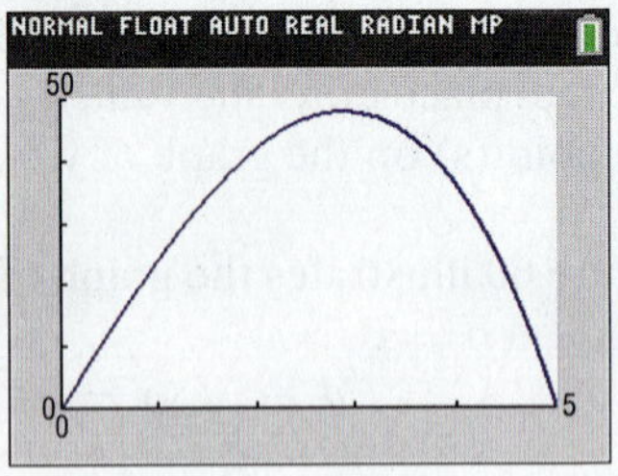

Figure 63 $A(x) = 25x - x^3$

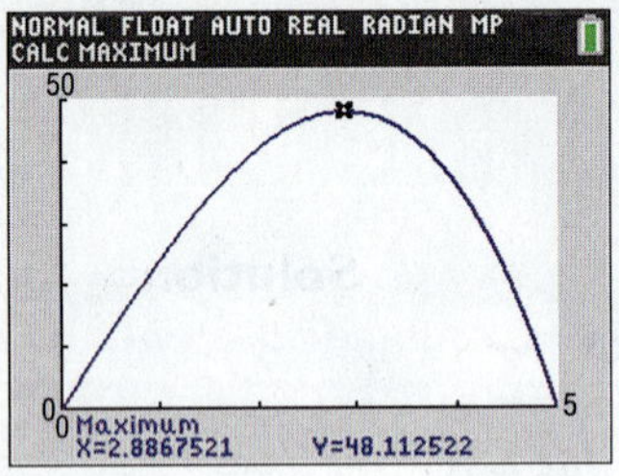

Figure 64

Now Work PROBLEM 7

EXAMPLE 3 Close Call?

Suppose two planes flying at the same altitude are headed toward each other. One plane is flying due south at a groundspeed of 400 miles per hour and is 600 miles from the potential intersection point of the planes. The other plane is flying due west with a groundspeed of 250 miles per hour and is 400 miles from the potential intersection point of the planes. See Figure 65.

(a) Build a model that expresses the distance d between the planes as a function of time t.

(b) Use a graphing utility to graph $d = d(t)$. How close do the planes come to each other? At what time are the planes closest?

Solution (a) Refer to Figure 65. The distance d between the two planes is the hypotenuse of a right triangle. At any time t, the length of the north/south leg of the triangle is $600 - 400t$. At any time t, the length of the east/west leg of the triangle is $400 - 250t$. Use the Pythagorean Theorem to find that the square of the distance between the two planes is

$$d^2 = (600 - 400t)^2 + (400 - 250t)^2$$

Therefore, the distance between the two planes as a function of time is given by the model

$$d(t) = \sqrt{(600 - 400t)^2 + (400 - 250t)^2}$$

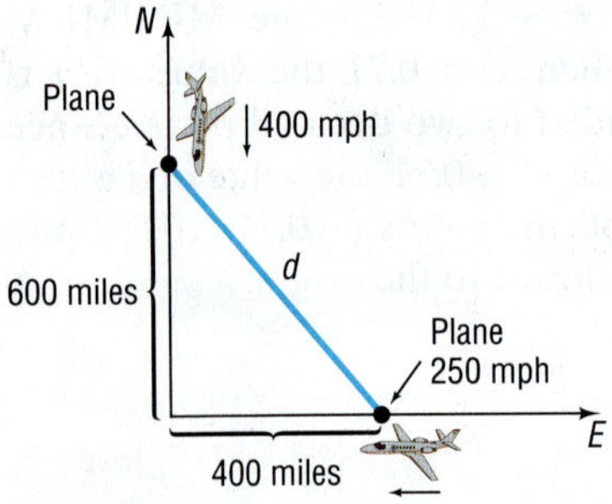

Figure 65

(b) Figure 66(a) shows the graph of $d = d(t)$. Using MINIMUM, the minimum distance between the planes is 21.20 miles, and the time at which the planes are closest is after 1.53 hours, each rounded to two decimal places. See Figure 66(b).

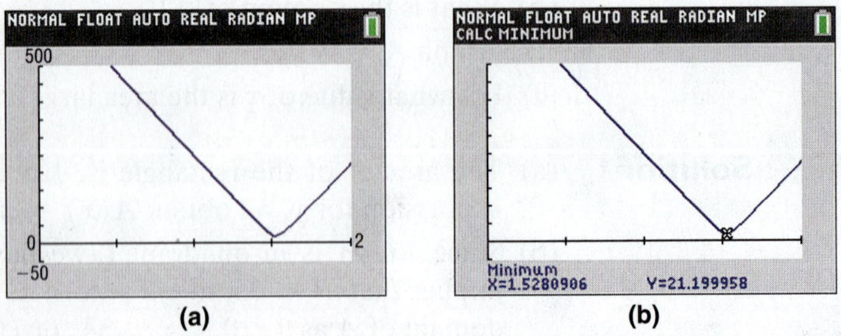

Figure 66

Now Work PROBLEM 19

3.6 Assess Your Understanding

Applications and Extensions

1. Let $P = (x, y)$ be a point on the graph of $y = x^2 - 8$.
   (a) Express the distance d from P to the origin as a function of x.
   (b) What is d if $x = 0$?
   (c) What is d if $x = 1$?
   (d) Use a graphing utility to graph $d = d(x)$.
   (e) For what values of x is d smallest?

2. Let $P = (x, y)$ be a point on the graph of $y = x^2 - 8$.
   (a) Express the distance d from P to the point $(0, -1)$ as a function of x.
   (b) What is d if $x = 0$?
   (c) What is d if $x = -1$?
   (d) Use a graphing utility to graph $d = d(x)$.
   (e) For what values of x is d smallest?

3. Let $P = (x, y)$ be a point on the graph of $y = \sqrt{x}$.
   (a) Express the distance d from P to the point $(1, 0)$ as a function of x.
   (b) Use a graphing utility to graph $d = d(x)$.
   (c) For what values of x is d smallest?

4. Let $P = (x, y)$ be a point on the graph of $y = \dfrac{1}{x}$.
   (a) Express the distance d from P to the origin as a function of x.
   (b) Use a graphing utility to graph $d = d(x)$.
   (c) For what values of x is d smallest?

5. A right triangle has one vertex on the graph of $y = x^3$, $x > 0$, at (x, y), another at the origin, and the third on the positive y-axis at $(0, y)$, as shown in the figure. Express the area A of the triangle as a function of x.

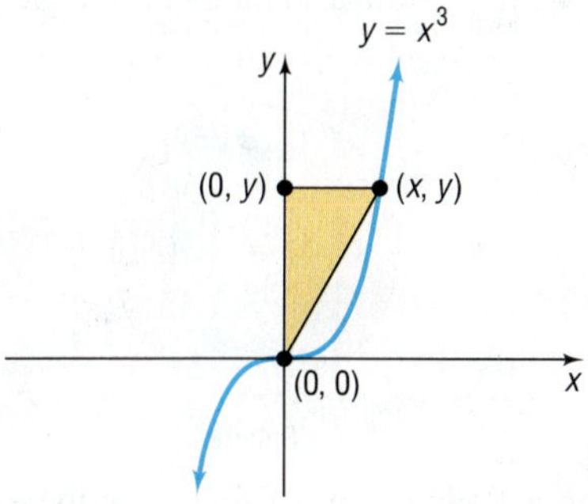

6. A right triangle has one vertex on the graph of $y = 9 - x^2$, $x > 0$, at (x, y), another at the origin, and the third on the positive x-axis at $(x, 0)$. Express the area A of the triangle as a function of x.

7. A rectangle has one corner in quadrant I on the graph of $y = 16 - x^2$, another at the origin, a third on the positive y-axis, and the fourth on the positive x-axis. See the figure (top, right).
   (a) Express the area A of the rectangle as a function of x.
   (b) What is the domain of A?
   (c) Graph $A = A(x)$. For what value of x is A largest?

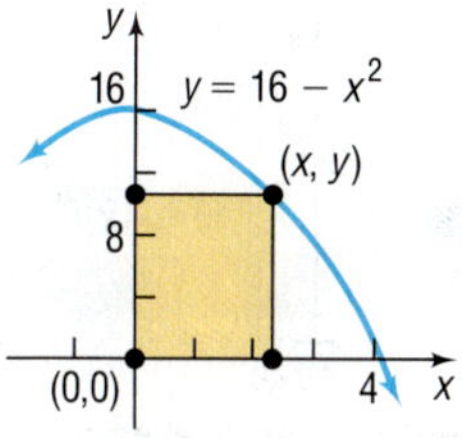

8. A rectangle is inscribed in a semicircle of radius 2. See the figure. Let $P = (x, y)$ be the point in quadrant I that is a vertex of the rectangle and is on the circle.

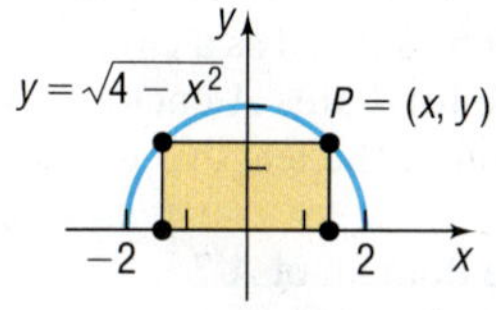

   (a) Express the area A of the rectangle as a function of x.
   (b) Express the perimeter p of the rectangle as a function of x.
   (c) Graph $A = A(x)$. For what value of x is A largest?
   (d) Graph $p = p(x)$. For what value of x is p largest?

9. A rectangle is inscribed in a circle of radius 2. See the figure. Let $P = (x, y)$ be the point in quadrant I that is a vertex of the rectangle and is on the circle.

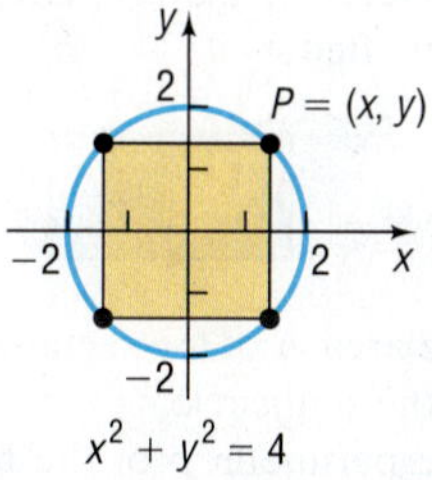

   (a) Express the area A of the rectangle as a function of x.
   (b) Express the perimeter p of the rectangle as a function of x.
   (c) Graph $A = A(x)$. For what value of x is A largest?
   (d) Graph $p = p(x)$. For what value of x is p largest?

10. A circle of radius r is inscribed in a square. See the figure.

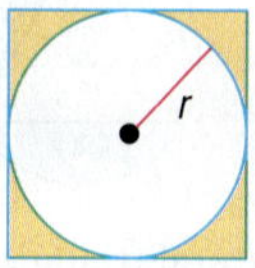

   (a) Express the area A of the square as a function of the radius r of the circle.
   (b) Express the perimeter p of the square as a function of r.

11. Geometry A wire 10 meters long is to be cut into two pieces. One piece will be shaped as a square, and the other piece will be shaped as a circle. See the figure.

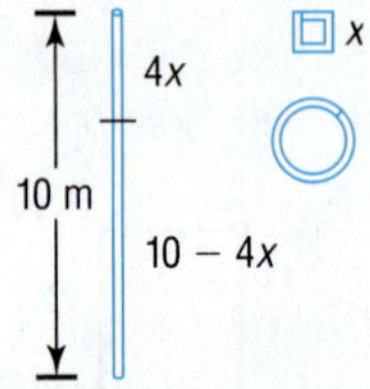

(a) Express the total area A enclosed by the pieces of wire as a function of the length x of a side of the square.
(b) What is the domain of A ?
(c) Graph $A = A(x)$. For what value of x is A smallest?

12. Geometry A wire 10 meters long is to be cut into two pieces. One piece will be shaped as an equilateral triangle, and the other piece will be shaped as a circle.
(a) Express the total area A enclosed by the pieces of wire as a function of the length x of a side of the equilateral triangle.
(b) What is the domain of A ?
(c) Graph $A = A(x)$. For what value of x is A smallest?

13. Geometry A wire of length x is bent into the shape of a circle.
(a) Express the circumference C of the circle as a function of x.
(b) Express the area A of the circle as a function of x.

14. Geometry A wire of length x is bent into the shape of a square.
(a) Express the perimeter p of the square as a function of x.
(b) Express the area A of the square as a function of x.

15. Geometry A semicircle of radius r is inscribed in a rectangle so that the diameter of the semicircle is the length of the rectangle. See the figure.

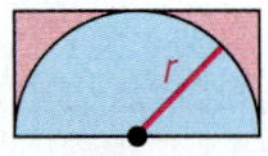

(a) Express the area A of the rectangle as a function of the radius r of the semicircle.
(b) Express the perimeter p of the rectangle as a function of r.

16. Geometry An equilateral triangle is inscribed in a circle of radius r. See the figure. Express the circumference C of the circle as a function of the length x of a side of the triangle.

[**Hint:** First show that $r^2 = \frac{x^2}{3}$.]

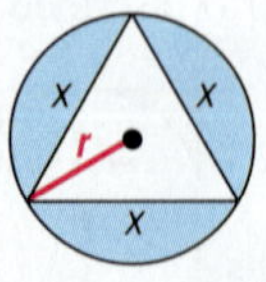

17. Geometry An equilateral triangle is inscribed in a circle of radius r. See the figure in Problem 16. Express the area A within the circle, but outside the triangle, as a function of the length x of a side of the triangle.

18. Uniform Motion Two cars leave an intersection at the same time. One is headed south at a constant speed of 30 miles per hour, and the other is headed west at a constant speed of 40 miles per hour (see the figure). Build a model that expresses the distance d between the cars as a function of the time t.

[**Hint:** At $t = 0$, the cars leave the intersection.]

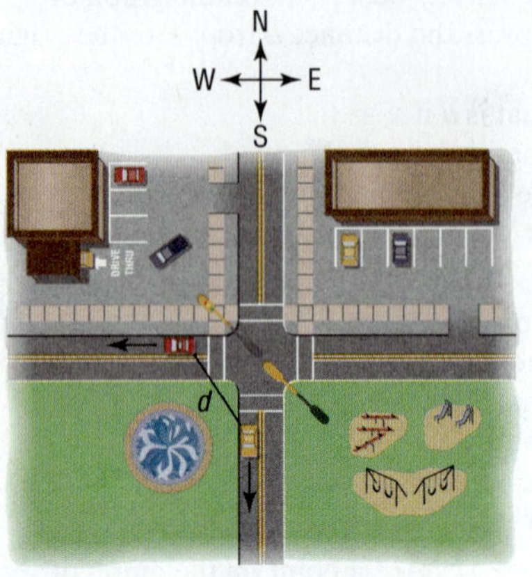

19. Uniform Motion Two cars are approaching an intersection. One is 2 miles south of the intersection and is moving at a constant speed of 30 miles per hour. At the same time, the other car is 3 miles east of the intersection and is moving at a constant speed of 40 miles per hour.
(a) Build a model that expresses the distance d between the cars as a function of time t.
[**Hint:** At $t = 0$, the cars are 2 miles south and 3 miles east of the intersection, respectively.]
(b) Use a graphing utility to graph $d = d(t)$. For what value of t is d smallest?

20. Inscribing a Cylinder in a Sphere Inscribe a right circular cylinder of height h and radius r in a sphere of fixed radius R. See the illustration. Express the volume V of the cylinder as a function of h.

[**Hint:** $V = \pi r^2 h$. Note also the right triangle.]

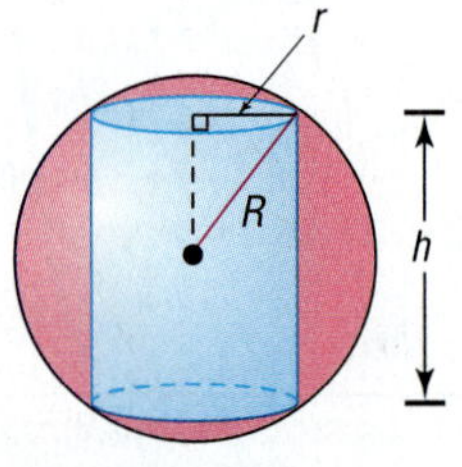

Sphere

21. Inscribing a Cylinder in a Cone Inscribe a right circular cylinder of height h and radius r in a cone of fixed radius R and fixed height H. See the illustration. Express the volume V of the cylinder as a function of r.

[**Hint:** $V = \pi r^2 h$. Note also the similar triangles.]

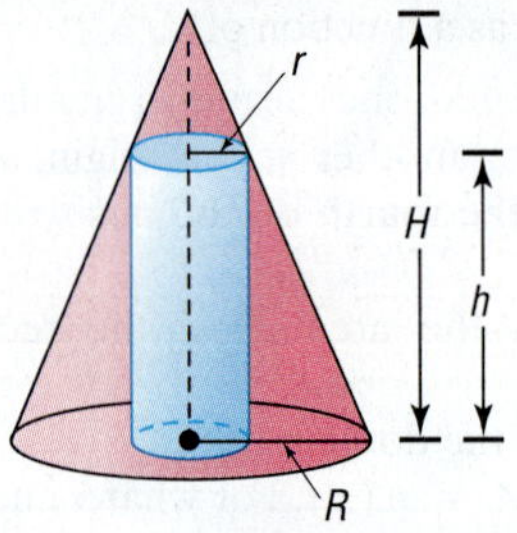

Cone

22. Installing Cable TV MetroMedia Cable is asked to provide service to a customer whose house is located 2 miles from the road along which the cable is buried. The nearest connection box for the cable is located 5 miles down the road. See the figure.

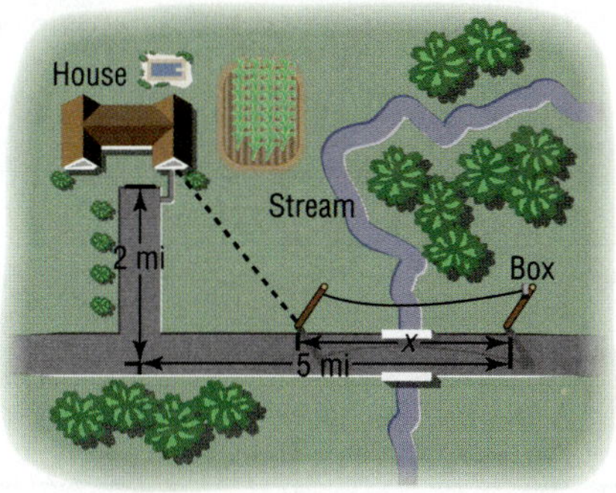

(a) If the installation cost is \$500 per mile along the road and \$700 per mile off the road, build a model that expresses the total cost C of installation as a function of the distance x (in miles) from the connection box to the point where the cable installation turns off the road. Find the domain of $C = C(x)$.

(b) Compute the cost if $x = 1$ mile.

(c) Compute the cost if $x = 3$ miles.

(d) Graph the function $C = C(x)$. Use TRACE to see how the cost C varies as x changes from 0 to 5.

(e) What value of x results in the least cost?

23. Time Required to Go from an Island to a Town An island is 2 miles from the nearest point P on a straight shoreline. A town is 12 miles down the shore from P. See the illustration.

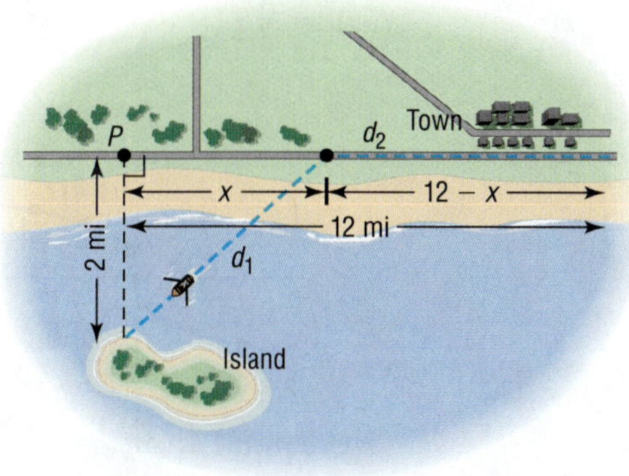

(a) If a person can row a boat at an average speed of 3 miles per hour and the same person can walk 5 miles per hour, build a model that expresses the time T that it takes to go from the island to town as a function of the distance x from P to where the person lands the boat.

(b) What is the domain of T?

(c) How long will it take to travel from the island to town if the person lands the boat 4 miles from P?

(d) How long will it take if the person lands the boat 8 miles from P?

24. Filling a Conical Tank Water is poured into a container in the shape of a right circular cone with radius 4 feet and height 16 feet. See the figure. Express the volume V of the water in the cone as a function of the height h of the water.

[**Hint:** The volume V of a cone of radius r and height h is $V = \frac{1}{3}\pi r^2 h$.]

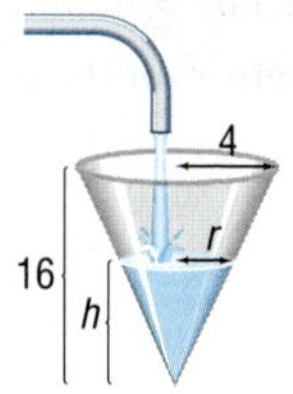

25. Constructing an Open Box An open box with a square base is to be made from a square piece of cardboard 24 inches on a side by cutting out a square from each corner and turning up the sides. See the figure.

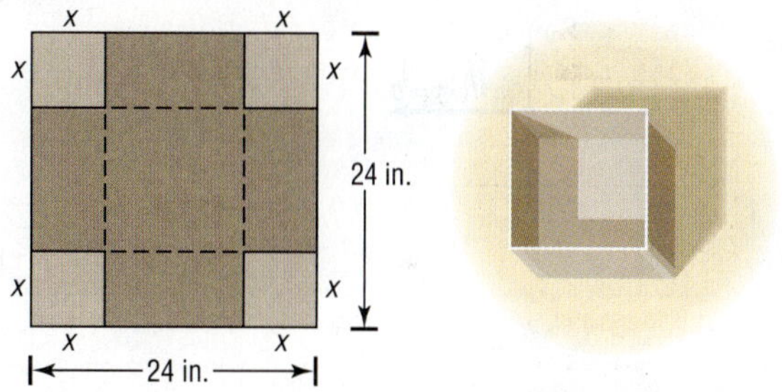

(a) Express the volume V of the box as a function of the length x of the side of the square cut from each corner.

(b) What is the volume if a 3-inch square is cut out?

(c) What is the volume if a 10-inch square is cut out?

(d) Graph $V = V(x)$. For what value of x is V largest?

26. Constructing an Open Box An open box with a square base is required to have a volume of 10 cubic feet.

(a) Express the amount A of material used to make such a box as a function of the length x of a side of the square base.

(b) How much material is required for a base 1 foot by 1 foot?

(c) How much material is required for a base 2 feet by 2 feet?

(d) Use a graphing utility to graph $A = A(x)$. For what value of x is A smallest?

Retain Your Knowledge

Problems 27–30 are based on material learned earlier in the course. The purpose of these problems is to keep the material fresh in your mind so that you are better prepared for the final exam.

27. Solve: $|2x - 3| - 5 = -2$

28. A 16-foot long Ford Fusion wants to pass a 50-foot truck traveling at 55 mi/h. How fast must the car travel to completely pass the truck in 5 seconds?

29. Find the slope of the line containing the points $(3, -2)$ and $(1, 6)$.

30. Find the missing length x for the given pair of similar triangles.

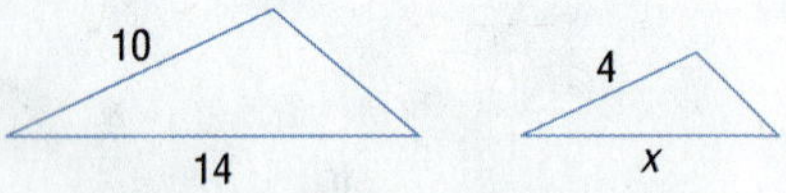

Chapter Review

Library of Functions

Constant function (p. 239)

$f(x) = b$

The graph is a horizontal line with y-intercept b.

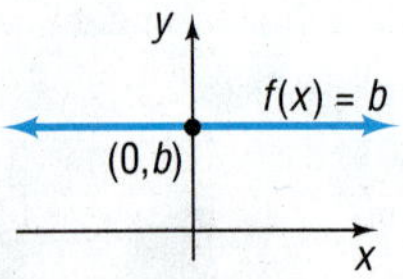

Identity function (p. 239)

$f(x) = x$

The graph is a line with slope 1 and y-intercept 0.

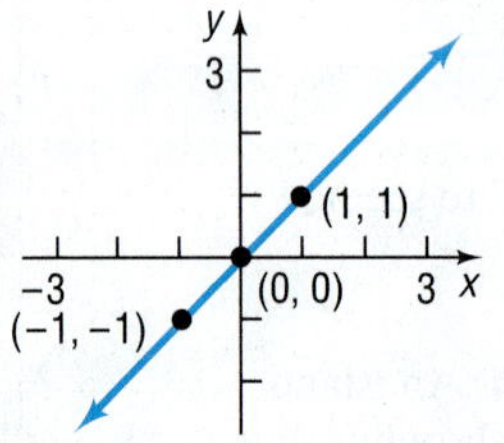

Square function (p. 239)

$f(x) = x^2$

The graph is a parabola with intercept at $(0, 0)$.

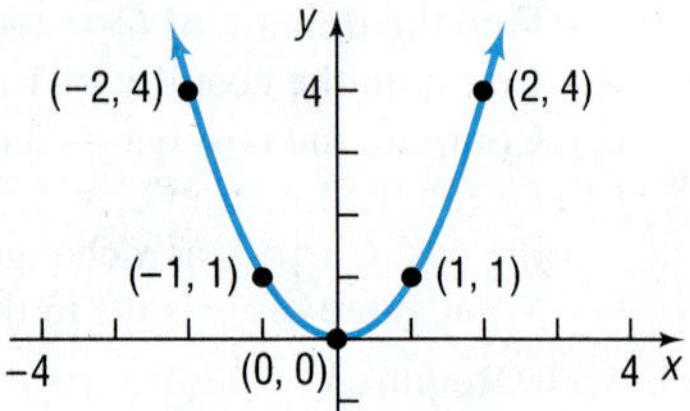

Cube function (p. 240)

$f(x) = x^3$

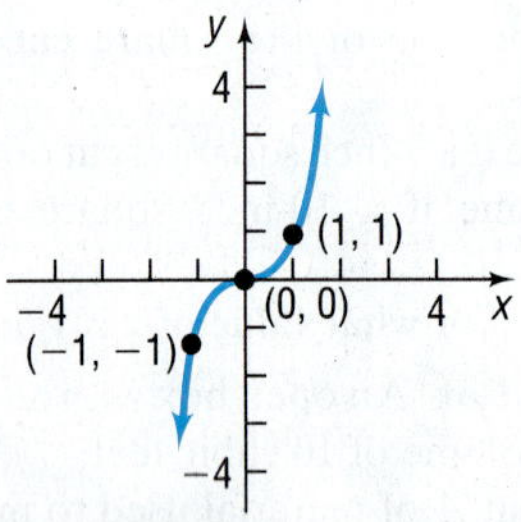

Square root function (p. 240)

$f(x) = \sqrt{x}$

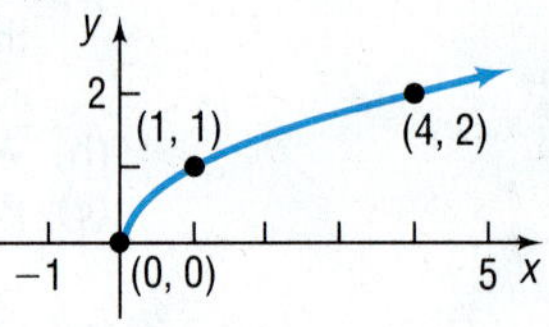

Cube root function (p. 240)

$f(x) = \sqrt[3]{x}$

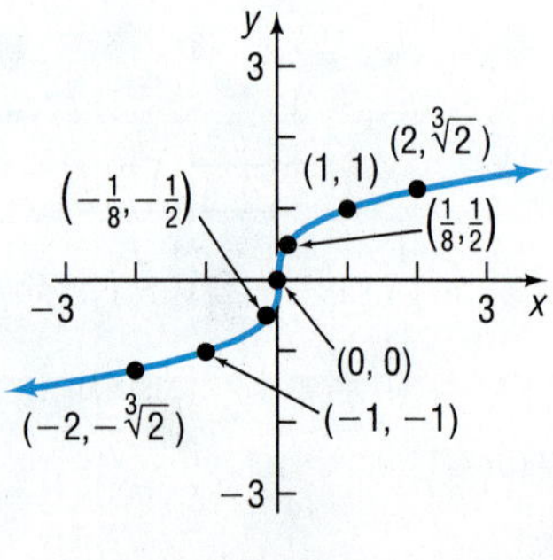

Reciprocal function (p. 240)

$f(x) = \dfrac{1}{x}$

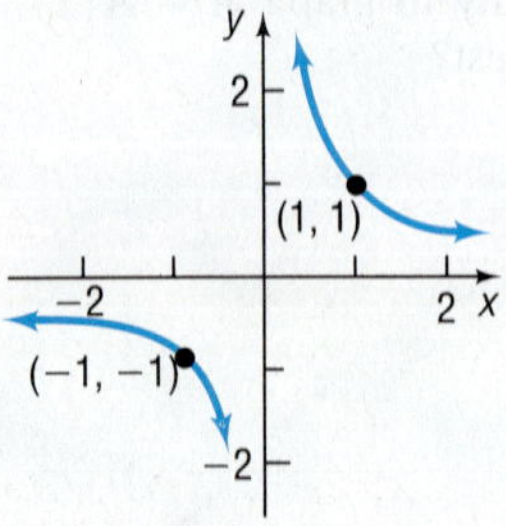

Absolute value function (p. 240)

$f(x) = |x|$

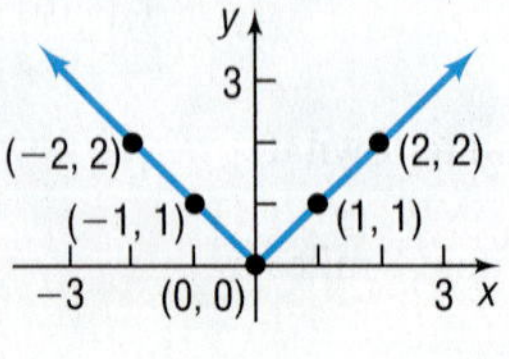

Greatest integer function (p. 241)

$f(x) = \text{int}(x)$

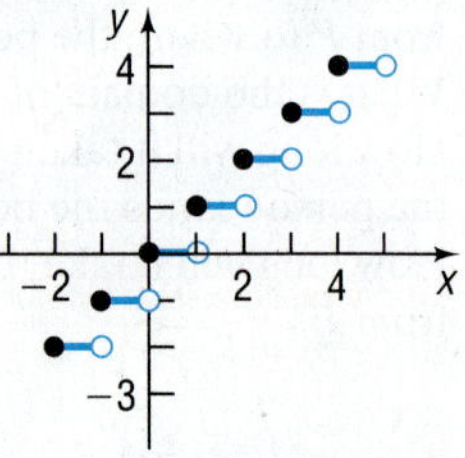

Things to Know

Function (pp. 199–202)

A relation between two sets so that each element x in the first set, the domain, has corresponding to it exactly one element y in the second set. The range is the set of y-values of the function for the x-values in the domain.

A function can also be characterized as a set of ordered pairs (x, y) in which no first element is paired with two different second elements.

Function notation (pp. 202–205)

$y = f(x)$

f is a symbol for the function.

x is the argument, or independent variable.

y is the dependent variable.

$f(x)$ is the value of the function at x, or the image of x.

A function f may be defined implicitly by an equation involving x and y or explicitly by writing $y = f(x)$.

Difference quotient of f (p. 205)

$$\frac{f(x+h) - f(x)}{h} \quad h \neq 0$$

Domain (pp. 206–208)

If unspecified, the domain of a function f defined by an equation is the largest set of real numbers for which $f(x)$ is a real number.

Vertical-line test (p. 214)

A set of points in the xy-plane is the graph of a function if and only if every vertical line intersects the graph in at most one point.

Even function f (p. 224)

$f(-x) = f(x)$ for every x in the domain ($-x$ must also be in the domain).

Odd function f (p. 224)

$f(-x) = -f(x)$ for every x in the domain ($-x$ must also be in the domain).

Increasing function (p. 226)

A function f is increasing on an open interval I if, for any choice of x_1 and x_2 in I, with $x_1 < x_2$, we have $f(x_1) < f(x_2)$.

Decreasing function (p. 226)

A function f is decreasing on an open interval I if, for any choice of x_1 and x_2 in I, with $x_1 < x_2$, we have $f(x_1) > f(x_2)$.

Constant function (p. 226)

A function f is constant on an open interval I if, for all choices of x in I, the values of $f(x)$ are equal.

Local maximum (p. 227)

A function f, defined on some interval I, has a local maximum at c if there is an open interval in I containing c such that, for all x in this open interval, $f(x) \leq f(c)$.

Local minimum (p. 227)

A function f, defined on some interval I, has a local minimum at c if there is an open interval in I containing c such that, for all x in this open interval, $f(x) \geq f(c)$.

Absolute maximum and Absolute minimum (p. 227)

Let f denote a function defined on some interval I.
If there is a number u in I for which $f(x) \leq f(u)$ for all x in I, then f has an absolute maximum at u, and the number $f(u)$ is the absolute maximum of f on I.

If there is a number v in I for which $f(x) \geq f(v)$, for all x in I, then f has an absolute minimum at v and the number $f(v)$ is the absolute minimum of f on I.

Average rate of change of a function (p. 230)

The average rate of change of f from a to b is

$$\frac{\Delta y}{\Delta x} = \frac{f(b) - f(a)}{b - a} \quad a \neq b$$

Objectives

Section	You should be able to . . .	Examples	Review Exercises
3.1	1 Determine whether a relation represents a function (p. 199)	1–5	1, 2
	2 Find the value of a function (p. 202)	6, 7	3–5, 39
	3 Find the difference quotient of a function (p. 205)	8	15
	4 Find the domain of a function defined by an equation (p. 206)	9, 10	6–11
	5 Form the sum, difference, product, and quotient of two functions (p. 208)	11	12–14
3.2	1 Identify the graph of a function (p. 214)	1	27, 28
	2 Obtain information from or about the graph of a function (p. 215)	2–4	16(a)–(e), 17(a), 17(e), 17(g)

Section		You should be able to ...	Examples	Review Exercises
3.3	1	Determine even and odd functions from a graph (p. 223)	1	17(f)
	2	Identify even and odd functions from an equation (p. 225)	2	18–21
	3	Use a graph to determine where a function is increasing, decreasing, or constant (p. 225)	3	17(b)
	4	Use a graph to locate local maxima and local minima (p. 226)	4	17(c)
	5	Use a graph to locate the absolute maximum and the absolute minimum (p. 227)	5	17(d)
	6	Use a graphing utility to approximate local maxima and local minima and to determine where a function is increasing or decreasing (p. 229)	6	22, 23, 40(d), 41(b)
	7	Find the average rate of change of a function (p. 230)	7, 8	24–26
3.4	1	Graph the functions listed in the library of functions (p. 237)	1, 2	29, 30
	2	Graph piecewise-defined functions (p. 242)	3, 4	37, 38
3.5	1	Graph functions using vertical and horizontal shifts (p. 248)	1–5, 11–13	16(f), 31, 33–36
	2	Graph functions using compressions and stretches (p. 251)	6–8, 12	16(g), 32, 36
	3	Graph functions using reflections about the x-axis and the y-axis (p. 253)	9, 10, 11, 13	16(h), 32, 34, 36
3.6	1	Build and analyze functions (p. 260)	1–3	40, 41

Review Exercises

In Problems 1 and 2, determine whether each relation represents a function. For each function, state the domain and range.

1. $\{(-1,0),(2,3),(4,0)\}$ **2.** $\{(4,-1),(2,1),(4,2)\}$

In Problems 3–5, find the following for each function:

(a) $f(2)$ (b) $f(-2)$ (c) $f(-x)$ (d) $-f(x)$ (e) $f(x-2)$ (f) $f(2x)$

3. $f(x)=\dfrac{3x}{x^2-1}$ **4.** $f(x)=\sqrt{x^2-4}$ **5.** $f(x)=\dfrac{x^2-4}{x^2}$

In Problems 6–11, find the domain of each function.

6. $f(x)=\dfrac{x}{x^2-9}$ **7.** $f(x)=\sqrt{2-x}$ **8.** $g(x)=\dfrac{|x|}{x}$

9. $f(x)=\dfrac{x}{x^2+2x-3}$ **10.** $f(x)=\dfrac{\sqrt{x+1}}{x^2-4}$ **11.** $g(x)=\dfrac{x}{\sqrt{x+8}}$

In Problems 12–14, find $f+g$, $f-g$, $f\cdot g$, and $\dfrac{f}{g}$ for each pair of functions. State the domain of each of these functions.

12. $f(x)=2-x;\ g(x)=3x+1$ **13.** $f(x)=3x^2+x+1;\ g(x)=3x$ **14.** $f(x)=\dfrac{x+1}{x-1};\ g(x)=\dfrac{1}{x}$

15. Find the difference quotient of $f(x)=-2x^2+x+1$; that is, find $\dfrac{f(x+h)-f(x)}{h}$, $h\neq 0$.

16. Consider the graph of the function f on the right.

(a) Find the domain and the range of f.
(b) List the intercepts.
(c) Find $f(-2)$.
(d) For what value of x does $f(x)=-3$?
(e) Solve $f(x)>0$.
(f) Graph $y=f(x-3)$.
(g) Graph $y=f\left(\frac{1}{2}x\right)$.
(h) Graph $y=-f(x)$.

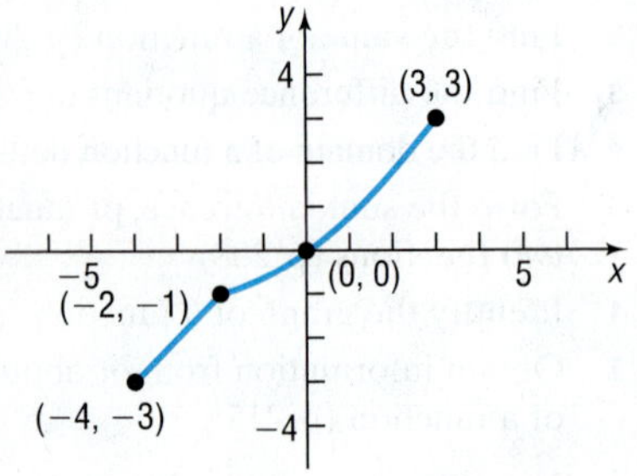

17. Use the graph of the function f shown to find:

(a) The domain and the range of f.
(b) The intervals on which f is increasing, decreasing, or constant.
(c) The local minimum values and local maximum values.
(d) The absolute maximum and absolute minimum.
(e) Whether the graph is symmetric with respect to the x-axis, the y-axis, or the origin.
(f) Whether the function is even, odd, or neither.
(g) The intercepts, if any.

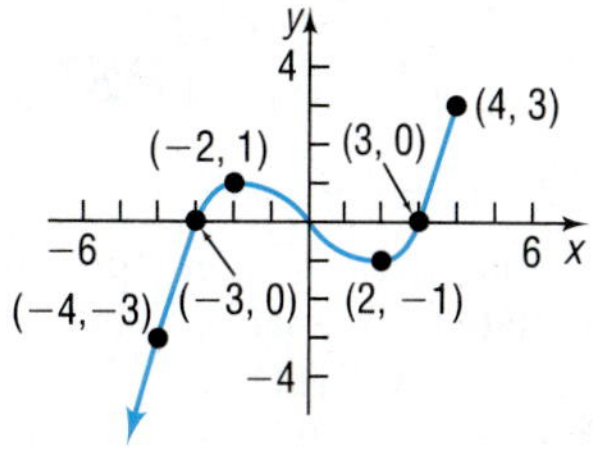

In Problems 18–21, determine (algebraically) whether the given function is even, odd, or neither.

18. $f(x) = x^3 - 4x$ **19.** $g(x) = \dfrac{4 + x^2}{1 + x^4}$ **20.** $G(x) = 1 - x + x^3$ **21.** $f(x) = \dfrac{x}{1 + x^2}$

In Problems 22 and 23, use a graphing utility to graph each function over the indicated interval. Approximate any local maximum values and local minimum values. Determine where the function is increasing and where it is decreasing.

22. $f(x) = 2x^3 - 5x + 1 \quad (-3, 3)$ **23.** $f(x) = 2x^4 - 5x^3 + 2x + 1 \quad (-2, 3)$

24. Find the average rate of change of $f(x) = 8x^2 - x$:
(a) From 1 to 2 (b) From 0 to 1 (c) From 2 to 4

In Problems 25 and 26, find the average rate of change from 2 to 3 for each function f. Be sure to simplify.

25. $f(x) = 2 - 5x$ **26.** $f(x) = 3x - 4x^2$

In Problems 27 and 28, is the graph shown the graph of a function?

27.

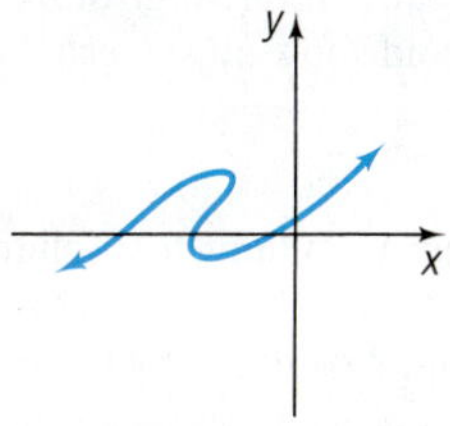

28.

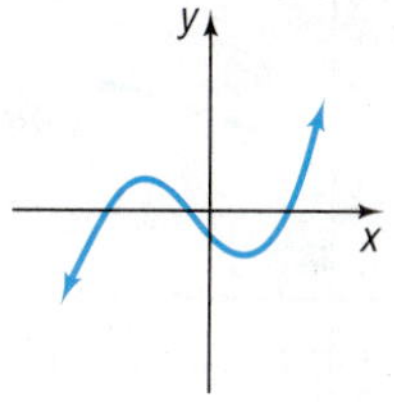

In Problems 29 and 30, graph each function. Be sure to label at least three points.

29. $f(x) = |x|$ **30.** $f(x) = \sqrt{x}$

In Problems 31–36, graph each function using the techniques of shifting, compressing or stretching, and reflections. Identify any intercepts of the graph. State the domain and, based on the graph, find the range.

31. $F(x) = |x| - 4$ **32.** $g(x) = -2|x|$ **33.** $h(x) = \sqrt{x - 1}$

34. $f(x) = \sqrt{1 - x}$ **35.** $h(x) = (x - 1)^2 + 2$ **36.** $g(x) = -2(x + 2)^3 - 8$

In Problems 37 and 38:
(a) Find the domain of each function. *(b) Locate any intercepts.* *(c) Graph each function.*
(d) Based on the graph, find the range. *(e) Is f continuous on its domain?*

37. $f(x) = \begin{cases} 3x & \text{if } -2 < x \le 1 \\ x + 1 & \text{if } x > 1 \end{cases}$ **38.** $f(x) = \begin{cases} x & \text{if } -4 \le x < 0 \\ 1 & \text{if } x = 0 \\ 3x & \text{if } x > 0 \end{cases}$

39. A function f is defined by

$$f(x) = \frac{Ax + 5}{6x - 2}$$

If $f(1) = 4$, find A.

40. Constructing a Closed Box A closed box with a square base is required to have a volume of 10 cubic feet.
(a) Build a model that expresses the amount A of material used to make such a box as a function of the length x of a side of the square base.
(b) How much material is required for a base 1 foot by 1 foot?
(c) How much material is required for a base 2 feet by 2 feet?
(d) Graph $A = A(x)$. For what value of x is A smallest?

41. Area of a Rectangle A rectangle has one vertex in quadrant I on the graph of $y = 10 - x^2$, another at the origin, one on the positive x-axis, and one on the positive y-axis.
(a) Express the area A of the rectangle as a function of x.
(b) Find the largest area A that can be enclosed by the rectangle.

Chapter Test

CHAPTER **Test Prep** VIDEOS

The Chapter Test Prep Videos are step-by-step solutions available in MyMathLab, or on this text's YouTube Channel. Flip back to the Resources for Success page for a link to this text's YouTube channel.

1. Determine whether each relation represents a function. For each function, state the domain and the range.

(a) $\{(2,5), (4,6), (6,7), (8,8)\}$

(b) $\{(1,3), (4,-2), (-3,5), (1,7)\}$

(c)

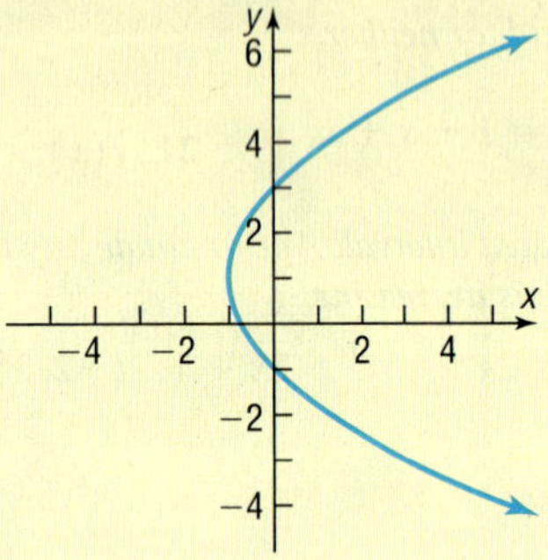

(d)

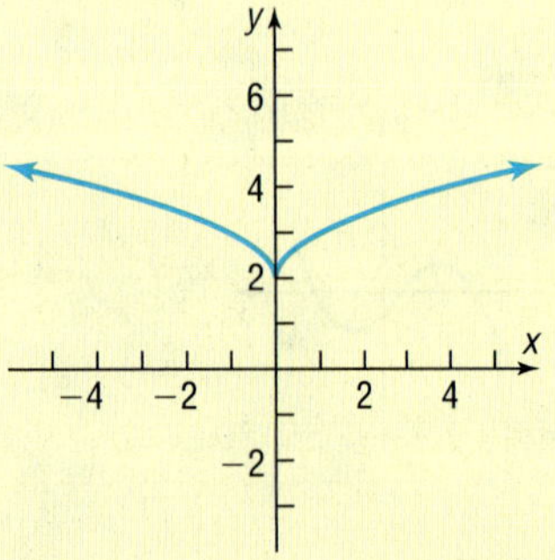

In Problems 2–4, find the domain of each function and evaluate each function at $x = -1$.

2. $f(x) = \sqrt{4 - 5x}$

3. $g(x) = \dfrac{x + 2}{|x + 2|}$

4. $h(x) = \dfrac{x - 4}{x^2 + 5x - 36}$

5. Consider the graph of the function f below.

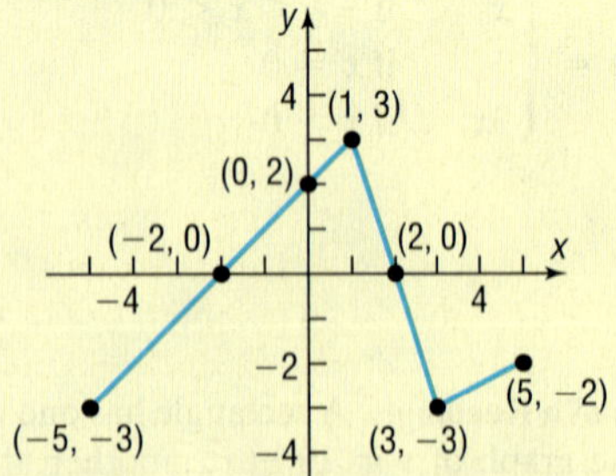

(a) Find the domain and the range of f.

(b) List the intercepts.

(c) Find $f(1)$.

(d) For what value(s) of x does $f(x) = -3$?

(e) Solve $f(x) < 0$.

6. Use a graphing utility to graph the function $f(x) = -x^4 + 2x^3 + 4x^2 - 2$ on the interval $(-5, 5)$. Then approximate any local maximum values and local minimum values rounded to two decimal places. Determine where the function is increasing and where it is decreasing.

7. Consider the function $g(x) = \begin{cases} 2x + 1 & \text{if } x < -1 \\ x - 4 & \text{if } x \geq -1 \end{cases}$

(a) Graph the function.

(b) List the intercepts.

(c) Find $g(-5)$.

(d) Find $g(2)$.

8. For the function $f(x) = 3x^2 - 2x + 4$, find the average rate of change of f from 3 to 4.

9. For the functions $f(x) = 2x^2 + 1$ and $g(x) = 3x - 2$, find the following and simplify.

(a) $(f - g)(x)$

(b) $(f \cdot g)(x)$

(c) $f(x + h) - f(x)$

10. Graph each function using the techniques of shifting, compressing or stretching, and reflecting. Start with the graph of the basic function and show all stages.

(a) $h(x) = -2(x + 1)^3 + 3$

(b) $g(x) = |x + 4| + 2$

11. The variable interest rate on a student loan changes each July 1 based on the bank prime loan rate. For the years 1992–2007, this rate can be approximated by the model

$$r(x) = -0.115x^2 + 1.183x + 5.623,$$

where x is the number of years since 1992 and r is the interest rate as a percent.

(a) Use a graphing utility to estimate the highest rate during this time period. During which year was the interest rate the highest?

(b) Use the model to estimate the rate in 2010. Does this value seem reasonable?

Source: U.S. Federal Reserve

12. A community skating rink is in the shape of a rectangle with semicircles attached at the ends. The length of the rectangle is 20 feet less than twice the width. The thickness of the ice is 0.75 inch.

(a) Build a model that expresses the ice volume, V, as a function of the width, x.

(b) How much ice is in the rink if the width is 90 feet?

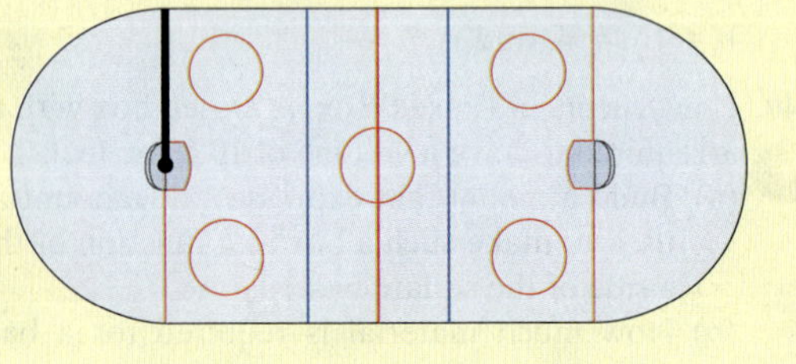

Cumulative Review

In Problems 1–6, find the real solutions of each equation.

1. $3x - 8 = 10$ **2.** $3x^2 - x = 0$

3. $x^2 - 8x - 9 = 0$ **4.** $6x^2 - 5x + 1 = 0$

5. $|2x + 3| = 4$ **6.** $\sqrt{2x + 3} = 2$

In Problems 7–9, solve each inequality. Graph the solution set.

7. $2 - 3x > 6$ **8.** $|2x - 5| < 3$ **9.** $|4x + 1| \geq 7$

10. (a) Find the distance from $P_1 = (-2, -3)$ to $P_2 = (3, -5)$.
(b) What is the midpoint of the line segment from P_1 to P_2?
(c) What is the slope of the line containing the points P_1 and P_2?

In Problems 11–14, graph each equation.

11. $3x - 2y = 12$ **12.** $x = y^2$

13. $x^2 + (y - 3)^2 = 16$ **14.** $y = \sqrt{x}$

15. For the equation $3x^2 - 4y = 12$, find the intercepts and check for symmetry.

16. Find the slope–intercept form of the equation of the line containing the points $(-2, 4)$ and $(6, 8)$.

In Problems 17–19, graph each function.

17. $f(x) = (x + 2)^2 - 3$

18. $f(x) = \dfrac{1}{x}$

19. $f(x) = \begin{cases} 2 - x & \text{if } x \leq 2 \\ |x| & \text{if } x > 2 \end{cases}$

Chapter Projects

Internet-based Project

I. Choosing a Wireless Data Plan Collect information from your family, friends, or consumer agencies such as Consumer Reports. Then decide on a cellular provider, choosing the company that you feel offers the best service. Once you have selected a service provider, research the various types of individual plans offered by the company by visiting the provider's website. Many providers offer family plans that include unlimited talk and text. The monthly cost is primarily determined by the amount of data used and the number of devices.

1. Suppose you expect to use 10 gigabytes of data for a single smartphone. What would be the monthly cost of each plan you are considering?
2. Suppose you expect to use 30 gigabytes of data and want a personal hotspot, but you still have only a single smartphone. What would be the monthly cost of each plan you are considering?
3. Suppose you expect to use 20 gigabytes of data with three smartphones sharing the data. What would be the monthly cost of each plan you are considering?
4. Suppose you expect to use 20 gigabytes of data with a single smartphone and a personal hotspot. What would be the monthly cost of each plan you are considering?
5. Build a model that describes the monthly cost C, in dollars, as a function of the number of data gigabytes used, g, assuming a single smartphone and a personal hotspot for each plan you are considering.
6. Graph each function from Problem 5.
7. Based on your particular usage, which plan is best for you?
8. Now, develop an Excel spreadsheet to analyze the various plans you are considering. Suppose you want a family plan with unlimited talk and text that offers 10 gigabytes of shared data and costs \$100 per month. Additional gigabytes of data cost \$15 per gigabyte, extra phones can be added to the plan for \$15 each per month, and each hotspot costs \$20 per month. Because wireless

data plans have a cost structure based on piecewise-defined functions, we need an "if/then" statement within Excel to analyze the cost of the plan. Use the accompanying Excel spreadsheet as a guide in developing your spreadsheet. Enter into your spreadsheet a variety of possible amounts of data and various numbers of additional phones and hotsposts.

	A	B	C	D	
1					
2	Monthly fee	$100			
3	Allotted data per month (GB)	10			
4	Data used (GB)	12			
5	Cost per additional GB of data	$15			
6					
7	Monthly cost of hotspot	$20			
8	Number of hotspots	1			
9	Monthly cost of additional phone	$15			
10	Number of additional phones	2			
11					
12	Cost of data	=IF(B4<B3,B2,B2+B5*(B4-B3))			
13	Cost of additional devices/hotspots	=B8*B7+B10*B9			
14					
15	Total Cost	=B12+B13			
16					

9. Write a paragraph supporting the choice in plans that best meets your needs.

10. How are "if/then" loops similar to a piecewise-defined function?

The following projects are available on the Instructor's Resource Center (IRC).

II. Project at Motorola: *Wireless Internet Service* Use functions and their graphs to analyze the total cost of various wireless Internet service plans.

III. Cost of Cable When government regulations and customer preference influence the path of a new cable line, the Pythagorean Theorem can be used to assess the cost of installation.

IV. Oil Spill Functions are used to analyze the size and spread of an oil spill from a leaking tanker.

Linear and Quadratic Functions

4

The Beta of a Stock

Investing in the stock market can be rewarding and fun, but how does one go about selecting which stocks to purchase? Financial investment firms hire thousands of analysts who track individual stocks (equities) and assess the value of the underlying company. One measure the analysts consider is the *beta* of the stock. **Beta** measures the risk of an individual company's equity relative to that of a market basket of stocks, such as the Standard & Poor's 500. But how is beta computed?

—See the Internet-based Chapter Project I—

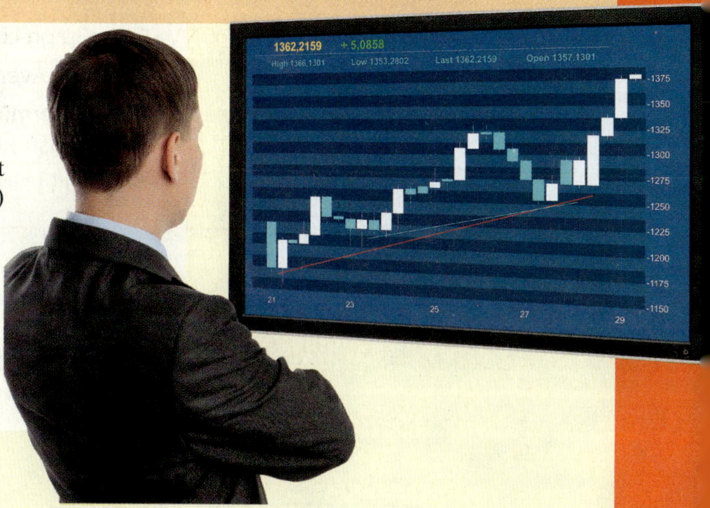

••• A Look Back

Up to now, our discussion has focused on graphs of equations and functions. We learned how to graph equations using the point-plotting method, intercepts, and the tests for symmetry. In addition, we learned what a function is and how to identify whether a relation represents a function. We also discussed properties of functions, such as domain/range, increasing/decreasing, even/odd, and average rate of change.

A Look Ahead •••

Going forward, we will look at classes of functions. This chapter focuses on linear and quadratic functions, their properties, and their applications.

Outline

4.1 Properties of Linear Functions and Linear Models

PREPARING FOR THIS SECTION *Before getting started, review the following:*

- Lines (Section 2.3, pp. 167–175)
- Graphs of Equations in Two Variables; Intercepts; Symmetry (Section 2.2, pp. 157–164)
- Linear Equations (Section 1.1, pp. 82–87)
- Functions (Section 3.1, pp. 199–208)
- The Graph of a Function (Section 3.2, pp. 214–217)
- Properties of Functions (Section 3.3, pp. 223–231)

Now Work the **'Are You Prepared?'** problems on page 280.

OBJECTIVES
1. Graph Linear Functions (p. 274)
2. Use Average Rate of Change to Identify Linear Functions (p. 274)
3. Determine Whether a Linear Function Is Increasing, Decreasing, or Constant (p. 277)
4. Build Linear Models from Verbal Descriptions (p. 278)

1 Graph Linear Functions

In Section 2.3 we discussed lines. In particular, for nonvertical lines we developed the slope–intercept form of the equation of a line $y = mx + b$. When the slope–intercept form of a line is written using function notation, the result is a *linear function*.

DEFINITION

A **linear function** is a function of the form

$$f(x) = mx + b$$

The graph of a linear function is a line with slope m and y-intercept b. Its domain is the set of all real numbers.

Functions that are not linear are said to be **nonlinear**.

EXAMPLE 1 **Graphing a Linear Function**

Graph the linear function $f(x) = -3x + 7$. What are the domain and the range of f?

Solution This is a linear function with slope $m = -3$ and y-intercept $b = 7$. To graph this function, plot the point $(0, 7)$, the y-intercept, and use the slope to find an additional point by moving right 1 unit and down 3 units. See Figure 1. The domain and the range of f are each the set of all real numbers. ●

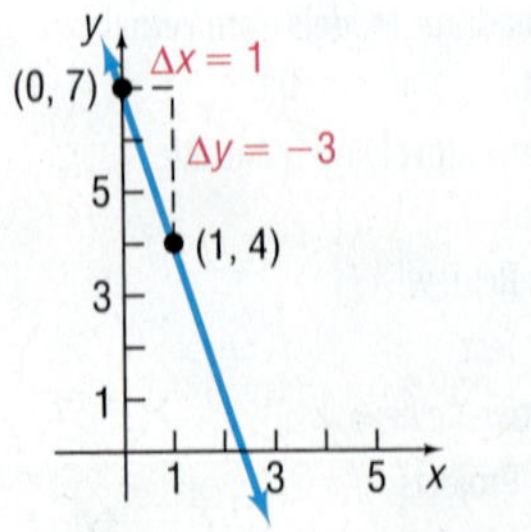

Figure 1 $f(x) = -3x + 7$

Alternatively, an additional point could have been found by evaluating the function at some $x \neq 0$. For $x = 1$, $f(1) = -3(1) + 7 = 4$ and the point $(1, 4)$ lies on the graph.

Now Work PROBLEMS 13(a) AND (b)

2 Use Average Rate of Change to Identify Linear Functions

Look at Table 1, which shows certain values of the independent variable x and corresponding values of the dependent variable y for the function $f(x) = -3x + 7$. Notice that as the value of the independent variable, x, increases by 1, the value of the dependent variable y decreases by 3. That is, the average rate of change of y with respect to x is a constant, -3.

Table 1

x	$y = f(x) = -3x + 7$	Average Rate of Change $= \dfrac{\Delta y}{\Delta x}$
−2	13	
		$\dfrac{10 - 13}{-1 - (-2)} = \dfrac{-3}{1} = -3$
−1	10	
		$\dfrac{7 - 10}{0 - (-1)} = \dfrac{-3}{1} = -3$
0	7	
		−3
1	4	
		−3
2	1	
		−3
3	−2	

It is not a coincidence that the average rate of change of the linear function $f(x) = -3x + 7$ is the slope of the linear function. That is, $\dfrac{\Delta y}{\Delta x} = m = -3$. The following theorem states this fact.

THEOREM

Average Rate of Change of a Linear Function

Linear functions have a constant average rate of change. That is, the average rate of change of a linear function $f(x) = mx + b$ is

$$\frac{\Delta y}{\Delta x} = m$$

Proof The average rate of change of $f(x) = mx + b$ from x_1 to x_2, $x_1 \neq x_2$, is

$$\frac{\Delta y}{\Delta x} = \frac{f(x_2) - f(x_1)}{x_2 - x_1} = \frac{(mx_2 + b) - (mx_1 + b)}{x_2 - x_1}$$

$$= \frac{mx_2 - mx_1}{x_2 - x_1} = \frac{m(x_2 - x_1)}{x_2 - x_1} = m$$ ■

Based on the theorem just proved, the average rate of change of the function $g(x) = -\frac{2}{5}x + 5$ is $-\frac{2}{5}$.

Now Work PROBLEM 13(c)

As it turns out, only linear functions have a constant average rate of change. Because of this, the average rate of change can be used to determine whether a function is linear. This is especially useful if the function is defined by a data set.

EXAMPLE 2 **Using the Average Rate of Change to Identify Linear Functions**

(a) A strain of *E. coli* known as Beu 397-recA441 is placed into a Petri dish at 30° Celsius and allowed to grow. The data shown in Table 2 on the next page are collected. The population is measured in grams and the time in hours. Plot the ordered pairs (x, y) in the Cartesian plane, and use the average rate of change to determine whether the function is linear.

(b) The data in Table 3 represent the maximum number of heartbeats that a healthy individual of different ages should have during a 15-second interval of time while exercising. Plot the ordered pairs (x, y) in the Cartesian plane, and use the average rate of change to determine whether the function is linear.

Table 2

Time (hours), *x*	Population (grams), *y*	(*x*, *y*)
0	0.09	(0, 0.09)
1	0.12	(1, 0.12)
2	0.16	(2, 0.16)
3	0.22	(3, 0.22)
4	0.29	(4, 0.29)
5	0.39	(5, 0.39)

Table 3

Age, *x*	Maximum Number of Heartbeats, *y*	(*x*, *y*)
20	50	(20, 50)
30	47.5	(30, 47.5)
40	45	(40, 45)
50	42.5	(50, 42.5)
60	40	(60, 40)
70	37.5	(70, 37.5)

Source: American Heart Association

Solution Compute the average rate of change of each function. If the average rate of change is constant, the function is linear. If the average rate of change is not constant, the function is nonlinear.

(a) Figure 2 shows the points listed in Table 2 plotted in the Cartesian plane. Note that it is impossible to draw a straight line that contains all the points. Table 4 displays the average rate of change of the population.

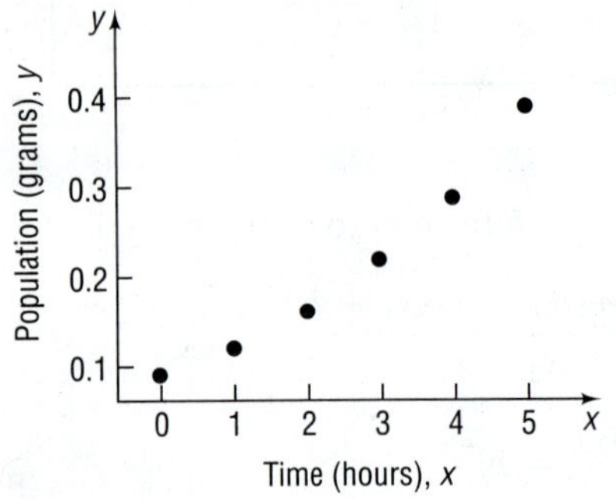

Figure 2

Table 4

Time (hours), *x*	Population (grams), *y*	Average Rate of Change $= \dfrac{\Delta y}{\Delta x}$
0	0.09	
		$\dfrac{0.12 - 0.09}{1 - 0} = 0.03$
1	0.12	
		0.04
2	0.16	
		0.06
3	0.22	
		0.07
4	0.29	
		0.10
5	0.39	

Because the average rate of change is not constant, the function is not linear. In fact, because the average rate of change is increasing as the value of the independent variable increases, the function is increasing at an increasing rate. So not only is the population increasing over time, but it is also growing more rapidly as time passes.

(b) Figure 3 shows the points listed in Table 3 plotted in the Cartesian plane. Note that the data in Figure 3 lie on a straight line. Table 5 displays the average rate of change of the maximum number of heartbeats. The average rate of change of the heartbeat data is constant, -0.25 beat per year, so the function is linear.

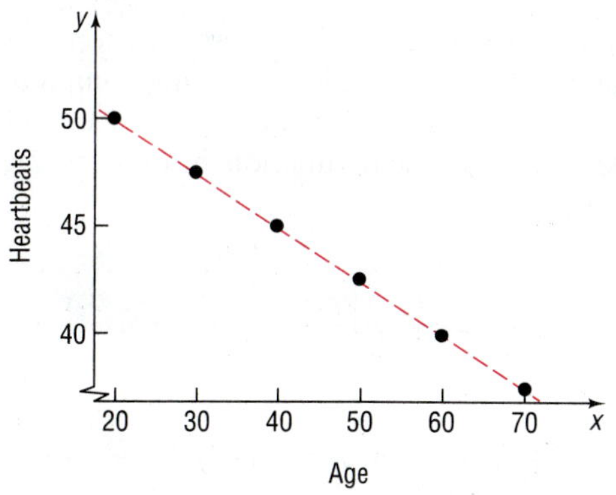

Figure 3

Table 5

Age, x	Maximum Number of Heartbeats, y	Average Rate of Change $= \frac{\Delta y}{\Delta x}$
20	50	
		$\frac{47.5 - 50}{30 - 20} = -0.25$
30	47.5	
		-0.25
40	45	
		-0.25
50	42.5	
		-0.25
60	40	
		-0.25
70	37.5	

Now Work PROBLEM 21

3 Determine Whether a Linear Function Is Increasing, Decreasing, or Constant

Look back at the Seeing the Concept on page 169. When the slope m of a linear function is positive ($m > 0$), the line slants upward from left to right. When the slope m of a linear function is negative ($m < 0$), the line slants downward from left to right. When the slope m of a linear function is zero ($m = 0$), the line is horizontal.

THEOREM

Increasing, Decreasing, and Constant Linear Functions

A linear function $f(x) = mx + b$ is increasing over its domain if its slope, m, is positive. It is decreasing over its domain if its slope, m, is negative. It is constant over its domain if its slope, m, is zero.

EXAMPLE 3 **Determining Whether a Linear Function Is Increasing, Decreasing, or Constant**

Determine whether the following linear functions are increasing, decreasing, or constant.

(a) $f(x) = 5x - 2$ (b) $g(x) = -2x + 8$

(c) $s(t) = \frac{3}{4}t - 4$ (d) $h(z) = 7$

Solution

(a) For the linear function $f(x) = 5x - 2$, the slope is 5, which is positive. The function f is increasing on the interval $(-\infty, \infty)$.

(b) For the linear function $g(x) = -2x + 8$, the slope is -2, which is negative. The function g is decreasing on the interval $(-\infty, \infty)$.

(c) For the linear function $s(t) = \frac{3}{4}t - 4$, the slope is $\frac{3}{4}$, which is positive. The function s is increasing on the interval $(-\infty, \infty)$.

(d) The linear function h can be written as $h(z) = 0z + 7$. Because the slope is 0, the function h is constant on the interval $(-\infty, \infty)$.

Now Work PROBLEM 13 (d)

4 Build Linear Models from Verbal Descriptions

When the average rate of change of a function is constant, a linear function can model the relation between the two variables. For example, if a recycling company pays \$0.52 per pound for aluminum cans, then the relation between the price paid p and the pounds recycled x can be modeled as the linear function $p(x) = 0.52x$, with slope $m = \dfrac{0.52 \text{ dollar}}{1 \text{ pound}}$.

Modeling with a Linear Function

If the average rate of change of a function is a constant m, a linear function f can be used to model the relation between the two variables as follows:

$$f(x) = mx + b$$

where b is the value of f at 0; that is, $b = f(0)$.

EXAMPLE 4 **Straight-line Depreciation**

Book value is the value of an asset that a company uses to create its balance sheet. Some companies depreciate assets using straight-line depreciation so that the value of the asset declines by a fixed amount each year. The amount of the decline depends on the useful life that the company assigns to the asset. Suppose a company just purchased a fleet of new cars for its sales force at a cost of \$31,500 per car. The company chooses to depreciate each vehicle using the straight-line method over 7 years. This means that each car will depreciate by $\dfrac{\$31{,}500}{7} = \$4500$ per year.

(a) Write a linear function that expresses the book value V of each car as a function of its age, x, in years.
(b) Graph the linear function.
(c) What is the book value of each car after 3 years?
(d) Interpret the slope.
(e) When will the book value of each car be \$9000?
[**Hint:** Solve the equation $V(x) = 9000$.]

Solution (a) If we let $V(x)$ represent the value of each car after x years, then $V(0)$ represents the original value of each car, so $V(0) = \$31{,}500$. The y-intercept of the linear function is \$31,500. Because each car depreciates by \$4500 per year, the slope of the linear function is -4500. The linear function that represents the book value V of each car after x years is

$$V(x) = -4500x + 31{,}500$$

(b) Figure 4 shows the graph of V.

(c) The book value of each car after 3 years is

$$V(3) = -4500(3) + 31{,}500$$
$$= \$18{,}000$$

(d) Since the slope of $V(x) = -4500x + 31{,}500$ is -4500, the average rate of change of the book value is $-\$4500$/year. So for each additional year that passes, the book value of the car decreases by \$4500.

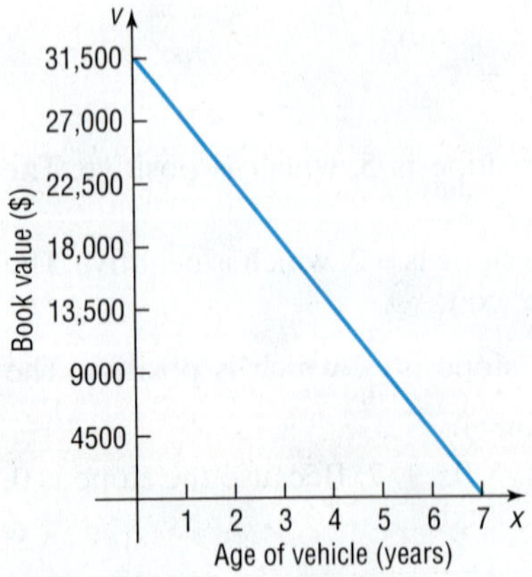

Figure 4 $V(x) = -4500x + 31{,}500$

(e) To find when the book value will be \$9000, solve the equation

$$
\begin{aligned}
V(x) &= 9000 \\
-4500x + 31{,}500 &= 9000 \\
-4500x &= -22{,}500 && \text{Subtract 31,500 from each side.} \\
x &= \frac{-22{,}500}{-4500} = 5 && \text{Divide by } -4500.
\end{aligned}
$$

The car will have a book value of \$9000 when it is 5 years old.

Now Work PROBLEM 45

EXAMPLE 5

Supply and Demand

The **quantity supplied** of a good is the amount of a product that a company is willing to make available for sale at a given price. The **quantity demanded** of a good is the amount of a product that consumers are willing to purchase at a given price. Suppose that the quantity supplied, S, and the quantity demanded, D, of cellular telephones each month are given by the following functions:

$$S(p) = 60p - 900$$
$$D(p) = -15p + 2850$$

where p is the price (in dollars) of the telephone.

(a) The **equilibrium price** of a product is defined as the price at which quantity supplied equals quantity demanded. That is, the equilibrium price is the price at which $S(p) = D(p)$. Find the equilibrium price of cellular telephones. What is the **equilibrium quantity**, the amount demanded (or supplied) at the equilibrium price?

(b) Determine the prices for which quantity supplied is greater than quantity demanded. That is, solve the inequality $S(p) > D(p)$.

(c) Graph $S = S(p)$ and $D = D(p)$, and label the **equilibrium point**, the point of intersection of S and D.

Solution

(a) To find the equilibrium price, solve the equation $S(p) = D(p)$.

$$
\begin{aligned}
60p - 900 &= -15p + 2850 && S(p) = 60p - 900;\ D(p) = -15p + 2850 \\
60p &= -15p + 3750 && \text{Add 900 to each side.} \\
75p &= 3750 && \text{Add } 15p \text{ to each side.} \\
p &= 50 && \text{Divide each side by 75.}
\end{aligned}
$$

The equilibrium price is \$50 per cellular phone. To find the equilibrium quantity, evaluate either $S(p)$ or $D(p)$ at $p = 50$.

$$S(50) = 60(50) - 900 = 2100$$

The equilibrium quantity is 2100 cellular phones. At a price of \$50 per phone, the company will produce and sell 2100 phones each month and have no shortages or excess inventory.

(b) The inequality $S(p) > D(p)$ is

$$
\begin{aligned}
60p - 900 &> -15p + 2850 && S(p) > D(p) \\
60p &> -15p + 3750 && \text{Add 900 to each side.} \\
75p &> 3750 && \text{Add } 15p \text{ to each side.} \\
p &> 50 && \text{Divide each side by 75.}
\end{aligned}
$$

If the company charges more than \$50 per phone, quantity supplied will exceed quantity demanded. In this case the company will have excess phones in inventory.

(c) Figure 5 shows the graphs of $S = S(p)$ and $D = D(p)$ with the equilibrium point labeled.

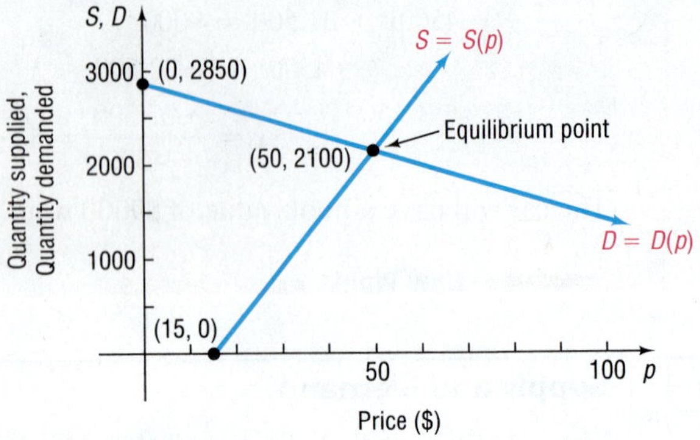

Figure 5 Supply and demand functions

Now Work PROBLEM 39

4.1 Assess Your Understanding

'Are You Prepared?' *Answers are given at the end of these exercises. If you get a wrong answer, read the pages listed in red.*

1. Graph $y = x^2 - 1$. (pp. 157–164)
2. Find the slope of the line joining the points $(2, 5)$ and $(-1, 3)$. (pp. 167–175)
3. Find the average rate of change of $f(x) = 3x^2 - 2$, from 2 to 4. (pp. 223–231)
4. Solve: $60x - 900 = -15x + 2850$. (pp. 82–87)
5. If $f(x) = x^2 - 4$, find $f(-2)$. (pp. 199–208)
6. ***True or False*** The graph of the function $f(x) = x^2$ is increasing on the interval $(0, \infty)$. (pp. 214–217)

Concepts and Vocabulary

7. For the graph of the linear function $f(x) = mx + b$, m is the ________ and b is the __________.
8. If the slope m of the graph of a linear function is __________, the function is increasing over its domain.
9. ***True or False*** The slope of a nonvertical line is the average rate of change of the linear function.
10. ***True or False*** The average rate of change of $f(x) = 2x + 8$ is 8.
11. What is the only type of function that has a constant average rate of change?
    (a) linear function (b) quadratic function
    (c) step function (d) absolute value function
12. A car has 12,500 miles on its odometer. Say the car is driven an average of 40 miles per day. Choose the model that expresses the number of miles N that will be on its odometer after x days.
    (a) $N(x) = -40x + 12{,}500$ (b) $N(x) = 40x - 12{,}500$
    (c) $N(x) = 12{,}500x + 40$ (d) $N(x) = 40x + 12{,}500$

Skill Building

In Problems 13–20, a linear function is given.
(a) Determine the slope and y-intercept of each function.
(b) Use the slope and y-intercept to graph the linear function.
(c) Determine the average rate of change of each function.
(d) Determine whether the linear function is increasing, decreasing, or constant.

13. $f(x) = 2x + 3$ 14. $g(x) = 5x - 4$ 15. $h(x) = -3x + 4$ 16. $p(x) = -x + 6$

17. $f(x) = \frac{1}{4}x - 3$ 18. $h(x) = -\frac{2}{3}x + 4$ 19. $F(x) = 4$ 20. $G(x) = -2$

In Problems 21–28, determine whether the given function is linear or nonlinear. If it is linear, determine the slope.

21.

x	y = f(x)
−2	4
−1	1
0	−2
1	−5
2	−8

22.

x	y = f(x)
−2	1/4
−1	1/2
0	1
1	2
2	4

23.

x	y = f(x)
−2	−8
−1	−3
0	0
1	1
2	0

24.

x	y = f(x)
−2	−4
−1	0
0	4
1	8
2	12

25.

x	$y = f(x)$
−2	−26
−1	−4
0	2
1	−2
2	−10

26.

x	$y = f(x)$
−2	−4
−1	−3.5
0	−3
1	−2.5
2	−2

27.

x	$y = f(x)$
−2	8
−1	8
0	8
1	8
2	8

28.

x	$y = f(x)$
−2	0
−1	1
0	4
1	9
2	16

Applications and Extensions

29. Suppose that $f(x) = 4x - 1$ and $g(x) = -2x + 5$.
(a) Solve $f(x) = 0$. (b) Solve $f(x) > 0$.
(c) Solve $f(x) = g(x)$. (d) Solve $f(x) \leq g(x)$.
(e) Graph $y = f(x)$ and $y = g(x)$ and label the point that represents the solution to the equation $f(x) = g(x)$.

30. Suppose that $f(x) = 3x + 5$ and $g(x) = -2x + 15$.
(a) Solve $f(x) = 0$. (b) Solve $f(x) < 0$.
(c) Solve $f(x) = g(x)$. (d) Solve $f(x) \geq g(x)$.
(e) Graph $y = f(x)$ and $y = g(x)$ and label the point that represents the solution to the equation $f(x) = g(x)$.

31. In parts (a)–(f), use the following figure.

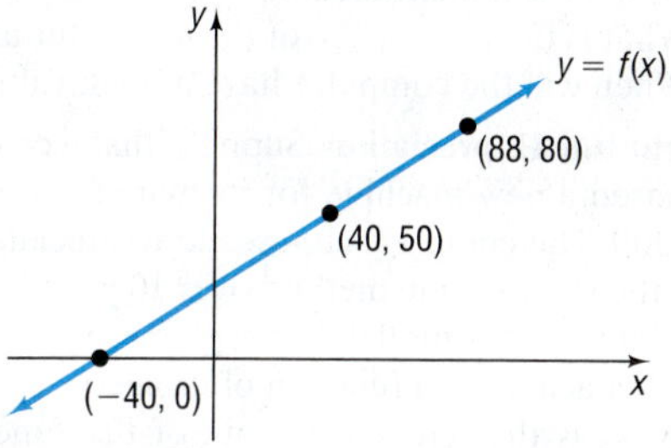

(a) Solve $f(x) = 50$. (b) Solve $f(x) = 80$.
(c) Solve $f(x) = 0$. (d) Solve $f(x) > 50$.
(e) Solve $f(x) \leq 80$. (f) Solve $0 < f(x) < 80$.

32. In parts (a)–(f), use the following figure.

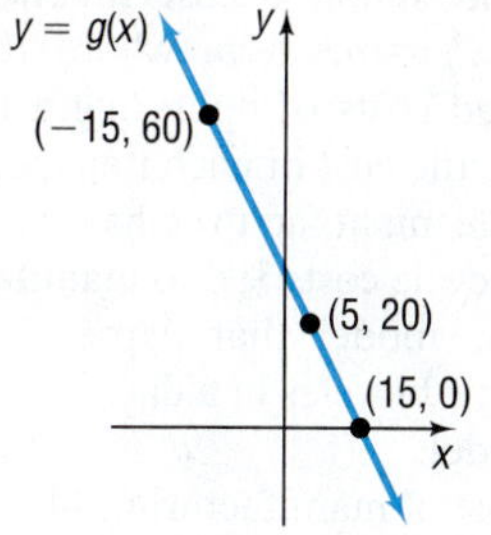

(a) Solve $g(x) = 20$. (b) Solve $g(x) = 60$.
(c) Solve $g(x) = 0$. (d) Solve $g(x) > 20$.
(e) Solve $g(x) \leq 60$. (f) Solve $0 < g(x) < 60$.

33. In parts (a) and (b), use the following figure.

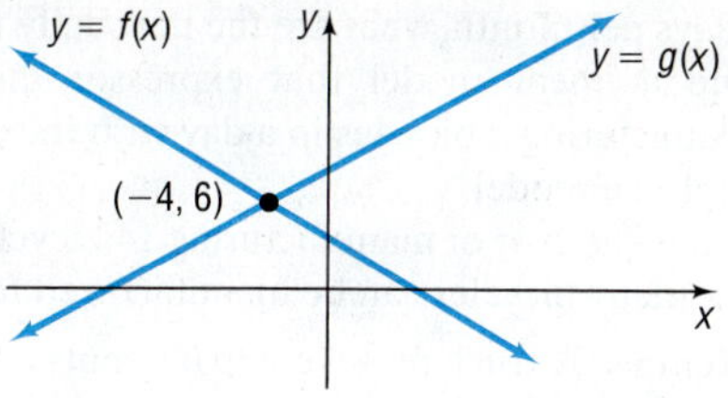

(a) Solve the equation: $f(x) = g(x)$.
(b) Solve the inequality: $f(x) > g(x)$.

34. In parts (a) and (b), use the following figure.

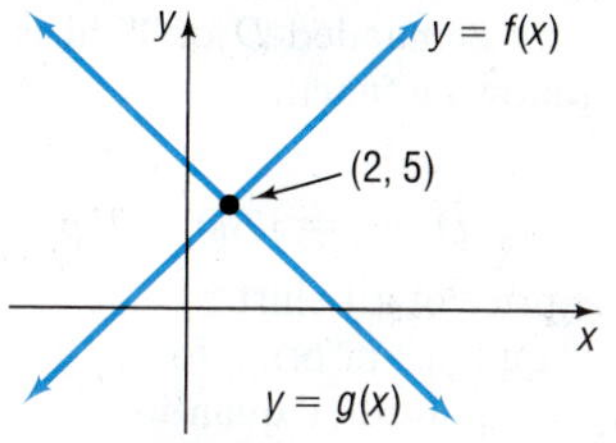

(a) Solve the equation: $f(x) = g(x)$.
(b) Solve the inequality: $f(x) \leq g(x)$.

35. In parts (a) and (b), use the following figure.

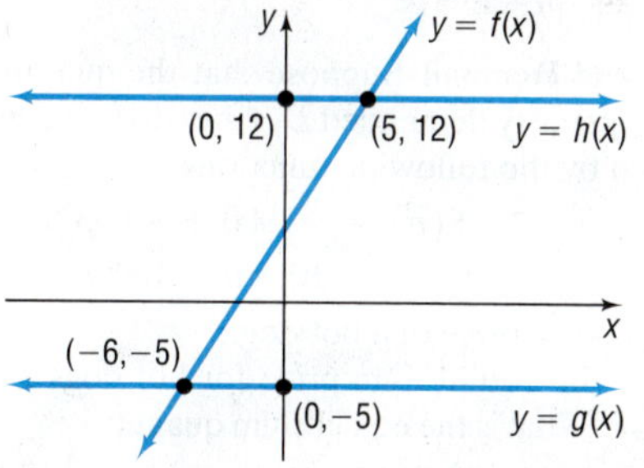

(a) Solve the equation: $f(x) = g(x)$.
(b) Solve the inequality: $g(x) \leq f(x) < h(x)$.

36. In parts (a) and (b), use the following figure.

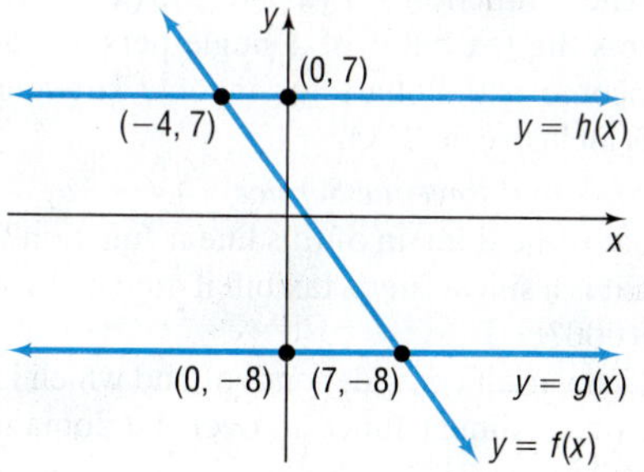

(a) Solve the equation: $f(x) = g(x)$.
(b) Solve the inequality: $g(x) < f(x) \leq h(x)$.

37. Truck Rentals The cost C, in dollars, of a one-day truck rental is modeled by the function $C(x) = 0.35x + 45$, where x is the number of miles driven.
(a) What is the cost if you drive $x = 40$ miles?
(b) If the cost of renting the truck is $108, how many miles did you drive?
(c) Suppose that you want the cost to be no more than $150. What is the maximum number of miles that you can drive?
(d) What is the implied domain of C?

38. Phone Charges The monthly cost C, in dollars, for calls from the United States to Germany on a certain phone plan is modeled by the function $C(x) = 0.26x + 5$, where x is the number of minutes used.

(a) What is the cost if you talk on the phone for $x = 50$ minutes?
(b) Suppose that your monthly bill is $21.64. How many minutes did you use the phone?
(c) Suppose that you budget yourself $50 per month for the phone. What is the maximum number of minutes that you can talk?
(d) What is the implied domain of C if there are 30 days in the month?

39. Supply and Demand Suppose that the quantity supplied S and the quantity demanded D of T-shirts at a concert are given by the following functions:

$$S(p) = -600 + 50p$$
$$D(p) = 1200 - 25p$$

where p is the price of a T-shirt.

(a) Find the equilibrium price for T-shirts at this concert. What is the equilibrium quantity?
(b) Determine the prices for which quantity demanded is greater than quantity supplied.
(c) What do you think will eventually happen to the price of T-shirts if quantity demanded is greater than quantity supplied?

40. Supply and Demand Suppose that the quantity supplied S and the quantity demanded D of hot dogs at a baseball game are given by the following functions:

$$S(p) = -2000 + 3000p$$
$$D(p) = 10{,}000 - 1000p$$

where p is the price of a hot dog.

(a) Find the equilibrium price for hot dogs at the baseball game. What is the equilibrium quantity?
(b) Determine the prices for which quantity demanded is less than quantity supplied.
(c) What do you think will eventually happen to the price of hot dogs if quantity demanded is less than quantity supplied?

41. Taxes The function $T(x) = 0.15(x - 9075) + 907.50$ represents the tax bill T of a single person whose adjusted gross income is x dollars for income between $9075 and $36,900, inclusive, in 2014.

Source: Internal Revenue Service

(a) What is the domain of this linear function?
(b) What is a single filer's tax bill if adjusted gross income is $20,000?
(c) Which variable is independent and which is dependent?
(d) Graph the linear function over the domain specified in part (a).
(e) What is a single filer's adjusted gross income if the tax bill is $3671.25?

42. Luxury Tax In 2011, major league baseball signed a labor agreement with the players. In this agreement, any team whose payroll exceeded $189 million in 2014 had to pay a luxury tax of 50%. The linear function $T(p) = 0.50(p - 189)$ describes the luxury tax T for a team whose payroll was p (in millions of dollars).

Source: Major League Baseball

(a) What is the implied domain of this linear function?
(b) What was the luxury tax for the New York Yankees, whose 2014 payroll was $203.4 million?
(c) Graph the linear function.
(d) What is the payroll of a team that pays a luxury tax of $15.7 million?

The point at which a company's profits equal zero is called the company's ***break-even point.*** *For Problems 43 and 44, let R represent a company's revenue, let C represent the company's costs, and let x represent the number of units produced and sold each day.*

(a) Find the firm's break-even point; that is, find x so that R = C.
(b) Find the values of x such that $R(x) > C(x)$. This represents the number of units that the company must sell to earn a profit.

43. $R(x) = 8x$
$C(x) = 4.5x + 17{,}500$

44. $R(x) = 12x$
$C(x) = 10x + 15{,}000$

45. Straight-line Depreciation Suppose that a company has just purchased a new computer for $3000. The company chooses to depreciate the computer using the straight-line method over 3 years.

(a) Write a linear model that expresses the book value V of the computer as a function of its age x.
(b) What is the implied domain of the function found in part (a)?
(c) Graph the linear function.
(d) What is the book value of the computer after 2 years?
(e) When will the computer have a book value of $2000?

46. Straight-line Depreciation Suppose that a company has just purchased a new machine for its manufacturing facility for $120,000. The company chooses to depreciate the machine using the straight-line method over 10 years.

(a) Write a linear model that expresses the book value V of the machine as a function of its age x.
(b) What is the implied domain of the function found in part (a)?
(c) Graph the linear function.
(d) What is the book value of the machine after 4 years?
(e) When will the machine have a book value of $72,000?

47. Cost Function The simplest cost function is the linear cost function, $C(x) = mx + b$, where the y-intercept b represents the fixed costs of operating a business and the slope m represents the cost of each item produced. Suppose that a small bicycle manufacturer has daily fixed costs of $1800, and each bicycle costs $90 to manufacture.

(a) Write a linear model that expresses the cost C of manufacturing x bicycles in a day.
(b) Graph the model.
(c) What is the cost of manufacturing 14 bicycles in a day?
(d) How many bicycles could be manufactured for $3780?

48. Cost Function Refer to Problem 47. Suppose that the landlord of the building increases the bicycle manufacturer's rent by $100 per month.

(a) Assuming that the manufacturer is open for business 20 days per month, what are the new daily fixed costs?
(b) Write a linear model that expresses the cost C of manufacturing x bicycles in a day with the higher rent.
(c) Graph the model.
(d) What is the cost of manufacturing 14 bicycles in a day?
(e) How many bicycles can be manufactured for $3780?

49. Truck Rentals A truck rental company rents a truck for one day by charging $31.95 plus $0.89 per mile.

(a) Write a linear model that relates the cost C, in dollars, of renting the truck to the number x of miles driven.

(b) What is the cost of renting the truck if the truck is driven 110 miles? 230 miles?

50. International Call Plan A cell phone company offers an international plan by charging \$30 for the first 80 minutes, plus \$0.50 for each minute over 80.

(a) Write a linear model that relates the cost C, in dollars, of talking x minutes, assuming $x \geq 80$.

(b) What is the cost of talking 105 minutes? 120 minutes?

Mixed Practice

51. Building a Linear Model from Data How many songs can an iPod hold? The following data represent the memory m and the number of songs, n.

Memory, *m* (gigabytes)	Number of Songs, *n*
8	1750
16	3500
32	7000
64	14,000

(a) Plot the ordered pairs (m, n) in a Cartesian plane.

(b) Show that the number of songs n is a linear function of the memory m.

(c) Determine the linear function that describes the relation between m and n.

(d) What is the implied domain of the linear function?

(e) Graph the linear function in the Cartesian plane drawn in part (a).

(f) Interpret the slope.

52. Building a Linear Model from Data The following data represent the various combinations of soda and hot dogs that Yolanda can buy at a baseball game with \$60.

Soda, *s*	Hot Dogs, *h*
20	0
15	3
10	6
5	9

(a) Plot the ordered pairs (s, h) in a Cartesian plane.

(b) Show that the number of hot dogs purchased h is a linear function of the number of sodas purchased s.

(c) Determine the linear function that describes the relation between s and h.

(d) What is the implied domain of the linear function?

(e) Graph the linear function in the Cartesian plane drawn in part (a).

(f) Interpret the slope.

(g) Interpret the values of the intercepts.

Explaining Concepts: Discussion and Writing

53. Which of the following functions might have the graph shown? (More than one answer is possible.)

(a) $f(x) = 2x - 7$

(b) $g(x) = -3x + 4$

(c) $H(x) = 5$

(d) $F(x) = 3x + 4$

(e) $G(x) = \frac{1}{2}x + 2$

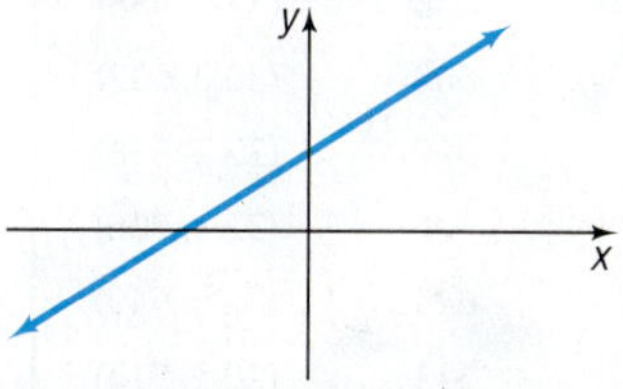

54. Which of the following functions might have the graph shown? (More than one answer is possible.)

(a) $f(x) = 3x + 1$

(b) $g(x) = -2x + 3$

(c) $H(x) = 3$

(d) $F(x) = -4x - 1$

(e) $G(x) = -\frac{2}{3}x + 3$

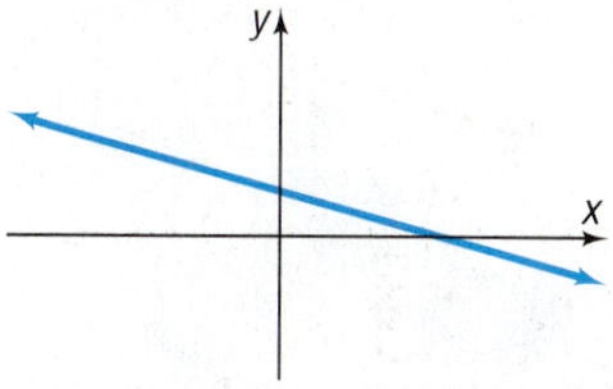

55. Under what circumstances is a linear function $f(x) = mx + b$ odd? Can a linear function ever be even?

56. Explain how the graph of $f(x) = mx + b$ can be used to solve $mx + b > 0$.

Retain Your Knowledge

Problems 57–60 are based on material learned earlier in the course. The purpose of these problems is to keep the material fresh in your mind so that you are better prepared for the final exam.

57. Graph $x^2 - 4x + y^2 + 10y - 7 = 0$.

58. If $f(x) = \dfrac{2x + B}{x - 3}$ and $f(5) = 8$, what is the value of B?

59. Find the average rate of change of $f(x) = 3x^2 - 5x$ from 1 to 3.

60. Graph $g(x) = \begin{cases} x^2 & x \leq 0 \\ \sqrt{x} + 1 & x > 0 \end{cases}$

'Are You Prepared?' Answers

1.

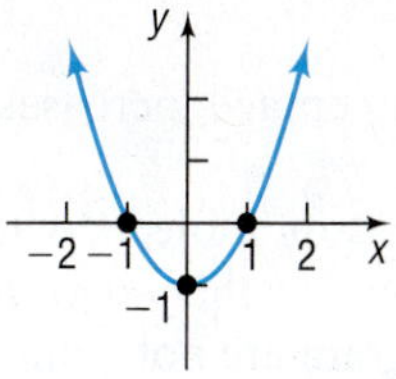

2. $\frac{2}{3}$ **3.** 18 **4.** $\{50\}$ **5.** 0 **6.** True

4.2 Building Linear Models from Data

PREPARING FOR THIS SECTION *Before getting started, review the following:*

- Rectangular Coordinates (Section 2.1, pp. 150–151)
- Functions (Section 3.1, pp. 199–208)
- Lines (Section 2.3, pp. 167–175)

Now Work the 'Are You Prepared?' problems on page 287.

OBJECTIVES
1 Draw and Interpret Scatter Diagrams (p. 284)
2 Distinguish between Linear and Nonlinear Relations (p. 285)
3 Use a Graphing Utility to Find the Line of Best Fit (p. 286)

1 Draw and Interpret Scatter Diagrams

In Section 4.1, we built linear models from verbal descriptions. Linear models can also be constructed by fitting a linear function to data. The first step is to plot the ordered pairs using rectangular coordinates. The resulting graph is a **scatter diagram**.

EXAMPLE 1 **Drawing and Interpreting a Scatter Diagram**

In baseball, the on-base percentage for a team represents the percentage of time that the players safely reach base. The data given in Table 6 represent the number of runs scored y and the on-base percentage x for teams in the National League during the 2013 baseball season.

Table 6

Team	On-Base Percentage, x	Runs Scored, y	(x, y)
Arizona	32.3	685	(32.3, 685)
Atlanta	32.1	688	(32.1, 688)
Chicago Cubs	30.0	602	(30.0, 602)
Cincinnati	32.7	698	(32.7, 698)
Colorado	32.3	706	(32.3, 706)
LA Dodgers	32.6	649	(32.6, 649)
Miami	29.3	513	(29.3, 513)
Milwaukee	31.1	640	(31.1, 640)
NY Mets	30.6	619	(30.6, 619)
Philadelphia	30.6	610	(30.6, 610)
Pittsburgh	31.3	634	(31.3, 634)
San Diego	30.8	618	(30.8, 618)
San Francisco	32.0	629	(32.0, 629)
St. Louis	33.2	783	(33.2, 783)
Washington	31.3	656	(31.3, 656)

Source: espn.go.com

(a) Draw a scatter diagram of the data, treating on-base percentage as the independent variable.
(b) Use a graphing utility to draw a scatter diagram.
(c) Describe what happens to runs scored as the on-base percentage increases.

Solution (a) To draw a scatter diagram, plot the ordered pairs listed in Table 6, with the on-base percentage as the x-coordinate and the runs scored as the y-coordinate. See Figure 6(a). Notice that the points in the scatter diagram are not connected.

(b) Figure 6(b) shows a scatter diagram using a TI-84 Plus C graphing calculator.

(c) The scatter diagrams show that as the on-base percentage increases, the number of runs scored also increases.

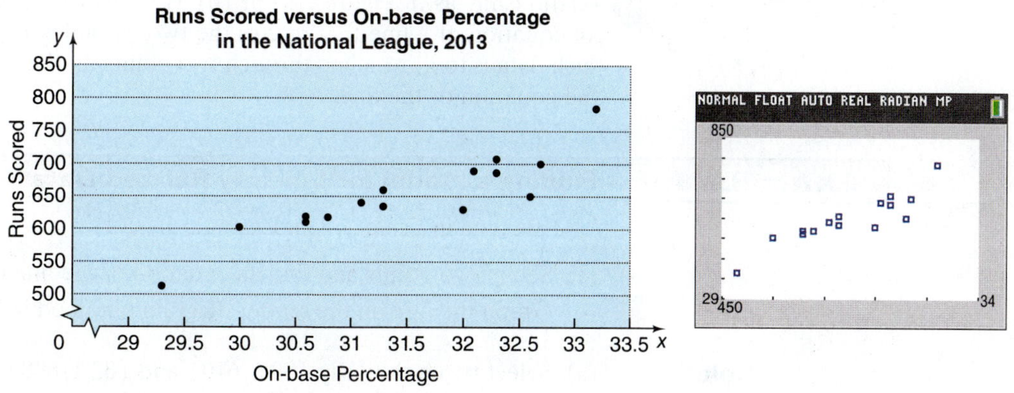

Figure 6 (a) (b)

Now Work PROBLEM 11(a)

2 Distinguish between Linear and Nonlinear Relations

Notice that the points in Figure 6 do not follow a perfect linear relation (as they do in Figure 3 in Section 4.1). However, the data do exhibit a linear pattern. There are numerous possible explanations why the data are not perfectly linear, but one easy explanation is the fact that other variables besides on-base percentage (such as number of home runs hit) play a role in determining runs scored.

Scatter diagrams are used to help us to see the type of relation that exists between two variables. In this text, we will discuss a variety of different relations that may exist between two variables. For now, we concentrate on distinguishing between linear and nonlinear relations. See Figure 7.

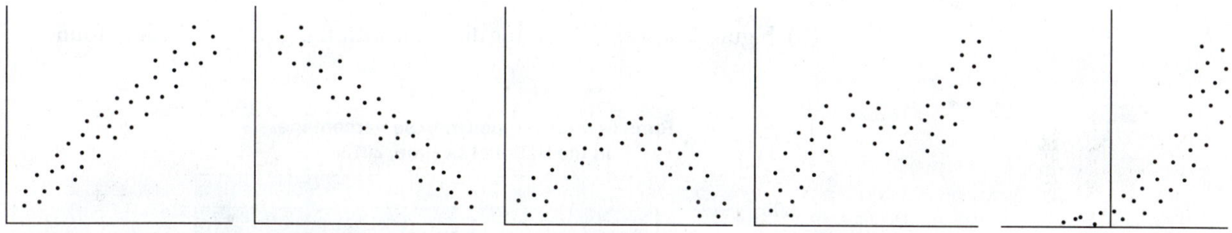

Figure 7 **(a)** Linear $y = mx + b, m > 0$ **(b)** Linear $y = mx + b, m < 0$ **(c)** Nonlinear **(d)** Nonlinear **(e)** Nonlinear

EXAMPLE 2 Distinguishing between Linear and Nonlinear Relations

Determine whether the relation between the two variables in each scatter diagram in Figure 8 is linear or nonlinear.

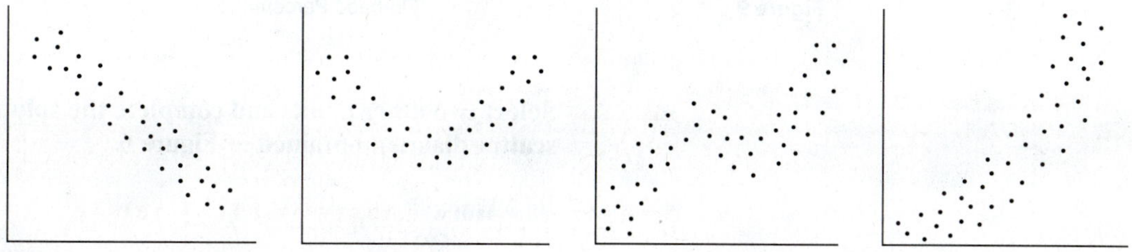

Figure 8 (a) (b) (c) (d)

Solution (a) Linear (b) Nonlinear (c) Nonlinear (d) Nonlinear

Now Work PROBLEM 5

This section considers data whose scatter diagrams suggest that a linear relation exists between the two variables.

Suppose that the scatter diagram of a set of data appears to indicate a linear relationship, as in Figure 7(a) or (b). We might want to model the data by finding an equation of a line that relates the two variables. One way to obtain a model for such data is to draw a line through two points on the scatter diagram and determine the equation of the line.

EXAMPLE 3 **Finding a Model for Linearly Related Data**

Use the data in Table 6 from Example 1.

(a) Select two points and find an equation of the line containing the points.

(b) Graph the line on the scatter diagram obtained in Example 1(a).

Solution (a) Select two points, say $(30.6, 610)$ and $(32.1, 688)$. The slope of the line joining the points $(30.6, 610)$ and $(32.1, 688)$ is

$$m = \frac{688 - 610}{32.1 - 30.6} = \frac{78}{1.5} = 52$$

The equation of the line that has slope 52 and passes through $(30.6, 610)$ is found using the point–slope form with $m = 52$, $x_1 = 30.6$, and $y_1 = 610$.

$$y - y_1 = m(x - x_1) \quad \text{Point–slope form of a line}$$
$$y - 610 = 52(x - 30.6) \quad x_1 = 30.6,\ y_1 = 610,\ m = 52$$
$$y - 610 = 52x - 1591.2$$
$$y = 52x - 981.2 \quad \text{The model}$$

(b) Figure 9 shows the scatter diagram with the graph of the line found in part (a).

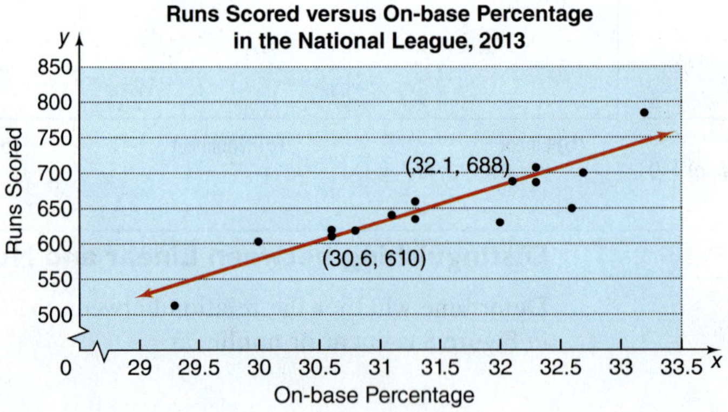

Figure 9

Select two other points and complete the solution. Graph the line on the scatter diagram obtained in Figure 6.

Now Work PROBLEMS 11(b) AND (c)

3 Use a Graphing Utility to Find the Line of Best Fit

The model obtained in Example 3 depends on the selection of points, which will vary from person to person. So the model that we found might be different from the model you found. Although the model in Example 3 appears to fit the data well, there may be a model that "fits them better." Do you think your model fits the data better? Is there a *line of best fit*? As it turns out, there is a method for finding a model that best fits linearly related data (called the **line of best fit**).*

EXAMPLE 4 Finding a Model for Linearly Related Data

Use the data in Table 6 from Example 1.

(a) Use a graphing utility to find the line of best fit that models the relation between on-base percentage and runs scored.

(b) Graph the line of best fit on the scatter diagram obtained in Example 1(b).

(c) Interpret the slope.

(d) Use the line of best fit to predict the number of runs a team will score if their on-base percentage is 33.1.

Solution

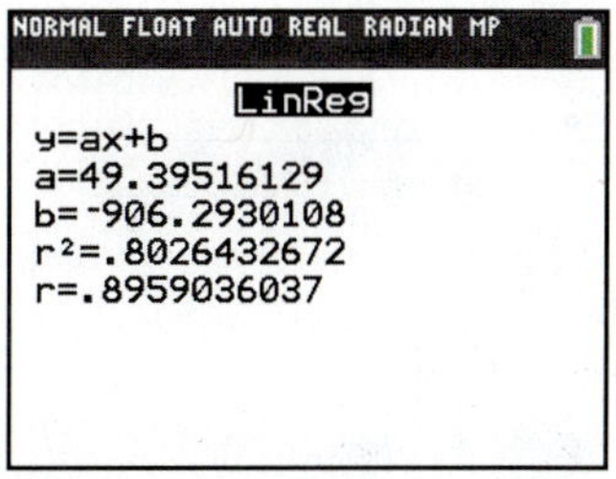

Figure 10

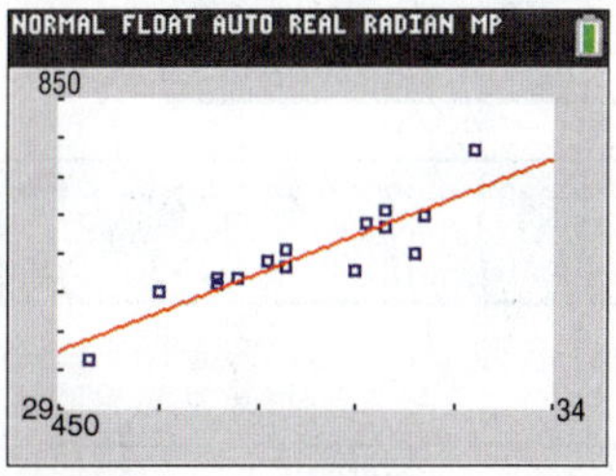

Figure 11

(a) Graphing utilities contain built-in programs that find the line of best fit for a collection of points in a scatter diagram. Executing the LINear REGression program provides the results shown in Figure 10. This output shows the equation $y = ax + b$, where a is the slope of the line and b is the y-intercept. The line of best fit that relates on-base percentage to runs scored may be expressed as the line

$$y = 49.40x - 906.29 \quad \text{The model}$$

(b) Figure 11 shows the graph of the line of best fit, along with the scatter diagram.

(c) The slope of the line of best fit is 49.40, which means that, for every 1 percent increase in the on-base percentage, runs scored increase by 49.40, on average.

(d) Letting $x = 33.1$ in the equation of the line of best fit, we obtain $y = 49.40(33.1) - 906.29 \approx 729$ runs. ●

Now Work PROBLEMS 11(d) AND (e)

Does the line of best fit appear to be a good fit? In other words, does it appear to accurately describe the relation between on-base percentage and runs scored?

And just how "good" is this line of best fit? Look again at Figure 10. The last line of output is $r = 0.896$. This number, called the **correlation coefficient**, r, $-1 \le r \le 1$, is a measure of the strength of the linear relation that exists between two variables. The closer $|r|$ is to 1, the more nearly perfect the linear relationship is. If r is close to 0, there is little or no linear relationship between the variables. A negative value of r, $r < 0$, indicates that as x increases, y decreases; a positive value of r, $r > 0$, indicates that as x increases, y does also. The data given in Table 6, which have a correlation coefficient of 0.896, are indicative of a linear relationship with positive slope.

4.2 Assess Your Understanding

'Are You Prepared?' *Answers are given at the end of these exercises. If you get a wrong answer, read the pages listed in red.*

1. Plot the points (1, 5), (2, 6), (3, 9), (1, 12) in the Cartesian plane. Is the relation $\{(1,5), (2,6), (3,9), (1,12)\}$ a function? Why? (pp. 150 and 199–208)

2. Find an equation of the line containing the points $(1,4)$ and $(3,8)$. (pp. 167–175)

Concepts and Vocabulary

3. A _______________ is used to help us to see what type of relation, if any, may exist between two variables.

4. ***True or False*** The correlation coefficient is a measure of the strength of a linear relation between two variables and must lie between -1 and 1, inclusive.

* We shall not discuss the underlying mathematics of lines of best fit in this text.

Skill Building

In Problems 5–10, examine the scatter diagram and determine whether the type of relation is linear or nonlinear.

5.

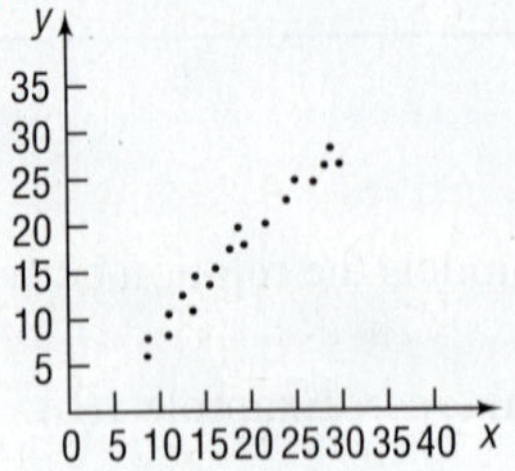

6.

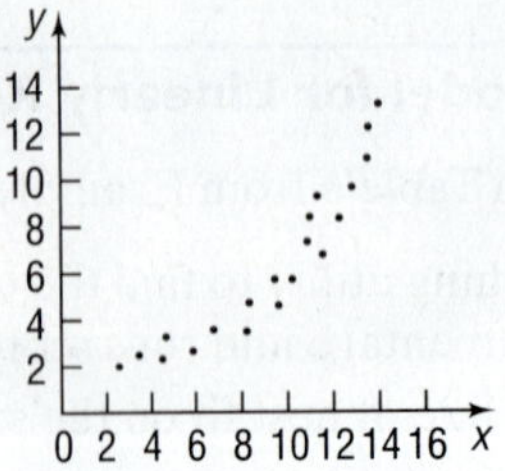

7.

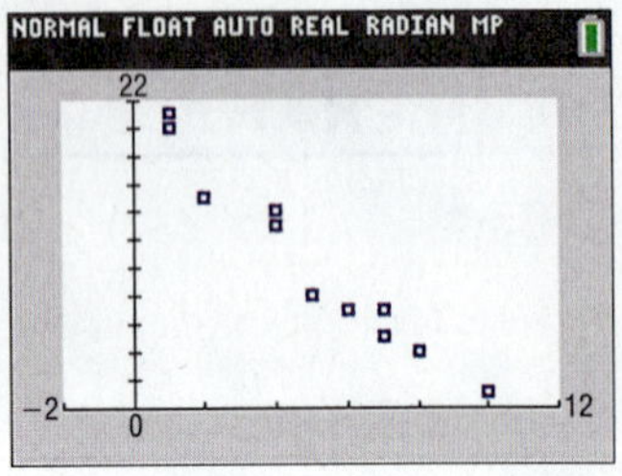

8.

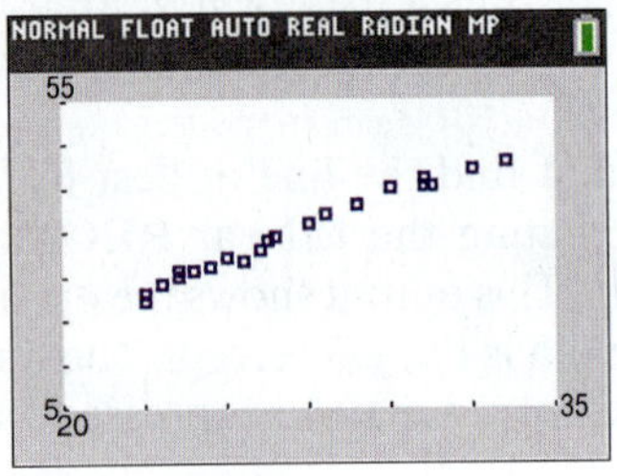

9.

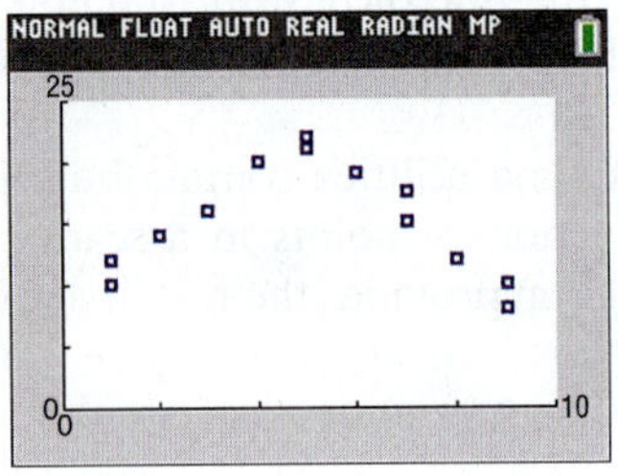

10.

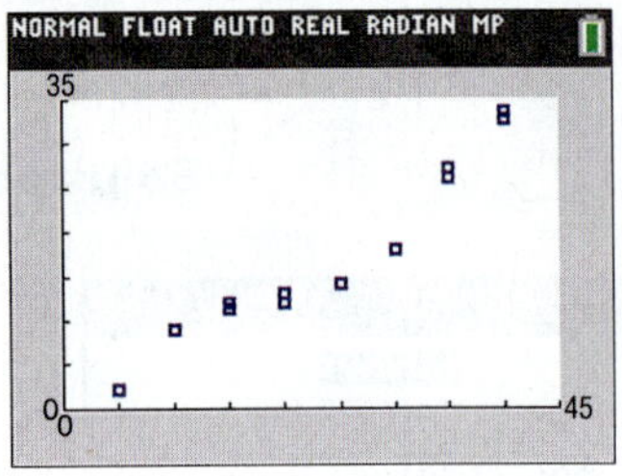

**In Problems 11–16:*

(a) Draw a scatter diagram.
(b) Select two points from the scatter diagram, and find the equation of the line containing the points selected.
(c) Graph the line found in part (b) on the scatter diagram.
(d) Use a graphing utility to find the line of best fit.
(e) Use a graphing utility to draw the scatter diagram and graph the line of best fit on it.

11.

x	3	4	5	6	7	8	9
y	4	6	7	10	12	14	16

12.

x	3	5	7	9	11	13
y	0	2	3	6	9	11

13.

x	−2	−1	0	1	2
y	−4	0	1	4	5

14.

x	−2	−1	0	1	2
y	7	6	3	2	0

15.

x	−20	−17	−15	−14	−10
y	100	120	118	130	140

16.

x	−30	−27	−25	−20	−14
y	10	12	13	13	18

Applications and Extensions

17. Candy The following data represent the weight (in grams) of various candy bars and the corresponding number of calories.

Candy Bar	Weight, x	Calories, y
Hershey's Milk Chocolate®	44.28	230
Nestle's Crunch®	44.84	230
Butterfinger®	61.30	270
Baby Ruth®	66.45	280
Almond Joy®	47.33	220
Twix® (with caramel)	58.00	280
Snickers®	61.12	280
Heath®	39.52	210

Source: Megan Pocius, student at Joliet Junior College

(a) Draw a scatter diagram of the data, treating weight as the independent variable.
(b) What type of relation appears to exist between the weight of a candy bar and the number of calories?
(c) Select two points and find a linear model that contains the points.
(d) Graph the line on the scatter diagram drawn in part (a).
(e) Use the linear model to predict the number of calories in a candy bar that weighs 62.3 grams.
(f) Interpret the slope of the line found in part (c).

18. Raisins The following data represent the weight (in grams) of a box of raisins and the number of raisins in the box.

Weight (grams), w	Number of Raisins, N
42.3	87
42.7	91
42.8	93
42.4	87
42.6	89
42.4	90
42.3	82
42.5	86
42.7	86
42.5	86

Source: Jennifer Maxwell, student at Joliet Junior College

(a) Draw a scatter diagram of the data, treating weight as the independent variable.
(b) What type of relation appears to exist between the weight of a box of raisins and the number of raisins?
(c) Select two points and find a linear model that contains the points.
(d) Graph the line on the scatter diagram drawn in part (b).
(e) Use the linear model to predict the number of raisins in a box that weighs 42.5 grams.
(f) Interpret the slope of the line found in part (c).

19. Video Games and Grade-Point Average Professor Grant Alexander wanted to find a linear model that relates the number of hours a student plays video games each week, h, to the cumulative grade-point average, G, of the student. He obtained a random sample of 10 full-time students at his college and asked each student to disclose the number of hours spent playing video games and the student's cumulative grade-point average.

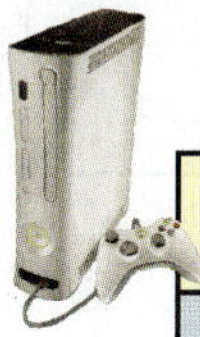

Hours of Video Games per Week, h	Grade-point Average, G
0	3.49
0	3.05
2	3.24
3	2.82
3	3.19
5	2.78
8	2.31
8	2.54
10	2.03
12	2.51

(a) Explain why the number of hours spent playing video games is the independent variable and cumulative grade-point average is the dependent variable.
(b) Use a graphing utility to draw a scatter diagram.
(c) Use a graphing utility to find the line of best fit that models the relation between number of hours of video game playing each week and grade-point average. Express the model using function notation.
(d) Interpret the slope.
(e) Predict the grade-point average of a student who plays video games for 8 hours each week.
(f) How many hours of video game playing do you think a student plays whose grade-point average is 2.40?

20. Height versus Head Circumference A pediatrician wanted to find a linear model that relates a child's height, H, to head circumference, C. She randomly selects nine children from her practice, measures their height and head circumference, and obtains the data shown. Let H represent the independent variable and C the dependent variable.

(a) Use a graphing utility to draw a scatter diagram.
(b) Use a graphing utility to find the line of best fit that models the relation between height and head circumference. Express the model using function notation.
(c) Interpret the slope.
(d) Predict the head circumference of a child who is 26 inches tall.
(e) What is the height of a child whose head circumference is 17.4 inches?

Height, H (inches)	Head Circumference, C (inches)
25.25	16.4
25.75	16.9
25	16.9
27.75	17.6
26.5	17.3
27	17.5
26.75	17.3
26.75	17.5
27.5	17.5

Source: Denise Slucki, student at Joliet Junior College

21. Flight Time and Ticket Price The following data represent nonstop flight time (in minutes) and one-way ticket price (in dollars) for flying from Chicago to various cities on Southwest Airlines.

City	Time (minutes), t	Price (dollars), P
Atlanta, GA	110	140
Los Angeles, CA	260	199
Nashville, TN	80	108
Oklahoma City, OK	125	130
Omaha, NE	85	89
Ft. Myers, FL	170	175
Phoenix, AZ	225	191
Houston, TX	155	147
Seattle, WA	265	196
St. Louis, MO	65	93

Source: Southwest.com, for midmorning flights in May 2014

(a) Use a graphing utility to draw a scatter diagram.
(b) Use a graphing utility to find the line of best fit that models the relation between flight time and airfare. Express the model using function notation.
(c) Interpret the slope.
(d) Predict the airfare for a flight from Chicago to Kansas City, Missouri, if the flight time is 85 minutes. Round to the nearest dollar.
(e) Predict the flight time from Chicago to Baltimore, Maryland if the airfare is $120. Round to the nearest minute.

Explaining Concepts: Discussion and Writing

22. **Maternal Age versus Down Syndrome** A biologist would like to know how the age of the mother affects the incidence of Down syndrome. The data to the right represent the age of the mother and the incidence of Down syndrome per 1000 pregnancies. Draw a scatter diagram treating age of the mother as the independent variable. Would it make sense to find the line of best fit for these data? Why or why not?

Age of Mother, x	Incidence of Down Syndrome, y
33	2.4
34	3.1
35	4
36	5
37	6.7
38	8.3
39	10
40	13.3
41	16.7
42	22.2
43	28.6
44	33.3
45	50

Source: Hook, E.B., *Journal of the American Medical Association*, 249, 2034-2038, 1983.

23. Find the line of best fit for the ordered pairs $(1, 5)$ and $(3, 8)$. What is the correlation coefficient for these data? Why is this result reasonable?
24. What does a correlation coefficient of 0 imply?
25. Explain why it does not make sense to interpret the y-intercept in Problem 17.
26. Refer to Problem 19. Solve $G(h) = 0$. Provide an interpretation of this result. Find $G(0)$. Provide an interpretation of this result.

Retain Your Knowledge

Problems 27–30 are based on material learned earlier in the course. The purpose of these problems is to keep the material fresh in your mind so that you are better prepared for the final exam.

27. Find an equation for the line containing the points $(-1, 5)$ and $(3, -3)$. Express your answer using either the general form or the slope-intercept form of the equation of a line, whichever you prefer.
28. Find the domain of $f(x) = \dfrac{x-1}{x^2-25}$.
29. For $f(x) = 5x - 8$ and $g(x) = x^2 - 3x + 4$, find $(g - f)(x)$.
30. Write the function whose graph is the graph of $y = x^2$, but shifted to the left 3 units and shifted down 4 units.

'Are You Prepared?' Answers

1. 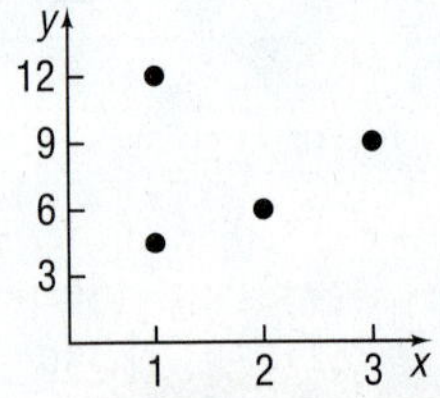

No, because the input, 1, corresponds to two different outputs.

2. $y = 2x + 2$

4.3 Quadratic Functions and Their Properties

PREPARING FOR THIS SECTION *Before getting started, review the following:*

- Intercepts (Section 2.2, pp. 159–160)
- Graphing Techniques: Transformations (Section 3.5, pp. 247–256)
- Completing the Square (Section R.5, p. 56)
- Quadratic Equations (Section 1.2, pp. 92–99)

Now Work the **'Are You Prepared?'** problems on page 299.

OBJECTIVES
1. Graph a Quadratic Function Using Transformations (p. 292)
2. Identify the Vertex and Axis of Symmetry of a Quadratic Function (p. 294)
3. Graph a Quadratic Function Using Its Vertex, Axis, and Intercepts (p. 294)
4. Find a Quadratic Function Given Its Vertex and One Other Point (p. 297)
5. Find the Maximum or Minimum Value of a Quadratic Function (p. 298)

Quadratic Functions

Here are some examples of quadratic functions.

$$F(x) = 3x^2 - 5x + 1 \quad g(x) = -6x^2 + 1 \quad H(x) = \frac{1}{2}x^2 + \frac{2}{3}x$$

DEFINITION

A **quadratic function** is a function of the form

$$f(x) = ax^2 + bx + c$$

where a, b, and c are real numbers and $a \neq 0$. The domain of a quadratic function is the set of all real numbers.

In Words

A quadratic function is a function defined by a second-degree polynomial in one variable.

Many applications require a knowledge of quadratic functions. For example, suppose that Texas Instruments collects the data shown in Table 7, which relate the number of calculators sold to the price p (in dollars) per calculator. Since the price of a product determines the quantity that will be purchased, we treat price as the independent variable. The relationship between the number x of calculators sold and the price p per calculator is given by the linear equation

$$x = 21{,}000 - 150p$$

Table 7

Price p per Calculator, (in dollars)	Number of Calculators, x
60	12,000
65	11,250
70	10,500
75	9,750
80	9,000
85	8,250
90	7,500

Then the revenue R derived from selling x calculators at the price p per calculator is equal to the unit selling price p of the calculator times the number x of units actually sold. That is,

$$R = xp$$
$$R(p) = (21{,}000 - 150p)p \quad x = 21{,}000 - 150p$$
$$= -150p^2 + 21{,}000p$$

So the revenue R is a quadratic function of the price p. Figure 12 illustrates the graph of this revenue function, whose domain is $0 \le p \le 140$, since both x and p must be nonnegative.

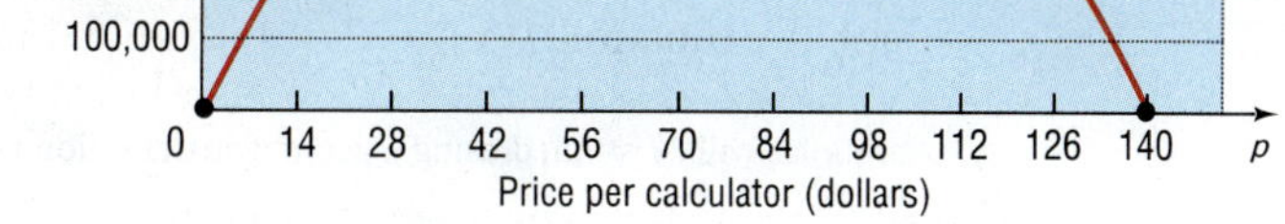

Figure 12
$R(p) = -150p^2 + 21{,}000p$

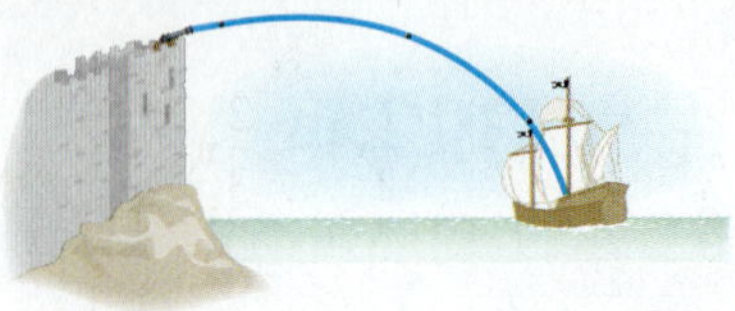

Figure 13
Path of a cannonball

A second situation in which a quadratic function appears involves the motion of a projectile. Based on Newton's Second Law of Motion (force equals mass times acceleration, $F = ma$), it can be shown that, ignoring air resistance, the path of a projectile propelled upward at an inclination to the horizontal is the graph of a quadratic function. See Figure 13 for an illustration.

1 Graph a Quadratic Function Using Transformations

We know how to graph the square function $f(x) = x^2$. Figure 14 shows the graph of three functions of the form $f(x) = ax^2, a > 0$, for $a = 1, a = \frac{1}{2}$, and $a = 3$. Note that the larger the value of a, the "narrower" the graph is, and the smaller the value of a, the "wider" the graph is.

Figure 15 shows the graphs of $f(x) = ax^2$ for $a < 0$. Notice that these graphs are reflections about the x-axis of the graphs in Figure 14. Based on the results of these two figures, general conclusions can be drawn about the graph of $f(x) = ax^2$. First, as $|a|$ increases, the graph is stretched vertically (becomes "taller"), and as $|a|$ gets closer to zero, the graph is compressed vertically (becomes "shorter"). Second, if a is positive, the graph opens "up," and if a is negative, the graph opens "down."

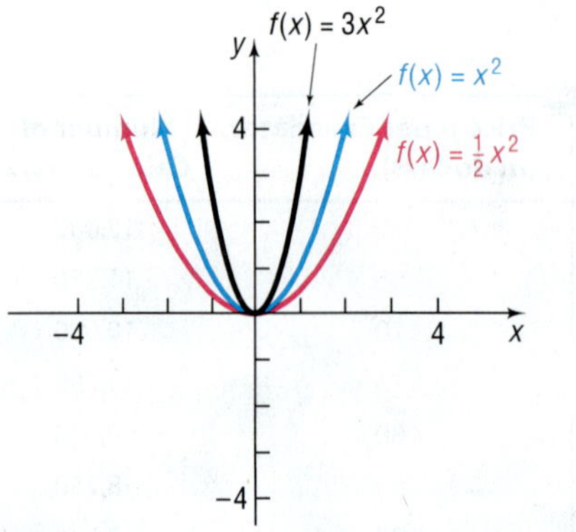

Figure 14

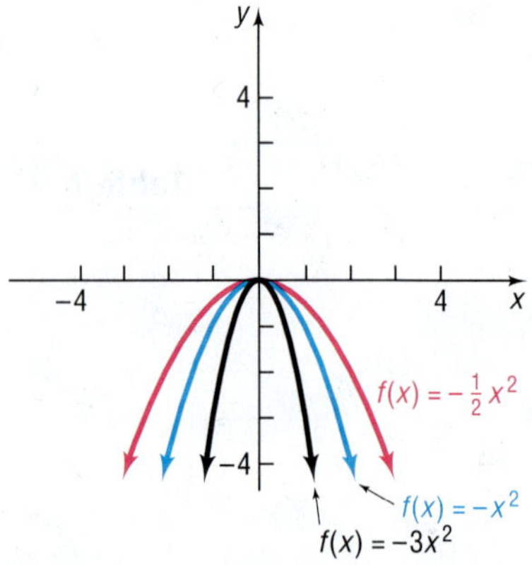

Figure 15

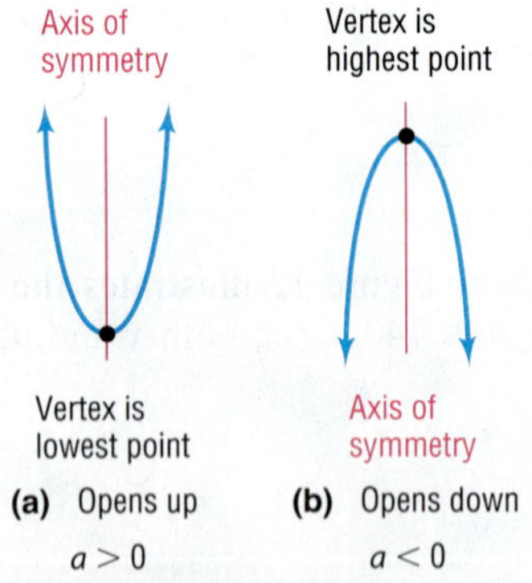

Figure 16
Graphs of a quadratic function, $f(x) = ax^2 + bx + c, a \neq 0$

The graphs in Figures 14 and 15 are typical of the graphs of all quadratic functions, which are called **parabolas.*** Refer to Figure 16, where two parabolas are pictured. The one on the left **opens up** and has a lowest point; the one on the right **opens down** and has a highest point. The lowest or highest point of a parabola is called the **vertex.** The vertical line passing through the vertex in each parabola in Figure 16 is called the **axis of symmetry** (usually abbreviated to **axis**) of the parabola. Because the parabola is symmetric about its axis, the axis of symmetry of a parabola can be used to find additional points on the parabola.

The parabolas shown in Figure 16 are the graphs of a quadratic function $f(x) = ax^2 + bx + c, a \neq 0$. Notice that the coordinate axes are not included in the figure. Depending on the values of a, b, and c, the axes could be placed anywhere. The important fact is that the shape of the graph of a quadratic function will look like one of the parabolas in Figure 16.

In the following example, techniques from Section 3.5 are used to graph a quadratic function $f(x) = ax^2 + bx + c, a \neq 0$. The method of completing the square is used to write the function f in the form $f(x) = a(x - h)^2 + k$.

EXAMPLE 1 **Graphing a Quadratic Function Using Transformations**

Graph the function $f(x) = 2x^2 + 8x + 5$. Find the vertex and axis of symmetry.

* Parabolas will be studied using a geometric definition later in this text.

Solution Begin by completing the square on the right side.

$$\begin{aligned} f(x) &= 2x^2 + 8x + 5 \\ &= 2(x^2 + 4x) + 5 && \text{Factor out the 2 from } 2x^2 + 8x. \\ &= 2(x^2 + 4x + 4) + 5 - 8 && \text{Complete the square of } x^2 + 4x \text{ by adding 4. Notice that the factor of 2} \\ &= 2(x + 2)^2 - 3 && \text{requires that 8 be added and subtracted.} \end{aligned}$$

The graph of f can be obtained from the graph of $y = x^2$ in three stages, as shown in Figure 17. Now compare this graph to the graph in Figure 16(a). The graph of $f(x) = 2x^2 + 8x + 5$ is a parabola that opens up and has its vertex (lowest point) at $(-2, -3)$. Its axis of symmetry is the line $x = -2$.

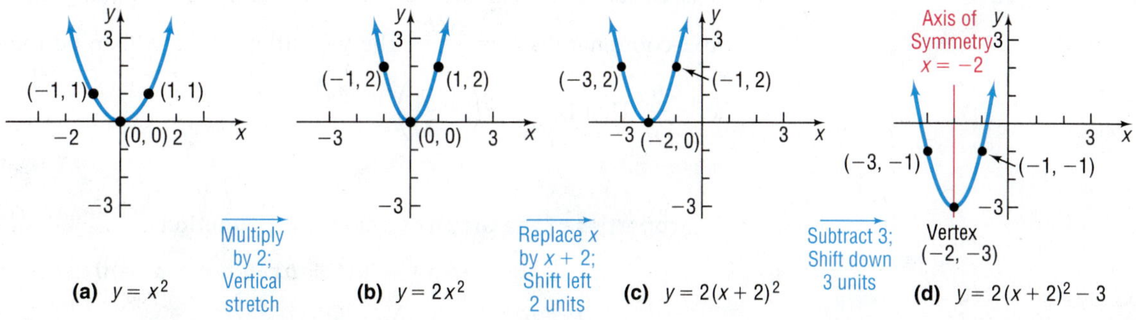

Figure 17 (a) $y = x^2$ (b) $y = 2x^2$ (c) $y = 2(x+2)^2$ (d) $y = 2(x+2)^2 - 3$

Now Work PROBLEM 25

The method used in Example 1 can be used to graph any quadratic function $f(x) = ax^2 + bx + c, a \neq 0$, as follows:

$$\begin{aligned} f(x) &= ax^2 + bx + c \\ &= a\left(x^2 + \frac{b}{a}x\right) + c && \text{Factor out } a \text{ from } ax^2 + bx. \\ &= a\left(x^2 + \frac{b}{a}x + \frac{b^2}{4a^2}\right) + c - a\left(\frac{b^2}{4a^2}\right) && \text{Complete the square by adding } \frac{b^2}{4a^2}. \text{ Look closely at this step!} \\ &= a\left(x + \frac{b}{2a}\right)^2 + c - \frac{b^2}{4a} && \text{Factor; simplify.} \\ &= a\left(x + \frac{b}{2a}\right)^2 + \frac{4ac - b^2}{4a} && c - \frac{b^2}{4a} = c \cdot \frac{4a}{4a} - \frac{b^2}{4a} = \frac{4ac - b^2}{4a} \end{aligned}$$

These results lead to the following conclusion:

If $h = -\dfrac{b}{2a}$ and $k = \dfrac{4ac - b^2}{4a}$, then

$$f(x) = ax^2 + bx + c = a(x - h)^2 + k \qquad (1)$$

The graph of $f(x) = a(x - h)^2 + k$ is the parabola $y = ax^2$ shifted horizontally h units (replace x by $x - h$) and vertically k units (add k). As a result, the vertex is at (h, k), and the graph opens up if $a > 0$ and down if $a < 0$. The axis of symmetry is the vertical line $x = h$.

For example, compare equation (1) with the solution given in Example 1.

$$\begin{aligned} f(x) &= 2(x+2)^2 - 3 \\ &= 2(x-(-2))^2 + (-3) \\ &= a(x-h)^2 + k \end{aligned}$$

Because $a = 2$, the graph opens up. Also, because $h = -2$ and $k = -3$, its vertex is at $(-2, -3)$.

2 Identify the Vertex and Axis of Symmetry of a Quadratic Function

We do not need to complete the square to obtain the vertex. In almost every case, it is easier to obtain the vertex of a quadratic function f by remembering that its x-coordinate is $h = -\frac{b}{2a}$. The y-coordinate k can then be found by evaluating f at $-\frac{b}{2a}$. That is, $k = f\left(-\frac{b}{2a}\right)$.

Properties of the Graph of a Quadratic Function

$$f(x) = ax^2 + bx + c \qquad a \neq 0$$

Vertex $= \left(-\frac{b}{2a}, f\left(-\frac{b}{2a}\right)\right)$ Axis of symmetry: the vertical line $x = -\frac{b}{2a}$ **(2)**

Parabola opens up if $a > 0$; the vertex is a minimum point.
Parabola opens down if $a < 0$; the vertex is a maximum point.

EXAMPLE 2 **Locating the Vertex without Graphing**

Without graphing, locate the vertex and axis of symmetry of the parabola defined by $f(x) = -3x^2 + 6x + 1$. Does it open up or down?

Solution For this quadratic function, $a = -3$, $b = 6$, and $c = 1$. The x-coordinate of the vertex is

$$h = -\frac{b}{2a} = -\frac{6}{2(-3)} = 1$$

The y-coordinate of the vertex is

$$k = f\left(-\frac{b}{2a}\right) = f(1) = -3(1)^2 + 6(1) + 1 = 4$$

The vertex is located at the point $(1, 4)$. The axis of symmetry is the line $x = 1$. Because $a = -3 < 0$, the parabola opens down. ●

3 Graph a Quadratic Function Using Its Vertex, Axis, and Intercepts

The location of the vertex and intercepts, along with knowledge of whether the graph opens up or down, usually provides enough information to graph $f(x) = ax^2 + bx + c$, $a \neq 0$.

The y-intercept is the value of f at $x = 0$; that is, the y-intercept is $f(0) = c$.
The x-intercepts, if there are any, are found by solving the quadratic equation

$$ax^2 + bx + c = 0$$

This equation has two, one, or no real solutions, depending on whether the discriminant $b^2 - 4ac$ is positive, 0, or negative. Depending on the value of the discriminant, the graph of f has x-intercepts, as follows:

The x-Intercepts of a Quadratic Function

1. If the discriminant $b^2 - 4ac > 0$, the graph of $f(x) = ax^2 + bx + c$ has two distinct x-intercepts so it crosses the x-axis in two places.
2. If the discriminant $b^2 - 4ac = 0$, the graph of $f(x) = ax^2 + bx + c$ has one x-intercept so it touches the x-axis at its vertex.
3. If the discriminant $b^2 - 4ac < 0$, the graph of $f(x) = ax^2 + bx + c$ has no x-intercepts so it does not cross or touch the x-axis.

Figure 18 illustrates these possibilities for parabolas that open up.

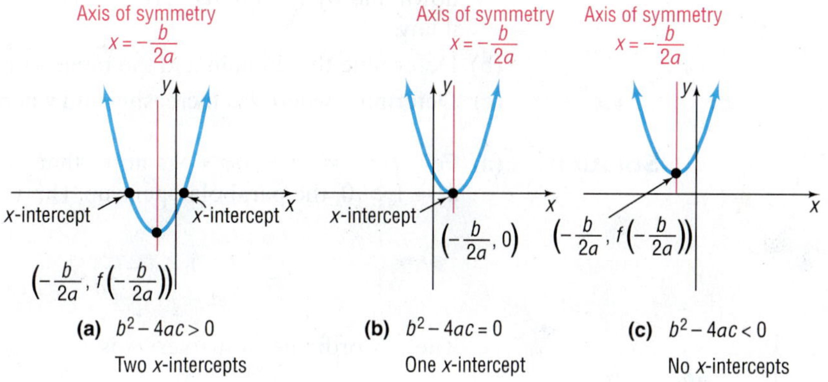

Figure 18 $f(x) = ax^2 + bx + c$, $a > 0$

EXAMPLE 3 Graphing a Quadratic Function Using Its Vertex, Axis, and Intercepts

(a) Use the information from Example 2 and the locations of the intercepts to graph $f(x) = -3x^2 + 6x + 1$.

(b) Determine the domain and the range of f.

(c) Determine where f is increasing and where it is decreasing.

Solution (a) In Example 2, we found the vertex to be at $(1, 4)$ and the axis of symmetry to be $x = 1$. The y-intercept is found by letting $x = 0$. The y-intercept is $f(0) = 1$. The x-intercepts are found by solving the equation $f(x) = 0$. This results in the equation

$$-3x^2 + 6x + 1 = 0 \quad a = -3, b = 6, c = 1$$

The discriminant $b^2 - 4ac = (6)^2 - 4(-3)(1) = 36 + 12 = 48 > 0$, so the equation has two real solutions and the graph has two x-intercepts. Use the quadratic formula to find that

$$x = \frac{-b + \sqrt{b^2 - 4ac}}{2a} = \frac{-6 + \sqrt{48}}{-6} = \frac{-6 + 4\sqrt{3}}{-6} \approx -0.15$$

and

$$x = \frac{-b - \sqrt{b^2 - 4ac}}{2a} = \frac{-6 - \sqrt{48}}{-6} = \frac{-6 - 4\sqrt{3}}{-6} \approx 2.15$$

The x-intercepts are approximately -0.15 and 2.15.

The graph is illustrated in Figure 19. Notice how we used the y-intercept and the axis of symmetry, $x = 1$, to obtain the additional point $(2, 1)$ on the graph.

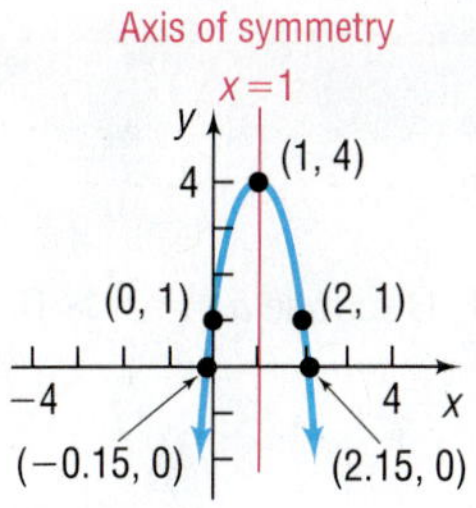

Figure 19 $f(x) = -3x^2 + 6x + 1$

(b) The domain of f is the set of all real numbers. Based on the graph, the range of f is the interval $(-\infty, 4]$.

(c) The function f is increasing on the interval $(-\infty, 1)$ and decreasing on the interval $(1, \infty)$. ●

Graph the function in Example 3 by completing the square and using transformations. Which method do you prefer?

Now Work PROBLEM 33

If the graph of a quadratic function has only one x-intercept or no x-intercepts, it is usually necessary to plot an additional point to obtain the graph.

EXAMPLE 4 Graphing a Quadratic Function Using Its Vertex, Axis, and Intercepts

(a) Graph $f(x) = x^2 - 6x + 9$ by determining whether the graph opens up or down and by finding its vertex, axis of symmetry, y-intercept, and x-intercepts, if any.

(b) Determine the domain and the range of f.

(c) Determine where f is increasing and where it is decreasing.

Solution (a) For $f(x) = x^2 - 6x + 9$, note that $a = 1, b = -6$, and $c = 9$. Because $a = 1 > 0$, the parabola opens up. The x-coordinate of the vertex is

$$h = -\frac{b}{2a} = -\frac{-6}{2(1)} = 3$$

The y-coordinate of the vertex is

$$k = f(3) = (3)^2 - 6(3) + 9 = 0$$

The vertex is at $(3, 0)$. The axis of symmetry is the line $x = 3$. The y-intercept is $f(0) = 9$. Since the vertex $(3, 0)$ lies on the x-axis, the graph touches the x-axis at the x-intercept. By using the axis of symmetry and the y-intercept at $(0, 9)$, we can locate the additional point $(6, 9)$ on the graph. See Figure 20.

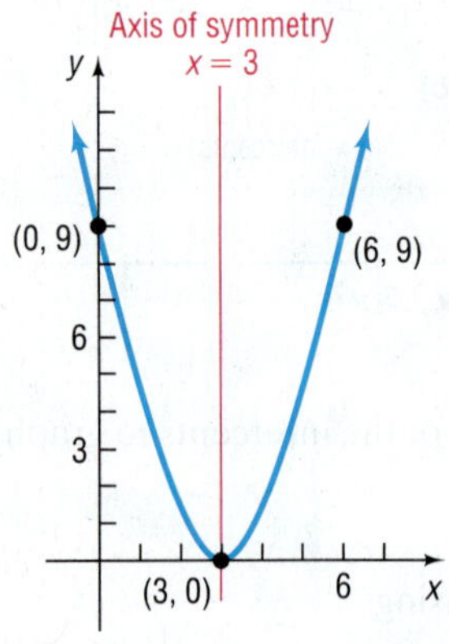

Figure 20 $f(x) = x^2 - 6x + 9$

(b) The domain of f is the set of all real numbers. Based on the graph, the range of f is the interval $[0, \infty)$.

(c) The function f is decreasing on the interval $(-\infty, 3)$ and increasing on the interval $(3, \infty)$. ●

Now Work PROBLEM 39

EXAMPLE 5 Graphing a Quadratic Function Using Its Vertex, Axis, and Intercepts

(a) Graph $f(x) = 2x^2 + x + 1$ by determining whether the graph opens up or down and by finding its vertex, axis of symmetry, y-intercept, and x-intercepts, if any.

(b) Determine the domain and the range of f.

(c) Determine where f is increasing and where it is decreasing.

Solution (a) For $f(x) = 2x^2 + x + 1$, we have $a = 2$, $b = 1$, and $c = 1$. Because $a = 2 > 0$, the parabola opens up. The x-coordinate of the vertex is

$$h = -\frac{b}{2a} = -\frac{1}{4}$$

NOTE In Example 5, since the vertex is above the x-axis and the parabola opens up, we can conclude that the graph of the quadratic function will have no x-intercepts. ■

The y-coordinate of the vertex is

$$k = f\left(-\frac{1}{4}\right) = 2\left(\frac{1}{16}\right) + \left(-\frac{1}{4}\right) + 1 = \frac{7}{8}$$

The vertex is at $\left(-\frac{1}{4}, \frac{7}{8}\right)$. The axis of symmetry is the line $x = -\frac{1}{4}$. The y-intercept is $f(0) = 1$. The x-intercept(s), if any, satisfy the equation $2x^2 + x + 1 = 0$. The discriminant $b^2 - 4ac = (1)^2 - 4(2)(1) = -7 < 0$. This equation has no real solutions, which means the graph has no x-intercepts. Use the point $(0, 1)$ and the axis of symmetry $x = -\frac{1}{4}$ to locate the additional point $\left(-\frac{1}{2}, 1\right)$ on the graph. See Figure 21.

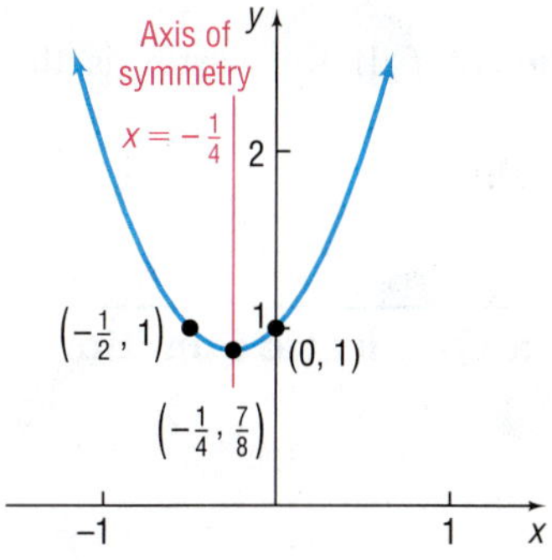

Figure 21 $f(x) = 2x^2 + x + 1$

(b) The domain of f is the set of all real numbers. Based on the graph, the range of f is the interval $\left[\frac{7}{8}, \infty\right)$.

(c) The function f is decreasing on the interval $\left(-\infty, -\frac{1}{4}\right)$ and is increasing on the interval $\left(-\frac{1}{4}, \infty\right)$. ●

Now Work PROBLEM 43

4 Find a Quadratic Function Given Its Vertex and One Other Point

If the vertex (h, k) and one additional point on the graph of a quadratic function $f(x) = ax^2 + bx + c, a \neq 0$, are known, then

$$f(x) = a(x - h)^2 + k \qquad \textbf{(3)}$$

can be used to obtain the quadratic function.

EXAMPLE 6 Finding the Quadratic Function Given Its Vertex and One Other Point

Determine the quadratic function whose vertex is $(1, -5)$ and whose y-intercept is -3.

Solution The vertex is $(1, -5)$, so $h = 1$ and $k = -5$. Substitute these values into equation (3).

$$f(x) = a(x - h)^2 + k \quad \text{Equation (3)}$$
$$f(x) = a(x - 1)^2 - 5 \quad h = 1, k = -5$$

To determine the value of a, use the fact that $f(0) = -3$ (the y-intercept).

$$f(x) = a(x - 1)^2 - 5$$
$$-3 = a(0 - 1)^2 - 5 \quad x = 0, y = f(0) = -3$$
$$-3 = a - 5$$
$$a = 2$$

The quadratic function we seek is

$$f(x) = a(x - h)^2 + k = 2(x - 1)^2 - 5 = 2x^2 - 4x - 3$$ ●

See Figure 22.

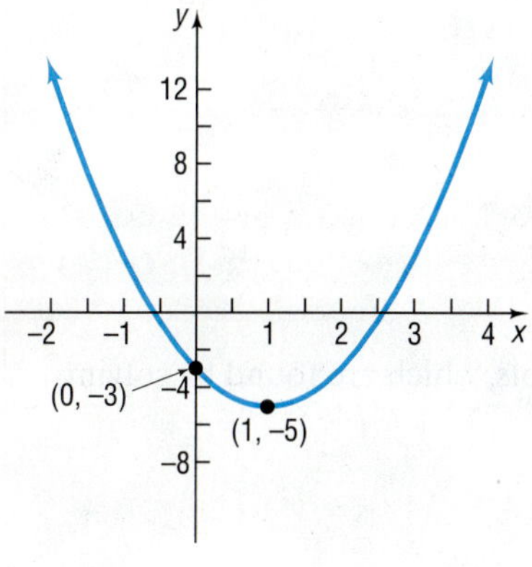

Figure 22 $f(x) = 2x^2 - 4x - 3$

Now Work PROBLEM 49

5 Find the Maximum or Minimum Value of a Quadratic Function

The graph of a quadratic function

$$f(x) = ax^2 + bx + c \qquad a \neq 0$$

is a parabola with vertex at $\left(-\frac{b}{2a}, f\left(-\frac{b}{2a}\right)\right)$. This vertex is the highest point on the graph if $a < 0$ and the lowest point on the graph if $a > 0$. If the vertex is the highest point $(a < 0)$, then $f\left(-\frac{b}{2a}\right)$ is the **maximum value** of f. If the vertex is the lowest point $(a > 0)$, then $f\left(-\frac{b}{2a}\right)$ is the **minimum value** of f.

EXAMPLE 7 Finding the Maximum or Minimum Value of a Quadratic Function

Determine whether the quadratic function

$$f(x) = x^2 - 4x - 5$$

has a maximum or a minimum value. Then find the maximum or minimum value.

Solution Compare $f(x) = x^2 - 4x - 5$ to $f(x) = ax^2 + bx + c$. Then $a = 1, b = -4$, and $c = -5$. Because $a > 0$, the graph of f opens up, which means the vertex is a minimum point. The minimum value occurs at

$$x = -\frac{b}{2a} = -\frac{-4}{2(1)} = \frac{4}{2} = 2$$

$\uparrow$ $a = 1, b = -4$

The minimum value is

$$f\left(-\frac{b}{2a}\right) = f(2) = 2^2 - 4(2) - 5 = 4 - 8 - 5 = -9$$

Now Work PROBLEM 57

SUMMARY

Steps for Graphing a Quadratic Function $f(x) = ax^2 + bx + c,\ a \neq 0$

Option 1

STEP 1: Complete the square in x to write the quadratic function in the form $f(x) = a(x - h)^2 + k$.

STEP 2: Graph the function in stages using transformations.

Option 2

STEP 1: Determine whether the parabola opens up $(a > 0)$ or down $(a < 0)$.

STEP 2: Determine the vertex $\left(-\frac{b}{2a}, f\left(-\frac{b}{2a}\right)\right)$.

STEP 3: Determine the axis of symmetry, $x = -\frac{b}{2a}$.

STEP 4: Determine the y-intercept, $f(0)$, and the x-intercepts, if any.

(a) If $b^2 - 4ac > 0$, the graph of the quadratic function has two x-intercepts, which are found by solving the equation $ax^2 + bx + c = 0$.

(b) If $b^2 - 4ac = 0$, the vertex is the x-intercept.

(c) If $b^2 - 4ac < 0$, there are no x-intercepts.

STEP 5: Determine an additional point by using the y-intercept and the axis of symmetry.

STEP 6: Plot the points and draw the graph.

4.3 Assess Your Understanding

'Are You Prepared?' *Answers are given at the end of these exercises. If you get a wrong answer, read the pages listed in red.*

1. List the intercepts of the equation $y = x^2 - 9$. (pp. 159–160)
2. Find the real solutions of the equation $2x^2 + 7x - 4 = 0$. (pp. 92–99)
3. To complete the square of $x^2 - 5x$, you add the number ______. (p. 56)
4. To graph $y = (x - 4)^2$, you shift the graph of $y = x^2$ to the ______ a distance of ____ units. (pp. 247–256)

Concepts and Vocabulary

5. The graph of a quadratic function is called a(n) ________.
6. The vertical line passing through the vertex of a parabola is called the _______.
7. The x-coordinate of the vertex of $f(x) = ax^2 + bx + c$, $a \neq 0$, is ________.
8. ***True or False*** The graph of $f(x) = 2x^2 + 3x - 4$ opens up.
9. ***True or False*** The y-coordinate of the vertex of $f(x) = -x^2 + 4x + 5$ is $f(2)$.
10. ***True or False*** If the discriminant $b^2 - 4ac = 0$, the graph of $f(x) = ax^2 + bx + c$, $a \neq 0$, will touch the x-axis at its vertex.
11. If $b^2 - 4ac > 0$, which of the following conclusions can be made about the graph of $f(x) = ax^2 + bx + c$, $a \neq 0$?
    (a) The graph has two distinct x-intercepts.
    (b) The graph has no x-intercepts.
    (c) The graph has three distinct x-intercepts.
    (d) The graph has one x-intercept.
12. If the graph of $f(x) = ax^2 + bx + c$, $a \neq 0$, has a maximum value at its vertex, which of the following conditions must be true?
    (a) $-\dfrac{b}{2a} > 0$ (b) $-\dfrac{b}{2a} < 0$
    (c) $a > 0$ (d) $a < 0$

Skill Building

In Problems 13–20, match each graph to one the following functions.

13. $f(x) = x^2 - 1$
14. $f(x) = -x^2 - 1$
15. $f(x) = x^2 - 2x + 1$
16. $f(x) = x^2 + 2x + 1$
17. $f(x) = x^2 - 2x + 2$
18. $f(x) = x^2 + 2x$
19. $f(x) = x^2 - 2x$
20. $f(x) = x^2 + 2x + 2$

A.
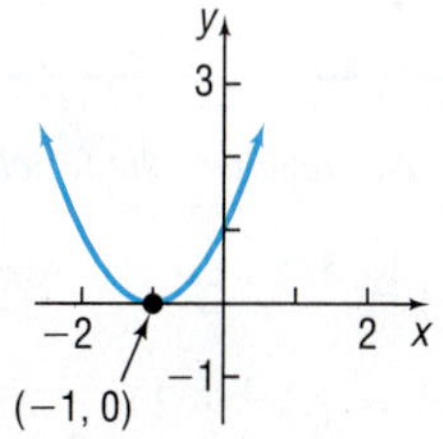

B.
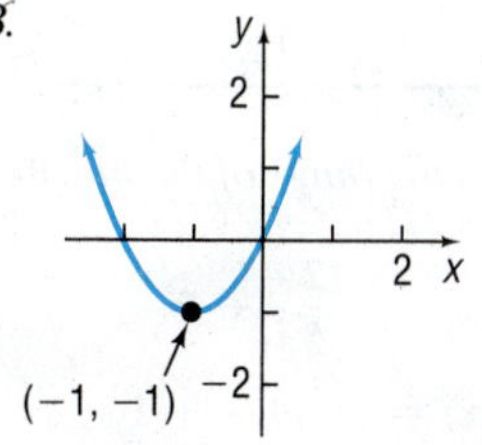

C.
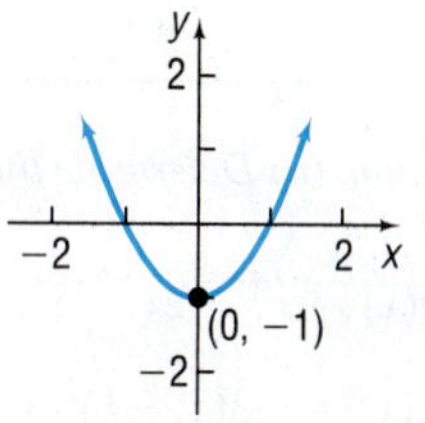

D.
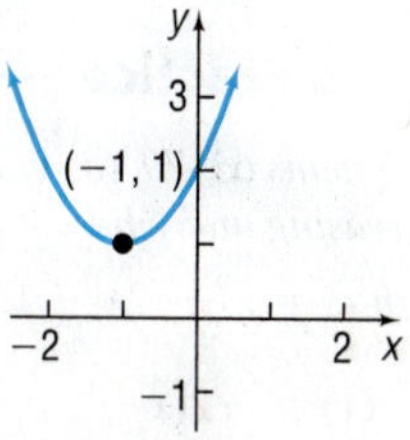

E.
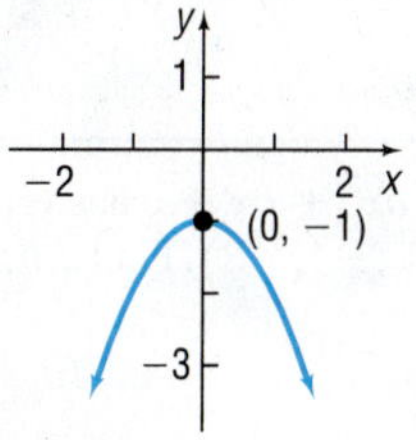

F.
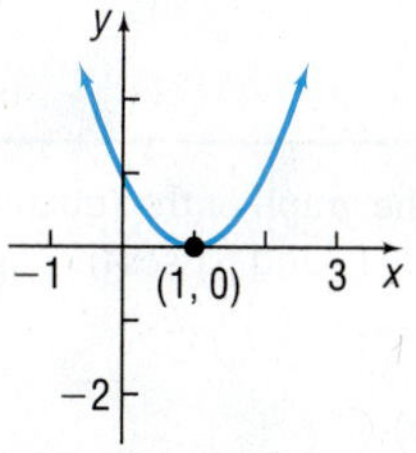

G.
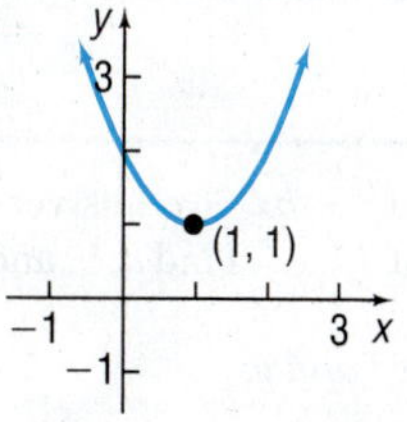

H.
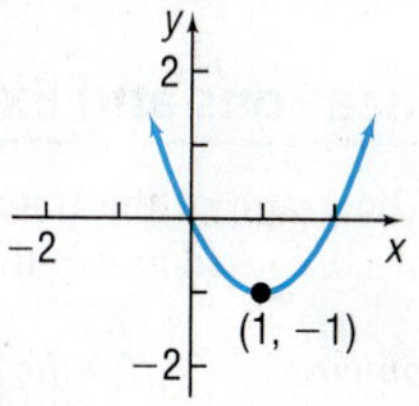

In Problems 21–32, graph the function f by starting with the graph of $y = x^2$ and using transformations (shifting, compressing, stretching, and/or reflecting).
[**Hint:** If necessary, write f in the form $f(x) = a(x - h)^2 + k$.]

21. $f(x) = \frac{1}{4}x^2$
22. $f(x) = 2x^2 + 4$
23. $f(x) = (x + 2)^2 - 2$
24. $f(x) = (x - 3)^2 - 10$
25. $f(x) = x^2 + 4x + 2$
26. $f(x) = x^2 - 6x - 1$
27. $f(x) = 2x^2 - 4x + 1$
28. $f(x) = 3x^2 + 6x$
29. $f(x) = -x^2 - 2x$
30. $f(x) = -2x^2 + 6x + 2$
31. $f(x) = \frac{1}{2}x^2 + x - 1$
32. $f(x) = \frac{2}{3}x^2 + \frac{4}{3}x - 1$

In Problems 33–48, (a) graph each quadratic function by determining whether its graph opens up or down and by finding its vertex, axis of symmetry, y-intercept, and x-intercepts, if any. (b) Determine the domain and the range of the function. (c) Determine where the function is increasing and where it is decreasing.

33. $f(x) = x^2 + 2x$
34. $f(x) = x^2 - 4x$
35. $f(x) = -x^2 - 6x$
36. $f(x) = -x^2 + 4x$
37. $f(x) = x^2 + 2x - 8$
38. $f(x) = x^2 - 2x - 3$
39. $f(x) = x^2 + 2x + 1$
40. $f(x) = x^2 + 6x + 9$

41. $f(x) = 2x^2 - x + 2$ **42.** $f(x) = 4x^2 - 2x + 1$ **43.** $f(x) = -2x^2 + 2x - 3$ **44.** $f(x) = -3x^2 + 3x - 2$

45. $f(x) = 3x^2 + 6x + 2$ **46.** $f(x) = 2x^2 + 5x + 3$ **47.** $f(x) = -4x^2 - 6x + 2$ **48.** $f(x) = 3x^2 - 8x + 2$

In Problems 49–54, determine the quadratic function whose graph is given.

49.

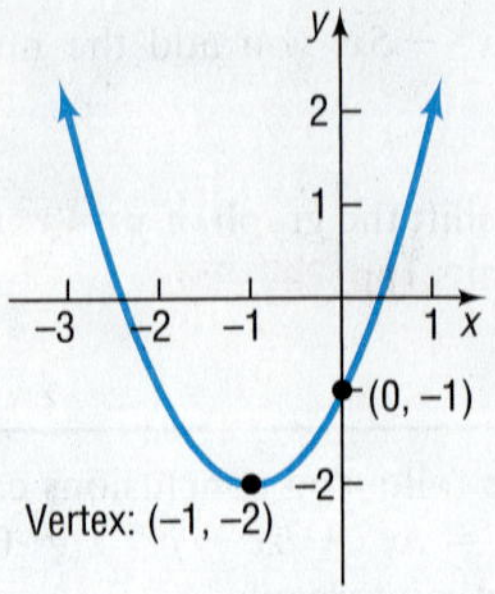

50.

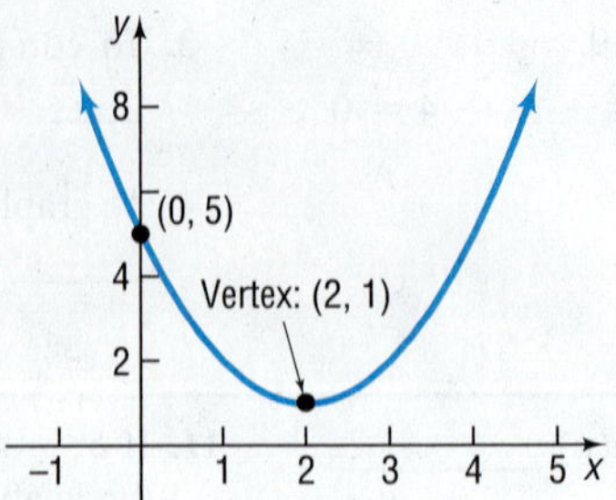

51.

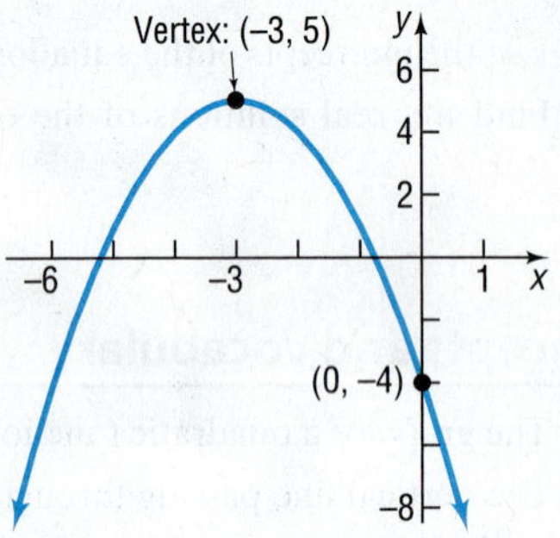

52.

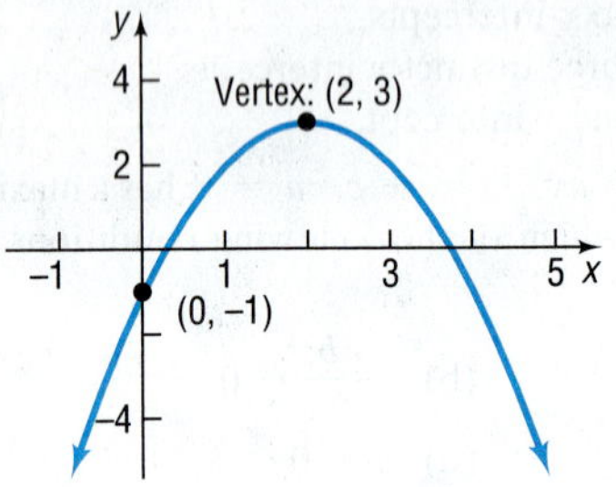

53.

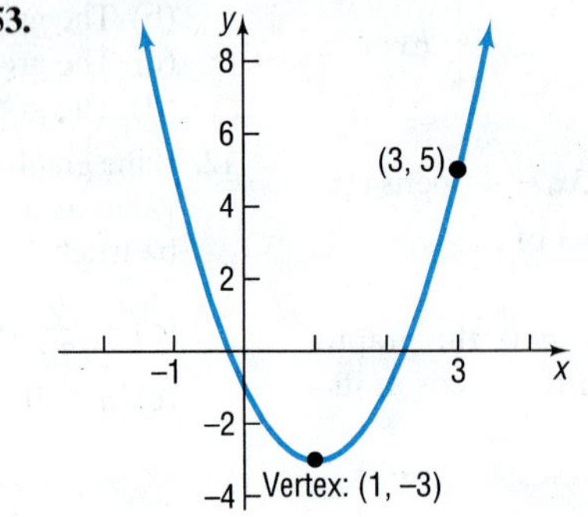

54.

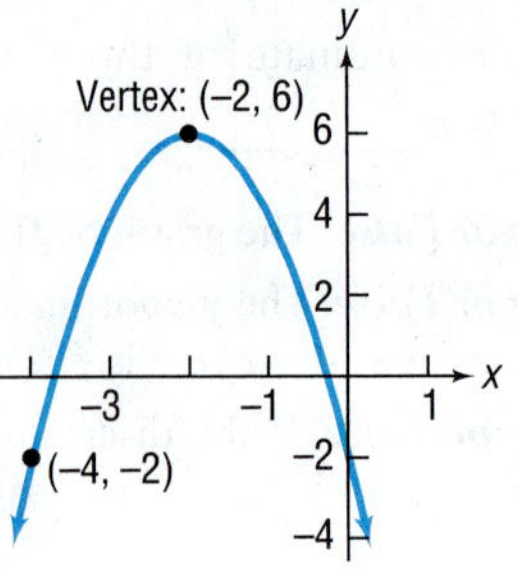

In Problems 55–62, determine, without graphing, whether the given quadratic function has a maximum value or a minimum value, and then find the value.

55. $f(x) = 2x^2 + 12x$ **56.** $f(x) = -2x^2 + 12x$ **57.** $f(x) = 2x^2 + 12x - 3$ **58.** $f(x) = 4x^2 - 8x + 3$

59. $f(x) = -x^2 + 10x - 4$ **60.** $f(x) = -2x^2 + 8x + 3$ **61.** $f(x) = -3x^2 + 12x + 1$ **62.** $f(x) = 4x^2 - 4x$

Mixed Practice

In Problems 63–70, (a) graph each function. (b) Determine the domain and the range of the function. (c) Determine where the function is increasing and where it is decreasing.

63. $f(x) = x^2 - 2x - 15$ **64.** $g(x) = x^2 - 2x - 8$ **65.** $h(x) = -\frac{2}{5}x + 4$ **66.** $f(x) = \frac{3}{2}x - 2$

67. $g(x) = -2(x - 3)^2 + 2$ **68.** $h(x) = -3(x + 1)^2 + 4$ **69.** $f(x) = 2x^2 + x + 1$ **70.** $F(x) = -4x^2 + 20x - 25$

Applications and Extensions

71. The graph of the function $f(x) = ax^2 + bx + c$ has vertex at $(0, 2)$ and passes through the point $(1, 8)$. Find a, b, and c.

72. The graph of the function $f(x) = ax^2 + bx + c$ has vertex at $(1, 4)$ and passes through the point $(-1, -8)$. Find a, b, and c.

In Problems 73–78, for the given functions f and g:
(a) Graph f and g on the same Cartesian plane. *(b) Solve $f(x) = g(x)$.*
(c) Use the result of part (b) to label the points of intersection of the graphs of f and g.
(d) Shade the region for which $f(x) > g(x)$—that is, the region below f and above g.

73. $f(x) = 2x - 1;\ g(x) = x^2 - 4$ **74.** $f(x) = -2x - 1;\ g(x) = x^2 - 9$

75. $f(x) = -x^2 + 4;\ g(x) = -2x + 1$ **76.** $f(x) = -x^2 + 9;\ g(x) = 2x + 1$

77. $f(x) = -x^2 + 5x;\ g(x) = x^2 + 3x - 4$ **78.** $f(x) = -x^2 + 7x - 6;\ g(x) = x^2 + x - 6$

Answer Problems 79 and 80 using the following: A quadratic function of the form $f(x) = ax^2 + bx + c$ with $b^2 - 4ac > 0$ may also be written in the form $f(x) = a(x - r_1)(x - r_2)$, where r_1 and r_2 are the x-intercepts of the graph of the quadratic function.

79. (a) Find a quadratic function whose x-intercepts are -3 and 1 with $a = 1$; $a = 2$; $a = -2$; $a = 5$.
(b) How does the value of a affect the intercepts?
(c) How does the value of a affect the axis of symmetry?
(d) How does the value of a affect the vertex?
(e) Compare the x-coordinate of the vertex with the midpoint of the x-intercepts. What might you conclude?

80. (a) Find a quadratic function whose x-intercepts are -5 and 3 with $a = 1$; $a = 2$; $a = -2$; $a = 5$.
(b) How does the value of a affect the intercepts?
(c) How does the value of a affect the axis of symmetry?
(d) How does the value of a affect the vertex?
(e) Compare the x-coordinate of the vertex with the midpoint of the x-intercepts. What might you conclude?

81. Suppose that $f(x) = x^2 + 4x - 21$.
(a) What is the vertex of f?
(b) What are the x-intercepts of the graph of f?
(c) Solve $f(x) = -21$ for x. What points are on the graph of f?
(d) Use the information obtained in parts (a)–(c) to graph $f(x) = x^2 + 4x - 21$.

82. Suppose that $f(x) = x^2 + 2x - 8$.
(a) What is the vertex of f?
(b) What are the x-intercepts of the graph of f?
(c) Solve $f(x) = -8$ for x. What points are on the graph of f?
(d) Use the information obtained in parts (a)–(c) to graph $f(x) = x^2 + 2x - 8$.

83. Find the point on the line $y = x$ that is closest to the point $(3, 1)$.
[**Hint:** Express the distance d from the point to the line as a function of x, and then find the minimum value of $[d(x)]^2$.

84. Find the point on the line $y = x + 1$ that is closest to the point $(4, 1)$.

85. Maximizing Revenue Suppose that the manufacturer of a gas clothes dryer has found that when the unit price is p dollars, the revenue R (in dollars) is

$$R(p) = -4p^2 + 4000p$$

What unit price should be established for the dryer to maximize revenue? What is the maximum revenue?

86. Maximizing Revenue The John Deere company has found that the revenue, in dollars, from sales of riding mowers is a function of the unit price p, in dollars, that it charges. If the revenue R is

$$R(p) = -\frac{1}{2}p^2 + 1900p$$

what unit price p should be charged to maximize revenue? What is the maximum revenue?

87. Minimizing Marginal Cost The **marginal cost** of a product can be thought of as the cost of producing one additional unit of output. For example, if the marginal cost of producing the 50th product is \$6.20, it cost \$6.20 to increase production from 49 to 50 units of output. Suppose the marginal cost C (in dollars) to produce x thousand digital music players is given by the function

$$C(x) = x^2 - 140x + 7400$$

(a) How many players should be produced to minimize the marginal cost?
(b) What is the minimum marginal cost?

88. Minimizing Marginal Cost (See Problem 87.) The marginal cost C (in dollars) of manufacturing x cell phones (in thousands) is given by

$$C(x) = 5x^2 - 200x + 4000$$

(a) How many cell phones should be manufactured to minimize the marginal cost?
(b) What is the minimum marginal cost?

89. Business The monthly revenue R achieved by selling x wristwatches is figured to be $R(x) = 75x - 0.2x^2$. The monthly cost C of selling x wristwatches is

$$C(x) = 32x+1750$$

(a) How many wristwatches must the firm sell to maximize revenue? What is the maximum revenue?
(b) Profit is given as $P(x) = R(x) - C(x)$. What is the profit function?
(c) How many wristwatches must the firm sell to maximize profit? What is the maximum profit?
(d) Provide a reasonable explanation as to why the answers found in parts (a) and (c) differ. Explain why a quadratic function is a reasonable model for revenue.

90. Business The daily revenue R achieved by selling x boxes of candy is figured to be $R(x) = 9.5x - 0.04x^2$. The daily cost C of selling x boxes of candy is $C(x) = 1.25x + 250$.
(a) How many boxes of candy must the firm sell to maximize revenue? What is the maximum revenue?
(b) Profit is given as $P(x) = R(x) - C(x)$. What is the profit function?
(c) How many boxes of candy must the firm sell to maximize profit? What is the maximum profit?
(d) Provide a reasonable explanation as to why the answers found in parts (a) and (c) differ. Explain why a quadratic function is a reasonable model for revenue.

91. Stopping Distance An accepted relationship between stopping distance, d (in feet), and the speed of a car, v (in mph), is $d = 1.1v + 0.06v^2$ on dry, level concrete.
(a) How many feet will it take a car traveling 45 mph to stop on dry, level concrete?
(b) If an accident occurs 200 feet ahead of you, what is the maximum speed you can be traveling to avoid being involved?
(c) What might the term $1.1v$ represent?

92. Birth Rate of Unmarried Women In the United States, the birth rate B of unmarried women (births per 1000 unmarried women) for women whose age is a is modeled by the function $B(a) = -0.30a^2 + 16.26a - 158.90$
(a) What is the age of unmarried women with the highest birth rate?
(b) What is the highest birth rate of unmarried women?
(c) Evaluate and interpret $B(40)$.

Source: National Vital Statistics System, 2013

93. Let $f(x) = ax^2 + bx + c$, where a, b, and c are odd integers. If x is an integer, show that $f(x)$ must be an odd integer.
[**Hint:** x is either an even integer or an odd integer.]

Explaining Concepts: Discussion and Writing

94. Make up a quadratic function that opens down and has only one x-intercept. Compare yours with others in the class. What are the similarities? What are the differences?

95. On one set of coordinate axes, graph the family of parabolas $f(x) = x^2 + 2x + c$ for $c = -3$, $c = 0$, and $c = 1$. Describe the characteristics of a member of this family.

96. On one set of coordinate axes, graph the family of parabolas $f(x) = x^2 + bx + 1$ for $b = -4$, $b = 0$, and $b = 4$. Describe the general characteristics of this family.

97. State the circumstances that cause the graph of a quadratic function $f(x) = ax^2 + bx + c$ to have no x-intercepts.

98. Why does the graph of a quadratic function open up if $a > 0$ and down if $a < 0$?

99. Can a quadratic function have a range of $(-\infty, \infty)$? Justify your answer.

100. What are the possibilities for the number of times the graphs of two different quadratic functions intersect?

Retain Your Knowledge

Problems 101–104 are based on material learned earlier in the course. The purpose of these problems is to keep the material fresh in your mind so that you are better prepared for the final exam.

101. Determine whether $x^2 + 4y^2 = 16$ is symmetric respect to the x-axis, the y-axis, and/or the origin.

102. Solve the inequality $27 - x \geq 5x + 3$. Write the solution in both set notation and interval notation.

103. Find the center and radius of the circle $x^2 + y^2 - 10x + 4y + 20 = 0$.

104. Write the function whose graph is the graph of $y = \sqrt{x}$, but reflected about the y-axis.

'Are You Prepared?' Answers

1. $(0, -9), (-3, 0), (3, 0)$ **2.** $\left\{-4, \frac{1}{2}\right\}$ **3.** $\frac{25}{4}$ **4.** right; 4

4.4 Build Quadratic Models from Verbal Descriptions and from Data

PREPARING FOR THIS SECTION *Before getting started, review the following:*

- Problem Solving (Section 1.7, pp. 134–140)
- Building Linear Models from Data (Section 4.2, pp. 284–287)

Now Work the **'Are You Prepared?'** problems on page 307.

OBJECTIVES **1** Build Quadratic Models from Verbal Descriptions (p. 302)
2 Build Quadratic Models from Data (p. 306)

In this section we will first discuss models in the form of a quadratic function when a verbal description of the problem is given. We end the section by fitting a quadratic function to data, which is another form of modeling.

When a mathematical model is in the form of a quadratic function, the properties of the graph of the function can provide important information about the model. In particular, we can use the quadratic function to determine the maximum or minimum value of the function. The fact that the graph of a quadratic function has a maximum or minimum value enables us to answer questions involving **optimization**—that is, finding the maximum or minimum values in models.

1 Build Quadratic Models from Verbal Descriptions

In economics, revenue R, in dollars, is defined as the amount of money received from the sale of an item and is equal to the unit selling price p, in dollars, of the item times the number x of units actually sold. That is,

$$R = xp$$

The Law of Demand states that p and x are related: As one increases, the other decreases. The equation that relates p and x is called the **demand equation**. When the demand equation is linear, the revenue model is a quadratic function.

EXAMPLE 1 **Maximizing Revenue**

The marketing department at Texas Instruments has found that when certain calculators are sold at a price of p dollars per unit, the number x of calculators sold is given by the demand equation

$$x = 21{,}000 - 150p$$

(a) Find a model that expresses the revenue R as a function of the price p.
(b) What is the domain of R?
(c) What unit price should be used to maximize revenue?
(d) If this price is charged, what is the maximum revenue?

(e) How many units are sold at this price?
(f) Graph R.
(g) What price should Texas Instruments charge to collect at least \$675,000 in revenue?

Solution

(a) The revenue is $R = xp$, where $x = 21{,}000 - 150p$.

$$R = xp = (21{,}000 - 150p)p = -150p^2 + 21{,}000p \quad \text{The model}$$

(b) Because x represents the number of calculators sold, we have $x \geq 0$, so $21{,}000 - 150p \geq 0$. Solving this linear inequality gives $p \leq 140$. In addition, Texas Instruments will charge only a positive price for the calculator, so $p > 0$. Combining these inequalities gives the domain of R, which is $\{p \mid 0 < p \leq 140\}$.

(c) The function R is a quadratic function with $a = -150$, $b = 21{,}000$, and $c = 0$. Because $a < 0$, the vertex is the highest point on the parabola. The revenue R is a maximum when the price p is

$$p = -\frac{b}{2a} \underset{a = -150,\, b = 21{,}000}{=} -\frac{21{,}000}{2(-150)} = \$70.00$$

(d) The maximum revenue R is

$$R(70) = -150(70)^2 + 21{,}000(70) = \$735{,}000$$

(e) The number of calculators sold is given by the demand equation $x = 21{,}000 - 150p$. At a price of $p = \$70$,

$$x = 21{,}000 - 150(70) = 10{,}500$$

calculators are sold.

(f) To graph R, plot the intercept $(140, 0)$ and the vertex $(70, 735\,000)$. See Figure 23 for the graph.

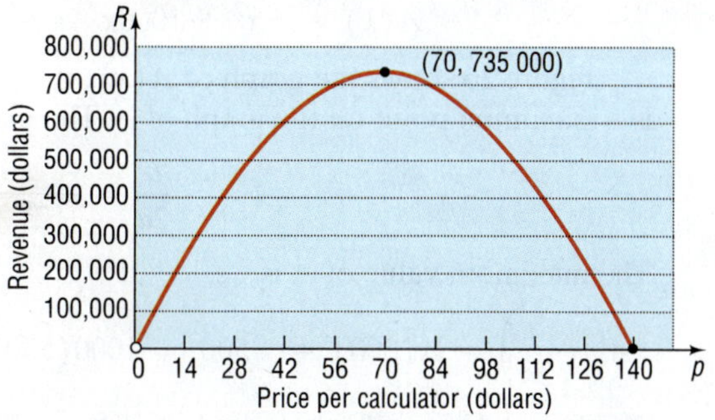

Figure 23

(g) Graph $R = 675{,}000$ and $R(p) = -150p^2 + 21{,}000p$ on the same Cartesian plane. See Figure 24. We find where the graphs intersect by solving

$$\begin{aligned} 675{,}000 &= -150p^2 + 21{,}000p \\ 150p^2 - 21{,}000p + 675{,}000 &= 0 && \text{Add } 150p^2 - 21{,}000p \text{ to both sides.} \\ p^2 - 140p + 4500 &= 0 && \text{Divide both sides by 150.} \\ (p - 50)(p - 90) &= 0 && \text{Factor.} \\ p = 50 \text{ or } p &= 90 && \text{Use the Zero-Product Property.} \end{aligned}$$

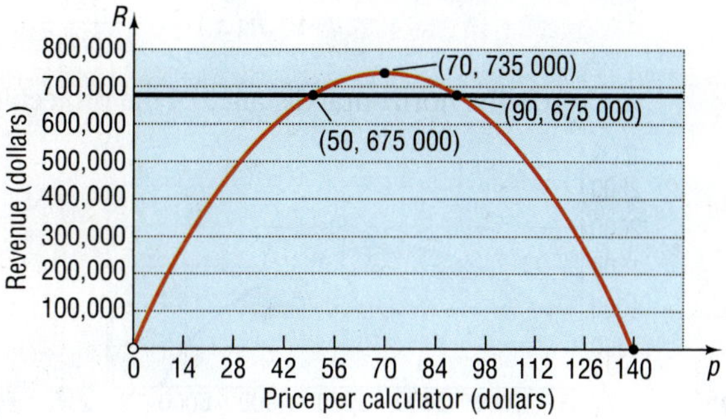

Figure 24

The graphs intersect at $(50, 675\,000)$ and $(90, 675\,000)$. Based on the graph in Figure 24, Texas Instruments should charge between \$50 and \$90 to earn at least \$675,000 in revenue. ●

Now Work PROBLEM 3

EXAMPLE 2 Maximizing the Area Enclosed by a Fence

A farmer has 2000 yards of fence to enclose a rectangular field. What are the dimensions of the rectangle that encloses the most area?

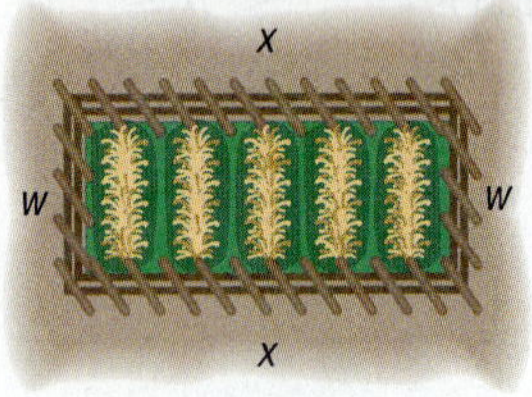

Figure 25

Solution Figure 25 illustrates the situation. The available fence represents the perimeter of the rectangle. If x is the length and w is the width, then

$$2x + 2w = 2000 \quad (1)$$

The area A of the rectangle is

$$A = xw$$

To express A in terms of a single variable, solve equation (1) for w and substitute the result in $A = xw$. Then A involves only the variable x. [You could also solve equation (1) for x and express A in terms of w alone. Try it!]

$$2x + 2w = 2000$$
$$2w = 2000 - 2x$$
$$w = \frac{2000 - 2x}{2} = 1000 - x$$

Then the area A is

$$A = xw = x(1000 - x) = -x^2 + 1000x$$

Now, A is a quadratic function of x.

$$A(x) = -x^2 + 1000x \quad a = -1, b = 1000, c = 0$$

Figure 26 shows the graph of $A(x) = -x^2 + 1000x$. Because $a < 0$, the vertex is a maximum point on the graph of A. The maximum value occurs at

$$x = -\frac{b}{2a} = -\frac{1000}{2(-1)} = 500$$

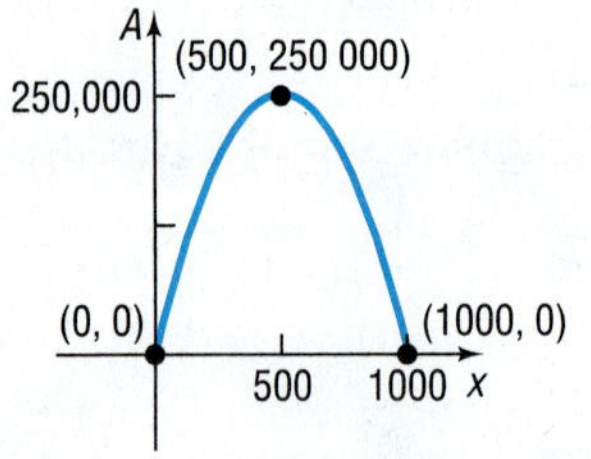

Figure 26 $A(x) = -x^2 + 1000x$

The maximum value of A is

$$A\left(-\frac{b}{2a}\right) = A(500) = -500^2 + 1000(500) = -250{,}000 + 500{,}000 = 250{,}000$$

The largest rectangle that can be enclosed by 2000 yards of fence has an area of 250,000 square yards. Its dimensions are 500 yards by 500 yards. ●

Now Work PROBLEM 7

EXAMPLE 3 Analyzing the Motion of a Projectile

A projectile is fired from a cliff 500 feet above the water at an inclination of 45° to the horizontal, with a muzzle velocity of 400 feet per second. From physics, the height h of the projectile above the water can be modeled by

$$h(x) = \frac{-32x^2}{(400)^2} + x + 500$$

where x is the horizontal distance of the projectile from the base of the cliff. See Figure 27.

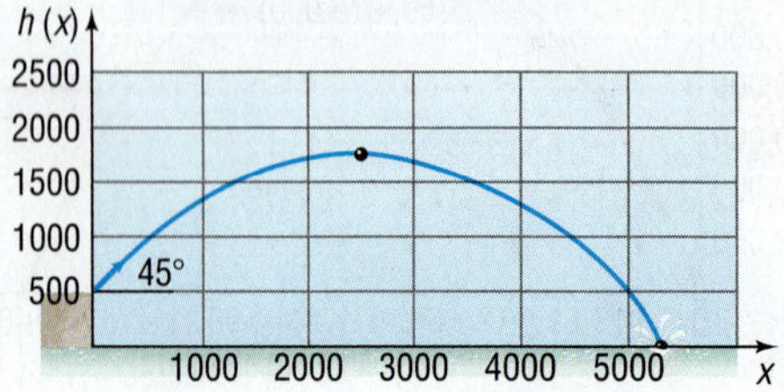

Figure 27

(a) Find the maximum height of the projectile.

(b) How far from the base of the cliff will the projectile strike the water?

Solution (a) The height of the projectile is given by a quadratic function.

$$h(x) = \frac{-32x^2}{(400)^2} + x + 500 = \frac{-1}{5000}x^2 + x + 500$$

We are looking for the maximum value of h. Because $a < 0$, the maximum value occurs at the vertex, whose x-coordinate is

$$x = -\frac{b}{2a} = -\frac{1}{2\left(\frac{-1}{5000}\right)} = \frac{5000}{2} = 2500$$

The maximum height of the projectile is

$$h(2500) = \frac{-1}{5000}(2500)^2 + 2500 + 500 = -1250 + 2500 + 500 = 1750 \text{ ft}$$

(b) The projectile will strike the water when the height is zero. To find the distance x traveled, solve the equation

$$h(x) = \frac{-1}{5000}x^2 + x + 500 = 0$$

The discriminant of this quadratic equation is

$$b^2 - 4ac = 1^2 - 4\left(\frac{-1}{5000}\right)(500) = 1.4$$

Then

$$x = \frac{-b \pm \sqrt{b^2 - 4ac}}{2a} = \frac{-1 \pm \sqrt{1.4}}{2\left(\frac{-1}{5000}\right)} \approx \begin{cases} -458 \\ 5458 \end{cases}$$

Discard the negative solution. The projectile will strike the water at a distance of about 5458 feet from the base of the cliff.

Seeing the Concept

Graph

$$h(x) = \frac{-1}{5000}x^2 + x + 500$$
$$0 \le x \le 5500$$

Use MAXIMUM to find the maximum height of the projectile, and use ROOT or ZERO to find the distance from the base of the cliff to where it strikes the water. Compare your results with those obtained in Example 3.

Now Work PROBLEM 11

EXAMPLE 4 The Golden Gate Bridge

The Golden Gate Bridge, a suspension bridge, spans the entrance to San Francisco Bay. Its 746-foot-tall towers are 4200 feet apart. The bridge is suspended from two huge cables more than 3 feet in diameter; the 90-foot-wide roadway is 220 feet above the water. The cables are parabolic in shape* and touch the road surface at the center of the bridge. Find the height of the cable above the road at a distance of 1000 feet from the center.

Solution See Figure 28 on page 306. Begin by choosing the placement of the coordinate axes so that the x-axis coincides with the road surface and the origin coincides with the center of the bridge. As a result, the 746-foot towers will be vertical (height $746 - 220 = 526$ feet above the road) and located 2100 feet from the center. Also, the cable, which has the shape of a parabola, will extend from the towers, open up, and have its vertex at $(0, 0)$. This choice of placement of the axes enables the equation of the parabola to have the form $y = ax^2$, $a > 0$. Note that the points $(-2100, 526)$ and $(2100, 526)$ are on the graph.

*A cable suspended from two towers is in the shape of a **catenary**, but when a horizontal roadway is suspended from the cable, the cable takes the shape of a parabola.

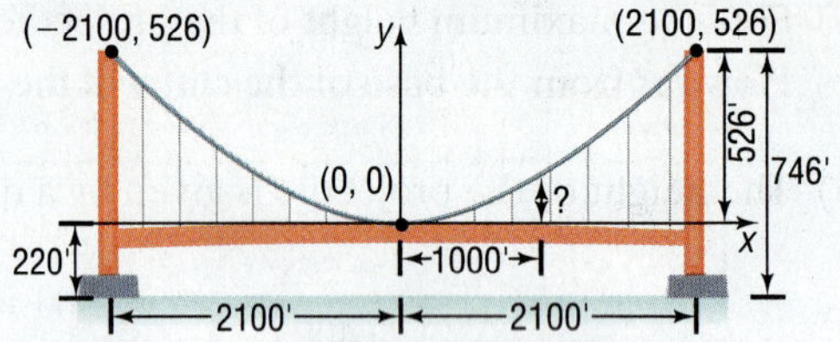

Figure 28

Use these facts to find the value of a in $y = ax^2$.

$$y = ax^2$$

$$526 = a(2100)^2 \quad x = 2100, y = 526$$

$$a = \frac{526}{(2100)^2}$$

The equation of the parabola is

$$y = \frac{526}{(2100)^2}x^2$$

When $x = 1000$, the height of the cable is

$$y = \frac{526}{(2100)^2}(1000)^2 \approx 119.3 \text{ feet}$$

The cable is 119.3 feet above the road at a distance of 1000 feet from the center of the bridge.

Now Work PROBLEM 13

2 Build Quadratic Models from Data

In Section 4.2, we found the line of best fit for data that appeared to be linearly related. It was noted that data may also follow a nonlinear relation. Figures 29(a) and (b) show scatter diagrams of data that follow a quadratic relation.

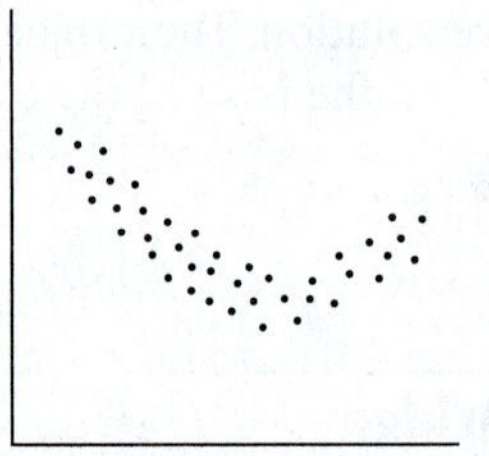

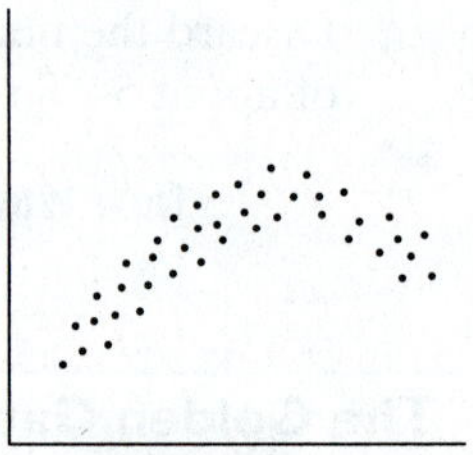

$y = ax^2 + bx + c, a > 0$ (a) $y = ax^2 + bx + c, a < 0$ (b)

Figure 29

EXAMPLE 5 Fitting a Quadratic Function to Data

The data in Table 8 on page 307 represent the percentage D of the population that is divorced for various ages x in 2012.

(a) Draw a scatter diagram of the data, treating age as the independent variable. Comment on the type of relation that may exist between age and percentage of the population divorced.

(b) Use a graphing utility to find the quadratic function of best fit that models the relation between age and percentage of the population divorced.

(c) Use the model found in part (b) to approximate the age at which the percentage of the population divorced is greatest.

(d) Use the model found in part (b) to approximate the highest percentage of the population that is divorced.

(e) Use a graphing utility to draw the quadratic function of best fit on the scatter diagram.

Table 8

Age, x	Percentage Divorced, D
22	0.9
27	3.6
32	7.4
37	10.4
42	12.7
50	15.7
60	16.2
70	13.1
80	6.5

Source: United States Statistical Abstract, 2012

Solution

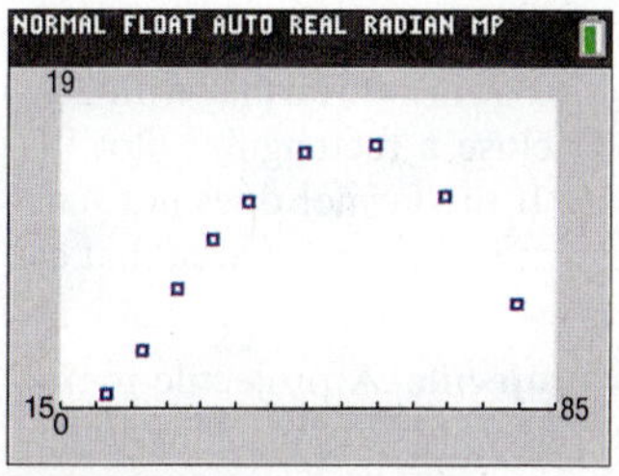

Figure 30

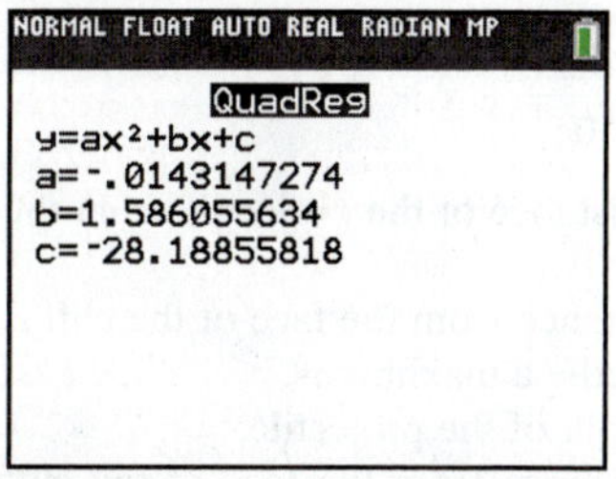

Figure 31

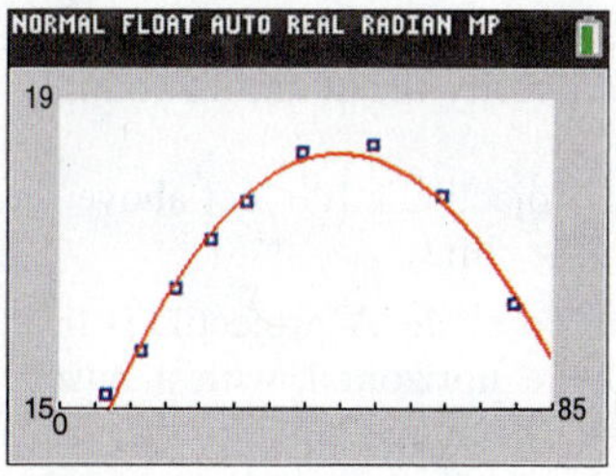

Figure 32

(a) Figure 30 shows the scatter diagram, from which it appears the data follow a quadratic relation, with $a < 0$.

(b) Execute the QUADratic REGression program to obtain the results shown in Figure 31. The output shows the equation $y = ax^2 + bx + c$. The quadratic function of best fit that models the relation between age and percentage divorced is

$$D(x) = -0.0143x^2 + 1.5861x - 28.1886 \quad \text{The model}$$

where x represents age and D represents the percentage divorced.

(c) Based on the quadratic function of best fit, the age with the greatest percentage divorced is

$$-\frac{b}{2a} = -\frac{1.5861}{2(-0.0143)} \approx 55 \text{ years}$$

(d) Evaluate the function $D(x)$ at $x = 55$.

$$D(55) = -0.0143(55)^2 + 1.5861(55) - 28.1886 \approx 15.8 \text{ percent}$$

According to the model, 55-year-olds have the highest percentage divorced at 15.8 percent.

(e) Figure 32 shows the graph of the quadratic function found in part (b) drawn on the scatter diagram.

Look again at Figure 31. Notice that the output given by the graphing calculator does not include r, the correlation coefficient. Recall that the correlation coefficient is a measure of the strength of a linear relation that exists between two variables. The graphing calculator does not provide an indication of how well the function fits the data in terms of r, since a quadratic function cannot be expressed as a linear function.

Now Work PROBLEM 25

4.4 Assess Your Understanding

'Are You Prepared?' *Answers are given at the end of these exercises. If you get a wrong answer, read the pages listed in red.*

1. Translate the following sentence into a mathematical equation: The total revenue R from selling x hot dogs is \$3 times the number of hot dogs sold. (pp. 134–140)

2. Use a graphing utility to find the line of best fit for the following data: (pp. 284–287)

x	3	5	5	6	7	8
y	10	13	12	15	16	19

Applications and Extensions

3. Maximizing Revenue The price p (in dollars) and the quantity x sold of a certain product obey the demand equation

$$p = -\frac{1}{6}x + 100$$

(a) Find a model that expresses the revenue R as a function of x. (Remember, $R = xp$.)
(b) What is the domain of R?
(c) What is the revenue if 200 units are sold?
(d) What quantity x maximizes revenue? What is the maximum revenue?
(e) What price should the company charge to maximize revenue?

4. Maximizing Revenue The price p (in dollars) and the quantity x sold of a certain product obey the demand equation

$$p = -\frac{1}{3}x + 100$$

(a) Find a model that expresses the revenue R as a function of x.
(b) What is the domain of R?
(c) What is the revenue if 100 units are sold?
(d) What quantity x maximizes revenue? What is the maximum revenue?
(e) What price should the company charge to maximize revenue?

5. Maximizing Revenue The price p (in dollars) and the quantity x sold of a certain product obey the demand equation

$$x = -5p + 100 \qquad 0 < p \le 20$$

(a) Express the revenue R as a function of x.
(b) What is the revenue if 15 units are sold?
(c) What quantity x maximizes revenue? What is the maximum revenue?
(d) What price should the company charge to maximize revenue?
(e) What price should the company charge to earn at least \$480 in revenue?

6. Maximizing Revenue The price p (in dollars) and the quantity x sold of a certain product obey the demand equation

$$x = -20p + 500 \qquad 0 < p \le 25$$

(a) Express the revenue R as a function of x.
(b) What is the revenue if 20 units are sold?
(c) What quantity x maximizes revenue? What is the maximum revenue?
(d) What price should the company charge to maximize revenue?
(e) What price should the company charge to earn at least \$3000 in revenue?

7. Enclosing a Rectangular Field David has 400 yards of fencing and wishes to enclose a rectangular area.
(a) Express the area A of the rectangle as a function of the width w of the rectangle.
(b) For what value of w is the area largest?
(c) What is the maximum area?

8. Enclosing a Rectangular Field Beth has 3000 feet of fencing available to enclose a rectangular field.
(a) Express the area A of the rectangle as a function of x, where x is the length of the rectangle.
(b) For what value of x is the area largest?
(c) What is the maximum area?

9. Enclosing the Most Area with a Fence A farmer with 4000 meters of fencing wants to enclose a rectangular plot that borders on a river. If the farmer does not fence the side along the river, what is the largest area that can be enclosed? (See the figure.)

10. Enclosing the Most Area with a Fence A farmer with 2000 meters of fencing wants to enclose a rectangular plot that borders on a straight highway. If the farmer does not fence the side along the highway, what is the largest area that can be enclosed?

11. Analyzing the Motion of a Projectile A projectile is fired from a cliff 200 feet above the water at an inclination of 45° to the horizontal, with a muzzle velocity of 50 feet per second. The height h of the projectile above the water is modeled by

$$h(x) = \frac{-32x^2}{(50)^2} + x + 200$$

where x is the horizontal distance of the projectile from the face of the cliff.
(a) At what horizontal distance from the face of the cliff is the height of the projectile a maximum?
(b) Find the maximum height of the projectile.
(c) At what horizontal distance from the face of the cliff will the projectile strike the water?
(d) Using a graphing utility, graph the function h, $0 \le x \le 200$.
(e) Use a graphing utility to verify the solutions found in parts (b) and (c).
(f) When the height of the projectile is 100 feet above the water, how far is it from the cliff?

12. Analyzing the Motion of a Projectile A projectile is fired at an inclination of 45° to the horizontal, with a muzzle velocity of 100 feet per second. The height h of the projectile is modeled by

$$h(x) = \frac{-32x^2}{(100)^2} + x$$

where x is the horizontal distance of the projectile from the firing point.
(a) At what horizontal distance from the firing point is the height of the projectile a maximum?
(b) Find the maximum height of the projectile.
(c) At what horizontal distance from the firing point will the projectile strike the ground?
(d) Using a graphing utility, graph the function h, $0 \le x \le 350$.

(e) Use a graphing utility to verify the results obtained in parts (b) and (c).

(f) When the height of the projectile is 50 feet above the ground, how far has it traveled horizontally?

13. Suspension Bridge A suspension bridge with weight uniformly distributed along its length has twin towers that extend 75 meters above the road surface and are 400 meters apart. The cables are parabolic in shape and are suspended from the tops of the towers. The cables touch the road surface at the center of the bridge. Find the height of the cables at a point 100 meters from the center. (Assume that the road is level.)

14. Architecture A parabolic arch has a span of 120 feet and a maximum height of 25 feet. Choose suitable rectangular coordinate axes and find the equation of the parabola. Then calculate the height of the arch at points 10 feet, 20 feet, and 40 feet from the center.

15. Constructing Rain Gutters A rain gutter is to be made of aluminum sheets that are 12 inches wide by turning up the edges 90°. See the illustration.

(a) What depth will provide maximum cross-sectional area and hence allow the most water to flow?

(b) What depths will allow at least 16 square inches of water to flow?

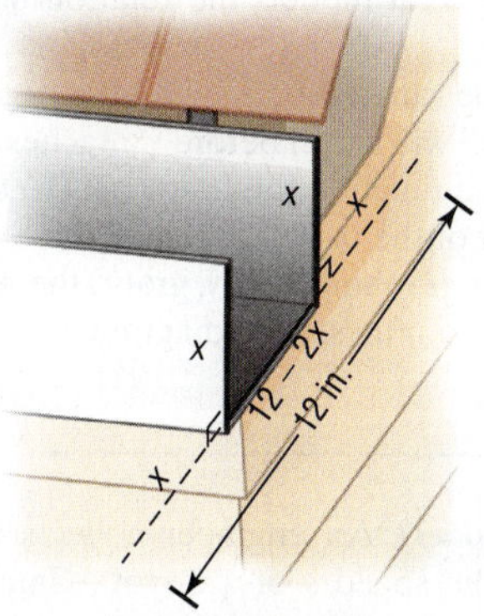

16. Norman Windows A **Norman window** has the shape of a rectangle surmounted by a semicircle of diameter equal to the width of the rectangle. See the figure. If the perimeter of the window is 20 feet, what dimensions will admit the most light (maximize the area)?

[**Hint:** Circumference of a circle $= 2\pi r$; area of a circle $= \pi r^2$, where r is the radius of the circle.]

17. Constructing a Stadium A track-and-field playing area is in the shape of a rectangle with semicircles at each end. See the figure (top, right). The inside perimeter of the track is to be 1500 meters. What should the dimensions of the rectangle be so that the area of the rectangle is a maximum?

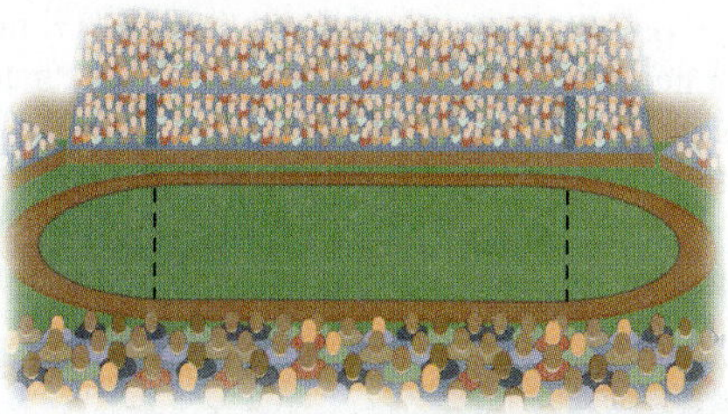

18. Architecture A special window has the shape of a rectangle surmounted by an equilateral triangle. See the figure. If the perimeter of the window is 16 feet, what dimensions will admit the most light?

[**Hint:** Area of an equilateral triangle $= \left(\frac{\sqrt{3}}{4}\right)x^2$, where x is the length of a side of the triangle.]

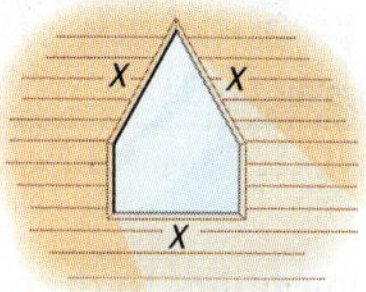

19. Chemical Reactions A self-catalytic chemical reaction results in the formation of a compound that causes the formation ratio to increase. If the reaction rate V is modeled by

$$V(x) = kx(a - x), \qquad 0 \le x \le a$$

where k is a positive constant, a is the initial amount of the compound, and x is the variable amount of the compound, for what value of x is the reaction rate a maximum?

20. Calculus: Simpson's Rule The figure shows the graph of $y = ax^2 + bx + c$. Suppose that the points $(-h, y_0)$, $(0, y_1)$, and (h, y_2) are on the graph. It can be shown that the area enclosed by the parabola, the x-axis, and the lines $x = -h$ and $x = h$ is

$$\text{Area} = \frac{h}{3}(2ah^2 + 6c)$$

Show that this area may also be given by

$$\text{Area} = \frac{h}{3}(y_0 + 4y_1 + y_2)$$

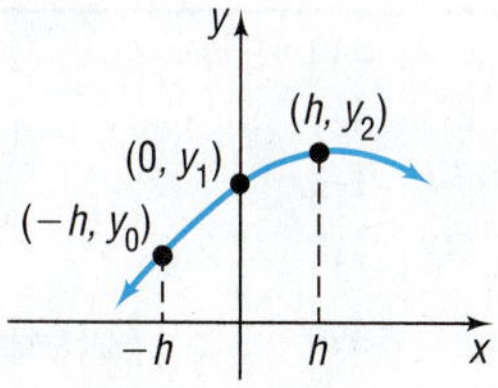

21. Use the result obtained in Problem 20 to find the area enclosed by $f(x) = -5x^2 + 8$, the x-axis, and the lines $x = -1$ and $x = 1$.

22. Use the result obtained in Problem 20 to find the area enclosed by $f(x) = 2x^2 + 8$, the x-axis, and the lines $x = -2$ and $x = 2$.

23. Use the result obtained in Problem 20 to find the area enclosed by $f(x) = x^2 + 3x + 5$, the x-axis, and the lines $x = -4$ and $x = 4$.

24. Use the result obtained in Problem 20 to find the area enclosed by $f(x) = -x^2 + x + 4$, the x-axis, and the lines $x = -1$ and $x = 1$.

25. Life Cycle Hypothesis An individual's income varies with his or her age. The following table shows the median income I of males of different age groups within the United States for 2012. For each age group, let the class midpoint represent the independent variable, x. For the class "65 years and older," we will assume that the class midpoint is 69.5.

Age	Class Midpoint, x	Median Income, I
15–24 years	19.5	$10,869
25–34 years	29.5	$34,113
35–44 years	39.5	$45,225
45–54 years	49.5	$46,466
55–64 years	59.5	$42,176
65 years and older	69.5	$27,612

Source: U.S. Census Bureau

(a) Use a graphing utility to draw a scatter diagram of the data. Comment on the type of relation that may exist between the two variables.
(b) Use a graphing utility to find the quadratic function of best fit that models the relation between age and median income.
(c) Use the function found in part (b) to determine the age at which an individual can expect to earn the most income.
(d) Use the function found in part (b) to predict the peak income earned.
(e) With a graphing utility, graph the quadratic function of best fit on the scatter diagram.

26. Height of a Ball A shot-putter throws a ball at an inclination of 45° to the horizontal. The following data represent the height of the ball h, in feet, at the instant that it has traveled x feet horizontally.

Distance, x	Height, h
20	25
40	40
60	55
80	65
100	71
120	77
140	77
160	75
180	71
200	64

(a) Use a graphing utility to draw a scatter diagram of the data. Comment on the type of relation that may exist between the two variables.
(b) Use a graphing utility to find the quadratic function of best fit that models the relation between distance and height.
(c) Use the function found in part (b) to determine how far the ball will travel before it reaches its maximum height.
(d) Use the function found in part (b) to find the maximum height of the ball.
(e) With a graphing utility, graph the quadratic function of best fit on the scatter diagram.

Mixed Practice

27. Which Model? The following data represent the square footage and rents (dollars per month) for apartments in the La Jolla area of San Diego, California.

Square Footage, x	Rent per Month, R
520	$1525
621	$1750
718	$1785
753	$1850
850	$1900
968	$2130
1020	$2180

Source: apartments.com, 2014

(a) Using a graphing utility, draw a scatter diagram of the data treating square footage as the independent variable. What type of relation appears to exist between square footage and rent?
(b) Based on your response to part (a), find either a linear or a quadratic model that describes the relation between square footage and rent.
(c) Use your model to predict the rent for an apartment in San Diego that is 875 square feet.

28. Which Model? An engineer collects the following data showing the speed s of a Toyota Camry and its average miles per gallon, M.

Speed, s	Miles per Gallon, M
30	18
35	20
40	23
40	25
45	25
50	28
55	30
60	29
65	26
65	25
70	25

(a) Using a graphing utility, draw a scatter diagram of the data, treating speed as the independent variable. What type of relation appears to exist between speed and miles per gallon?

(b) Based on your response to part (a), find either a linear model or a quadratic model that describes the relation between speed and miles per gallon.

(c) Use your model to predict the miles per gallon for a Camry that is traveling 63 miles per hour.

29. Which Model? The following data represent the birth rate (births per 1000 population) for women whose age is *a*, in 2012.

Age, *a*	Birth Rate, *B*
16	14.1
19	51.4
22	83.1
27	106.5
32	97.3
37	48.3
42	10.4

Source: National Vital Statistics System, 2013

(a) Using a graphing utility, draw a scatter diagram of the data, treating age as the independent variable. What type of relation appears to exist between age and birth rate?

(b) Based on your response to part (a), find either a linear or a quadratic model that describes the relation between age and birth rate.

(c) Use your model to predict the birth rate for 35-year-old women.

30. Which Model? A cricket makes a chirping noise by sliding its wings together rapidly. Perhaps you have noticed that the number of chirps seems to increase with the temperature. The following data list the temperature (in degrees Fahrenheit) and the number of chirps per second for the striped ground cricket.

Temperature (°F), *x*	Chirps per Second, *C*
88.6	20.0
93.3	19.8
80.6	17.1
69.7	14.7
69.4	15.4
79.6	15.0
80.6	16.0
76.3	14.4
75.2	15.5

Source: Pierce, George W. *The Songs of Insects*. Cambridge, MA Harvard University Press, 1949, pp. 12 – 21

(a) Using a graphing utility, draw a scatter diagram of the data, treating temperature as the independent variable. What type of relation appears to exist between temperature and chirps per second?

(b) Based on your response to part (a), find either a linear or a quadratic model that best describes the relation between temperature and chirps per second.

(c) Use your model to predict the chirps per second if the temperature is 80°F.

Explaining Concepts: Discussion and Writing

31. Refer to Example 1 in this section. Notice that if the price charged for the calculators is \$0 or \$140, then the revenue is \$0. It is easy to explain why revenue would be \$0 if the price charged were \$0, but how can revenue be \$0 if the price charged is \$140?

Retain Your Knowledge

Problems 32–35 are based on material learned earlier in the course. The purpose of these problems is to keep the material fresh in your mind so that you are better prepared for the final exam.

32. Express as a complex number: $\sqrt{-225}$

33. Find the distance between the points $P_1 = (4, -7)$ and $P_2 = (-1, 5)$.

34. Find the equation of the circle with center $(-6, 0)$ and radius $r = \sqrt{7}$.

35. Solve: $5x^2 + 8x - 3 = 0$

'Are You Prepared?' Answers

1. $R = 3x$ **2.** $y = 1.7826x + 4.0652$

4.5 Inequalities Involving Quadratic Functions

PREPARING FOR THIS SECTION *Before getting started, review the following:*

- Solve Inequalities (Section 1.5, pp. 123–126)
- Use Interval Notation (Section 1.5, pp. 120–121)

Now Work the **'Are You Prepared?'** problems on page 314.

OBJECTIVE 1 Solve Inequalities Involving a Quadratic Function (p. 312)

1 Solve Inequalities Involving a Quadratic Function

In this section we solve inequalities that involve quadratic functions. We will accomplish this by using their graphs. For example, to solve the inequality

$$ax^2 + bx + c > 0 \qquad a \neq 0$$

graph the function $f(x) = ax^2 + bx + c$ and, from the graph, determine where it is above the x-axis—that is, where $f(x) > 0$. To solve the inequality $ax^2 + bx + c < 0$, $a \neq 0$, graph the function $f(x) = ax^2 + bx + c$ and determine where the graph is below the x-axis. If the inequality is not strict, include the x-intercepts, if any, in the solution.

EXAMPLE 1 **Solving an Inequality**

Solve the inequality $x^2 - 4x - 12 \leq 0$ and graph the solution set.

Solution Graph the function $f(x) = x^2 - 4x - 12$.

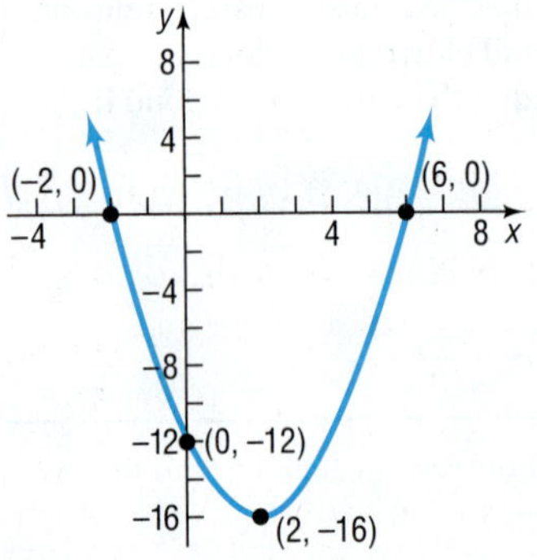

Figure 33 $f(x) = x^2 - 4x - 12$

y-intercept: $f(0) = -12$ Evaluate f at 0.

x-intercepts (if any): $x^2 - 4x - 12 = 0$ Solve $f(x) = 0$.

$(x - 6)(x + 2) = 0$ Factor.

$x - 6 = 0$ or $x + 2 = 0$ Apply the Zero-Product Property.

$x = 6$ or $x = -2$

The y-intercept is -12; the x-intercepts are -2 and 6.

The vertex is at $x = -\dfrac{b}{2a} = -\dfrac{-4}{2} = 2$. Because $f(2) = -16$, the vertex is at $(2, -16)$.

See Figure 33 for the graph.

The graph is below the x-axis for $-2 < x < 6$. Because the original inequality is not strict, include the x-intercepts. The solution set is $\{x \mid -2 \leq x \leq 6\}$ or, using interval notation, $[-2, 6]$. See Figure 34 for the graph of the solution set. ●

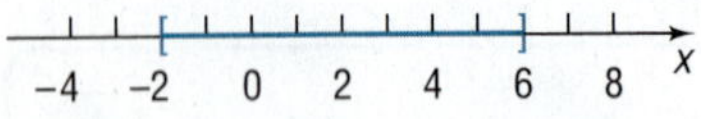

Figure 34

Now Work PROBLEM 9

EXAMPLE 2 **Solving an Inequality**

Solve the inequality $2x^2 < x + 10$ and graph the solution set.

Solution *Option 1* Rearrange the inequality so that 0 is on the right side.

$$2x^2 < x + 10$$

$$2x^2 - x - 10 < 0 \qquad \text{Subtract } x + 10 \text{ from both sides.}$$

This inequality is equivalent to the original inequality.

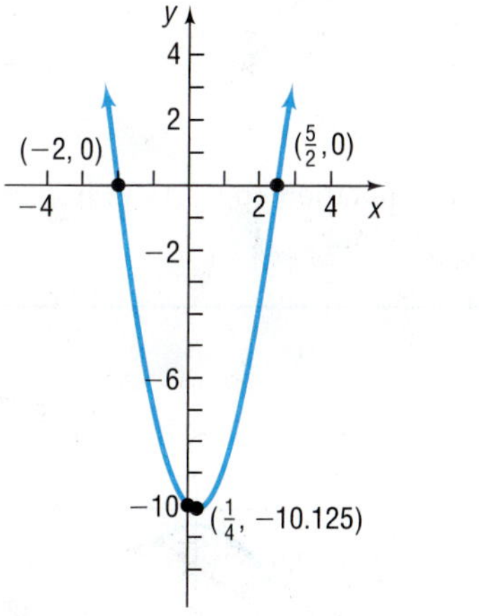

Figure 35 $f(x) = 2x^2 - x - 10$

Next graph the function $f(x) = 2x^2 - x - 10$ to find where $f(x) < 0$.

y-intercept: $f(0) = -10$ — Evaluate f at 0.

x-intercepts (if any): $2x^2 - x - 10 = 0$ — Solve $f(x) = 0$.

$$(2x - 5)(x + 2) = 0 \quad \text{Factor.}$$

$$2x - 5 = 0 \quad \text{or} \quad x + 2 = 0 \quad \text{Apply the Zero-Product Property.}$$

$$x = \frac{5}{2} \quad \text{or} \quad x = -2$$

The y-intercept is -10; the x-intercepts are -2 and $\frac{5}{2}$.

The vertex is at $x = -\frac{b}{2a} = -\frac{-1}{4} = \frac{1}{4}$. Because $f\left(\frac{1}{4}\right) = -10.125$, the vertex is $\left(\frac{1}{4}, -10.125\right)$. See Figure 35 for the graph.

The graph is below the x-axis ($f(x) < 0$) between $x = -2$ and $x = \frac{5}{2}$. Because the inequality is strict, the solution set is $\left\{x \middle| -2 < x < \frac{5}{2}\right\}$ or, using interval notation, $\left(-2, \frac{5}{2}\right)$.

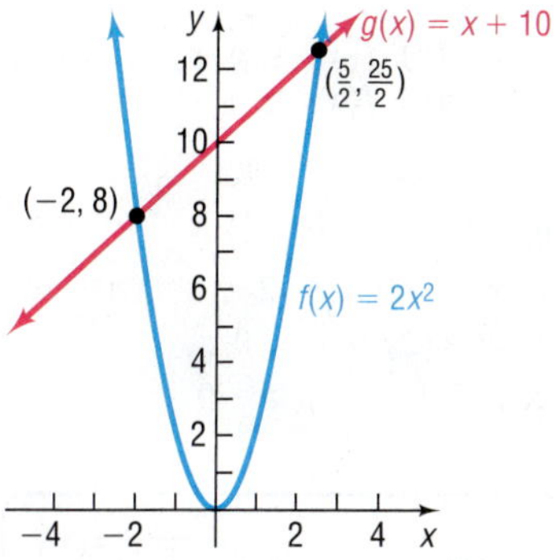

Figure 36

Option 2 If $f(x) = 2x^2$ and $g(x) = x + 10$, then the inequality to be solved is $f(x) < g(x)$. Graph the functions $f(x) = 2x^2$ and $g(x) = x + 10$. See Figure 36. The graphs intersect where $f(x) = g(x)$. Then

$$2x^2 = x + 10 \quad f(x) = g(x)$$

$$2x^2 - x - 10 = 0$$

$$(2x - 5)(x + 2) = 0 \quad \text{Factor.}$$

$$2x - 5 = 0 \quad \text{or} \quad x + 2 = 0 \quad \text{Apply the Zero-Product Property.}$$

$$x = \frac{5}{2} \quad \text{or} \quad x = -2$$

The graphs intersect at the points $(-2, 8)$ and $\left(\frac{5}{2}, \frac{25}{2}\right)$. To solve $f(x) < g(x)$, find where the graph of f is below the graph of g. This happens between the points of intersection. Because the inequality is strict, the solution set is $\left\{x \middle| -2 < x < \frac{5}{2}\right\}$ or, using interval notation, $\left(-2, \frac{5}{2}\right)$.

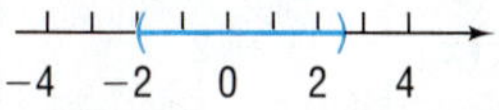

Figure 37

See Figure 37 for the graph of the solution set.

Now Work PROBLEMS 5 AND 13

EXAMPLE 3 Solving an Inequality

Solve the inequality $x^2 + x + 1 > 0$ and graph the solution set.

Solution

Graph the function $f(x) = x^2 + x + 1$. The y-intercept is 1; there are no x-intercepts (Do you see why? Check the discriminant). The vertex is at $x = -\frac{b}{2a} = -\frac{1}{2}$. Since $f\left(-\frac{1}{2}\right) = \frac{3}{4}$, the vertex is at $\left(-\frac{1}{2}, \frac{3}{4}\right)$. The points $(1, 3)$ and $(-1, 1)$ are also on the graph. See Figure 38.

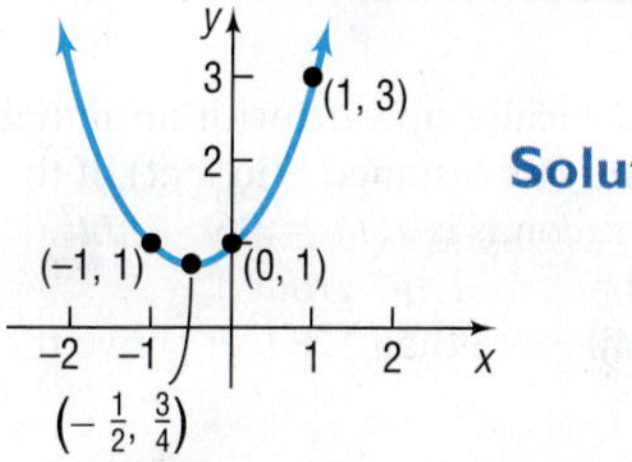

Figure 38 $f(x) = x^2 + x + 1$

The graph of f lies above the x-axis for all x. The solution set is the set of all real numbers, or $(-\infty, \infty)$. See Figure 39.

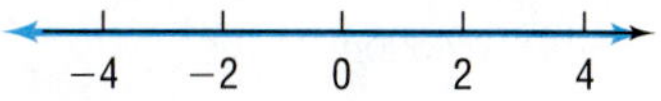

Figure 39

Now Work PROBLEM 17

4.5 Assess Your Understanding

'Are You Prepared?' *Answers are given at the end of these exercises. If you get a wrong answer, read the pages listed in red.*

1. Solve the inequality $-3x - 2 < 7$ (pp. 123–126)

2. Write $(-2, 7]$ using inequality notation. (pp. 120–121)

Skill Building

In Problems 3–6, use the figure to solve each inequality.

3.

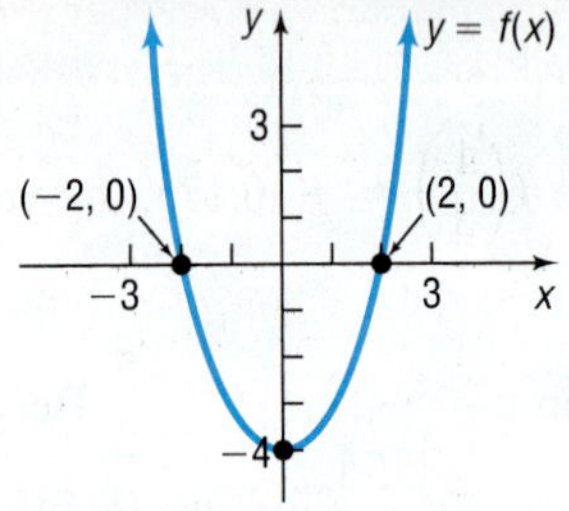

(a) $f(x) > 0$
(b) $f(x) \le 0$

4.

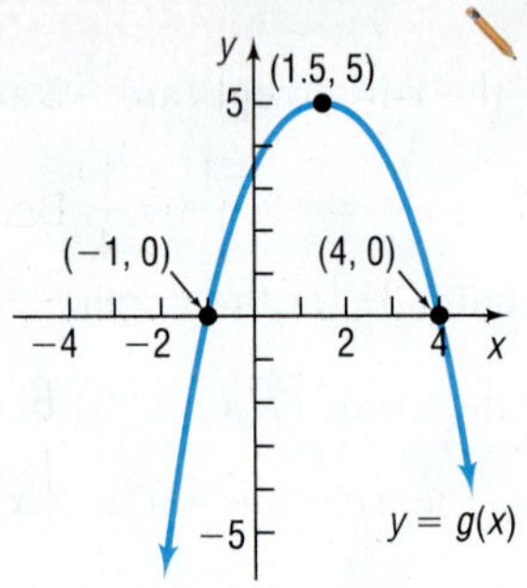

(a) $g(x) < 0$
(b) $g(x) \ge 0$

5.

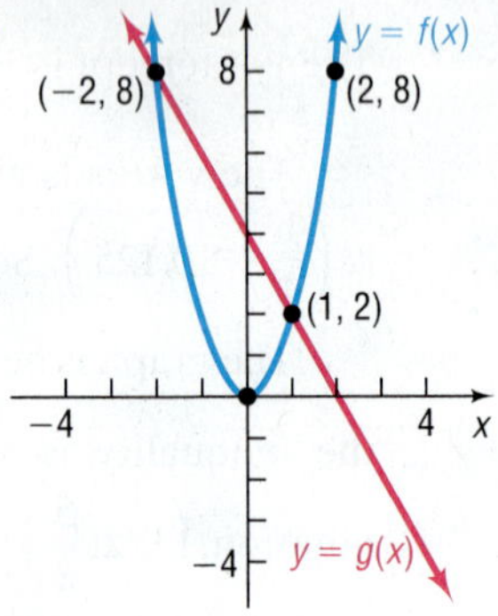

(a) $g(x) \ge f(x)$
(b) $f(x) > g(x)$

6.

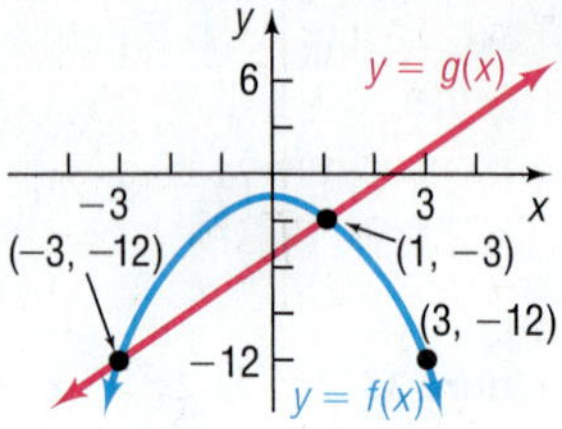

(a) $f(x) < g(x)$
(b) $f(x) \ge g(x)$

In Problems 7–22, solve each inequality.

7. $x^2 - 3x - 10 < 0$
8. $x^2 + 3x - 10 > 0$
9. $x^2 - 4x > 0$
10. $x^2 + 8x > 0$
11. $x^2 - 9 < 0$
12. $x^2 - 1 < 0$
13. $x^2 + x > 12$
14. $x^2 + 7x < -12$
15. $2x^2 < 5x + 3$
16. $6x^2 < 6 + 5x$
17. $x^2 - x + 1 \le 0$
18. $x^2 + 2x + 4 > 0$
19. $4x^2 + 9 < 6x$
20. $25x^2 + 16 < 40x$
21. $6(x^2 - 1) > 5x$
22. $2(2x^2 - 3x) > -9$

Mixed Practice

23. What is the domain of the function $f(x) = \sqrt{x^2 - 16}$?

24. What is the domain of the function $f(x) = \sqrt{x - 3x^2}$?

In Problems 25–32, use the given functions f and g.

(a) Solve $f(x) = 0$.
(b) Solve $g(x) = 0$.
(c) Solve $f(x) = g(x)$.
(d) Solve $f(x) > 0$.
(e) Solve $g(x) \le 0$.
(f) Solve $f(x) > g(x)$.
(g) Solve $f(x) \ge 1$.

25. $f(x) = x^2 - 1$; $g(x) = 3x + 3$
26. $f(x) = -x^2 + 3$; $g(x) = -3x + 3$
27. $f(x) = -x^2 + 1$; $g(x) = 4x + 1$
28. $f(x) = -x^2 + 4$; $g(x) = -x - 2$
29. $f(x) = x^2 - 4$; $g(x) = -x^2 + 4$
30. $f(x) = x^2 - 2x + 1$; $g(x) = -x^2 + 1$
31. $f(x) = x^2 - x - 2$; $g(x) = x^2 + x - 2$
32. $f(x) = -x^2 - x + 1$; $g(x) = -x^2 + x + 6$

Applications and Extensions

33. Physics A ball is thrown vertically upward with an initial velocity of 80 feet per second. The distance s (in feet) of the ball from the ground after t seconds is $s(t) = 80t - 16t^2$.

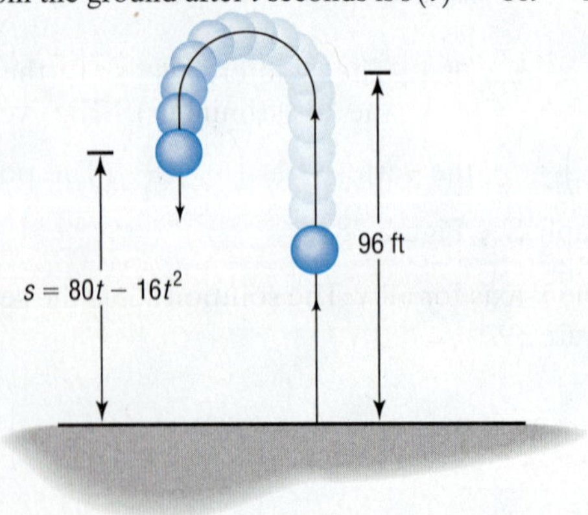

(a) At what time t will the ball strike the ground?
(b) For what time t is the ball more than 96 feet above the ground?

34. Physics A ball is thrown vertically upward with an initial velocity of 96 feet per second. The distance s (in feet) of the ball from the ground after t seconds is $s(t) = 96t - 16t^2$.
(a) At what time t will the ball strike the ground?
(b) For what time t is the ball more than 128 feet above the ground?

35. Revenue Suppose that the manufacturer of a gas clothes dryer has found that when the unit price is p dollars, the revenue R (in dollars) is

$$R(p) = -4p^2 + 4000p$$

(a) At what prices p is revenue zero?
(b) For what range of prices will revenue exceed \$800,000?

36. Revenue The John Deere company has found that the revenue from sales of heavy-duty tractors is a function of the unit price p, in dollars, that it charges. The revenue R, in dollars, is given by

$$R(p) = -\frac{1}{2}p^2 + 1900p$$

(a) At what prices p is revenue zero?
(b) For what range of prices will revenue exceed \$1,200,000?

37. Artillery A projectile fired from the point $(0, 0)$ at an angle to the positive x-axis has a trajectory given by

$$y = cx - (1 + c^2)\left(\frac{g}{2}\right)\left(\frac{x}{v}\right)^2$$

where

x = horizontal distance in meters
y = height in meters
v = initial muzzle velocity in meters per second (m/sec)
g = acceleration due to gravity = 9.81 meters per second squared (m/sec^2)
$c > 0$ is a constant determined by the angle of elevation.

A howitzer fires an artillery round with a muzzle velocity of 897 m/sec.

(a) If the round must clear a hill 200 meters high at a distance of 2000 meters in front of the howitzer, what c values are permitted in the trajectory equation?
(b) If the goal in part (a) is to hit a target on the ground 75 kilometers away, is it possible to do so? If so, for what values of c? If not, what is the maximum distance the round will travel?

Source: *www.answers.com*

38. Runaway Car Using Hooke's Law, we can show that the *work* done in compressing a spring a distance of x feet from its at-rest position is $W = \frac{1}{2}kx^2$, where k is a stiffness constant depending on the spring. It can also be shown that the work done by a body in motion before it comes to rest is given by $\widetilde{W} = \frac{w}{2g}v^2$, where w = weight of the object (in lbs), g = acceleration due to gravity (32.2 ft/sec^2), and v = object's velocity (in ft/sec). A parking garage has a spring shock absorber at the end of a ramp to stop runaway cars. The spring has a stiffness constant $k = 9450$ lb/ft and must be able to stop a 4000-lb car traveling at 25 mph. What is the least compression required of the spring? Express your answer using feet to the nearest tenth.
[**Hint:** Solve $W > \widetilde{W}, x \geq 0$].

Source: *www.sciforums.com*

Explaining Concepts: Discussion and Writing

39. Show that the inequality $(x - 4)^2 \leq 0$ has exactly one solution.
40. Show that the inequality $(x - 2)^2 > 0$ has one real number that is not a solution.
41. Explain why the inequality $x^2 + x + 1 > 0$ has all real numbers as the solution set.
42. Explain why the inequality $x^2 - x + 1 < 0$ has the empty set as the solution set.
43. Explain the circumstances under which the x-intercepts of the graph of a quadratic function are included in the solution set of a quadratic inequality.

Retain Your Knowledge

Problems 44–47 are based on material learned earlier in the course. The purpose of these problems is to keep the material fresh in your mind so that you are better prepared for the final exam.

44. Determine the domain of $f(x) = \sqrt{10 - 2x}$.

45. Determine algebraically whether $f(x) = \frac{-x}{x^2 + 9}$ is even, odd, or neither.

46. Consider the linear function $f(x) = \frac{2}{3}x - 6$.
(a) Find the intercepts of the graph of f.
(b) Graph f.

47. Multiply $(6 - 5i)(4 - i)$. Write the answer in the form $a + bi$.

'Are You Prepared?' Answers

1. $\{x | x > -3\}$ or $(-3, \infty)$ **2.** $-2 < x \leq 7$

Chapter Review

Things to Know

Linear function (p. 274)

$f(x) = mx + b$ Average rate of change $= m$

The graph is a line with slope m and y-intercept b.

Quadratic function (pp. 291–295)

$f(x) = ax^2 + bx + c, a \neq 0$ — The graph is a parabola that opens up if $a > 0$ and opens down if $a < 0$.

Vertex: $\left(-\frac{b}{2a}, f\left(-\frac{b}{2a}\right)\right)$

Axis of symmetry: $x = -\frac{b}{2a}$

y-intercept: $f(0) = c$

x-intercept(s): If any, found by finding the real solutions of the equation $ax^2 + bx + c = 0$

Objectives

Section		You should be able to...	Examples	Review Exercises
4.1	1	Graph linear functions (p. 274)	1	1(a)–3(a), 1(c)–3(c)
	2	Use average rate of change to identify linear functions (p. 274)	2	1(b)–3(b), 4, 5
	3	Determine whether a linear function is increasing, decreasing, or constant (p. 277)	3	1(d)–3(d)
	4	Build linear models from verbal descriptions (p. 278)	4, 5	21
4.2	1	Draw and interpret scatter diagrams (p. 284)	1	29(a), 30(a)
	2	Distinguish between linear and nonlinear relations (p. 285)	2	29(b), 30(a)
	3	Use a graphing utility to find the line of best fit (p. 286)	4	29(c)
4.3	1	Graph a quadratic function using transformations (p. 292)	1	6–8
	2	Identify the vertex and axis of symmetry of a quadratic function (p. 294)	2	9–13
	3	Graph a quadratic function using its vertex, axis, and intercepts (p. 294)	3–5	9–13
	4	Find a quadratic function given its vertex and one other point (p. 297)	6	19, 20
	5	Find the maximum or minimum value of a quadratic function (p. 298)	7	14–16, 22–27
4.4	1	Build quadratic models from verbal descriptions (p. 302)	1–4	22–28
	2	Build quadratic models from data (p. 306)	5	30
4.5	3	Solve inequalities involving a quadratic function (p. 312)	1–3	17, 18

Review Exercises

In Problems 1–3:

(a) Determine the slope and y-intercept of each linear function.
(b) Find the average rate of change of each function.
(c) Graph each function. Label the intercepts.
(d) Determine whether the function is increasing, decreasing, or constant.

1. $f(x) = 2x - 5$ **2.** $h(x) = \frac{4}{5}x - 6$ **3.** $G(x) = 4$

In Problems 4 and 5, determine whether the function is linear or nonlinear. If the function is linear, state its slope.

4.

x	y = f(x)
−1	−2
0	3
1	8
2	13
3	18

5.

x	y = g(x)
−1	−3
0	4
1	7
2	6
3	1

In Problems 6–8, graph each quadratic function using transformations (shifting, compressing, stretching, and/or reflecting).

6. $f(x) = (x + 1)^2 - 4$ **7.** $f(x) = -(x - 4)^2$ **8.** $f(x) = -3(x + 2)^2 + 1$

In Problems 9–13, (a) graph each quadratic function by determining whether its graph opens up or down and by finding its vertex, axis of symmetry, y-intercept, and x-intercepts, if any. (b) Determine the domain and the range of the function. (c) Determine where the function is increasing and where it is decreasing.

9. $f(x) = (x-2)^2 + 2$

10. $f(x) = \frac{1}{4}x^2 - 16$

11. $f(x) = -4x^2 + 4x$

12. $f(x) = \frac{9}{2}x^2 + 3x + 1$

13. $f(x) = 3x^2 + 4x - 1$

In Problems 14–16, determine whether the given quadratic function has a maximum value or a minimum value, and then find the value.

14. $f(x) = 3x^2 - 6x + 4$

15. $f(x) = -x^2 + 8x - 4$

16. $f(x) = -3x^2 + 12x + 4$

In Problems 17 and 18, solve each quadratic inequality.

17. $x^2 + 6x - 16 < 0$

18. $3x^2 \geq 14x + 5$

In Problems 19 and 20, find the quadratic function for which:

19. Vertex is (2, –4); y-intercept is –16

20. Vertex is $(-1, 2)$; contains the point $(1, 6)$

21. Sales Commissions Bill was just offered a sales position for a computer company. His salary would be \$25,000 per year plus 1% of his total annual sales.
(a) Find a linear function that relates Bill's annual salary, S, to his total annual sales, x.
(b) If Bill's total annual sales were \$1,000,000, what would be Bill's salary?
(c) What would Bill have to sell to earn \$100,000?
(d) Determine the sales required of Bill for his salary to exceed \$150,000.

22. Demand Equation The price p (in dollars) and the quantity x sold of a certain product obey the demand equation

$$p = -\frac{1}{10}x + 150 \qquad 0 \leq x \leq 1500$$

(a) Express the revenue R as a function of x.
(b) What is the revenue if 100 units are sold?
(c) What quantity x maximizes revenue? What is the maximum revenue?
(d) What price should the company charge to maximize revenue?

23. Landscaping A landscape engineer has 200 feet of border to enclose a rectangular pond. What dimensions will result in the largest pond?

24. Enclosing the Most Area with a Fence A farmer with 10,000 meters of fencing wants to enclose a rectangular field and then divide it into two plots with a fence parallel to one of the sides. See the figure. What is the largest area that can be enclosed?

25. Architecture A special window in the shape of a rectangle with semicircles at each end is to be constructed so that the outside perimeter is 100 feet. See the illustration. Find the dimensions of the rectangle that maximizes its area.

26. Minimizing Marginal Cost Callaway Golf Company has determined that the marginal cost C of manufacturing x Big Bertha golf clubs may be expressed by the quadratic function

$$C(x) = 4.9x^2 - 617.4x + 19{,}600$$

(a) How many clubs should be manufactured to minimize the marginal cost?
(b) At this level of production, what is the marginal cost?

27. Maximizing Area A rectangle has one vertex on the line $y = 10 - x$, $x > 0$, another at the origin, one on the positive x-axis, and one on the positive y-axis. Express the area A of the rectangle as a function of x. Find the largest area A that can be enclosed by the rectangle.

28. Parabolic Arch Bridge A horizontal bridge is in the shape of a parabolic arch. Given the information shown in the figure, what is the height h of the arch 2 feet from shore?

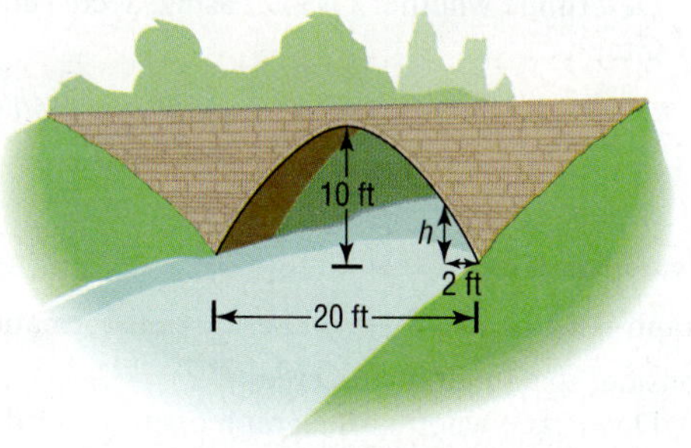

29. Bone Length Research performed at NASA, led by Dr. Emily R. Morey-Holton, measured the lengths of the right humerus and right tibia in 11 rats that were sent to space on Spacelab Life Sciences 2. The data on page 318 were collected.
(a) Draw a scatter diagram of the data, treating length of the right humerus as the independent variable.
(b) Based on the scatter diagram, do you think that there is a linear relation between the length of the right humerus and the length of the right tibia?

(c) Use a graphing utility to find the line of best fit relating length of the right humerus and length of the right tibia.

(d) Predict the length of the right tibia on a rat whose right humerus is 26.5 millimeters (mm).

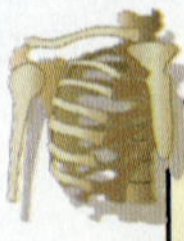

Right Humerus (mm), x	Right Tibia (mm), y
24.80	36.05
24.59	35.57
24.59	35.57
24.29	34.58
23.81	34.20
24.87	34.73
25.90	37.38
26.11	37.96
26.63	37.46
26.31	37.75
26.84	38.50

Source: NASA Life Sciences Data Archive

30. Advertising A small manufacturing firm collected the following data on advertising expenditures A (in thousands of dollars) and total revenue R (in thousands of dollars).

(a) Draw a scatter diagram of the data. Comment on the type of relation that may exist between the two variables.

Advertising Expenditures ($1000s)	Total Revenue ($1000s)
20	6101
22	6222
25	6350
25	6378
27	6453
28	6423
29	6360
31	6231

(b) The quadratic function of best fit to these data is

$$R(A) = -7.76A^2 + 411.88A + 942.72$$

Use this function to determine the optimal level of advertising.

(c) Use the function to predict the total revenue when the optimal level of advertising is spent.

(d) Use a graphing utility to verify that the function given in part (b) is the quadratic function of best fit.

(e) Use a graphing utility to draw a scatter diagram of the data, and then graph the quadratic function of best fit on the scatter diagram.

Chapter Test

CHAPTER **Test Prep** VIDEOS

The Chapter Test Prep Videos are step-by-step solutions available in MyMathLab, or on this text's YouTube Channel. Flip back to the Resources for Success page for a link to this text's YouTube channel.

1. Consider the linear function $f(x) = -4x + 3$:

(a) Find the slope and y-intercept.

(b) What is the average rate of change of f?

(c) Determine whether f is increasing, decreasing, or constant.

(d) Graph f.

In Problems 2 and 3, find the intercepts, if any, of each quadratic function.

2. $f(x) = 3x^2 - 2x - 8$

3. $G(x) = -2x^2 + 4x + 1$

4. Given that $f(x) = x^2 + 3x$ and $g(x) = 5x + 3$, solve $f(x) = g(x)$. Graph each function and label the points of intersection.

5. Graph $f(x) = (x - 3)^2 - 2$ using transformations.

6. Consider the quadratic function $f(x) = 3x^2 - 12x + 4$:

(a) Determine whether the graph opens up or down.

(b) Determine the vertex.

(c) Determine the axis of symmetry.

(d) Determine the intercepts.

(e) Use the information from parts (a)–(d) to graph f.

7. Determine whether $f(x) = -2x^2 + 12x + 3$ has a maximum or a minimum. Then find the maximum or minimum value.

8. Solve $x^2 - 10x + 24 \geq 0$.

9. RV Rental The weekly rental cost of a 20-foot recreational vehicle is \$129.50 plus \$0.15 per mile.

(a) Find a linear function that expresses the cost C as a function of miles driven m.

(b) What is the rental cost if 860 miles are driven?

(c) How many miles were driven if the rental cost is \$213.80?

Cumulative Review

1. Find the distance between the points $P = (-1, 3)$ and $Q = (4, -2)$. Find the midpoint of the line segment P to Q.
2. Which of the following points are on the graph of $y = x^3 - 3x + 1$?
   (a) $(-2, -1)$ (b) $(2, 3)$ (c) $(3, 1)$
3. Solve the inequality $5x + 3 \geq 0$ and graph the solution set.
4. Find the equation of the line containing the points $(-1, 4)$ and $(2, -2)$. Express your answer in slope–intercept form and graph the line.
5. Find the equation of the line perpendicular to the line $y = 2x + 1$ and containing the point $(3, 5)$. Express your answer in slope–intercept form and graph both lines.
6. Graph the equation $x^2 + y^2 - 4x + 8y - 5 = 0$.
7. Does the following relation represent a function? $\{(-3, 8), (1, 3), (2, 5), (3, 8)\}$.
8. For the function f defined by $f(x) = x^2 - 4x + 1$, find:
   (a) $f(2)$ (b) $f(x) + f(2)$
   (c) $f(-x)$ (d) $-f(x)$
   (e) $f(x + 2)$ (f) $\dfrac{f(x + h) - f(x)}{h}$, $h \neq 0$
9. Find the domain of $h(z) = \dfrac{3z - 1}{6z - 7}$.
10. Is the following graph the graph of a function?

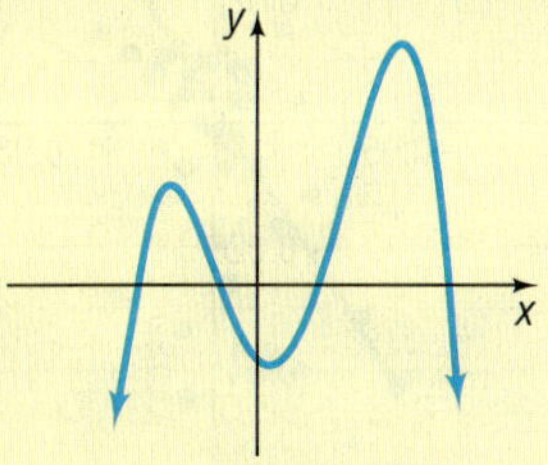

11. Consider the function $f(x) = \dfrac{x}{x + 4}$.
    (a) Is the point $\left(1, \frac{1}{4}\right)$ on the graph of f?
    (b) If $x = -2$, what is $f(x)$? What point is on the graph of f?
    (c) If $f(x) = 2$, what is x? What point is on the graph of f?
12. Is the function $f(x) = \dfrac{x^2}{2x + 1}$ even, odd, or neither?
13. Approximate the local maximum values and local minimum values of $f(x) = x^3 - 5x + 1$ on $(-4, 4)$. Determine where the function is increasing and where it is decreasing.
14. If $f(x) = 3x + 5$ and $g(x) = 2x + 1$:
    (a) Solve $f(x) = g(x)$.
    (b) Solve $f(x) > g(x)$.
15. Consider the graph below of the function f.
    (a) Find the domain and the range of f.
    (b) Find the intercepts.
    (c) Is the graph of f symmetric with respect to the x-axis, the y-axis, or the origin?
    (d) Find $f(2)$.
    (e) For what value(s) of x is $f(x) = 3$?
    (f) Solve $f(x) < 0$.
    (g) Graph $y = f(x) + 2$.
    (h) Graph $y = f(-x)$.
    (i) Graph $y = 2f(x)$.
    (j) Is f even, odd, or neither?
    (k) Find the interval(s) on which f is increasing.

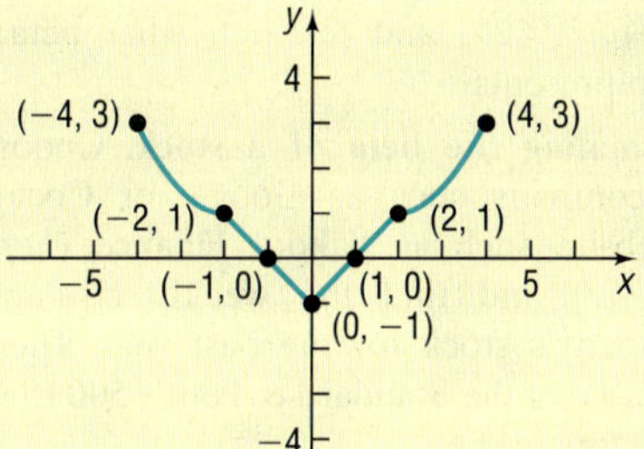

Chapter Projects

Internet-based Project

I. The Beta of a Stock You want to invest in the stock market but are not sure which stock to purchase. Information is the key to making an informed investment decision. One piece of information that many stock analysts use is the beta of the stock. Go to Wikipedia (*http://en.wikipedia.org/wiki/Beta_%28finance%29*) and research what beta measures and what it represents.

1. **Approximating the beta of a stock.** Choose a well-known company such as Google or Coca-Cola. Go to a website such as Yahoo! Finance (*http://finance.yahoo.com/*) and find the weekly closing price of the company's stock for the past year. Then find the closing price of the Standard & Poor's 500 (S&P500) for the same time period.

   To get the historical prices in Yahoo! Finance, select Historical Prices from the left menu. Choose the appropriate time period. Select Weekly and Get Prices. Finally, select Download to Spreadsheet. Repeat this for the S&P500, and copy the data into the same spreadsheet. Finally, rearrange the data in chronological order. Be sure to expand the selection to sort all the data. Now, using the adjusted close price, compute the percentage change in price for each week, using the formula % change $= \frac{P_1 - P_o}{P_o}$. For example, if week 1 price is in cell D1 and week 2 price is in cell D2, then % change $= \frac{D2 - D1}{D1}$. Repeat this for the S&P500 data.

2. **Using Excel to draw a scatter diagram.** Treat the percentage change in the S&P500 as the independent variable and the percentage change in the stock you chose as the dependent variable. The easiest way to draw a scatter diagram in Excel is to place the two columns of data next to each other (for example, have the percentage change in the S&P500 in column F and the percentage change in the stock you chose in column G). Then highlight the data and select the Scatter Diagram icon under Insert. Comment on the type of relation that appears to exist between the two variables.

3. **Finding beta.** To find beta requires that we find the line of best fit using least-squares regression. The easiest approach is to click inside the scatter diagram. Select the Chart Elements icon (+). Check the box for Trendline, select the arrow to the right, and choose More Options. Select Linear and check the box for Display Equation on chart. The line of best fit appears on the scatter diagram. See below.

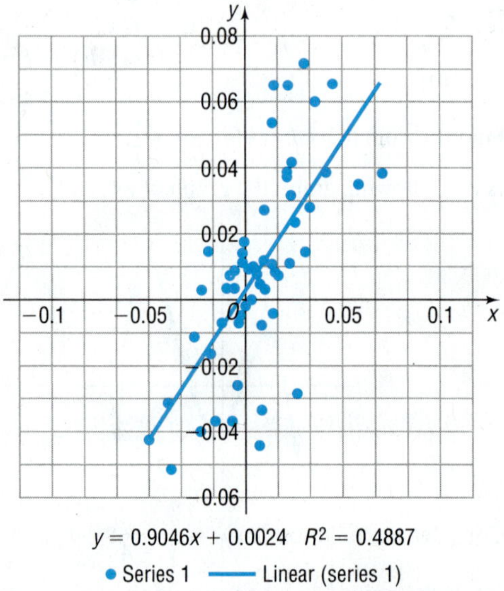

The line of best fit for this data is $y = 0.9046x + 0.0024$. You may click on Chart Title or either axis title and insert the appropriate names. The beta is the slope of the line of best fit, 0.9046. We interpret this by saying, "If the S&P500 increases by 1%, then this stock will increase by 0.9%, on average." Find the beta of your stock and provide an interpretation. NOTE: Another way to use Excel to find the line of best fit requires using the Data Analysis Tool Pack under add-ins.

The following projects are available on the Instructor's Resource Center (IRC):

II. Cannons A battery commander uses the weight of a missile, its initial velocity, and the position of its gun to determine where the missile will travel.

III. First and Second Differences Finite differences provide a numerical method that is used to estimate the graph of an unknown function.

IV. CBL Experiment Computer simulation is used to study the physical properties of a bouncing ball.

Polynomial and Rational Functions

5

Day Length

Day length is the length of time each day from the moment the upper limb of the sun's disk appears above the horizon during sunrise to the moment when the upper limb disappears below the horizon during sunset. The length of a day depends on the day of the year as well as the latitude of the location. Latitude gives the location of a point on Earth north or south of the equator. In the Internet Project at the end of this chapter, we use information from the chapter to investigate the relation between day length and latitude for a specific day of the year.

—See the Internet-based Chapter Project I—

••• A Look Back

In Chapter 3, we began our discussion of functions. We defined domain, range, and independent and dependent variables, found the value of a function, and graphed functions. We continued our study of functions by listing the properties that a function might have, such as being even or odd, and created a library of functions, naming key functions and listing their properties, including their graphs.

In Chapter 4, we discussed linear functions and quadratic functions, which belong to the class of *polynomial functions*.

A Look Ahead •••

In this chapter, we look at two general classes of functions, polynomial functions and rational functions, and examine their properties. Polynomial functions are arguably the simplest expressions in algebra. For this reason, they are often used to approximate other, more complicated functions. Rational functions are ratios of polynomial functions.

Outline

5.1 Polynomial Functions and Models

PREPARING FOR THIS SECTION *Before getting started, review the following:*

- Polynomials (Chapter R, Section R.4, pp. 39–47)
- Using a Graphing Utility to Approximate Local Maxima and Local Minima (Section 3.3, p. 229)
- Intercepts of a Function (Section 3.2, pp. 215–217)
- Graphing Techniques: Transformations (Section 3.5, pp. 247–256)
- Intercepts (Section 2.2, pp. 159–160)

Now Work the 'Are You Prepared?' problems on page 338.

OBJECTIVES
1. Identify Polynomial Functions and Their Degree (p. 322)
2. Graph Polynomial Functions Using Transformations (p. 326)
3. Identify the Real Zeros of a Polynomial Function and Their Multiplicity (p. 327)
4. Analyze the Graph of a Polynomial Function (p. 334)
5. Build Cubic Models from Data (p. 336)

1 Identify Polynomial Functions and Their Degree

In Chapter 4, we studied the linear function $f(x) = mx + b$, which can be written as

$$f(x) = a_1x + a_0$$

and the quadratic function $f(x) = ax^2 + bx + c, a \neq 0$, which can be written as

$$f(x) = a_2x^2 + a_1x + a_0 \qquad a_2 \neq 0$$

Each of these functions is an example of a *polynomial function*.

DEFINITION

A **polynomial function** in one variable is a function of the form

$$f(x) = a_nx^n + a_{n-1}x^{n-1} + \cdots + a_1x + a_0 \qquad \textbf{(1)}$$

where $a_n, a_{n-1}, \ldots, a_1, a_0$ are constants, called the **coefficients** of the polynomial, $n \geq 0$ is an integer, and x is the variable. If $a_n \neq 0$, it is called the **leading coefficient**, and n is the **degree** of the polynomial.

The domain of a polynomial function is the set of all real numbers.

In Words
A polynomial function is a sum of monomials.

The monomials that make up a polynomial are called its **terms**. If $a_n \neq 0$, a_nx^n is called the **leading term**; a_0 is called the **constant term**. If all of the coefficients are 0, the polynomial is called the **zero polynomial**, which has no degree.

Polynomials are usually written in **standard form**, beginning with the nonzero term of highest degree and continuing with terms in descending order according to degree. If a power of x is missing, it is because its coefficient is zero.

Polynomial functions are among the simplest expressions in algebra. They are easy to evaluate: only addition and repeated multiplication are required. Because of this, they are often used to approximate other, more complicated functions. In this section, we investigate properties of this important class of functions.

EXAMPLE 1 **Identifying Polynomial Functions**

Determine which of the following are polynomial functions. For those that are, state the degree; for those that are not, tell why not. Write each polynomial in standard form, and then identify the leading term and the constant term.

(a) $p(x) = 5x^3 - \frac{1}{4}x^2 - 9$ (b) $f(x) = x + 2 - 3x^4$ (c) $g(x) = \sqrt{x}$

(d) $h(x) = \dfrac{x^2 - 2}{x^3 - 1}$ (e) $G(x) = 8$ (f) $H(x) = -2x^3(x - 1)^2$

Solution

(a) p is a polynomial function of degree 3, and it is already in standard form. The leading term is $5x^3$, and the constant term is -9.

(b) f is a polynomial function of degree 4. Its standard form is $f(x) = -3x^4 + x + 2$. The leading term is $-3x^4$, and the constant term is 2.

(c) g is not a polynomial function because $g(x) = \sqrt{x} = x^{\frac{1}{2}}$, so the variable x is raised to the $\frac{1}{2}$ power, which is not a nonnegative integer.

(d) h is not a polynomial function. It is the ratio of two distinct polynomials, and the polynomial in the denominator is of positive degree.

(e) G is a nonzero constant polynomial function so it is of degree 0. The polynomial is in standard form. The leading term and constant term are both 8.

(f) $H(x) = -2x^3(x-1)^2 = -2x^3(x^2 - 2x + 1) = -2x^5 + 4x^4 - 2x^3$. So, H is a polynomial function of degree 5. Because $H(x) = -2x^5 + 4x^4 - 2x^3$, the leading term is $-2x^5$. Since no constant term is shown, the constant term is 0. Do you see a way to find the degree of H without multiplying it out?

Now Work PROBLEMS 17 AND 21

We have already discussed in detail polynomial functions of degrees 0, 1, and 2. See Table 1 for a summary of the properties of the graphs of these polynomial functions.

Table 1

Degree	Form	Name	Graph
No degree	$f(x) = 0$	Zero function	The x-axis
0	$f(x) = a_0, \quad a_0 \neq 0$	Constant function	Horizontal line with y-intercept a_0
1	$f(x) = a_1x + a_0, \quad a_1 \neq 0$	Linear function	Nonvertical, nonhorizontal line with slope a_1 and y-intercept a_0
2	$f(x) = a_2x^2 + a_1x + a_0, \quad a_2 \neq 0$	Quadratic function	Parabola: graph opens up if $a_2 > 0$; graph opens down if $a_2 < 0$

One objective of this section is to analyze the graph of a polynomial function. If you take a course in calculus, you will learn that the graph of every polynomial function is both smooth and continuous. By **smooth**, we mean that the graph contains no sharp corners or cusps; by **continuous**, we mean that the graph has no gaps or holes and can be drawn without lifting your pencil from the paper. See Figures 1(a) and (b).

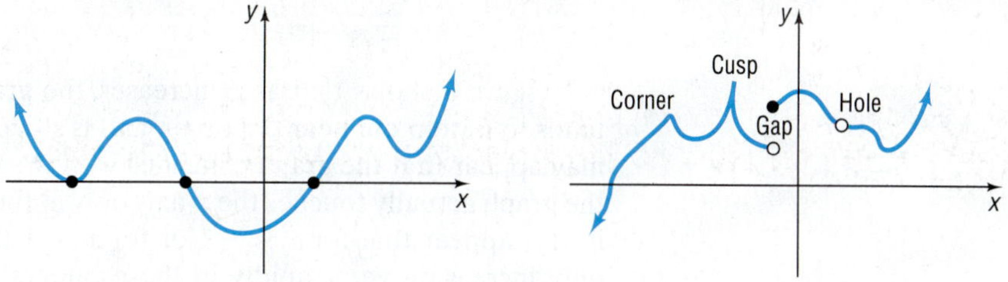

Figure 1 **(a)** Graph of a polynomial function: smooth, continuous **(b)** Cannot be the graph of a polynomial function

Power Functions

We begin the analysis of the graph of a polynomial function by discussing *power functions,* a special kind of polynomial function.

DEFINITION

A **power function of degree n** is a monomial function of the form

$$f(x) = ax^n \tag{2}$$

where a is a real number, $a \neq 0$, and $n > 0$ is an integer.

In Words
A power function is defined by a single monomial.

Examples of power functions are

$$f(x) = 3x \qquad f(x) = -5x^2 \qquad f(x) = 8x^3 \qquad f(x) = -5x^4$$
degree 1   degree 2   degree 3   degree 4

The graph of a power function of degree 1, $f(x) = ax$, is a straight line, with slope a, that passes through the origin. The graph of a power function of degree 2, $f(x) = ax^2$, is a parabola, with vertex at the origin, that opens up if $a > 0$ and opens down if $a < 0$.

If we know how to graph a power function of the form $f(x) = x^n$, a compression or stretch and, perhaps, a reflection about the x-axis will enable us to obtain the graph of $g(x) = ax^n$. Consequently, we shall concentrate on graphing power functions of the form $f(x) = x^n$.

We begin with power functions of even degree of the form $f(x) = x^n$, $n \geq 2$ and n even. The domain of f is the set of all real numbers, and the range is the set of nonnegative real numbers. Such a power function is an even function. (Do you see why?). Its graph is symmetric with respect to the y-axis. Its graph always contains the origin and the points $(-1, 1)$ and $(1, 1)$.

If $n = 2$, the graph is the familiar parabola $y = x^2$ that opens up, with vertex at the origin. If $n \geq 4$, the graph of $f(x) = x^n$, n even, will be closer to the x-axis than the parabola $y = x^2$ if $-1 < x < 1, x \neq 0$, and farther from the x-axis than the parabola $y = x^2$ if $x < -1$ or if $x > 1$. Figure 2(a) illustrates this conclusion. Figure 2(b) shows the graphs of $y = x^4$ and $y = x^8$ for comparison.

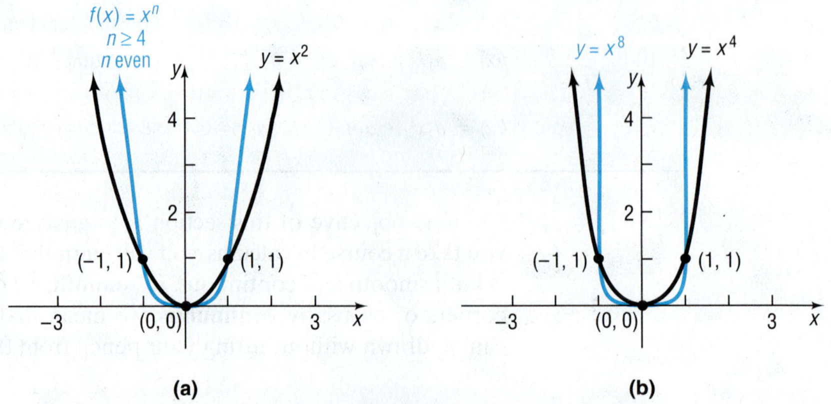

Figure 2

Figure 2 shows that as n increases, the graph of $f(x) = x^n$, $n \geq 2$ and n even, tends to flatten out near the origin and is steeper when x is far from 0. For large n, it may appear that the graph coincides with the x-axis near the origin, but it does not; the graph actually touches the x-axis only at the origin (see Table 2). Also, for large n, it may appear that for $x < -1$ or for $x > 1$ the graph is vertical, but it is not; it is only increasing very rapidly in these intervals. If the graphs were enlarged many times, these distinctions would be clear.

Table 2

	$x = 0.1$	$x = 0.3$	$x = 0.5$
$f(x) = x^8$	10^{-8}	0.0000656	0.0039063
$f(x) = x^{20}$	10^{-20}	$3.487 \cdot 10^{-11}$	0.000001
$f(x) = x^{40}$	10^{-40}	$1.216 \cdot 10^{-21}$	$9.095 \cdot 10^{-13}$

Seeing the Concept

Graph $Y_1 = x^4$, $Y_2 = x^8$, and $Y_3 = x^{12}$ using the viewing rectangle $-2 \leq x \leq 2, -4 \leq y \leq 16$. Then graph each again using the viewing rectangle $-1 \leq x \leq 1, 0 \leq y \leq 1$. See Figure 3. TRACE along one of the graphs to confirm that for x close to 0 the graph is above the x-axis and that for $x > 0$ the graph is increasing.

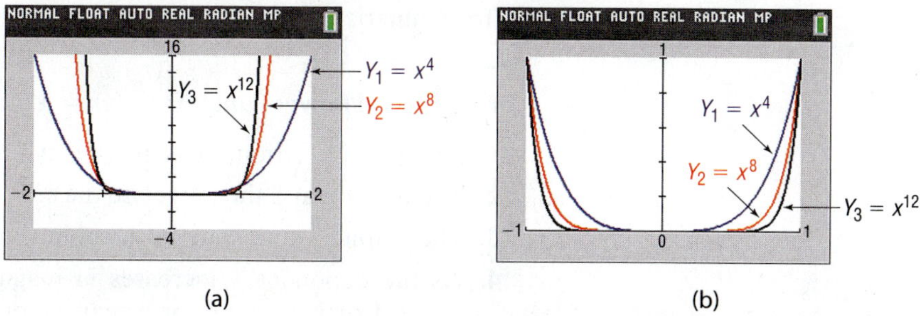

Figure 3

Properties of Power Functions, $f(x) = x^n$, n Is a Positive Even Integer

1. f is an even function, so its graph is symmetric with respect to the y-axis.
2. The domain is the set of all real numbers. The range is the set of nonnegative real numbers.
3. The graph always contains the points $(-1, 1)$, $(0, 0)$, and $(1, 1)$.
4. As the exponent n increases in magnitude, the graph is steeper when $x < -1$ or $x > 1$; but for x near the origin, the graph tends to flatten out and lie closer to the x-axis.

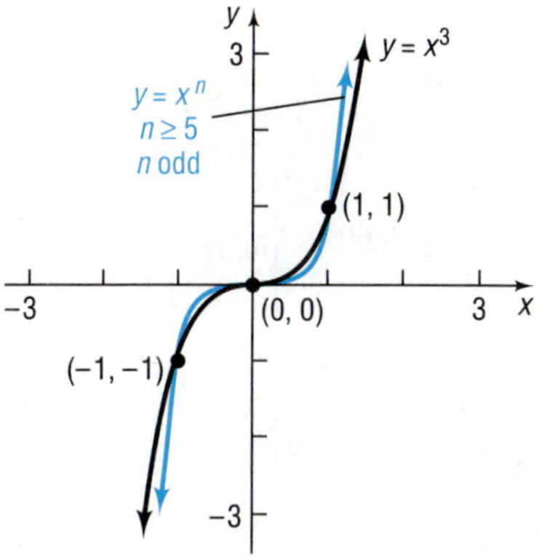

Figure 4

Now we consider power functions of odd degree of the form $f(x) = x^n$, $n \geq 3$ and n odd. The domain and the range of f are the set of real numbers. Such a power function is an odd function. (Do you see why?). Its graph is symmetric with respect to the origin. Its graph always contains the origin and the points $(-1, -1)$ and $(1, 1)$.

The graph of $f(x) = x^n$ when $n = 3$ has been shown several times and is repeated in Figure 4. If $n \geq 5$, the graph of $f(x) = x^n$, n odd, will be closer to the x-axis than that of $y = x^3$ if $-1 < x < 1$ and farther from the x-axis than that of $y = x^3$ if $x < -1$ or if $x > 1$. Figure 4 illustrates this conclusion. Figure 5 shows the graphs of $y = x^5$ and $y = x^9$ for further comparison.

It appears that each graph coincides with the x-axis near the origin, but it does not; each graph actually crosses the x-axis at the origin. Also, it appears that as x increases the graphs become vertical, but they do not; each graph is just increasing very rapidly.

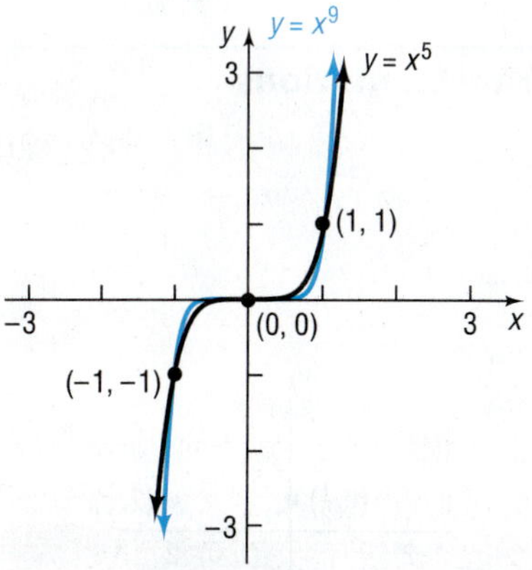

Figure 5

Seeing the Concept

Graph $Y_1 = x^3$, $Y_2 = x^7$, and $Y_3 = x^{11}$ using the viewing rectangle $-2 \leq x \leq 2, -16 \leq y \leq 16$. Then graph each again using the viewing rectangle $-1 \leq x \leq 1, -1 \leq y \leq 1$. See Figure 6. TRACE along one of the graphs to confirm that the graph is increasing and crosses the x-axis at the origin.

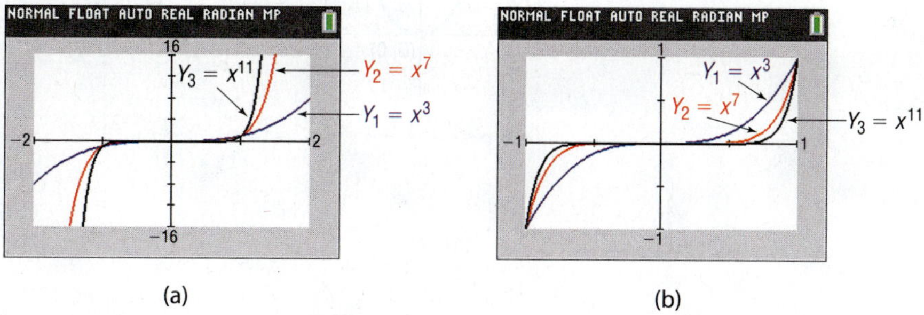

Figure 6

To summarize:

Properties of Power Functions, $f(x) = x^n$, n Is a Positive Odd Integer

1. f is an odd function, so its graph is symmetric with respect to the origin.
2. The domain and the range are the set of all real numbers.
3. The graph always contains the points $(-1, -1)$, $(0, 0)$, and $(1, 1)$.
4. As the exponent n increases in magnitude, the graph is steeper when $x < -1$ or $x > 1$; but for x near the origin, the graph tends to flatten out and lie closer to the x-axis.

2 Graph Polynomial Functions Using Transformations

The methods of shifting, compression, stretching, and reflection (studied in Section 3.5), when used with the facts just presented, will enable us to graph polynomial functions that are transformations of power functions.

EXAMPLE 2 **Graphing a Polynomial Function Using Transformations**

Graph: $f(x) = 1 - x^5$

Solution It is helpful to rewrite f as $f(x) = -x^5 + 1$. Figure 7 shows the required stages.

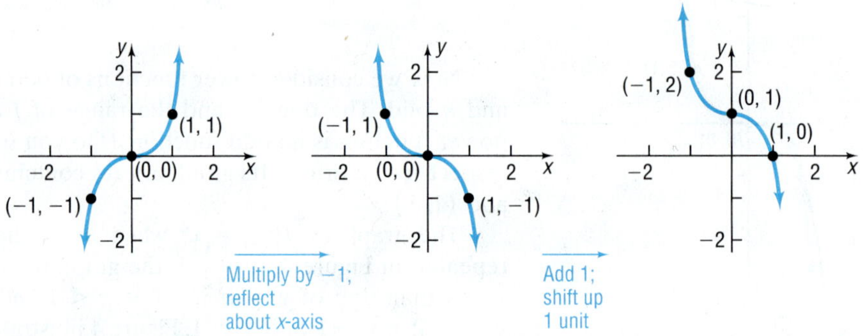

Figure 7 (a) $y = x^5$ (b) $y = -x^5$ (c) $y = -x^5 + 1 = 1 - x^5$

EXAMPLE 3 **Graphing a Polynomial Function Using Transformations**

Graph: $f(x) = \frac{1}{2}(x - 1)^4$

Solution Figure 8 shows the required stages.

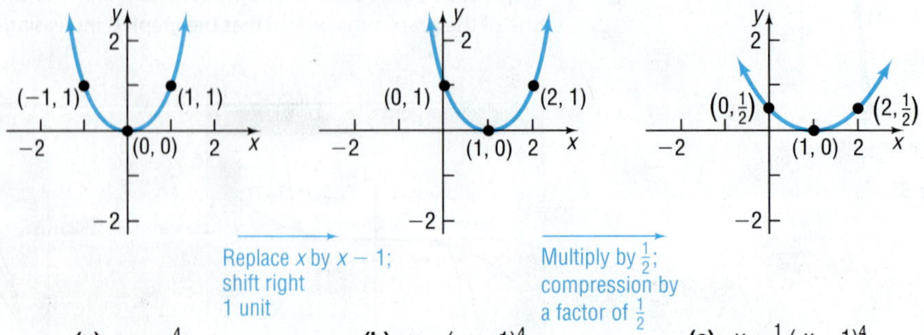

Figure 8 (a) $y = x^4$ (b) $y = (x - 1)^4$ (c) $y = \frac{1}{2}(x - 1)^4$

Now Work PROBLEMS 29 AND 35

3 Identify the Real Zeros of a Polynomial Function and Their Multiplicity

Figure 9 shows the graph of a polynomial function with four x-intercepts. Notice that at the x-intercepts, the graph must either cross the x-axis or touch the x-axis. Consequently, between consecutive x-intercepts the graph is either above the x-axis or below the x-axis.

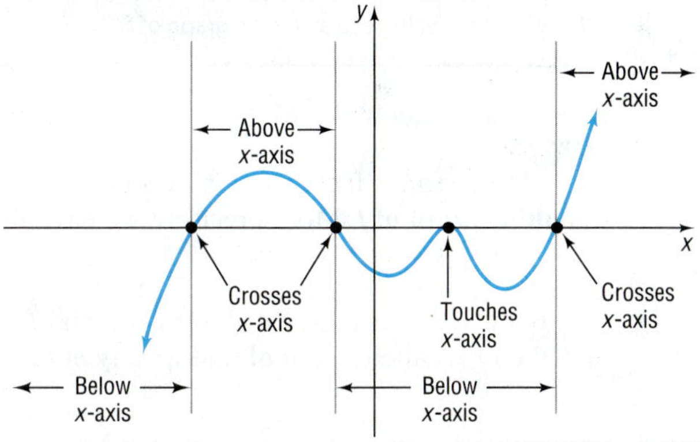

Figure 9 Graph of a polynomial function

If a polynomial function f is factored completely, it is easy to locate the x-intercepts of the graph by solving the equation $f(x) = 0$ and using the Zero-Product Property. For example, if $f(x) = (x - 1)^2(x + 3)$, then the solutions of the equation

$$f(x) = (x - 1)^2(x + 3) = 0$$

are identified as 1 and -3. That is, $f(1) = 0$ and $f(-3) = 0$.

DEFINITION

If f is a function and r is a real number for which $f(r) = 0$, then r is called a **real zero** of f.

As a consequence of this definition, the following statements are equivalent.

1. r is a real zero of a polynomial function f.
2. r is an x-intercept of the graph of f.
3. $x - r$ is a factor of f.
4. r is a solution to the equation $f(x) = 0$.

So the real zeros of a polynomial function are the x-intercepts of its graph, and they are found by solving the equation $f(x) = 0$.

EXAMPLE 4 Finding a Polynomial Function from Its Zeros

(a) Find a polynomial function of degree 3 whose zeros are -3, 2, and 5.

(b) Use a graphing utility to graph the polynomial found in part (a) to verify your result.

Solution

(a) If r is a real zero of a polynomial function f, then $x - r$ is a factor of f. This means that $x - (-3) = x + 3$, $x - 2$, and $x - 5$ are factors of f. As a result, any polynomial function of the form

$$f(x) = a(x + 3)(x - 2)(x - 5)$$

where a is a nonzero real number, qualifies. The value of a causes a stretch, compression, or reflection, but it does not affect the x-intercepts of the graph. Do you know why?

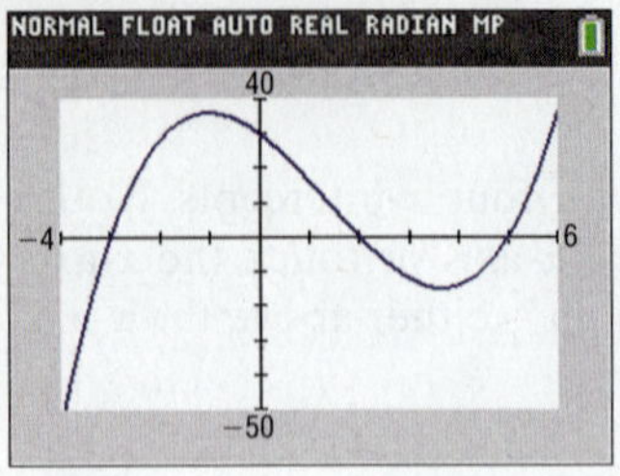

Figure 10 $f(x) = x^3 - 4x^2 - 11x + 30$

(b) We choose to graph f with $a = 1$. Then

$$f(x) = (x + 3)(x - 2)(x - 5) = x^3 - 4x^2 - 11x + 30$$

Figure 10 shows the graph of f. Notice that the x-intercepts are -3, 2, and 5. ●

Seeing the Concept

Graph the function found in Example 4 for $a = 2$ and $a = -1$. Does the value of a affect the zeros of f? How does the value of a affect the graph of f?

Now Work PROBLEM 43

If the same factor $x - r$ occurs more than once, r is called a **repeated**, or **multiple, zero of *f***. More precisely, we have the following definition.

DEFINITION

If $(x - r)^m$ is a factor of a polynomial f and $(x - r)^{m+1}$ is not a factor of f, then r is called a **zero of multiplicity *m* of *f***.*

EXAMPLE 5 Identifying Zeros and Their Multiplicities

For the polynomial

$$f(x) = 5x^2(x + 2)\left(x - \frac{1}{2}\right)^4$$

- -2 is a zero of multiplicity 1 because the exponent on the factor $x + 2$ is 1.
- 0 is a zero of multiplicity 2 because the exponent on the factor x is 2.
- $\frac{1}{2}$ is a zero of multiplicity 4 because the exponent on the factor $x - \frac{1}{2}$ is 4. ●

In Words
The multiplicity of a zero is the number of times its corresponding factor occurs.

Now Work PROBLEM 57(a)

Suppose that it is possible to completely factor a polynomial function and, as a result, locate all the x-intercepts of its graph (the real zeros of the function). These x-intercepts then divide the x-axis into open intervals and, on each such interval, the graph of the polynomial will be either above or below the x-axis over the entire interval. Let's look at an example.

EXAMPLE 6 Graphing a Polynomial Using Its x-Intercepts

Consider the following polynomial: $f(x) = (x + 1)^2(x - 2)$

(a) Find the x- and y-intercepts of the graph of f.

(b) Use the x-intercepts to find the intervals on which the graph of f is above the x-axis and the intervals on which the graph of f is below the x-axis.

(c) Locate other points on the graph, and connect all the points plotted with a smooth, continuous curve.

Solution (a) The y-intercept is $f(0) = (0 + 1)^2(0 - 2) = -2$. The x-intercepts satisfy the equation

$$f(x) = (x + 1)^2(x - 2) = 0$$

from which we find

$$(x + 1)^2 = 0 \quad \text{or} \quad x - 2 = 0$$
$$x = -1 \quad \text{or} \quad x = 2$$

The x-intercepts are -1 and 2.

*Some books use the terms **multiple root** and **root of multiplicity *m***.

(b) The two x-intercepts divide the x-axis into three intervals:

$$(-\infty, -1) \qquad (-1, 2) \qquad (2, \infty)$$

Since the graph of f crosses or touches the x-axis only at $x = -1$ and $x = 2$, it follows that the graph of f is either above the x-axis $[f(x) > 0]$ or below the x-axis $[f(x) < 0]$ on each of these three intervals. To see where the graph lies, we need only pick a number in each interval, evaluate f there, and see whether the value is positive (above the x-axis) or negative (below the x-axis). See Table 3.

(c) In constructing Table 3, we obtained three additional points on the graph: $(-2, -4)$, $(1, -4)$ and $(3, 16)$. Figure 11 illustrates these points, the intercepts, and a smooth, continuous curve (the graph of f) connecting them.

Table 3

	−1	2	→ x
Interval	$(-\infty, -1)$	$(-1, 2)$	$(2, \infty)$
Number chosen	-2	1	3
Value of f	$f(-2) = -4$	$f(1) = -4$	$f(3) = 16$
Location of graph	Below x-axis	Below x-axis	Above x-axis
Point on graph	$(-2, -4)$	$(1, -4)$	$(3, 16)$

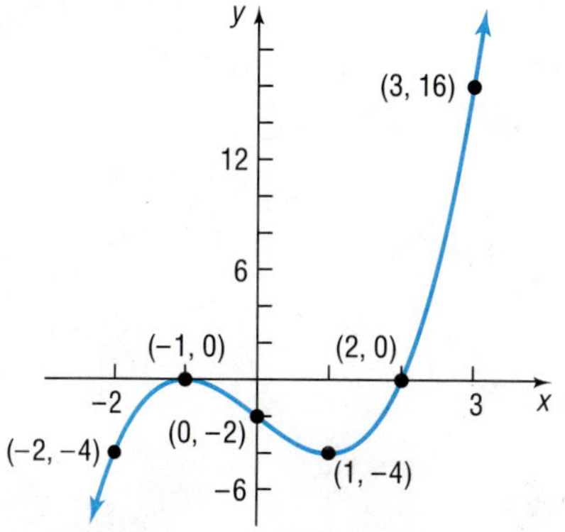

Figure 11 $f(x) = (x + 1)^2(x - 2)$

Look again at Table 3. Since the graph of $f(x) = (x + 1)^2(x - 2)$ is below the x-axis on both sides of -1, the graph of f *touches* the x-axis at $x = -1$, a *zero of multiplicity 2*. Since the graph of f is below the x-axis for $x < 2$ and above the x-axis for $x > 2$, the graph of f *crosses* the x-axis at $x = 2$, a *zero of multiplicity 1*.

This suggests the following results:

If r Is a Zero of Even Multiplicity

Numerically: The sign of $f(x)$ does not change from one side to the other side of r.

Graphically: The graph of f **touches** the x-axis at r.

If r Is a Zero of Odd Multiplicity

Numerically: The sign of $f(x)$ changes from one side to the other side of r.

Graphically: The graph of f **crosses** the x-axis at r.

Now Work PROBLEM 57(b)

Turning Points

Look again at Figure 11 above. We cannot be sure just how low the graph actually goes between $x = -1$ and $x = 2$. But we do know that somewhere in the interval $(-1, 2)$ the graph of f must change direction (from decreasing to increasing). The points at which a graph changes direction are called **turning points**.* Each turning point yields either a **local maximum** or a **local minimum** (see Section 3.3). The following result from calculus tells us the maximum number of turning points that the graph of a polynomial function can have.

*Graphing utilities can be used to approximate turning points. For most polynomials, calculus is needed to find the exact turning points.

THEOREM

Turning Points

If f is a polynomial function of degree n, then the graph of f has at most $n - 1$ turning points.

If the graph of a polynomial function f has $n - 1$ turning points, then the degree of f is at least n.

Based on the first part of the theorem, a polynomial function of degree 5 will have at most $5 - 1 = 4$ turning points. Based on the second part of the theorem, if the graph of a polynomial function has three turning points, then the degree of the function must be at least 4.

Exploration

A graphing utility can be used to locate the turning points of a graph. Graph $Y_1 = (x + 1)^2(x - 2)$. Use MINIMUM to find the location of the turning point for $0 < x < 2$. See Figure 12.

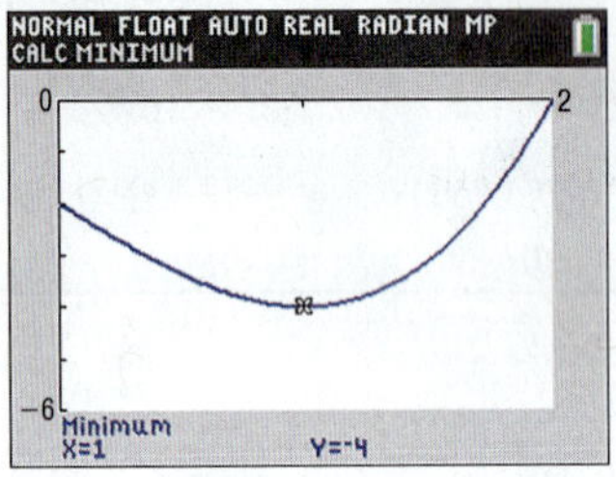

Figure 12 $Y_1 = (x + 1)^2(x - 2)$

Now Work PROBLEM 57(c)

EXAMPLE 7

Identifying the Graph of a Polynomial Function

Which of the graphs in Figure 13 could be the graph of a polynomial function? For those that could, list the real zeros and state the least degree the polynomial can have. For those that could not, say why not.

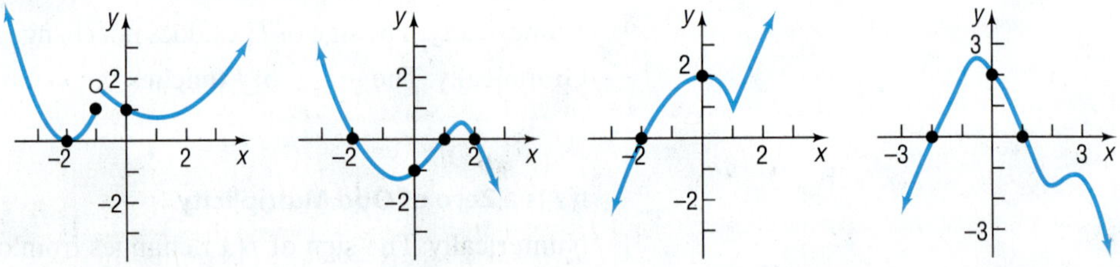

Figure 13 (a) (b) (c) (d)

Solution

(a) The graph in Figure 13(a) cannot be the graph of a polynomial function because of the gap that occurs at $x = -1$. Remember, the graph of a polynomial function is continuous—no gaps or holes. (See Figure 1.)

(b) The graph in Figure 13(b) could be the graph of a polynomial function because the graph is smooth and continuous. It has three real zeros: $-2, 1,$ and 2. Since the graph has two turning points, the degree of the polynomial function must be at least 3.

(c) The graph in Figure 13(c) cannot be the graph of a polynomial function because of the cusp at $x = 1$. Remember, the graph of a polynomial function is smooth.

(d) The graph in Figure 13(d) could be the graph of a polynomial function. It has two real zeros: -2 and 1. Since the graph has three turning points, the degree of the polynomial function is at least 4.

Now Work PROBLEM 69

End Behavior

One last remark about Figure 11. For very large values of x, either positive or negative, the graph of $f(x) = (x+1)^2(x-2)$ looks like the graph of $y = x^3$. To see why, we write f in the form

$$f(x) = (x+1)^2(x-2) = x^3 - 3x - 2 = x^3\left(1 - \frac{3}{x^2} - \frac{2}{x^3}\right)$$

Now, for large values of x, either positive or negative, the terms $\frac{3}{x^2}$ and $\frac{2}{x^3}$ are close to 0, so for large values of x,

$$f(x) = x^3 - 3x - 2 = x^3\left(1 - \frac{3}{x^2} - \frac{2}{x^3}\right) \approx x^3$$

The behavior of the graph of a function for large values of x, either positive or negative, is referred to as its **end behavior**.

THEOREM

End Behavior

For large values of x, either positive or negative, the graph of the polynomial function

$$f(x) = a_nx^n + a_{n-1}x^{n-1} + \cdots + a_1x + a_0 \quad a_n \neq 0$$

resembles the graph of the power function

$$y = a_nx^n$$

In Words

The end behavior of a polynomial function resembles that of its leading term.

For example, if $f(x) = -2x^3 + 5x^2 + x - 4$, then the graph of f will behave like the graph of $y = -2x^3$ for very large values of x, either positive or negative. We can see that the graphs of f and $y = -2x^3$ "behave" the same by considering Table 4 and Figure 14.

Table 4

x	$f(x)$	$y = -2x^3$
10	−1,494	−2,000
100	−1,949,904	−2,000,000
500	−248,749,504	−250,000,000
1,000	−1,994,999,004	−2,000,000,000

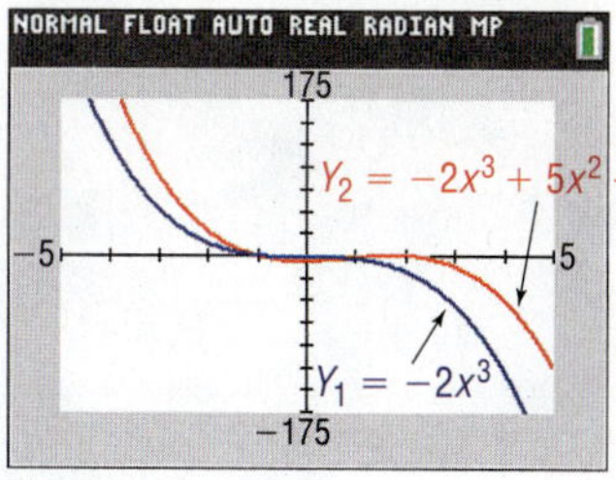

Figure 14

NOTE Infinity (∞) and negative infinity ($-\infty$) are not numbers. Rather, they are symbols that represent unboundedness. ■

Notice that as x becomes a larger and larger positive number, the values of f become larger and larger negative numbers. When this happens, we say that f is **unbounded in the negative direction**. Rather than using words to describe the behavior of the graph of the function, we explain its behavior using notation. We can symbolize "the value of f becomes a larger and larger negative number as x becomes a larger and larger positive number" by writing $f(x) \to -\infty$ as $x \to \infty$ (read "the values of f approach negative infinity as x approaches infinity"). In calculus, **limits** are used to convey these ideas. There we use the symbolism $\lim_{x \to \infty} f(x) = -\infty$, read "the limit of $f(x)$ as x approaches infinity equals negative infinity," to mean that $f(x) \to -\infty$ as $x \to \infty$.

When we say that the value of a limit equals infinity (or negative infinity), we mean that the values of the function are unbounded in the positive (or negative) direction and call the limit an **infinite limit**. When we discuss limits as x becomes unbounded in the negative direction or unbounded in the positive direction, we are discussing **limits at infinity**.

Look back at Figures 2 and 4. Based on the preceding theorem and the previous discussion on power functions, the end behavior of a polynomial function can be of only four types. See Figure 15.

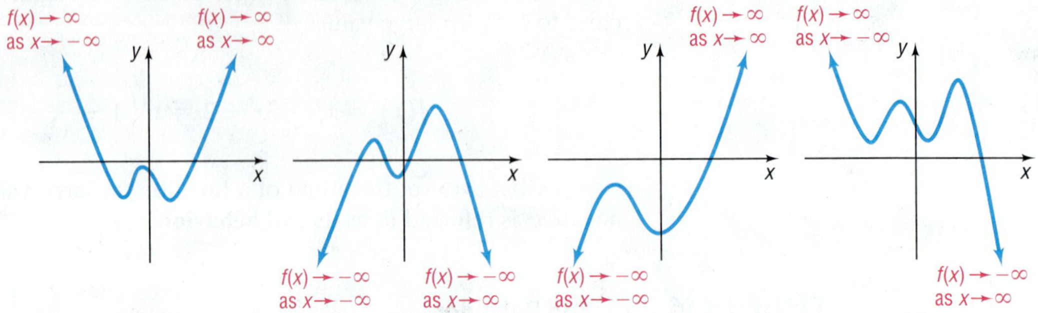

(a) $n \geq 2$ even; $a_n > 0$ | (b) $n \geq 2$ even; $a_n < 0$ | (c) $n \geq 3$ odd; $a_n > 0$ | (d) $n \geq 3$ odd; $a_n < 0$

Figure 15 End behavior of $f(x) = a_nx^n + a_{n-1}x^{n-1} + \cdots + a_1x + a_0$

For example, if $f(x) = -2x^4 + x^3 + 4x^2 - 7x + 1$, the graph of f will resemble the graph of the power function $y = -2x^4$ for large $|x|$. The graph of f will behave like Figure 15(b) for large $|x|$.

Now Work PROBLEM 57(d)

EXAMPLE 8

Identifying the Graph of a Polynomial Function

Which of the graphs in Figure 16 could be the graph of

$$f(x) = x^4 + ax^3 + bx^2 - 5x - 6$$

where $a > 0, b > 0$?

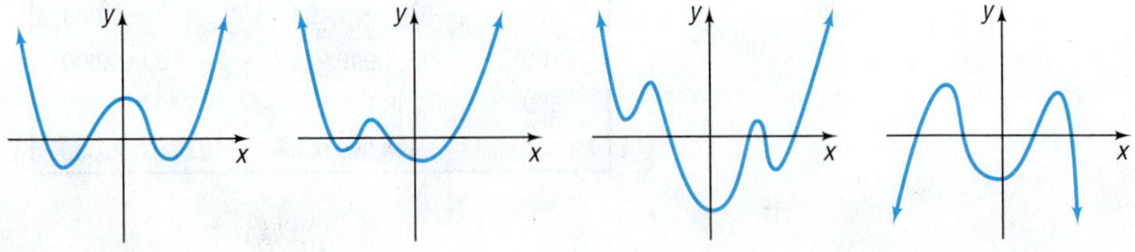

(a) (b) (c) (d)

Figure 16

Solution The y-intercept of f is $f(0) = -6$. We can eliminate the graph in Figure 16(a), whose y-intercept is positive.

We are not able to solve $f(x) = 0$ to find the x-intercepts of f, so we move on to investigate the turning points of each graph. Since f is of degree 4, the graph of f has at most 3 turning points. We eliminate the graph in Figure 16(c) because that graph has 5 turning points.

Now we look at end behavior. For large values of x, the graph of f will behave like the graph of $y = x^4$. This eliminates the graph in Figure 16(d), whose end behavior is like the graph of $y = -x^4$.

Only the graph in Figure 16(b) could be the graph of

$$f(x) = x^4 + ax^3 + bx^2 - 5x - 6$$

where $a > 0, b > 0$.

EXAMPLE 9 **Writing a Polynomial Function from Its Graph**

Write a polynomial function whose graph is shown in Figure 17 (use the smallest degree possible).

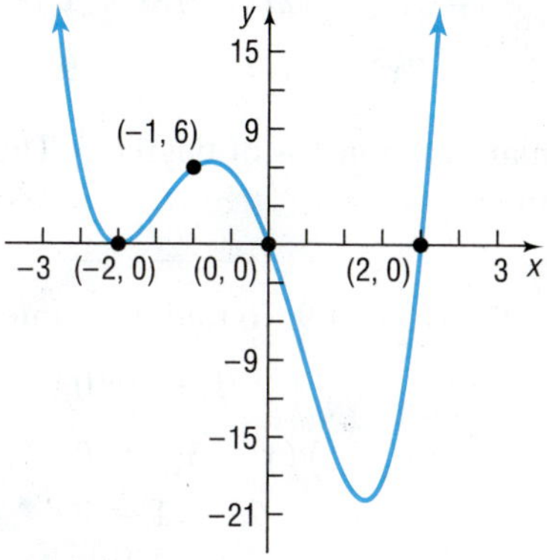

Figure 17

Solution The x-intercepts are -2, 0, and 2. Therefore, the polynomial must have the factors $(x + 2)$, x, and $(x - 2)$, respectively. There are three turning points, so the degree of the polynomial must be at least 4. The graph touches the x-axis at $x = -2$, so -2 must have an even multiplicity. The graph crosses the x-axis at $x = 0$ and $x = 2$, so 0 and 2 must have odd multiplicities. Using the smallest degree possible (1 for odd multiplicity and 2 for even multiplicity), we can write

$$f(x) = ax(x + 2)^2(x - 2)$$

All that remains is to find the leading coefficient, a. From Figure 17, the point $(-1, 6)$ must lie on the graph.

$$\begin{aligned} 6 &= a(-1)(-1 + 2)^2(-1 - 2) \quad && f(-1) = 6 \\ 6 &= 3a \\ 2 &= a \end{aligned}$$

The polynomial function $f(x) = 2x(x + 2)^2(x - 2)$ would have the graph in Figure 17.

Check: Graph $Y_1 = 2x(x + 2)^2(x - 2)$ using a graphing utility to verify this result.

Now Work PROBLEMS 73 AND 77

SUMMARY

Graph of a Polynomial Function $f(x) = a_n x^n + a_{n-1}x^{n-1} + \cdots + a_1 x + a_0 \quad a_n \neq 0$

Degree of the polynomial function f: n
y-intercept: $f(0) = a_0$.
Graph is smooth and continuous.
Maximum number of turning points: $n - 1$
At a zero of even multiplicity: The graph of f touches the x-axis.
At a zero of odd multiplicity: The graph of f crosses the x-axis.
Between zeros, the graph of f is either above or below the x-axis.
End behavior: For large $|x|$, the graph of f behaves like the graph of $y = a_n x^n$.

4 Analyze the Graph of a Polynomial Function

EXAMPLE 10 **How to Analyze the Graph of a Polynomial Function**

Analyze the factored form of the polynomial function $f(x) = (2x + 1)(x - 3)^2$.

Step-by-Step Solution

Step 1: Determine the end behavior of the graph of the function.

Expand the polynomial:

$$f(x) = (2x + 1)(x - 3)^2 = (2x + 1)(x^2 - 6x + 9)$$
$$= 2x^3 - 12x^2 + 18x + x^2 - 6x + 9 \quad \text{Multiply.}$$
$$= 2x^3 - 11x^2 + 12x + 9 \quad \text{Combine like terms.}$$

The polynomial function f is of degree 3. The graph of f behaves like $y = 2x^3$ for large values of $|x|$.

Step 2: Find the x- and y-intercepts of the graph of the function.

The y-intercept is $f(0) = 9$. To find the x-intercepts, solve $f(x) = 0$.

$$f(x) = 0$$
$$(2x + 1)(x - 3)^2 = 0$$
$$2x + 1 = 0 \quad \text{or} \quad (x - 3)^2 = 0$$
$$x = -\frac{1}{2} \quad \text{or} \quad x = 3$$

The x-intercepts are $-\frac{1}{2}$ and 3.

Step 3: Determine the zeros of the function and their multiplicity. Use this information to determine whether the graph crosses or touches the x-axis at each x-intercept.

The zeros of f are $-\frac{1}{2}$ and 3. The zero $-\frac{1}{2}$ is a zero of multiplicity 1, so the graph of f crosses the x-axis at $x = -\frac{1}{2}$. The zero 3 is a zero of multiplicity 2, so the graph of f touches the x-axis at $x = 3$.

Step 4: Determine the maximum number of turning points on the graph of the function.

Because the polynomial function is of degree 3 (Step 1), the graph of the function will have at most $3 - 1 = 2$ turning points.

Step 5: Put all the information from Steps 1 through 4 together to obtain the graph of f. To help establish the y-axis scale, find additional points on the graph on each side of any x-intercept.

Figure 18(a) illustrates the information obtained from Steps 1 through 4. We evaluate f at -1, 1, and 4 to help establish the scale on the y-axis.

We find that $f(-1) = -16$, $f(1) = 12$, and $f(4) = 9$, so we plot the points $(-1, -16)$, $(1, 12)$, and $(4, 9)$. The graph of f is given in Figure 18(b).

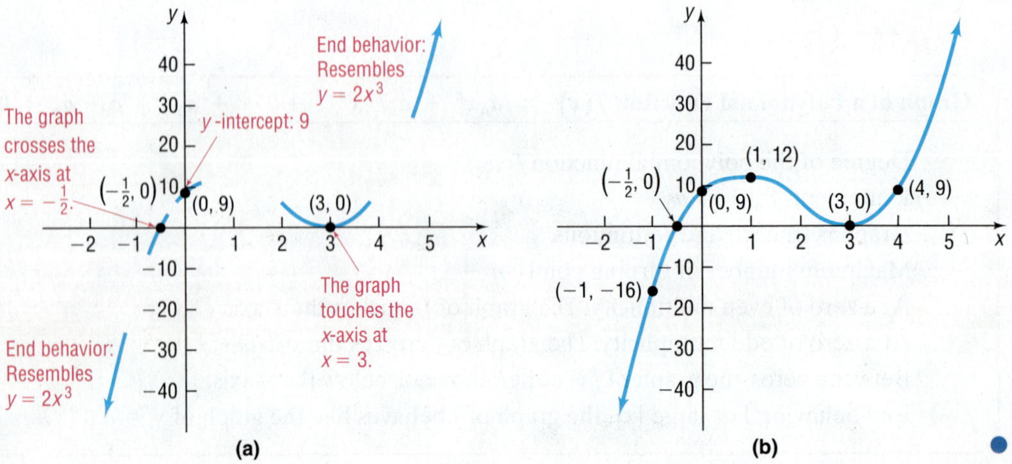

Figure 18

SUMMARY

Analyzing the Graph of a Polynomial Function

STEP 1: Determine the end behavior of the graph of the function.

STEP 2: Find the x- and y-intercepts of the graph of the function.

STEP 3: Determine the zeros of the function and their multiplicity. Use this information to determine whether the graph crosses or touches the x-axis at each x-intercept.

STEP 4: Determine the maximum number of turning points on the graph of the function.

STEP 5: Use the information in Steps 1 through 4 to draw a complete graph of the function. To help establish the y-axis scale, find additional points on the graph on each side of any x-intercept.

Now Work PROBLEM 81

For polynomial functions that have noninteger coefficients and for polynomials that are not easily factored, we use a graphing utility early in the analysis. This is because the amount of information that can be obtained from algebraic analysis is limited.

EXAMPLE 11 How to Use a Graphing Utility to Analyze the Graph of a Polynomial Function

Analyze the graph of the polynomial function

$$f(x) = x^3 + 2.48x^2 - 4.3155x + 2.484406$$

Step-by-Step Solution

Step 1: Determine the end behavior of the graph of the function.

The polynomial function f is of degree 3. The graph of f behaves like $y = x^3$ for large values of $|x|$.

Step 2: Graph the function using a graphing utility.

See Figure 19 for the graph of f.

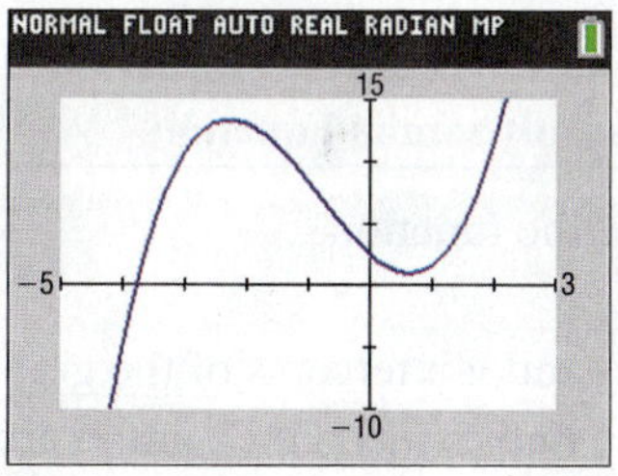

Figure 19 $f(x) = x^3 + 2.48x^2 - 4.3155x + 2.484406$

Step 3: Use a graphing utility to approximate the x- and y-intercepts of the graph.

The y-intercept is $f(0) = 2.484406$.

In Example 10, the polynomial function was factored, so it was easy to find the x-intercepts algebraically. However, it is not readily apparent how to factor f in this example. Therefore, we use a graphing utility's ZERO (or ROOT or SOLVE) feature and find the lone x-intercept to be -3.79, rounded to two decimal places.

Step 4: Use a graphing utility to create a TABLE to find points on the graph around each *x*-intercept.

Table 5 below shows values of x on each side of the x-intercept. The points $(-4, -4.57)$ and $(-2, 13.04)$ are on the graph.

Table 5

NORMAL FLOAT AUTO REAL RADIAN MP
PRESS ENTER TO EDIT

X	Y1
-4	-4.574
-2	13.035

Y1=X³+2.48X²-4.3155X+2.4844

Step 5: Approximate the turning points of the graph.

From the graph of f shown in Figure 19, we can see that f has two turning points. Using MAXIMUM reveals one turning point is at $(-2.28, 13.36)$, rounded to two decimal places. Using MINIMUM shows that the other turning point is at $(0.63, 1)$, rounded to two decimal places.

Step 6: Use the information in Steps 1 through 5 to draw a complete graph of the function by hand.

Figure 20 shows a graph of f drawn by hand using the information in Steps 1 through 5.

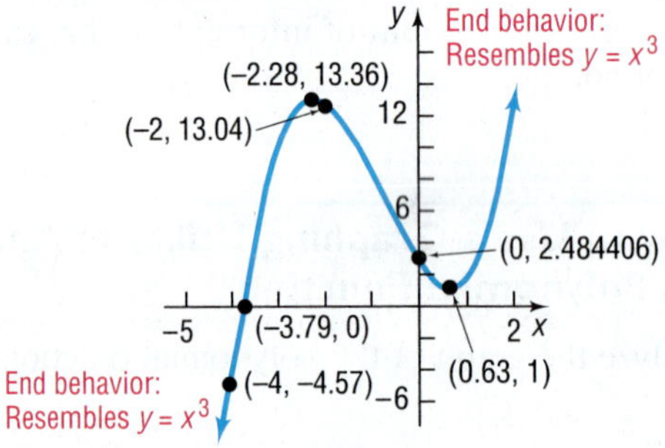

Figure 20

Step 7: Find the domain and the range of the function.

The domain and the range of f are the set of all real numbers.

Step 8: Use the graph to determine where the function is increasing and where it is decreasing.

Based on the graph, f is increasing on the intervals $(-\infty, -2.28)$ and $(0.63, \infty)$. Also, f is decreasing on the interval $(-2.28, 0.63)$.

SUMMARY

Using a Graphing Utility to Analyze the Graph of a Polynomial Function

STEP 1: Determine the end behavior of the graph of the function.
STEP 2: Graph the function using a graphing utility.
STEP 3: Use a graphing utility to approximate the x- and y-intercepts of the graph.
STEP 4: Use a graphing utility to create a TABLE to find points on the graph around each x-intercept.
STEP 5: Approximate the turning points of the graph.
STEP 6: Use the information in Steps 1 through 5 to draw a complete graph of the function by hand.
STEP 7: Find the domain and the range of the function.
STEP 8: Use the graph to determine where the function is increasing and where it is decreasing.

Now Work PROBLEM 99

5 Build Cubic Models from Data

In Section 4.2 we found the line of best fit from data, and in Section 4.4 we found the quadratic function of best fit. It is also possible to find polynomial functions of best fit. However, most statisticians do not recommend finding polynomials of best fit of degree higher than 3.

Data that follow a cubic relation should look like Figure 21(a) or (b).

$y = ax^3 + bx^2 + cx + d, a > 0$ (a) $y = ax^3 + bx^2 + cx + d, a < 0$ (b)

Figure 21 Cubic relation

EXAMPLE 12 A Cubic Function of Best Fit

The data in Table 6 represent the weekly cost C (in thousands of dollars) of printing x thousand textbooks.

Table 6

Number x of Textbooks, (thousands)	Cost, C ($1000s)
0	100
5	128.1
10	144
13	153.5
17	161.2
18	162.6
20	166.3
23	178.9
25	190.2
27	221.8

(a) Draw a scatter diagram of the data using x as the independent variable and C as the dependent variable. Comment on the type of relation that may exist between the two variables x and C.

(b) Using a graphing utility, find the cubic function of best fit $C = C(x)$ that models the relation between number of texts and cost.

(c) Graph the cubic function of best fit on your scatter diagram.

(d) Use the function found in part (b) to predict the cost of printing 22 thousand texts per week.

Solution

(a) Figure 22 shows the scatter diagram. A cubic relation may exist between the two variables.

(b) Upon executing the CUBIC REGression program, we obtain the results shown in Figure 23. The output that the utility provides shows us the equation $y = ax^3 + bx^2 + cx + d$. The cubic function of best fit to the data is $C(x) = 0.0155x^3 - 0.5951x^2 + 9.1502x + 98.4327$.

(c) Figure 24 shows the graph of the cubic function of best fit on the scatter diagram. The function fits the data reasonably well.

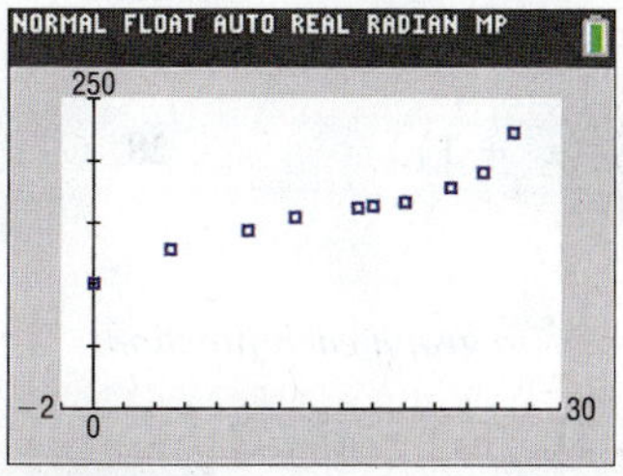

Figure 22

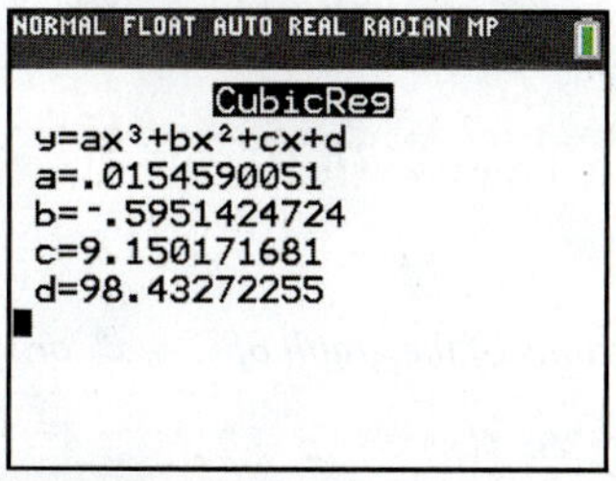

Figure 23

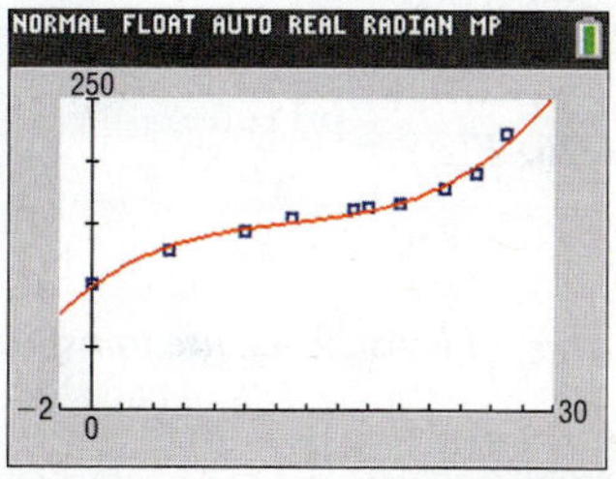

Figure 24

(d) Evaluate the function $C(x)$ at $x = 22$.

$$C(22) = 0.0155(22)^3 - 0.5951(22)^2 + 9.1502(22) + 98.4327 \approx 176.8$$

The model predicts that the cost of printing 22 thousand textbooks in a week will be 176.8 thousand dollars—that is, $176,800.

5.1 Assess Your Understanding

'Are You Prepared?' *Answers are given at the end of these exercises. If you get a wrong answer, read the pages listed in red.*

1. The intercepts of the equation $9x^2 + 4y = 36$ are ______. (pp. 159–160)

2. Is the expression $4x^3 - 3.6x^2 - \sqrt{2}$ a polynomial? If so, what is its degree? (pp. 39–47)

3. To graph $y = x^2 - 4$, you would shift the graph of $y = x^2$ ______ a distance of ____ units. (pp. 247–256)

4. Use a graphing utility to approximate (rounded to two decimal places) the local maximum value and local minimum value of $f(x) = x^3 - 2x^2 - 4x + 5$, for $-3 < x < 3$. (p. 229)

5. ***True or False*** The x-intercepts of the graph of a function $y = f(x)$ are the real solutions of the equation $f(x) = 0$. (pp. 215–217)

6. If $g(5) = 0$, what point is on the graph of g? What is the corresponding x-intercept of the graph of g? (pp. 215–217)

Concepts and Vocabulary

7. The graph of every polynomial function is both ______ and ______.

8. If r is a real zero of even multiplicity of a function f, then the graph of f ______ (crosses/touches) the x-axis at r.

9. The graphs of power functions of the form $f(x) = x^n$, where n is an even integer, always contain the points ______, ______, and ______.

10. If r is a solution to the equation $f(x) = 0$, name three additional statements that can be made about f and r, assuming f is a polynomial function.

11. The points at which a graph changes direction (from increasing to decreasing or decreasing to increasing) are called ______.

12. The graph of the function $f(x) = 3x^4 - x^3 + 5x^2 - 2x - 7$ will behave like the graph of ______ for large values of $|x|$.

13. If $f(x) = -2x^5 + x^3 - 5x^2 + 7$, then $\lim_{x\to-\infty} f(x) =$ ____ and $\lim_{x\to\infty} f(x) =$ ______.

14. Explain what the notation $\lim_{x\to\infty} f(x) = -\infty$ means.

15. The ______ of a zero is the number of times its corresponding factor occurs.
(a) degree (b) multiplicity (c) turning point (d) limit

16. The graph of $y = 5x^6 - 3x^4 + 2x - 9$ has at most how many turning points?
(a) -9 (b) 14 (c) 6 (d) 5

Skill Building

In Problems 17–28, determine which functions are polynomial functions. For those that are, state the degree. For those that are not, tell why not. Write each polynomial in standard form. Then identify the leading term and the constant term.

17. $f(x) = 4x + x^3$
18. $f(x) = 5x^2 + 4x^4$
19. $g(x) = \dfrac{1 - x^2}{2}$
20. $h(x) = 3 - \dfrac{1}{2}x$
21. $f(x) = 1 - \dfrac{1}{x}$
22. $f(x) = x(x - 1)$
23. $g(x) = x^{3/2} - x^2 + 2$
24. $h(x) = \sqrt{x}(\sqrt{x} - 1)$
25. $F(x) = 5x^4 - \pi x^3 + \dfrac{1}{2}$
26. $F(x) = \dfrac{x^2 - 5}{x^3}$
27. $G(x) = 2(x - 1)^2(x^2 + 1)$
28. $G(x) = -3x^2(x + 2)^3$

In Problems 29–42, use transformations of the graph of $y = x^4$ or $y = x^5$ to graph each function.

29. $f(x) = (x + 1)^4$
30. $f(x) = (x - 2)^5$
31. $f(x) = x^5 - 3$
32. $f(x) = x^4 + 2$
33. $f(x) = \dfrac{1}{2}x^4$
34. $f(x) = 3x^5$
35. $f(x) = -x^5$
36. $f(x) = -x^4$
37. $f(x) = (x - 1)^5 + 2$
38. $f(x) = (x + 2)^4 - 3$
39. $f(x) = 2(x + 1)^4 + 1$
40. $f(x) = \dfrac{1}{2}(x - 1)^5 - 2$
41. $f(x) = 4 - (x - 2)^5$
42. $f(x) = 3 - (x + 2)^4$

In Problems 43–50, form a polynomial function whose real zeros and degree are given. Answers will vary depending on the choice of a leading coefficient.

43. Zeros: $-1, 1, 3$; degree 3

44. Zeros: $-2, 2, 3$; degree 3

45. Zeros: $-3, 0, 4$; degree 3

46. Zeros: $-4, 0, 2$; degree 3

47. Zeros: $-4, -1, 2, 3$; degree 4

48. Zeros: $-3, -1, 2, 5$; degree 4

49. Zeros: -1, multiplicity 1; 3, multiplicity 2; degree 3

50. Zeros: -2, multiplicity 2; 4, multiplicity 1; degree 3

In Problems 51–56, find the polynomial function with the given zeros whose graph passes through the given point.

51. Zeros: $-3, 1, 4$
Point: $(6, 180)$

52. Zeros: $-2, 0, 2$
Point: $(-4, 16)$

53. Zeros: $-1, 0, 2, 4$
Point: $\left(\frac{1}{2}, 63\right)$

54. Zeros: $-5, -1, 2, 6$
Point: $\left(\frac{5}{2}, 15\right)$

55. Zeros: -1 (multiplicity 2), 1 (multiplicity 2)
Point: $(-2, 45)$

56. Zeros: -1 (multiplicity 2), 0, 3 (multiplicity 2)
Point: $(1, -48)$

In Problems 57–68, for each polynomial function:
(a) List each real zero and its multiplicity.
(b) Determine whether the graph crosses or touches the x-axis at each x-intercept.
(c) Determine the maximum number of turning points on the graph.
(d) Determine the end behavior; that is, find the power function that the graph of f resembles for large values of $|x|$.

57. $f(x) = 3(x - 7)(x + 3)^2$

58. $f(x) = 4(x + 4)(x + 3)^3$

59. $f(x) = 4(x^2 + 1)(x - 2)^3$

60. $f(x) = 2(x - 3)(x^2 + 4)^3$

61. $f(x) = -2\left(x + \frac{1}{2}\right)^2(x + 4)^3$

62. $f(x) = \left(x - \frac{1}{3}\right)^2(x - 1)^3$

63. $f(x) = (x - 5)^3(x + 4)^2$

64. $f(x) = (x + \sqrt{3})^2(x - 2)^4$

65. $f(x) = 3(x^2 + 8)(x^2 + 9)^2$

66. $f(x) = -2(x^2 + 3)^3$

67. $f(x) = -2x^2(x^2 - 2)$

68. $f(x) = 4x(x^2 - 3)$

In Problems 69–72, identify which of the graphs could be the graph of a polynomial function. For those that could, list the real zeros and state the least degree the polynomial can have. For those that could not, say why not.

69.

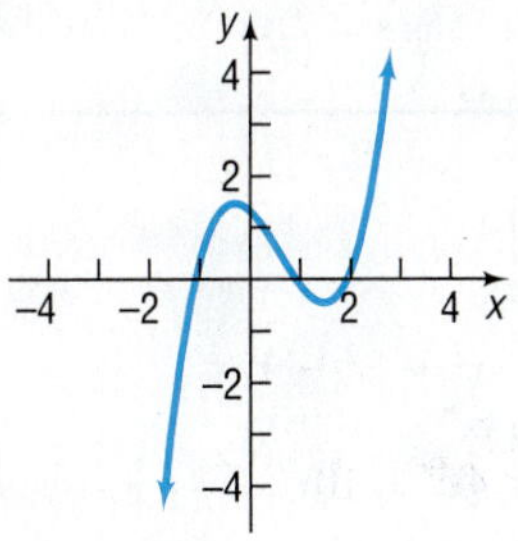

70.

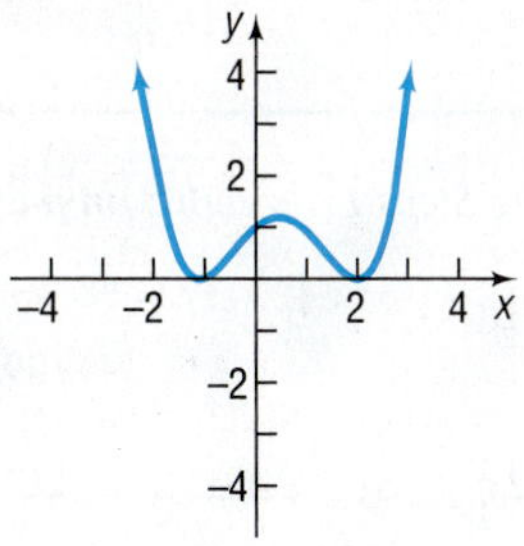

71.

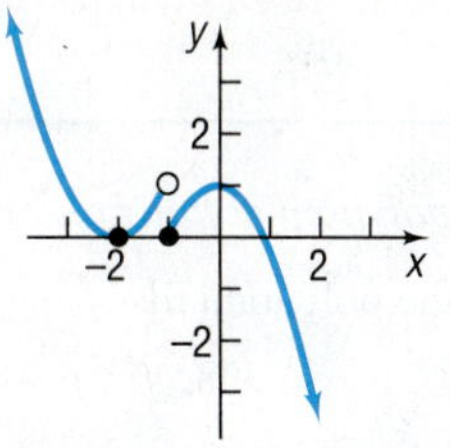

72.

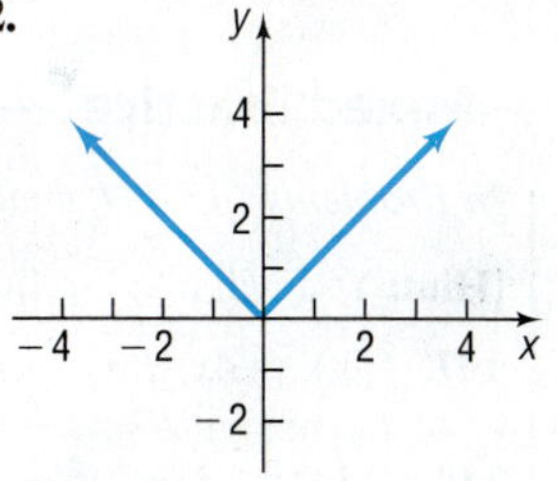

In Problems 73–76, construct a polynomial function that might have the given graph. (More than one answer may be possible.)

73.

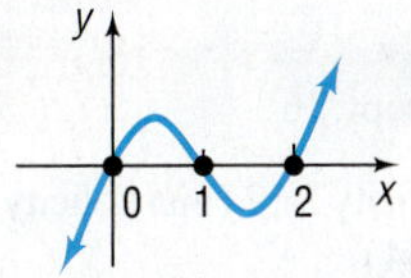

74.

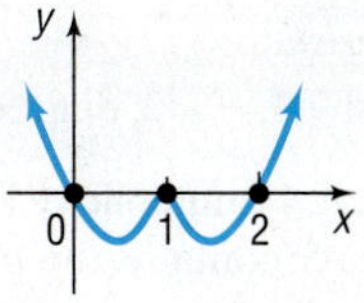

75.

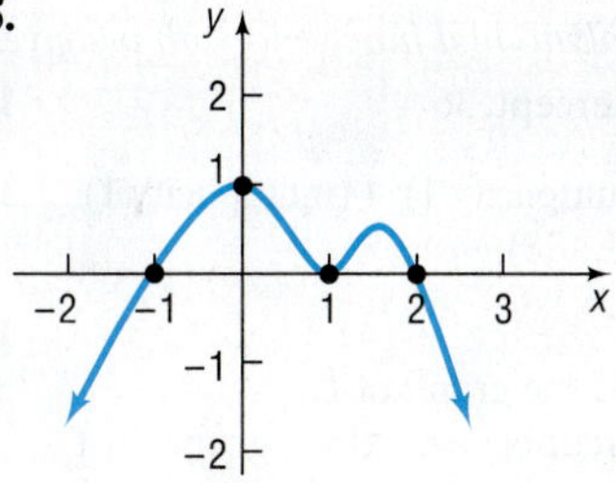

76.

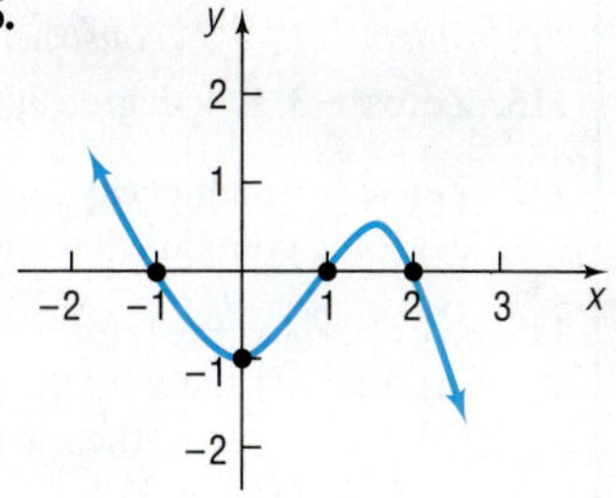

In Problems 77–80, write a polynomial function whose graph is shown (use the smallest degree possible).

77.

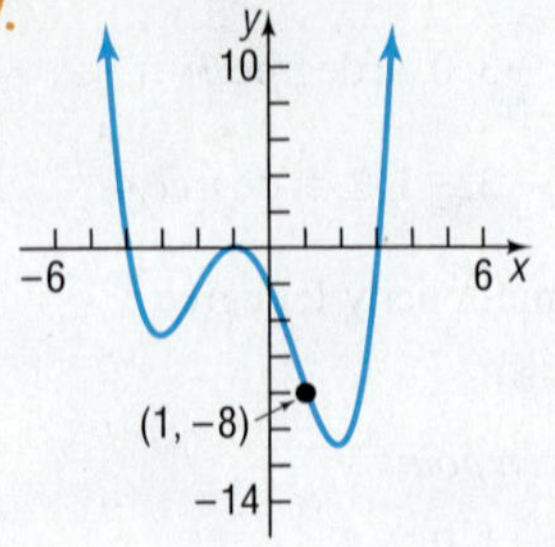

78.

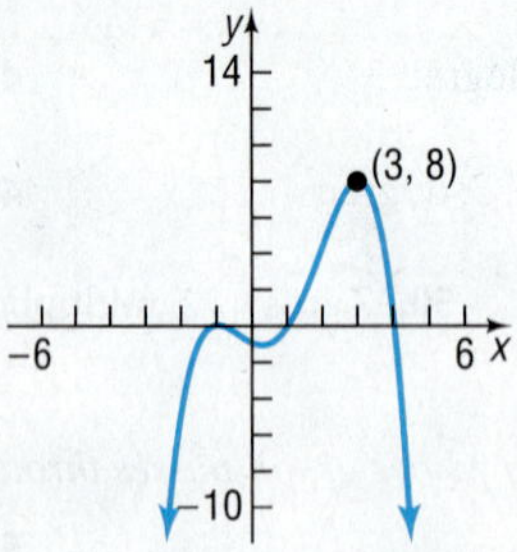

79.

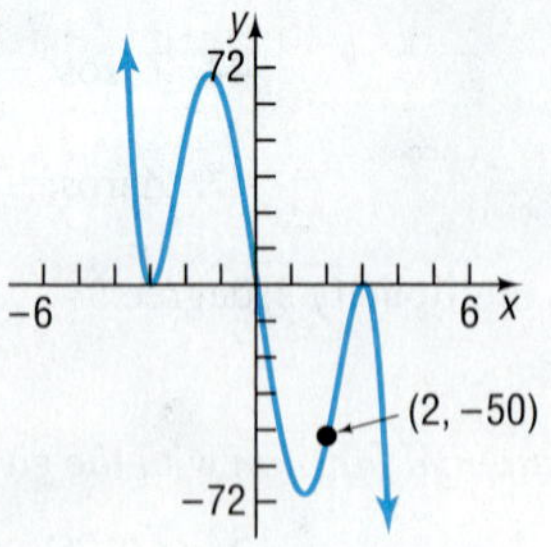

80.

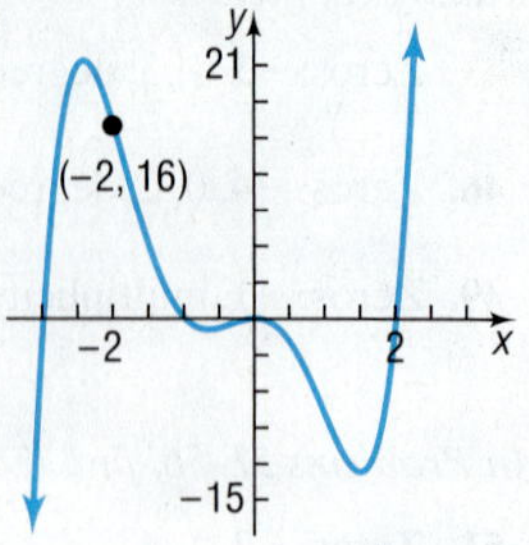

In Problems 81–98, analyze each polynomial function by following Steps 1 through 5 on page 335.

81. $f(x) = x^2(x - 3)$

82. $f(x) = x(x + 2)^2$

83. $f(x) = (x + 4)^2(1 - x)$

84. $f(x) = (x - 1)(x + 3)^2$

85. $f(x) = -2(x + 2)(x - 2)^3$

86. $f(x) = -\frac{1}{2}(x + 4)(x - 1)^3$

87. $f(x) = (x + 1)(x - 2)(x + 4)$

88. $f(x) = (x - 1)(x + 4)(x - 3)$

89. $f(x) = x^2(x - 2)(x + 2)$

90. $f(x) = x^2(x - 3)(x + 4)$

91. $f(x) = (x + 1)^2(x - 2)^2$

92. $f(x) = (x - 4)^2(x + 2)^2$

93. $f(x) = x^2(x + 3)(x + 1)$

94. $f(x) = x^2(x - 3)(x - 1)$

95. $f(x) = 5x(x^2 - 4)(x + 3)$

96. $f(x) = (x - 2)^2(x + 2)(x + 4)$

97. $f(x) = x^2(x - 2)(x^2 + 3)$

98. $f(x) = x^2(x^2 + 1)(x + 4)$

In Problems 99–106, analyze each polynomial function f by following Steps 1 through 8 on page 336.

99. $f(x) = x^3 + 0.2x^2 - 1.5876x - 0.31752$

100. $f(x) = x^3 - 0.8x^2 - 4.6656x + 3.73248$

101. $f(x) = x^3 + 2.56x^2 - 3.31x + 0.89$

102. $f(x) = x^3 - 2.91x^2 - 7.668x - 3.8151$

103. $f(x) = x^4 - 2.5x^2 + 0.5625$

104. $f(x) = x^4 - 18.5x^2 + 50.2619$

105. $f(x) = 2x^4 - \pi x^3 + \sqrt{5}x - 4$

106. $f(x) = -1.2x^4 + 0.5x^2 - \sqrt{3}x + 2$

Mixed Practice

In Problems 107–114, analyze each polynomial function by following Steps 1 through 5 on page 335.

[**Hint:** You will need to first factor the polynomial].

107. $f(x) = 4x - x^3$

108. $f(x) = x - x^3$

109. $f(x) = x^3 + x^2 - 12x$

110. $f(x) = x^3 + 2x^2 - 8x$

111. $f(x) = 2x^4 + 12x^3 - 8x^2 - 48x$

112. $f(x) = 4x^3 + 10x^2 - 4x - 10$

113. $f(x) = -x^5 - x^4 + x^3 + x^2$

114. $f(x) = -x^5 + 5x^4 + 4x^3 - 20x^2$

In Problems 115–118, construct a polynomial function f with the given characteristics.

115. Zeros: $-3, 1, 4$; degree 3; y-intercept: 36

116. Zeros: $-4, -1, 2$; degree 3; y-intercept: 16

117. Zeros: -5(multiplicity 2); 2 (multiplicity 1); 4 (multiplicity 1); degree 4; contains the point $(3, 128)$

118. Zeros: -4 (multiplicity 1); 0 (multiplicity 3); 2 (multiplicity 1); degree 5; contains the point $(-2, 64)$

119. $G(x) = (x + 3)^2(x - 2)$
(a) Identify the x-intercepts of the graph of G.
(b) What are the x-intercepts of the graph of $y = G(x + 3)$?

120. $h(x) = (x + 2)(x - 4)^3$
(a) Identify the x-intercepts of the graph of h.
(b) What are the x-intercepts of the graph of $y = h(x - 2)$?

Applications and Extensions

121. Hurricanes In 2012, Hurricane Sandy struck the East Coast of the United States, killing 147 people and causing an estimated $75 billion in damage. With a gale diameter of about 1000 miles, it was the largest ever to form over the Atlantic Basin. The accompanying data represent the number of major hurricane strikes in the Atlantic Basin (category 3, 4, or 5) each decade from 1921 to 2010.

Decade, x	Major Hurricanes Striking Atlantic Basin, H
1921–1930, 1	17
1931–1940, 2	16
1941–1950, 3	29
1951–1960, 4	33
1961–1970, 5	27
1971–1980, 6	16
1981–1990, 7	16
1991–2000, 8	27
2001–2010, 9	33

Source: National Oceanic & Atmospheric Administration

(a) Draw a scatter diagram of the data. Comment on the type of relation that may exist between the two variables.

(b) Use a graphing utility to find the cubic function of best fit that models the relation between decade and number of major hurricanes.

(c) Use the model found in part (b) to predict the number of major hurricanes that struck the Atlantic Basin between 1961 and 1970.

(d) With a graphing utility, draw a scatter diagram of the data and then graph the cubic function of best fit on the scatter diagram.

(e) Concern has risen about the increase in the number and intensity of hurricanes, but some scientists believe this is just a natural fluctuation that could last another decade or two. Use your model to predict the number of major hurricanes that will strike the Atlantic Basin between 2011 and 2020. Is your result reasonable?

122. Poverty Rates The following data represent the percentage of families in the United States whose income is below the poverty level.

(a) With a graphing utility, draw a scatter diagram of the data. Comment on the type of relation that appears to exist between the two variables.

(b) Decide on a function of best fit to these data (linear, quadratic, or cubic), and use this function to predict the percentage of U.S. families that were below the poverty level in 2012 ($t = 23$). Compare your prediction to the actual value of 15.9.

(c) Draw the function of best fit on the scatter diagram drawn in part (a).

Year, t	Percent below Poverty Level, p	Year, t	Percent below Poverty Level, p
1990, 1	13.5	2001, 12	11.7
1991, 2	14.2	2002, 13	12.1
1992, 3	14.8	2003, 14	12.5
1993, 4	15.1	2004, 15	12.7
1994, 5	14.5	2005, 16	12.6
1995, 6	13.8	2006, 17	12.3
1996, 7	13.7	2007, 18	12.5
1997, 8	13.3	2008, 19	13.2
1998, 9	12.7	2009, 20	14.3
1999, 10	11.9	2010, 21	15.3
2000, 11	11.3	2011, 22	15.9

Source: U.S. Census Bureau

123. Temperature The following data represent the temperature T (°Fahrenheit) in Kansas City, Missouri, x hours after midnight on April 15, 2014.

Hours after Midnight, x	Temperature (°F), T
3	36.1
6	32.0
9	39.0
12	46.2
15	52.0
18	55.0
21	52.0
24	48.9

Source: The Weather Underground

(a) Draw a scatter diagram of the data. Comment on the type of relation that may exist between the two variables.

(b) Find the average rate of change in temperature from 9 AM to 12 noon.

(c) What is the average rate of change in temperature from 3 PM to 6 PM ?

(d) Decide on a function of best fit to these data (linear, quadratic, or cubic) and use this function to predict the temperature at 5 PM.

(e) With a graphing utility, draw a scatter diagram of the data and then graph the function of best fit on the scatter diagram.

(f) Interpret the y-intercept.

124. Future Value of Money Suppose that you make deposits of $500 at the beginning of every year into an Individual Retirement Account (IRA) earning interest r (expressed as a decimal). At the beginning of the first year, the value of the account will be $500; at the beginning of the second year, the value of the account, will be

$$\underbrace{\$500 + \$500r}_{\text{Value of 1st deposit}} + \underbrace{\$500}_{\text{Value of 2nd deposit}} = \$500(1 + r) + \$500 = 500r + 1000$$

(a) Verify that the value of the account at the beginning of the third year is $T(r) = 500r^2 + 1500r + 1500$.

(b) The account value at the beginning of the fourth year is $F(r) = 500r^3 + 2000r^2 + 3000r + 2000$. If the annual rate of interest is $5\% = 0.05$, what will be the value of the account at the beginning of the fourth year?

125. A Geometric Series In calculus, you will learn that certain functions can be approximated by polynomial functions. We will explore one such function now.

(a) Using a graphing utility, create a table of values with $Y_1 = f(x) = \dfrac{1}{1-x}$ and $Y_2 = g_2(x) = 1 + x + x^2 + x^3$ for $-1 < x < 1$ with $\Delta\text{Tbl} = 0.1$.

(b) Using a graphing utility, create a table of values with $Y_1 = f(x) = \dfrac{1}{1-x}$ and $Y_2 = g_3(x) = 1 + x + x^2 + x^3 + x^4$ for $-1 < x < 1$ with $\Delta\text{Tbl} = 0.1$.

(c) Using a graphing utility, create a table of values with $Y_1 = f(x) = \dfrac{1}{1-x}$ and $Y_2 = g_4(x) = 1 + x + x^2 + x^3 + x^4 + x^5$ for $-1 < x < 1$ with $\Delta\text{Tbl} = 0.1$.

(d) What do you notice about the values of the function as more terms are added to the polynomial? Are there some values of x for which the approximations are better?

Explaining Concepts: Discussion and Writing

126. Can the graph of a polynomial function have no y-intercept? Can it have no x-intercepts? Explain.

127. Write a few paragraphs that provide a general strategy for graphing a polynomial function. Be sure to mention the following: degree, intercepts, end behavior, and turning points.

128. Make up a polynomial that has the following characteristics: crosses the x-axis at -1 and 4, touches the x-axis at 0 and 2, and is above the x-axis between 0 and 2. Give your polynomial to a fellow classmate and ask for a written critique.

129. Make up two polynomials, not of the same degree, with the following characteristics: crosses the x-axis at -2, touches the x-axis at 1, and is above the x-axis between -2 and 1. Give your polynomials to a fellow classmate and ask for a written critique.

130. The graph of a polynomial function is always smooth and continuous. Name a function studied earlier that is smooth but not continuous. Name one that is continuous but not smooth.

131. Which of the following statements are true regarding the graph of the cubic polynomial $f(x) = x^3 + bx^2 + cx + d$? (Give reasons for your conclusions.)

(a) It intersects the y-axis in one and only one point.
(b) It intersects the x-axis in at most three points.
(c) It intersects the x-axis at least once.
(d) For $|x|$ very large, it behaves like the graph of $y = x^3$.
(e) It is symmetric with respect to the origin.
(f) It passes through the origin.

132. The illustration shows the graph of a polynomial function.

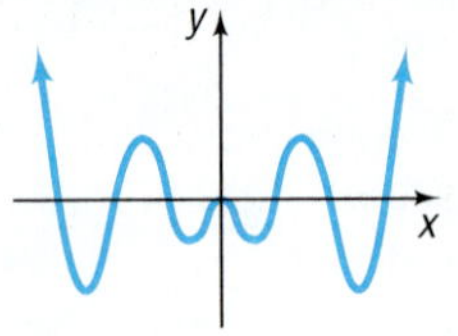

(a) Is the degree of the polynomial even or odd?
(b) Is the leading coefficient positive or negative?
(c) Is the function even, odd, or neither?
(d) Why is x^2 necessarily a factor of the polynomial?
(e) What is the minimum degree of the polynomial?
(f) Formulate five different polynomials whose graphs could look like the one shown. Compare yours to those of other students. What similarities do you see? What differences?

133. Design a polynomial function with the following characteristics: degree 6; four distinct real zeros, one of multiplicity 3; y-intercept 3; behaves like $y = -5x^6$ for large values of $|x|$. Is this polynomial unique? Compare your polynomial with those of other students. What terms will be the same as everyone else's? Add some more characteristics, such as symmetry or naming the real zeros. How does this modify the polynomial?

Retain Your Knowledge

Problems 134–137 are based on material learned earlier in the course. The purpose of these problems is to keep the material fresh in your mind so that you are better prepared for the final exam.

134. Find the equation of the line that contains the point $(2, -3)$ and is perpendicular to the line $5x - 2y = 6$.

135. Find the domain of the function $h(x) = \dfrac{x-3}{x+5}$.

136. Use the quadratic formula to find the zeros of the function $f(x) = 4x^2 + 8x - 3$.

137. Solve: $|5x - 3| = 7$.

'Are You Prepared?' Answers

1. $(-2, 0), (2, 0), (0, 9)$ **2.** Yes; 3 **3.** Down; 4 **4.** Local maximum value 6.48 at $x = -0.67$; local minimum value -3 at $x = 2$
5. True **6.** $(5, 0)$; 5

5.2 Properties of Rational Functions

PREPARING FOR THIS SECTION *Before getting started, review the following:*

- Rational Expressions (Chapter R, Section R.7, pp. 62–69)
- Polynomial Division (Chapter R, Section R.4, pp. 44–47)
- Graph of $f(x) = \frac{1}{x}$ (Section 2.2, Example 12, p. 164)
- Graphing Techniques: Transformations (Section 3.5, pp. 247–256)

Now Work the **'Are You Prepared?'** problems on page 350.

OBJECTIVES
1. Find the Domain of a Rational Function (p. 343)
2. Find the Vertical Asymptotes of a Rational Function (p. 346)
3. Find the Horizontal or Oblique Asymptote of a Rational Function (p. 348)

Ratios of integers are called *rational numbers*. Similarly, ratios of polynomial functions are called *rational functions*. Examples of rational functions are

$$R(x) = \frac{x^2 - 4}{x^2 + x + 1} \qquad F(x) = \frac{x^3}{x^2 - 4} \qquad G(x) = \frac{3x^2}{x^4 - 1}$$

DEFINITION

A **rational function** is a function of the form

$$R(x) = \frac{p(x)}{q(x)}$$

where p and q are polynomial functions and q is not the zero polynomial. The domain of a rational function is the set of all real numbers except those for which the denominator q is 0.

1 Find the Domain of a Rational Function

EXAMPLE 1 **Finding the Domain of a Rational Function**

(a) The domain of $R(x) = \frac{2x^2 - 4}{x + 5}$ is the set of all real numbers x except -5; that is, the domain is $\{x \mid x \neq -5\}$.

(b) The domain of $R(x) = \frac{1}{x^2 - 4} = \frac{1}{(x + 2)(x - 2)}$ is the set of all real numbers x except -2 and 2; that is, the domain is $\{x \mid x \neq -2, x \neq 2\}$.

(c) The domain of $R(x) = \frac{x^3}{x^2 + 1}$ is the set of all real numbers.

(d) The domain of $R(x) = \frac{x^2 - 1}{x - 1}$ is the set of all real numbers x except 1; that is, the domain is $\{x \mid x \neq 1\}$. ●

Although $\frac{x^2 - 1}{x - 1}$ reduces to $x + 1$, it is important to observe that the functions

$$R(x) = \frac{x^2 - 1}{x - 1} \quad \text{and} \quad f(x) = x + 1$$

are not equal, since the domain of R is $\{x \mid x \neq 1\}$ and the domain of f is the set of all real numbers.

Now Work PROBLEM 17

WARNING The domain of a rational function must be found *before* writing the function in lowest terms. ■

If $R(x) = \dfrac{p(x)}{q(x)}$ is a rational function, and if p and q have no common factors, then the rational function R is said to be in **lowest terms.** For a rational function $R(x) = \dfrac{p(x)}{q(x)}$ in lowest terms, the real zeros, if any, of the numerator in the domain of R are the x-intercepts of the graph of R and so will play a major role in the graph of R. The real zeros of the denominator of R [that is, the numbers x, if any, for which $q(x) = 0$], although not in the domain of R, also play a major role in the graph of R.

We have already discussed the properties of the rational function $y = \dfrac{1}{x}$. (Refer to Example 12, page 164). The next rational function that we take up is $H(x) = \dfrac{1}{x^2}$.

EXAMPLE 2 Graphing $y = \dfrac{1}{x^2}$

Analyze the graph of $H(x) = \dfrac{1}{x^2}$.

Solution The domain of $H(x) = \dfrac{1}{x^2}$ is the set of all real numbers x except 0. The graph has no y-intercept, because x can never equal 0. The graph has no x-intercept because the equation $H(x) = 0$ has no solution. Therefore, the graph of H will not cross or touch either of the coordinate axes. Because

$$H(-x) = \frac{1}{(-x)^2} = \frac{1}{x^2} = H(x)$$

H is an even function, so its graph is symmetric with respect to the y-axis.

Table 7 shows the behavior of $H(x) = \dfrac{1}{x^2}$ for selected positive numbers x. (We use symmetry to obtain the graph of H when $x < 0$.) From the first three rows of Table 7, we see that as the values of x approach (get closer to) 0, the values of $H(x)$ become larger and larger positive numbers, so H is unbounded in the positive direction. In calculus we use limit notation, $\lim_{x \to 0} H(x) = \infty$, which is read "the limit of $H(x)$ as x approaches zero equals infinity," to mean that $H(x) \to \infty$ as $x \to 0$.

Look at the last four rows of Table 7. As $x \to \infty$, the values of $H(x)$ approach 0 (the end behavior of the graph). In calculus, this is expressed by writing $\lim_{x \to \infty} H(x) = 0$. Figure 25 shows the graph. Notice the use of red dashed lines to convey the ideas discussed above.

Table 7

x	$H(x) = \dfrac{1}{x^2}$
$\frac{1}{2}$	4
$\frac{1}{100}$	10,000
$\frac{1}{10,000}$	100,000,000
1	1
2	$\frac{1}{4}$
100	$\frac{1}{10,000}$
10,000	$\frac{1}{100,000,000}$

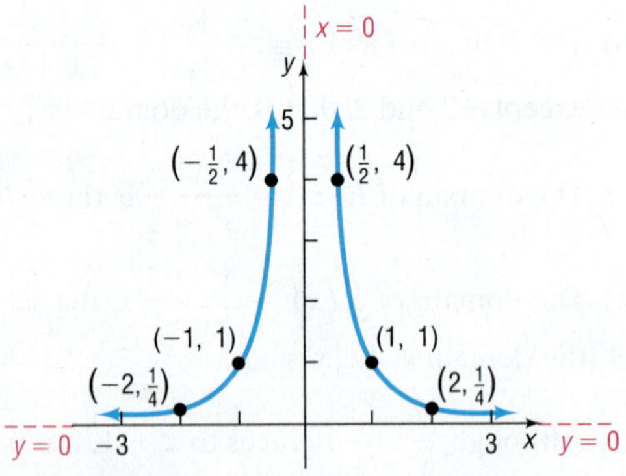

Figure 25 $H(x) = \dfrac{1}{x^2}$ ●

EXAMPLE 3 Using Transformations to Graph a Rational Function

Graph the rational function: $R(x) = \dfrac{1}{(x-2)^2} + 1$

Solution The domain of R is the set of all real numbers except $x = 2$. To graph R, start with the graph of $y = \frac{1}{x^2}$. See Figure 26 for the steps.

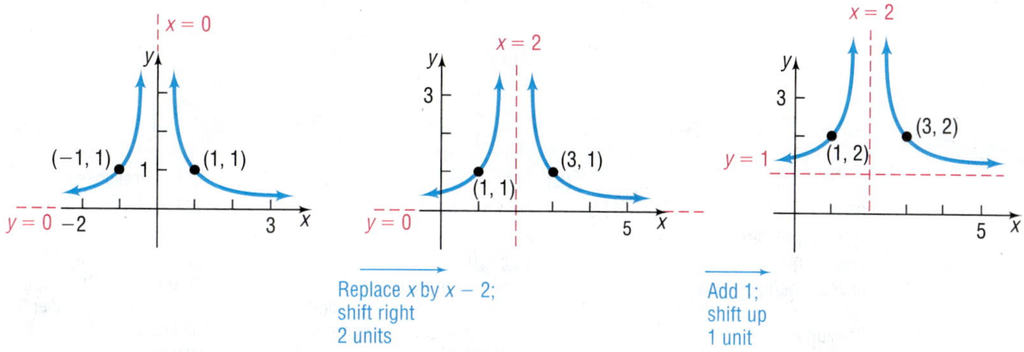

Figure 26 (a) $y = \frac{1}{x^2}$ (b) $y = \frac{1}{(x-2)^2}$ (c) $y = \frac{1}{(x-2)^2} + 1$

Now Work PROBLEMS 35(a) AND 35(b)

Asymptotes

Let's investigate the roles of the vertical line $x = 2$ and the horizontal line $y = 1$ in Figure 26(c).

First, we look at the end behavior of $R(x) = \frac{1}{(x-2)^2} + 1$. Table 8(a) shows the values of R at $x = 10, 100, 1000$, and $10{,}000$. Note that as x becomes unbounded in the positive direction, the values of R approach 1, so $\lim_{x\to\infty} R(x) = 1$. From Table 8(b) we see that as x becomes unbounded in the negative direction, the values of R also approach 1, so $\lim_{x\to-\infty} R(x) = 1$.

Even though $x = 2$ is not in the domain of R, the behavior of the graph of R near $x = 2$ is important. Table 8(c) shows the values of R at $x = 1.5, 1.9, 1.99, 1.999$, and 1.9999. We see that as x approaches 2 for $x < 2$, denoted $x \to 2^-$, the values of R are increasing without bound, so $\lim_{x\to 2^-} R(x) = \infty$. From Table 8(d), we see that as x approaches 2 for $x > 2$, denoted $x \to 2^+$, the values of R are also increasing without bound, so $\lim_{x\to 2^+} R(x) = \infty$.

Table 8

x	R(x)
10	1.0156
100	1.0001
1000	1.000001
10,000	1.00000001

(a)

x	R(x)
−10	1.0069
−100	1.0001
−1000	1.000001
−10,000	1.00000001

(b)

x	R(x)
1.5	5
1.9	101
1.99	10,001
1.999	1,000,001
1.9999	100,000,001

(c)

x	R(x)
2.5	5
2.1	101
2.01	10,001
2.001	1,000,001
2.0001	100,000,001

(d)

The vertical line $x = 2$ and the horizontal line $y = 1$ are called *asymptotes* of the graph of R.

DEFINITION

Let R denote a function.

If, as $x \to -\infty$ or as $x \to \infty$, the values of $R(x)$ approach some fixed number L, then the line $y = L$ is a **horizontal asymptote** of the graph of R. [Refer to Figures 27(a) and (b) on page 346.]

If, as x approaches some number c, the values $|R(x)| \to \infty$ [that is, $R(x) \to -\infty$ or $R(x) \to \infty$], then the line $x = c$ is a **vertical asymptote** of the graph of R. [Refer to Figures 27(c) and (d).]

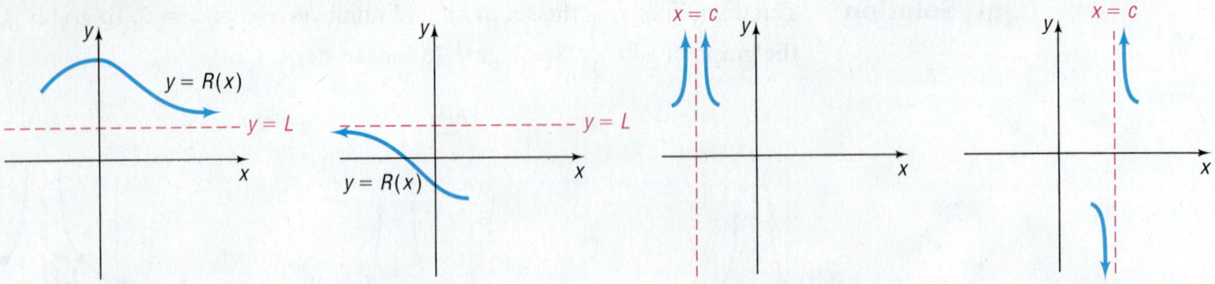

(a) End behavior: As $x \to \infty$, the values of $R(x)$ approach L [$\lim_{x\to\infty} R(x) = L$]. That is, the points on the graph of R are getting closer to the line $y = L$; $y = L$ is a horizontal asymptote.

(b) End behavior: As $x \to -\infty$, the values of $R(x)$ approach L [$\lim_{x\to-\infty} R(x) = L$]. That is, the points on the graph of R are getting closer to the line $y = L$; $y = L$ is a horizontal asymptote.

(c) As x approaches c, the values of $|R(x)| \to \infty$ [$\lim_{x\to c^-} R(x) = \infty$; $\lim_{x\to c^+} R(x) = \infty$]. That is, the points on the graph of R are getting closer to the line $x = c$; $x = c$ is a vertical asymptote.

(d) As x approaches c, the values of $|R(x)| \to \infty$ [$\lim_{x\to c^-} R(x) = -\infty$; $\lim_{x\to c^+} R(x) = \infty$]. That is, the points on the graph of R are getting closer to the line $x = c$; $x = c$ is a vertical asymptote.

Figure 27

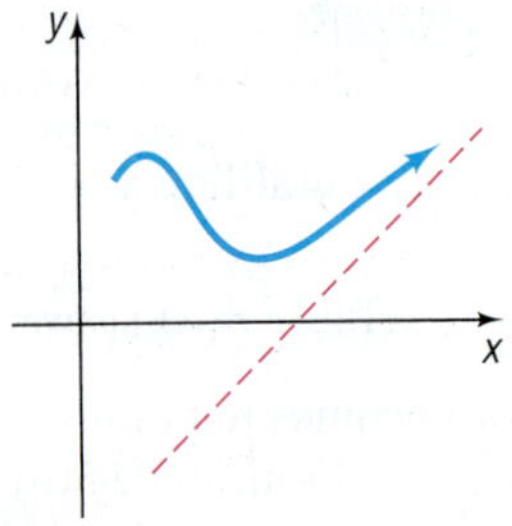

Figure 28 Oblique asymptote

A horizontal asymptote, when it occurs, describes the **end behavior** of the graph as $x \to \infty$ or as $x \to -\infty$. **The graph of a function may intersect a horizontal asymptote.**

A vertical asymptote, when it occurs, describes the behavior of the graph when x is close to some number c. **The graph of a rational function will never intersect a vertical asymptote.**

There is a third possibility. If, as $x \to -\infty$ or as $x \to \infty$, the value of a rational function $R(x)$ approaches a linear expression $ax + b$, $a \neq 0$, then the line $y = ax + b$, $a \neq 0$, is an **oblique (or slant) asymptote** of R. Figure 28 shows an oblique asymptote. An oblique asymptote, when it occurs, describes the end behavior of the graph. **The graph of a function may intersect an oblique asymptote.**

Now Work PROBLEMS 27 AND 35(c)

2 Find the Vertical Asymptotes of a Rational Function

The vertical asymptotes of a rational function $R(x) = \dfrac{p(x)}{q(x)}$, in lowest terms, are located at the real zeros of the denominator $q(x)$. Suppose that r is a real zero of q, so $x - r$ is a factor of q. As x approaches r, symbolized as $x \to r$, the values of $x - r$ approach 0, causing the ratio to become unbounded; that is, $|R(x)| \to \infty$. Based on the definition, we conclude that the line $x = r$ is a vertical asymptote.

THEOREM

Locating Vertical Asymptotes

A rational function $R(x) = \dfrac{p(x)}{q(x)}$, *in lowest terms*, will have a vertical asymptote $x = r$ if r is a real zero of the *denominator* q. That is, if $x - r$ is a factor of the denominator q of a rational function $R(x) = \dfrac{p(x)}{q(x)}$, in lowest terms, R will have the vertical asymptote $x = r$.

WARNING If a rational function is not in lowest terms, an application of this theorem may result in an incorrect listing of vertical asymptotes. ■

EXAMPLE 4 **Finding Vertical Asymptotes**

Find the vertical asymptotes, if any, of the graph of each rational function.

(a) $F(x) = \dfrac{x + 3}{x - 1}$ (b) $R(x) = \dfrac{x}{x^2 - 4}$

(c) $H(x) = \dfrac{x^2}{x^2 + 1}$ (d) $G(x) = \dfrac{x^2 - 9}{x^2 + 4x - 21}$

Solution

WARNING In Example 4(a), the vertical asymptote is $x = 1$. Do not say that the vertical asymptote is 1. ■

(a) F is in lowest terms, and the only zero of the denominator is 1. The line $x = 1$ is the vertical asymptote of the graph of F.

(b) R is in lowest terms, and the zeros of the denominator $x^2 - 4$ are -2 and 2. The lines $x = -2$ and $x = 2$ are the vertical asymptotes of the graph of R.

(c) H is in lowest terms, and the denominator has no real zeros because the equation $x^2 + 1 = 0$ has no real solutions. The graph of H has no vertical asymptotes.

(d) Factor the numerator and denominator of $G(x)$ to determine whether it is in lowest terms.

$$G(x) = \frac{x^2 - 9}{x^2 + 4x - 21} = \frac{(x + 3)(x - 3)}{(x + 7)(x - 3)} = \frac{x + 3}{x + 7} \qquad x \neq 3$$

The only zero of the denominator of $G(x)$ in lowest terms is -7. The line $x = -7$ is the only vertical asymptote of the graph of G. ●

As Example 4 points out, rational functions can have no vertical asymptotes, one vertical asymptote, or more than one vertical asymptote.

Multiplicity and Vertical Asymptotes

Recall from Figure 15 in Section 5.1 that the end behavior of a polynomial function is always one of four types. For polynomials of odd degree, the ends of the graph go in opposite directions (one up and one down), whereas for polynomials of even degree, the ends go in the same direction (both up or both down).

For a rational function in lowest terms, the multiplicities of the zeros in the denominator can be used in a similar fashion to determine the behavior of the graph around each vertical asymptote. Consider the following four functions, each with a single vertical asymptote, $x = 2$.

$$R_1(x) = \frac{1}{x - 2} \quad R_2(x) = -\frac{1}{x - 2} \quad R_3(x) = \frac{1}{(x - 2)^2} \quad R_4(x) = -\frac{1}{(x - 2)^2}$$

Figure 29 shows the graphs of each function. The graphs of R_1 and R_2 are transformations of the graph of $y = \frac{1}{x}$, and the graphs of R_3 and R_4 are transformations of the graph of $y = \frac{1}{x^2}$.

Based on Figure 29, we can make the following conclusions:

- If the multiplicity of the zero that gives rise to a vertical asymptote is odd, the graph approaches ∞ on one side of the vertical asymptote and approaches $-\infty$ on the other side.
- If the multiplicity of the zero that gives rise to the vertical asymptote is even, the graph approaches either ∞ or $-\infty$ on both sides of the vertical asymptote.

These results are true in general and will be helpful when graphing rational functions in the next section.

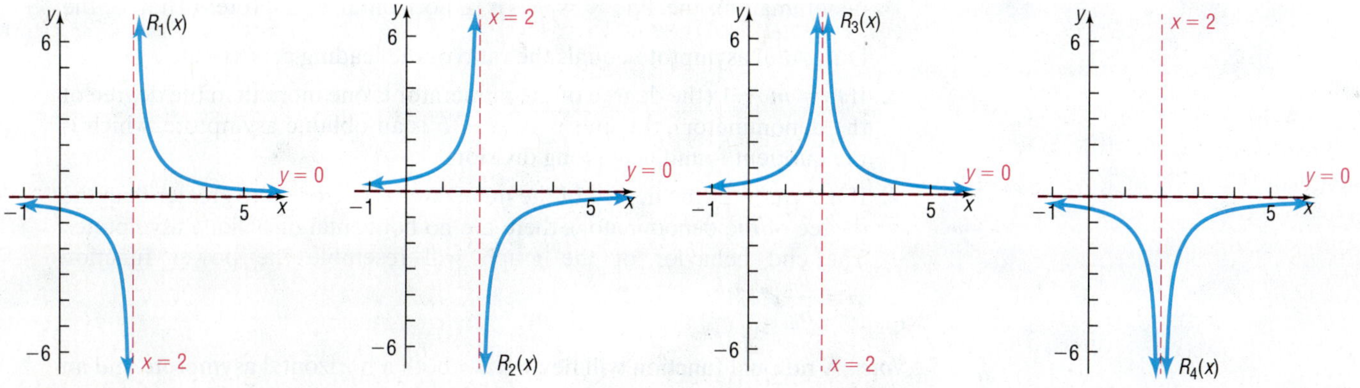

(a) Odd multiplicity $\lim_{x \to 2^-} R_1(x) = -\infty$ $\lim_{x \to 2^+} R_1(x) = \infty$

(b) Odd multiplicity $\lim_{x \to 2^-} R_2(x) = \infty$ $\lim_{x \to 2^+} R_2(x) = -\infty$

(c) Even multiplicity $\lim_{x \to 2^-} R_3(x) = \infty$ $\lim_{x \to 2^+} R_3(x) = \infty$

(d) Even multiplicity $\lim_{x \to 2^-} R_4(x) = -\infty$ $\lim_{x \to 2^+} R_4(x) = -\infty$

Figure 29

3 Find the Horizontal or Oblique Asymptote of a Rational Function

To find horizontal or oblique asymptotes, we need to know how the value of the function behaves as $x \to -\infty$ or as $x \to \infty$. That is, we need to determine the end behavior of the function. This can be done by examining the degrees of the numerator and denominator, and the respective power functions that each resembles. For example, consider the rational function

$$R(x) = \frac{3x - 2}{5x^2 - 7x + 1}$$

The degree of the numerator, 1, is less than the degree of the denominator, 2. When $|x|$ is very large, the numerator of R can be approximated by the power function $y = 3x$, and the denominator can be approximated by the power function $y = 5x^2$. This means

$$R(x) = \frac{3x - 2}{5x^2 - 7x + 1} \underset{\uparrow}{\approx} \frac{3x}{5x^2} = \frac{3}{5x} \underset{\uparrow}{\to} 0$$

For $|x|$ very large   As $x \to -\infty$ or $x \to \infty$

which shows that the line $y = 0$ is a horizontal asymptote. This result is true for all rational functions that are **proper** (that is, the degree of the numerator is less than the degree of the denominator). If a rational function is **improper** (that is, if the degree of the numerator is greater than or equal to the degree of the denominator), there could be a horizontal asymptote, an oblique asymptote, or neither. The following summary details how to find horizontal or oblique asymptotes.

Finding a Horizontal or Oblique Asymptote of a Rational Function

Consider the rational function

$$R(x) = \frac{p(x)}{q(x)} = \frac{a_n x^n + a_{n-1}x^{n-1} + \cdots + a_1 x + a_0}{b_m x^m + b_{m-1}x^{m-1} + \cdots + b_1 x + b_0}$$

in which the degree of the numerator is n and the degree of the denominator is m.

1. If $n < m$ (the degree of the numerator is less than the degree of the denominator), the line $y = 0$ is a horizontal asymptote.
2. If $n = m$ (the degree of the numerator equals the degree of the denominator), the line $y = \frac{a_n}{b_m}$ is a horizontal asymptote. (That is, the horizontal asymptote equals the ratio of the leading coefficients.)
3. If $n = m + 1$ (the degree of the numerator is one more than the degree of the denominator), the line $y = ax + b$ is an oblique asymptote, which is the quotient found using long division.
4. If $n \geq m + 2$ (the degree of the numerator is two or more greater than the degree of the denominator), there are no horizontal or oblique asymptotes. The end behavior of the graph will resemble the power function $y = \frac{a_n}{b_m}x^{n-m}$.

Note: A rational function will never have both a horizontal asymptote and an oblique asymptote. A rational function may have neither a horizontal nor an oblique asymptote.

We illustrate each of the possibilities in Examples 5 through 8.

EXAMPLE 5

Finding a Horizontal Asymptote

Find the horizontal asymptote, if one exists, of the graph of

$$R(x) = \frac{4x^3 - 5x + 2}{7x^5 + 2x^4 - 3x}$$

Solution Since the degree of the numerator, 3, is less than the degree of the denominator, 5, the rational function R is proper. The line $y = 0$ is a horizontal asymptote of the graph of R. ●

EXAMPLE 6

Finding a Horizontal or Oblique Asymptote

Find the horizontal or oblique asymptote, if one exists, of the graph of

$$H(x) = \frac{3x^4 - x^2}{x^3 - x^2 + 1}$$

Solution Since the degree of the numerator, 4, is exactly one greater than the degree of the denominator, 3, the rational function H has an oblique asymptote. Find the asymptote by using long division.

$$\begin{array}{r} 3x + 3 \\ x^3 - x^2 + 1 \overline{\big) 3x^4 - x^2 } \\ \underline{3x^4 - 3x^3 + 3x} \\ 3x^3 - x^2 - 3x \\ \underline{3x^3 - 3x^2 + 3} \\ 2x^2 - 3x - 3 \end{array}$$

As a result,

$$H(x) = \frac{3x^4 - x^2}{x^3 - x^2 + 1} = 3x + 3 + \frac{2x^2 - 3x - 3}{x^3 - x^2 + 1}$$

As $x \to -\infty$ or as $x \to \infty$,

$$\frac{2x^2 - 3x - 3}{x^3 - x^2 + 1} \approx \frac{2x^2}{x^3} = \frac{2}{x} \to 0$$

As $x \to -\infty$ or as $x \to \infty$, we have $H(x) \to 3x + 3$. The graph of the rational function H has an oblique asymptote $y = 3x + 3$. Put another way, as $x \to \pm\infty$, the graph of H will behave like the graph of $y = 3x + 3$. ●

EXAMPLE 7

Finding a Horizontal or Oblique Asymptote

Find the horizontal or oblique asymptote, if one exists, of the graph of

$$R(x) = \frac{8x^2 - x + 2}{4x^2 - 1}$$

Solution Since the degree of the numerator, 2, equals the degree of the denominator, 2, the rational function R has a horizontal asymptote equal to the ratio of the leading coefficients.

$$y = \frac{a_n}{b_m} = \frac{8}{4} = 2$$

To see why the horizontal asymptote equals the ratio of the leading coefficients, investigate the behavior of R as $x \to -\infty$ or as $x \to \infty$. When $|x|$ is very large, the numerator of R can be approximated by the power function $y = 8x^2$, and the denominator can be approximated by the power function $y = 4x^2$. This means that as $x \to -\infty$ or as $x \to \infty$,

$$R(x) = \frac{8x^2 - x + 2}{4x^2 - 1} \approx \frac{8x^2}{4x^2} = \frac{8}{4} = 2$$

The graph of the rational function R has a horizontal asymptote $y = 2$. The graph of R will behave like $y = 2$ as $x \to \pm\infty$.

EXAMPLE 8 Finding a Horizontal or Oblique Asymptote

Find the horizontal or oblique asymptote, if one exists, of the graph of

$$G(x) = \frac{2x^5 - x^3 + 2}{x^3 - 1}$$

Solution Since the degree of the numerator, 5, is greater than the degree of the denominator, 3, by more than one, the rational function G has no horizontal or oblique asymptote. The end behavior of the graph will resemble the power function $y = 2x^{5-3} = 2x^2$.

To see why this is the case, investigate the behavior of G as $x \to -\infty$ or as $x \to \infty$. When $|x|$ is very large, the numerator of G can be approximated by the power function $y = 2x^5$, and the denominator can be approximated by the power function $y = x^3$. This means as $x \to -\infty$ or as $x \to \infty$,

$$G(x) = \frac{2x^5 - x^3 + 2}{x^3 - 1} \approx \frac{2x^5}{x^3} = 2x^{5-3} = 2x^2$$

Since this is not linear, the graph of G has no horizontal or oblique asymptote. The graph of G will behave like $y = 2x^2$ as $x \to \pm\infty$.

Now Work PROBLEMS 45, 47, AND 49

5.2 Assess Your Understanding

'Are You Prepared?' *Answers are given at the end of these exercises. If you get a wrong answer, read the pages listed in red.*

1. ***True or False*** The quotient of two polynomial expressions is a rational expression. (pp. 62–69)
2. What are the quotient and remainder when $3x^4 - x^2$ is divided by $x^3 - x^2 + 1$. (pp. 44–47)
3. Graph $y = \frac{1}{x}$. (p. 164)
4. Graph $y = 2(x + 1)^2 - 3$ using transformations. (pp. 247–256)

Concepts and Vocabulary

5. ***True or False*** The domain of every rational function is the set of all real numbers.
6. If, as $x \to -\infty$ or as $x \to \infty$, the values of $R(x)$ approach some fixed number L, then the line $y = L$ is a ________ ________ of the graph of R.
7. If, as x approaches some number c, the values of $|R(x)| \to \infty$, then the line $x = c$ is a ________ ________ of the graph of R.
8. For a rational function R, if the degree of the numerator is less than the degree of the denominator, then R is ________.
9. ***True or False*** The graph of a rational function may intersect a horizontal asymptote.
10. ***True or False*** The graph of a rational function may intersect a vertical asymptote.
11. If a rational function is proper, then ________ is a horizontal asymptote.
12. ***True or False*** If the degree of the numerator of a rational function equals the degree of the denominator, then the ratio of the leading coefficients gives rise to the horizontal asymptote.
13. If $R(x) = \frac{p(x)}{q(x)}$ is a rational function and if p and q have no common factors, then R is ________.
    (a) improper (b) proper
    (c) undefined (d) in lowest terms
14. Which type of asymptote, when it occurs, describes the behavior of a graph when x is close to some number?
    (a) vertical (b) horizontal (c) oblique (d) all of these

Skill Building

In Problems 15–26, find the domain of each rational function.

15. $R(x) = \dfrac{4x}{x-3}$ **16.** $R(x) = \dfrac{5x^2}{3+x}$ **17.** $H(x) = \dfrac{-4x^2}{(x-2)(x+4)}$

18. $G(x) = \dfrac{6}{(x+3)(4-x)}$ **19.** $F(x) = \dfrac{3x(x-1)}{2x^2-5x-3}$ **20.** $Q(x) = \dfrac{-x(1-x)}{3x^2+5x-2}$

21. $R(x) = \dfrac{x}{x^3-8}$ **22.** $R(x) = \dfrac{x}{x^4-1}$ **23.** $H(x) = \dfrac{3x^2+x}{x^2+4}$

24. $G(x) = \dfrac{x-3}{x^4+1}$ **25.** $R(x) = \dfrac{3(x^2-x-6)}{4(x^2-9)}$ **26.** $F(x) = \dfrac{-2(x^2-4)}{3(x^2+4x+4)}$

In Problems 27–32, use the graph shown to find
(a) The domain and range of each function *(b) The intercepts, if any* *(c) Horizontal asymptotes, if any*
(d) Vertical asymptotes, if any *(e) Oblique asymptotes, if any*

27.

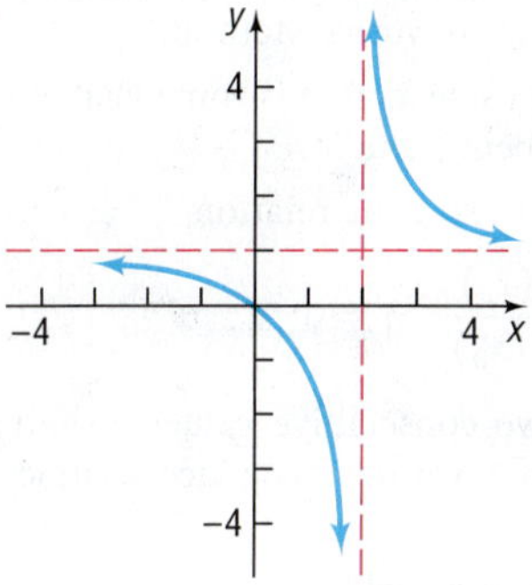

28.

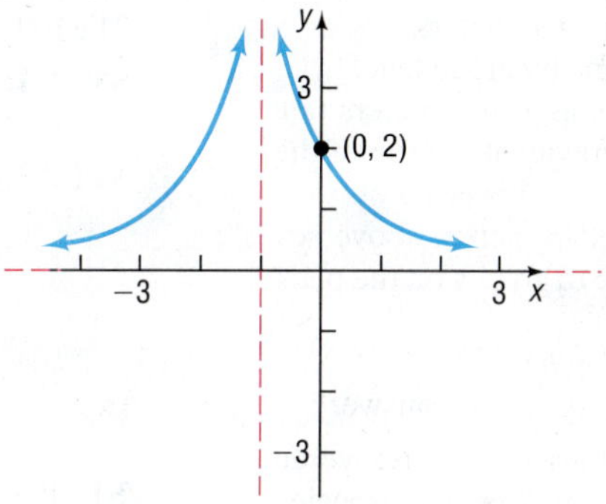

29.

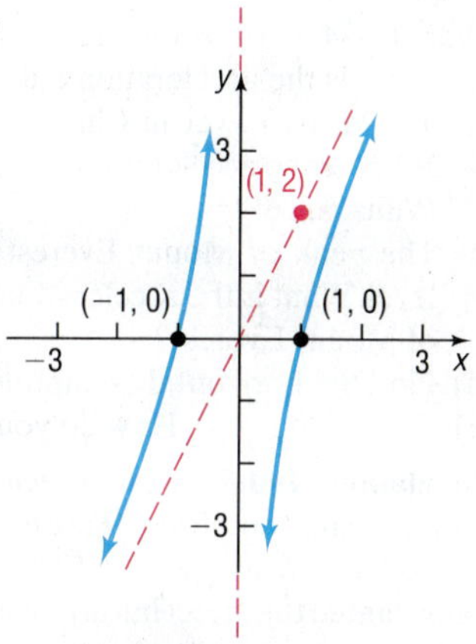

30.

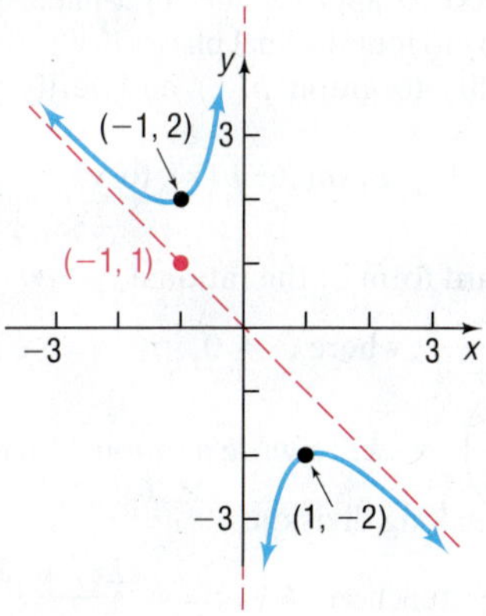

31.

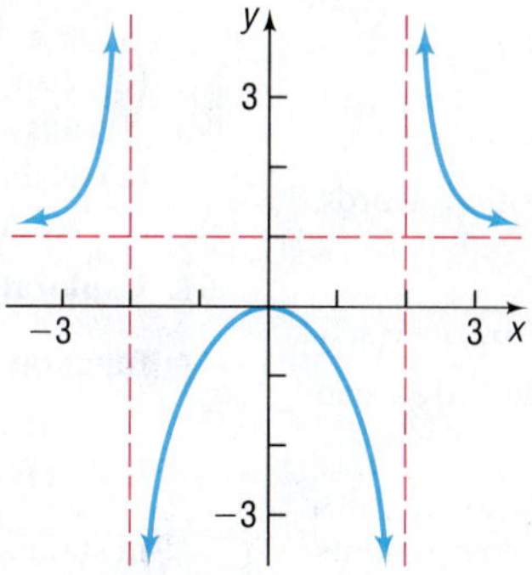

32.

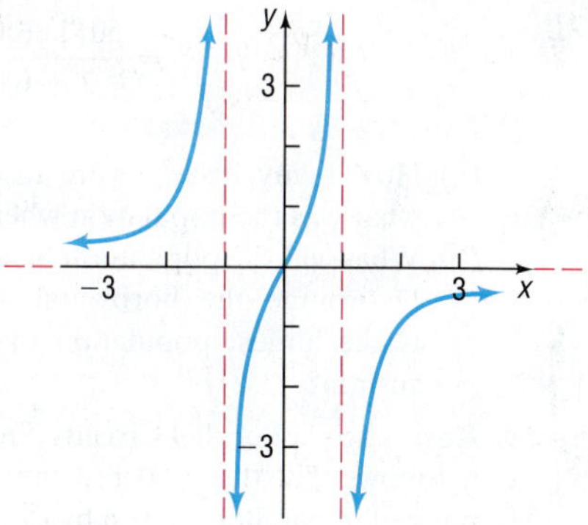

In Problems 33–44, (a) graph the rational function using transformations, (b) use the final graph to find the domain and range, and (c) use the final graph to list any vertical, horizontal, or oblique asymptotes.

33. $F(x) = 2 + \dfrac{1}{x}$ **34.** $Q(x) = 3 + \dfrac{1}{x^2}$ **35.** $R(x) = \dfrac{1}{(x-1)^2}$ **36.** $R(x) = \dfrac{3}{x}$

37. $H(x) = \dfrac{-2}{x+1}$ **38.** $G(x) = \dfrac{2}{(x+2)^2}$ **39.** $R(x) = \dfrac{-1}{x^2+4x+4}$ **40.** $R(x) = \dfrac{1}{x-1} + 1$

41. $G(x) = 1 + \dfrac{2}{(x-3)^2}$ **42.** $F(x) = 2 - \dfrac{1}{x+1}$ **43.** $R(x) = \dfrac{x^2-4}{x^2}$ **44.** $R(x) = \dfrac{x-4}{x}$

In Problems 45–56, find the vertical, horizontal, and oblique asymptotes, if any, of each rational function.

45. $R(x) = \dfrac{3x}{x + 4}$

46. $R(x) = \dfrac{3x + 5}{x - 6}$

47. $H(x) = \dfrac{x^3 - 8}{x^2 - 5x + 6}$

48. $G(x) = \dfrac{x^3 + 1}{x^2 - 5x - 14}$

49. $T(x) = \dfrac{x^3}{x^4 - 1}$

50. $P(x) = \dfrac{4x^2}{x^3 - 1}$

51. $Q(x) = \dfrac{2x^2 - 5x - 12}{3x^2 - 11x - 4}$

52. $F(x) = \dfrac{x^2 + 6x + 5}{2x^2 + 7x + 5}$

53. $R(x) = \dfrac{6x^2 + 7x - 5}{3x + 5}$

54. $R(x) = \dfrac{8x^2 + 26x - 7}{4x - 1}$

55. $G(x) = \dfrac{x^4 - 1}{x^2 - x}$

56. $F(x) = \dfrac{x^4 - 16}{x^2 - 2x}$

Applications and Extensions

57. Gravity In physics, it is established that the acceleration due to gravity, g (in meters/sec^2), at a height h meters above sea level is given by

$$g(h) = \frac{3.99 \times 10^{14}}{(6.374 \times 10^6 + h)^2}$$

where 6.374×10^6 is the radius of Earth in meters.

(a) What is the acceleration due to gravity at sea level?

(b) The Willis Tower in Chicago, Illinois, is 443 meters tall. What is the acceleration due to gravity at the top of the Willis Tower?

(c) The peak of Mount Everest is 8848 meters above sea level. What is the acceleration due to gravity on the peak of Mount Everest?

(d) Find the horizontal asymptote of $g(h)$.

(e) Solve $g(h) = 0$. How do you interpret your answer?

58. Population Model A rare species of insect was discovered in the Amazon Rain Forest. To protect the species, environmentalists declared the insect endangered and transplanted the insect into a protected area. The population P of the insect t months after being transplanted is

$$P(t) = \frac{50(1 + 0.5t)}{2 + 0.01t}$$

(a) How many insects were discovered? In other words, what was the population when $t = 0$?

(b) What will the population be after 5 years?

(c) Determine the horizontal asymptote of $P(t)$. What is the largest population that the protected area can sustain?

59. Resistance in Parallel Circuits From Ohm's Law for circuits, it follows that the total resistance R_{tot} of two components hooked in parallel is given by the equation

$$R_{tot} = \frac{R_1 R_2}{R_1 + R_2}$$

where R_1 and R_2 are the individual resistances.

(a) Let $R_1 = 10$ ohms, and graph R_{tot} as a function of R_2.

(b) Find and interpret any asymptotes of the graph obtained in part (a).

(c) If $R_2 = 2\sqrt{R_1}$, what value of R_1 will yield an R_{tot} of 17 ohms?

60. Newton's Method In calculus you will learn that if

$$p(x) = a_n x^n + a_{n-1}x^{n-1} + \cdots + a_1 x + a_0$$

is a polynomial function, then the *derivative* of $p(x)$ is

$$p'(x) = na_n x^{n-1} + (n - 1)a_{n-1}x^{n-2} + \cdots + 2a_2 x + a_1$$

Newton's Method is an efficient method for approximating the x-intercepts (or real zeros) of a function, such as $p(x)$. The following steps outline Newton's Method.

STEP 1: Select an initial value x_0 that is somewhat close to the x-intercept being sought.

STEP 2: Find values for x using the relation

$$x_{n+1} = x_n - \frac{p(x_n)}{p'(x_n)} \qquad n = 1, 2, \ldots$$

until you get two consecutive values x_n and x_{n+1} that agree to whatever decimal place accuracy you desire.

STEP 3: The approximate zero will be x_{n+1}.

Consider the polynomial $p(x) = x^3 - 7x - 40$.

(a) Evaluate $p(5)$ and $p(-3)$.

(b) What might we conclude about a zero of p? Explain.

(c) Use Newton's Method to approximate an x-intercept, r, $-3 < r < 5$, of $p(x)$ to four decimal places.

(d) Use a graphing utility to graph $p(x)$ and verify your answer in part (c).

(e) Using a graphing utility, evaluate $p(r)$ to verify your result.

61. Exploration The standard form of the rational function $R(x) = \dfrac{mx + b}{cx + d}$, where $c \neq 0$, is $R(x) = a\left(\dfrac{1}{x - h}\right) + k$. To write a rational function in standard form requires long division.

(a) Write the rational function $R(x) = \dfrac{2x + 3}{x - 1}$ in standard form by writing R in the form

$$\text{Quotient} + \frac{\text{remainder}}{\text{divisor}}$$

(b) Graph R using transformations.

(c) Determine the vertical asymptote and the horizontal asymptote of R.

62. Exploration Repeat Problem 61 for the rational function $R(x) = \dfrac{-6x + 16}{2x - 7}$.

Explaining Concepts: Discussion and Writing

63. If the graph of a rational function R has the vertical asymptote $x = 4$, the factor $x - 4$ must be present in the denominator of R. Explain why.

64. If the graph of a rational function R has the horizontal asymptote $y = 2$, the degree of the numerator of R equals the degree of the denominator of R. Explain why.

65. The graph of a rational function cannot have both a horizontal and an oblique asymptote? Explain why.

66. Make up a rational function that has $y = 2x + 1$ as an oblique asymptote. Explain the methodology that you used.

Retain Your Knowledge

Problems 67–70 are based on material learned earlier in the course. The purpose of these problems is to keep the material fresh in your mind so that you are better prepared for the final exam.

67. Find the equation of a vertical line passing through the point $(5, -3)$.

68. Solve: $\frac{2}{5}(3x - 7) + 1 = \frac{x}{4} - 2$

69. Determine whether the graph of the equation $2x^3 - xy^2 = 4$ is symmetric with respect to the x-axis, the y-axis, the origin, or none of these.

70. What are the points of intersection of the graphs of the functions $f(x) = -3x + 2$ and $g(x) = x^2 - 2x - 4$?

'Are You Prepared?' Answers

1. True

2. Quotient: $3x + 3$; remainder: $2x^2 - 3x - 3$

3.

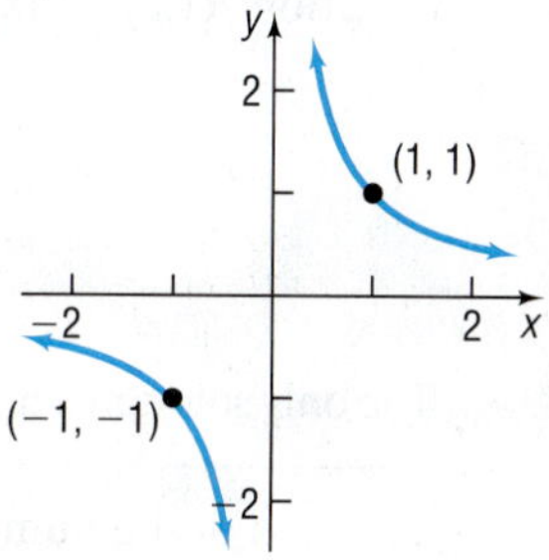

4.

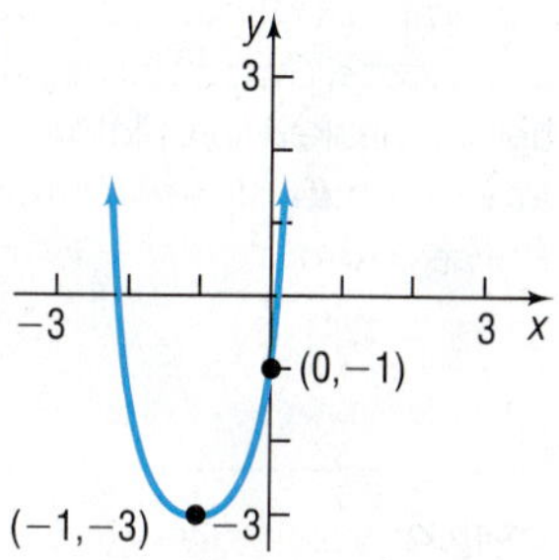

5.3 The Graph of a Rational Function

PREPARING FOR THIS SECTION *Before getting started, review the following:*

- Intercepts Section 2.2, pp. 159–160

Now Work the **'Are You Prepared?'** problem on page 365.

OBJECTIVES **1** Analyze the Graph of a Rational Function (p. 353)
2 Solve Applied Problems Involving Rational Functions (p. 364)

1 Analyze the Graph of a Rational Function

We commented earlier that calculus provides the tools required to graph a polynomial function accurately. The same holds true for rational functions. However, we can gather together quite a bit of information about their graphs to get an idea of the general shape and position of the graph.

EXAMPLE 1 How to Analyze the Graph of a Rational Function

Analyze the graph of the rational function: $R(x) = \frac{x - 1}{x^2 - 4}$

Step-by-Step Solution

Step 1: Factor the numerator and denominator of R. Find the domain of the rational function.

$$R(x) = \frac{x - 1}{x^2 - 4} = \frac{x - 1}{(x + 2)(x - 2)}$$

The domain of R is $\{x \mid x \neq -2, x \neq 2\}$.

Step 2: Write R in lowest terms.

Because there are no common factors between the numerator and denominator, R is in lowest terms.

Step 3: Find and plot the intercepts of the graph. Use multiplicity to determine the behavior of the graph of R at each x-intercept.

Since 0 is in the domain of R, the y-intercept is $R(0) = \frac{1}{4}$. Plot the point $\left(0, \frac{1}{4}\right)$. The x-intercepts are found by determining the real zeros of the numerator of R written in lowest terms. By solving $x - 1 = 0$, we find that the only real zero of the numerator is 1, so the only x-intercept of the graph of R is 1. Plot the point $(1, 0)$. The multiplicity of 1 is odd, so the graph will cross the x-axis at $x = 1$.

Step 4: Find the vertical asymptotes. Graph each vertical asymptote using a dashed line. Determine the behavior of the graph on either side of each vertical asymptote.

The vertical asymptotes are the zeros of the denominator with the rational function in lowest terms. With R written in lowest terms, we find that the graph of R has two vertical asymptotes: the lines $x = -2$ and $x = 2$. See Figure 30(a). The multiplicities of the zeros that give rise to the vertical asymptotes are both odd. Therefore, the graph will approach ∞ on one side of each vertical asymptote, and will approach $-\infty$ on the other side.

Step 5: Find the horizontal or oblique asymptote, if one exists. Find points, if any, at which the graph of R intersects this asymptote. Graph the asymptote using a dashed line. Plot any points at which the graph of R intersects the asymptote.

Because the degree of the numerator is less than the degree of the denominator, R is proper and the line $y = 0$ (the x-axis) is a horizontal asymptote of the graph. To determine whether the graph of R intersects the horizontal asymptote, solve the equation $R(x) = 0$:

$$\frac{x-1}{x^2-4} = 0$$

$$x - 1 = 0$$

$$x = 1$$

The only solution is $x = 1$, so the graph of R intersects the horizontal asymptote at $(1, 0)$.

Step 6: Use the zeros of the numerator and denominator of R to divide the x-axis into intervals. Determine where the graph of R is above or below the x-axis by choosing a number in each interval and evaluating R there. Plot the points found.

The zero of the numerator, 1, and the zeros of the denominator, -2 and 2, divide the x-axis into four intervals:

$$(-\infty, -2) \qquad (-2, 1) \qquad (1, 2) \qquad (2, \infty)$$

Now construct Table 9.

Table 9

Number line: -2, 1, 2, x

Interval	$(-\infty, -2)$	$(-2, 1)$	$(1, 2)$	$(2, \infty)$
Number chosen	-3	-1	$\frac{3}{2}$	3
Value of R	$R(-3) = -0.8$	$R(-1) = \frac{2}{3}$	$R\left(\frac{3}{2}\right) = -\frac{2}{7}$	$R(3) = 0.4$
Location of graph	Below x-axis	Above x-axis	Below x-axis	Above x-axis
Point on graph	$(-3, -0.8)$	$\left(-1, \frac{2}{3}\right)$	$\left(\frac{3}{2}, -\frac{2}{7}\right)$	$(3, 0.4)$

Figure 30(a) shows the asymptotes, the points from Table 9, the y-intercept, and the x-intercept.

Step 7: Use the results obtained in Steps 1 through 6 to graph R.

- The graph crosses the x-axis at $x = 1$, changing from being above the x-axis for $x < 1$ to below it for $x > 1$. Indicate this on the graph. See Figure 30(b).
- Since $y = 0$ (the x-axis) is a horizontal asymptote and the graph lies below the x-axis for $x < -2$, we can sketch a portion of the graph by placing a small arrow to the far left and under the x-axis.

- Since the line $x = -2$ is a vertical asymptote and the graph lies below the x-axis for $x < -2$, we place an arrow well below the x-axis and approaching the line $x = -2$ from the left $\left(\lim_{x \to -2^-} R(x) = -\infty\right)$.
- Since the graph approaches $-\infty$ on one side of $x = -2$, and -2 is a zero of odd multiplicity, the graph will approach ∞ on the other side of $x = -2$. That is, $\lim_{x \to -2^+} R(x) = \infty$. Similar analysis leads to $\lim_{x \to 2^-} R(x) = -\infty$ and $\lim_{x \to 2^+} R(x) = \infty$. Finally, $\lim_{x \to -\infty} R(x) = 0$ and $\lim_{x \to \infty} R(x) = 0$.

Figure 30(b) illustrates these conclusions and Figure 30(c) shows the graph of R.

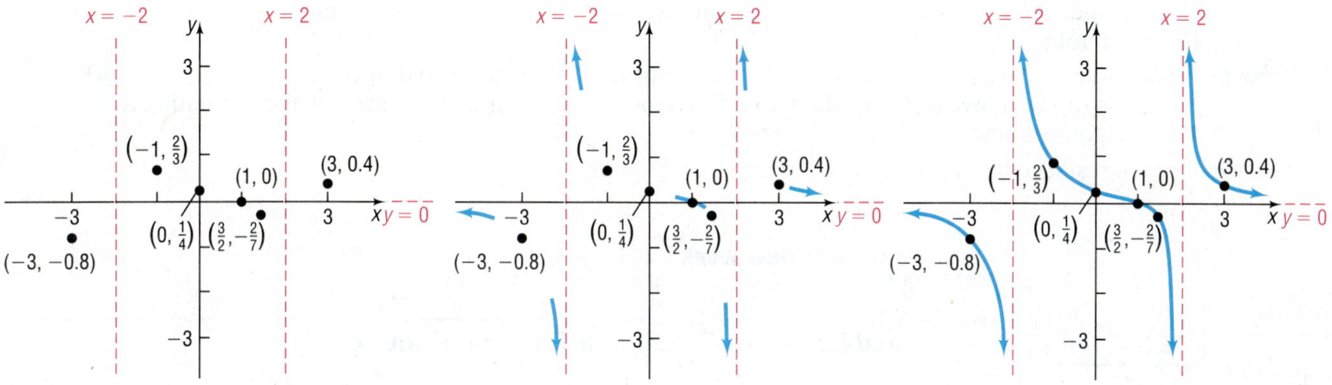

Figure 30 (a) (b) (c)

Exploration

Graph the rational function: $R(x) = \dfrac{x - 1}{x^2 - 4}$

Result The analysis just completed in Example 1 helps us to set the viewing rectangle to obtain a complete graph. Figure 31(a) shows the graph of $R(x) = \dfrac{x-1}{x^2-4}$ in connected mode, and Figure 31(b) shows it in dot mode. Notice in Figure 31(a) that the graph has vertical lines at $x = -2$ and $x = 2$. This is due to the fact that when a graphing utility is in connected mode, some will connect the dots between consecutive pixels, and vertical lines may occur. We know that the graph of R does not cross the lines $x = -2$ and $x = 2$, since R is not defined at $x = -2$ or $x = 2$. So, when graphing rational functions, use dot mode if extraneous vertical lines are present in connected mode. Newer graphing utilities may not have extraneous vertical lines in connected mode. See Figure 31(c).

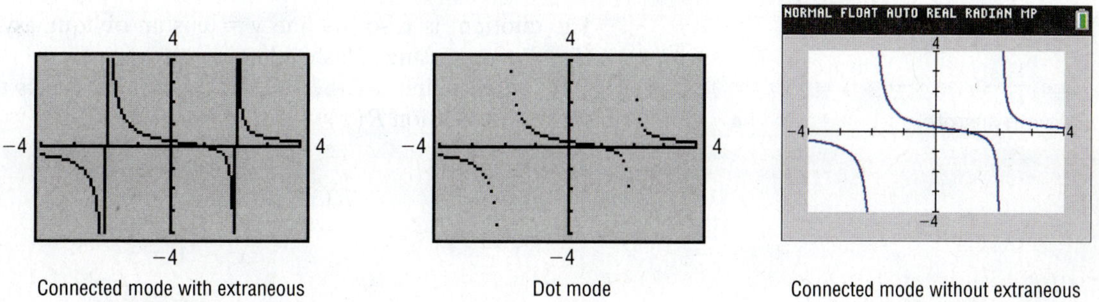

Figure 31

(a) Connected mode with extraneous vertical lines

(b) Dot mode

(c) Connected mode without extraneous vertical lines

SUMMARY

Analyzing the Graph of a Rational Function R

STEP 1: Factor the numerator and denominator of R. Find the domain of the rational function.

STEP 2: Write R in lowest terms.

STEP 3: Find and plot the intercepts of the graph. Use multiplicity to determine the behavior of the graph of R at each x-intercept.

STEP 4: Find the vertical asymptotes. Graph each vertical asymptote using a dashed line. Determine the behavior of the graph of R on either side of each vertical asymptote.

STEP 5: Find the horizontal or oblique asymptote, if one exists. Find points, if any, at which the graph of R intersects this asymptote. Graph the asymptote using a dashed line. Plot any points at which the graph of R intersects the asymptote.

STEP 6: Use the zeros of the numerator and denominator of R to divide the x-axis into intervals. Determine where the graph of R is above or below the x-axis by choosing a number in each interval and evaluating R there. Plot the points found.

STEP 7: Use the results obtained in Steps 1 through 6 to graph R.

Now Work PROBLEM 7

EXAMPLE 2 Analyzing the Graph of a Rational Function

Analyze the graph of the rational function: $R(x) = \dfrac{x^2 - 1}{x}$

Solution

STEP 1: $R(x) = \dfrac{(x + 1)(x - 1)}{x}$. The domain of R is $\{x \mid x \neq 0\}$.

STEP 2: R is in lowest terms.

STEP 3: Because x cannot equal 0, there is no y-intercept. The graph has two x-intercepts, -1 and 1, each with odd multiplicity. Plot the points $(-1, 0)$ and $(1, 0)$. The graph will cross the x-axis at both points.

STEP 4: The real zero of the denominator with R in lowest terms is 0, so the graph of R has the line $x = 0$ (the y-axis) as a vertical asymptote. Graph $x = 0$ using a dashed line. The multiplicity of 0 is odd, so the graph will approach ∞ on one side of the asymptote $x = 0$, and $-\infty$ on the other side.

STEP 5: Since the degree of the numerator, 2, is one greater than the degree of the denominator, 1, the rational function will have an oblique asymptote. To find the oblique asymptote, use long division.

$$\begin{array}{r} x \\ x\overline{)x^2 - 1} \\ \underline{x^2 } \\ -1 \end{array}$$

NOTE Because the denominator of the rational function is a monomial, we can also find the oblique asymptote as follows:

$$\frac{x^2 - 1}{x} = \frac{x^2}{x} - \frac{1}{x} = x - \frac{1}{x}$$

Since $\frac{1}{x} \to 0$ as $x \to \infty$, $y = x$ is the oblique asymptote. ■

The quotient is x, so the line $y = x$ is an oblique asymptote of the graph. Graph $y = x$ using a dashed line.

To determine whether the graph of R intersects the asymptote $y = x$, solve the equation $R(x) = x$.

$$R(x) = \frac{x^2 - 1}{x} = x$$

$$x^2 - 1 = x^2$$

$$-1 = 0 \quad \text{Impossible}$$

The equation $\dfrac{x^2 - 1}{x} = x$ has no solution, so the graph of R does not intersect the line $y = x$.

STEP 6: The zeros of the numerator are -1 and 1; the zero of the denominator is 0. Use these values to divide the x-axis into four intervals:

$$(-\infty, -1) \quad (-1, 0) \quad (0, 1) \quad (1, \infty)$$

Now construct Table 10. Plot the points from Table 10. You should now have Figure 32(a).

Table 10

	-1	0	1	x
Interval	$(-\infty, -1)$	$(-1, 0)$	$(0, 1)$	$(1, \infty)$
Number chosen	-2	$-\frac{1}{2}$	$\frac{1}{2}$	2
Value of R	$R(-2) = -\frac{3}{2}$	$R\left(-\frac{1}{2}\right) = \frac{3}{2}$	$R\left(\frac{1}{2}\right) = -\frac{3}{2}$	$R(2) = \frac{3}{2}$
Location of graph	Below x-axis	Above x-axis	Below x-axis	Above x-axis
Point on graph	$\left(-2, -\frac{3}{2}\right)$	$\left(-\frac{1}{2}, \frac{3}{2}\right)$	$\left(\frac{1}{2}, -\frac{3}{2}\right)$	$\left(2, \frac{3}{2}\right)$

STEP 7: The graph crosses the x-axis at $x = -1$ and $x = 1$, changing from being below the x-axis to being above it in both cases.

Since the graph of R is below the x-axis for $x < -1$ and is above the x-axis for $x > 1$, and since the graph of R does not intersect the oblique asymptote $y = x$, the graph of R will approach the line $y = x$ as shown in Figure 32(b).

NOTE Notice that R in Example 2 is an odd function. Do you see the symmetry about the origin in the graph of R in Figure 32(c)? ■

Since the graph of R is above the x-axis for $-1 < x < 0$, the graph of R will approach the vertical asymptote $x = 0$ at the top to the left of $x = 0$ $[\lim_{x \to 0^-} R(x) = \infty]$; since the graph of R approaches ∞ on one side of the asymptote and $-\infty$ on the other, the graph of R will approach the vertical asymptote $x = 0$ at the bottom to the right of $x = 0$ $[\lim_{x \to 0^+} R(x) = -\infty]$. See Figure 32(b).

The complete graph is given in Figure 32(c).

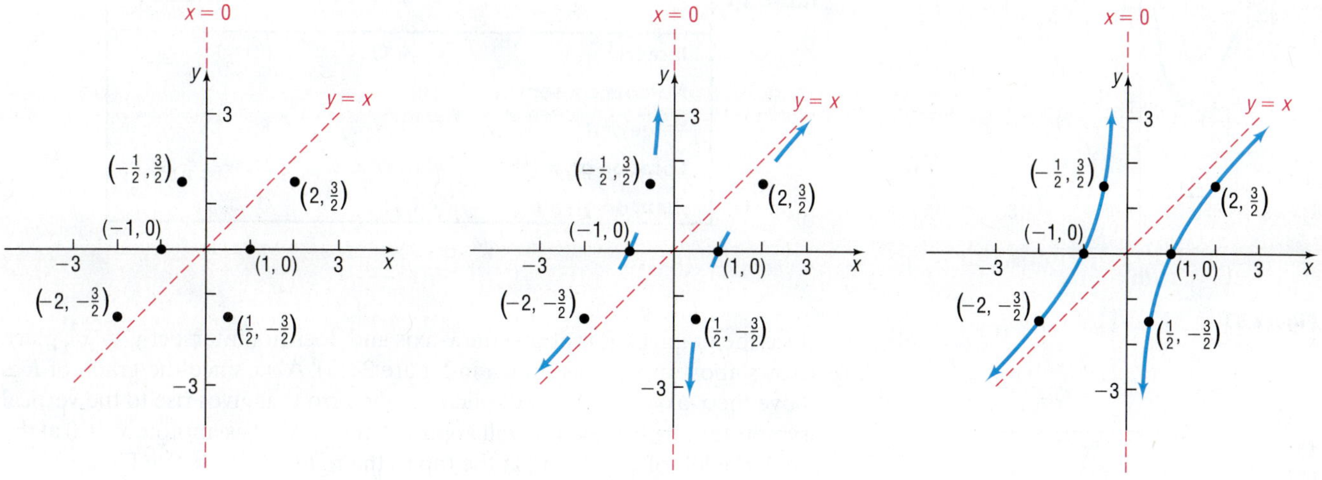

Figure 32 (a) (b) (c)

Seeing the Concept

Graph $R(x) = \dfrac{x^2 - 1}{x}$ and compare what you see with Figure 32(c). Could you have predicted from the graph that $y = x$ is an oblique asymptote? Graph $y = x$ and ZOOM-OUT. What do you observe?

Now Work PROBLEM 15

EXAMPLE 3 Analyzing the Graph of a Rational Function

Analyze the graph of the rational function: $R(x) = \dfrac{x^4 + 1}{x^2}$

Solution

Step 1: R is completely factored. The domain of R is $\{x \mid x \neq 0\}$.

Step 2: R is in lowest terms.

Step 3: There is no y-intercept. Since $x^4 + 1 = 0$ has no real solutions, there are no x-intercepts.

Step 4: R is in lowest terms, so $x = 0$ (the y-axis) is a vertical asymptote of R. Graph the line $x = 0$ using dashes. The multiplicity of 0 is even, so the graph will approach either ∞ or $-\infty$ on both sides of the asymptote.

Step 5: Since the degree of the numerator, 4, is two more than the degree of the denominator, 2, the rational function will not have a horizontal or oblique asymptote. Find the end behavior of R. As $|x| \to \infty$,

$$R(x) = \frac{x^4 + 1}{x^2} \approx \frac{x^4}{x^2} = x^2$$

The graph of R will approach the graph of $y = x^2$ as $x \to -\infty$ and as $x \to \infty$. The graph of R does not intersect $y = x^2$. Do you know why? Graph $y = x^2$ using dashes.

Step 6: The numerator has no real zeros, and the denominator has one real zero at 0. Divide the x-axis into the two intervals

$$(-\infty, 0) \qquad (0, \infty)$$

and construct Table 11.

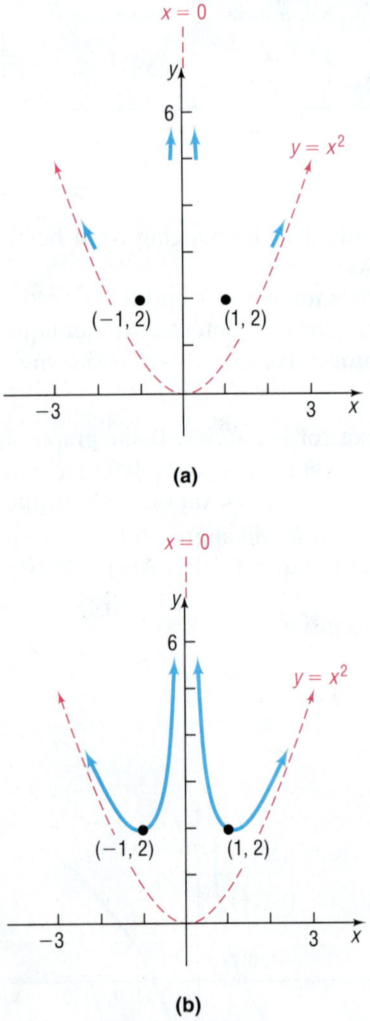

Figure 33

Table 11

	0 → x	
Interval	$(-\infty, 0)$	$(0, \infty)$
Number chosen	-1	1
Value of R	$R(-1) = 2$	$R(1) = 2$
Location of graph	Above x-axis	Above x-axis
Point on graph	$(-1, 2)$	$(1, 2)$

Step 7: Since the graph of R is above the x-axis and does not intersect $y = x^2$, place arrows above $y = x^2$ as shown in Figure 33(a). Also, since the graph of R is above the x-axis, and the multiplicity of the zero that gives rise to the vertical asymptote, $x = 0$, is even, it will approach the vertical asymptote $x = 0$ at the top to the left of $x = 0$ and at the top to the right of $x = 0$. See Figure 33(a).

Figure 33(b) shows the complete graph.

NOTE Notice that R in Example 3 is an even function. Do you see the symmetry about the y-axis in the graph of R?

Seeing the Concept

Graph $R(x) = \dfrac{x^4 + 1}{x^2}$ and compare what you see with Figure 33(b). Use MINIMUM to find the two turning points. Enter $Y_2 = x^2$ and ZOOM-OUT. What do you see?

Now Work PROBLEM 13

EXAMPLE 4 Analyzing the Graph of a Rational Function

Analyze the graph of the rational function: $R(x) = \dfrac{3x^2 - 3x}{x^2 + x - 12}$

Solution **STEP 1:** Factor R to get

$$R(x) = \frac{3x(x - 1)}{(x + 4)(x - 3)}$$

The domain of R is $\{x \mid x \neq -4, x \neq 3\}$.

STEP 2: R is in lowest terms.

STEP 3: The y-intercept is $R(0) = 0$. Plot the point $(0, 0)$. Since the real solutions of the equation $3x(x - 1) = 0$ are $x = 0$ and $x = 1$, the graph has two x-intercepts, 0 and 1, each with odd multiplicity. Plot the points $(0, 0)$ and $(1, 0)$; the graph will cross the x-axis at both points.

STEP 4: R is in lowest terms. The real solutions of the equation $(x + 4)(x - 3) = 0$ are $x = -4$ and $x = 3$, so the graph of R has two vertical asymptotes, the lines $x = -4$ and $x = 3$. Graph these lines using dashes. The multiplicities that give rise to the vertical asymptotes are both odd, so the graph will approach ∞ on one side of each vertical asymptote and $-\infty$ on the other side.

STEP 5: Since the degree of the numerator equals the degree of the denominator, the graph has a horizontal asymptote. To find it, form the quotient of the leading coefficient of the numerator, 3, and the leading coefficient of the denominator, 1. The graph of R has the horizontal asymptote $y = 3$.

To find out whether the graph of R intersects the asymptote, solve the equation $R(x) = 3$.

$$\begin{aligned} R(x) = \frac{3x^2 - 3x}{x^2 + x - 12} &= 3 \\ 3x^2 - 3x &= 3x^2 + 3x - 36 \\ -6x &= -36 \\ x &= 6 \end{aligned}$$

The graph intersects the line $y = 3$ at $x = 6$, and $(6, 3)$ is a point on the graph of R. Plot the point $(6, 3)$ and graph the line $y = 3$ using dashes.

STEP 6: The real zeros of the numerator, 0 and 1, and the real zeros of the denominator, -4 and 3, divide the x-axis into five intervals:

$$(-\infty, -4) \quad (-4, 0) \quad (0, 1) \quad (1, 3) \quad (3, \infty)$$

Construct Table 12. See page 360. Plot the points from Table 12. Figure 34(a) shows the graph so far.

STEP 7: Since the graph of R is above the x-axis for $x < -4$ and only crosses the line $y = 3$ at $(6, 3)$, as x approaches $-\infty$ the graph of R will approach the horizontal asymptote $y = 3$ from above $\left(\lim_{x \to -\infty} R(x) = 3\right)$. The graph of R will approach the vertical asymptote $x = -4$ at the top to the left of $x = -4$ $\left(\lim_{x \to -4^-} R(x) = \infty\right)$ and at the bottom to the right of $x = -4$ $\left(\lim_{x \to -4^+} R(x) = -\infty\right)$. The graph of R will approach the vertical asymptote

Table 12

Interval	$(-\infty, -4)$	$(-4, 0)$	$(0, 1)$	$(1, 3)$	$(3, \infty)$
Number chosen	-5	-2	$\frac{1}{2}$	2	4
Value of *R*	$R(-5) = 11.25$	$R(-2) = -1.8$	$R\left(\frac{1}{2}\right) = \frac{1}{15}$	$R(2) = -1$	$R(4) = 4.5$
Location of graph	Above x-axis	Below x-axis	Above x-axis	Below x-axis	Above x-axis
Point on graph	$(-5, 11.25)$	$(-2, -1.8)$	$\left(\frac{1}{2}, \frac{1}{15}\right)$	$(2, -1)$	$(4, 4.5)$

$x = 3$ at the bottom to the left of $x = 3$ $\left(\lim_{x\to 3^-} R(x) = -\infty\right)$ and at the top to the right of $x = 3$ $\left(\lim_{x\to 3^+} R(x) = \infty\right)$.

We do not know whether the graph of R crosses or touches the line $y = 3$ at $(6, 3)$. To see whether the graph, in fact, crosses or touches the line $y = 3$, plot an additional point to the right of $(6, 3)$. We use $x = 7$ to find $R(7) = \frac{63}{22} < 3$. The graph crosses $y = 3$ at $x = 6$. Because $(6, 3)$ is the only point where the graph of R intersects the asymptote $y = 3$, the graph must approach the line $y = 3$ from below as $x \to \infty$ $\left(\lim_{x\to\infty} R(x) = 3\right)$.

The graph crosses the x-axis at $x = 0$, changing from being below the x-axis to being above. The graph also crosses the x-axis at $x = 1$, changing from being above the x-axis to being below. See Figure 34(b). The complete graph is shown in Figure 34(c).

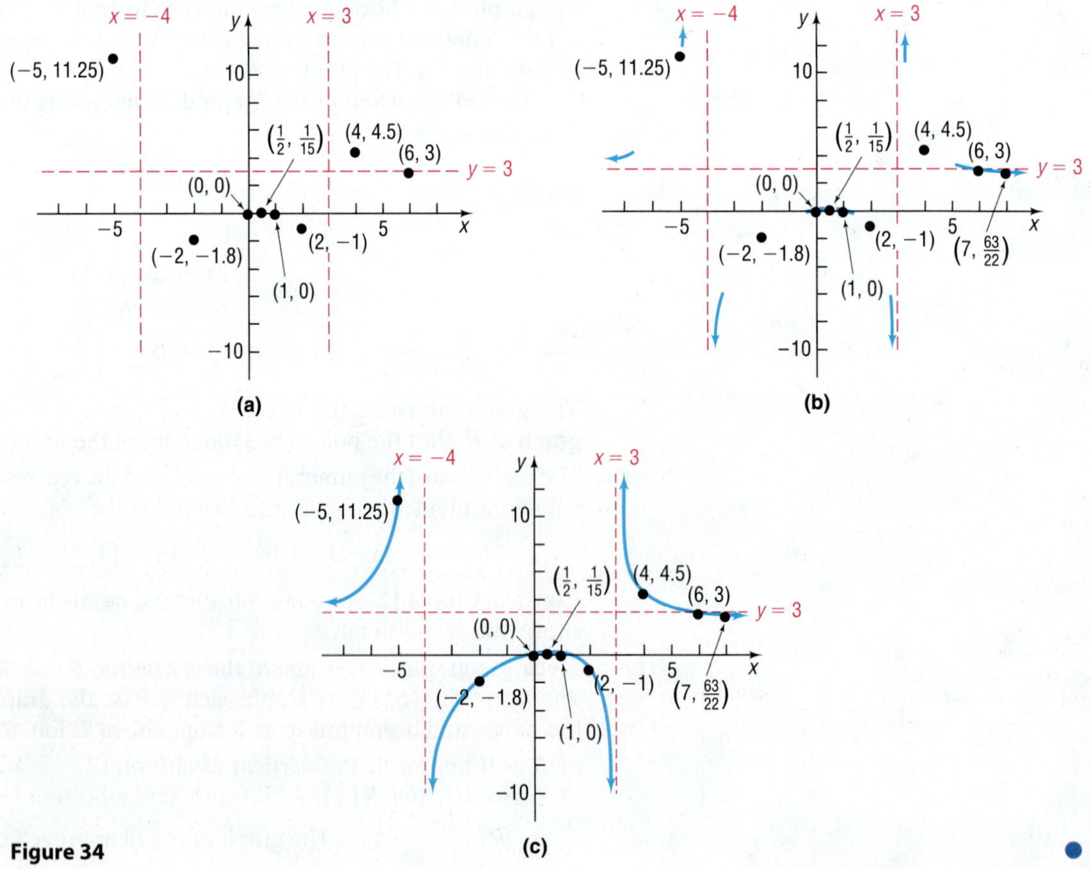

Figure 34

Exploration

Graph the rational function: $R(x) = \dfrac{3x^2 - 3x}{x^2 + x - 12}$

Result Figure 35(a) shows the graph. The graph does not clearly display the behavior of the function between the two x-intercepts, 0 and 1. Nor does it clearly display the fact that the graph crosses the horizontal asymptote at (6, 3). To see these parts better, graph R for $-1 \le x \le 2$ [Figure 35(b)] and for $4 \le x \le 60$ [Figure 35(c)].

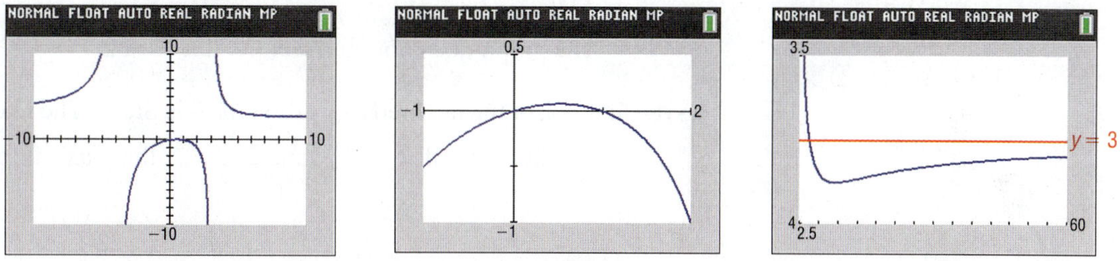

Figure 35 (a) (b) (c)

The new graphs reflect the behavior produced by the analysis. Furthermore, we observe two turning points, one between 0 and 1 and the other to the right of 6. Rounded to two decimal places, these turning points are (0.52, 0.07) and (11.48, 2.75).

Now Work PROBLEM 31

EXAMPLE 5 Analyzing the Graph of a Rational Function with a Hole

Analyze the graph of the rational function: $R(x) = \dfrac{2x^2 - 5x + 2}{x^2 - 4}$

Solution

STEP 1: Factor R and obtain

$$R(x) = \frac{(2x - 1)(x - 2)}{(x + 2)(x - 2)}$$

The domain of R is $\{x \mid x \neq -2, x \neq 2\}$.

STEP 2: In lowest terms,

$$R(x) = \frac{2x - 1}{x + 2} \qquad x \neq -2, x \neq 2$$

STEP 3: The y-intercept is $R(0) = -\frac{1}{2}$. Plot the point $\left(0, -\frac{1}{2}\right)$. The graph has one x-intercept, $\frac{1}{2}$, with odd multiplicity. Plot the point $\left(\frac{1}{2}, 0\right)$. The graph will cross the x-axis at $x = \frac{1}{2}$. See Figure 36(a) on page 363.

STEP 4: Since $x + 2$ is the only factor of the denominator of $R(x)$ *in lowest terms*, the graph has one vertical asymptote, $x = -2$. However, the rational function is undefined at both $x = 2$ and $x = -2$. Graph the line $x = -2$ using dashes. The multiplicity of -2 is odd, so the graph will approach ∞ on one side of the vertical asymptote and $-\infty$ on the other side.

STEP 5: Since the degree of the numerator equals the degree of the denominator, the graph has a horizontal asymptote. To find it, form the quotient of the leading coefficient of the numerator, 2, and the leading coefficient of the denominator, 1. The graph of R has the horizontal asymptote $y = 2$. Graph the line $y = 2$ using dashes.

To find out whether the graph of R intersects the horizontal asymptote $y = 2$, solve the equation $R(x) = 2$.

$$\begin{aligned} R(x) = \frac{2x-1}{x+2} &= 2 \\ 2x - 1 &= 2(x+2) \\ 2x - 1 &= 2x + 4 \\ -1 &= 4 \quad \text{Impossible} \end{aligned}$$

The graph does not intersect the line $y = 2$.

STEP 6: Look at the factored expression for R in Step 1. The real zeros of the numerator and denominator, -2, $\frac{1}{2}$, and 2, divide the x-axis into four intervals:

$$(-\infty, -2) \quad \left(-2, \frac{1}{2}\right) \quad \left(\frac{1}{2}, 2\right) \quad (2, \infty)$$

Construct Table 13. Plot the points in Table 13.

Table 13

	-2	$1/2$	2	x
Interval	$(-\infty, -2)$	$\left(-2, \frac{1}{2}\right)$	$\left(\frac{1}{2}, 2\right)$	$(2, \infty)$
Number chosen	-3	-1	1	3
Value of R	$R(-3) = 7$	$R(-1) = -3$	$R(1) = \frac{1}{3}$	$R(3) = 1$
Location of graph	Above x-axis	Below x-axis	Above x-axis	Above x-axis
Point on graph	$(-3, 7)$	$(-1, -3)$	$\left(1, \frac{1}{3}\right)$	$(3, 1)$

STEP 7: From Table 13 we know that the graph of R is above the x-axis for $x < -2$. From Step 5 we know that the graph of R does not intersect the asymptote $y = 2$. Therefore, the graph of R will approach $y = 2$ from above as $x \to -\infty$ and will approach the vertical asymptote $x = -2$ at the top from the left.

Since the graph of R is below the x-axis for $-2 < x < \frac{1}{2}$, the graph will approach $x = -2$ at the bottom from the right. Finally, since the graph of R is above the x-axis for $x > \frac{1}{2}$ and does not intersect the horizontal asymptote $y = 2$, the graph of R will approach $y = 2$ from below as $x \to \infty$. The graph crosses the x-axis at $x = \frac{1}{2}$, changing from being below the x-axis to being above. See Figure 36(a).

See Figure 36(b) for the complete graph. Since R is not defined at 2, there is a hole at the point $\left(2, \frac{3}{4}\right)$.

NOTE The coordinates of the hole were obtained by evaluating R in lowest terms at $x = 2$. R in lowest terms is $\frac{2x-1}{x+2}$, which, at $x = 2$, is $\frac{2(2)-1}{2+2} = \frac{3}{4}$. ■

Exploration

Graph $R(x) = \frac{2x^2 - 5x + 2}{x^2 - 4}$. Do you see the hole at $\left(2, \frac{3}{4}\right)$? TRACE along the graph. Did you obtain an ERROR at $x = 2$? Are you convinced that an algebraic analysis of a rational function is required in order to accurately interpret the graph obtained with a graphing utility?

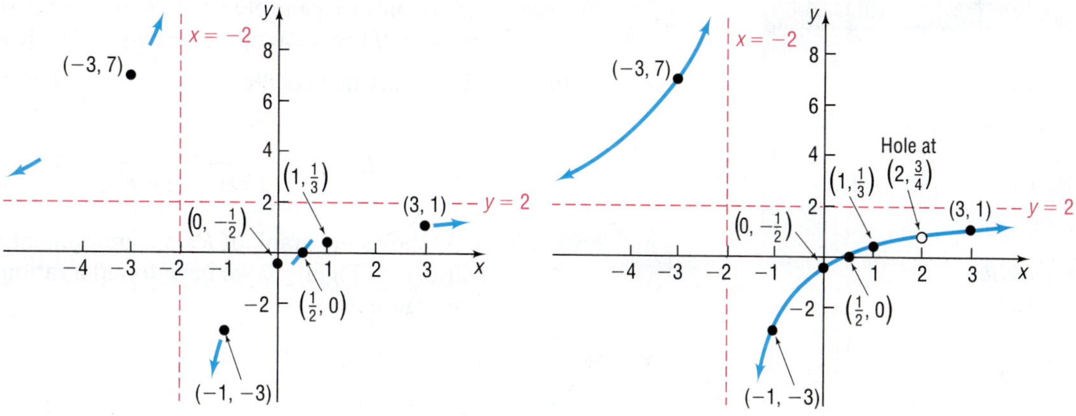

Figure 36 (a) (b)

As Example 5 shows, **the zeros of the denominator of a rational function give rise to either vertical asymptotes or holes on the graph.**

Now Work PROBLEM 33

EXAMPLE 6 Constructing a Rational Function from Its Graph

Find a rational function that might have the graph shown in Figure 37.

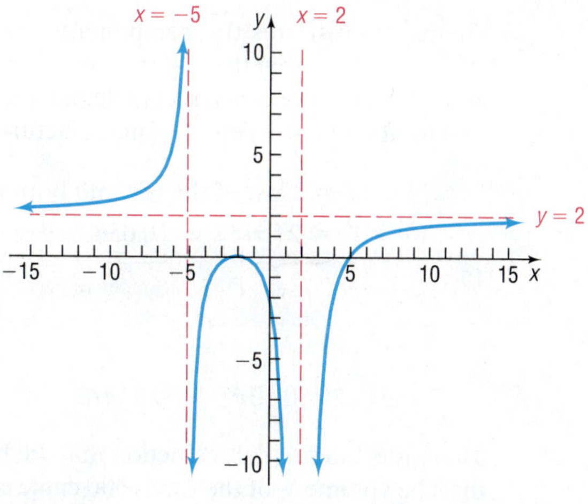

Figure 37

Solution The numerator of a rational function $R(x) = \dfrac{p(x)}{q(x)}$ in lowest terms determines the x-intercepts of its graph. The graph shown in Figure 37 has x-intercepts -2 (even multiplicity; graph touches the x-axis) and 5 (odd multiplicity; graph crosses the x-axis). So one possibility for the numerator is $p(x) = (x + 2)^2(x - 5)$.

The denominator of a rational function in lowest terms determines the vertical asymptotes of its graph. The vertical asymptotes of the graph are $x = -5$ and $x = 2$. Since $R(x)$ approaches ∞ to the left of $x = -5$ and $R(x)$ approaches $-\infty$ to the right of $x = -5$, we know that $(x + 5)$ is a factor of odd multiplicity in $q(x)$. Also, $R(x)$ approaches $-\infty$ on both sides of $x = 2$, so $(x - 2)$ is a factor of even multiplicity in $q(x)$. A possibility for the denominator is $q(x) = (x + 5)(x - 2)^2$.

So far we have $R(x) = \dfrac{(x + 2)^2(x - 5)}{(x + 5)(x - 2)^2}$.

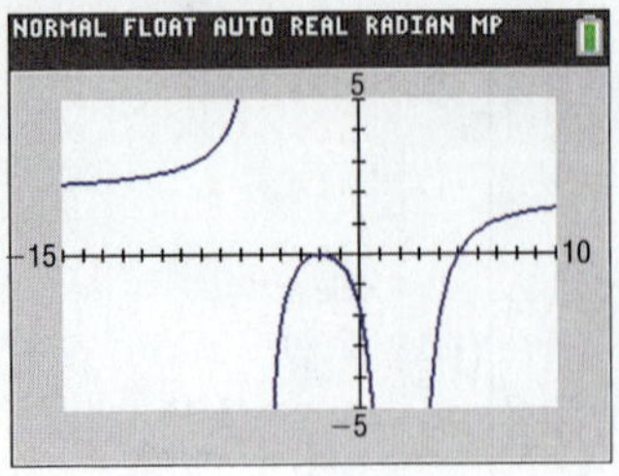

Figure 38

However, the horizontal asymptote of the graph given in Figure 37 is $y = 2$, so we know that the degree of the numerator must equal the degree of the denominator and that the quotient of leading coefficients must be $\frac{2}{1}$. This leads to

$$R(x) = \frac{2(x+2)^2(x-5)}{(x+5)(x-2)^2}$$

Check: Figure 38 shows the graph of R on a graphing utility. Since Figure 38 looks similar to Figure 37, we have found a rational function R for the graph in Figure 37.

Now Work PROBLEM 51

2 Solve Applied Problems Involving Rational Functions

EXAMPLE 7 **Finding the Least Cost of a Can**

Reynolds Metal Company manufactures aluminum cans in the shape of a cylinder with a capacity of 500 cubic centimeters $\left(\frac{1}{2}\text{ liter}\right)$. The top and bottom of the can are made of a special aluminum alloy that costs 0.05¢ per square centimeter. The sides of the can are made of material that costs 0.02¢ per square centimeter.

(a) Express the cost of material for the can as a function of the radius r of the can.
(b) Use a graphing utility to graph the function $C = C(r)$.
(c) What value of r will result in the least cost?
(d) What is this least cost?

Solution

(a) Figure 39 illustrates the components of a can in the shape of a right circular cylinder. Notice that the material required to produce a cylindrical can of height h and radius r consists of a rectangle of area $2\pi rh$ and two circles, each of area πr^2. The total cost C (in cents) of manufacturing the can is therefore

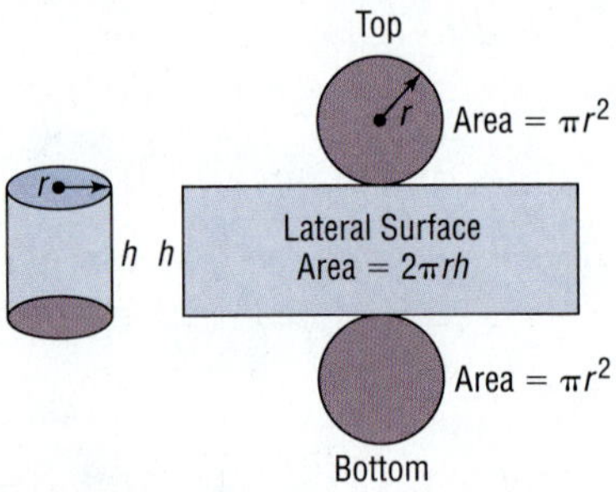

Figure 39

$$\begin{aligned} C &= \text{Cost of the top and bottom} + \text{Cost of the side} \\ &= \underbrace{2(\pi r^2)}_{\substack{\text{Total area}\\ \text{of top and}\\ \text{bottom}}} \cdot \underbrace{(0.05)}_{\substack{\text{Cost/unit}\\ \text{area}}} + \underbrace{(2\pi rh)}_{\substack{\text{Total}\\ \text{area of}\\ \text{side}}} \cdot \underbrace{(0.02)}_{\substack{\text{Cost/unit}\\ \text{area}}} \\ &= 0.10\pi r^2 + 0.04\pi rh \end{aligned}$$

There is an additional restriction that the height h and radius r must be chosen so that the volume V of the can is 500 cubic centimeters. Since $V = \pi r^2 h$, we have

$$500 = \pi r^2 h \quad \text{so} \quad h = \frac{500}{\pi r^2}$$

Substituting this expression for h, we find that the cost C, in cents, as a function of the radius r is

$$C(r) = 0.10\pi r^2 + 0.04\pi r \cdot \frac{500}{\pi r^2} = 0.10\pi r^2 + \frac{20}{r} = \frac{0.10\pi r^3 + 20}{r}$$

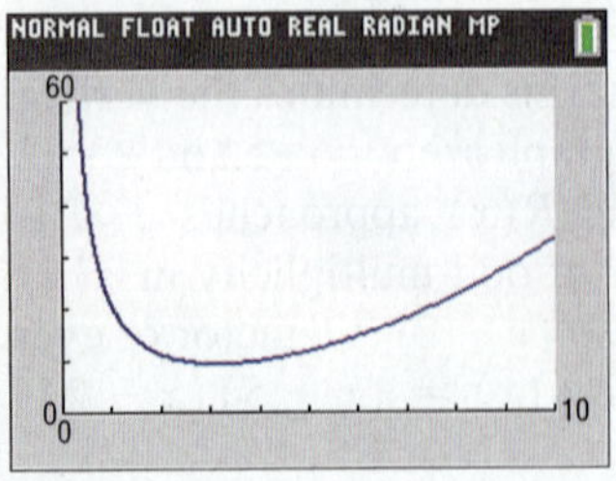

Figure 40

(b) See Figure 40 for the graph of $C = C(r)$.
(c) Using the MINIMUM command, the cost is least for a radius of about 3.17 centimeters.
(d) The least cost is $C(3.17) \approx 9.47$¢.

Now Work PROBLEM 61

5.3 Assess Your Understanding

'Are You Prepared?' *The answer is given at the end of these exercises. If you get a wrong answer, read the pages listed in red.*

1. Find the intercepts of the graph of the equation $y = \dfrac{x^2 - 1}{x^2 - 4}$. (pp. 159–160)

Concepts and Vocabulary

2. ***True or False*** Every rational function has at least one asymptote.

3. Which type of asymptote will never intersect the graph of a rational function?
(a) horizontal (b) oblique (c) vertical (d) all of these

4. Identify the y-intercept of the graph of $R(x) = \dfrac{6(x - 1)}{(x + 1)(x + 2)}$.
(a) -3 (b) -2 (c) -1 (d) 1

5. $R(x) = \dfrac{x(x - 2)^2}{x - 2}$
(a) Find the domain of R.
(b) Find the x-intercepts of R.

6. ***True or False*** The graph of a rational function sometimes has a hole.

Skill Building

In Problems 7–50, follow Steps 1 through 7 on page 356 to analyze the graph of each function.

7. $R(x) = \dfrac{x + 1}{x(x + 4)}$

8. $R(x) = \dfrac{x}{(x - 1)(x + 2)}$

9. $R(x) = \dfrac{3x + 3}{2x + 4}$

10. $R(x) = \dfrac{2x + 4}{x - 1}$

11. $R(x) = \dfrac{3}{x^2 - 4}$

12. $R(x) = \dfrac{6}{x^2 - x - 6}$

13. $P(x) = \dfrac{x^4 + x^2 + 1}{x^2 - 1}$

14. $Q(x) = \dfrac{x^4 - 1}{x^2 - 4}$

15. $H(x) = \dfrac{x^3 - 1}{x^2 - 9}$

16. $G(x) = \dfrac{x^3 + 1}{x^2 + 2x}$

17. $R(x) = \dfrac{x^2}{x^2 + x - 6}$

18. $R(x) = \dfrac{x^2 + x - 12}{x^2 - 4}$

19. $G(x) = \dfrac{x}{x^2 - 4}$

20. $G(x) = \dfrac{3x}{x^2 - 1}$

21. $R(x) = \dfrac{3}{(x - 1)(x^2 - 4)}$

22. $R(x) = \dfrac{-4}{(x + 1)(x^2 - 9)}$

23. $H(x) = \dfrac{x^2 - 1}{x^4 - 16}$

24. $H(x) = \dfrac{x^2 + 4}{x^4 - 1}$

25. $F(x) = \dfrac{x^2 - 3x - 4}{x + 2}$

26. $F(x) = \dfrac{x^2 + 3x + 2}{x - 1}$

27. $R(x) = \dfrac{x^2 + x - 12}{x - 4}$

28. $R(x) = \dfrac{x^2 - x - 12}{x + 5}$

29. $F(x) = \dfrac{x^2 + x - 12}{x + 2}$

30. $G(x) = \dfrac{x^2 - x - 12}{x + 1}$

31. $R(x) = \dfrac{x(x - 1)^2}{(x + 3)^3}$

32. $R(x) = \dfrac{(x - 1)(x + 2)(x - 3)}{x(x - 4)^2}$

33. $R(x) = \dfrac{x^2 + x - 12}{x^2 - x - 6}$

34. $R(x) = \dfrac{x^2 + 3x - 10}{x^2 + 8x + 15}$

35. $R(x) = \dfrac{6x^2 - 7x - 3}{2x^2 - 7x + 6}$

36. $R(x) = \dfrac{8x^2 + 26x + 15}{2x^2 - x - 15}$

37. $R(x) = \dfrac{x^2 + 5x + 6}{x + 3}$

38. $R(x) = \dfrac{x^2 + x - 30}{x + 6}$

39. $H(x) = \dfrac{3x - 6}{4 - x^2}$

40. $H(x) = \dfrac{2 - 2x}{x^2 - 1}$

41. $F(x) = \dfrac{x^2 - 5x + 4}{x^2 - 2x + 1}$

42. $F(x) = \dfrac{x^2 - 2x - 15}{x^2 + 6x + 9}$

43. $G(x) = \dfrac{x}{(x + 2)^2}$

44. $G(x) = \dfrac{2 - x}{(x - 1)^2}$

45. $f(x) = x + \dfrac{1}{x}$

46. $f(x) = 2x + \dfrac{9}{x}$

47. $f(x) = x^2 + \dfrac{1}{x}$

48. $f(x) = 2x^2 + \dfrac{16}{x}$

49. $f(x) = x + \dfrac{1}{x^3}$

50. $f(x) = 2x + \dfrac{9}{x^3}$

In Problems 51–54, find a rational function that might have the given graph. (More than one answer might be possible.)

51.

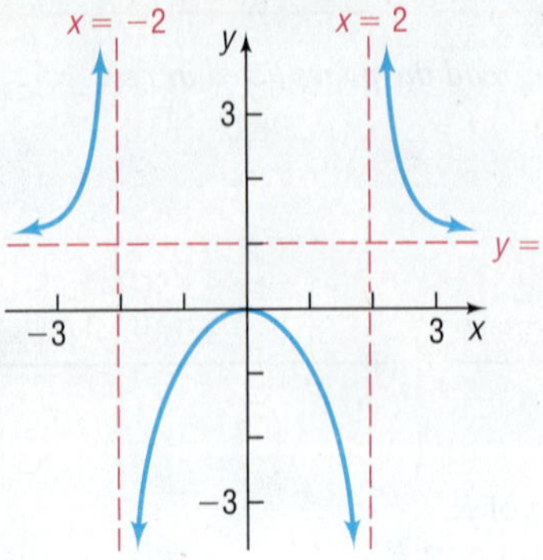

52.

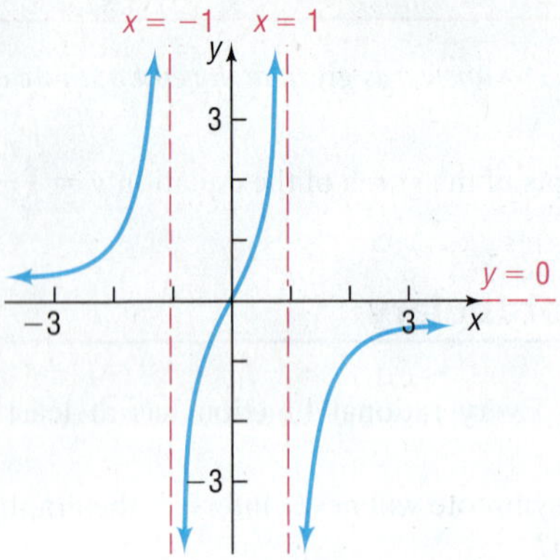

53.

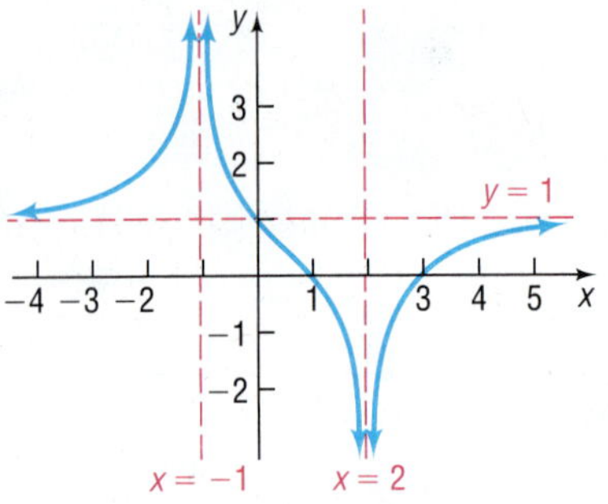

54.

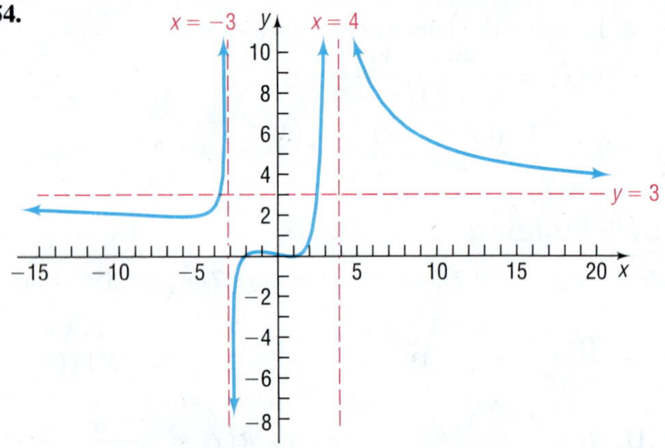

Applications and Extensions

55. Drug Concentration The concentration C of a certain drug in a patient's bloodstream t hours after injection is given by

$$C(t) = \frac{t}{2t^2 + 1}$$

(a) Find the horizontal asymptote of $C(t)$. What happens to the concentration of the drug as t increases?
(b) Using your graphing utility, graph $C = C(t)$.
(c) Determine the time at which the concentration is highest.

56. Drug Concentration The concentration C of a certain drug in a patient's bloodstream t minutes after injection is given by

$$C(t) = \frac{50t}{t^2 + 25}$$

(a) Find the horizontal asymptote of $C(t)$. What happens to the concentration of the drug as t increases?
(b) Using your graphing utility, graph $C = C(t)$.
(c) Determine the time at which the concentration is highest.

57. Minimum Cost A rectangular area adjacent to a river is to be fenced in; no fence is needed on the river side. The enclosed area is to be 1000 square feet. Fencing for the side parallel to the river is \$5 per linear foot, and fencing for the other two sides is \$8 per linear foot; the four corner posts are \$25 apiece. Let x be the length of one of the sides perpendicular to the river.

(a) Write a function $C(x)$ that describes the cost of the project.
(b) What is the domain of C?
(c) Use a graphing utility to graph $C = C(x)$.
(d) Find the dimensions of the cheapest enclosure.

Source: http://dl.uncw.edu/digilib/mathematics/algebra/mat111hb/pandr/rational/rational.html

58. Doppler Effect The Doppler effect (named after Christian Doppler) is the change in the pitch (frequency) of the sound from a source (s) as heard by an observer (o) when one or both are in motion. If we assume both the source and the observer are moving in the same direction, the relationship is

$$f' = f_a\left(\frac{v - v_o}{v - v_s}\right)$$

where f' = perceived pitch by the observer
f_a = actual pitch of the source
v = speed of sound in air (assume 772.4 mph)
v_o = speed of the observer
v_s = speed of the source

Suppose that you are traveling down the road at 45 mph and you hear an ambulance (with siren) coming toward you from the rear. The actual pitch of the siren is 600 hertz (Hz).

(a) Write a function $f'(v_s)$ that describes this scenario.
(b) If $f' = 620$ Hz, find the speed of the ambulance.
(c) Use a graphing utility to graph the function.
(d) Verify your answer from part (b).

Source: www.acs.psu.edu/drussell/

59. Minimizing Surface Area United Parcel Service has contracted you to design a closed box with a square base that has a volume of 10,000 cubic inches. See the illustration.

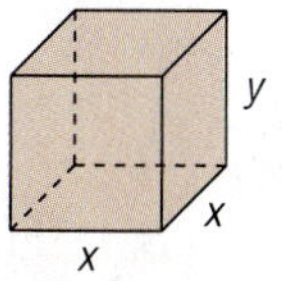

(a) Express the surface area S of the box as a function of x.
(b) Using a graphing utility, graph the function found in part (a).
(c) What is the minimum amount of cardboard that can be used to construct the box?
(d) What are the dimensions of the box that minimize the surface area?
(e) Why might UPS be interested in designing a box that minimizes the surface area?

60. Minimizing Surface Area United Parcel Service has contracted you to design an open box with a square base that has a volume of 5000 cubic inches. See the illustration.

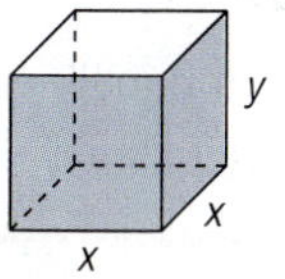

(a) Express the surface area S of the box as a function of x.
(b) Using a graphing utility, graph the function found in part (a).
(c) What is the minimum amount of cardboard that can be used to construct the box?
(d) What are the dimensions of the box that minimize the surface area?
(e) Why might UPS be interested in designing a box that minimizes the surface area?

61. Cost of a Can A can in the shape of a right circular cylinder is required to have a volume of 500 cubic centimeters. The top and bottom are made of material that costs 6¢ per square centimeter, while the sides are made of material that costs 4¢ per square centimeter.
(a) Express the total cost C of the material as a function of the radius r of the cylinder. (Refer to Figure 39.)
(b) Graph $C = C(r)$. For what value of r is the cost C a minimum?

62. Material Needed to Make a Drum A steel drum in the shape of a right circular cylinder is required to have a volume of 100 cubic feet.

(a) Express the amount A of material required to make the drum as a function of the radius r of the cylinder.
(b) How much material is required if the drum's radius is 3 feet?
(c) How much material is required if the drum's radius is 4 feet?
(d) How much material is required if the drum's radius is 5 feet?
(e) Graph $A = A(r)$. For what value of r is A smallest?

Discussion and Writing

63. Graph each of the following functions:

$$y = \frac{x^2 - 1}{x - 1} \qquad y = \frac{x^3 - 1}{x - 1}$$
$$y = \frac{x^4 - 1}{x - 1} \qquad y = \frac{x^5 - 1}{x - 1}$$

Is $x = 1$ a vertical asymptote? Why not? What is happening for $x = 1$? What do you conjecture about $y = \dfrac{x^n - 1}{x - 1}$, $n \geq 1$ an integer, for $x = 1$?

64. Graph each of the following functions:

$$y = \frac{x^2}{x - 1} \quad y = \frac{x^4}{x - 1} \quad y = \frac{x^6}{x - 1} \quad y = \frac{x^8}{x - 1}$$

What similarities do you see? What differences?

65. Write a few paragraphs that provide a general strategy for graphing a rational function. Be sure to mention the following: proper, improper, intercepts, and asymptotes.

66. Create a rational function that has the following characteristics: crosses the x-axis at 2; touches the x-axis at -1; one vertical asymptote at $x = -5$ and another at $x = 6$; and one horizontal asymptote, $y = 3$. Compare your function to a fellow classmate's. How do they differ? What are their similarities?

67. Create a rational function that has the following characteristics: crosses the x-axis at 3; touches the x-axis at -2; one vertical asymptote, $x = 1$; and one horizontal asymptote, $y = 2$. Give your rational function to a fellow classmate and ask for a written critique of your rational function.

68. Create a rational function with the following characteristics: three real zeros, one of multiplicity 2; y-intercept 1; vertical asymptotes, $x = -2$ and $x = 3$; oblique asymptote, $y = 2x + 1$. Is this rational function unique? Compare your function with those of other students. What will be the same as everyone else's? Add some more characteristics, such as symmetry or naming the real zeros. How does this modify the rational function?

69. Explain the circumstances under which the graph of a rational function will have a hole.

Retain Your Knowledge

Problems 70–73 are based on material learned earlier in the course. The purpose of these problems is to keep the material fresh in your mind so that you are better prepared for the final exam.

70. Subtract: $(4x^3 - 7x + 1) - (5x^2 - 9x + 3)$

71. Solve: $\dfrac{3x}{3x + 1} = \dfrac{x - 2}{x + 5}$

72. Find the maximum value of $f(x) = -\dfrac{2}{3}x^2 + 6x - 5$.

73. Approximate $\dfrac{\sqrt{5} - 3}{\sqrt{7} + 2}$. Round your answer to three decimal places.

'Are You Prepared?' Answers

1. $\left(0, \frac{1}{4}\right)$, $(1, 0)$, $(-1, 0)$

5.4 Polynomial and Rational Inequalities

PREPARING FOR THIS SECTION *Before getting started, review the following:*

- Solving Linear Inequalities (Section 1.5, pp. 123–125)
- Solving Quadratic Inequalities (Section 4.5, pp. 312–313)

Now Work the 'Are You Prepared?' problems on page 372.

OBJECTIVES 1 Solve Polynomial Inequalities (p. 368)
2 Solve Rational Inequalities (p. 370)

1 Solve Polynomial Inequalities

In this section we solve inequalities that involve polynomials of degree 3 and higher, along with inequalities that involve rational functions. To help understand the algebraic procedure for solving such inequalities, we use the information obtained in the previous three sections about the graphs of polynomial and rational functions. The approach follows the same methodology that we used to solve inequalities involving quadratic functions.

EXAMPLE 1 Solving a Polynomial Inequality Using Its Graph

Solve $(x + 3)(x - 1)^2 > 0$ by graphing $f(x) = (x + 3)(x - 1)^2$.

Solution Graph $f(x) = (x + 3)(x - 1)^2$ and determine the intervals of x for which the graph is above the x-axis. These values of x result in $f(x)$ being positive. Using Steps 1 through 5 on page 335, we obtain the graph shown in Figure 41.

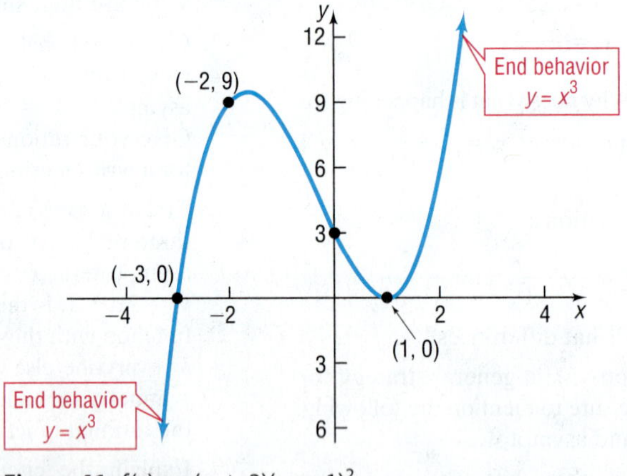

Figure 41 $f(x) = (x + 3)(x - 1)^2$

From the graph, we can see that $f(x) > 0$ for $-3 < x < 1$ or $x > 1$. The solution set is $\{x \mid -3 < x < 1 \text{ or } x > 1\}$ or, using interval notation, $(-3, 1) \cup (1, \infty)$. ●

Now Work PROBLEM 9

The results of Example 1 lead to the following approach to solving polynomial and rational inequalities algebraically. Suppose that the polynomial or rational inequality is in one of the forms

$$f(x) < 0 \qquad f(x) > 0 \qquad f(x) \leq 0 \qquad f(x) \geq 0$$

Locate the zeros of f if f is a polynomial function, and locate the zeros of the numerator and the denominator if f is a rational function. If we use these zeros to divide the real number line into intervals, we know that on each interval, the graph of f is either above the x-axis $[f(x) > 0]$ or below the x-axis $[f(x) < 0]$. This will enable us to identify the solution of the inequality.

EXAMPLE 2 How to Solve a Polynomial Inequality Algebraically

Solve the inequality $x^4 > x$ algebraically, and graph the solution set.

Step-by-Step Solution

Step 1: Write the inequality so that a polynomial expression f is on the left side and zero is on the right side.

Rearrange the inequality so that 0 is on the right side.

$$x^4 > x$$

$$x^4 - x > 0 \quad \text{Subtract } x \text{ from both sides of the inequality.}$$

This inequality is equivalent to the one we wish to solve.

Step 2: Determine the real zeros (x-intercepts of the graph) of f.

Find the real zeros of $f(x) = x^4 - x$ by solving $x^4 - x = 0$.

$$x^4 - x = 0$$

$$x(x^3 - 1) = 0 \quad \text{Factor out } x.$$

$$x(x-1)(x^2 + x + 1) = 0 \quad \text{Factor the difference of two cubes.}$$

$$x = 0 \quad \text{or} \quad x - 1 = 0 \quad \text{or} \quad x^2 + x + 1 = 0 \quad \text{Set each factor equal to zero and solve.}$$

$$x = 0 \quad \text{or} \quad x = 1$$

The equation $x^2 + x + 1 = 0$ has no real solutions. Do you see why?

Step 3: Use the zeros found in Step 2 to divide the real number line into intervals.

Use the real zeros to separate the real number line into three intervals:

$$(-\infty, 0) \qquad (0, 1) \qquad (1, \infty)$$

Step 4: Select a number in each interval, evaluate f at the number, and determine whether f is positive or negative. If f is positive, all values of f in the interval are positive. If f is negative, all values of f in the interval are negative.

Select a test number in each interval found in Step 3 and evaluate $f(x) = x^4 - x$ at each number to determine whether $f(x)$ is positive or negative. See Table 14.

Table 14

	0	1	x
Interval	$(-\infty, 0)$	$(0, 1)$	$(1, \infty)$
Number chosen	-1	$\frac{1}{2}$	2
Value of f	$f(-1) = 2$	$f\left(\frac{1}{2}\right) = -\frac{7}{16}$	$f(2) = 14$
Conclusion	Positive	Negative	Positive

Since we want to know where $f(x)$ is positive, conclude that $f(x) > 0$ for all numbers x for which $x < 0$ or $x > 1$. Because the original inequality is strict, numbers x that satisfy the equation $x^4 = x$ are not solutions. The solution set of the inequality $x^4 > x$ is $\{x \mid x < 0 \text{ or } x > 1\}$ or, using interval notation, $(-\infty, 0) \cup (1, \infty)$.

Figure 42 shows the graph of the solution set. ●

NOTE If the inequality is not strict (that is, if it is $\leq$ or $\geq$), include the solutions of $f(x) = 0$ in the solution set. ■

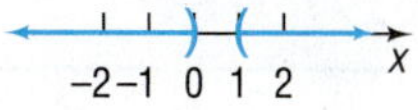

Figure 42

The Role of Multiplicity in Solving Polynomial Inequalities

In Example 2, we used the number -1 and found that f is positive for all $x < 0$. Because the "cut point" of 0 is the result of a zero of odd multiplicity (x is a factor to the first power), we know that the sign of f will change on either side of 0, so for

$0 < x < 1$, f will be negative. Similarly, we know that f will be positive for $x > 1$, since the multiplicity of the zero 1 is odd. Therefore, the solution set of $x^4 > x$ is $\{x | x < 0 \text{ or } x > 1\}$ or, using interval notation, $(-\infty, 0) \cup (1, \infty)$.

Now Work PROBLEM 21

2 Solve Rational Inequalities

Just as we presented a graphical approach to help us understand the algebraic procedure for solving inequalities involving polynomials, we present a graphical approach to help us understand the algebraic procedure for solving inequalities involving rational expressions.

EXAMPLE 3 Solving a Rational Inequality Using Its Graph

Solve $\dfrac{x-1}{x^2-4} \geq 0$ by graphing $R(x) = \dfrac{x-1}{x^2-4}$.

Solution Graph $R(x) = \dfrac{x-1}{x^2-4}$ and determine the intervals of x such that the graph is above or on the x-axis. These values of x result in $R(x)$ being positive or zero. We graphed $R(x) = \dfrac{x-1}{x^2-4}$ in Example 1, Section 5.3 (pp. 353–355). We reproduce the graph in Figure 43.

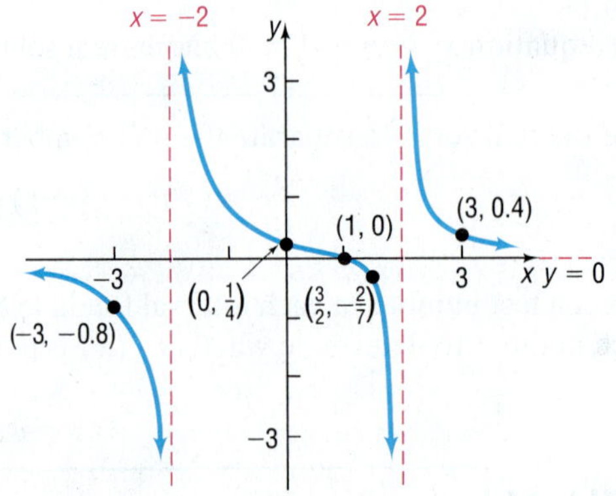

Figure 43 $R(x) = \dfrac{x-1}{x^2-4}$

From the graph, we can see that $R(x) \geq 0$ for $-2 < x \leq 1$ or $x > 2$. The solution set is $\{x | -2 < x \leq 1 \text{ or } x > 2\}$ or, using interval notation, $(-2, 1] \cup (2, \infty)$. ●

Now Work PROBLEM 15

To solve a rational inequality algebraically, we follow the same approach that we used to solve a polynomial inequality algebraically. However, we must also identify the zeros of the denominator of the rational function, because the sign of a rational function may change on either side of a vertical asymptote. Convince yourself of this by looking at Figure 43. Notice that the function values are negative for $x < -2$ and are positive for $x > -2$ (but less than 1).

EXAMPLE 4 How to Solve a Rational Inequality Algebraically

Solve the inequality $\dfrac{3x^2 + 13x + 9}{(x+2)^2} \leq 3$ algebraically, and graph the solution set.

Step-by-Step Solution

Step 1: Write the inequality so that a rational expression f is on the left side and zero is on the right side.

Rearrange the inequality so that 0 is on the right side.

$$\frac{3x^2 + 13x + 9}{(x + 2)^2} \leq 3$$

$$\frac{3x^2 + 13x + 9}{x^2 + 4x + 4} - 3 \leq 0 \quad \text{Subtract 3 from both sides of the inequality; Expand } (x + 2)^2$$

$$\frac{3x^2 + 13x + 9}{x^2 + 4x + 4} - 3 \cdot \frac{x^2 + 4x + 4}{x^2 + 4x + 4} \leq 0 \quad \text{Multiply 3 by } \frac{x^2 + 4x + 4}{x^2 + 4x + 4}.$$

$$\frac{3x^2 + 13x + 9 - 3x^2 - 12x - 12}{x^2 + 4x + 4} \leq 0 \quad \text{Write as a single quotient.}$$

$$\frac{x - 3}{(x + 2)^2} \leq 0 \quad \text{Combine like terms.}$$

Step 2: Determine the real zeros (x-intercepts of the graph) of f and the real numbers for which f is undefined.

The zero of $f(x) = \dfrac{x - 3}{(x + 2)^2}$ is 3. Also, f is undefined for $x = -2$.

Step 3: Use the zeros and undefined values found in Step 2 to divide the real number line into intervals.

Use the zero and the undefined value to separate the real number line into three intervals:

$$(-\infty, -2) \quad (-2, 3) \quad (3, \infty)$$

Step 4: Select a number in each interval, evaluate f at the number, and determine whether $f(x)$ is positive or negative. If $f(x)$ is positive, all values of f in the interval are positive. If $f(x)$ is negative, all values of f in the interval are negative.

Select a test number in each interval from Step 3, and evaluate f at each number to determine whether $f(x)$ is positive or negative. See Table 15.

Table 15

	-2	3	x
Interval	$(-\infty, -2)$	$(-2, 3)$	$(3, \infty)$
Number chosen	-3	0	4
Value of f	$f(-3) = -6$	$f(0) = -\frac{3}{4}$	$f(4) = \frac{1}{36}$
Conclusion	Negative	Negative	Positive

NOTE If the inequality is not strict ($\leq$ or $\geq$), include the solutions of $f(x) = 0$ in the solution set. ■

Since we want to know where $f(x)$ is negative or zero, we conclude that $f(x) \leq 0$ for all numbers for which $x < -2$ or $-2 < x \leq 3$. Notice that we do not include -2 in the solution because -2 is not in the domain of f. The solution set of the inequality $\dfrac{3x^2 + 13x + 9}{(x + 2)^2} \leq 3$ is $\{x \mid x < -2 \text{ or } -2 < x \leq 3\}$ or, using interval notation, $(-\infty, -2) \cup (-2, 3]$. Figure 44 shows the graph of the solution set. ●

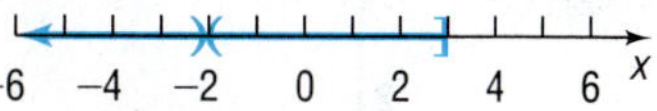

Figure 44

The Role of Multiplicity in Solving Rational Inequalities

In Example 4, we used the number -3 and found that R is negative for all $x < -2$. Because the "cut point" of -2 is the result of a zero of even multiplicity, we know the sign of R will not change on either side of -2, so for $-2 < x < 3$, R will also be negative. Because the "cut point" of 3 is the result of a zero of odd multiplicity, the sign of R will change on either side of 3, so for $x > 3$, R will be positive. Therefore, the solution set of $\dfrac{3x^2 + 13x + 9}{(x + 2)^2} \leq 3$ is $\{x \mid x < -2 \text{ or } -2 < x \leq 3\}$ or, using interval notation, $(-\infty, -2) \cup (-2, 3]$.

Now Work PROBLEMS **33 and 39**

SUMMARY

Steps for Solving Polynomial and Rational Inequalities Algebraically

STEP 1: Write the inequality so that a polynomial or rational expression f is on the left side and zero is on the right side in one of the following forms:

$$f(x) > 0 \quad f(x) \geq 0 \quad f(x) < 0 \quad f(x) \leq 0$$

For rational expressions, be sure that the left side is written as a single quotient, and find the domain of f.

STEP 2: Determine the real numbers at which the expression f equals zero and, if the expression is rational, the real numbers at which the expression f is undefined.

STEP 3: Use the numbers found in Step 2 to separate the real number line into intervals.

STEP 4: Select a number in each interval and evaluate f at the number.

(a) If the value of f is positive, then $f(x) > 0$ for all numbers x in the interval.

(b) If the value of f is negative, then $f(x) < 0$ for all numbers x in the interval.

If the inequality is not strict ($\geq$ or $\leq$), include the solutions of $f(x) = 0$ that are in the domain of f in the solution set. Be careful to exclude values of x where f is undefined.

5.4 Assess Your Understanding

'Are You Prepared?' *Answers are given at the end of these exercises. If you get a wrong answer, read the pages listed in red.*

1. Solve the inequality $3 - 4x > 5$. Graph the solution set. (pp. 123–125)

2. Solve the inequality $x^2 - 5x \leq 24$. Graph the solution set. (pp. 312–313)

Concepts and Vocabulary

3. Which of the following could be a test number for the interval $-2 < x < 5$?
(a) -3 (b) -2 (c) 4 (d) 7

4. ***True or False*** The graph of $f(x) = \dfrac{x}{x-3}$ is above the x-axis for $x < 0$ or $x > 3$, so the solution set of the inequality $\dfrac{x}{x-3} \geq 0$ is $\{x \mid x \leq 0 \text{ or } x \geq 3\}$.

Skill Building

In Problems 5–8, use the graph of the function f to solve the inequality.

5. (a) $f(x) > 0$
(b) $f(x) \leq 0$

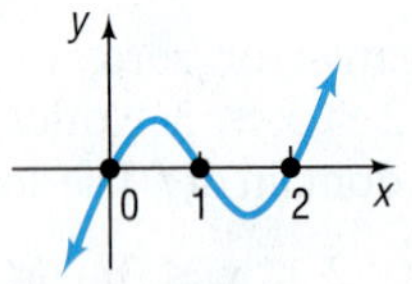

6. (a) $f(x) < 0$
(b) $f(x) \geq 0$

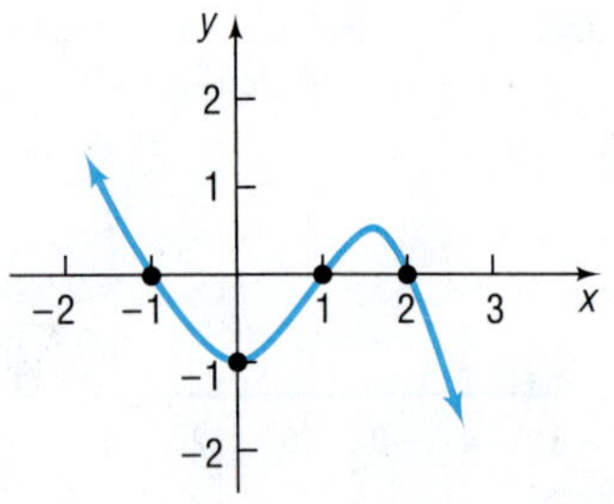

7. (a) $f(x) < 0$
(b) $f(x) \geq 0$

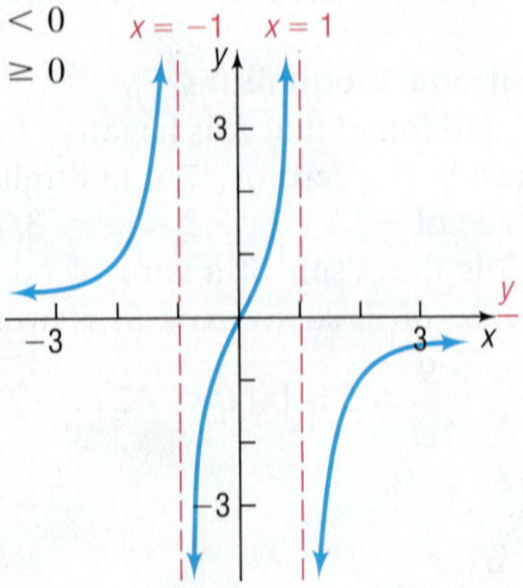

8. (a) $f(x) > 0$
(b) $f(x) \leq 0$

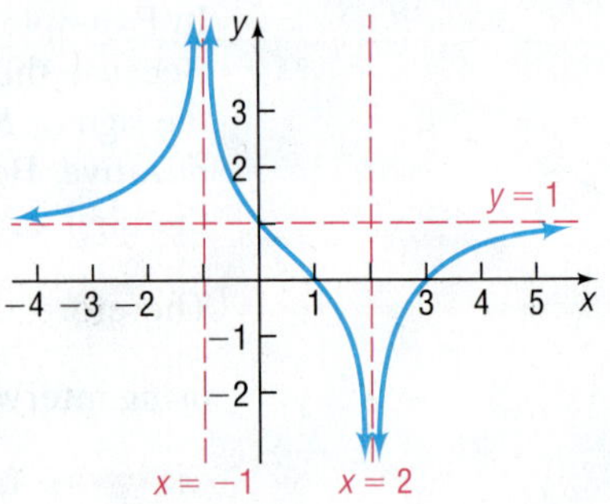

In Problems 9–14, solve the inequality by using the graph of the function.
[**Hint:** The graphs were drawn in Problems 81–86 of Section 5.1.]

9. Solve $f(x) < 0$, where $f(x) = x^2(x - 3)$.

10. Solve $f(x) \leq 0$, where $f(x) = x(x + 2)^2$.

11. Solve $f(x) \geq 0$, where $f(x) = (x + 4)^2(1 - x)$.

12. Solve $f(x) > 0$, where $f(x) = (x - 1)(x + 3)^2$.

13. Solve $f(x) \leq 0$, where $f(x) = -2(x + 2)(x - 2)^3$.

14. Solve $f(x) < 0$, where $f(x) = -\frac{1}{2}(x + 4)(x - 1)^3$.

In Problems 15–18, solve the inequality by using the graph of the function.
[**Hint:** The graphs were drawn in Problems 7–10 of Section 5.3.]

15. Solve $R(x) > 0$, where $R(x) = \dfrac{x + 1}{x(x + 4)}$.

16. Solve $R(x) < 0$, where $R(x) = \dfrac{x}{(x - 1)(x + 2)}$.

17. Solve $R(x) \leq 0$, where $R(x) = \dfrac{3x + 3}{2x + 4}$.

18. Solve $R(x) \geq 0$, where $R(x) = \dfrac{2x + 4}{x - 1}$.

In Problems 19–48, solve each inequality algebraically.

19. $(x - 5)^2(x + 2) < 0$

20. $(x - 5)(x + 2)^2 > 0$

21. $x^3 - 4x^2 > 0$

22. $x^3 + 8x^2 < 0$

23. $2x^3 > -8x^2$

24. $3x^3 < -15x^2$

25. $(x - 1)(x - 2)(x - 3) \leq 0$

26. $(x + 1)(x + 2)(x + 3) \leq 0$

27. $x^3 - 2x^2 - 3x > 0$

28. $x^3 + 2x^2 - 3x > 0$

29. $x^4 > x^2$

30. $x^4 < 9x^2$

31. $x^4 > 1$

32. $x^3 > 1$

33. $\dfrac{x + 1}{x - 1} > 0$

34. $\dfrac{x - 3}{x + 1} > 0$

35. $\dfrac{(x - 1)(x + 1)}{x} \leq 0$

36. $\dfrac{(x - 3)(x + 2)}{x - 1} \leq 0$

37. $\dfrac{(x - 2)^2}{x^2 - 1} \geq 0$

38. $\dfrac{(x + 5)^2}{x^2 - 4} \geq 0$

39. $\dfrac{x + 4}{x - 2} \leq 1$

40. $\dfrac{x + 2}{x - 4} \geq 1$

41. $\dfrac{3x - 5}{x + 2} \leq 2$

42. $\dfrac{x - 4}{2x + 4} \geq 1$

43. $\dfrac{1}{x - 2} < \dfrac{2}{3x - 9}$

44. $\dfrac{5}{x - 3} > \dfrac{3}{x + 1}$

45. $\dfrac{x^2(3 + x)(x + 4)}{(x + 5)(x - 1)} \geq 0$

46. $\dfrac{x(x^2 + 1)(x - 2)}{(x - 1)(x + 1)} \geq 0$

47. $\dfrac{(3 - x)^3(2x + 1)}{x^3 - 1} < 0$

48. $\dfrac{(2 - x)^3(3x - 2)}{x^3 + 1} < 0$

Mixed Practice

In Problems 49–60, solve each inequality algebraically.

49. $(x + 1)(x - 3)(x - 5) > 0$

50. $(2x - 1)(x + 2)(x + 5) < 0$

51. $7x - 4 \geq -2x^2$

52. $x^2 + 3x \geq 10$

53. $\dfrac{x + 1}{x - 3} \leq 2$

54. $\dfrac{x - 1}{x + 2} \geq -2$

55. $3(x^2 - 2) < 2(x - 1)^2 + x^2$

56. $(x - 3)(x + 2) < x^2 + 3x + 5$

57. $6x - 5 < \dfrac{6}{x}$

58. $x + \dfrac{12}{x} < 7$

59. $x^3 - 9x \leq 0$

60. $x^3 - x \geq 0$

In Problems 61–64, (a) graph each function by hand, and (b) solve $f(x) \geq 0$.

61. $f(x) = \dfrac{x^2 + 5x - 6}{x^2 - 4x + 4}$

62. $f(x) = \dfrac{2x^2 + 9x + 9}{x^2 - 4}$

63. $f(x) = \dfrac{x^3 + 2x^2 - 11x - 12}{x^2 - x - 6}$

64. $f(x) = \dfrac{x^3 - 6x^2 + 9x - 4}{x^2 + x - 20}$

Applications and Extensions

65. For what positive numbers will the cube of a number exceed four times its square?

66. For what positive numbers will the cube of a number be less than the number?

67. What is the domain of the function $f(x) = \sqrt{x^4 - 16}$?

68. What is the domain of the function $f(x) = \sqrt{x^3 - 3x^2}$?

69. What is the domain of the function $f(x) = \sqrt{\dfrac{x-2}{x+4}}$?

70. What is the domain of the function $f(x) = \sqrt{\dfrac{x-1}{x+4}}$?

In Problems 71–74, determine where the graph of f is below the graph of g by solving the inequality $f(x) \le g(x)$. Graph f and g together.

71. $f(x) = x^4 - 1$
$g(x) = -2x^2 + 2$

72. $f(x) = x^4 - 1$
$g(x) = x - 1$

73. $f(x) = x^4 - 4$
$g(x) = 3x^2$

74. $f(x) = x^4$
$g(x) = 2 - x^2$

75. Average Cost Suppose that the daily cost C of manufacturing bicycles is given by $C(x) = 80x + 5000$. Then the average daily cost $\overline{C}$ is given by $\overline{C}(x) = \dfrac{80x + 5000}{x}$. How many bicycles must be produced each day for the average cost to be no more than \$100?

76. Average Cost See Problem 75. Suppose that the government imposes a \$1000-per-day tax on the bicycle manufacturer so that the daily cost C of manufacturing x bicycles is now given by $C(x) = 80x + 6000$. Now the average daily cost $\overline{C}$ is given by $\overline{C}(x) = \dfrac{80x + 6000}{x}$. How many bicycles must be produced each day for the average cost to be no more than \$100?

77. Bungee Jumping Originating on Pentecost Island in the Pacific, the practice of a person jumping from a high place harnessed to a flexible attachment was introduced to Western culture in 1979 by the Oxford University Dangerous Sport Club. One important parameter to know before attempting a bungee jump is the amount the cord will stretch at the bottom of the fall. The stiffness of the cord is related to the amount of stretch by the equation

$$K = \frac{2W(S + L)}{S^2}$$

where W = weight of the jumper (pounds)
K = cord's stiffness (pounds per foot)
L = free length of the cord (feet)
S = stretch (feet)

(a) A 150-pound person plans to jump off a ledge attached to a cord of length 42 feet. If the stiffness of the cord is no less than 16 pounds per foot, how much will the cord stretch?

(b) If safety requirements will not permit the jumper to get any closer than 3 feet to the ground, what is the minimum height required for the ledge in part (a)?

Source: American Institute of Physics, Physics News Update, No. 150, November 5, 1993.

78. Gravitational Force According to Newton's Law of Universal Gravitation, the attractive force F between two bodies is given by

$$F = G\frac{m_1 m_2}{r^2}$$

where m_1, m_2 = the masses of the two bodies
r = distance between the two bodies
G = gravitational constant = 6.6742×10^{-11} newtons · meter^2 · kilogram^{-2}

Suppose an object is traveling directly from Earth to the moon. The mass of Earth is 5.9742×10^{24} kilograms, the mass of the moon is 7.349×10^{22} kilograms, and the mean distance from Earth to the moon is 384,400 kilometers. For an object between Earth and the moon, how far from Earth is the force on the object due to the moon greater than the force on the object due to Earth?

Source: www.solarviews.com;en.wikipedia.org

79. Field Trip Mrs. West has decided to take her fifth grade class to a play. The manager of the theater agreed to discount the regular \$40 price of the ticket by \$0.20 for each ticket sold. The cost of the bus, \$500, will be split equally among the students. How many students must attend to keep the cost per student at or below \$40?

Explaining Concepts: Discussion and Writing

80. Make up an inequality that has no solution. Make up one that has exactly one solution.

81. The inequality $x^4 + 1 < -5$ has no solution. Explain why.

82. A student attempted to solve the inequality $\dfrac{x+4}{x-3} \le 0$ by multiplying both sides of the inequality by $x - 3$ to get $x + 4 \le 0$. This led to a solution of $\{x \mid x \le -4\}$. Is the student correct? Explain.

83. Write a rational inequality whose solution set is $\{x \mid -3 < x \le 5\}$.

Retain Your Knowledge

Problems 84–87 are based on material learned earlier in the course. The purpose of these problems is to keep the material fresh in your mind so that you are better prepared for the final exam.

84. Solve: $9 - 2x \le 4x + 1$

85. Given $f(x) = x^2 + 3x - 2$, find $f(x - 2)$.

86. Factor completely: $6x^4y^4 + 3x^3y^5 - 18x^2y^6$

87. Suppose y varies directly with $\sqrt{x}$. Write a general formula to describe the variation if $y = 2$ when $x = 9$.

'Are You Prepared?' Answers

1. $\left\{x \,\middle|\, x < -\frac{1}{2}\right\}$ or $\left(-\infty, -\frac{1}{2}\right)$

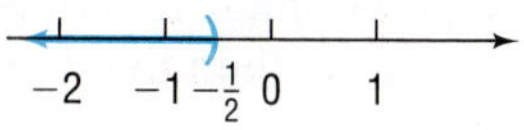

2. $\{x \mid -3 \le x \le 8\}$ or $[-3, 8]$

5.5 The Real Zeros of a Polynomial Function

PREPARING FOR THIS SECTION *Before getting started, review the following:*

- Evaluating Functions (Section 3.1, pp. 202–204)
- Factoring Polynomials (Chapter R, Section R.5, pp. 49–55)
- Synthetic Division (Chapter R, Section R.6, pp. 58–61)
- Polynomial Division (Chapter R, Section R.4, pp. 44–47)
- Solve a Quadratic Equation (Section 1.2, pp. 92–99)

Now Work the 'Are You Prepared?' problems on page 386.

OBJECTIVES
1 Use the Remainder and Factor Theorems (p. 375)
2 Use Descartes' Rule of Signs to Determine the Number of Positive and the Number of Negative Real Zeros of a Polynomial Function (p. 378)
3 Use the Rational Zeros Theorem to List the Potential Rational Zeros of a Polynomial Function (p. 379)
4 Find the Real Zeros of a Polynomial Function (p. 380)
5 Solve Polynomial Equations (p. 382)
6 Use the Theorem for Bounds on Zeros (p. 383)
7 Use the Intermediate Value Theorem (p. 384)

In Section 5.1, we were able to identify the real zeros of a polynomial function because either the polynomial function was in factored form or it could be easily factored. But how do we find the real zeros of a polynomial function if it is not factored or cannot be easily factored?

Recall that if r is a real zero of a polynomial function f, then $f(r) = 0$, r is an x-intercept of the graph of f, $x - r$ is a factor of f, and r is a solution of the equation $f(x) = 0$. For example, if $x - 4$ is a factor of f, then 4 is a real zero of f and 4 is a solution to the equation $f(x) = 0$. For polynomial functions, we have seen the importance of the real zeros for graphing. In most cases, however, the real zeros of a polynomial function are difficult to find using algebraic methods. No nice formulas like the quadratic formula are available to help us find zeros for polynomials of degree 3 or higher. Formulas do exist for solving any third- or fourth-degree polynomial equation, but they are somewhat complicated. No general formulas exist for polynomial equations of degree 5 or higher. Refer to the Historical Feature at the end of this section for more information.

1 Use the Remainder and Factor Theorems

When one polynomial (the dividend) is divided by another (the divisor), a quotient polynomial and a remainder are obtained, the remainder being either the zero polynomial or a polynomial whose degree is less than the degree of the divisor. To check, verify that

$$(\text{Quotient})(\text{Divisor}) + \text{Remainder} = \text{Dividend}$$

This checking routine is the basis for a famous theorem called the **division algorithm*** **for polynomials**, which we now state without proof.

*A systematic process in which certain steps are repeated a finite number of times is called an **algorithm**. For example, long division is an algorithm.

THEOREM

Division Algorithm for Polynomials

If $f(x)$ and $g(x)$ denote polynomial functions and if $g(x)$ is a polynomial whose degree is greater than zero, then there are unique polynomial functions $q(x)$ and $r(x)$ such that

$$\frac{f(x)}{g(x)} = q(x) + \frac{r(x)}{g(x)} \quad \text{or} \quad \underset{\text{dividend}}{f(x)} = \underset{\text{quotient}}{q(x)}\,\underset{\text{divisor}}{g(x)} + \underset{\text{remainder}}{r(x)} \qquad \textbf{(1)}$$

where $r(x)$ is either the zero polynomial or a polynomial of degree less than that of $g(x)$.

In equation (1), $f(x)$ is the **dividend**, $g(x)$ is the **divisor**, $q(x)$ is the **quotient**, and $r(x)$ is the **remainder**.

If the divisor $g(x)$ is a first-degree polynomial of the form

$$g(x) = x - c \qquad c \text{ a real number}$$

then the remainder $r(x)$ is either the zero polynomial or a polynomial of degree 0. As a result, for such divisors, the remainder is some number, say R, and we may write

$$f(x) = (x - c)q(x) + R \qquad \textbf{(2)}$$

This equation is an identity in x and is true for all real numbers x. Suppose that $x = c$. Then equation (2) becomes

$$f(c) = (c - c)q(c) + R$$
$$f(c) = R$$

Substitute $f(c)$ for R in equation (2) to obtain

$$f(x) = (x - c)q(x) + f(c) \qquad \textbf{(3)}$$

which proves the **Remainder Theorem**.

REMAINDER THEOREM

Let f be a polynomial function. If $f(x)$ is divided by $x - c$, then the remainder is $f(c)$.

EXAMPLE 1 **Using the Remainder Theorem**

Find the remainder when $f(x) = x^3 - 4x^2 - 5$ is divided by

(a) $x - 3$ (b) $x + 2$

Solution (a) Either long division or synthetic division could be used, but it is easier to use the Remainder Theorem, which says that the remainder is $f(3)$.

$$f(3) = (3)^3 - 4(3)^2 - 5 = 27 - 36 - 5 = -14$$

The remainder is -14.

(b) To find the remainder when $f(x)$ is divided by $x + 2 = x - (-2)$, find $f(-2)$.

$$f(-2) = (-2)^3 - 4(-2)^2 - 5 = -8 - 16 - 5 = -29$$

The remainder is -29.

Compare the method used in Example 1(a) with the method used in Example 1 of Chapter R, Section R.6. Which method do you prefer?

COMMENT A graphing utility provides another way to find the value of a function using the eVALUEate feature. Consult your manual for details. Then check the results of Example 1. ■

An important and useful consequence of the Remainder Theorem is the **Factor Theorem**.

FACTOR THEOREM

Let f be a polynomial function. Then $x - c$ is a factor of $f(x)$ if and only if $f(c) = 0$.

The Factor Theorem actually consists of two separate statements:

1. If $f(c) = 0$, then $x - c$ is a factor of $f(x)$.
2. If $x - c$ is a factor of $f(x)$, then $f(c) = 0$.

The proof requires two parts.

Proof

1. Suppose that $f(c) = 0$. Then, by equation (3), we have
$$f(x) = (x - c)q(x)$$
for some polynomial $q(x)$. That is, $x - c$ is a factor of $f(x)$.
2. Suppose that $x - c$ is a factor of $f(x)$. Then there is a polynomial function q such that
$$f(x) = (x - c)q(x)$$
Replacing x by c, we find that
$$f(c) = (c - c)q(c) = 0 \cdot q(c) = 0$$
This completes the proof. ■

One use of the Factor Theorem is to determine whether a polynomial has a particular factor.

EXAMPLE 2 Using the Factor Theorem

Use the Factor Theorem to determine whether the function
$$f(x) = 2x^3 - x^2 + 2x - 3$$
has the factor

(a) $x - 1$ (b) $x + 2$

Solution The Factor Theorem states that if $f(c) = 0$, then $x - c$ is a factor.

(a) Because $x - 1$ is of the form $x - c$ with $c = 1$, find the value of $f(1)$. We choose to use substitution.
$$f(1) = 2(1)^3 - (1)^2 + 2(1) - 3 = 2 - 1 + 2 - 3 = 0$$
By the Factor Theorem, $x - 1$ is a factor of $f(x)$.

(b) To test the factor $x + 2$, first write it in the form $x - c$. Since $x + 2 = x - (-2)$, find the value of $f(-2)$. We choose to use synthetic division.

$$\begin{array}{r|rrrr} -2 & 2 & -1 & 2 & -3 \\ & & -4 & 10 & -24 \\ \hline & 2 & -5 & 12 & -27 \end{array}$$

Because $f(-2) = -27 \neq 0$, conclude from the Factor Theorem that $x - (-2) = x + 2$ is not a factor of $f(x)$. ●

Now Work PROBLEM 11

In Example 2(a), $x - 1$ was found to be a factor of f. To write f in factored form, use long division or synthetic division.

$$\begin{array}{r|rrrr} 1 & 2 & -1 & 2 & -3 \\ & & 2 & 1 & 3 \\ \hline & 2 & 1 & 3 & 0 \end{array}$$

The quotient is $q(x) = 2x^2 + x + 3$ with a remainder of 0, as expected. Write f in factored form as

$$f(x) = 2x^3 - x^2 + 2x - 3 = (x - 1)(2x^2 + x + 3)$$

The next theorem concerns the number of real zeros that a polynomial function may have. In counting the zeros of a polynomial, count each zero as many times as its multiplicity.

THEOREM

Number of Real Zeros

A polynomial function cannot have more real zeros than its degree.

Proof The proof is based on the Factor Theorem. If r is a real zero of a polynomial function f, then $f(r) = 0$, and $x - r$ is a factor of $f(x)$. Each real zero corresponds to a factor of degree 1. Because f cannot have more first-degree factors than its degree, the result follows. ■

2 Use Descartes' Rule of Signs to Determine the Number of Positive and the Number of Negative Real Zeros of a Polynomial Function

Descartes' Rule of Signs provides information about the number and location of the real zeros of a polynomial function written in standard form (omitting terms with a 0 coefficient). It utilizes the number of variations in the sign of the coefficients of $f(x)$ and $f(-x)$.

For example, the following polynomial function has two variations in the signs of the coefficients.

$$f(x) = -3x^7 + 4x^4 + 3x^2 - 2x - 1$$

− to +   + to −

Replacing x by $-x$ gives

$$\begin{aligned} f(-x) &= -3(-x)^7 + 4(-x)^4 + 3(-x)^2 - 2(-x) - 1 \\ &= 3x^7 + 4x^4 + 3x^2 + 2x - 1 \end{aligned}$$

+ to −

which has one variation in sign.

THEOREM

Descartes' Rule of Signs

Let f denote a polynomial function written in standard form.

The number of positive real zeros of f either equals the number of variations in the sign of the nonzero coefficients of $f(x)$ or else equals that number less an even integer.

The number of negative real zeros of f either equals the number of variations in the sign of the nonzero coefficients of $f(-x)$ or else equals that number less an even integer.

We shall not prove Descartes' Rule of Signs. Let's see how it is used.

EXAMPLE 3 **Using the Number of Real Zeros Theorem and Descartes' Rule of Signs**

Discuss the real zeros of $f(x) = 3x^7 - 4x^4 + 3x^3 + 2x^2 - x - 3$.

Solution Because the polynomial is of degree 7, by the Number of Real Zeros Theorem there are at most seven real zeros. Since there are three variations in the sign of the nonzero coefficients of $f(x)$, by Descartes' Rule of Signs we expect either three positive real zeros or one positive real zero. To continue, look at $f(-x)$.

$$f(-x) = -3x^7 - 4x^4 - 3x^3 + 2x^2 + x - 3$$

There are two variations in sign, so we expect either two negative real zeros or no negative real zeros. Equivalently, we now know that the graph of f has either three positive x-intercepts or one positive x-intercept and two negative x intercepts or no negative x-intercepts. ●

Now Work PROBLEM 21

3 Use the Rational Zeros Theorem to List the Potential Rational Zeros of a Polynomial Function

The next result, called the **Rational Zeros Theorem**, provides information about the rational zeros of a polynomial *with integer coefficients*.

THEOREM

Rational Zeros Theorem

Let f be a polynomial function of degree 1 or higher of the form

$$f(x) = a_nx^n + a_{n-1}x^{n-1} + \cdots + a_1x + a_0 \qquad a_n \neq 0 \quad a_0 \neq 0$$

where each coefficient is an integer. If $\frac{p}{q}$, in lowest terms, is a rational zero of f, then p must be a factor of a_0, and q must be a factor of a_n.

EXAMPLE 4 **Listing Potential Rational Zeros**

List the potential rational zeros of

$$f(x) = 2x^3 + 11x^2 - 7x - 6$$

Solution Because f has integer coefficients, the Rational Zeros Theorem may be used. First, list all the integers p that are factors of the constant term $a_0 = -6$ and all the integers q that are factors of the leading coefficient $a_3 = 2$.

p: $\pm 1, \pm 2, \pm 3, \pm 6$ Factors of -6

q: $\pm 1, \pm 2$ Factors of 2

Now form all possible ratios $\frac{p}{q}$.

$$\frac{p}{q}: \quad \pm\frac{1}{1}, \pm\frac{2}{1}, \pm\frac{3}{1}, \pm\frac{6}{1}, \pm\frac{1}{2}, \pm\frac{2}{2}, \pm\frac{3}{2}, \pm\frac{6}{2}$$

which simplify to

$$\frac{p}{q}: \quad \pm 1, \pm 2, \pm 3, \pm 6, \pm\frac{1}{2}, \pm\frac{3}{2}$$

If f has a rational zero, it will be found in this list, which contains 12 possibilities. ●

In Words
For the polynomial function $f(x) = 2x^3 + 11x^2 - 7x - 6$, we know 5 is not a zero, because 5 is not in the list of potential rational zeros. However, -1 may or may not be a zero.

Now Work PROBLEM 33

Be sure that you understand what the Rational Zeros Theorem says: For a polynomial with integer coefficients, *if* there is a rational zero, it is one of those listed. It may be the case that the function does not have any rational zeros.

Long division, synthetic division, or substitution can be used to test each potential rational zero to determine whether it is indeed a zero. To make the work easier, integers are usually tested first.

4 Find the Real Zeros of a Polynomial Function

EXAMPLE 5

How to Find the Real Zeros of a Polynomial Function

Find the real zeros of the polynomial function $f(x) = 2x^3 + 11x^2 - 7x - 6$. Write f in factored form.

Step-by-Step Solution

Step 1: Use the degree of the polynomial to determine the maximum number of zeros.

Since f is a polynomial of degree 3, there are at most three real zeros.

Step 2: Use Descartes' Rule of Signs to determine the possible number of positive zeros and negative zeros.

By Descartes' Rule of Signs, there is one positive real zero. Also, because

$$f(-x) = -2x^3 + 11x^2 + 7x - 6$$

there are two negative zeros or no negative zeros.

Step 3: If the polynomial has integer coefficients, use the Rational Zeros Theorem to identify those rational numbers that potentially can be zeros. Use the Factor Theorem to determine whether each potential rational zero is a zero. If it is, use synthetic division or long division to factor the polynomial function. Repeat Step 3 until all the rational zeros of the polynomial function have been identified.

List the potential rational zeros obtained in Example 4:

$$\pm 1, \pm 2, \pm 3, \pm 6, \pm \frac{1}{2}, \pm \frac{3}{2}$$

From our list of potential rational zeros, we will test 6 to determine whether it is a zero of f. Because $f(6) = 780 \neq 0$, we know that 6 is not a zero of f. Now, let's test whether -6 is a zero. Because $f(-6) = 0$, we know that -6 is a zero and $x - (-6) = x + 6$ is a factor of f. Use long division or synthetic division to factor f. (We will not show the division here, but you are encouraged to verify the results shown.) After dividing f by $x + 6$, the quotient is $2x^2 - x - 1$, so

$$\begin{aligned} f(x) &= 2x^3 + 11x^2 - 7x - 6 \\ &= (x + 6)(2x^2 - x - 1) \end{aligned}$$

Now any solution of the equation $2x^2 - x - 1 = 0$ will be a zero of f. The equation $2x^2 - x - 1 = 0$ is called a **depressed equation** of f. Because any solution to the equation $2x^2 - x - 1 = 0$ is a zero of f, work with the depressed equation to find the remaining zeros of f.

The depressed equation $2x^2 - x - 1 = 0$ is a quadratic equation with discriminant $b^2 - 4ac = (-1)^2 - 4(2)(-1) = 9 > 0$. The equation has two real solutions, which can be found by factoring.

$$2x^2 - x - 1 = (2x + 1)(x - 1) = 0$$

$$2x + 1 = 0 \quad \text{or} \quad x - 1 = 0$$

$$x = -\frac{1}{2} \quad \text{or} \quad x = 1$$

The zeros of f are -6, $-\frac{1}{2}$, and 1.

Factor f completely as follows:

$$\begin{aligned} f(x) = 2x^3 + 11x^2 - 7x - 6 &= (x + 6)(2x^2 - x - 1) \\ &= (x + 6)(2x + 1)(x - 1) \end{aligned}$$

Notice that all three zeros of f are in the list of potential rational zeros. Also note that the zeros confirm what was expected from Descartes' Rule of Signs. ●

SUMMARY

Steps for Finding the Real Zeros of a Polynomial Function

STEP 1: Use the degree of the polynomial to determine the maximum number of real zeros.

STEP 2: Use Descartes' Rule of Signs to determine the possible number of positive zeros and negative zeros.

STEP 3: (a) If the polynomial has integer coefficients, use the Rational Zeros Theorem to identify those rational numbers that potentially could be zeros.

(b) Use substitution, synthetic division, or long division to test each potential rational zero. Each time that a zero (and thus a factor) is found, repeat Step 3 on the depressed equation.

In attempting to find the zeros, remember to use (if possible) the factoring techniques that you already know (special products, factoring by grouping, and so on).

EXAMPLE 6 Finding the Real Zeros of a Polynomial Function

Find the real zeros of $f(x) = x^5 - 7x^4 + 19x^3 - 37x^2 + 60x - 36$. Write f in factored form.

Solution

STEP 1: Because f is a polynomial of degree 5, there are at most five real zeros.

STEP 2: By Descartes' Rule of Signs, there are five, three, or one positive zeros. There are no negative zeros because

$$f(-x) = -x^5 - 7x^4 - 19x^3 - 37x^2 - 60x - 36$$

has no sign variation.

STEP 3: Because the leading coefficient $a_5 = 1$ and there are no negative zeros, the potential rational zeros are limited to the positive integers 1, 2, 3, 4, 6, 9, 12, 18, and 36 (the positive factors of the constant term, 36). Test the potential rational zero 1 first, using synthetic division.

$$\begin{array}{r|rrrrrr} 1 & 1 & -7 & 19 & -37 & 60 & -36 \\ & & 1 & -6 & 13 & -24 & 36 \\ \hline & 1 & -6 & 13 & -24 & 36 & 0 \end{array}$$

The remainder is $f(1) = 0$, so 1 is a zero and $x - 1$ is a factor of f. Use the entries in the bottom row of the synthetic division to begin factoring f.

$$\begin{aligned} f(x) &= x^5 - 7x^4 + 19x^3 - 37x^2 + 60x - 36 \\ &= (x - 1)(x^4 - 6x^3 + 13x^2 - 24x + 36) \end{aligned}$$

Continue the process using the first depressed equation:

$$q_1(x) = x^4 - 6x^3 + 13x^2 - 24x + 36 = 0$$

REPEAT STEP 3: The potential rational zeros of q_1 are still 1, 2, 3, 4, 6, 9, 12, 18, and 36. Test 1 again, since it may be a repeated zero of f.

$$\begin{array}{r|rrrrr} 1 & 1 & -6 & 13 & -24 & 36 \\ & & 1 & -5 & 8 & -16 \\ \hline & 1 & -5 & 8 & -16 & 20 \end{array}$$

Since the remainder is 20, 1 is not a repeated zero. Try 2 next.

$$\begin{array}{r|rrrrr} 2 & 1 & -6 & 13 & -24 & 36 \\ & & 2 & -8 & 10 & -28 \\ \hline & 1 & -4 & 5 & -14 & 8 \end{array}$$

Since the remainder is 8, 2 is not a zero. Try 3 next.

$$\begin{array}{r|rrrrr} 3 & 1 & -6 & 13 & -24 & 36 \\ & & 3 & -9 & 12 & -36 \\ \hline & 1 & -3 & 4 & -12 & 0 \end{array}$$

The remainder is $f(3) = 0$, so 3 is a zero and $x - 3$ is a factor of f. Use the bottom row of the synthetic division to continue the factoring of f.

$$f(x) = x^5 - 7x^4 + 19x^3 - 37x^2 + 60x - 36$$
$$= (x - 1)(x - 3)(x^3 - 3x^2 + 4x - 12)$$

The remaining zeros satisfy the new depressed equation

$$q_2(x) = x^3 - 3x^2 + 4x - 12 = 0$$

Notice that $q_2(x)$ can be factored by grouping. (Alternatively, Step 3 could be repeated to again check the potential rational zero 3. The potential rational zeros 1 or 2 would no longer be checked, because they have already been eliminated from further consideration.) Then

$$x^3 - 3x^2 + 4x - 12 = 0$$
$$x^2(x - 3) + 4(x - 3) = 0$$
$$(x^2 + 4)(x - 3) = 0$$
$$x^2 + 4 = 0 \text{ or } x - 3 = 0$$
$$x = 3$$

Since $x^2 + 4 = 0$ has no real solutions, the real zeros of f are 1 and 3, with 3 being a repeated zero of multiplicity 2. The factored form of f is

$$f(x) = x^5 - 7x^4 + 19x^3 - 37x^2 + 60x - 36$$
$$= (x - 1)(x - 3)^2(x^2 + 4)$$

Now Work PROBLEM 45

5 Solve Polynomial Equations

EXAMPLE 7 **Solving a Polynomial Equation**

Find the real solutions of the equation: $x^5 - 7x^4 + 19x^3 - 37x^2 + 60x - 36 = 0$

Solution The real solutions of this equation are the real zeros of the polynomial function

$$f(x) = x^5 - 7x^4 + 19x^3 - 37x^2 + 60x - 36$$

Using the result of Example 6, the real zeros of f are 1 and 3. The real solutions of the equation $x^5 - 7x^4 + 19x^3 - 37x^2 + 60x - 36 = 0$ are 1 and 3.

Now Work PROBLEM 57

In Example 6, the quadratic factor $x^2 + 4$ that appears in the factored form of f is called *irreducible*, because the polynomial $x^2 + 4$ cannot be factored over the real numbers. In general, a quadratic factor $ax^2 + bx + c$ is **irreducible** if it cannot be factored over the real numbers—that is, if it is prime over the real numbers.

Refer to Examples 5 and 6. The polynomial function of Example 5 has three real zeros, and its factored form contains three linear factors. The polynomial function of Example 6 has two distinct real zeros, and its factored form contains two distinct linear factors and one irreducible quadratic factor.

THEOREM

Every polynomial function (with real coefficients) can be uniquely factored into a product of linear factors and/or irreducible quadratic factors.

We prove this result in Section 5.6, and in fact, we shall draw several additional conclusions about the zeros of a polynomial function. One conclusion is worth noting now. If a polynomial with real coefficients is of odd degree, it must contain at least one linear factor. (Do you see why? Consider the end behavior of polynomial functions of odd degree.) This means that it must have at least one real zero.

THEOREM

A polynomial function (with real coefficients) of odd degree has at least one real zero.

6 Use the Theorem for Bounds on Zeros

The work involved in finding the zeros of a polynomial function can be reduced somewhat if upper and lower bounds to the zeros can be found. A number M is an **upper bound** to the zeros of a polynomial f if no zero of f is greater than M. The number m is a **lower bound** if no zero of f is less than m. Accordingly, if m is a lower bound and M is an upper bound to the zeros of a polynomial function f, then

$$m \leq \text{ any zero of } f \leq M$$

For polynomials with integer coefficients, knowing the values of a lower bound m and an upper bound M may enable you to eliminate some potential rational zeros—that is, any zeros outside of the interval $[m, M]$.

COMMENT The bounds on the zeros of a polynomial provide good choices for setting Xmin and Xmax of the viewing rectangle. With these choices, all the x-intercepts of the graph can be seen. ■

THEOREM

Bounds on Zeros

Let f denote a polynomial function whose leading coefficient is positive.

- If $M > 0$ is a real number and if the third row in the process of synthetic division of f by $x - M$ contains only numbers that are positive or zero, then M is an upper bound to the zeros of f.
- If $m < 0$ is a real number and if the third row in the process of synthetic division of f by $x - m$ contains numbers that alternate positive (or 0) and negative (or 0), then m is a lower bound to the zeros of f.

NOTE When finding a lower bound, remember that a 0 can be treated as either positive or negative, but not both. For example, 3, 0, 5 would be considered to alternate sign, whereas 3, 0, −5 would not. ■

Proof (Outline) We give only an outline of the proof of the first part of the theorem. Suppose that M is a positive real number, and the third row in the process of synthetic division of the polynomial f by $x - M$ contains only numbers that are positive or 0. Then there are a quotient q and a remainder R such that

$$f(x) = (x - M)q(x) + R$$

where the coefficients of $q(x)$ are positive or 0 and the remainder $R \geq 0$. Then, for any $x > M$, we must have $x - M > 0$, $q(x) > 0$, and $R \geq 0$, so that $f(x) > 0$. That is, there is no zero of f larger than M. The proof of the second part follows similar reasoning. ■

In finding bounds, it is preferable to find the smallest upper bound and largest lower bound. This will require repeated synthetic division until a desired pattern is observed. For simplicity, we will consider only potential rational zeros that are integers. If a bound is not found using these values, continue checking positive and/or negative integers until you find both an upper and a lower bound.

EXAMPLE 8 **Finding Upper and Lower Bounds of Zeros**

For the polynomial function $f(x) = 2x^3 + 11x^2 - 7x - 6$, use the Bounds on Zeros Theorem to find integer upper and lower bounds to the zeros of f.

Solution From Example 4, the potential rational zeros of f are $\pm 1, \pm 2, \pm 3, \pm 6, \pm \frac{1}{2}, \pm \frac{3}{2}$. To find an upper bound, start with the smallest positive integer that is a potential rational zero, which is 1. Continue checking 2, 3, and 6 (and then subsequent positive integers), if necessary, until an upper bound is found. To find a lower bound, start with the largest negative integer that is a potential rational zero, which is −1. Continue checking −2, −3, and −6 (and then subsequent negative integers), if necessary, until a lower bound is found. Table 16 summarizes the results of doing repeated synthetic divisions by showing only the third row of each division. For example, the first row of the table shows the result of dividing $f(x)$ by $x - 1$.

$$\begin{array}{r|rrrr} 1 & 2 & 11 & -7 & -6 \\ & & 2 & 13 & 6 \\ \hline & 2 & 13 & 6 & 0 \end{array}$$

Table 16 Synthetic Division Summary

	r	Coefficients of *q*(*x*)			Remainder	
Upper bound →	1	2	13	6	0	All nonnegative
	−1	2	9	−16	10	
	−2	2	7	−21	36	
	−3	2	5	−22	60	
	−6	2	−1	−1	0	
Lower bound →	−7	2	−3	14	−104	Alternating Signs

NOTE Keep track of any zeros that are found when looking for bounds. ■

For $r = 1$, the third row of synthetic division contains only numbers that are positive or 0, so we know there are no zeros greater than 1. Since the third row of synthetic division for $r = -7$ results in alternating positive (or 0) and negative (or 0) values, we know that -7 is a lower bound. There are no zeros less than -7. Notice that in looking for bounds, two zeros were discovered. These zeros are 1 and -6. ●

Now Work PROBLEM 69

If the leading coefficient of f is negative, the upper and lower bounds can still be found by first multiplying the polynomial by -1. Since $-f(x) = (-1)f(x)$, the zeros of $-f(x)$ are the same as the zeros of $f(x)$.

7 Use the Intermediate Value Theorem

The next result, called the **Intermediate Value Theorem**, is based on the fact that the graph of a polynomial function is continuous; that is, it contains no "holes" or "gaps." Although the proof of this result requires advanced methods in calculus, it is easy to "see" why the result is true. Look at Figure 45.

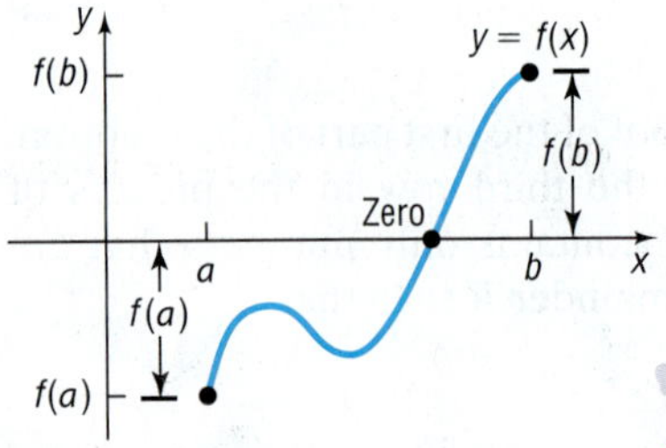

Figure 45 If $f(a) < 0$ and $f(b) > 0$, and if f is continuous, there is a zero between a and b.

THEOREM

Intermediate Value Theorem

Let f denote a polynomial function. If $a < b$ and if $f(a)$ and $f(b)$ are of opposite sign, there is at least one real zero of f between a and b.

EXAMPLE 9 **Using the Intermediate Value Theorem to Locate a Real Zero**

Show that $f(x) = x^5 - x^3 - 1$ has a zero between 1 and 2.

Solution Evaluate f at 1 and at 2.

$$f(1) = -1 \quad \text{and} \quad f(2) = 23$$

Because $f(1) < 0$ and $f(2) > 0$, it follows from the Intermediate Value Theorem that the polynomial function f has at least one zero between 1 and 2. ●

Now Work PROBLEM 79

Let's look at the polynomial f of Example 9 more closely. Based on Descartes' Rule of Signs, f has exactly one positive real zero. Based on the Rational Zeros Theorem, 1 is the only potential positive rational zero. Since $f(1) \neq 0$, the zero between 1 and 2 is irrational. The Intermediate Value Theorem can be used to approximate it.

Approximating the Real Zeros of a Polynomial Function

Step 1: Find two consecutive integers a and $a + 1$ such that f has a zero between them.

Step 2: Divide the interval $[a, a + 1]$ into 10 equal subintervals.

Step 3: Evaluate f at each endpoint of the subintervals until the Intermediate Value Theorem applies; this interval then contains a zero.

Step 4: Now divide the new interval into 10 equal subintervals and repeat Step 3.

Step 5: Continue with Steps 3 and 4 until the desired accuracy is achieved.

EXAMPLE 10 Approximating a Real Zero of a Polynomial Function

Find the positive zero of $f(x) = x^5 - x^3 - 1$ correct to two decimal places.

Solution From Example 9 we know that the positive zero is between 1 and 2. Divide the interval $[1, 2]$ into 10 equal subintervals: $[1, 1.1]$, $[1.1, 1.2]$, $[1.2, 1.3]$, $[1.3, 1.4]$, $[1.4, 1.5]$, $[1.5, 1.6]$, $[1.6, 1.7]$, $[1.7, 1.8]$, $[1.8, 1.9]$, $[1.9, 2]$. Now find the value of f at each endpoint until the Intermediate Value Theorem applies.

$$f(x) = x^5 - x^3 - 1$$

$$f(1.0) = -1 \qquad f(1.2) = -0.23968$$

$$f(1.1) = -0.72049 \qquad f(1.3) = 0.51593$$

We can stop here and conclude that the zero is between 1.2 and 1.3. Now divide the interval $[1.2, 1.3]$ into 10 equal subintervals and proceed to evaluate f at each endpoint.

$$f(1.20) = -0.23968 \qquad f(1.23) \approx -0.0455613$$

$$f(1.21) \approx -0.1778185 \qquad f(1.24) \approx 0.025001$$

$$f(1.22) \approx -0.1131398$$

The zero lies between 1.23 and 1.24, and so, correct to two decimal places, the zero is 1.23. ●

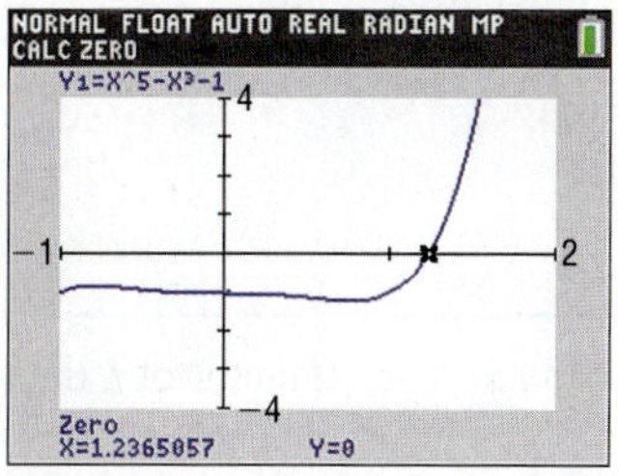

Figure 46

Exploration

We examine the polynomial function f given in Example 10. The Theorem on Bounds of Zeros tells us that every zero is between -1 and 2. If we graph f using $-1 \leq x \leq 2$ (see Figure 46), we see that f has exactly one x-intercept. Using ZERO or ROOT, we find this zero to be 1.24 rounded to two decimal places. Correct to two decimal places, the zero is 1.23.

Now Work PROBLEM 91

COMMENT The TABLE feature of a graphing calculator makes the computations in the solution to Example 10 a lot easier. ■

There are many other numerical techniques for approximating the zeros of a polynomial. The one outlined in Example 10 (a variation of the *bisection method*) has the advantages that it will always work, it can be programmed rather easily on a computer, and each time it is used another decimal place of accuracy is achieved. See Problem 119 for the bisection method, which places the zero in a succession of intervals, with each new interval being half the length of the preceding one.

Historical Feature

Formulas for the solution of third- and fourth-degree polynomial equations exist, and although they are not very practical, they do have an interesting history.

In the 1500s in Italy, mathematical contests were a popular pastime, and people who possessed methods for solving problems kept them secret. (Solutions that were published were already common knowledge.) Niccolo of Brescia (1499–1557), commonly referred to as Tartaglia ("the stammerer"), had the secret for solving cubic (third-degree) equations, which gave him a decided advantage in the contests. Girolamo Cardano (1501–1576) found out that Tartaglia had the secret, and, being interested in cubics, he requested it from Tartaglia. The reluctant Tartaglia hesitated for some time, but finally, swearing Cardano to secrecy with midnight oaths by candlelight, told him the secret. Cardano then published the solution in his book *Ars Magna* (1545), giving Tartaglia the credit but rather compromising the secrecy. Tartaglia exploded into bitter recriminations, and each wrote pamphlets that reflected on the other's mathematics, moral character, and ancestry.

The quartic (fourth-degree) equation was solved by Cardano's student Lodovico Ferrari, and this solution also was included, with credit and this time with permission, in the *Ars Magna*.

Attempts were made to solve the fifth-degree equation in similar ways, all of which failed. In the early 1800s, P. Ruffini, Niels Abel, and Evariste Galois all found ways to show that it is not possible to solve fifth-degree equations by formula, but the proofs required the introduction of new methods. Galois's methods eventually developed into a large part of modern algebra.

Historical Problems

Problems 1–8 develop the Tartaglia–Cardano solution of the cubic equation and show why it is not altogether practical.

1. Show that the general cubic equation $y^3 + by^2 + cy + d = 0$ can be transformed into an equation of the form $x^3 + px + q = 0$ by using the substitution $y = x - \frac{b}{3}$.
2. In the equation $x^3 + px + q = 0$, replace x by $H + K$. Let $3HK = -p$, and show that $H^3 + K^3 = -q$.
3. Based on Problem 2, we have the two equations
$$3HK = -p \quad \text{and} \quad H^3 + K^3 = -q$$
Solve for K in $3HK = -p$ and substitute into $H^3 + K^3 = -q$. Then show that
$$H = \sqrt[3]{\frac{-q}{2} + \sqrt{\frac{q^2}{4} + \frac{p^3}{27}}}$$
[**Hint:** Look for an equation that is quadratic in form.]
4. Use the solution for H from Problem 3 and the equation $H^3 + K^3 = -q$ to show that
$$K = \sqrt[3]{\frac{-q}{2} - \sqrt{\frac{q^2}{4} + \frac{p^3}{27}}}$$
5. Use the results from Problems 2 to 4 to show that the solution of $x^3 + px + q = 0$ is
$$x = \sqrt[3]{\frac{-q}{2} + \sqrt{\frac{q^2}{4} + \frac{p^3}{27}}} + \sqrt[3]{\frac{-q}{2} - \sqrt{\frac{q^2}{4} + \frac{p^3}{27}}}$$
6. Use the result of Problem 5 to solve the equation $x^3 - 6x - 9 = 0$.
7. Use a calculator and the result of Problem 5 to solve the equation $x^3 + 3x - 14 = 0$.
8. Use the methods of this section to solve the equation $x^3 + 3x - 14 = 0$.

5.5 Assess Your Understanding

'Are You Prepared?' *Answers are given at the end of these exercises. If you get a wrong answer, read the pages listed in red.*

1. Find $f(-1)$ if $f(x) = 2x^2 - x$. (pp. 202–204)
2. Factor the expression $6x^2 + x - 2$. (pp. 49–55)
3. Find the quotient and remainder if $3x^4 - 5x^3 + 7x - 4$ is divided by $x - 3$. (pp. 44–47 or 58–61)
4. Solve $x^2 + x - 3 = 0$. (pp. 92–99)

Concepts and Vocabulary

5. If $f(x) = q(x)g(x) + r(x)$, the function $r(x)$ is called the______.
(a) remainder (b) dividend (c) quotient (d) divisor
6. When a polynomial function f is divided by $x - c$, the remainder is ________.
7. Given $f(x) = 3x^4 - 2x^3 + 7x - 2$, how many sign changes are there in the coefficients of $f(-x)$?
(a) 0 (b) 1 (c) 2 (d) 3
8. ***True or False*** Every polynomial function of degree 3 with real coefficients has exactly three real zeros.
9. If f is a polynomial function and $x - 4$ is a factor of f, then $f(4) =$ ___.
10. ***True or False*** If f is a polynomial function of degree 4 and if $f(2) = 5$, then
$$\frac{f(x)}{x-2} = p(x) + \frac{5}{x-2}$$
where $p(x)$ is a polynomial of degree 3.

Skill Building

In Problems 11–20, use the Remainder Theorem to find the remainder when $f(x)$ is divided by $x - c$. Then use the Factor Theorem to determine whether $x - c$ is a factor of $f(x)$.

11. $f(x) = 4x^3 - 3x^2 - 8x + 4; x - 2$
12. $f(x) = -4x^3 + 5x^2 + 8; x + 3$
13. $f(x) = 3x^4 - 6x^3 - 5x + 10; x - 2$
14. $f(x) = 4x^4 - 15x^2 - 4; x - 2$
15. $f(x) = 3x^6 + 82x^3 + 27; x + 3$
16. $f(x) = 2x^6 - 18x^4 + x^2 - 9; x + 3$
17. $f(x) = 4x^6 - 64x^4 + x^2 - 15; x + 4$
18. $f(x) = x^6 - 16x^4 + x^2 - 16; x + 4$
19. $f(x) = 2x^4 - x^3 + 2x - 1; x - \frac{1}{2}$
20. $f(x) = 3x^4 + x^3 - 3x + 1; x + \frac{1}{3}$

In Problems 21–32, tell the maximum number of real zeros that each polynomial function may have. Then use Descartes' Rule of Signs to determine how many positive and how many negative zeros each polynomial function may have. Do not attempt to find the zeros.

21. $f(x) = -4x^7 + x^3 - x^2 + 2$
22. $f(x) = 5x^4 + 2x^2 - 6x - 5$
23. $f(x) = 2x^6 - 3x^2 - x + 1$
24. $f(x) = -3x^5 + 4x^4 + 2$
25. $f(x) = 3x^3 - 2x^2 + x + 2$
26. $f(x) = -x^3 - x^2 + x + 1$
27. $f(x) = -x^4 + x^2 - 1$
28. $f(x) = x^4 + 5x^3 - 2$
29. $f(x) = x^5 + x^4 + x^2 + x + 1$
30. $f(x) = x^5 - x^4 + x^3 - x^2 + x - 1$
31. $f(x) = x^6 - 1$
32. $f(x) = x^6 + 1$

In Problems 33–44, list the potential rational zeros of each polynomial function. Do not attempt to find the zeros.

33. $f(x) = 3x^4 - 3x^3 + x^2 - x + 1$
34. $f(x) = x^5 - x^4 + 2x^2 + 3$
35. $f(x) = x^5 - 6x^2 + 9x - 3$
36. $f(x) = 2x^5 - x^4 - x^2 + 1$
37. $f(x) = -4x^3 - x^2 + x + 2$
38. $f(x) = 6x^4 - x^2 + 2$
39. $f(x) = 6x^4 - x^2 + 9$
40. $f(x) = -4x^3 + x^2 + x + 6$
41. $f(x) = 2x^5 - x^3 + 2x^2 + 12$
42. $f(x) = 3x^5 - x^2 + 2x + 18$
43. $f(x) = 6x^4 + 2x^3 - x^2 + 20$
44. $f(x) = -6x^3 - x^2 + x + 10$

In Problems 45–56, use the Rational Zeros Theorem to find all the real zeros of each polynomial function. Use the zeros to factor f over the real numbers.

45. $f(x) = x^3 + 2x^2 - 5x - 6$
46. $f(x) = x^3 + 8x^2 + 11x - 20$
47. $f(x) = 2x^3 - x^2 + 2x - 1$
48. $f(x) = 2x^3 + x^2 + 2x + 1$
49. $f(x) = 2x^3 - 4x^2 - 10x + 20$
50. $f(x) = 3x^3 + 6x^2 - 15x - 30$
51. $f(x) = 2x^4 + x^3 - 7x^2 - 3x + 3$
52. $f(x) = 2x^4 - x^3 - 5x^2 + 2x + 2$
53. $f(x) = x^4 + x^3 - 3x^2 - x + 2$
54. $f(x) = x^4 - x^3 - 6x^2 + 4x + 8$
55. $f(x) = 4x^4 + 5x^3 + 9x^2 + 10x + 2$
56. $f(x) = 3x^4 + 4x^3 + 7x^2 + 8x + 2$

In Problems 57–68, solve each equation in the real number system.

57. $x^4 - x^3 + 2x^2 - 4x - 8 = 0$
58. $2x^3 + 3x^2 + 2x + 3 = 0$
59. $3x^3 + 4x^2 - 7x + 2 = 0$
60. $2x^3 - 3x^2 - 3x - 5 = 0$

61. $3x^3 - x^2 - 15x + 5 = 0$

62. $2x^3 - 11x^2 + 10x + 8 = 0$

63. $x^4 + 4x^3 + 2x^2 - x + 6 = 0$

64. $x^4 - 2x^3 + 10x^2 - 18x + 9 = 0$

65. $x^3 - \frac{2}{3}x^2 + \frac{8}{3}x + 1 = 0$

66. $x^3 + \frac{3}{2}x^2 + 3x - 2 = 0$

67. $2x^4 - 19x^3 + 57x^2 - 64x + 20 = 0$

68. $2x^4 + x^3 - 24x^2 + 20x + 16 = 0$

In Problems 69–78, find bounds on the real zeros of each polynomial function.

69. $f(x) = x^4 - 3x^2 - 4$

70. $f(x) = x^4 - 5x^2 - 36$

71. $f(x) = x^4 + x^3 - x - 1$

72. $f(x) = x^4 - x^3 + x - 1$

73. $f(x) = 3x^4 + 3x^3 - x^2 - 12x - 12$

74. $f(x) = 3x^4 - 3x^3 - 5x^2 + 27x - 36$

75. $f(x) = 4x^5 - x^4 + 2x^3 - 2x^2 + x - 1$

76. $f(x) = 4x^5 + x^4 + x^3 + x^2 - 2x - 2$

77. $f(x) = -x^4 + 3x^3 - 4x^2 - 2x + 9$

78. $f(x) = -4x^5 + 5x^3 + 9x^2 + 3x - 12$

In Problems 79–84, use the Intermediate Value Theorem to show that each polynomial function has a zero in the given interval.

79. $f(x) = 8x^4 - 2x^2 + 5x - 1;\ [0, 1]$

80. $f(x) = x^4 + 8x^3 - x^2 + 2;\ [-1, 0]$

81. $f(x) = 2x^3 + 6x^2 - 8x + 2;\ [-5, -4]$

82. $f(x) = 3x^3 - 10x + 9;\ [-3, -2]$

83. $f(x) = x^5 - x^4 + 7x^3 - 7x^2 - 18x + 18;\ [1.4, 1.5]$

84. $f(x) = x^5 - 3x^4 - 2x^3 + 6x^2 + x + 2;\ [1.7, 1.8]$

In Problems 85–88, each equation has a solution r in the interval indicated. Use the method of Example 10 to approximate this solution correct to two decimal places.

85. $8x^4 - 2x^2 + 5x - 1 = 0;\ 0 \le r \le 1$

86. $x^4 + 8x^3 - x^2 + 2 = 0;\ -1 \le r \le 0$

87. $2x^3 + 6x^2 - 8x + 2 = 0;\ -5 \le r \le -4$

88. $3x^3 - 10x + 9 = 0;\ -3 \le r \le -2$

In Problems 89–92, each polynomial function has exactly one positive zero. Use the method of Example 10 to approximate the zero correct to two decimal places.

89. $f(x) = x^3 + x^2 + x - 4$

90. $f(x) = 2x^4 + x^2 - 1$

91. $f(x) = 2x^4 - 3x^3 - 4x^2 - 8$

92. $f(x) = 3x^3 - 2x^2 - 20$

Mixed Practice

In Problems 93–104, graph each polynomial function.

93. $f(x) = x^3 + 2x^2 - 5x - 6$

94. $f(x) = x^3 + 8x^2 + 11x - 20$

95. $f(x) = 2x^3 - x^2 + 2x - 1$

96. $f(x) = 2x^3 + x^2 + 2x + 1$

97. $f(x) = x^4 + x^2 - 2$

98. $f(x) = x^4 - 3x^2 - 4$

99. $f(x) = 4x^4 + 7x^2 - 2$

100. $f(x) = 4x^4 + 15x^2 - 4$

101. $f(x) = x^4 + x^3 - 3x^2 - x + 2$

102. $f(x) = x^4 - x^3 - 6x^2 + 4x + 8$

103. $f(x) = 4x^5 - 8x^4 - x + 2$

104. $f(x) = 4x^5 + 12x^4 - x - 3$

105. Suppose that $f(x) = 3x^3 + 16x^2 + 3x - 10$. Find the zeros of $f(x + 3)$.

106. Suppose that $f(x) = 4x^3 - 11x^2 - 26x + 24$. Find the zeros of $f(x - 2)$.

Applications and Extensions

107. Find k such that $f(x) = x^3 - kx^2 + kx + 2$ has the factor $x - 2$.

108. Find k such that $f(x) = x^4 - kx^3 + kx^2 + 1$ has the factor $x + 2$.

109. What is the remainder when $f(x) = 2x^{20} - 8x^{10} + x - 2$ is divided by $x - 1$?

110. What is the remainder when $f(x) = -3x^{17} + x^9 - x^5 + 2x$ is divided by $x + 1$?

111. Use the Factor Theorem to prove that $x - c$ is a factor of $x^n - c^n$ for any positive integer n.

112. Use the Factor Theorem to prove that $x + c$ is a factor of $x^n + c^n$ if $n \geq 1$ is an odd integer.

113. One solution of the equation $x^3 - 8x^2 + 16x - 3 = 0$ is 3. Find the sum of the remaining solutions.

114. One solution of the equation $x^3 + 5x^2 + 5x - 2 = 0$ is -2. Find the sum of the remaining solutions.

115. Geometry What is the length of the edge of a cube if, after a slice 1 inch thick is cut from one side, the volume remaining is 294 cubic inches?

116. Geometry What is the length of the edge of a cube if its volume could be doubled by an increase of 6 centimeters in one edge, an increase of 12 centimeters in a second edge, and a decrease of 4 centimeters in the third edge?

117. Let $f(x)$ be a polynomial function whose coefficients are integers. Suppose that r is a real zero of f and that the leading coefficient of f is 1. Use the Rational Zeros Theorem to show that r is either an integer or an irrational number.

118. Prove the Rational Zeros Theorem.

[**Hint:** Let $\frac{p}{q}$, where p and q have no common factors except 1 and -1, be a zero of the polynomial function

$$f(x) = a_n x^n + a_{n-1}x^{n-1} + \cdots + a_1 x + a_0$$

whose coefficients are all integers. Show that

$$a_n p^n + a_{n-1}p^{n-1}q + \cdots + a_1 pq^{n-1} + a_0 q^n = 0$$

Now, because p is a factor of the first n terms of this equation, p must also be a factor of the term $a_0 q^n$. Since p is not a factor of q (why?), p must be a factor of a_0. Similarly, q must be a factor of a_n.]

119. Bisection Method for Approximating Zeros of a Function f We begin with two consecutive integers, a and $a + 1$, such that $f(a)$ and $f(a + 1)$ are of opposite sign. Evaluate f at the midpoint m_1 of a and $a + 1$. If $f(m_1) = 0$, then m_1 is the zero of f, and we are finished. Otherwise, $f(m_1)$ is of opposite sign to either $f(a)$ or $f(a + 1)$. Suppose that it is $f(a)$ and $f(m_1)$ that are of opposite sign. Now evaluate f at the midpoint m_2 of a and m_1. Repeat this process until the desired degree of accuracy is obtained. Note that each iteration places the zero in an interval whose length is half that of the previous interval. Use the bisection method to approximate the zero of $f(x) = 8x^4 - 2x^2 + 5x - 1$ in the interval $[0, 1]$ correct to three decimal places.

[**Hint:** The process ends when both endpoints agree to the desired number of decimal places.]

Discussion and Writing

120. Is $\frac{1}{3}$ a zero of $f(x) = 2x^3 + 3x^2 - 6x + 7$? Explain.

121. Is $\frac{1}{3}$ a zero of $f(x) = 4x^3 - 5x^2 - 3x + 1$? Explain.

122. Is $\frac{3}{5}$ a zero of $f(x) = 2x^6 - 5x^4 + x^3 - x + 1$? Explain.

123. Is $\frac{2}{3}$ a zero of $f(x) = x^7 + 6x^5 - x^4 + x + 2$? Explain.

Retain Your Knowledge

Problems 124–127 are based on material learned earlier in the course. The purpose of these problems is to keep the material fresh in your mind so that you are better prepared for the final exam.

124. Solve $2x - 5y = 3$ for y.

125. Express the inequality $3 \leq x < 8$ using interval notation.

126. Find the intercepts of the graph of the equation $3x + y^2 = 12$.

127. Use the figure to determine the interval(s) on which the function is increasing.

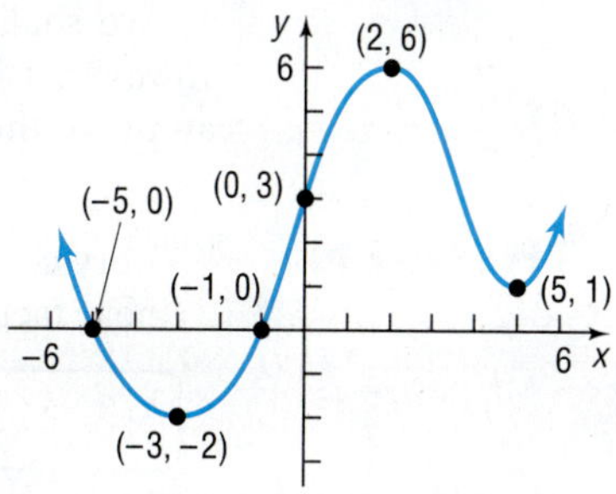

'Are You Prepared?' Answers

1. 3 **2.** $(3x + 2)(2x - 1)$ **3.** Quotient: $3x^3 + 4x^2 + 12x + 43$; Remainder: 125 **4.** $\frac{-1 - \sqrt{13}}{2}, \frac{-1 + \sqrt{13}}{2}$

5.6 Complex Zeros; Fundamental Theorem of Algebra

PREPARING FOR THIS SECTION *Before getting started, review the following:*

- Complex Numbers (Section 1.3, pp. 104–109)
- Complex Solutions of a Quadratic Equation (Section 1.3, pp. 109–111)

Now Work the 'Are You Prepared?' problems on page 394.

OBJECTIVES
1 Use the Conjugate Pairs Theorem (p. 391)
2 Find a Polynomial Function with Specified Zeros (p. 392)
3 Find the Complex Zeros of a Polynomial Function (p. 393)

In Section 1.2, we found the real solutions of a quadratic equation. That is, we found the real zeros of a polynomial function of degree 2. Then, in Section 1.3 we found the complex solutions of a quadratic equation. That is, we found the complex zeros of a polynomial function of degree 2.

In Section 5.5, we found the real zeros of polynomial functions of degree 3 or higher. In this section we will find the *complex zeros* of polynomial functions of degree 3 or higher.

DEFINITION

A variable in the complex number system is referred to as a **complex variable.** A **complex polynomial function** f of degree n is a function of the form

$$f(x) = a_n x^n + a_{n-1}x^{n-1} + \cdots + a_1 x + a_0 \qquad \textbf{(1)}$$

where $a_n, a_{n-1}, \ldots, a_1, a_0$ are complex numbers, $a_n \neq 0$, n is a nonnegative integer, and x is a complex variable. As before, a_n is called the **leading coefficient** of f. A complex number r is called a **complex zero** of f if $f(r) = 0$.

In most of our work, the coefficients in (1) will be real numbers.

We have learned that some quadratic equations have no real solutions, but that in the complex number system every quadratic equation has a solution, either real or complex. The next result, proved by Karl Friedrich Gauss (1777–1855) when he was 22 years old,* gives an extension to complex polynomials. In fact, this result is so important and useful that it has become known as the **Fundamental Theorem of Algebra**.

FUNDAMENTAL THEOREM OF ALGEBRA

Every complex polynomial function f of degree $n \geq 1$ has at least one complex zero.

We shall not prove this result, as the proof is beyond the scope of this text. However, using the Fundamental Theorem of Algebra and the Factor Theorem, we can prove the following result:

THEOREM

Every complex polynomial function f of degree $n \geq 1$ can be factored into n linear factors (not necessarily distinct) of the form

$$f(x) = a_n(x - r_1)(x - r_2) \cdot \cdots \cdot (x - r_n) \qquad \textbf{(2)}$$

where $a_n, r_1, r_2, \ldots, r_n$ are complex numbers. That is, every complex polynomial function of degree $n \geq 1$ has exactly n complex zeros, some of which may repeat.

*In all, Gauss gave four different proofs of this theorem, the first one in 1799 being the subject of his doctoral dissertation.

Proof Let

$$f(x) = a_n x^n + a_{n-1}x^{n-1} + \cdots + a_1 x + a_0$$

By the Fundamental Theorem of Algebra, f has at least one zero, say r_1. Then, by the Factor Theorem, $x - r_1$ is a factor, and

$$f(x) = (x - r_1)q_1(x)$$

where $q_1(x)$ is a complex polynomial of degree $n - 1$ whose leading coefficient is a_n. Repeating this argument n times, we arrive at

$$f(x) = (x - r_1)(x - r_2) \cdot \cdots \cdot (x - r_n)q_n(x)$$

where $q_n(x)$ is a complex polynomial of degree $n - n = 0$ whose leading coefficient is also a_n. That is, $q_n(x) = a_n x^0 = a_n$, and so

$$f(x) = a_n(x - r_1)(x - r_2) \cdot \cdots \cdot (x - r_n)$$

We conclude that every complex polynomial function f of degree $n \geq 1$ has exactly n (not necessarily distinct) zeros. ■

1 Use the Conjugate Pairs Theorem

The Fundamental Theorem of Algebra can be used to obtain valuable information about the complex zeros of polynomial functions whose coefficients are real numbers.

CONJUGATE PAIRS THEOREM

Let f be a polynomial function whose coefficients are real numbers. If $r = a + bi$ is a zero of f, the complex conjugate $\bar{r} = a - bi$ is also a zero of f.

In other words, for polynomial functions whose coefficients are real numbers, the complex zeros occur in conjugate pairs. This result should not be all that surprising since the complex zeros of a quadratic function occurred in conjugate pairs.

Proof Let

$$f(x) = a_n x^n + a_{n-1}x^{n-1} + \cdots + a_1 x + a_0$$

where $a_n, a_{n-1}, \ldots, a_1, a_0$ are real numbers and $a_n \neq 0$. If $r = a + bi$ is a zero of f, then $f(r) = f(a + bi) = 0$, so

$$a_n r^n + a_{n-1}r^{n-1} + \cdots + a_1 r + a_0 = 0$$

Take the conjugate of both sides to get

$$\overline{a_n r^n + a_{n-1}r^{n-1} + \cdots + a_1 r + a_0} = \bar{0}$$

$$\overline{a_n r^n} + \overline{a_{n-1}r^{n-1}} + \cdots + \overline{a_1 r} + \overline{a_0} = \bar{0} \quad \text{The conjugate of a sum equals the sum of the conjugates (see Section 1.3).}$$

$$\overline{a_n}(\bar{r})^n + \overline{a_{n-1}}(\bar{r})^{n-1} + \cdots + \overline{a_1}\bar{r} + \overline{a_0} = \bar{0} \quad \text{The conjugate of a product equals the product of the conjugates.}$$

$$a_n(\bar{r})^n + a_{n-1}(\bar{r})^{n-1} + \cdots + a_1\bar{r} + a_0 = 0 \quad \text{The conjugate of a real number equals the real number.}$$

This last equation states that $f(\bar{r}) = 0$; that is, $\bar{r} = a - bi$ is a zero of f. ■

The importance of this result should be clear. Once we know that, say, $3 + 4i$ is a zero of a polynomial function with real coefficients, then we know that $3 - 4i$ is also a zero. This result has an important corollary.

COROLLARY

A polynomial function f of odd degree with real coefficients has at least one real zero.

Proof Because complex zeros occur as conjugate pairs in a polynomial function with real coefficients, there will always be an even number of zeros that are not real numbers. Consequently, since f is of odd degree, one of its zeros has to be a real number. ■

For example, the polynomial function $f(x) = x^5 - 3x^4 + 4x^3 - 5$ has at least one zero that is a real number, since f is of degree 5 (odd) and has real coefficients.

EXAMPLE 1 **Using the Conjugate Pairs Theorem**

A polynomial function f of degree 5 whose coefficients are real numbers has the zeros $1, 5i$, and $1 + i$. Find the remaining two zeros.

Solution Since f has coefficients that are real numbers, complex zeros appear as conjugate pairs. It follows that $-5i$, the conjugate of $5i$, and $1 - i$, the conjugate of $1 + i$, are the two remaining zeros.

Now Work PROBLEM 7

2 Find a Polynomial Function with Specified Zeros

EXAMPLE 2 **Finding a Polynomial Function Whose Zeros Are Given**

Find a polynomial function f of degree 4 whose coefficients are real numbers that has the zeros $1, 1$, and $-4 + i$.

Solution Since $-4 + i$ is a zero, by the Conjugate Pairs Theorem, $-4 - i$ must also be a zero of f. Because of the Factor Theorem, if $f(c) = 0$, then $x - c$ is a factor of $f(x)$. So f can now be written as

$$f(x) = a(x - 1)(x - 1)[x - (-4 + i)][x - (-4 - i)]$$

where a is any real number. Then

$$\begin{aligned} f(x) &= a(x - 1)(x - 1)[x - (-4 + i)][x - (-4 - i)] \\ &= a(x^2 - 2x + 1)[(x + 4) - i][(x + 4) + i] \\ &= a(x^2 - 2x + 1)[(x + 4)^2 - i^2] \\ &= a(x^2 - 2x + 1)[x^2 + 8x + 16 - (-1)] \\ &= a(x^2 - 2x + 1)(x^2 + 8x + 17) \\ &= a(x^4 + 8x^3 + 17x^2 - 2x^3 - 16x^2 - 34x + x^2 + 8x + 17) \\ &= a(x^4 + 6x^3 + 2x^2 - 26x + 17) \end{aligned}$$

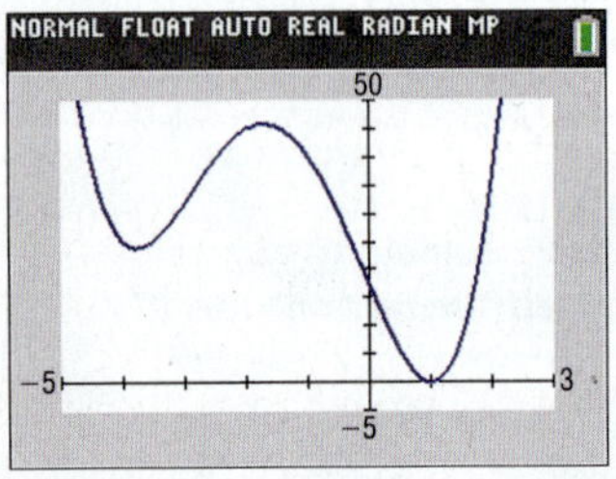

Figure 47
$f(x) = x^4 + 6x^3 + 2x^2 - 26x + 17$

Exploration

Graph the function f found in Example 2 for $a = 1$. Does the value of a affect the zeros of f? How does the value of a affect the graph of f? What information about f is sufficient to uniquely determine a?

Result A quick analysis of the polynomial function f tells us what to expect:

At most three turning points.

For large $|x|$, the graph will behave like $y = x^4$.

A repeated real zero at 1, so the graph will touch the x-axis at 1.

The only x-intercept is 1; the y-intercept is 17.

Figure 47 shows the complete graph. (Do you see why? The graph has exactly three turning points.) The value of a causes a stretch or compression; a reflection also occurs if $a < 0$. The zeros are not affected.

If any point other than an x-intercept on the graph of f is known, then a can be determined. For example, if (2, 3) is on the graph, then $f(2) = 3 = a(37)$, so $a = 3/37$. Why won't an x-intercept work?

Now we can prove the theorem we conjectured in Section 5.5.

THEOREM

Every polynomial function with real coefficients can be uniquely factored over the real numbers into a product of linear factors and/or irreducible quadratic factors.

Proof Every complex polynomial function f of degree n has exactly n zeros and can be factored into a product of n linear factors. If its coefficients are real, those zeros that are complex numbers will always occur as conjugate pairs. As a result, if $r = a + bi$ is a complex zero, then so is $\bar{r} = a - bi$. Consequently, when the linear factors $x - r$ and $x - \bar{r}$ of f are multiplied, we have

$$(x - r)(x - \bar{r}) = x^2 - (r + \bar{r})x + r\bar{r} = x^2 - 2ax + a^2 + b^2$$

This second-degree polynomial has real coefficients and is irreducible (over the real numbers). So, the factors of f are either linear or irreducible quadratic factors. ■

Now Work PROBLEM 17

3 Find the Complex Zeros of a Polynomial Function

The steps for finding the complex zeros of a polynomial function are the same as those for finding the real zeros.

EXAMPLE 3

Finding the Complex Zeros of a Polynomial Function

Find the complex zeros of the polynomial function

$$f(x) = 3x^4 + 5x^3 + 25x^2 + 45x - 18$$

Write f in factored form.

Solution

STEP 1: The degree of f is 4, so f will have four complex zeros.

STEP 2: Since the coefficients of f are real numbers, Descartes' Rule of Signs can be used to obtain information about the real zeros. For this polynomial function, there is one positive real zero. There are three negative real zeros or one negative real zero, because

$$f(-x) = 3x^4 - 5x^3 + 25x^2 - 45x - 18$$

has three variations in sign.

STEP 3: Since the coefficients of f are integers, the Rational Zeros Theorem can be used to obtain information about the potential rational zeros of f. The potential rational zeros are

$$\pm\frac{1}{3}, \pm\frac{2}{3}, \pm 1, \pm 2, \pm 3, \pm 6, \pm 9, \pm 18$$

Table 17 summarizes some results of synthetic division.

Table 17

r	Coefficients of $q(x)$				Remainder	
1	3	8	33	78	60	1 is not a zero.
−1	3	2	23	22	−40	−1 is not a zero.
2	3	11	47	139	260	2 is not a zero.
−2	3	−1	27	−9	0	−2 is a zero.

Since $f(-2) = 0$, then -2 is a zero and $x + 2$ is a factor of f. The depressed equation is

$$3x^3 - x^2 + 27x - 9 = 0$$

REPEAT STEP 3: We factor the depressed equation by grouping.

$$3x^3 - x^2 + 27x - 9 = 0$$

$$x^2(3x - 1) + 9(3x - 1) = 0 \quad \text{Factor } x^2 \text{ from } 3x^3 - x^2 \text{ and 9 from } 27x - 9.$$

$$(x^2 + 9)(3x - 1) = 0 \quad \text{Factor out the common factor } 3x-1.$$

$$x^2 + 9 = 0 \quad \text{or} \quad 3x - 1 = 0 \quad \text{Apply the Zero-Product Property.}$$

$$x^2 = -9 \quad \text{or} \quad x = \frac{1}{3}$$

$$x = -3i, \; x = 3i \quad \text{or} \quad x = \frac{1}{3}$$

The four complex zeros of f are $\left\{-3i, 3i, -2, \frac{1}{3}\right\}$.

The factored form of f is

$$\begin{aligned} f(x) &= 3x^4 + 5x^3 + 25x^2 + 45x - 18 \\ &= 3(x + 3i)(x - 3i)(x + 2)\left(x - \frac{1}{3}\right) \end{aligned}$$

Now Work PROBLEM 33

5.6 Assess Your Understanding

'Are You Prepared?' *Answers are given at the end of these exercises. If you get the wrong answer, read the pages listed in red.*

1. Find the sum and the product of the complex numbers $3 - 2i$ and $-3 + 5i$. (pp. 104–109)
2. In the complex number system, find the complex zeros of $f(x) = x^2 + 2x + 2$. (pp. 109–111)

Concepts and Vocabulary

3. Every polynomial function of odd degree with real coefficients will have at least ________ real zero(s).
4. If $3 + 4i$ is a zero of a polynomial function of degree 5 with real coefficients, then so is ________.
5. ***True or False*** A polynomial function of degree n with real coefficients has exactly n complex zeros. At most n of them are real zeros.
6. ***True or False*** A polynomial function of degree 4 with real coefficients could have $-3, 2 + i, 2 - i$, and $-3 + 5i$ as its zeros.

Skill Building

In Problems 7–16, information is given about a polynomial function f whose coefficients are real numbers. Find the remaining zeros of f.

7. Degree 3; zeros: $3, 4 - i$
8. Degree 3; zeros: $4, 3 + i$
9. Degree 4; zeros: $i, 1 + i$
10. Degree 4; zeros: $1, 2, 2 + i$
11. Degree 5; zeros: $1, i, 2i$
12. Degree 5; zeros: $0, 1, 2, i$
13. Degree 4; zeros: $i, 2, -2$
14. Degree 4; zeros: $2 - i, -i$
15. Degree 6; zeros: $2, 2 + i, -3 - i, 0$
16. Degree 6; zeros: $i, 3 - 2i, -2 + i$

In Problems 17–22, find a polynomial function f with real coefficients having the given degree and zeros. Answers will vary depending on the choice of leading coefficient.

17. Degree 4; zeros: $3 + 2i$; 4, multiplicity 2
18. Degree 4; zeros: $i, 1 + 2i$
19. Degree 5; zeros: $2; -i; 1 + i$
20. Degree 6; zeros: $i, 4 - i; 2 + i$
21. Degree 4; zeros: 3, multiplicity 2; $-i$
22. Degree 5; zeros: 1, multiplicity 3; $1 + i$

In Problems 23–30, use the given zero to find the remaining zeros of each polynomial function.

23. $f(x) = x^3 - 4x^2 + 4x - 16$; zero: $2i$
24. $g(x) = x^3 + 3x^2 + 25x + 75$; zero: $-5i$
25. $f(x) = 2x^4 + 5x^3 + 5x^2 + 20x - 12$; zero: $-2i$
26. $h(x) = 3x^4 + 5x^3 + 25x^2 + 45x - 18$; zero: $3i$
27. $h(x) = x^4 - 9x^3 + 21x^2 + 21x - 130$; zero: $3 - 2i$
28. $f(x) = x^4 - 7x^3 + 14x^2 - 38x - 60$; zero: $1 + 3i$
29. $h(x) = 3x^5 + 2x^4 + 15x^3 + 10x^2 - 528x - 352$; zero: $-4i$
30. $g(x) = 2x^5 - 3x^4 - 5x^3 - 15x^2 - 207x + 108$; zero: $3i$

In Problems 31–40, find the complex zeros of each polynomial function. Write f in factored form.

31. $f(x) = x^3 - 1$

32. $f(x) = x^4 - 1$

33. $f(x) = x^3 - 8x^2 + 25x - 26$

34. $f(x) = x^3 + 13x^2 + 57x + 85$

35. $f(x) = x^4 + 5x^2 + 4$

36. $f(x) = x^4 + 13x^2 + 36$

37. $f(x) = x^4 + 2x^3 + 22x^2 + 50x - 75$

38. $f(x) = x^4 + 3x^3 - 19x^2 + 27x - 252$

39. $f(x) = 3x^4 - x^3 - 9x^2 + 159x - 52$

40. $f(x) = 2x^4 + x^3 - 35x^2 - 113x + 65$

Mixed Practice

41. Given $f(x) = 2x^3 - 14x^2 + bx - 3$ with $f(2) = 0$, $g(x) = x^3 + cx^2 - 8x + 30$, with the zero $x = 3 - i$, and b and c real numbers, find $(f \cdot g)(1)$.†

42. Let f be the polynomial function of degree 4 with real coefficients, leading coefficient 1, and zeros $x = 3 + i, 2, -2$. Let g be the polynomial function of degree 4 with intercept $(0, -4)$ and zeros $x = i, 2i$. Find $(f + g)(1)$.†

43. The complex zeros of $f(x) = x^4 + 1$ For the function $f(x) = x^4 + 1$:
(a) Factor f into the product of two irreducible quadratics. (***Hint:*** complete the square by adding and subtracting $2x^2$.)
(b) Find the zeros of f by finding the zeros of each irreducible quadratic.

†Courtesy of the Joliet Junior College Mathematics Department

Discussion and Writing

In Problems 44 and 45, explain why the facts given are contradictory.

44. f is a polynomial function of degree 3 whose coefficients are real numbers; its zeros are $2, i$, and $3 + i$.

45. f is a polynomial function of degree 3 whose coefficients are real numbers; its zeros are $4 + i, 4 - i$, and $2 + i$.

46. f is a polynomial function of degree 4 whose coefficients are real numbers; two of its zeros are -3 and $4 - i$. Explain why one of the remaining zeros must be a real number. Write down one of the missing zeros.

47. f is a polynomial function of degree 4 whose coefficients are real numbers; three of its zeros are 2, $1 + 2i$, and $1 - 2i$. Explain why the remaining zero must be a real number.

48. For the polynomial function $f(x) = x^2 + 2ix - 10$:
(a) Verify that $3 - i$ is a zero of f.
(b) Verify that $3 + i$ is not a zero of f.
(c) Explain why these results do not contradict the Conjugate Pairs Theorem.

Retain Your Knowledge

Problems 49–52 are based on material learned earlier in the course. The purpose of these problems is to keep the material fresh in your mind so that you are better prepared for the final exam.

49. Draw a scatter diagram for the given data.

x	−1	1	2	5	8	10
y	−4	0	3	1	5	7

50. Solve: $\sqrt{3 - x} = 5$

51. Multiply: $(2x - 5)(3x^2 + x - 4)$

52. Find the area and circumference of a circle with a diameter of 6 feet.

'Are You Prepared?' Answers

1. Sum: $3i$; product: $1 + 21i$ **2.** $-1 - i, \quad -1 + i$

Chapter Review

Things to Know

Power function (pp. 323–326)

$f(x) = x^n, n \geq 2$ even — Domain: all real numbers Range: nonnegative real numbers
Passes through $(-1, 1)$, $(0, 0)$, $(1, 1)$
Even function
Decreasing on $(-\infty, 0)$, increasing on $(0, \infty)$

$f(x) = x^n, n \geq 3$ odd — Domain: all real numbers Range: all real numbers
Passes through $(-1, -1)$, $(0, 0)$, $(1, 1)$
Odd function
Increasing on $(-\infty, \infty)$

Polynomial function (pp. 322, 327–333)

$f(x) = a_n x^n + a_{n-1}x^{n-1} + \cdots + a_1 x + a_0,\ a_n \neq 0$ — Domain: all real numbers
At most $n - 1$ turning points
End behavior: Behaves like $y = a_n x^n$ for large $|x|$

Real zeros of a polynomial function *f* (p. 327) — Real numbers for which $f(x) = 0$; the real zeros of f are the x-intercepts of the graph of f.

Rational function (pp. 343–350)

$R(x) = \dfrac{p(x)}{q(x)}$ — Domain: $\{x \mid q(x) \neq 0\}$

p, q are polynomial functions and q is not the zero polynomial. — Vertical asymptotes: With $R(x)$ in lowest terms, if $q(r) = 0$ for some real number, then $x = r$ is a vertical asymptote.
Horizontal or oblique asymptote: See the summary on page 348.

Remainder Theorem (p. 376) — If a polynomial function $f(x)$ is divided by $x - c$, then the remainder is $f(c)$.

Factor Theorem (p. 377) — $x - c$ is a factor of a polynomial function $f(x)$ if and only if $f(c) = 0$.

Descartes' Rule of Signs (p. 378) — Let f denote a polynomial function written in standard form. The number of positive zeros of f either equals the number of variations in sign of the nonzero coefficients of $f(x)$ or else equals that number less some even integer. The number of negative zeros of f either equals the number of variations in sign of the nonzero coefficients of $f(-x)$ or else equals that number less some even integer.

Rational Zeros Theorem (p. 379) — Let f be a polynomial function of degree 1 or higher of the form

$$f(x) = a_n x^n + a_{n-1}x^{n-1} + \cdots + a_1 x + a_0 \quad a_n \neq 0, a_0 \neq 0$$

where each coefficient is an integer. If $\dfrac{p}{q}$, in lowest terms, is a rational zero of f, then p must be a factor of a_0, and q must be a factor of a_n.

Intermediate Value Theorem (p. 384) — Let f be a polynomial function. If $a < b$ and $f(a)$ and $f(b)$ are of opposite sign, then there is at least one real zero of f between a and b.

Fundamental Theorem of Algebra (p. 390) — Every complex polynomial function f of degree $n \geq 1$ has at least one complex zero.

Conjugate Pairs Theorem (p. 391) — Let f be a polynomial function whose coefficients are real numbers. If $r = a + bi$ is a zero of f, then its complex conjugate $\bar{r} = a - bi$ is also a zero of f.

Objectives

Section		You should be able to ...	Example(s)	Review Exercises
5.1	1	Identify polynomial functions and their degree (p. 322)	1	1–4
	2	Graph polynomial functions using transformations (p. 326)	2, 3	5–7
	3	Identify the real zeros of a polynomial function and their multiplicity (p. 327)	4–9	8–11
	4	Analyze the graph of a polynomial function (p. 334)	10, 11	8–11
	5	Build cubic models from data (p. 336)	12	50
5.2	1	Find the domain of a rational function (p. 343)	1–3	12–14
	2	Find the vertical asymptotes of a rational function (p. 346)	4	12–14
	3	Find the horizontal or oblique asymptote of a rational function (p. 348)	5–8	12–14
5.3	1	Analyze the graph of a rational function (p. 353)	1–6	15–20
	2	Solve applied problems involving rational functions (p. 364)	7	49
5.4	1	Solve polynomial inequalities (p. 368)	1, 2	21, 22
	2	Solve rational inequalities (p. 370)	3, 4	23–25
5.5	1	Use the Remainder and Factor Theorems (p. 375)	1, 2	26–28
	2	Use Descartes' Rule of Signs to determine the number of positive and the number of negative real zeros of a polynomial function (p. 378)	3	29, 30
	3	Use the Rational Zeros Theorem to list the potential rational zeros of a polynomial function (p. 379)	4	31–34
	4	Find the real zeros of a polynomial function (p. 380)	5, 6	32–34
	5	Solve polynomial equations (p. 382)	7	35, 36
	6	Use the Theorem for Bounds on Zeros (p. 383)	8	37, 38
	7	Use the Intermediate Value Theorem (p. 384)	9, 10	39–42
5.6	1	Use the Conjugate Pairs Theorem (p. 391)	1	43, 44
	2	Find a polynomial function with specified zeros (p. 392)	2	43, 44
	3	Find the complex zeros of a polynomial function (p. 393)	3	45–48

Review Exercises

In Problems 1–4, determine whether the function is a polynomial function, a rational function, or neither. For those that are polynomial functions, state the degree. For those that are not polynomial functions, tell why not.

1. $f(x) = 4x^5 - 3x^2 + 5x - 2$ **2.** $f(x) = \dfrac{3x^5}{2x + 1}$ **3.** $f(x) = 3x^2 + 5x^{1/2} - 1$ **4.** $f(x) = 3$

In Problems 5–7, graph each function using transformations (shifting, compressing, stretching, and reflecting). Show all the stages.

5. $f(x) = (x + 2)^3$ **6.** $f(x) = -(x - 1)^4$ **7.** $f(x) = (x - 1)^4 + 2$

In Problems 8–11, analyze the graph of each polynomial function by following Steps 1 through 5 on page 335.

8. $f(x) = x(x + 2)(x + 4)$ **9.** $f(x) = (x - 2)^2(x + 4)$

10. $f(x) = -2x^3 + 4x^2$ **11.** $f(x) = (x - 1)^2(x + 3)(x + 1)$

In Problems 12–14, find the domain of each rational function. Find any horizontal, vertical, or oblique asymptotes.

12. $R(x) = \dfrac{x + 2}{x^2 - 9}$ **13.** $R(x) = \dfrac{x^2 + 4}{x - 2}$ **14.** $R(x) = \dfrac{x^2 + 3x + 2}{(x + 2)^2}$

In Problems 15–20, analyze the graph of each rational function following the seven steps given on page 356.

15. $R(x) = \dfrac{2x - 6}{x}$ **16.** $H(x) = \dfrac{x + 2}{x(x - 2)}$ **17.** $R(x) = \dfrac{x^2 + x - 6}{x^2 - x - 6}$

18. $F(x) = \dfrac{x^3}{x^2 - 4}$ **19.** $R(x) = \dfrac{2x^4}{(x - 1)^2}$ **20.** $G(x) = \dfrac{x^2 - 4}{x^2 - x - 2}$

In Problems 21–25, solve each inequality. Graph the solution set.

21. $x^3 + x^2 < 4x + 4$ **22.** $x^3 + 4x^2 \geq x + 4$ **23.** $\frac{2x - 6}{1 - x} < 2$ **24.** $\frac{(x - 2)(x - 1)}{x - 3} \geq 0$ **25.** $\frac{x^2 - 8x + 12}{x^2 - 16} > 0$

In Problems 26 and 27, find the remainder R when $f(x)$ is divided by $g(x)$. Is g a factor of f?

26. $f(x) = 8x^3 - 3x^2 + x + 4;\quad g(x) = x - 1$ **27.** $f(x) = x^4 - 2x^3 + 15x - 2;\quad g(x) = x + 2$

28. Use synthetic division to find the value of $f(x) = 12x^6 - 8x^4 + 1$ at $x = 4$.

In Problems 29 and 30, use Descartes' Rule of Signs to determine how many positive and negative zeros each polynomial function may have. Do not attempt to find the zeros.

29. $f(x) = 12x^8 - x^7 + 8x^4 - 2x^3 + x + 3$

30. $f(x) = -6x^5 + x^4 + 5x^3 + x + 1$

31. List all the potential rational zeros of $f(x) = 12x^8 - x^7 + 6x^4 - x^3 + x - 3$.

In Problems 32–34, use the Rational Zeros Theorem to find all the real zeros of each polynomial function. Use the zeros to factor f over the real numbers.

32. $f(x) = x^3 - 3x^2 - 6x + 8$ **33.** $f(x) = 4x^3 + 4x^2 - 7x + 2$ **34.** $f(x) = x^4 - 4x^3 + 9x^2 - 20x + 20$

In Problems 35 and 36, solve each equation in the real number system.

35. $2x^4 + 2x^3 - 11x^2 + x - 6 = 0$ **36.** $2x^4 + 7x^3 + x^2 - 7x - 3 = 0$

In Problems 37 and 38, find bounds to the real zeros of each polynomial function.

37. $f(x) = x^3 - x^2 - 4x + 2$ **38.** $f(x) = 2x^3 - 7x^2 - 10x + 35$

In Problems 39 and 40, use the Intermediate Value Theorem to show that each polynomial function has a zero in the given interval.

39. $f(x) = 3x^3 - x - 1;\ [0, 1]$ **40.** $f(x) = 8x^4 - 4x^3 - 2x - 1;\ [0, 1]$

In Problems 41 and 42, each polynomial function has exactly one positive zero. Approximate the zero correct to two decimal places.

41. $f(x) = x^3 - x - 2$ **42.** $f(x) = 8x^4 - 4x^3 - 2x - 1$

In Problems 43 and 44, information is given about a complex polynomial f whose coefficients are real numbers. Find the remaining zeros of f. Then find a polynomial function with real coefficients that has the zeros.

43. Degree 3; zeros: $4 + i, 6$ **44.** Degree 4; zeros: $i, 1 + i$

In Problems 45–48, find the complex zeros of each polynomial function f. Write f in factored form.

45. $f(x) = x^3 - 3x^2 - 6x + 8$ **46.** $f(x) = 4x^3 + 4x^2 - 7x + 2$

47. $f(x) = x^4 - 4x^3 + 9x^2 - 20x + 20$ **48.** $f(x) = 2x^4 + 2x^3 - 11x^2 + x - 6$

---

49. Making a Can A can in the shape of a right circular cylinder is required to have a volume of 250 cubic centimeters.

(a) Express the amount A of material needed to make the can as a function of the radius r of the cylinder.

(b) How much material is required if the can is of radius 3 centimeters?

(c) How much material is required if the can is of radius 5 centimeters?

(d) Graph $A = A(r)$. For what value of r is A smallest?

50. Housing Prices The data in the table on the right represent the January median new-home prices in the United States for the years shown.

(a) With a graphing utility, draw a scatter diagram of the data. Comment on the type of relation that appears to exist between the two variables.

(b) Decide on the function of best fit to these data (linear, quadratic, or cubic), and use this function to predict the median new-home price in the United States for January 2020 ($t = 9$).

(c) Draw the function of best fit on the scatter diagram obtained in part (a).

Year, *t*	Median Price, *P* ($1000s)
2004, 1	209.5
2006, 2	244.9
2008, 3	232.4
2010, 4	218.2
2012, 5	221.7
2014, 6	262.7

Chapter Test

CHAPTER Test Prep VIDEOS The Chapter Test Prep Videos are step-by-step solutions available in MyMathLab, or on this text's YouTube Channel. Flip back to the Resources for Success page for a link to this text's YouTube channel.

1. Graph $f(x) = (x-3)^4 - 2$ using transformations.
2. For the polynomial function $g(x) = 2x^3 + 5x^2 - 28x - 15$,
   (a) Determine the maximum number of real zeros that the function may have.
   (b) List the potential rational zeros.
   (c) Determine the real zeros of g. Factor g over the reals.
   (d) Find the x- and y-intercepts of the graph of g.
   (e) Determine whether the graph crosses or touches the x-axis at each x-intercept.
   (f) Find the power function that the graph of g resembles for large values of $|x|$.
   (g) Put all the information together to obtain the graph of g.
3. Find the complex zeros of $f(x) = x^3 - 4x^2 + 25x - 100$.
4. Solve $3x^3 + 2x - 1 = 8x^2 - 4$ in the complex number system.

In Problems 5 and 6, find the domain of each function. Find any horizontal, vertical, or oblique asymptotes.

5. $g(x) = \dfrac{2x^2 - 14x + 24}{x^2 + 6x - 40}$
6. $r(x) = \dfrac{x^2 + 2x - 3}{x + 1}$
7. Sketch the graph of the function in Problem 6. Label all intercepts, vertical asymptotes, horizontal asymptotes, and oblique asymptotes.

In Problems 8 and 9, write a function that meets the given conditions.

8. Fourth-degree polynomial with real coefficients; zeros: $-2, 0, 3 + i$.
9. Rational function; asymptotes: $y = 2, x = 4$; domain: $\{x | x \neq 4, x \neq 9\}$
10. Use the Intermediate Value Theorem to show that the function $f(x) = -2x^2 - 3x + 8$ has at least one real zero on the interval $[0, 4]$.
11. Solve: $\dfrac{x+2}{x-3} < 2$

Cumulative Review

1. Find the distance between the points $P = (1, 3)$ and $Q = (-4, 2)$.
2. Solve the inequality $x^2 \geq x$ and graph the solution set.
3. Solve the inequality $x^2 - 3x < 4$ and graph the solution set.
4. Find a linear function with slope -3 that contains the point $(-1, 4)$. Graph the function.
5. Find the equation of the line parallel to the line $y = 2x + 1$ and containing the point $(3, 5)$. Express your answer in slope–intercept form, and graph the line.
6. Graph the equation $y = x^3$.
7. Does the relation $\{(3, 6), (1, 3), (2, 5), (3, 8)\}$ represent a function? Why or why not?
8. Solve the equation $x^3 - 6x^2 + 8x = 0$.
9. Solve the inequality $3x + 2 \leq 5x - 1$ and graph the solution set.
10. Find the center and the radius of the circle $x^2 + 4x + y^2 - 2y - 4 = 0$. Graph the circle.
11. For the equation $y = x^3 - 9x$, determine the intercepts and test for symmetry.
12. Find an equation of the line perpendicular to $3x - 2y = 7$ that contains the point $(1, 5)$.
13. Is the following the graph of a function? Why or why not?

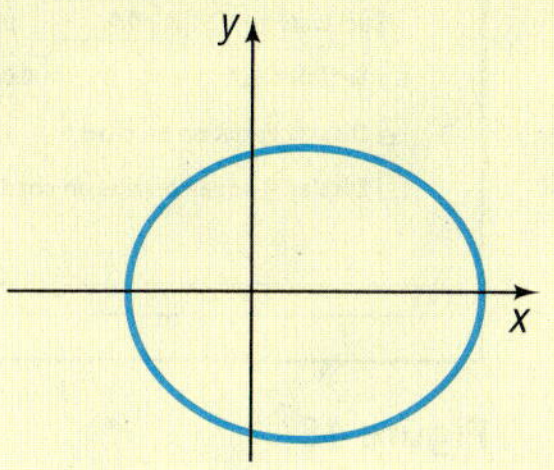

14. For the function $f(x) = x^2 + 5x - 2$, find
    (a) $f(3)$
    (b) $f(-x)$
    (c) $-f(x)$
    (d) $f(3x)$
    (e) $\dfrac{f(x+h) - f(x)}{h}$ $h \neq 0$
15. Answer the following questions regarding the function
$$f(x) = \frac{x+5}{x-1}$$
    (a) What is the domain of f?
    (b) Is the point $(2, 6)$ on the graph of f?
    (c) If $x = 3$, what is $f(x)$? What point is on the graph of f?
    (d) If $f(x) = 9$, what is x? What point is on the graph of f?
    (e) Is f a polynomial or a rational function?
16. Graph the function $f(x) = -3x + 7$.
17. Graph $f(x) = 2x^2 - 4x + 1$ by determining whether its graph opens up or down and by finding its vertex, axis of symmetry, y-intercept, and x-intercepts, if any.
18. Find the average rate of change of $f(x) = x^2 + 3x + 1$ from 1 to 2. Use this result to find the equation of the secant line containing $(1, f(1))$ and $(2, f(2))$.
19. In parts (a) to (f) on page 400, use the following graph.

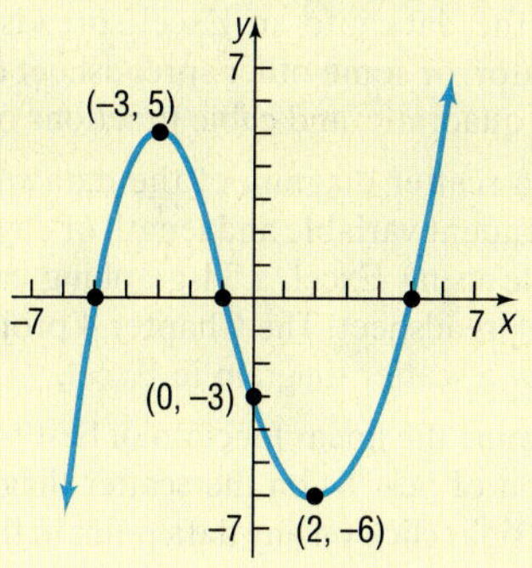

(a) Determine the intercepts.
(b) Based on the graph, tell whether the graph is symmetric with respect to the x-axis, the y-axis, and/or the origin.
(c) Based on the graph, tell whether the function is even, odd, or neither.
(d) List the intervals on which f is increasing. List the intervals on which f is decreasing.
(e) List the numbers, if any, at which f has a local maximum. What are the local maxima values?
(f) List the numbers, if any, at which f has a local minimum. What are the local minima values?

20. Determine algebraically whether the function

$$f(x) = \frac{5x}{x^2 - 9}$$

is even, odd, or neither.

21. For the function $f(x) = \begin{cases} 2x + 1 & \text{if } -3 < x < 2 \\ -3x + 4 & \text{if } x \geq 2 \end{cases}$

(a) Find the domain of f.
(b) Locate any intercepts.
(c) Graph the function.
(d) Based on the graph, find the range.

22. Graph the function $f(x) = -3(x + 1)^2 + 5$ using transformations.

23. Suppose that $f(x) = x^2 - 5x + 1$ and $g(x) = -4x - 7$.
(a) Find $f + g$ and state its domain.
(b) Find $\frac{f}{g}$ and state its domain.

24. Demand Equation The price p (in dollars) and the quantity x sold of a certain product obey the demand equation

$$p = -\frac{1}{10}x + 150, \quad 0 \leq x \leq 1500$$

(a) Express the revenue R as a function of x.
(b) What is the revenue if 100 units are sold?
(c) What quantity x maximizes revenue? What is the maximum revenue?
(d) What price should the company charge to maximize revenue?

Chapter Projects

Internet-based Project

I. Length of Day Go to *http://en.wikipedia.org/wiki/Latitude* and read about latitude through the subhead "Effect of Latitude." Now go to *http://www.orchidculture.com/COD/daylength.html#0N*.

1. For a particular day of the year, record in a table the length of day for the equator (0°N), 5°N, 10°N, . . ., 60°N. Enter the data into an Excel spreadsheet, TI-graphing calculator, or some other spreadsheet capable of finding linear, quadratic, and cubic functions of best fit.
2. Draw a scatter diagram of the data with latitude as the independent variable and length of day as the dependent variable using Excel, a TI-graphing calculator, or some other spreadsheet. The Chapter 4 project describes how to draw a scatter diagram in Excel.
3. Determine the linear function of best fit. Graph the linear function of best fit on the scatter diagram. To do this in Excel, right click on any data point in the scatter diagram. Now click the Add Chart Element menu, select Trendline, and then select More Trendline Options. Select the Linear radio button and select Display Equation on Chart. See Figure 48. Move the Trendline Options window off to the side and you will see the linear function of best fit displayed on the scatter diagram. Do you think the function accurately describes the relation between latitude and length of day?

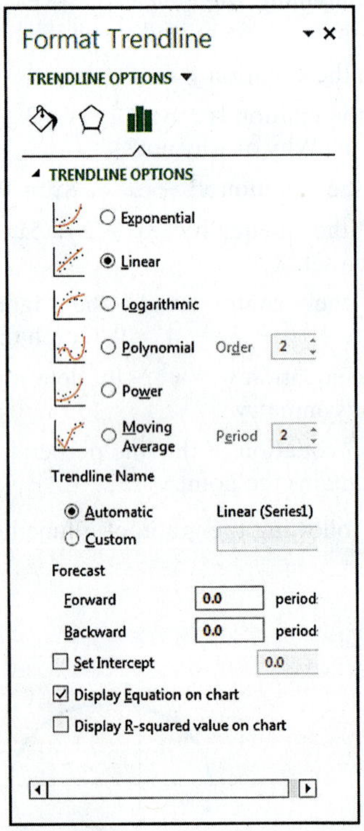

Figure 48

4. Determine the quadratic function of best fit. Graph the quadratic function of best fit on the scatter diagram. To do this in Excel, click on any data point in the scatter diagram. Now click the Add Chart Element menu, select Trendline, and then select More Trendline Options. Select the Polynomial radio button with Order set to 2. Select Display Equation on chart. Move the Trendline Options window off to the side and you will see the quadratic function of best fit displayed on the scatter diagram. Do you think the function accurately describes the relation between latitude and length of day?

5. Determine the cubic function of best fit. Graph the cubic function of best fit on the scatter diagram. To do this in Excel, click on any data point in the scatter diagram. Now click the Add Chart Element menu, select Trendline, and then select More Trendline Options. Select the Polynomial radio button with Order set to 3. Select Display Equation on chart. Move the Trendline Options window off to the side and you will see the cubic function of best fit displayed on the scatter diagram. Do you think the function accurately describes the relation between latitude and length of day?

6. Which of the three models seems to fit the data best? Explain your reasoning.

7. Use your model to predict the hours of daylight on the day you selected for Chicago (41.85 degrees north latitude). Go to the Old Farmer's Almanac or another website to determine the hours of daylight in Chicago for the day you selected. How do the two compare?

The following project is available at the Instructor's Resource Center (IRC):

II. Theory of Equations The coefficients of a polynomial function can be found if its zeros are known, an advantage of using polynomials in modeling.

6 Exponential and Logarithmic Functions

Depreciation of Cars

You are ready to buy that first new car. You know that cars lose value over time due to depreciation and that different cars have different rates of depreciation. So you will research the depreciation rates for the cars you are thinking of buying. After all, for cars that sell for about the same price, the lower the depreciation rate, the more the car will be worth each year.

—See the Internet-based Chapter Project I—

Outline

••• A Look Back

Until now, our study of functions has concentrated on polynomial and rational functions. These functions belong to the class of **algebraic functions**—that is, functions that can be expressed in terms of sums, differences, products, quotients, powers, or roots of polynomials. Functions that are not algebraic are termed **transcendental** (they transcend, or go beyond, algebraic functions).

A Look Ahead •••

In this chapter, we study two transcendental functions: the exponential function and the logarithmic function. These functions occur frequently in a wide variety of applications, such as biology, chemistry, economics, and psychology.

The chapter begins with a discussion of composite, one-to-one, and inverse functions—concepts that are needed to explain the relationship between exponential and logarithmic functions.

6.1 Composite Functions

PREPARING FOR THIS SECTION *Before getting started, review the following:*

- Find the Value of a Function (Section 3.1, pp. 202–206)
- Domain of a Function (Section 3.1, pp. 206–208)

Now Work the **'Are You Prepared?'** problems on page 408.

OBJECTIVES 1 Form a Composite Function (p. 403)
2 Find the Domain of a Composite Function (p. 404)

1 Form a Composite Function

Suppose that an oil tanker is leaking oil and you want to determine the area of the circular oil patch around the ship. See Figure 1. It is determined that the oil is leaking from the tanker in such a way that the radius of the circular patch of oil around the ship is increasing at a rate of 3 feet per minute. Therefore, the radius r of the oil patch at any time t, in minutes, is given by $r(t) = 3t$. So after 20 minutes, the radius of the oil patch is $r(20) = 3(20) = 60$ feet.

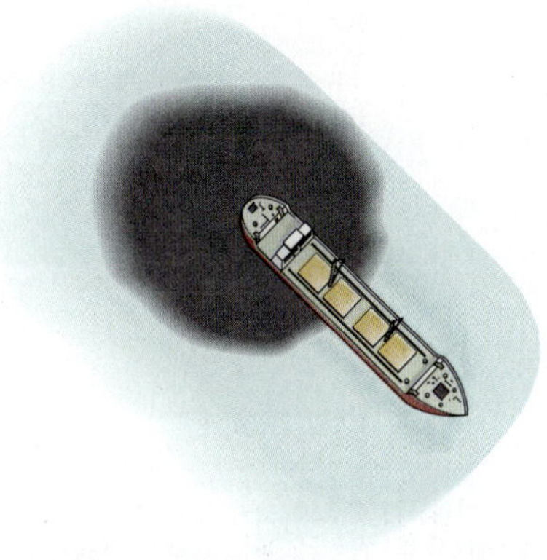

Figure 1

The area A of a circle as a function of the radius r is given by $A(r) = \pi r^2$. The area of the circular patch of oil after 20 minutes is $A(60) = \pi(60)^2 = 3600\pi$ square feet. Note that $60 = r(20)$, so $A(60) = A(r(20))$. The argument of the function A is the output of the function r!

In general, the area of the oil patch can be expressed as a function of time t by evaluating $A(r(t))$ and obtaining $A(r(t)) = A(3t) = \pi(3t)^2 = 9\pi t^2$. The function $A(r(t))$ is a special type of function called a *composite function.*

As another example, consider the function $y = (2x + 3)^2$. Let $y = f(u) = u^2$ and $u = g(x) = 2x + 3$. Then by a substitution process, the original function is obtained as follows: $y = f(u) = f(g(x)) = (2x + 3)^2$.

In general, suppose that f and g are two functions and that x is a number in the domain of g. Evaluating g at x yields $g(x)$. If $g(x)$ is in the domain of f, then evaluating f at $g(x)$ yields the expression $f(g(x))$. The correspondence from x to $f(g(x))$ is called a *composite function* $f \circ g$.

DEFINITION

Given two functions f and g, the **composite function**, denoted by $f \circ g$ (read as "f composed with g"), is defined by

$$(f \circ g)(x) = f(g(x))$$

The domain of $f \circ g$ is the set of all numbers x in the domain of g such that $g(x)$ is in the domain of f.

Look carefully at Figure 2. Only those values of x in the domain of g for which $g(x)$ is in the domain of f can be in the domain of $f \circ g$. The reason is that if $g(x)$ is not in the domain of f, then $f(g(x))$ is not defined. Because of this, the domain of $f \circ g$ is a subset of the domain of g; the range of $f \circ g$ is a subset of the range of f.

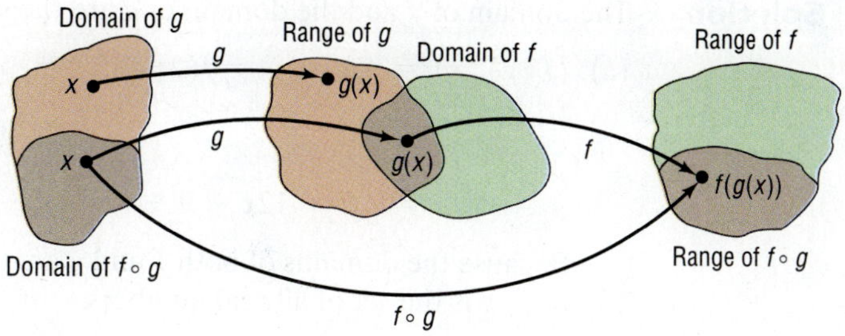

Figure 2

Figure 3 provides a second illustration of the definition. Here x is the input to the function g, yielding $g(x)$. Then $g(x)$ is the input to the function f, yielding $f(g(x))$. Note that the "inside" function g in $f(g(x))$ is "processed" first.

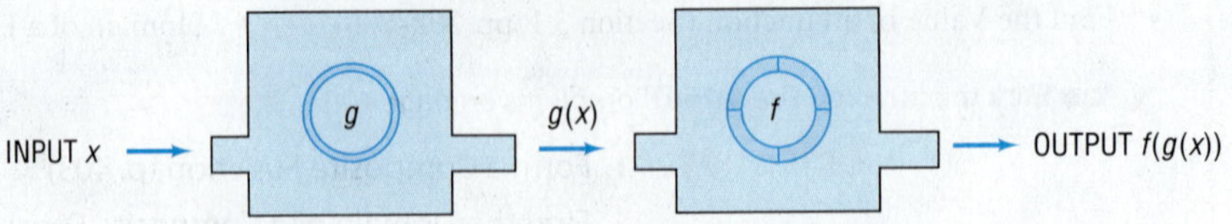

Figure 3

EXAMPLE 1

Evaluating a Composite Function

Suppose that $f(x) = 2x^2 - 3$ and $g(x) = 4x$. Find:

(a) $(f \circ g)(1)$ (b) $(g \circ f)(1)$ (c) $(f \circ f)(-2)$ (d) $(g \circ g)(-1)$

Solution (a) $(f \circ g)(1) = f(g(1)) = f(4) = 2 \cdot 4^2 - 3 = 29$

$g(x) = 4x$, $g(1) = 4$; $f(x) = 2x^2 - 3$

(b) $(g \circ f)(1) = g(f(1)) = g(-1) = 4 \cdot (-1) = -4$

$f(x) = 2x^2 - 3$, $f(1) = -1$; $g(x) = 4x$

(c) $(f \circ f)(-2) = f(f(-2)) = f(5) = 2 \cdot 5^2 - 3 = 47$

$f(-2) = 2(-2)^2 - 3 = 5$

(d) $(g \circ g)(-1) = g(g(-1)) = g(-4) = 4 \cdot (-4) = -16$

$g(-1) = -4$

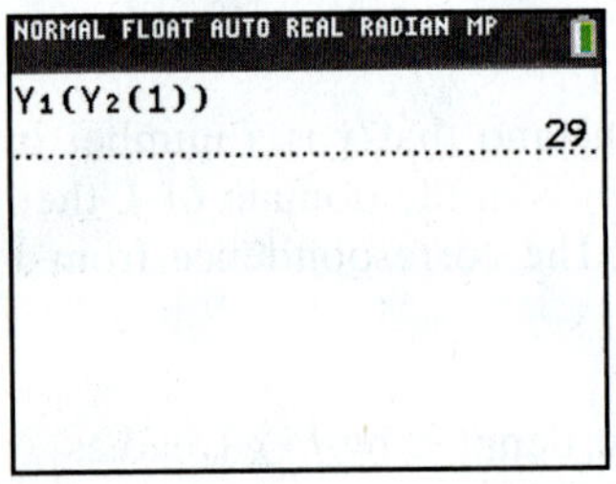

Figure 4

COMMENT Graphing calculators can be used to evaluate composite functions.* Let $Y_1 = f(x) = 2x^2 - 3$ and $Y_2 = g(x) = 4x$. Then, using a TI-84 Plus C graphing calculator, find $(f \circ g)(1)$ as shown in Figure 4. Note that this is the result obtained in Example 1(a).

Now Work PROBLEM 13

2 Find the Domain of a Composite Function

EXAMPLE 2

Finding a Composite Function and Its Domain

Suppose that $f(x) = x^2 + 3x - 1$ and $g(x) = 2x + 3$.

Find: (a) $f \circ g$ (b) $g \circ f$

Then find the domain of each composite function.

Solution The domain of f and the domain of g are the set of all real numbers.

(a) $(f \circ g)(x) = f(g(x)) = f(2x + 3) = (2x + 3)^2 + 3(2x + 3) - 1$

$f(x) = x^2 + 3x - 1$

$$= 4x^2 + 12x + 9 + 6x + 9 - 1 = 4x^2 + 18x + 17$$

Because the domains of both f and g are the set of all real numbers, the domain of $f \circ g$ is the set of all real numbers.

* Consult your owner's manual for the appropriate keystrokes.

(b) $(g \circ f)(x) = g(f(x)) = g(x^2 + 3x - 1) = 2(x^2 + 3x - 1) + 3$

$\uparrow$ $g(x) = 2x + 3$

$$= 2x^2 + 6x - 2 + 3 = 2x^2 + 6x + 1$$

Because the domains of both f and g are the set of all real numbers, the domain of $g \circ f$ is the set of all real numbers.

Example 2 illustrates that, in general, $f \circ g \neq g \circ f$. Sometimes $f \circ g$ does equal $g \circ f$, as we shall see in Example 5.

Look back at Figure 2 on page 403. In determining the domain of the composite function $(f \circ g)(x) = f(g(x))$, keep the following two thoughts in mind about the input x.

1. Any x not in the domain of g must be excluded.
2. Any x for which $g(x)$ is not in the domain of f must be excluded.

EXAMPLE 3 Finding the Domain of $f \circ g$

Find the domain of $f \circ g$ if $f(x) = \dfrac{1}{x + 2}$ and $g(x) = \dfrac{4}{x - 1}$.

Solution For $(f \circ g)(x) = f(g(x))$, first note that the domain of g is $\{x \mid x \neq 1\}$, so 1 is excluded from the domain of $f \circ g$. Next note that the domain of f is $\{x \mid x \neq -2\}$, which means that $g(x)$ cannot equal -2. Solve the equation $g(x) = -2$ to determine what additional value(s) of x to exclude.

$$\frac{4}{x - 1} = -2 \qquad g(x) = -2$$

$$4 = -2(x - 1) \qquad \text{Multiply both sides by } x - 1.$$

$$4 = -2x + 2 \qquad \text{Apply the Distributive Property.}$$

$$2x = -2 \qquad \text{Add } 2x \text{ to both sides. Subtract 4 from both sides.}$$

$$x = -1 \qquad \text{Divide both sides by 2.}$$

Also exclude -1 from the domain of $f \circ g$.

The domain of $f \circ g$ is $\{x \mid x \neq -1, x \neq 1\}$.

✓**Check:** For $x = 1$, $g(x) = \dfrac{4}{x - 1}$ is not defined, so $(f \circ g)(x) = f(g(x))$ is not defined.

For $x = -1$, $g(-1) = -2$, and $(f \circ g)(-1) = f(g(-1)) = f(-2)$ is not defined.

EXAMPLE 4 Finding a Composite Function and Its Domain

Suppose that $f(x) = \dfrac{1}{x + 2}$ and $g(x) = \dfrac{4}{x - 1}$.

Find: (a) $f \circ g$ (b) $f \circ f$

Then find the domain of each composite function.

Solution The domain of f is $\{x \mid x \neq -2\}$ and the domain of g is $\{x \mid x \neq 1\}$.

(a) $$(f \circ g)(x) = f(g(x)) = f\left(\frac{4}{x - 1}\right) = \frac{1}{\dfrac{4}{x - 1} + 2} = \frac{x - 1}{4 + 2(x - 1)} = \frac{x - 1}{2x + 2} = \frac{x - 1}{2(x + 1)}$$

$\uparrow$ $f(x) = \dfrac{1}{x + 2}$ $\qquad$ $\uparrow$ Multiply by $\dfrac{x - 1}{x - 1}$.

In Example 3, the domain of $f \circ g$ was found to be $\{x \mid x \neq -1, x \neq 1\}$.

The domain of $f \circ g$ also can be found by first looking at the domain of g: $\{x \mid x \neq 1\}$. Exclude 1 from the domain of $f \circ g$ as a result. Then look at $f \circ g$ and note that x cannot equal -1, because $x = -1$ results in division by 0. So exclude -1 from the domain of $f \circ g$. Therefore, the domain of $f \circ g$ is $\{x \mid x \neq -1, x \neq 1\}$.

(b) $$(f \circ f)(x) = f(f(x)) = f\left(\frac{1}{x+2}\right) = \frac{1}{\dfrac{1}{x+2} + 2} = \frac{x+2}{1 + 2(x+2)} = \frac{x+2}{2x+5}$$

$f(x) = \dfrac{1}{x+2}$ ↑   Multiply by $\dfrac{x+2}{x+2}$. ↑

The domain of $f \circ f$ consists of all values of x in the domain of f, $\{x \mid x \neq -2\}$, for which

$$f(x) = \frac{1}{x+2} \neq -2 \qquad \frac{1}{x+2} = -2$$

$$1 = -2(x+2)$$
$$1 = -2x - 4$$
$$2x = -5$$
$$x = -\frac{5}{2}$$

or, equivalently,

$$x \neq -\frac{5}{2}$$

The domain of $f \circ f$ is $\left\{x \,\middle|\, x \neq -\dfrac{5}{2}, x \neq -2\right\}$.

The domain of $f \circ f$ also can be found by recognizing that -2 is not in the domain of f and so should be excluded from the domain of $f \circ f$. Then, looking at $f \circ f$, note that x cannot equal $-\dfrac{5}{2}$. Do you see why? Therefore, the domain of $f \circ f$ is $\left\{x \,\middle|\, x \neq -\dfrac{5}{2}, x \neq -2\right\}$. ●

Now Work PROBLEMS 27 AND 29

EXAMPLE 5 Showing That Two Composite Functions Are Equal

If $f(x) = 3x - 4$ and $g(x) = \dfrac{1}{3}(x+4)$, show that

$$(f \circ g)(x) = (g \circ f)(x) = x$$

for every x in the domain of $f \circ g$ and $g \circ f$.

Solution $$(f \circ g)(x) = f(g(x))$$

$$= f\left(\frac{x+4}{3}\right) \qquad g(x) = \frac{1}{3}(x+4) = \frac{x+4}{3}$$

$$= 3\left(\frac{x+4}{3}\right) - 4 \qquad f(x) = 3x - 4.$$

$$= x + 4 - 4 = x$$

Seeing the Concept

Using a graphing calculator, let

$Y_1 = f(x) = 3x - 4$

$Y_2 = g(x) = \frac{1}{3}(x + 4)$

$Y_3 = f \circ g,\ Y_4 = g \circ f$

Using the viewing window $-3 \le x \le 3$, $-2 \le y \le 2$, graph only Y_3 and Y_4. What do you see? TRACE to verify that $Y_3 = Y_4$.

$$\begin{aligned}(g \circ f)(x) &= g(f(x))\\ &= g(3x - 4) && f(x) = 3x - 4\\ &= \frac{1}{3}[(3x - 4) + 4] && g(x) = \frac{1}{3}(x + 4).\\ &= \frac{1}{3}(3x) = x\end{aligned}$$

We conclude that $(f \circ g)(x) = (g \circ f)(x) = x$. ●

In Section 6.2, we shall see that there is an important relationship between functions f and g for which $(f \circ g)(x) = (g \circ f)(x) = x$.

Now Work PROBLEM 39

Calculus Application

Some techniques in calculus require the ability to determine the components of a composite function. For example, the function $H(x) = \sqrt{x + 1}$ is the composition of the functions f and g, where $f(x) = \sqrt{x}$ and $g(x) = x + 1$, because $H(x) = (f \circ g)(x) = f(g(x)) = f(x + 1) = \sqrt{x + 1}$.

EXAMPLE 6 Finding the Components of a Composite Function

Find functions f and g such that $f \circ g = H$ if $H(x) = (x^2 + 1)^{50}$.

Solution The function H takes $x^2 + 1$ and raises it to the power 50. A natural way to decompose H is to raise the function $g(x) = x^2 + 1$ to the power 50. Let $f(x) = x^{50}$ and $g(x) = x^2 + 1$. Then

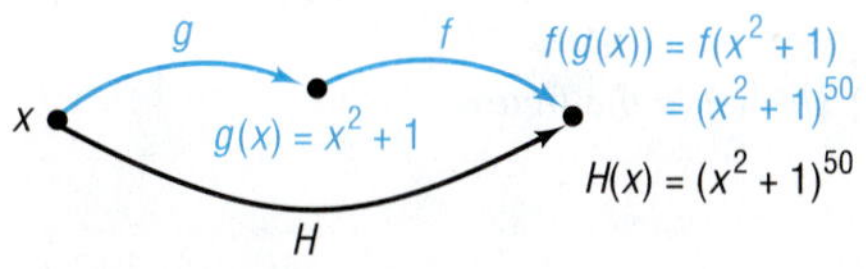

Figure 5

$$\begin{aligned}(f \circ g)(x) &= f(g(x))\\ &= f(x^2 + 1)\\ &= (x^2 + 1)^{50} = H(x)\end{aligned}$$

See Figure 5. ●

Other functions f and g may be found for which $f \circ g = H$ in Example 6. For instance, if $f(x) = x^2$ and $g(x) = (x^2 + 1)^{25}$, then

$$(f \circ g)(x) = f(g(x)) = f((x^2 + 1)^{25}) = [(x^2 + 1)^{25}]^2 = (x^2 + 1)^{50}$$

Although the functions f and g found as a solution to Example 6 are not unique, there is usually a "natural" selection for f and g that comes to mind first.

EXAMPLE 7 Finding the Components of a Composite Function

Find functions f and g such that $f \circ g = H$ if $H(x) = \dfrac{1}{x + 1}$.

Solution Here H is the reciprocal of $g(x) = x + 1$. Let $f(x) = \dfrac{1}{x}$ and $g(x) = x + 1$. Then

$$(f \circ g)(x) = f(g(x)) = f(x + 1) = \frac{1}{x + 1} = H(x)$$ ●

Now Work PROBLEM 47

6.1 Assess Your Understanding

'Are You Prepared?' *Answers are given at the end of these exercises. If you get a wrong answer, read the pages listed in red.*

1. Find $f(3)$ if $f(x) = -4x^2 + 5x$. (pp. 202–206)
2. Find $f(3x)$ if $f(x) = 4 - 2x^2$. (pp. 202–206)
3. Find the domain of the function $f(x) = \dfrac{x^2 - 1}{x^2 - 25}$. (pp. 206–208)

Concepts and Vocabulary

4. Given two functions f and g, the ________ ________, denoted $f \circ g$, is defined by $(f \circ g)(x) =$ ________.
5. ***True or False*** If $f(x) = x^2$ and $g(x) = \sqrt{x + 9}$, then $(f \circ g)(4) = 5$.
6. If $f(x) = \sqrt{x + 2}$ and $g(x) = \dfrac{3}{x}$, which of the following does $(f \circ g)(x)$ equal?

   (a) $\dfrac{3}{\sqrt{x + 2}}$ (b) $\dfrac{3}{\sqrt{x}} + 2$ (c) $\sqrt{\dfrac{3}{x} + 2}$ (d) $\sqrt{\dfrac{3}{x + 2}}$
7. If $H = f \circ g$ and $H(x) = \sqrt{25 - 4x^2}$, which of the following cannot be the component functions f and g?

   (a) $f(x) = \sqrt{25 - x^2}$; $g(x) = 4x$
   (b) $f(x) = \sqrt{x}$; $g(x) = 25 - 4x^2$
   (c) $f(x) = \sqrt{25 - x}$; $g(x) = 4x^2$
   (d) $f(x) = \sqrt{25 - 4x}$; $g(x) = x^2$
8. ***True or False*** The domain of the composite function $(f \circ g)(x)$ is the same as the domain of $g(x)$.

Skill Building

In Problems 9 and 10, evaluate each expression using the values given in the table.

9.

x	−3	−2	−1	0	1	2	3
f(x)	−7	−5	−3	−1	3	5	7
g(x)	8	3	0	−1	0	3	8

(a) $(f \circ g)(1)$ (b) $(f \circ g)(-1)$
(c) $(g \circ f)(-1)$ (d) $(g \circ f)(0)$
(e) $(g \circ g)(-2)$ (f) $(f \circ f)(-1)$

10.

x	−3	−2	−1	0	1	2	3
f(x)	11	9	7	5	3	1	−1
g(x)	−8	−3	0	1	0	−3	−8

(a) $(f \circ g)(1)$ (b) $(f \circ g)(2)$
(c) $(g \circ f)(2)$ (d) $(g \circ f)(3)$
(e) $(g \circ g)(1)$ (f) $(f \circ f)(3)$

In Problems 11 and 12, evaluate each expression using the graphs of $y = f(x)$ and $y = g(x)$ shown in the figure.

11. (a) $(g \circ f)(-1)$ (b) $(g \circ f)(0)$
    (c) $(f \circ g)(-1)$ (d) $(f \circ g)(4)$
12. (a) $(g \circ f)(1)$ (b) $(g \circ f)(5)$
    (c) $(f \circ g)(0)$ (d) $(f \circ g)(2)$

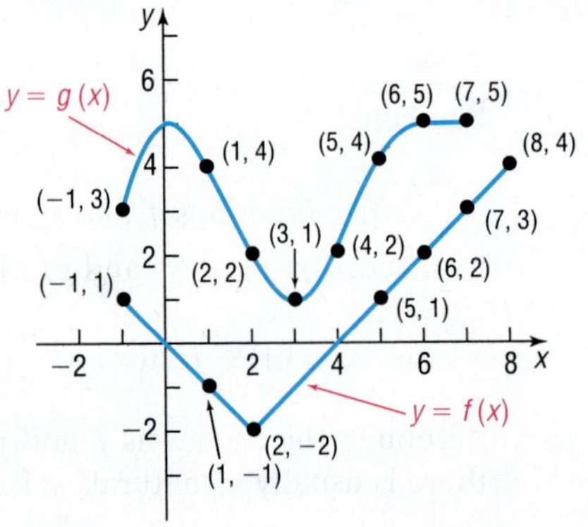

In Problems 13–22, for the given functions f and g, find:

(a) $(f \circ g)(4)$ *(b)* $(g \circ f)(2)$ *(c)* $(f \circ f)(1)$ *(d)* $(g \circ g)(0)$

13. $f(x) = 2x$; $g(x) = 3x^2 + 1$
14. $f(x) = 3x + 2$; $g(x) = 2x^2 - 1$
15. $f(x) = 4x^2 - 3$; $g(x) = 3 - \dfrac{1}{2}x^2$
16. $f(x) = 2x^2$; $g(x) = 1 - 3x^2$
17. $f(x) = \sqrt{x}$; $g(x) = 2x$
18. $f(x) = \sqrt{x + 1}$; $g(x) = 3x$
19. $f(x) = |x|$; $g(x) = \dfrac{1}{x^2 + 1}$
20. $f(x) = |x - 2|$; $g(x) = \dfrac{3}{x^2 + 2}$
21. $f(x) = \dfrac{3}{x + 1}$; $g(x) = \sqrt[3]{x}$
22. $f(x) = x^{3/2}$; $g(x) = \dfrac{2}{x + 1}$

In Problems 23–38, for the given functions f and g, find:
(a) $f \circ g$ *(b)* $g \circ f$ *(c)* $f \circ f$ *(d)* $g \circ g$
State the domain of each composite function.

23. $f(x) = 2x + 3;\ g(x) = 3x$

24. $f(x) = -x;\ g(x) = 2x - 4$

25. $f(x) = 3x + 1;\ g(x) = x^2$

26. $f(x) = x + 1;\ g(x) = x^2 + 4$

27. $f(x) = x^2;\ g(x) = x^2 + 4$

28. $f(x) = x^2 + 1;\ g(x) = 2x^2 + 3$

29. $f(x) = \dfrac{3}{x-1};\ g(x) = \dfrac{2}{x}$

30. $f(x) = \dfrac{1}{x+3};\ g(x) = -\dfrac{2}{x}$

31. $f(x) = \dfrac{x}{x-1};\ g(x) = -\dfrac{4}{x}$

32. $f(x) = \dfrac{x}{x+3};\ g(x) = \dfrac{2}{x}$

33. $f(x) = \sqrt{x};\ g(x) = 2x + 3$

34. $f(x) = \sqrt{x-2};\ g(x) = 1 - 2x$

35. $f(x) = x^2 + 1;\ g(x) = \sqrt{x-1}$

36. $f(x) = x^2 + 4;\ g(x) = \sqrt{x-2}$

37. $f(x) = \dfrac{x-5}{x+1};\ g(x) = \dfrac{x+2}{x-3}$

38. $f(x) = \dfrac{2x-1}{x-2};\ g(x) = \dfrac{x+4}{2x-5}$

In Problems 39–46, show that $(f \circ g)(x) = (g \circ f)(x) = x$.

39. $f(x) = 2x;\ g(x) = \dfrac{1}{2}x$

40. $f(x) = 4x;\ g(x) = \dfrac{1}{4}x$

41. $f(x) = x^3;\ g(x) = \sqrt[3]{x}$

42. $f(x) = x + 5;\ g(x) = x - 5$

43. $f(x) = 2x - 6;\ g(x) = \dfrac{1}{2}(x + 6)$

44. $f(x) = 4 - 3x;\ g(x) = \dfrac{1}{3}(4 - x)$

45. $f(x) = ax + b;\ g(x) = \dfrac{1}{a}(x - b)\quad a \neq 0$

46. $f(x) = \dfrac{1}{x};\ g(x) = \dfrac{1}{x}$

In Problems 47–52, find functions f and g so that $f \circ g = H$.

47. $H(x) = (2x + 3)^4$

48. $H(x) = (1 + x^2)^3$

49. $H(x) = \sqrt{x^2 + 1}$

50. $H(x) = \sqrt{1 - x^2}$

51. $H(x) = |2x + 1|$

52. $H(x) = |2x^2 + 3|$

Applications and Extensions

53. If $f(x) = 2x^3 - 3x^2 + 4x - 1$ and $g(x) = 2$, find $(f \circ g)(x)$ and $(g \circ f)(x)$.

54. If $f(x) = \dfrac{x+1}{x-1}$, find $(f \circ f)(x)$.

55. If $f(x) = 2x^2 + 5$ and $g(x) = 3x + a$, find a so that the graph of $f \circ g$ crosses the y-axis at 23.

56. If $f(x) = 3x^2 - 7$ and $g(x) = 2x + a$, find a so that the graph of $f \circ g$ crosses the y-axis at 68.

In Problems 57 and 58, use the functions f and g to find:
(a) $f \circ g$ *(b)* $g \circ f$
(c) the domain of $f \circ g$ *and of* $g \circ f$
(d) the conditions for which $f \circ g = g \circ f$

57. $f(x) = ax + b\quad g(x) = cx + d$

58. $f(x) = \dfrac{ax+b}{cx+d}\quad g(x) = mx$

59. Surface Area of a Balloon The surface area S (in square meters) of a hot-air balloon is given by

$$S(r) = 4\pi r^2$$

where r is the radius of the balloon (in meters). If the radius r is increasing with time t (in seconds) according to the formula $r(t) = \dfrac{2}{3}t^3, t \geq 0$, find the surface area S of the balloon as a function of the time t.

60. Volume of a Balloon The volume V (in cubic meters) of the hot-air balloon described in Problem 59 is given by $V(r) = \dfrac{4}{3}\pi r^3$. If the radius r is the same function of t as in Problem 59, find the volume V as a function of the time t.

61. Automobile Production The number N of cars produced at a certain factory in one day after t hours of operation is given by $N(t) = 100t - 5t^2, 0 \leq t \leq 10$. If the cost C (in dollars) of producing N cars is $C(N) = 15{,}000 + 8000N$, find the cost C as a function of the time t of operation of the factory.

62. Environmental Concerns The spread of oil leaking from a tanker is in the shape of a circle. If the radius r (in feet) of the spread after t hours is $r(t) = 200\sqrt{t}$, find the area A of the oil slick as a function of the time t.

63. Production Cost The price p, in dollars, of a certain product and the quantity x sold obey the demand equation

$$p = -\frac{1}{4}x + 100\quad 0 \leq x \leq 400$$

Suppose that the cost C, in dollars, of producing x units is

$$C = \frac{\sqrt{x}}{25} + 600$$

Assuming that all items produced are sold, find the cost C as a function of the price p.

[**Hint:** Solve for x in the demand equation and then form the composite function.]

64. Cost of a Commodity The price p, in dollars, of a certain commodity and the quantity x sold obey the demand equation

$$p = -\frac{1}{5}x + 200 \quad 0 \le x \le 1000$$

Suppose that the cost C, in dollars, of producing x units is

$$C = \frac{\sqrt{x}}{10} + 400$$

Assuming that all items produced are sold, find the cost C as a function of the price p.

65. Volume of a Cylinder The volume V of a right circular cylinder of height h and radius r is $V = \pi r^2 h$. If the height is twice the radius, express the volume V as a function of r.

66. Volume of a Cone The volume V of a right circular cone is $V = \frac{1}{3}\pi r^2 h$. If the height is twice the radius, express the volume V as a function of r.

67. Foreign Exchange Traders often buy foreign currency in the hope of making money when the currency's value changes. For example, on April 28, 2014, one U.S. dollar could purchase 0.7235 euro, and one euro could purchase 141.119 yen. Let $f(x)$ represent the number of euros you can buy with x dollars, and let $g(x)$ represent the number of yen you can buy with x euros.

(a) Find a function that relates dollars to euros.

(b) Find a function that relates euros to yen.

(c) Use the results of parts (a) and (b) to find a function that relates dollars to yen. That is, find

$$(g \circ f)(x) = g(f(x)).$$

(d) What is $g(f(1000))$?

68. Temperature Conversion The function $C(F) = \frac{5}{9}(F - 32)$ converts a temperature in degrees Fahrenheit, F, to a temperature in degrees Celsius, C. The function $K(C) = C + 273$, converts a temperature in degrees Celsius to a temperature in kelvins, K.

(a) Find a function that converts a temperature in degrees Fahrenheit to a temperature in kelvins.

(b) Determine 80 degrees Fahrenheit in kelvins.

69. Discounts The manufacturer of a computer is offering two discounts on last year's model computer. The first discount is a \$200 rebate and the second discount is 20% off the regular price, p.

(a) Write a function f that represents the sale price if only the rebate applies.

(b) Write a function g that represents the sale price if only the 20% discount applies.

(c) Find $f \circ g$ and $g \circ f$. What does each of these functions represent? Which combination of discounts represents a better deal for the consumer? Why?

70. Taxes Suppose that you work for \$15 per hour. Write a function that represents gross salary G as a function of hours worked h. Your employer is required to withhold taxes (federal income tax, Social Security, Medicare) from your paycheck. Suppose your employer withholds 20% of your income for taxes. Write a function that represents net salary N as a function of gross salary G. Find and interpret $N \circ G$.

71. Let $f(x) = ax + b$ and $g(x) = bx + a$, where a and b are integers. If $f(1) = 8$ and $f(g(20)) - g(f(20)) = -14$, find the product of a and b.*

72. If f and g are odd functions, show that the composite function $f \circ g$ is also odd.

73. If f is an odd function and g is an even function, show that the composite functions $f \circ g$ and $g \circ f$ are both even.

* Courtesy of the Joliet Junior College Mathematics Department

Retain Your Knowledge

Problems 74–77 are based on material learned earlier in the course. The purpose of these problems is to keep the material fresh in your mind so that you are better prepared for the final exam.

74. Given $f(x) = 3x + 8$ and $g(x) = x - 5$, find $(f + g)(x)$, $(f - g)(x)$, $(f \cdot g)(x)$, and $\left(\frac{f}{g}\right)(x)$. State the domain of each.

75. Find the real zeros of $f(x) = 2x - 5\sqrt{x} + 2$.

76. Use a graphing utility to graph $f(x) = -x^3 + 4x - 2$ over the interval $(-3, 3)$. Approximate any local maxima and local minima. Determine where the function is increasing and where it is decreasing.

77. Find the domain of $R(x) = \frac{x^2 + 6x + 5}{x - 3}$. Find any horizontal, vertical, or oblique asymptotes.

'Are You Prepared?' Answers

1. -21 **2.** $4 - 18x^2$ **3.** $\{x \mid x \neq -5, x \neq 5\}$

6.2 One-to-One Functions; Inverse Functions

PREPARING FOR THIS SECTION *Before getting started, review the following:*

- Functions (Section 3.1, pp. 199–208)
- Increasing/Decreasing Functions (Section 3.3, pp. 225–226)
- Rational Expressions (Chapter R, Section R.7, pp. 62–69)

Now Work the **'Are You Prepared?'** problems on page 419.

OBJECTIVES
1. Determine Whether a Function Is One-to-One (p. 411)
2. Determine the Inverse of a Function Defined by a Map or a Set of Ordered Pairs (p. 413)
3. Obtain the Graph of the Inverse Function from the Graph of the Function (p. 416)
4. Find the Inverse of a Function Defined by an Equation (p. 417)

1 Determine Whether a Function Is One-to-One

Section 3.1 presented four different ways to represent a function: (1) a map, (2) a set of ordered pairs, (3) a graph, and (4) an equation. For example, Figures 6 and 7 illustrate two different functions represented as mappings. The function in Figure 6 shows the correspondence between states and their populations (in millions). The function in Figure 7 shows a correspondence between animals and life expectancies (in years).

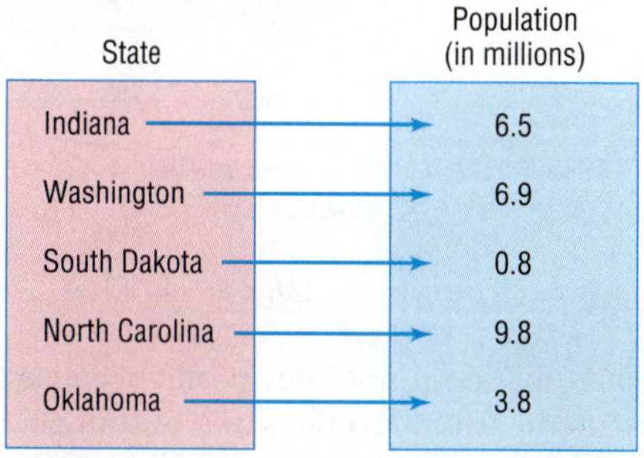

Figure 6

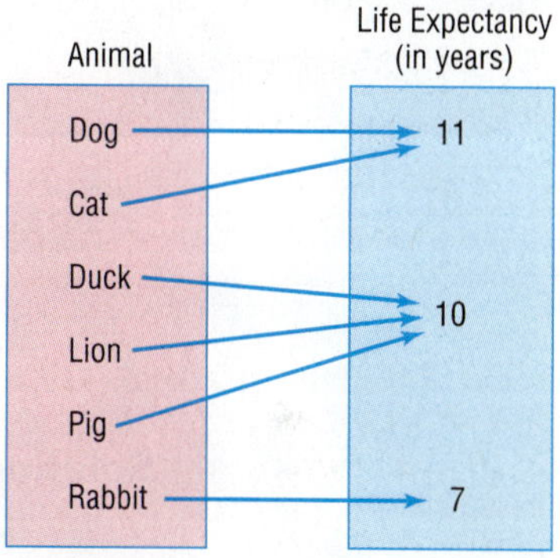

Figure 7

Suppose several people are asked to name a state that has a population of 0.8 million based on the function in Figure 6. Everyone will respond "South Dakota." Now, if the same people are asked to name an animal whose life expectancy is 11 years based on the function in Figure 7, some may respond "dog," while others may respond "cat." What is the difference between the functions in Figures 6 and 7? In Figure 6, no two elements in the domain correspond to the same element in the range. In Figure 7, this is not the case: Different elements in the domain correspond to the same element in the range. Functions such as the one in Figure 6 are given a special name.

DEFINITION

A function is **one-to-one** if any two different inputs in the domain correspond to two different outputs in the range. That is, if x_1 and x_2 are two different inputs of a function f, then f is one-to-one if $f(x_1) \neq f(x_2)$.

In Words
A function is not one-to-one if two different inputs correspond to the same output.

Put another way, a function f is one-to-one if no y in the range is the image of more than one x in the domain. A function is not one-to-one if any two (or more) different elements in the domain correspond to the same element in the range. So the function in Figure 7 is not one-to-one because two different elements in the domain, *dog* and *cat*, both correspond to 11 (and also because three different elements in the domain correspond to 10). Figure 8 illustrates the distinction among one-to-one functions, functions that are not one-to-one, and relations that are not functions.

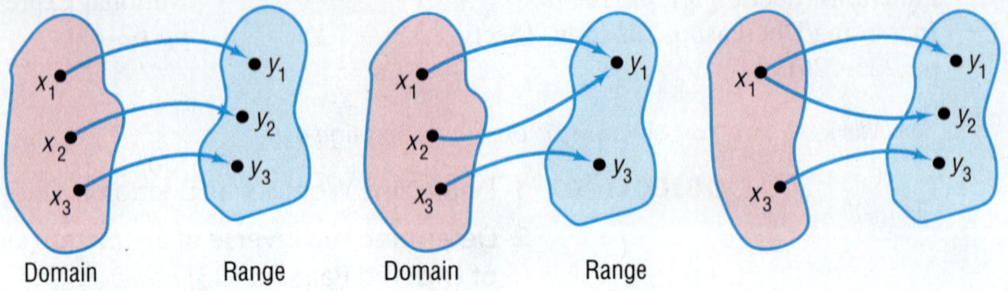

Figure 8 **(a)** One-to-one function: Each x in the domain has one and only one image in the range. **(b)** Not a one-to-one function: y_1 is the image of both x_1 and x_2. **(c)** Not a function: x_1 has two images, y_1 and y_2.

EXAMPLE 1 Determining Whether a Function Is One-to-One

Determine whether the following functions are one-to-one.

(a) For the following function, the domain represents the ages of five males, and the range represents their HDL (good) cholesterol scores (mg/dL).

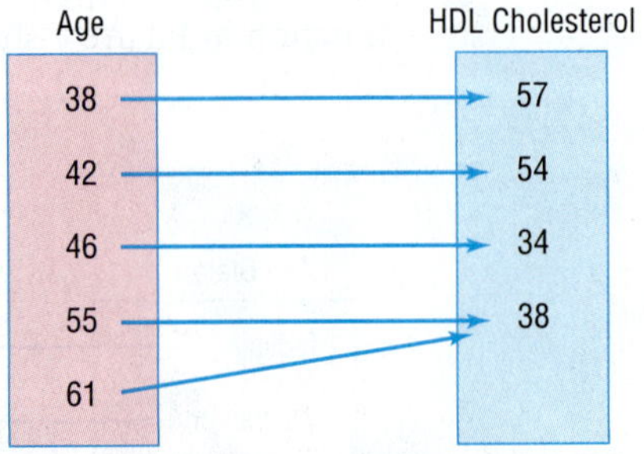

(b) $\{(-2,6), (-1,3), (0,2), (1,5), (2,8)\}$

Solution (a) The function is not one-to-one because there are two different inputs, 55 and 61, that correspond to the same output, 38.

(b) The function is one-to-one because no two distinct inputs correspond to the same output.

Now Work PROBLEMS 13 AND 17

For functions defined by an equation $y = f(x)$ and for which the graph of f is known, there is a simple test, called the **horizontal-line test**, to determine whether f is one-to-one.

THEOREM **Horizontal-line Test**

If every horizontal line intersects the graph of a function f in at most one point, then f is one-to-one.

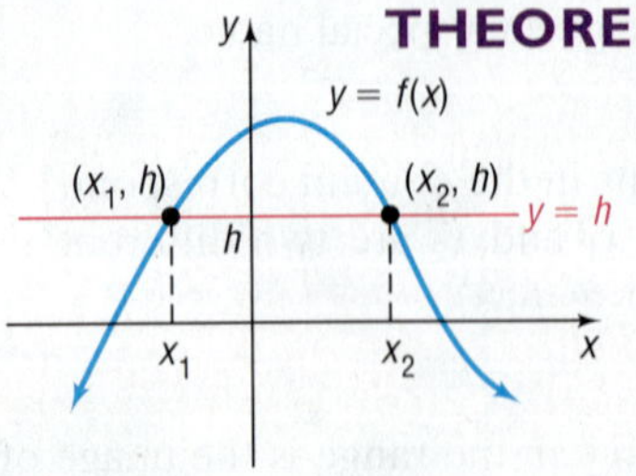

Figure 9 $f(x_1) = f(x_2) = h$ and $x_1 \neq x_2$; f is not a one-to-one function.

The reason why this test works can be seen in Figure 9, where the horizontal line $y = h$ intersects the graph at two distinct points, (x_1, h) and (x_2, h). Since h is the image of both x_1 and x_2 and $x_1 \neq x_2$, f is not one-to-one. Based on Figure 9, we can state the horizontal-line test in another way: If the graph of any horizontal line intersects the graph of a function f at more than one point, then f is not one-to-one.

EXAMPLE 2 **Using the Horizontal-line Test**

For each function, use its graph to determine whether the function is one-to-one.

(a) $f(x) = x^2$ (b) $g(x) = x^3$

Solution

(a) Figure 10(a) illustrates the horizontal-line test for $f(x) = x^2$. The horizontal line $y = 1$ intersects the graph of f twice, at $(1, 1)$ and at $(-1, 1)$, so f is not one-to-one.

(b) Figure 10(b) illustrates the horizontal-line test for $g(x) = x^3$. Because every horizontal line intersects the graph of g exactly once, it follows that g is one-to-one.

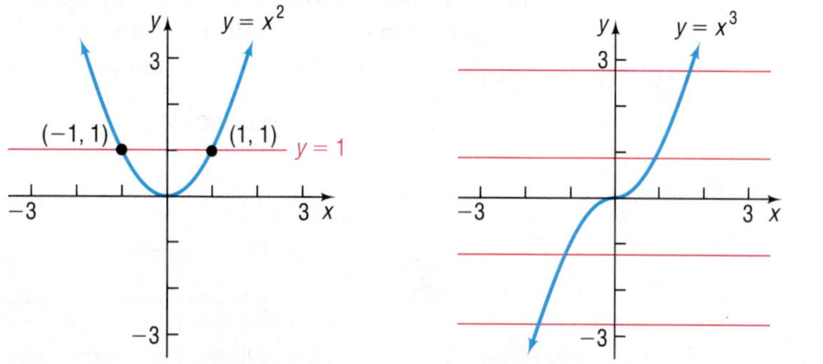

Figure 10 **(a)** A horizontal line intersects the graph twice; f is not one-to-one. **(b)** Every horizontal line intersects the graph exactly once; g is one-to-one.

Now Work PROBLEM 21

Look more closely at the one-to-one function $g(x) = x^3$. This function is an increasing function. Because an increasing (or decreasing) function will always have different y-values for unequal x-values, it follows that a function that is increasing (or decreasing) over its domain is also a one-to-one function.

THEOREM

A function that is increasing on an interval I is a one-to-one function on I.
A function that is decreasing on an interval I is a one-to-one function on I.

2 Determine the Inverse of a Function Defined by a Map or a Set of Ordered Pairs

DEFINITION

Suppose that f is a one-to-one function. Then, corresponding to each x in the domain of f, there is exactly one y in the range (because f is a function); and corresponding to each y in the range of f, there is exactly one x in the domain (because f is one-to-one). The correspondence from the range of f back to the domain of f is called the **inverse function of** f. The symbol f^{-1} is used to denote the inverse function of f.

In Words

Suppose that f is a one-to-one function so that the input 5 corresponds to the output 10. In the inverse function f^{-1}, the input 10 will correspond to the output 5.

We will discuss how to find inverses for all four representations of functions: (1) maps, (2) sets of ordered pairs, (3) graphs, and (4) equations. We begin with finding inverses of functions represented by maps or sets of ordered pairs.

EXAMPLE 3 **Finding the Inverse of a Function Defined by a Map**

Find the inverse of the function defined by the map on the next page. Let the domain of the function represent certain states, and let the range represent the states' populations (in millions). Find the domain and the range of the inverse function.

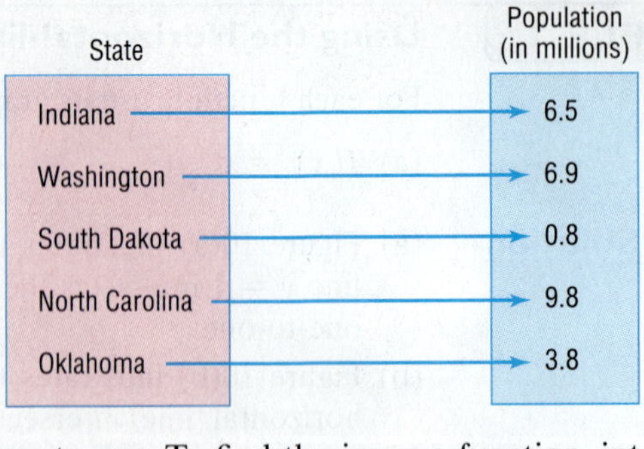

Solution The function is one-to-one. To find the inverse function, interchange the elements in the domain with the elements in the range. For example, the function receives as input Indiana and outputs 6.5 million. So the inverse receives as input 6.5 million and outputs Indiana. The inverse function is shown next.

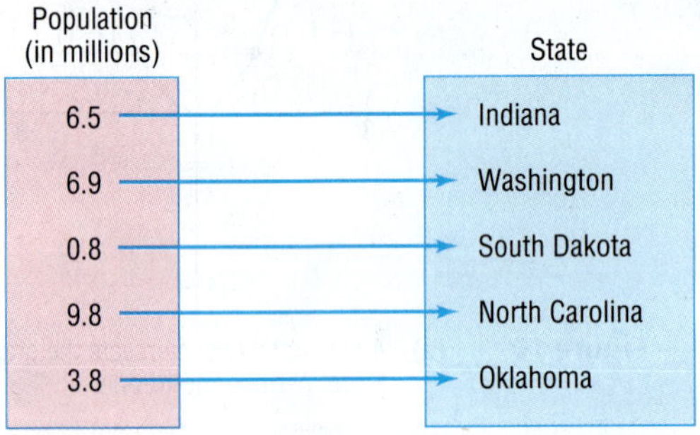

The domain of the inverse function is $\{6.5, 6.9, 0.8, 9.8, 3.8\}$. The range of the inverse function is {Indiana, Washington, South Dakota, North Carolina, Oklahoma}. ●

If the function f is a set of ordered pairs (x, y), then the inverse function of f, denoted f^{-1}, is the set of ordered pairs (y, x).

EXAMPLE 4 Finding the Inverse of a Function Defined by a Set of Ordered Pairs

Find the inverse of the following one-to-one function:

$$\{(-3, -27), (-2, -8), (-1, -1), (0, 0), (1, 1), (2, 8), (3, 27)\}$$

State the domain and the range of the function and its inverse.

Solution The inverse of the given function is found by interchanging the entries in each ordered pair and so is given by

$$\{(-27, -3), (-8, -2), (-1, -1), (0, 0), (1, 1), (8, 2), (27, 3)\}$$

The domain of the function is $\{-3, -2, -1, 0, 1, 2, 3\}$. The range of the function is $\{-27, -8, -1, 0, 1, 8, 27\}$. The domain of the inverse function is $\{-27, -8, -1, 0, 1, 8, 27\}$. The range of the inverse function is $\{-3, -2, -1, 0, 1, 2, 3\}$. ●

Now Work PROBLEMS 27 AND 31

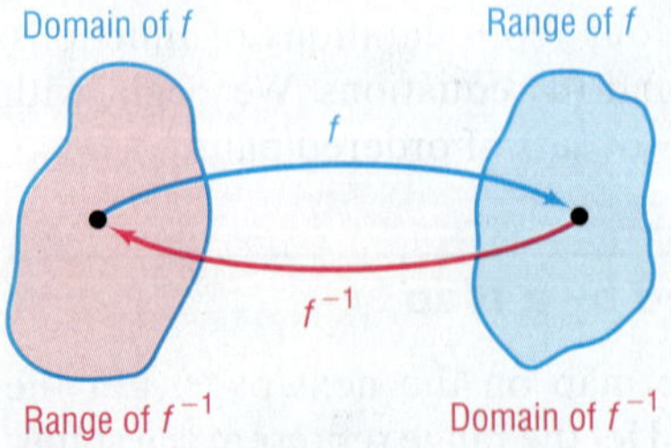

Figure 11

Remember, if f is a one-to-one function, it has an inverse function, f^{-1}. See Figure 11.

Based on the results of Example 4 and Figure 11, two facts are now apparent about a one-to-one function f and its inverse f^{-1}.

Domain of f = Range of f^{-1} Range of f = Domain of f^{-1}

Look again at Figure 11 to visualize the relationship. Starting with x, applying f, and then applying f^{-1} gets x back again. Starting with x, applying f^{-1}, and then applying f

WARNING *Be careful!* f^{-1} is a symbol for the inverse function of f. The -1 used in f^{-1} is not an exponent. That is, f^{-1} does *not* mean the reciprocal of f; $f^{-1}(x)$ is not equal to $\dfrac{1}{f(x)}$. ■

gets the number x back again. To put it simply, what f does, f^{-1} undoes, and vice versa. See the illustration that follows.

Input x from domain of f	Apply f $\longrightarrow$	$f(x)$	Apply f^{-1} $\longrightarrow$	$f^{-1}(f(x)) = x$
Input x from domain of f^{-1}	Apply f^{-1} $\longrightarrow$	$f^{-1}(x)$	Apply f $\longrightarrow$	$f(f^{-1}(x)) = x$

In other words,

$$f^{-1}(f(x)) = x \quad \text{where } x \text{ is in the domain of } f$$
$$f(f^{-1}(x)) = x \quad \text{where } x \text{ is in the domain of } f^{-1}$$

Consider the function $f(x) = 2x$, which multiplies the argument x by 2. The inverse function f^{-1} undoes whatever f does. So the inverse function of f is $f^{-1}(x) = \frac{1}{2}x$, which divides the argument by 2. For example, $f(3) = 2(3) = 6$ and $f^{-1}(6) = \frac{1}{2}(6) = 3$, so f^{-1} undoes what f did. This is verified by showing that

$$f^{-1}(f(x)) = f^{-1}(2x) = \frac{1}{2}(2x) = x \quad \text{and} \quad f(f^{-1}(x)) = f\left(\frac{1}{2}x\right) = 2\left(\frac{1}{2}x\right) = x$$

See Figure 12.

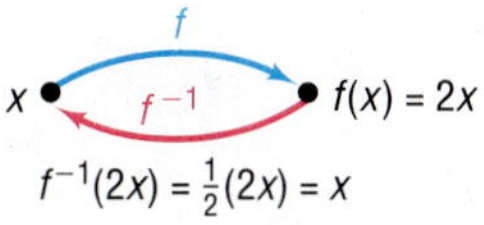

Figure 12

EXAMPLE 5 Verifying Inverse Functions

(a) Verify that the inverse of $g(x) = x^3$ is $g^{-1}(x) = \sqrt[3]{x}$.

(b) Verify that the inverse of $f(x) = 2x + 3$ is $f^{-1}(x) = \frac{1}{2}(x - 3)$.

Solution

(a) $g^{-1}(g(x)) = g^{-1}(x^3) = \sqrt[3]{x^3} = x$ for all x in the domain of g

$g(g^{-1}(x)) = g(\sqrt[3]{x}) = (\sqrt[3]{x})^3 = x$ for all x in the domain of g^{-1}

(b) $f^{-1}(f(x)) = f^{-1}(2x + 3) = \frac{1}{2}[(2x + 3) - 3] = \frac{1}{2}(2x) = x$ for all x in the domain of f

$f(f^{-1}(x)) = f\left(\frac{1}{2}(x - 3)\right) = 2\left[\frac{1}{2}(x - 3)\right] + 3 = (x - 3) + 3 = x$ for all x in the domain of f^{-1} ●

EXAMPLE 6 Verifying Inverse Functions

Verify that the inverse of $f(x) = \dfrac{1}{x - 1}$ is $f^{-1}(x) = \dfrac{1}{x} + 1$. For what values of x is $f^{-1}(f(x)) = x$? For what values of x is $f(f^{-1}(x)) = x$?

Solution The domain of f is $\{x \mid x \neq 1\}$ and the domain of f^{-1} is $\{x \mid x \neq 0\}$. Now

$$f^{-1}(f(x)) = f^{-1}\left(\frac{1}{x - 1}\right) = \frac{1}{\frac{1}{x - 1}} + 1 = x - 1 + 1 = x \quad \text{provided } x \neq 1$$

$$f(f^{-1}(x)) = f\left(\frac{1}{x} + 1\right) = \frac{1}{\left(\frac{1}{x} + 1\right) - 1} = \frac{1}{\frac{1}{x}} = x \quad \text{provided } x \neq 0$$ ●

Now Work PROBLEMS 35 AND 39

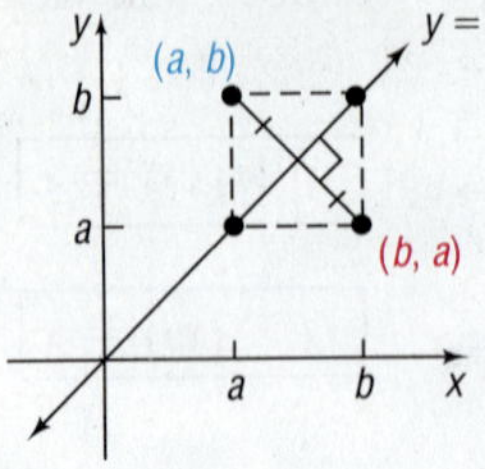

Figure 13

3 Obtain the Graph of the Inverse Function from the Graph of the Function

Suppose that (a, b) is a point on the graph of a one-to-one function f defined by $y = f(x)$. Then $b = f(a)$. This means that $a = f^{-1}(b)$, so (b, a) is a point on the graph of the inverse function f^{-1}. The relationship between the point (a, b) on f and the point (b, a) on f^{-1} is shown in Figure 13. The line segment with endpoints (a, b) and (b, a) is perpendicular to the line $y = x$ and is bisected by the line $y = x$. (Do you see why?) It follows that the point (b, a) on f^{-1} is the reflection about the line $y = x$ of the point (a, b) on f.

THEOREM

The graph of a one-to-one function f and the graph of its inverse function f^{-1} are symmetric with respect to the line $y = x$.

Figure 14 illustrates this result. Once the graph of f is known, the graph of f^{-1} may be obtained by reflecting the graph of f about the line $y = x$.

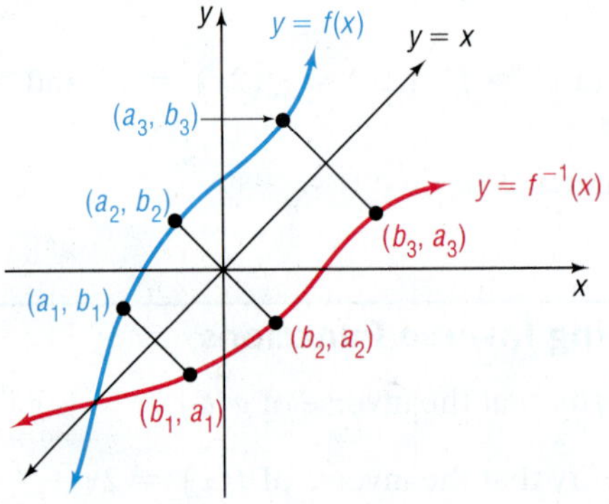

Figure 14

EXAMPLE 7 **Graphing the Inverse Function**

The graph in Figure 15(a) is that of a one-to-one function $y = f(x)$. Draw the graph of its inverse.

Solution Begin by adding the graph of $y = x$ to Figure 15(a). Since the points $(-2, -1)$, $(-1, 0)$, and $(2, 1)$ are on the graph of f, the points $(-1, -2)$, $(0, -1)$, and $(1, 2)$ must be on the graph of f^{-1}. Keeping in mind that the graph of f^{-1} is the reflection about the line $y = x$ of the graph of f, draw the graph of f^{-1}. See Figure 15(b).

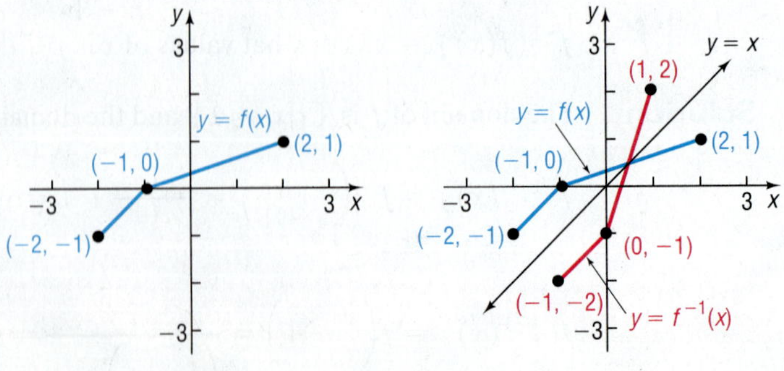

Figure 15 (a) (b)

Now Work PROBLEM 45

4 Find the Inverse of a Function Defined by an Equation

The fact that the graphs of a one-to-one function f and its inverse function f^{-1} are symmetric with respect to the line $y = x$ tells us more. It says that we can obtain f^{-1} by interchanging the roles of x and y in f. Look again at Figure 14. If f is defined by the equation

$$y = f(x)$$

then f^{-1} is defined by the equation

$$x = f(y)$$

The equation $x = f(y)$ defines f^{-1} *implicitly*. If we can solve this equation for y, we will have the *explicit* form of f^{-1}, that is,

$$y = f^{-1}(x)$$

Let's use this procedure to find the inverse of $f(x) = 2x + 3$. (Because f is a linear function and is increasing, f is one-to-one and so has an inverse function.)

EXAMPLE 8 How to Find the Inverse Function

Find the inverse of $f(x) = 2x + 3$. Graph f and f^{-1} on the same coordinate axes.

Step-by-Step Solution

Step 1: Replace $f(x)$ with y. In $y = f(x)$, interchange the variables x and y to obtain $x = f(y)$. This equation defines the inverse function f^{-1} implicitly.

Replace $f(x)$ with y in $f(x) = 2x + 3$ and obtain $y = 2x + 3$. Now interchange the variables x and y to obtain

$$x = 2y + 3$$

This equation defines the inverse function f^{-1} implicitly.

Step 2: If possible, solve the implicit equation for y in terms of x to obtain the explicit form of f^{-1}, $y = f^{-1}(x)$.

To find the explicit form of the inverse, solve $x = 2y + 3$ for y.

$$x = 2y + 3$$

$$2y + 3 = x \quad \text{Reflexive Property; If } a = b, \text{ then } b = a.$$

$$2y = x - 3 \quad \text{Subtract 3 from both sides.}$$

$$y = \frac{1}{2}(x - 3) \quad \text{Multiply both sides by } \frac{1}{2}.$$

The explicit form of the inverse function f^{-1} is

$$f^{-1}(x) = \frac{1}{2}(x - 3)$$

Step 3: Check the result by showing that $f^{-1}(f(x)) = x$ and $f(f^{-1}(x)) = x$.

We verified that f and f^{-1} are inverses in Example 5(b).

The graphs of $f(x) = 2x + 3$ and its inverse $f^{-1}(x) = \frac{1}{2}(x - 3)$ are shown in Figure 16. Note the symmetry of the graphs with respect to the line $y = x$. ●

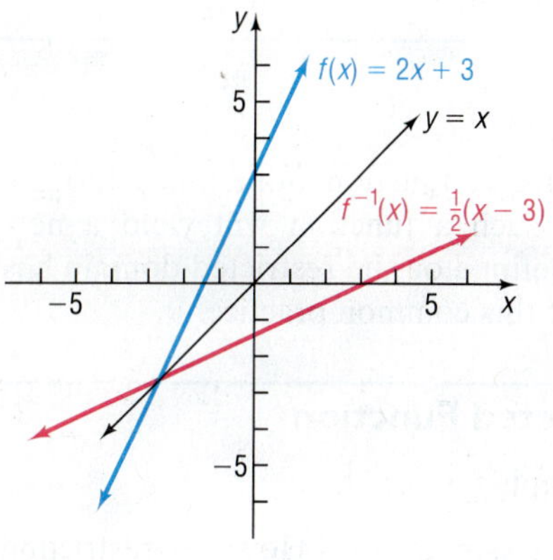

Figure 16

Procedure for Finding the Inverse of a One-to-One Function

STEP 1: In $y = f(x)$, interchange the variables x and y to obtain

$$x = f(y)$$

This equation defines the inverse function f^{-1} implicitly.

STEP 2: If possible, solve the implicit equation for y in terms of x to obtain the explicit form of f^{-1}:

$$y = f^{-1}(x)$$

STEP 3: Check the result by showing that

$$f^{-1}(f(x)) = x \quad \text{and} \quad f(f^{-1}(x)) = x$$

EXAMPLE 9 **Finding the Inverse Function**

The function

$$f(x) = \frac{2x+1}{x-1} \quad x \neq 1$$

is one-to-one. Find its inverse function and check the result.

Solution **Step 1:** Replace $f(x)$ with y and interchange the variables x and y in

$$y = \frac{2x+1}{x-1}$$

to obtain

$$x = \frac{2y+1}{y-1}$$

Step 2: Solve for y.

$$x = \frac{2y+1}{y-1}$$

$$x(y-1) = 2y+1 \quad \text{Multiply both sides by } y-1.$$

$$xy - x = 2y + 1 \quad \text{Apply the Distributive Property.}$$

$$xy - 2y = x + 1 \quad \text{Subtract } 2y \text{ from both sides; add } x \text{ to both sides.}$$

$$(x-2)y = x+1 \quad \text{Factor.}$$

$$y = \frac{x+1}{x-2} \quad \text{Divide by } x-2.$$

The inverse function is

$$f^{-1}(x) = \frac{x+1}{x-2} \quad x \neq 2 \quad \text{Replace } y \text{ by } f^{-1}(x).$$

Step 3: ✓**Check:**

$$f^{-1}(f(x)) = f^{-1}\left(\frac{2x+1}{x-1}\right) = \frac{\frac{2x+1}{x-1}+1}{\frac{2x+1}{x-1}-2} = \frac{2x+1+x-1}{2x+1-2(x-1)} = \frac{3x}{3} = x, \quad x \neq 1$$

$$f(f^{-1}(x)) = f\left(\frac{x+1}{x-2}\right) = \frac{2\left(\frac{x+1}{x-2}\right)+1}{\frac{x+1}{x-2}-1} = \frac{2(x+1)+x-2}{x+1-(x-2)} = \frac{3x}{3} = x, \quad x \neq 2$$

●

Exploration

In Example 9, we found that if $f(x) = \frac{2x+1}{x-1}$, then $f^{-1}(x) = \frac{x+1}{x-2}$. Compare the vertical and horizontal asymptotes of f and f^{-1}.

Result The vertical asymptote of f is $x = 1$, and the horizontal asymptote is $y = 2$. The vertical asymptote of f^{-1} is $x = 2$, and the horizontal asymptote is $y = 1$.

Now Work PROBLEMS 53 AND 67

If a function is not one-to-one, it has no inverse function. Sometimes, though, an appropriate restriction on the domain of such a function will yield a new function that *is* one-to-one. Then the function defined on the restricted domain has an inverse function. Let's look at an example of this common practice.

EXAMPLE 10 **Finding the Inverse of a Domain-restricted Function**

Find the inverse of $y = f(x) = x^2$ if $x \geq 0$. Graph f and f^{-1}.

Solution The function $y = x^2$ is not one-to-one. [Refer to Example 2(a).] However, restricting the domain of this function to $x \geq 0$, as indicated, results in a new function that is increasing and therefore is one-to-one. Consequently, the function defined by $y = f(x) = x^2, x \geq 0$, has an inverse function, f^{-1}.

Follow the steps given previously to find f^{-1}.

Step 1: In the equation $y = x^2, x \geq 0$, interchange the variables x and y. The result is

$$x = y^2 \qquad y \geq 0$$

This equation defines the inverse function implicitly.

Step 2: Solve for y to get the explicit form of the inverse. Because $y \geq 0$, only one solution for y is obtained: $y = \sqrt{x}$. So $f^{-1}(x) = \sqrt{x}$.

Step 3: ✓**Check:** $f^{-1}(f(x)) = f^{-1}(x^2) = \sqrt{x^2} = |x| = x$ because $x \geq 0$

$$f(f^{-1}(x)) = f(\sqrt{x}) = (\sqrt{x})^2 = x$$

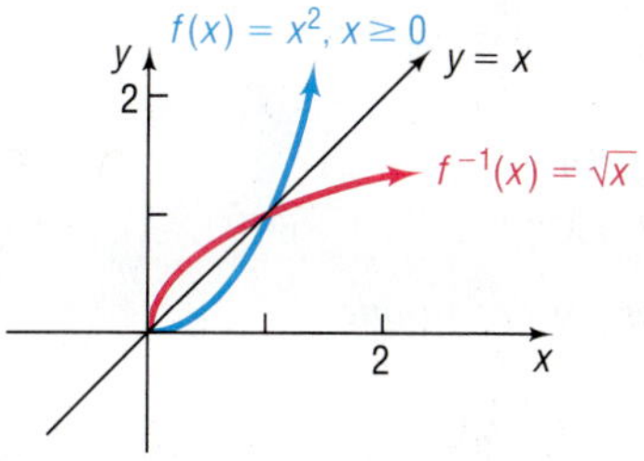

Figure 17

Figure 17 illustrates the graphs of $f(x) = x^2, x \geq 0$, and $f^{-1}(x) = \sqrt{x}$. ●

SUMMARY

1. If a function f is one-to-one, then it has an inverse function f^{-1}.
2. Domain of f = Range of f^{-1}; Range of f = Domain of f^{-1}.
3. To verify that f^{-1} is the inverse of f, show that $f^{-1}(f(x)) = x$ for every x in the domain of f and that $f(f^{-1}(x)) = x$ for every x in the domain of f^{-1}.
4. The graphs of f and f^{-1} are symmetric with respect to the line $y = x$.

6.2 Assess Your Understanding

'Are You Prepared?' *Answers are given at the end of these exercises. If you get a wrong answer, read the pages listed in red.*

1. Is the set of ordered pairs $\{(1,3), (2,3), (-1,2)\}$ a function? Why or why not? (pp. 199–208)
2. Where is the function $f(x) = x^2$ increasing? Where is it decreasing? (pp. 225–226)
3. What is the domain of $f(x) = \dfrac{x+5}{x^2 + 3x - 18}$? (pp. 199–208)
4. Simplify: $\dfrac{\frac{1}{x} + 1}{\frac{1}{x^2} - 1}$ (pp. 62–69)

Concepts and Vocabulary

5. If x_1 and x_2 are two different inputs of a function f, then f is one-to-one if ______________.
6. If every horizontal line intersects the graph of a function f at no more than one point, then f is a(n) ___________ function.
7. If f is a one-to-one function and $f(3) = 8$, then $f^{-1}(8) =$ ________.
8. If f^{-1} denotes the inverse of a function f, then the graphs of f and f^{-1} are symmetric with respect to the line ________.
9. If the domain of a one-to-one function f is $[4, \infty)$, then the range of its inverse function f^{-1} is ______________.
10. ***True or False*** If f and g are inverse functions, then the domain of f is the same as the range of g.
11. If $(-2, 3)$ is a point on the graph of a one-to-one function f, which of the following points is on the graph of f^{-1}?
(a) $(3, -2)$ (b) $(2, -3)$ (c) $(-3, 2)$ (d) $(-2, -3)$
12. Suppose f is a one-to-one function with a domain of $\{x | x \neq 3\}$ and a range of $\left\{x \middle| x \neq \frac{2}{3}\right\}$. Which of the following is the domain of f^{-1}?
(a) $\{x | x \neq 3\}$ (b) All real numbers
(c) $\left\{x \middle| x \neq \frac{2}{3}, x \neq 3\right\}$ (d) $\left\{x \middle| x \neq \frac{2}{3}\right\}$

Skill Building

In Problems 13–20, determine whether the function is one-to-one.

13.
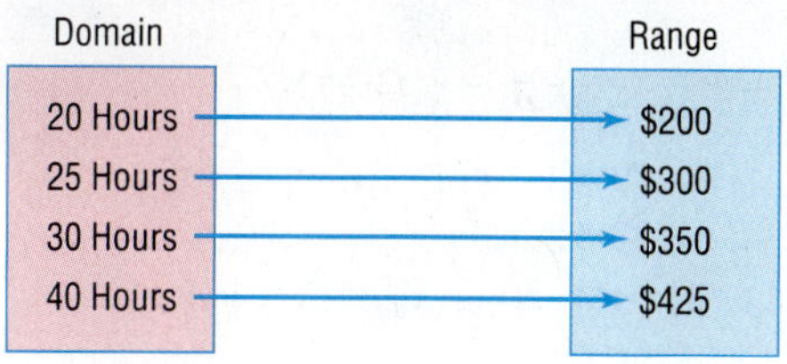

14.
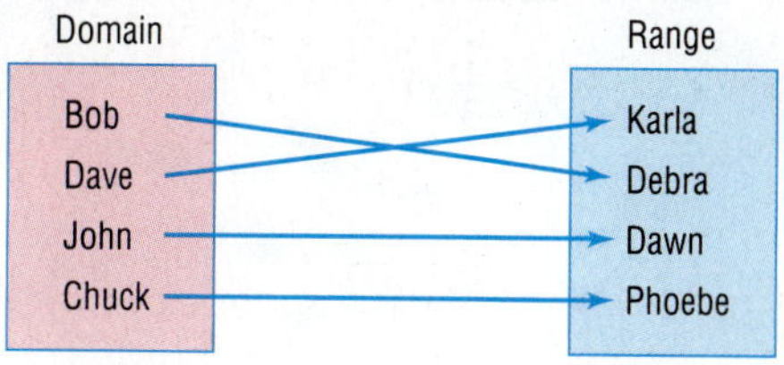

15.

16.

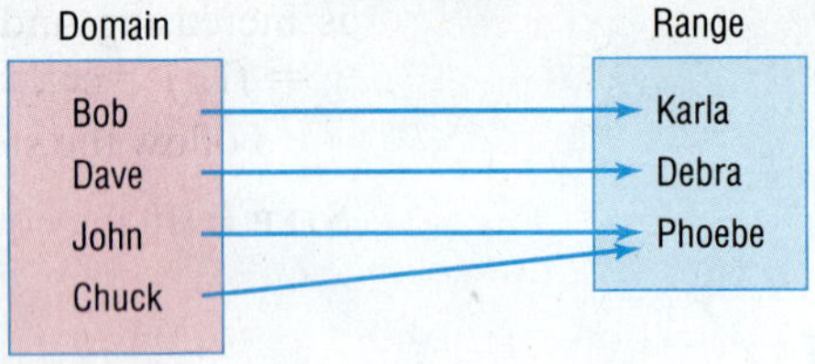

17. $\{(2,6), (-3,6), (4,9), (1,10)\}$

18. $\{(-2,5), (-1,3), (3,7), (4,12)\}$

19. $\{(0,0), (1,1), (2,16), (3,81)\}$

20. $\{(1,2), (2,8), (3,18), (4,32)\}$

In Problems 21–26, the graph of a function f is given. Use the horizontal-line test to determine whether f is one-to-one.

21.

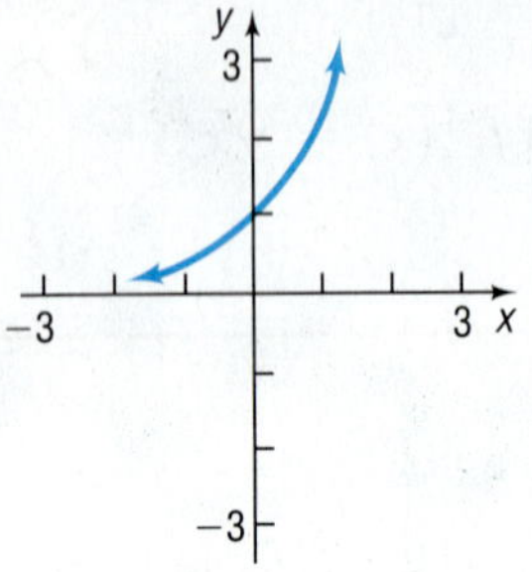

22.

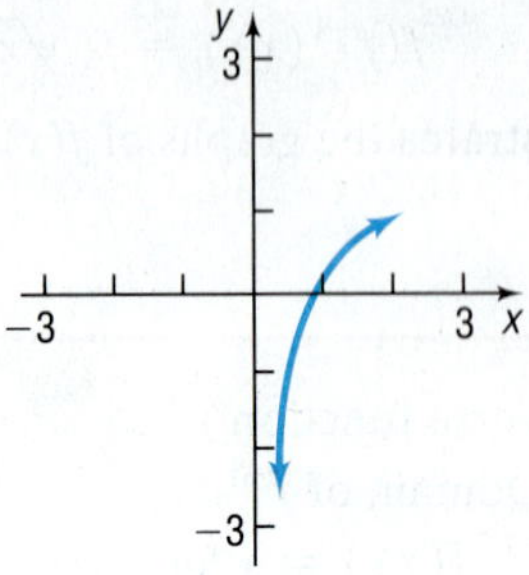

23.

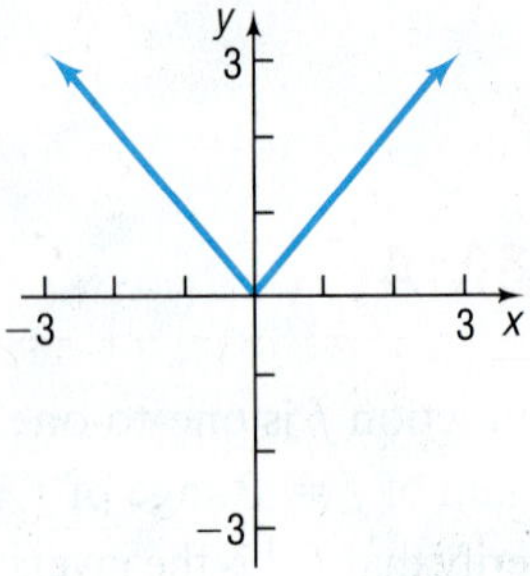

24.

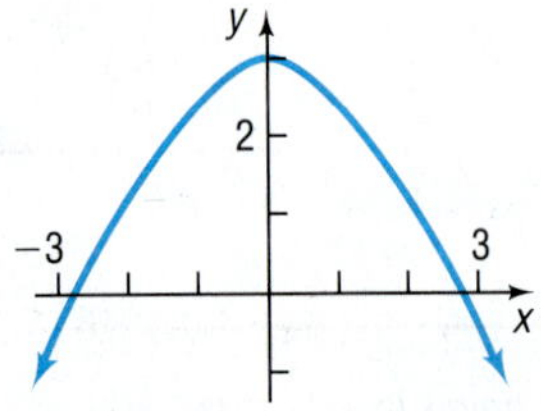

25.

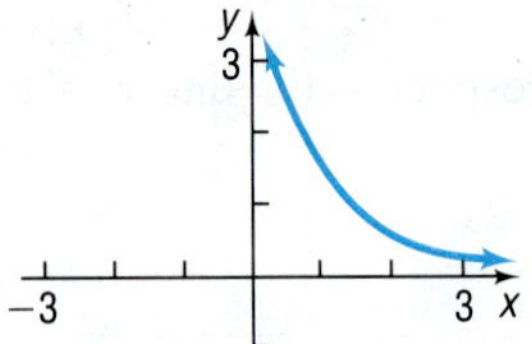

26.

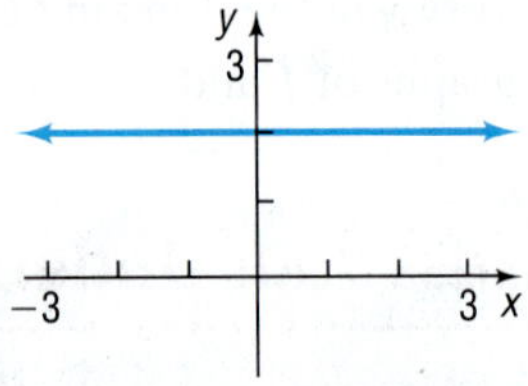

In Problems 27–34, find the inverse of each one-to-one function. State the domain and the range of each inverse function.

27.

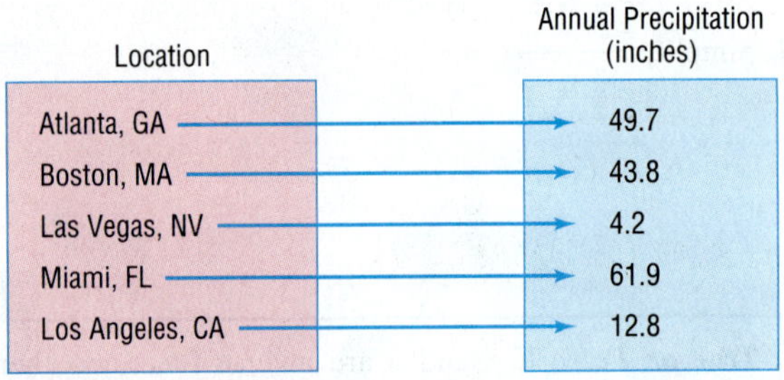

Source: currentresults.com

28.

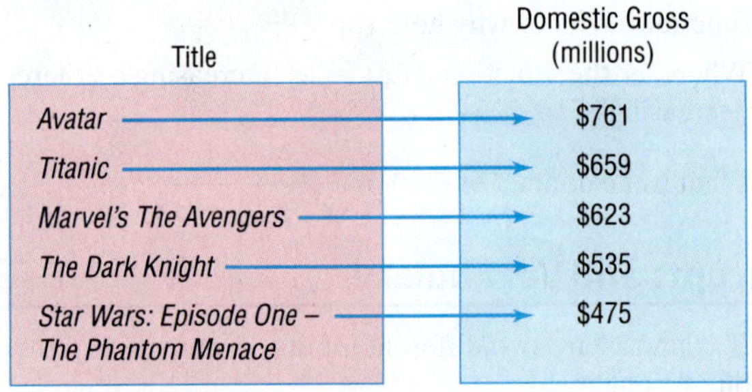

Source: boxofficemojo.com

29.

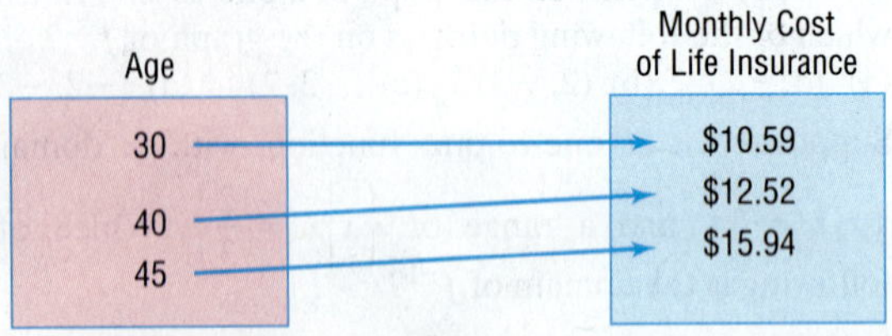

Source: tiaa-cref.org

30.

Source: United States Bureau of Labor Statistics, March 2014

31. $\{(-3,5), (-2,9), (-1,2), (0,11), (1,-5)\}$

32. $\{(-2,2), (-1,6), (0,8), (1,-3), (2,9)\}$

33. $\{(-2,1), (-3,2), (-10,0), (1,9), (2,4)\}$

34. $\{(-2,-8), (-1,-1), (0,0), (1,1), (2,8)\}$

In Problems 35–44, verify that the functions f and g are inverses of each other by showing that $f(g(x)) = x$ *and* $g(f(x)) = x$*. Give any values of x that need to be excluded from the domain of f and the domain of g.*

35. $f(x) = 3x + 4; \quad g(x) = \frac{1}{3}(x - 4)$

36. $f(x) = 3 - 2x; \quad g(x) = -\frac{1}{2}(x - 3)$

37. $f(x) = 4x - 8; \quad g(x) = \frac{x}{4} + 2$

38. $f(x) = 2x + 6; \quad g(x) = \frac{1}{2}x - 3$

39. $f(x) = x^3 - 8; \quad g(x) = \sqrt[3]{x + 8}$

40. $f(x) = (x - 2)^2, x \geq 2; \quad g(x) = \sqrt{x} + 2$

41. $f(x) = \frac{1}{x};\ g(x) = \frac{1}{x}$

42. $f(x) = x;\ g(x) = x$

43. $f(x) = \frac{2x+3}{x+4};\ g(x) = \frac{4x-3}{2-x}$

44. $f(x) = \frac{x-5}{2x+3};\ g(x) = \frac{3x+5}{1-2x}$

In Problems 45–50, the graph of a one-to-one function f is given. Draw the graph of the inverse function f^{-1}.

45.

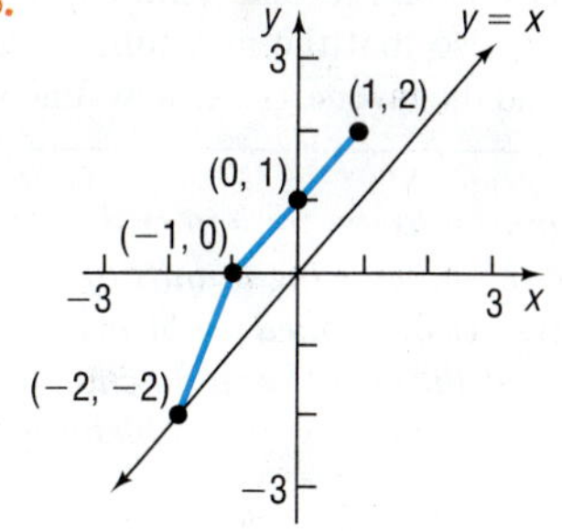

46.

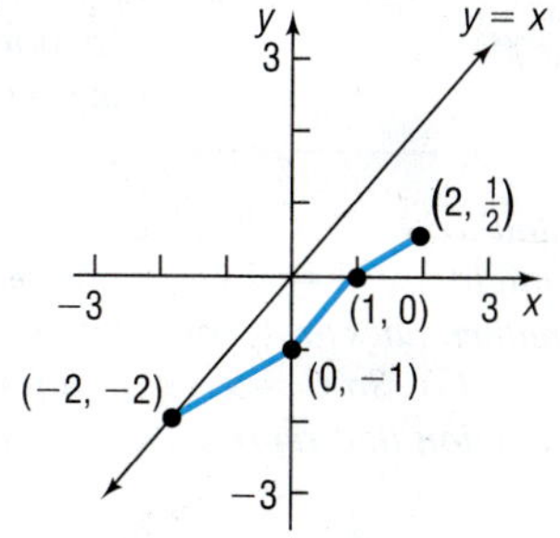

47.

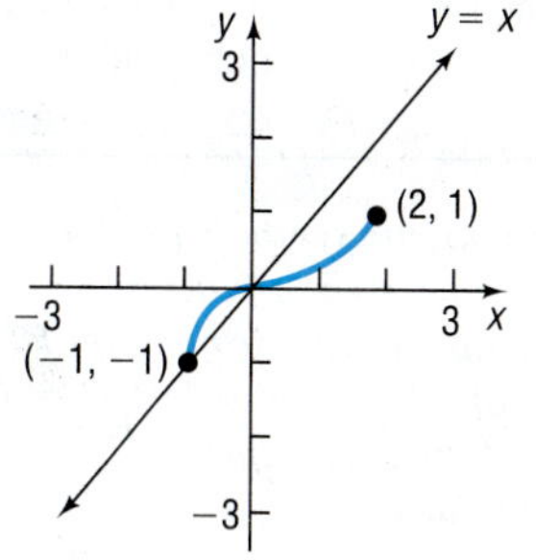

48.

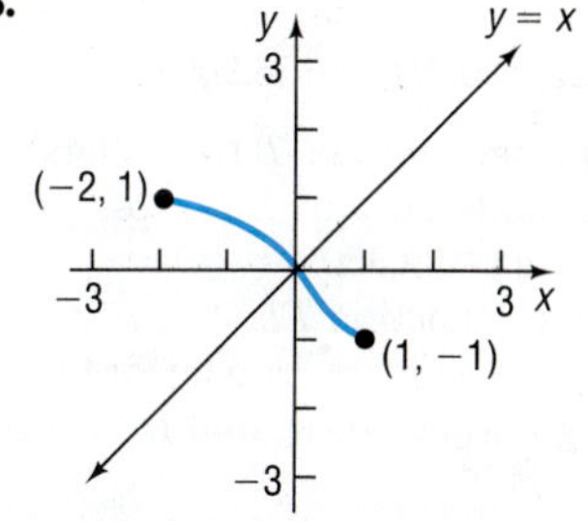

49.

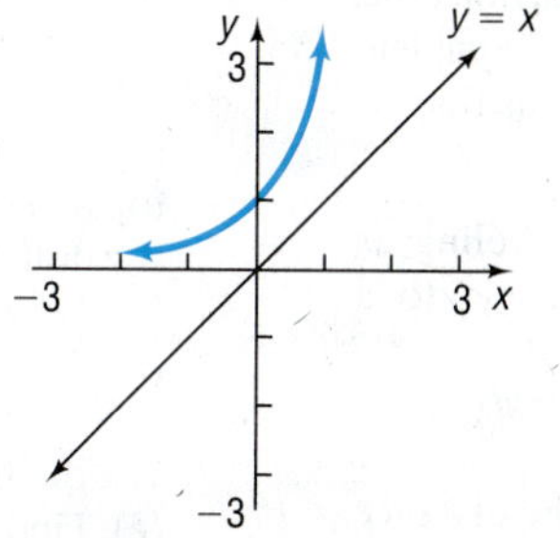

50.

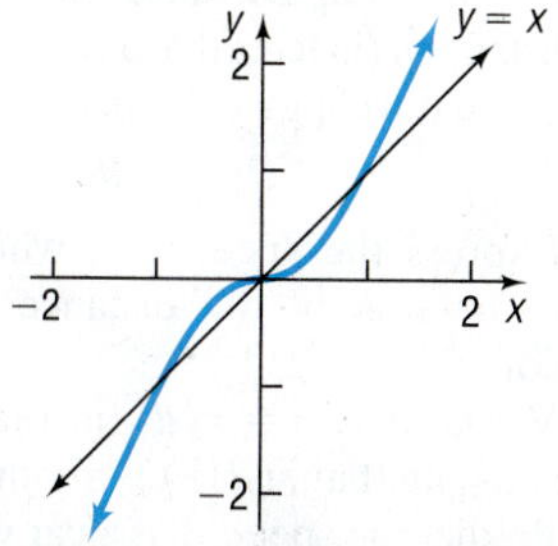

In Problems 51–62, the function f is one-to-one. (a) Find its inverse function f^{-1} and check your answer. (b) Find the domain and the range of f and f^{-1}. (c) Graph f, f^{-1}, and $y = x$ on the same coordinate axes.

51. $f(x) = 3x$

52. $f(x) = -4x$

53. $f(x) = 4x + 2$

54. $f(x) = 1 - 3x$

55. $f(x) = x^3 - 1$

56. $f(x) = x^3 + 1$

57. $f(x) = x^2 + 4,\quad x \ge 0$

58. $f(x) = x^2 + 9,\quad x \ge 0$

59. $f(x) = \frac{4}{x}$

60. $f(x) = -\frac{3}{x}$

61. $f(x) = \frac{1}{x-2}$

62. $f(x) = \frac{4}{x+2}$

In Problems 63–74, the function f is one-to-one. (a) Find its inverse function f^{-1} and check your answer. (b) Find the domain and the range of f and f^{-1}.

63. $f(x) = \frac{2}{3+x}$

64. $f(x) = \frac{4}{2-x}$

65. $f(x) = \frac{3x}{x+2}$

66. $f(x) = -\frac{2x}{x-1}$

67. $f(x) = \frac{2x}{3x-1}$

68. $f(x) = -\frac{3x+1}{x}$

69. $f(x) = \frac{3x+4}{2x-3}$

70. $f(x) = \frac{2x-3}{x+4}$

71. $f(x) = \frac{2x+3}{x+2}$

72. $f(x) = \frac{-3x-4}{x-2}$

73. $f(x) = \frac{x^2-4}{2x^2},\quad x > 0$

74. $f(x) = \frac{x^2+3}{3x^2}\quad x > 0$

Applications and Extensions

75. Use the graph of $y = f(x)$ given in Problem 45 to evaluate the following:
(a) $f(-1)$ (b) $f(1)$ (c) $f^{-1}(1)$ (d) $f^{-1}(2)$

76. Use the graph of $y = f(x)$ given in Problem 46 to evaluate the following:
(a) $f(2)$ (b) $f(1)$ (c) $f^{-1}(0)$ (d) $f^{-1}(-1)$

77. If $f(7) = 13$ and f is one-to-one, what is $f^{-1}(13)$?

78. If $g(-5) = 3$ and g is one-to-one, what is $g^{-1}(3)$?

79. The domain of a one-to-one function f is $[5, \infty)$, and its range is $[-2, \infty)$. State the domain and the range of f^{-1}.

80. The domain of a one-to-one function f is $[0, \infty)$, and its range is $[5, \infty)$. State the domain and the range of f^{-1}.

81. The domain of a one-to-one function g is $(-\infty, 0]$, and its range is $[0, \infty)$. State the domain and the range of g^{-1}.

82. The domain of a one-to-one function g is $[0, 15]$, and its range is $(0, 8)$. State the domain and the range of g^{-1}.

83. A function $y = f(x)$ is increasing on the interval $(0, 5)$. What conclusions can you draw about the graph of $y = f^{-1}(x)$?

84. A function $y = f(x)$ is decreasing on the interval $(0, 5)$. What conclusions can you draw about the graph of $y = f^{-1}(x)$?

85. Find the inverse of the linear function

$$f(x) = mx + b, \quad m \neq 0$$

86. Find the inverse of the function

$$f(x) = \sqrt{r^2 - x^2}, \quad 0 \le x \le r$$

87. A function f has an inverse function f^{-1}. If the graph of f lies in quadrant I, in which quadrant does the graph of f^{-1} lie?

88. A function f has an inverse function f^{-1}. If the graph of f lies in quadrant II, in which quadrant does the graph of f^{-1} lie?

89. The function $f(x) = |x|$ is not one-to-one. Find a suitable restriction on the domain of f so that the new function that results is one-to-one. Then find the inverse of the new function.

90. The function $f(x) = x^4$ is not one-to-one. Find a suitable restriction on the domain of f so that the new function that results is one-to-one. Then find the inverse of the new function.

In applications, the symbols used for the independent and dependent variables are often based on common usage. So, rather than using $y = f(x)$ to represent a function, an applied problem might use $C = C(q)$ to represent the cost C of manufacturing q units of a good. Because of this, the inverse notation f^{-1} used in a pure mathematics problem is not used when finding inverses of applied problems. Rather, the inverse of a function such as $C = C(q)$ will be $q = q(C)$. So $C = C(q)$ is a function that represents the cost C as a function of the number q of units manufactured, and $q = q(C)$ is a function that represents the number q as a function of the cost C. Problems 91–94 illustrate this idea.

91. Vehicle Stopping Distance Taking into account reaction time, the distance d (in feet) that a car requires to come to a complete stop while traveling r miles per hour is given by the function

$$d(r) = 6.97r - 90.39$$

(a) Express the speed r at which the car is traveling as a function of the distance d required to come to a complete stop.
(b) Verify that $r = r(d)$ is the inverse of $d = d(r)$ by showing that $r(d(r)) = r$ and $d(r(d)) = d$.
(c) Predict the speed that a car was traveling if the distance required to stop was 300 feet.

92. Height and Head Circumference The head circumference C of a child is related to the height H of the child (both in inches) through the function

$$H(C) = 2.15C - 10.53$$

(a) Express the head circumference C as a function of height H.
(b) Verify that $C = C(H)$ is the inverse of $H = H(C)$ by showing that $H(C(H)) = H$ and $C(H(C)) = C$.
(c) Predict the head circumference of a child who is 26 inches tall.

93. Ideal Body Weight One model for the ideal body weight W for men (in kilograms) as a function of height h (in inches) is given by the function

$$W(h) = 50 + 2.3(h - 60)$$

(a) What is the ideal weight of a 6-foot male?
(b) Express the height h as a function of weight W.
(c) Verify that $h = h(W)$ is the inverse of $W = W(h)$ by showing that $h(W(h)) = h$ and $W(h(W)) = W$.
(d) What is the height of a male who is at his ideal weight of 80 kilograms?

[**Note:** The ideal body weight W for women (in kilograms) as a function of height h (in inches) is given by $W(h) = 45.5 + 2.3(h - 60)$.]

94. Temperature Conversion The function $F(C) = \frac{9}{5}C + 32$ converts a temperature from C degrees Celsius to F degrees Fahrenheit.

(a) Express the temperature in degrees Celsius C as a function of the temperature in degrees Fahrenheit F.
(b) Verify that $C = C(F)$ is the inverse of $F = F(C)$ by showing that $C(F(C)) = C$ and $F(C(F)) = F$.
(c) What is the temperature in degrees Celsius if it is 70 degrees Fahrenheit?

95. Income Taxes The function

$$T(g) = 5081.25 + 0.25(g - 36{,}900)$$

represents the 2014 federal income tax T (in dollars) due for a "single" filer whose modified adjusted gross income is g dollars, where $36{,}900 \le g \le 89{,}350$.

(a) What is the domain of the function T?
(b) Given that the tax due T is an increasing linear function of modified adjusted gross income g, find the range of the function T.
(c) Find adjusted gross income g as a function of federal income tax T. What are the domain and the range of this function?

96. Income Taxes The function

$$T(g) = 1815 + 0.15(g - 18{,}150)$$

represents the 2014 federal income tax T (in dollars) due for a "married filing jointly" filer whose modified adjusted gross income is g dollars, where $18{,}150 \le g \le 73{,}800$.

(a) What is the domain of the function T?
(b) Given that the tax due T is an increasing linear function of modified adjusted gross income g, find the range of the function T.
(c) Find adjusted gross income g as a function of federal income tax T. What are the domain and the range of this function?

97. Gravity on Earth If a rock falls from a height of 100 meters on Earth, the height H (in meters) after t seconds is approximately

$$H(t) = 100 - 4.9t^2$$

(a) In general, quadratic functions are not one-to-one. However, the function H is one-to-one. Why?
(b) Find the inverse of H and verify your result.
(c) How long will it take a rock to fall 80 meters?

98. Period of a Pendulum The period T (in seconds) of a simple pendulum as a function of its length l (in feet) is given by

$$T(l) = 2\pi\sqrt{\frac{l}{32.2}}$$

(a) Express the length l as a function of the period T.
(b) How long is a pendulum whose period is 3 seconds?

99. Given

$$f(x) = \frac{ax + b}{cx + d}$$

find $f^{-1}(x)$. If $c \neq 0$, under what conditions on $a, b, c,$ and d is $f = f^{-1}$?

Explaining Concepts: Discussion and Writing

100. Can a one-to-one function and its inverse be equal? What must be true about the graph of f for this to happen? Give some examples to support your conclusion.
101. Draw the graph of a one-to-one function that contains the points $(-2, -3)$, $(0, 0)$, and $(1, 5)$. Now draw the graph of its inverse. Compare your graph to those of other students. Discuss any similarities. What differences do you see?
102. Give an example of a function whose domain is the set of real numbers and that is neither increasing nor decreasing on its domain, but is one-to-one.
[**Hint:** Use a piecewise-defined function.]
103. Is every odd function one-to-one? Explain.
104. Suppose that $C(g)$ represents the cost C, in dollars, of manufacturing g cars. Explain what $C^{-1}(800{,}000)$ represents.
105. Explain why the horizontal-line test can be used to identify one-to-one functions from a graph.
106. Explain why a function must be one-to-one in order to have an inverse that is a function. Use the function $y = x^2$ to support your explanation.

Retain Your Knowledge

Problems 107–110 are based on material learned earlier in the course. The purpose of these problems is to keep the material fresh in your mind so that you are better prepared for the final exam.

107. If $f(x) = 3x^2 - 7x$, find $f(x + h) - f(x)$.
108. Use the techniques of shifting, compressing or stretching, and reflections to graph $f(x) = -|x + 2| + 3$.
109. Find the zeros of the quadratic function $f(x) = 3x^2 + 5x + 1$. What are the x-intercepts, if any, of the graph of the function?
110. Find the domain of $R(x) = \dfrac{6x^2 - 11x - 2}{2x^2 - x - 6}$. Find any horizontal, vertical, or oblique asymptotes.

'Are You Prepared?' Answers

1. Yes; for each input x there is one output y.
2. Increasing on $(0, \infty)$; decreasing on $(-\infty, 0)$
3. $\{x \mid x \neq -6, x \neq 3\}$
4. $\dfrac{x}{1 - x}$, $x \neq 0, x \neq -1$

6.3 Exponential Functions

PREPARING FOR THIS SECTION *Before getting started, review the following:*

- Exponents (Chapter R, Section R.2, pp. 21–24, and Section R.8, pp. 73–77)
- Graphing Techniques: Transformations (Section 3.5, pp. 247–256)
- Solving Equations (Section 1.1, pp. 82–87 and Section 1.2, pp. 92–99)
- Average Rate of Change (Section 3.3, pp. 230–231)
- Quadratic Functions (Section 4.3, pp. 290–298)
- Linear Functions (Section 4.1, pp. 274–277)
- Horizontal Asymptotes (Section 5.2, pp. 348–350)

Now Work the 'Are You Prepared?' problems on page 434.

OBJECTIVES
1 Evaluate Exponential Functions (p. 423)
2 Graph Exponential Functions (p. 427)
3 Define the Number e (p. 430)
4 Solve Exponential Equations (p. 432)

1 Evaluate Exponential Functions

Chapter R, Section R.8 gives a definition for raising a real number a to a rational power. That discussion provides meaning to expressions of the form

$$a^r$$

where the base a is a positive real number and the exponent r is a rational number.

But what is the meaning of a^x, where the base a is a positive real number and the exponent x is an irrational number? Although a rigorous definition requires methods discussed in calculus, the basis for the definition is easy to follow: Select a rational number r that is formed by truncating (removing) all but a finite number of digits from the irrational number x. Then it is reasonable to expect that

$$a^x \approx a^r$$

For example, take the irrational number $\pi = 3.14159\ldots$. Then an approximation to a^π is

$$a^\pi \approx a^{3.14}$$

where the digits after the hundredths position have been removed from the value for π. A better approximation would be

$$a^\pi \approx a^{3.14159}$$

where the digits after the hundred-thousandths position have been removed. Continuing in this way, we can obtain approximations to a^π to any desired degree of accuracy.

Most calculators have an $\boxed{x^y}$ key or a caret key $\boxed{\wedge}$ for working with exponents. To evaluate expressions of the form a^x, enter the base a, then press the $\boxed{x^y}$ key (or the $\boxed{\wedge}$ key), enter the exponent x, and press $\boxed{=}$ (or $\boxed{\text{ENTER}}$).

EXAMPLE 1

Using a Calculator to Evaluate Powers of 2

Using a calculator, evaluate:

(a) $2^{1.4}$ (b) $2^{1.41}$ (c) $2^{1.414}$ (d) $2^{1.4142}$ (e) $2^{\sqrt{2}}$

Solution

(a) $2^{1.4} \approx 2.639015822$ (b) $2^{1.41} \approx 2.657371628$

(c) $2^{1.414} \approx 2.66474965$ (d) $2^{1.4142} \approx 2.665119089$

(d) $2^{\sqrt{2}} \approx 2.665144143$

Now Work PROBLEM 15

It can be shown that the familiar laws for rational exponents hold for real exponents.

THEOREM

Laws of Exponents

If $s, t, a,$ and b are real numbers with $a > 0$ and $b > 0$, then

$$a^s \cdot a^t = a^{s+t} \qquad (a^s)^t = a^{st} \qquad (ab)^s = a^s \cdot b^s$$
$$1^s = 1 \qquad a^{-s} = \frac{1}{a^s} = \left(\frac{1}{a}\right)^s \qquad a^0 = 1 \tag{1}$$

Introduction to Exponential Growth

Suppose a function f has the following two properties:

1. The value of f doubles with every 1-unit increase in the independent variable x.
2. The value of f at $x = 0$ is 5, so $f(0) = 5$.

Table 1 shows values of the function f for $x = 0, 1, 2, 3,$ and 4.

Let's find an equation $y = f(x)$ that describes this function f. The key fact is that the value of f doubles for every 1-unit increase in x.

Table 1

x	$f(x)$
0	5
1	10
2	20
3	40
4	80

$$f(0) = 5$$
$$f(1) = 2f(0) = 2\cdot 5 = 5\cdot 2^1 \quad \text{Double the value of } f \text{ at 0 to get the value at 1.}$$
$$f(2) = 2f(1) = 2(5\cdot 2) = 5\cdot 2^2 \quad \text{Double the value of } f \text{ at 1 to get the value at 2.}$$
$$f(3) = 2f(2) = 2(5\cdot 2^2) = 5\cdot 2^3$$
$$f(4) = 2f(3) = 2(5\cdot 2^3) = 5\cdot 2^4$$

The pattern leads to

$$f(x) = 2f(x-1) = 2(5\cdot 2^{x-1}) = 5\cdot 2^x$$

DEFINITION

An **exponential function** is a function of the form

$$f(x) = Ca^x$$

where a is a positive real number $(a > 0)$, $a \neq 1$, and $C \neq 0$ is a real number. The domain of f is the set of all real numbers. The base a is the **growth factor**, and, because $f(0) = Ca^0 = C$, C is called the **initial value.**

WARNING It is important to distinguish a power function, $g(x) = ax^n$, $n \geq 2$ an integer, from an exponential function, $f(x) = C \cdot a^x$, $a \neq 1$, $a > 0$. In a power function, the base is a variable and the exponent is a constant. In an exponential function, the base is a constant and the exponent is a variable. ■

In the definition of an exponential function, the base $a = 1$ is excluded because this function is simply the constant function $f(x) = C \cdot 1^x = C$. Bases that are negative are also excluded; otherwise, many values of x would have to be excluded from the domain, such as $x = \frac{1}{2}$ and $x = \frac{3}{4}$. [Recall that $(-2)^{1/2} = \sqrt{-2}$, $(-3)^{3/4} = \sqrt[4]{(-3)^3} = \sqrt[4]{-27}$, and so on, are not defined in the set of real numbers.]

Transformations (vertical shifts, horizontal shifts, reflections, and so on) of a function of the form $f(x) = Ca^x$ also represent exponential functions. Some examples of exponential functions are

$$f(x) = 2^x \qquad F(x) = \left(\frac{1}{3}\right)^x + 5 \qquad G(x) = 2 \cdot 3^{x-3}$$

For each function, note that the base of the exponential expression is a constant and the exponent contains a variable.

In the function $f(x) = 5 \cdot 2^x$, notice that the ratio of consecutive outputs is constant for 1-unit increases in the input. This ratio equals the constant 2, the base of the exponential function. In other words,

$$\frac{f(1)}{f(0)} = \frac{5 \cdot 2^1}{5} = 2 \qquad \frac{f(2)}{f(1)} = \frac{5 \cdot 2^2}{5 \cdot 2^1} = 2 \qquad \frac{f(3)}{f(2)} = \frac{5 \cdot 2^3}{5 \cdot 2^2} = 2 \quad \text{and so on}$$

This leads to the following result.

THEOREM

For an exponential function $f(x) = Ca^x$, $a > 0$, $a \neq 1$, and $C \neq 0$, if x is any real number, then

$$\frac{f(x+1)}{f(x)} = a \quad \text{or} \quad f(x+1) = af(x)$$

In Words

For 1-unit changes in the input x of an exponential function $f(x) = C \cdot a^x$, the ratio of consecutive outputs is the constant a.

Proof

$$\frac{f(x+1)}{f(x)} = \frac{Ca^{x+1}}{Ca^x} = a^{x+1-x} = a^1 = a$$ ■

EXAMPLE 2 **Identifying Linear or Exponential Functions**

Determine whether the given function is linear, exponential, or neither. For those that are linear, find a linear function that models the data. For those that are exponential, find an exponential function that models the data.

(a)

x	y
−1	5
0	2
1	−1
2	−4
3	−7

(b)

x	y
−1	32
0	16
1	8
2	4
3	2

(c)

x	y
−1	2
0	4
1	7
2	11
3	16

Solution For each function, compute the average rate of change of y with respect to x and the ratio of consecutive outputs. If the average rate of change is constant, then the function is linear. If the ratio of consecutive outputs is constant, then the function is exponential.

Table 2

x	y	Average Rate of Change	Ratio of Consecutive Outputs
-1	5		
		$\frac{\Delta y}{\Delta x} = \frac{2-5}{0-(-1)} = -3$	$\frac{2}{5}$
0	2		
		$\frac{-1-2}{1-0} = -3$	$\frac{-1}{2} = -\frac{1}{2}$
1	-1		
		$\frac{-4-(-1)}{2-1} = -3$	$\frac{-4}{-1} = 4$
2	-4		
		$\frac{-7-(-4)}{3-2} = -3$	$\frac{-7}{-4} = \frac{7}{4}$
3	-7		

(a)

x	y	Average Rate of Change	Ratio of Consecutive Outputs
-1	32		
		$\frac{\Delta y}{\Delta x} = \frac{16-32}{0-(-1)} = -16$	$\frac{16}{32} = \frac{1}{2}$
0	16		
		-8	$\frac{8}{16} = \frac{1}{2}$
1	8		
		-4	$\frac{4}{8} = \frac{1}{2}$
2	4		
		-2	$\frac{2}{4} = \frac{1}{2}$
3	2		

(b)

x	y	Average Rate of Change	Ratio of Consecutive Outputs
-1	2		
		$\frac{\Delta y}{\Delta x} = \frac{4-2}{0-(-1)} = 2$	2
0	4		
		3	$\frac{7}{4}$
1	7		
		4	$\frac{11}{7}$
2	11		
		5	$\frac{16}{11}$
3	16		

(c)

(a) See Table 2(a). The average rate of change for every 1-unit increase in x is -3. Therefore, the function is a linear function. In a linear function the average rate of change is the slope m, so $m = -3$. The y-intercept b is the value of the function at $x = 0$, so $b = 2$. The linear function that models the data is $f(x) = mx + b = -3x + 2$.

(b) See Table 2(b). For this function, the average rate of change from -1 to 0 is -16, and the average rate of change from 0 to 1 is -8. Because the average rate of change is not constant, the function is not a linear function. The ratio of consecutive outputs for a 1-unit increase in the inputs is a constant, $\frac{1}{2}$. Because the ratio of consecutive outputs is constant, the function is an exponential function with growth factor $a = \frac{1}{2}$. The initial value C of the exponential function is $C = 16$, the value of the function at 0. Therefore, the exponential function that models the data is $g(x) = Ca^x = 16 \cdot \left(\frac{1}{2}\right)^x$.

(c) See Table 2(c). For this function, the average rate of change from -1 to 0 is 2, and the average rate of change from 0 to 1 is 3. Because the average rate of change is not constant, the function is not a linear function. The ratio of consecutive outputs from -1 to 0 is 2, and the ratio of consecutive outputs from 0 to 1 is $\frac{7}{4}$. Because the ratio of consecutive outputs is not a constant, the function is not an exponential function.

Now Work PROBLEM 27

2 Graph Exponential Functions

If we know how to graph an exponential function of the form $f(x) = a^x$, then we can use transformations (shifting, stretching, and so on) to obtain the graph of any exponential function.

First, let's graph the exponential function $f(x) = 2^x$.

EXAMPLE 3 Graphing an Exponential Function

Graph the exponential function: $f(x) = 2^x$

Solution The domain of $f(x) = 2^x$ is the set of all real numbers. Begin by locating some points on the graph of $f(x) = 2^x$, as listed in Table 3.

Because $2^x > 0$ for all x, the range of f is $(0, \infty)$. Therefore, the graph has no x-intercepts, and in fact the graph will lie above the x-axis for all x. As Table 3 indicates, the y-intercept is 1. Table 3 also indicates that as $x \to -\infty$, the value of $f(x) = 2^x$ gets closer and closer to 0. Therefore, the x-axis $(y = 0)$ is a horizontal asymptote to the graph as $x \to -\infty$. This provides the end behavior for x large and negative.

To determine the end behavior for x large and positive, look again at Table 3. As $x \to \infty$, $f(x) = 2^x$ grows very quickly, causing the graph of $f(x) = 2^x$ to rise very rapidly. It is apparent that f is an increasing function and so is one-to-one.

Using all this information, plot some of the points from Table 3 and connect them with a smooth, continuous curve, as shown in Figure 18.

Table 3

x	$f(x) = 2^x$
-10	$2^{-10} \approx 0.00098$
-3	$2^{-3} = \frac{1}{8}$
-2	$2^{-2} = \frac{1}{4}$
-1	$2^{-1} = \frac{1}{2}$
0	$2^0 = 1$
1	$2^1 = 2$
2	$2^2 = 4$
3	$2^3 = 8$
10	$2^{10} = 1024$

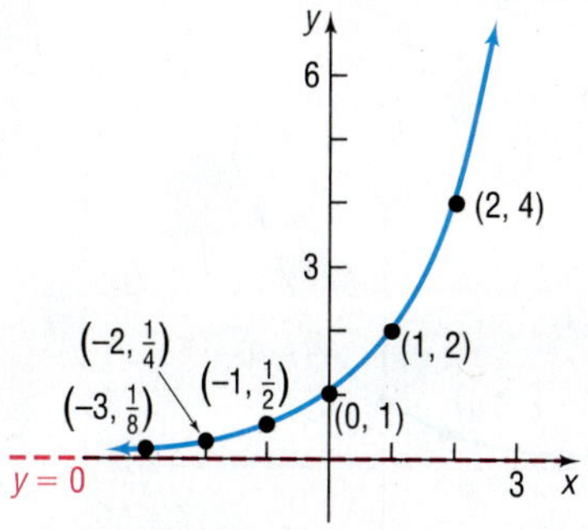

Figure 18 $f(x) = 2^x$

Graphs that look like the one in Figure 18 occur very frequently in a variety of situations. For example, the graph in Figure 19 illustrates the number of Facebook subscribers by year from 2004 to 2013. One might conclude from this graph that the number of Facebook subscribers is growing *exponentially*.

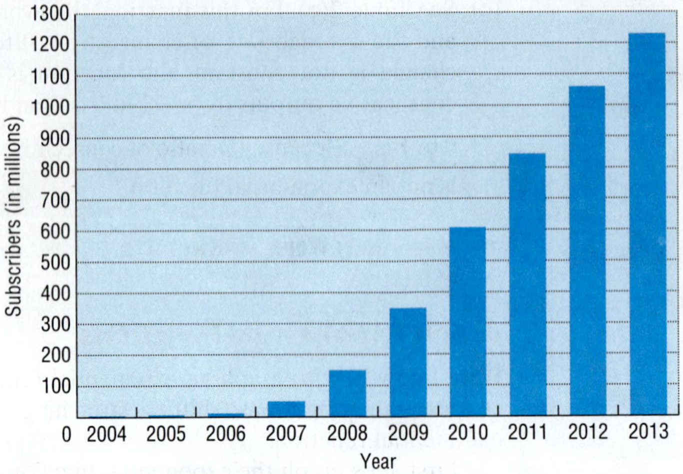

Source: Facebook Newsroom

Figure 19

Later in this chapter, more will be said about situations that lead to exponential growth. For now, let's continue to explore properties of exponential functions.

The graph of $f(x) = 2^x$ in Figure 18 is typical of all exponential functions of the form $f(x) = a^x$ with $a > 1$. Such functions are increasing functions and hence are one-to-one. Their graphs lie above the x-axis, pass through the point $(0, 1)$, and thereafter rise rapidly as $x \to \infty$. As $x \to -\infty$, the x-axis $(y = 0)$ is a horizontal asymptote. There are no vertical asymptotes. Finally, the graphs are smooth and continuous with no corners or gaps.

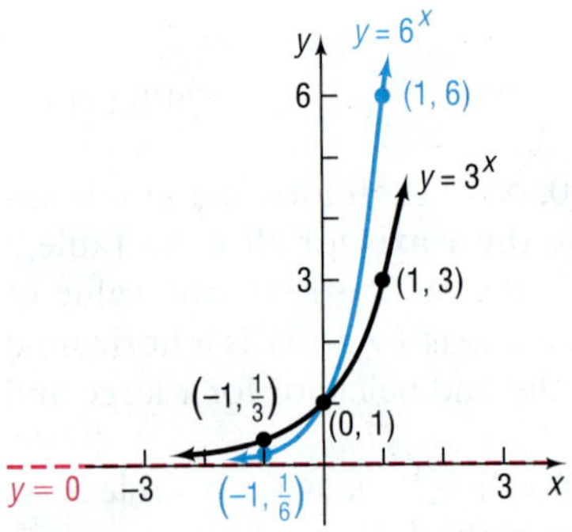

Figure 20

Figure 20 illustrates the graphs of two more exponential functions whose bases are larger than 1. Notice that the larger the base, the steeper the graph is when $x > 0$, and when $x < 0$, the larger the base, the closer the graph of the equation is to the x-axis.

Seeing the Concept

Graph $Y_1 = 2^x$ and compare what you see to Figure 18. Clear the screen, graph $Y_1 = 3^x$ and $Y_2 = 6^x$, and compare what you see to Figure 20. Clear the screen and graph $Y_1 = 10^x$ and $Y_2 = 100^x$.

Properties of the Exponential Function $f(x) = a^x, a > 1$

1. The domain is the set of all real numbers, or $(-\infty, \infty)$ using interval notation; the range is the set of positive real numbers, or $(0, \infty)$ using interval notation.
2. There are no x-intercepts; the y-intercept is 1.
3. The x-axis $(y = 0)$ is a horizontal asymptote as $x \to -\infty$. $\left[\lim_{x \to -\infty} a^x = 0\right]$.
4. $f(x) = a^x, a > 1$, is an increasing function and is one-to-one.
5. The graph of f contains the points $\left(-1, \frac{1}{a}\right)$, $(0, 1)$, and $(1, a)$.
6. The graph of f is smooth and continuous, with no corners or gaps. See Figure 21.

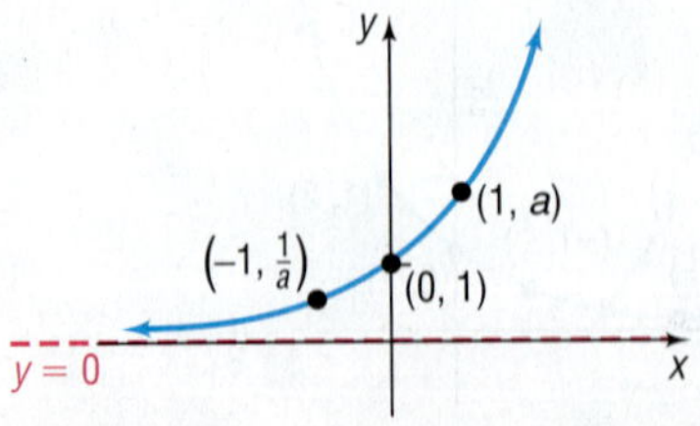

Figure 21 $f(x) = a^x, a > 1$

Now consider $f(x) = a^x$ when $0 < a < 1$.

EXAMPLE 4 Graphing an Exponential Function

Graph the exponential function: $f(x) = \left(\frac{1}{2}\right)^x$

Solution The domain of $f(x) = \left(\frac{1}{2}\right)^x$ consists of all real numbers. As before, locate some points on the graph as shown in Table 4. Because $\left(\frac{1}{2}\right)^x > 0$ for all x, the range of f is the interval $(0, \infty)$. The graph lies above the x-axis and has no x-intercepts. The y-intercept is 1. As $x \to -\infty$, $f(x) = \left(\frac{1}{2}\right)^x$ grows very quickly. As $x \to \infty$, the values of $f(x)$ approach 0. The x-axis $(y = 0)$ is a horizontal asymptote as $x \to \infty$. It is apparent that f is a decreasing function and so is one-to-one. Figure 22 illustrates the graph.

Table 4

x	$f(x) = \left(\frac{1}{2}\right)^x$
-10	$\left(\frac{1}{2}\right)^{-10} = 1024$
-3	$\left(\frac{1}{2}\right)^{-3} = 8$
-2	$\left(\frac{1}{2}\right)^{-2} = 4$
-1	$\left(\frac{1}{2}\right)^{-1} = 2$
0	$\left(\frac{1}{2}\right)^{0} = 1$
1	$\left(\frac{1}{2}\right)^{1} = \frac{1}{2}$
2	$\left(\frac{1}{2}\right)^{2} = \frac{1}{4}$
3	$\left(\frac{1}{2}\right)^{3} = \frac{1}{8}$
10	$\left(\frac{1}{2}\right)^{10} \approx 0.00098$

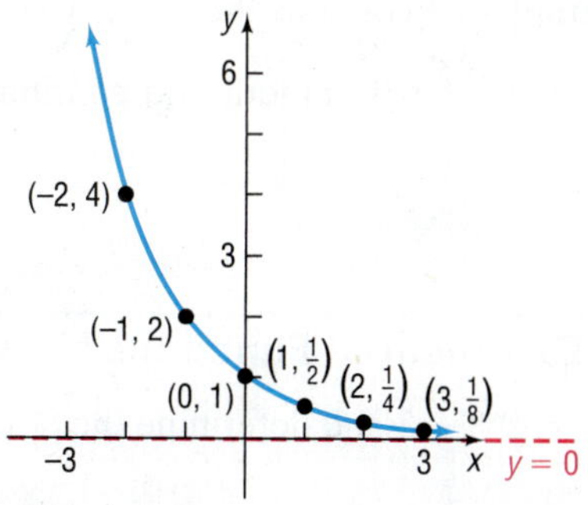

Figure 22 $f(x) = \left(\frac{1}{2}\right)^x$

The graph of $y = \left(\frac{1}{2}\right)^x$ also can be obtained from the graph of $y = 2^x$ using transformations. The graph of $y = \left(\frac{1}{2}\right)^x = 2^{-x}$ is a reflection about the y-axis of the graph of $y = 2^x$ (replace x by $-x$). See Figures 23(a) and (b).

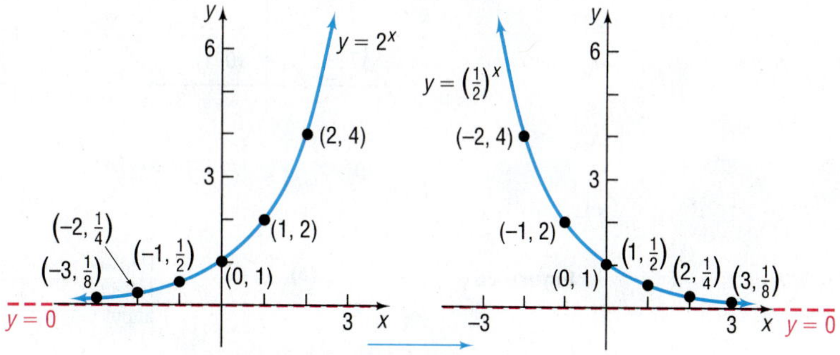

Figure 23 **(a)** $y = 2^x$ **(b)** $y = 2^{-x} = \left(\frac{1}{2}\right)^x$

Seeing the Concept

Using a graphing utility, simultaneously graph:

(a) $Y_1 = 3^x, Y_2 = \left(\frac{1}{3}\right)^x$

(b) $Y_1 = 6^x, Y_2 = \left(\frac{1}{6}\right)^x$

Conclude that the graph of $Y_2 = \left(\frac{1}{a}\right)^x$, for $a > 0$, is the reflection about the y-axis of the graph of $Y_1 = a^x$.

The graph of $f(x) = \left(\frac{1}{2}\right)^x$ in Figure 22 is typical of all exponential functions of the form $f(x) = a^x$ with $0 < a < 1$. Such functions are decreasing and one-to-one. Their graphs lie above the x-axis and pass through the point $(0, 1)$. The graphs rise rapidly as $x \to -\infty$. As $x \to \infty$, the x-axis $(y = 0)$ is a horizontal asymptote. There are no vertical asymptotes. Finally, the graphs are smooth and continuous, with no corners or gaps.

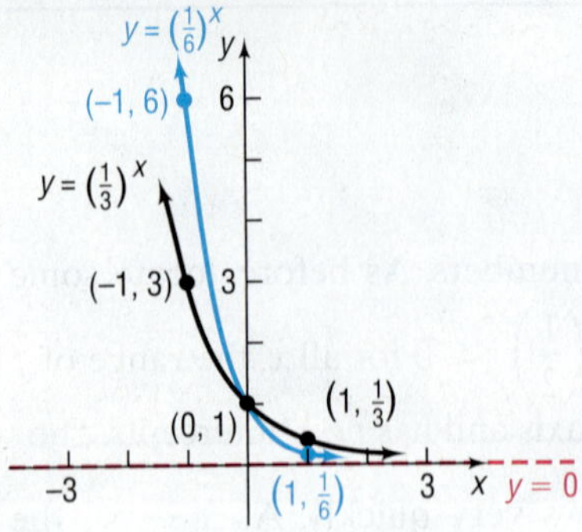

Figure 24

Figure 24 illustrates the graphs of two more exponential functions whose bases are between 0 and 1. Notice that the smaller base results in a graph that is steeper when $x < 0$. When $x > 0$, the graph of the equation with the smaller base is closer to the x-axis.

Properties of the Exponential Function $f(x) = a^x, 0 < a < 1$

1. The domain is the set of all real numbers, or $(-\infty, \infty)$ using interval notation; the range is the set of positive real numbers, or $(0, \infty)$ using interval notation.
2. There are no x-intercepts; the y-intercept is 1.
3. The x-axis $(y = 0)$ is a horizontal asymptote as $x \to \infty$ $\left[\lim_{x\to\infty} a^x = 0\right]$.
4. $f(x) = a^x, 0 < a < 1$, is a decreasing function and is one-to-one.
5. The graph of f contains the points $\left(-1, \frac{1}{a}\right)$, $(0, 1)$, and $(1, a)$.
6. The graph of f is smooth and continuous, with no corners or gaps. See Figure 25.

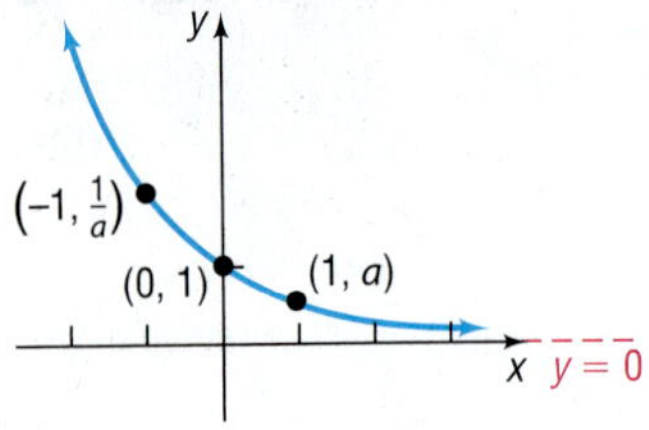

Figure 25 $f(x) = a^x, 0 < a < 1$

EXAMPLE 5

Graphing Exponential Functions Using Transformations

Graph $f(x) = 2^{-x} - 3$ and determine the domain, range, and horizontal asymptote of f.

Solution Begin with the graph of $y = 2^x$. Figure 26 shows the stages.

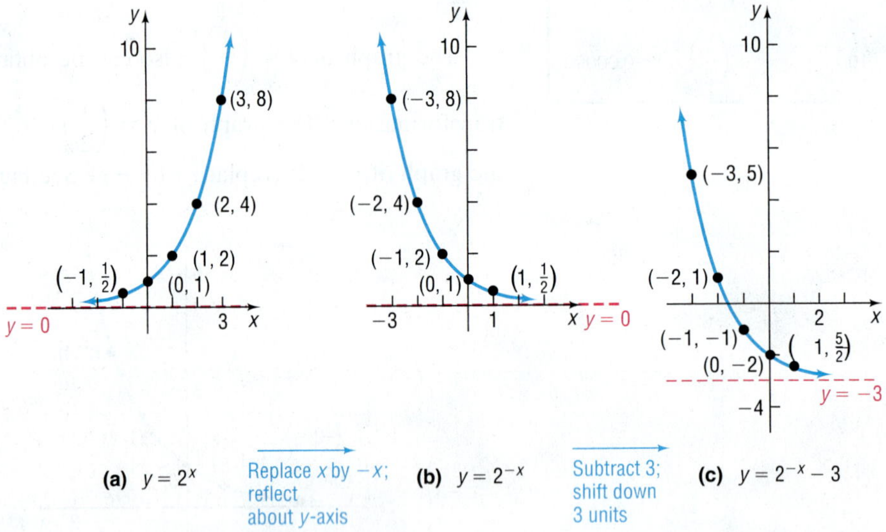

Figure 26

As Figure 26(c) illustrates, the domain of $f(x) = 2^{-x} - 3$ is the interval $(-\infty, \infty)$ and the range is the interval $(-3, \infty)$. The horizontal asymptote of f is the line $y = -3$.

Now Work PROBLEM 43

3 Define the Number e

Many problems that occur in nature require the use of an exponential function whose base is a certain irrational number, symbolized by the letter e.

One way of arriving at this important number e is given next.

DEFINITION

The **number e** is defined as the number that the expression

$$\left(1 + \frac{1}{n}\right)^n \qquad (2)$$

approaches as $n \to \infty$. In calculus, this is expressed, using limit notation, as

$$e = \lim_{n\to\infty}\left(1 + \frac{1}{n}\right)^n$$

Table 5 illustrates what happens to the defining expression (2) as n takes on increasingly large values. The last number in the right column in the table approximates e correct to nine decimal places. That is, $e = 2.718281827\ldots$. Remember, the three dots indicate that the decimal places continue. Because these decimal places continue but do not repeat, e is an irrational number. The number e is often expressed as a decimal rounded to a specific number of places. For example, $e \approx 2.71828$ is rounded to five decimal places.

Table 5

n	$\frac{1}{n}$	$1 + \frac{1}{n}$	$\left(1 + \frac{1}{n}\right)^n$
1	1	2	2
2	0.5	1.5	2.25
5	0.2	1.2	2.48832
10	0.1	1.1	2.59374246
100	0.01	1.01	2.704813829
1,000	0.001	1.001	2.716923932
10,000	0.0001	1.0001	2.718145927
100,000	0.00001	1.00001	2.718268237
1,000,000	0.000001	1.000001	2.718280469
1,000,000,000	10^{-9}	$1 + 10^{-9}$	2.718281827

Table 6

x	e^x
-2	$e^{-2} \approx 0.14$
-1	$e^{-1} \approx 0.37$
0	$e^0 = 1$
1	$e^1 \approx 2.72$
2	$e^2 \approx 7.39$

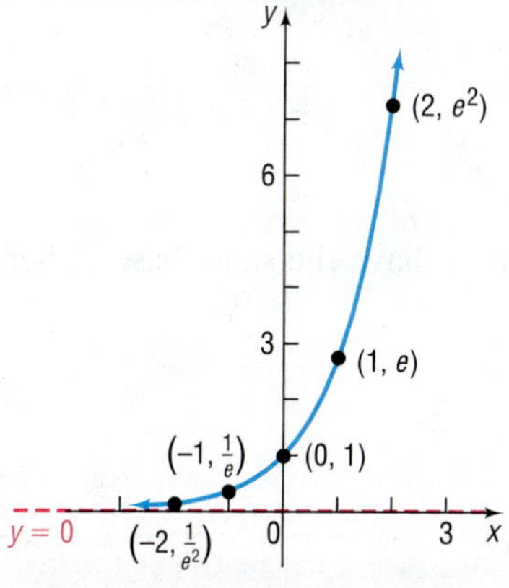

Figure 27 $y = e^x$

The exponential function $f(x) = e^x$, whose base is the number e, occurs with such frequency in applications that it is usually referred to as *the* exponential function. Indeed, most calculators have the key $\boxed{e^x}$ or $\boxed{\exp(x)}$, which may be used to evaluate the exponential function for a given value of x.*

Now use your calculator to approximate e^x for $x = -2, x = -1, x = 0, x = 1$, and $x = 2$. See Table 6. The graph of the exponential function $f(x) = e^x$ is given in Figure 27. Since $2 < e < 3$, the graph of $y = e^x$ lies between the graphs of $y = 2^x$ and $y = 3^x$. Do you see why? (Refer to Figures 18 and 20.)

Seeing the Concept

Graph $Y_1 = e^x$ and compare what you see to Figure 27. Use eVALUEate or TABLE to verify the points on the graph shown in Figure 27. Now graph $Y_2 = 2^x$ and $Y_3 = 3^x$ on the same screen as $Y_1 = e^x$. Notice that the graph of $Y_1 = e^x$ lies between these two graphs.

EXAMPLE 6 **Graphing Exponential Functions Using Transformations**

Graph $f(x) = -e^{x-3}$ and determine the domain, range, and horizontal asymptote of f.

*If your calculator does not have one of these keys, refer to your owner's manual.

Solution Begin with the graph of $y = e^x$. Figure 28 shows the stages.

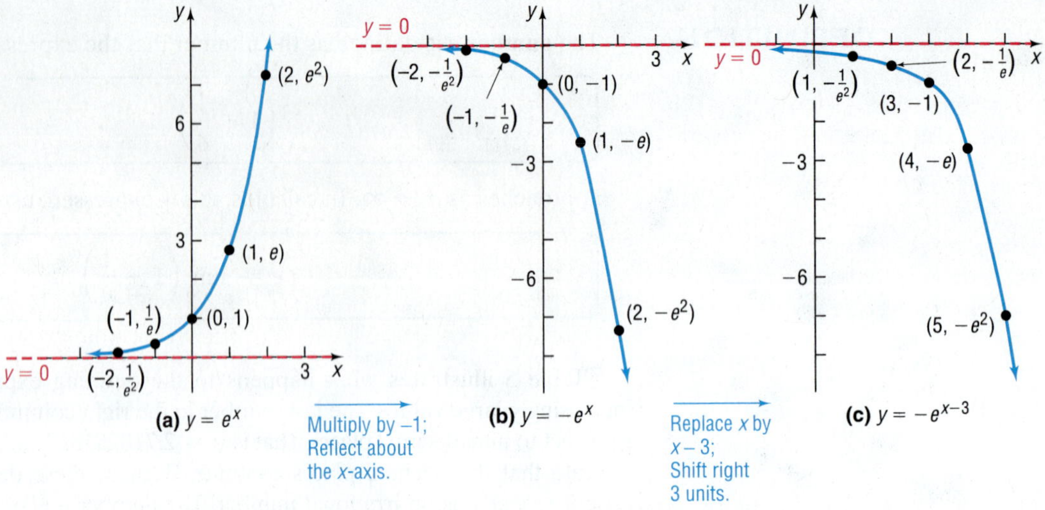

Figure 28

As Figure 28(c) illustrates, the domain of $f(x) = -e^{x-3}$ is the interval $(-\infty, \infty)$, and the range is the interval $(-\infty, 0)$. The horizontal asymptote is the line $y = 0$.

Now Work PROBLEM 55

4 Solve Exponential Equations

Equations that involve terms of the form a^x, where $a > 0$ and $a \neq 1$, are referred to as **exponential equations**. Such equations can sometimes be solved by appropriately applying the Laws of Exponents and property (3):

In Words
When two exponential expressions with the same base are equal, then their exponents are equal.

$$\text{If} \quad a^u = a^v, \quad \text{then} \quad u = v. \qquad \textbf{(3)}$$

Property (3) is a consequence of the fact that exponential functions are one-to-one. To use property (3), each side of the equality must be written with the same base.

EXAMPLE 7 **Solving Exponential Equations**

Solve each exponential equation.

(a) $3^{x+1} = 81$ (a) $4^{2x-1} = 8^{x+3}$

Solution (a) Since $81 = 3^4$, write the equation as

$$3^{x+1} = 81 = 3^4$$

Now the expressions on both sides of the equation have the same base, 3. Set the exponents equal to each other to obtain

$$x + 1 = 4$$
$$x = 3$$

The solution set is $\{3\}$.

(b)
$$4^{2x-1} = 8^{x+3}$$
$$(2^2)^{(2x-1)} = (2^3)^{(x+3)} \quad 4 = 2^2; 8 = 2^3$$
$$2^{2(2x-1)} = 2^{3(x+3)} \quad (a^r)^s = a^{rs}$$
$$2(2x - 1) = 3(x + 3) \quad \text{If } a^u = a^v, \text{ then } u = v.$$
$$4x - 2 = 3x + 9$$
$$x = 11$$

The solution set is $\{11\}$.

Now Work PROBLEMS 65 AND 75

EXAMPLE 8 Solving an Exponential Equation

Solve: $e^{-x^2} = (e^x)^2 \cdot \frac{1}{e^3}$

Solution Use the Laws of Exponents first to get a single expression with the base e on the right side.

$$(e^x)^2 \cdot \frac{1}{e^3} = e^{2x} \cdot e^{-3} = e^{2x-3}$$

As a result,

$$\begin{aligned} e^{-x^2} &= e^{2x-3} \\ -x^2 &= 2x - 3 && \text{Apply property (3).} \\ x^2 + 2x - 3 &= 0 && \text{Place the quadratic equation in standard form.} \\ (x+3)(x-1) &= 0 && \text{Factor.} \\ x = -3 \quad \text{or} \quad x &= 1 && \text{Use the Zero-Product Property.} \end{aligned}$$

The solution set is $\{-3, 1\}$.

Now Work PROBLEM 81

EXAMPLE 9 Exponential Probability

Between 9:00 PM and 10:00 PM, cars arrive at Burger King's drive-thru at the rate of 12 cars per hour (0.2 car per minute). The following formula from statistics can be used to determine the probability that a car will arrive within t minutes of 9:00 PM.

$$F(t) = 1 - e^{-0.2t}$$

(a) Determine the probability that a car will arrive within 5 minutes of 9 PM (that is, before 9:05 PM).

(b) Determine the probability that a car will arrive within 30 minutes of 9 PM (before 9:30 PM).

(c) Graph F using your graphing utility.

(d) What value does F approach as t increases without bound in the positive direction?

Solution (a) The probability that a car will arrive within 5 minutes is found by evaluating $F(t)$ at $t = 5$.

$$F(5) = 1 - e^{-0.2(5)} \approx 0.63212$$

↑ Use a calculator.

There is a 63% probability that a car will arrive within 5 minutes.

(b) The probability that a car will arrive within 30 minutes is found by evaluating $F(t)$ at $t = 30$.

$$F(30) = 1 - e^{-0.2(30)} \approx 0.9975$$

↑ Use a calculator.

There is a 99.75% probability that a car will arrive within 30 minutes.

(c) See Figure 29 for the graph of F.

(d) As time passes, the probability that a car will arrive increases. The value that F approaches can be found by letting $t \to \infty$. Since $e^{-0.2t} = \frac{1}{e^{0.2t}}$, it follows that $e^{-0.2t} \to 0$ as $t \to \infty$. Therefore, F approaches 1 as t gets large. The algebraic analysis is confirmed by Figure 29.

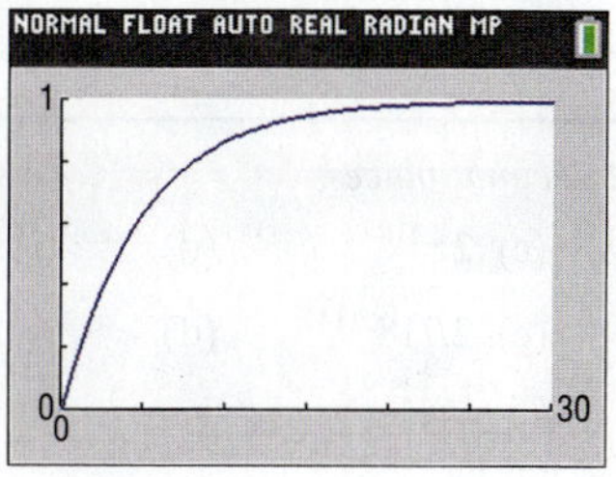

Figure 29 $F(t) = 1 - e^{-0.2t}$

Now Work PROBLEM 113

SUMMARY

Properties of the Exponential Function

$f(x) = a^x, \quad a > 1$	Domain: the interval $(-\infty, \infty)$; range: the interval $(0, \infty)$ x-intercepts: none; y-intercept: 1 Horizontal asymptote: x-axis $(y = 0)$ as $x \to -\infty$ Increasing; one-to-one; smooth; continuous See Figure 21 for a typical graph.
$f(x) = a^x, \quad 0 < a < 1$	Domain: the interval $(-\infty, \infty)$; range: the interval $(0, \infty)$ x-intercepts: none; y-intercept: 1 Horizontal asymptote: x-axis $(y = 0)$ as $x \to \infty$ Decreasing; one-to-one; smooth; continuous See Figure 25 for a typical graph.
If $a^u = a^v$, then $u = v$.	

6.3 Assess Your Understanding

'Are You Prepared?' *Answers are given at the end of these exercises. If you get a wrong answer, read the pages listed in red.*

1. $4^3 =$ _____; $8^{2/3} =$ _____; $3^{-2} =$ _____. (pp. 21–24 and pp. 73–77)

2. Solve: $x^2 + 3x = 4$ (pp. 92–99)

3. ***True or False*** To graph $y = (x - 2)^3$, shift the graph of $y = x^3$ to the left 2 units. (pp. 247–256)

4. Find the average rate of change of $f(x) = 3x - 5$ from $x = 0$ to $x = 4$. (pp. 230–231)

5. ***True or False*** The function $f(x) = \dfrac{2x}{x - 3}$ has $y = 2$ as a horizontal asymptote. (pp. 348–350)

Concepts and Vocabulary

6. A(n) ________ ________ is a function of the form $f(x) = Ca^x$, where $a > 0, a \neq 1$, and $C \neq 0$ are real numbers. The base a is the ______ ______ and C is the ______ ______.

7. For an exponential function $f(x) = Ca^x$, $\dfrac{f(x + 1)}{f(x)} =$ ____.

8. ***True or False*** The domain of the exponential function $f(x) = a^x$, where $a > 0$ and $a \neq 1$, is the set of all real numbers.

9. ***True or False*** The graph of the exponential function $f(x) = a^x$, where $a > 0$ and $a \neq 1$, has no x-intercept.

10. The graph of every exponential function $f(x) = a^x$, where $a > 0$ and $a \neq 1$, passes through three points: ____, ____, and ____.

11. If $3^x = 3^4$, then $x =$ ____.

12. ***True or False*** The graphs of $y = 3^x$ and $y = \left(\dfrac{1}{3}\right)^x$ are identical.

13. Which of the following exponential functions is an increasing function?

(a) $f(x) = 0.5^x$ (b) $f(x) = \left(\dfrac{5}{2}\right)^x$
(c) $f(x) = \left(\dfrac{2}{3}\right)^x$ (d) $f(x) = 0.9^x$

14. Which of the following is the range of the exponential function $f(x) = a^x, a > 0$ and $a \neq 1$?

(a) $(-\infty, \infty)$ (b) $(-\infty, 0)$
(c) $(0, \infty)$ (d) $(-\infty, 0) \cup (0, \infty)$

Skill Building

In Problems 15–26, approximate each number using a calculator. Express your answer rounded to three decimal places.

15. (a) $2^{3.14}$ (b) $2^{3.141}$ (c) $2^{3.1415}$ (d) 2^{π}

16. (a) $2^{2.7}$ (b) $2^{2.71}$ (c) $2^{2.718}$ (d) 2^{e}

17. (a) $3.1^{2.7}$ (b) $3.14^{2.71}$ (c) $3.141^{2.718}$ (d) π^{e}

18. (a) $2.7^{3.1}$ (b) $2.71^{3.14}$ (c) $2.718^{3.141}$ (d) e^{π}

19. $(1 + 0.04)^6$

20. $\left(1 + \dfrac{0.09}{12}\right)^{24}$

21. $8.4\left(\dfrac{1}{3}\right)^{2.9}$

22. $158\left(\dfrac{5}{6}\right)^{8.63}$

23. $e^{1.2}$

24. $e^{-1.3}$

25. $125e^{0.026(7)}$

26. $83.6e^{-0.157(9.5)}$

In Problems 27–34, determine whether the given function is linear, exponential, or neither. For those that are linear functions, find a linear function that models the data; for those that are exponential, find an exponential function that models the data.

27.

x	$f(x)$
−1	3
0	6
1	12
2	18
3	30

28.

x	$g(x)$
−1	2
0	5
1	8
2	11
3	14

29.

x	$H(x)$
−1	$\frac{1}{4}$
0	1
1	4
2	16
3	64

30.

x	$F(x)$
−1	$\frac{2}{3}$
0	1
1	$\frac{3}{2}$
2	$\frac{9}{4}$
3	$\frac{27}{8}$

31.

x	$f(x)$
−1	$\frac{3}{2}$
0	3
1	6
2	12
3	24

32.

x	$g(x)$
−1	6
0	1
1	0
2	3
3	10

33.

x	$H(x)$
−1	2
0	4
1	6
2	8
3	10

34.

x	$F(x)$
−1	$\frac{1}{2}$
0	$\frac{1}{4}$
1	$\frac{1}{8}$
2	$\frac{1}{16}$
3	$\frac{1}{32}$

In Problems 35–42, the graph of an exponential function is given. Match each graph to one of the following functions.

(A) $y = 3^x$ (B) $y = 3^{-x}$ (C) $y = -3^x$ (D) $y = -3^{-x}$
(E) $y = 3^x - 1$ (F) $y = 3^{x-1}$ (G) $y = 3^{1-x}$ (H) $y = 1 - 3^x$

35.
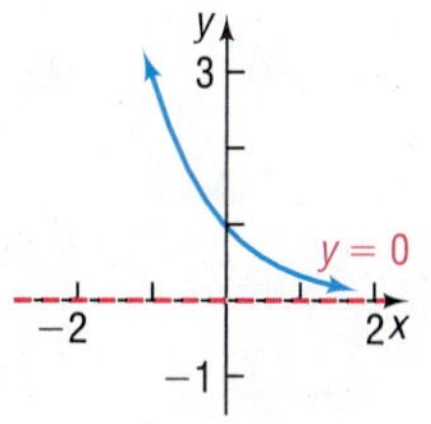

36.
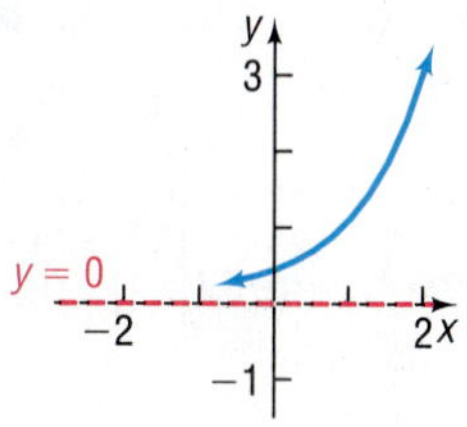

37.
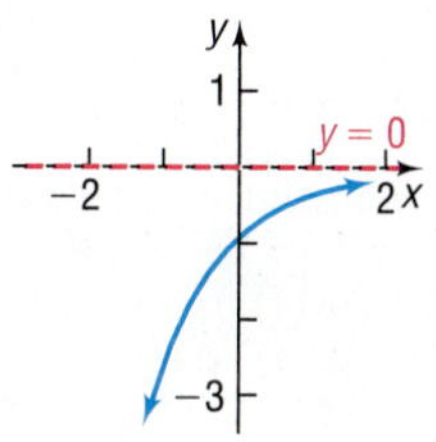

38.
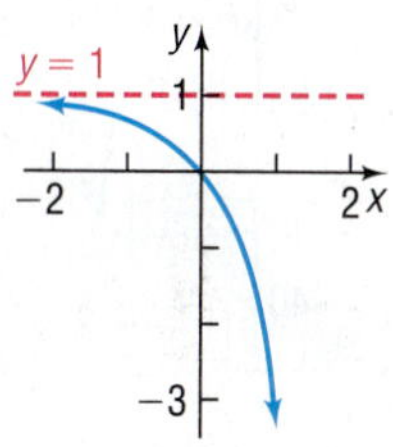

39.
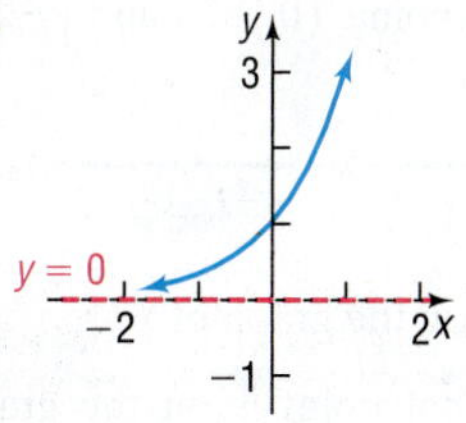

40.
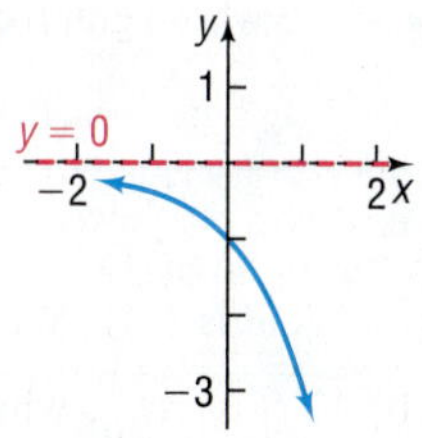

41.
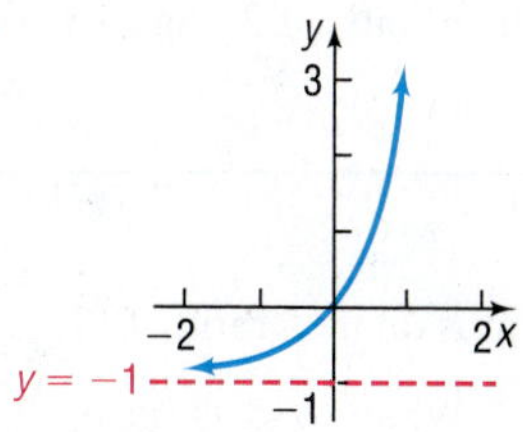

42.
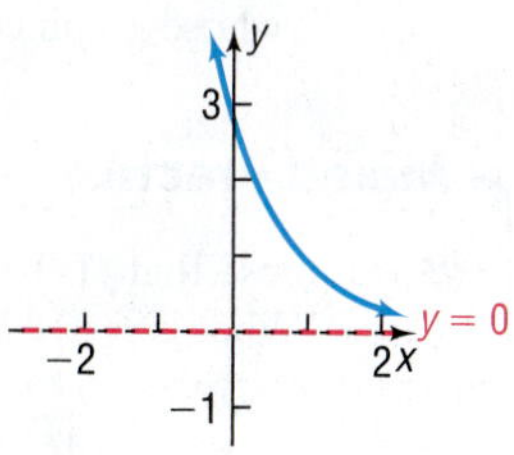

In Problems 43–54, use transformations to graph each function. Determine the domain, range, and horizontal asymptote of each function.

43. $f(x) = 2^x + 1$ **44.** $f(x) = 3^x - 2$ **45.** $f(x) = 3^{x-1}$ **46.** $f(x) = 2^{x+2}$

47. $f(x) = 3 \cdot \left(\frac{1}{2}\right)^x$ **48.** $f(x) = 4 \cdot \left(\frac{1}{3}\right)^x$ **49.** $f(x) = 3^{-x} - 2$ **50.** $f(x) = -3^x + 1$

51. $f(x) = 2 + 4^{x-1}$ **52.** $f(x) = 1 - 2^{x+3}$ **53.** $f(x) = 2 + 3^{x/2}$ **54.** $f(x) = 1 - 2^{-x/3}$

In Problems 55–62, begin with the graph of $y = e^x$ (Figure 27) and use transformations to graph each function. Determine the domain, range, and horizontal asymptote of each function.

55. $f(x) = e^{-x}$ **56.** $f(x) = -e^x$ **57.** $f(x) = e^{x+2}$ **58.** $f(x) = e^x - 1$

59. $f(x) = 5 - e^{-x}$ **60.** $f(x) = 9 - 3e^{-x}$ **61.** $f(x) = 2 - e^{-x/2}$ **62.** $f(x) = 7 - 3e^{2x}$

In Problems 63–82, solve each equation.

63. $7^x = 7^3$ **64.** $5^x = 5^{-6}$ **65.** $2^{-x} = 16$ **66.** $3^{-x} = 81$

67. $\left(\frac{1}{5}\right)^x = \frac{1}{25}$ **68.** $\left(\frac{1}{4}\right)^x = \frac{1}{64}$ **69.** $2^{2x-1} = 4$ **70.** $5^{x+3} = \frac{1}{5}$

71. $3^{x^3} = 9^x$ **72.** $4^{x^2} = 2^x$ **73.** $8^{-x+14} = 16^x$ **74.** $9^{-x+15} = 27^x$

75. $3^{x^2-7} = 27^{2x}$ **76.** $5^{x^2+8} = 125^{2x}$ **77.** $4^x \cdot 2^{x^2} = 16^2$ **78.** $9^{2x} \cdot 27^{x^2} = 3^{-1}$

79. $e^x = e^{3x+8}$ **80.** $e^{3x} = e^{2-x}$ **81.** $e^{x^2} = e^{3x} \cdot \frac{1}{e^2}$ **82.** $(e^4)^x \cdot e^{x^2} = e^{12}$

83. If $4^x = 7$, what does 4^{-2x} equal?

84. If $2^x = 3$, what does 4^{-x} equal?

85. If $3^{-x} = 2$, what does 3^{2x} equal?

86. If $5^{-x} = 3$, what does 5^{3x} equal?

87. If $9^x = 25$, what does 3^x equal?

88. If $2^{-3x} = \frac{1}{1000}$, what does 2^x equal?

In Problems 89–92, determine the exponential function whose graph is given.

89.

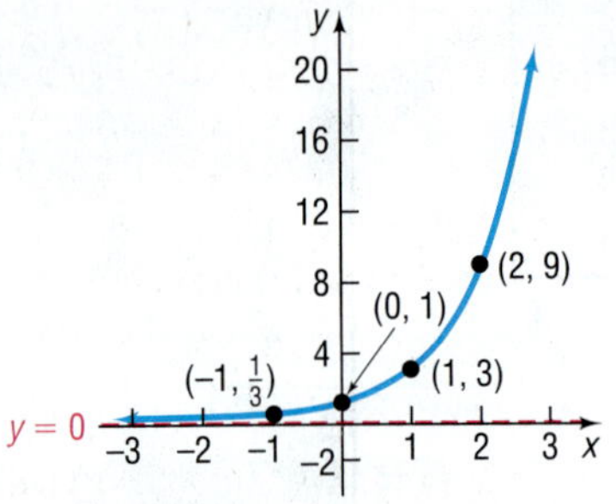

90.

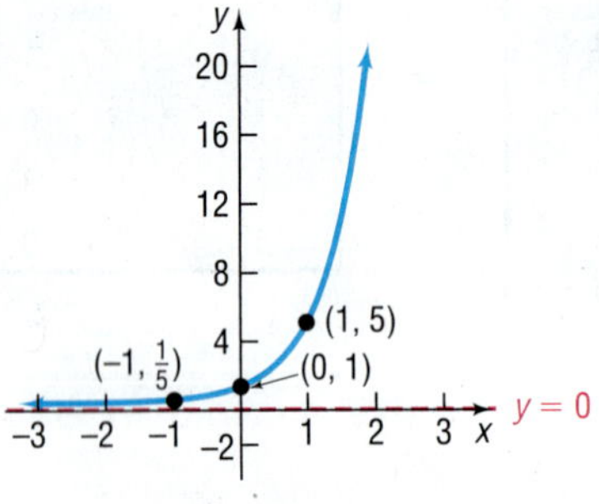

91.

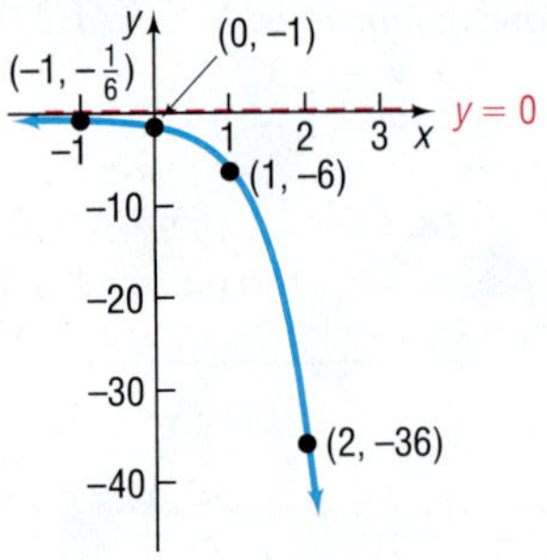

92.

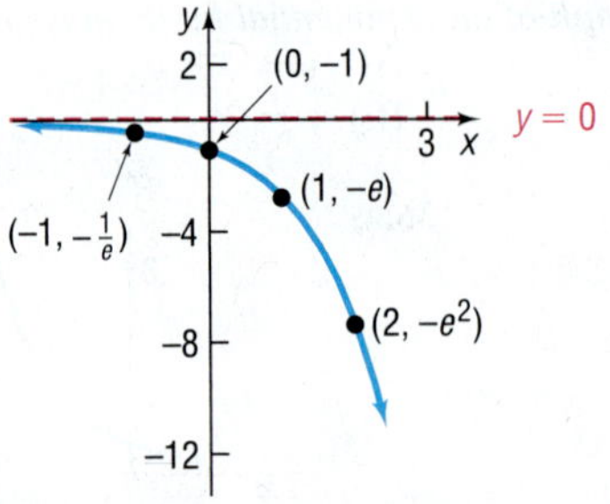

93. Find an exponential function with horizontal asymptote $y = 2$ whose graph contains the points $(0, 3)$ and $(1, 5)$.

94. Find an exponential function with horizontal asymptote $y = -3$ whose graph contains the points $(0, -2)$ and $(-2, 1)$.

Mixed Practice

95. Suppose that $f(x) = 2^x$.
(a) What is $f(4)$? What point is on the graph of f?
(b) If $f(x) = \frac{1}{16}$, what is x? What point is on the graph of f?

96. Suppose that $f(x) = 3^x$.
(a) What is $f(4)$? What point is on the graph of f?
(b) If $f(x) = \frac{1}{9}$, what is x? What point is on the graph of f?

97. Suppose that $g(x) = 4^x + 2$.
(a) What is $g(-1)$? What point is on the graph of g?
(b) If $g(x) = 66$, what is x? What point is on the graph of g?

98. Suppose that $g(x) = 5^x - 3$.
(a) What is $g(-1)$? What point is on the graph of g?
(b) If $g(x) = 122$, what is x? What point is on the graph of g?

99. Suppose that $H(x) = \left(\frac{1}{2}\right)^x - 4$.
(a) What is $H(-6)$? What point is on the graph of H?
(b) If $H(x) = 12$, what is x? What point is on the graph of H?
(c) Find the zero of H.

100. Suppose that $F(x) = \left(\frac{1}{3}\right)^x - 3$.
(a) What is $F(-5)$? What point is on the graph of F?
(b) If $F(x) = 24$, what is x? What point is on the graph of F?
(c) Find the zero of F.

In Problems 101–104, graph each function. Based on the graph, state the domain and the range, and find any intercepts.

101. $f(x) = \begin{cases} e^{-x} & \text{if } x < 0 \\ e^{x} & \text{if } x \geq 0 \end{cases}$

102. $f(x) = \begin{cases} e^{x} & \text{if } x < 0 \\ e^{-x} & \text{if } x \geq 0 \end{cases}$

103. $f(x) = \begin{cases} -e^{x} & \text{if } x < 0 \\ -e^{-x} & \text{if } x \geq 0 \end{cases}$

104. $f(x) = \begin{cases} -e^{-x} & \text{if } x < 0 \\ -e^{x} & \text{if } x \geq 0 \end{cases}$

Applications and Extensions

105. Optics If a single pane of glass obliterates 3% of the light passing through it, the percent p of light that passes through n successive panes is given approximately by the function

$$p(n) = 100(0.97)^n$$

(a) What percent of light will pass through 10 panes?
(b) What percent of light will pass through 25 panes?
(c) Explain the meaning of the base 0.97 in this problem.

106. Atmospheric Pressure The atmospheric pressure p on a balloon or airplane decreases with increasing height. This pressure, measured in millimeters of mercury, is related to the height h (in kilometers) above sea level by the function

$$p(h) = 760e^{-0.145h}$$

(a) Find the atmospheric pressure at a height of 2 km (over a mile).
(b) What is it at a height of 10 kilometers (over 30,000 feet)?

107. Depreciation The price p, in dollars, of a Honda Civic EX-L sedan that is x years old is modeled by

$$p(x) = 22{,}265(0.90)^x$$

(a) How much should a 3-year-old Civic EX-L sedan cost?
(b) How much should a 9-year-old Civic EX-L sedan cost?
(c) Explain the meaning of the base 0.90 in this problem.

108. Healing of Wounds The normal healing of wounds can be modeled by an exponential function. If A_0 represents the original area of the wound and if A equals the area of the wound, then the function

$$A(n) = A_0e^{-0.35n}$$

describes the area of a wound after n days following an injury when no infection is present to retard the healing. Suppose that a wound initially had an area of 100 square millimeters.
(a) If healing is taking place, how large will the area of the wound be after 3 days?
(b) How large will it be after 10 days?

109. Advanced-Stage Pancreatic Cancer The percentage of patients P who have survived t years after initial diagnosis of advanced-stage pancreatic cancer is modeled by the function

$$P(t) = 100(0.3)^t$$

Source: *Cancer Treatment Centers of America*
(a) According to the model, what percent of patients survive 1 year after initial diagnosis?
(b) What percent of patients survive 2 years after initial diagnosis?
(c) Explain the meaning of the base 0.3 in the context of this problem.

110. Endangered Species In a protected environment, the population P of a certain endangered species recovers over time t (in years) according to the model

$$P(t) = 30(1.149)^t$$

(a) What is the size of the initial population of the species?
(b) According to the model, what will be the population of the species in 5 years?
(c) According to the model, what will be the population of the species in 10 years?
(d) According to the model, what will be the population of the species in 15 years?
(e) What is happening to the population every 5 years?

111. Drug Medication The function

$$D(h) = 5e^{-0.4h}$$

can be used to find the number of milligrams D of a certain drug that is in a patient's bloodstream h hours after the drug has been administered. How many milligrams will be present after 1 hour? After 6 hours?

112. Spreading of Rumors A model for the number N of people in a college community who have heard a certain rumor is

$$N = P(1 - e^{-0.15d})$$

where P is the total population of the community and d is the number of days that have elapsed since the rumor began. In a community of 1000 students, how many students will have heard the rumor after 3 days?

113. Exponential Probability Between 12:00 PM and 1:00 PM, cars arrive at Citibank's drive-thru at the rate of 6 cars per hour (0.1 car per minute). The following formula from probability can be used to determine the probability that a car will arrive within t minutes of 12:00 PM.

$$F(t) = 1 - e^{-0.1t}$$

(a) Determine the probability that a car will arrive within 10 minutes of 12:00 PM (that is, before 12:10 PM).
(b) Determine the probability that a car will arrive within 40 minutes of 12:00 PM (before 12:40 PM).
(c) What value does F approach as t becomes unbounded in the positive direction?
(d) Graph F using a graphing utility.
(e) Using INTERSECT, determine how many minutes are needed for the probability to reach 50%.

114. Exponential Probability Between 5:00 PM and 6:00 PM, cars arrive at Jiffy Lube at the rate of 9 cars per hour (0.15 car per minute). This formula from probability can be used to determine the probability that a car will arrive within t minutes of 5:00 PM:

$$F(t) = 1 - e^{-0.15t}$$

(a) Determine the probability that a car will arrive within 15 minutes of 5:00 PM (that is, before 5:15 PM).
(b) Determine the probability that a car will arrive within 30 minutes of 5:00 PM (before 5:30 PM).
(c) What value does F approach as t becomes unbounded in the positive direction?
(d) Graph F using a graphing utility.
(e) Using INTERSECT, determine how many minutes are needed for the probability to reach 60%.

115. Poisson Probability Between 5:00 PM and 6:00 PM, cars arrive at a McDonald's drive-thru at the rate of 20 cars per hour. The following formula from probability can be used to determine the probability that x cars will arrive between 5:00 PM and 6:00 PM.

$$P(x) = \frac{20^x e^{-20}}{x!}$$

where

$$x! = x \cdot (x-1) \cdot (x-2) \cdot \cdots \cdot 3 \cdot 2 \cdot 1$$

(a) Determine the probability that $x = 15$ cars will arrive between 5:00 PM and 6:00 PM.
(b) Determine the probability that $x = 20$ cars will arrive between 5:00 PM and 6:00 PM.

116. Poisson Probability People enter a line for the *Demon Roller Coaster* at the rate of 4 per minute. The following formula from probability can be used to determine the probability that x people will arrive within the next minute.

$$P(x) = \frac{4^x e^{-4}}{x!}$$

where

$$x! = x \cdot (x-1) \cdot (x-2) \cdot \cdots \cdot 3 \cdot 2 \cdot 1$$

(a) Determine the probability that $x = 5$ people will arrive within the next minute.
(b) Determine the probability that $x = 8$ people will arrive within the next minute.

117. Relative Humidity The relative humidity is the ratio (expressed as a percent) of the amount of water vapor in the air to the maximum amount that the air can hold at a specific temperature. The relative humidity, R, is found using the following formula:

$$R = 10^{\left(\frac{4221}{T+459.4} - \frac{4221}{D+459.4} + 2\right)}$$

where T is the air temperature (in °F) and D is the dew point temperature (in °F).

(a) Determine the relative humidity if the air temperature is 50° Fahrenheit and the dew point temperature is 41° Fahrenheit.
(b) Determine the relative humidity if the air temperature is 68° Fahrenheit and the dew point temperature is 59° Fahrenheit.
(c) What is the relative humidity if the air temperature and the dew point temperature are the same?

118. Learning Curve Suppose that a student has 500 vocabulary words to learn. If the student learns 15 words after 5 minutes, the function

$$L(t) = 500(1 - e^{-0.0061t})$$

approximates the number of words L that the student will have learned after t minutes.

(a) How many words will the student have learned after 30 minutes?
(b) How many words will the student have learned after 60 minutes?

119. Current in an RL Circuit The equation governing the amount of current I (in amperes) after time t (in seconds) in a single RL circuit consisting of a resistance R (in ohms), an inductance L (in henrys), and an electromotive force E (in volts) is

$$I = \frac{E}{R}\left[1 - e^{-(R/L)t}\right]$$

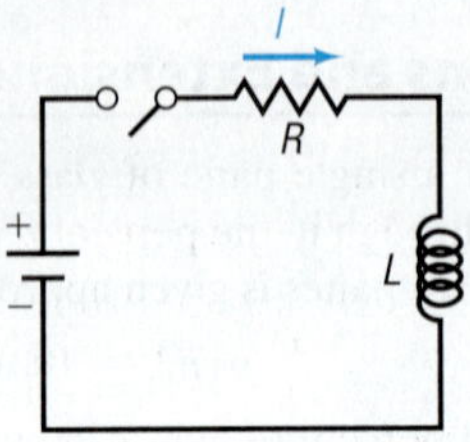

(a) If $E = 120$ volts, $R = 10$ ohms, and $L = 5$ henrys, how much current I_1 is flowing after 0.3 second? After 0.5 second? After 1 second?
(b) What is the maximum current?
(c) Graph this function $I = I_1(t)$, measuring I along the y-axis and t along the x-axis.
(d) If $E = 120$ volts, $R = 5$ ohms, and $L = 10$ henrys, how much current I_2 is flowing after 0.3 second? After 0.5 second? After 1 second?
(e) What is the maximum current?
(f) Graph the function $I = I_2(t)$ on the same coordinate axes as $I_1(t)$.

120. Current in an RC Circuit The equation governing the amount of current I (in amperes) after time t (in microseconds) in a single RC circuit consisting of a resistance R (in ohms), a capacitance C (in microfarads), and an electromotive force E (in volts) is

$$I = \frac{E}{R} e^{-t/(RC)}$$

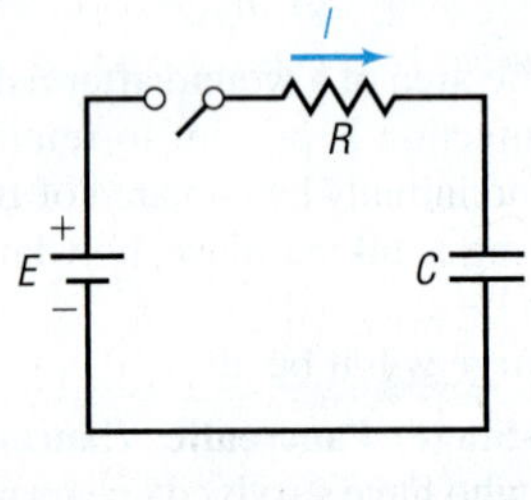

(a) If $E = 120$ volts, $R = 2000$ ohms, and $C = 1.0$ microfarad, how much current I_1 is flowing initially ($t = 0$)? After 1000 microseconds? After 3000 microseconds?
(b) What is the maximum current?
(c) Graph the function $I = I_1(t)$, measuring I along the y-axis and t along the x-axis.
(d) If $E = 120$ volts, $R = 1000$ ohms, and $C = 2.0$ microfarads, how much current I_2 is flowing initially? After 1000 microseconds? After 3000 microseconds?
(e) What is the maximum current?
(f) Graph the function $I = I_2(t)$ on the same coordinate axes as $I_1(t)$.

121. If f is an exponential function of the form $f(x) = Ca^x$ with growth factor 3, and if $f(6) = 12$, what is $f(7)$?

122. Another Formula for e Use a calculator to compute the values of

$$2 + \frac{1}{2!} + \frac{1}{3!} + \cdots + \frac{1}{n!}$$

for $n = 4, 6, 8$, and 10. Compare each result with e.
[**Hint:** $1! = 1, 2! = 2 \cdot 1, 3! = 3 \cdot 2 \cdot 1,$
$n! = n(n-1) \cdot \cdots \cdot (3)(2)(1)$.]

123. Another Formula for e Use a calculator to compute the various values of the expression. Compare the values to e.

$$2 + \cfrac{1}{1 + \cfrac{1}{2 + \cfrac{2}{3 + \cfrac{3}{4 + \cfrac{4}{\text{etc.}}}}}}$$

124. Difference Quotient If $f(x) = a^x$, show that

$$\frac{f(x+h) - f(x)}{h} = a^x \cdot \frac{a^h - 1}{h} \qquad h \neq 0$$

125. If $f(x) = a^x$, show that $f(A + B) = f(A) \cdot f(B)$.

126. If $f(x) = a^x$, show that $f(-x) = \dfrac{1}{f(x)}$.

127. If $f(x) = a^x$, show that $f(\alpha x) = [f(x)]^\alpha$.

Problems 128 and 129 provide definitions for two other transcendental functions.

128. The **hyperbolic sine function,** designated by $\sinh x$, is defined as

$$\sinh x = \frac{1}{2}(e^x - e^{-x})$$

(a) Show that $f(x) = \sinh x$ is an odd function.
(b) Graph $f(x) = \sinh x$ using a graphing utility.

129. The **hyperbolic cosine function,** designated by $\cosh x$, is defined as

$$\cosh x = \frac{1}{2}(e^x + e^{-x})$$

(a) Show that $f(x) = \cosh x$ is an even function.
(b) Graph $f(x) = \cosh x$ using a graphing utility.
(c) Refer to Problem 128. Show that, for every x,

$$(\cosh x)^2 - (\sinh x)^2 = 1$$

130. Historical Problem Pierre de Fermat (1601–1665) conjectured that the function

$$f(x) = 2^{(2^x)} + 1$$

for $x = 1, 2, 3, \ldots$, would always have a value equal to a prime number. But Leonhard Euler (1707–1783) showed that this formula fails for $x = 5$. Use a calculator to determine the prime numbers produced by f for $x = 1, 2, 3, 4$. Then show that $f(5) = 641 \times 6{,}700{,}417$, which is not prime.

Explaining Concepts: Discussion and Writing

131. The bacteria in a 4-liter container double every minute. After 60 minutes the container is full. How long did it take to fill half the container?

132. Explain in your own words what the number e is. Provide at least two applications that use this number.

133. Do you think that there is a power function that increases more rapidly than an exponential function whose base is greater than 1? Explain.

134. As the base a of an exponential function $f(x) = a^x$, where $a > 1$, increases, what happens to the behavior of its graph for $x > 0$? What happens to the behavior of its graph for $x < 0$?

135. The graphs of $y = a^{-x}$ and $y = \left(\dfrac{1}{a}\right)^x$ are identical. Why?

Retain Your Knowledge

Problems 136–139 are based on material learned earlier in the course. The purpose of these problems is to keep the material fresh in your mind so that you are better prepared for the final exam.

136. Solve the inequality: $x^3 + 5x^2 \leq 4x + 20$.

137. Solve the inequality: $\dfrac{x+1}{x-2} \geq 1$.

138. Find the equation of the quadratic function f that has its vertex at $(3, 5)$ and contains the point $(2, 3)$.

139. Consider the quadratic function $f(x) = x^2 + 2x - 3$.
(a) Graph f by determining whether its graph opens up or down and by finding its vertex, axis of symmetry, y-intercept, and x-intercepts, if any.
(b) Determine the domain and range of f.
(c) Determine where f is increasing and where it is decreasing.

'Are You Prepared?' Answers

1. $64; 4; \frac{1}{9}$ **2.** $\{-4, 1\}$ **3.** False **4.** 3 **5.** True

6.4 Logarithmic Functions

PREPARING FOR THIS SECTION *Before getting started, review the following:*

- Solving Inequalities (Section 1.5, pp. 119–126)
- Quadratic Inequalities (Section 4.5, pp. 312–313)
- Polynomial and Rational Inequalities (Section 5.4, pp. 370–372)
- Solve Linear Equations (Section 1.1, pp. 82–87)

Now Work the 'Are You Prepared?' problems on page 448.

OBJECTIVES
1. Change Exponential Statements to Logarithmic Statements and Logarithmic Statements to Exponential Statements (p. 440)
2. Evaluate Logarithmic Expressions (p. 441)
3. Determine the Domain of a Logarithmic Function (p. 441)
4. Graph Logarithmic Functions (p. 442)
5. Solve Logarithmic Equations (p. 446)

Recall that a one-to-one function $y = f(x)$ has an inverse function that is defined implicitly by the equation $x = f(y)$. In particular, the exponential function $y = f(x) = a^x$, where $a > 0$ and $a \neq 1$, is one-to-one and hence has an inverse function that is defined implicitly by the equation

$$x = a^y \qquad a > 0 \qquad a \neq 1$$

This inverse function is so important that it is given a name, the *logarithmic function*.

DEFINITION

The **logarithmic function with base *a***, where $a > 0$ and $a \neq 1$, is denoted by $y = \log_a x$ (read as "y is the logarithm with base a of x") and is defined by

$$y = \log_a x \quad \text{if and only if} \quad x = a^y$$

The domain of the logarithmic function $y = \log_a x$ is $x > 0$.

In Words

When you need to evaluate $\log_a x$, think to yourself "a raised to what power gives me x?"

As this definition illustrates, **a logarithm is a name for a certain exponent**. So $\log_a x$ represents the exponent to which a must be raised to obtain x.

EXAMPLE 1 **Relating Logarithms to Exponents**

(a) If $y = \log_3 x$, then $x = 3^y$. For example, the logarithmic statement $4 = \log_3 81$ is equivalent to the exponential statement $81 = 3^4$.

(b) If $y = \log_5 x$, then $x = 5^y$. For example, $-1 = \log_5\left(\frac{1}{5}\right)$ is equivalent to $\frac{1}{5} = 5^{-1}$.

1 Change Exponential Statements to Logarithmic Statements and Logarithmic Statements to Exponential Statements

The definition of a logarithm can be used to convert from exponential form to logarithmic form, and vice versa, as the following two examples illustrate.

EXAMPLE 2 **Changing Exponential Statements to Logarithmic Statements**

Change each exponential statement to an equivalent statement involving a logarithm.

(a) $1.2^3 = m$ (b) $e^b = 9$ (c) $a^4 = 24$

Solution Use the fact that $y = \log_a x$ and $x = a^y$, where $a > 0$ and $a \neq 1$, are equivalent.

(a) If $1.2^3 = m$, then $3 = \log_{1.2} m$. (b) If $e^b = 9$, then $b = \log_e 9$.
(c) If $a^4 = 24$, then $4 = \log_a 24$.

Now Work PROBLEM 11

EXAMPLE 3 **Changing Logarithmic Statements to Exponential Statements**

Change each logarithmic statement to an equivalent statement involving an exponent.

(a) $\log_a 4 = 5$ (b) $\log_e b = -3$ (c) $\log_3 5 = c$

Solution (a) If $\log_a 4 = 5$, then $a^5 = 4$.
(b) If $\log_e b = -3$, then $e^{-3} = b$.
(c) If $\log_3 5 = c$, then $3^c = 5$.

Now Work PROBLEM 19

2 Evaluate Logarithmic Expressions

To find the exact value of a logarithm, write the logarithm in exponential notation using the fact that $y = \log_a x$ is equivalent to $a^y = x$, and use the fact that if $a^u = a^v$, then $u = v$.

EXAMPLE 4 **Finding the Exact Value of a Logarithmic Expression**

Find the exact value of:

(a) $\log_2 16$ (b) $\log_3 \frac{1}{27}$

Solution

(a) To evaluate $\log_2 16$, think "2 raised to what power yields 16?" Then,

$y = \log_2 16$

$2^y = 16$ Change to exponential form.

$2^y = 2^4$ $16 = 2^4$

$y = 4$ Equate exponents.

Therefore, $\log_2 16 = 4$.

(b) To evaluate $\log_3 \frac{1}{27}$, think "3 raised to what power yields $\frac{1}{27}$?" Then,

$y = \log_3 \frac{1}{27}$

$3^y = \frac{1}{27}$ Change to exponential form.

$3^y = 3^{-3}$ $\frac{1}{27} = \frac{1}{3^3} = 3^{-3}$

$y = -3$ Equate exponents.

Therefore, $\log_3 \frac{1}{27} = -3$.

Now Work PROBLEM 27

3 Determine the Domain of a Logarithmic Function

The logarithmic function $y = \log_a x$ has been defined as the inverse of the exponential function $y = a^x$. That is, if $f(x) = a^x$, then $f^{-1}(x) = \log_a x$. Based on the discussion in Section 6.2 on inverse functions, for a function f and its inverse f^{-1},

$$\text{Domain of } f^{-1} = \text{Range of } f \quad \text{and} \quad \text{Range of } f^{-1} = \text{Domain of } f$$

Consequently, it follows that

> Domain of the logarithmic function = Range of the exponential function = $(0, \infty)$
> Range of the logarithmic function = Domain of the exponential function = $(-\infty, \infty)$

The next box summarizes some properties of the logarithmic function.

> $y = \log_a x$ (defining equation: $x = a^y$)
> Domain: $0 < x < \infty$ Range: $-\infty < y < \infty$

The domain of a logarithmic function consists of the *positive* real numbers, so the argument of a logarithmic function must be greater than zero.

EXAMPLE 5 Finding the Domain of a Logarithmic Function

Find the domain of each logarithmic function.

(a) $F(x) = \log_2(x + 3)$ (b) $g(x) = \log_5\left(\dfrac{1 + x}{1 - x}\right)$ (c) $h(x) = \log_{1/2}|x|$

Solution (a) The domain of F consists of all x for which $x + 3 > 0$, that is, $x > -3$. Using interval notation, the domain of F is $(-3, \infty)$.

(b) The domain of g is restricted to

$$\frac{1 + x}{1 - x} > 0$$

Solve this inequality to find that the domain of g consists of all x between -1 and 1, that is, $-1 < x < 1$, or, using interval notation, $(-1, 1)$.

(c) Since $|x| > 0$, provided that $x \neq 0$, the domain of h consists of all real numbers except zero, or, using interval notation, $(-\infty, 0) \cup (0, \infty)$. ●

Now Work PROBLEMS 41 AND 47

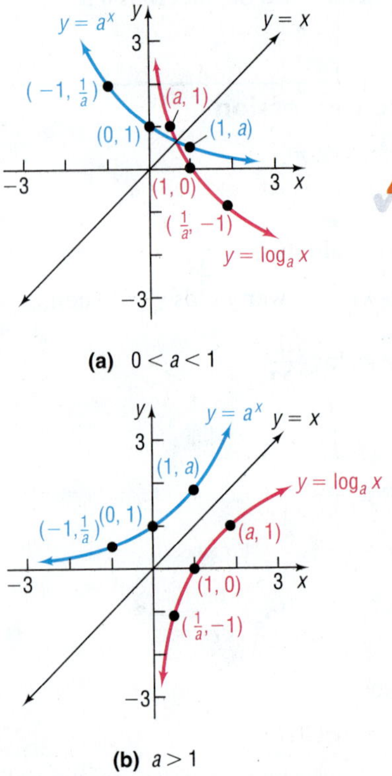

Figure 30

4 Graph Logarithmic Functions

Because exponential functions and logarithmic functions are inverses of each other, the graph of the logarithmic function $y = \log_a x$ is the reflection about the line $y = x$ of the graph of the exponential function $y = a^x$, as shown in Figure 30.

For example, to graph $y = \log_2 x$, graph $y = 2^x$ and reflect it about the line $y = x$. See Figure 31. To graph $y = \log_{1/3} x$, graph $y = \left(\dfrac{1}{3}\right)^x$ and reflect it about the line $y = x$. See Figure 32.

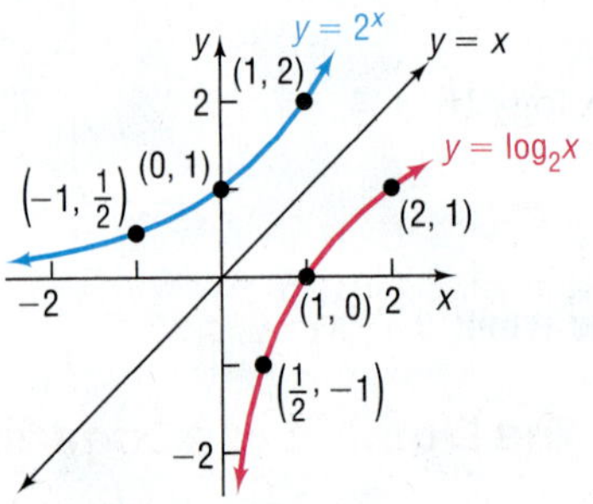

Figure 31

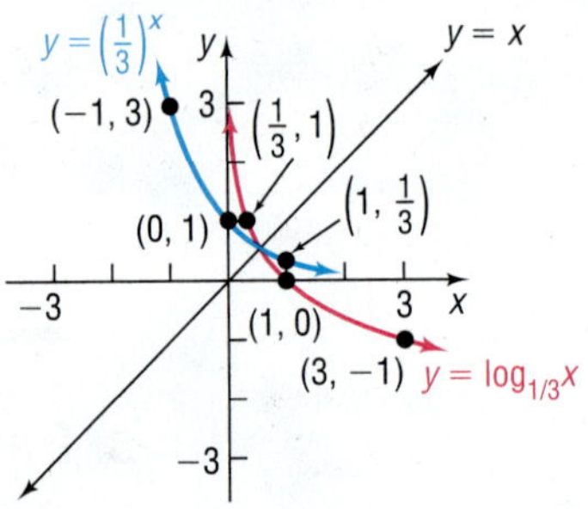

Figure 32

Now Work PROBLEM 61

The graphs of $y = \log_a x$ in Figures 30(a) and (b) lead to the following properties.

Properties of the Logarithmic Function $f(x) = \log_a x$; $a > 0$, $a \neq 1$

1. The domain is the set of positive real numbers, or $(0, \infty)$ using interval notation; the range is the set of all real numbers, or $(-\infty, \infty)$ using interval notation.
2. The x-intercept of the graph is 1. There is no y-intercept.
3. The y-axis $(x = 0)$ is a vertical asymptote of the graph.
4. A logarithmic function is decreasing if $0 < a < 1$ and is increasing if $a > 1$.
5. The graph of f contains the points $(1, 0)$, $(a, 1)$, and $\left(\dfrac{1}{a}, -1\right)$.
6. The graph is smooth and continuous, with no corners or gaps.

If the base of a logarithmic function is the number e, the result is the **natural logarithm function**. This function occurs so frequently in applications that it is given a special symbol, **ln** (from the Latin, *logarithmus naturalis*). That is,

> **In Words**
> $y = \log_e x$ is written $y = \ln x$.

$$y = \ln x \quad \text{if and only if} \quad x = e^y \qquad (1)$$

Because $y = \ln x$ and the exponential function $y = e^x$ are inverse functions, the graph of $y = \ln x$ can be obtained by reflecting the graph of $y = e^x$ about the line $y = x$. See Figure 33.

Using a calculator with an [ln] key, we can obtain other points on the graph of $f(x) = \ln x$. See Table 7.

Seeing the Concept

Graph $Y_1 = e^x$ and $Y_2 = \ln x$ on the same square screen. Use eVALUEate to verify the points on the graph given in Figure 33. Do you see the symmetry of the two graphs with respect to the line $y = x$?

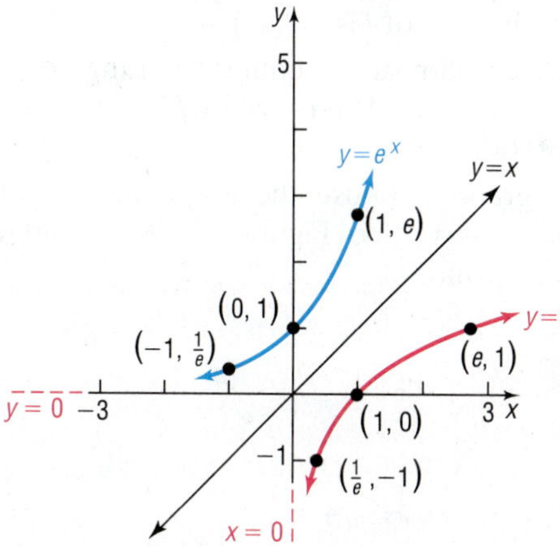

Figure 33

Table 7

x	$\ln x$
$\frac{1}{2}$	−0.69
2	0.69
3	1.10

EXAMPLE 6 Graphing a Logarithmic Function and Its Inverse

(a) Find the domain of the logarithmic function $f(x) = -\ln(x - 2)$.
(b) Graph f.
(c) From the graph, determine the range and vertical asymptote of f.
(d) Find f^{-1}, the inverse of f.
(e) Find the domain and the range of f^{-1}.
(f) Graph f^{-1}.

Solution

(a) The domain of f consists of all x for which $x - 2 > 0$, or equivalently, $x > 2$. The domain of f is $\{x \mid x > 2\}$, or $(2, \infty)$ in interval notation.

(b) To obtain the graph of $y = -\ln(x - 2)$, begin with the graph of $y = \ln x$ and use transformations. See Figure 34.

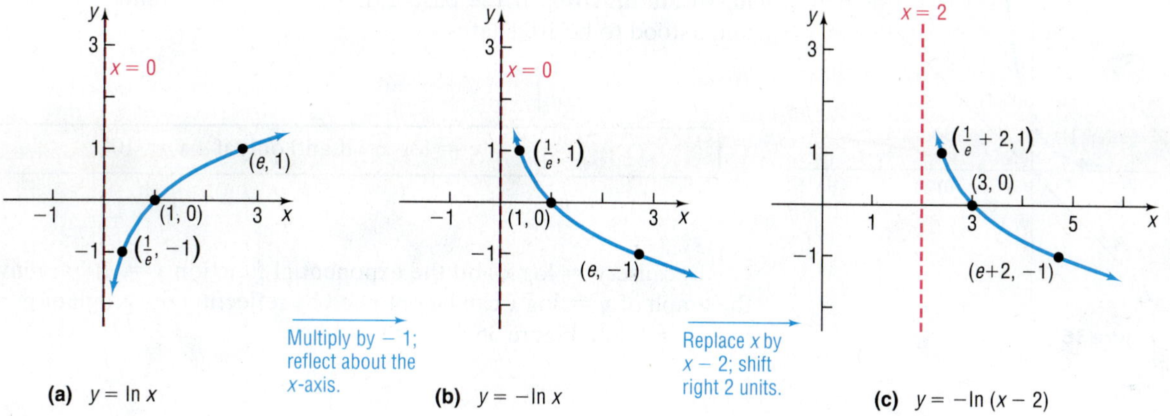

Figure 34 (a) $y = \ln x$ (b) $y = -\ln x$ (c) $y = -\ln(x - 2)$

(c) The range of $f(x) = -\ln(x - 2)$ is the set of all real numbers. The vertical asymptote is $x = 2$. [Do you see why? The original asymptote $(x = 0)$ is shifted to the right 2 units.]

(d) To find f^{-1}, begin with $y = -\ln(x - 2)$. The inverse function is defined implicitly by the equation

$$x = -\ln(y - 2)$$

Now solve for y.

$$-x = \ln(y - 2) \quad \text{Isolate the logarithm.}$$
$$e^{-x} = y - 2 \quad \text{Change to exponential form.}$$
$$y = e^{-x} + 2 \quad \text{Solve for } y.$$

The inverse of f is $f^{-1}(x) = e^{-x} + 2$.

(e) The domain of f^{-1} equals the range of f, which is the set of all real numbers, from part (c). The range of f^{-1} is the domain of f, which is $(2, \infty)$ in interval notation.

(f) To graph f^{-1}, use the graph of f in Figure 34(c) and reflect it about the line $y = x$. See Figure 35. We could also graph $f^{-1}(x) = e^{-x} + 2$ using transformations.

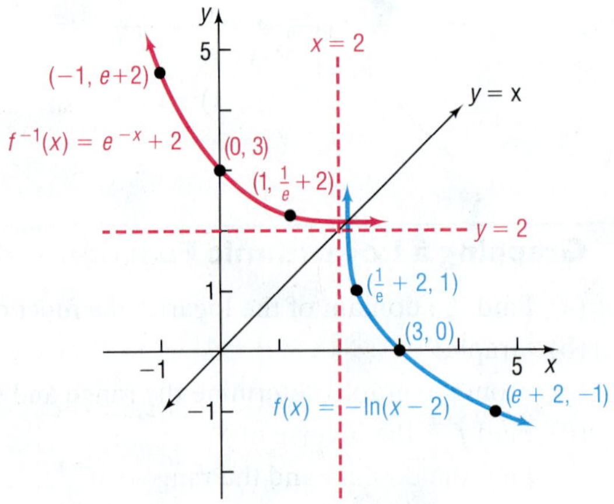

Figure 35

Now Work PROBLEM 73

If the base of a logarithmic function is the number 10, the result is the **common logarithm function**. If the base a of the logarithmic function is not indicated, it is understood to be 10. That is,

$$y = \log x \quad \text{if and only if} \quad x = 10^y$$

Because $y = \log x$ and the exponential function $y = 10^x$ are inverse functions, the graph of $y = \log x$ can be obtained by reflecting the graph of $y = 10^x$ about the line $y = x$. See Figure 36.

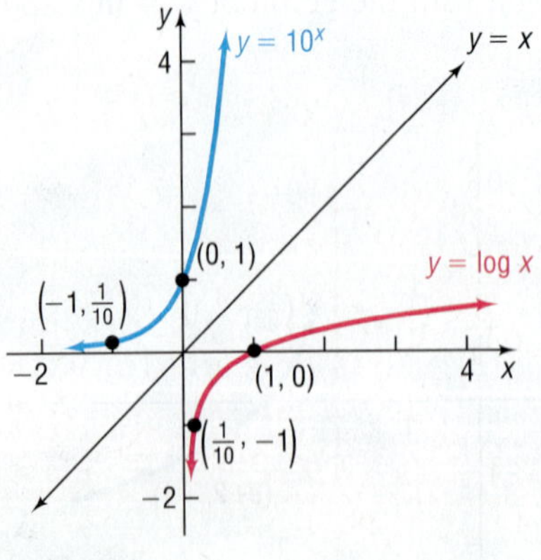

Figure 36

EXAMPLE 7 **Graphing a Logarithmic Function and Its Inverse**

(a) Find the domain of the logarithmic function $f(x) = 3\log(x - 1)$.
(b) Graph f.
(c) From the graph, determine the range and vertical asymptote of f.
(d) Find f^{-1}, the inverse of f.
(e) Find the domain and the range of f^{-1}.
(f) Graph f^{-1}.

Solution (a) The domain of f consists of all x for which $x - 1 > 0$, or equivalently, $x > 1$. The domain of f is $\{x|x > 1\}$, or $(1, \infty)$ in interval notation.

(b) To obtain the graph of $y = 3\log(x - 1)$, begin with the graph of $y = \log x$ and use transformations. See Figure 37.

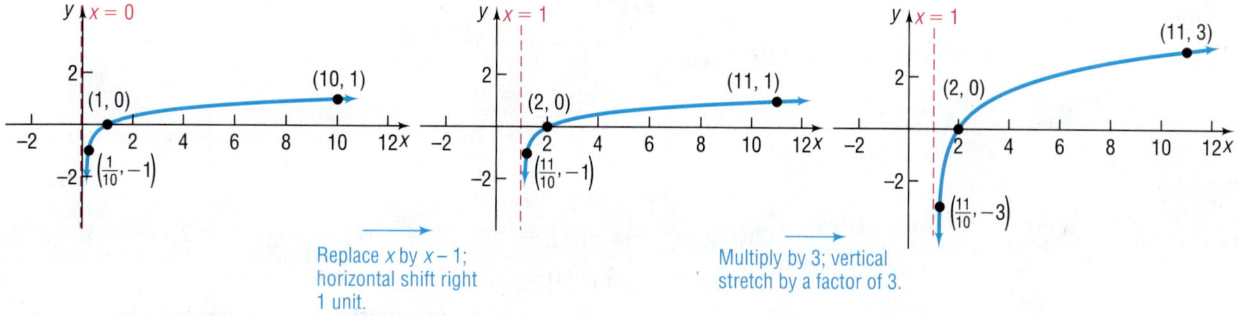

Figure 37 **(a)** $y = \log x$ **(b)** $y = \log(x - 1)$ **(c)** $y = 3\log(x - 1)$

(c) The range of $f(x) = 3\log(x - 1)$ is the set of all real numbers. The vertical asymptote is $x = 1$.

(d) Begin with $y = 3\log(x - 1)$. The inverse function is defined implicitly by the equation

$$x = 3\log(y - 1)$$

Proceed to solve for y.

$$\frac{x}{3} = \log(y - 1) \quad \text{Isolate the logarithm.}$$

$$10^{x/3} = y - 1 \quad \text{Change to exponential form.}$$

$$y = 10^{x/3} + 1 \quad \text{Solve for } y.$$

The inverse of f is $f^{-1}(x) = 10^{x/3} + 1$.

(e) The domain of f^{-1} is the range of f, which is the set of all real numbers, from part (c). The range of f^{-1} is the domain of f, which is $(1, \infty)$ in interval notation.

(f) To graph f^{-1}, use the graph of f in Figure 37(c) and reflect it about the line $y = x$. See Figure 38. We could also graph $f^{-1}(x) = 10^{x/3} + 1$ using transformations.

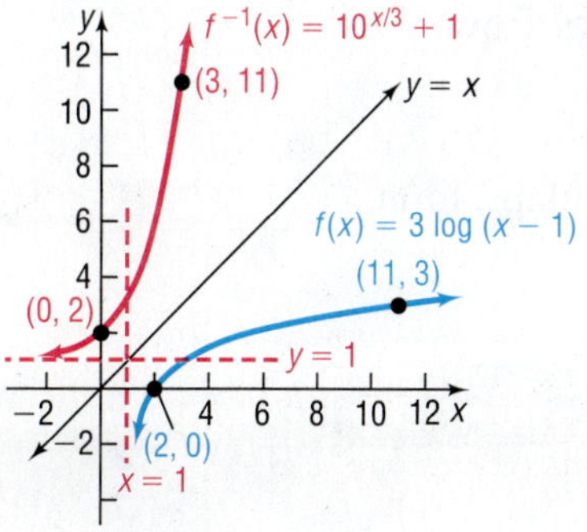

Figure 38

Now Work PROBLEM 81

5 Solve Logarithmic Equations

Equations that contain logarithms are called **logarithmic equations**. Care must be taken when solving logarithmic equations algebraically. In the expression $\log_a M$, remember that a and M are positive and $a \neq 1$. Be sure to check each apparent solution in the original equation and discard any that are extraneous.

Some logarithmic equations can be solved by changing the logarithmic equation to exponential form using the fact that $y = \log_a x$ means $a^y = x$.

EXAMPLE 8 **Solving Logarithmic Equations**

Solve:

(a) $\log_3(4x - 7) = 2$ (a) $\log_x 64 = 2$

Solution (a) To solve, change the logarithmic equation to exponential form.

$$\log_3(4x - 7) = 2$$
$$4x - 7 = 3^2 \quad \text{Change to exponential form.}$$
$$4x - 7 = 9$$
$$4x = 16$$
$$x = 4$$

Check: $\log_3(4x - 7) = \log_3(4 \cdot 4 - 7) = \log_3 9 = 2$ $3^2 = 9$

The solution set is $\{4\}$.

(b) To solve, change the logarithmic equation to exponential form.

$$\log_x 64 = 2$$
$$x^2 = 64 \quad \text{Change to exponential form.}$$
$$x = \pm\sqrt{64} = \pm 8 \quad \text{Square Root Method}$$

Because the base of a logarithm must be positive, discard -8. Check the potential solution 8.

Check: $\log_8 64 = 2$ $8^2 = 64$

The solution set is $\{8\}$.

EXAMPLE 9 **Using Logarithms to Solve an Exponential Equation**

Solve: $e^{2x} = 5$

Solution To solve, change the exponential equation to logarithmic form.

$$e^{2x} = 5$$
$$\ln 5 = 2x \quad \text{Change to logarithmic form.}$$
$$x = \frac{\ln 5}{2} \quad \text{Exact solution}$$
$$\approx 0.805 \quad \text{Approximate solution}$$

The solution set is $\left\{\frac{\ln 5}{2}\right\}$.

Now Work PROBLEMS 89 AND 101

EXAMPLE 10 Alcohol and Driving

Blood alcohol concentration (BAC) is a measure of the amount of alcohol in a person's bloodstream. A BAC of 0.04% means that a person has 4 parts alcohol per 10,000 parts blood in the body. Relative risk is defined as the likelihood of one event occurring divided by the likelihood of a second event occurring. For example, if an individual with a BAC of 0.02% is 1.4 times as likely to have a car accident as an individual who has not been drinking, the relative risk of an accident with a BAC of 0.02% is 1.4. Recent medical research suggests that the relative risk R of having an accident while driving a car can be modeled by an equation of the form

$$R = e^{kx}$$

where x is the percent concentration of alcohol in the bloodstream and k is a constant.

(a) Research indicates that the relative risk of a person having an accident with a BAC of 0.02% is 1.4. Find the constant k in the equation.

(b) Using this value of k, what is the relative risk if the concentration is 0.17%?

(c) Using this same value of k, what BAC corresponds to a relative risk of 100?

(d) If the law asserts that anyone with a relative risk of 4 or more should not have driving privileges, at what concentration of alcohol in the bloodstream should a driver be arrested and charged with DUI (driving under the influence)?

Solution

(a) For a concentration of alcohol in the blood of 0.02% and a relative risk of 1.4, let $x = 0.02$ and $R = 1.4$ in the equation and solve for k.

$$R = e^{kx}$$

$$1.4 = e^{k(0.02)} \qquad R = 1.4;\ x = 0.02$$

$$0.02k = \ln 1.4 \qquad \text{Change to a logarithmic statement.}$$

$$k = \frac{\ln 1.4}{0.02} \approx 16.82 \qquad \text{Solve for } k.$$

(b) A concentration of 0.17% means $x = 0.17$. Use $k = 16.82$ in the equation to find the relative risk R:

$$R = e^{kx} = e^{(16.82)(0.17)} \approx 17.5$$

For a concentration of alcohol in the blood of 0.17%, the relative risk of an accident is about 17.5. That is, a person with a BAC of 0.17% is 17.5 times as likely to have a car accident as a person with no alcohol in the bloodstream.

(c) A relative risk of 100 means $R = 100$. Use $k = 16.82$ in the equation $R = e^{kx}$. The concentration x of alcohol in the blood obeys

$$100 = e^{16.82x} \qquad R = e^{kx},\ R = 100,\ k = 16.82$$

$$16.82x = \ln 100 \qquad \text{Change to a logarithmic statement.}$$

$$x = \frac{\ln 100}{16.82} \approx 0.27 \qquad \text{Solve for } x.$$

For a concentration of alcohol in the blood of 0.27%, the relative risk of an accident is 100.

NOTE A BAC of 0.30% results in a loss of consciousness in most people. ■

(d) A relative risk of 4 means $R = 4$. Use $k = 16.82$ in the equation $R = e^{kx}$. The concentration x of alcohol in the bloodstream obeys

$$4 = e^{16.82x}$$

$$16.82x = \ln 4$$

$$x = \frac{\ln 4}{16.82} \approx 0.082$$

A driver with a BAC of 0.082% or more should be arrested and charged with DUI. ●

NOTE In most states, the blood alcohol content at which a DUI citation is given is 0.08%. ■

SUMMARY

Properties of the Logarithmic Function

$f(x) = \log_a x, \quad a > 1$ ($y = \log_a x$ means $x = a^y$)	Domain: the interval $(0, \infty)$; Range: the interval $(-\infty, \infty)$ x-intercept: 1; y-intercept: none; vertical asymptote: $x = 0$ (y-axis); increasing; one-to-one See Figure 39(a) for a typical graph.
$f(x) = \log_a x, \quad 0 < a < 1$ ($y = \log_a x$ means $x = a^y$)	Domain: the interval $(0, \infty)$; Range: the interval $(-\infty, \infty)$ x-intercept: 1; y-intercept: none; vertical asymptote: $x = 0$ (y-axis); decreasing; one-to-one See Figure 39(b) for a typical graph.

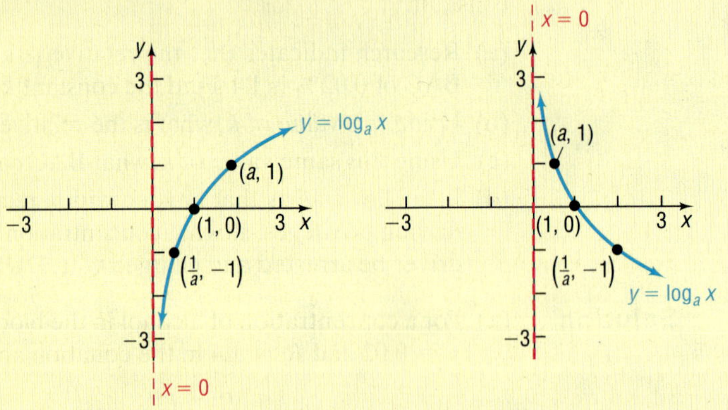

Figure 39 (a) $a > 1$ (b) $0 < a < 1$

6.4 Assess Your Understanding

'Are You Prepared?' *Answers are given at the end of these exercises. If you get a wrong answer, read the pages listed in red.*

1. Solve each inequality:
   (a) $3x - 7 \le 8 - 2x$ (pp. 119–126)
   (b) $x^2 - x - 6 > 0$ (pp. 312–313)
2. Solve the inequality: $\frac{x-1}{x+4} > 0$ (pp. 370–372)
3. Solve: $2x + 3 = 9$ (pp. 82–87)

Concepts and Vocabulary

4. The domain of the logarithmic function $f(x) = \log_a x$ is ______.
5. The graph of every logarithmic function $f(x) = \log_a x$, where $a > 0$ and $a \ne 1$, passes through three points: ______, ______, and ______.
6. If the graph of a logarithmic function $f(x) = \log_a x$, where $a > 0$ and $a \ne 1$, is increasing, then its base must be larger than ______.
7. ***True or False*** If $y = \log_a x$, then $y = a^x$.
8. ***True or False*** The graph of $f(x) = \log_a x$, where $a > 0$ and $a \ne 1$, has an x-intercept equal to 1 and no y-intercept.
9. Select the answer that completes the statement: $y = \ln x$ if and only if ______.
   (a) $x = e^y$ (b) $y = e^x$ (c) $x = 10^y$ (d) $y = 10^x$
10. Choose the domain of $f(x) = \log_3(x + 2)$.
   (a) $(-\infty, \infty)$ (b) $(2, \infty)$ (c) $(-2, \infty)$ (d) $(0, \infty)$

Skill Building

In Problems 11–18, change each exponential statement to an equivalent statement involving a logarithm.

11. $9 = 3^2$	12. $16 = 4^2$	13. $a^2 = 1.6$	14. $a^3 = 2.1$
15. $2^x = 7.2$	16. $3^x = 4.6$	17. $e^x = 8$	18. $e^{2.2} = M$

In Problems 19–26, change each logarithmic statement to an equivalent statement involving an exponent.

19. $\log_2 8 = 3$	20. $\log_3\left(\frac{1}{9}\right) = -2$	21. $\log_a 3 = 6$	22. $\log_b 4 = 2$
23. $\log_3 2 = x$	24. $\log_2 6 = x$	25. $\ln 4 = x$	26. $\ln x = 4$

In Problems 27–38, find the exact value of each logarithm without using a calculator.

27. $\log_2 1$ **28.** $\log_8 8$ **29.** $\log_5 25$ **30.** $\log_3\left(\frac{1}{9}\right)$

31. $\log_{1/2} 16$ **32.** $\log_{1/3} 9$ **33.** $\log_{10} \sqrt{10}$ **34.** $\log_5 \sqrt[3]{25}$

35. $\log_{\sqrt{2}} 4$ **36.** $\log_{\sqrt{3}} 9$ **37.** $\ln\sqrt{e}$ **38.** $\ln e^3$

In Problems 39–50, find the domain of each function.

39. $f(x) = \ln(x - 3)$ **40.** $g(x) = \ln(x - 1)$ **41.** $F(x) = \log_2 x^2$

42. $H(x) = \log_5 x^3$ **43.** $f(x) = 3 - 2\log_4\left(\frac{x}{2} - 5\right)$ **44.** $g(x) = 8 + 5\ln(2x + 3)$

45. $f(x) = \ln\left(\frac{1}{x + 1}\right)$ **46.** $g(x) = \ln\left(\frac{1}{x - 5}\right)$ **47.** $g(x) = \log_5\left(\frac{x + 1}{x}\right)$

48. $h(x) = \log_3\left(\frac{x}{x - 1}\right)$ **49.** $f(x) = \sqrt{\ln x}$ **50.** $g(x) = \frac{1}{\ln x}$

In Problems 51–58, use a calculator to evaluate each expression. Round your answer to three decimal places.

51. $\ln\frac{5}{3}$ **52.** $\frac{\ln 5}{3}$ **53.** $\frac{\ln\frac{10}{3}}{0.04}$ **54.** $\frac{\ln\frac{2}{3}}{-0.1}$

55. $\frac{\ln 4 + \ln 2}{\log 4 + \log 2}$ **56.** $\frac{\log 15 + \log 20}{\ln 15 + \ln 20}$ **57.** $\frac{2\ln 5 + \log 50}{\log 4 - \ln 2}$ **58.** $\frac{3\log 80 - \ln 5}{\log 5 + \ln 20}$

59. Find a so that the graph of $f(x) = \log_a x$ contains the point $(2, 2)$.

60. Find a so that the graph of $f(x) = \log_a x$ contains the point $\left(\frac{1}{2}, -4\right)$.

In Problems 61–64, graph each function and its inverse on the same set of axes.

61. $f(x) = 3^x;\ f^{-1}(x) = \log_3 x$ **62.** $f(x) = 4^x;\ f^{-1}(x) = \log_4 x$

63. $f(x) = \left(\frac{1}{2}\right)^x;\ f^{-1}(x) = \log_{1/2} x$ **64.** $f(x) = \left(\frac{1}{3}\right)^x;\ f^{-1}(x) = \log_{1/3} x$

In Problems 65–72, the graph of a logarithmic function is given. Match each graph to one of the following functions:

(A) $y = \log_3 x$ (B) $y = \log_3(-x)$ (C) $y = -\log_3 x$ (D) $y = -\log_3(-x)$
(E) $y = \log_3 x - 1$ (F) $y = \log_3(x - 1)$ (G) $y = \log_3(1 - x)$ (H) $y = 1 - \log_3 x$

65.
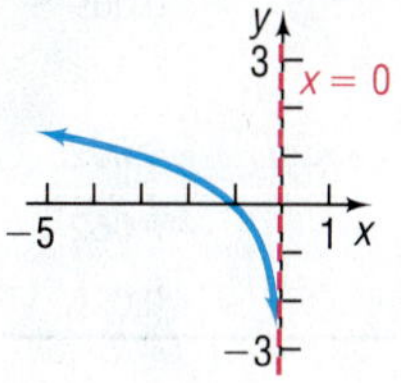

66.
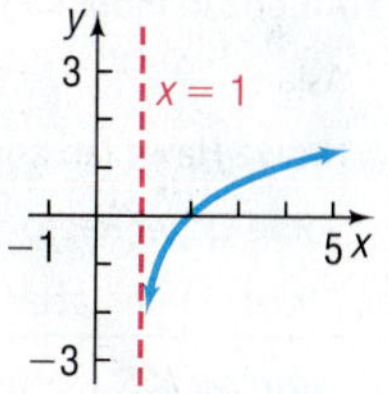

67.
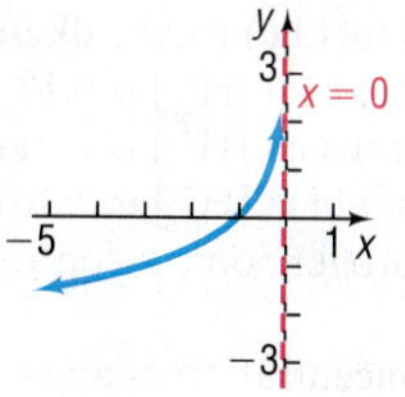

68.
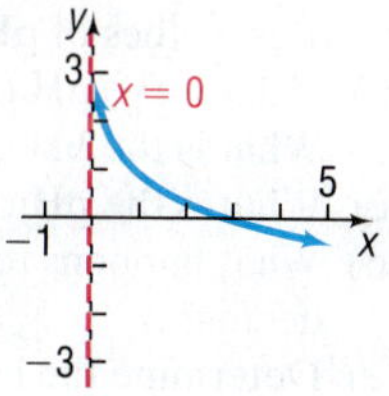

69.
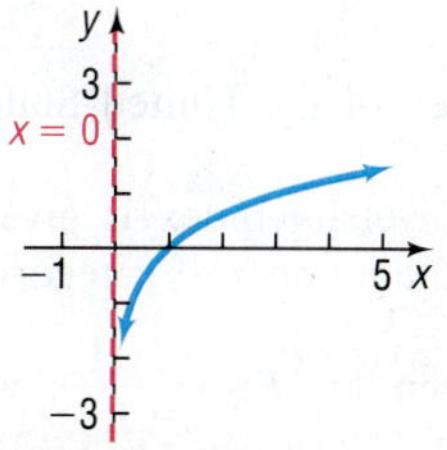

70.
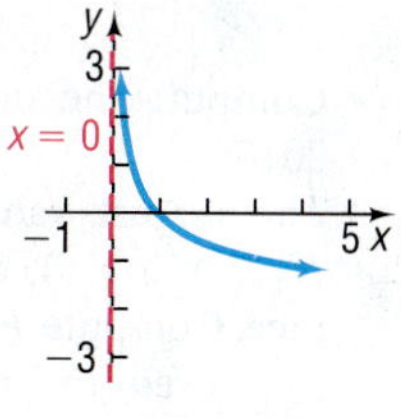

71.
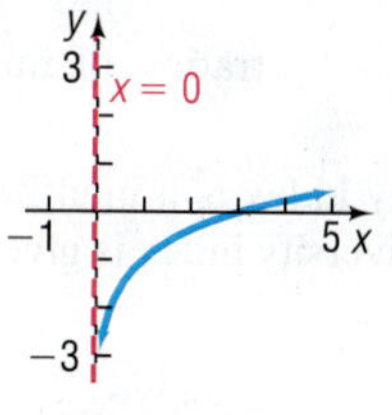

72.
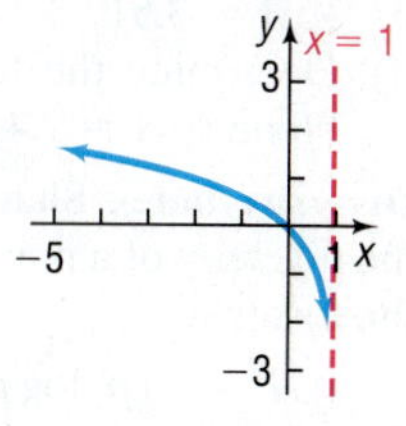

In Problems 73–88, use the given function f.
(a) Find the domain of f. (b) Graph f. (c) From the graph, determine the range and any asymptotes of f.
(d) Find f^{-1}, the inverse of f. (e) Find the domain and the range of f^{-1}. (f) Graph f^{-1}.

73. $f(x) = \ln(x + 4)$ **74.** $f(x) = \ln(x - 3)$ **75.** $f(x) = 2 + \ln x$ **76.** $f(x) = -\ln(-x)$

77. $f(x) = \ln(2x) - 3$ **78.** $f(x) = -2\ln(x + 1)$ **79.** $f(x) = \log(x - 4) + 2$ **80.** $f(x) = \frac{1}{2}\log x - 5$

81. $f(x) = \frac{1}{2}\log(2x)$ **82.** $f(x) = \log(-2x)$ **83.** $f(x) = 3 + \log_3(x + 2)$ **84.** $f(x) = 2 - \log_3(x + 1)$

85. $f(x) = e^{x+2} - 3$ **86.** $f(x) = 3e^x + 2$ **87.** $f(x) = 2^{x/3} + 4$ **88.** $f(x) = -3^{x+1}$

In Problems 89–112, solve each equation.

89. $\log_3 x = 2$
90. $\log_5 x = 3$
91. $\log_2(2x + 1) = 3$
92. $\log_3(3x - 2) = 2$
93. $\log_x 4 = 2$
94. $\log_x\left(\frac{1}{8}\right) = 3$
95. $\ln e^x = 5$
96. $\ln e^{-2x} = 8$
97. $\log_4 64 = x$
98. $\log_5 625 = x$
99. $\log_3 243 = 2x + 1$
100. $\log_6 36 = 5x + 3$
101. $e^{3x} = 10$
102. $e^{-2x} = \frac{1}{3}$
103. $e^{2x+5} = 8$
104. $e^{-2x+1} = 13$
105. $\log_3(x^2 + 1) = 2$
106. $\log_5(x^2 + x + 4) = 2$
107. $\log_2 8^x = -3$
108. $\log_3 3^x = -1$
109. $5e^{0.2x} = 7$
110. $8 \cdot 10^{2x-7} = 3$
111. $2 \cdot 10^{2-x} = 5$
112. $4e^{x+1} = 5$

Mixed Practice

113. Suppose that $G(x) = \log_3(2x + 1) - 2$.
(a) What is the domain of G?
(b) What is $G(40)$? What point is on the graph of G?
(c) If $G(x) = 3$, what is x? What point is on the graph of G?
(d) What is the zero of G?

114. Suppose that $F(x) = \log_2(x + 1) - 3$.
(a) What is the domain of F?
(b) What is $F(7)$? What point is on the graph of F?
(c) If $F(x) = -1$, what is x? What point is on the graph of F?
(d) What is the zero of F?

In Problems 115–118, graph each function. Based on the graph, state the domain and the range, and find any intercepts.

115. $f(x) = \begin{cases} \ln(-x) & \text{if } x < 0 \\ \ln x & \text{if } x > 0 \end{cases}$

116. $f(x) = \begin{cases} \ln(-x) & \text{if } x \le -1 \\ -\ln(-x) & \text{if } -1 < x < 0 \end{cases}$

117. $f(x) = \begin{cases} -\ln x & \text{if } 0 < x < 1 \\ \ln x & \text{if } x \ge 1 \end{cases}$

118. $f(x) = \begin{cases} \ln x & \text{if } 0 < x < 1 \\ -\ln x & \text{if } x \ge 1 \end{cases}$

Applications and Extensions

119. **Chemistry** The pH of a chemical solution is given by the formula

$$\text{pH} = -\log_{10}[H^+]$$

where $[H^+]$ is the concentration of hydrogen ions in moles per liter. Values of pH range from 0 (acidic) to 14 (alkaline).
(a) What is the pH of a solution for which $[H^+]$ is 0.1?
(b) What is the pH of a solution for which $[H^+]$ is 0.01?
(c) What is the pH of a solution for which $[H^+]$ is 0.001?
(d) What happens to pH as the hydrogen ion concentration decreases?
(e) Determine the hydrogen ion concentration of an orange (pH = 3.5).
(f) Determine the hydrogen ion concentration of human blood (pH = 7.4).

120. **Diversity Index Shannon's diversity index** is a measure of the diversity of a population. The diversity index is given by the formula

$$H = -(p_1 \log p_1 + p_2 \log p_2 + \cdots + p_n \log p_n)$$

where p_1 is the proportion of the population that is species 1, p_2 is the proportion of the population that is species 2, and so on. In this problem, the population is people in the United States and the species is race.
(a) According to the U.S. Census Bureau, the distribution of race in the United States in 2010 was as follows:

Race	Proportion
White	0.724
Black or African American	0.126
American Indian and Alaska Native	0.009
Asian	0.048
Native Hawaiian and Other Pacific Islander	0.002
Some Other Race	0.062
Two or More Races	0.029

Source: U.S. Census Bureau

Compute the diversity index of the United States in 2010.
(b) The largest value of the diversity index is given by $H_{max} = \log(S)$, where S is the number of categories of race. Compute H_{max}.
(c) The **evenness ratio** is given by $E_H = \frac{H}{H_{max}}$, where $0 \le E_H \le 1$. If $E_H = 1$, there is complete evenness. Compute the evenness ratio for the United States.
(d) Obtain the distribution of race for the United States in 2000 from the Census Bureau. Compute Shannon's diversity index. Is the United States becoming more diverse? Why?

121. Atmospheric Pressure The atmospheric pressure p on an object decreases with increasing height. This pressure, measured in millimeters of mercury, is related to the height h (in kilometers) above sea level by the function

$$p(h) = 760e^{-0.145h}$$

(a) Find the height of an aircraft if the atmospheric pressure is 320 millimeters of mercury.
(b) Find the height of a mountain if the atmospheric pressure is 667 millimeters of mercury.

122. Healing of Wounds The normal healing of wounds can be modeled by an exponential function. If A_0 represents the original area of the wound, and if A equals the area of the wound, then the function

$$A(n) = A_0e^{-0.35n}$$

describes the area of a wound after n days following an injury when no infection is present to retard the healing. Suppose that a wound initially had an area of 100 square millimeters.

(a) If healing is taking place, after how many days will the wound be one-half its original size?
(b) How long before the wound is 10% of its original size?

123. Exponential Probability Between 12:00 PM and 1:00 PM, cars arrive at Citibank's drive-thru at the rate of 6 cars per hour (0.1 car per minute). The following formula from statistics can be used to determine the probability that a car will arrive within t minutes of 12:00 PM.

$$F(t) = 1 - e^{-0.1t}$$

(a) Determine how many minutes are needed for the probability to reach 50%.
(b) Determine how many minutes are needed for the probability to reach 80%.
(c) Is it possible for the probability to equal 100%? Explain.

124. Exponential Probability Between 5:00 PM and 6:00 PM, cars arrive at Jiffy Lube at the rate of 9 cars per hour (0.15 car per minute). The following formula from statistics can be used to determine the probability that a car will arrive within t minutes of 5:00 PM.

$$F(t) = 1 - e^{-0.15t}$$

(a) Determine how many minutes are needed for the probability to reach 50%.
(b) Determine how many minutes are needed for the probability to reach 80%.

125. Drug Medication The formula

$$D = 5e^{-0.4h}$$

can be used to find the number of milligrams D of a certain drug that is in a patient's bloodstream h hours after the drug was administered. When the number of milligrams reaches 2, the drug is to be administered again. What is the time between injections?

126. Spreading of Rumors A model for the number N of people in a college community who have heard a certain rumor is

$$N = P(1 - e^{-0.15d})$$

where P is the total population of the community and d is the number of days that have elapsed since the rumor began. In a community of 1000 students, how many days will elapse before 450 students have heard the rumor?

127. Current in an RL Circuit The equation governing the amount of current I (in amperes) after time t (in seconds) in a simple RL circuit consisting of a resistance R (in ohms), an inductance L (in henrys), and an electromotive force E (in volts) is

$$I = \frac{E}{R}[1 - e^{-(R/L)t}]$$

If $E = 12$ volts, $R = 10$ ohms, and $L = 5$ henrys, how long does it take to obtain a current of 0.5 ampere? Of 1.0 ampere? Graph the equation.

128. Learning Curve Psychologists sometimes use the function

$$L(t) = A(1 - e^{-kt})$$

to measure the amount L learned at time t. Here A represents the amount to be learned, and the number k measures the rate of learning. Suppose that a student has an amount A of 200 vocabulary words to learn. A psychologist determines that the student has learned 20 vocabulary words after 5 minutes.

(a) Determine the rate of learning k.
(b) Approximately how many words will the student have learned after 10 minutes?
(c) After 15 minutes?
(d) How long does it take for the student to learn 180 words?

Loudness of Sound *Problems 129–132 use the following discussion: The* ***loudness*** $L(x)$, *measured in decibels (dB), of a sound of intensity x, measured in watts per square meter, is defined as* $L(x) = 10\log\frac{x}{I_0}$, *where* $I_0 = 10^{-12}$ *watt per square meter is the least intense sound that a human ear can detect. Determine the loudness, in decibels, of each of the following sounds.*

129. Normal conversation: intensity of $x = 10^{-7}$ watt per square meter.

130. Amplified rock music: intensity of 10^{-1} watt per square meter.

131. Heavy city traffic: intensity of $x = 10^{-3}$ watt per square meter.

132. Diesel truck traveling 40 miles per hour 50 feet away: intensity 10 times that of a passenger car traveling 50 miles per hour 50 feet away, whose loudness is 70 decibels.

The Richter Scale *Problems 133 and 134 on the next page use the following discussion: The* ***Richter scale*** *is one way of converting seismographic readings into numbers that provide an easy reference for measuring the magnitude M of an earthquake. All earthquakes are compared to a* ***zero-level earthquake*** *whose seismographic reading measures 0.001 millimeter at a distance of 100 kilometers from the epicenter. An earthquake whose seismographic reading measures x millimeters has* ***magnitude*** $M(x)$, *given by*

$$M(x) = \log\left(\frac{x}{x_0}\right)$$

where $x_0 = 10^{-3}$ *is the reading of a zero-level earthquake the same distance from its epicenter. In Problems 133 and 134, determine the magnitude of each earthquake.*

133. Magnitude of an Earthquake Mexico City in 1985: seismographic reading of 125,892 millimeters 100 kilometers from the center

134. Magnitude of an Earthquake San Francisco in 1906: seismographic reading of 50,119 millimeters 100 kilometers from the center

135. Alcohol and Driving The concentration of alcohol in a person's bloodstream is measurable. Suppose that the relative risk R of having an accident while driving a car can be modeled by an equation of the form

$$R = e^{kx}$$

where x is the percent concentration of alcohol in the bloodstream and k is a constant.

(a) Suppose that a concentration of alcohol in the bloodstream of 0.03 percent results in a relative risk of an accident of 1.4. Find the constant k in the equation.

(b) Using this value of k, what is the relative risk if the concentration is 0.17 percent?

(c) Using the same value of k, what concentration of alcohol corresponds to a relative risk of 100?

(d) If the law asserts that anyone with a relative risk of having an accident of 5 or more should not have driving privileges, at what concentration of alcohol in the bloodstream should a driver be arrested and charged with a DUI?

(e) Compare this situation with that of Example 10. If you were a lawmaker, which situation would you support? Give your reasons.

Explaining Concepts: Discussion and Writing

136. Is there any function of the form $y = x^{\alpha}, 0 < \alpha < 1$, that increases more slowly than a logarithmic function whose base is greater than 1? Explain.

137. In the definition of the logarithmic function, the base a is not allowed to equal 1. Why?

138. Critical Thinking In buying a new car, one consideration might be how well the price of the car holds up over time. Different makes of cars have different depreciation rates. One way to compute a depreciation rate for a car is given here. Suppose that the current prices of a certain automobile are as shown in the table.

	Age in Years				
New	1	2	3	4	5
\$38,000	\$36,600	\$32,400	\$28,750	\$25,400	\$21,200

Use the formula New $=$ Old(e^{Rt}) to find R, the annual depreciation rate, for a specific time t. When might be the best time to trade in the car? Consult the NADA ("blue") book and compare two like models that you are interested in. Which has the better depreciation rate?

Retain Your Knowledge

Problems 139–142 are based on material learned earlier in the course. The purpose of these problems is to keep the material fresh in your mind so that you are better prepared for the final exam.

139. Find the real zeros of $g(x) = 4x^4 - 37x^2 + 9$. What are the x-intercepts of the graph of g?

140. Find the average rate of change of $f(x) = 9^x$ from $\frac{1}{2}$ to 1.

141. Use the Intermediate Value Theorem to show that the function $f(x) = 4x^3 - 2x^2 - 7$ has a real zero in the interval $[1, 2]$.

142. A complex polynomial function f of degree 4 with real coefficients has the zeros $-1, 2$, and $3 - i$. Find the remaining zero(s) of f. Then find a polynomial function that has the zeros.

'Are You Prepared?' Answers

1. (a) $x \leq 3$ (b) $x < -2$ or $x > 3$ **2.** $x < -4$ or $x > 1$ **3.** $\{3\}$

6.5 Properties of Logarithms

1 Work with the Properties of Logarithms

Logarithms have some very useful properties that can be derived directly from the definition and the laws of exponents.

EXAMPLE 1 **Establishing Properties of Logarithms**

(a) Show that $\log_a 1 = 0$. (a) Show that $\log_a a = 1$.

Solution (a) This fact was established when we graphed $y = \log_a x$ (see Figure 30 on page 442). To show the result algebraically, let $y = \log_a 1$. Then

$$
\begin{aligned}
y &= \log_a 1 \\
a^y &= 1 && \text{Change to exponential form.} \\
a^y &= a^0 && a^0 = 1 \text{ since } a > 0, a \neq 1 \\
y &= 0 && \text{Equate exponents.} \\
\log_a 1 &= 0 && y = \log_a 1
\end{aligned}
$$

(b) Let $y = \log_a a$. Then

$$
\begin{aligned}
y &= \log_a a \\
a^y &= a && \text{Change to exponential form.} \\
a^y &= a^1 && a = a^1 \\
y &= 1 && \text{Equate exponents.} \\
\log_a a &= 1 && y = \log_a a
\end{aligned}
$$

To summarize:

$$\log_a 1 = 0 \qquad \log_a a = 1$$

THEOREM

Properties of Logarithms

In the properties given next, M and a are positive real numbers, $a \neq 1$, and r is any real number.

The number $\log_a M$ is the exponent to which a must be raised to obtain M. That is,

$$a^{\log_a M} = M \qquad \textbf{(1)}$$

The logarithm with base a of a raised to a power equals that power. That is,

$$\log_a a^r = r \qquad \textbf{(2)}$$

The proof uses the fact that $y = a^x$ and $y = \log_a x$ are inverse functions.

Proof of Property (1) For inverse functions,

$$f(f^{-1}(x)) = x \quad \text{for all } x \text{ in the domain of } f^{-1}$$

Use $f(x) = a^x$ and $f^{-1}(x) = \log_a x$ to find

$$f(f^{-1}(x)) = a^{\log_a x} = x \quad \text{for } x > 0$$

Now let $x = M$ to obtain $a^{\log_a M} = M$, where $M > 0$. ∎

Proof of Property (2) For inverse functions,

$$f^{-1}(f(x)) = x \quad \text{for all } x \text{ in the domain of } f$$

Use $f(x) = a^x$ and $f^{-1}(x) = \log_a x$ to find

$$f^{-1}(f(x)) = \log_a a^x = x \quad \text{for all real numbers } x$$

Now let $x = r$ to obtain $\log_a a^r = r$, where r is any real number. ∎

EXAMPLE 2 **Using Properties (1) and (2)**

(a) $2^{\log_2 \pi} = \pi$ (b) $\log_{0.2} 0.2^{-\sqrt{2}} = -\sqrt{2}$ (c) $\ln e^{kt} = kt$

Now Work PROBLEM 15

Other useful properties of logarithms are given next.

THEOREM

Properties of Logarithms

In the following properties, M, N, and a are positive real numbers, $a \neq 1$, and r is any real number.

The Log of a Product Equals the Sum of the Logs

$$\log_a(MN) = \log_a M + \log_a N \quad (3)$$

The Log of a Quotient Equals the Difference of the Logs

$$\log_a\left(\frac{M}{N}\right) = \log_a M - \log_a N \quad (4)$$

The Log of a Power Equals the Product of the Power and the Log

$$\log_a M^r = r \log_a M \quad (5)$$

$$a^r = e^{r \ln a} \quad (6)$$

We shall derive properties (3), (5), and (6) and leave the derivation of property (4) as an exercise (see Problem 109).

Proof of Property (3) Let $A = \log_a M$ and let $B = \log_a N$. These expressions are equivalent to the exponential expressions

$$a^A = M \quad \text{and} \quad a^B = N$$

Now

$$\begin{aligned} \log_a(MN) = \log_a(a^A a^B) &= \log_a a^{A+B} && \text{Law of Exponents} \\ &= A + B && \text{Property (2) of logarithms} \\ &= \log_a M + \log_a N \end{aligned}$$

■

Proof of Property (5) Let $A = \log_a M$. This expression is equivalent to

$$a^A = M$$

Now

$$\begin{aligned} \log_a M^r = \log_a(a^A)^r &= \log_a a^{rA} && \text{Law of Exponents} \\ &= rA && \text{Property (2) of logarithms} \\ &= r \log_a M \end{aligned}$$

■

Proof of Property (6) Property (1), with $a = e$, gives

$$e^{\ln M} = M$$

Now let $M = a^r$ and apply property (5).

$$e^{\ln a^r} = e^{r \ln a} = a^r$$

■

Now Work PROBLEM 19

2 Write a Logarithmic Expression as a Sum or Difference of Logarithms

Logarithms can be used to transform products into sums, quotients into differences, and powers into factors. Such transformations prove useful in certain types of calculus problems.

EXAMPLE 3 **Writing a Logarithmic Expression as a Sum of Logarithms**

Write $\log_a(x\sqrt{x^2+1})$, $x > 0$, as a sum of logarithms. Express all powers as factors.

Solution

$$\begin{aligned}\log_a(x\sqrt{x^2+1}) &= \log_a x + \log_a\sqrt{x^2+1} && \log_a(M\cdot N) = \log_a M + \log_a N\\ &= \log_a x + \log_a(x^2+1)^{1/2}\\ &= \log_a x + \frac{1}{2}\log_a(x^2+1) && \log_a M^r = r\log_a M\end{aligned}$$

EXAMPLE 4 **Writing a Logarithmic Expression as a Difference of Logarithms**

Write

$$\ln\frac{x^2}{(x-1)^3} \qquad x > 1$$

as a difference of logarithms. Express all powers as factors.

Solution

$$\ln\frac{x^2}{(x-1)^3} = \ln x^2 - \ln(x-1)^3 = 2\ln x - 3\ln(x-1)$$

$\log_a\left(\frac{M}{N}\right) = \log_a M - \log_a N$ $\qquad$ $\log_a M^r = r\log_a M$

EXAMPLE 5 **Writing a Logarithmic Expression as a Sum and Difference of Logarithms**

Write

$$\log_a\frac{\sqrt{x^2+1}}{x^3(x+1)^4} \qquad x > 0$$

as a sum and difference of logarithms. Express all powers as factors.

Solution

$$\begin{aligned}\log_a\frac{\sqrt{x^2+1}}{x^3(x+1)^4} &= \log_a\sqrt{x^2+1} - \log_a[x^3(x+1)^4] && \text{Property (4)}\\ &= \log_a\sqrt{x^2+1} - [\log_a x^3 + \log_a(x+1)^4] && \text{Property (3)}\\ &= \log_a(x^2+1)^{1/2} - \log_a x^3 - \log_a(x+1)^4\\ &= \frac{1}{2}\log_a(x^2+1) - 3\log_a x - 4\log_a(x+1) && \text{Property (5)}\end{aligned}$$

WARNING In using properties (3) through (5), be careful about the values that the variable may assume. For example, the domain of the variable for $\log_a x$ is $x > 0$ and for $\log_a(x-1)$ is $x > 1$. If these functions are added, the domain is $x > 1$. That is, the equality

$$\log_a x + \log_a(x-1) = \log_a[x(x-1)]$$

is true only for $x > 1$.

Now Work PROBLEM 51

3 Write a Logarithmic Expression as a Single Logarithm

Another use of properties (3) through (5) is to write sums and/or differences of logarithms with the same base as a single logarithm. This skill will be needed to solve certain logarithmic equations discussed in the next section.

EXAMPLE 6 **Writing Expressions as a Single Logarithm**

Write each of the following as a single logarithm.

(a) $\log_a 7 + 4\log_a 3$ $\qquad$ (b) $\frac{2}{3}\ln 8 - \ln(5^2-1)$

(c) $\log_a x + \log_a 9 + \log_a(x^2+1) - \log_a 5$

Solution (a) $\log_a 7 + 4\log_a 3 = \log_a 7 + \log_a 3^4$ $\quad r\log_a M = \log_a M^r$

$= \log_a 7 + \log_a 81$

$= \log_a(7\cdot 81)$ $\quad \log_a M + \log_a N = \log_a(M\cdot N)$

$= \log_a 567$

(b) $\frac{2}{3}\ln 8 - \ln(5^2 - 1) = \ln 8^{2/3} - \ln(25 - 1)$ $\quad r\log_a M = \log_a M^r$

$= \ln 4 - \ln 24$ $\quad 8^{2/3} = (\sqrt[3]{8})^2 = 2^2 = 4$

$= \ln\left(\frac{4}{24}\right)$ $\quad \log_a M - \log_a N = \log_a\left(\frac{M}{N}\right)$

$= \ln\left(\frac{1}{6}\right)$

$= \ln 1 - \ln 6$

$= -\ln 6$ $\quad \ln 1 = 0$

(c) $\log_a x + \log_a 9 + \log_a(x^2 + 1) - \log_a 5 = \log_a(9x) + \log_a(x^2 + 1) - \log_a 5$

$= \log_a[9x(x^2 + 1)] - \log_a 5$

$= \log_a\left[\frac{9x(x^2 + 1)}{5}\right]$ ●

WARNING A common error that some students make is to express the logarithm of a sum as the sum of logarithms.

$\log_a(M + N)$ is not equal to $\log_a M + \log_a N$

$\log_a(MN) = \log_a M + \log_a N$

Another common error is to express the difference of logarithms as the quotient of logarithms.

$\log_a M - \log_a N$ is not equal to $\frac{\log_a M}{\log_a N}$

$\log_a M - \log_a N = \log_a\left(\frac{M}{N}\right)$

A third common error is to express a logarithm raised to a power as the product of the power times the logarithm.

$(\log_a M)^r$ is not equal to $r\log_a M$

$\log_a M^r = r\log_a M$ ■

Now Work PROBLEMS 57 AND 63

Two other important properties of logarithms are consequences of the fact that the logarithmic function $y = \log_a x$ is a one-to-one function.

THEOREM

Properties of Logarithms

In the following properties, M, N, and a are positive real numbers, $a \neq 1$.

If $M = N$, then $\log_a M = \log_a N$. **(7)**

If $\log_a M = \log_a N$, then $M = N$. **(8)**

Property (7) is used as follows: Starting with the equation $M = N$, "take the logarithm of both sides" to obtain $\log_a M = \log_a N$.

Properties (7) and (8) are useful for solving *exponential and logarithmic equations*, a topic discussed in the next section.

4 Evaluate Logarithms Whose Base Is Neither 10 Nor *e*

Logarithms with base 10—common logarithms—were used to facilitate arithmetic computations before the widespread use of calculators. (See the Historical Feature at the end of this section.) Natural logarithms—that is, logarithms whose base is the number e—remain very important because they arise frequently in the study of natural phenomena.

Common logarithms are usually abbreviated by writing **log**, with the base understood to be 10, just as natural logarithms are abbreviated by **ln**, with the base understood to be e.

Most calculators have both [log] and [ln] keys to calculate the common logarithm and the natural logarithm of a number, respectively. Let's look at an example to see how to approximate logarithms having a base other than 10 or e.

EXAMPLE 7 **Approximating a Logarithm Whose Base Is Neither 10 Nor e**

Approximate $\log_2 7$. Round the answer to four decimal places.

Solution Remember, evaluating $\log_2 7$ means answering the question "2 raised to what exponent equals 7?" Let $y = \log_2 7$. Then $2^y = 7$. Because $2^2 = 4$ and $2^3 = 8$, the value of $\log_2 7$ is between 2 and 3.

$$2^y = 7$$

$$\ln 2^y = \ln 7 \quad \text{Property (7)}$$

$$y \ln 2 = \ln 7 \quad \text{Property (5)}$$

$$y = \frac{\ln 7}{\ln 2} \quad \text{Exact value}$$

$$y \approx 2.8074 \quad \text{Approximate value rounded to four decimal places}$$

Example 7 shows how to approximate a logarithm whose base is 2 by changing to logarithms involving the base e. In general, the **Change-of-Base Formula** is used.

THEOREM

Change-of-Base Formula

If $a \neq 1$, $b \neq 1$, and M are positive real numbers, then

$$\log_a M = \frac{\log_b M}{\log_b a} \tag{9}$$

Proof Let $y = \log_a M$. Then

$$a^y = M$$

$$\log_b a^y = \log_b M \quad \text{Property (7)}$$

$$y \log_b a = \log_b M \quad \text{Property (5)}$$

$$y = \frac{\log_b M}{\log_b a} \quad \text{Solve for } y.$$

$$\log_a M = \frac{\log_b M}{\log_b a} \quad y = \log_a M$$

■

Because calculators have keys only for [log] and [ln], in practice, the Change-of-Base Formula uses either $b = 10$ or $b = e$. That is,

$$\log_a M = \frac{\log M}{\log a} \quad \text{and} \quad \log_a M = \frac{\ln M}{\ln a} \tag{10}$$

EXAMPLE 8 Using the Change-of-Base Formula

Approximate:

(a) $\log_5 89$ (b) $\log_{\sqrt{2}} \sqrt{5}$

Round answers to four decimal places.

Solution

(a) $\log_5 89 = \dfrac{\log 89}{\log 5} \approx \dfrac{1.949390007}{0.6989700043}$
≈ 2.7889

or

$\log_5 89 = \dfrac{\ln 89}{\ln 5} \approx \dfrac{4.48863637}{1.609437912}$
≈ 2.7889

(b) $\log_{\sqrt{2}} \sqrt{5} = \dfrac{\log \sqrt{5}}{\log \sqrt{2}} = \dfrac{\frac{1}{2}\log 5}{\frac{1}{2}\log 2}$
$= \dfrac{\log 5}{\log 2} \approx 2.3219$

or

$\log_{\sqrt{2}} \sqrt{5} = \dfrac{\ln \sqrt{5}}{\ln \sqrt{2}} = \dfrac{\frac{1}{2}\ln 5}{\frac{1}{2}\ln 2}$
$= \dfrac{\ln 5}{\ln 2} \approx 2.3219$

Now Work PROBLEMS 23 AND 71

COMMENT To graph logarithmic functions when the base is different from e or 10 requires the Change-of-Base Formula. For example, to graph $y = \log_2 x$, graph either $y = \dfrac{\ln x}{\ln 2}$ or $y = \dfrac{\log x}{\log 2}$.

Now Work PROBLEM 79

SUMMARY

Properties of Logarithms

In the list that follows, a, b, M, N, and r are real numbers. Also, $a > 0, a \neq 1, b > 0, b \neq 1, M > 0$, and $N > 0$.

Definition	$y = \log_a x$ means $x = a^y$		
Properties of logarithms	$\log_a 1 = 0$	$\log_a a = 1$	$\log_a M^r = r \log_a M$
	$a^{\log_a M} = M$	$\log_a a^r = r$	$a^r = e^{r \ln a}$
	$\log_a(MN) = \log_a M + \log_a N$		$\log_a\left(\dfrac{M}{N}\right) = \log_a M - \log_a N$
	If $M = N$, then $\log_a M = \log_a N$.		If $\log_a M = \log_a N$, then $M = N$.
Change-of-Base Formula	$\log_a M = \dfrac{\log_b M}{\log_b a}$		

Historical Feature

John Napier (1550–1617)

Logarithms were invented about 1590 by John Napier (1550–1617) and Joost Bürgi (1552–1632), working independently. Napier, whose work had the greater influence, was a Scottish lord, a secretive man whose neighbors were inclined to believe him to be in league with the devil. His approach to logarithms was very different from ours; it was based on the relationship between arithmetic and geometric sequences, discussed in a later chapter, and not on the inverse function relationship of logarithms to exponential functions (described in Section 6.4). Napier's tables, published in 1614, listed what would now be called *natural logarithms* of sines and were rather difficult to use. A London professor, Henry Briggs, became interested in the tables and visited Napier. In their conversations, they developed the idea of common logarithms, which were published in 1617. The importance of this tool for calculation was immediately recognized, and by 1650 common logarithms were being printed as far away as China. They remained an important calculation tool until the advent of the inexpensive handheld calculator about 1972, which has decreased their calculational—but not their theoretical—importance.

A side effect of the invention of logarithms was the popularization of the decimal system of notation for real numbers.

6.5 Assess Your Understanding

Concepts and Vocabulary

1. $\log_a 1 =$ ____
2. $a^{\log_a M} =$ ____
3. $\log_a a^r =$ ____
4. $\log_a (MN) =$ ____ + ____
5. $\log_a\left(\frac{M}{N}\right) =$ ____ − ____
6. $\log_a M^r =$ ____
7. If $\log_8 M = \frac{\log_5 7}{\log_5 8}$, then $M =$ ____.
8. ***True or False*** $\ln(x+3) - \ln(2x) = \frac{\ln(x+3)}{\ln(2x)}$
9. ***True or False*** $\log_2(3x^4) = 4\log_2(3x)$
10. ***True or False*** $\log\left(\frac{2}{3}\right) = \frac{\log 2}{\log 3}$
11. Choose the expression equivalent to 2^x.
    (a) e^{2x} (b) $e^{x \ln 2}$ (c) $e^{\log_2 x}$ (d) $e^{2 \ln x}$
12. Writing $\log_a x - \log_a y + 2\log_a z$ as a single logarithm results in which of the following?
    (a) $\log_a(x - y + 2z)$ (b) $\log_a\left(\frac{xz^2}{y}\right)$
    (c) $\log_a\left(\frac{2xz}{y}\right)$ (d) $\log_a\left(\frac{x}{yz^2}\right)$

Skill Building

In Problems 13–28, use properties of logarithms to find the exact value of each expression. Do not use a calculator.

13. $\log_3 3^{71}$
14. $\log_2 2^{-13}$
15. $\ln e^{-4}$
16. $\ln e^{\sqrt{2}}$
17. $2^{\log_2 7}$
18. $e^{\ln 8}$
19. $\log_8 2 + \log_8 4$
20. $\log_6 9 + \log_6 4$
21. $\log_6 18 - \log_6 3$
22. $\log_8 16 - \log_8 2$
23. $\log_2 6 \cdot \log_6 8$
24. $\log_3 8 \cdot \log_8 9$
25. $3^{\log_3 5 - \log_3 4}$
26. $5^{\log_5 6 + \log_5 7}$
27. $e^{\log_{e^2} 16}$
28. $e^{\log_{e^2} 9}$

In Problems 29–36, suppose that $\ln 2 = a$ and $\ln 3 = b$. Use properties of logarithms to write each logarithm in terms of a and b.

29. $\ln 6$
30. $\ln \frac{2}{3}$
31. $\ln 1.5$
32. $\ln 0.5$
33. $\ln 8$
34. $\ln 27$
35. $\ln \sqrt[5]{6}$
36. $\ln \sqrt[4]{\frac{2}{3}}$

In Problems 37–56, write each expression as a sum and/or difference of logarithms. Express powers as factors.

37. $\log_5(25x)$
38. $\log_3 \frac{x}{9}$
39. $\log_2 z^3$
40. $\log_7 x^5$
41. $\ln(ex)$
42. $\ln \frac{e}{x}$
43. $\ln \frac{x}{e^x}$
44. $\ln(xe^x)$
45. $\log_a(u^2v^3)$ $u > 0, v > 0$
46. $\log_2\left(\frac{a}{b^2}\right)$ $a > 0, b > 0$
47. $\ln(x^2\sqrt{1-x})$ $0 < x < 1$
48. $\ln(x\sqrt{1+x^2})$ $x > 0$
49. $\log_2\left(\frac{x^3}{x-3}\right)$ $x > 3$
50. $\log_5\left(\frac{\sqrt[3]{x^2+1}}{x^2-1}\right)$ $x > 1$
51. $\log\left[\frac{x(x+2)}{(x+3)^2}\right]$ $x > 0$
52. $\log\left[\frac{x^3\sqrt{x+1}}{(x-2)^2}\right]$ $x > 2$
53. $\ln\left[\frac{x^2-x-2}{(x+4)^2}\right]^{1/3}$ $x > 2$
54. $\ln\left[\frac{(x-4)^2}{x^2-1}\right]^{2/3}$ $x > 4$
55. $\ln \frac{5x\sqrt{1+3x}}{(x-4)^3}$ $x > 4$
56. $\ln\left[\frac{5x^2\sqrt[3]{1-x}}{4(x+1)^2}\right]$ $0 < x < 1$

In Problems 57–70, write each expression as a single logarithm.

57. $3\log_5 u + 4\log_5 v$
58. $2\log_3 u - \log_3 v$
59. $\log_3\sqrt{x} - \log_3 x^3$
60. $\log_2\left(\frac{1}{x}\right) + \log_2\left(\frac{1}{x^2}\right)$
61. $\log_4(x^2-1) - 5\log_4(x+1)$
62. $\log(x^2+3x+2) - 2\log(x+1)$
63. $\ln\left(\frac{x}{x-1}\right) + \ln\left(\frac{x+1}{x}\right) - \ln(x^2-1)$
64. $\log\left(\frac{x^2+2x-3}{x^2-4}\right) - \log\left(\frac{x^2+7x+6}{x+2}\right)$
65. $8\log_2\sqrt{3x-2} - \log_2\left(\frac{4}{x}\right) + \log_2 4$
66. $21\log_3\sqrt[3]{x} + \log_3(9x^2) - \log_3 9$
67. $2\log_a(5x^3) - \frac{1}{2}\log_a(2x+3)$
68. $\frac{1}{3}\log(x^3+1) + \frac{1}{2}\log(x^2+1)$
69. $2\log_2(x+1) - \log_2(x+3) - \log_2(x-1)$
70. $3\log_5(3x+1) - 2\log_5(2x-1) - \log_5 x$

In Problems 71–78, use the Change-of-Base Formula and a calculator to evaluate each logarithm. Round your answer to three decimal places.

71. $\log_3 21$ **72.** $\log_5 18$ **73.** $\log_{1/3} 71$ **74.** $\log_{1/2} 15$

75. $\log_{\sqrt{2}} 7$ **76.** $\log_{\sqrt{5}} 8$ **77.** $\log_{\pi} e$ **78.** $\log_{\pi} \sqrt{2}$

In Problems 79–84, graph each function using a graphing utility and the Change-of-Base Formula.

79. $y = \log_4 x$ **80.** $y = \log_5 x$ **81.** $y = \log_2(x+2)$

82. $y = \log_4(x-3)$ **83.** $y = \log_{x-1}(x+1)$ **84.** $y = \log_{x+2}(x-2)$

Mixed Practice

85. If $f(x) = \ln x$, $g(x) = e^x$, and $h(x) = x^2$, find:
(a) $(f \circ g)(x)$. What is the domain of $f \circ g$?
(b) $(g \circ f)(x)$. What is the domain of $g \circ f$?
(c) $(f \circ g)(5)$
(d) $(f \circ h)(x)$. What is the domain of $f \circ h$?
(e) $(f \circ h)(e)$

86. If $f(x) = \log_2 x$, $g(x) = 2^x$, and $h(x) = 4x$, find:
(a) $(f \circ g)(x)$. What is the domain of $f \circ g$?
(b) $(g \circ f)(x)$. What is the domain of $g \circ f$?
(c) $(f \circ g)(3)$
(d) $(f \circ h)(x)$. What is the domain of $f \circ h$?
(e) $(f \circ h)(8)$

Applications and Extensions

In Problems 87–96, express y as a function of x. The constant C is a positive number.

87. $\ln y = \ln x + \ln C$

88. $\ln y = \ln(x + C)$

89. $\ln y = \ln x + \ln(x+1) + \ln C$

90. $\ln y = 2\ln x - \ln(x+1) + \ln C$

91. $\ln y = 3x + \ln C$

92. $\ln y = -2x + \ln C$

93. $\ln(y-3) = -4x + \ln C$

94. $\ln(y+4) = 5x + \ln C$

95. $3\ln y = \frac{1}{2}\ln(2x+1) - \frac{1}{3}\ln(x+4) + \ln C$

96. $2\ln y = -\frac{1}{2}\ln x + \frac{1}{3}\ln(x^2+1) + \ln C$

97. Find the value of $\log_2 3 \cdot \log_3 4 \cdot \log_4 5 \cdot \log_5 6 \cdot \log_6 7 \cdot \log_7 8$.

98. Find the value of $\log_2 4 \cdot \log_4 6 \cdot \log_6 8$.

99. Find the value of $\log_2 3 \cdot \log_3 4 \cdot \cdots \cdot \log_n(n+1) \cdot \log_{n+1} 2$.

100. Find the value of $\log_2 2 \cdot \log_2 4 \cdot \cdots \cdot \log_2 2^n$.

101. Show that $\log_a(x + \sqrt{x^2-1}) + \log_a(x - \sqrt{x^2-1}) = 0$.

102. Show that $\log_a(\sqrt{x} + \sqrt{x-1}) + \log_a(\sqrt{x} - \sqrt{x-1}) = 0$.

103. Show that $\ln(1 + e^{2x}) = 2x + \ln(1 + e^{-2x})$.

104. Difference Quotient If $f(x) = \log_a x$, show that $\frac{f(x+h) - f(x)}{h} = \log_a\left(1 + \frac{h}{x}\right)^{1/h}$, $h \neq 0$.

105. If $f(x) = \log_a x$, show that $-f(x) = \log_{1/a} x$.

106. If $f(x) = \log_a x$, show that $f(AB) = f(A) + f(B)$.

107. If $f(x) = \log_a x$, show that $f\left(\frac{1}{x}\right) = -f(x)$.

108. If $f(x) = \log_a x$, show that $f(x^\alpha) = \alpha f(x)$.

109. Show that $\log_a\left(\frac{M}{N}\right) = \log_a M - \log_a N$, where a, M, and N are positive real numbers and $a \neq 1$.

110. Show that $\log_a\left(\frac{1}{N}\right) = -\log_a N$, where a and N are positive real numbers and $a \neq 1$.

Explaining Concepts: Discussion and Writing

111. Graph $Y_1 = \log(x^2)$ and $Y_2 = 2\log(x)$ using a graphing utility. Are they equivalent? What might account for any differences in the two functions?

112. Write an example that illustrates why $(\log_a x)^r \neq r\log_a x$.

113. Write an example that illustrates why $\log_2(x+y) \neq \log_2 x + \log_2 y$.

114. Does $3^{\log_3(-5)} = -5$? Why or why not?

Retain Your Knowledge

Problems 115–118 are based on material learned earlier in the course. The purpose of these problems is to keep the material fresh in your mind so that you are better prepared for the final exam.

115. Use a graphing utility to solve $x^3 - 3x^2 - 4x + 8 = 0$. Round answers to two decimal places.

116. Without solving, determine the character of the solution of the quadratic equation $4x^2 - 28x + 49 = 0$ in the complex number system.

117. Find the real zeros of $f(x) = 5x^5 + 44x^4 + 116x^3 + 95x^2 - 4x - 4$

118. Graph $f(x) = \sqrt{2 - x}$ using the techniques of shifting, compressing or stretching, and reflecting. State the domain and the range of f.

6.6 Logarithmic and Exponential Equations

PREPARING FOR THIS SECTION *Before getting started, review the following:*

- Solving Equations Using a Graphing Utility (Appendix, Section 4, pp. A8–A10)
- Solving Quadratic Equations (Section 1.2, pp. 92–99)
- Solving Equations Quadratic in Form (Section 1.4, pp. 114–116)

Now Work the **'Are You Prepared?'** problems on page 465.

OBJECTIVES
1 Solve Logarithmic Equations (p. 461)
2 Solve Exponential Equations (p. 463)
3 Solve Logarithmic and Exponential Equations Using a Graphing Utility (p. 464)

1 Solve Logarithmic Equations

In Section 6.4 we solved logarithmic equations by changing a logarithmic expression to an exponential expression. That is, we used the definition of a logarithm:

$$y = \log_a x \quad \text{is equivalent to} \quad x = a^y \qquad a > 0 \quad a \neq 1$$

For example, to solve the equation $\log_2(1 - 2x) = 3$, write the logarithmic equation as an equivalent exponential equation $1 - 2x = 2^3$ and solve for x.

$$\log_2(1 - 2x) = 3$$

$$1 - 2x = 2^3 \quad \text{Change to exponential form.}$$

$$-2x = 7 \quad \text{Simplify.}$$

$$x = -\frac{7}{2} \quad \text{Solve.}$$

You should check this solution for yourself.

For most logarithmic equations, some manipulation of the equation (usually using properties of logarithms) is required to obtain a solution. Also, to avoid extraneous solutions with logarithmic equations, determine the domain of the variable first.

Let's begin with an example of a logarithmic equation that requires using the fact that a logarithmic function is a one-to-one function:

$$\text{If } \log_a M = \log_a N, \text{ then } M = N \qquad M, N, \text{ and } a \text{ are positive and } a \neq 1$$

EXAMPLE 1 **Solving a Logarithmic Equation**

Solve: $2 \log_5 x = \log_5 9$

Solution The domain of the variable in this equation is $x > 0$. Note that each logarithm has the same base, 5. Then find the exact solution as follows:

$$2 \log_5 x = \log_5 9$$

$$\log_5 x^2 = \log_5 9 \quad r\log_a M = \log_a M^r$$

$$x^2 = 9 \quad \text{If } \log_a M = \log_a N, \text{ then } M = N.$$

$$x = 3 \quad \text{or} \quad x = -3$$

Recall that the domain of the variable is $x > 0$. Therefore, -3 is extraneous and must be discarded.

✓**Check:** $2\log_5 3 \stackrel{?}{=} \log_5 9$

$\log_5 3^2 \stackrel{?}{=} \log_5 9$ $r\log_a M = \log_a M^r$

$\log_5 9 = \log_5 9$

The solution set is $\{3\}$.

Now Work PROBLEM 13

Often one or more properties of logarithms are needed to rewrite the equation as a single logarithm. In the next example, the log of a product property is used.

EXAMPLE 2 Solving a Logarithmic Equation

Solve: $\log_5(x+6) + \log_5(x+2) = 1$

Solution The domain of the variable requires that $x + 6 > 0$ and $x + 2 > 0$, so $x > -6$ and $x > -2$. This means any solution must satisfy $x > -2$. To obtain an exact solution, first express the left side as a single logarithm. Then change the equation to an equivalent exponential equation.

$$\log_5(x+6) + \log_5(x+2) = 1$$

$$\log_5[(x+6)(x+2)] = 1 \quad \log_a M + \log_a N = \log_a(MN)$$

$$(x+6)(x+2) = 5^1 = 5 \quad \text{Change to exponential form.}$$

$$x^2 + 8x + 12 = 5 \quad \text{Multiply out.}$$

$$x^2 + 8x + 7 = 0 \quad \text{Place the quadratic equation in standard form.}$$

$$(x+7)(x+1) = 0 \quad \text{Factor.}$$

$$x = -7 \quad \text{or} \quad x = -1 \quad \text{Zero-Product Property}$$

Only $x = -1$ satisfies the restriction that $x > -2$, so $x = -7$ is extraneous. The solution set is $\{-1\}$, which you should check.

WARNING A negative solution is not automatically extraneous. You must determine whether the potential solution causes the argument of any logarithmic expression in the equation to be negative or 0.

Now Work PROBLEM 21

EXAMPLE 3 Solving a Logarithmic Equation

Solve: $\ln x = \ln(x+6) - \ln(x-4)$

Solution The domain of the variable requires that $x > 0$, $x + 6 > 0$, and $x - 4 > 0$. As a result, the domain of the variable here is $x > 4$. Begin the solution using the log of a difference property.

$$\ln x = \ln(x+6) - \ln(x-4)$$

$$\ln x = \ln\left(\frac{x+6}{x-4}\right) \quad \ln M - \ln N = \ln\left(\frac{M}{N}\right)$$

$$x = \frac{x+6}{x-4} \quad \text{If } \ln M = \ln N \text{, then } M = N.$$

$$x(x-4) = x+6 \quad \text{Multiply both sides by } x - 4.$$

$$x^2 - 4x = x + 6 \quad \text{Multiply out.}$$

$$x^2 - 5x - 6 = 0 \quad \text{Place the quadratic equation in standard form.}$$

$$(x-6)(x+1) = 0 \quad \text{Factor.}$$

$$x = 6 \quad \text{or} \quad x = -1 \quad \text{Zero-Product Property}$$

Because the domain of the variable is $x > 4$, discard -1 as extraneous. The solution set is $\{6\}$, which you should check.

WARNING In using properties of logarithms to solve logarithmic equations, avoid using the property $\log_a x^r = r\log_a x$, when r is even. The reason can be seen in this example:

Solve: $\log_3 x^2 = 4$

Solution: The domain of the variable x is all real numbers except 0.

(a) $\log_3 x^2 = 4$

$x^2 = 3^4 = 81$

$x = -9$ or $x = 9$

(b) $\log_3 x^2 = 4$

$2\log_3 x = 4$ $\quad x > 0$

$\log_3 x = 2$

$x = 9$

Both -9 and 9 are solutions of $\log_3 x^2 = 4$ (as you can verify). The solution in part (b) does not find the solution -9 because the domain of the variable was further restricted due to the application of the property $\log_a x^r = r\log_a x$. ■

Now Work PROBLEM 31

2 Solve Exponential Equations

In Sections 6.3 and 6.4, we solved exponential equations algebraically by expressing each side of the equation using the same base. That is, we used the one-to-one property of the exponential function:

$$\text{If } a^u = a^v, \text{ then } u = v \qquad a > 0 \quad a \neq 1$$

For example, to solve the exponential equation $4^{2x+1} = 16$, notice that $16 = 4^2$ and apply the property above to obtain the equation $2x + 1 = 2$, from which we find $x = \frac{1}{2}$.

Not all exponential equations can be readily expressed so that each side of the equation has the same base. For such equations, algebraic techniques often can be used to obtain exact solutions.

EXAMPLE 4 **Solving Exponential Equations**

Solve: (a) $2^x = 5$ (b) $8\cdot 3^x = 5$

Solution (a) Because 5 cannot be written as an integer power of 2 ($2^2 = 4$ and $2^3 = 8$), write the exponential equation as the equivalent logarithmic equation.

$$2^x = 5$$

$$x = \log_2 5 = \frac{\ln 5}{\ln 2}$$

↑ Change-of-Base Formula (10), Section 6.5

Alternatively, the equation $2^x = 5$ can be solved by taking the natural logarithm (or common logarithm) of each side.

$$2^x = 5$$

$$\ln 2^x = \ln 5 \qquad \text{If } M = N, \text{ then } \ln M = \ln N.$$

$$x\ln 2 = \ln 5 \qquad \ln M^r = r\ln M$$

$$x = \frac{\ln 5}{\ln 2} \qquad \text{Exact solution}$$

$$\approx 2.322 \qquad \text{Approximate solution}$$

The solution set is $\left\{\frac{\ln 5}{\ln 2}\right\}$.

(b) $8\cdot 3^x = 5$

$$3^x = \frac{5}{8} \qquad \text{Solve for } 3^x.$$

$$x = \log_3\left(\frac{5}{8}\right) = \frac{\ln\left(\frac{5}{8}\right)}{\ln 3} \quad \text{Exact solution}$$

$$\approx -0.428 \quad \text{Approximate solution}$$

The solution set is $\left\{\dfrac{\ln\left(\frac{5}{8}\right)}{\ln 3}\right\}$.

Now Work PROBLEM 43

EXAMPLE 5 **Solving an Exponential Equation**

Solve: $5^{x-2} = 3^{3x+2}$

Solution Because the bases are different, first apply property (7), Section 6.5 (take the natural logarithm of each side), and then use a property of logarithms. The result is an equation in x that can be solved.

$$\begin{aligned}
5^{x-2} &= 3^{3x+2} \\
\ln 5^{x-2} &= \ln 3^{3x+2} && \text{If } M = N, \ln M = \ln N. \\
(x-2)\ln 5 &= (3x+2)\ln 3 && \ln M^r = r\ln M \\
(\ln 5)x - 2\ln 5 &= (3\ln 3)x + 2\ln 3 && \text{Distribute.} \\
(\ln 5)x - (3\ln 3)x &= 2\ln 3 + 2\ln 5 && \text{Place terms involving } x \text{ on the left.} \\
(\ln 5 - 3\ln 3)x &= 2(\ln 3 + \ln 5) && \text{Factor.} \\
x &= \frac{2(\ln 3 + \ln 5)}{\ln 5 - 3\ln 3} && \text{Exact solution} \\
&\approx -3.212 && \text{Approximate solution}
\end{aligned}$$

The solution set is $\left\{\dfrac{2(\ln 3 + \ln 5)}{\ln 5 - 3\ln 3}\right\}$.

NOTE: Because of the properties of logarithms, exact solutions involving logarithms often can be expressed in multiple ways. For example, the solution to $5^{x-2} = 3^{3x+2}$ from Example 5 can be expressed equivalently as $\dfrac{2\ln 15}{\ln 5 - \ln 27}$ or as $\dfrac{\ln 225}{\ln(5/27)}$, among others. Do you see why?

Now Work PROBLEM 53

EXAMPLE 6 **Solving an Exponential Equation That Is Quadratic in Form**

Solve: $4^x - 2^x - 12 = 0$

Solution Note that $4^x = (2^2)^x = 2^{(2x)} = (2^x)^2$, so the equation is quadratic in form and can be written as

$$(2^x)^2 - 2^x - 12 = 0 \quad \text{Let } u = 2^x\text{; then } u^2 - u - 12 = 0.$$

Now factor as usual.

$$\begin{aligned}
(2^x - 4)(2^x + 3) &= 0 && (u-4)(u+3) = 0 \\
2^x - 4 = 0 \quad \text{or} \quad 2^x + 3 &= 0 && u - 4 = 0 \quad \text{or} \quad u + 3 = 0 \\
2^x = 4 \qquad\qquad 2^x &= -3 && u = 2^x = 4 \qquad u = 2^x = -3
\end{aligned}$$

The equation on the left has the solution $x = 2$, since $2^x = 4 = 2^2$; the equation on the right has no solution, since $2^x > 0$ for all x. The only solution is 2. The solution set is $\{2\}$.

Now Work PROBLEM 61

3 Solve Logarithmic and Exponential Equations Using a Graphing Utility

The algebraic techniques introduced in this section to obtain exact solutions apply only to certain types of logarithmic and exponential equations. Solutions for other types are generally studied in calculus, using numerical methods. For such types, we can use a graphing utility to approximate the solution.

EXAMPLE 7 **Solving Equations Using a Graphing Utility**

Solve: $x + e^x = 2$

Express the solution(s) rounded to two decimal places.

Solution The solution is found by graphing $Y_1 = x + e^x$ and $Y_2 = 2$. Since Y_1 is an increasing function (do you know why?), there is only one point of intersection for Y_1 and Y_2. Figure 40 shows the graphs of Y_1 and Y_2. Using the INTERSECT command reveals that the solution is 0.44, rounded to two decimal places. ●

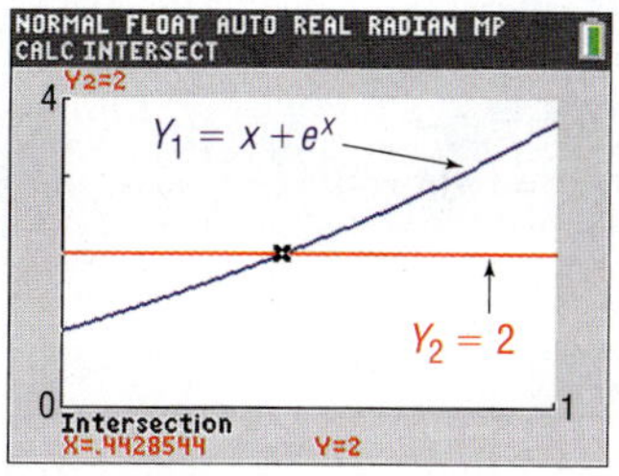

Figure 40

Now Work PROBLEM 71

6.6 Assess Your Understanding

'Are You Prepared?' *Answers are given at the end of these exercises. If you get a wrong answer, read the pages listed in red.*

1. Solve $x^2 - 7x - 30 = 0$. (pp. 92–99)
2. Solve $(x + 3)^2 - 4(x + 3) + 3 = 0$. (pp. 114–116)
3. Approximate the solution(s) to $x^3 = x^2 - 5$ using a graphing utility. (pp. A8–A10)
4. Approximate the solution(s) to $x^3 - 2x + 2 = 0$ using a graphing utility. (pp. A8–A10)

Skill Building

In Problems 5–40, solve each logarithmic equation. Express irrational solutions in exact form and as a decimal rounded to three decimal places.

5. $\log_4 x = 2$
6. $\log (x + 6) = 1$
7. $\log_2 (5x) = 4$
8. $\log_3 (3x - 1) = 2$
9. $\log_4 (x + 2) = \log_4 8$
10. $\log_5 (2x + 3) = \log_5 3$
11. $\frac{1}{2} \log_3 x = 2 \log_3 2$
12. $-2 \log_4 x = \log_4 9$
13. $3 \log_2 x = -\log_2 27$
14. $2 \log_5 x = 3 \log_5 4$
15. $3 \log_2 (x - 1) + \log_2 4 = 5$
16. $2 \log_3 (x + 4) - \log_3 9 = 2$
17. $\log x + \log (x + 15) = 2$
18. $\log x + \log (x - 21) = 2$
19. $\log (2x + 1) = 1 + \log (x - 2)$
20. $\log (2x) - \log (x - 3) = 1$
21. $\log_2 (x + 7) + \log_2 (x + 8) = 1$
22. $\log_6 (x + 4) + \log_6 (x + 3) = 1$
23. $\log_8 (x + 6) = 1 - \log_8 (x + 4)$
24. $\log_5 (x + 3) = 1 - \log_5 (x - 1)$
25. $\ln x + \ln (x + 2) = 4$
26. $\ln (x + 1) - \ln x = 2$
27. $\log_3 (x + 1) + \log_3 (x + 4) = 2$
28. $\log_2 (x + 1) + \log_2 (x + 7) = 3$
29. $\log_{1/3} (x^2 + x) - \log_{1/3} (x^2 - x) = -1$
30. $\log_4 (x^2 - 9) - \log_4 (x + 3) = 3$
31. $\log_a (x - 1) - \log_a (x + 6) = \log_a (x - 2) - \log_a (x + 3)$
32. $\log_a x + \log_a (x - 2) = \log_a (x + 4)$
33. $2 \log_5 (x - 3) - \log_5 8 = \log_5 2$
34. $\log_3 x - 2 \log_3 5 = \log_3 (x + 1) - 2 \log_3 10$
35. $2 \log_6 (x + 2) = 3 \log_6 2 + \log_6 4$
36. $3(\log_7 x - \log_7 2) = 2 \log_7 4$
37. $2 \log_{13} (x + 2) = \log_{13} (4x + 7)$
38. $\log (x - 1) = \frac{1}{3} \log 2$
39. $(\log_3 x)^2 - 5(\log_3 x) = 6$
40. $\ln x - 3\sqrt{\ln x} + 2 = 0$

In Problems 41–68, solve each exponential equation. Express irrational solutions in exact form and as a decimal rounded to three decimal places.

41. $2^{x-5} = 8$
42. $5^{-x} = 25$
43. $2^x = 10$
44. $3^x = 14$
45. $8^{-x} = 1.2$
46. $2^{-x} = 1.5$
47. $5(2^{3x}) = 8$
48. $0.3(4^{0.2x}) = 0.2$
49. $3^{1-2x} = 4^x$
50. $2^{x+1} = 5^{1-2x}$
51. $\left(\frac{3}{5}\right)^x = 7^{1-x}$
52. $\left(\frac{4}{3}\right)^{1-x} = 5^x$
53. $1.2^x = (0.5)^{-x}$
54. $0.3^{1+x} = 1.7^{2x-1}$
55. $\pi^{1-x} = e^x$
56. $e^{x+3} = \pi^x$

57. $2^{2x} + 2^x - 12 = 0$
58. $3^{2x} + 3^x - 2 = 0$
59. $3^{2x} + 3^{x+1} - 4 = 0$
60. $2^{2x} + 2^{x+2} - 12 = 0$
61. $16^x + 4^{x+1} - 3 = 0$
62. $9^x - 3^{x+1} + 1 = 0$
63. $25^x - 8 \cdot 5^x = -16$
64. $36^x - 6 \cdot 6^x = -9$
65. $3 \cdot 4^x + 4 \cdot 2^x + 8 = 0$
66. $2 \cdot 49^x + 11 \cdot 7^x + 5 = 0$
67. $4^x - 10 \cdot 4^{-x} = 3$
68. $3^x - 14 \cdot 3^{-x} = 5$

In Problems 69–82, use a graphing utility to solve each equation. Express your answer rounded to two decimal places.

69. $\log_5(x + 1) - \log_4(x - 2) = 1$
70. $\log_2(x - 1) - \log_6(x + 2) = 2$
71. $e^x = -x$
72. $e^{2x} = x + 2$
73. $e^x = x^2$
74. $e^x = x^3$
75. $\ln x = -x$
76. $\ln(2x) = -x + 2$
77. $\ln x = x^3 - 1$
78. $\ln x = -x^2$
79. $e^x + \ln x = 4$
80. $e^x - \ln x = 4$
81. $e^{-x} = \ln x$
82. $e^{-x} = -\ln x$

Mixed Practice

In Problems 83–94, solve each equation. Express irrational solutions in exact form and as a decimal rounded to three decimal places.

83. $\log_2(x + 1) - \log_4 x = 1$
[**Hint:** Change $\log_4 x$ to base 2.]

84. $\log_2(3x + 2) - \log_4 x = 3$

85. $\log_{16} x + \log_4 x + \log_2 x = 7$

86. $\log_9 x + 3\log_3 x = 14$

87. $(\sqrt[3]{2})^{2-x} = 2^{x^2}$

88. $\log_2 x^{\log_2 x} = 4$

89. $\dfrac{e^x + e^{-x}}{2} = 1$
[**Hint:** Multiply each side by e^x.]

90. $\dfrac{e^x + e^{-x}}{2} = 3$

91. $\dfrac{e^x - e^{-x}}{2} = 2$

92. $\dfrac{e^x - e^{-x}}{2} = -2$

93. $\log_5 x + \log_3 x = 1$
[**Hint:** Use the Change-of-Base Formula.]

94. $\log_2 x + \log_6 x = 3$

---

95. $f(x) = \log_2(x + 3)$ and $g(x) = \log_2(3x + 1)$.
(a) Solve $f(x) = 3$. What point is on the graph of f?
(b) Solve $g(x) = 4$. What point is on the graph of g?
(c) Solve $f(x) = g(x)$. Do the graphs of f and g intersect? If so, where?
(d) Solve $(f + g)(x) = 7$.
(e) Solve $(f - g)(x) = 2$.

96. $f(x) = \log_3(x + 5)$ and $g(x) = \log_3(x - 1)$.
(a) Solve $f(x) = 2$. What point is on the graph of f?
(b) Solve $g(x) = 3$. What point is on the graph of g?
(c) Solve $f(x) = g(x)$. Do the graphs of f and g intersect? If so, where?
(d) Solve $(f + g)(x) = 3$.
(e) Solve $(f - g)(x) = 2$.

97. (a) If $f(x) = 3^{x+1}$ and $g(x) = 2^{x+2}$, graph f and g on the same Cartesian plane.
(b) Find the point(s) of intersection of the graphs of f and g by solving $f(x) = g(x)$. Round answers to three decimal places. Label any intersection points on the graph drawn in part (a).
(c) Based on the graph, solve $f(x) > g(x)$.

98. (a) If $f(x) = 5^{x-1}$ and $g(x) = 2^{x+1}$, graph f and g on the same Cartesian plane.
(b) Find the point(s) of intersection of the graphs of f and g by solving $f(x) = g(x)$. Label any intersection points on the graph drawn in part (a).
(c) Based on the graph, solve $f(x) > g(x)$.

99. (a) Graph $f(x) = 3^x$ and $g(x) = 10$ on the same Cartesian plane.
(b) Shade the region bounded by the y-axis, $f(x) = 3^x$, and $g(x) = 10$ on the graph drawn in part (a).
(c) Solve $f(x) = g(x)$ and label the point of intersection on the graph drawn in part (a).

100. (a) Graph $f(x) = 2^x$ and $g(x) = 12$ on the same Cartesian plane.
(b) Shade the region bounded by the y-axis, $f(x) = 2^x$, and $g(x) = 12$ on the graph drawn in part (a).
(c) Solve $f(x) = g(x)$ and label the point of intersection on the graph drawn in part (a).

101. (a) Graph $f(x) = 2^{x+1}$ and $g(x) = 2^{-x+2}$ on the same Cartesian plane.
(b) Shade the region bounded by the y-axis, $f(x) = 2^{x+1}$, and $g(x) = 2^{-x+2}$ on the graph drawn in part (a).
(c) Solve $f(x) = g(x)$ and label the point of intersection on the graph drawn in part (a).

102. (a) Graph $f(x) = 3^{-x+1}$ and $g(x) = 3^{x-2}$ on the same Cartesian plane.
(b) Shade the region bounded by the y-axis, $f(x) = 3^{-x+1}$, and $g(x) = 3^{x-2}$ on the graph drawn in part (a).
(c) Solve $f(x) = g(x)$ and label the point of intersection on the graph drawn in part (a).

103. (a) Graph $f(x) = 2^x - 4$.
(b) Find the zero of f.
(c) Based on the graph, solve $f(x) < 0$.

104. (a) Graph $g(x) = 3^x - 9$.
(b) Find the zero of g.
(c) Based on the graph, solve $g(x) > 0$.

Applications and Extensions

105. A Population Model The resident population of the United States in 2014 was 317 million people and was growing at a rate of 0.7% per year. Assuming that this growth rate continues, the model $P(t) = 317(1.007)^{t-2014}$ represents the population P (in millions of people) in year t.

(a) According to this model, when will the population of the United States be 400 million people?

(b) According to this model, when will the population of the United States be 435 million people?

Source: U.S. Census Bureau

106. A Population Model The population of the world in 2014 was 7.14 billion people and was growing at a rate of 1.1% per year. Assuming that this growth rate continues, the model $P(t) = 7.14(1.011)^{t-2014}$ represents the population P (in billions of people) in year t.

(a) According to this model, when will the population of the world be 9 billion people?

(b) According to this model, when will the population of the world be 12.5 billion people?

Source: U.S. Census Bureau

107. Depreciation The value V of a Chevy Cruze that is t years old can be modeled by $V(t) = 18{,}700(0.84)^t$.

(a) According to the model, when will the car be worth \$9000?

(b) According to the model, when will the car be worth \$6000?

(c) According to the model, when will the car be worth \$2000?

Source: Kelley Blue Book

108. Depreciation The value V of a Honda Civic LX that is t years old can be modeled by $V(t) = 18{,}955(0.905)^t$.

(a) According to the model, when will the car be worth \$16,000?

(b) According to the model, when will the car be worth \$10,000?

(c) According to the model, when will the car be worth \$7500?

Source: Kelley Blue Book

Explaining Concepts: Discussion and Writing

109. Fill in the reason for each step in the following two solutions.

Solve: $\log_3(x-1)^2 = 2$

Solution A

$\log_3(x-1)^2 = 2$

$(x-1)^2 = 3^2 = 9$ ______

$(x-1) = \pm 3$ ______

$x - 1 = -3$ or $x - 1 = 3$ ______

$x = -2$ or $x = 4$ ______

Solution B

$\log_3(x-1)^2 = 2$

$2\log_3(x-1) = 2$ ______

$\log_3(x-1) = 1$ ______

$x - 1 = 3^1 = 3$ ______

$x = 4$ ______

Both solutions given in Solution A check. Explain what caused the solution $x = -2$ to be lost in Solution B.

Retain Your Knowledge

Problems 110–113 are based on material learned earlier in the course. The purpose of these problems is to keep the material fresh in your mind so that you are better prepared for the final exam.

110. Solve: $4x^3 + 3x^2 - 25x + 6 = 0$

111. Determine whether the function $\{(0, -4), (2, -2), (4, 0), (6, 2)\}$ is one-to-one.

112. For $f(x) = \dfrac{x}{x-2}$ and $g(x) = \dfrac{x+5}{x-3}$, find $f \circ g$. Then find the domain of $f \circ g$.

113. Find the domain of $f(x) = \sqrt{x+3} + \sqrt{x-1}$.

'Are You Prepared?' Answers

1. $\{-3, 10\}$ **2.** $\{-2, 0\}$ **3.** $\{-1.43\}$ **4.** $\{-1.77\}$

6.7 Financial Models

PREPARING FOR THIS SECTION *Before getting started, review the following:*

- Simple Interest (Section 1.7, p. 136)

Now Work the 'Are You Prepared?' problems on page 474.

OBJECTIVES
1. Determine the Future Value of a Lump Sum of Money (p. 468)
2. Calculate Effective Rates of Return (p. 471)
3. Determine the Present Value of a Lump Sum of Money (p. 472)
4. Determine the Rate of Interest or the Time Required to Double a Lump Sum of Money (p. 473)

1 Determine the Future Value of a Lump Sum of Money

Interest is money paid for the use of money. The total amount borrowed (whether by an individual from a bank in the form of a loan or by a bank from an individual in the form of a savings account) is called the **principal**. The **rate of interest**, expressed as a percent, is the amount charged for the use of the principal for a given period of time, usually on a yearly (that is, per annum) basis.

THEOREM

Simple Interest Formula

If a principal of P dollars is borrowed for a period of t years at a per annum interest rate r, expressed as a decimal, the interest I charged is

$$I = Prt \qquad (1)$$

Interest charged according to formula (1) is called **simple interest**.

In problems involving interest, the term **payment period** is defined as follows.

Annually:	Once per year	**Monthly:**	12 times per year
Semiannually:	Twice per year	**Daily:**	365 times per year*
Quarterly:	Four times per year		

When the interest due at the end of a payment period is added to the principal so that the interest computed at the end of the next payment period is based on this new principal amount (old principal + interest), the interest is said to have been **compounded**. **Compound interest** is interest paid on the principal and on previously earned interest.

EXAMPLE 1 **Computing Compound Interest**

A credit union pays interest of 2% per annum compounded quarterly on a certain savings plan. If \$1000 is deposited in such a plan and the interest is left to accumulate, how much is in the account after 1 year?

Solution Use the simple interest formula, $I = Prt$. The principal P is \$1000 and the rate of interest is $2\% = 0.02$. After the first quarter of a year, the time t is $\frac{1}{4}$ year, so the interest earned is

$$I = Prt = (\$1000)(0.02)\left(\frac{1}{4}\right) = \$5$$

* Most banks use a 360-day "year." Why do you think they do?

The new principal is $P + I = \$1000 + \$5 = \$1005$. At the end of the second quarter, the interest on this principal is

$$I = (\$1005)(0.02)\left(\frac{1}{4}\right) = \$5.03$$

At the end of the third quarter, the interest on the new principal of $\$1005 + \$5.03 = \$1010.03$ is

$$I = (\$1010.03)(0.02)\left(\frac{1}{4}\right) = \$5.05$$

Finally, after the fourth quarter, the interest is

$$I = (\$1015.08)(0.02)\left(\frac{1}{4}\right) = \$5.08$$

After 1 year the account contains $\$1015.08 + \$5.08 = \$1020.16$.

The pattern of the calculations performed in Example 1 leads to a general formula for compound interest. For this purpose, let P represent the principal to be invested at a per annum interest rate r that is compounded n times per year, so the time of each compounding period is $\frac{1}{n}$ years. (For computing purposes, r is expressed as a decimal.) The interest earned after each compounding period is given by formula (1).

$$\text{Interest} = \text{principal} \times \text{rate} \times \text{time} = P \cdot r \cdot \frac{1}{n} = P \cdot \left(\frac{r}{n}\right)$$

The amount A after one compounding period is

$$A = P + P \cdot \left(\frac{r}{n}\right) = P \cdot \left(1 + \frac{r}{n}\right)$$

After two compounding periods, the amount A, based on the new principal $P \cdot \left(1 + \frac{r}{n}\right)$, is

$$A = \underbrace{P \cdot \left(1 + \frac{r}{n}\right)}_{\text{New principal}} + \underbrace{P \cdot \left(1 + \frac{r}{n}\right)\left(\frac{r}{n}\right)}_{\text{Interest on new principal}} \underset{\uparrow \text{ Factor out } P\cdot\left(1+\frac{r}{n}\right)}{=} P \cdot \left(1 + \frac{r}{n}\right) \cdot \left(1 + \frac{r}{n}\right) = P \cdot \left(1 + \frac{r}{n}\right)^2$$

After three compounding periods, the amount A is

$$A = P \cdot \left(1 + \frac{r}{n}\right)^2 + P \cdot \left(1 + \frac{r}{n}\right)^2\left(\frac{r}{n}\right) = P \cdot \left(1 + \frac{r}{n}\right)^2 \cdot \left(1 + \frac{r}{n}\right) = P \cdot \left(1 + \frac{r}{n}\right)^3$$

Continuing this way, after n compounding periods (1 year), the amount A is

$$A = P \cdot \left(1 + \frac{r}{n}\right)^n$$

Because t years will contain $n \cdot t$ compounding periods, the amount after t years is

$$A = P \cdot \left(1 + \frac{r}{n}\right)^{nt}$$

THEOREM

Compound Interest Formula

The amount A after t years due to a principal P invested at an annual interest rate r, expressed as a decimal, compounded n times per year is

$$A = P \cdot \left(1 + \frac{r}{n}\right)^{nt} \qquad \textbf{(2)}$$

Exploration

To observe the effects of compounding interest monthly on an initial deposit of \$1, graph $Y_1 = \left(1 + \frac{r}{12}\right)^{12x}$ with $r = 0.06$ and $r = 0.12$ for $0 \le x \le 30$. What is the future value of \$1 in 30 years when the interest rate per annum is $r = 0.06$ (6%)? What is the future value of \$1 in 30 years when the interest rate per annum is $r = 0.12$ (12%)? Does doubling the interest rate double the future value?

For example, to rework Example 1, use $P = \$1000$, $r = 0.02$, $n = 4$ (quarterly compounding), and $t = 1$ year to obtain

$$A = P \cdot \left(1 + \frac{r}{n}\right)^{nt} = 1000\left(1 + \frac{0.02}{4}\right)^{4 \cdot 1} = \$1020.16$$

In equation (2), the amount A is typically referred to as the **future value** of the account, and P is called the **present value.**

Now Work PROBLEM 7

EXAMPLE 2 **Comparing Investments Using Different Compounding Periods**

Investing \$1000 at an annual rate of 10% compounded annually, semiannually, quarterly, monthly, and daily will yield the following amounts after 1 year:

Annual compounding $(n = 1)$:
$$A = P \cdot (1 + r) = (\$1000)(1 + 0.10) = \$1100.00$$

Semiannual compounding $(n = 2)$:
$$A = P \cdot \left(1 + \frac{r}{2}\right)^2 = (\$1000)(1 + 0.05)^2 = \$1102.50$$

Quarterly compounding $(n = 4)$:
$$A = P \cdot \left(1 + \frac{r}{4}\right)^4 = (\$1000)(1 + 0.025)^4 = \$1103.81$$

Monthly compounding $(n = 12)$:
$$A = P \cdot \left(1 + \frac{r}{12}\right)^{12} = (\$1000)\left(1 + \frac{0.10}{12}\right)^{12} = \$1104.71$$

Daily compounding $(n = 365)$:
$$A = P \cdot \left(1 + \frac{r}{365}\right)^{365} = (\$1000)\left(1 + \frac{0.10}{365}\right)^{365} = \$1105.16$$

From Example 2, note that the effect of compounding more frequently is that the amount after 1 year is higher: \$1000 compounded 4 times a year at 10% results in \$1103.81, \$1000 compounded 12 times a year at 10% results in \$1104.71, and \$1000 compounded 365 times a year at 10% results in \$1105.16. This leads to the following question: What would happen to the amount after 1 year if the number of times that the interest is compounded were increased without bound?

Let's find the answer. Suppose that P is the principal, r is the per annum interest rate, and n is the number of times that the interest is compounded each year. The amount A after 1 year is

$$A = P \cdot \left(1 + \frac{r}{n}\right)^n$$

Rewrite this expression as follows:

$$A = P \cdot \left(1 + \frac{r}{n}\right)^n = P \cdot \left(1 + \frac{1}{\frac{n}{r}}\right)^n = P \cdot \left[\left(1 + \frac{1}{\frac{n}{r}}\right)^{n/r}\right]^r = P \cdot \left[\left(1 + \frac{1}{h}\right)^h\right]^r \quad \textbf{(3)}$$

$$\uparrow \; h = \frac{n}{r}$$

Now suppose that the number n of times that the interest is compounded per year gets larger and larger; that is, suppose that $n \to \infty$. Then $h = \frac{n}{r} \to \infty$, and the expression in brackets in equation (3) equals e. That is, $A \to Pe^r$.

Table 8 compares $\left(1 + \frac{r}{n}\right)^n$, for large values of n, to e^r for $r = 0.05$, $r = 0.10$, $r = 0.15$, and $r = 1$. The larger that n gets, the closer $\left(1 + \frac{r}{n}\right)^n$ gets to e^r. No matter how frequent the compounding, the amount after 1 year has the definite ceiling Pe^r.

Table 8

	$(1 + \frac{r}{n})^n$			
	$n = 100$	$n = 1000$	$n = 10{,}000$	e^r
$r = 0.05$	1.0512580	1.0512698	1.051271	1.0512711
$r = 0.10$	1.1051157	1.1051654	1.1051704	1.1051709
$r = 0.15$	1.1617037	1.1618212	1.1618329	1.1618342
$r = 1$	2.7048138	2.7169239	2.7181459	2.7182818

When interest is compounded so that the amount after 1 year is Pe^r, the interest is said to be **compounded continuously**.

THEOREM

Continuous Compounding

The amount A after t years due to a principal P invested at an annual interest rate r compounded continuously is

$$A = Pe^{rt} \qquad \textbf{(4)}$$

EXAMPLE 3 **Using Continuous Compounding**

The amount A that results from investing a principal P of \$1000 at an annual rate r of 10% compounded continuously for a time t of 1 year is

$$A = \$1000e^{0.10} = (\$1000)(1.10517) = \$1105.17$$

Now Work PROBLEM 13

2 Calculate Effective Rates of Return

Suppose that you have \$1000 and a bank offers to pay you 3% annual interest on a savings account with interest compounded monthly. What annual interest rate must be earned for you to have the same amount at the end of the year as if the interest had been compounded annually (once per year)? To answer this question, first determine the value of the \$1000 in the account that earns 3% compounded monthly.

$$A = \$1000\left(1 + \frac{0.03}{12}\right)^{12} \qquad \text{Use } A = P\left(1 + \frac{r}{n}\right)^n \text{ with } P = \$1000, r = 0.03, n = 12.$$
$$= \$1030.42$$

So the interest earned is \$30.42. Using $I = Prt$ with $t = 1$, $I = \$30.42$, and $P = \$1000$, the annual simple interest rate is $0.03042 = 3.042\%$. This interest rate is known as the *effective rate of interest*.

The **effective rate of interest** is the annual simple interest rate that would yield the same amount as compounding n times per year, or continuously, after 1 year.

THEOREM

Effective Rate of Interest

The effective rate of interest r_e of an investment earning an annual interest rate r is given by

$$\text{Compounding } n \text{ times per year: } r_e = \left(1 + \frac{r}{n}\right)^n - 1$$

$$\text{Continuous compounding: } r_e = e^r - 1$$

EXAMPLE 4 **Computing the Effective Rate of Interest—Which Is the Best Deal?**

Suppose you want to buy a 5-year certificate of deposit (CD). You visit three banks to determine their CD rates. American Express offers you 2.15% annual interest compounded monthly, and First Internet Bank offers you 2.20% compounded quarterly. Discover offers 2.12% compounded daily. Determine which bank is offering the best deal.

Solution The bank that offers the best deal is the one with the highest effective interest rate.

American Express	First Internet Bank	Discover
$r_e = \left(1 + \frac{0.0215}{12}\right)^{12} - 1$	$r_e = \left(1 + \frac{0.022}{4}\right)^{4} - 1$	$r_e = \left(1 + \frac{0.0212}{365}\right)^{365} - 1$
$\approx 1.02171 - 1$	$\approx 1.02218 - 1$	$\approx 1.02143 - 1$
$= 0.02171$	$= 0.02218$	$= 0.02143$
$= 2.171\%$	$= 2.218\%$	$= 2.143\%$

The effective rate of interest is highest for First Internet Bank, so First Internet Bank is offering the best deal. ●

Now Work PROBLEM 23

3 Determine the Present Value of a Lump Sum of Money

When people in finance speak of the "time value of money," they are usually referring to the *present value* of money. The **present value** of A dollars to be received at a future date is the principal that you would need to invest now so that it will grow to A dollars in the specified time period. The present value of money to be received at a future date is always less than the amount to be received, since the amount to be received will equal the present value (money invested now) *plus* the interest accrued over the time period.

The compound interest formula (2) is used to develop a formula for present value. If P is the present value of A dollars to be received after t years at a per annum interest rate r compounded n times per year, then, by formula (2),

$$A = P\cdot\left(1 + \frac{r}{n}\right)^{nt}$$

To solve for P, divide both sides by $\left(1 + \frac{r}{n}\right)^{nt}$. The result is

$$\frac{A}{\left(1 + \frac{r}{n}\right)^{nt}} = P \quad \text{or} \quad P = A\cdot\left(1 + \frac{r}{n}\right)^{-nt}$$

THEOREM

Present Value Formulas

The present value P of A dollars to be received after t years, assuming a per annum interest rate r compounded n times per year, is

$$P = A \cdot \left(1 + \frac{r}{n}\right)^{-nt} \quad (5)$$

If the interest is compounded continuously, then

$$P = Ae^{-rt} \quad (6)$$

To derive (6), solve formula (4) for P.

EXAMPLE 5 **Computing the Value of a Zero-Coupon Bond**

A zero-coupon (noninterest-bearing) bond can be redeemed in 10 years for \$1000. How much should you be willing to pay for it now if you want a return of

(a) 8% compounded monthly? (b) 7% compounded continuously?

Solution (a) To find the present value of \$1000, use formula (5) with $A = \$1000$, $n = 12$, $r = 0.08$, and $t = 10$.

$$P = A \cdot \left(1 + \frac{r}{n}\right)^{-nt} = \$1000\left(1 + \frac{0.08}{12}\right)^{-12(10)} = \$450.52$$

For a return of 8% compounded monthly, pay \$450.52 for the bond.

(b) Here use formula (6) with $A = \$1000$, $r = 0.07$, and $t = 10$.

$$P = Ae^{-rt} = \$1000e^{-(0.07)(10)} = \$496.59$$

For a return of 7% compounded continuously, pay \$496.59 for the bond. ●

Now Work PROBLEM 15

4 Determine the Rate of Interest or the Time Required to Double a Lump Sum of Money

EXAMPLE 6 **Rate of Interest Required to Double an Investment**

What annual rate of interest compounded annually is needed in order to double an investment in 5 years?

Solution If P is the principal and P is to double, then the amount A will be $2P$. Use the compound interest formula with $n = 1$ and $t = 5$ to find r.

$$A = P \cdot \left(1 + \frac{r}{n}\right)^{nt}$$

$$2P = P \cdot (1 + r)^5 \qquad A = 2P, n = 1, t = 5$$

$$2 = (1 + r)^5 \qquad \text{Divide both sides by } P.$$

$$1 + r = \sqrt[5]{2} \qquad \text{Take the fifth root of each side.}$$

$$r = \sqrt[5]{2} - 1 \approx 1.148698 - 1 = 0.148698$$

The annual rate of interest needed to double the principal in 5 years is 14.87%. ●

Now Work PROBLEM 31

EXAMPLE 7 **Time Required to Double or Triple an Investment**

(a) How long will it take for an investment to double in value if it earns 5% compounded continuously?

(b) How long will it take to triple at this rate?

Solution (a) If P is the initial investment and P is to double, then the amount A will be $2P$. Use formula (4) for continuously compounded interest with $r = 0.05$.

$$\begin{aligned} A &= Pe^{rt} \\ 2P &= Pe^{0.05t} && A = 2P, r = 0.05 \\ 2 &= e^{0.05t} && \text{Divide out the } P\text{'s.} \\ 0.05t &= \ln 2 && \text{Rewrite as a logarithm.} \\ t &= \frac{\ln 2}{0.05} \approx 13.86 && \text{Solve for } t. \end{aligned}$$

It will take about 14 years to double the investment.

(b) To triple the investment, let $A = 3P$ in formula (4).

$$\begin{aligned} A &= Pe^{rt} \\ 3P &= Pe^{0.05t} && A = 3P, r = 0.05 \\ 3 &= e^{0.05t} && \text{Divide out the } P\text{'s.} \\ 0.05t &= \ln 3 && \text{Rewrite as a logarithm.} \\ t &= \frac{\ln 3}{0.05} \approx 21.97 && \text{Solve for } t. \end{aligned}$$

It will take about 22 years to triple the investment.

Now Work PROBLEM 35

6.7 Assess Your Understanding

'Are You Prepared?' *Answers are given at the end of these exercises. If you get a wrong answer, read the pages listed in red.*

1. What is the interest due when \$500 is borrowed for 6 months at a simple interest rate of 6% per annum? (p. 136)

2. If you borrow \$5000 and, after 9 months, pay off the loan in the amount of \$5500, what per annum rate of interest was charged? (p. 136)

Concepts and Vocabulary

3. The total amount borrowed (whether by an individual from a bank in the form of a loan or by a bank from an individual in the form of a savings account) is called the ________.

4. If a principal of P dollars is borrowed for a period of t years at a per annum interest rate r, expressed as a decimal, the interest I charged is _____ = ______. Interest charged according to this formula is called ______ ______.

5. In problems involving interest, if the payment period of the interest is quarterly, then interest is paid _____ times per year.

6. The ________ ____ __ ______ is the annual simple interest rate that would yield the same amount as compounding n times per year, or continuously, after 1 year.

Skill Building

In Problems 7–14, find the amount that results from each investment.

7. \$100 invested at 4% compounded quarterly after a period of 2 years

8. \$50 invested at 6% compounded monthly after a period of 3 years

9. \$500 invested at 8% compounded quarterly after a period of $2\frac{1}{2}$ years

10. \$300 invested at 12% compounded monthly after a period of $1\frac{1}{2}$ years

11. \$600 invested at 5% compounded daily after a period of 3 years

12. \$700 invested at 6% compounded daily after a period of 2 years

13. \$1000 invested at 11% compounded continuously after a period of 2 years

14. \$400 invested at 7% compounded continuously after a period of 3 years

In Problems 15–22, find the principal needed now to get each amount; that is, find the present value.

15. To get \$100 after 2 years at 6% compounded monthly

16. To get \$75 after 3 years at 8% compounded quarterly

17. To get \$1000 after $2\frac{1}{2}$ years at 6% compounded daily

18. To get \$800 after $3\frac{1}{2}$ years at 7% compounded monthly

19. To get \$600 after 2 years at 4% compounded quarterly

20. To get \$300 after 4 years at 3% compounded daily

21. To get \$80 after $3\frac{1}{4}$ years at 9% compounded continuously

22. To get \$800 after $2\frac{1}{2}$ years at 8% compounded continuously

In Problems 23–26, find the effective rate of interest.

23. For 5% compounded quarterly

24. For 6% compounded monthly

25. For 5% compounded continuously

26. For 6% compounded continuously

In Problems 27–30, determine the rate that represents the better deal.

27. 6% compounded quarterly or $6\frac{1}{4}$% compounded annually

28. 9% compounded quarterly or $9\frac{1}{4}$% compounded annually

29. 9% compounded monthly or 8.8% compounded daily

30. 8% compounded semiannually or 7.9% compounded daily

31. What rate of interest compounded annually is required to double an investment in 3 years?

32. What rate of interest compounded annually is required to double an investment in 6 years?

33. What rate of interest compounded annually is required to triple an investment in 5 years?

34. What rate of interest compounded annually is required to triple an investment in 10 years?

35. (a) How long does it take for an investment to double in value if it is invested at 8% compounded monthly?
(b) How long does it take if the interest is compounded continuously?

36. (a) How long does it take for an investment to triple in value if it is invested at 6% compounded monthly?
(b) How long does it take if the interest is compounded continuously?

37. What rate of interest compounded quarterly will yield an effective interest rate of 7%?

38. What rate of interest compounded continuously will yield an effective interest rate of 6%?

Applications and Extensions

39. **Time Required to Reach a Goal** If Tanisha has \$100 to invest at 4% per annum compounded monthly, how long will it be before she has \$150? If the compounding is continuous, how long will it be?

40. **Time Required to Reach a Goal** If Angela has \$100 to invest at 2.5% per annum compounded monthly, how long will it be before she has \$175? If the compounding is continuous, how long will it be?

41. **Time Required to Reach a Goal** How many years will it take for an initial investment of \$10,000 to grow to \$25,000? Assume a rate of interest of 6% compounded continuously.

42. **Time Required to Reach a Goal** How many years will it take for an initial investment of \$25,000 to grow to \$80,000? Assume a rate of interest of 7% compounded continuously.

43. **Price Appreciation of Homes** What will a \$90,000 condominium cost 5 years from now if the price appreciation for condos over that period averages 3% compounded annually?

44. **Credit Card Interest** A department store charges 1.25% per month on the unpaid balance for customers with charge accounts (interest is compounded monthly). A customer charges \$200 and does not pay her bill for 6 months. What is the bill at that time?

45. **Saving for a Car** Jerome will be buying a used car for \$15,000 in 3 years. How much money should he ask his parents for now so that, if he invests it at 5% compounded continuously, he will have enough to buy the car?

46. **Paying off a Loan** John requires \$3000 in 6 months to pay off a loan that has no prepayment privileges. If he has the \$3000 now, how much of it should he save in an account paying 3% compounded monthly so that in 6 months he will have exactly \$3000?

47. **Return on a Stock** George contemplates the purchase of 100 shares of a stock selling for \$15 per share. The stock pays no dividends. The history of the stock indicates that it should grow at an annual rate of 15% per year. How much should the 100 shares of stock be worth in 5 years?

48. **Return on an Investment** A business purchased for \$650,000 in 2010 is sold in 2013 for \$850,000. What is the annual rate of return for this investment?

49. **Comparing Savings Plans** Jim places \$1000 in a bank account that pays 5.6% compounded continuously. After 1 year, will he have enough money to buy a computer system that costs \$1060? If another bank will pay Jim 5.9% compounded monthly, is this a better deal?

50. **Savings Plans** On January 1, Kim places \$1000 in a certificate of deposit that pays 6.8% compounded continuously and matures in 3 months. Then Kim places the \$1000 and the interest in a passbook account that pays 5.25% compounded monthly. How much does Kim have in the passbook account on May 1?

51. **Comparing IRA Investments** Will invests \$2000 of the money in his IRA in a bond trust that pays 9% interest compounded semiannually. His friend Henry invests \$2000 in his IRA in a certificate of deposit that pays $8\frac{1}{2}\%$ compounded continuously. Who has more money after 20 years, Will or Henry?

52. **Comparing Two Alternatives** Suppose that April has access to an investment that will pay 10% interest compounded continuously. Which is better: to be given \$1000 now so that she can take advantage of this investment opportunity or to be given \$1325 after 3 years?

53. **College Costs** The average annual cost of college at 4-year private colleges was \$30,094 in the 2013–2014 academic year. This was a 3.8% increase from the previous year.
    (a) If the cost of college increases by 3.8% each year, what will be the average cost of college at a 4-year private college for the 2033–2034 academic year?
    (b) College savings plans, such as a 529 plan, allow individuals to put money aside now to help pay for college later. If one such plan offers a rate of 2% compounded continuously, how much should be put in a college savings plan in 2015 to pay for 1 year of the cost of college at a 4-year private college for an incoming freshman in 2033?

    Source: *The College Board*

54. **Analyzing Interest Rates on a Mortgage** Colleen and Bill have just purchased a house for \$650,000, with the seller holding a second mortgage of \$100,000. They promise to pay the seller \$100,000 plus all accrued interest 5 years from now. The seller offers them three interest options on the second mortgage:
    (a) Simple interest at 6% per annum
    (b) 5.5% interest compounded monthly
    (c) 5.25% interest compounded continuously

    Which option is best? That is, which results in paying the least interest on the loan?

55. **2009 Federal Stimulus Package** In February 2009, President Obama signed into law a \$787 billion federal stimulus package. At that time, 20-year Series EE bonds had a fixed rate of 1.3% compounded semiannually. If the federal government financed the stimulus through EE bonds, how much would it have to pay back in 2029? How much interest was paid to finance the stimulus?

    Source: *U.S. Treasury Department*

56. **Per Capita Federal Debt** In 2014, the federal debt was about \$17.5 trillion. In 2014, the U.S. population was about 317 million. Assuming that the federal debt is increasing about 6.4% per year and the U.S. population is increasing about 0.7% per year, determine the per capita debt (total debt divided by population) in 2030.

Problems 57–62 require the following discussion. ***Inflation*** *is a term used to describe the erosion of the purchasing power of money. For example, if the annual inflation rate is 3%, then \$1000 worth of purchasing power now will have only \$970 worth of purchasing power in 1 year because 3% of the original \$1000 (0.03 × 1000 = 30) has been eroded due to inflation. In general, if the rate of inflation averages r% per annum over n years, the amount A that \$P will purchase after n years is*

$$A = P \cdot (1 - r)^n$$

where r is expressed as a decimal.

57. **Inflation** If the inflation rate averages 3%, what will be the purchasing power of \$1000 in 2 years?

58. **Inflation** If the inflation rate averages 2%, what will be the purchasing power of \$1000 in 3 years?

59. **Inflation** If the purchasing power of \$1000 is only \$950 after 2 years, what was the average inflation rate?

60. **Inflation** If the purchasing power of \$1000 is only \$930 after 2 years, what was the average inflation rate?

61. **Inflation** If the average inflation rate is 2%, how long is it until purchasing power is cut in half?

62. **Inflation** If the average inflation rate is 4%, how long is it until purchasing power is cut in half?

Problems 63–66 involve zero-coupon bonds. A ***zero-coupon bond*** *is a bond that is sold now at a discount and will pay its face value at the time when it matures; no interest payments are made.*

63. **Zero-Coupon Bonds** A zero-coupon bond can be redeemed in 20 years for \$10,000. How much should you be willing to pay for it now if you want a return of:
    (a) 5% compounded monthly?
    (b) 5% compounded continuously?

64. **Zero-Coupon Bonds** A child's grandparents are considering buying an \$80,000 face-value, zero-coupon bond at her birth so that she will have enough money for her college education 17 years later. If they want a rate of return of 6% compounded annually, what should they pay for the bond?

65. **Zero-Coupon Bonds** How much should a \$10,000 face-value, zero-coupon bond, maturing in 10 years, be sold for now if its rate of return is to be 4.5% compounded annually?

66. **Zero-Coupon Bonds** If Pat pays \$15,334.65 for a \$25,000 face-value, zero-coupon bond that matures in 8 years, what is his annual rate of return?

67. **Time to Double or Triple an Investment** The formula

$$t = \frac{\ln m}{n \ln\left(1 + \frac{r}{n}\right)}$$

can be used to find the number of years t required to multiply an investment m times when r is the per annum interest rate compounded n times a year.
    (a) How many years will it take to double the value of an IRA that compounds annually at the rate of 6%?
    (b) How many years will it take to triple the value of a savings account that compounds quarterly at an annual rate of 5%?
    (c) Give a derivation of this formula.

68. Time to Reach an Investment Goal The formula

$$t = \frac{\ln A - \ln P}{r}$$

can be used to find the number of years t required for an investment P to grow to a value A when compounded continuously at an annual rate r.

(a) How long will it take to increase an initial investment of \$1000 to \$4500 at an annual rate of 5.75%?
(b) What annual rate is required to increase the value of a \$2000 IRA to \$30,000 in 35 years?
(c) Give a derivation of this formula.

Problems 69–72 require the following discussion. The **consumer price index (CPI)** *indicates the relative change in price over time for a fixed basket of goods and services. It is a cost-of-living index that helps measure the effect of inflation on the cost of goods and services. The CPI uses the base period 1982–1984 for comparison (the CPI for this period is 100). The CPI for March 2014 was 236.29. This means that \$100 in the period 1982–1984 had the same purchasing power as \$236.29 in March 2014. In general, if the rate of inflation averages r% per annum over n years, then the CPI index after n years is*

$$\text{CPI} = \text{CPI}_0\left(1 + \frac{r}{100}\right)^n$$

where CPI_0 is the CPI index at the beginning of the n-year period.

Source: U.S. Bureau of Labor Statistics

69. Consumer Price Index
(a) The CPI was 215.3 for 2008 and 233.0 for 2013. Assuming that annual inflation remained constant for this time period, determine the average annual inflation rate.
(b) Using the inflation rate from part (a), in what year will the CPI reach 300?

70. Consumer Price Index If the current CPI is 234.2 and the average annual inflation rate is 2.8%, what will be the CPI in 5 years?

71. Consumer Price Index If the average annual inflation rate is 3.1%, how long will it take for the CPI index to double? (A doubling of the CPI index means purchasing power is cut in half.)

72. Consumer Price Index The base period for the CPI changed in 1998. Under the previous weight and item structure, the CPI for 1995 was 456.5. If the average annual inflation rate was 5.57%, what year was used as the base period for the CPI?

Explaining Concepts: Discussion and Writing

73. Explain in your own words what the term *compound interest* means. What does *continuous compounding* mean?

74. Explain in your own words the meaning of *present value.*

75. **Critical Thinking** You have just contracted to buy a house and will seek financing in the amount of \$100,000. You go to several banks. Bank 1 will lend you \$100,000 at the rate of 4.125% amortized over 30 years with a loan origination fee of 0.45%. Bank 2 will lend you \$100,000 at the rate of 3.375% amortized over 15 years with a loan origination fee of 0.95%. Bank 3 will lend you \$100,000 at the rate of 4.25% amortized over 30 years with no loan origination fee. Bank 4 will lend you \$100,000 at the rate of 3.625% amortized over 15 years with no loan origination fee. Which loan would you take? Why? Be sure to have sound reasons for your choice. Use the information in the table to assist you. If the amount of the monthly payment does not matter to you, which loan would you take? Again, have sound reasons for your choice. Compare your final decision with others in the class. Discuss.

	Monthly Payment	Loan Origination Fee
Bank 1	\$485	\$450
Bank 2	\$709	\$950
Bank 3	\$492	\$0
Bank 4	\$721	\$0

Retain Your Knowledge

Problems 76–79 are based on material learned earlier in the course. The purpose of these problems is to keep the material fresh in your mind so that you are better prepared for the final exam.

76. Find the remainder R when $f(x) = 6x^3 + 3x^2 + 2x - 11$ is divided by $g(x) = x - 1$. Is g a factor of f?

77. The function $f(x) = \dfrac{x}{x - 2}$ is one-to-one. Find f^{-1}.

78. Find the real zeros of

$$f(x) = x^5 - x^4 - 15x^3 - 21x^2 - 16x - 20.$$

Then write f in factored form.

79. Solve: $\log_2(x + 3) = 2\log_2(x - 3)$

'Are You Prepared?' Answers

1. \$15 **2.** $13\frac{1}{3}\%$

6.8 Exponential Growth and Decay Models; Newton's Law; Logistic Growth and Decay Models

OBJECTIVES
1 Find Equations of Populations That Obey the Law of Uninhibited Growth (p. 478)
2 Find Equations of Populations That Obey the Law of Decay (p. 480)
3 Use Newton's Law of Cooling (p. 481)
4 Use Logistic Models (p. 483)

1 Find Equations of Populations That Obey the Law of Uninhibited Growth

Many natural phenomena have been found to follow the law that an amount A varies with time t according to the function

$$A(t) = A_0 e^{kt} \quad (1)$$

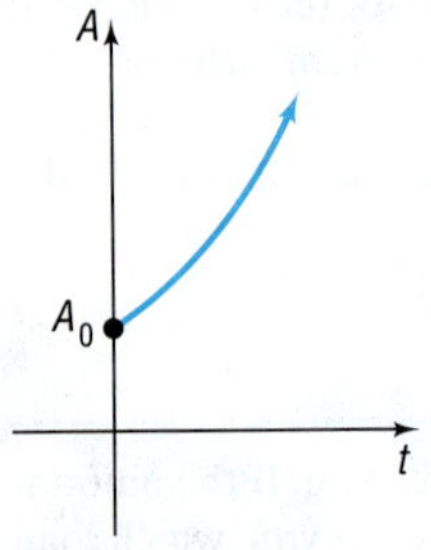

(a) $A(t) = A_0 e^{kt}$, $k > 0$
Exponential growth

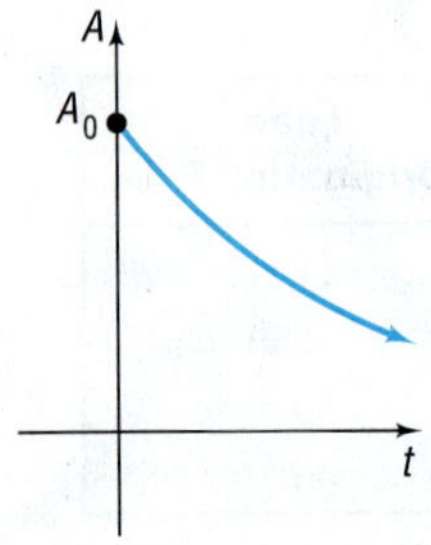

(b) $A(t) = A_0 e^{kt}$, $k < 0$
Exponential decay

Figure 41

Here A_0 is the original amount $(t = 0)$ and $k \neq 0$ is a constant.

If $k > 0$, then equation (1) states that the amount A is increasing over time; if $k < 0$, the amount A is decreasing over time. In either case, when an amount A varies over time according to equation (1), it is said to follow the **exponential law**, or the **law of uninhibited growth** $(k > 0)$ **or decay** $(k < 0)$. See Figure 41.

For example, in Section 6.7, continuously compounded interest was shown to follow the law of uninhibited growth. In this section we shall look at some additional phenomena that follow the exponential law.

Cell division is the growth process of many living organisms, such as amoebas, plants, and human skin cells. Based on an ideal situation in which no cells die and no by-products are produced, the number of cells present at a given time follows the law of uninhibited growth. Actually, however, after enough time has passed, growth at an exponential rate will cease as a consequence of factors such as lack of living space and dwindling food supply. The law of uninhibited growth accurately models only the early stages of the cell division process.

The cell division process begins with a culture containing N_0 cells. Each cell in the culture grows for a certain period of time and then divides into two identical cells. Assume that the time needed for each cell to divide in two is constant and does not change as the number of cells increases. These new cells then grow, and eventually each divides in two, and so on.

Uninhibited Growth of Cells

A model that gives the number N of cells in a culture after a time t has passed (in the early stages of growth) is

$$N(t) = N_0 e^{kt} \qquad k > 0 \quad (2)$$

where N_0 is the initial number of cells and k is a positive constant that represents the growth rate of the cells.

Using formula (2) to model the growth of cells employs a function that yields positive real numbers, even though the number of cells being counted must be an integer. This is a common practice in many applications.

EXAMPLE 1

Bacterial Growth

A colony of bacteria that grows according to the law of uninhibited growth is modeled by the function $N(t) = 100e^{0.045t}$, where N is measured in grams and t is measured in days.

(a) Determine the initial amount of bacteria.
(b) What is the growth rate of the bacteria?
(c) What is the population after 5 days?
(d) How long will it take for the population to reach 140 grams?
(e) What is the doubling time for the population?

Solution

(a) The initial amount of bacteria, N_0, is obtained when $t = 0$, so

$$N_0 = N(0) = 100e^{0.045(0)} = 100 \text{ grams}$$

(b) Compare $N(t) = 100e^{0.045t}$ to $N(t) = N_0 e^{kt}$. The value of k, 0.045, indicates a growth rate of 4.5%.

(c) The population after 5 days is $N(5) = 100e^{0.045(5)} \approx 125.2$ grams.

(d) To find how long it takes for the population to reach 140 grams, solve the equation $N(t) = 140$.

$$100e^{0.045t} = 140$$

$$e^{0.045t} = 1.4 \quad \text{Divide both sides of the equation by 100.}$$

$$0.045t = \ln 1.4 \quad \text{Rewrite as a logarithm.}$$

$$t = \frac{\ln 1.4}{0.045} \quad \text{Divide both sides of the equation by 0.045.}$$

$$\approx 7.5 \text{ days}$$

The population reaches 140 grams in about 7.5 days.

(e) The population doubles when $N(t) = 200$ grams, so the doubling time is found by solving the equation $200 = 100e^{0.045t}$ for t.

$$200 = 100e^{0.045t}$$

$$2 = e^{0.045t} \quad \text{Divide both sides of the equation by 100.}$$

$$\ln 2 = 0.045t \quad \text{Rewrite as a logarithm.}$$

$$t = \frac{\ln 2}{0.045} \quad \text{Divide both sides of the equation by 0.045.}$$

$$\approx 15.4 \text{ days}$$

The population doubles approximately every 15.4 days. ●

Now Work PROBLEM 1

EXAMPLE 2

Bacterial Growth

A colony of bacteria increases according to the law of uninhibited growth.

(a) If N is the number of cells and t is the time in hours, express N as a function of t.
(b) If the number of bacteria doubles in 3 hours, find the function that gives the number of cells in the culture.
(c) How long will it take for the size of the colony to triple?
(d) How long will it take for the population to double a second time (that is, to increase four times)?

Solution

(a) Using formula (2), the number N of cells at time t is

$$N(t) = N_0 e^{kt}$$

where N_0 is the initial number of bacteria present and k is a positive number.

(b) To find the growth rate k, note that the number of cells doubles in 3 hours, so

$$N(3) = 2N_0$$

But $N(3) = N_0 e^{k(3)}$, so

$$N_0 e^{k(3)} = 2N_0$$

$$e^{3k} = 2 \quad \text{Divide both sides by } N_0.$$

$$3k = \ln 2 \quad \text{Write the exponential equation as a logarithm.}$$

$$k = \frac{1}{3}\ln 2 \approx 0.23105$$

The function that models this growth process is therefore

$$N(t) = N_0\, e^{0.23105t}$$

(c) The time t needed for the size of the colony to triple requires that $N = 3N_0$. Substitute $3N_0$ for N to get

$$3N_0 = N_0\, e^{0.23105t}$$

$$3 = e^{0.23105t}$$

$$0.23105t = \ln 3$$

$$t = \frac{\ln 3}{0.23105} \approx 4.755 \text{ hours}$$

It will take about 4.755 hours, or 4 hours and 45 minutes, for the size of the colony to triple.

(d) If a population doubles in 3 hours, it will double a second time in 3 more hours, for a total time of 6 hours.

2 Find Equations of Populations That Obey the Law of Decay

Radioactive materials follow the law of uninhibited decay.

Uninhibited Radioactive Decay

The amount A of a radioactive material present at time t is given by

$$A(t) = A_0 e^{kt} \qquad k < 0 \qquad \textbf{(3)}$$

where A_0 is the original amount of radioactive material and k is a negative number that represents the rate of decay.

All radioactive substances have a specific **half-life,** which is the time required for half of the radioactive substance to decay. **Carbon dating** uses the fact that all living organisms contain two kinds of carbon, carbon-12 (a stable carbon) and carbon-14 (a radioactive carbon with a half-life of 5730 years). While an organism is living, the ratio of carbon-12 to carbon-14 is constant. But when an organism dies, the original amount of carbon-12 present remains unchanged, whereas the amount of carbon-14 begins to decrease. This change in the amount of carbon-14 present relative to the amount of carbon-12 present makes it possible to calculate when the organism died.

EXAMPLE 3 **Estimating the Age of Ancient Tools**

Traces of burned wood along with ancient stone tools in an archeological dig in Chile were found to contain approximately 1.67% of the original amount of carbon-14. If the half-life of carbon-14 is 5730 years, approximately when was the tree cut and burned?

Solution Using formula (3), the amount A of carbon-14 present at time t is

$$A(t) = A_0 e^{kt}$$

where A_0 is the original amount of carbon-14 present and k is a negative number. We first seek the number k. To find it, we use the fact that after 5730 years, half of the original amount of carbon-14 remains, so $A(5730) = \frac{1}{2}A_0$. Then

$$\frac{1}{2}A_0 = A_0 e^{k(5730)}$$

$$\frac{1}{2} = e^{5730k}$$ Divide both sides of the equation by A_0.

$$5730k = \ln\frac{1}{2}$$ Rewrite as a logarithm.

$$k = \frac{1}{5730}\ln\frac{1}{2} \approx -0.000120968$$

Formula (3) therefore becomes

$$A(t) = A_0\, e^{-0.000120968t}$$

If the amount A of carbon-14 now present is 1.67% of the original amount, it follows that

$$0.0167A_0 = A_0\, e^{-0.000120968t}$$

$$0.0167 = e^{-0.000120968t}$$ Divide both sides of the equation by A_0.

$$-0.000120968t = \ln 0.0167$$ Rewrite as a logarithm.

$$t = \frac{\ln 0.0167}{-0.000120968} \approx 33{,}830 \text{ years}$$

The tree was cut and burned about 33,830 years ago. Some archeologists use this conclusion to argue that humans lived in the Americas nearly 34,000 years ago, much earlier than is generally accepted. ●

Now Work PROBLEM 3

3 Use Newton's Law of Cooling

Newton's Law of Cooling* states that the temperature of a heated object decreases exponentially over time toward the temperature of the surrounding medium.

Newton's Law of Cooling

The temperature u of a heated object at a given time t can be modeled by the following function:

$$u(t) = T + (u_0 - T)e^{kt} \qquad k < 0 \tag{4}$$

where T is the constant temperature of the surrounding medium, u_0 is the initial temperature of the heated object, and k is a negative constant.

EXAMPLE 4 **Using Newton's Law of Cooling**

An object is heated to 100°C (degrees Celsius) and is then allowed to cool in a room whose air temperature is 30°C.

(a) If the temperature of the object is 80°C after 5 minutes, when will its temperature be 50°C?

(b) Determine the elapsed time before the temperature of the object is 35°C.

(c) What do you notice about the temperature as time passes?

*Named after Sir Isaac Newton (1643–1727), one of the cofounders of calculus.

Solution (a) Using formula (4) with $T = 30$ and $u_0 = 100$, the temperature $u(t)$ (in degrees Celsius) of the object at time t (in minutes) is

$$u(t) = 30 + (100 - 30)e^{kt} = 30 + 70e^{kt} \qquad \textbf{(5)}$$

where k is a negative constant. To find k, use the fact that $u = 80$ when $t = 5$. Then

$$\begin{aligned} u(t) &= 30 + 70e^{kt} \\ 80 &= 30 + 70e^{k(5)} && u(5) = 80 \\ 50 &= 70e^{5k} && \text{Simplify.} \\ e^{5k} &= \frac{50}{70} && \text{Solve for } e^{5k}. \\ 5k &= \ln\frac{5}{7} && \text{Rewrite as a logarithm.} \\ k &= \frac{1}{5}\ln\frac{5}{7} \approx -0.0673 && \text{Solve for } k. \end{aligned}$$

Formula (5) therefore becomes

$$u(t) = 30 + 70e^{-0.0673t} \qquad \textbf{(6)}$$

To find t when $u = 50°\text{C}$, solve the equation

$$\begin{aligned} 50 &= 30 + 70e^{-0.0673t} \\ 20 &= 70e^{-0.0673t} && \text{Simplify.} \\ e^{-0.0673t} &= \frac{20}{70} \\ -0.0673t &= \ln\frac{2}{7} && \text{Rewrite as a logarithm.} \\ t &= \frac{\ln\frac{2}{7}}{-0.0673} \approx 18.6 \text{ minutes} && \text{Solve for } t. \end{aligned}$$

The temperature of the object will be 50°C after about 18.6 minutes, or 18 minutes, 36 seconds.

(b) Use equation (6) to find t when $u = 35°\text{C}$.

$$\begin{aligned} 35 &= 30 + 70e^{-0.0673t} \\ 5 &= 70e^{-0.0673t} && \text{Simplify.} \\ e^{-0.0673t} &= \frac{5}{70} \\ -0.0673t &= \ln\frac{5}{70} && \text{Rewrite as a logarithm.} \\ t &= \frac{\ln\frac{5}{70}}{-0.0673} \approx 39.2 \text{ minutes} && \text{Solve for } t. \end{aligned}$$

The object will reach a temperature of 35°C after about 39.2 minutes.

(c) Look at equation (6). As t increases, the exponent $-0.0673t$ becomes unbounded in the negative direction. As a result, the value of $e^{-0.0673t}$ approaches zero, so the value of u, the temperature of the object, approaches 30°C, the air temperature of the room.

Now Work PROBLEM 13

4 Use Logistic Models

The exponential growth model $A(t) = A_0e^{kt}$, $k > 0$, assumes uninhibited growth, meaning that the value of the function grows without limit. Recall that cell division could be modeled using this function, assuming that no cells die and no by-products are produced. However, cell division eventually is limited by factors such as living space and food supply. The **logistic model**, given next, can describe situations where the growth or decay of the dependent variable is limited.

Logistic Model

In a logistic model, the population P after time t is given by the function

$$P(t) = \frac{c}{1 + ae^{-bt}} \qquad (7)$$

where a, b, and c are constants with $a > 0$ and $c > 0$. The model is a growth model if $b > 0$; the model is a decay model if $b < 0$.

The number c is called the **carrying capacity** (for growth models) because the value $P(t)$ approaches c as t approaches infinity; that is, $\lim_{t\to\infty} P(t) = c$. The number $|b|$ is the growth rate for $b > 0$ and the decay rate for $b < 0$. Figure 42(a) shows the graph of a typical logistic growth function, and Figure 42(b) shows the graph of a typical logistic decay function.

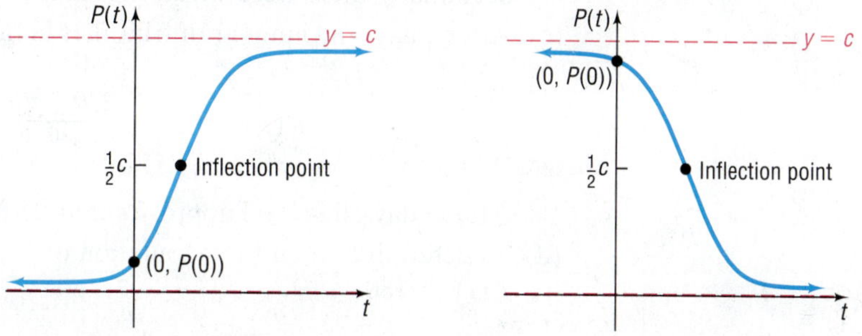

Figure 42

(a) $P(t) = \dfrac{c}{1 + ae^{-bt}}$, $b > 0$
Logistic growth

(b) $P(t) = \dfrac{c}{1 + ae^{-bt}}$, $b < 0$
Logistic decay

Based on the figures, the following properties of logistic functions emerge.

Properties of the Logistic Model, Equation (7)

1. The domain is the set of all real numbers. The range is the interval $(0, c)$, where c is the carrying capacity.
2. There are no x-intercepts; the y-intercept is $P(0)$.
3. There are two horizontal asymptotes: $y = 0$ and $y = c$.
4. $P(t)$ is an increasing function if $b > 0$ and a decreasing function if $b < 0$.
5. There is an **inflection point** where $P(t)$ equals $\frac{1}{2}$ of the carrying capacity. The inflection point is the point on the graph where the graph changes from being curved upward to being curved downward for growth functions, and the point where the graph changes from being curved downward to being curved upward for decay functions.
6. The graph is smooth and continuous, with no corners or gaps.

EXAMPLE 5 Fruit Fly Population

Fruit flies are placed in a half-pint milk bottle with a banana (for food) and yeast plants (for food and to provide a stimulus to lay eggs). Suppose that the fruit fly population after t days is given by

$$P(t) = \frac{230}{1 + 56.5e^{-0.37t}}$$

(a) State the carrying capacity and the growth rate.
(b) Determine the initial population.
(c) What is the population after 5 days?
(d) How long does it take for the population to reach 180?
(e) Use a graphing utility to determine how long it takes for the population to reach one-half of the carrying capacity.

Solution

(a) As $t \to \infty$, $e^{-0.37t} \to 0$ and $P(t) \to \frac{230}{1}$. The carrying capacity of the half-pint bottle is 230 fruit flies. The growth rate is $|b| = |0.37| = 37\%$ per day.

(b) To find the initial number of fruit flies in the half-pint bottle, evaluate $P(0)$.

$$P(0) = \frac{230}{1 + 56.5e^{-0.37(0)}} = \frac{230}{1 + 56.5} = 4$$

So, initially, there were 4 fruit flies in the half-pint bottle.

(c) After 5 days the number of fruit flies in the half-pint bottle is

$$P(5) = \frac{230}{1 + 56.5e^{-0.37(5)}} \approx 23 \text{ fruit flies}$$

After 5 days, there are approximately 23 fruit flies in the bottle.

(d) To determine when the population of fruit flies will be 180, solve the equation $P(t) = 180$.

$$\frac{230}{1 + 56.5e^{-0.37t}} = 180$$

$$230 = 180(1 + 56.5e^{-0.37t})$$

$$1.2778 = 1 + 56.5e^{-0.37t} \quad \text{Divide both sides by 180.}$$

$$0.2778 = 56.5e^{-0.37t} \quad \text{Subtract 1 from both sides.}$$

$$0.0049 = e^{-0.37t} \quad \text{Divide both sides by 56.5.}$$

$$\ln(0.0049) = -0.37t \quad \text{Rewrite as a logarithmic expression.}$$

$$t \approx 14.4 \text{ days} \quad \text{Divide both sides by } -0.37.$$

It will take approximately 14.4 days (14 days, 10 hours) for the population to reach 180 fruit flies.

(e) One-half of the carrying capacity is 115 fruit flies. Solve $P(t) = 115$ by graphing $Y_1 = \frac{230}{1 + 56.5e^{-0.37t}}$ and $Y_2 = 115$ and using INTERSECT. See Figure 43. The population will reach one-half of the carrying capacity in about 10.9 days (10 days, 22 hours).

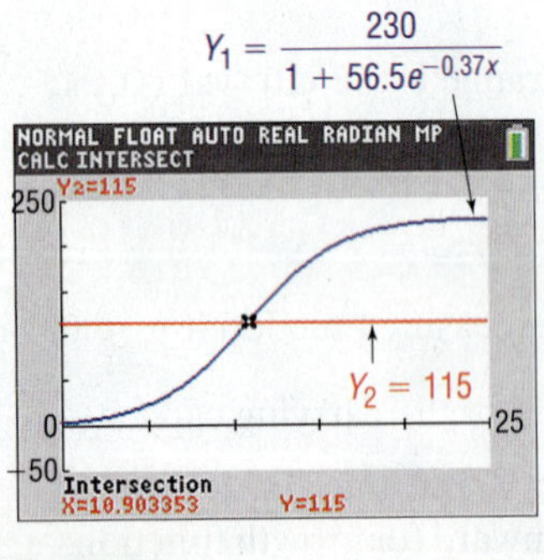

Figure 43

Look at Figure 43. Notice the point where the graph reaches 115 fruit flies (one-half of the carrying capacity): The graph changes from being curved upward to being curved downward. Using the language of calculus, we say the graph changes from increasing at an increasing rate to increasing at a decreasing rate. For any logistic growth function, when the population reaches one-half the carrying capacity, the population growth starts to slow down.

Now Work PROBLEM 23

Exploration

On the same viewing rectangle, graph

$$Y_1 = \frac{500}{1 + 24e^{-0.03t}} \quad \text{and} \quad Y_2 = \frac{500}{1 + 24e^{-0.08t}}$$

What effect does the growth rate $|b|$ have on the logistic growth function?

EXAMPLE 6 Wood Products

The EFISCEN wood product model classifies wood products according to their life-span. There are four classifications: short (1 year), medium short (4 years), medium long (16 years), and long (50 years). Based on data obtained from the European Forest Institute, the percentage of remaining wood products after t years for wood products with long life-spans (such as those used in the building industry) is given by

$$P(t) = \frac{100.3952}{1 + 0.0316e^{0.0581t}}$$

(a) What is the decay rate?
(b) What is the percentage of remaining wood products after 10 years?
(c) How long does it take for the percentage of remaining wood products to reach 50%?
(d) Explain why the numerator given in the model is reasonable.

Solution

(a) The decay rate is $|b| = |-0.0581| = 5.81\%$ per year.
(b) Evaluate $P(10)$.

$$P(10) = \frac{100.3952}{1 + 0.0316e^{0.0581(10)}} \approx 95.0$$

So 95% of long-life-span wood products remain after 10 years.

(c) Solve the equation $P(t) = 50$.

$$\frac{100.3952}{1 + 0.0316e^{0.0581t}} = 50$$

$$100.3952 = 50(1 + 0.0316e^{0.0581t})$$

$$2.0079 = 1 + 0.0316e^{0.0581t} \quad \text{Divide both sides by 50.}$$

$$1.0079 = 0.0316e^{0.0581t} \quad \text{Subtract 1 from both sides.}$$

$$31.8956 = e^{0.0581t} \quad \text{Divide both sides by 0.0316.}$$

$$\ln(31.8956) = 0.0581t \quad \text{Rewrite as a logarithmic expression.}$$

$$t \approx 59.6 \text{ years} \quad \text{Divide both sides by 0.0581.}$$

It will take approximately 59.6 years for the percentage of long-life-span wood products remaining to reach 50%.

(d) The numerator of 100.3952 is reasonable because the maximum percentage of wood products remaining that is possible is 100%.

6.8 Assess Your Understanding

Applications and Extensions

1. **Growth of an Insect Population** The size P of a certain insect population at time t (in days) obeys the law of uninhibited growth $P(t) = 500e^{0.02t}$.
   (a) Determine the number of insects at $t = 0$ days.
   (b) What is the growth rate of the insect population?
   (c) What is the population after 10 days?
   (d) When will the insect population reach 800?
   (e) When will the insect population double?

2. **Growth of Bacteria** The number N of bacteria present in a culture at time t (in hours) obeys the law of uninhibited growth $N(t) = 1000e^{0.01t}$.
   (a) Determine the number of bacteria at $t = 0$ hours.
   (b) What is the growth rate of the bacteria?
   (c) What is the population after 4 hours?
   (d) When will the number of bacteria reach 1700?
   (e) When will the number of bacteria double?

3. **Radioactive Decay** Strontium-90 is a radioactive material that decays according to the function $A(t) = A_0e^{-0.0244t}$, where A_0 is the initial amount present and A is the amount present at time t (in years). Assume that a scientist has a sample of 500 grams of strontium-90.
   (a) What is the decay rate of strontium-90?
   (b) How much strontium-90 is left after 10 years?
   (c) When will 400 grams of strontium-90 be left?
   (d) What is the half-life of strontium-90?

4. **Radioactive Decay** Iodine-131 is a radioactive material that decays according to the function $A(t) = A_0e^{-0.087t}$, where A_0 is the initial amount present and A is the amount present at time t (in days). Assume that a scientist has a sample of 100 grams of iodine-131.
   (a) What is the decay rate of iodine-131?
   (b) How much iodine-131 is left after 9 days?
   (c) When will 70 grams of iodine-131 be left?
   (d) What is the half-life of iodine-131?

5. **Growth of a Colony of Mosquitoes** The population of a colony of mosquitoes obeys the law of uninhibited growth.
   (a) If N is the population of the colony and t is the time in days, express N as a function of t.
   (b) If there are 1000 mosquitoes initially and there are 1800 after 1 day, what is the size of the colony after 3 days?
   (c) How long is it until there are 10,000 mosquitoes?

6. **Bacterial Growth** A culture of bacteria obeys the law of uninhibited growth.
   (a) If N is the number of bacteria in the culture and t is the time in hours, express N as a function of t.
   (b) If 500 bacteria are present initially and there are 800 after 1 hour, how many will be present in the culture after 5 hours?
   (c) How long is it until there are 20,000 bacteria?

7. **Population Growth** The population of a southern city follows the exponential law.
   (a) If N is the population of the city and t is the time in years, express N as a function of t.
   (b) If the population doubled in size over an 18-month period and the current population is 10,000, what will the population be 2 years from now?

8. **Population Decline** The population of a midwestern city follows the exponential law.
   (a) If N is the population of the city and t is the time in years, express N as a function of t.
   (b) If the population decreased from 900,000 to 800,000 from 2008 to 2010, what will the population be in 2012?

9. **Radioactive Decay** The half-life of radium is 1690 years. If 10 grams is present now, how much will be present in 50 years?

10. **Radioactive Decay** The half-life of radioactive potassium is 1.3 billion years. If 10 grams is present now, how much will be present in 100 years? In 1000 years?

11. **Estimating the Age of a Tree** A piece of charcoal is found to contain 30% of the carbon-14 that it originally had. When did the tree die from which the charcoal came? Use 5730 years as the half-life of carbon-14.

12. **Estimating the Age of a Fossil** A fossilized leaf contains 70% of its normal amount of carbon-14. How old is the fossil?

13. **Cooling Time of a Pizza Pan** A pizza pan is removed at 5:00 PM from an oven whose temperature is fixed at 450°F into a room that is a constant 70°F. After 5 minutes, the temperature of the pan is 300°F.
    (a) At what time is the temperature of the pan 135°F?
    (b) Determine the time that needs to elapse before the temperature of the pan is 160°F.
    (c) What do you notice about the temperature as time passes?

14. **Newton's Law of Cooling** A thermometer reading 72°F is placed in a refrigerator where the temperature is a constant 38°F.
    (a) If the thermometer reads 60°F after 2 minutes, what will it read after 7 minutes?
    (b) How long will it take before the thermometer reads 39°F?
    (c) Determine the time that must elapse before the thermometer reads 45°F.
    (d) What do you notice about the temperature as time passes?

15. **Newton's Law of Heating** A thermometer reading 8°C is brought into a room with a constant temperature of 35°C. If the thermometer reads 15°C after 3 minutes, what will it read after being in the room for 5 minutes? For 10 minutes?
    [**Hint:** You need to construct a formula similar to equation (4).]

16. **Warming Time of a Beer Stein** A beer stein has a temperature of 28°F. It is placed in a room with a constant temperature of 70°F. After 10 minutes, the temperature of the stein has risen to 35°F. What will the temperature of the stein be after 30 minutes? How long will it take the stein to reach a temperature of 45°F? (See the hint given for Problem 15.)

17. **Decomposition of Chlorine in a Pool** Under certain water conditions, the free chlorine (hypochlorous acid, HOCl) in a swimming pool decomposes according to the law of uninhibited decay. After shocking his pool, Ben tested the water and found the amount of free chlorine to be 2.5 parts per million (ppm). Twenty-four hours later, Ben tested the water again and found the amount of free chlorine to be 2.2 ppm. What will be the reading after 3 days (that is, 72 hours)? When the chlorine level reaches 1.0 ppm, Ben must shock the pool again. How long can Ben go before he must shock the pool again?

18. **Decomposition of Dinitrogen Pentoxide** At 45°C, dinitrogen pentoxide (N_2O_5) decomposes into nitrous dioxide (NO_2) and oxygen (O_2) according to the law of uninhibited decay. An initial amount of 0.25 M N_2O_5 (M is a measure of concentration known as molarity) decomposes to 0.15 M N_2O_5 in 17 minutes. What concentration of N_2O_5 will remain after 30 minutes? How long will it take until only 0.01 M N_2O_5 remains?

19. **Decomposition of Sucrose** Reacting with water in an acidic solution at 35°C, sucrose ($C_{12}H_{22}O_{11}$) decomposes into glucose ($C_6H_{12}O_6$) and fructose ($C_6H_{12}O_6$)* according to the law of uninhibited decay. An initial concentration of 0.40 M of sucrose decomposes to 0.36 M sucrose in 30 minutes. What concentration of sucrose will remain after 2 hours? How long will it take until only 0.10 M sucrose remains?

20. **Decomposition of Salt in Water** Salt (NaCl) decomposes in water into sodium (Na^+) and chloride (Cl^-) ions according to the law of uninhibited decay. If the initial amount of salt is 25 kilograms and, after 10 hours, 15 kilograms of salt is left, how much salt is left after 1 day? How long does it take until $\frac{1}{2}$ kilogram of salt is left?

21. **Radioactivity from Chernobyl** After the release of radioactive material into the atmosphere from a nuclear power plant at Chernobyl (Ukraine) in 1986, the hay in Austria was contaminated by iodine 131 (half-life 8 days). If it is safe to feed the hay to cows when 10% of the iodine 131 remains, how long did the farmers need to wait to use this hay?

22. **Word Users** According to a survey by Olsten Staffing Services, the percentage of companies reporting usage of Microsoft Word t years since 1984 is given by

$$P(t) = \frac{99.744}{1 + 3.014e^{-0.799t}}$$

(a) What is the growth rate in the percentage of Microsoft Word users?
(b) Use a graphing utility to graph $P = P(t)$.
(c) What was the percentage of Microsoft Word users in 1990?
(d) During what year did the percentage of Microsoft Word users reach 90%?
(e) Explain why the numerator given in the model is reasonable. What does it imply?

23. **Home Computers** The logistic model

$$P(t) = \frac{95.4993}{1 + 0.0405e^{0.1968t}}$$

represents the percentage of households that do not own a personal computer t years since 1984.
(a) Evaluate and interpret $P(0)$.
(b) Use a graphing utility to graph $P = P(t)$.
(c) What percentage of households did not own a personal computer in 1995?
(d) In what year did the percentage of households that do not own a personal computer reach 10%?

Source: *U.S. Department of Commerce*

24. **Farmers** The logistic model

$$W(t) = \frac{14{,}656{,}248}{1 + 0.059e^{0.057t}}$$

represents the number of farm workers in the United States t years after 1910.
(a) Evaluate and interpret $W(0)$.
(b) Use a graphing utility to graph $W = W(t)$.
(c) How many farm workers were there in the United States in 2010?
(d) When did the number of farm workers in the United States reach 10,000,000?
(e) According to this model, what happens to the number of farm workers in the United States as t approaches ∞? Based on this result, do you think that it is reasonable to use this model to predict the number of farm workers in the United States in 2060? Why?

Source: *U.S. Department of Agriculture*

25. **Birthdays** The logistic model

$$P(n) = \frac{113.3198}{1 + 0.115e^{0.0912n}}$$

models the probability that, in a room of n people, no two people share the same birthday.
(a) Use a graphing utility to graph $P = P(n)$.
(b) In a room of $n = 15$ people, what is the probability that no two share the same birthday?
(c) How many people must be in a room before the probability that no two people share the same birthday falls below 10%?
(d) What happens to the probability as n increases? Explain what this result means.

26. **Population of an Endangered Species** Environmentalists often capture an endangered species and transport the species to a controlled environment where the species can produce offspring and regenerate its population. Suppose that six American bald eagles are captured, transported to Montana, and set free. Based on experience, the environmentalists expect the population to grow according to the model

$$P(t) = \frac{500}{1 + 83.33e^{-0.162t}}$$

where t is measured in years. (continued on the next page)

*Author's Note: Surprisingly, the chemical formulas for glucose and fructose are the same: This is not a typo.

(a) Determine the carrying capacity of the environment.
(b) What is the growth rate of the bald eagle?
(c) What is the population after 3 years?
(d) When will the population be 300 eagles?
(e) How long does it take for the population to reach one-half of the carrying capacity?

27. The *Challenger* Disaster After the *Challenger* disaster in 1986, a study was made of the 23 launches that preceded the fatal flight. A mathematical model was developed involving the relationship between the Fahrenheit temperature x around the O-rings and the number y of eroded or leaky primary O-rings. The model stated that

$$y = \frac{6}{1 + e^{-(5.085 - 0.1156x)}}$$

where the number 6 indicates the 6 primary O-rings on the spacecraft.
(a) What is the predicted number of eroded or leaky primary O-rings at a temperature of 100°F?
(b) What is the predicted number of eroded or leaky primary O-rings at a temperature of 60°F?
(c) What is the predicted number of eroded or leaky primary O-rings at a temperature of 30°F?
(d) Graph the equation using a graphing utility. At what temperature is the predicted number of eroded or leaky O-rings 1? 3? 5?

Source: *Linda Tappin, "Analyzing Data Relating to the* Challenger *Disaster,"* Mathematics Teacher, *Vol. 87, No. 6, September 1994, pp. 423–426.*

Problems 28 and 29 use the following discussion: Uninhibited growth can be modeled by exponential functions other than $A(t) = A_0e^{kt}$. For example, if an initial population P_0 requires n units of time to double, then the function $P(t) = P_0 \cdot 2^{t/n}$ models the size of the population at time t. Likewise, a population requiring n units of time to triple can be modeled by $P(t) = P_0 \cdot 3^{t/n}$.

28. Growth of a Human Population The population of a town is growing exponentially.
(a) If its population doubled in size over an 8-year period and the current population is 25,000, write an exponential function of the form $P(t) = P_0 \cdot 2^{t/n}$ that models the population.
(b) What will the population be in 3 years?
(c) When will the population reach 80,000?
(d) Express the model from part (a) in the form $A(t) = A_0e^{kt}$.

29. Growth of an Insect Population An insect population grows exponentially.
(a) If the population triples in 20 days, and 50 insects are present initially, write an exponential function of the form $P(t) = P_0 \cdot 3^{t/n}$ that models the population.
(b) What will the population be in 47 days?
(c) When will the population reach 700?
(d) Express the model from part (a) in the form $A(t) = A_0e^{kt}$.

Retain Your Knowledge

Problems 30–33 are based on material learned earlier in the course. The purpose of these problems is to keep the material fresh in your mind so that you are better prepared for the final exam.

30. Find the equation of the linear function f that passes through the points (4, 1) and (8, −5).

31. Determine whether the graphs of the linear functions $f(x) = 5x - 1$ and $g(x) = \frac{1}{5}x + 1$ are parallel, perpendicular, or neither.

32. Write the logarithmic expression $\ln\left(\frac{x^2\sqrt{y}}{z}\right)$ as the sum and/or difference of logarithms. Express powers as factors.

33. Rationalize the denominator of $\frac{10}{\sqrt[3]{25}}$.

6.9 Building Exponential, Logarithmic, and Logistic Models from Data

PREPARING FOR THIS SECTION *Before getting started, review the following:*

- Building Linear Models from Data (Section 4.2, pp. 284–287)
- Building Cubic Models from Data (Section 5.1, pp. 336–337)
- Building Quadratic Models from Data (Section 4.4, pp. 306–307)

OBJECTIVES
1 Build an Exponential Model from Data (p. 489)
2 Build a Logarithmic Model from Data (p. 491)
3 Build a Logistic Model from Data (p. 491)

In Section 4.2 we discussed how to find the linear function of best fit $(y = ax + b)$, in Section 4.4 we discussed how to find the quadratic function of best fit $(y = ax^2 + bx + c)$, and in Section 5.1 we discussed how to find the cubic function of best fit $(y = ax^3 + bx^2 + cx + d)$.

In this section we discuss how to use a graphing utility to find equations of best fit that describe the relation between two variables when the relation is thought to be exponential $(y = ab^x)$, logarithmic $(y = a + b\ln x)$, or logistic $\left(y = \dfrac{c}{1 + ae^{-bx}}\right)$. As before, we draw a scatter diagram of the data to help to determine the appropriate model to use.

Figure 44 shows scatter diagrams that will typically be observed for the three models. Below each scatter diagram are any restrictions on the values of the parameters.

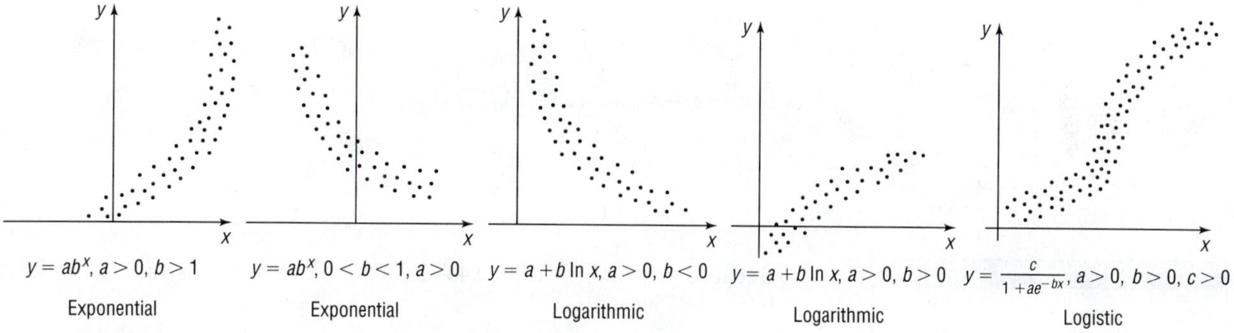

Figure 44

Most graphing utilities have REGression options that fit data to a specific type of curve. Once the data have been entered and a scatter diagram obtained, the type of curve that you want to fit to the data is selected. Then that REGression option is used to obtain the curve of *best fit* of the type selected.

The correlation coefficient r will appear only if the model can be written as a linear expression. As it turns out, r will appear for the linear, power, exponential, and logarithmic models, since these models can be written as a linear expression. Remember, the closer $|r|$ is to 1, the better the fit.

1 Build an Exponential Model from Data

We saw in Section 6.7 that the future value of money behaves exponentially, and we saw in Section 6.8 that growth and decay models also behave exponentially. The next example shows how data can lead to an exponential model.

EXAMPLE 1 Fitting an Exponential Function to Data

Mariah deposited \$20,000 in a well-diversified mutual fund 6 years ago. The data in Table 9 represent the value of the account at the beginning of each year for the last 7 years.

Table 9

Year, x	Account Value, y
0	20,000
1	21,516
2	23,355
3	24,885
4	27,484
5	30,053
6	32,622

(a) Using a graphing utility, draw a scatter diagram with year as the independent variable.

(b) Using a graphing utility, build an exponential model from the data.

(c) Express the function found in part (b) in the form $A = A_0 e^{kt}$.

(d) Graph the exponential function found in part (b) or (c) on the scatter diagram.

(e) Using the solution to part (b) or (c), predict the value of the account after 10 years.

(f) Interpret the value of k found in part (c).

Solution

(a) Enter the data into the graphing utility and draw the scatter diagram as shown in Figure 45.

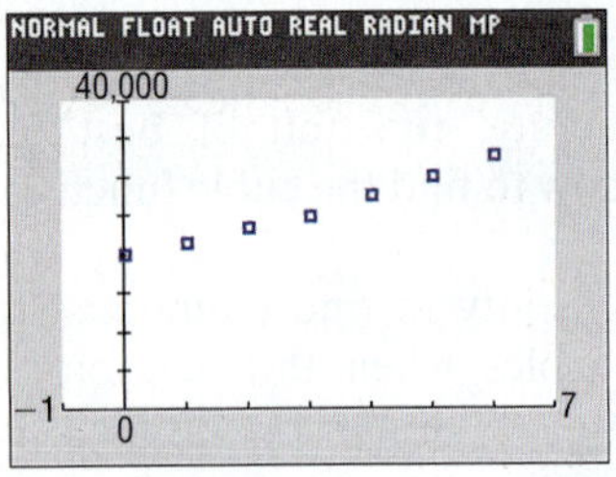

Figure 45

(b) A graphing utility fits the data in Table 9 to an exponential model of the form $y = ab^x$ using the EXPonential REGression option. Figure 46 shows that $y = ab^x = 19{,}820.43(1.085568)^x$. Notice that $|r| = 0.999$, which is close to 1, indicating a good fit.

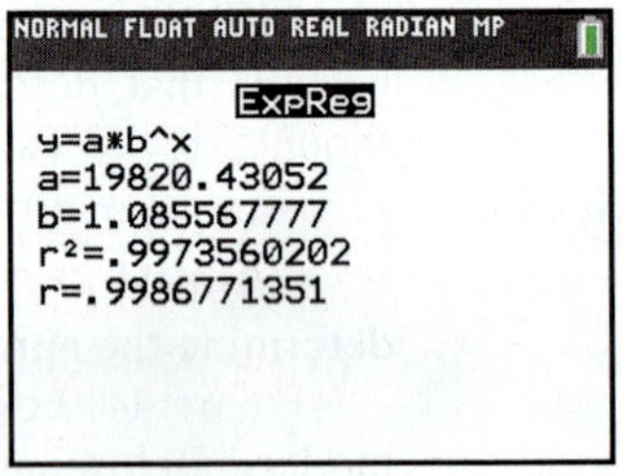

Figure 46

(c) To express $y = ab^x$ in the form $A = A_0 e^{kt}$, where $x = t$ and $y = A$, proceed as follows:

$$ab^x = A_0 e^{kt}$$

If $x = t = 0$, then $a = A_0$. This leads to

$$b^x = e^{kt}$$
$$b^x = (e^k)^t$$
$$b = e^k \qquad x = t$$

Because $y = ab^x = 19{,}820.43(1.085568)^x$, this means that $a = 19{,}820.43$ and $b = 1.085568$.

$$a = A_0 = 19{,}820.43 \quad \text{and} \quad b = e^k = 1.085568$$

To find k, rewrite $e^k = 1.085568$ as a logarithm to obtain

$$k = \ln(1.085568) \approx 0.08210$$

As a result, $A = A_0 e^{kt} = 19{,}820.43e^{0.08210t}$.

(d) See Figure 47 for the graph of the exponential function of best fit.

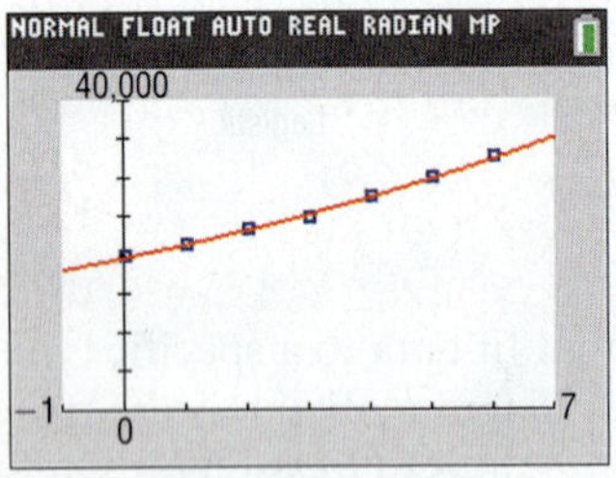

Figure 47

(e) Let $t = 10$ in the function found in part (c). The predicted value of the account after 10 years is

$$A = A_0 e^{kt} = 19{,}820.43e^{0.08210(10)} \approx \$45{,}047$$

(f) The value of $k = 0.08210 = 8.210\%$ represents the annual growth rate of the account. It represents the rate of interest earned, assuming the account is growing continuously. ●

Now Work PROBLEM 1

2 Build a Logarithmic Model from Data

Some relations between variables follow a logarithmic model.

EXAMPLE 2

Fitting a Logarithmic Function to Data

Jodi, a meteorologist, is interested in finding a function that explains the relation between the height of a weather balloon (in kilometers) and the atmospheric pressure (measured in millimeters of mercury) on the balloon. She collects the data shown in Table 10.

(a) Using a graphing utility, draw a scatter diagram of the data with atmospheric pressure as the independent variable.

(b) It is known that the relation between atmospheric pressure and height follows a logarithmic model. Using a graphing utility, build a logarithmic model from the data.

(c) Draw the logarithmic function found in part (b) on the scatter diagram.

(d) Use the function found in part (b) to predict the height of the weather balloon if the atmospheric pressure is 560 millimeters of mercury.

Table 10

Atmospheric Pressure, p	Height, h
760	0
740	0.184
725	0.328
700	0.565
650	1.079
630	1.291
600	1.634
580	1.862
550	2.235

Solution

(a) Enter the data into the graphing utility, and draw the scatter diagram. See Figure 48.

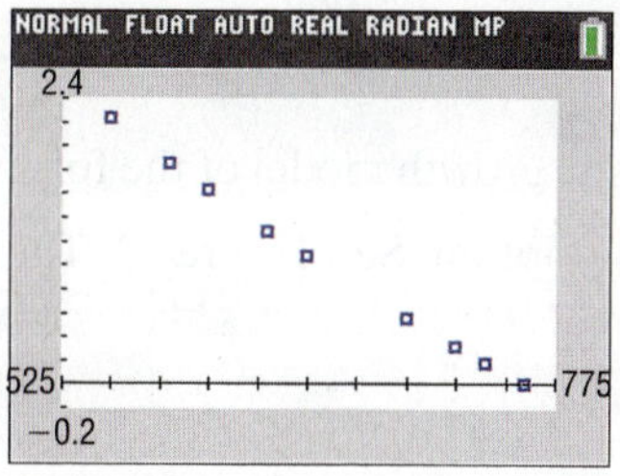

Figure 48

(b) A graphing utility fits the data in Table 10 to a logarithmic function of the form $y = a + b \ln x$ by using the LOGarithm REGression option. See Figure 49. The logarithmic model from the data is

$$h(p) = 45.7863 - 6.9025 \ln p$$

where h is the height of the weather balloon and p is the atmospheric pressure. Notice that $|r|$ is close to 1, indicating a good fit.

(c) Figure 50 shows the graph of $h(p) = 45.7863 - 6.9025 \ln p$ on the scatter diagram.

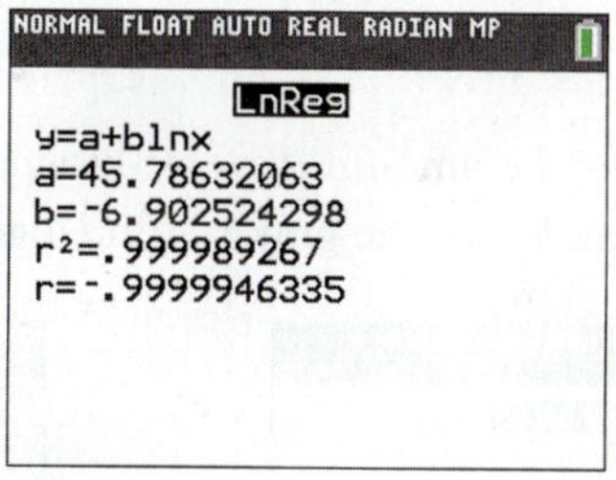

Figure 49

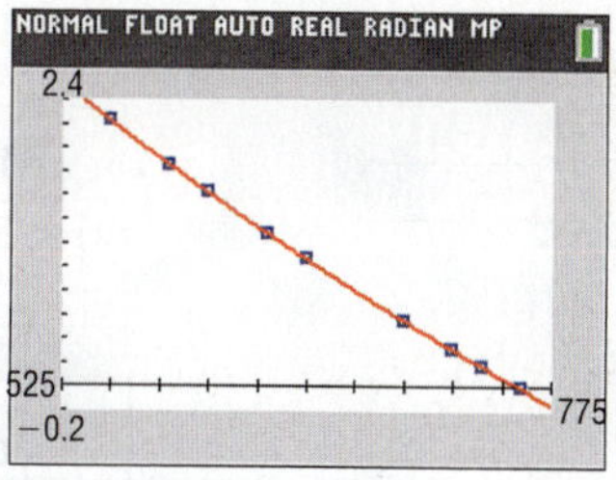

Figure 50

(d) Using the function found in part (b), Jodi predicts the height of the weather balloon when the atmospheric pressure is 560 to be

$$h(560) = 45.7863 - 6.9025 \ln 560$$

$$\approx 2.108 \text{ kilometers}$$

Now Work PROBLEM 5

3 Build a Logistic Model from Data

Logistic growth models can be used to model situations for which the value of the dependent variable is limited. Many real-world situations conform to this scenario. For example, the population of the human race is limited by the availability of natural resources such as food and shelter. When the value of the dependent variable is limited, a logistic growth model is often appropriate.

EXAMPLE 3 Fitting a Logistic Function to Data

The data in Table 11 represent the amount of yeast biomass in a culture after t hours.

Table 11

Time (hours)	Yeast Biomass	Time (hours)	Yeast Biomass	Time (hours)	Yeast Biomass
0	9.6	7	257.3	14	640.8
1	18.3	8	350.7	15	651.1
2	29.0	9	441.0	16	655.9
3	47.2	10	513.3	17	659.6
4	71.1	11	559.7	18	661.8
5	119.1	12	594.8		
6	174.6	13	629.4		

Source: Tor Carlson (Über Geschwindigkeit and Grösse der Hefevermehrung in Würze, Biochemische Zeitschrift, Bd. 57, pp. 313–334, 1913)

(a) Using a graphing utility, draw a scatter diagram of the data with time as the independent variable.

(b) Using a graphing utility, build a logistic model from the data.

(c) Using a graphing utility, graph the function found in part (b) on the scatter diagram.

(d) What is the predicted carrying capacity of the culture?

(e) Use the function found in part (b) to predict the population of the culture at $t = 19$ hours.

Solution

(a) See Figure 51 for a scatter diagram of the data.

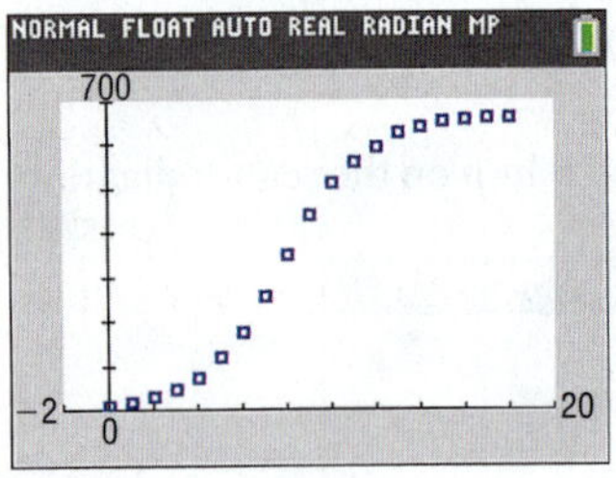

Figure 51

(b) A graphing utility fits the data in Table 11 to a logistic growth model of the form $y = \dfrac{c}{1 + ae^{-bx}}$ by using the LOGISTIC regression option. See Figure 52. The logistic model from the data is

$$y = \frac{663.0}{1 + 71.6e^{-0.5470x}}$$

where y is the amount of yeast biomass in the culture and x is the time.

(c) See Figure 53 for the graph of the logistic model.

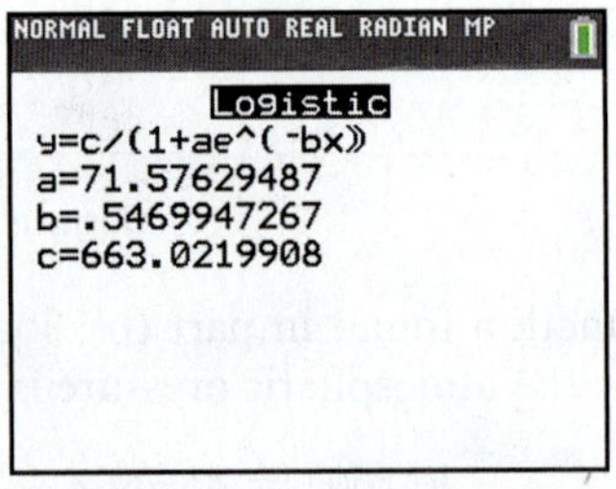

Figure 52

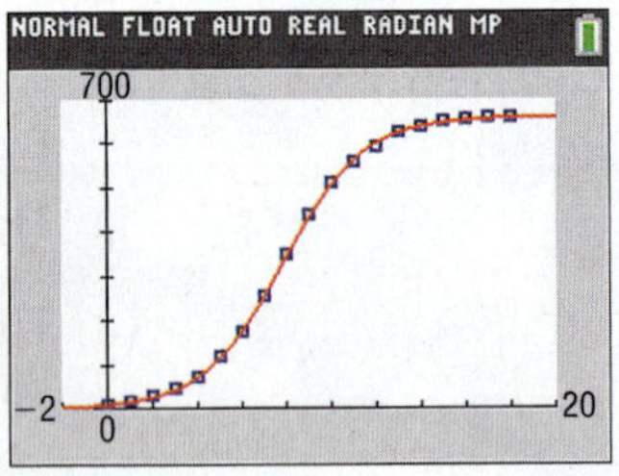

Figure 53

(d) Based on the logistic growth model found in part (b), the carrying capacity of the culture is 663.

(e) Using the logistic growth model found in part (b), the predicted amount of yeast biomass at $t = 19$ hours is

$$y = \frac{663.0}{1 + 71.6e^{-0.5470(19)}} \approx 661.5$$

Now Work PROBLEM 7

6.9 Assess Your Understanding

Applications and Extensions

1. **Biology** A strain of E-coli Beu 397-recA441 is placed into a nutrient broth at 30° Celsius and allowed to grow. The following data are collected. Theory states that the number of bacteria in the petri dish will initially grow according to the law of uninhibited growth. The population is measured using an optical device in which the amount of light that passes through the petri dish is measured.

Time (hours), x	Population, y
0	0.09
2.5	0.18
3.5	0.26
4.5	0.35
6	0.50

Source: Dr. Polly Lavery, Joliet Junior College

(a) Draw a scatter diagram treating time as the independent variable.
(b) Using a graphing utility, build an exponential model from the data.
(c) Express the function found in part (b) in the form $N(t) = N_0e^{kt}$.
(d) Graph the exponential function found in part (b) or (c) on the scatter diagram.
(e) Use the exponential function from part (b) or (c) to predict the population at $x = 7$ hours.
(f) Use the exponential function from part (b) or (c) to predict when the population will reach 0.75.

2. **Ethanol Production** The data in the table below represent ethanol production (in billions of gallons) in the United States from 2000 to 2013.

Year	Ethanol Produced (billion gallons)	Year	Ethanol Produced (billion gallons)
2000 ($x = 0$)	1.6	2007 ($x = 7$)	6.5
2001 ($x = 1$)	1.8	2008 ($x = 8$)	9.3
2002 ($x = 2$)	2.1	2009 ($x = 9$)	10.9
2003 ($x = 3$)	2.8	2010 ($x = 10$)	13.3
2004 ($x = 4$)	3.4	2011 ($x = 11$)	13.9
2005 ($x = 5$)	3.9	2012 ($x = 12$)	13.2
2006 ($x = 6$)	4.9	2013 ($x = 13$)	13.3

Source: Renewable Fuels Association, 2014

(a) Using a graphing utility, draw a scatter diagram of the data using 0 for 2000, 1 for 2001, and so on, as the independent variable.
(b) Using a graphing utility, build an exponential model from the data.
(c) Express the function found in part (b) in the form $A(t) = A_0e^{kt}$.
(d) Graph the exponential function found in part (b) or (c) on the scatter diagram
(e) Use the model to predict the amount of ethanol that will be produced in 2015.
(f) Interpret the meaning of k in the function found in part (c).

3. **Advanced-Stage Breast Cancer** The data in the table below represents the percentage of patients who have survived after diagnosis of advanced-stage breast cancer at 6-month intervals of time.

Time after Diagnosis (years)	Percentage Surviving
0.5	95.7
1	83.6
1.5	74.0
2	58.6
2.5	47.4
3	41.9
3.5	33.6

Source: Cancer Treatment Centers of America

(a) Using a graphing utility, draw a scatter diagram of the data with time after diagnosis as the independent variable.
(b) Using a graphing utility, build an exponential model from the data.
(c) Express the function found in part (b) in the form $A(t) = A_0e^{kt}$.
(d) Graph the exponential function found in part (b) or (c) on the scatter diagram.
(e) What percentage of patients diagnosed with advanced-stage cancer are expected to survive for 4 years after initial diagnosis?
(f) Interpret the meaning of k in the function found in part (c).

4. **Chemistry** A chemist has a 100-gram sample of a radioactive material. He records the amount of radioactive material every week for 7 weeks and obtains the following data: (continued on the next page)

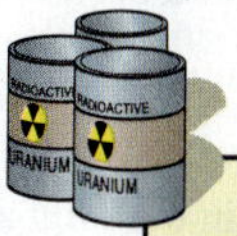

Week	Weight (in grams)
0	100.0
1	88.3
2	75.9
3	69.4
4	59.1
5	51.8
6	45.5

(a) Using a graphing utility, draw a scatter diagram with week as the independent variable.
(b) Using a graphing utility, build an exponential model from the data.
(c) Express the function found in part (b) in the form $A(t) = A_0e^{kt}$.
(d) Graph the exponential function found in part (b) or (c) on the scatter diagram.
(e) From the result found in part (b), determine the half-life of the radioactive material.
(f) How much radioactive material will be left after 50 weeks?
(g) When will there be 20 grams of radioactive material?

5. **Milk Production** The data in the table below represent the number of dairy farms (in thousands) and the amount of milk produced (in billions of pounds) in the United States for various years.

Year	Dairy Farms (thousands)	Milk Produced (billion pounds)
1980	334	128
1985	269	143
1990	193	148
1995	140	155
2000	105	167
2005	78	177
2010	63	193

Source: *Statistical Abstract of the United States*, 2012

(a) Using a graphing utility, draw a scatter diagram of the data with the number of dairy farms as the independent variable.
(b) Using a graphing utility, build a logarithmic model from the data.
(c) Graph the logarithmic function found in part (b) on the scatter diagram.
(d) In 2008, there were 67 thousand dairy farms in the United States. Use the function in part (b) to predict the amount of milk produced in 2008.
(e) The actual amount of milk produced in 2008 was 190 billion pounds. How does your prediction in part (d) compare to this?

6. **Cable Rates** The data (top, right) represent the average monthly rate charged for expanded basic cable television in the United States from 1995 to 2012. A market researcher believes that external factors, such as the growth of satellite television and internet programming, have affected the cost of basic cable. She is interested in building a model that will describe the average monthly cost of basic cable.
(a) Using a graphing utility, draw a scatter diagram of the data using 0 for 1995, 1 for 1996, and so on, as the independent variable and average monthly rate as the dependent variable.
(b) Using a graphing utility, build a logistic model from the data.
(c) Graph the logistic function found in part (b) on the scatter diagram.
(d) Based on the model found in part (b), what is the maximum possible average monthly rate for basic cable?
(e) Use the model to predict the average rate for basic cable in 2017.

Year	Average Monthly Rate (dollars)
1995 ($x = 0$)	22.35
1996 ($x = 1$)	24.28
1997 ($x = 2$)	26.31
1998 ($x = 3$)	27.88
1999 ($x = 4$)	28.94
2000 ($x = 5$)	31.22
2001 ($x = 6$)	33.75
2002 ($x = 7$)	36.47
2003 ($x = 8$)	38.95
2004 ($x = 9$)	41.04
2005 ($x = 10$)	43.04
2006 ($x = 11$)	45.26
2007 ($x = 12$)	47.27
2008 ($x = 13$)	49.65
2009 ($x = 14$)	52.37
2010 ($x = 15$)	54.44
2011 ($x = 16$)	57.46
2012 ($x = 17$)	61.63

Source: Federal Communications Commission, 2013

7. **Population Model** The following data represent the population of the United States. An ecologist is interested in building a model that describes the population of the United States.

Year	Population
1900	76,212,168
1910	92,228,496
1920	106,021,537
1930	123,202,624
1940	132,164,569
1950	151,325,798
1960	179,323,175
1970	203,302,031
1980	226,542,203
1990	248,709,873
2000	281,421,906
2010	308,745,538

Source: U.S. Census Bureau

(a) Using a graphing utility, draw a scatter diagram of the data using years since 1900 as the independent variable and population as the dependent variable.
(b) Using a graphing utility, build a logistic model from the data.
(c) Using a graphing utility, draw the function found in part (b) on the scatter diagram.
(d) Based on the function found in part (b), what is the carrying capacity of the United States?
(e) Use the function found in part (b) to predict the population of the United States in 2012.
(f) When will the United States population be 350,000,000?
(g) Compare actual U.S. Census figures to the predictions found in parts (e) and (f). Discuss any differences.

8. **Population Model** The data on the right represent the world population. An ecologist is interested in building a model that describes the world population.

(a) Using a graphing utility, draw a scatter diagram of the data using years since 2000 as the independent variable and population as the dependent variable.
(b) Using a graphing utility, build a logistic model from the data.
(c) Using a graphing utility, draw the function found in part (b) on the scatter diagram.
(d) Based on the function found in part (b), what is the carrying capacity of the world?
(e) Use the function found in part (b) to predict the populationof the world in 2020.
(f) When will world population be 10 billion?

Year	Population (billions)	Year	Population (billions)
2001	6.17	2008	6.71
2002	6.24	2009	6.79
2003	6.32	2010	6.86
2004	6.40	2011	6.94
2005	6.47	2012	7.02
2006	6.55	2013	7.10
2007	6.63		

Source: U.S. Census Bureau

9. **Cell Phone Towers** The following data represent the number of cell sites in service in the United States from 1985 to 2012 at the end of June each year.

Year	Cell Sites (thousands)	Year	Cell Sites (thousands)	Year	Cell Sites (thousands)
1985 ($x = 1$)	0.6	1995 ($x = 11$)	19.8	2004 ($x = 20$)	174.4
1986 ($x = 2$)	1.2	1996 ($x = 12$)	24.8	2005 ($x = 21$)	178.0
1987 ($x = 3$)	1.7	1997 ($x = 13$)	38.7	2006 ($x = 22$)	197.6
1988 ($x = 4$)	2.8	1998 ($x = 14$)	57.7	2007 ($x = 23$)	210.4
1989 ($x = 5$)	3.6	1999 ($x = 15$)	74.2	2008 ($x = 24$)	220.5
1990 ($x = 6$)	4.8	2000 ($x = 16$)	95.7	2009 ($x = 25$)	245.9
1991 ($x = 7$)	6.7	2001 ($x = 17$)	114.1	2010 ($x = 26$)	251.6
1992 ($x = 8$)	8.9	2002 ($x = 18$)	131.4	2011 ($x = 27$)	256.9
1993 ($x = 9$)	11.6	2003 ($x = 19$)	147.7	2012 ($x = 28$)	285.6
1994 ($x = 10$)	14.7				

Source: ©2013 CTIA-The Wireless Association®. All Rights Reserved.

(a) Using a graphing utility, draw a scatter diagram of the data using 1 for 1985, 2 for 1986, and so on, as the independent variable and number of cell sites as the dependent variable.
(b) Using a graphing utility, build a logistic model from the data.
(c) Graph the logistic function found in part (b) on the scatter diagram.
(d) What is the predicted carrying capacity for cell sites in the United States?
(e) Use the model to predict the number of cell sites in the United States at the end of June 2017.

Mixed Practice

10. **Age versus Total Cholesterol** The following data represent the age and average total cholesterol for adult males at various ages.

Age	Total Cholesterol
27	189
40	205
50	215
60	210
70	210
80	194

(a) Using a graphing utility, draw a scatter diagram of the data using age, x, as the independent variable and total cholesterol, y, as the dependent variable.
(b) Based on the scatter diagram drawn in part (a), decide on a model (linear, quadratic, cubic, exponential, logarithmic, or logistic) that you think best describes the relation between age and total cholesterol. Be sure to justify your choice of model.
(c) Using a graphing utility, find the model of best fit.
(d) Using a graphing utility, draw the model of best fit on the scatter diagram drawn in part (a).
(e) Use your model to predict the total cholesterol of a 35-year-old male.

11. Golfing The data below represent the expected percentage of putts that will be made by professional golfers on the PGA Tour depending on distance. For example, it is expected that 99.3% of 2-foot putts will be made.

Distance (feet)	Expected Percentage	Distance (feet)	Expected Percentage
2	99.3	14	25.0
3	94.8	15	22.0
4	85.8	16	20.0
5	74.7	17	19.0
6	64.7	18	17.0
7	55.6	19	16.0
8	48.5	20	14.0
9	43.4	21	13.0
10	38.3	22	12.0
11	34.2	23	11.0
12	30.1	24	11.0
13	27.0	25	10.0

Source: TheSandTrap.com

(a) Using a graphing utility, draw a scatter diagram of the data with distance as the independent variable.
(b) Based on the scatter diagram drawn in part (a), decide on a model (linear, quadratic, cubic, exponential, logarithmic, or logistic) that you think best describes the relation between distance and expected percentage. Be sure to justify your choice of model.
(c) Using a graphing utility, find the model of best fit.
(d) Graph the function found in part (c) on the scatter diagram.
(e) Use the function found in part (c) to predict what percentage of 30-foot putts will be made.

12. Income versus Crime Rate The following data represent property crime rate against individuals (crimes per 1000 households) and their household income (in dollars) in the United States in 2009.

Income Level	Property Crime Rate
5000	201.1
11,250	157.0
20,000	141.6
30,000	134.1
42,500	139.7
62,500	120.0

Source: Statistical Abstract of the United States, 2012

(a) Using a graphing utility, draw a scatter diagram of the data using income, x, as the independent variable and crime rate, y, as the dependent variable.
(b) Based on the scatter diagram drawn in part (a), decide on a model (linear, quadratic, cubic, exponential, logarithmic, or logistic) that you think best describes the relation between income and crime rate. Be sure to justify your choice of model.
(c) Using a graphing utility, find the model of best fit.
(d) Using a graphing utility, draw the model of best fit on the scatter diagram you drew in part (a).
(e) Use your model to predict the crime rate of a household whose income is $55,000.

Retain Your Knowledge

Problems 13–16 are based on material learned earlier in the course. The purpose of these problems is to keep the material fresh in your mind so that you are better prepared for the final exam.

13. Construct a polynomial function that might have the graph shown. (More than one answer is possible.)

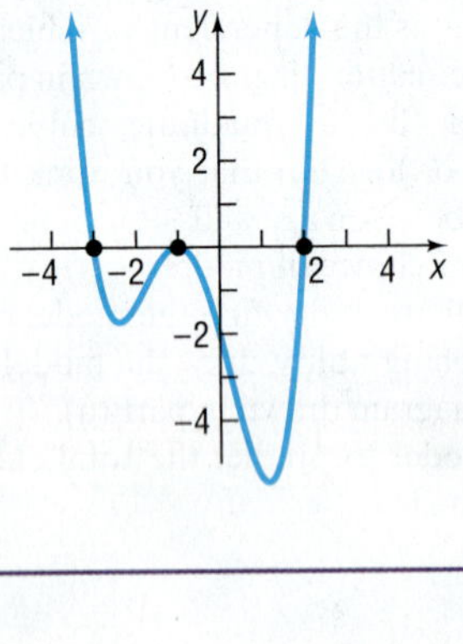

14. Rationalize the denominator of $\dfrac{3}{\sqrt{2}}$.

15. Use the Pythagorean Theorem to find the exact length of the unlabeled side in the given right triangle.

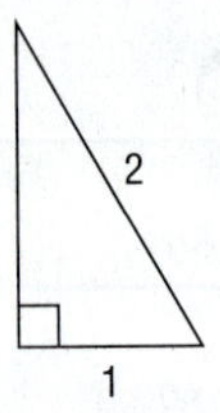

16. Graph the equation $(x - 3)^2 + y^2 = 25$.

Chapter Review

Things to Know

Composite function (p. 403)
$(f \circ g)(x) = f(g(x))$ The domain of $f \circ g$ is the set of all numbers x in the domain of g for which $g(x)$ is in the domain of f.

One-to-one function f (p. 411)
A function for which any two different inputs in the domain correspond to two different outputs in the range
For any choice of elements x_1, x_2 in the domain of f, if $x_1 \neq x_2$, then $f(x_1) \neq f(x_2)$.

Horizontal-line test (p. 412)
If every horizontal line intersects the graph of a function f in at most one point, f is one-to-one.

Inverse function f^{-1} of f (pp. 413–416)
Domain of f = range of f^{-1}; range of f = domain of f^{-1}
$f^{-1}(f(x)) = x$ for all x in the domain of f
$f(f^{-1}(x)) = x$ for all x in the domain of f^{-1}
The graphs of f and f^{-1} are symmetric with respect to the line $y = x$.

Properties of the exponential function (pp. 425, 428, 430)

$f(x) = Ca^x \quad a > 1, C > 0$
- Domain: the interval $(-\infty, \infty)$
- Range: the interval $(0, \infty)$
- x-intercepts: none; y-intercept: C
- Horizontal asymptote: x-axis $(y = 0)$ as $x \to -\infty$
- Increasing; one-to-one; smooth; continuous
- See Figure 21 for a typical graph.

$f(x) = Ca^x \quad 0 < a < 1, C > 0$
- Domain: the interval $(-\infty, \infty)$
- Range: the interval $(0, \infty)$
- x-intercepts: none; y-intercept: C
- Horizontal asymptote: x-axis $(y = 0)$ as $x \to \infty$
- Decreasing; one-to-one; smooth; continuous
- See Figure 25 for a typical graph.

Number e (p. 431)
Number approached by the expression $\left(1 + \frac{1}{n}\right)^n$ as $n \to \infty$; that is, $\lim_{n\to\infty}\left(1 + \frac{1}{n}\right)^n = e$.

Property of exponents (p. 432)
If $a^u = a^v$, then $u = v$.

Properties of the logarithmic function (pp. 440–442)

$f(x) = \log_a x \quad a > 1$
$(y = \log_a x$ means $x = a^y)$
- Domain: the interval $(0, \infty)$
- Range: the interval $(-\infty, \infty)$
- x-intercept: 1; y-intercept: none
- Vertical asymptote: $x = 0$ (y-axis)
- Increasing; one-to-one; smooth; continuous
- See Figure 39(a) for a typical graph.

$f(x) = \log_a x \quad 0 < a < 1$
$(y = \log_a x$ means $x = a^y)$
- Domain: the interval $(0, \infty)$
- Range: the interval $(-\infty, \infty)$
- x-intercept: 1; y-intercept: none
- Vertical asymptote: $x = 0$ (y-axis)
- Decreasing; one-to-one; smooth; continuous
- See Figure 39(b) for a typical graph.

Natural logarithm (p. 443)
$y = \ln x$ means $x = e^y$.

Properties of logarithms (pp. 453–454, 456)
$\log_a 1 = 0 \qquad \log_a a = 1 \qquad a^{\log_a M} = M \qquad \log_a a^r = r \qquad a^r = e^{r \ln a}$

$\log_a(MN) = \log_a M + \log_a N \qquad \log_a\left(\frac{M}{N}\right) = \log_a M - \log_a N$
$\log_a M^r = r \log_a M$
If $M = N$, then $\log_a M = \log_a N$.
If $\log_a M = \log_a N$, then $M = N$.

Formulas

Change-of-Base Formula (p. 457)	$\log_a M = \dfrac{\log_b M}{\log_b a}$
Compound Interest Formula (p. 469)	$A = P \cdot \left(1 + \dfrac{r}{n}\right)^{nt}$
Continuous compounding (p. 471)	$A = Pe^{rt}$
Effective rate of interest (p. 472)	Compounding n times per year: $r_e = \left(1 + \dfrac{r}{n}\right)^n - 1$ Continuous compounding: $r_e = e^r - 1$
Present Value Formulas (p. 473)	$P = A \cdot \left(1 + \dfrac{r}{n}\right)^{-nt}$ or $P = Ae^{-rt}$
Uninhibited Growth and decay (pp. 478, 480)	$A(t) = A_0 e^{kt}$
Newton's Law of Cooling (p. 481)	$u(t) = T + (u_0 - T)e^{kt} \quad k < 0$
Logistic model (p. 483)	$P(t) = \dfrac{c}{1 + ae^{-bt}}$

Objectives

Section		You should be able to . . .	Example(s)	Review Exercises
6.1	1	Form a composite function (p. 403)	1, 2, 4, 5	1–6
	2	Find the domain of a composite function (p. 404)	2–4	4–6
6.2	1	Determine whether a function is one-to-one (p. 411)	1, 2	7(a), 8
	2	Determine the inverse of a function defined by a map or a set of ordered pairs (p. 413)	3, 4	7(b)
	3	Obtain the graph of the inverse function from the graph of the function (p. 416)	7	8
	4	Find the inverse of a function defined by an equation (p. 417)	8, 9, 10	9–12
6.3	1	Evaluate exponential functions (p. 423)	1	13(a), (c), 46(a)
	2	Graph exponential functions (p. 427)	3–6	30–32
	3	Define the number e (p. 430)	p. 431	
	4	Solve exponential equations (p. 432)	7, 8	34, 35, 38, 40, 46(b)
6.4	1	Change exponential statements to logarithmic statements and logarithmic statements to exponential statements (p. 440)	2, 3	14, 15
	2	Evaluate logarithmic expressions (p. 441)	4	13(b), (d), 18, 45(b), 47(a), 48
	3	Determine the domain of a logarithmic function (p. 441)	5	16, 17, 33(a)
	4	Graph logarithmic functions (p. 442)	6, 7	29, 33(b), 45(a)
	5	Solve logarithmic equations (p. 446)	8, 9	36, 39, 45(c), 47(b)
6.5	1	Work with the properties of logarithms (p. 452)	1, 2	19, 20
	2	Write a logarithmic expression as a sum or difference of logarithms (p. 454)	3–5	21–24
	3	Write a logarithmic expression as a single logarithm (p. 455)	6	25–27
	4	Evaluate logarithms whose base is neither 10 nor e (p. 457)	7, 8	28
6.6	1	Solve logarithmic equations (p. 461)	1–3	36, 42
	2	Solve exponential equations (p. 463)	4–6	37, 41, 43, 44
	3	Solve logarithmic and exponential equations using a graphing utility (p. 464)	7	34–44
6.7	1	Determine the future value of a lump sum of money (p. 468)	1–3	49
	2	Calculate effective rates of return (p. 471)	4	49
	3	Determine the present value of a lump sum of money (p. 472)	5	50
	4	Determine the rate of interest or the time required to double a lump sum of money (p. 473)	6, 7	49

Section	You should be able to . . .	Example(s)	Review Exercises
6.8	1 Find equations of populations that obey the law of uninhibited growth (p. 478)	1, 2	53
	2 Find equations of populations that obey the law of decay (p. 480)	3	51, 54
	3 Use Newton's Law of Cooling (p. 481)	4	52
	4 Use logistic models (p. 483)	5, 6	55
6.9	1 Build an exponential model from data (p. 489)	1	56
	2 Build a logarithmic model from data (p. 491)	2	57
	3 Build a logistic model from data (p. 491)	3	58

Review Exercises

In Problems 1–3, for the given functions f and g, find:
(a) $(f \circ g)(2)$ (b) $(g \circ f)(-2)$ (c) $(f \circ f)(4)$ (d) $(g \circ g)(-1)$

1. $f(x) = 3x - 5;\; g(x) = 1 - 2x^2$ **2.** $f(x) = \sqrt{x+2};\; g(x) = 2x^2 + 1$ **3.** $f(x) = e^x;\; g(x) = 3x - 2$

In Problems 4–6, find $f \circ g$, $g \circ f$, $f \circ f$, and $g \circ g$ for each pair of functions. State the domain of each composite function.

4. $f(x) = 2 - x;\; g(x) = 3x + 1$ **5.** $f(x) = \sqrt{3x};\; g(x) = 1 + x + x^2$ **6.** $f(x) = \dfrac{x+1}{x-1};\; g(x) = \dfrac{1}{x}$

7. *(a) Verify that the function below is one-to-one; (b) find its inverse.*
$\{(1,2), (3,5), (5,8), (6,10)\}$

8. *State why the graph of the function is one-to-one. Then draw the graph of the inverse function f^{-1}.*

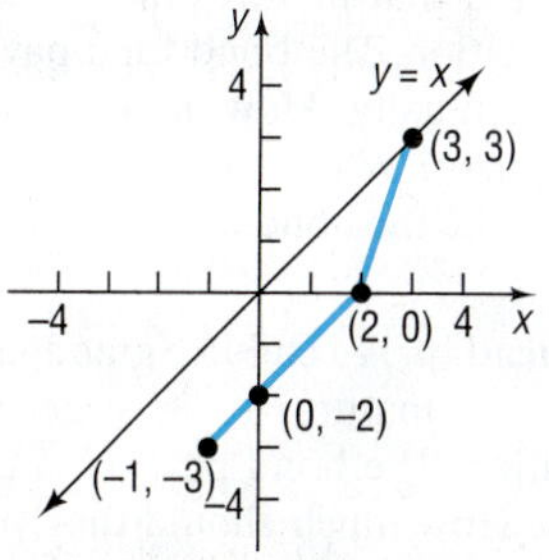

In Problems 9–12, each function is one-to-one. Find the inverse of each function and check your answer.

9. $f(x) = \dfrac{2x+3}{5x-2}$ **10.** $f(x) = \dfrac{1}{x-1}$

11. $f(x) = \sqrt{x-2}$ **12.** $f(x) = x^{1/3} + 1$

In Problem 13, $f(x) = 3^x$ and $g(x) = \log_3 x$.

13. Evaluate: (a) $f(4)$ (b) $g(9)$ (c) $f(-2)$ (d) $g\left(\dfrac{1}{27}\right)$

14. Change $5^2 = z$ to an equivalent statement involving a logarithm.

15. Change $\log_5 u = 13$ to an equivalent statement involving an exponent.

In Problems 16 and 17, find the domain of each logarithmic function.

16. $f(x) = \log(3x - 2)$ **17.** $H(x) = \log_2(x^2 - 3x + 2)$

In Problems 18–20, find the exact value of each expression. Do not use a calculator.

18. $\log_2\left(\dfrac{1}{8}\right)$ **19.** $\ln e^{\sqrt{2}}$ **20.** $2^{\log_2 0.4}$

In Problems 21–24, write each expression as the sum and/or difference of logarithms. Express powers as factors.

21. $\log_3\left(\dfrac{uv^2}{w}\right)\quad u > 0, v > 0, w > 0$ **22.** $\log_2\left(a^2\sqrt{b}\right)^4\quad a > 0, b > 0$

23. $\log\left(x^2\sqrt{x^3+1}\right)\quad x > 0$ **24.** $\ln\left(\dfrac{2x+3}{x^2-3x+2}\right)^2\quad x > 2$

In Problems 25–27, write each expression as a single logarithm.

25. $3\log_4 x^2 + \frac{1}{2}\log_4 \sqrt{x}$

26. $\ln\left(\frac{x-1}{x}\right) + \ln\left(\frac{x}{x+1}\right) - \ln(x^2 - 1)$

27. $\frac{1}{2}\ln(x^2+1) - 4\ln\frac{1}{2} - \frac{1}{2}[\ln(x-4) + \ln x]$

28. Use the Change-of-Base Formula and a calculator to evaluate $\log_4 19$. Round your answer to three decimal places.

29. Graph $y = \log_3 x$ using a graphing utility and the Change-of-Base Formula.

In Problems 30–33, use the given function f to:
(a) Find the domain of f. (b) Graph f. (c) From the graph, determine the range and any asymptotes of f.
(d) Find f^{-1}, the inverse function of f. (e) Find the domain and the range of f^{-1}. (f) Graph f^{-1}.

30. $f(x) = 2^{x-3}$ **31.** $f(x) = 1 + 3^{-x}$ **32.** $f(x) = 3e^{x-2}$ **33.** $f(x) = \frac{1}{2}\ln(x+3)$

In Problems 34–44, solve each equation. Express any irrational solution in exact form and as a decimal rounded to three decimal places.

34. $8^{6+3x} = 4$ **35.** $3^{x^2+x} = \sqrt{3}$ **36.** $\log_x 64 = -3$ **37.** $5^x = 3^{x+2}$

38. $25^{2x} = 5^{x^2-12}$ **39.** $\log_3 \sqrt{x-2} = 2$ **40.** $8 = 4^{x^2} \cdot 2^{5x}$ **41.** $2^x \cdot 5 = 10^x$

42. $\log_6(x+3) + \log_6(x+4) = 1$ **43.** $e^{1-x} = 5$ **44.** $9^x + 4 \cdot 3^x - 3 = 0$

45. Suppose that $f(x) = \log_2(x-2) + 1$.
(a) Graph f.
(b) What is $f(6)$? What point is on the graph of f?
(c) Solve $f(x) = 4$. What point is on the graph of f?
(d) Based on the graph drawn in part (a), solve $f(x) > 0$.
(e) Find $f^{-1}(x)$. Graph f^{-1} on the same Cartesian plane as f.

46. Amplifying Sound An amplifier's power output P (in watts) is related to its decibel voltage gain d by the formula

$$P = 25e^{0.1d}$$

(a) Find the power output for a decibel voltage gain of 4 decibels.
(b) For a power output of 50 watts, what is the decibel voltage gain?

47. Limiting Magnitude of a Telescope A telescope is limited in its usefulness by the brightness of the star that it is aimed at and by the diameter of its lens. One measure of a star's brightness is its *magnitude;* the dimmer the star, the larger its magnitude. A formula for the limiting magnitude L of a telescope—that is, the magnitude of the dimmest star that it can be used to view—is given by

$$L = 9 + 5.1\log d$$

where d is the diameter (in inches) of the lens.
(a) What is the limiting magnitude of a 3.5-inch telescope?
(b) What diameter is required to view a star of magnitude 14?

48. Salvage Value The number of years n for a piece of machinery to depreciate to a known salvage value can be found using the formula

$$n = \frac{\log s - \log i}{\log(1-d)}$$

where s is the salvage value of the machinery, i is its initial value, and d is the annual rate of depreciation.
(a) How many years will it take for a piece of machinery to decline in value from \$90,000 to \$10,000 if the annual rate of depreciation is 0.20 (20%)?
(b) How many years will it take for a piece of machinery to lose half of its value if the annual rate of depreciation is 15%?

49. Funding a College Education A child's grandparents purchase a \$10,000 bond fund that matures in 18 years to be used for her college education. The bond fund pays 4% interest compounded semiannually. How much will the bond fund be worth at maturity? What is the effective rate of interest? How long will it take the bond to double in value under these terms?

50. Funding a College Education A child's grandparents wish to purchase a bond that matures in 18 years to be used for her college education. The bond pays 4% interest compounded semiannually. How much should they pay so that the bond will be worth \$85,000 at maturity?

51. Estimating the Date When a Prehistoric Man Died The bones of a prehistoric man found in the desert of New Mexico contain approximately 5% of the original amount of carbon-14. If the half-life of carbon-14 is 5730 years, approximately how long ago did the man die?

52. Temperature of a Skillet A skillet is removed from an oven where the temperature is 450°F and placed in a room where the temperature is 70°F. After 5 minutes, the temperature of the skillet is 400°F. How long will it be until its temperature is 150°F?

53. World Population The annual growth rate of the world's population in 2014 was $k = 1.1\% = 0.011$. The population of the world in 2014 was 7,137,577,750. Letting $t = 0$ represent 2014, use the uninhibited growth model to predict the world's population in the year 2024.

Source: U.S. Census Bureau

54. Radioactive Decay The half-life of radioactive cobalt is 5.27 years. If 100 grams of radioactive cobalt is present now, how much will be present in 20 years? In 40 years?

55. Logistic Growth The logistic growth model

$$P(t) = \frac{0.8}{1 + 1.67e^{-0.16t}}$$

represents the proportion of new cars with a global positioning system (GPS). Let $t = 0$ represent 2006, $t = 1$ represent 2007, and so on.

(a) What proportion of new cars in 2006 had a GPS?
(b) Determine the maximum proportion of new cars that have a GPS.
(c) Using a graphing utility, graph $P = P(t)$.
(d) When will 75% of new cars have a GPS?

56. Rising Tuition The following data represent the average in-state tuition and fees (in 2013 dollars) at public four-year colleges and universities in the United States from the academic year 1983–84 to the academic year 2013–14.

Academic Year	Tuition and Fees (2013 dollars)
1983–84 ($x = 0$)	2684
1988–89 ($x = 5$)	3111
1993–94 ($x = 10$)	4101
1998–99 ($x = 15$)	4648
2003–04 ($x = 20$)	5900
2008–09 ($x = 25$)	7008
2013–14 ($x = 30$)	8893

Source: The College Board

(a) Using a graphing utility, draw a scatter diagram with academic year as the independent variable.
(b) Using a graphing utility, build an exponential model from the data.
(c) Express the function found in part (b) in the form $A(t) = A_0e^{kt}$.
(d) Graph the exponential function found in part (b) or (c) on the scatter diagram.
(e) Predict the academic year when the average tuition will reach $12,000.

57. Wind Chill Factor The data (top, right) represent the wind speed (mph) and the wind chill factor at an air temperature of 15°F.

(a) Using a graphing utility, draw a scatter diagram with wind speed as the independent variable.
(b) Using a graphing utility, build a logarithmic model from the data.
(c) Using a graphing utility, draw the logarithmic function found in part (b) on the scatter diagram.
(d) Use the function found in part (b) to predict the wind chill factor if the air temperature is 15°F and the wind speed is 23 mph.

Wind Speed (mph)	Wind Chill Factor (°F)
5	7
10	3
15	0
20	−2
25	−4
30	−5
35	−7

Source: U.S. National Weather Service

58. Spreading of a Disease Jack and Diane live in a small town of 50 people. Unfortunately, both Jack and Diane have a cold. Those who come in contact with someone who has this cold will themselves catch the cold. The following data represent the number of people in the small town who have caught the cold after *t* days.

Days, t	Number of People with Cold, C
0	2
1	4
2	8
3	14
4	22
5	30
6	37
7	42
8	44

(a) Using a graphing utility, draw a scatter diagram of the data. Comment on the type of relation that appears to exist between the number of days that have passed and the number of people with a cold.
(b) Using a graphing utility, build a logistic model from the data.
(c) Graph the function found in part (b) on the scatter diagram.
(d) According to the function found in part (b), what is the maximum number of people who will catch the cold? In reality, what is the maximum number of people who could catch the cold?
(e) Sometime between the second day and the third day, 10 people in the town had a cold. According to the model found in part (b), when did 10 people have a cold?
(f) How long will it take for 46 people to catch the cold?

Chapter Test

CHAPTER **Test Prep** VIDEOS

The Chapter Test Prep Videos are step-by-step solutions available in MyMathLab, or on this text's YouTube Channel. Flip back to the Resources for Success page for a link to this text's YouTube channel.

1. Given $f(x) = \dfrac{x+2}{x-2}$ and $g(x) = 2x + 5$, find:
(a) $f \circ g$ and state its domain
(b) $(g \circ f)(-2)$
(c) $(f \circ g)(-2)$

2. Determine whether the function is one-to-one.
(a) $y = 4x^2 + 3$ (b) $y = \sqrt{x+3} - 5$

3. Find the inverse of $f(x) = \dfrac{2}{3x-5}$ and check your answer. State the domain and the range of f and f^{-1}.

4. If the point $(3, -5)$ is on the graph of a one-to-one function f, what point must be on the graph of f^{-1}?

In Problems 5–7, solve each equation.

5. $3^x = 243$ **6.** $\log_b 16 = 2$
7. $\log_5 x = 4$

In Problems 8–11, use a calculator to evaluate each expression. Round your answer to three decimal places.

8. $e^3 + 2$ **9.** $\log 20$
10. $\log_3 21$ **11.** $\ln 133$

In Problems 12 and 13, use the given function f.
(a) Find the domain of f.
(b) Graph f.
(c) From the graph, determine the range and any asymptotes of f.
(d) Find f^{-1}, the inverse of f.
(e) Find the domain and the range of f^{-1}.
(f) Graph f^{-1}.

12. $f(x) = 4^{x+1} - 2$
13. $f(x) = 1 - \log_5(x - 2)$

In Problems 14–19, solve each equation.

14. $5^{x+2} = 125$ **15.** $\log(x + 9) = 2$
16. $8 - 2e^{-x} = 4$ **17.** $\log(x^2 + 3) = \log(x + 6)$
18. $7^{x+3} = e^x$ **19.** $\log_2(x - 4) + \log_2(x + 4) = 3$

20. Write $\log_2\left(\dfrac{4x^3}{x^2 - 3x - 18}\right)$ as the sum and/or difference of logarithms. Express powers as factors.

21. A 50-mg sample of a radioactive substance decays to 34 mg after 30 days. How long will it take for there to be 2 mg remaining?

22. (a) If \$1000 is invested at 5% compounded monthly, how much is there after 8 months?
(b) If you want to have \$1000 in 9 months, how much do you need to place in a savings account now that pays 5% compounded quarterly?
(c) How long does it take to double your money if you can invest it at 6% compounded annually?

23. The decibel level, D, of sound is given by the equation $D = 10\log\left(\dfrac{I}{I_0}\right)$, where I is the intensity of the sound and $I_0 = 10^{-12}$ watt per square meter.
(a) If the shout of a single person measures 80 decibels, how loud would the sound be if two people shouted at the same time? That is, how loud would the sound be if the intensity doubled?
(b) The pain threshold for sound is 125 decibels. If the Athens Olympic Stadium 2004 (Olympiako Stadio Athinas 'Spyros Louis') can seat 74,400 people, how many people in the crowd need to shout at the same time for the resulting sound level to meet or exceed the pain threshold? (Ignore any possible sound dampening.)

Cumulative Review

1. Is the following graph the graph of a function? If it is, is the function one-to-one?

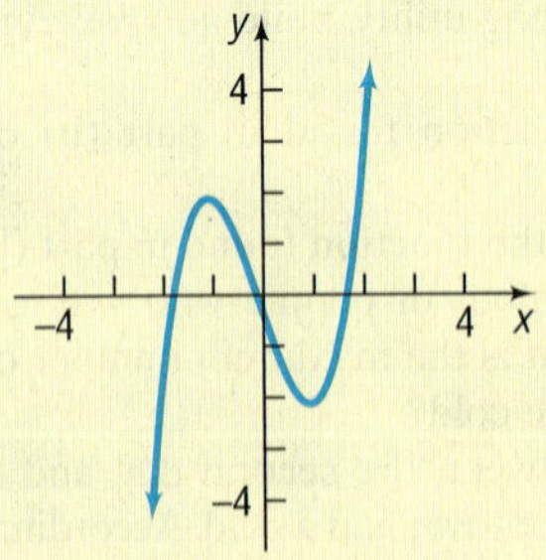

2. For the function $f(x) = 2x^2 - 3x + 1$, find the following:
(a) $f(3)$ (b) $f(-x)$ (c) $f(x + h)$

3. Determine which of the following points are on the graph of $x^2 + y^2 = 1$.
(a) $\left(\dfrac{1}{2}, \dfrac{1}{2}\right)$ (b) $\left(\dfrac{1}{2}, \dfrac{\sqrt{3}}{2}\right)$

4. Solve the equation $3(x - 2) = 4(x + 5)$.

5. Graph the line $2x - 4y = 16$.

6. (a) Graph the quadratic function $f(x) = -x^2 + 2x - 3$ by determining whether its graph opens up or down and by finding its vertex, axis of symmetry, y-intercept, and x-intercept(s), if any.
(b) Solve $f(x) \le 0$.

7. Determine the quadratic function whose graph is given in the figure.

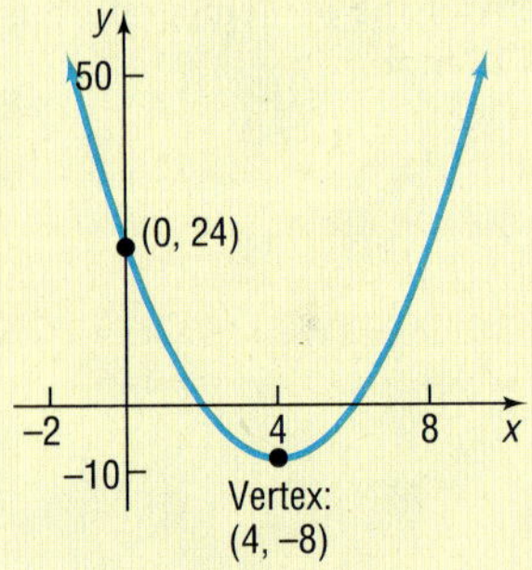

8. Graph $f(x) = 3(x+1)^3 - 2$ using transformations.

9. Given that $f(x) = x^2 + 2$ and $g(x) = \dfrac{2}{x-3}$, find $(f \circ g)(x)$ and state its domain. What is $(f \circ g)(5)$?

10. For the polynomial function $f(x) = 4x^3 + 9x^2 - 30x - 8$:
    (a) Find the real zeros of f.
    (b) Determine the intercepts of the graph of f.
    (c) Use a graphing utility to approximate the local maxima and local minima.
    (d) Draw a complete graph of f. Be sure to label the intercepts and turning points.

11. For the function $g(x) = 3^x + 2$:
    (a) Graph g using transformations. State the domain, range, and horizontal asymptote of g.
    (b) Determine the inverse of g. State the domain, range, and vertical asymptote of g^{-1}.
    (c) On the same graph as g, graph g^{-1}.

12. Solve the equation: $4^{x-3} = 8^{2x}$

13. Solve the equation: $\log_3(x+1) + \log_3(2x-3) = \log_9 9$

14. Suppose that $f(x) = \log_3(x+2)$. Solve:
    (a) $f(x) = 0$ (b) $f(x) > 0$
    (c) $f(x) = 3$

15. **Data Analysis** The following data represent the percent of all drivers by age who have been stopped by the police for any reason within the past year. The median age represents the midpoint of the upper and lower limit for the age range.

Age Range	Median Age, x	Percent Stopped, y
16–19	17.5	18.2
20–29	24.5	16.8
30–39	34.5	11.3
40–49	44.5	9.4
50–59	54.5	7.7
≥60	69.5	3.8

(a) Using your graphing utility, draw a scatter diagram of the data treating median age, x, as the independent variable.
(b) What type of model do you feel best describes the relation between median age and percent stopped? You may choose from among linear, quadratic, cubic, exponential, logarithmic, and logistic models.
(c) Provide a justification for the model that you selected in part (b).

Chapter Projects

Internet-based Project

I. Depreciation of Cars Kelley Blue Book is a guide that provides the current retail price of cars. You can access the Kelley Blue Book at your library or online at www.kbb.com.

1. Identify three cars that you are considering purchasing, and find the Kelley Blue Book value of the cars for 0 (brand new), 1, 2, 3, 4, and 5 years of age. Online, the value of the car can be found by selecting What should I pay for a used car? Enter the year, make, and model of the car you are selecting. To be consistent, we will assume the cars will be driven 12,000 miles per year, so a 1-year-old car will have 12,000 miles, a 2-year-old car will have 24,000 miles, and so on. Choose the same options for each year, and finally determine the suggested retail price for cars that are in Excellent, Good, and Fair shape. You should have a total of 16 observations (1 for a brand new car, 3 for a 1-year-old car, 3 for a 2-year-old car, and so on).
2. Draw a scatter diagram of the data with age as the independent variable and value as the dependent variable using Excel, a TI-graphing calculator, or some other spreadsheet. The Chapter 4 project describes how to draw a scatter diagram in Excel.
3. Determine the exponential function of best fit. Graph the exponential function of best fit on the scatter diagram. To do this in Excel, click on any data point in the scatter diagram. Now select the Chart Element icon (+). Check the box for Trendline, select the arrow to the right, and choose More Options. Select the Exponential radio button and select Display Equation on Chart. See Figure 54 on page 504. Move the Trendline Options window off to the side, and you will see the exponential function of best fit displayed on the scatter diagram. Do you think the function accurately describes the relation between age of the car and suggested retail price?

4. The exponential function of best fit is of the form $y = Ce^{rx}$, where y is the suggested retail value of the car and x is the age of the car (in years). What does the value of C represent? What does the value of r represent? What is the depreciation rate for each car that you are considering?
5. Write a report detailing which car you would purchase based on the depreciation rate you found for each car.

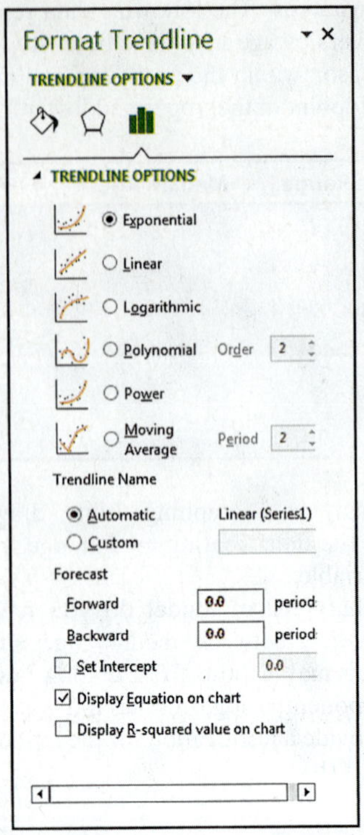

Figure 54

Citation: Excel © 2013 Microsoft Corporation. Used with permission from Microsoft.

The following projects are available on the Instructor's Resource Center (IRC):

II. Hot Coffee A fast-food restaurant wants a special container to hold coffee. The restaurant wishes the container to quickly cool the coffee from 200° to 130°F and keep the liquid between 110° and 130°F as long as possible. The restaurant has three containers to select from. Which one should be purchased?

III. Project at Motorola *Thermal Fatigue of Solder Connections* Product reliability is a major concern of a manufacturer. Here a logarithmic transformation is used to simplify the analysis of a cell phone's ability to withstand temperature change.

Systems of Equations and Inequalities

8

Economic Outcomes

Annual Earnings of Young Adults

For both males and females, earnings increase with education: full-time workers with at least a bachelor's degree have higher median earnings than those with less education. For example, in 2012, male college graduates ages 25–34 earned 67% more than male high school completers ages 25–34. In the same age range, females with a bachelor's or higher degree earned 72% more than female high school completers. Males and females who dropped out of high school earned 25% and 28% less, respectively, than male and female high school completers.

The median earnings of young adults who had at least a bachelor's degree declined in the 1970s relative to their counterparts who were high school completers, before increasing between 1980 and 2012. Males with a bachelor's degree or higher had earnings 19% higher than male high school completers in 1980 and had earnings 67% higher in 2012. Among females, those with at least a bachelor's degree had earnings 34% higher than female high school completers in 1980, compared with earnings 72% higher in 2012.

—See Chapter Project I—

••• A Look Back

In Chapters 1, 4, 5, and 6, we solved various kinds of equations and inequalities involving a single variable.

A Look Ahead •••

In this chapter we take up the problem of solving equations and inequalities containing two or more variables. There are various ways to solve such problems.

The *method of substitution* for solving equations in several variables dates back to ancient times.

The *method of elimination*, although it had existed for centuries, was put into systematic order by Karl Friedrich Gauss (1777–1855) and by Camille Jordan (1838–1922).

The theory of *matrices* was developed in 1857 by Arthur Cayley (1821–1895), although only later were matrices used as we use them in this chapter. Matrices have become a very flexible instrument, useful in almost all areas of mathematics.

The method of *determinants* was invented by Takakazu Seki Kôwa (1642–1708) in 1683 in Japan and by Gottfried Wilhelm von Leibniz (1646–1716) in 1693 in Germany. *Cramer's Rule* is named after Gabriel Cramer (1704–1752) of Switzerland, who popularized the use of determinants for solving linear systems.

Section 8.5, on *partial fraction decomposition*, provides an application of systems of equations. This particular application is one that is used in integral calculus.

Section 8.8 introduces *linear programming*, a modern application of linear inequalities. This topic is particularly useful for students interested in operations research.

Outline

8.1 Systems of Linear Equations: Substitution and Elimination

PREPARING FOR THIS SECTION *Before getting started, review the following:*

- Linear Equations (Section 1.1, pp. 82–85)
- Lines (Section 2.3, pp. 167–177)

Now Work the **'Are You Prepared?'** problems on page 554.

OBJECTIVES
1. Solve Systems of Equations by Substitution (p. 546)
2. Solve Systems of Equations by Elimination (p. 547)
3. Identify Inconsistent Systems of Equations Containing Two Variables (p. 549)
4. Express the Solution of a System of Dependent Equations Containing Two Variables (p. 549)
5. Solve Systems of Three Equations Containing Three Variables (p. 550)
6. Identify Inconsistent Systems of Equations Containing Three Variables (p. 552)
7. Express the Solution of a System of Dependent Equations Containing Three Variables (p. 553)

EXAMPLE 1 **Movie Theater Ticket Sales**

A movie theater sells tickets for \$8.00 each, with seniors receiving a discount of \$2.00. One evening the theater took in \$3580 in revenue. If x represents the number of tickets sold at \$8.00 and y the number of tickets sold at the discounted price of \$6.00, write an equation that relates these variables.

Solution Each nondiscounted ticket brings in \$8.00, so x tickets will bring in $8x$ dollars. Similarly, y discounted tickets bring in $6y$ dollars. Because the total brought in is \$3580, we must have

$$8x + 6y = 3580$$

In Example 1, suppose that we also know that 525 tickets were sold that evening. Then we have another equation relating the variables x and y:

$$x + y = 525$$

The two equations

$$\begin{cases} 8x + 6y = 3580 \\ x + y = 525 \end{cases}$$

form a *system* of equations.

In general, a **system of equations** is a collection of two or more equations, each containing one or more variables. Example 2 gives some illustrations of systems of equations.

EXAMPLE 2 **Examples of Systems of Equations**

(a) $\begin{cases} 2x + y = 5 & (1) \\ -4x + 6y = -2 & (2) \end{cases}$ Two equations containing two variables, x and y

(b) $\begin{cases} x + y^2 = 5 & (1) \\ 2x + y = 4 & (2) \end{cases}$ Two equations containing two variables, x and y

(c) $\begin{cases} x + y + z = 6 & (1) \\ 3x - 2y + 4z = 9 & (2) \\ x - y - z = 0 & (3) \end{cases}$ Three equations containing three variables, x, y, and z

(d) $\begin{cases} x + y + z = 5 & (1) \\ x - y = 2 & (2) \end{cases}$ Two equations containing three variables, x, y, and z

(e) $\begin{cases} x + y + z = 6 & (1) \\ 2x + 2z = 4 & (2) \\ y + z = 2 & (3) \\ x = 4 & (4) \end{cases}$ Four equations containing three variables, x, y, and z

We use a brace to remind us that we are dealing with a system of equations, and we number each equation in the system for convenient reference.

A **solution** of a system of equations consists of values for the variables that are solutions of each equation of the system. To **solve** a system of equations means to find all solutions of the system.

For example, $x = 2, y = 1$ is a solution of the system in Example 2(a), because

$$\begin{cases} 2x + y = 5 & (1) \\ -4x + 6y = -2 & (2) \end{cases} \qquad \begin{cases} 2(2) + 1 = 4 + 1 = 5 \\ -4(2) + 6(1) = -8 + 6 = -2 \end{cases}$$

This solution may also be written as the ordered pair $(2, 1)$.

A solution of the system in Example 2(b) is $x = 1, y = 2$, because

$$\begin{cases} x + y^2 = 5 & (1) \\ 2x + y = 4 & (2) \end{cases} \qquad \begin{cases} 1 + 2^2 = 1 + 4 = 5 \\ 2(1) + 2 = 2 + 2 = 4 \end{cases}$$

Another solution of the system in Example 2(b) is $x = \frac{11}{4}, y = -\frac{3}{2}$, which you can check for yourself.

A solution of the system in Example 2(c) is $x = 3, y = 2, z = 1$, because

$$\begin{cases} x + y + z = 6 & (1) \\ 3x - 2y + 4z = 9 & (2) \\ x - y - z = 0 & (3) \end{cases} \qquad \begin{cases} 3 + 2 + 1 = 6 \\ 3(3) - 2(2) + 4(1) = 9 - 4 + 4 = 9 \\ 3 - 2 - 1 = 0 \end{cases}$$

This solution may also be written as the ordered triplet $(3, 2, 1)$.

Note that $x = 3, y = 3, z = 0$ is not a solution of the system in Example 2(c).

$$\begin{cases} x + y + z = 6 & (1) \\ 3x - 2y + 4z = 9 & (2) \\ x - y - z = 0 & (3) \end{cases} \qquad \begin{cases} 3 + 3 + 0 = 6 \\ 3(3) - 2(3) + 4(0) = 3 \neq 9 \\ 3 - 3 - 0 = 0 \end{cases}$$

Although $x = 3, y = 3$, and $z = 0$ satisfy equations (1) and (3), they do not satisfy equation (2). Any solution of the system must satisfy *each* equation of the system.

Now Work PROBLEM 11

When a system of equations has at least one solution, it is said to be **consistent**. When a system of equations has no solution, it is called **inconsistent**.

An equation in n variables is said to be **linear** if it is equivalent to an equation of the form

$$a_1x_1 + a_2x_2 + \cdots + a_nx_n = b$$

where $x_1, x_2, \ldots, x_n$ are n distinct variables, $a_1, a_2, \ldots, a_n, b$ are constants, and at least one of the a's is not 0.

Some examples of linear equations are

$$2x + 3y = 2 \qquad 5x - 2y + 3z = 10 \qquad 8x + 8y - 2z + 5w = 0$$

If each equation in a system of equations is linear, we have a **system of linear equations**. The systems in Examples 2(a), (c), (d), and (e) are linear, whereas the system in Example 2(b) is nonlinear. In this chapter we shall solve linear systems in Sections 8.1 to 8.3. Nonlinear systems are discussed in Section 8.6.

We begin by discussing a system of two linear equations containing two variables. The problem of solving such a system can be viewed as a geometry problem. The graph of each equation in such a system is a line. So a system of two linear equations containing two variables represents a pair of lines. The lines may (1) intersect, or (2) be parallel, or (3) be **coincident** (that is, identical).

1. If the lines intersect, the system of equations has one solution, given by the point of intersection. The system is **consistent** and the equations are **independent**. See Figure 1(a).
2. If the lines are parallel, the system of equations has no solution, because the lines never intersect. The system is **inconsistent**. See Figure 1(b).
3. If the lines are coincident (the lines lie on top of each other), the system of equations has infinitely many solutions, represented by the totality of points on the line. The system is **consistent** and the equations are **dependent**. See Figure 1(c).

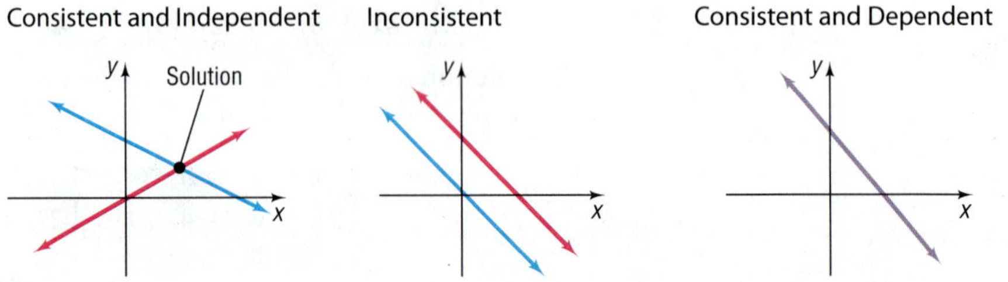

Figure 1 **(a)** Intersecting lines; system has one solution **(b)** Parallel lines; system has no solution **(c)** Coincident lines; system has infinitely many solutions

EXAMPLE 3 Graphing a System of Linear Equations

Graph the system: $\begin{cases} 2x + y = -1 & (1) \\ -4x + 6y = 42 & (2) \end{cases}$

Solution First, solve each equation for y. This is equivalent to writing each equation in slope–intercept form. Equation (1) in slope–intercept form is $y = -2x - 1$, which has slope -2 and y-intercept -1. Equation (2) in slope–intercept form is $y = \frac{2}{3}x + 7$, which has slope $\frac{2}{3}$ and y-intercept 7. Figure 2 shows their graphs. ●

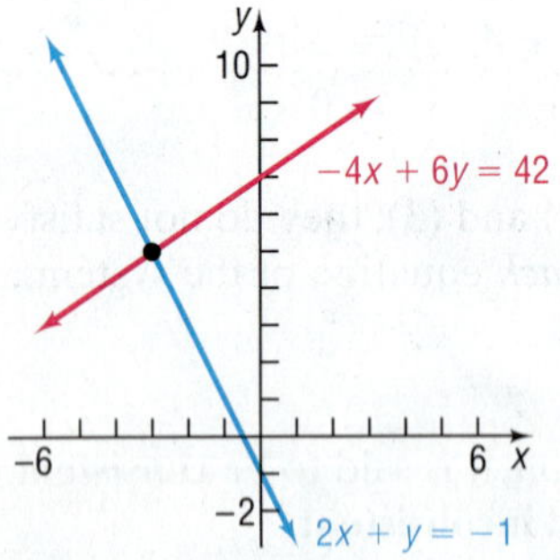

Figure 2

From the graph in Figure 2, we see that the lines intersect, so the system given in Example 3 is consistent. The graph can also be used as a means of approximating the solution. For this system, the solution appears to be the point $(-3, 5)$.

Seeing the Concept

Graph the lines $2x + y = -1$ $(Y_1 = -2x - 1)$ and $-4x + 6y = 42$ $\left(Y_2 = \frac{2}{3}x + 7\right)$, and compare what you see with Figure 2. Use INTERSECT to verify that the point of intersection is $(-3, 5)$.

1 Solve Systems of Equations by Substitution

Most of the time we must use algebraic methods to obtain exact solutions. A number of methods are available for solving systems of linear equations algebraically. In this section, we introduce two methods: *substitution* and *elimination*. We illustrate the **method of substitution** by solving the system given in Example 3.

EXAMPLE 4

How to Solve a System of Linear Equations by Substitution

Solve: $\begin{cases} 2x + y = -1 & (1) \\ -4x + 6y = 42 & (2) \end{cases}$

Step-by-Step Solution

Step 1: Pick one of the equations, and solve for one of the variables in terms of the remaining variable(s).

Solve equation (1) for y.

$$2x + y = -1 \quad \text{Equation (1)}$$
$$y = -2x - 1 \quad \text{Subtract } 2x \text{ from each side of (1).}$$

Step 2: Substitute the result into the remaining equation(s).

Substitute $-2x - 1$ for y in equation (2). The result is an equation containing just the variable x, which we can solve.

$$-4x + 6y = 42 \quad \text{Equation (2)}$$
$$-4x + 6(-2x - 1) = 42 \quad \text{Substitute } -2x - 1 \text{ for } y \text{ in (2).}$$

Step 3: If one equation in one variable results, solve this equation. Otherwise, repeat Steps 1 and 2 until a single equation with one variable remains.

$$-4x - 12x - 6 = 42 \quad \text{Distribute.}$$
$$-16x - 6 = 42 \quad \text{Combine like terms.}$$
$$-16x = 48 \quad \text{Add 6 to both sides.}$$
$$x = \frac{48}{-16} \quad \text{Divide both sides by } -16.$$
$$x = -3 \quad \text{Simplify.}$$

Step 4: Find the values of the remaining variables by back-substitution.

Because we know that $x = -3$, we can find the value of y by **back-substitution**, that is, by substituting -3 for x in one of the original equations. Equation (1) seems easier to work with, so we will back-substitute into equation (1).

$$2x + y = -1 \quad \text{Equation (1)}$$
$$2(-3) + y = -1 \quad \text{Substitute } -3 \text{ for } x \text{ in equation (1).}$$
$$-6 + y = -1 \quad \text{Simplify.}$$
$$y = -1 + 6 \quad \text{Add 6 to both sides.}$$
$$y = 5 \quad \text{Simplify.}$$

Step 5: Check the solution found.

We have $x = -3$ and $y = 5$. Verify that both equations are satisfied (true) for these values.

$$\begin{cases} 2x + y = -1 & 2(-3) + 5 = -6 + 5 = -1 \\ -4x + 6y = 42 & -4(-3) + 6(5) = 12 + 30 = 42 \end{cases}$$

The solution of the system is $x = -3$ and $y = 5$. The solution can also be written as the ordered pair $(-3, 5)$.

Now Use Substitution to Work PROBLEM 21

2 Solve Systems of Equations by Elimination

A second method for solving a system of linear equations is the *method of elimination*. This method is usually preferred over substitution if substitution leads to fractions or if the system contains more than two variables. Elimination also provides the necessary motivation for solving systems using matrices (the subject of Section 8.2).

The idea behind the **method of elimination** is to replace the original system of equations by an equivalent system so that adding two of the equations eliminates a variable. The rules for obtaining equivalent equations are the same as those studied earlier. However, we may also interchange any two equations of the system and/or replace any equation in the system by the sum (or difference) of that equation and a nonzero multiple of any other equation in the system.

In Words

When using elimination, get the coefficients of one of the variables to be negatives of each other.

Rules for Obtaining an Equivalent System of Equations

1. Interchange any two equations of the system.
2. Multiply (or divide) each side of an equation by the same nonzero constant.
3. Replace any equation in the system by the sum (or difference) of that equation and a nonzero multiple of any other equation in the system.

An example will give you the idea. As you work through the example, pay particular attention to the pattern being followed.

EXAMPLE 5

How to Solve a System of Linear Equations by Elimination

Solve: $\begin{cases} 2x + 3y = 1 & (1) \\ -x + y = -3 & (2) \end{cases}$

Step-by-Step Solution

Step 1: Multiply both sides of one or both equations by a nonzero constant so that the coefficients of one of the variables are additive inverses.

Multiply both sides of equation (2) by 2 so that the coefficients of x in the two equations are additive inverses.

$$\begin{cases} 2x + 3y = 1 & (1) \\ -x + y = -3 & (2) \end{cases}$$

$$\begin{cases} 2x + 3y = 1 & (1) \\ 2(-x + y) = 2(-3) & (2) \quad \text{Multiply by 2.} \end{cases}$$

$$\begin{cases} 2x + 3y = 1 & (1) \\ -2x + 2y = -6 & (2) \end{cases}$$

Step 2: Add the equations to eliminate the variable. Solve the resulting equation for the remaining unknown.

$$\begin{cases} 2x + 3y = 1 & (1) \\ -2x + 2y = -6 & (2) \end{cases}$$

$$5y = -5 \quad \text{Add equations (1) and (2).}$$

$$y = -1 \quad \text{Divide both sides by 5.}$$

Step 3: Back-substitute the value of the variable found in Step 2 into one of the *original* equations to find the value of the remaining variable.

Back-substitute $y = -1$ into equation (1) and solve for x.

$$\begin{aligned} 2x + 3y &= 1 && \text{Equation (1)} \\ 2x + 3(-1) &= 1 && \text{Substitute } y = -1 \text{ into equation (1).} \\ 2x - 3 &= 1 && \text{Simplify.} \\ 2x &= 4 && \text{Add 3 to both sides.} \\ x &= 2 && \text{Divide both sides by 2.} \end{aligned}$$

Step 4: Check the solution found.

The check is left to you.

The solution of the system is $x = 2$ and $y = -1$. The solution also can be written as the ordered pair $(2, -1)$. ●

Now Use Elimination to Work PROBLEM 21

EXAMPLE 6

Movie Theater Ticket Sales

A movie theater sells tickets for \$8.00 each, with seniors receiving a discount of \$2.00. One evening the theater sold 525 tickets and took in \$3580 in revenue. How many of each type of ticket were sold?

Solution If x represents the number of tickets sold at \$8.00 and y the number of tickets sold at the discounted price of \$6.00, then the given information results in the system of equations

$$\begin{cases} 8x + 6y = 3580 & (1) \\ x + y = 525 & (2) \end{cases}$$

Using the method of elimination, first multiply the second equation by -6, and then add the equations.

$$\begin{cases} 8x + 6y = 3580 \\ -6x - 6y = -3150 \end{cases}$$
$$2x = 430 \quad \text{Add the equations.}$$
$$x = 215$$

Since $x + y = 525$, then $y = 525 - x = 525 - 215 = 310$. So 215 nondiscounted tickets and 310 senior discount tickets were sold.

3 Identify Inconsistent Systems of Equations Containing Two Variables

The previous examples dealt with consistent systems of equations that had a single solution. The next two examples deal with two other possibilities that may occur, the first being a system that has no solution.

EXAMPLE 7 An Inconsistent System of Linear Equations

Solve: $\begin{cases} 2x + y = 5 & (1) \\ 4x + 2y = 8 & (2) \end{cases}$

Solution We choose to use the method of substitution and solve equation (1) for y.

$$2x + y = 5 \quad (1)$$
$$y = -2x + 5 \quad \text{Subtract } 2x \text{ from each side.}$$

Now substitute $-2x + 5$ for y in equation (2) and solve for x.

$$4x + 2y = 8 \quad (2)$$
$$4x + 2(-2x + 5) = 8 \quad \text{Substitute } y = -2x + 5 \text{ in (2).}$$
$$4x - 4x + 10 = 8 \quad \text{Multiply out.}$$
$$0 = -2 \quad \text{Subtract 10 from both sides.}$$

This statement is false. Conclude that the system has no solution and is therefore inconsistent.

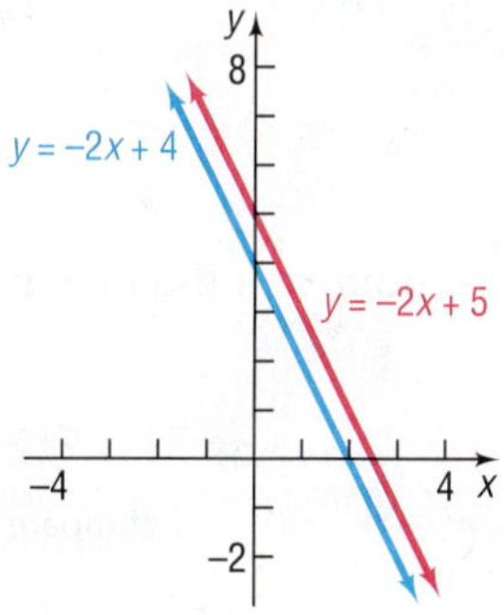

Figure 3

Figure 3 illustrates the pair of lines whose equations form the system in Example 7. Notice that the graphs of the two equations are lines, each with slope -2; one has a y-intercept of 5, the other a y-intercept of 4. The lines are parallel and have no point of intersection. This geometric statement is equivalent to the algebraic statement that the system has no solution.

Seeing the Concept

Graph the lines $2x + y = 5$ ($Y_1 = -2x + 5$) and $4x + 2y = 8$ ($Y_2 = -2x + 4$), and compare what you see with Figure 3. How can you be sure that the lines are parallel?

4 Express the Solution of a System of Dependent Equations Containing Two Variables

EXAMPLE 8 Solving a System of Dependent Equations

Solve: $\begin{cases} 2x + y = 4 & (1) \\ -6x - 3y = -12 & (2) \end{cases}$

Solution We choose to use the method of elimination.

$$\begin{cases} 2x + y = 4 & (1) \\ -6x - 3y = -12 & (2) \end{cases}$$

$$\begin{cases} 6x + 3y = 12 & (1) \quad \text{Multiply each side of equation (1) by 3.} \\ -6x - 3y = -12 & (2) \end{cases}$$

$$0 = 0 \quad \text{Add equations (1) and (2).}$$

The statement $0 = 0$ is true. This means the equation $6x + 3y = 12$ is equivalent to $-6x - 3y = -12$. Therefore, the original system is equivalent to a system containing one equation, so the equations are dependent. This means that any values of x and y that satisfy $6x + 3y = 12$ or, equivalently, $2x + y = 4$ are solutions. For example, $x = 2, y = 0$; $x = 0, y = 4$; $x = -2, y = 8$; $x = 4, y = -4$; and so on, are solutions. There are, in fact, infinitely many values of x and y for which $2x + y = 4$, so the original system has infinitely many solutions. We will write the solution of the original system either as

$$y = -2x + 4, \text{ where } x \text{ can be any real number}$$

or as

$$x = -\frac{1}{2}y + 2, \text{ where } y \text{ can be any real number.}$$

The solution can also be expressed as $\{(x, y) \mid y = -2x + 4, x \text{ is any real number}\}$ or as $\left\{(x, y) \,\middle|\, x = -\frac{1}{2}y + 2, y \text{ is any real number}\right\}$.

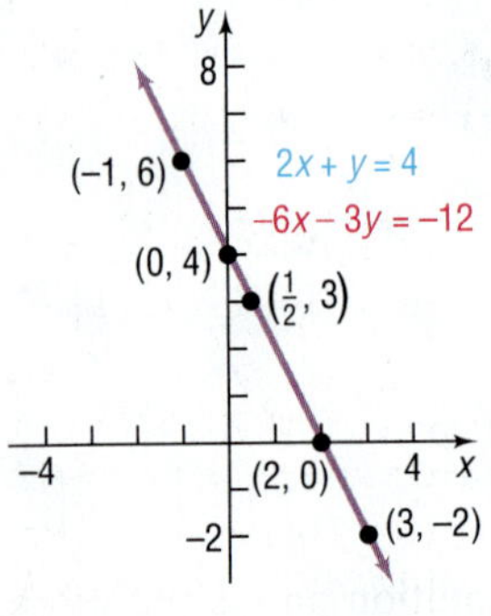

Figure 4 $y = -2x + 4$

Figure 4 illustrates the system given in Example 8. Notice that the graphs of the two equations are lines, each with slope -2 and each with y-intercept 4. The lines are coincident. Notice also that equation (2) in the original system is -3 times equation (1), indicating that the two equations are dependent.

For the system in Example 8, some of the infinite number of solutions can be written down by assigning values to x and then finding $y = -2x + 4$.

If $x = -1$, then $y = -2(-1) + 4 = 6$.

If $x = 0$, then $y = 4$.

If $x = 2$, then $y = 0$.

The ordered pairs $(-1, 6)$, $(0, 4)$, and $(2, 0)$ are three of the points on the line in Figure 4.

Seeing the Concept

Graph the lines $2x + y = 4$ ($Y_1 = -2x + 4$) and $-6x - 3y = -12$ ($Y_2 = -2x + 4$), and compare what you see with Figure 4. How can you be sure that the lines are coincident?

Now Work PROBLEMS 27 AND 31

5 Solve Systems of Three Equations Containing Three Variables

Just like a system of two linear equations containing two variables, a system of three linear equations containing three variables has (1) exactly one solution (a consistent system with independent equations), or (2) no solution (an inconsistent system), or (3) infinitely many solutions (a consistent system with dependent equations).

The problem of solving a system of three linear equations containing three variables can be viewed as a geometry problem. The graph of each equation in such a system is a plane in space. A system of three linear equations containing three variables represents three planes in space. Figure 5 illustrates some of the possibilities.

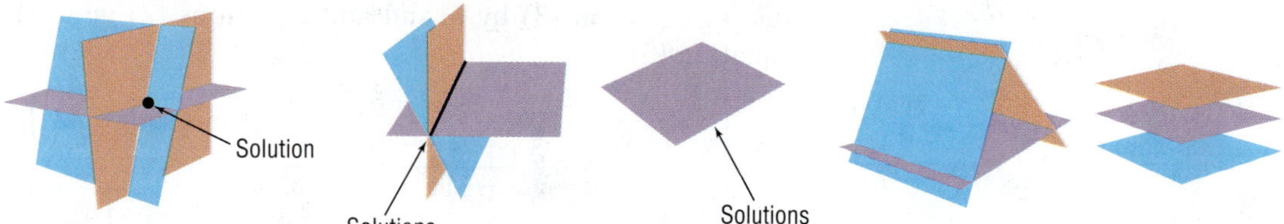

Figure 5 **(a)** Consistent system; one solution **(b)** Consistent system; infinite number of solutions **(c)** Inconsistent system; no solution

Recall that a **solution** to a system of equations consists of values for the variables that are solutions of each equation of the system. For example, $x = 3$, $y = -1$, $z = -5$, or, using an ordered triplet, $(3, -1, -5)$, is a solution to the system of equations

$$\begin{cases} x + y + z = -3 & (1) \\ 2x - 3y + 6z = -21 & (2) \\ -3x + 5y = -14 & (3) \end{cases} \qquad \begin{aligned} &3 + (-1) + (-5) = -3 \\ &2(3) - 3(-1) + 6(-5) = 6 + 3 - 30 = -21 \\ &-3(3) + 5(-1) = -9 - 5 = -14 \end{aligned}$$

because these values of the variables are solutions of each equation.

Typically, when solving a system of three linear equations containing three variables, we use the method of elimination. Recall that the idea behind the method of elimination is to form equivalent equations so that adding two of the equations eliminates a variable.

EXAMPLE 9 Solving a System of Three Linear Equations with Three Variables

Use the method of elimination to solve the system of equations.

$$\begin{cases} x + y - z = -1 & (1) \\ 4x - 3y + 2z = 16 & (2) \\ 2x - 2y - 3z = 5 & (3) \end{cases}$$

Solution For a system of three equations, attempt to eliminate one variable at a time, using pairs of equations, until an equation with a single variable remains. Our strategy for solving this system will be to use equation (1) to eliminate the variable x from equations (2) and (3). We can then treat the new equations (2) and (3) as a system with two unknowns. Alternatively, we could use equation (1) to eliminate either y or z from equations (2) and (3). Try one of these approaches for yourself.

Begin by multiplying each side of equation (1) by -4 and adding the result to equation (2). (Do you see why? The coefficients of x are now negatives of one another.) We also multiply equation (1) by -2 and add the result to equation (3). Notice that these two procedures result in the elimination of the variable x from equations (2) and (3).

$$\begin{cases} x + y - z = -1 & (1) \quad \text{Multiply by } -4. \\ 4x - 3y + 2z = 16 & (2) \end{cases} \qquad \begin{cases} -4x - 4y + 4z = 4 & (1) \\ 4x - 3y + 2z = 16 & (2) \end{cases}$$
$$-7y + 6z = 20 \quad \text{Add}$$

$$\begin{cases} x + y - z = -1 & (1) \quad \text{Multiply by } -2. \\ 2x - 2y - 3z = 5 & (3) \end{cases} \qquad \begin{cases} -2x - 2y + 2z = 2 & (1) \\ 2x - 2y - 3z = 5 & (3) \end{cases}$$
$$-4y - z = 7 \quad \text{Add}$$

$$\begin{cases} x + y - z = -1 & (1) \\ -7y + 6z = 20 & (2) \\ -4y - z = 7 & (3) \end{cases}$$

Now concentrate on the new equations (2) and (3), treating them as a system of two equations containing two variables. It is easiest to eliminate z. Multiply each side of equation (3) by 6, and add equations (2) and (3). The result is the new equation (3).

$$\begin{array}{ll} -7y + 6z = 20 & (2) \\ -4y - z = 7 & (3) \end{array} \quad \text{Multiply by 6.} \quad \begin{array}{ll} -7y + 6z = 20 & (2) \\ -24y - 6z = 42 & (3) \\ \hline -31y \quad = 62 & \end{array} \quad \text{Add} \rightarrow \begin{cases} x + y - z = -1 & (1) \\ -7y + 6z = 20 & (2) \\ -31y = 62 & (3) \end{cases}$$

Now solve equation (3) for y by dividing both sides of the equation by -31.

$$\begin{cases} x + y - z = -1 & (1) \\ -7y + 6z = 20 & (2) \\ y = -2 & (3) \end{cases}$$

Back-substitute $y = -2$ in equation (2) and solve for z.

$$\begin{aligned} -7y + 6z &= 20 && (2) \\ -7(-2) + 6z &= 20 && \text{Substitute } y = -2 \text{ in (2).} \\ 6z &= 6 && \text{Subtract 14 from both sides of the equation.} \\ z &= 1 && \text{Divide both sides of the equation by 6.} \end{aligned}$$

Finally, back-substitute $y = -2$ and $z = 1$ in equation (1) and solve for x.

$$\begin{aligned} x + y - z &= -1 && (1) \\ x + (-2) - 1 &= -1 && \text{Substitute } y = -2 \text{ and } z = 1 \text{ in (1).} \\ x - 3 &= -1 && \text{Simplify.} \\ x &= 2 && \text{Add 3 to both sides.} \end{aligned}$$

The solution of the original system is $x = 2, y = -2, z = 1$, or, using an ordered triplet, $(2, -2, 1)$. You should check this solution. ●

Look back over the solution given in Example 9. Note the pattern of removing one of the variables from two of the equations, followed by solving this system of two equations and two unknowns. Although which variables to remove is your choice, the methodology remains the same for all systems.

Now Work PROBLEM 45

6 Identify Inconsistent Systems of Equations Containing Three Variables

EXAMPLE 10 **Identify an Inconsistent System of Linear Equations**

Solve: $\begin{cases} 2x + y - z = -2 & (1) \\ x + 2y - z = -9 & (2) \\ x - 4y + z = 1 & (3) \end{cases}$

Solution Our strategy is the same as in Example 9. However, in this system, it seems easiest to eliminate the variable z first. Do you see why?

Multiply each side of equation (1) by -1, and add the result to equation (2). Also, add equations (2) and (3). Here equation (2) is used twice and becomes equation (1) in the new system.

$$\begin{array}{ll} 2x + y - z = -2 & (1) \\ x + 2y - z = -9 & (2) \end{array} \quad \text{Multiply by } -1. \quad \begin{array}{ll} -2x - y + z = 2 & (1) \\ x + 2y - z = -9 & (2) \\ \hline -x + y \quad = -7 & \text{Add} \end{array}$$

$$\begin{array}{ll} x + 2y - z = -9 & (2) \\ x - 4y + z = 1 & (3) \\ \hline 2x - 2y \quad = -8 & \text{Add} \end{array}$$

$$\begin{cases} x + 2y - z = -9 & (1) \\ -x + y = -7 & (2) \\ 2x - 2y = -8 & (3) \end{cases}$$

Now concentrate on the new equations (2) and (3), treating them as a system of two equations containing two variables. Multiply each side of equation (2) by 2, and add the result to equation (3).

$$\begin{aligned} -x + y &= -7 \quad (2) \\ 2x - 2y &= -8 \quad (3) \end{aligned} \quad \xrightarrow{\text{Multiply by 2.}} \quad \begin{aligned} -2x + 2y &= -14 \quad (2) \\ 2x - 2y &= -8 \quad (3) \\ \hline 0 &= -22 \end{aligned} \quad \xrightarrow{\text{Add}} \quad \begin{cases} x + 2y - z = -9 & (1) \\ -x + y = -7 & (2) \\ 0 = -22 & (3) \end{cases}$$

Equation (3) has no solution, so the system is inconsistent. ●

7 Express the Solution of a System of Dependent Equations Containing Three Variables

EXAMPLE 11 Solving a System of Dependent Equations

$$\text{Solve: } \begin{cases} x - 2y - z = 8 & (1) \\ 2x - 3y + z = 23 & (2) \\ 4x - 5y + 5z = 53 & (3) \end{cases}$$

Solution Our plan is to eliminate x from equations (2) and (3). Multiply each side of equation (1) by -2, and add the result to equation (2). Also, multiply each side of equation (1) by -4, and add the result to equation (3).

$$\begin{aligned} x - 2y - z &= 8 \quad (1) \\ 2x - 3y + z &= 23 \quad (2) \end{aligned} \quad \xrightarrow{\text{Multiply by } -2.} \quad \begin{aligned} -2x + 4y + 2z &= -16 \quad (1) \\ 2x - 3y + z &= 23 \quad (2) \\ \hline y + 3z &= 7 \quad \text{Add} \end{aligned}$$

$$\begin{aligned} x - 2y - z &= 8 \quad (1) \\ 4x - 5y + 5z &= 53 \quad (3) \end{aligned} \quad \xrightarrow{\text{Multiply by } -4.} \quad \begin{aligned} -4x + 8y + 4z &= -32 \quad (1) \\ 4x - 5y + 5z &= 53 \quad (3) \\ \hline 3y + 9z &= 21 \quad \text{Add} \end{aligned}$$

$$\begin{cases} x - 2y - z = 8 & (1) \\ y + 3z = 7 & (2) \\ 3y + 9z = 21 & (3) \end{cases}$$

Treat equations (2) and (3) as a system of two equations containing two variables, and eliminate the variable y by multiplying both sides of equation (2) by -3 and adding the result to equation (3).

$$\begin{aligned} y + 3z &= 7 \quad (2) \\ 3y + 9z &= 21 \quad (3) \end{aligned} \quad \xrightarrow{\text{Multiply by } -3.} \quad \begin{aligned} -3y - 9z &= -21 \\ 3y + 9z &= 21 \\ \hline 0 &= 0 \end{aligned} \quad \xrightarrow{\text{Add}} \quad \begin{cases} x - 2y - z = 8 & (1) \\ y + 3z = 7 & (2) \\ 0 = 0 & (3) \end{cases}$$

The original system is equivalent to a system containing two equations, so the equations are dependent and the system has infinitely many solutions. If we solve equation (2) for y, we can express y in terms of z as $y = -3z + 7$. Substitute this expression into equation (1) to determine x in terms of z.

$$\begin{aligned} x - 2y - z &= 8 && (1) \\ x - 2(-3z + 7) - z &= 8 && \text{Substitute } y = -3z + 7 \text{ in (1).} \\ x + 6z - 14 - z &= 8 && \text{Multiply out.} \\ x + 5z &= 22 && \text{Combine like terms.} \\ x &= -5z + 22 && \text{Solve for } x. \end{aligned}$$

We will write the solution to the system as

$$\begin{cases} x = -5z + 22 \\ y = -3z + 7 \end{cases} \quad \text{where } z \text{ can be any real number.}$$

This way of writing the solution makes it easier to find specific solutions of the system. To find specific solutions, choose any value of z and use the equations $x = -5z + 22$ and $y = -3z + 7$ to determine x and y. For example, if $z = 0$, then $x = 22$ and $y = 7$, and if $z = 1$, then $x = 17$ and $y = 4$.

Using ordered triplets, the solution is

$$\{(x, y, z) \mid x = -5z + 22, y = -3z + 7, z \text{ is any real number}\}$$

Now Work PROBLEM 47

Two distinct points in the Cartesian plane determine a unique line. Given three noncollinear points, we can find the unique quadratic function whose graph contains these three points.

EXAMPLE 12 **Curve Fitting**

Find real numbers a, b, and c so that the graph of the quadratic function $y = ax^2 + bx + c$ contains the points $(-1, -4)$, $(1, 6)$, and $(3, 0)$.

Solution The three points must satisfy the equation $y = ax^2 + bx + c$.

For the point $(-1, -4)$ we have: $-4 = a(-1)^2 + b(-1) + c \quad -4 = a - b + c$

For the point $(1, 6)$ we have: $6 = a(1)^2 + b(1) + c \quad 6 = a + b + c$

For the point $(3, 0)$ we have: $0 = a(3)^2 + b(3) + c \quad 0 = 9a + 3b + c$

Determine a, b, and c so that each equation is satisfied. That is, solve the following system of three equations containing three variables:

$$\begin{cases} a - b + c = -4 & (1) \\ a + b + c = 6 & (2) \\ 9a + 3b + c = 0 & (3) \end{cases}$$

Solving this system of equations, we obtain $a = -2$, $b = 5$, and $c = 3$. So the quadratic function whose graph contains the points $(-1, -4)$, $(1, 6)$, and $(3, 0)$ is

$$y = -2x^2 + 5x + 3 \quad y = ax^2 + bx + c, \quad a = -2, b = 5, c = 3$$

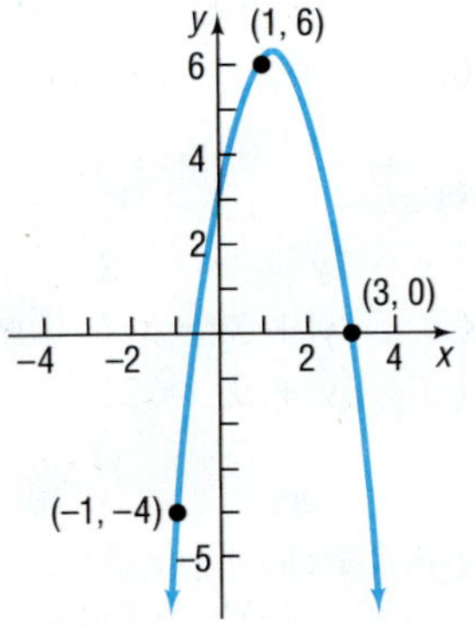

Figure 6 $y = -2x^2 + 5x + 3$

Figure 6 shows the graph of the function, along with the three points.

Now Work PROBLEM 73

8.1 Assess Your Understanding

'Are You Prepared?' *Answers are given at the end of these exercises. If you get a wrong answer, read the pages listed in red.*

1. Solve the equation: $3x + 4 = 8 - x$. (pp. 82–85)

2. (a) Graph the line: $3x + 4y = 12$.
(b) What is the slope of a line parallel to this line? (pp. 167–177)

Concepts and Vocabulary

3. If a system of equations has no solution, it is said to be ________.

4. If a system of equations has one solution, the system is ________ and the equations are ________.

5. If the only solution to a system of two linear equations containing two unknowns is $x = 3, y = -2$, then the lines intersect at the point ________.

6. If the lines that make up a system of two linear equations are coincident, then the system is ________ and the equations are ________.

7. If a system of two linear equations in two variables is inconsistent, which of the following best describes the graphs of the lines in the system?
(a) intersecting (b) parallel
(c) coincident (d) perpendicular

8. If a system of dependent equations containing three variables has the general solution
$\{(x, y, z) \mid x = -z + 4, y = -2z + 5, z \text{ is any real number}\}$
which of the following is one of the infinite number of solutions of the system?
(a) $(1, -1, 3)$ (b) $(0, 4, 5)$ (c) $(4, -3, 0)$ (d) $(-1, 5, 7)$

Skill Building

In Problems 9–18, verify that the values of the variables listed are solutions of the system of equations.

9. $\begin{cases} 2x - y = 5 \\ 5x + 2y = 8 \end{cases}$
$x = 2, y = -1; (2, -1)$

10. $\begin{cases} 3x + 2y = 2 \\ x - 7y = -30 \end{cases}$
$x = -2, y = 4; (-2, 4)$

11. $\begin{cases} 3x - 4y = 4 \\ \frac{1}{2}x - 3y = -\frac{1}{2} \end{cases}$
$x = 2, y = \frac{1}{2}; \left(2, \frac{1}{2}\right)$

12. $\begin{cases} 2x + \frac{1}{2}y = 0 \\ 3x - 4y = -\frac{19}{2} \end{cases}$
$x = -\frac{1}{2}, y = 2; \left(-\frac{1}{2}, 2\right)$

13. $\begin{cases} x - y = 3 \\ \frac{1}{2}x + y = 3 \end{cases}$
$x = 4, y = 1; (4, 1)$

14. $\begin{cases} x - y = 3 \\ -3x + y = 1 \end{cases}$
$x = -2, y = -5; (-2, -5)$

15. $\begin{cases} 3x + 3y + 2z = 4 \\ x - y - z = 0 \\ 2y - 3z = -8 \end{cases}$
$x = 1, y = -1, z = 2;$ $(1, -1, 2)$

16. $\begin{cases} 4x - z = 7 \\ 8x + 5y - z = 0 \\ -x - y + 5z = 6 \end{cases}$
$x = 2, y = -3, z = 1;$ $(2, -3, 1)$

17. $\begin{cases} 3x + 3y + 2z = 4 \\ x - 3y + z = 10 \\ 5x - 2y - 3z = 8 \end{cases}$
$x = 2, y = -2, z = 2; (2, -2, 2)$

18. $\begin{cases} 4x - 5z = 6 \\ 5y - z = -17 \\ -x - 6y + 5z = 24 \end{cases}$
$x = 4, y = -3, z = 2; (4, -3, 2)$

In Problems 19–56, solve each system of equations. If the system has no solution, say that it is inconsistent. For Problems 19–30, graph the lines of the system.

19. $\begin{cases} x + y = 8 \\ x - y = 4 \end{cases}$

20. $\begin{cases} x + 2y = -7 \\ x + y = -3 \end{cases}$

21. $\begin{cases} 5x - y = 21 \\ 2x + 3y = -12 \end{cases}$

22. $\begin{cases} x + 3y = 5 \\ 2x - 3y = -8 \end{cases}$

23. $\begin{cases} 3x = 24 \\ x + 2y = 0 \end{cases}$

24. $\begin{cases} 4x + 5y = -3 \\ -2y = -8 \end{cases}$

25. $\begin{cases} 3x - 6y = 2 \\ 5x + 4y = 1 \end{cases}$

26. $\begin{cases} 2x + 4y = \frac{2}{3} \\ 3x - 5y = -10 \end{cases}$

27. $\begin{cases} 2x + y = 1 \\ 4x + 2y = 3 \end{cases}$

28. $\begin{cases} x - y = 5 \\ -3x + 3y = 2 \end{cases}$

29. $\begin{cases} 2x - y = 0 \\ 4x + 2y = 12 \end{cases}$

30. $\begin{cases} 3x + 3y = -1 \\ 4x + y = \frac{8}{3} \end{cases}$

31. $\begin{cases} x + 2y = 4 \\ 2x + 4y = 8 \end{cases}$

32. $\begin{cases} 3x - y = 7 \\ 9x - 3y = 21 \end{cases}$

33. $\begin{cases} 2x - 3y = -1 \\ 10x + y = 11 \end{cases}$

34. $\begin{cases} 3x - 2y = 0 \\ 5x + 10y = 4 \end{cases}$

35. $\begin{cases} 2x + 3y = 6 \\ x - y = \frac{1}{2} \end{cases}$

36. $\begin{cases} \frac{1}{2}x + y = -2 \\ x - 2y = 8 \end{cases}$

37. $\begin{cases} \frac{1}{2}x + \frac{1}{3}y = 3 \\ \frac{1}{4}x - \frac{2}{3}y = -1 \end{cases}$

38. $\begin{cases} \frac{1}{3}x - \frac{3}{2}y = -5 \\ \frac{3}{4}x + \frac{1}{3}y = 11 \end{cases}$

39. $\begin{cases} 3x - 5y = 3 \\ 15x + 5y = 21 \end{cases}$

40. $\begin{cases} 2x - y = -1 \\ x + \frac{1}{2}y = \frac{3}{2} \end{cases}$

41. $\begin{cases} \frac{1}{x} + \frac{1}{y} = 8 \\ \frac{3}{x} - \frac{5}{y} = 0 \end{cases}$

42. $\begin{cases} \frac{4}{x} - \frac{3}{y} = 0 \\ \frac{6}{x} + \frac{3}{2y} = 2 \end{cases}$

[**Hint:** Let $u = \frac{1}{x}$ and $v = \frac{1}{y}$, and solve for u and v. Then $x = \frac{1}{u}$ and $y = \frac{1}{v}$.]

43. $\begin{cases} x - y = 6 \\ 2x - 3z = 16 \\ 2y + z = 4 \end{cases}$

44. $\begin{cases} 2x + y = -4 \\ -2y + 4z = 0 \\ 3x - 2z = -11 \end{cases}$

45. $\begin{cases} x - 2y + 3z = 7 \\ 2x + y + z = 4 \\ -3x + 2y - 2z = -10 \end{cases}$

46. $\begin{cases} 2x + y - 3z = 0 \\ -2x + 2y + z = -7 \\ 3x - 4y - 3z = 7 \end{cases}$

47. $\begin{cases} x - y - z = 1 \\ 2x + 3y + z = 2 \\ 3x + 2y = 0 \end{cases}$

48. $\begin{cases} 2x - 3y - z = 0 \\ -x + 2y + z = 5 \\ 3x - 4y - z = 1 \end{cases}$

49. $\begin{cases} x - y - z = 1 \\ -x + 2y - 3z = -4 \\ 3x - 2y - 7z = 0 \end{cases}$

50. $\begin{cases} 2x - 3y - z = 0 \\ 3x + 2y + 2z = 2 \\ x + 5y + 3z = 2 \end{cases}$

51. $\begin{cases} 2x - 2y + 3z = 6 \\ 4x - 3y + 2z = 0 \\ -2x + 3y - 7z = 1 \end{cases}$

52. $\begin{cases} 3x - 2y + 2z = 6 \\ 7x - 3y + 2z = -1 \\ 2x - 3y + 4z = 0 \end{cases}$

53. $\begin{cases} x + y - z = 6 \\ 3x - 2y + z = -5 \\ x + 3y - 2z = 14 \end{cases}$

54. $\begin{cases} x - y + z = -4 \\ 2x - 3y + 4z = -15 \\ 5x + y - 2z = 12 \end{cases}$

55. $\begin{cases} x + 2y - z = -3 \\ 2x - 4y + z = -7 \\ -2x + 2y - 3z = 4 \end{cases}$

56. $\begin{cases} x + 4y - 3z = -8 \\ 3x - y + 3z = 12 \\ x + y + 6z = 1 \end{cases}$

Applications and Extensions

57. The perimeter of a rectangular floor is 90 feet. Find the dimensions of the floor if the length is twice the width.

58. The length of fence required to enclose a rectangular field is 3000 meters. What are the dimensions of the field if it is known that the difference between its length and width is 50 meters?

59. Orbital Launches In 2013 there was a total of 81 commercial and noncommercial orbital launches worldwide. In addition, the number of noncommercial orbital launches was twelve more than twice the number of commercial orbital launches. Determine the number of commercial and noncommercial orbital launches in 2013.

Source: Federal Aviation Administration

60. Movie Theater Tickets A movie theater charges \$9.00 for adults and \$7.00 for senior citizens. On a day when 325 people paid for admission, the total receipts were \$2495. How many who paid were adults? How many were seniors?

61. Mixing Nuts A store sells cashews for \$5.00 per pound and peanuts for \$1.50 per pound. The manager decides to mix 30 pounds of peanuts with some cashews and sell the mixture for \$3.00 per pound. How many pounds of cashews should be mixed with the peanuts so that the mixture will produce the same revenue as selling the nuts separately?

62. Mixing a Solution A chemist wants to make 14 liters of a 40% acid solution. She has solutions that are 30% acid and 65% acid. How much of each must she mix?

63. Presale Order A wireless store owner takes presale orders for a new smartphone and tablet. He gets 340 preorders for the smartphone and 250 preorders for the tablet. The combined value of the preorders is \$270,500. If the price of a smartphone and tablet together is \$965, how much does each device cost?

64. Financial Planning A recently retired couple needs \$12,000 per year to supplement their Social Security. They have \$150,000 to invest to obtain this income. They have decided on two investment options: AA bonds yielding 10% per annum and a Bank Certificate yielding 5%.

(a) How much should be invested in each to realize exactly \$12,000?

(b) If, after 2 years, the couple requires \$14,000 per year in income, how should they reallocate their investment to achieve the new amount?

65. Computing Wind Speed With a tail wind, a small Piper aircraft can fly 600 miles in 3 hours. Against this same wind, the Piper can fly the same distance in 4 hours. Find the average wind speed and the average airspeed of the Piper.

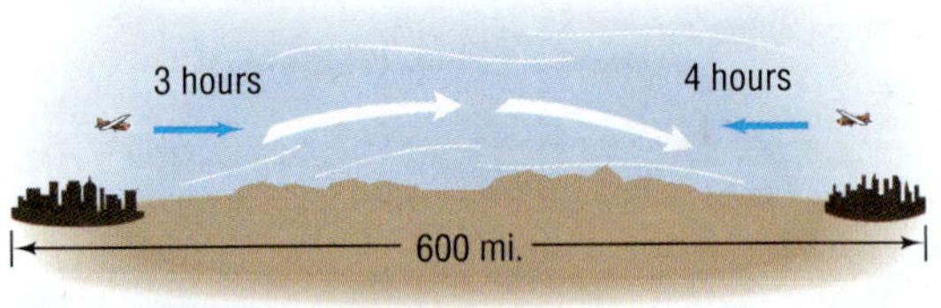

66. Computing Wind Speed The average airspeed of a single-engine aircraft is 150 miles per hour. If the aircraft flew the same distance in 2 hours with the wind as it flew in 3 hours against the wind, what was the wind speed?

67. Restaurant Management A restaurant manager wants to purchase 200 sets of dishes. One design costs \$25 per set, and another costs \$45 per set. If she has only \$7400 to spend, how many sets of each design should she order?

68. Cost of Fast Food One group of people purchased 10 hot dogs and 5 soft drinks at a cost of \$35.00. A second bought 7 hot dogs and 4 soft drinks at a cost of \$25.25. What is the cost of a single hot dog? A single soft drink?

69. Computing a Refund The grocery store we use does not mark prices on its goods. My wife went to this store, bought three 1-pound packages of bacon and two cartons of eggs, and paid a total of \$13.45. Not knowing that she went to the store, I also went to the same store, purchased two 1-pound packages of bacon and three cartons of eggs, and paid a total of \$11.45. Now we want to return two 1-pound packages of bacon and two cartons of eggs. How much will be refunded?

70. Finding the Current of a Stream Pamela requires 3 hours to swim 15 miles downstream on the Illinois River. The return trip upstream takes 5 hours. Find Pamela's average speed in still water. How fast is the current? (Assume that Pamela's speed is the same in each direction.)

71. Pharmacy A doctor's prescription calls for a daily intake containing 40 milligrams (mg) of vitamin C and 30 mg of vitamin D. Your pharmacy stocks two liquids that can be used: One contains 20% vitamin C and 30% vitamin D, the other 40% vitamin C and 20% vitamin D. How many milligrams of each compound should be mixed to fill the prescription?

72. Pharmacy A doctor's prescription calls for the creation of pills that contain 12 units of vitamin B_{12} and 12 units of vitamin E. Your pharmacy stocks two powders that can be used to make these pills: One contains 20% vitamin B_{12} and 30% vitamin E, the other 40% vitamin B_{12} and 20% vitamin E. How many units of each powder should be mixed in each pill?

73. Curve Fitting Find real numbers a, b, and c so that the graph of the function $y = ax^2 + bx + c$ contains the points $(-1, 4)$, $(2, 3)$, and $(0, 1)$.

74. Curve Fitting Find real numbers a, b, and c so that the graph of the function $y = ax^2 + bx + c$ contains the points $(-1, -2)$, $(1, -4)$, and $(2, 4)$.

75. IS–LM Model in Economics In economics, the IS curve is a linear equation that represents all combinations of income Y and interest rates r that maintain an equilibrium in the market for goods in the economy. The LM curve is a linear equation that represents all combinations of income Y and interest rates r that maintain an equilibrium in the market for money in the economy. In an economy, suppose that the equilibrium level of income (in millions of dollars) and interest rates satisfy the system of equations

$$\begin{cases} 0.06Y - 5000r = 240 \\ 0.06Y + 6000r = 900 \end{cases}$$

Find the equilibrium level of income and interest rates.

76. IS–LM Model in Economics In economics, the IS curve is a linear equation that represents all combinations of income Y and interest rates r that maintain an equilibrium in the market for goods in the economy. The LM curve is a linear equation that represents all combinations of income Y and interest rates r that maintain an equilibrium in the market for money in the economy. In an economy, suppose that the equilibrium level of income (in millions of dollars) and interest rates satisfy the system of equations

$$\begin{cases} 0.05Y - 1000r = 10 \\ 0.05Y + 800r = 100 \end{cases}$$

Find the equilibrium level of income and interest rates.

77. Electricity: Kirchhoff's Rules An application of Kirchhoff's Rules to the circuit shown results in the following system of equations:

$$\begin{cases} I_2 = I_1 + I_3 \\ 5 - 3I_1 - 5I_2 = 0 \\ 10 - 5I_2 - 7I_3 = 0 \end{cases}$$

Find the currents I_1, I_2, and I_3.

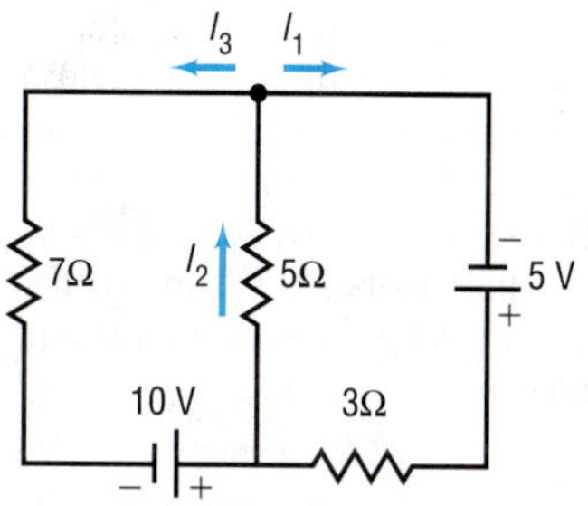

Source: *Physics for Scientists & Engineers*, 9th ed., by Serway. © 2013 Cengage Learning.

78. Electricity: Kirchhoff's Rules An application of Kirchhoff's Rules to the circuit shown below results in the following system of equations:

$$\begin{cases} I_3 = I_1 + I_2 \\ 8 = 4I_3 + 6I_2 \\ 8I_1 = 4 + 6I_2 \end{cases}$$

Find the currents I_1, I_2, and I_3.

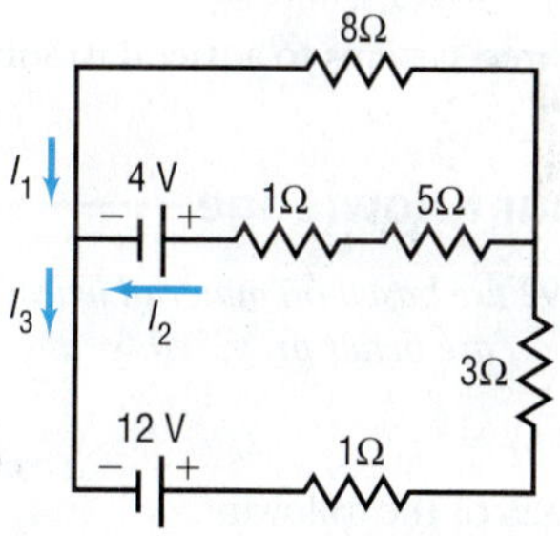

Source: *Physics for Scientists & Engineers*, 9th ed., by Serway. © 2013 Cengage Learning.

79. Theater Revenues A Broadway theater has 500 seats, divided into orchestra, main, and balcony seating. Orchestra seats sell for \$150, main seats for \$135, and balcony seats for \$110. If all the seats are sold, the gross revenue to the theater is \$64,250. If all the main and balcony seats are sold, but only half the orchestra seats are sold, the gross revenue is \$56,750. How many of each kind of seat are there?

80. Theater Revenues A movie theater charges \$11.00 for adults, \$6.50 for children, and \$9.00 for senior citizens. One day the theater sold 405 tickets and collected \$3315 in receipts. Twice as many children's tickets were sold as adult tickets. How many adults, children, and senior citizens went to the theater that day?

81. Nutrition A dietitian wishes a patient to have a meal that has 66 grams (g) of protein, 94.5 g of carbohydrates, and 910 milligrams (mg) of calcium. The hospital food service tells the dietitian that the dinner for today is chicken, corn, and 2% milk. Each serving of chicken has 30 g of protein, 35 g of carbohydrates, and 200 mg of calcium. Each serving of corn has 3 g of protein, 16 g of carbohydrates, and 10 mg of calcium. Each glass of 2% milk has 9 g of protein, 13 g of carbohydrates, and 300 mg of calcium. How many servings of each food should the dietitian provide for the patient?

82. Investments Kelly has \$20,000 to invest. As her financial planner, you recommend that she diversify into three investments: Treasury bills that yield 5% simple interest, Treasury bonds that yield 7% simple interest, and corporate bonds that yield 10% simple interest. Kelly wishes to earn \$1390 per year in income. Also, Kelly wants her investment in Treasury bills to be \$3000 more than her investment in corporate bonds. How much money should Kelly place in each investment?

83. Prices of Fast Food One group of customers bought 8 deluxe hamburgers, 6 orders of large fries, and 6 large colas for \$26.10. A second group ordered 10 deluxe hamburgers, 6 large fries, and 8 large colas and paid \$31.60. Is there sufficient information to determine the price of each food item? If not, construct a table showing the various possibilities. Assume that the hamburgers cost between \$1.75 and \$2.25, the fries between \$0.75 and \$1.00, and the colas between \$0.60 and \$0.90.

84. Prices of Fast Food Use the information given in Problem 83. Suppose that a third group purchased 3 deluxe hamburgers, 2 large fries, and 4 large colas for \$10.95. Now is there sufficient information to determine the price of each food item? If so, determine each price.

85. Painting a House Three painters (Beth, Bill, and Edie), working together, can paint the exterior of a home in 10 hours (h). Bill and Edie together have painted a similar house in 15 h. One day, all three worked on this same kind of house for 4 h, after which Edie left. Beth and Bill required 8 more h to finish. Assuming no gain or loss in efficiency, how long should it take each person to complete such a job alone?

Explaining Concepts: Discussion and Writing

86. Make up a system of three linear equations containing three variables that has:
   (a) No solution
   (b) Exactly one solution
   (c) Infinitely many solutions

   Give the three systems to a friend to solve and critique.

87. Write a brief paragraph outlining your strategy for solving a system of two linear equations containing two variables.

88. Do you prefer the method of substitution or the method of elimination for solving a system of two linear equations containing two variables? Give your reasons.

Retain Your Knowledge

Problems 89–92 are based on material learned earlier in the course. The purpose of these problems is to keep the material fresh in your mind so that you are better prepared for the final exam.

89. Graph $f(x) = -3^{1-x} + 2$.

90. Factor each of the following:

(a) $4(2x-3)^3\cdot 2\cdot (x^3+5)^2 + 2(x^3+5)\cdot 3x^2\cdot (2x-3)^4$

(b) $\frac{1}{2}(3x-5)^{-\frac{1}{2}}\cdot 3\cdot (x+3)^{-\frac{1}{2}} - \frac{1}{2}(x+3)^{-\frac{3}{2}}(3x-5)^{\frac{1}{2}}$

91. Solve: $|3x-2| + 5 \le 9$.

92. The exponential function $f(x) = 1 + 2^x$ is one-to-one. Find f^{-1}.

'Are You Prepared?' Answers

1. $\{1\}$ **2.** (a)

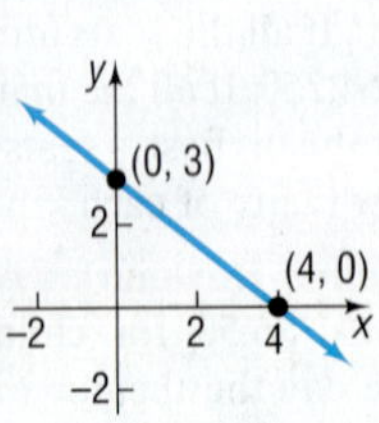

(b) $-\frac{3}{4}$

Appendix
Graphing Utilities

Outline

1 The Viewing Rectangle

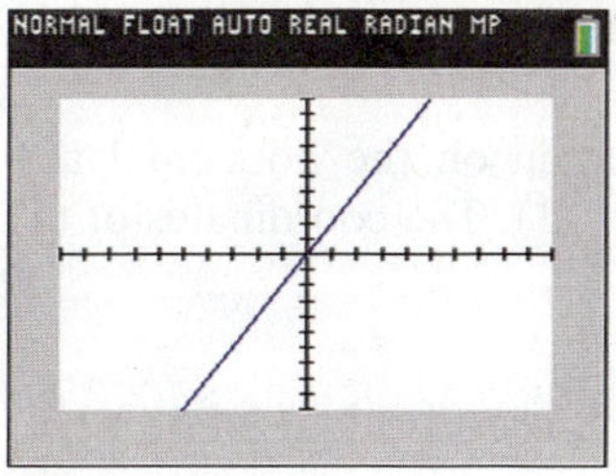

Figure 1 $y = 2x$

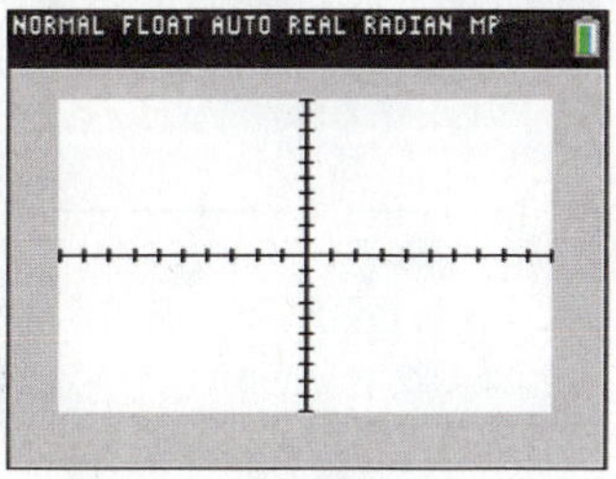

Figure 2 Viewing window

All graphing utilities (that is, all graphing calculators and all computer software graphing packages) graph equations by plotting points on a screen. The screen itself actually consists of small rectangles called **pixels.** The more pixels the screen has, the better the resolution. Most graphing calculators have 50 to 100 pixels per inch; most smartphones have 300 to 450 pixels per inch. When a point to be plotted lies inside a pixel, the pixel is turned on (lights up). The graph of an equation is a collection of pixels. Figure 1 shows how the graph of $y = 2x$ looks on a TI-84 Plus C graphing calculator.

The screen of a graphing utility will display the coordinate axes of a rectangular coordinate system. However, the scale must be set on each axis. The smallest and largest values of x and y to be included in the graph must also be set. This is called **setting the viewing rectangle** or **viewing window.** Figure 2 shows a typical viewing window.

To select the viewing window, values must be given to the following expressions:

Xmin:	the smallest value of x
Xmax:	the largest value of x
Xscl:	the number of units per tick mark on the x-axis
Ymin:	the smallest value of y
Ymax:	the largest value of y
Yscl:	the number of units per tick mark on the y-axis

Figure 3 illustrates these settings and their relation to the Cartesian coordinate system.

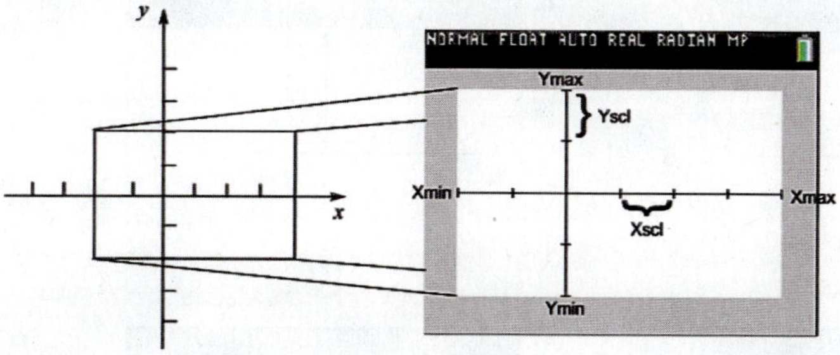

Figure 3

If the scale used on each axis is known, the minimum and maximum values of x and y shown on the screen can be determined by counting the tick marks. Look again at Figure 2. For a scale of 1 on each axis, the minimum and maximum values of x are -10 and 10, respectively; the minimum and maximum values of y are

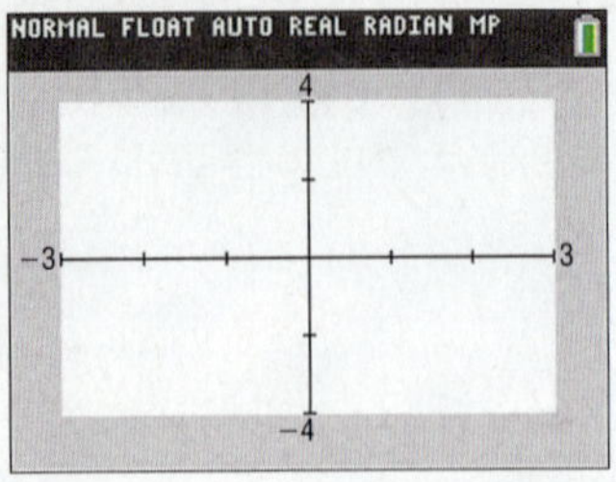

Figure 4

also -10 and 10. If the scale is 2 on each axis, then the minimum and maximum values of x are -20 and 20, respectively; and the minimum and maximum values of y are -20 and 20, respectively.

Conversely, if the minimum and maximum values of x and y are known, the scales can be determined by counting the tick marks displayed. This text follows the practice of showing the minimum and maximum values of x and y in illustrations so that the reader will know how the viewing window was set. See Figure 4. The numbers outside of the viewing window stand for

$$\text{Xmin} = -3 \qquad \text{Ymin} = -4$$
$$\text{Xmax} = 3 \qquad \text{Ymax} = 4$$
$$\text{Xscl} = 1 \qquad \text{Yscl} = 2$$

EXAMPLE 1

Finding the Coordinates of a Point Shown on a Graphing Utility Screen

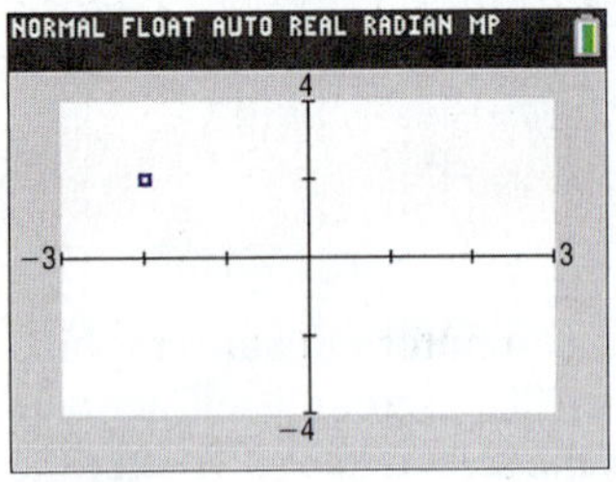

Figure 5

Find the coordinates of the point shown in Figure 5. Assume that the coordinates are integers.

Solution First note that the viewing window used in Figure 5 is

$$\text{Xmin} = -3 \qquad \text{Ymin} = -4$$
$$\text{Xmax} = 3 \qquad \text{Ymax} = 4$$
$$\text{Xscl} = 1 \qquad \text{Yscl} = 2$$

The point shown is 2 tick units to the left of the origin on the horizontal axis $(\text{scale} = 1)$ and 1 tick up on the vertical axis $(\text{scale} = 2)$. The coordinates of the point shown are $(-2, 2)$.

1 Exercises

In Problems 1–4, determine the coordinates of the points shown. Tell in which quadrant each point lies. Assume that the coordinates are integers.

1.
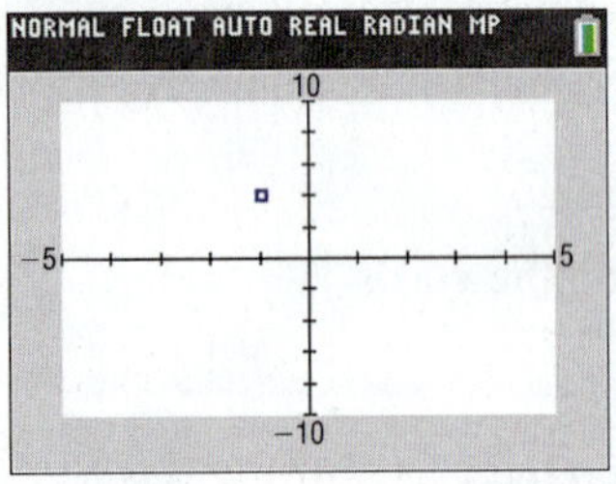

2.
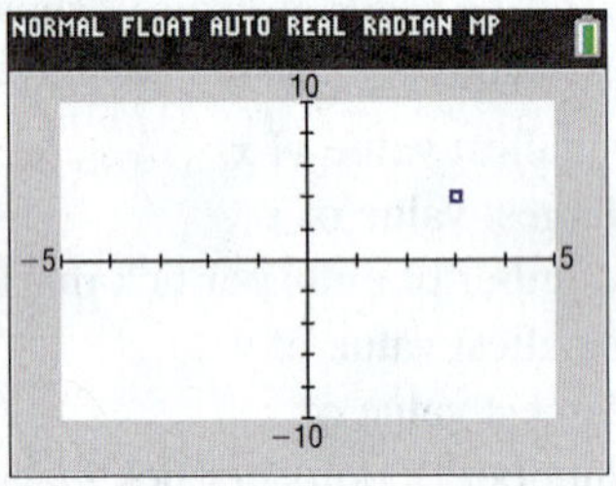

3.
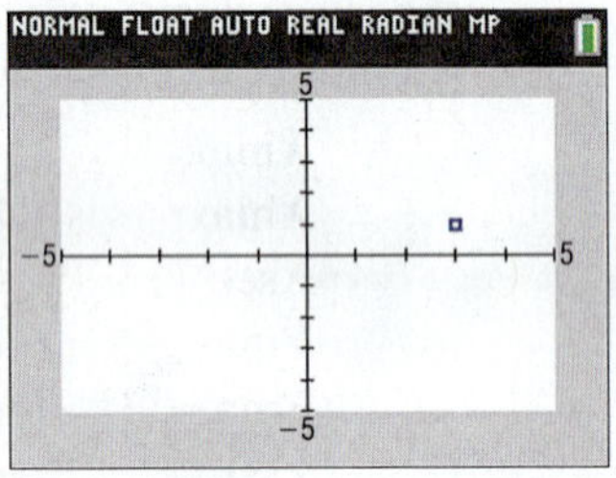

4.
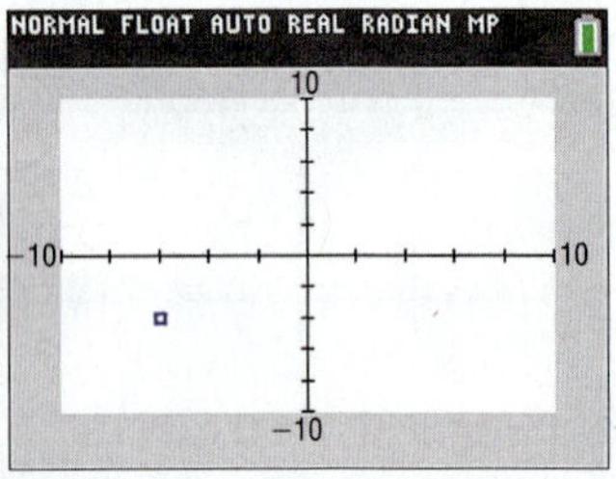

In Problems 5–10, determine the viewing window used.

5.
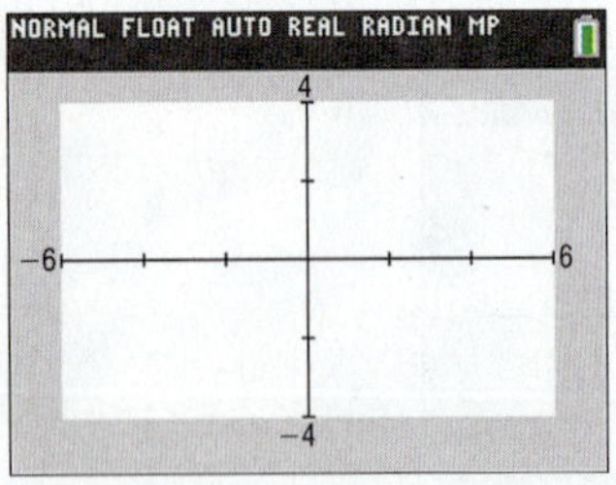

6.
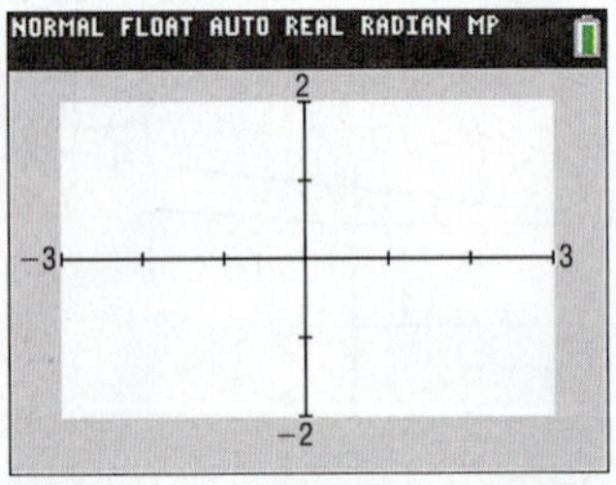

7.
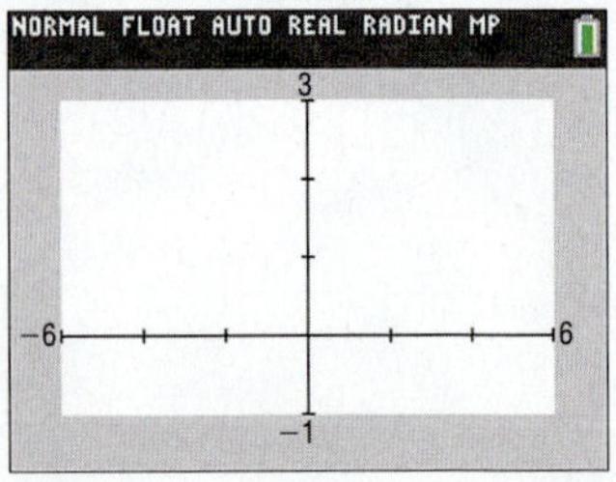

8.
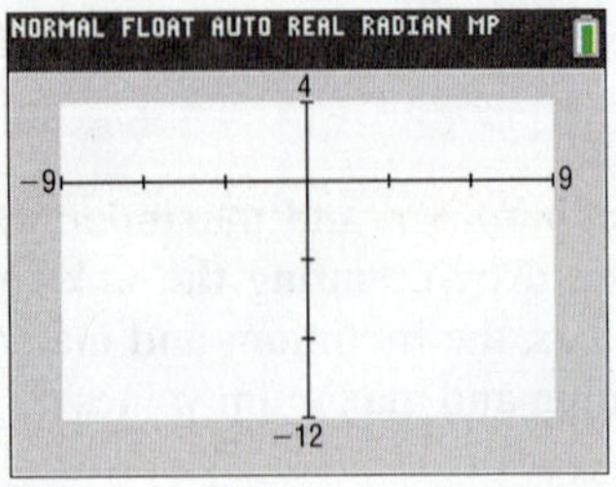

9.
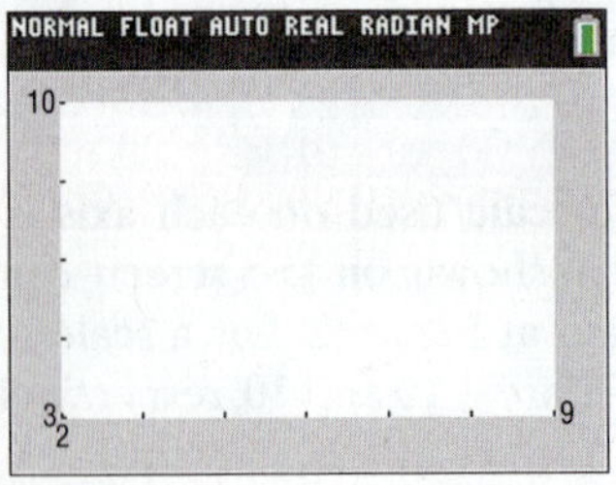

10.
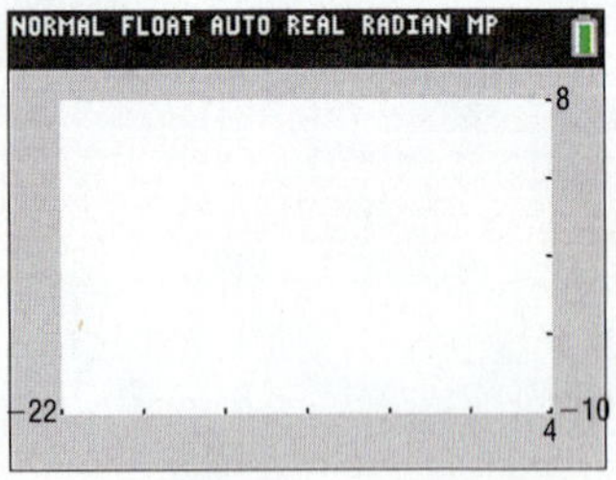

In Problems 11–16, select a setting so that each of the given points will lie within the viewing rectangle.

11. $(-10, 5), (3, -2), (4, -1)$ **12.** $(5, 0), (6, 8), (-2, -3)$ **13.** $(40, 20), (-20, -80), (10, 40)$

14. $(-80, 60), (20, -30), (-20, -40)$ **15.** $(0, 0), (100, 5), (5, 150)$ **16.** $(0, -1), (100, 50), (-10, 30)$

2 Using a Graphing Utility to Graph Equations

From Examples 2 and 3 in Section 2.2, recall that a graph can be obtained by plotting points in a rectangular coordinate system and connecting them. Graphing utilities perform these same steps when graphing an equation. For example, the TI-84 Plus C determines 265 evenly spaced input values,* starting at *X*min and ending at *X*max; uses the equation to determine the output values; plots these points on the screen; and finally (if in the connected mode) draws a line between consecutive points.

To graph an equation in two variables x and y using a graphing utility requires that the equation be written in the form $y = \{expression\ in\ x\}$. If the original equation is not in this form, replace it by equivalent equations until the form $y = \{expression\ in\ x\}$ is obtained.

Steps for Graphing an Equation Using a Graphing Utility

STEP 1: Solve the equation for y in terms of x.

STEP 2: Get into the graphing mode of the graphing utility. The screen will usually display $Y_1 =$, prompting you to enter the expression involving x found in Step 1. (Consult your manual for the correct way to enter the expression; for example, $y = x^2$ might be entered as x^2 or as x*x or as x x^y 2).

STEP 3: Select the viewing window. Without prior knowledge about the behavior of the graph of the equation, it is common to select the **standard viewing window**** initially. The viewing window is then adjusted based on the graph that appears. In this text the standard viewing window is

$$X\text{min} = -10 \qquad Y\text{min} = -10$$
$$X\text{max} = 10 \qquad Y\text{max} = 10$$
$$X\text{scl} = 1 \qquad Y\text{scl} = 1$$

STEP 4: Graph.

STEP 5: Adjust the viewing window until a complete graph is obtained.

EXAMPLE 1

Graphing an Equation on a Graphing Utility

Graph the equation: $6x^2 + 3y = 36$

Solution **STEP 1:** Solve for y in terms of x.

$$6x^2 + 3y = 36$$
$$3y = -6x^2 + 36 \quad \text{Subtract } 6x^2 \text{ from both sides of the equation.}$$
$$y = -2x^2 + 12 \quad \text{Divide both sides of the equation by 3 and simplify.}$$

*These input values depend on the values of *X*min and *X*max. For example, if $X\text{min} = -10$ and $X\text{max} = 10$, then the first input value will be -10 and the next input value will be $-10 + \dfrac{10 - (-10)}{264} = -9.9242$, and so on.

**Some graphing utilities have a ZOOM-STANDARD feature that automatically sets the viewing window to the standard viewing window and graphs the equation.

STEP 2: From the $Y_1 =$ screen, enter the expression $-2x^2 + 12$ after the prompt.

STEP 3: Set the viewing window to the standard viewing window.

STEP 4: Graph. The screen should look like Figure 6.

STEP 5: The graph of $y = -2x^2 + 12$ is not complete. The value of Ymax must be increased so that the top portion of the graph is visible. After increasing the value of Ymax to 12, we obtain the graph in Figure 7. The graph is now complete.

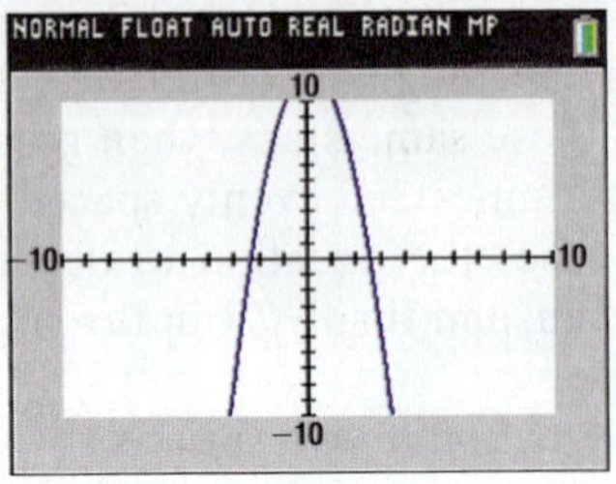

Figure 6 $y = -2x^2 + 12$

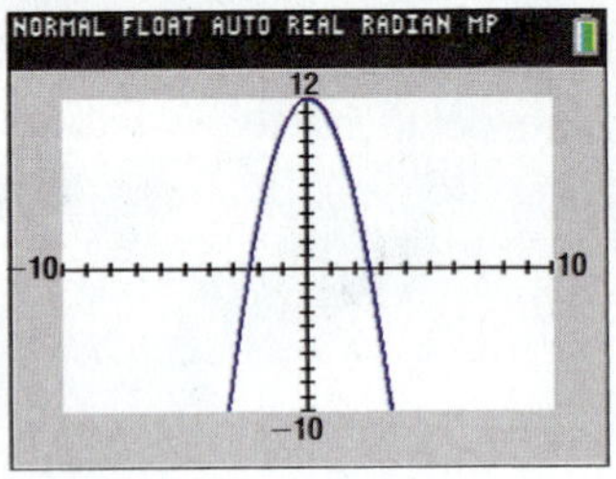

Figure 7 $y = -2x^2 + 12$

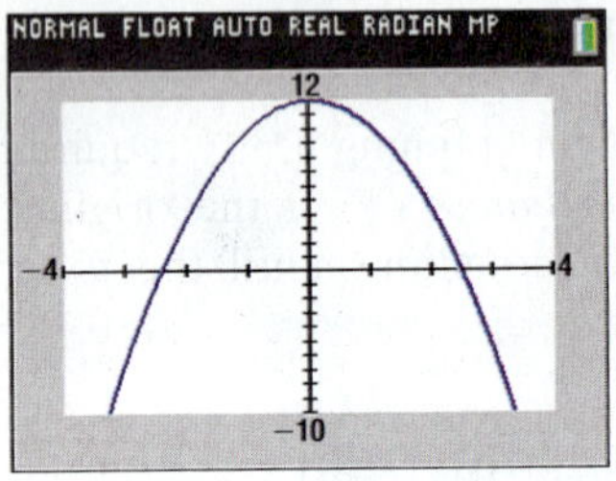

Figure 8 $y = -2x^2 + 12$

Look again at Figure 7. Although a complete graph is shown, the graph might be improved by adjusting the values of Xmin and Xmax. Figure 8 shows the graph of $y = -2x^2 + 12$ using Xmin $= -4$ and Xmax $= 4$. Do you think this is a better choice for the viewing window?

EXAMPLE 2 **Creating a Table and Graphing an Equation**

Create a table and graph the equation: $y = x^3$

Solution Most graphing utilities have the capability of creating a table of values for an equation. (Check your manual to see if your graphing utility has this capability.) Table 1 illustrates a table of values for $y = x^3$ on a TI-84 Plus C. See Figure 9 for the graph.

Table 1

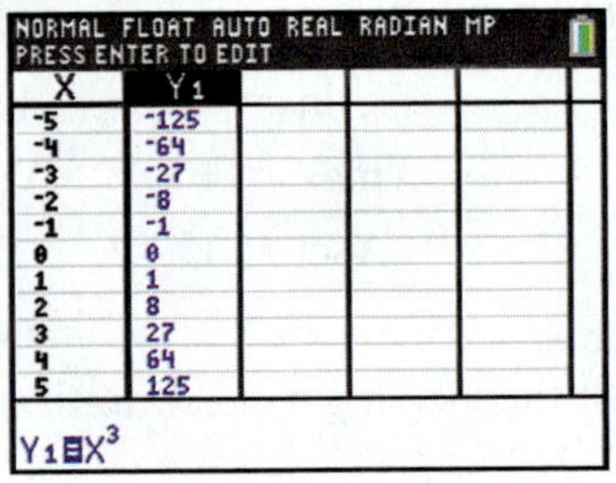

NORMAL FLOAT AUTO REAL RADIAN MP
PRESS ENTER TO EDIT

X	Y₁
-5	-125
-4	-64
-3	-27
-2	-8
-1	-1
0	0
1	1
2	8
3	27
4	64
5	125

$Y_1 = X^3$

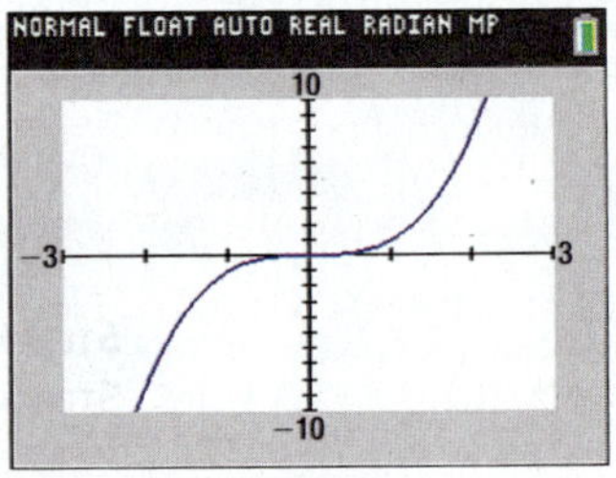

Figure 9 $y = x^3$

2 Exercises

In Problems 1–16, graph each equation using the following viewing windows:

(a) Xmin $= -5$
Xmax $= 5$
Xscl $= 1$
Ymin $= -4$
Ymax $= 4$
Yscl $= 1$

(b) Xmin $= -10$
Xmax $= 10$
Xscl $= 2$
Ymin $= -8$
Ymax $= 8$
Yscl $= 2$

1. $y = x + 2$
2. $y = x - 2$
3. $y = -x + 2$
4. $y = -x - 2$
5. $y = 2x + 2$
6. $y = 2x - 2$
7. $y = -2x + 2$
8. $y = -2x - 2$
9. $y = x^2 + 2$
10. $y = x^2 - 2$
11. $y = -x^2 + 2$
12. $y = -x^2 - 2$
13. $3x + 2y = 6$
14. $3x - 2y = 6$
15. $-3x + 2y = 6$
16. $-3x - 2y = 6$

17–32. *For each of the above equations, create a table, $-5 \le x \le 5$, and list points on the graph.*

3 Using a Graphing Utility to Locate Intercepts and Check for Symmetry

Value and Zero (or Root)

Most graphing utilities have an eVALUEate feature that, given a value of x, determines the value of y for an equation. This feature can be used to evaluate an equation at $x = 0$ to determine the y-intercept. Most graphing utilities also have a ZERO (or ROOT) feature that can be used to determine the x-intercept(s) of an equation.

EXAMPLE 1 **Finding Intercepts Using a Graphing Utility**

Use a graphing utility to find the intercepts of the equation $y = x^3 - 8$.

Solution Figure 10(a) shows the graph of $y = x^3 - 8$.

The eVALUEate feature of a TI-84 Plus C graphing calculator accepts as input a value of x and determines the value of y. Letting $x = 0$, we find that the y-intercept is -8. See Figure 10(b).

The ZERO feature of a TI-84 Plus C is used to find the x-intercept(s). See Figure 10(c). The x-intercept is 2.

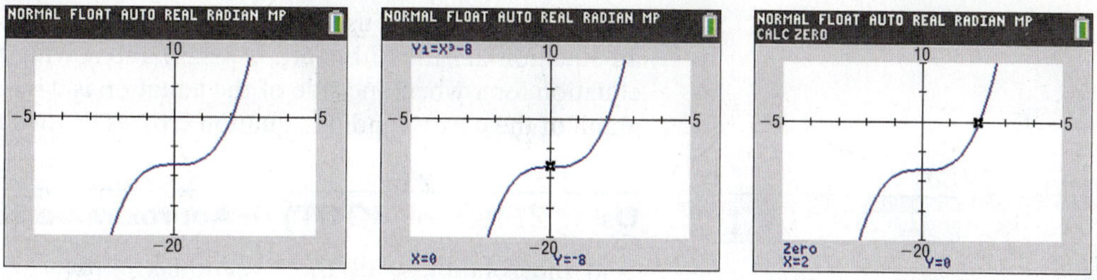

Figure 10 (a) (b) (c)

EXAMPLE 2 **Graphing the Equation $y = \frac{1}{x}$**

Graph the equation $y = \frac{1}{x}$. Based on the graph, infer information about intercepts and symmetry.

Solution Figure 11 shows the graph. Infer from the graph that there are no intercepts; also infer that symmetry with respect to the origin is a possibility. The TABLE feature on a graphing utility can provide further evidence of symmetry with respect to the origin. Using a TABLE, observe that for any ordered pair (x, y), the ordered pair $(-x, -y)$ is also a point on the graph.

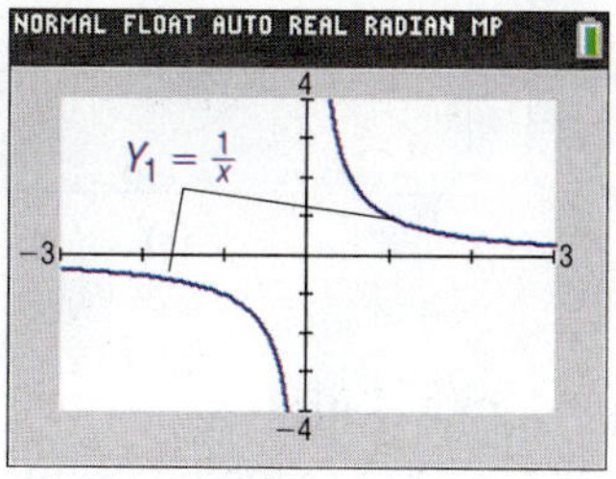

Figure 11

3 Exercises

In Problems 1–6, use ZERO (or ROOT) to approximate the smaller of the two x-intercepts of each equation. Express the answer rounded to two decimal places.

1. $y = x^2 + 4x + 2$

2. $y = x^2 + 4x - 3$

3. $y = 2x^2 + 4x + 1$

4. $y = 3x^2 + 5x + 1$

5. $y = 2x^2 - 3x - 1$

6. $y = 2x^2 - 4x - 1$

In Problems 7–12, use ZERO (or ROOT) to approximate the ***positive*** *x-intercepts of each equation. Express each answer rounded to two decimal places.*

7. $y = x^3 + 3.2x^2 - 16.83x - 5.31$

8. $y = x^3 + 3.2x^2 - 7.25x - 6.3$

9. $y = x^4 - 1.4x^3 - 33.71x^2 + 23.94x + 292.41$

10. $y = x^4 + 1.2x^3 - 7.46x^2 - 4.692x + 15.2881$

11. $y = x^3 + 19.5x^2 - 1021x + 1000.5$

12. $y = x^3 + 14.2x^2 - 4.8x - 12.4$

4 Using a Graphing Utility to Solve Equations

For many equations, there are no algebraic techniques that lead to a solution. For such equations, a graphing utility can often be used to investigate possible solutions. When a graphing utility is used to solve an equation, *approximate* solutions usually are obtained. Unless otherwise stated, this text follows the practice of giving approximate solutions *rounded to two decimal places*.

The ZERO (or ROOT) feature of a graphing utility can be used to find the solutions of an equation when one side of the equation is 0. In using this feature to solve equations, make use of the fact that the x-intercepts (or zeros) of the graph of an equation are found by letting $y = 0$ and solving the equation for x. Solving an equation for x when one side of the equation is 0 is equivalent to finding where the graph of the corresponding equation crosses or touches the x-axis.

EXAMPLE 1 Using ZERO (or ROOT) to Approximate Solutions of an Equation

Find the solution(s) of the equation $x^2 - 6x + 7 = 0$. Round answers to two decimal places.

Solution The solutions of the equation $x^2 - 6x + 7 = 0$ are the same as the x-intercepts of the graph of $Y_1 = x^2 - 6x + 7$. Begin by graphing the equation. See Figure 12(a).

From the graph there appear to be two x-intercepts (solutions to the equation): one between 1 and 2, the other between 4 and 5.

Using the ZERO (or ROOT) feature of the graphing utility, determine that the x-intercepts, and thus the solutions to the equation, are $x = 1.59$ and $x = 4.41$, rounded to two decimal places. See Figures 12(b) and (c).

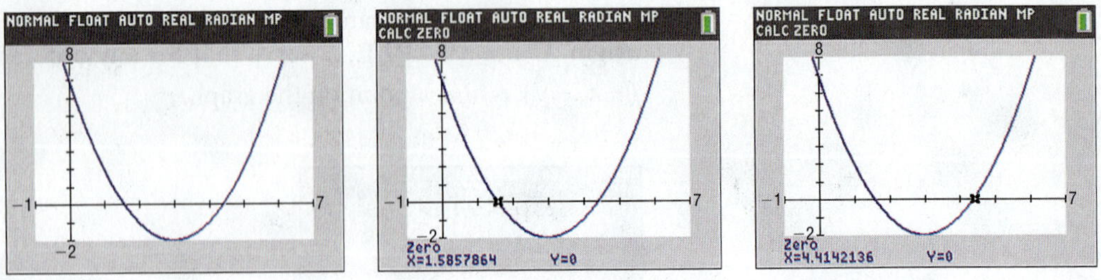

Figure 12 (a) (b) (c)

A second method for solving equations using a graphing utility involves the INTERSECT feature of the graphing utility. This feature is used most effectively when one side of the equation is not 0.

EXAMPLE 2 **Using INTERSECT to Approximate Solutions of an Equation**

Find the solution(s) of the equation $3(x-2) = 5(x-1)$.

Solution Begin by graphing each side of the equation as follows: graph $Y_1 = 3(x-2)$ and $Y_2 = 5(x-1)$. See Figure 13(a).

At the point of intersection of the graphs, the value of the y-coordinate is the same. Conclude that the x-coordinate of the point of intersection represents the solution of the equation. Do you see why? The INTERSECT feature on a graphing utility determines the point of intersection of the graphs. Using this feature, find that the graphs intersect at $(-0.5, -7.5)$. See Figure 13(b). The solution of the equation is therefore $x = -0.5$.

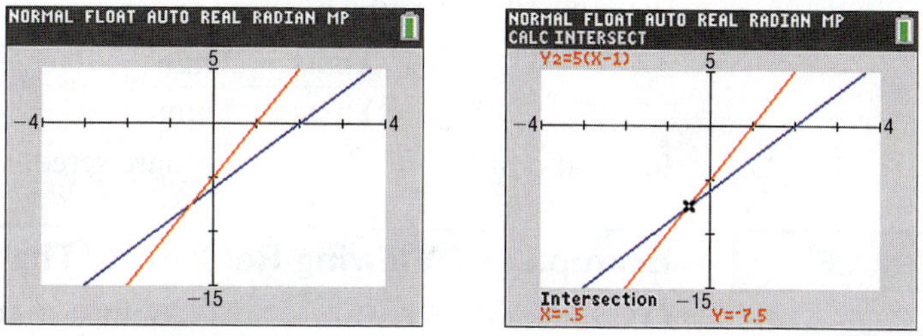

(a) (b)

Figure 13

SUMMARY

The following steps can be used for approximating solutions of equations.

Steps for Approximating Solutions of Equations Using ZERO (or ROOT)

STEP 1: Write the equation in the form $\{expression\ in\ x\} = 0$.

STEP 2: Graph $Y_1 = \{expression\ in\ x\}$.

Be sure that the graph is complete. That is, be sure that all the intercepts are shown on the screen.

STEP 3: Use ZERO (or ROOT) to determine each x-intercept of the graph.

Steps for Approximating Solutions of Equations Using INTERSECT

STEP 1: Graph $Y_1 = \{expression\ in\ x\ on\ the\ left\ side\ of\ the\ equation\}$.

Graph $Y_2 = \{expression\ in\ x\ on\ the\ right\ side\ of\ the\ equation\}$.

STEP 2: Use INTERSECT to determine each x-coordinate of the point(s) of intersection, if any.

Be sure that the graphs are complete. That is, be sure that all the points of intersection are shown on the screen.

EXAMPLE 3 **Solving a Radical Equation**

Find the real solutions of the equation $\sqrt[3]{2x-4} - 2 = 0$.

Solution Figure 14 shows the graph of the equation $Y_1 = \sqrt[3]{2x-4} - 2$. From the graph, there is one x-intercept near 6. Using ZERO (or ROOT), find that the x-intercept is 6. The only solution is $x = 6$.

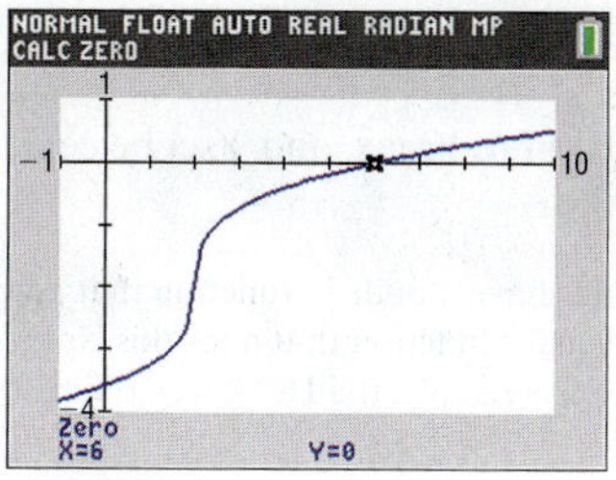

Figure 14

5 Square Screens

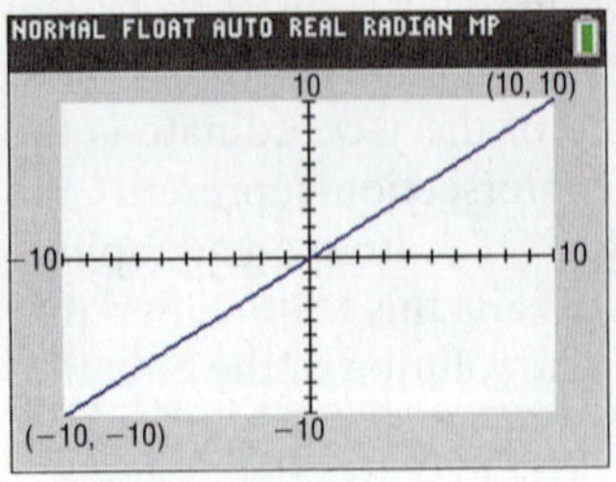

Figure 15 $y = x$

Most graphing utilities have a rectangular screen. Because of this, using the same settings for both x and y will result in a distorted view. For example, Figure 15 shows the graph of the line $y = x$ connecting the points $(-10, -10)$ and $(10, 10)$.

We expect the line to bisect the first and third quadrants, but it doesn't. The selections for Xmin, Xmax, Ymin, and Ymax must be adjusted so that a **square screen** results. On the TI-84 Plus C, this is accomplished by setting the ratio of x to y at 8:5.* For example, if

$$Xmin = -16 \qquad Ymin = -10$$
$$Xmax = 16 \qquad Ymax = 10$$

then the ratio of x to y is

$$\frac{Xmax - Xmin}{Ymax - Ymin} = \frac{16 - (-16)}{10 - (-10)} = \frac{32}{20} = \frac{8}{5}$$

for a ratio of 8:5, resulting in a square screen.

EXAMPLE 1 **Examples of Viewing Rectangles That Result in Square Screens**

(a)		(b)		(c)	
Xmin =	−8	Xmin =	−16	Xmin =	−24
Xmax =	8	Xmax =	16	Xmax =	24
Xscl =	1	Xscl =	2	Xscl =	3
Ymin =	−5	Ymin =	−10	Ymin =	−15
Ymax =	5	Ymax =	10	Ymax =	15
Yscl =	1	Yscl =	1	Yscl =	3

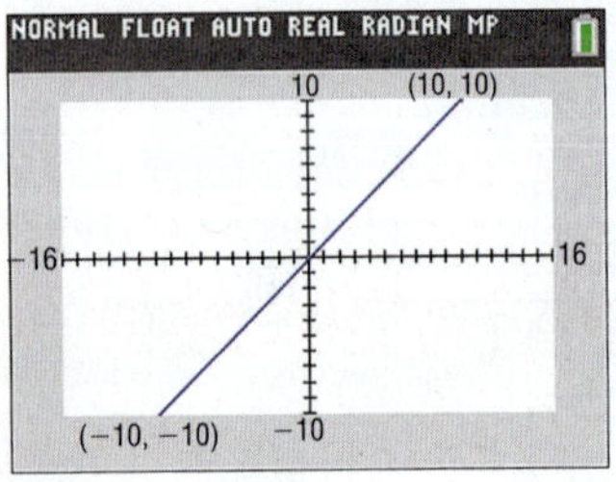

Figure 16 $y = x$

Figure 16 shows the graph of the line $y = x$ on a square screen using the viewing rectangle given in part (b). Notice that the line now bisects the first and third quadrants. Compare this illustration to Figure 15.

5 Exercises

In Problems 1–8, determine which of the given viewing rectangles result in a square screen.

	1.	2.	3.	4.	5.	6.	7.	8.
Xmin =	−8	−5	0	−16	−6	−8	−3	−10
Xmax =	8	5	16	16	6	8	5	14
Xscl =	2	1	4	8	1	4	1	2
Ymin =	−5	−4	−2	−10	−2	−5	−2	−7
Ymax =	5	4	8	10	2	5	3	8
Yscl =	2	1	2	5	0.5	1	1	3

9. If Xmin = −4, Xmax = 12, and Xscl = 1, how should Ymin, Ymax, and Yscl be selected so that the viewing rectangle contains the point $(4, 8)$ and the screen is square?

10. If Xmin = −6, Xmax = 10, and Xscl = 2, how should Ymin, Ymax, and Yscl be selected so that the viewing rectangle contains the point $(4, 8)$ and the screen is square?

*Some graphing utilities have a built-in function that automatically squares the screen. For example, the TI-84 Plus C has a ZSquare function that does this. Some graphing utilities require a ratio other than 8:5 to square the screen. For example, the HP 48G requires the ratio of x to y to be 2:1 for a square screen. Consult your manual.

6 Using a Graphing Utility to Graph Inequalities

EXAMPLE 1 **Graphing an Inequality Using a Graphing Utility**

Use a graphing utility to graph: $3x + y - 6 \leq 0$

Solution Begin by graphing the equation $3x + y - 6 = 0$ $(Y_1 = -3x + 6)$. See Figure 17.

As with graphing by hand, select test points from each region and determine whether they satisfy the inequality. To test the point $(-1, 2)$, for example, enter $3(-1) + 2 - 6 \leq 0$. See Figure 18(a). The 1 that appears indicates that the statement entered (the inequality) is true. When the point $(5, 5)$ is tested, a 0 appears, indicating that the statement entered is false. Thus, $(-1, 2)$ is a part of the graph of the inequality, and $(5, 5)$ is not. Figure 18(b) shows the graph of the inequality on a TI-84 Plus C.*

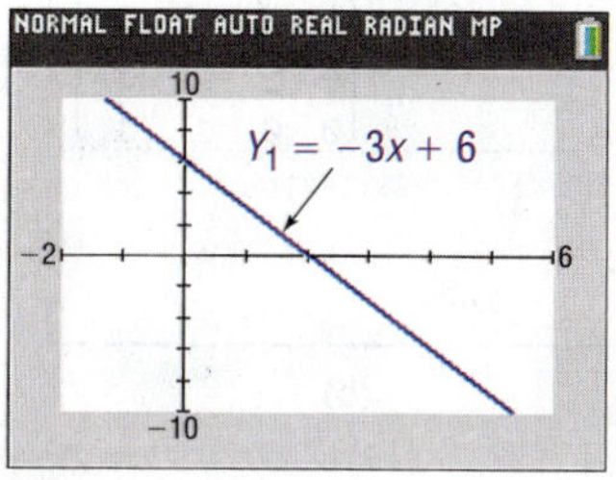

Figure 17

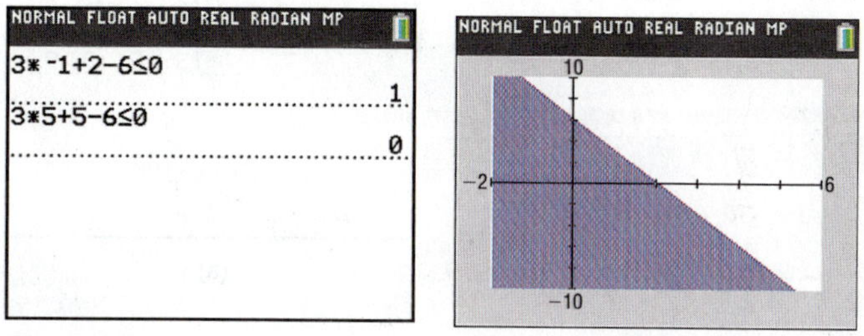

Figure 18 (a) (b)

Steps for Graphing an Inequality Using a Graphing Utility

STEP 1: Replace the inequality symbol by an equal sign, solve the equation for y, and graph the equation.

STEP 2: In each region, select a test point P and determine whether the coordinates of P satisfy the inequality.

(a) If the test point satisfies the inequality, then so do all the points in the region. Indicate this by using the graphing utility to shade the region.

(b) If the coordinates of P do not satisfy the inequality, then neither will any of the other points in that region.

7 Using a Graphing Utility to Solve Systems of Linear Equations

Most graphing utilities have the capability to put the augmented matrix of a system of linear equations in row echelon form. The next example, Example 6 from Section 8.2, page 564, demonstrates this feature using a TI-84 Plus C graphing calculator.

*Consult your owner's manual for shading techniques.

EXAMPLE 1

Solving a System of Linear Equations Using a Graphing Utility

Solve: $\begin{cases} x - y + z = 8 & (1) \\ 2x + 3y - z = -2 & (2) \\ 3x - 2y - 9z = 9 & (3) \end{cases}$

Solution The augmented matrix of the system is

$$\left[\begin{array}{rrr|r} 1 & -1 & 1 & 8 \\ 2 & 3 & -1 & -2 \\ 3 & -2 & -9 & 9 \end{array}\right]$$

Enter this matrix into a graphing utility and name it A. See Figure 19(a). Using the REF (row echelon form) command on matrix A, the results shown in Figure 19(b) are obtained. If the entire matrix does not fit on the screen, scroll right to see the rest of it.

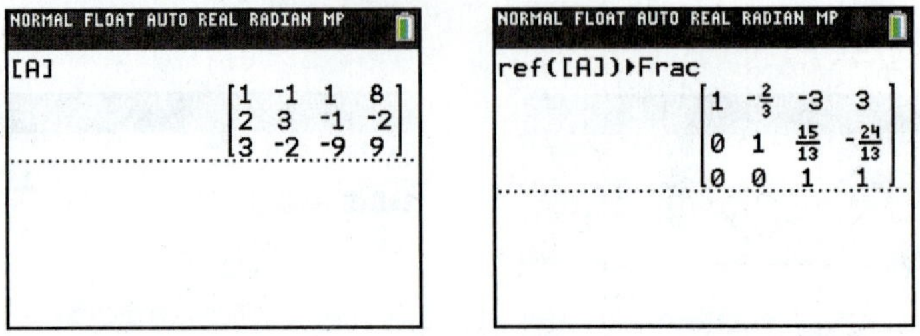

Figure 19 (a) (b)

The system of equations represented by the matrix in row echelon form is

$$\left[\begin{array}{rrr|r} 1 & -\frac{2}{3} & -3 & 3 \\ 0 & 1 & \frac{15}{13} & -\frac{24}{13} \\ 0 & 0 & 1 & 1 \end{array}\right] \quad \begin{cases} x - \frac{2}{3}y - 3z = 3 & (1) \\ y + \frac{15}{13}z = -\frac{24}{13} & (2) \\ z = 1 & (3) \end{cases}$$

Using $z = 1$, back-substitute to get

$$\begin{cases} x - \frac{2}{3}y - 3(1) = 3 & (1) \\ y + \frac{15}{13}(1) = -\frac{24}{13} & (2) \end{cases} \xrightarrow{\text{Simplify.}} \begin{cases} x - \frac{2}{3}y = 6 & (1) \\ y = -\frac{39}{13} = -3 & (2) \end{cases}$$

Solve the second equation for y to find that $y = -3$. Back-substitute $y = -3$ into $x - \frac{2}{3}y = 6$ to find that $x = 4$. The solution of the system is $x = 4$, $y = -3$, $z = 1$. ●

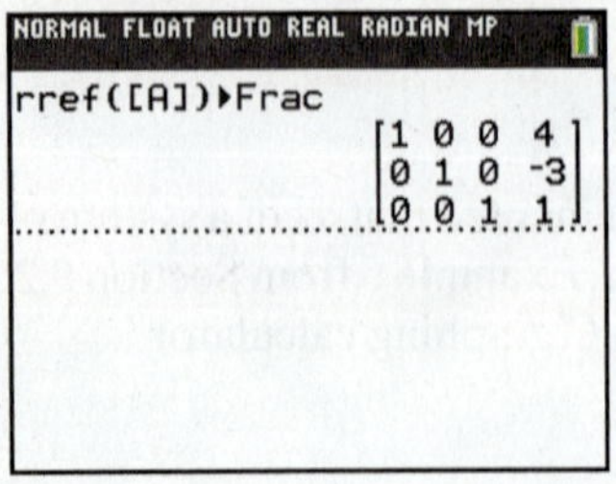

Figure 20

Notice that the row echelon form of the augmented matrix using the graphing utility differs from the row echelon form obtained in Chapter 8 (p. 565), yet both matrices provide the same solution! This is because the two solutions used different row operations to obtain the row echelon form. In all likelihood, the two solutions parted ways in Step 4 of the algebraic solution, where fractions were avoided by interchanging rows 2 and 3.

Most graphing utilities also have the ability to put a matrix in reduced row echelon form. Figure 20 shows the reduced row echelon form of the augmented matrix from Example 1 using the **rref** command on a TI-84 Plus C graphing calculator. Using this command, note that the solution of the system is $x = 4$, $y = -3$, $z = 1$.

Answers

CHAPTER R Review

R.1 Assess Your Understanding *(page 15)*

1. rational **2.** 31 **3.** Distributive **4.** $5(x+3) = 6$ **5.** a **6.** b **7.** T **8.** F **9.** F **10.** T **11.** $\{1, 2, 3, 4, 5, 6, 7, 8, 9\}$ **13.** $\{4\}$ **15.** $\{1, 3, 4, 6\}$ **17.** $\{0, 2, 6, 7, 8\}$ **19.** $\{0, 1, 2, 3, 5, 6, 7, 8, 9\}$ **21.** $\{0, 1, 2, 3, 5, 6, 7, 8, 9\}$ **23.** **(a)** $\{2, 5\}$ **(b)** $\{-6, 2, 5\}$ **(c)** $\left\{-6, \frac{1}{2}, -1.333\cdots, 2, 5\right\}$ **(d)** $\{\pi\}$ **(e)** $\left\{-6, \frac{1}{2}, -1.333..., \pi, 2, 5\right\}$ **25.** **(a)** $\{1\}$ **(b)** $\{0, 1\}$ **(c)** $\left\{0, 1, \frac{1}{2}, \frac{1}{3}, \frac{1}{4}\right\}$ **(d)** None **(e)** $\left\{0, 1, \frac{1}{2}, \frac{1}{3}, \frac{1}{4}\right\}$ **27.** **(a)** None **(b)** None **(c)** None **(d)** $\left\{\sqrt{2}, \pi, \sqrt{2}+1, \pi+\frac{1}{2}\right\}$ **(e)** $\left\{\sqrt{2}, \pi, \sqrt{2}+1, \pi+\frac{1}{2}\right\}$ **29.** **(a)** 18.953 **(b)** 18.952 **31.** **(a)** 28.653 **(b)** 28.653 **33.** **(a)** 0.063 **(b)** 0.062 **35.** **(a)** 9.999 **(b)** 9.998 **37.** **(a)** 0.429 **(b)** 0.428 **39.** **(a)** 34.733 **(b)** 34.733 **41.** $3 + 2 = 5$ **43.** $x + 2 = 3 \cdot 4$ **45.** $3y = 1 + 2$ **47.** $x - 2 = 6$ **49.** $\frac{x}{2} = 6$ **51.** 7 **53.** 6 **55.** 1 **57.** $\frac{13}{3}$ **59.** -11 **61.** 11 **63.** -4 **65.** 1 **67.** 6 **69.** $\frac{2}{7}$ **71.** $\frac{4}{45}$ **73.** $\frac{23}{20}$ **75.** $\frac{79}{30}$ **77.** $\frac{13}{36}$ **79.** $-\frac{16}{45}$ **81.** $\frac{1}{60}$ **83.** $\frac{15}{22}$ **85.** 1 **87.** $\frac{15}{8}$ **89.** $6x + 24$ **91.** $x^2 - 4x$ **93.** $\frac{3}{2}x - 1$ **95.** $x^2 + 6x + 8$ **97.** $x^2 - x - 2$ **99.** $x^2 - 10x + 16$ **101.** $2x + 3x = (2+3)x = 5x$ **103.** $2(3 \cdot 4) = 2 \cdot 12 = 24$; $(2 \cdot 3) \cdot (2 \cdot 4) = 6 \cdot 8 = 48$ **105.** No; $2 - 3 \neq 3 - 2$ **107.** No; $\frac{2}{3} \neq \frac{3}{2}$ **109.** Symmetric Property **111.** No; no

R.2 Assess Your Understanding *(page 26)*

1. variable **2.** origin **3.** strict **4.** base; exponent, or power **5.** 1.2345678×10^3 **6.** d **7.** b **8.** T **9.** T **10.** F **11.** F **12.** F **13.**

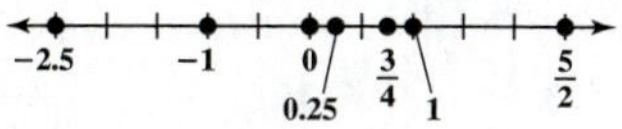

15. $>$ **17.** $>$ **19.** $>$ **21.** $=$ **23.** $<$ **25.** $x > 0$ **27.** $x < 2$ **29.** $x \le 1$

31.

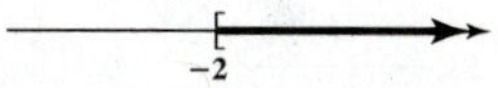

33.

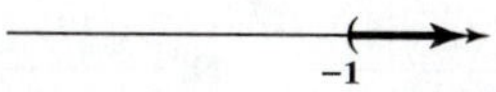

35. 1 **37.** 2 **39.** 6 **41.** 4 **43.** -28 **45.** $\frac{4}{5}$ **47.** 0 **49.** 1 **51.** 5 **53.** 1 **55.** 22 **57.** 2 **59.** $x = 0$ **61.** $x = 3$ **63.** None **65.** $x = 0, x = 1, x = -1$ **67.** $\{x \mid x \neq 5\}$ **69.** $\{x \mid x \neq -4\}$ **71.** 0°C **73.** 25°C **75.** 16 **77.** $\frac{1}{16}$ **79.** $\frac{1}{9}$ **81.** 9 **83.** 5 **85.** 4 **87.** $64x^6$ **89.** $\frac{x^4}{y^2}$ **91.** $\frac{x}{y}$ **93.** $-\frac{8x^3z}{9y}$ **95.** $\frac{16x^2}{9y^2}$ **97.** -4 **99.** 5 **101.** 4 **103.** 2 **105.** $\sqrt{5}$ **107.** $\frac{1}{2}$ **109.** 10; 0 **111.** 81 **113.** 304,006.671 **115.** 0.004 **117.** 481.890 **119.** 0.000 **121.** 4.542×10^2 **123.** 1.3×10^{-2} **125.** 3.2155×10^4 **127.** 4.23×10^{-4} **129.** 61,500 **131.** 0.001214 **133.** 110,000,000 **135.** 0.081 **137.** $A = lw$ **139.** $C = \pi d$ **141.** $A = \frac{\sqrt{3}}{4}x^2$ **143.** $V = \frac{4}{3}\pi r^3$ **145.** $V = x^3$ **147.** **(a)** \$6000 **(b)** \$8000 **149.** $|x - 4| \ge 6$ **151.** **(a)** $2 \le 5$ **(b)** $6 > 5$ **153.** **(a)** Yes **(b)** No **155.** 400,000,000 m **157.** 0.0000005 m **159.** 5×10^{-4} in. **161.** 5.865696×10^{12} mi **163.** No; $\frac{1}{3}$ is larger; 0.000333... **165.** No

R.3 Assess Your Understanding *(page 36)*

1. right; hypotenuse **2.** $A = \frac{1}{2}bh$ **3.** $C = 2\pi r$ **4.** similar **5.** c **6.** b **7.** T **8.** T **9.** F **10.** T **11.** T **12.** F **13.** 13 **15.** 26 **17.** 25 **19.** Right triangle; 5 **21.** Not a right triangle **23.** Right triangle; 25 **25.** Not a right triangle **27.** 8 in.^{2} **29.** 4 in.^{2} **31.** $A = 25\pi$ m^2; $C = 10\pi$ m **33.** $V = 224$ ft^3; $S = 232$ ft^2 **35.** $V = \frac{256}{3}\pi$ cm^3; $S = 64\pi$ cm^2 **37.** $V = 648\pi$ in.^{3}; $S = 306\pi$ in.^{2} **39.** π square units **41.** 2π square units **43.** $x = 4$ units; $A = 90°$; $B = 60°$; $C = 30°$ **45.** $x = 67.5$ units; $A = 60°$; $B = 95°$; $C = 25°$ **47.** About 16.8 ft **49.** 64 ft^2 **51.** $24 + 2\pi \approx 30.28$ ft^2; $16 + 2\pi \approx 22.28$ ft **53.** 160 paces **55.** About 5.477 mi **57.** From 100 ft: 12.2 mi; From 150 ft: 15.0 mi

R.4 Assess Your Understanding *(page 47)*

1. 4; 3 **2.** $x^4 - 16$ **3.** $x^3 - 8$ **4.** a **5.** c **6.** F **7.** T **8.** F **9.** Monomial; variable: x; coefficient: 2; degree: 3 **11.** Not a monomial; the exponent of the variable is not a nonnegative integer **13.** Monomial; variables: x, y; coefficient: -2; degree: 3 **15.** Not a monomial; the exponent of one of the variables is not a nonnegative integer **17.** Not a monomial; it has more than one term **19.** Yes; 2 **21.** Yes; 0 **23.** No; the exponent of the variable of one of the terms is not a nonnegative integer **25.** Yes; 3 **27.** No; the polynomial of the denominator has a degree greater than 0 **29.** $x^2 + 7x + 2$ **31.** $x^3 - 4x^2 + 9x + 7$ **33.** $6x^5 + 5x^4 + 3x^2 + x$ **35.** $7x^2 - x - 7$ **37.** $-2x^3 + 18x^2 - 18$ **39.** $2x^2 - 4x + 6$ **41.** $15y^2 - 27y + 30$ **43.** $x^3 + x^2 - 4x$ **45.** $-8x^5 - 10x^2$ **47.** $x^3 + 3x^2 - 2x - 4$ **49.** $x^2 + 6x + 8$ **51.** $2x^2 + 9x + 10$ **53.** $x^2 - 2x - 8$ **55.** $x^2 - 5x + 6$ **57.** $2x^2 - x - 6$ **59.** $-2x^2 + 11x - 12$ **61.** $2x^2 + 8x + 8$ **63.** $x^2 - xy - 2y^2$ **65.** $-6x^2 - 13xy - 6y^2$ **67.** $x^2 - 49$ **69.** $4x^2 - 9$ **71.** $x^2 + 8x + 16$ **73.** $x^2 - 8x + 16$ **75.** $9x^2 - 16$ **77.** $4x^2 - 12x + 9$ **79.** $x^2 - y^2$ **81.** $9x^2 - y^2$ **83.** $x^2 + 2xy + y^2$ **85.** $x^2 - 4xy + 4y^2$ **87.** $x^3 - 6x^2 + 12x - 8$ **89.** $8x^3 + 12x^2 + 6x + 1$ **91.** $4x^2 - 11x + 23$; remainder -45 **93.** $4x - 3$; remainder $x + 1$ **95.** $5x^2 - 13$; remainder $x + 27$ **97.** $2x^2$; remainder $-x^2 + x + 1$ **99.** $x^2 - 2x + \frac{1}{2}$; remainder $\frac{5}{2}x + \frac{1}{2}$ **101.** $-4x^2 - 3x - 3$; remainder -7 **103.** $x^2 - x - 1$; remainder $2x + 2$ **105.** $x^2 + ax + a^2$; remainder 0

R.5 Assess Your Understanding *(page 57)*

1. $3x(x-2)(x+2)$ **2.** prime **3.** c **4.** b **5.** d **6.** c **7.** T **8.** F **9.** $3(x+2)$ **11.** $a(x^2+1)$ **13.** $x(x^2+x+1)$ **15.** $2x(x-1)$ **17.** $3xy(x-2y+4)$ **19.** $(x+1)(x-1)$ **21.** $(2x+1)(2x-1)$ **23.** $(x+4)(x-4)$ **25.** $(5x+2)(5x-2)$ **27.** $(x+1)^2$ **29.** $(x+2)^2$ **31.** $(x-5)^2$ **33.** $(2x+1)^2$ **35.** $(4x+1)^2$ **37.** $(x-3)(x^2+3x+9)$ **39.** $(x+3)(x^2-3x+9)$ **41.** $(2x+3)(4x^2-6x+9)$ **43.** $(x+2)(x+3)$ **45.** $(x+6)(x+1)$ **47.** $(x+5)(x+2)$ **49.** $(x-8)(x-2)$ **51.** $(x-8)(x+1)$ **53.** $(x+8)(x-1)$ **55.** $(x+2)(2x+3)$ **57.** $(x-2)(2x+1)$ **59.** $(2x+3)(3x+2)$ **61.** $(3x+1)(x+1)$ **63.** $(z+1)(2z+3)$ **65.** $(x+2)(3x-4)$ **67.** $(x-2)(3x+4)$ **69.** $(x+4)(3x+2)$ **71.** $(x+4)(3x-2)$ **73.** $25;\ (x+5)^2$ **75.** $9;\ (y-3)^2$ **77.** $\frac{1}{16};\ \left(x-\frac{1}{4}\right)^2$ **79.** $(x+6)(x-6)$ **81.** $2(1+2x)(1-2x)$ **83.** $(x+1)(x+10)$ **85.** $(x-7)(x-3)$ **87.** $4(x^2-2x+8)$ **89.** Prime **91.** $-(x-5)(x+3)$ **93.** $3(x+2)(x-6)$ **95.** $y^2(y+5)(y+6)$ **97.** $(2x+3)^2$ **99.** $2(3x+1)(x+1)$ **101.** $(x-3)(x+3)(x^2+9)$ **103.** $(x-1)^2(x^2+x+1)^2$ **105.** $x^5(x-1)(x+1)$ **107.** $(4x+3)^2$ **109.** $-(4x-5)(4x+1)$ **111.** $(2y-5)(2y-3)$ **113.** $-(3x-1)(3x+1)(x^2+1)$ **115.** $(x+3)(x-6)$ **117.** $(x+2)(x-3)$ **119.** $(3x-5)(9x^2-3x+7)$ **121.** $(x+5)(3x+11)$ **123.** $(x-1)(x+1)(x+2)$ **125.** $(x-1)(x+1)(x^2-x+1)$ **127.** $2(3x+4)(9x+13)$ **129.** $2x(3x+5)$ **131.** $5(x+3)(x-2)^2(x+1)$ **133.** $3(4x-3)(4x-1)$ **135.** $6(3x-5)(2x+1)^2(5x-4)$ **137.** The possibilities are $(x\pm 1)(x\pm 4)=x^2\pm 5x+4$ or $(x\pm 2)(x\pm 2)=x^2\pm 4x+4$, none of which equals x^2+4.

R.6 Assess Your Understanding *(page 61)*

1. quotient; divisor; remainder **2.** $-3\overline{)2\ 0\ -5\ 1}$ **3.** d **4.** a **5.** T **6.** T **7.** x^2+x+4; remainder 12 **9.** $3x^2+11x+32$; remainder 99 **11.** $x^4-3x^3+5x^2-15x+46$; remainder -138 **13.** $4x^5+4x^4+x^3+x^2+2x+2$; remainder 7 **15.** $0.1x^2-0.11x+0.321$; remainder -0.3531 **17.** $x^4+x^3+x^2+x+1$; remainder 0 **19.** No **21.** Yes **23.** Yes **25.** No **27.** Yes **29.** -9

R.7 Assess Your Understanding *(page 70)*

1. lowest terms **2.** least common multiple **3.** d **4.** a **5.** T **6.** F **7.** $\frac{3}{x-3}$ **9.** $\frac{x}{3}$ **11.** $\frac{4x}{2x-1}$ **13.** $\frac{y+5}{2(y+1)}$ **15.** $\frac{x+5}{x-1}$ **17.** $-(x+7)$ **19.** $\frac{3}{5x(x-2)}$ **21.** $\frac{2x(x^2+4x+16)}{x+4}$ **23.** $\frac{8}{3x}$ **25.** $\frac{x-3}{x+7}$ **27.** $\frac{4x}{(x-2)(x-3)}$ **29.** $\frac{4}{5(x-1)}$ **31.** $-\frac{(x-4)^2}{4x}$ **33.** $\frac{(x+3)^2}{(x-3)^2}$ **35.** $\frac{(x-4)(x+3)}{(x-1)(2x+1)}$ **37.** $\frac{x+5}{2}$ **39.** $\frac{(x-2)(x+2)}{2x-3}$ **41.** $\frac{3x-2}{x-3}$ **43.** $\frac{x+9}{2x-1}$ **45.** $\frac{4-x}{x-2}$ **47.** $\frac{2(x+5)}{(x-1)(x+2)}$ **49.** $\frac{3x^2-2x-3}{(x+1)(x-1)}$ **51.** $\frac{-(11x+2)}{(x+2)(x-2)}$ **53.** $\frac{2(x^2-2)}{x(x-2)(x+2)}$ **55.** $(x-2)(x+2)(x+1)$ **57.** $x(x-1)(x+1)$ **59.** $x^3(2x-1)^2$ **61.** $x(x-1)^2(x+1)(x^2+x+1)$ **63.** $\frac{5x}{(x-6)(x-1)(x+4)}$ **65.** $\frac{2(2x^2+5x-2)}{(x-2)(x+2)(x+3)}$ **67.** $\frac{5x+1}{(x-1)^2(x+1)^2}$ **69.** $\frac{-x^2+3x+13}{(x-2)(x+1)(x+4)}$ **71.** $\frac{x^3-2x^2+4x+3}{x^2(x+1)(x-1)}$ **73.** $\frac{-1}{x(x+h)}$ **75.** $\frac{x+1}{x-1}$ **77.** $\frac{(x-1)(x+1)}{2x(2x+1)}$ **79.** $\frac{2(5x-1)}{(x-2)(x+1)^2}$ **81.** $\frac{-2x(x^2-2)}{(x+2)(x^2-x-3)}$ **83.** $\frac{-1}{x-1}$ **85.** $\frac{3x-1}{2x+1}$ **87.** $\frac{19}{(3x-5)^2}$ **89.** $\frac{(x+1)(x-1)}{(x^2+1)^2}$ **91.** $\frac{x(3x+2)}{(3x+1)^2}$ **93.** $-\frac{(x+3)(3x-1)}{(x^2+1)^2}$ **95.** $f=\frac{R_1\cdot R_2}{(n-1)(R_1+R_2)};\ \frac{2}{15}$ m

R.8 Assess Your Understanding *(page 78)*

3. index **4.** cube root **5.** b **6.** d **7.** c **8.** c **9.** T **10.** F **11.** 3 **13.** -2 **15.** $2\sqrt{2}$ **17.** $-2x\sqrt[3]{x}$ **19.** x^3y^2 **21.** x^2y **23.** $6\sqrt{x}$ **25.** $3x^2y^3\sqrt[4]{2x}$ **27.** $6x\sqrt{x}$ **29.** $15\sqrt[3]{3}$ **31.** $12\sqrt{3}$ **33.** $7\sqrt{2}$ **35.** $\sqrt{2}$ **37.** $2\sqrt{3}$ **39.** $-\sqrt[3]{2}$ **41.** $x-2\sqrt{x}+1$ **43.** $(2x-1)\sqrt[3]{2x}$ **45.** $(2x-15)\sqrt{2x}$ **47.** $-(x+5y)\sqrt[3]{2xy}$ **49.** $\frac{\sqrt{2}}{2}$ **51.** $-\frac{\sqrt{15}}{5}$ **53.** $\frac{(5+\sqrt{2})\sqrt{3}}{23}$ **55.** $\frac{8\sqrt{5}-19}{41}$ **57.** $5\sqrt{2}+5$ **59.** $\frac{5\sqrt[3]{4}}{2}$ **61.** $\frac{2x+h-2\sqrt{x^2+xh}}{h}$ **63.** 4 **65.** -3 **67.** 64 **69.** $\frac{1}{27}$ **71.** $\frac{27\sqrt{2}}{32}$ **73.** $\frac{27\sqrt{2}}{32}$ **75.** $-\frac{1}{10}$ **77.** $\frac{25}{16}$ **79.** $x^{7/12}$ **81.** xy^2 **83.** $x^{2/3}y$ **85.** $\frac{8x^{5/4}}{y^{3/4}}$ **87.** $\frac{3x+2}{(1+x)^{1/2}}$ **89.** $\frac{x(3x^2+2)}{(x^2+1)^{1/2}}$ **91.** $\frac{22x+5}{10\sqrt{x-5}\sqrt{4x+3}}$ **93.** $\frac{2+x}{2(1+x)^{3/2}}$ **95.** $\frac{4-x}{(x+4)^{3/2}}$ **97.** $\frac{1}{x^2(x^2-1)^{1/2}}$ **99.** $\frac{1-3x^2}{2\sqrt{x}(1+x^2)^2}$ **101.** $\frac{1}{2}(5x+2)(x+1)^{1/2}$ **103.** $2x^{1/2}(3x-4)(x+1)$ **105.** $(x^2+4)^{1/3}(11x^2+12)$ **107.** $(3x+5)^{1/3}(2x+3)^{1/2}(17x+27)$ **109.** $\frac{3(x+2)}{2x^{1/2}}$ **111.** 1.41 **113.** 1.59 **115.** 4.89 **117.** 2.15 **119.** (a) 15,660.4 gal **(b)** 390.7 gal **121.** $2\sqrt{2}\pi\approx 8.89$ sec

CHAPTER 1 Equations and Inequalities

1.1 Assess Your Understanding *(page 90)*

4. F **5.** identity **6.** linear; first-degree **7.** F **8.** T **9.** b **10.** d **11.** {3} **13.** {−5} **15.** $\left\{\frac{3}{2}\right\}$ **17.** $\left\{\frac{5}{4}\right\}$ **19.** {−2} **21.** {3} **23.** {−1} **25.** {−2} **27.** {−18} **29.** {−4} **31.** $\left\{-\frac{3}{4}\right\}$ **33.** {−20} **35.** {2} **37.** {0.5} **39.** $\left\{\frac{29}{10}\right\}$ **41.** {2} **43.** {8} **45.** {2} **47.** {−1} **49.** {3} **51.** No solution **53.** No solution **55.** {−6} **57.** {34} **59.** $\left\{-\frac{20}{39}\right\}$ **61.** {−1} **63.** $\left\{-\frac{11}{6}\right\}$ **65.** {−6} **67.** {5.91} **69.** {0.41} **71.** $x=\frac{b+c}{a}$ **73.** $x=\frac{abc}{a+b}$ **75.** $a=3$ **77.** $R=\frac{R_1R_2}{R_1+R_2}$ **79.** $R=\frac{mv^2}{F}$ **81.** $r=\frac{S-a}{S}$ **83.** \$11,500 will be invested in bonds and \$8500 in CDs.

85. The regular hourly rate is \$10.50. **87.** Brooke needs a score of 85. **89.** The original price was \$500,000; purchasing the model saves \$75,000. **91.** The theater paid \$0.80 for the candy. **93.** There were 2187 adults. **95.** The length is 19 ft; the width is 11 ft. **97.** breakfast: 675 cal; lunch: 550 cal; dinner: 800 cal **99.** Judy pays \$10.80 and Tom, \$7.20.

Historical Problems *(page 100)*

1. The area of each shaded square is 9, so the larger square will have area $85 + 4(9) = 121$. The area of the larger square is also given by the expression $(x + 6)^2$ so $(x + 6)^2 = 121$. Taking the positive square root of each side, $x + 6 = 11$ or $x = 5$.
2. Let $z = -6$, so $z^2 + 12z - 85 = -121$. We get the equation $u^2 - 121 = 0$ or $u^2 = 121$. Thus $u = \pm 11$, so $x = \pm 11 - 6$. $x = -17$ or $x = 5$.

3.
$$\left(x + \frac{b}{2a}\right)^2 = \left(\frac{\sqrt{b^2 - 4ac}}{2a}\right)^2$$
$$\left(x + \frac{b}{2a}\right)^2 - \left(\frac{\sqrt{b^2 - 4ac}}{2a}\right)^2 = 0$$
$$\left(x + \frac{b}{2a} - \frac{\sqrt{b^2 - 4ac}}{2a}\right)\left(x + \frac{b}{2a} + \frac{\sqrt{b^2 - 4ac}}{2a}\right) = 0$$
$$\left(x + \frac{b - \sqrt{b^2 - 4ac}}{2a}\right)\left(x + \frac{b + \sqrt{b^2 - 4ac}}{2a}\right) = 0$$
$$x = \frac{-b + \sqrt{b^2 - 4ac}}{2a} \text{ or } x = \frac{-b - \sqrt{b^2 - 4ac}}{2a}$$

1.2 Assess Your Understanding *(page 101)*

6. discriminant; negative **7.** F **8.** F **9.** b **10.** d **11.** $\{0, 9\}$ **13.** $\{-5, 5\}$ **15.** $\{-3, 2\}$ **17.** $\left\{-\frac{1}{2}, 3\right\}$ **19.** $\{-4, 4\}$ **21.** $\{2, 6\}$ **23.** $\left\{\frac{3}{2}\right\}$ **25.** $\left\{-\frac{2}{3}, \frac{3}{2}\right\}$ **27.** $\left\{-\frac{2}{3}, \frac{3}{2}\right\}$ **29.** $\left\{-\frac{3}{4}, 2\right\}$ **31.** $\{-5, 5\}$ **33.** $\{-1, 3\}$ **35.** $\{-3, 0\}$ **37.** $\{-7, 3\}$ **39.** $\left\{-\frac{1}{4}, \frac{3}{4}\right\}$ **41.** $\left\{\frac{-1 - \sqrt{7}}{6}, \frac{-1 + \sqrt{7}}{6}\right\}$ **43.** $\{2 - \sqrt{2}, 2 + \sqrt{2}\}$ **45.** $\{2 - \sqrt{5}, 2 + \sqrt{5}\}$ **47.** $\left\{1, \frac{3}{2}\right\}$ **49.** No real solution **51.** $\left\{\frac{-1 - \sqrt{5}}{4}, \frac{-1 + \sqrt{5}}{4}\right\}$ **53.** $\left\{0, \frac{9}{4}\right\}$ **55.** $\left\{\frac{1}{3}\right\}$ **57.** $\left\{-\frac{2}{3}, 1\right\}$ **59.** $\left\{\frac{3 - \sqrt{29}}{10}, \frac{3 + \sqrt{29}}{10}\right\}$ **61.** $\left\{\frac{-2 - \sqrt{10}}{2}, \frac{-2 + \sqrt{10}}{2}\right\}$ **63.** $\left\{\frac{1 - \sqrt{33}}{8}, \frac{1 + \sqrt{33}}{8}\right\}$ **65.** $\left\{\frac{9 - \sqrt{73}}{2}, \frac{9 + \sqrt{73}}{2}\right\}$ **67.** $\{0.63, 3.47\}$ **69.** $\{-2.80, 1.07\}$ **71.** $\{-0.85, 1.17\}$ **73.** No real solution **75.** Repeated real solution **77.** Two unequal real solutions **79.** $\{-\sqrt{5}, \sqrt{5}\}$ **81.** $\left\{\frac{1}{4}\right\}$ **83.** $\left\{-\frac{3}{5}, \frac{5}{2}\right\}$ **85.** $\left\{-\frac{1}{2}, \frac{2}{3}\right\}$ **87.** $\left\{\frac{-\sqrt{2} + 2}{2}, \frac{-\sqrt{2} - 2}{2}\right\}$ **89.** $\left\{\frac{-1 - \sqrt{17}}{2}, \frac{-1 + \sqrt{17}}{2}\right\}$ **91.** $\{5\}$
93. 2; 5 meters, 12 meters, 13 meters; 20 meters, 21 meters; 29 meters **95.** The dimensions are 11 ft by 13 ft. **97.** The dimensions are 5 m by 8 m. **99.** The dimensions should be 4 ft by 4 ft. **101. a.** The ball strikes the ground after 6 sec. **b.** The ball passes the top of the building on its way down after 5 sec. **103.** The dimensions should be 11.55 cm by 6.55 cm by 3 cm. **105.** The border will be 2.71 ft wide. **107.** The border will be 2.56 ft wide. **109.** The screen of an iPad Air in 4:3 format has an area of 45.16 square inches; the screen of the Google Nexus in 16:10 format has an area of 44.94 square inches. The iPad Air has a larger screen. **111.** 1.1 ft **113.** 29 hr **115.** 37 consecutive integers must be added.
117. $\frac{-b + \sqrt{b^2 - 4ac}}{2a} + \frac{-b - \sqrt{b^2 - 4ac}}{2a} = \frac{-2b}{2a} = -\frac{b}{a}$ **119.** $k = \frac{1}{2}$ or $k = -\frac{1}{2}$
121. $ax^2 + bx + c = 0, x = \frac{-b \pm \sqrt{b^2 - 4ac}}{2a}$; $ax^2 - bx + c = 0, x = \frac{b \pm \sqrt{(-b)^2 - 4ac}}{2a} = \frac{b \pm \sqrt{b^2 - 4ac}}{2a} = -\frac{-b \pm \sqrt{b^2 - 4ac}}{2a}$ **123.** b

1.3 Assess Your Understanding *(page 111)*

4. real; imaginary; imaginary unit **5.** F **6.** T **7.** F **8.** b **9.** a **10.** c **11.** $8 + 5i$ **13.** $-7 + 6i$ **15.** $-6 - 11i$ **17.** $6 - 18i$ **19.** $6 + 4i$ **21.** $10 - 5i$ **23.** 37 **25.** $\frac{6}{5} + \frac{8}{5}i$ **27.** $1 - 2i$ **29.** $\frac{5}{2} - \frac{7}{2}i$ **31.** $-\frac{1}{2} + \frac{\sqrt{3}}{2}i$ **33.** $2i$ **35.** $-i$ **37.** i **39.** -6 **41.** $-10i$ **43.** $-2 + 2i$ **45.** 0 **47.** 0 **49.** $2i$ **51.** $5i$ **53.** $5i$ **55.** $\{-2i, 2i\}$ **57.** $\{-4, 4\}$ **59.** $\{3 - 2i, 3 + 2i\}$ **61.** $\{3 - i, 3 + i\}$ **63.** $\left\{\frac{1}{4} - \frac{1}{4}i, \frac{1}{4} + \frac{1}{4}i\right\}$ **65.** $\left\{\frac{1}{5} - \frac{2}{5}i, \frac{1}{5} + \frac{2}{5}i\right\}$ **67.** $\left\{-\frac{1}{2} - \frac{\sqrt{3}}{2}i, -\frac{1}{2} + \frac{\sqrt{3}}{2}i\right\}$ **69.** $\{2, -1 - \sqrt{3}i, -1 + \sqrt{3}i\}$ **71.** $\{-2, 2, -2i, 2i\}$ **73.** $\{-3i, -2i, 2i, 3i\}$
75. Two complex solutions that are conjugates of each other **77.** Two unequal real solutions **79.** A repeated real solution
81. $2 - 3i$ **83.** 6 **85.** 25 **87.** $2 + 3i$ ohms **89.** $z + \bar{z} = (a + bi) + (a - bi) = 2a$; $z - \bar{z} = (a + bi) - (a - bi) = 2bi$
91. $\overline{z + w} = \overline{(a + bi) + (c + di)} = \overline{(a + c) + (b + d)i} = (a + c) - (b + d)i = (a - bi) + (c - di) = \bar{z} + \bar{w}$

1.4 Assess Your Understanding *(page 117)*

4. F **5.** quadratic in form **6.** T **7.** a **8.** c **9.** $\{1\}$ **11.** No real solution **13.** $\{-13\}$ **15.** $\{-1\}$ **17.** $\{0, 64\}$ **19.** $\{3\}$ **21.** $\{2\}$ **23.** $\left\{-\frac{8}{5}\right\}$ **25.** $\{8\}$ **27.** $\{2\}$ **29.** $\{-1, 3\}$ **31.** $\{1, 5\}$ **33.** $\{1\}$ **35.** $\{5\}$ **37.** $\{2\}$ **39.** $\{-4, 4\}$ **41.** $\{0, 3\}$ **43.** $\{-2, -1, 1, 2\}$ **45.** $\{-1, 1\}$ **47.** $\{-2, 1\}$ **49.** $\{-6, -5\}$ **51.** $\left\{-\frac{1}{3}\right\}$ **53.** $\left\{-\frac{3}{2}, 2\right\}$ **55.** $\left\{0, \frac{1}{16}\right\}$ **57.** $\{16\}$ **59.** $\{1\}$ **61.** $\left\{\left(\frac{9 - \sqrt{17}}{8}\right)^4, \left(\frac{9 + \sqrt{17}}{8}\right)^4\right\}$ **63.** $\{\sqrt{2}, \sqrt{3}\}$ **65.** $\{-4, 1\}$ **67.** $\left\{-2, -\frac{1}{2}\right\}$ **69.** $\left\{-\frac{3}{2}, \frac{1}{3}\right\}$ **71.** $\left\{-\frac{1}{8}, 27\right\}$ **73.** $\left\{-2, -\frac{4}{5}\right\}$ **75.** $\{-3, 0, 3\}$ **77.** $\left\{0, \frac{3}{4}\right\}$ **79.** $\{-5, 0, 4\}$ **81.** $\{-1, 1\}$ **83.** $\{-2, 2, 3\}$ **85.** $\left\{-2, \frac{1}{2}, 2\right\}$

87. $\left\{\frac{2}{5}\right\}$ **89.** $\left\{0, \frac{5}{2}, 3\right\}$ **91.** {0.34, 11.66} **93.** {−1.03, 1.03} **95.** {−1.85, 0.17} **97.** $\left\{\frac{3}{2}, 5\right\}$ **99.** The depth of the well is 229.94 ft. **101.** 220.7 ft
105. −1 is extraneous.

1.5 Assess Your Understanding *(page 127)*

3. closed interval **4.** Multiplication Properties **5.** T **6.** T **7.** T **8.** F **9.** T **10.** F **11.** b **12.** c **13.** $[0, 2]$; $0 \le x \le 2$ **15.** $[2, \infty)$; $x \ge 2$
17. $[0, 3)$; $0 \le x < 3$ **19. (a)** $6 < 8$ **(b)** $-2 < 0$ **(c)** $9 < 15$ **(d)** $-6 > -10$ **21. (a)** $7 > 0$ **(b)** $-1 > -8$ **(c)** $12 > -9$ **(d)** $-8 < 6$
23. (a) $2x + 4 < 5$ **(b)** $2x - 4 < -3$ **(c)** $6x + 3 < 6$ **(d)** $-4x - 2 > -4$

25. $[0, 4]$

27. $[4, 6)$

29. $[4, \infty)$
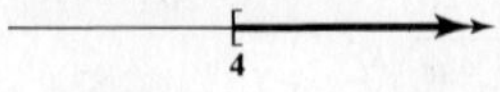

31. $(-\infty, -4)$
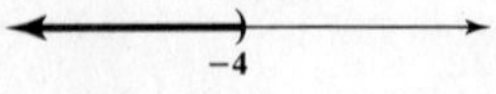

33. $2 \le x \le 5$

35. $-3 < x < -2$

37. $x \ge 4$
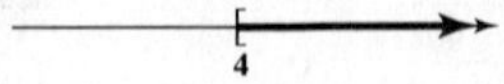

39. $x < -3$
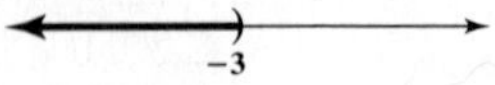

41. $<$ **43.** $>$ **45.** $\ge$ **47.** $<$ **49.** $\le$ **51.** $>$ **53.** $\ge$

55. $\{x \mid x < 4\}$ or $(-\infty, 4)$
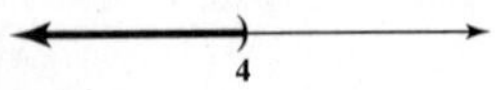

57. $\{x \mid x \ge -1\}$ or $[-1, \infty)$
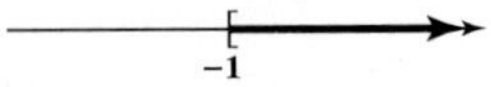

59. $\{x \mid x > 3\}$ or $(3, \infty)$
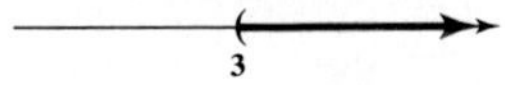

61. $\{x \mid x \ge 2\}$ or $[2, \infty)$
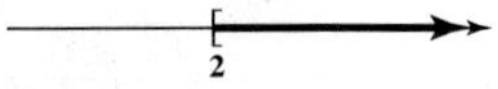

63. $\{x \mid x > -7\}$ or $(-7, \infty)$

65. $\left\{x \middle| x \le \frac{2}{3}\right\}$ or $\left(-\infty, \frac{2}{3}\right]$
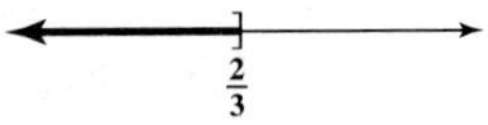

67. $\{x \mid x < -20\}$ or $(-\infty, -20)$
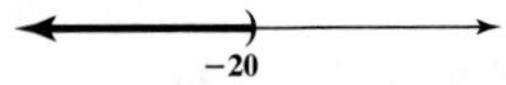

69. $\left\{x \middle| x \ge \frac{4}{3}\right\}$ or $\left[\frac{4}{3}, \infty\right)$
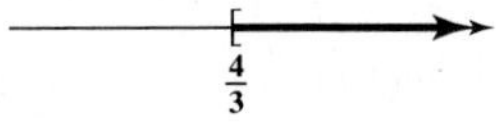

71. $\{x \mid 3 \le x \le 5\}$ or $[3, 5]$

73. $\left\{x \middle| \frac{2}{3} \le x \le 3\right\}$ or $\left[\frac{2}{3}, 3\right]$

75. $\left\{x \middle| -\frac{11}{2} < x < \frac{1}{2}\right\}$ or $\left(-\frac{11}{2}, \frac{1}{2}\right)$
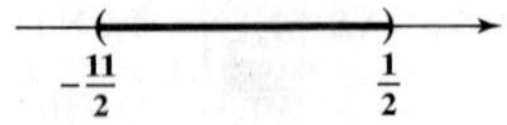

77. $\{x \mid -6 < x < 0\}$ or $(-6, 0)$

79. $\{x \mid x < -5\}$ or $(-\infty, -5)$
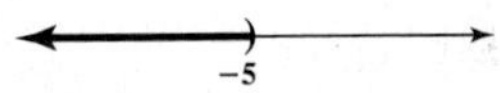

81. $\{x \mid x \ge -1\}$ or $[-1, \infty)$
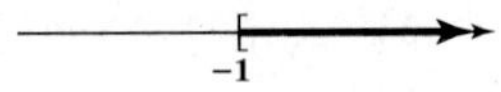

83. $\left\{x \middle| \frac{1}{2} \le x < \frac{5}{4}\right\}$ or $\left[\frac{1}{2}, \frac{5}{4}\right)$
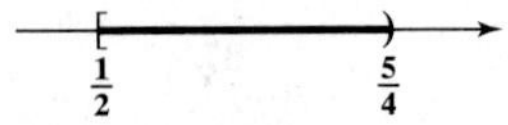

85. $\left\{x \middle| x < -\frac{1}{2}\right\}$ or $\left(-\infty, -\frac{1}{2}\right)$
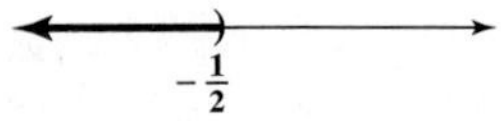

87. $\left\{x \middle| \frac{9}{35} \le x < \frac{2}{7}\right\}$ or $\left[\frac{9}{35}, \frac{2}{7}\right)$
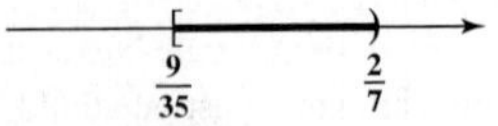

89. $\left\{x \middle| x > \frac{10}{3}\right\}$ or $\left(\frac{10}{3}, \infty\right)$
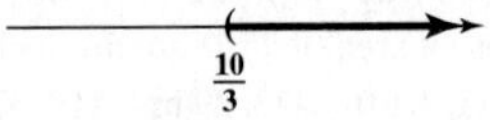

91. $\{x \mid x > 3\}$ or $(3, \infty)$
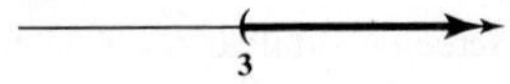

93. $a = 3, b = 5$ **95.** $a = -12, b = -8$ **97.** $a = 3, b = 11$ **99.** $a = \frac{1}{4}, b = 1$ **101.** $a = 4, b = 16$ **103.** $\{x \mid x \ge -2\}$ **105.** $21 < \text{Age} < 30$
107. (a) Male $\ge$ 81.9 years **(b)** Female $\ge$ 85.6 years **(c)** A female could expect to live 3.7 years longer.
109. The agent's commission ranges from \$45,000 to \$95,000, inclusive. As a percent of selling price, the commission ranges from 5% to 8.6%, inclusive.
111. The amount withheld varies from \$134.50 to \$184.50, inclusive. **113.** The usage varied from 700 kW · hr to 2700 kW · hr, inclusive.
115. The dealer's cost varies from \$15,254.24 to \$16,071.43, inclusive.
117. (a) You need at least a 74 on the fifth test. **(b)** You need at least a 77 on the fifth test.
119. $\frac{a+b}{2} - a = \frac{a+b-2a}{2} = \frac{b-a}{2} > 0$; therefore, $a < \frac{a+b}{2}$. $b - \frac{a+b}{2} = \frac{2b-a-b}{2} = \frac{b-a}{2} > 0$; therefore, $b > \frac{a+b}{2}$.
121. $(\sqrt{ab})^2 - a^2 = ab - a^2 = a(b - a) > 0$; thus $(\sqrt{ab})^2 > a^2$ and $\sqrt{ab} > a$.
$b^2 - (\sqrt{ab})^2 = b^2 - ab = b(b - a) > 0$; thus $b^2 > (\sqrt{ab})^2$ and $b > \sqrt{ab}$.
123. $h - a = \frac{2ab}{a+b} - a = \frac{ab - a^2}{a+b} = \frac{a(b-a)}{a+b} > 0$; thus $h > a$. $b - h = b - \frac{2ab}{a+b} = \frac{b^2 - ab}{a+b} = \frac{b(b-a)}{a+b} > 0$; thus $h < b$.
125. Since $0 < a < b$, then $a - b < 0$ and $\frac{a-b}{ab} < 0$. So $\frac{a}{ab} - \frac{b}{ab} < 0$, or $\frac{1}{b} - \frac{1}{a} < 0$. Therefore, $\frac{1}{b} < \frac{1}{a}$. And $0 < \frac{1}{b}$ because $b > 0$.

1.6 Assess Your Understanding *(page 132)*

3. $\{-5, 5\}$ **4.** $\{x \mid -5 < x < 5\}$ **5.** T **6.** T **7.** d **8.** a **9.** $\{-3, 3\}$ **11.** $\{-4, 1\}$ **13.** $\left\{-1, \frac{3}{2}\right\}$ **15.** $\{-4, 4\}$ **17.** $\{2\}$ **19.** $\left\{-\frac{27}{2}, \frac{27}{2}\right\}$ **21.** $\left\{-\frac{36}{5}, \frac{24}{5}\right\}$
23. No solution **25.** $\left\{-\frac{1}{2}, \frac{1}{2}\right\}$ **27.** $\{-3, 3\}$ **29.** $\{-1, 3\}$ **31.** $\{-2, -1, 0, 1\}$ **33.** $\left\{\frac{8}{7}, 4\right\}$ **35.** $\left\{-\frac{1}{2}, 0\right\}$

37. $\{x \mid -4 < x < 4\}$; $(-4, 4)$

39. $\{x \mid x < -4$ or $x > 4\}$; $(-\infty, -4) \cup (4, \infty)$
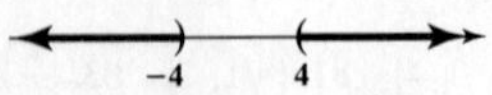

41. $\{x \mid 1 < x < 3\}$; $(1, 3)$

43. $\left\{t \middle| -\frac{2}{3} \le t \le 2\right\}$; $\left[-\frac{2}{3}, 2\right]$
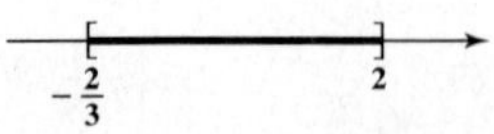

45. $\left\{x \middle| x \le \frac{1}{2} \text{ or } x \ge \frac{5}{2}\right\}; \left(-\infty, \frac{1}{2}\right] \cup \left[\frac{5}{2}, \infty\right)$

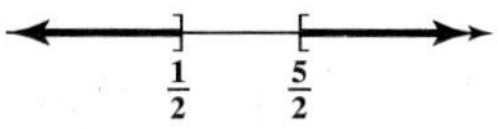

47. $\left\{x \middle| -1 < x < \frac{3}{2}\right\}; \left(-1, \frac{3}{2}\right)$

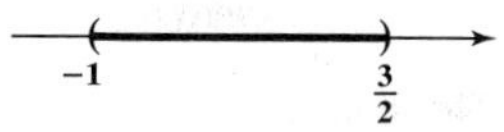

49. $\{x|x < -1 \text{ or } x > 2\}; (-\infty, -1) \text{ or } (2, \infty)$

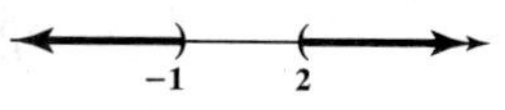

51. No solution

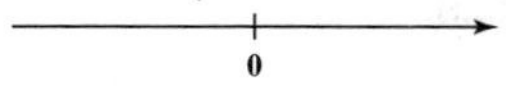

53. $\left\{x \middle| x < -\frac{3}{2} \text{ or } x > \frac{3}{2}\right\}; \left(-\infty, -\frac{3}{2}\right) \cup \left(\frac{3}{2}, \infty\right)$

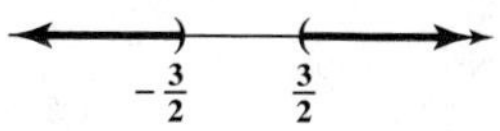

55. $\{x| -1 \le x \le 2\}; [-1, 2]$

57. No solution

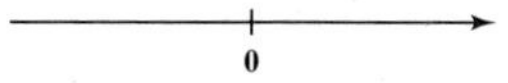

59. All real numbers; $(-\infty, \infty)$

61. $\left\{x \middle| -\frac{9}{4} < x < \frac{3}{4}\right\}; \left(-\frac{9}{4}, \frac{3}{4}\right)$

63. All real numbers; $(-\infty, \infty)$

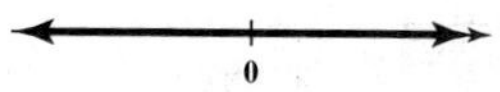

65. $|x - 98.6| \ge 1.5; x \le 97.1$ or $x \ge 101.1$ **67.** $|x - 13.6| < 1.8$; between 11.8 and 15.4 books per year are read. **69.** $|x - 3| < \frac{1}{2}; \frac{5}{2} < x < \frac{7}{2}$
71. $|x + 3| > 2; x < -5$ or $x > -1$ **73.** $a = 2, b = 8$ **75.** $a = -15, b = -7$ **77.** $a = -1, b = -\frac{1}{15}$
79. $b - a = (\sqrt{b} - \sqrt{a})(\sqrt{b} + \sqrt{a})$. Since $\sqrt{b} - \sqrt{a} > 0$, $\sqrt{a} > 0$, $\sqrt{b} > 0$, then $b - a > 0$, so $a < b$.
81. $(a + b)^2 = a^2 + 2ab + b^2 \le |a|^2 + 2|a||b| + |b|^2 = (|a| + |b|)^2$; thus, $|a + b| \le |a| + |b|$.
83. $x^2 - a < 0; (x - \sqrt{a})(x + \sqrt{a}) < 0$; therefore, $-\sqrt{a} < x < \sqrt{a}$. **85.** $\{x| -1 < x < 1\}$ **87.** $\{x|x \le -3 \text{ or } x \ge 3\}$
89. $\{x| -4 \le x \le 4\}$ **91.** $\{x|x < -2 \text{ or } x > 2\}$ **93.** $\{-1, 5\}$

1.7 Assess Your Understanding *(page 140)*

1. mathematical modeling **2.** interest **3.** uniform motion **4.** F **5.** T **6.** a **7.** b **8.** c **9.** $A = \pi r^2$; r = radius, A = area
11. $A = s^2$; A = area, s = length of a side **13.** $F = ma$; F = force, m = mass, a = acceleration **15.** $W = Fd$; W = work, F = force, d = distance
17. $C = 150x$; C = total variable cost, x = number of dishwashers **19.** Invest \$31,250 in bonds and \$18,750 in CDs.
21. \$11,600 was loaned out at 8%. **23.** Mix 75 lb of Earl Grey tea with 25 lb of Orange Pekoe tea. **25.** Mix 160 lb of cashews with the almonds.
27. The speed of the current is 2.286 mi/h. **29.** The speed of the current is 5 mi/h. **31.** Karen walked at 4.05 ft/sec. **33.** A doubles tennis court is 78 feet long and 36 feet wide. **35.** Working together, it takes 12 min. **37. (a)** The dimensions are 10 ft by 5 ft. **(b)** The area is 50 sq ft. **(c)** The dimensions would be 7.5 ft by 7.5 ft. **(d)** The area would be 56.25 sq ft. **39.** The defensive back catches up to the tight end at the tight end's 45-yd line.
41. Add $\frac{2}{3}$ gal of water. **43.** Evaporate 10.67 oz of water. **45.** 40 g of 12-karat gold should be mixed with 20 g of pure gold.
47. Mike passes Dan $\frac{1}{3}$ mile from the start, 2 min from the time Mike started to run. **49.** Start the auxiliary pump at 9:45 AM. **51.** The tub will fill in 1 hr.
53. Run: 12 miles; bicycle: 75 miles **55.** Bolt would beat Burke by 19.25 m. **57.** Set the original price at \$40. At 50% off, there will be no profit.
61. The tail wind was 91.47 knots.

Review Exercises *(page 145)*

1. $\{-18\}$ **2.** $\{6\}$ **3.** $\left\{\frac{1}{5}\right\}$ **4.** $\{6\}$ **5.** No real solution **6.** $\{-3, 2\}$ **7.** $\left\{\frac{11}{8}\right\}$ **8.** $\left\{-\frac{27}{13}\right\}$ **9.** $\left\{-2, \frac{3}{2}\right\}$ **10.** $\left\{\frac{1 - \sqrt{13}}{4}, \frac{1 + \sqrt{13}}{4}\right\}$

11. $\{-3, 3\}$ **12.** $\{2\}$ **13.** No real solution **14.** $\{-2, -1, 1, 2\}$ **15.** $\{2\}$ **16.** $\left\{\frac{13}{2}\right\}$ **17.** $\left\{\frac{\sqrt{5}}{2}\right\}$ **18.** $\{5, 41\}$ **19.** $\left\{\frac{9}{4}\right\}$ **20.** $\left\{-1, \frac{1}{2}\right\}$

21. $\left\{\frac{m}{1 - n}, \frac{m}{1 + n}\right\}$ **22.** $\left\{-\frac{9b}{5a}, \frac{2b}{a}\right\}$ **23.** $\left\{-\frac{9}{5}\right\}$ **24.** $\{-5, 2\}$ **25.** $\left\{-\frac{5}{3}, 3\right\}$ **26.** $\left\{0, \frac{3}{2}\right\}$ **27.** $\left\{-\frac{5}{2}, -2, 2\right\}$

28. $\{x|x \ge 14\}; [14, \infty)$

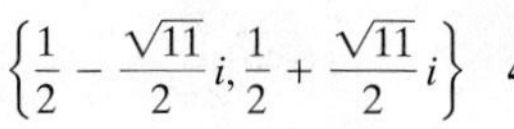

29. $\left\{x \middle| -\frac{31}{2} \le x \le \frac{33}{2}\right\}; \left[-\frac{31}{2}, \frac{33}{2}\right]$

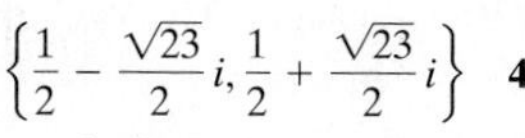

30. $\{x| -23 < x < -7\}; (-23, -7)$

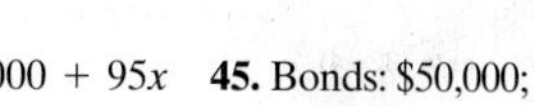

31. $\left\{x \middle| -\frac{3}{2} < x < -\frac{7}{6}\right\}; \left(-\frac{3}{2}, -\frac{7}{6}\right)$

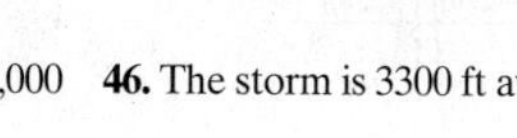

32. $\{x|x \le -2 \text{ or } x \ge 7\}; (-\infty, -2] \cup [7, \infty)$

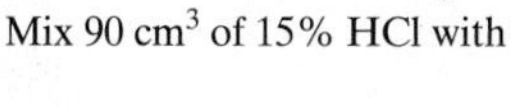

33. $\left\{x \middle| 0 \le x \le \frac{4}{3}\right\}; \left[0, \frac{4}{3}\right]$

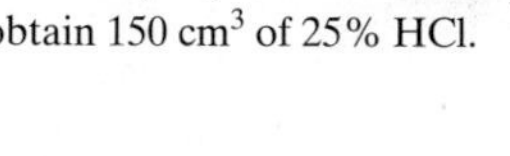

34. $\left\{x \middle| x < -1 \text{ or } x > \frac{7}{3}\right\}; (-\infty, -1) \cup \left(\frac{7}{3}, \infty\right)$

35. $4 + 7i$ **36.** $-3 + 2i$ **37.** $\frac{9}{10} - \frac{3}{10}i$ **38.** -1 **39.** $-46 + 9i$ **40.** $\left\{-\frac{1}{2} - \frac{\sqrt{3}}{2}i, -\frac{1}{2} + \frac{\sqrt{3}}{2}i\right\}$ **41.** $\left\{\frac{-1 - \sqrt{17}}{4}, \frac{-1 + \sqrt{17}}{4}\right\}$

42. $\left\{\frac{1}{2} - \frac{\sqrt{11}}{2}i, \frac{1}{2} + \frac{\sqrt{11}}{2}i\right\}$ **43.** $\left\{\frac{1}{2} - \frac{\sqrt{23}}{2}i, \frac{1}{2} + \frac{\sqrt{23}}{2}i\right\}$ **44.** $C = 50{,}000 + 95x$ **45.** Bonds: \$50,000; CD: \$20,000 **46.** The storm is 3300 ft away.
47. From 0.5 m to 0.75 m **48.** The search plane can go as far as 616 miles. **49.** The helicopter will reach the life raft in a little less than 1 hr 35 min.
50. (a) In 8 sec **(b)** The height is 896 ft. **51.** It takes Clarissa 10 days by herself. **52.** 12 hr 23 min **53.** Add 256 oz of water. **54.** 6 in. by 8 in.
55. Mix 90 cm^3 of 15% HCl with the 60 cm^3 of 40% HCl to obtain 150 cm^3 of 25% HCl.

56. **(a)** 6.5 in. by 6.5 in.; 12.5 in. by 12.5 in. **(b)** $8\frac{2}{3}$ in. by $4\frac{1}{3}$ in.; $14\frac{2}{3}$ in. by $10\frac{1}{3}$ in.

57. Scott receives \$400,000, Alice receives \$300,000, and Tricia receives \$200,000. **58.** It would take the older copier 180 min or 3 hr.

59. **(a)** No **(b)** Todd **(c)** $\frac{1}{4}$ m **(d)** 5.26 m **(e)** Yes **60.** The freight train is 190.67 feet long.

Chapter Test *(page 147)*

1. $\left\{\frac{5}{2}\right\}$ **2.** $\{-2, 3\}$ **3.** $\{-2, 2\}$ **4.** $\left\{\frac{9}{2}\right\}$ **5.** $\{0, 3\}$ **6.** $\left\{-2, -\frac{2}{3}, 2\right\}$ **7.** No real solution **8.** $\left[-\frac{2}{3}, \frac{16}{3}\right]$

9. $\left(-4, \frac{4}{3}\right)$

10. $(-\infty, -1] \cup [6, \infty)$

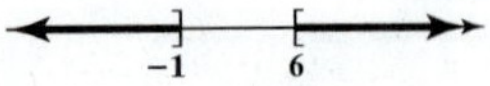

11. $-\frac{3}{5} - \frac{1}{5}i$ **12.** $\left\{\frac{1}{2} - i, \frac{1}{2} + i\right\}$ **13.** Add $6\frac{2}{3}$ lb of \$8/lb coffee to get $26\frac{2}{3}$ lb of \$5/lb coffee.

CHAPTER 2 Graphs

2.1 Assess Your Understanding *(page 154)*

7. *x*-coordinate or abscissa; *y*-coordinate or ordinate **8.** quadrants **9.** midpoint **10.** F **11.** F **12.** T **13.** b **14.** a

15. **(a)** Quadrant II **(b)** *x*-axis **(c)** Quadrant III **(d)** Quadrant I **(e)** *y*-axis **(f)** Quadrant IV

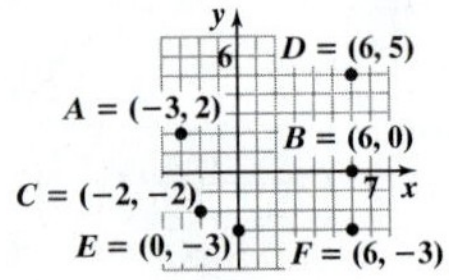

17. The points will be on a vertical line that is 2 units to the right of the *y*-axis.

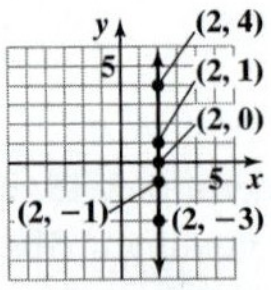

19. $\sqrt{5}$
21. $\sqrt{10}$
23. $2\sqrt{17}$
25. $\sqrt{85}$
27. $\sqrt{53}$
29. $\sqrt{a^2 + b^2}$

31. $d(A, B) = \sqrt{13}$
$d(B, C) = \sqrt{13}$
$d(A, C) = \sqrt{26}$
$(\sqrt{13})^2 + (\sqrt{13})^2 = (\sqrt{26})^2$
Area $= \frac{13}{2}$ square units

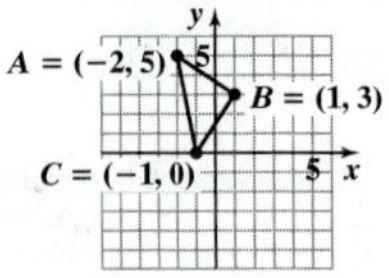

33. $d(A, B) = \sqrt{130}$
$d(B, C) = \sqrt{26}$
$d(A, C) = 2\sqrt{26}$
$(\sqrt{26})^2 + (2\sqrt{26})^2 = (\sqrt{130})^2$
Area $= 26$ square units

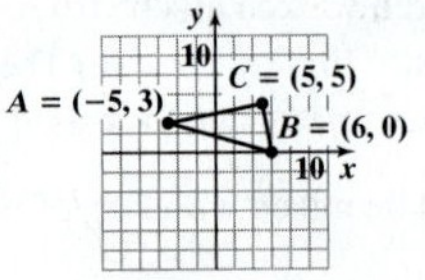

35. $d(A, B) = 4$
$d(B, C) = \sqrt{41}$
$d(A, C) = 5$
$4^2 + 5^2 = (\sqrt{41})^2$
Area $= 10$ square units

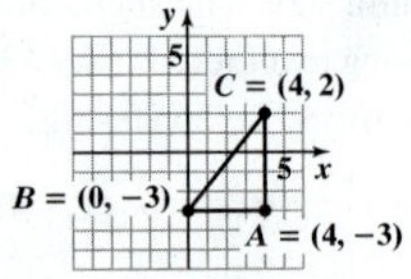

37. $(4, 0)$ **39.** $\left(\frac{3}{2}, 1\right)$ **41.** $(5, -1)$ **43.** $\left(\frac{a}{2}, \frac{b}{2}\right)$ **45.** $(5, 3)$ **47.** $(3, -13), (3, 11)$

49. $(4 + 3\sqrt{3}, 0); (4 - 3\sqrt{3}, 0)$ **51.** **(a)** $(-1, 1)$ **(b)** $(0, 13)$ **53.** $(1, 2)$ **55.** $\sqrt{17}; 2\sqrt{5}; \sqrt{29}$ **57.** $\left(\frac{s}{2}, \frac{s}{2}\right)$

59. $d(P_1, P_2) = 6$; $d(P_2, P_3) = 4$; $d(P_1, P_3) = 2\sqrt{13}$; right triangle **61.** $d(P_1, P_2) = 2\sqrt{17}$; $d(P_2, P_3) = \sqrt{34}$; $d(P_1, P_3) = \sqrt{34}$; isosceles right triangle

63. $90\sqrt{2} \approx 127.28$ ft **65.** **(a)** $(90, 0), (90, 90), (0, 90)$ **(b)** $5\sqrt{2161} \approx 232.43$ ft **(c)** $30\sqrt{149} \approx 366.20$ ft

67. $d = 50t$ mi **69.** **(a)** $(2.65, 1.6)$ **(b)** Approximately 1.285 units

71. \$21,142; a slight underestimate **73.** $\left\{x \,\middle|\, x \neq \frac{5}{2}\right\}$ **74.** $\left\{-\frac{5}{3}, 4\right\}$ **75.** $13 - 11i$ **76.** $(-\infty, 2]$

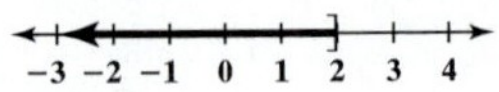

2.2 Assess Your Understanding *(page 164)*

3. intercepts **4.** $y = 0$ **5.** *y*-axis **6.** 4 **7.** $(-3, 4)$ **8.** T **9.** F **10.** F **11.** d **12.** c **13.** $(0, 0)$ is on the graph. **15.** $(0, 3)$ is on the graph.

17. $(0, 2)$ and $(\sqrt{2}, \sqrt{2})$ are on the graph.

19. $(-2, 0), (0, 2)$

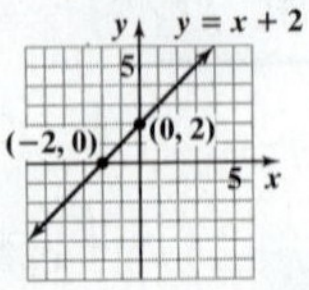

21. $(-4, 0), (0, 8)$

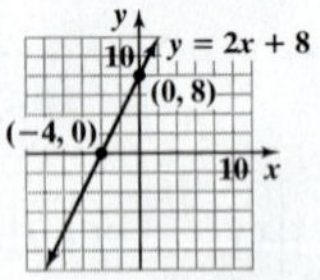

23. $(-1, 0), (1, 0), (0, -1)$

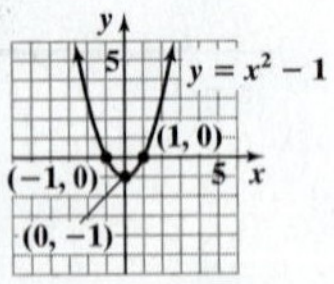

25. $(-2, 0), (2, 0), (0, 4)$

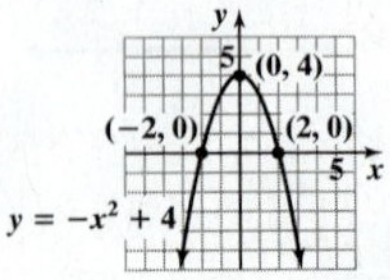

27. $(3, 0), (0, 2)$

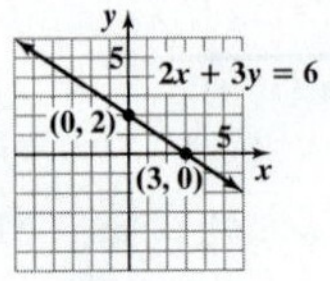

29. $(-2, 0), (2, 0), (0, 9)$

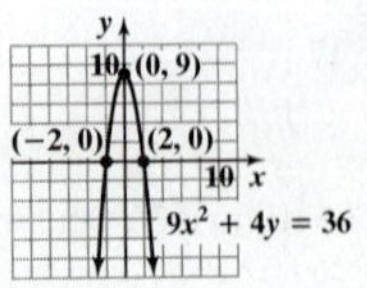

31.

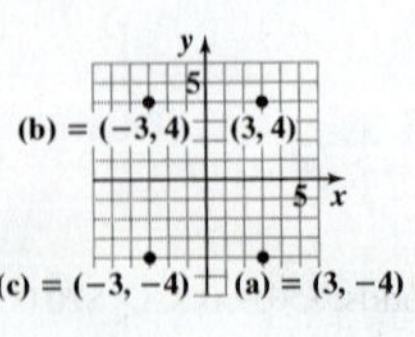

33.

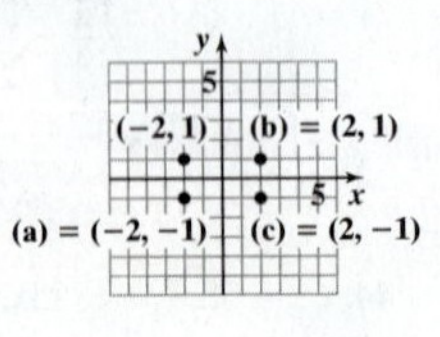

35.

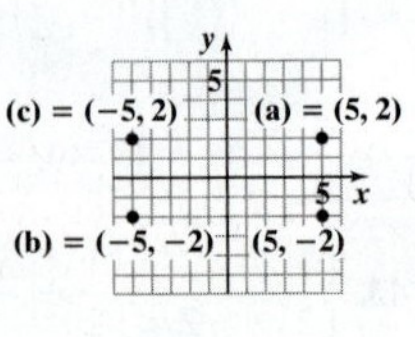

37.

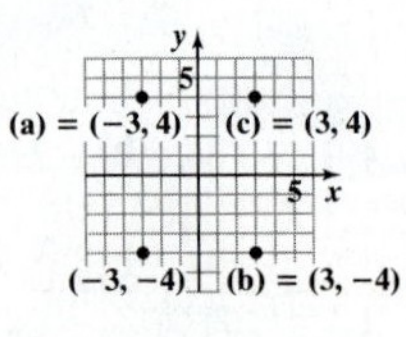

39.

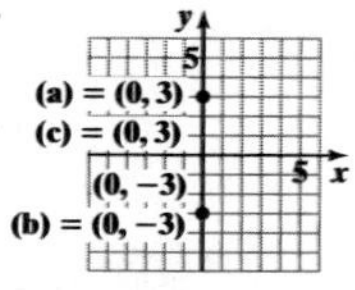

41. (a) $(-1, 0), (1, 0)$ **(b)** Symmetric with respect to the x-axis, the y-axis, and the origin **43. (a)** $\left(-\frac{\pi}{2}, 0\right), (0, 1), \left(\frac{\pi}{2}, 0\right)$ **(b)** Symmetric with respect to the y-axis **45. (a)** $(0, 0)$ **(b)** Symmetric with respect to the x-axis **47. (a)** $(-2, 0), (0, 0), (2, 0)$ **(b)** Symmetric with respect to the origin **49. (a)** $(x, 0), -2 \le x \le 1$ **(b)** No symmetry **51. (a)** No intercepts **(b)** Symmetric with respect to the origin

53.

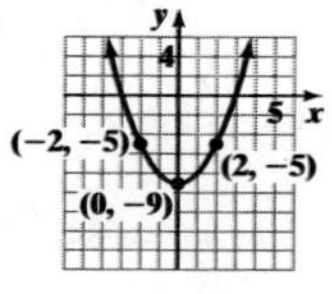

55.

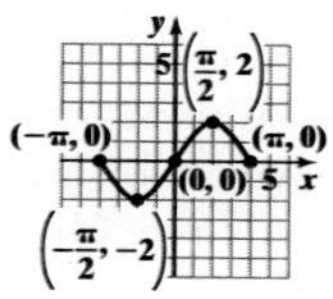

57. $(-4, 0), (0, -2), (0, 2)$; symmetric with respect to the x-axis **59.** $(0, 0)$; symmetric with respect to the origin **61.** $(0, 9), (3, 0), (-3, 0)$; symmetric with respect to the y-axis **63.** $(-2, 0), (2, 0), (0, -3), (0, 3)$; symmetric with respect to the x-axis, y-axis, and origin **65.** $(0, -27), (3, 0)$; no symmetry **67.** $(0, -4), (4, 0), (-1, 0)$; no symmetry **69.** $(0, 0)$; symmetric with respect to the origin **71.** $(0, 0)$; symmetric with respect to the origin

73.

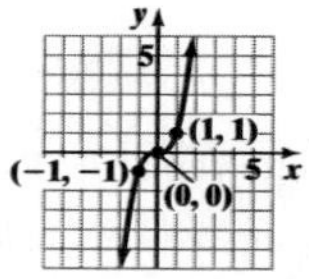

75.

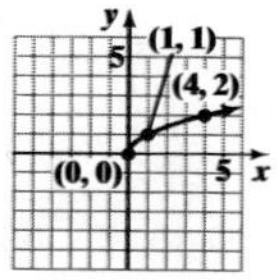

77. $a = -4$ or $a = 1$ **79.** $(-1, -2)$ **81.** 4 **83. (a)** $(0, 0), (2, 0), (0, 1), (0, -1)$ **(b)** x-axis symmetry **85. (a)** $y = \sqrt{x^2}$ and $y = |x|$ have the same graph. **(b)** $\sqrt{x^2} = |x|$ **(c)** $x \ge 0$ for $y = (\sqrt{x})^2$, while x can be any real number for $y = x$. **(d)** $y \ge 0$ for $y = \sqrt{x^2}$ **92.** $\frac{1}{2}$ **93.** $3(x - 5)^2$ **94.** $14i$ **95.** $\{4 - 2\sqrt{3}, 4 + 2\sqrt{3}\}$

2.3 Assess Your Understanding *(page 178)*

1. undefined; 0 **2.** 3; 2 **3.** T **4.** F **5.** T **6.** $m_1 = m_2$; y-intercepts; $m_1 m_2 = -1$ **7.** 2 **8.** $-\frac{1}{2}$ **9.** F **10.** d **11.** c **12.** b

13. (a) Slope $= \frac{1}{2}$ **(b)** If x increases by 2 units, y will increase by 1 unit. **15. (a)** Slope $= -\frac{1}{3}$ **(b)** If x increases by 3 units, y will decrease by 1 unit.

17. Slope $= -\frac{3}{2}$

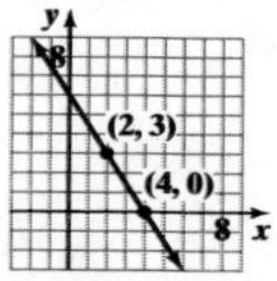

19. Slope $= -\frac{1}{2}$

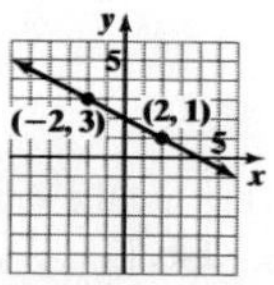

21. Slope $= 0$

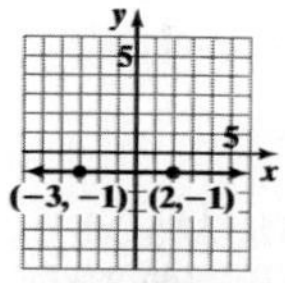

23. Slope undefined

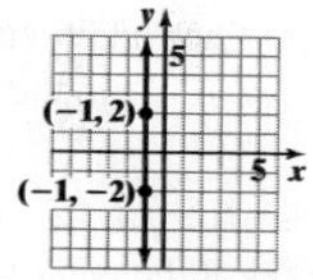

25.

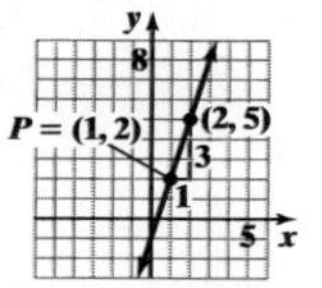

$y - 2 = 3(x - 1)$

27.

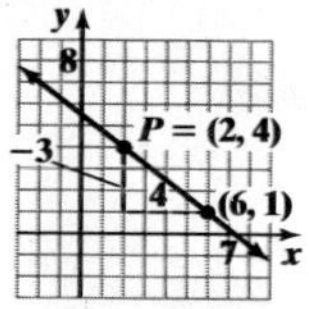

$y - 4 = -\frac{3}{4}(x - 2)$

29.

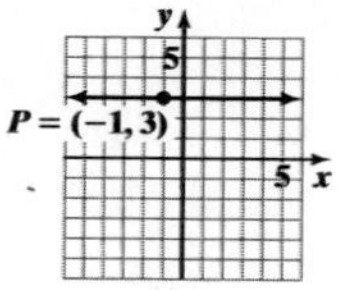

$y - 3 = 0$

31.

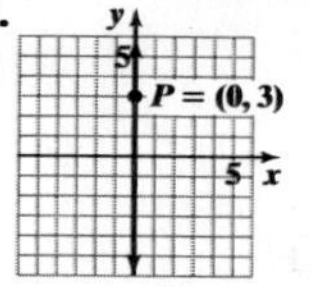

$x = 0$

33. $(2, 6); (3, 10); (4, 14)$ **35.** $(4, -7); (6, -10); (8, -13)$ **37.** $(-1, -5); (0, -7); (1, -9)$ **39.** $x - 2y = 0$ or $y = \frac{1}{2}x$

41. $x + y = 2$ or $y = -x + 2$ **43.** $2x - y = 3$ or $y = 2x - 3$ **45.** $x + 2y = 5$ or $y = -\frac{1}{2}x + \frac{5}{2}$ **47.** $3x - y = -9$ or $y = 3x + 9$ **49.** $2x + 3y = -1$ or $y = -\frac{2}{3}x - \frac{1}{3}$ **51.** $x - 2y = -5$ or $y = \frac{1}{2}x + \frac{5}{2}$ **53.** $3x + y = 3$ or $y = -3x + 3$ **55.** $x - 2y = 2$ or $y = \frac{1}{2}x - 1$ **57.** $x = 2$; no slope–intercept form **59.** $y = 2$ **61.** $2x - y = -4$ or $y = 2x + 4$ **63.** $2x - y = 0$ or $y = 2x$ **65.** $x = 4$; no slope–intercept form **67.** $2x + y = 0$ or $y = -2x$ **69.** $x - 2y = -3$ or $y = \frac{1}{2}x + \frac{3}{2}$ **71.** $y = 4$

73. Slope $= 2$; y-intercept $= 3$

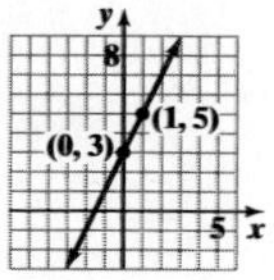

75. Slope $= 2$; y-intercept $= -2$

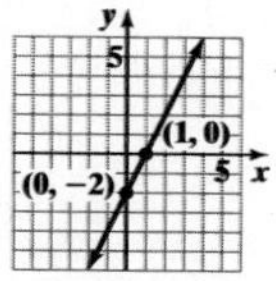

77. Slope $= \frac{1}{2}$; y-intercept $= 2$

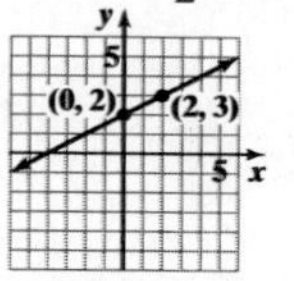

79. Slope $= -\frac{1}{2}$; y-intercept $= 2$

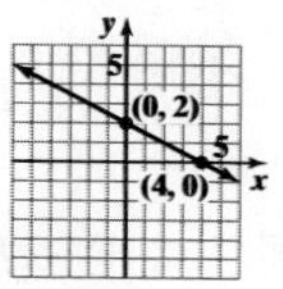

81. Slope $= \frac{2}{3}$; y-intercept $= -2$

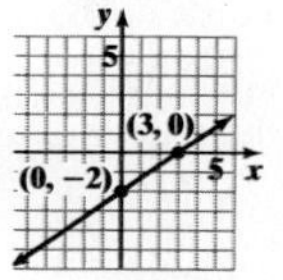

83. Slope $= -1$; y-intercept $= 1$

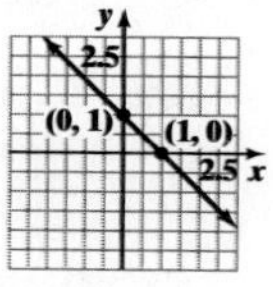

85. Slope undefined; no y-intercept

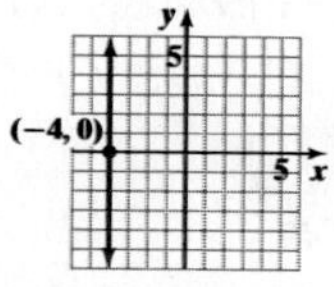

87. Slope $= 0$; y-intercept $= 5$

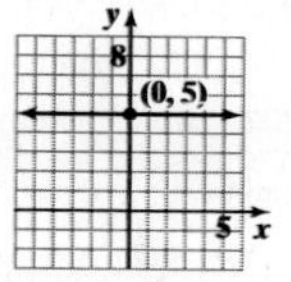

89. Slope $= 1$; y-intercept $= 0$

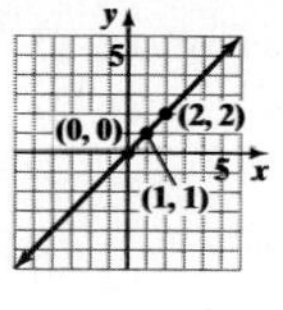

91. Slope $= \frac{3}{2}$; y-intercept $= 0$

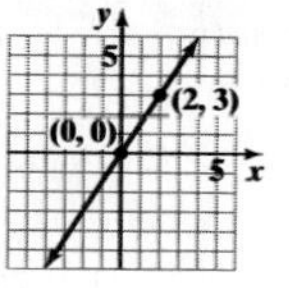

93. (a) x-intercept: 3; y-intercept: 2 **(b)**

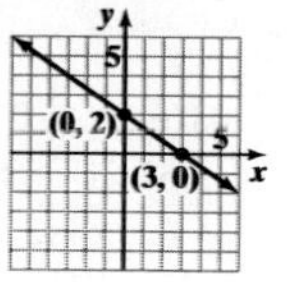

95. (a) x-intercept: -10; y-intercept: 8 **(b)**

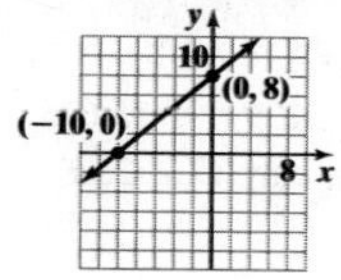

97. (a) x-intercept: 3; y-intercept: $\frac{21}{2}$ **(b)**

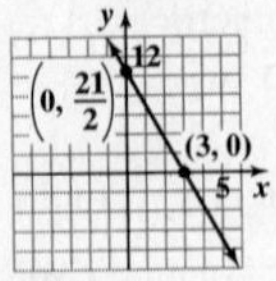

99. (a) x-intercept: 2; y-intercept: 3 **(b)**

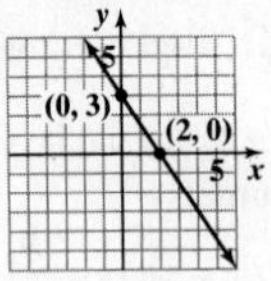

101. (a) x-intercept: 5; y-intercept: -2 **(b)**

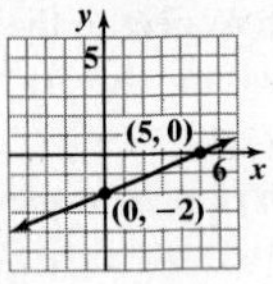

103. $y = 0$ **105.** Parallel **107.** Neither **109.** $x - y = -2$ or $y = x + 2$ **111.** $x + 3y = 3$ or $y = -\frac{1}{3}x + 1$

113. $P_1 = (-2, 5)$, $P_2 = (1, 3)$, $m_1 = -\frac{2}{3}$; $P_2 = (1, 3)$, $P_3 = (-1, 0)$, $m_2 = \frac{3}{2}$; because $m_1m_2 = -1$, the lines are perpendicular and the points $(-2, 5)$, $(1, 3)$, and $(-1, 0)$ are the vertices of a right triangle; thus, the points P_1, P_2, and P_3 are the vertices of a right triangle.
115. $P_1 = (-1, 0)$, $P_2 = (2, 3)$, $m = 1$; $P_3 = (1, -2)$, $P_4 = (4, 1)$, $m = 1$; $P_1 = (-1, 0)$, $P_3 = (1, -2)$, $m = -1$; $P_2 = (2, 3)$, $P_4 = (4, 1)$, $m = -1$; opposite sides are parallel, and adjacent sides are perpendicular; the points are the vertices of a rectangle.

117. $C = 0.60x + 39$; \$105.00; \$177.00 **119.** $C = 0.17x + 4462$
121. (a) $C = 0.0821x + 15.37, 0 \le x \le 800$ **(b)**

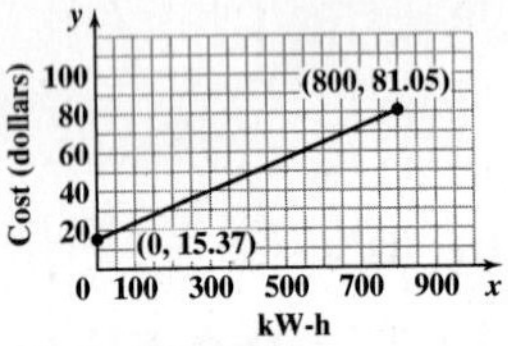

(c) \$31.79 **(d)** \$56.42
(e) Each additional kW-h used adds \$0.0821 to the bill.

123. °C $= \frac{5}{9}$(°F − 32); approximately 21.1°C **125. (a)** $y = -\frac{2}{25}x + 30$ **(b)** x-intercept: 375; The ramp meets the floor 375 in. (31.25 ft) from the base of the platform. **(c)** The ramp does not meet design requirements. It has a run of 31.25 ft. **(d)** The only slope possible for the ramp to comply with the requirement is for it to drop 1 in. for every 12-in. run.
127. (a) $A = \frac{1}{5}x + 20{,}000$ **(b)** \$80,000 **(c)** Each additional box sold requires an additional \$0.20 in advertising. **129.** All have the same slope, 2; the lines are parallel.

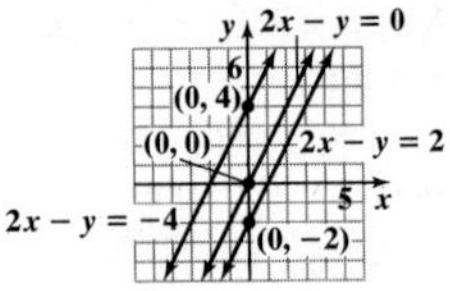

131. (b), (c), (e), (g) **133.** (c) **139.** No; no **141.** They are the same line.
143. Yes, if the y-intercept is 0. **146.** x^4y^{16} **147.** 17 **148.** $\{3 - 2\sqrt{6}, 3 + 2\sqrt{6}\}$ **149.** $\{x|1 < x < 4\}$ or $(1, 4)$

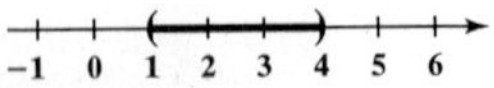

2.4 Assess Your Understanding *(page 186)*

3. F **4.** radius **5.** T **6.** F **7.** d **8.** a **9.** Center (2, 1); radius = 2; $(x - 2)^2 + (y - 1)^2 = 4$
11. Center $\left(\frac{5}{2}, 2\right)$; radius $= \frac{3}{2}$; $\left(x - \frac{5}{2}\right)^2 + (y - 2)^2 = \frac{9}{4}$

13. $x^2 + y^2 = 4$; $x^2 + y^2 - 4 = 0$

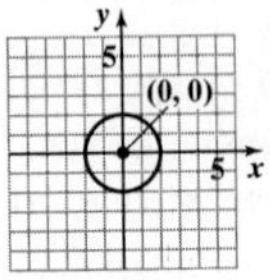

15. $x^2 + (y - 2)^2 = 4$; $x^2 + y^2 - 4y = 0$

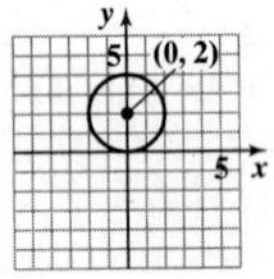

17. $(x - 4)^2 + (y + 3)^2 = 25$; $x^2 + y^2 - 8x + 6y = 0$

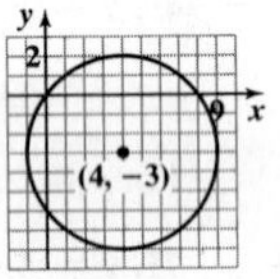

19. $(x + 2)^2 + (y - 1)^2 = 16$; $x^2 + y^2 + 4x - 2y - 11 = 0$

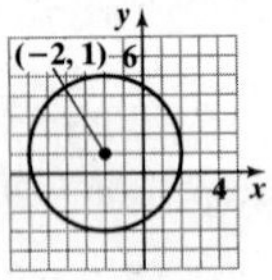

21. $\left(x - \frac{1}{2}\right)^2 + y^2 = \frac{1}{4}$; $x^2 + y^2 - x = 0$

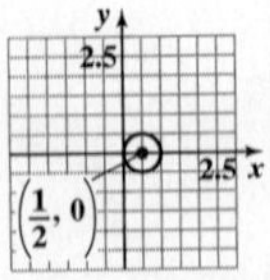

23. (a) $(h, k) = (0, 0)$; $r = 2$ **(b)**

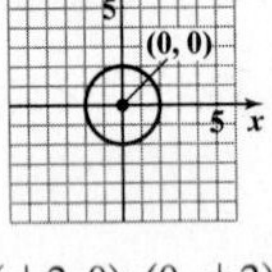

(c) $(\pm 2, 0)$; $(0, \pm 2)$

25. (a) $(h, k) = (3, 0)$; $r = 2$ **(b)**

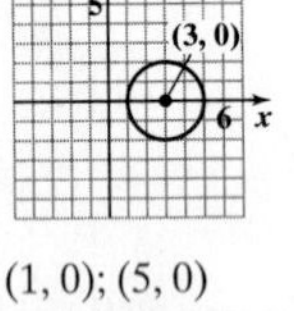

(c) (1, 0); (5, 0)

27. (a) $(h, k) = (1, 2)$; $r = 3$ **(b)**

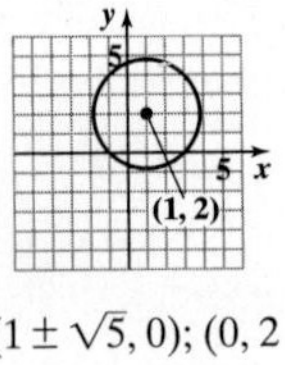

(c) $(1 \pm \sqrt{5}, 0)$; $(0, 2 \pm 2\sqrt{2})$

29. (a) $(h, k) = (-2, 2)$; $r = 3$ **(b)**

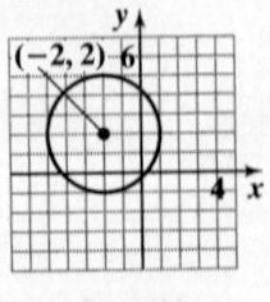

(c) $(-2 \pm \sqrt{5}, 0)$; $(0, 2 \pm \sqrt{5})$

31. (a) $(h, k) = \left(\frac{1}{2}, -1\right)$; $r = \frac{1}{2}$ **(b)**

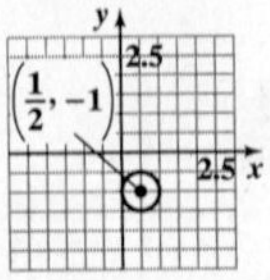

(c) $(0, -1)$

33. (a) $(h, k) = (3, -2)$; $r = 5$ **(b)**

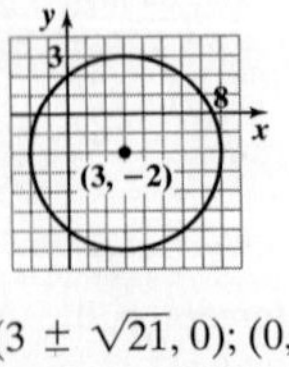

(c) $(3 \pm \sqrt{21}, 0)$; $(0, -6)$, $(0, 2)$

35. (a) $(h, k) = (-2, 0)$; $r = 2$ **(b)**

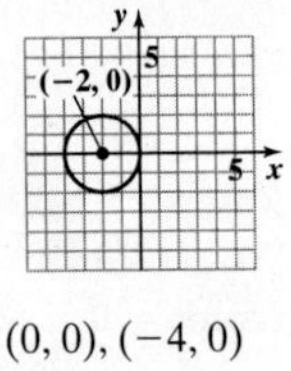

(c) (0, 0), (−4, 0)

37. $x^2 + y^2 = 13$ **39.** $(x - 2)^2 + (y - 3)^2 = 9$ **41.** $(x + 1)^2 + (y - 3)^2 = 5$ **43.** $(x + 1)^2 + (y - 3)^2 = 1$ **45.** (c) **47.** (b) **49.** 18 units^2
51. $x^2 + (y - 139)^2 = 15{,}625$ **53.** $x^2 + y^2 + 2x + 4y - 4168.16 = 0$ **55.** $\sqrt{2}x + 4y - 9\sqrt{2} = 0$ **57.** (1, 0) **59.** $y = 2$ **61.** (b), (c), (e), (g)
65. $A = 169\pi$ cm^2; $C = 26\pi$ cm **66.** $3x^3 - 8x^2 + 13x - 6$ **67.** {1} **68.** 12.32 minutes

2.5 Assess Your Understanding *(page 191)*

1. $y = kx$ **2.** F **3.** b **4.** c **5.** $y = \frac{1}{5}x$ **7.** $A = \pi x^2$ **9.** $F = \frac{250}{d^2}$ **11.** $z = \frac{1}{5}(x^2 + y^2)$ **13.** $M = \frac{9d^2}{2\sqrt{x}}$ **15.** $T^2 = \frac{8a^3}{d^2}$ **17.** $V = \frac{4\pi}{3}r^3$ **19.** $A = \frac{1}{2}bh$ **21.** $F = 6.67 \times 10^{-11}\left(\frac{mM}{d^2}\right)$ **23.** $p = 0.00649B$; \$941.05 **25.** 144 ft; 2 sec **27.** 2.25 **29.** $R = 3.95g$; \$41.48 **31. (a)** $D = \frac{429}{p}$ **(b)** 143 bags **33.** 450 cm^3 **35.** 124.76 lb **37.** $V = \pi r^2 h$ **39.** 0.012 foot-candle **41.** $\sqrt[3]{6} \approx 1.82$ in. **43.** 2812.5 joules **45.** 384 psi **51.** $(3x + 25)(x + 2)(x - 2)$ **52.** $\frac{6}{x + 4}$ **53.** $\frac{8}{125}$ **54.** $\sqrt{7} + 2$

Review Exercises *(page 195)*

1. (a) $2\sqrt{5}$ **(b)** $(2, 1)$ **(c)** $\frac{1}{2}$ **(d)** For each run of 2, there is a rise of 1. **2. (a)** 5 **(b)** $\left(-\frac{1}{2}, 1\right)$ **(c)** $-\frac{4}{3}$ **(d)** For each run of 3, there is a rise of -4. **3. (a)** 12 **(b)** $(4, 2)$ **(c)** undefined **(d)** no change in x

4.

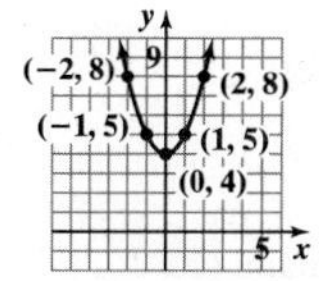

5. $(-4, 0), (0, 2), (0, 0), (0, -2), (2, 0)$ **6.** $(0, 0)$; symmetric with respect to the x-axis **7.** $(\pm 4, 0), (0, \pm 2)$; symmetric with respect to the x-axis, y-axis, and origin **8.** $(0, 1)$; symmetric with respect to the y-axis **9.** $(0, 0), (\pm 1, 0)$; symmetric with respect to the origin **10.** $(0, 0), (-1, 0), (0, -2)$; no symmetry **11.** $(x + 2)^2 + (y - 3)^2 = 16$ **12.** $(x + 1)^2 + (y + 2)^2 = 1$

13. Center $(0, 1)$; radius $= 2$

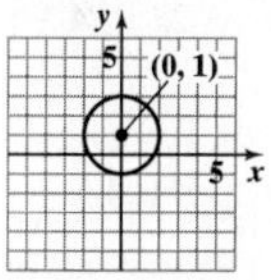

Intercepts: $(-\sqrt{3}, 0), (\sqrt{3}, 0), (0, -1), (0, 3)$

14. Center $(1, -2)$; radius $= 3$

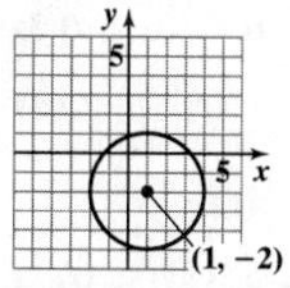

Intercepts: $(1 - \sqrt{5}, 0), (1 + \sqrt{5}, 0), (0, -2 - 2\sqrt{2}), (0, -2 + 2\sqrt{2})$

15. Center $(1, -2)$; radius $= \sqrt{5}$

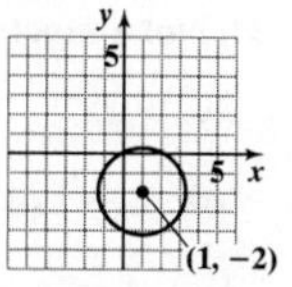

Intercepts: $(0, 0), (2, 0), (0, -4)$

16. $2x + y = 5$ or $y = -2x + 5$
17. $x = -3$; no slope–intercept form
18. $x + 5y = -10$ or $y = -\frac{1}{5}x - 2$
19. $5x + y = 11$ or $y = -5x + 11$
20. $2x - 3y = -19$ or $y = \frac{2}{3}x + \frac{19}{3}$
21. $x - y = 7$ or $y = x - 7$

22. Slope $= \frac{4}{5}$; y-intercept $= 4$

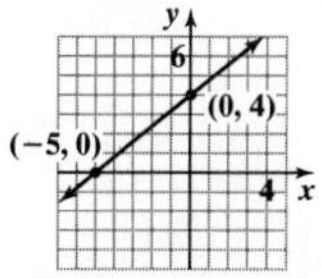

23. Slope $= \frac{3}{2}$; y-intercept $= \frac{1}{2}$

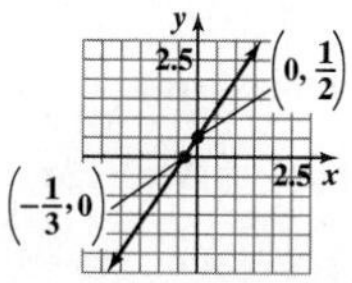

24. Intercepts: $(6, 0), (0, -4)$

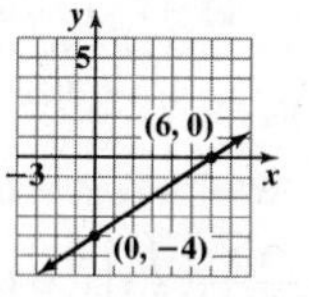

25. Intercepts: $(4, 0), (0, 6)$

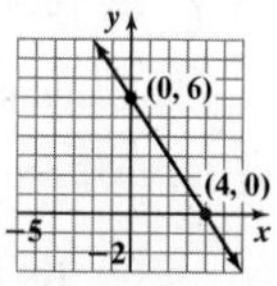

26.

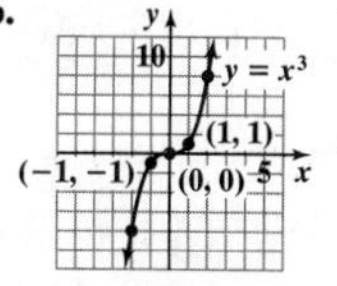

27.

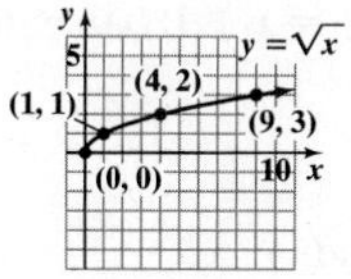

28.

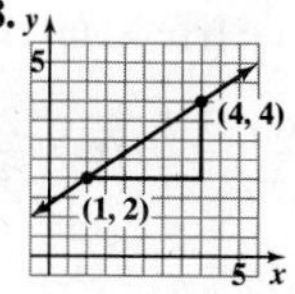

29. $d(A, B) = \sqrt{(1 - 3)^2 + (1 - 4)^2} = \sqrt{13}$ and $d(B, C) = \sqrt{(-2 - 1)^2 + (3 - 1)^2} = \sqrt{13}$
30. (a) $d(A, B) = 2\sqrt{5}$; $d(B, C) = \sqrt{145}$; $d(A, C) = 5\sqrt{5}$; $[d(A, B)]^2 + [d(A, C)]^2 = (2\sqrt{5})^2 + (5\sqrt{5})^2 = 20 + 125 = 145 = [d(B, C)]^2$
(b) Slope from A to B is -2; slope from A to C is $\frac{1}{2}$. Since $(-2)\left(\frac{1}{2}\right) = -1$, the lines are perpendicular.
31. Center $(1, -2)$; Radius $= 4\sqrt{2}$; $(x - 1)^2 + (y + 2)^2 = 32$
32. Slope from A to B is -1; slope from A to C is -1.
33. $p = \frac{427}{65{,}000}B$; \$1083.92 **34.** 199.9 lb **35.** 189 Btu

Chapter Test *(page 196)*

1. $d = 2\sqrt{13}$ **2.** $(2, 1)$ **3. (a)** $m = -\frac{2}{3}$ **(b)** For every 3-unit change in x, y will change by -2 units.

4.

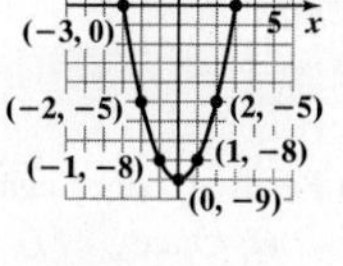

5.

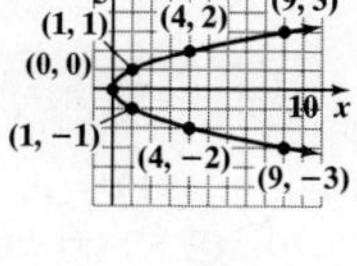

6. Intercepts: $(-3, 0), (3, 0), (0, 9)$; symmetric with respect to the y-axis

7. $y = -2x + 2$

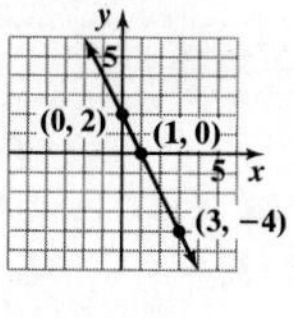

8. $x^2 + y^2 - 8x + 6y = 0$

9. Center: $(-2, 1)$; radius: 3

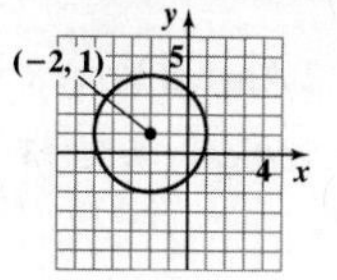

10. Parallel line: $y = -\frac{2}{3}x - \frac{1}{3}$; perpendicular line: $y = \frac{3}{2}x + 3$ **11.** 14.69 ohms

Cumulative Review *(page 196)*

1. $\left\{\frac{5}{3}\right\}$ **2.** $\{-3, 4\}$ **3.** $\left\{-\frac{1}{2}, 3\right\}$ **4.** $\{1 - \sqrt{3}, 1 + \sqrt{3}\}$ **5.** No real solution **6.** $\{4\}$ **7.** $\{1, 3\}$ **8.** $\{-2 - 2\sqrt{2}, -2 + 2\sqrt{2}\}$ **9.** $\{-3i, 3i\}$
10. $\{1 - 2i, 1 + 2i\}$ **11.** $\{x|x \le 5\}$ or $(-\infty, 5]$; 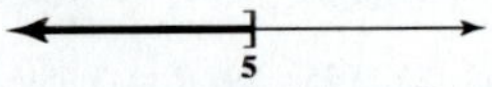
12. $\{x|-5 < x < 1\}$ or $(-5, 1)$; 
13. $\{x|1 \le x \le 3\}$ or $[1, 3]$; 
14. $\{x|x < -5$ or $x > 1\}$ or $(-\infty, -5) \cup (1, \infty)$; 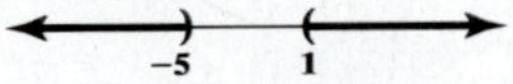
15. $5\sqrt{2}$; $\left(\frac{3}{2}, \frac{1}{2}\right)$ **16.** (a),(b) **17.**

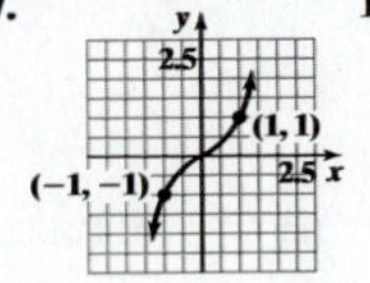

18. $y = -2x + 2$ **19.** $y = -\frac{1}{2}x + \frac{13}{2}$;

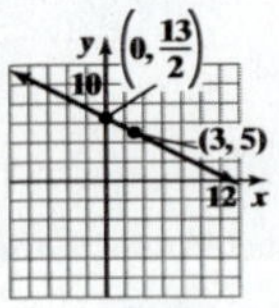

20.

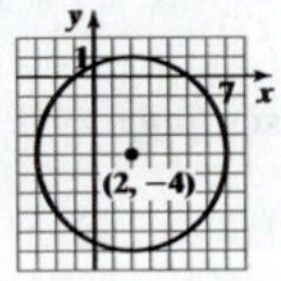

CHAPTER 3 Functions and Their Graphs

3.1 Assess Your Understanding *(page 210)*

7. independent; dependent **8.** $[0, 5]$ **9.** $\ne$; f; g **10.** $(g - f)(x)$ **11.** F **12.** T **13.** F **14.** F **15.** a **16.** c **17.** d **18.** a **19.** Function; Domain: {Elvis, Colleen, Kaleigh, Marissa}; Range: {January 8, March 15, September 17} **21.** Not a function; Domain: $\{20, 30, 40\}$; Range: $\{200, 300, 350, 425\}$ **23.** Not a function; Domain: $\{-3, 2, 4\}$; Range: $\{6, 9, 10\}$ **25.** Function; Domain: $\{1, 2, 3, 4\}$; Range: $\{3\}$ **27.** Not a function; Domain: $\{-2, 0, 3\}$; Range: $\{3, 4, 6, 7\}$ **29.** Function; Domain: $\{-2, -1, 0, 1\}$; Range: $\{0, 1, 4\}$ **31.** Function **33.** Function **35.** Not a function **37.** Not a function **39.** Function **41.** Not a function **43.** **(a)** -4 **(b)** 1 **(c)** -3 **(d)** $3x^2 - 2x - 4$ **(e)** $-3x^2 - 2x + 4$ **(f)** $3x^2 + 8x + 1$ **(g)** $12x^2 + 4x - 4$ **(h)** $3x^2 + 6xh + 3h^2 + 2x + 2h - 4$ **45.** **(a)** 0 **(b)** $\frac{1}{2}$ **(c)** $-\frac{1}{2}$ **(d)** $\frac{-x}{x^2 + 1}$ **(e)** $\frac{-x}{x^2 + 1}$ **(f)** $\frac{x + 1}{x^2 + 2x + 2}$ **(g)** $\frac{2x}{4x^2 + 1}$ **(h)** $\frac{x + h}{x^2 + 2xh + h^2 + 1}$ **47.** **(a)** 4 **(b)** 5 **(c)** 5 **(d)** $|x| + 4$ **(e)** $-|x| - 4$ **(f)** $|x + 1| + 4$ **(g)** $2|x| + 4$ **(h)** $|x + h| + 4$ **49.** **(a)** $-\frac{1}{5}$ **(b)** $-\frac{3}{2}$ **(c)** $\frac{1}{8}$ **(d)** $\frac{2x - 1}{3x + 5}$ **(e)** $\frac{-2x - 1}{3x - 5}$ **(f)** $\frac{2x + 3}{3x - 2}$ **(g)** $\frac{4x + 1}{6x - 5}$ **(h)** $\frac{2x + 2h + 1}{3x + 3h - 5}$ **51.** All real numbers **53.** All real numbers **55.** $\{x|x \ne -4, x \ne 4\}$ **57.** $\{x|x \ne 0\}$ **59.** $\{x|x \ge 4\}$ **61.** $\{x|x > 1\}$ **63.** $\{x|x > 4\}$ **65.** $\{t|t \ge 4, t \ne 7\}$ **67.** **(a)** $(f + g)(x) = 5x + 1$; All real numbers **(b)** $(f - g)(x) = x + 7$; All real numbers **(c)** $(f \cdot g)(x) = 6x^2 - x - 12$; All real numbers **(d)** $\left(\frac{f}{g}\right)(x) = \frac{3x + 4}{2x - 3}$; $\left\{x \middle| x \ne \frac{3}{2}\right\}$ **(e)** 16 **(f)** 11 **(g)** 10 **(h)** -7 **69.** **(a)** $(f + g)(x) = 2x^2 + x - 1$; All real numbers **(b)** $(f - g)(x) = -2x^2 + x - 1$; All real numbers **(c)** $(f \cdot g)(x) = 2x^3 - 2x^2$; All real numbers **(d)** $\left(\frac{f}{g}\right)(x) = \frac{x - 1}{2x^2}$; $\{x|x \ne 0\}$ **(e)** 20 **(f)** -29 **(g)** 8 **(h)** 0 **71.** **(a)** $(f + g)(x) = \sqrt{x} + 3x - 5$; $\{x|x \ge 0\}$ **(b)** $(f - g)(x) = \sqrt{x} - 3x + 5$; $\{x|x \ge 0\}$ **(c)** $(f \cdot g)(x) = 3x\sqrt{x} - 5\sqrt{x}$; $\{x|x \ge 0\}$ **(d)** $\left(\frac{f}{g}\right)(x) = \frac{\sqrt{x}}{3x - 5}$; $\left\{x \middle| x \ge 0, x \ne \frac{5}{3}\right\}$ **(e)** $\sqrt{3} + 4$ **(f)** -5 **(g)** $\sqrt{2}$ **(h)** $-\frac{1}{2}$ **73.** **(a)** $(f + g)(x) = 1 + \frac{2}{x}$; $\{x|x \ne 0\}$ **(b)** $(f - g)(x) = 1$; $\{x|x \ne 0\}$ **(c)** $(f \cdot g)(x) = \frac{1}{x} + \frac{1}{x^2}$; $\{x|x \ne 0\}$ **(d)** $\left(\frac{f}{g}\right)(x) = x + 1$; $\{x|x \ne 0\}$ **(e)** $\frac{5}{3}$ **(f)** 1 **(g)** $\frac{3}{4}$ **(h)** 2 **75.** **(a)** $(f + g)(x) = \frac{6x + 3}{3x - 2}$; $\left\{x \middle| x \ne \frac{2}{3}\right\}$ **(b)** $(f - g)(x) = \frac{-2x + 3}{3x - 2}$; $\left\{x \middle| x \ne \frac{2}{3}\right\}$ **(c)** $(f \cdot g)(x) = \frac{8x^2 + 12x}{(3x - 2)^2}$; $\left\{x \middle| x \ne \frac{2}{3}\right\}$ **(d)** $\left(\frac{f}{g}\right)(x) = \frac{2x + 3}{4x}$; $\left\{x \middle| x \ne 0, x \ne \frac{2}{3}\right\}$ **(e)** 3 **(f)** $-\frac{1}{2}$ **(g)** $\frac{7}{2}$ **(h)** $\frac{5}{4}$ **77.** $g(x) = 5 - \frac{7}{2}x$ **79.** 4 **81.** $2x + h$ **83.** $2x + h - 1$ **85.** $\frac{-(2x + h)}{x^2(x + h)^2}$ **87.** $\frac{6}{(x + 3)(x + h + 3)}$ **89.** $\frac{1}{\sqrt{x + h - 2} + \sqrt{x - 2}}$ **91.** $\{-2, 4\}$ **93.** $A = -\frac{7}{2}$ **95.** $A = -4$ **97.** $A(x) = \frac{1}{2}x^2$ **99.** $G(x) = 14x$ **101.** **(a)** P is the dependent variable; a is the independent variable. **(b)** $P(20) = 231.427$ million; In 2012, there were 231.427 million people 20 years of age or older. **(c)** $P(0) = 327.287$ million; In 2012, there were 327.287 million people. **103.** **(a)** 15.1 m, 14.071 m, 12.944 m, 11.719 m **(b)** 1.01 sec, 1.43 sec, 1.75 sec **(c)** 2.02 sec **105.** **(a)** \$222 **(b)** \$225 **(c)** \$220 **(d)** \$230 **107.** $R(x) = \frac{L(x)}{P(x)}$ **109.** $H(x) = P(x) \cdot I(x)$ **111.** **(a)** $P(x) = -0.05x^3 + 0.8x^2 + 155x - 500$ **(b)** $P(15) = \$1836.25$ **(c)** When 15 hundred cellphones are sold, the profit is \$1836.25. **113.** **(a)** $D(v) = 0.05v^2 + 2.6v - 15$ **(b)** 321 feet **(c)** The car will need 321 feet to stop once the impediment is observed. **115.** No; domain of f is all real numbers; domain of g is $\{x|x \ne -1\}$ **117.** $H(x) = \frac{3x - x^3}{\text{age}}$ **118.** Intercepts: $(-16, 0)$, $(-8, 0)$; x-axis symmetry **119.** $(4, 32)$ **120.** 7.5 lb **121.** $\{-3, 2, 3\}$

3.2 Assess Your Understanding *(page 218)*

3. vertical **4.** 5; -3 **5.** $a = -2$ **6.** F **7.** F **8.** T **9.** c **10.** a **11.** **(a)** $f(0) = 3$; $f(-6) = -3$ **(b)** $f(6) = 0$; $f(11) = 1$ **(c)** Positive **(d)** Negative **(e)** -3, 6, and 10 **(f)** $-3 < x < 6$; $10 < x \le 11$ **(g)** $\{x|-6 \le x \le 11\}$ **(h)** $\{y|-3 \le y \le 4\}$ **(i)** -3, 6, 10 **(j)** 3 **(k)** 3 times **(l)** Once **(m)** 0, 4 **(n)** -5, 8 **13.** Not a function **15.** Function **(a)** Domain: $\{x|-\pi \le x \le \pi\}$; Range: $\{y|-1 \le y \le 1\}$ **(b)** $\left(-\frac{\pi}{2}, 0\right)$, $\left(\frac{\pi}{2}, 0\right)$, $(0, 1)$ **(c)** y-axis **17.** Not a function **19.** Function **(a)** Domain: $\{x|0 < x < 3\}$; Range: $\{y|y < 2\}$ **(b)** $(1, 0)$ **(c)** None **21.** Function **(a)** Domain: all real numbers; Range: $\{y|y \le 2\}$ **(b)** $(-3, 0)$, $(3, 0)$, $(0, 2)$ **(c)** y-axis **23.** Function **(a)** Domain: all real numbers; Range: $\{y|y \ge -3\}$ **(b)** $(1, 0)$, $(3, 0)$, $(0, 9)$ **(c)** None **25.** **(a)** Yes **(b)** $f(-2) = 9$; $(-2, 9)$ **(c)** $0, \frac{1}{2}$; $(0, -1)$, $\left(\frac{1}{2}, -1\right)$ **(d)** All real numbers

(e) $-\frac{1}{2}$, 1 **(f)** -1 **27. (a)** No **(b)** $f(4) = -3$; $(4, -3)$ **(c)** 14; $(14, 2)$ **(d)** $\{x \mid x \neq 6\}$ **(e)** -2 **(f)** $-\frac{1}{3}$ **29. (a)** Yes **(b)** $f(2) = \frac{8}{17}$; $\left(2, \frac{8}{17}\right)$ **(c)** $-1, 1$; $(-1, 1)$, $(1, 1)$ **(d)** All real numbers **(e)** 0 **(f)** 0 **31. (a)** 3 **(b)** -2 **(c)** -1 **(d)** 1 **(e)** 2 **(f)** $-\frac{1}{3}$

33. (a) Approximately 10.4 ft high
(b) Approximately 9.9 ft high
(c)

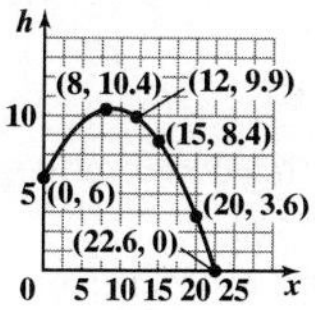

(d) The ball will not go through the hoop; $h(15) \approx 8.4$ ft. If $v = 30$ ft/sec, $h(15) = 10$ ft.

35. (a) About 81.07 ft **(b)** About 129.59 ft **(c)** About 26.63 ft
(d) About 528.13 ft
(e)

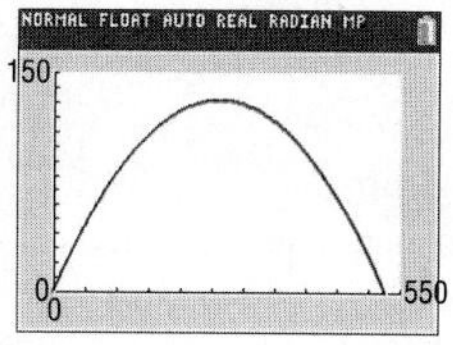

(f) About 115.07 ft and 413.05 ft **(g)** 275 ft; maximum height shown in the table is 131.8 ft **(h)** 264 ft

37. (a) \$223; \$220 **(b)** $\{x \mid x > 0\}$
(c)

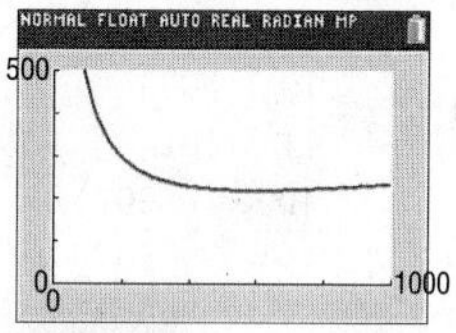

(d)

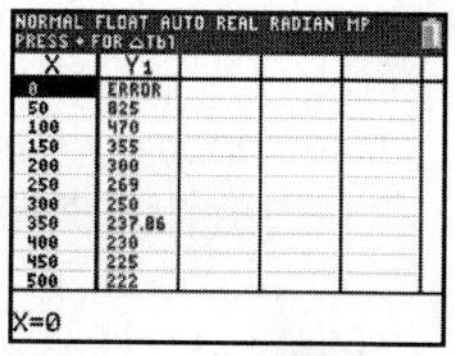
NORMAL FLOAT AUTO REAL RADIAN MP
PRESS + FOR ΔTbl

X	Y₁
0	ERROR
50	825
100	470
150	355
200	300
250	269
300	250
350	237.86
400	230
450	225
500	222

X=0

(e) 600 mi/h

39. (a) \$30; It costs \$30 if you use 0 gigabytes. **(b)** \$30; It costs \$30 if you use 5 gigabytes. **(c)** \$90; It costs \$90 if you use 15 gigabytes. **(d)** $\{g \mid 0 \le g \le 60\}$. There are at most 60 gigabytes in a month. **41.** The x-intercepts can number anywhere from 0 to infinitely many. There is at most one y-intercept.
43. (a) III **(b)** IV **(c)** I **(d)** V **(e)** II

45.

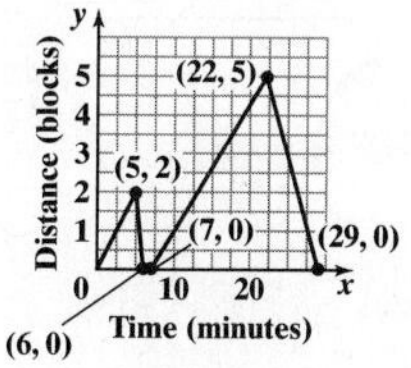

47. (a) 2 hr elapsed during which Kevin was between 0 and 3 mi from home **(b)** 0.5 hr elapsed during which Kevin was 3 mi from home **(c)** 0.3 hr elapsed during which Kevin was between 0 and 3 mi from home **(d)** 0.2 hr elapsed during which Kevin was 0 mi from home **(e)** 0.9 hr elapsed during which Kevin was between 0 and 2.8 mi from home **(f)** 0.3 hr elapsed during which Kevin was 2.8 mi from home **(g)** 1.1 hr elapsed during which Kevin was between 0 and 2.8 mi from home **(h)** 3 mi **(i)** Twice **49.** No points whose x-coordinate is 5 or whose y-coordinate is 0 can be on the graph.
52. $2(2x + 3)^2$ **53.** $2\sqrt{10}$ **54.** $y = \frac{2}{3}x + 8$ **55.** $4x^3 - 8x^2 - 5x + 4$

3.3 Assess Your Understanding *(page 232)*

6. increasing **7.** even; odd **8.** T **9.** T **10.** F **11.** c **12.** d **13.** Yes **15.** No **17.** $(-8, -2)$; $(0, 2)$; $(5, 7)$ **19.** Yes; 10 **21.** $-2, 2$; $6, 10$ **23.** $f(-8) = -4$ **25. (a)** $(-2, 0)$, $(0, 3)$, $(2, 0)$ **(b)** Domain: $\{x \mid -4 \le x \le 4\}$ or $[-4, 4]$; Range: $\{y \mid 0 \le y \le 3\}$ or $[0, 3]$ **(c)** Increasing on $(-2, 0)$ and $(2, 4)$; Decreasing on $(-4, -2)$ and $(0, 2)$ **(d)** Even **27. (a)** $(0, 1)$ **(b)** Domain: all real numbers; Range: $\{y \mid y > 0\}$ or $(0, \infty)$ **(c)** Increasing on $(-\infty, \infty)$ **(d)** Neither **29. (a)** $(-\pi, 0)$, $(0, 0)$, $(\pi, 0)$ **(b)** Domain: $\{x \mid -\pi \le x \le \pi\}$ or $[-\pi, \pi]$; Range: $\{y \mid -1 \le y \le 1\}$ or $[-1, 1]$ **(c)** Increasing on $\left(-\frac{\pi}{2}, \frac{\pi}{2}\right)$; Decreasing on $\left(-\pi, -\frac{\pi}{2}\right)$ and $\left(\frac{\pi}{2}, \pi\right)$ **(d)** Odd **31. (a)** $\left(0, \frac{1}{2}\right)$, $\left(\frac{1}{3}, 0\right)$, $\left(\frac{5}{2}, 0\right)$ **(b)** Domain: $\{x \mid -3 \le x \le 3\}$ or $[-3, 3]$; Range: $\{y \mid -1 \le y \le 2\}$ or $[-1, 2]$ **(c)** Increasing on $(2, 3)$; Decreasing on $(-1, 1)$; Constant on $(-3, -1)$ and $(1, 2)$ **(d)** Neither **33. (a)** 0; 3 **(b)** $-2, 2$; 0, 0 **35. (a)** $\frac{\pi}{2}$; 1 **(b)** $-\frac{\pi}{2}$; -1 **37.** Odd **39.** Even **41.** Odd **43.** Neither **45.** Even **47.** Odd **49.** Absolute maximum: $f(1) = 4$; absolute minimum: $f(5) = 1$; local maximum: $f(3) = 3$; local minimum: $f(2) = 2$ **51.** Absolute maximum: $f(3) = 4$; absolute minimum: $f(1) = 1$; local maximum: $f(3) = 4$; local minimum: $f(1) = 1$ **53.** Absolute maximum: none; absolute minimum: $f(0) = 0$: local maximum: $f(2) = 3$; local minimum: $f(0) = 0$ and $f(3) = 2$ **55.** Absolute maximum: none; absolute minimum: none: local maximum: none; local minimum: none

57.

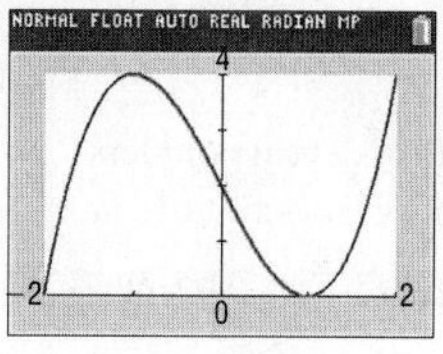

Increasing: $(-2, -1)$, $(1, 2)$
Decreasing: $(-1, 1)$
Local maximum: $f(-1) = 4$
Local minimum: $f(1) = 0$

59.

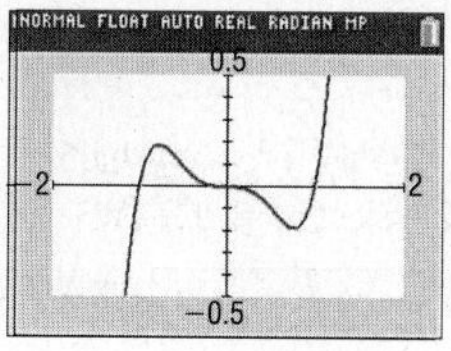

Increasing: $(-2, -0.77)$, $(0.77, 2)$
Decreasing: $(-0.77, 0.77)$
Local maximum: $f(-0.77) = 0.19$
Local minimum: $f(0.77) = -0.19$

61.

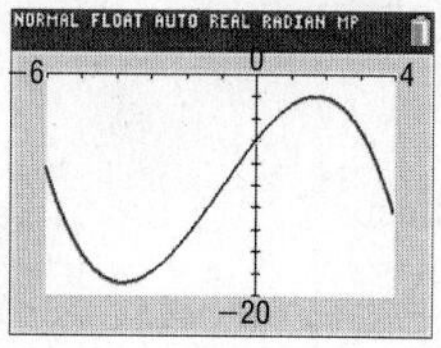

Increasing: $(-3.77, 1.77)$
Decreasing: $(-6, -3.77)$, $(1.77, 4)$
Local maximum: $f(1.77) = -1.91$
Local minimum: $f(-3.77) = -18.89$

63.

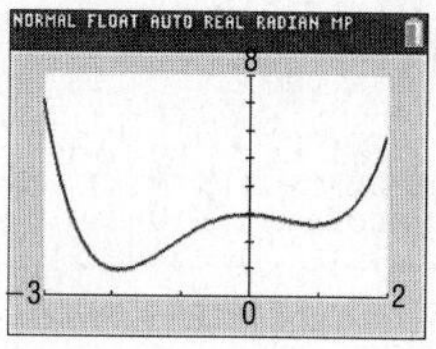

Increasing: $(-1.87, 0)$, $(0.97, 2)$
Decreasing: $(-3, -1.87)$, $(0, 0.97)$
Local maximum: $f(0) = 3$
Local minima: $f(-1.87) = 0.95$, $f(0.97) = 2.65$

65. (a) -4 **(b)** -8 **(c)** -10 **67. (a)** 17 **(b)** -1 **(c)** 11 **69. (a)** 5 **(b)** $y = 5x - 2$ **71. (a)** -1 **(b)** $y = -x$ **73. (a)** 4 **(b)** $y = 4x - 8$ **75. (a)** Odd **(b)** Local maximum value: 54 at $x = -3$ **77. (a)** Even **(b)** Local maximum value: 24 at $x = -2$ **(c)** 47.4 sq. units

79. (a)

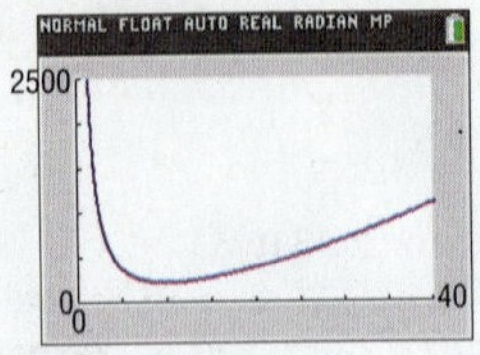

(b) 10 riding lawn mowers
(c) \$239/mower

81. (a), (b)

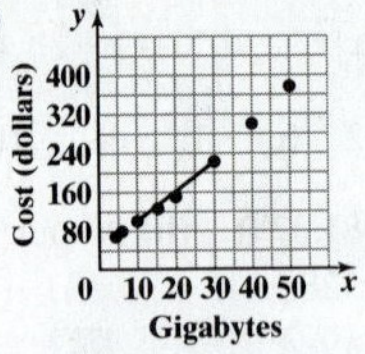

(c) \$5/gigabyte
(d) \$6.25/gigabyte
(e) \$7.50/gigabyte
(f) The average rate of change is increasing as the number of gigabytes increases.

83. (a) On average, the population is increasing at a rate of 0.036 g/h from 0 to 2.5 h. **(b)** On average, from 4.5 to 6 h, the population is increasing at a rate of 0.1 g/h. **(c)** The average rate of change is increasing over time.

85. (a) 1 **(b)** 0.5 **(c)** 0.1 **(d)** 0.01
(e) 0.001
(f)

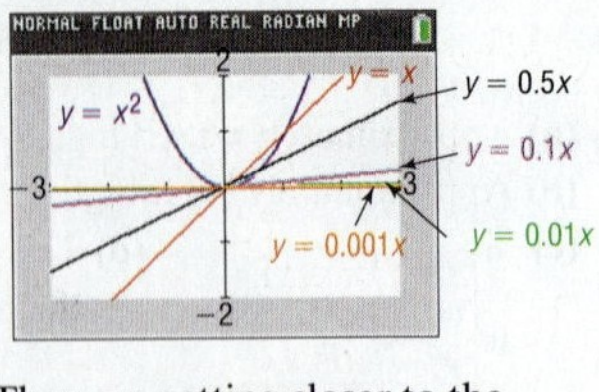

(g) They are getting closer to the tangent line at (0, 0).
(h) They are getting closer to 0.

87. (a) 2
(b) 2; 2; 2; 2
(c) $y = 2x + 5$
(d)

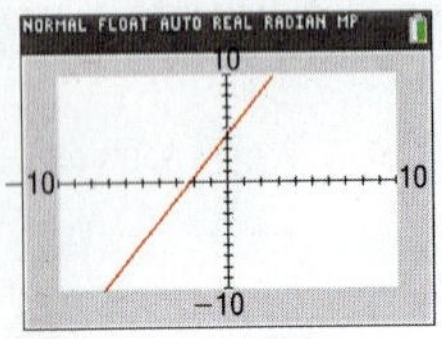

89. (a) $2x + h + 2$
(b) 4.5; 4.1; 4.01; 4
(c) $y = 4.01x - 1.01$
(d)

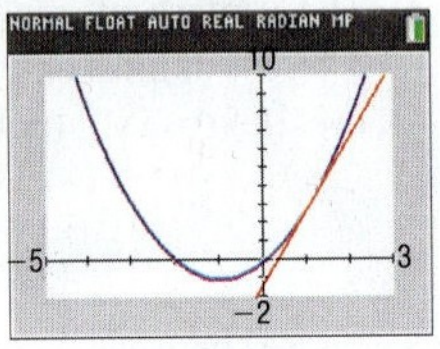

91. (a) $4x + 2h - 3$
(b) 2; 1.2; 1.02; 1
(c) $y = 1.02x - 1.02$
(d)

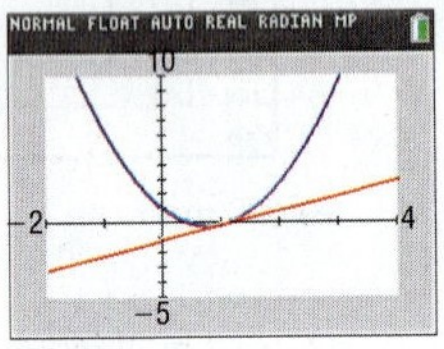

93. (a) $-\dfrac{1}{(x + h)x}$
(b) $-\dfrac{2}{3}; -\dfrac{10}{11}; -\dfrac{100}{101}; -1$
(c) $y = -\dfrac{100}{101}x + \dfrac{201}{101}$
(d)

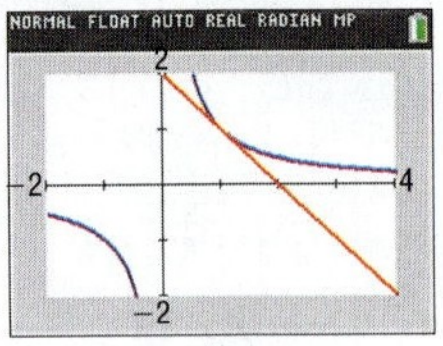

97. At most one **99.** Yes; the function $f(x) = 0$ is both even and odd. **101.** Not necessarily. It just means $f(5) > f(2)$.
103. (a) 7.01×10^{-6} **(b)** 2.305×10^{9} **104.** $6\sqrt{15}$ **105.** $[-8, -3)$ **106.** 8.25 days

3.4 Assess Your Understanding *(page 244)*

4. $(-\infty, 0)$ **5.** piecewise-defined **6.** T **7.** F **8.** F **9.** b **10.** a **11.** C **13.** E **15.** B **17.** F

19.

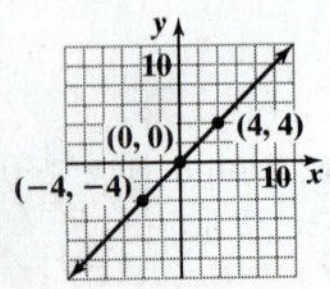

21.

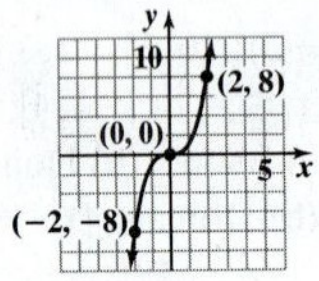

23.

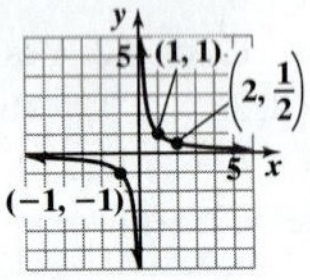

25.

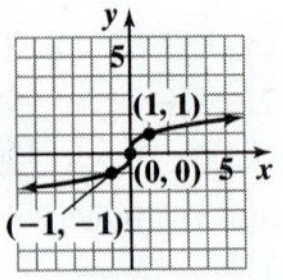

27. (a) 4 **(b)** 2 **(c)** 5 **29. (a)** −4 **(b)** −2 **(c)** 0 **(d)** 25

31. (a) All real numbers
(b) (0, 1)
(c)

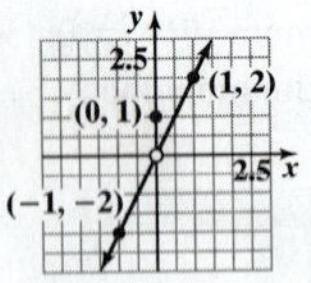

(d) $\{y \mid y \neq 0\}$; $(-\infty, 0) \cup (0, \infty)$
(e) Discontinuous at $x = 0$

33. (a) All real numbers
(b) (0, 3)
(c)

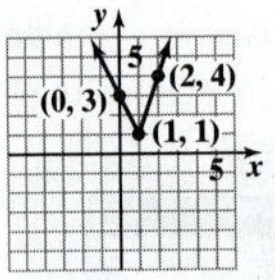

(d) $\{y \mid y \geq 1\}$; $[1, \infty)$
(e) Continuous

35. (a) $\{x \mid x \geq -2\}$; $[-2, \infty)$
(b) (0, 3), (2, 0)
(c)

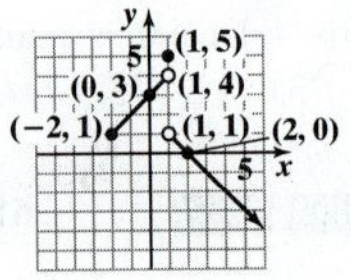

(d) $\{y \mid y < 4, y = 5\}$; $(-\infty, 4) \cup \{5\}$
(e) Discontinuous at $x = 1$

37. (a) All real numbers
(b) $(-1, 0)$, $(0, 0)$
(c)

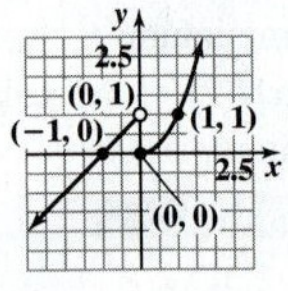

(d) All real numbers
(e) Discontinuous at $x = 0$

39. (a) $\{x \mid x \geq -2, x \neq 0\}$; $[-2, 0) \cup (0, \infty)$
(b) No intercepts
(c)

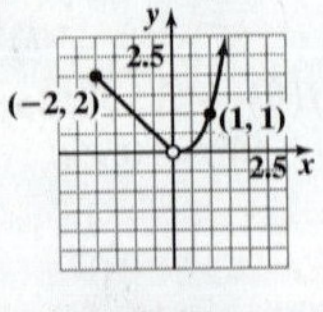

(d) $\{y \mid y > 0\}$; $(0, \infty)$
(e) Discontinuous at $x = 0$

41. (a) All real numbers
(b) $(x, 0)$ for $0 \leq x < 1$
(c)

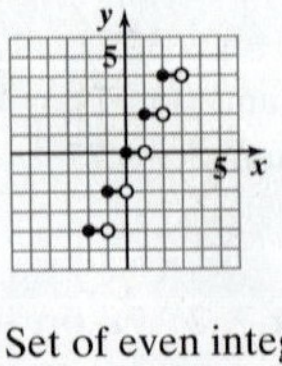

(d) Set of even integers
(e) Discontinuous at $\{x \mid x \text{ is an integer}\}$

43. $f(x) = \begin{cases} -x & \text{if } -1 \leq x \leq 0 \\ \frac{1}{2}x & \text{if } \ 0 < x \leq 2 \end{cases}$ (Other answers are possible.)

45. $f(x) = \begin{cases} -x & \text{if } x \leq 0 \\ -x + 2 & \text{if } 0 < x \leq 2 \end{cases}$ (Other answers are possible.)

47. (a) 2 **(b)** 3 **(c)** −4 **49. (a)** \$34.99 **(b)** \$64.99 **(c)** \$184.99

51. (a) \$44.46 **(b)** \$126.03

(c) $C(x) = \begin{cases} 1.24816x + 19.50 & \text{if } 0 \leq x \leq 30 \\ 0.5757x + 39.6738 & \text{if } x > 30 \end{cases}$

(d)

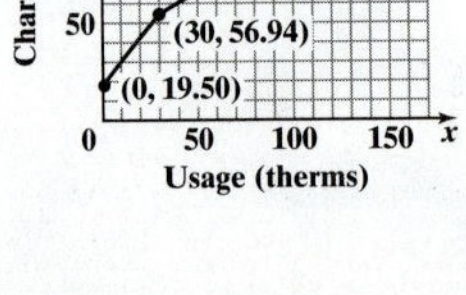

53. $f(x) = \begin{cases} 0.10x & \text{if} \quad 0 < x \le 9075 \\ 907.50 + 0.15(x - 9075) & \text{if} \quad 9075 < x \le 36{,}900 \\ 5081.25 + 0.25(x - 36{,}900) & \text{if} \quad 36{,}900 < x \le 89{,}350 \\ 18{,}193.75 + 0.28(x - 89{,}350) & \text{if} \quad 89{,}350 < x \le 186{,}350 \\ 45{,}353.75 + 0.33(x - 186{,}350) & \text{if} \quad 186{,}350 < x \le 405{,}100 \\ 117{,}541.25 + 0.35(x - 405{,}100) & \text{if} \quad 405{,}100 < x \le 406{,}750 \\ 118{,}188.75 + 0.396(x - 406{,}750) & \text{if} \quad x > 406{,}750 \end{cases}$

55. (a)

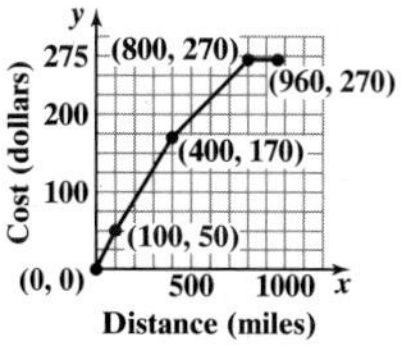

(b) $C(x) = 10 + 0.4x$ **(c)** $C(x) = 70 + 0.25x$

57. (a) $C(s) = \begin{cases} 9000 & \text{if} \quad s \le 659 \\ 7500 & \text{if} \quad 660 \le s \le 679 \\ 5250 & \text{if} \quad 680 \le s \le 699 \\ 3000 & \text{if} \quad 700 \le s \le 719 \\ 1500 & \text{if} \quad 720 \le s \le 739 \\ 750 & \text{if} \quad s \ge 740 \end{cases}$ **(b)** \$1500 **(c)** \$7500

59. (a) 10°C **(b)** 4°C **(c)** −3°C **(d)** −4°C
(e) The wind chill is equal to the air temperature.
(f) At wind speed greater than 20 m/s, the wind chill factor depends only on the air temperature.

61. $C(x) = \begin{cases} 0.98 & \text{if} \quad 0 < x \le 1 \\ 1.19 & \text{if} \quad 1 < x \le 2 \\ 1.40 & \text{if} \quad 2 < x \le 3 \\ 1.61 & \text{if} \quad 3 < x \le 4 \\ 1.82 & \text{if} \quad 4 < x \le 5 \\ 2.03 & \text{if} \quad 5 < x \le 6 \\ 2.24 & \text{if} \quad 6 < x \le 7 \\ 2.45 & \text{if} \quad 7 < x \le 8 \\ 2.66 & \text{if} \quad 8 < x \le 9 \\ 2.87 & \text{if} \quad 9 < x \le 10 \\ 3.08 & \text{if} \quad 10 < x \le 11 \\ 3.29 & \text{if} \quad 11 < x \le 12 \\ 3.50 & \text{if} \quad 12 < x \le 13 \end{cases}$

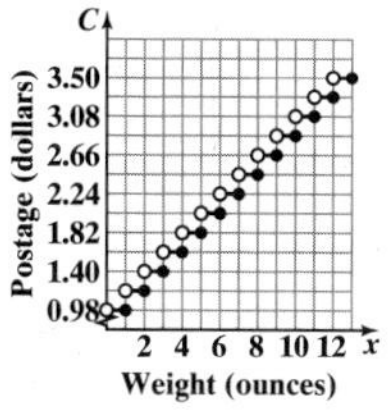

63. Each graph is that of $y = x^2$, but shifted horizontally. If $y = (x - k)^2$, $k > 0$, the shift is right k units; if $y = (x + k)^2$, $k > 0$, the shift is left k units. **65.** The graph of $y = -f(x)$ is the reflection about the x-axis of the graph of $y = f(x)$. **67.** Yes. The graph of $y = (x - 1)^3 + 2$ is the graph of $y = x^3$ shifted right 1 unit and up 2 units. **69.** They all have the same general shape. All three go through the points $(-1, -1)$, $(0, 0)$, and $(1, 1)$. As the exponent increases, the steepness of the curve increases (except near $x = 0$). **72.** $22 - 7i$ **73.** $(h, k) = (0, 3)$; $r = 5$ **74.** $\{-8\}$ **75.** CD: \$22,000; Mutual fund: \$38,000

3.5 Assess Your Understanding *(page 256)*

1. horizontal; right **2.** y **3.** F **4.** T **5.** d **6.** a **7.** B **9.** H **11.** I **13.** L **15.** F **17.** G **19.** $y = (x - 4)^3$ **21.** $y = x^3 + 4$ **23.** $y = -x^3$ **25.** $y = 4x^3$ **27.** $y = -(\sqrt{-x} + 2)$ **29.** $y = -\sqrt{x + 3} + 2$ **31.** (c) **33.** (c) **35. (a)** −7 and 1 **(b)** −3 and 5 **(c)** −5 and 3 **(d)** −3 and 5 **37. (a)** $(-3, 3)$ **(b)** $(4, 10)$ **(c)** Decreasing on $(-1, 5)$ **(d)** Decreasing on $(-5, 1)$

39.

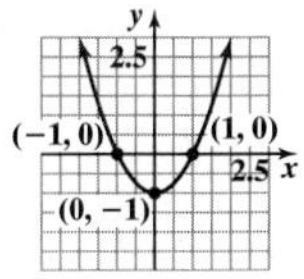

Domain: $(-\infty, \infty)$; Range: $[-1, \infty)$

41.

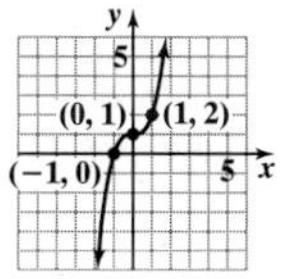

Domain: $(-\infty, \infty)$; Range: $(-\infty, \infty)$

43

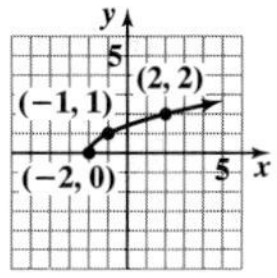

Domain: $[-2, \infty)$; Range: $[0, \infty)$

45.

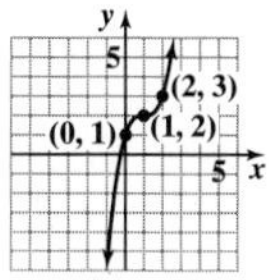

Domain: $(-\infty, \infty)$; Range: $(-\infty, \infty)$

47.

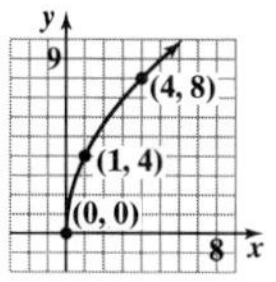

Domain: $[0, \infty)$; Range: $[0, \infty)$

49.

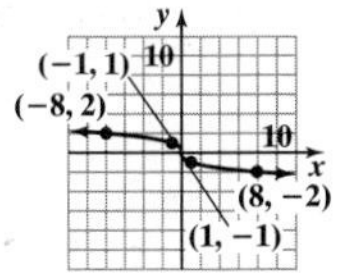

Domain: $(-\infty, \infty)$; Range: $(-\infty, \infty)$

51.

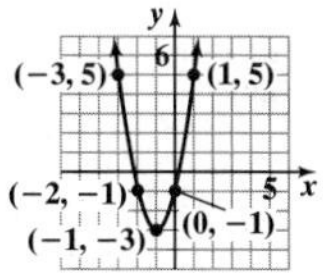

Domain: $(-\infty, \infty)$; Range: $[-3, \infty)$

53.

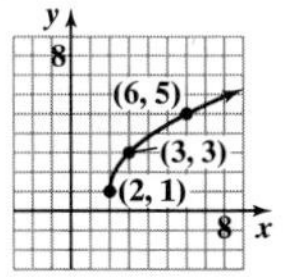

Domain: $[2, \infty)$; Range: $[1, \infty)$

55.

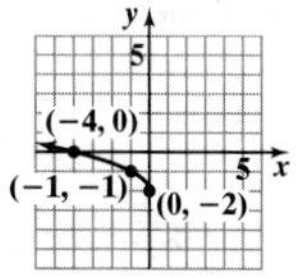

Domain: $(-\infty, 0]$; Range: $[-2, \infty)$

57.

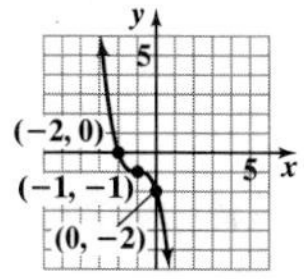

Domain: $(-\infty, \infty)$; Range: $(-\infty, \infty)$

59.

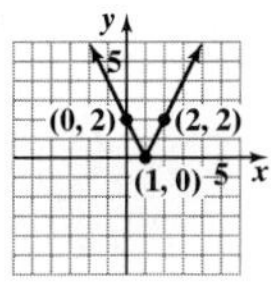

Domain: $(-\infty, \infty)$; Range: $[0, \infty)$

61.

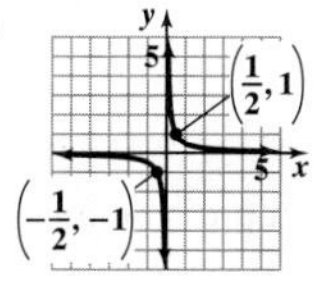

Domain: $(-\infty, 0) \cup (0, \infty)$; Range: $(-\infty, 0) \cup (0, \infty)$

63. (a) $F(x) = f(x) + 3$

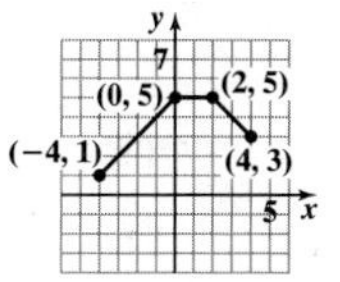

(b) $G(x) = f(x + 2)$

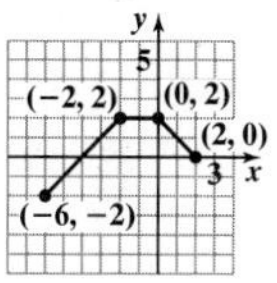

(c) $P(x) = -f(x)$

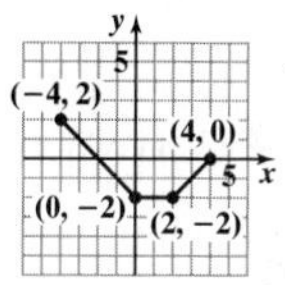

(d) $H(x) = f(x + 1) - 2$

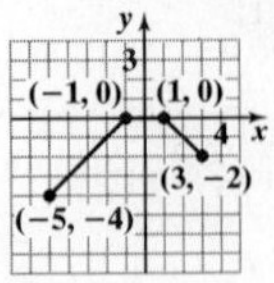

(e) $Q(x) = \frac{1}{2}f(x)$

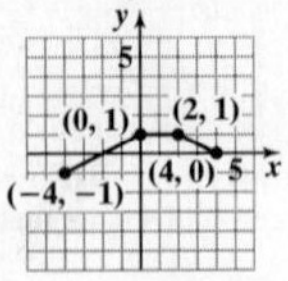

(f) $g(x) = f(-x)$

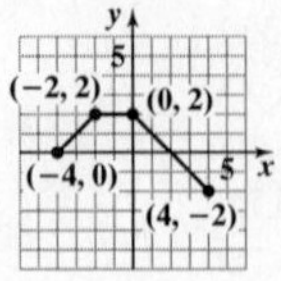

(g) $h(x) = f(2x)$

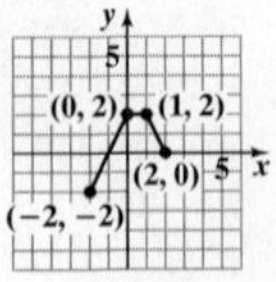

65. (a) $F(x) = f(x) + 3$

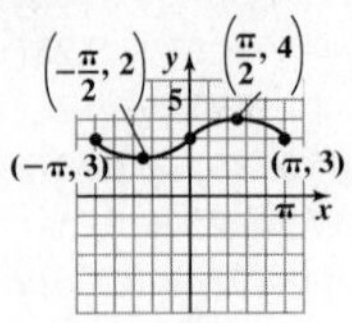

(b) $G(x) = f(x + 2)$

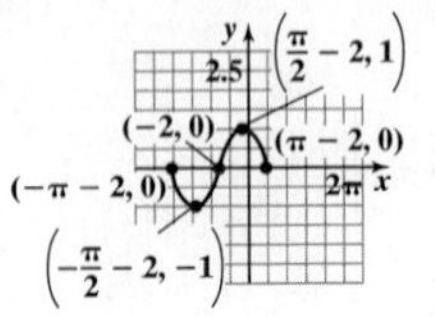

(c) $P(x) = -f(x)$

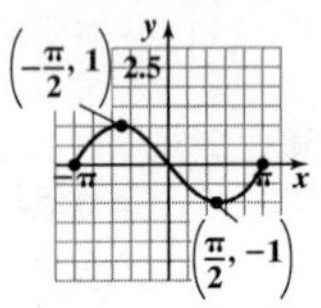

(d) $H(x) = f(x + 1) - 2$

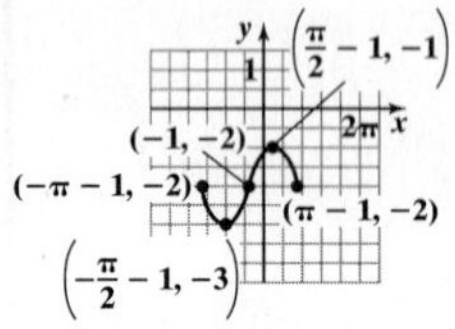

(e) $Q(x) = \frac{1}{2}f(x)$

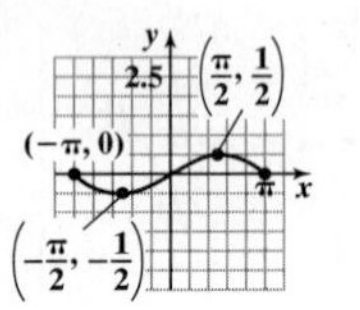

(f) $g(x) = f(-x)$

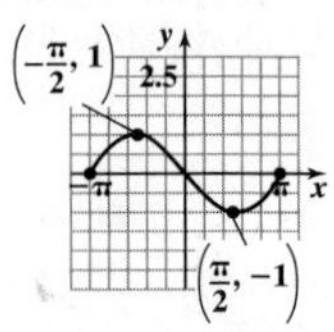

(g) $h(x) = f(2x)$

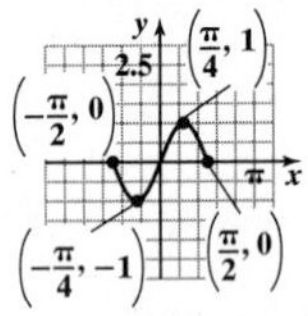

67. $f(x) = (x + 1)^2 - 1$

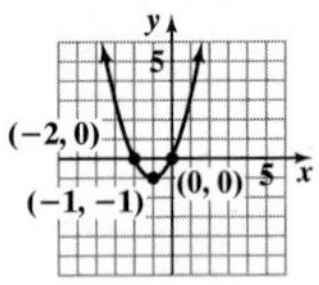

69. $f(x) = (x - 4)^2 - 15$

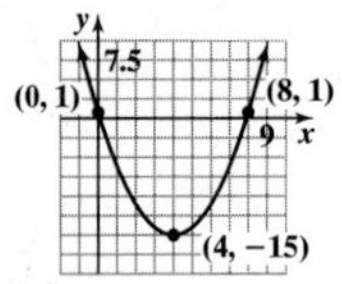

71. $f(x) = 2(x - 3)^2 + 1$

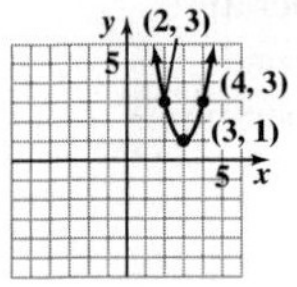

73. $f(x) = -3(x + 2)^2 - 5$

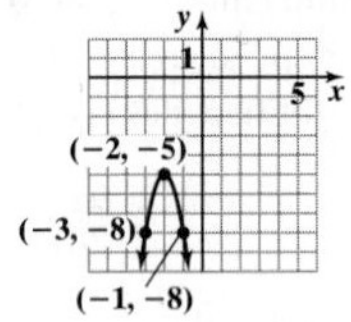

75. (a)

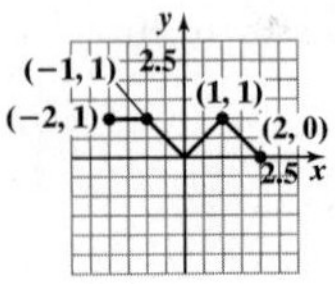

(b)

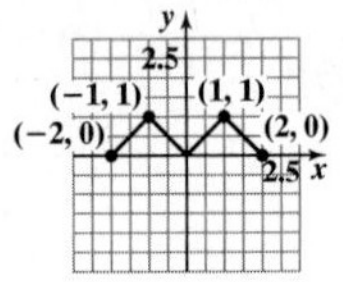

77. (a) $(-2, 2)$ **(b)** $(3, -5)$ **(c)** $(-1, 3)$

79. (a)

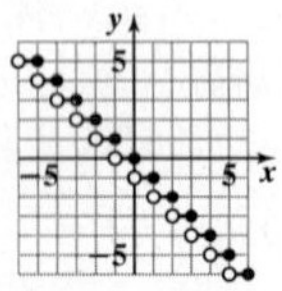

(b)

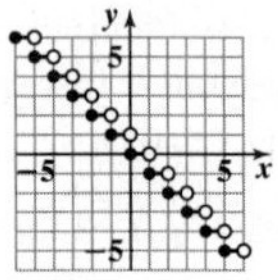

81. (a)

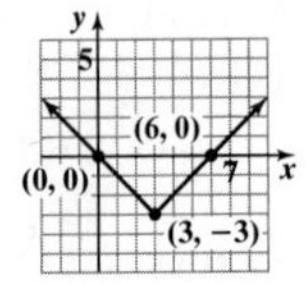

(b) 9 square units

83. (a) 72°F; 65°F **(b)** The temperature decreases by 2° to 70°F during the day and 63°F overnight.

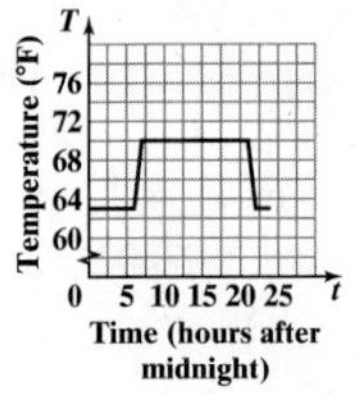

(c) The time at which the temperature adjusts between the daytime and overnight settings is moved to 1 hr sooner. It begins warming up at 5:00 AM instead of 6:00 AM, and it begins cooling down at 8:00 PM instead of 9:00 PM.

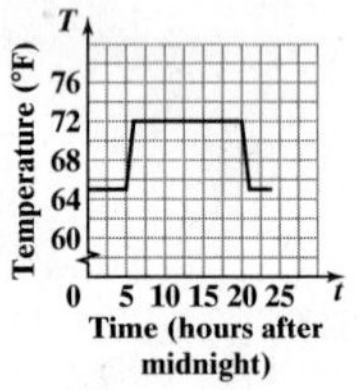

85.

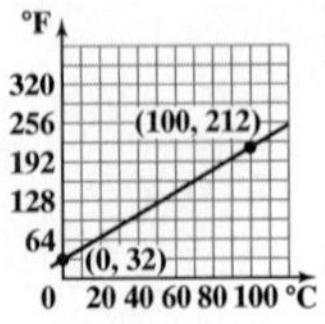

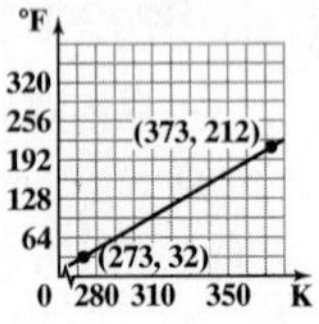

87.

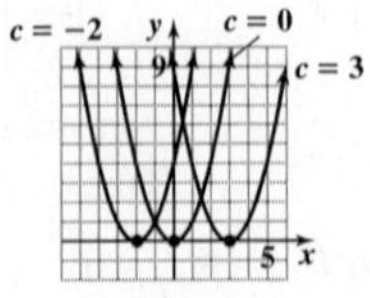

89. The graph of $y = 4f(x)$ is a vertical stretch by a factor of 4. The graph of $y = f(4x)$ is a horizontal compression by a factor of $\frac{1}{4}$. **91.** $\frac{16}{3}$ sq. units **93.** The domain of $g(x) = \sqrt{x}$ is $[0, \infty)$. The graph of $g(x - k)$ is the graph of g shifted k units to the right, so the domain of $g(x - k)$ is $[k, \infty)$. **94.** $m = \frac{3}{5}$; $b = -6$ **95.** $\frac{y^2}{x^4}$ **96.** 15.75 gal **97.** Intercepts: $(0, -2)$, $(0, 2)$, $(-4, 0)$; x-axis symmetry

3.6 Assess Your Understanding *(page 263)*

1. (a) $d(x) = \sqrt{x^4 - 15x^2 + 64}$ **(b)** $d(0) = 8$ **(c)** $d(1) = \sqrt{50} \approx 7.07$ **(d)**

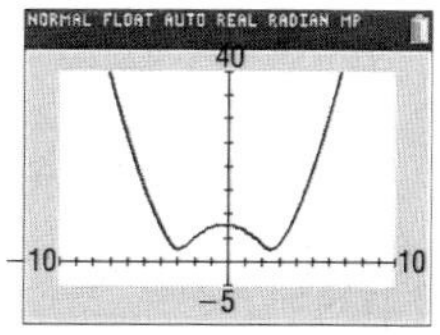

(e) d is smallest when $x \approx -2.74$ or $x \approx 2.74$

3. (a) $d(x) = \sqrt{x^2 - x + 1}$ **(b)**

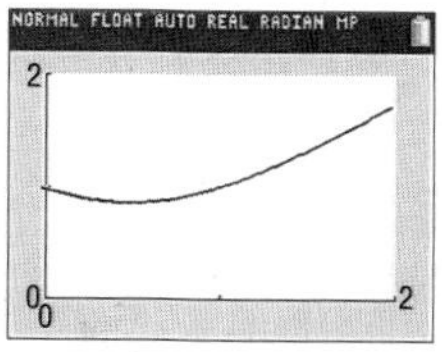

(c) d is smallest when $x = 0.5$.

5. $A(x) = \frac{1}{2}x^4$

7. (a) $A(x) = x(16 - x^2)$ **(b)** Domain: $\{x \mid 0 < x < 4\}$ **(c)** The area is largest when $x \approx 2.31$.

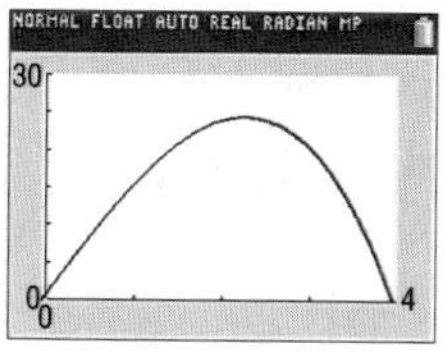

9. (a) $A(x) = 4x\sqrt{4 - x^2}$ **(b)** $p(x) = 4x + 4\sqrt{4 - x^2}$ **(c)** A is largest when $x \approx 1.41$. **(d)** p is largest when $x \approx 1.41$.

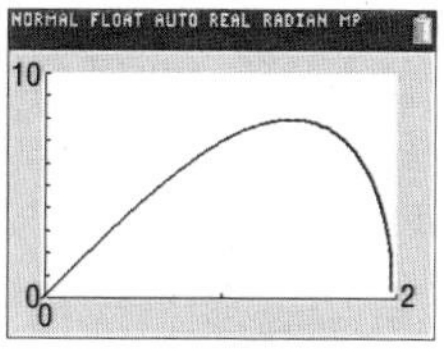

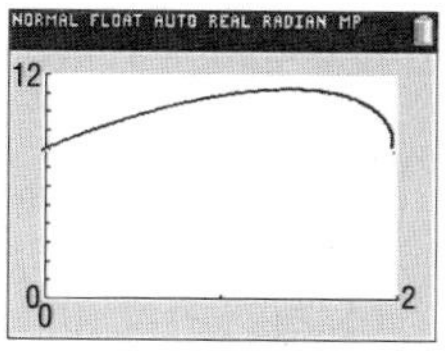

11. (a) $A(x) = x^2 + \dfrac{25 - 20x + 4x^2}{\pi}$ **(b)** Domain: $\{x \mid 0 < x < 2.5\}$ **(c)** A is smallest when $x \approx 1.40$ m.

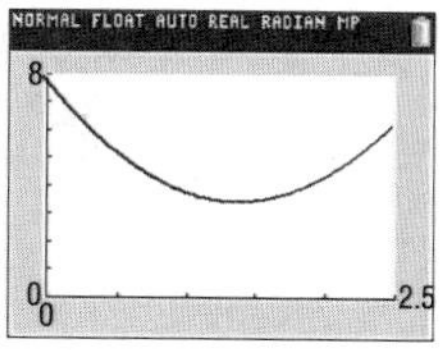

13. (a) $C(x) = x$ **(b)** $A(x) = \dfrac{x^2}{4\pi}$ **15. (a)** $A(r) = 2r^2$ **(b)** $p(r) = 6r$ **17.** $A(x) = \left(\dfrac{\pi}{3} - \dfrac{\sqrt{3}}{4}\right)x^2$

19. (a) $d(t) = \sqrt{2500t^2 - 360t + 13}$ **(b)** d is smallest when $t \approx 0.07$ hr.

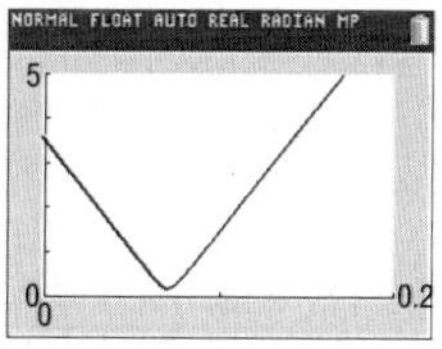

21. $V(r) = \dfrac{\pi H(R - r)r^2}{R}$

23. (a) $T(x) = \dfrac{12 - x}{5} + \dfrac{\sqrt{x^2 + 4}}{3}$ **(b)** $\{x \mid 0 \le x \le 12\}$ **(c)** 3.09 hr **(d)** 3.55 hr

25. (a) $V(x) = x(24 - 2x)^2$ **(b)** 972 in.^{3} **(c)** 160 in.^{3} **(d)** V is largest when $x = 4$.

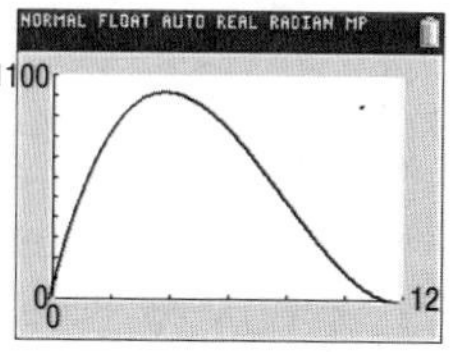

27. $\{0, 3\}$ **28.** 64 mi/h **29.** $m = -4$ **30.** 5.6

Review Exercises *(page 268)*

1. Function; domain $\{-1, 2, 4\}$, range $\{0, 3\}$ **2.** Not a function **3. (a)** 2 **(b)** -2 **(c)** $-\dfrac{3x}{x^2 - 1}$ **(d)** $-\dfrac{3x}{x^2 - 1}$ **(e)** $\dfrac{3(x - 2)}{x^2 - 4x + 3}$ **(f)** $\dfrac{6x}{4x^2 - 1}$ **4. (a)** 0 **(b)** 0 **(c)** $\sqrt{x^2 - 4}$ **(d)** $-\sqrt{x^2 - 4}$ **(e)** $\sqrt{x^2 - 4x}$ **(f)** $2\sqrt{x^2 - 1}$ **5. (a)** 0 **(b)** 0 **(c)** $\dfrac{x^2 - 4}{x^2}$ **(d)** $-\dfrac{x^2 - 4}{x^2}$ **(e)** $\dfrac{x(x - 4)}{(x - 2)^2}$ **(f)** $\dfrac{x^2 - 1}{x^2}$

6. $\{x \mid x \ne -3, x \ne 3\}$ **7.** $\{x \mid x \le 2\}$ **8.** $\{x \mid x \ne 0\}$ **9.** $\{x \mid x \ne -3, x \ne 1\}$ **10.** $[-1, 2) \cup (2, \infty)$ **11.** $\{x \mid x > -8\}$, or $(-8, \infty)$

12. $(f + g)(x) = 2x + 3$; Domain: all real numbers
$(f - g)(x) = -4x + 1$; Domain: all real numbers
$(f \cdot g)(x) = -3x^2 + 5x + 2$; Domain: all real numbers
$\left(\dfrac{f}{g}\right)(x) = \dfrac{2 - x}{3x + 1}$; Domain: $\left\{x \,\middle|\, x \ne -\dfrac{1}{3}\right\}$

13. $(f + g)(x) = 3x^2 + 4x + 1$; Domain: all real numbers
$(f - g)(x) = 3x^2 - 2x + 1$; Domain: all real numbers
$(f \cdot g)(x) = 9x^3 + 3x^2 + 3x$; Domain: all real numbers
$\left(\dfrac{f}{g}\right)(x) = \dfrac{3x^2 + x + 1}{3x}$; Domain: $\{x \mid x \ne 0\}$

14. $(f + g)(x) = \dfrac{x^2 + 2x - 1}{x(x - 1)}$; Domain: $\{x \mid x \ne 0, x \ne 1\}$
$(f - g)(x) = \dfrac{x^2 + 1}{x(x - 1)}$; Domain: $\{x \mid x \ne 0, x \ne 1\}$
$(f \cdot g)(x) = \dfrac{x + 1}{x(x - 1)}$; Domain: $\{x \mid x \ne 0, x \ne 1\}$
$\left(\dfrac{f}{g}\right)(x) = \dfrac{x(x + 1)}{x - 1}$; Domain: $\{x \mid x \ne 0, x \ne 1\}$

15. $-4x + 1 - 2h$ **16. (a)** Domain: $\{x \mid -4 \le x \le 3\}$; Range: $\{y \mid -3 \le y \le 3\}$ **(b)** $(0, 0)$ **(c)** -1 **(d)** -4 **(e)** $\{x \mid 0 < x \le 3\}$

(f)

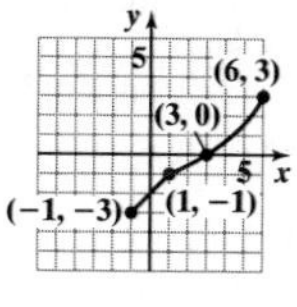

(g)

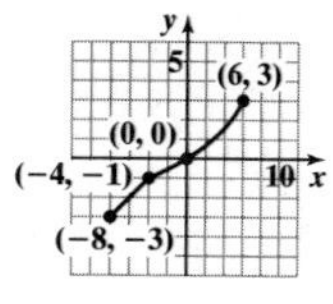

(h)

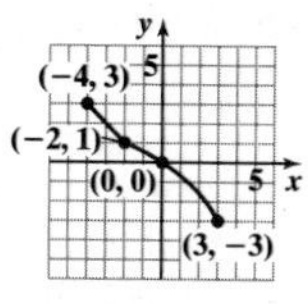

17. (a) Domain: $\{x|x \leq 4\}$ or $(-\infty, 4]$
Range: $\{y|y \leq 3\}$ or $(-\infty, 3]$
(b) Increasing on $(-\infty, -2)$ and $(2, 4)$; Decreasing on $(-2, 2)$
(c) Local maximum value is 1 and occurs at $x = -2$.
Local minimum value is -1 and occurs at $x = 2$.
(d) Absolute maximum: $f(4) = 3$
Absolute minimum: none
(e) No symmetry
(f) Neither
(g) x-intercepts: $-3, 0, 3$ y-intercept: 0

18. Odd **19.** Even **20.** Neither **21.** Odd

22.

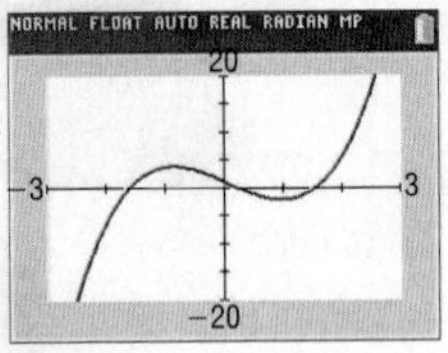

Local maximum value: 4.04 at $x = -0.91$
Local minimum value: -2.04 at $x = 0.91$
Increasing: $(-3, -0.91)$; $(0.91, 3)$
Decreasing: $(-0.91, 0.91)$

23.

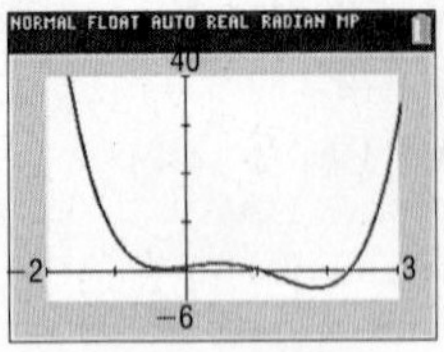

Local maximum value: 1.53 at $x = 0.41$
Local minima values: 0.54 at
$x = -0.34$ and -3.56 at $x = 1.80$
Increasing: $(-0.34, 0.41)$; $(1.80, 3)$
Decreasing: $(-2, -0.34)$; $(0.41, 1.80)$

24. (a) 23 **(b)** 7 **(c)** 47 **25.** -5 **26.** -17 **27.** No **28.** Yes

29.

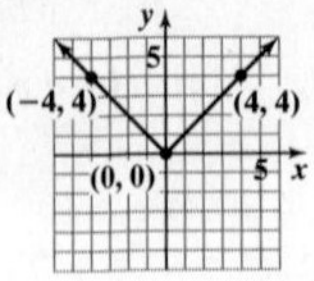

30.

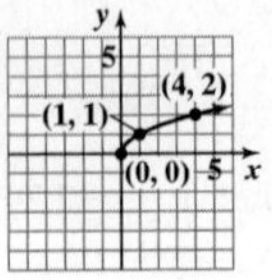

31.

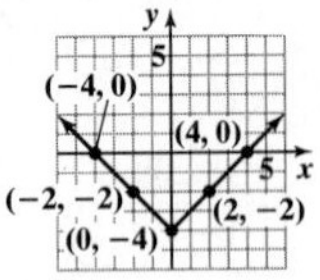

Intercepts: $(-4, 0)$, $(4, 0)$, $(0, -4)$
Domain: all real numbers
Range: $\{y|y \geq -4\}$ or $[-4, \infty)$

32.

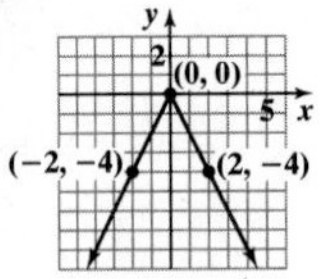

Intercept: (0, 0)
Domain: all real numbers
Range: $\{y|y \leq 0\}$ or $(-\infty, 0]$

33.

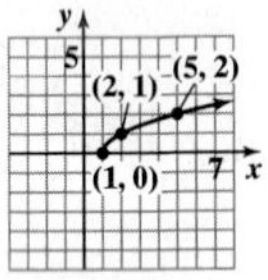

Intercept: (1, 0)
Domain: $\{x|x \geq 1\}$ or $[1, \infty)$
Range: $\{y|y \geq 0\}$ or $[0, \infty)$

34.

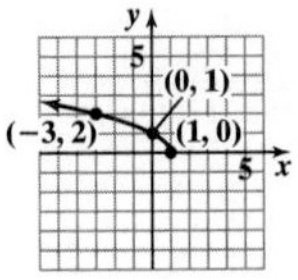

Intercepts: (0, 1), (1, 0)
Domain: $\{x|x \leq 1\}$ or $(-\infty, 1]$
Range: $\{y|y \geq 0\}$ or $[0, \infty)$

35.

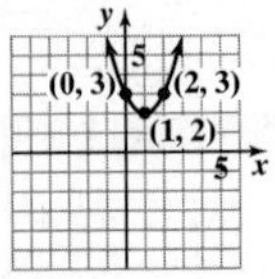

Intercept: (0, 3)
Domain: all real numbers
Range: $\{y|y \geq 2\}$ or $[2, \infty)$

36.

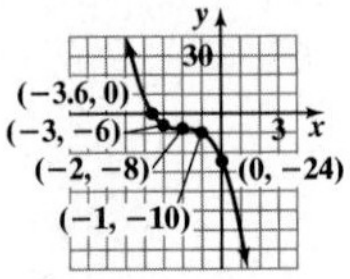

Intercepts: (0, –24),
$(-2 - \sqrt[3]{4}, 0)$ or about
$(-3.6, 0)$
Domain: all real numbers
Range: all real numbers

37. (a) $\{x|x > -2\}$ or $(-2, \infty)$
(b) (0, 0)
(c)

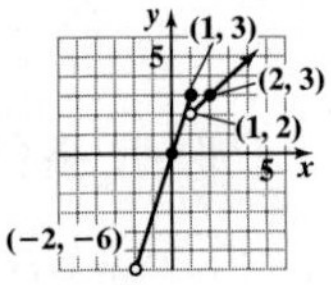

(d) $\{y|y > -6\}$ or $(-6, \infty)$
(e) Discontinuous at $x = 1$

38. (a) $\{x|x \geq -4\}$ or $[-4, \infty)$
(b) (0, 1)
(c)

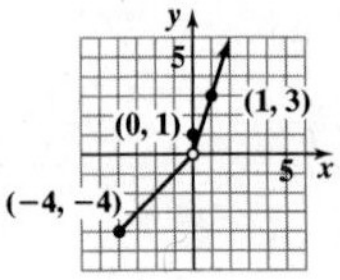

(d) $\{y|-4 \leq y < 0 \text{ or } y > 0\}$ or $[-4, 0) \cup (0, \infty)$
(e) Discontinuous at $x = 0$

39. $A = 11$

40. (a) $A(x) = 2x^2 + \dfrac{40}{x}$
(b) 42 ft^2 **(c)** 28 ft^2
(d)

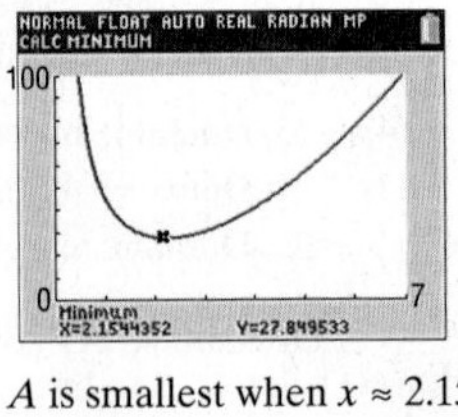

A is smallest when $x \approx 2.15$ ft.

41. (a) $A(x) = 10x - x^3$
(b) The largest area that can be enclosed by the rectangle is approximately 12.17 square units.

Chapter Test *(page 270)*

1. (a) Function; domain: $\{2, 4, 6, 8\}$; range: $\{5, 6, 7, 8\}$ **(b)** Not a function **(c)** Not a function **(d)** Function; domain; all real numbers; range: $\{y|y \geq 2\}$ **2.** Domain: $\left\{x \middle| x \leq \frac{4}{5}\right\}$; $f(-1) = 3$ **3.** Domain: $\{x|x \neq -2\}$; $g(-1) = 1$ **4.** Domain: $\{x|x \neq -9, x \neq 4\}$; $h(-1) = \frac{1}{8}$
5. (a) Domain: $\{x|-5 \leq x \leq 5\}$; range: $\{y|-3 \leq y \leq 3\}$ **(b)** $(0, 2)$, $(-2, 0)$, and $(2, 0)$ **(c)** $f(1) = 3$ **(d)** $x = -5$ and $x = 3$
(e) $\{x|-5 \leq x < -2 \text{ or } 2 < x \leq 5\}$ or $[-5, -2) \cup (2, 5]$ **6.** Local maxima values: $f(-0.85) \approx -0.86$; $f(2.35) \approx 15.55$; local minimum value: $f(0) = -2$; the function is increasing on the intervals $(-5, -0.85)$ and $(0, 2.35)$ and decreasing on the intervals $(-0.85, 0)$ and $(2.35, 5)$.

7. (a)

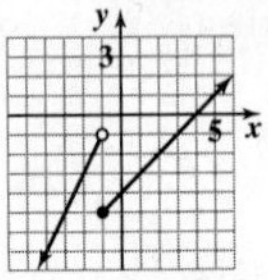

(b) $(0, -4)$, $(4, 0)$
(c) $g(-5) = -9$
(d) $g(2) = -2$

8. 19

9. (a) $(f - g)(x) = 2x^2 - 3x + 3$
(b) $(f \cdot g)(x) = 6x^3 - 4x^2 + 3x - 2$
(c) $f(x + h) - f(x) = 4xh + 2h^2$

10. (a)

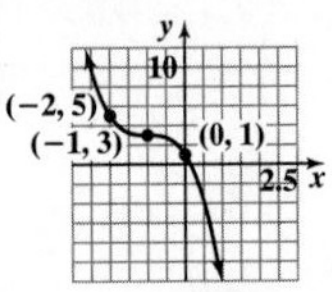

(b)

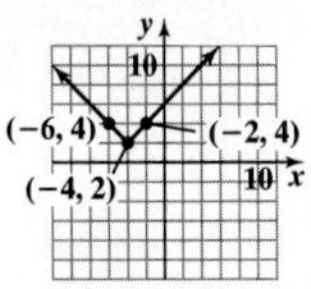

11. (a) 8.67% occurring in 1997 $(x \approx 5)$ **(b)** The model predicts that the interest rate will be -10.343%. This is not reasonable.

12. (a) $V(x) = \frac{x^2}{8} - \frac{5x}{4} + \frac{\pi x^2}{64}$ **(b)** 1297.61 ft^3

Cumulative Review *(page 271)*

1. $\{6\}$ **2.** $\left\{0, \frac{1}{3}\right\}$ **3.** $\{-1, 9\}$ **4.** $\left\{\frac{1}{3}, \frac{1}{2}\right\}$ **5.** $\left\{-\frac{7}{2}, \frac{1}{2}\right\}$ **6.** $\left\{\frac{1}{2}\right\}$

7. $\left\{x \middle| x < -\frac{4}{3}\right\}$; $\left(-\infty, -\frac{4}{3}\right)$

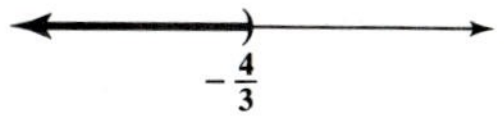

8. $\{x|1 < x < 4\}$; $(1, 4)$

9. $\left\{x \middle| x \le -2 \text{ or } x \ge \frac{3}{2}\right\}$; $(-\infty, -2] \cup \left[\frac{3}{2}, \infty\right)$

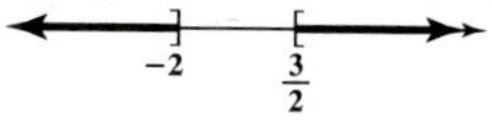

10. (a) distance: $\sqrt{29}$ **(b)** midpoint: $\left(\frac{1}{2}, -4\right)$ **(c)** slope: $-\frac{2}{5}$

11.

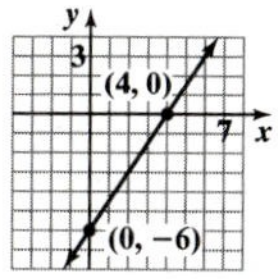

12.

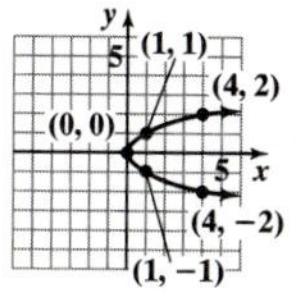

13.

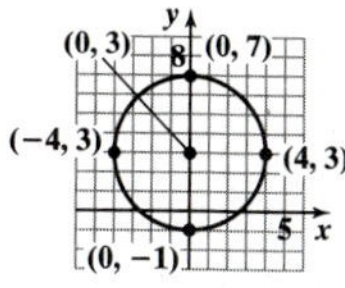

14.

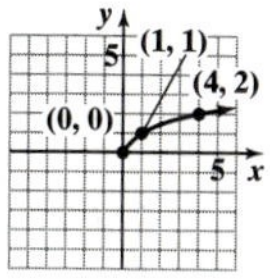

15. Intercepts: $(0, -3)$, $(-2, 0)$, $(2, 0)$; symmetry with respect to the y-axis

16. $y = \frac{1}{2}x + 5$

17.

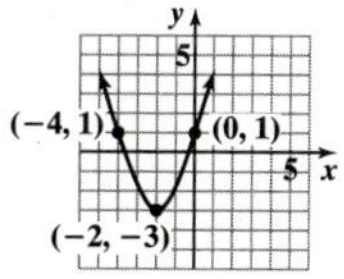

18.

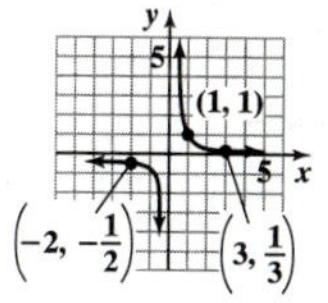

19.

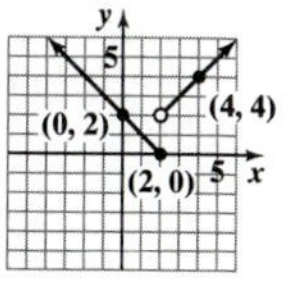

CHAPTER 4 Linear and Quadratic Functions

4.1 Assess Your Understanding *(page 280)*

7. slope; y-intercept **8.** positive **9.** T **10.** F **11.** a **12.** d

13. (a) $m = 2; b = 3$
(b)

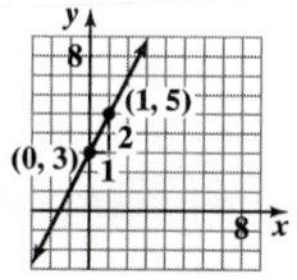

(c) 2 **(d)** Increasing

15. (a) $m = -3; b = 4$
(b)

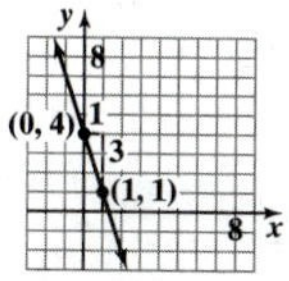

(c) -3 **(d)** Decreasing

17. (a) $m = \frac{1}{4}; b = -3$
(b)

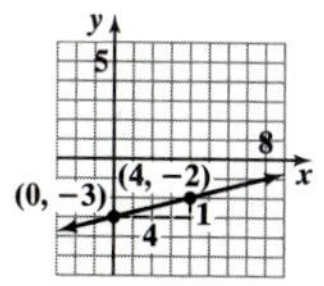

(c) $\frac{1}{4}$ **(d)** Increasing

19. (a) $m = 0; b = 4$
(b)

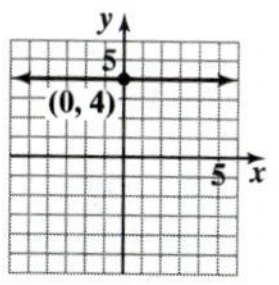

(c) 0 **(d)** Constant

21. Linear; -3 **23.** Nonlinear **25.** Nonlinear **27.** Linear; 0

29. (a) $\frac{1}{4}$ **(b)** $\left\{x \middle| x > \frac{1}{4}\right\}$ or $\left(\frac{1}{4}, \infty\right)$ **(c)** 1 **(d)** $\{x|x \le 1\}$ or $(-\infty, 1]$
(e)

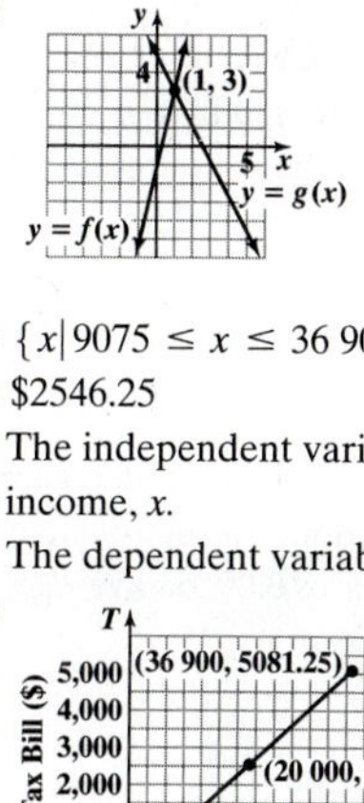

31. (a) 40 **(b)** 88 **(c)** -40 **(d)** $\{x|x > 40\}$ or $(40, \infty)$ **(e)** $\{x|x \le 88\}$ or $(-\infty, 88]$ **(f)** $\{x|-40 < x < 88\}$ or $(-40, 88)$ **33. (a)** -4 **(b)** $\{x|x < -4\}$ or $(-\infty, -4)$ **35. (a)** -6 **(b)** $\{x|-6 \le x < 5\}$ or $[-6, 5)$ **37. (a)** \$59 **(b)** 180 mi **(c)** 300 mi **(d)** $\{x|x \ge 0\}$ or $[0, \infty)$ **39. (a)** \$24; 600 T-shirts **(b)** $\$0 \le p < \$24$ **(c)** The price will increase.

41. (a) $\{x|9075 \le x \le 36\,900\}$ or $[9075, 36900]$
(b) \$2546.25
(c) The independent variable is adjusted gross income, x.
The dependent variable is the tax bill, T.
(d)

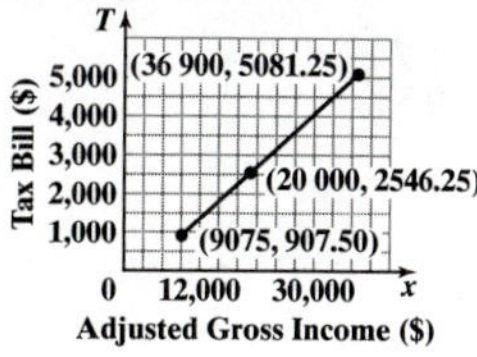

(e) \$27,500

43. (a) $x = 5000$
(b) $x > 5000$

45. (a) $V(x) = -1000x + 3000$
(b) $\{x|0 \le x \le 3\}$ or $[0, 3]$
(c)

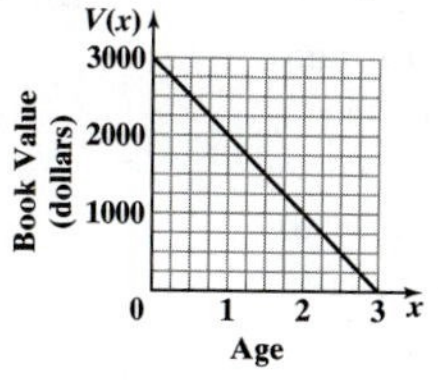

(d) \$1000
(e) After 1 year

47. (a) $C(x) = 90x + 1800$
(b)

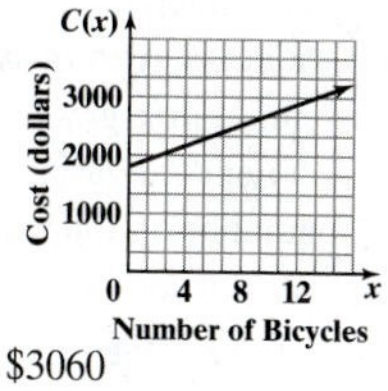

(c) \$3060
(d) 22 bicycles

49. (a) $C(x) = 0.89x + 31.95$ **(b)** \$129.85; \$236.65

51. (a)

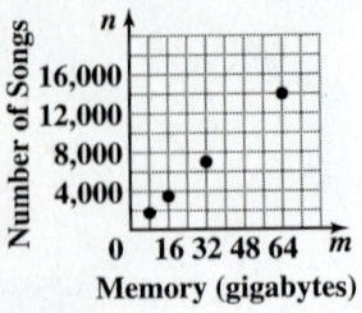

(b) Since each input (memory) corresponds to a single output (number of songs), we know that number of songs is a function of memory. Also, because the average rate of change is a constant 218.75 songs per gigabyte, the function is linear.

(c) $n(m) = 218.75\,m$ **(d)** $\{m \mid m \geq 0\}$ or $[0, \infty)$

(e)

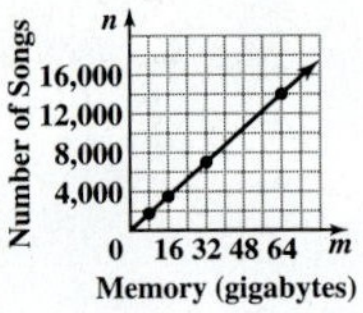

(f) If memory increases by 1 GB, then the number of songs increases by 218.75.

53. (d), (e) **55.** $b = 0$; yes, $f(x) = b$

57.

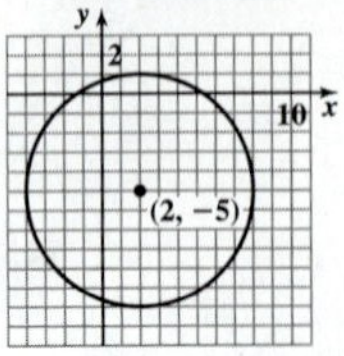

58. 6

59. 7

60.

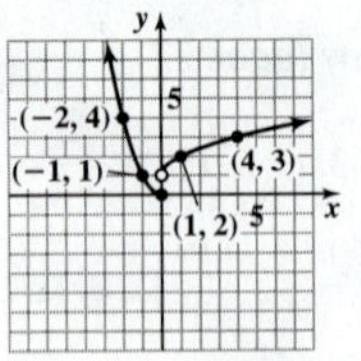

4.2 Assess Your Understanding *(page 287)*

3. scatter diagram **4.** T **5.** Linear relation, $m > 0$ **7.** Linear relation, $m < 0$ **9.** Nonlinear relation

11. (a)

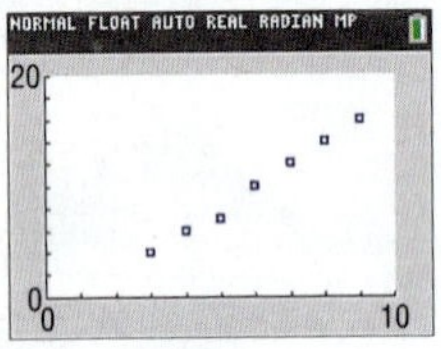

(b) Answers will vary. Using (4, 6) and (8, 14), $y = 2x - 2$.

(c)

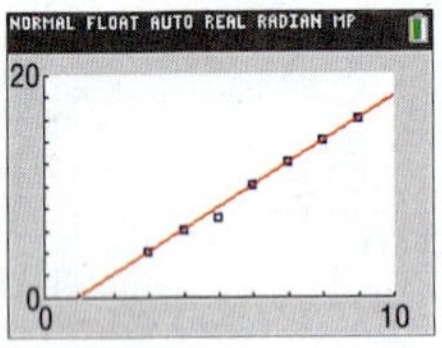

(d) $y = 2.0357x - 2.3571$

(e)

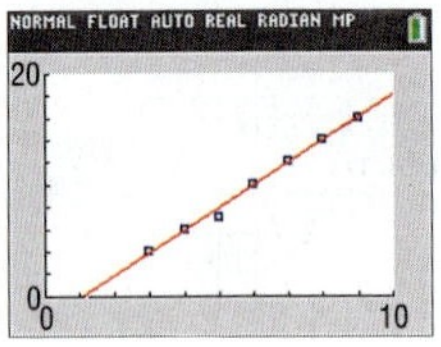

13. (a)

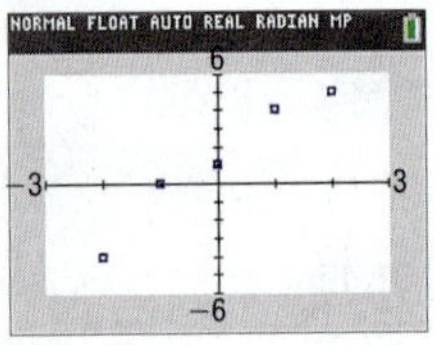

(b) Answers will vary. Using (−2, −4) and (2, 5), $y = \frac{9}{4}x + \frac{1}{2}$.

(c)

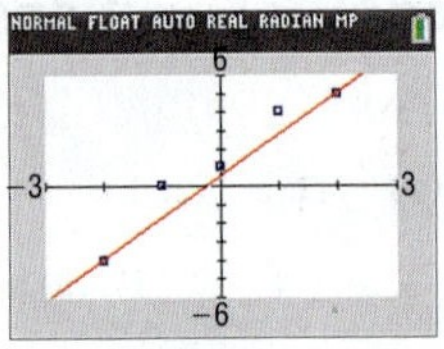

(d) $y = 2.2x + 1.2$

(e)

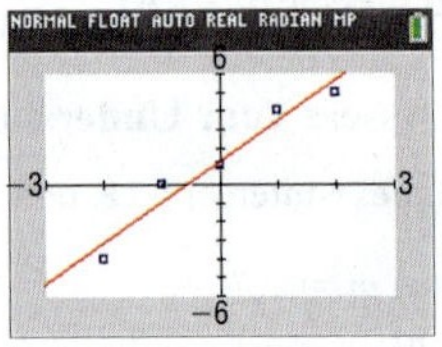

15. (a)

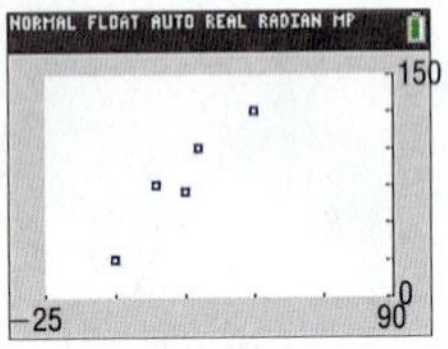

(b) Answers will vary. Using (−20, 100) and (−10, 140), $y = 4x + 180$.

(c)

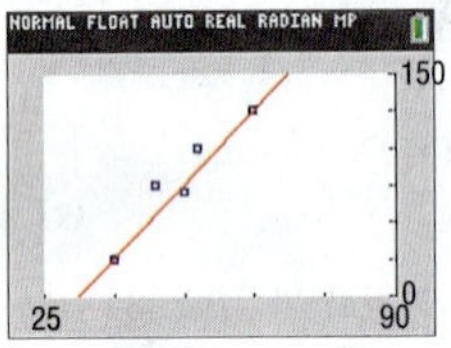

(d) $y = 3.8613x + 180.2920$

(e)

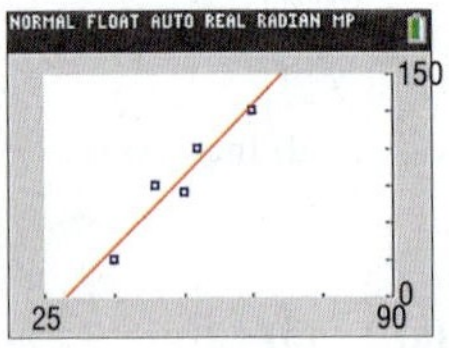

17. (a)

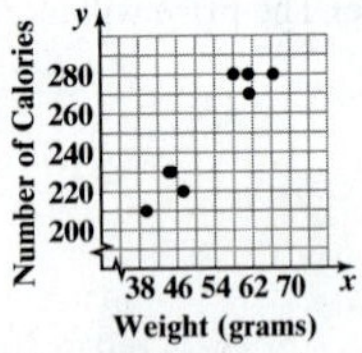

(b) Linear with positive slope

(c) Answers will vary. Using the points (39.52, 210) and (66.45, 280), $y = 2.599x + 107.288$.

(d)

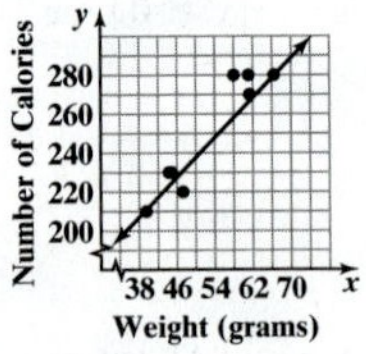

(e) 269 calories

(f) If the weight of a candy bar is increased by 1 gram, the number of calories will increase by 2.599, on average.

19. (a) The independent variable is the number of hours spent playing video games, and cumulative grade-point average is the dependent variable, because we are using number of hours playing video games to predict (or explain) cumulative grade-point average.

(b)

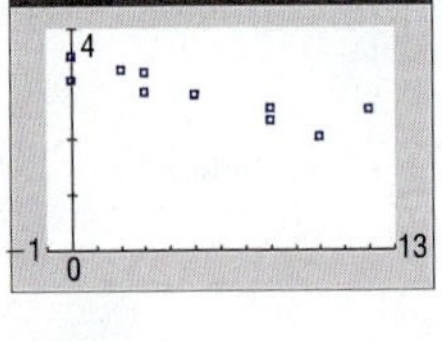

(c) $G(h) = -0.0942h + 3.2763$

(d) If the number of hours playing video games in a week increases by 1 hour, the cumulative grade-point average decreases 0.09, on average.

(e) 2.52

(f) Approximately 9.3 hours

21. (a)

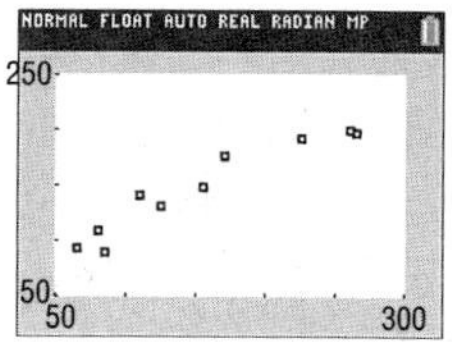

(b) $y = 0.5426x + 63.2336$
(c) If the flight time increases by 1 minute, then the ticket price increases by about $0.5426, on average.
(d) $109 **(e)** 105 minutes

23. $y = 1.5x + 3.5$; $r = 1$; The linear relationship between two points is perfect.
25. A value of $x = 0$ does not make sense.
27. $2x + y = 3$ or $y = -2x + 3$
28. $\{x \mid x \neq -5, x \neq 5\}$
29. $(g - f)(x) = x^2 - 8x + 12$
30. $y = (x + 3)^2 - 4$

4.3 Assess Your Understanding *(page 299)*

5. parabola **6.** axis or axis of symmetry **7.** $-\dfrac{b}{2a}$ **8.** T **9.** T **10.** T **11.** a **12.** d **13.** *C* **15.** *F* **17.** *G* **19.** *H*

21.

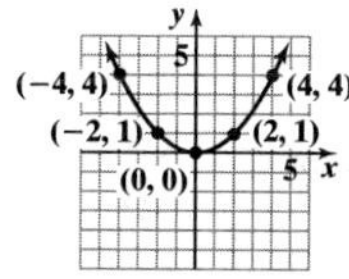

23.

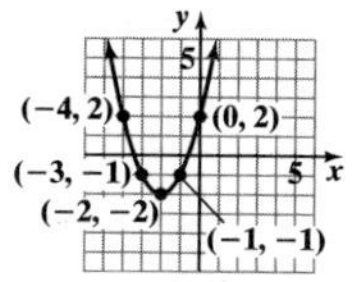

25. $f(x) = (x + 2)^2 - 2$

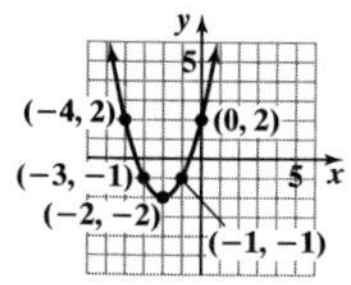

27. $f(x) = 2(x - 1)^2 - 1$

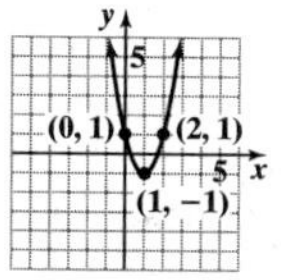

29. $f(x) = -(x + 1)^2 + 1$

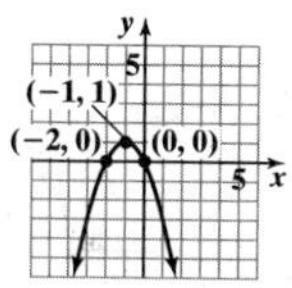

31. $f(x) = \dfrac{1}{2}(x + 1)^2 - \dfrac{3}{2}$

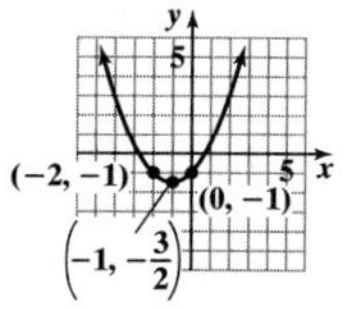

33. (a)

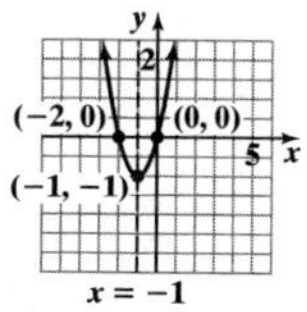

(b) Domain: $(-\infty, \infty)$
Range: $[-1, \infty)$
(c) Decreasing: $(-\infty, -1)$
Increasing: $(-1, \infty)$

35. (a)

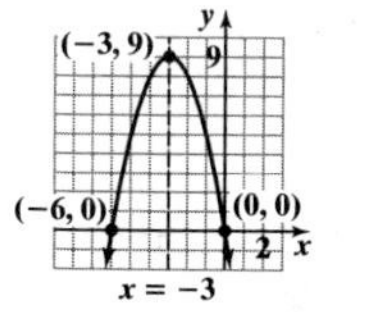

(b) Domain: $(-\infty, \infty)$
Range: $(-\infty, 9]$
(c) Increasing: $(-\infty, -3)$
Decreasing: $(-3, \infty)$

37. (a)

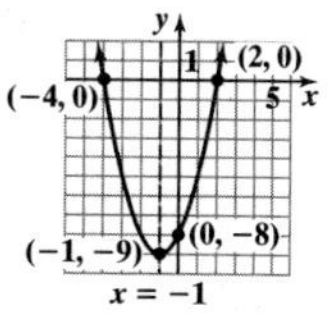

(b) Domain: $(-\infty, \infty)$
Range: $[-9, \infty)$
(c) Decreasing: $(-\infty, -1)$
Increasing: $(-1, \infty)$

39. (a)

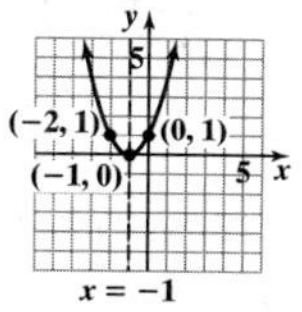

(b) Domain: $(-\infty, \infty)$
Range: $[0, \infty)$
(c) Decreasing: $(-\infty, -1)$
Increasing: $(-1, \infty)$

41. (a)

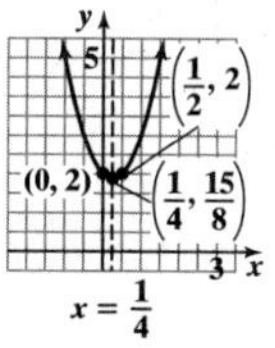

(b) Domain: $(-\infty, \infty)$
Range: $\left[\dfrac{15}{8}, \infty\right)$
(c) Decreasing: $\left(-\infty, \dfrac{1}{4}\right)$
Increasing: $\left(\dfrac{1}{4}, \infty\right)$

43. (a)

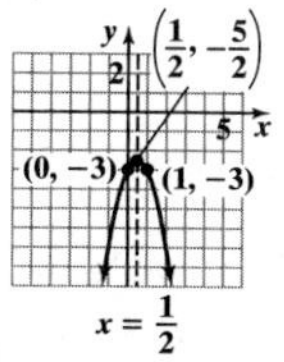

(b) Domain: $(-\infty, \infty)$
Range: $\left(-\infty, -\dfrac{5}{2}\right]$
(c) Increasing: $\left(-\infty, \dfrac{1}{2}\right)$
Decreasing: $\left(\dfrac{1}{2}, \infty\right)$

45. (a)

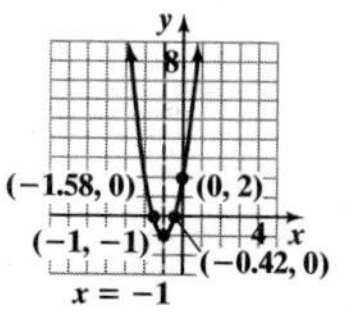

(b) Domain: $(-\infty, \infty)$
Range: $[-1, \infty)$
(c) Decreasing: $(-\infty, -1)$
Increasing: $(-1, \infty)$

47. (a)

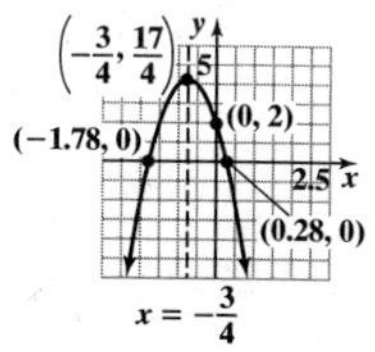

(b) Domain: $(-\infty, \infty)$
Range: $\left(-\infty, \dfrac{17}{4}\right]$
(c) Increasing: $\left(-\infty, -\dfrac{3}{4}\right)$
Decreasing: $\left(-\dfrac{3}{4}, \infty\right)$

49. $f(x) = (x + 1)^2 - 2 = x^2 + 2x - 1$ **51.** $f(x) = -(x + 3)^2 + 5 = -x^2 - 6x - 4$ **53.** $f(x) = 2(x - 1)^2 - 3 = 2x^2 - 4x - 1$
55. Minimum value; −18 **57.** Minimum value; −21 **59.** Maximum value; 21 **61.** Maximum value; 13

63. (a)

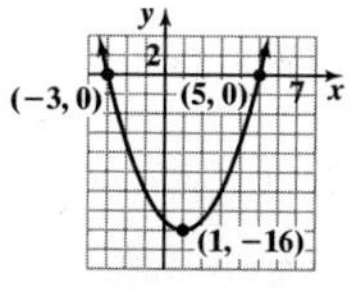

(b) Domain: $(-\infty, \infty)$
Range: $[-16, \infty)$
(c) Decreasing: $(-\infty, 1)$
Increasing: $(1, \infty)$

65. (a)

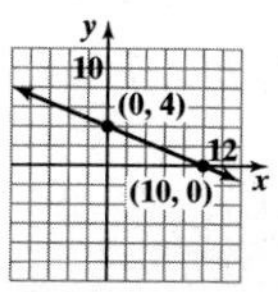

(b) Domain: $(-\infty, \infty)$
Range: $(-\infty, \infty)$
(c) Decreasing: $(-\infty, \infty)$

67. (a)

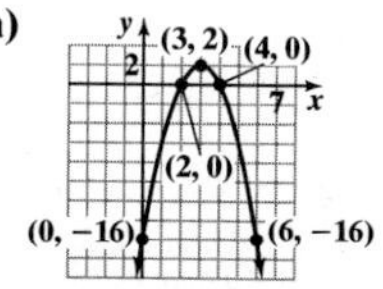

(b) Domain: $(-\infty, \infty)$
Range: $(-\infty, 2]$
(c) Increasing: $(-\infty, 3)$
Decreasing: $(3, \infty)$

69. (a)

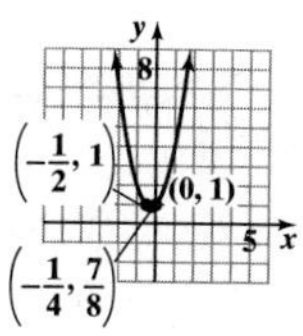

(b) Domain: $(-\infty, \infty)$
Range: $\left[\dfrac{7}{8}, \infty\right)$
(c) Decreasing: $\left(-\infty, -\dfrac{1}{4}\right)$
Increasing: $\left(-\dfrac{1}{4}, \infty\right)$

71. $a = 6, b = 0, c = 2$ **73. (a), (c), (d)**

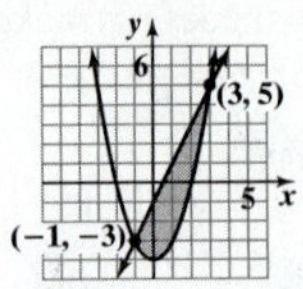

(b) $\{-1, 3\}$

75. (a), (c), (d)

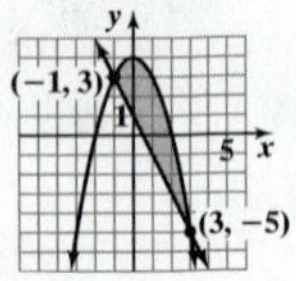

(b) $\{-1, 3\}$

77. (a), (c), (d)

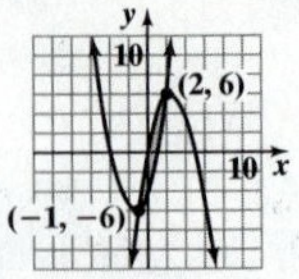

(b) $\{-1, 2\}$

79. (a) $a = 1: f(x) = (x + 3)(x - 1) = x^2 + 2x - 3$
$a = 2: f(x) = 2(x + 3)(x - 1) = 2x^2 + 4x - 6$
$a = -2: f(x) = -2(x + 3)(x - 1) = -2x^2 - 4x + 6$
$a = 5: f(x) = 5(x + 3)(x - 1) = 5x^2 + 10x - 15$
(b) The value of a does not affect the x-intercepts, but it changes the y-intercept by a factor of a.
(c) The value of a does not affect the axis of symmetry. It is $x = -1$ for all values of a.
(d) The value of a does not affect the x-coordinate of the vertex. However, the y-coordinate of the vertex is multiplied by a.
(e) The mean of the x-intercepts is the x-coordinate of the vertex.

81. (a) $(-2, -25)$
(b) $-7, 3$
(c) $-4, 0$; $(-4, -21)$, $(0, -21)$
(d)

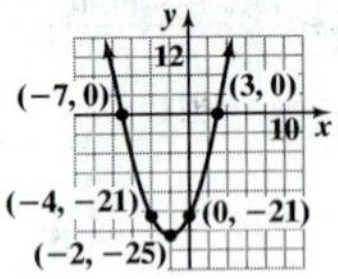

83. (2, 2) **85.** \$500; \$1,000,000 **87. (a)** 70,000 players **(b)** \$2500 **89. (a)** 187 or 188 watches; \$7031.20 **(b)** $P(x) = -0.2x^2 + 43x - 1750$ **(c)** 107 or 108 watches; \$561.20 **91. (a)** 171 ft **(b)** 49 mph **(c)** Reaction time **93.** If x is even, then ax^2 and bx are even and $ax^2 + bx$ is even, which means that $ax^2 + bx + c$ is odd. If x is odd, then ax^2 and bx are odd and $ax^2 + bx$ is even, which means that $ax^2 + bx + c$ is odd. In either case, $f(x)$ is odd.
95.

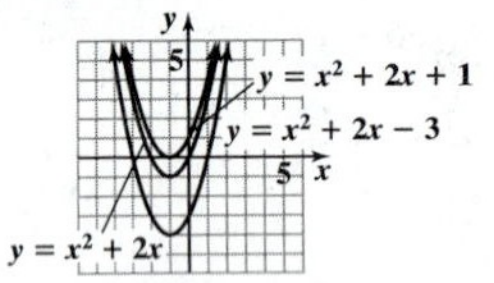

97. $b^2 - 4ac < 0$ **99.** No
101. Symmetric with respect to the x-axis, the y-axis, and the origin
102. $\{x | x \leq 4\}$ or $(-\infty, 4]$
103. Center $(5, -2)$; radius $= 3$
104. $y = \sqrt{-x}$

4.4 Assess Your Understanding *(page 307)*

3. (a) $R(x) = -\frac{1}{6}x^2 + 100x$ **(b)** $\{x | 0 \leq x \leq 600\}$ **(c)** \$13,333.33 **(d)** 300; \$15,000 **(e)** \$50

5. (a) $R(x) = -\frac{1}{5}x^2 + 20x$ **(b)** \$255 **(c)** 50; \$500 **(d)** \$10 **(e)** Between \$8 and \$12

7. (a) $A(w) = -w^2 + 200w$ **(b)** A is largest when $w = 100$ yd. **(c)** 10,000 yd^2 **9.** 2,000,000 m^2

11. (a) $\frac{625}{16} \approx 39$ ft **(b)** $\frac{7025}{32} \approx 219.5$ ft **(c)** About 170 ft

(d)

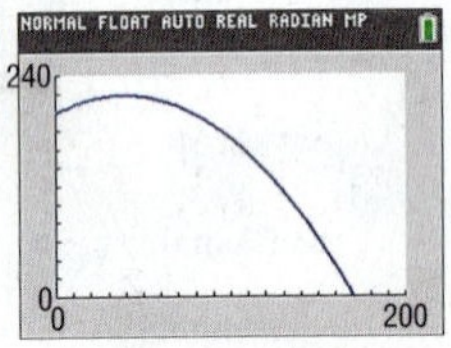

(f) When the height is 100 ft, the projectile is about 135.7 ft from the cliff.

13. 18.75 m **15. (a)** 3 in. **(b)** Between 2 in. and 4 in.
17. $\frac{750}{\pi} \approx 238.73$ m by 375 m
19. $x = \frac{a}{2}$ **21.** $\frac{38}{3}$ **23.** $\frac{248}{3}$

25. (a)

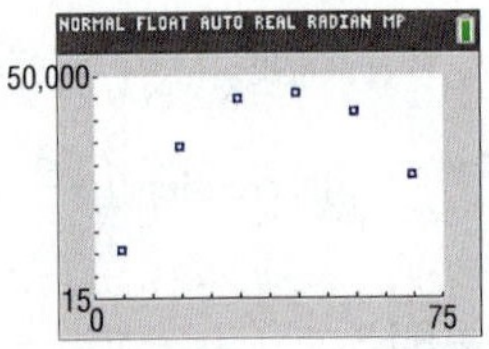

The data appear to follow a quadratic relation with $a < 0$.

(b) $I(x) = -44.759x^2 + 4295.356x - 55{,}045.418$
(c) About 48.0 years of age
(d) Approximately \$48,007

(e)

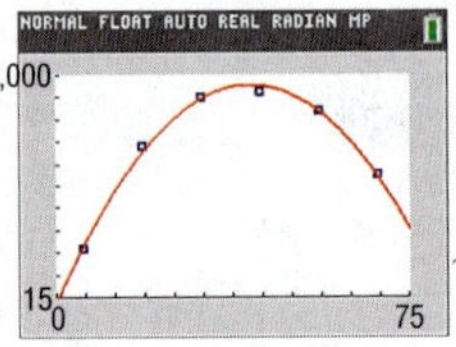

27. (a)

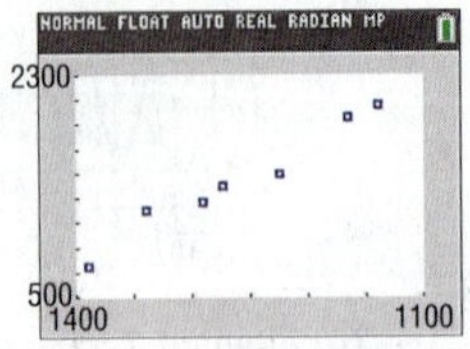

The data appear to be linearly related with positive slope.
(b) $R(x) = 1.229x + 917.385$ **(c)** \$1993

29. (a)

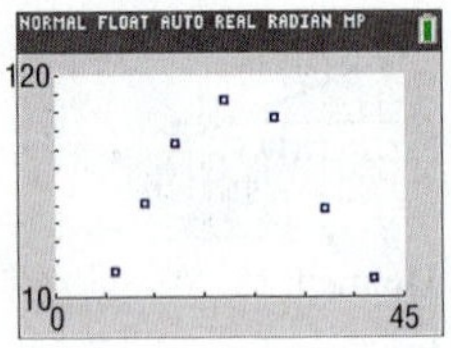

The data appear to follow a quadratic relation with $a < 0$.
(b) $B(a) = -0.547a^2 + 31.190a - 342.218$ **(c)** 79.357

32. $15i$ **33.** 13 **34.** $(x + 6)^2 + y^2 = 7$ **35.** $\left\{\frac{-4 - \sqrt{31}}{5}, \frac{-4 + \sqrt{31}}{5}\right\}$

4.5 Assess Your Understanding *(page 314)*

3. (a) $\{x \mid x < -2 \text{ or } x > 2\}$; $(-\infty, -2) \cup (2, \infty)$ **(b)** $\{x \mid -2 \le x \le 2\}$; $[-2, 2]$
5. (a) $\{x \mid -2 \le x \le 1\}$; $[-2, 1]$ **(b)** $\{x \mid x < -2 \text{ or } x > 1\}$; $(-\infty, -2) \cup (1, \infty)$
7. $\{x \mid -2 < x < 5\}$; $(-2, 5)$ **9.** $\{x \mid x < 0 \text{ or } x > 4\}$; $(-\infty, 0) \cup (4, \infty)$ **11.** $\{x \mid -3 < x < 3\}$; $(-3, 3)$
13. $\{x \mid x < -4 \text{ or } x > 3\}$; $(-\infty, -4) \cup (3, \infty)$ **15.** $\left\{x \middle| -\frac{1}{2} < x < 3\right\}$; $\left(-\frac{1}{2}, 3\right)$ **17.** No real solution **19.** No real solution
21. $\left\{x \middle| x < -\frac{2}{3} \text{ or } x > \frac{3}{2}\right\}$; $\left(-\infty, -\frac{2}{3}\right) \cup \left(\frac{3}{2}, \infty\right)$ **23.** $\{x \mid x \le -4 \text{ or } x \ge 4\}$; $(-\infty, -4] \cup [4, \infty)$ **25. (a)** $\{-1, 1\}$ **(b)** $\{-1\}$ **(c)** $\{-1, 4\}$
(d) $\{x \mid x < -1 \text{ or } x > 1\}$; $(-\infty, -1) \cup (1, \infty)$ **(e)** $\{x \mid x \le -1\}$; $(-\infty, -1]$ **(f)** $\{x \mid x < -1 \text{ or } x > 4\}$; $(-\infty, -1) \cup (4, \infty)$
(g) $\{x \mid x \le -\sqrt{2} \text{ or } x \ge \sqrt{2}\}$; $(-\infty, -\sqrt{2}] \cup [\sqrt{2}, \infty)$
27. (a) $\{-1, 1\}$ **(b)** $\left\{-\frac{1}{4}\right\}$ **(c)** $\{-4, 0\}$ **(d)** $\{x \mid -1 < x < 1\}$; $(-1, 1)$ **(e)** $\left\{x \middle| x \le -\frac{1}{4}\right\}$; $\left(-\infty, -\frac{1}{4}\right]$ **(f)** $\{x \mid -4 < x < 0\}$; $(-4, 0)$ **(g)** $\{0\}$
29. (a) $\{-2, 2\}$ **(b)** $\{-2, 2\}$ **(c)** $\{-2, 2\}$ **(d)** $\{x \mid x < -2 \text{ or } x > 2\}$; $(-\infty, -2) \cup (2, \infty)$ **(e)** $\{x \mid x \le -2 \text{ or } x \ge 2\}$; $(-\infty, -2] \cup [2, \infty)$
(f) $\{x \mid x < -2 \text{ or } x > 2\}$; $(-\infty, -2) \cup (2, \infty)$ **(g)** $\{x \mid x \le -\sqrt{5} \text{ or } x \ge \sqrt{5}\}$; $(-\infty, -\sqrt{5}] \cup [\sqrt{5}, \infty)$
31. (a) $\{-1, 2\}$ **(b)** $\{-2, 1\}$ **(c)** $\{0\}$ **(d)** $\{x \mid x < -1 \text{ or } x > 2\}$; $(-\infty, -1) \cup (2, \infty)$ **(e)** $\{x \mid -2 \le x \le 1\}$; $[-2, 1]$
(f) $\{x \mid x < 0\}$; $(-\infty, 0)$ **(g)** $\left\{x \middle| x \le \frac{1 - \sqrt{13}}{2} \text{ or } x \ge \frac{1 + \sqrt{13}}{2}\right\}$; $\left(-\infty, \frac{1 - \sqrt{13}}{2}\right] \cup \left[\frac{1 + \sqrt{13}}{2}, \infty\right)$
33. (a) 5 sec **(b)** The ball is more than 96 ft above the ground for time t between 2 and 3 sec, $2 < t < 3$.
35. (a) \$0, \$1000 **(b)** The revenue is more than \$800,000 for prices between \$276.39 and \$723.61, $\$276.39 < p < \$723.61$.
37. (a) $\{c \mid 0.112 < c < 81.907\}$; $(0.112, 81.907)$ **(b)** It is possible to hit a target 75 km away if $c = 0.651$ or $c = 1.536$.
44. $\{x \mid x \le 5\}$ **45.** Odd **46. (a)** $(9, 0)$, $(0, -6)$ **(b)**

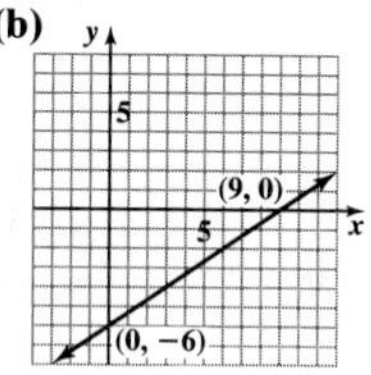

47. $19 - 26i$

Review Exercises *(page 316)*

1. (a) $m = 2$; $b = -5$ **(b)** 2
(c)

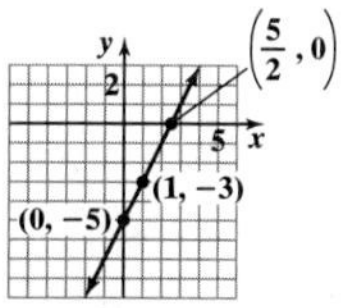

(d) Increasing

2. (a) $m = \frac{4}{5}$; $b = -6$ **(b)** $\frac{4}{5}$
(c)

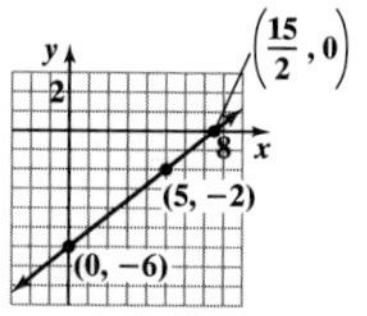

(d) Increasing

3. (a) $m = 0$; $b = 4$ **(b)** 0
(c)

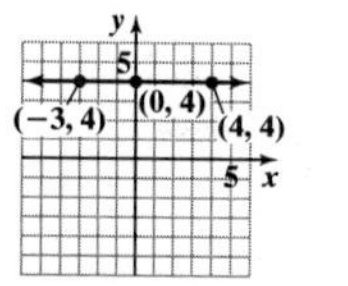

(d) Constant

4. Linear; Slope: 5
5. Nonlinear

6.

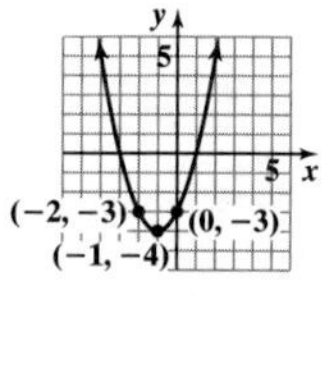

7.

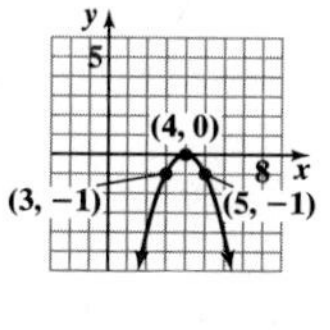

8.

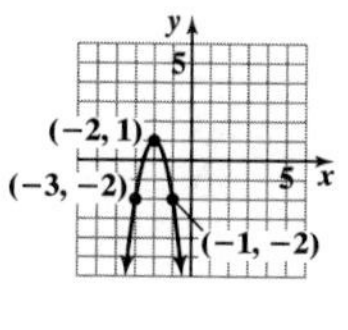

9. (a)

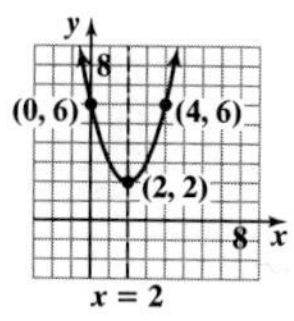

(b) Domain: $(-\infty, \infty)$
Range: $[2, \infty)$
(c) Decreasing: $(-\infty, 2)$
Increasing: $(2, \infty)$

10. (a)

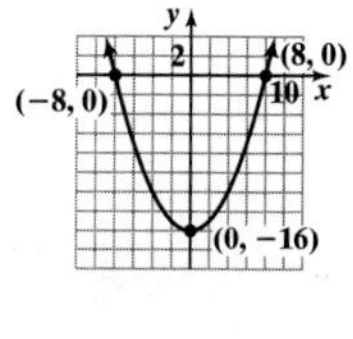

(b) Domain: $(-\infty, \infty)$
Range: $[-16, \infty)$
(c) Decreasing: $(-\infty, 0)$
Increasing: $(0, \infty)$

11. (a)

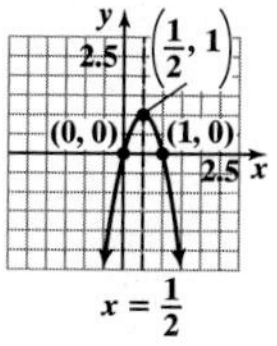

(b) Domain: $(-\infty, \infty)$
Range: $(-\infty, 1]$
(c) Increasing: $\left(-\infty, \frac{1}{2}\right)$
Decreasing: $\left(\frac{1}{2}, \infty\right)$

12. (a)

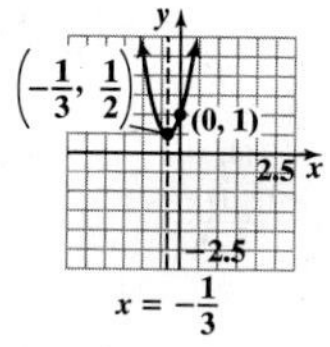

(b) Domain: $(-\infty, \infty)$
Range: $\left[\frac{1}{2}, \infty\right)$
(c) Decreasing: $\left(-\infty, -\frac{1}{3}\right)$
Increasing: $\left(-\frac{1}{3}, \infty\right)$

13. (a)

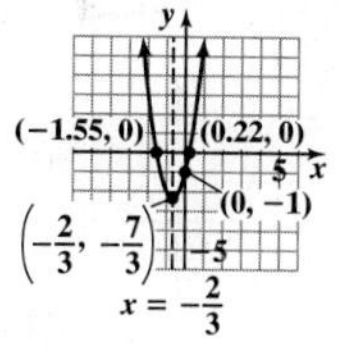

(b) Domain: $(-\infty, \infty)$
Range: $\left[-\frac{7}{3}, \infty\right)$
(c) Decreasing: $\left(-\infty, -\frac{2}{3}\right)$
Increasing: $\left(-\frac{2}{3}, \infty\right)$

14. Minimum value; 1 **15.** Maximum value; 12 **16.** Maximum value; 16 **17.** $\{x \mid -8 < x < 2\}$; $(-8, 2)$

18. $\left\{x \,\middle|\, x \le -\frac{1}{3} \text{ or } x \ge 5\right\}$; $\left(-\infty, -\frac{1}{3}\right] \cup [5, \infty)$ **19.** $y = -3x^2 + 12x - 16$ **20.** $y = x^2 + 2x + 3$

21. (a) $S(x) = 0.01x + 25{,}000$ **(b)** \$35,000 **(c)** \$7,500,000 **(d)** $x >$ \$12,500,000

22. (a) $R(x) = -\frac{1}{10}x^2 + 150x$ **(b)** \$14,000 **(c)** 750; \$56,250 **(d)** \$75 **23.** 50 ft by 50 ft **24.** 4,166,666.7 m^2

25. The side with the semicircles should be $\frac{50}{\pi}$ ft; the other side should be 25 ft. **26. (a)** 63 clubs **(b)** \$151.90 **27.** $A(x) = -x^2 + 10x$; 25 sq. units

28. 3.6 ft **29. (a)**

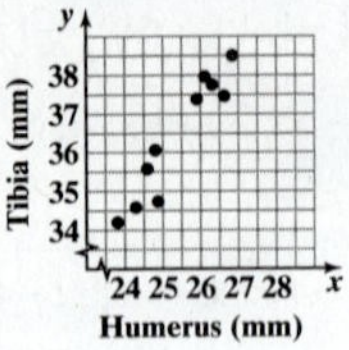

(b) yes **(c)** $y = 1.3902x + 1.1140$ **(d)** 37.95 mm

30. (a) Quadratic, $a < 0$

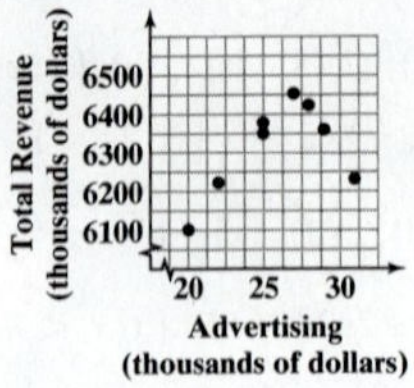

(b) About \$26.5 thousand **(c)** \$6408 thousand **(e)**

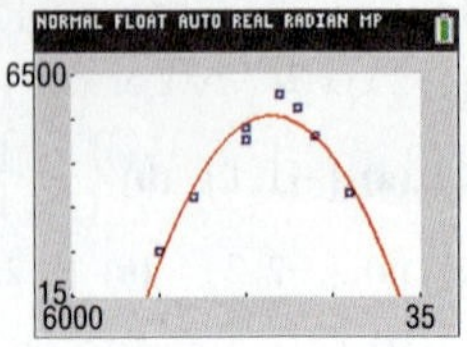

Chapter Test *(page 318)*

1. (a) Slope: -4; y-intercept: 3 **(b)** -4 **(c)** Decreasing **(d)**

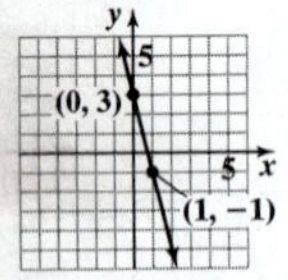

2. $\left(-\frac{4}{3}, 0\right)$, $(2, 0)$, $(0, -8)$

3. $\left(\frac{2 - \sqrt{6}}{2}, 0\right)$, $\left(\frac{2 + \sqrt{6}}{2}, 0\right)$, $(0, 1)$

4. $\{-1, 3\}$

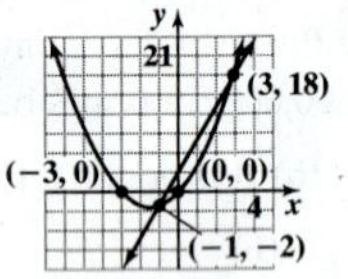

5.

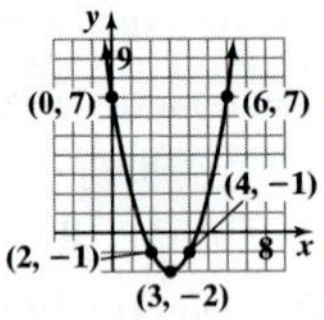

6. (a) Opens up **(b)** $(2, -8)$ **(c)** $x = 2$ **(d)** x-intercepts: $\frac{6 - 2\sqrt{6}}{3}$, $\frac{6 + 2\sqrt{6}}{3}$; y-intercept: 4

(e)

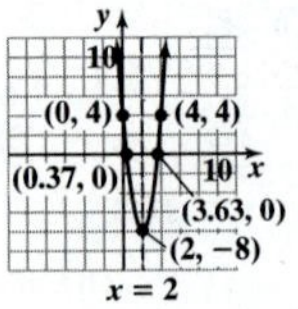

7. Maximum value; 21
8. $\{x \mid x \le 4 \text{ or } x \ge 6\}$; $(-\infty, 4] \cup [6, \infty)$
9. (a) $C(m) = 0.15m + 129.50$ **(b)** \$258.50 **(c)** 562 miles

Cumulative Review *(page 319)*

1. $5\sqrt{2}$; $\left(\frac{3}{2}, \frac{1}{2}\right)$ **2.** $(-2, -1)$ and $(2, 3)$ are on the graph.

3. $\left\{x \,\middle|\, x \ge -\frac{3}{5}\right\}$; $\left[-\frac{3}{5}, \infty\right)$

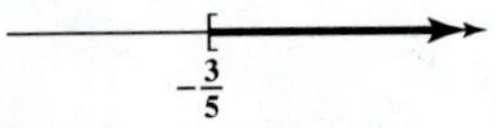

4. $y = -2x + 2$

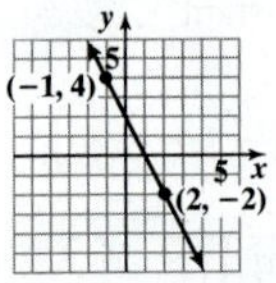

5. $y = -\frac{1}{2}x + \frac{13}{2}$

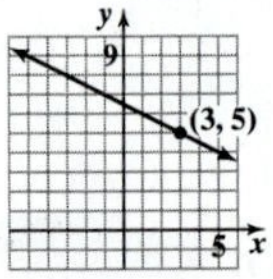

6. $(x - 2)^2 + (y + 4)^2 = 25$

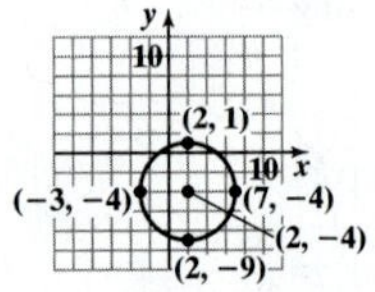

7. Yes **8. (a)** -3 **(b)** $x^2 - 4x - 2$ **(c)** $x^2 + 4x + 1$ **(d)** $-x^2 + 4x - 1$ **(e)** $x^2 - 3$ **(f)** $2x + h - 4$ **9.** $\left\{z \,\middle|\, z \ne \frac{7}{6}\right\}$

10. Yes **11. (a)** No **(b)** -1; $(-2, -1)$ is on the graph. **(c)** -8; $(-8, 2)$ is on the graph. **12.** Neither **13.** Local maximum value is 5.30 and occurs at $x = -1.29$. Local minimum value is -3.30 and occurs at $x = 1.29$. Increasing: $(-4, -1.29)$ and $(1.29, 4)$; Decreasing: $(-1.29, 1.29)$

14. (a) -4 **(b)** $\{x \mid x > -4\}$ or $(-4, \infty)$

15. (a) Domain: $\{x \mid -4 \le x \le 4\}$; Range: $\{y \mid -1 \le y \le 3\}$ **(b)** $(-1, 0)$, $(0, -1)$, $(1, 0)$ **(c)** y-axis **(d)** 1 **(e)** -4 and 4 **(f)** $\{x \mid -1 < x < 1\}$

(g)

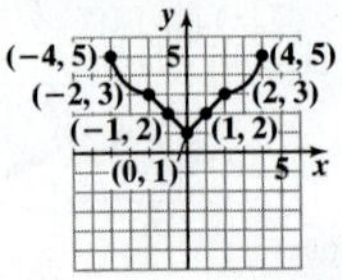

(h)

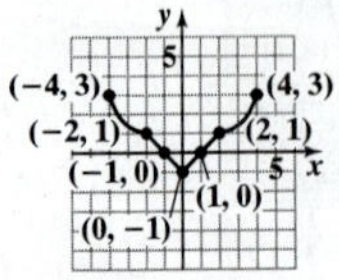

(i)

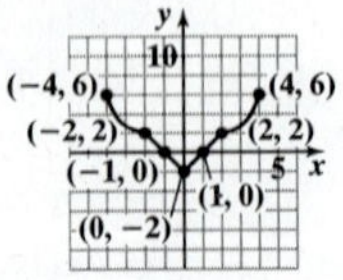

(j) Even **(k)** $(0, 4)$

CHAPTER 5 Polynomial and Rational Functions

5.1 Assess Your Understanding *(page 338)*

7. smooth; continuous **8.** touches **9.** $(-1, 1)$; $(0, 0)$; $(1, 1)$ **10.** r is a real zero of f; r is an x-intercept of the graph of f; $x - r$ is a factor of f. **11.** turning points **12.** $y = 3x^4$ **13.** ∞; $-\infty$ **14.** As x increases in the positive direction, $f(x)$ decreases without bound. **15.** b **16.** d

17. Yes; degree 3; $f(x) = x^3 + 4x$; leading term: x^3; constant term: 0 **19.** Yes; degree 2; $g(x) = -\frac{1}{2}x^2 + \frac{1}{2}$; leading term: $-\frac{1}{2}x^2$; constant term: $\frac{1}{2}$

21. No; x is raised to the -1 power **23.** No; x is raised to the $\frac{3}{2}$ power **25.** Yes; degree 4; $F(x) = 5x^4 - \pi x^3 + \frac{1}{2}$; leading term: $5x^4$; constant term: $\frac{1}{2}$

27. Yes; degree 4; $G(x) = 2x^4 - 4x^3 + 4x^2 - 4x + 2$; leading term: $2x^4$; constant term: 2

29.

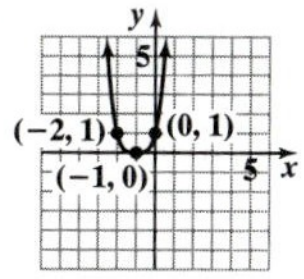

31.

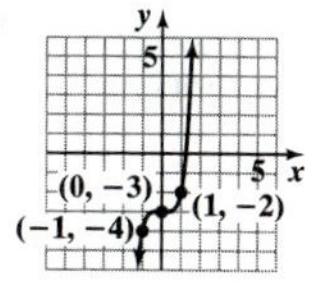

33.

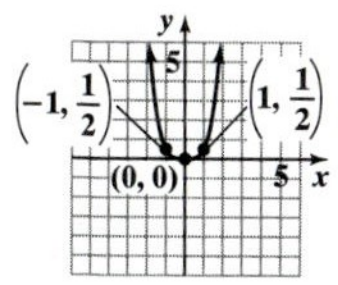

35.

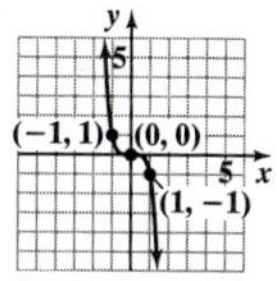

37.

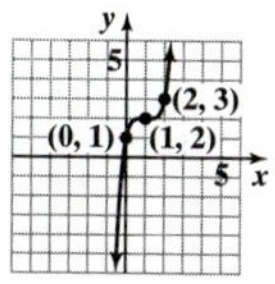

39.

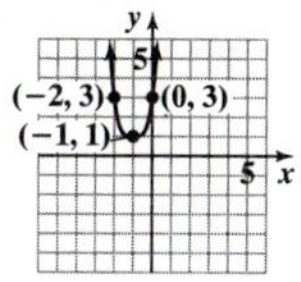

41.

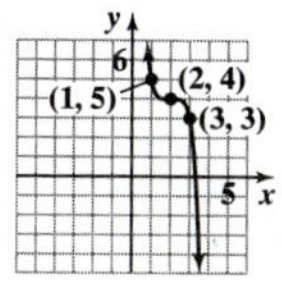

43. $f(x) = x^3 - 3x^2 - x + 3$ for $a = 1$ **45.** $f(x) = x^3 - x^2 - 12x$ for $a = 1$ **47.** $f(x) = x^4 - 15x^2 + 10x + 24$ for $a = 1$ **49.** $f(x) = x^3 - 5x^2 + 3x + 9$ for $a = 1$ **51.** $f(x) = 2x^3 - 4x^2 - 22x + 24$ **53.** $f(x) = 16x^4 - 80x^3 + 32x^2 + 128x$ **55.** $f(x) = 5x^4 - 10x^2 + 5$ **57. (a)** 7, multiplicity 1; -3, multiplicity 2 **(b)** Graph touches the x-axis at -3 and crosses it at 7. **(c)** 2 **(d)** $y = 3x^3$

59. (a) 2, multiplicity 3 **(b)** Graph crosses the x-axis at 2. **(c)** 4 **(d)** $y = 4x^5$ **61. (a)** $-\frac{1}{2}$, multiplicity 2; -4, multiplicity 3 **(b)** Graph touches the x-axis at $-\frac{1}{2}$ and crosses at -4. **(c)** 4 **(d)** $y = -2x^5$ **63. (a)** 5, multiplicity 3; -4, multiplicity 2 **(b)** Graph touches the x-axis at -4 and crosses it at 5. **(c)** 4 **(d)** $y = x^5$ **65. (a)** No real zeros **(b)** Graph neither crosses nor touches the x-axis. **(c)** 5 **(d)** $y = 3x^6$ **67. (a)** 0, multiplicity 2; $-\sqrt{2}$, $\sqrt{2}$, multiplicity 1 **(b)** Graph touches the x-axis at 0 and crosses at $-\sqrt{2}$ and $\sqrt{2}$. **(c)** 3 **(d)** $y = -2x^4$ **69.** Could be; zeros: $-1, 1, 2$; Least degree is 3. **71.** Cannot be the graph of a polynomial; gap at $x = -1$ **73.** $f(x) = x(x-1)(x-2)$

75. $f(x) = -\frac{1}{2}(x+1)(x-1)^2(x-2)$ **77.** $f(x) = 0.2(x+4)(x+1)^2(x-3)$ **79.** $f(x) = -x(x+3)^2(x-3)^2$

81. Step 1: $y = x^3$
Step 2: x-intercepts: 0, 3; y-intercept: 0
Step 3: 0: multiplicity 2, touches; 3: multiplicity 1, crosses
Step 4: At most 2 turning points
Step 5: $f(-1) = -4$; $f(2) = -4$; $f(4) = 16$

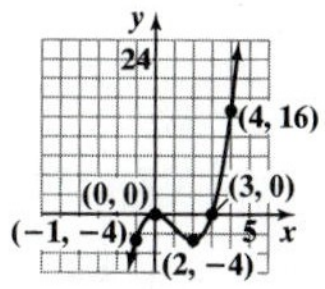

83. Step 1: $y = -x^3$
Step 2: x-intercepts: -4, 1; y-intercept: 16
Step 3: -4: multiplicity 2, touches; 1: multiplicity 1, crosses
Step 4: At most 2 turning points
Step 5: $f(-5) = 6$; $f(-2) = 12$; $f(2) = -36$

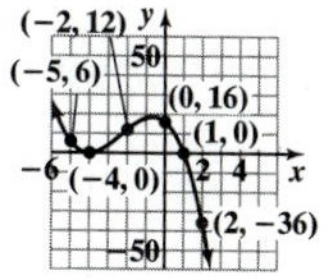

85. Step 1: $y = -2x^4$
Step 2: x-intercepts: -2, 2; y-intercept: 32
Step 3: -2: multiplicity 1, crosses; 2: multiplicity 3, crosses
Step 4: At most 3 turning points
Step 5: $f(-3) = -250$; $f(-1) = 54$; $f(3) = -10$

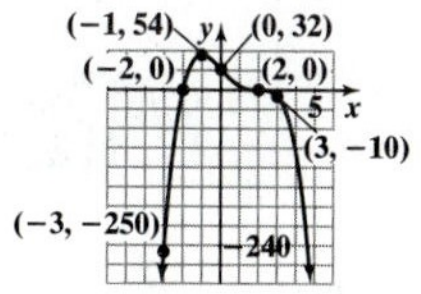

87. Step 1: $y = x^3$
Step 2: x-intercepts: $-4, -1, 2$; y-intercept: -8
Step 3: $-4, -1, 2$: multiplicity 1, crosses
Step 4: At most 2 turning points
Step 5: $f(-5) = -28$; $f(-2) = 8$; $f(1) = -10$; $f(3) = 28$

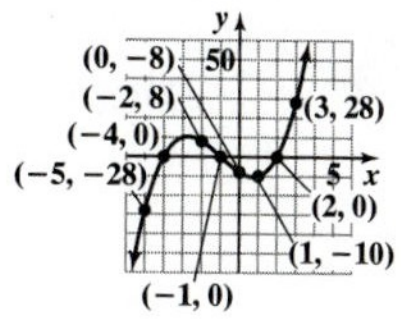

89. Step 1: $y = x^4$
Step 2: x-intercepts: $-2, 0, 2$; y-intercept: 0
Step 3: $-2, 2$: multiplicity 1, crosses; 0: multiplicity 2, touches
Step 4: At most 3 turning points
Step 5: $f(-3) = 45$; $f(-1) = -3$; $f(1) = -3$; $f(3) = 45$

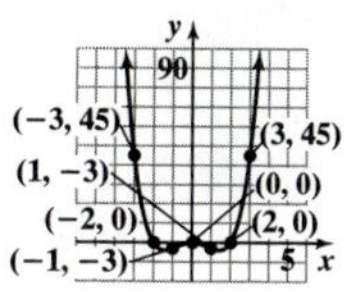

91. Step 1: $y = x^4$
Step 2: x-intercepts: -1, 2; y-intercept: 4
Step 3: -1, 2: multiplicity 2, touches
Step 4: At most 3 turning points
Step 5: $f(-2) = 16$; $f(1) = 4$; $f(3) = 16$

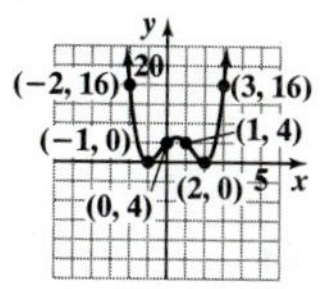

93. Step 1: $y = x^4$

Step 2: x-intercepts: $-3, -1, 0$; y-intercept: 3

Step 3: -3: multiplicity 1, crosses; -1: multiplicity 1, crosses; 0: multiplicity 2, touches

Step 4: At most 3 turning points

Step 5: $f(-4) = 48;\ f(-2) = -4;\ f\left(-\frac{1}{2}\right) = \frac{5}{16};\ f(1) = 8$

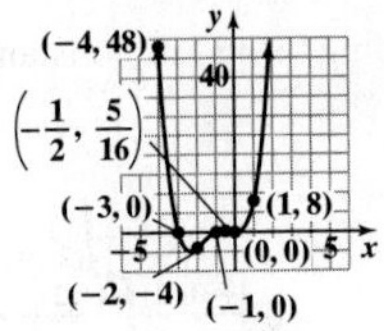

95. Step 1: $y = 5x^4$

Step 2: x-intercepts: $-3, -2, 0, 2$; y-intercept: 0

Step 3: $-3, -2, 0, 2$: multiplicity 1, crosses

Step 4: At most 3 turning points

Step 5: $f(-4) = 240;\ f(-2.5) \approx -14.1;$ $f(-1) = 30;\ f(1) = -60;\ f(3) = 450$

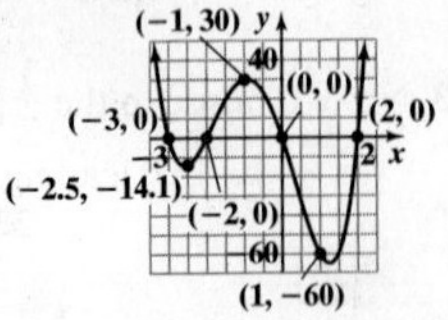

97. Step 1: $y = x^5$

Step 2: x-intercepts: 0, 2; y-intercept: 0

Step 3: 2: multiplicity 1, crosses;
0: multiplicity 2, touches

Step 4: At most 4 turning points

Step 5: $f(-1) = -12; f(1) = -4; f(3) = 108$

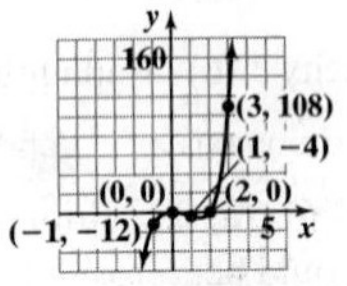

99. Step 1: $y = x^3$

Step 2:

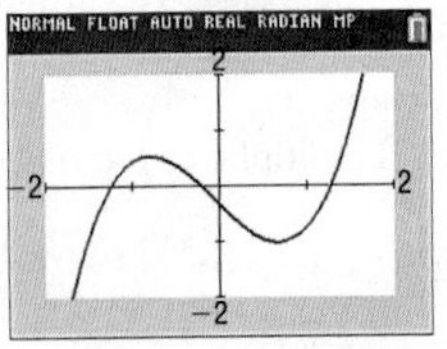

Step 3: x-intercepts: $-1.26, -0.20, 1.26$
y-intercept: -0.31752

Step 4:

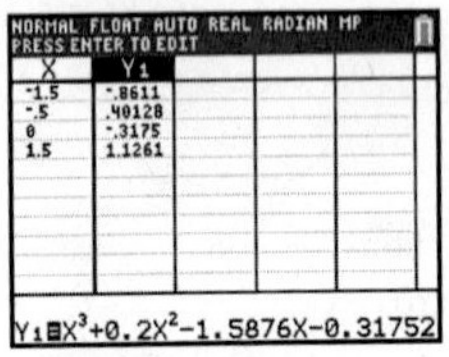
NORMAL FLOAT AUTO REAL RADIAN MP
PRESS ENTER TO EDIT

X	Y1
-1.5	-.8611
-.5	.40128
0	-.3175
1.5	1.1261

Y1=X³+0.2X²−1.5876X−0.31752

Step 5: $(-0.80, 0.57); (0.66, -0.99)$

Step 6:

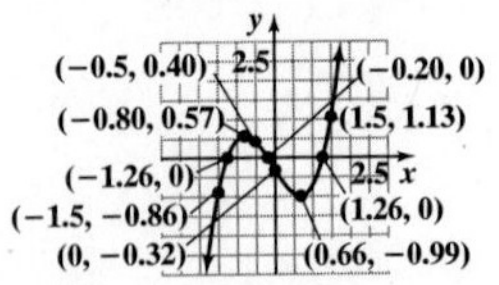

Step 7: Domain: $(-\infty, \infty)$;
Range: $(-\infty, \infty)$

Step 8: Increasing on $(-\infty, -0.80)$ and $(0.66, \infty)$
Decreasing on $(-0.80, 0.66)$

101. Step 1: $y = x^3$

Step 2:

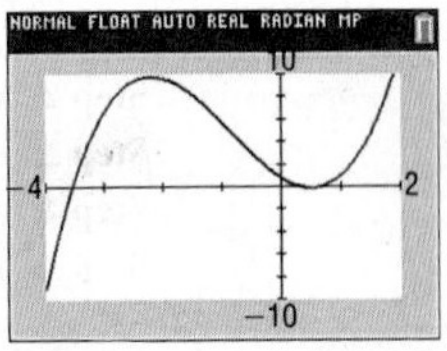

Step 3: x-intercepts: $-3.56, 0.50$
y-intercept: 0.89

Step 4:

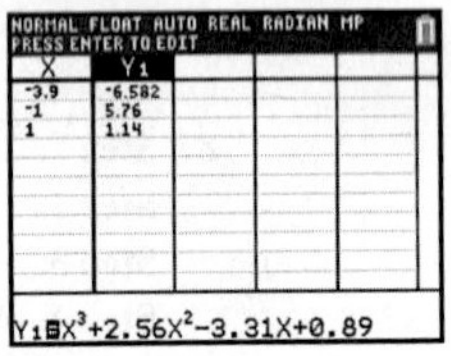
NORMAL FLOAT AUTO REAL RADIAN MP
PRESS ENTER TO EDIT

X	Y1
-3.9	-6.582
-1	5.76
1	1.14

Y1=X³+2.56X²−3.31X+0.89

Step 5: $(-2.21, 9.91); (0.50, 0)$

Step 6:

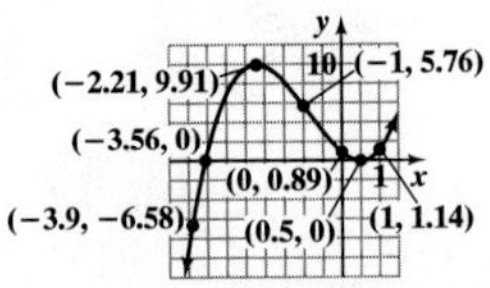

Step 7: Domain: $(-\infty, \infty)$;
Range: $(-\infty, \infty)$

Step 8: Increasing on $(-\infty, -2.21)$ and $(0.50, \infty)$
Decreasing on $(-2.21, 0.50)$

103. Step 1: $y = x^4$

Step 2:

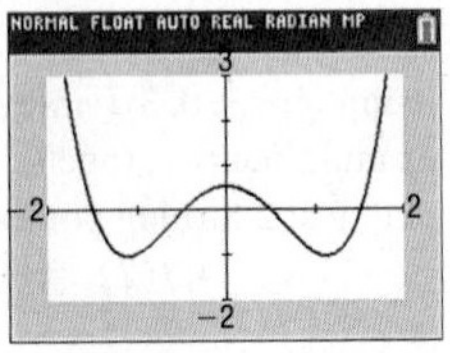

Step 3: x-intercepts: $-1.5, -0.5, 0.5, 1.5$
y-intercept: 0.5625

Step 4:

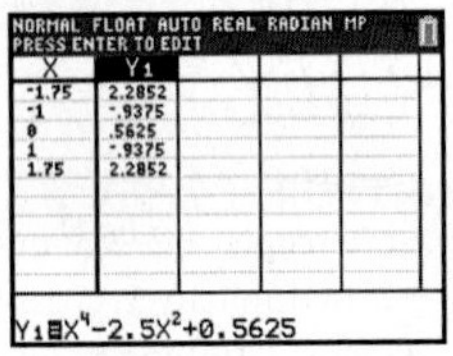
NORMAL FLOAT AUTO REAL RADIAN MP
PRESS ENTER TO EDIT

X	Y1
-1.75	2.2852
-1	-.9375
0	.5625
1	-.9375
1.75	2.2852

Y1=X⁴−2.5X²+0.5625

Step 5: $(-1.12, -1); (1.12, -1), (0, 0.56)$

Step 6:

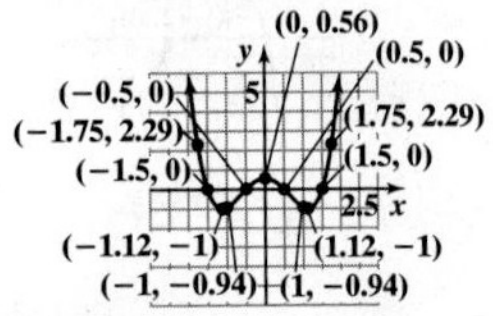

Step 7: Domain: $(-\infty, \infty)$; Range: $[-1, \infty)$

Step 8: Increasing on $(-1.12, 0)$ and $(1.12, \infty)$
Decreasing on $(-\infty, -1.12)$ and $(0, 1.12)$

105. Step 1: $y = 2x^4$

Step 2:

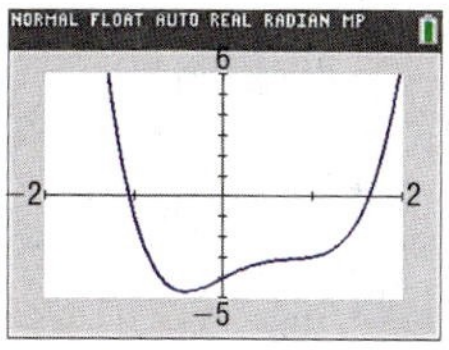

Step 3: x-intercepts: $-1.07, 1.62$; y-intercept: -4

Step 4:

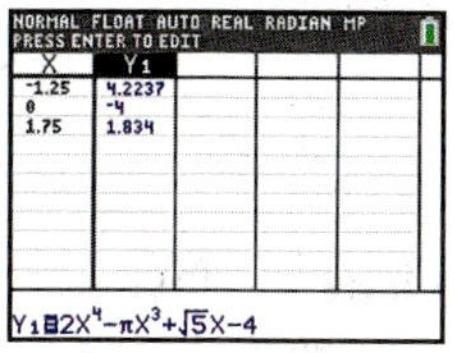

X	Y1
-1.25	4.2237
0	-4
1.75	1.834

Step 5: $(-0.42, -4.64)$

Step 6:

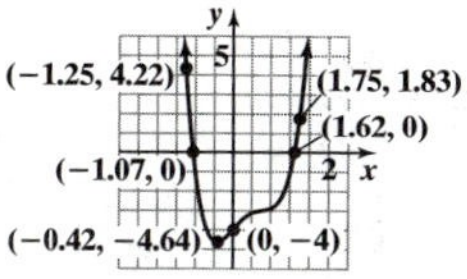

Step 7: Domain: $(-\infty, \infty)$; Range: $[-4.64, \infty)$

Step 8: Increasing on $(-0.42, \infty)$
Decreasing on $(-\infty, -0.42)$

107. $f(x) = -x(x + 2)(x - 2)$

Step 1: $y = -x^3$

Step 2: x-intercepts: $-2, 0, 2$; y-intercept: 0

Step 3: $-2, 0, 2$: multiplicity 1, crosses

Step 4: At most 2 turning points

Step 5: $f(-3) = 15$; $f(-1) = -3$; $f(1) = 3$; $f(3) = -15$

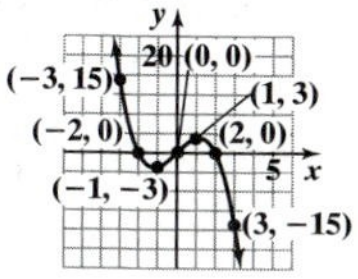

109. $f(x) = x(x + 4)(x - 3)$

Step 1: $y = x^3$

Step 2: x-intercepts: $-4, 0, 3$; y-intercept: 0

Step 3: $-4, 0, 3$: multiplicity 1, crosses

Step 4: At most 2 turning points

Step 5: $f(-5) = -40$; $f(-2) = 20$; $f(2) = -12$; $f(4) = 32$

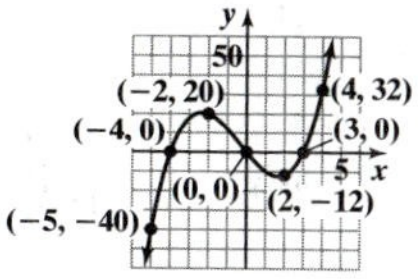

111. $f(x) = 2x(x + 6)(x - 2)(x + 2)$

Step 1: $y = 2x^4$

Step 2: x-intercepts: $-6, -2, 0, 2$; y-intercept: 0

Step 3: $-6, -2, 0, 2$: multiplicity 1, crosses

Step 4: At most 3 turning points

Step 5: $f(-7) = 630$; $f(-4) = -192$; $f(-1) = 30$; $f(1) = -42$; $f(3) = 270$

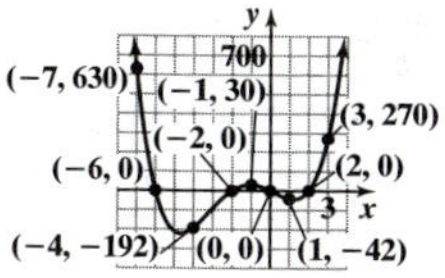

113. $f(x) = -x^2(x + 1)^2(x - 1)$

Step 1: $y = -x^5$

Step 2: x-intercepts: $-1, 0, 1$; y-intercept: 0

Step 3: 1: multiplicity 1, crosses; $-1, 0$: multiplicity 2, touches

Step 4: At most 4 turning points

Step 5: $f(-1.5) = 1.40625$; $f(-0.54) = 0.10$; $f(0.74) = 0.43$; $f(1.2) = -1.39392$

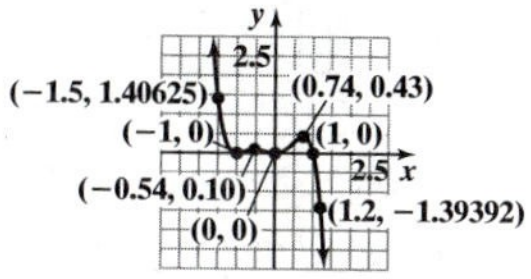

115. $f(x) = 3(x + 3)(x - 1)(x - 4)$ **117.** $f(x) = -2(x + 5)^2(x - 2)(x - 4)$ **119. (a)** $-3, 2$ **(b)** $-6, -1$

121. (a)

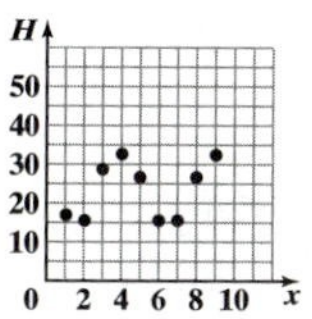

The relation appears to be cubic.

(b) $H(x) = 0.3948x^3 - 5.9563x^2 + 26.1965x - 7.4127$

(c) ≈ 24

(d)

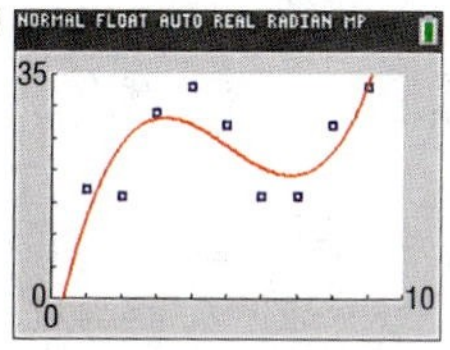

(e) ≈ 54; no

123. (a)

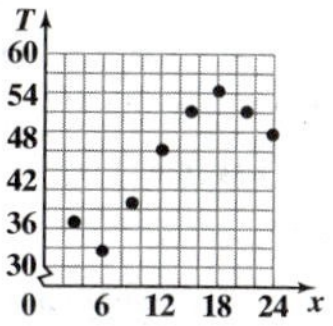

The relation appears to be cubic.

(b) 2.4°/h **(c)** 1°/h

(d) $T(x) = -0.01308x^3 + 0.4674x^2 - 3.4159x + 41.1929$; 53.9°F

(e)

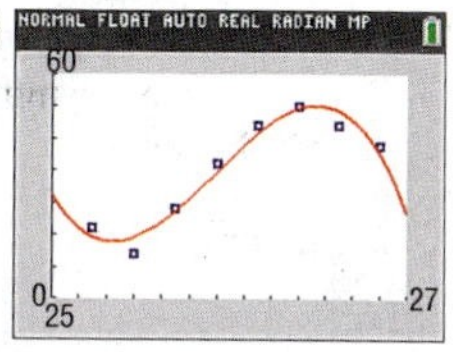

(f) The predicted temperature at midnight is 41.2°F.

125. (a)

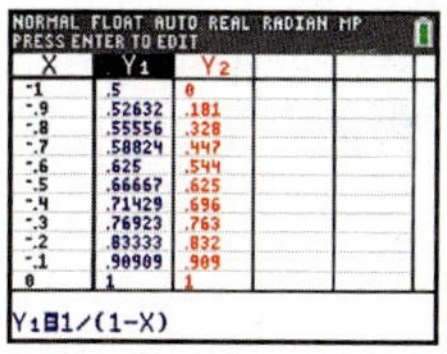

X	Y1	Y2
-1	.5	0
-.9	.52632	.181
-.8	.55556	.328
-.7	.58824	.447
-.6	.625	.544
-.5	.66667	.625
-.4	.71429	.696
-.3	.76923	.763
-.2	.83333	.832
-.1	.90909	.909
0	1	1

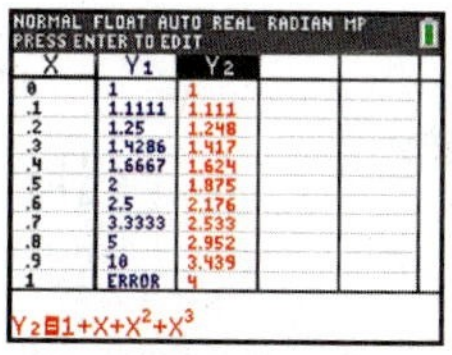

X	Y1	Y2
0	1	1
.1	1.1111	1.111
.2	1.25	1.248
.3	1.4286	1.417
.4	1.6667	1.624
.5	2	1.875
.6	2.5	2.176
.7	3.3333	2.533
.8	5	2.952
.9	10	3.439
1	ERROR	4

(b)

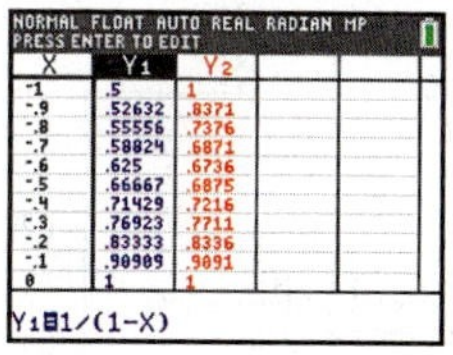

X	Y1	Y2
-1	.5	1
-.9	.52632	.8371
-.8	.55556	.7376
-.7	.58824	.6871
-.6	.625	.6736
-.5	.66667	.6875
-.4	.71429	.7216
-.3	.76923	.7711
-.2	.83333	.8336
-.1	.90909	.9091
0	1	1

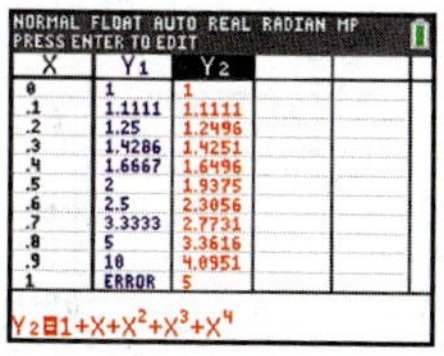

X	Y1	Y2
0	1	1
.1	1.1111	1.1111
.2	1.25	1.2496
.3	1.4286	1.4251
.4	1.6667	1.6496
.5	2	1.9375
.6	2.5	2.3056
.7	3.3333	2.7731
.8	5	3.3616
.9	10	4.0951
1	ERROR	5

(c)

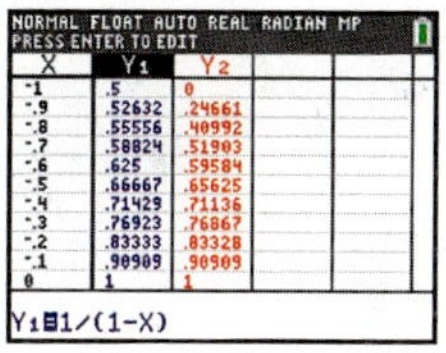

X	Y1	Y2
-1	.5	0
-.9	.52632	.24661
-.8	.55556	.40992
-.7	.58824	.51903
-.6	.625	.59584
-.5	.66667	.65625
-.4	.71429	.71136
-.3	.76923	.76867
-.2	.83333	.83328
-.1	.90909	.90909
0	1	1

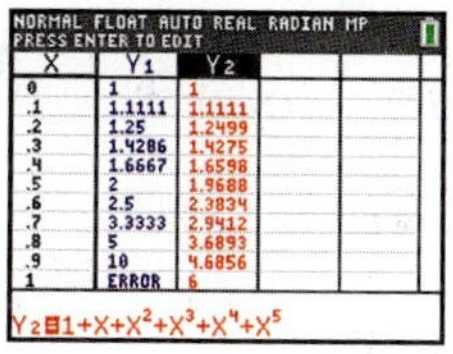

X	Y1	Y2
0	1	1
.1	1.1111	1.1111
.2	1.25	1.2499
.3	1.4286	1.4275
.4	1.6667	1.6598
.5	2	1.9688
.6	2.5	2.3834
.7	3.3333	2.9412
.8	5	3.6893
.9	10	4.6856
1	ERROR	6

(d) As more terms are added, the values of the polynomial function get closer to the values of f. The approximations near 0 are better than those near -1 or 1.

131. (a)–(d) **134.** $y = -\frac{2}{5}x - \frac{11}{5}$ **135.** $\{x | x \neq -5\}$ **136.** $\frac{-2-\sqrt{7}}{2}, \frac{-2+\sqrt{7}}{2}$ **137.** $\left\{-\frac{4}{5}, 2\right\}$

5.2 Assess Your Understanding *(page 350)*

5. F **6.** horizontal asymptote **7.** vertical asymptote **8.** proper **9.** T **10.** F **11.** $y = 0$ **12.** T **13.** d **14.** a **15.** All real numbers except 3; $\{x|x \neq 3\}$ **17.** All real numbers except 2 and -4; $\{x|x \neq 2, x \neq -4\}$ **19.** All real numbers except $-\frac{1}{2}$ and 3; $\left\{x \middle| x \neq -\frac{1}{2}, x \neq 3\right\}$
21. All real numbers except 2; $\{x|x \neq 2\}$ **23.** All real numbers **25.** All real numbers except -3 and 3; $\{x|x \neq -3, x \neq 3\}$
27. **(a)** Domain: $\{x|x \neq 2\}$; range: $\{y|y \neq 1\}$ **(b)** $(0, 0)$ **(c)** $y = 1$ **(d)** $x = 2$ **(e)** None
29. **(a)** Domain: $\{x|x \neq 0\}$; range: all real numbers **(b)** $(-1, 0), (1, 0)$ **(c)** None **(d)** $x = 0$ **(e)** $y = 2x$
31. **(a)** Domain: $\{x|x \neq -2, x \neq 2\}$; range: $\{y|y \leq 0, y > 1\}$ **(b)** $(0, 0)$ **(c)** $y = 1$ **(d)** $x = -2, x = 2$ **(e)** None

33. **(a)**

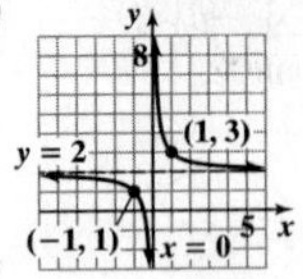

(b) Domain: $\{x|x \neq 0\}$; range: $\{y|y \neq 2\}$
(c) Vertical asymptote: $x = 0$; horizontal asymptote: $y = 2$

35. **(a)**

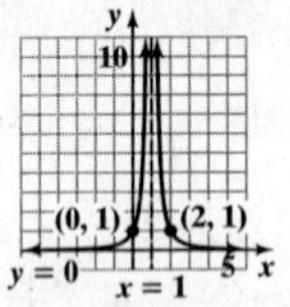

(b) Domain: $\{x|x \neq 1\}$; range: $\{y|y > 0\}$
(c) Vertical asymptote: $x = 1$; horizontal asymptote: $y = 0$

37. **(a)**

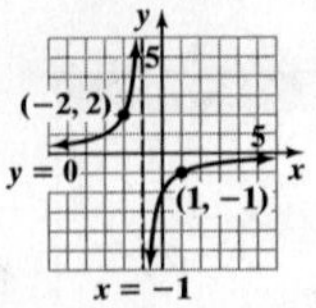

(b) Domain: $\{x|x \neq -1\}$; range: $\{y|y \neq 0\}$
(c) Vertical asymptote: $x = -1$; horizontal asymptote: $y = 0$

39. **(a)**

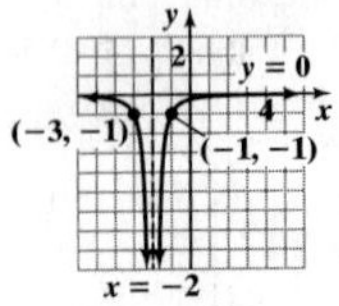

(b) Domain: $\{x|x \neq -2\}$; range: $\{y|y < 0\}$
(c) Vertical asymptote: $x = -2$; horizontal asymptote: $y = 0$

41. **(a)**

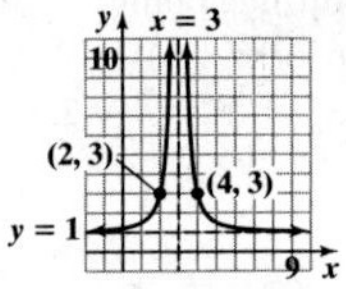

(b) Domain: $\{x|x \neq 3\}$; range: $\{y|y > 1\}$
(c) Vertical asymptote: $x = 3$; horizontal asymptote: $y = 1$

43. **(a)**

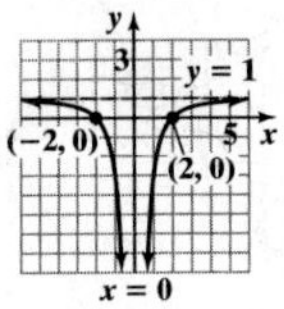

(b) Domain: $\{x|x \neq 0\}$; range: $\{y|y < 1\}$
(c) Vertical asymptote: $x = 0$; horizontal asymptote: $y = 1$

45. Vertical asymptote: $x = -4$; horizontal asymptote: $y = 3$ **47.** Vertical asymptote: $x = 3$; oblique asymptote: $y = x + 5$
49. Vertical asymptotes: $x = 1, x = -1$; horizontal asymptote: $y = 0$ **51.** Vertical asymptote: $x = -\frac{1}{3}$; horizontal asymptote: $y = \frac{2}{3}$
53. Vertical asymptote: none; oblique asymptote: $y = 2x - 1$ **55.** Vertical asymptote: $x = 0$; no horizontal or oblique asymptote
57. **(a)** 9.8208 m/sec^2 **(b)** 9.8195 m/sec^2 **(c)** 9.7936 m/sec^2 **(d)** h-axis **(e)** $\varnothing$

59. **(a)**

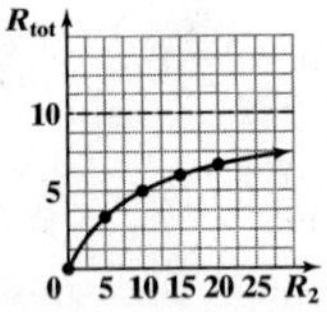

(b) Horizontal: $R_{tot} = 10$; as the resistance of R_2 increases without bound, the total resistance approaches 10 ohms, the resistance R_1.
(c) $R_1 \approx 103.5$ ohms

61. **(a)** $R(x) = 2 + \dfrac{5}{x - 1} = 5\left(\dfrac{1}{x - 1}\right) + 2$
(b)

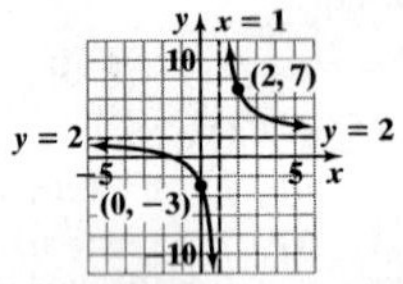

(c) Vertical asymptote: $x = 1$; horizontal asymptote: $y = 2$

67. $x = 5$ **68.** $\left\{-\frac{4}{19}\right\}$
69. x-axis symmetry
70. $(-3, 11), (2, -4)$

5.3 Assess Your Understanding *(page 365)*

2. False **3.** c **4.** a **5.** **(a)** $\{x|x \neq 2\}$ **(b)** 0 **6.** True
7. **1.** Domain: $\{x\,|\,x \neq 0, x \neq -4\}$ **2.** R is in lowest terms **3.** no y-intercept; x-intercept: -1
4. R is in lowest terms; vertical asymptotes: $x = 0, x = -4$ **5.** Horizontal asymptote: $y = 0$, intersected at $(-1, 0)$

6. Number line: -4, -1, 0

Interval	$(-\infty, -4)$	$(-4, -1)$	$(-1, 0)$	$(0, \infty)$
Number Chosen	-5	-2	$-\frac{1}{2}$	1
Value of R	$R(-5) = -\frac{4}{5}$	$R(-2) = \frac{1}{4}$	$R\left(-\frac{1}{2}\right) = -\frac{2}{7}$	$R(1) = \frac{2}{5}$
Location of Graph	Below x-axis	Above x-axis	Below x-axis	Above x-axis
Point on Graph	$\left(-5, -\frac{4}{5}\right)$	$\left(-2, \frac{1}{4}\right)$	$\left(-\frac{1}{2}, -\frac{2}{7}\right)$	$\left(1, \frac{2}{5}\right)$

7.

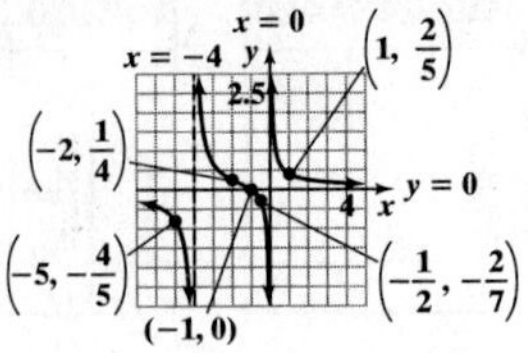

9. **1.** $R(x) = \dfrac{3(x + 1)}{2(x + 2)}$; Domain: $\{x|x \neq -2\}$ **2.** R is in lowest terms **3.** y-intercept: $\frac{3}{4}$; x-intercept: -1
4. R is in lowest terms; vertical asymptote: $x = -2$ **5.** Horizontal asymptote: $y = \frac{3}{2}$, not intersected

6.

Interval	$(-\infty, -2)$	$(-2, -1)$	$(-1, \infty)$
Number Chosen	-3	$-\frac{3}{2}$	0
Value of R	$R(-3) = 3$	$R\left(-\frac{3}{2}\right) = -\frac{3}{2}$	$R(0) = \frac{3}{4}$
Location of Graph	Above x-axis	Below x-axis	Above x-axis
Point on Graph	$(-3, 3)$	$\left(-\frac{3}{2}, -\frac{3}{2}\right)$	$\left(0, \frac{3}{4}\right)$

7.

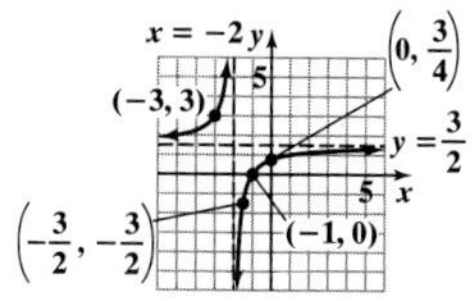

11. 1. $R(x) = \dfrac{3}{(x+2)(x-2)}$; Domain: $\{x \mid x \neq -2, x \neq 2\}$ **2.** R is in lowest terms **3.** y-intercept: $-\dfrac{3}{4}$; no x-intercept
4. R is in lowest terms; vertical asymptotes: $x = 2, x = -2$ **5.** Horizontal asymptote: $y = 0$, not intersected

6.

Interval	$(-\infty, -2)$	$(-2, 2)$	$(2, \infty)$
Number Chosen	-3	0	3
Value of R	$R(-3) = \frac{3}{5}$	$R(0) = -\frac{3}{4}$	$R(3) = \frac{3}{5}$
Location of Graph	Above x-axis	Below x-axis	Above x-axis
Point on Graph	$\left(-3, \frac{3}{5}\right)$	$\left(0, -\frac{3}{4}\right)$	$\left(3, \frac{3}{5}\right)$

7.

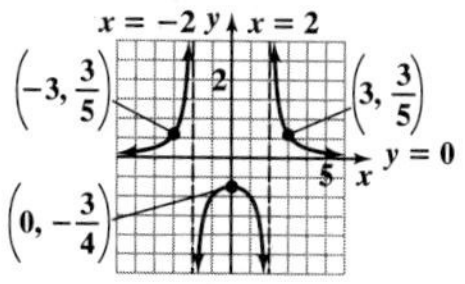

13. 1. $P(x) = \dfrac{(x^2+x+1)(x^2-x+1)}{(x+1)(x-1)}$; Domain: $\{x \mid x \neq -1, x \neq 1\}$ **2.** P is in lowest terms **3.** y-intercept: -1; no x-intercept
4. P is in lowest terms; vertical asymptotes: $x = -1, x = 1$ **5.** No horizontal or oblique asymptote

6.

Interval	$(-\infty, -1)$	$(-1, 1)$	$(1, \infty)$
Number Chosen	-2	0	2
Value of P	$P(-2) = 7$	$P(0) = -1$	$P(2) = 7$
Location of Graph	Above x-axis	Below x-axis	Above x-axis
Point on Graph	$(-2, 7)$	$(0, -1)$	$(2, 7)$

7.

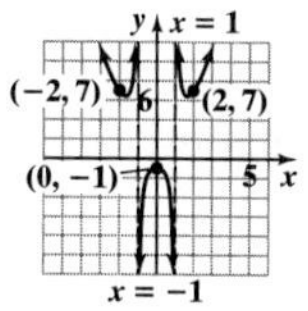

15. 1. $H(x) = \dfrac{(x-1)(x^2+x+1)}{(x+3)(x-3)}$; Domain: $\{x \mid x \neq -3, x \neq 3\}$ **2.** H is in lowest terms **3.** y-intercept: $\dfrac{1}{9}$; x-intercept: 1
4. H is in lowest terms; vertical asymptotes: $x = 3, x = -3$ **5.** Oblique asymptote: $y = x$, intersected at $\left(\dfrac{1}{9}, \dfrac{1}{9}\right)$

6.

Interval	$(-\infty, -3)$	$(-3, 1)$	$(1, 3)$	$(3, \infty)$
Number Chosen	-4	0	2	4
Value of H	$H(-4) \approx -9.3$	$H(0) = \frac{1}{9}$	$H(2) = -1.4$	$H(4) = 9$
Location of Graph	Below x-axis	Above x-axis	Below x-axis	Above x-axis
Point on Graph	$(-4, -9.3)$	$\left(0, \frac{1}{9}\right)$	$(2, -1.4)$	$(4, 9)$

7.

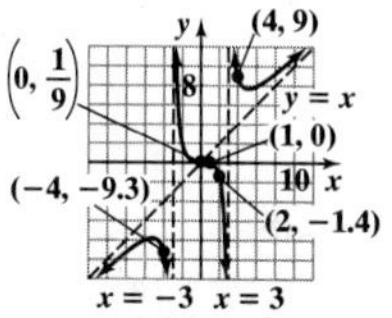

17. 1. $R(x) = \dfrac{x^2}{(x+3)(x-2)}$; Domain: $\{x \neq -3, x \neq 2\}$ **2.** R is in lowest terms **3.** y-intercept: 0; x-intercept: 0
4. R is in lowest terms; vertical asymptotes: $x = 2, x = -3$ **5.** Horizontal asymptote: $y = 1$, intersected at $(6, 1)$

6.

Interval	$(-\infty, -3)$	$(-3, 0)$	$(0, 2)$	$(2, \infty)$
Number Chosen	-6	-1	1	3
Value of R	$R(-6) = 1.5$	$R(-1) = -\frac{1}{6}$	$R(1) = -0.25$	$R(3) = 1.5$
Location of Graph	Above x-axis	Below x-axis	Below x-axis	Above x-axis
Point on Graph	$(-6, 1.5)$	$\left(-1, -\frac{1}{6}\right)$	$(1, -0.25)$	$(3, 1.5)$

7.

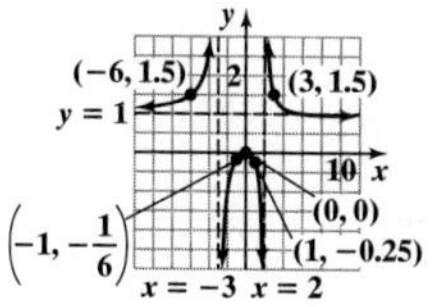

19. 1. $G(x) = \dfrac{x}{(x+2)(x-2)}$; Domain: $\{x \mid x \neq -2, x \neq 2\}$ **2.** G is in lowest terms **3.** y-intercept: 0; x-intercept: 0
4. G is in lowest terms; vertical asymptotes: $x = -2, x = 2$ **5.** Horizontal asymptote: $y = 0$, intersected at $(0, 0)$

6.

Interval	$(-\infty, -2)$	$(-2, 0)$	$(0, 2)$	$(2, \infty)$
Number Chosen	-3	-1	1	3
Value of G	$G(-3) = -\frac{3}{5}$	$G(-1) = \frac{1}{3}$	$G(1) = -\frac{1}{3}$	$G(3) = \frac{3}{5}$
Location of Graph	Below x-axis	Above x-axis	Below x-axis	Above x-axis
Point on Graph	$\left(-3, -\frac{3}{5}\right)$	$\left(-1, \frac{1}{3}\right)$	$\left(1, -\frac{1}{3}\right)$	$\left(3, \frac{3}{5}\right)$

7.

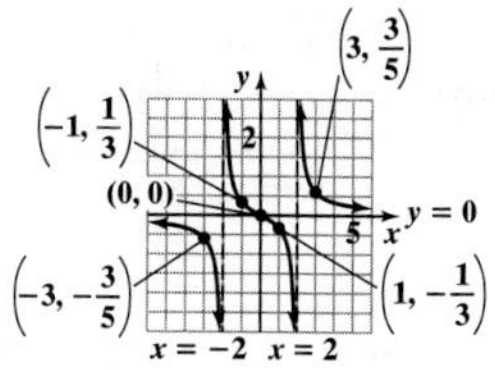

21. 1. $R(x) = \dfrac{3}{(x-1)(x+2)(x-2)}$; Domain: $\{x \mid x \neq 1, x \neq -2, x \neq 2\}$ **2.** R is in lowest terms **3.** y-intercept: $\dfrac{3}{4}$; no x-intercept

4. R is in lowest terms; vertical asymptotes: $x = -2, x = 1, x = 2$ **5.** Horizontal asymptote: $y = 0$, not intersected

6.

Number line: -2, 1, 2

Interval	$(-\infty, -2)$	$(-2, 1)$	$(1, 2)$	$(2, \infty)$
Number Chosen	-3	0	1.5	3
Value of R	$R(-3) = -\frac{3}{20}$	$R(0) = \frac{3}{4}$	$R(1.5) = -\frac{24}{7}$	$R(3) = \frac{3}{10}$
Location of Graph	Below x-axis	Above x-axis	Below x-axis	Above x-axis
Point on Graph	$\left(-3, -\frac{3}{20}\right)$	$\left(0, \frac{3}{4}\right)$	$\left(1.5, -\frac{24}{7}\right)$	$\left(3, \frac{3}{10}\right)$

7.

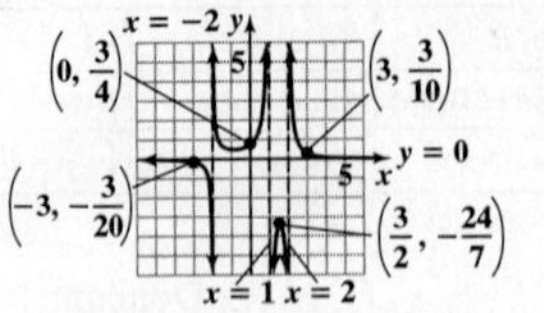

23. 1. $H(x) = \dfrac{(x+1)(x-1)}{(x^2+4)(x+2)(x-2)}$; Domain: $\{x \mid x \neq -2, x \neq 2\}$ **2.** H is in lowest terms **3.** y-intercept: $\dfrac{1}{16}$; x-intercepts: $-1, 1$

4. H is in lowest terms; vertical asymptotes: $x = -2, x = 2$ **5.** Horizontal asymptote: $y = 0$, intersected at $(-1, 0)$ and $(1, 0)$

6.

Number line: -2, -1, 1, 2

Interval	$(-\infty, -2)$	$(-2, -1)$	$(-1, 1)$	$(1, 2)$	$(2, \infty)$
Number Chosen	-3	-1.5	0	1.5	3
Value of H	$H(-3) \approx 0.12$	$H(-1.5) \approx -0.11$	$H(0) = \frac{1}{16}$	$H(1.5) \approx -0.11$	$H(3) \approx 0.12$
Location of Graph	Above x-axis	Below x-axis	Above x-axis	Below x-axis	Above x-axis
Point on Graph	$(-3, 0.12)$	$(-1.5, -0.11)$	$\left(0, \frac{1}{16}\right)$	$(1.5, -0.11)$	$(3, 0.12)$

7.

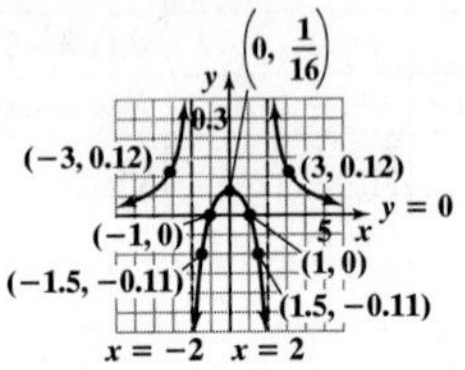

25. 1. $F(x) = \dfrac{(x+1)(x-4)}{x+2}$; Domain: $\{x \mid x \neq -2\}$ **2.** F is in lowest terms **3.** y-intercept: -2; x-intercept: $-1, 4$

4. F is in lowest terms; vertical asymptote: $x = -2$ **5.** Oblique asymptote: $y = x - 5$, not intersected

6.

Number line: -2, -1, 4

Interval	$(-\infty, -2)$	$(-2, -1)$	$(-1, 4)$	$(4, \infty)$
Number Chosen	-3	-1.5	0	5
Value of F	$F(-3) = -14$	$F(-1.5) = 5.5$	$F(0) = -2$	$F(5) \approx 0.86$
Location of Graph	Below x-axis	Above x-axis	Below x-axis	Above x-axis
Point on Graph	$(-3, -14)$	$(-1.5, 5.5)$	$(0, -2)$	$(5, 0.86)$

7.

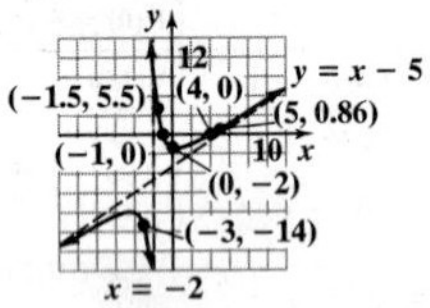

27. 1. $R(x) = \dfrac{(x+4)(x-3)}{x-4}$; Domain: $\{x \mid x \neq 4\}$ **2.** R is in lowest terms **3.** y-intercept: 3; x-intercepts: $-4, 3$

4. R is in lowest terms; vertical asymptote: $x = 4$ **5.** Oblique asymptote: $y = x + 5$, not intersected

6.

Number line: -4, 3, 4

Interval	$(-\infty, -4)$	$(-4, 3)$	$(3, 4)$	$(4, \infty)$
Number Chosen	-5	0	3.5	5
Value of R	$R(-5) = -\frac{8}{9}$	$R(0) = 3$	$R(3.5) = -7.5$	$R(5) = 18$
Location of Graph	Below x-axis	Above x-axis	Below x-axis	Above x-axis
Point on Graph	$\left(-5, -\frac{8}{9}\right)$	$(0, 3)$	$(3.5, -7.5)$	$(5, 18)$

7.

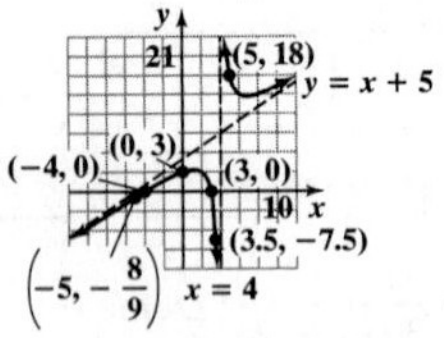

29. 1. $F(x) = \dfrac{(x+4)(x-3)}{x+2}$; Domain: $\{x \mid x \neq -2\}$ **2.** F is in lowest terms **3.** y-intercept: -6; x-intercepts: $-4, 3$

4. F is in lowest terms; vertical asymptote: $x = -2$ **5.** Oblique asymptote: $y = x - 1$, not intersected

6.

Number line: -4, -2, 3

Interval	$(-\infty, -4)$	$(-4, -2)$	$(-2, 3)$	$(3, \infty)$
Number Chosen	-5	-3	0	4
Value of F	$F(-5) = -\frac{8}{3}$	$F(-3) = 6$	$F(0) = -6$	$F(4) = \frac{4}{3}$
Location of Graph	Below x-axis	Above x-axis	Below x-axis	Above x-axis
Point on Graph	$\left(-5, -\frac{8}{3}\right)$	$(-3, 6)$	$(0, -6)$	$\left(4, \frac{4}{3}\right)$

7.

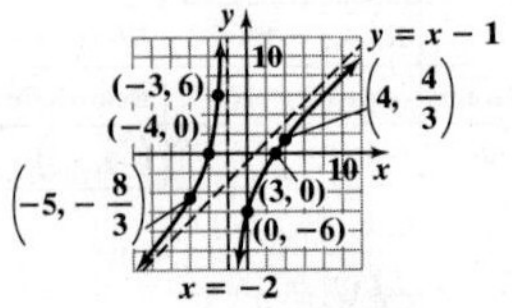

31. 1. Domain: $\{x \mid x \neq -3\}$ **2.** R is in lowest terms **3.** y-intercept: 0; x-intercepts: 0, 1 **4.** Vertical asymptote: $x = -3$

5. Horizontal asymptote: $y = 1$, not intersected

6.

Number line: -3, 0, 1

Interval	$(-\infty, -3)$	$(-3, 0)$	$(0, 1)$	$(1, \infty)$
Number Chosen	-4	-1	$\frac{1}{2}$	2
Value of R	$R(-4) = 100$	$R(-1) = -0.5$	$R\left(\frac{1}{2}\right) \approx 0.003$	$R(2) = 0.016$
Location of Graph	Above x-axis	Below x-axis	Above x-axis	Above x-axis
Point on Graph	$(-4, 100)$	$(-1, -0.5)$	$\left(\frac{1}{2}, 0.003\right)$	$(2, 0.016)$

7.

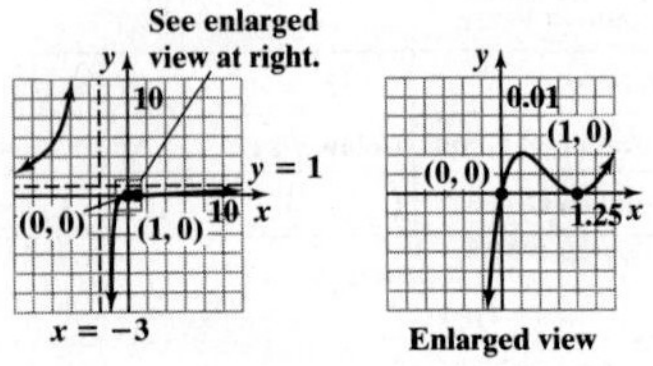

33. 1. $R(x) = \dfrac{(x+4)(x-3)}{(x+2)(x-3)}$; Domain: $\{x \mid x \neq -2, x \neq 3\}$ **2.** In lowest terms, $R(x) = \dfrac{x+4}{x+2}$ **3.** y-intercept: 2; x-intercept: -4

4. Vertical asymptote: $x = -2$; hole at $\left(3, \dfrac{7}{5}\right)$ **5.** Horizontal asymptote: $y = 1$, not intersected

6. $-4 \quad -2 \quad 3$

Interval	$(-\infty, -4)$	$(-4, -2)$	$(-2, 3)$	$(3, \infty)$
Number Chosen	-5	-3	0	4
Value of R	$R(-5) = \frac{1}{3}$	$R(-3) = -1$	$R(0) = 2$	$R(4) = \frac{4}{3}$
Location of Graph	Above x-axis	Below x-axis	Above x-axis	Above x-axis
Point on Graph	$\left(-5, \frac{1}{3}\right)$	$(-3, -1)$	$(0, 2)$	$\left(4, \frac{4}{3}\right)$

7.

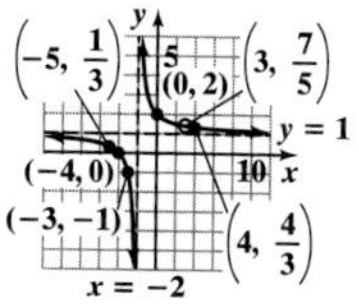

35. 1. $R(x) = \dfrac{(3x+1)(2x-3)}{(x-2)(2x-3)}$; Domain: $\left\{x \,\middle|\, x \neq \dfrac{3}{2}, x \neq 2\right\}$ **2.** In lowest terms, $R(x) = \dfrac{3x+1}{x-2}$ **3.** y-intercept: $-\dfrac{1}{2}$; x-intercept: $-\dfrac{1}{3}$

4. Vertical asymptote: $x = 2$; hole at $\left(\dfrac{3}{2}, -11\right)$ **5.** Horizontal asymptote: $y = 3$, not intersected

6. $-\frac{1}{3} \quad \frac{3}{2} \quad 2$

Interval	$\left(-\infty, -\frac{1}{3}\right)$	$\left(-\frac{1}{3}, \frac{3}{2}\right)$	$\left(\frac{3}{2}, 2\right)$	$(2, \infty)$
Number Chosen	-1	0	1.7	6
Value of R	$R(-1) = \frac{2}{3}$	$R(0) = -\frac{1}{2}$	$R(1.7) \approx -20.3$	$R(6) = 4.75$
Location of Graph	Above x-axis	Below x-axis	Below x-axis	Above x-axis
Point on Graph	$\left(-1, \frac{2}{3}\right)$	$\left(0, -\frac{1}{2}\right)$	$(1.7, -20.3)$	$(6, 4.75)$

7.

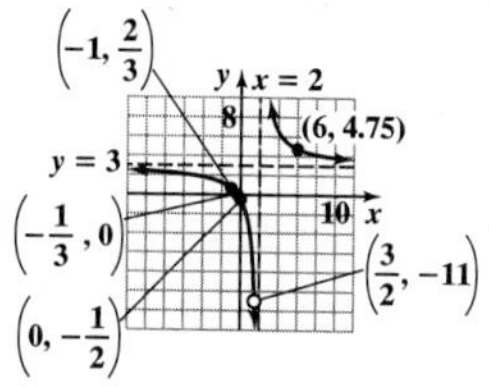

37. 1. $R(x) = \dfrac{(x+3)(x+2)}{x+3}$; Domain: $\{x \mid x \neq -3\}$ **2.** In lowest terms, $R(x) = x + 2$ **3.** y-intercept: 2; x-intercept: -2

4. Vertical asymptote: none; hole at $(-3, -1)$ **5.** Oblique asymptote: $y = x + 2$ intersected at all points except $x = -3$

6. $-3 \quad -2$

Interval	$(-\infty, -3)$	$(-3, -2)$	$(-2, \infty)$
Number Chosen	-4	$-\frac{5}{2}$	0
Value of R	$R(-4) = -2$	$R\left(-\frac{5}{2}\right) = -\frac{1}{2}$	$R(0) = 2$
Location of Graph	Below x-axis	Below x-axis	Above x-axis
Point on Graph	$(-4, -2)$	$\left(-\frac{5}{2}, -\frac{1}{2}\right)$	$(0, 2)$

7.

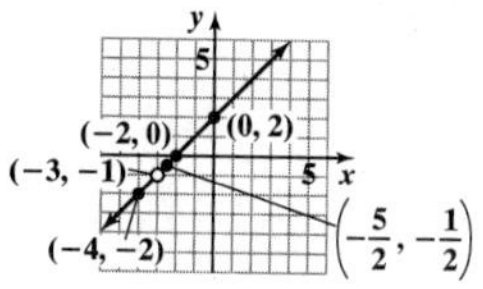

39. 1. $H(x) = \dfrac{-3(x-2)}{(x-2)(x+2)}$; Domain: $\{x \mid x \neq -2, x \neq 2\}$ **2.** In lowest terms, $H(x) = \dfrac{-3}{x+2}$ **3.** y-intercept: $-\dfrac{3}{2}$; no x-intercept

4. Vertical asymptote: $x = -2$; hole at $\left(2, -\dfrac{3}{4}\right)$ **5.** Horizontal asymptote: $y = 0$; not intersected

6. $-2 \quad 2$

Interval	$(-\infty, -2)$	$(-2, 2)$	$(2, \infty)$
Number Chosen	-3	0	3
Value of H	$H(-3) = 3$	$H(0) = -\frac{3}{2}$	$H(3) = -\frac{3}{5}$
Location of Graph	Above x-axis	Below x-axis	Below x-axis
Point on Graph	$(-3, 3)$	$\left(0, -\frac{3}{2}\right)$	$\left(3, -\frac{3}{5}\right)$

7

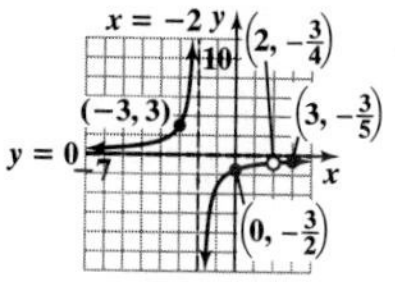

41. 1. $F(x) = \dfrac{(x-1)(x-4)}{(x-1)^2}$; Domain: $\{x \mid x \neq 1\}$ **2.** In lowest terms, $F(x) = \dfrac{x-4}{x-1}$ **3.** y-intercept: 4; x-intercept: 4

4. Vertical asymptote: $x = 1$ **5.** Horizontal asymptote: $y = 1$; not intersected

6. $1 \quad 4$

Interval	$(-\infty, 1)$	$(1, 4)$	$(4, \infty)$
Number Chosen	0	2	5
Value of F	$F(0) = 4$	$F(2) = -2$	$F(5) = \frac{1}{4}$
Location of Graph	Above x-axis	Below x-axis	Above x-axis
Point on Graph	$(0, 4)$	$(2, -2)$	$\left(5, \frac{1}{4}\right)$

7.

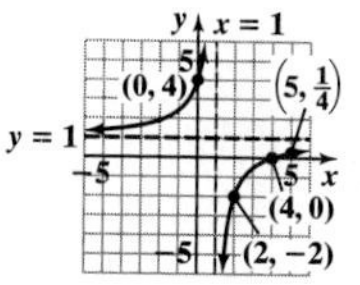

43. 1. $G(x) = \dfrac{x}{(x+2)^2}$; Domain: $\{x \mid x \neq -2\}$ **2.** G is in lowest terms **3.** y-intercept: 0; x-intercept: 0

4. Vertical asymptote: $x = -2$ **5.** Horizontal asymptote: $y = 0$; intersected at $(0, 0)$

6. $-2 \quad 0$

Interval	$(-\infty, -2)$	$(-2, 0)$	$(0, \infty)$
Number Chosen	-3	-1	1
Value of G	$G(-3) = -3$	$G(-1) = -1$	$G(1) = \frac{1}{9}$
Location of Graph	Below x-axis	Below x-axis	Above x-axis
Point on Graph	$(-3, -3)$	$(-1, -1)$	$\left(1, \frac{1}{9}\right)$

7.

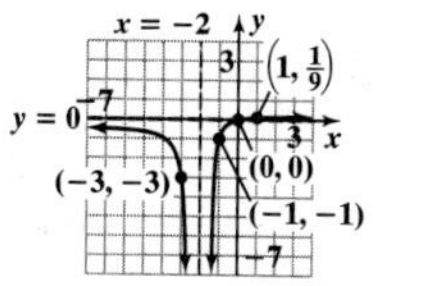

45. 1. $f(x) = \frac{x^2 + 1}{x}$; Domain: $\{x \mid x \neq 0\}$ **2.** f is in lowest terms **3.** no y-intercept; no x-intercepts
4. f is in lowest terms; vertical asymptote: $x = 0$ **5.** Oblique asymptote: $y = x$, not intersected
6.

Interval	$(-\infty, 0)$	$(0, \infty)$
Number Chosen	-1	1
Value of f	$f(-1) = -2$	$f(1) = 2$
Location of Graph	Below x-axis	Above x-axis
Point on Graph	$(-1, -2)$	$(1, 2)$

7.
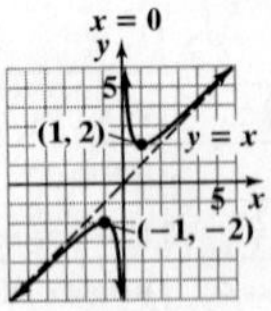

47. 1. $f(x) = \frac{x^3 + 1}{x} = \frac{(x + 1)(x^2 - x + 1)}{x}$; Domain: $\{x \mid x \neq 0\}$ **2.** f is in lowest terms **3.** no y-intercept; x-intercept: -1
4. f is in lowest terms; vertical asymptote: $x = 0$ **5.** No horizontal or oblique asymptote
6.

Interval	$(-\infty, -1)$	$(-1, 0)$	$(0, \infty)$
Number Chosen	-2	$-\frac{1}{2}$	1
Value of f	$f(-2) = \frac{7}{2}$	$f\left(-\frac{1}{2}\right) = -\frac{7}{4}$	$f(1) = 2$
Location of Graph	Above x-axis	Below x-axis	Above x-axis
Point on Graph	$\left(-2, \frac{7}{2}\right)$	$\left(-\frac{1}{2}, -\frac{7}{4}\right)$	$(1, 2)$

7.
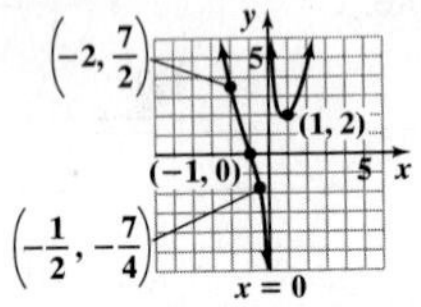

49. 1. $f(x) = \frac{x^4 + 1}{x^3}$; Domain: $\{x \mid x \neq 0\}$ **2.** f is in lowest terms **3.** no y-intercept; no x-intercepts
4. f is in lowest terms; vertical asymptote: $x = 0$ **5.** Oblique asymptote: $y = x$, not intersected
6.

Interval	$(-\infty, 0)$	$(0, \infty)$
Number Chosen	-1	1
Value of f	$f(-1) = -2$	$f(1) = 2$
Location of Graph	Below x-axis	Above x-axis
Point on Graph	$(-1, -2)$	$(1, 2)$

7.
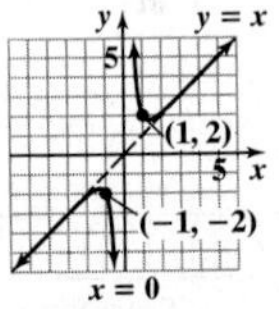

51. One possibility: $R(x) = \frac{x^2}{x^2 - 4}$ **53.** One possibility: $R(x) = \frac{(x - 1)(x - 3)\left(x^2 + \frac{4}{3}\right)}{(x + 1)^2 (x - 2)^2}$

55. (a) t-axis; $C(t) \to 0$
(b)
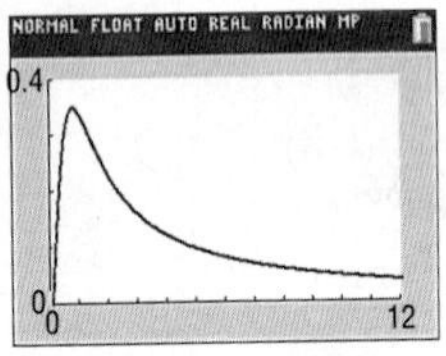

(c) 0.71 h after injection

57. (a) $C(x) = 16x + \frac{5000}{x} + 100$
(b) $x > 0$
(c)
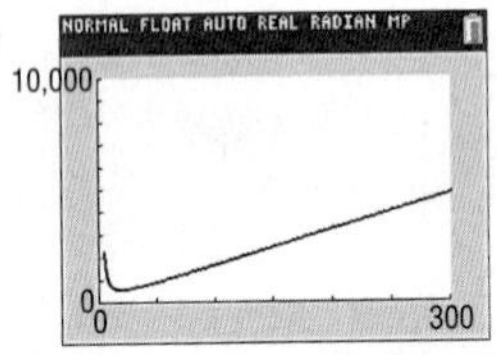

(d) Approximately 17.7 ft by 56.6 ft (longer side parallel to river)

59. (a) $S(x) = 2x^2 + \frac{40{,}000}{x}$
(b)
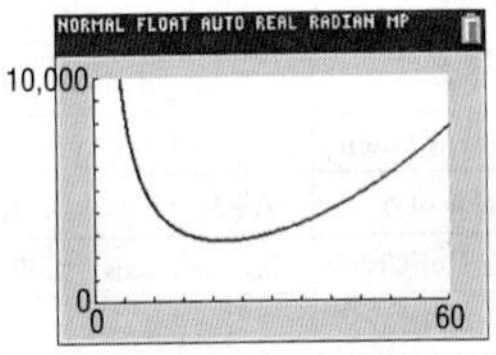

(c) 2784.95 in.^{2}
(d) 21.54 in. × 21.54 in. × 21.54 in.
(e) To minimize the cost of materials needed for construction

61. (a) $C(r) = 12\pi r^2 + \frac{4000}{r}$
(b)
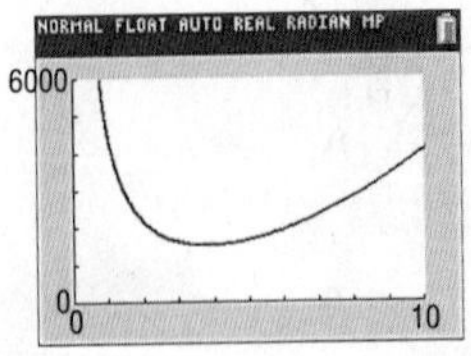

The cost is smallest when $r = 3.76$ cm.

63. No. Each function is a quotient of polynomials, but it is not written in lowest terms. Each function is undefined for $x = 1$; each graph has a hole at $x = 1$.
69. If there is a common factor between the numerator and the denominator, and the factor yields a real zero, then the graph will have a hole.
70. $4x^3 - 5x^2 + 2x - 2$ **71.** $\left\{-\frac{1}{10}\right\}$ **72.** $\frac{17}{2}$ **73.** ≈ -0.164

5.4 Assess Your Understanding *(page 372)*

3. c **4.** F **5. (a)** $\{x \mid 0 < x < 1 \text{ or } x > 2\}$; $(0, 1) \cup (2, \infty)$ **(b)** $\{x \mid x \leq 0 \text{ or } 1 \leq x \leq 2\}$; $(-\infty, 0] \cup [1, 2]$
7. (a) $\{x \mid -1 < x < 0 \text{ or } x > 1\}$; $(-1, 0) \cup (1, \infty)$ **(b)** $\{x \mid x < -1 \text{ or } 0 \leq x < 1\}$; $(-\infty, -1) \cup [0, 1)$
9. $\{x \mid x < 0 \text{ or } 0 < x < 3\}$; $(-\infty, 0) \cup (0, 3)$ **11.** $\{x \mid x \leq 1\}$; $(-\infty, 1]$ **13.** $\{x \mid x \leq -2 \text{ or } x \geq 2\}$; $(-\infty, -2] \cup [2, \infty)$

15. $\{x \mid -4 < x < -1 \text{ or } x > 0\}$; $(-4,-1) \cup (0,\infty)$ **17.** $\{x \mid -2 < x \le -1\}$; $(-2,-1]$ **19.** $\{x \mid x < -2\}$; $(-\infty,-2)$ **21.** $\{x \mid x > 4\}$; $(4,\infty)$ **23.** $\{x \mid -4 < x < 0 \text{ or } x > 0\}$; $(-4,0) \cup (0,\infty)$ **25.** $\{x \mid x \le 1 \text{ or } 2 \le x \le 3\}$; $(-\infty,1] \cup [2,3]$ **27.** $\{x \mid -1 < x < 0 \text{ or } x > 3\}$; $(-1,0) \cup (3,\infty)$ **29.** $\{x \mid x < -1 \text{ or } x > 1\}$; $(-\infty,-1) \cup (1,\infty)$ **31.** $\{x \mid x < -1 \text{ or } x > 1\}$; $(-\infty,-1) \cup (1,\infty)$ **33.** $\{x \mid x < -1 \text{ or } x > 1\}$; $(-\infty,-1) \cup (1,\infty)$ **35.** $\{x \mid x \le -1 \text{ or } 0 < x \le 1\}$; $(-\infty,-1] \cup (0,1]$ **37.** $\{x \mid x < -1 \text{ or } x > 1\}$; $(-\infty,-1) \cup (1,\infty)$ **39.** $\{x \mid x < 2\}$; $(-\infty,2)$ **41.** $\{x \mid -2 < x \le 9\}$; $(-2,9]$ **43.** $\{x \mid x < 2 \text{ or } 3 < x < 5\}$; $(-\infty,2) \cup (3,5)$ **45.** $\{x \mid x < -5 \text{ or } -4 \le x \le -3 \text{ or } x = 0 \text{ or } x > 1\}$; $(-\infty,-5) \cup [-4,-3] \cup \{0\} \cup (1,\infty)$ **47.** $\left\{x \,\middle|\, -\frac{1}{2} < x < 1 \text{ or } x > 3\right\}$; $\left(-\frac{1}{2},1\right) \cup (3,\infty)$ **49.** $\{x \mid -1 < x < 3 \text{ or } x > 5\}$; $(-1,3) \cup (5,\infty)$

51. $\left\{x \,\middle|\, x \le -4 \text{ or } x \ge \frac{1}{2}\right\}$; $(-\infty,-4] \cup \left[\frac{1}{2},\infty\right)$ **53.** $\{x \mid x < 3 \text{ or } x \ge 7\}$; $(-\infty,3) \cup [7,\infty)$ **55.** $\{x \mid x < 2\}$; $(-\infty,2)$

57. $\left\{x \,\middle|\, x < -\frac{2}{3} \text{ or } 0 < x < \frac{3}{2}\right\}$; $\left(-\infty,-\frac{2}{3}\right) \cup \left(0,\frac{3}{2}\right)$ **59.** $\{x \mid x \le -3 \text{ or } 0 \le x \le 3\}$; $(-\infty,-3] \cup [0,3]$

61. (a)

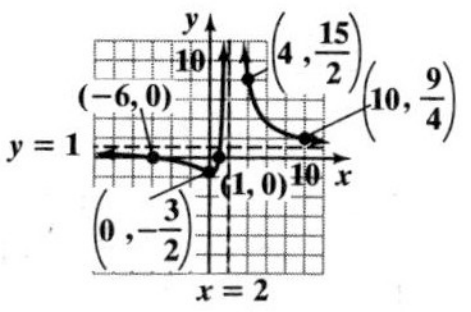

(b) $(-\infty,-6] \cup [1,2) \cup (2,\infty)$

63. (a)

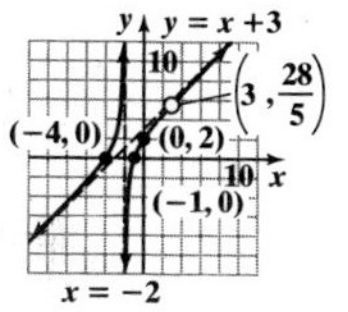

(b) $[-4,-2) \cup [-1,3) \cup (3,\infty)$

65. $\{x \mid x > 4\}$; $(4,\infty)$
67. $\{x \mid x \le -2 \text{ or } x \ge 2\}$; $(-\infty,-2] \cup [2,\infty)$
69. $\{x \mid x < -4 \text{ or } x \ge 2\}$; $(-\infty,-4) \cup [2,\infty)$

71.

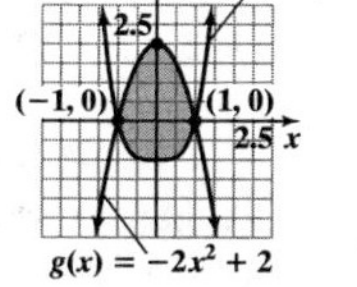

$f(x) \le g(x)$ if $-1 \le x \le 1$

73.

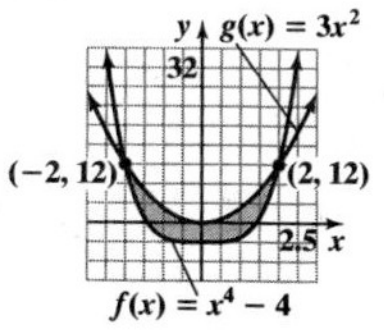

$f(x) \le g(x)$ if $-2 \le x \le 2$

75. Produce at least 250 bicycles
77. (a) The stretch is less than 39 ft.
(b) The ledge should be at least 84 ft above the ground for a 150-lb jumper.
79. At least 50 students must attend. **84.** $\left[\frac{4}{3},\infty\right)$
85. $x^2 - x - 4$ **86.** $3x^2y^4(x+2y)(2x-3y)$
87. $y = \frac{2}{3}\sqrt{x}$

Historical Problems *(page 386)*

1.

$$\left(x-\frac{b}{3}\right)^3 + b\left(x-\frac{b}{3}\right)^2 + c\left(x-\frac{b}{3}\right) + d = 0$$
$$x^3 - bx^2 + \frac{b^2x}{3} - \frac{b^3}{27} + bx^2 - \frac{2b^2x}{3} + \frac{b^3}{9} + cx - \frac{bc}{3} + d = 0$$
$$x^3 + \left(c - \frac{b^2}{3}\right)x + \left(\frac{2b^3}{27} - \frac{bc}{3} + d\right) = 0$$

Let $p = c - \frac{b^2}{3}$ and $q = \frac{2b^3}{27} - \frac{bc}{3} + d$. Then $x^3 + px + q = 0$.

2.

$$(H+K)^3 + p(H+K) + q = 0$$
$$H^3 + 3H^2K + 3HK^2 + K^3 + pH + pK + q = 0$$

Let $3HK = -p$.

$$H^3 - pH - pK + K^3 + pH + pK + q = 0,\ H^3 + K^3 = -q$$

3.

$$3HK = -p$$
$$K = -\frac{p}{3H}$$
$$H^3 + \left(-\frac{p}{3H}\right)^3 = -q$$
$$H^3 - \frac{p^3}{27H^3} = -q$$
$$27H^6 - p^3 = -27qH^3$$
$$27H^6 + 27qH^3 - p^3 = 0$$
$$H^3 = \frac{-27q \pm \sqrt{(27q)^2 - 4(27)(-p^3)}}{2 \cdot 27}$$
$$H^3 = \frac{-q}{2} \pm \sqrt{\frac{27^2q^2}{2^2(27^2)} + \frac{4(27)p^3}{2^2(27^2)}}$$
$$H^3 = \frac{-q}{2} \pm \sqrt{\frac{q^2}{4} + \frac{p^3}{27}}$$
$$H = \sqrt[3]{\frac{-q}{2} + \sqrt{\frac{q^2}{4} + \frac{p^3}{27}}}$$

Choose the positive root for now.

4.

$$H^3 + K^3 = -q$$
$$K^3 = -q - H^3$$
$$K^3 = -q - \left[\frac{-q}{2} + \sqrt{\frac{q^2}{4} + \frac{p^3}{27}}\right]$$
$$K^3 = \frac{-q}{2} - \sqrt{\frac{q^2}{4} + \frac{p^3}{27}}$$
$$K = \sqrt[3]{\frac{-q}{2} - \sqrt{\frac{q^2}{4} + \frac{p^3}{27}}}$$

5. $x = H + K$

$$x = \sqrt[3]{\frac{-q}{2} + \sqrt{\frac{q^2}{4} + \frac{p^3}{27}}} + \sqrt[3]{\frac{-q}{2} - \sqrt{\frac{q^2}{4} + \frac{p^3}{27}}}$$

(Note that if we had used the negative root in 3, the result would have been the same.)

6. $x = 3$ **7.** $x = 2$ **8.** $x = 2$

5.5 Assess Your Understanding *(page 386)*

5. a **6.** $f(c)$ **7.** b **8.** F **9.** 0 **10.** T **11.** $R = f(2) = 8$; no **13.** $R = f(2) = 0$; yes **15.** $R = f(-3) = 0$; yes **17.** $R = f(-4) = 1$; no **19.** $R = f\left(\frac{1}{2}\right) = 0$; yes **21.** 7; 3 or 1 positive; 2 or 0 negative **23.** 6; 2 or 0 positive; 2 or 0 negative **25.** 3; 2 or 0 positive; 1 negative **27.** 4; 2 or 0 positive; 2 or 0 negative **29.** 5; 0 positive; 3 or 1 negative **31.** 6; 1 positive; 1 negative

33. $\pm 1, \pm\frac{1}{3}$ **35.** $\pm 1, \pm 3$ **37.** $\pm 1, \pm 2, \pm\frac{1}{4}, \pm\frac{1}{2}$ **39.** $\pm 1, \pm 3, \pm 9, \pm\frac{1}{2}, \pm\frac{1}{3}, \pm\frac{1}{6}, \pm\frac{3}{2}, \pm\frac{9}{2}$

41. $\pm 1, \pm 2, \pm 3, \pm 4, \pm 6, \pm 12, \pm\frac{1}{2}, \pm\frac{3}{2}$ **43.** $\pm 1, \pm 2, \pm 4, \pm 5, \pm 10, \pm 20, \pm\frac{1}{2}, \pm\frac{5}{2}, \pm\frac{1}{3}, \pm\frac{2}{3}, \pm\frac{4}{3}, \pm\frac{5}{3}, \pm\frac{10}{3}, \pm\frac{20}{3}, \pm\frac{1}{6}, \pm\frac{5}{6}$

45. $-3, -1, 2; f(x) = (x+3)(x+1)(x-2)$ **47.** $\frac{1}{2}; f(x) = 2\left(x - \frac{1}{2}\right)(x^2+1)$ **49.** $2, \sqrt{5}, -\sqrt{5}; f(x) = 2(x-2)(x-\sqrt{5})(x+\sqrt{5})$

51. $-1, \frac{1}{2}, \sqrt{3}, -\sqrt{3}; f(x) = 2(x+1)\left(x-\frac{1}{2}\right)(x-\sqrt{3})(x+\sqrt{3})$ **53.** 1, multiplicity 2; $-2, -1; f(x) = (x+2)(x+1)(x-1)^2$

55. $-1, -\frac{1}{4}; f(x) = 4(x+1)\left(x+\frac{1}{4}\right)(x^2+2)$ **57.** $\{-1, 2\}$ **59.** $\left\{\frac{2}{3}, -1+\sqrt{2}, -1-\sqrt{2}\right\}$ **61.** $\left\{\frac{1}{3}, \sqrt{5}, -\sqrt{5}\right\}$ **63.** $\{-3, -2\}$

65. $\left\{-\frac{1}{3}\right\}$ **67.** $\left\{\frac{1}{2}, 2, 5\right\}$ **69.** LB $= -2$; UB $= 2$ **71.** LB $= -1$; UB $= 1$ **73.** LB $= -2$; UB $= 2$ **75.** LB $= -1$; UB $= 1$

77. LB $= -2$; UB $= 3$ **79.** $f(0) = -1; f(1) = 10$ **81.** $f(-5) = -58; f(-4) = 2$ **83.** $f(1.4) = -0.17536; f(1.5) = 1.40625$

85. 0.21 **87.** -4.04 **89.** 1.15 **91.** 2.53

93.

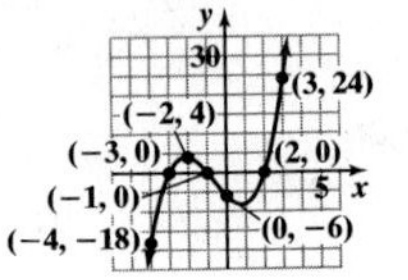

95.

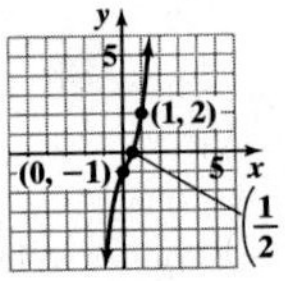

97.

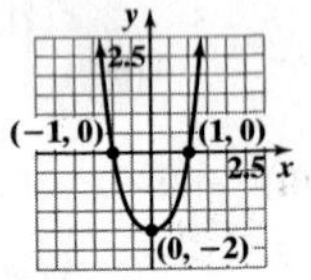

99.

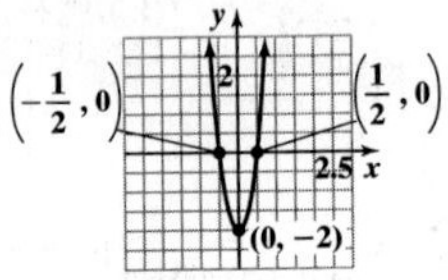

101.

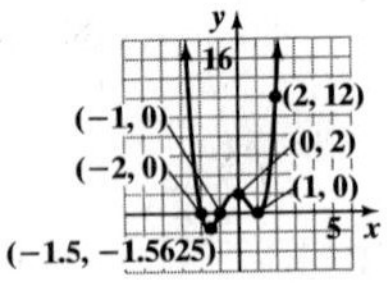

103.

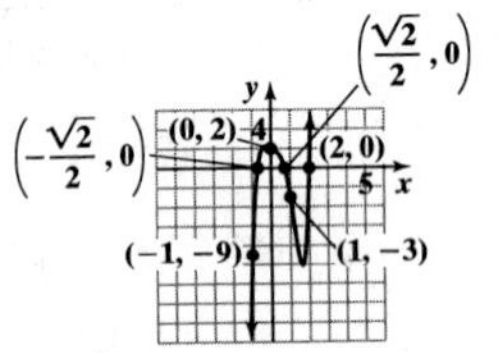

105. $-8, -4, -\frac{7}{3}$ **107.** $k = 5$ **109.** -7

111. If $f(x) = x^n - c^n$, then $f(c) = c^n - c^n = 0$, so $x - c$ is a factor of f.

113. 5 **115.** 7 in. **117.** All the potential rational zeros are integers, so r either is an integer or is not a rational zero (and is therefore irrational).

119. 0.215

121. No; by the Rational Zeros Theorem, $\frac{1}{3}$ is not a potential rational zero. **123.** No; by the Rational Zeros Theorem, $\frac{2}{3}$ is not a potential rational zero.

124. $y = \frac{2}{5}x - \frac{3}{5}$ **125.** $[3, 8)$ **126.** $(0, -2\sqrt{3}), (0, 2\sqrt{3}), (4, 0)$ **127.** $(-3, 2)$ and $(5, \infty)$

5.6 Assess Your Understanding *(page 394)*

3. one **4.** $3 - 4i$ **5.** T **6.** F **7.** $4 + i$ **9.** $-i, 1 - i$ **11.** $-i, -2i$ **13.** $-i$ **15.** $2 - i, -3 + i$

17. $f(x) = x^4 - 14x^3 + 77x^2 - 200x + 208; a = 1$ **19.** $f(x) = x^5 - 4x^4 + 7x^3 - 8x^2 + 6x - 4; a = 1$

21. $f(x) = x^4 - 6x^3 + 10x^2 - 6x + 9; a = 1$ **23.** $-2i, 4$ **25.** $2i, -3, \frac{1}{2}$ **27.** $3 + 2i, -2, 5$ **29.** $4i, -\sqrt{11}, \sqrt{11}, -\frac{2}{3}$

31. $1, -\frac{1}{2} - \frac{\sqrt{3}}{2}i, -\frac{1}{2} + \frac{\sqrt{3}}{2}i; f(x) = (x-1)\left(x + \frac{1}{2} + \frac{\sqrt{3}}{2}i\right)\left(x + \frac{1}{2} - \frac{\sqrt{3}}{2}i\right)$

33. $2, 3 - 2i, 3 + 2i; f(x) = (x-2)(x-3+2i)(x-3-2i)$ **35.** $-i, i, -2i, 2i; f(x) = (x+i)(x-i)(x+2i)(x-2i)$

37. $-5i, 5i, -3, 1; f(x) = (x+5i)(x-5i)(x+3)(x-1)$ **39.** $-4, \frac{1}{3}, 2 - 3i, 2 + 3i; f(x) = 3(x+4)\left(x - \frac{1}{3}\right)(x - 2 + 3i)(x - 2 - 3i)$

41. 130 **43. (a)** $f(x) = (x^2 - \sqrt{2}x + 1)(x^2 + \sqrt{2}x + 1)$ **(b)** $-\frac{\sqrt{2}}{2} - \frac{\sqrt{2}}{2}i, -\frac{\sqrt{2}}{2} + \frac{\sqrt{2}}{2}i, \frac{\sqrt{2}}{2} - \frac{\sqrt{2}}{2}i, \frac{\sqrt{2}}{2} + \frac{\sqrt{2}}{2}i$

45. Zeros that are complex numbers must occur in conjugate pairs; or a polynomial with real coefficients of odd degree must have at least one real zero.

47. If the remaining zero were a complex number, its conjugate would also be a zero, creating a polynomial of degree 5.

49.

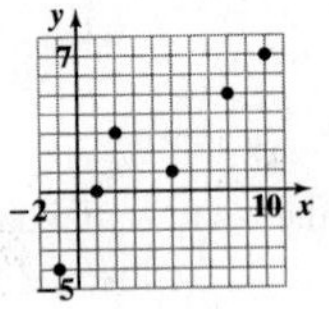

50. $\{-22\}$

51. $6x^3 - 13x^2 - 13x + 20$

52. $A = 9\pi$ ft^2 (≈ 28.274 ft^2); $C = 6\pi$ ft (≈ 18.850 ft)

Review Exercises *(page 397)*

1. Polynomial of degree 5 **2.** Rational **3.** Neither **4.** Polynomial of degree 0

5.

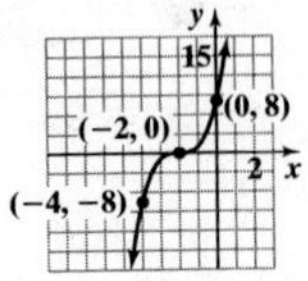

6.

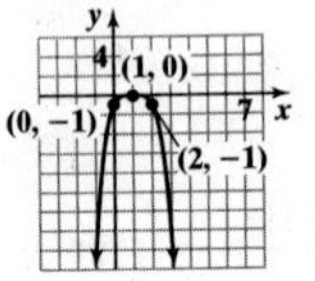

7.

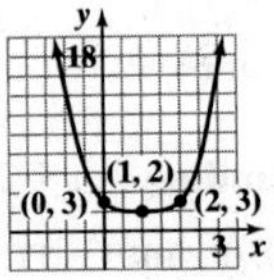

8. 1. $y = x^3$
2. x-intercepts: $-4, -2, 0$; y-intercept: 0
3. $-4, -2, 0$ (all multiplicity 1), crosses
4. 2
5. $f(-5) = -15; f(-3) = 3;$
$f(-1) = -3; f(1) = 15$

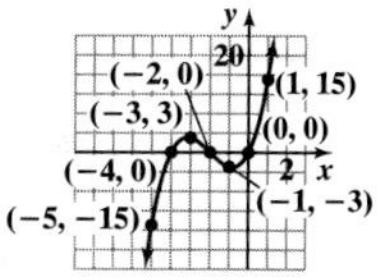

9. 1. $y = x^3$
2. x-intercepts: $-4, 2$; y-intercept: 16
3. -4, multiplicity 1, crosses; 2, multiplicity 2, touches
4. 2
5. $f(-5) = -49; f(-2) = 32; f(3) = 7$

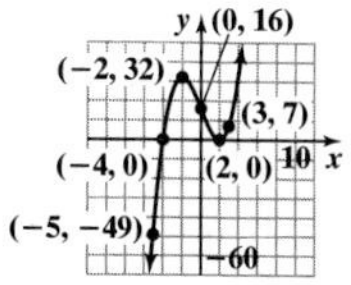

10. 1. $y = -2x^3$
2. $f(x) = -2x^2(x - 2)$
x-intercepts: 0, 2; y-intercept: 0
3. 0, multiplicity 2, touches; 2, multiplicity 1, crosses
4. 2
5. $f(-1) = 6; f(1) = 2; f(3) = -18$

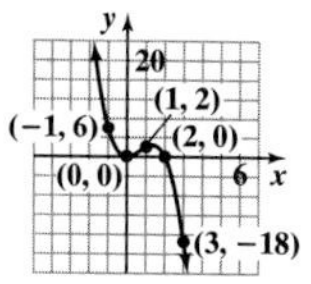

11. 1. $y = x^4$
2. x-intercepts: $-3, -1, 1$; y-intercept: 3
3. $-3, -1$ (both multiplicity 1), crosses; 1, multiplicity 2, touches
4. 3
5. $f(-4) = 75; f(-2) = -9; f(2) = 15$

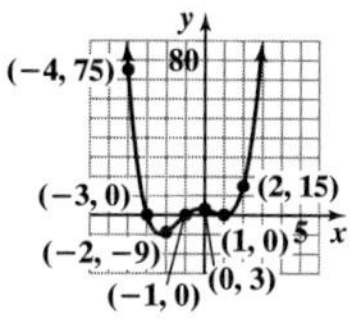

12. Domain: $\{x|x \neq -3, x \neq 3\}$; horizontal asymptote: $y = 0$; vertical asymptotes: $x = -3, x = 3$

13. Domain: $\{x|x \neq 2\}$; oblique asymptote: $y = x + 2$; vertical asymptote: $x = 2$

14. Domain: $\{x|x \neq -2\}$; horizontal asymptote: $y = 1$; vertical asymptote: $x = -2$

15. 1. $R(x) = \dfrac{2(x-3)}{x}$; domain: $\{x|x \neq 0\}$ **2.** R is in lowest terms **3.** no y-intercept; x-intercept: 3
4. R is in lowest terms; vertical asymptote: $x = 0$ **5.** Horizontal asymptote: $y = 2$; not intersected
6.

Interval	$(-\infty, 0)$	$(0, 3)$	$(3, \infty)$
Number Chosen	-2	1	4
Value of R	$R(-2) = 5$	$R(1) = -4$	$R(4) = \frac{1}{2}$
Location of Graph	Above x-axis	Below x-axis	Above x-axis
Point on Graph	$(-2, 5)$	$(1, -4)$	$\left(4, \frac{1}{2}\right)$

7.

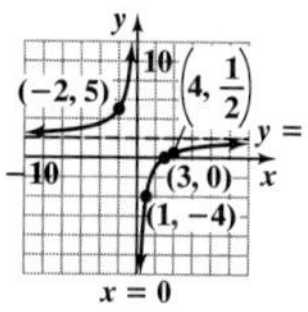

16. 1. Domain: $\{x|x \neq 0, x \neq 2\}$ **2.** H is in lowest terms **3.** no y-intercept; x-intercept: -2
4. H is in lowest terms; vertical asymptotes: $x = 0, x = 2$ **5.** Horizontal asymptote: $y = 0$; intersected at $(-2, 0)$
6.

Interval	$(-\infty, -2)$	$(-2, 0)$	$(0, 2)$	$(2, \infty)$
Number Chosen	-3	-1	1	3
Value of H	$H(-3) = -\frac{1}{15}$	$H(-1) = \frac{1}{3}$	$H(1) = -3$	$H(3) = \frac{5}{3}$
Location of Graph	Below x-axis	Above x-axis	Below x-axis	Above x-axis
Point on Graph	$\left(-3, -\frac{1}{15}\right)$	$\left(-1, \frac{1}{3}\right)$	$(1, -3)$	$\left(3, \frac{5}{3}\right)$

7.

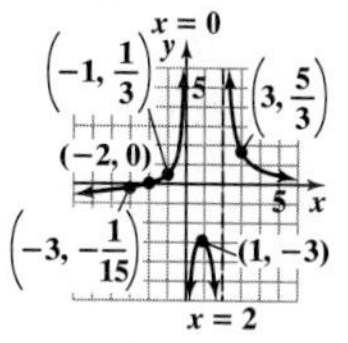

17. 1. $R(x) = \dfrac{(x+3)(x-2)}{(x-3)(x+2)}$; domain: $\{x|x \neq -2, x \neq 3\}$ **2.** R is in lowest terms **3.** y-intercept: 1; x-intercepts: $-3, 2$
4. R is in lowest terms; vertical asymptotes: $x = -2, x = 3$ **5.** Horizontal asymptote: $y = 1$; intersected at $(0, 1)$
6.

Interval	$(-\infty, -3)$	$(-3, -2)$	$(-2, 2)$	$(2, 3)$	$(3, \infty)$
Number Chosen	-4	$-\frac{5}{2}$	0	$\frac{5}{2}$	4
Value of R	$R(-4) = \frac{3}{7}$	$R\left(-\frac{5}{2}\right) = -\frac{9}{11}$	$R(0) = 1$	$R\left(\frac{5}{2}\right) = -\frac{11}{9}$	$R(4) = \frac{7}{3}$
Location of Graph	Above x-axis	Below x-axis	Above x-axis	Below x-axis	Above x-axis
Point on Graph	$\left(-4, \frac{3}{7}\right)$	$\left(-\frac{5}{2}, -\frac{9}{11}\right)$	$(0, 1)$	$\left(\frac{5}{2}, -\frac{11}{9}\right)$	$\left(4, \frac{7}{3}\right)$

7.

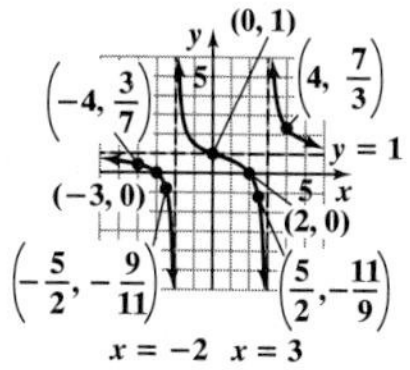

18. 1. $F(x) = \dfrac{x^3}{(x+2)(x-2)}$; domain: $\{x|x \neq -2, x \neq 2\}$ **2.** F is in lowest terms **3.** y-intercept: 0; x-intercept: 0
4. F is in lowest terms; vertical asymptotes: $x = -2, x = 2$ **5.** Oblique asymptote: $y = x$; intersected at $(0, 0)$
6.

Interval	$(-\infty, -2)$	$(-2, 0)$	$(0, 2)$	$(2, \infty)$
Number Chosen	-3	-1	1	3
Value of F	$F(-3) = -\frac{27}{5}$	$F(-1) = \frac{1}{3}$	$F(1) = -\frac{1}{3}$	$F(3) = \frac{27}{5}$
Location of Graph	Below x-axis	Above x-axis	Below x-axis	Above x-axis
Point on Graph	$\left(-3, -\frac{27}{5}\right)$	$\left(-1, \frac{1}{3}\right)$	$\left(1, -\frac{1}{3}\right)$	$\left(3, \frac{27}{5}\right)$

7.

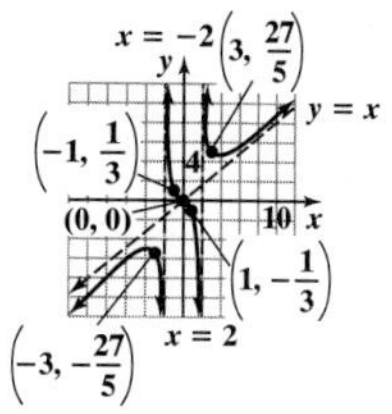

19. 1. Domain: $\{x \mid x \neq 1\}$ **2.** R is in lowest terms **3.** y-intercept: 0; x-intercept: 0
4. R is in lowest terms; vertical asymptote: $x = 1$ **5.** No oblique or horizontal asymptote
6.

Interval	$(-\infty, 0)$	$(0, 1)$	$(1, \infty)$
Number Chosen	-2	$\frac{1}{2}$	2
Value of R	$R(-2) = \frac{32}{9}$	$R\left(\frac{1}{2}\right) = \frac{1}{2}$	$R(2) = 32$
Location of Graph	Above x-axis	Above x-axis	Above x-axis
Point on Graph	$\left(-2, \frac{32}{9}\right)$	$\left(\frac{1}{2}, \frac{1}{2}\right)$	$(2, 32)$

7.
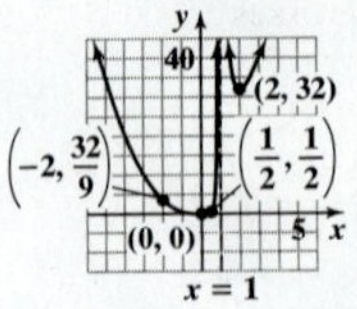

20. 1. $G(x) = \dfrac{(x + 2)(x - 2)}{(x + 1)(x - 2)}$; domain: $\{x \mid x \neq -1, x \neq 2\}$ **2.** In lowest terms, $G(x) = \dfrac{x + 2}{x + 1}$ **3.** y-intercept: 2; x-intercept: -2
4. Vertical asymptote: $x = -1$; hole at $\left(2, \dfrac{4}{3}\right)$ **5.** Horizontal asymptote: $y = 1$, not intersected
6.

Interval	$(-\infty, -2)$	$(-2, -1)$	$(-1, 2)$	$(2, \infty)$
Number Chosen	-3	$-\frac{3}{2}$	0	3
Value of G	$G(-3) = \frac{1}{2}$	$G\left(-\frac{3}{2}\right) = -1$	$G(0) = 2$	$G(3) = \frac{5}{4}$
Location of Graph	Above x-axis	Below x-axis	Above x-axis	Above x-axis
Point on Graph	$\left(-3, \frac{1}{2}\right)$	$\left(-\frac{3}{2}, -1\right)$	$(0, 2)$	$\left(3, \frac{5}{4}\right)$

7.
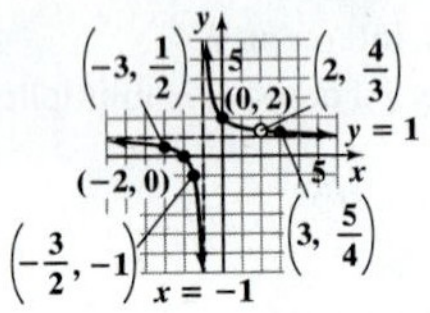

21. $\{x \mid x < -2 \text{ or } -1 < x < 2\}$; $(-\infty, -2) \cup (-1, 2)$

22. $\{x \mid -4 \leq x \leq -1 \text{ or } x \geq 1\}$ $[-4, -1] \cup [1, \infty)$
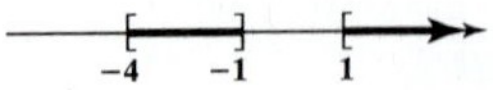

23. $\{x \mid x < 1 \text{ or } x > 2\}$; $(-\infty, 1) \cup (2, \infty)$
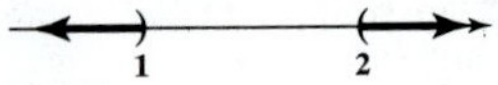

24. $\{x \mid 1 \leq x \leq 2 \text{ or } x > 3\}$; $[1, 2] \cup (3, \infty)$
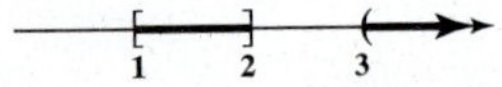

25. $\{x \mid x < -4 \text{ or } 2 < x < 4 \text{ or } x > 6\}$; $(-\infty, -4) \cup (2, 4) \cup (6, \infty)$
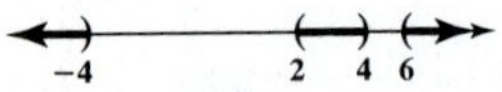

26. $R = 10$; g is not a factor of f. **27.** $R = 0$; g is a factor of f. **28.** $f(4) = 47{,}105$ **29.** 4, 2, or 0 positive; 2 or 0 negative **30.** 1 positive; 2 or 0 negative
31. $\pm 1, \pm 3, \pm\frac{1}{2}, \pm\frac{3}{2}, \pm\frac{1}{3}, \pm\frac{1}{4}, \pm\frac{3}{4}, \pm\frac{1}{6}, \pm\frac{1}{12}$ **32.** $-2, 1, 4; f(x) = (x + 2)(x - 1)(x - 4)$
33. $\frac{1}{2}$, multiplicity 2; -2; $f(x) = 4\left(x - \frac{1}{2}\right)^2 (x + 2)$ **34.** 2, multiplicity 2; $f(x) = (x - 2)^2 (x^2 + 5)$ **35.** $\{-3, 2\}$ **36.** $\left\{-3, -1, -\frac{1}{2}, 1\right\}$
37. lb: -2; ub: 3 **38.** lb: -3; ub: 5 **39.** $f(0) = -1$; $f(1) = 1$ **40.** $f(0) = -1$; $f(1) = 1$ **41.** 1.52 **42.** 0.93
43. $4 - i; f(x) = x^3 - 14x^2 + 65x - 102$ **44.** $-i, 1 - i; f(x) = x^4 - 2x^3 + 3x^2 - 2x + 2$ **45.** $-2, 1, 4; f(x) = (x + 2)(x - 1)(x - 4)$
46. $-2, \frac{1}{2}$ (multiplicity 2); $f(x) = 4(x + 2)\left(x - \frac{1}{2}\right)^2$ **47.** 2 (multiplicity 2), $-\sqrt{5}i, \sqrt{5}i; f(x) = (x + \sqrt{5}i)(x - \sqrt{5}i)(x - 2)^2$
48. $-3, 2, -\frac{\sqrt{2}}{2}i, \frac{\sqrt{2}}{2}i; f(x) = 2(x + 3)(x - 2)\left(x + \frac{\sqrt{2}}{2}i\right)\left(x - \frac{\sqrt{2}}{2}i\right)$

49. (a) $A(r) = 2\pi r^2 + \dfrac{500}{r}$
(b) 223.22 cm^2
(c) 257.08 cm^2
(d)
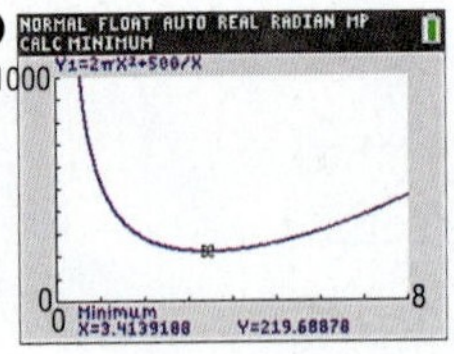

A is smallest when $r \approx 3.41$ cm.

50. (a)
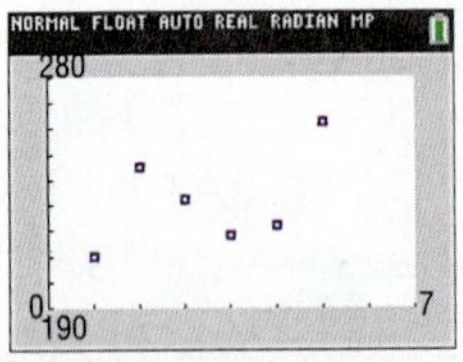

The relation appears to be cubic.
(b) $P(t) = 4.4926t^3 - 45.5294t^2 + 136.1209t + 115.4667$; $\approx$ \$928,000.
(c)
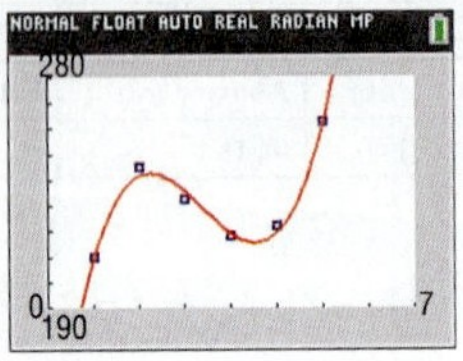

Chapter Test *(page 399)*

1.

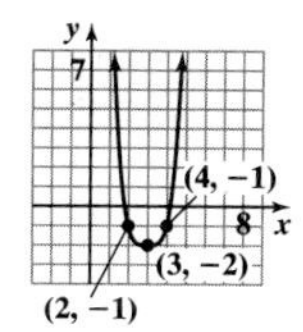

2. (a) 3
(b) $\frac{p}{q}: \pm\frac{1}{2}, \pm 1, \pm\frac{3}{2}, \pm\frac{5}{2}, \pm 3, \pm 5, \pm\frac{15}{2}, \pm 15$
(c) $-5, -\frac{1}{2}, 3; g(x) = (x + 5)(2x + 1)(x - 3)$
(d) y-intercept: -15; x-intercepts: $-5, -\frac{1}{2}, 3$
(e) Crosses at $-5, -\frac{1}{2}, 3$
(f) $y = 2x^3$
(g)

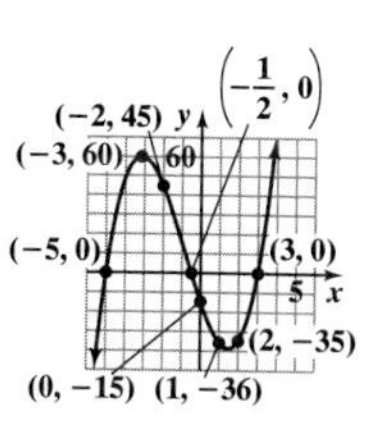

3. $4, -5i, 5i$ **4.** $\left\{1, \frac{5 - \sqrt{61}}{6}, \frac{5 + \sqrt{61}}{6}\right\}$ **5.** Domain: $\{x \mid x \neq -10, x \neq 4\}$; asymptotes: $x = -10, y = 2$

6. Domain: $\{x \mid x \neq -1\}$; asymptotes: $x = -1, y = x + 1$

7.

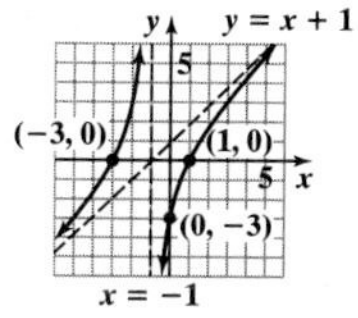

8. Answers may vary. One possibility is $f(x) = x^4 - 4x^3 - 2x^2 + 20x$.

9. Answers may vary. One possibility is $r(x) = \frac{2(x - 9)(x - 1)}{(x - 4)(x - 9)}$.

10. $f(0) = 8; f(4) = -36$
Since $f(0) = 8 > 0$ and $f(4) = -36 < 0$, the Intermediate Value Theorem guarantees that there is at least one real zero between 0 and 4.

11. $\{x \mid x < 3 \text{ or } x > 8\}; (-\infty, 3) \cup (8, \infty)$

Cumulative Review *(page 399)*

1. $\sqrt{26}$ **2.** $\{x \mid x \leq 0 \text{ or } x \geq 1\}$; $(-\infty, 0]$ or $[1, \infty)$

3. $\{x \mid -1 < x < 4\}; (-1, 4)$

4. $f(x) = -3x + 1$

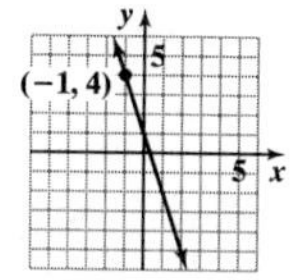

5. $y = 2x - 1$

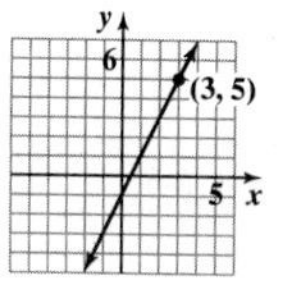

6.

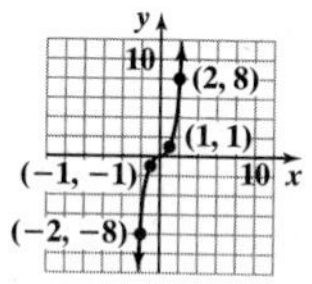

7. Not a function; 3 has two images. **8.** $\{0, 2, 4\}$ **9.** $\left\{x \middle| x \geq \frac{3}{2}\right\}; \left[\frac{3}{2}, \infty\right)$

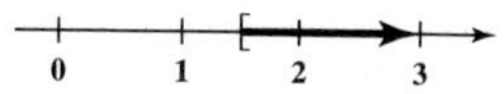

10. Center: $(-2, 1)$; radius: 3

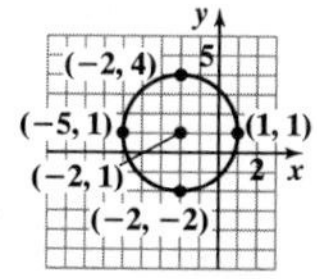

11. x-intercepts: $-3, 0, 3$; y-intercept: 0; symmetric with respect to the origin

12. $y = -\frac{2}{3}x + \frac{17}{3}$ **13.** Not a function; it fails the vertical-line test.

14. (a) 22 **(b)** $x^2 - 5x - 2$ **(c)** $-x^2 - 5x + 2$ **(d)** $9x^2 + 15x - 2$ **(e)** $2x + h + 5$

15. (a) $\{x \mid x \neq 1\}$ **(b)** No; $(2, 7)$ is on the graph. **(c)** 4; $(3, 4)$ is on the graph. **(d)** $\frac{7}{4}; \left(\frac{7}{4}, 9\right)$ is on the graph.
(e) Rational

16.

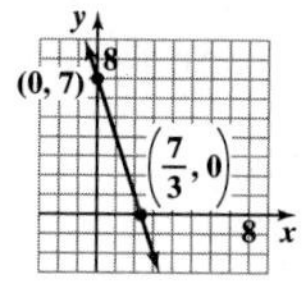

17.

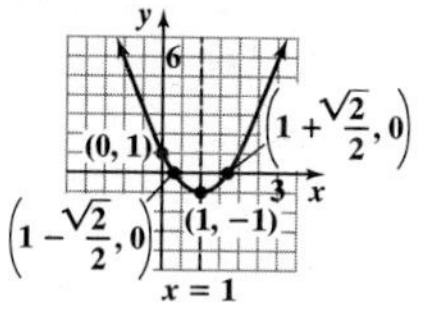

18. 6; $y = 6x - 1$

19. (a) x-intercepts: $-5, -1, 5$; y-intercept: -3
(b) No symmetry
(c) Neither
(d) Increasing: $(-\infty, -3)$ and $(2, \infty)$; decreasing: $(-3, 2)$
(e) A local maximum value of 5 occurs at $x = -3$.
(f) A local minimum value of -6 occurs at $x = 2$.

20. Odd

21. (a) Domain: $\{x \mid x > -3\}$ or $(-3, \infty)$
(b) x-intercept: $-\frac{1}{2}$; y-intercept: 1
(c)

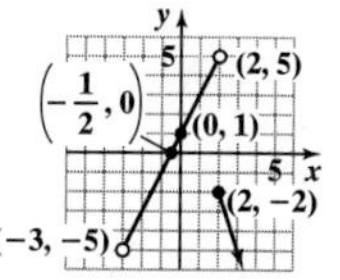

(d) Range: $\{y \mid y < 5\}$ or $(-\infty, 5)$

22.

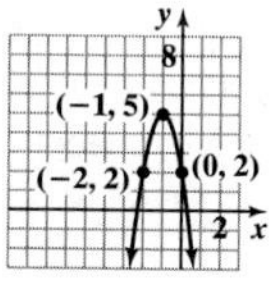

23. (a) $(f + g)(x) = x^2 - 9x - 6$; domain: all real numbers
(b) $\left(\frac{f}{g}\right)(x) = \frac{x^2 - 5x + 1}{-4x - 7}$; domain: $\left\{x \middle| x \neq -\frac{7}{4}\right\}$

24. (a) $R(x) = -\frac{1}{10}x^2 + 150x$
(b) \$14,000
(c) 750, \$56,250
(d) \$75

CHAPTER 6 Exponential and Logarithmic Functions

6.1 Assess Your Understanding *(page 408)*

4. composite function; $f(g(x))$ **5.** F **6.** c **7.** a **8.** *F* **9. (a)** -1 **(b)** -1 **(c)** 8 **(d)** 0 **(e)** 8 **(f)** -7 **11. (a)** 4 **(b)** 5 **(c)** -1 **(d)** -2
13. (a) 98 **(b)** 49 **(c)** 4 **(d)** 4 **15. (a)** 97 **(b)** $-\frac{163}{2}$ **(c)** 1 **(d)** $-\frac{3}{2}$ **17. (a)** $2\sqrt{2}$ **(b)** $2\sqrt{2}$ **(c)** 1 **(d)** 0 **19. (a)** $\frac{1}{17}$ **(b)** $\frac{1}{5}$ **(c)** 1 **(d)** $\frac{1}{2}$
21. (a) $\frac{3}{\sqrt[3]{4}+1}$ **(b)** 1 **(c)** $\frac{6}{5}$ **(d)** 0 **23. (a)** $(f \circ g)(x) = 6x + 3$; all real numbers **(b)** $(g \circ f)(x) = 6x + 9$; all real numbers
(c) $(f \circ f)(x) = 4x + 9$; all real numbers **(d)** $(g \circ g)(x) = 9x$; all real numbers **25. (a)** $(f \circ g)(x) = 3x^2 + 1$; all real numbers
(b) $(g \circ f)(x) = 9x^2 + 6x + 1$; all real numbers **(c)** $(f \circ f)(x) = 9x + 4$; all real numbers **(d)** $(g \circ g)(x) = x^4$; all real numbers
27. (a) $(f \circ g)(x) = x^4 + 8x^2 + 16$; all real numbers **(b)** $(g \circ f)(x) = x^4 + 4$; all real numbers **(c)** $(f \circ f)(x) = x^4$; all real numbers
(d) $(g \circ g)(x) = x^4 + 8x^2 + 20$; all real numbers **29. (a)** $(f \circ g)(x) = \frac{3x}{2-x}$; $\{x \mid x \neq 0, x \neq 2\}$ **(b)** $(g \circ f)(x) = \frac{2(x-1)}{3}$; $\{x \mid x \neq 1\}$
(c) $(f \circ f)(x) = \frac{3(x-1)}{4-x}$; $\{x \mid x \neq 1, x \neq 4\}$ **(d)** $(g \circ g)(x) = x$; $\{x \mid x \neq 0\}$ **31. (a)** $(f \circ g)(x) = \frac{4}{4+x}$; $\{x \mid x \neq -4, x \neq 0\}$
(b) $(g \circ f)(x) = \frac{-4(x-1)}{x}$; $\{x \mid x \neq 0, x \neq 1\}$ **(c)** $(f \circ f)(x) = x$; $\{x \mid x \neq 1\}$ **(d)** $(g \circ g)(x) = x$; $\{x \mid x \neq 0\}$
33. (a) $(f \circ g)(x) = \sqrt{2x+3}$; $\left\{x \mid x \geq -\frac{3}{2}\right\}$ **(b)** $(g \circ f)(x) = 2\sqrt{x} + 3$; $\{x \mid x \geq 0\}$ **(c)** $(f \circ f)(x) = \sqrt[4]{x}$; $\{x \mid x \geq 0\}$
(d) $(g \circ g)(x) = 4x + 9$; all real numbers **35. (a)** $(f \circ g)(x) = x$; $\{x \mid x \geq 1\}$ **(b)** $(g \circ f)(x) = |x|$; all real numbers
(c) $(f \circ f)(x) = x^4 + 2x^2 + 2$; all real numbers **(d)** $(g \circ g)(x) = \sqrt{\sqrt{x-1}-1}$; $\{x \mid x \geq 2\}$ **37. (a)** $(f \circ g)(x) = -\frac{4x-17}{2x-1}$; $\left\{x \,\middle|\, x \neq 3; x \neq \frac{1}{2}\right\}$
(b) $(g \circ f)(x) = -\frac{3x-3}{2x+8}$; $\{x \mid x \neq -4; x \neq -1\}$ **(c)** $(f \circ f)(x) = -\frac{2x+5}{x-2}$; $\{x \mid x \neq -1; x \neq 2\}$ **(d)** $(g \circ g)(x) = -\frac{3x-4}{2x-11}$; $\left\{x \,\middle|\, x \neq \frac{11}{2}; x \neq 3\right\}$
39. $(f \circ g)(x) = f(g(x)) = f\left(\frac{1}{2}x\right) = 2\left(\frac{1}{2}x\right) = x$; $(g \circ f)(x) = g(f(x)) = g(2x) = \frac{1}{2}(2x) = x$
41. $(f \circ g)(x) = f(g(x)) = f(\sqrt[3]{x}) = (\sqrt[3]{x})^3 = x$; $(g \circ f)(x) = g(f(x)) = g(x^3) = \sqrt[3]{x^3} = x$
43. $(f \circ g)(x) = f(g(x)) = f\left(\frac{1}{2}(x+6)\right) = 2\left[\frac{1}{2}(x+6)\right] - 6 = x + 6 - 6 = x$; $(g \circ f)(x) = g(f(x)) = g(2x-6) = \frac{1}{2}(2x - 6 + 6) = \frac{1}{2}(2x) = x$
45. $(f \circ g)(x) = f(g(x)) = f\left(\frac{1}{a}(x-b)\right) = a\left[\frac{1}{a}(x-b)\right] + b = x$; $(g \circ f)(x) = g(f(x)) = g(ax+b) = \frac{1}{a}(ax + b - b) = x$
47. $f(x) = x^4$; $g(x) = 2x + 3$ (Other answers are possible.) **49.** $f(x) = \sqrt{x}$; $g(x) = x^2 + 1$ (Other answers are possible.)
51. $f(x) = |x|$; $g(x) = 2x + 1$ (Other answers are possible.) **53.** $(f \circ g)(x) = 11$; $(g \circ f)(x) = 2$ **55.** $-3, 3$ **57. (a)** $(f \circ g)(x) = acx + ad + b$
(b) $(g \circ f)(x) = acx + bc + d$ **(c)** The domains of both $f \circ g$ and $g \circ f$ are all real numbers. **(d)** $f \circ g = g \circ f$ when $ad + b = bc + d$
59. $S(t) = \frac{16}{9}\pi t^6$ **61.** $C(t) = 15{,}000 + 800{,}000t - 40{,}000t^2$ **63.** $C(p) = \frac{2\sqrt{100-p}}{25} + 600, 0 \leq p \leq 100$ **65.** $V(r) = 2\pi r^3$
67. (a) $f(x) = 0.7235x$ **(b)** $g(x) = 141.119x$ **(c)** $g(f(x)) = g(0.7235x) = 102.0995965x$ **(d)** 102,099.5965 yen **69. (a)** $f(p) = p - 200$
(b) $g(p) = 0.8p$ **(c)** $(f \circ g)(p) = 0.8p - 200$; $(g \circ f)(p) = 0.8p - 160$; The 20% discount followed by the \$200 rebate is the better deal. **71.** 15
73. f is an odd function, so $f(-x) = -f(x)$. g is an even function, so $g(-x) = g(x)$. Then $(f \circ g)(-x) = f(g(-x)) = f(g(x)) = (f \circ g)(x)$. So $f \circ g$ is even.
Also, $(g \circ f)(-x) = g(f(-x)) = g(-f(x)) = g(f(x)) = (g \circ f)(x)$, so $g \circ f$ is even.

74. $(f + g)(x) = 4x + 3$; Domain: all real numbers
$(f - g)(x) = 2x + 13$; Domain: all real numbers
$(f \circ g)(x) = 3x^2 - 7x - 40$; Domain: all real numbers
$\left(\frac{f}{g}\right)(x) = \frac{3x+8}{x-5}$; Domain: $\{x \mid x \neq 5\}$

75. $\frac{1}{4}, 4$

76.
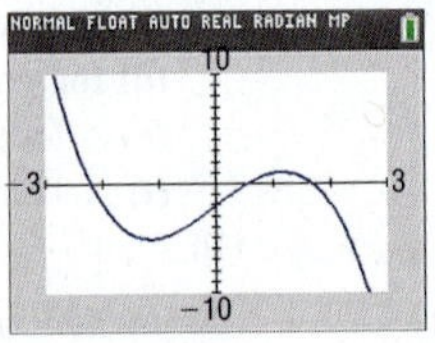

Local minimum: -5.08 at $x = -1.15$
Local maximum: 1.08 at $x = 1.15$
Decreasing: $(-3, -1.15)$; $(1.15, 3)$
Increasing: $(-1.15, 1.15)$

77. Domain: $\{x \mid x \neq 3\}$
Vertical asymptote: $x = 3$
Oblique asymptote: $y = x + 9$

6.2 Assess Your Understanding *(page 419)*

5. $f(x_1) \neq f(x_2)$ **6.** one-to-one **7.** 3 **8.** $y = x$ **9.** $[4, \infty)$ **10.** T **11.** a **12.** d **13.** one-to-one **15.** not one-to-one
17. not one-to-one **19.** one-to-one **21.** one-to-one **23.** not one-to-one **25.** one-to-one

27.

Annual Rainfall (inches)	Location
49.7	Atlanta, Georgia
43.8	Boston, Massachusetts
4.2	Las Vegas, Nevada
61.9	Miami, Florida
12.8	Los Angeles, California

Domain: {49.7, 43.8, 4.2, 61.9, 12.8}
Range: {Atlanta, Boston, Las Vegas, Miami, Los Angeles}

29.

Monthly Cost of Life Insurance	Age
\$10.59	30
\$12.52	40
\$15.94	45

Domain: {\$10.59, \$12.52, \$15.94}
Range: {30, 40, 45}

31. $\{(5,-3),(9,-2),(2,-1),(11,0),(-5,1)\}$
Domain: $\{5, 9, 2, 11, -5\}$
Range: $\{-3, -2, -1, 0, 1\}$

33. $\{(1,-2),(2,-3),(0,-10),(9,1),(4,2)\}$
Domain: $\{1, 2, 0, 9, 4\}$
Range: $\{-2, -3, -10, 1, 2\}$

35. $f(g(x)) = f\left(\frac{1}{3}(x-4)\right) = 3\left[\frac{1}{3}(x-4)\right] + 4 = (x-4) + 4 = x$

$g(f(x)) = g(3x+4) = \frac{1}{3}[(3x+4)-4] = \frac{1}{3}(3x) = x$

37. $f(g(x)) = f\left(\frac{x}{4}+2\right) = 4\left[\frac{x}{4}+2\right] - 8 = (x+8) - 8 = x$

$g(f(x)) = g(4x-8) = \frac{4x-8}{4} + 2 = (x-2) + 2 = x$

39. $f(g(x)) = f(\sqrt[3]{x+8}) = (\sqrt[3]{x+8})^3 - 8 = (x+8) - 8 = x$

$g(f(x)) = g(x^3-8) = \sqrt[3]{(x^3-8)+8} = \sqrt[3]{x^3} = x$

41. $f(g(x)) = f\left(\frac{1}{x}\right) = \frac{1}{\left(\frac{1}{x}\right)} = x; x \neq 0,\ g(f(x)) = g\left(\frac{1}{x}\right) = \frac{1}{\left(\frac{1}{x}\right)} = x,\ x \neq 0$

43. $f(g(x)) = f\left(\frac{4x-3}{2-x}\right) = \dfrac{2\left(\frac{4x-3}{2-x}\right)+3}{\frac{4x-3}{2-x}+4}$

$= \dfrac{2(4x-3)+3(2-x)}{4x-3+4(2-x)} = \dfrac{5x}{5} = x, x \neq 2$

$g(f(x)) = g\left(\frac{2x+3}{x+4}\right) = \dfrac{4\left(\frac{2x+3}{x+4}\right)-3}{2-\frac{2x+3}{x+4}}$

$= \dfrac{4(2x+3)-3(x+4)}{2(x+4)-(2x+3)} = \dfrac{5x}{5} = x, x \neq -4$

45.

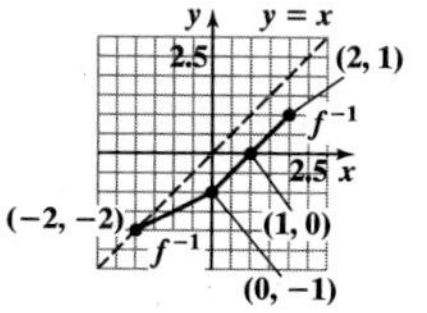

47.

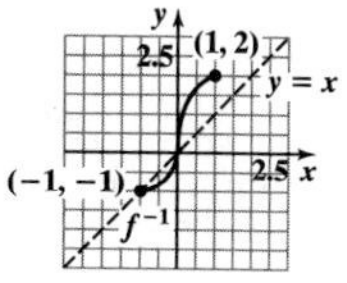

49.

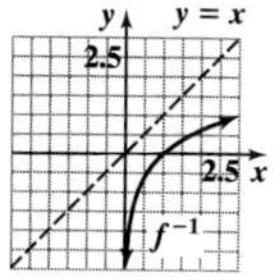

51. (a) $f^{-1}(x) = \frac{1}{3}x$

$f(f^{-1}(x)) = f\left(\frac{1}{3}x\right) = 3\left(\frac{1}{3}x\right) = x$

$f^{-1}(f(x)) = f^{-1}(3x) = \frac{1}{3}(3x) = x$

(b) Domain of f = Range of f^{-1} = All real numbers;
Range of f = Domain of f^{-1} = All real numbers

(c)

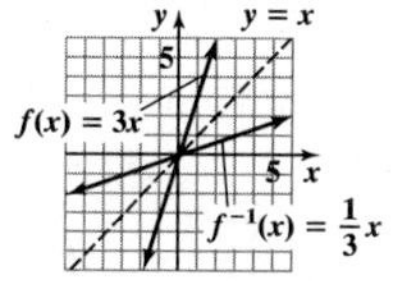

53. (a) $f^{-1}(x) = \frac{x}{4} - \frac{1}{2}$

$f(f^{-1}(x)) = f\left(\frac{x}{4}-\frac{1}{2}\right) = 4\left(\frac{x}{4}-\frac{1}{2}\right) + 2$

$= (x-2) + 2 = x$

$f^{-1}(f(x)) = f^{-1}(4x+2) = \frac{4x+2}{4} - \frac{1}{2}$

$= \left(x+\frac{1}{2}\right) - \frac{1}{2} = x$

(b) Domain of f = Range of f^{-1} = All real numbers;
Range of f = Domain of f^{-1} = All real numbers

(c)

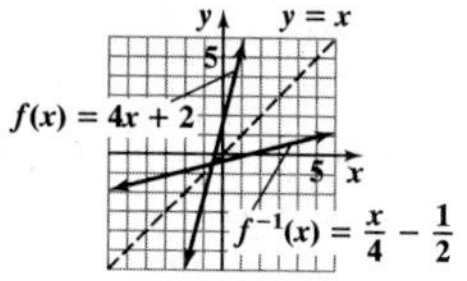

55. (a) $f^{-1}(x) = \sqrt[3]{x+1}$

$f(f^{-1}(x)) = f(\sqrt[3]{x+1})$

$= (\sqrt[3]{x+1})^3 - 1 = x$

$f^{-1}(f(x)) = f^{-1}(x^3-1)$

$= \sqrt[3]{(x^3-1)+1} = x$

(b) Domain of f = Range of f^{-1} = All real numbers;
Range of f = Domain of f^{-1} = All real numbers

(c)

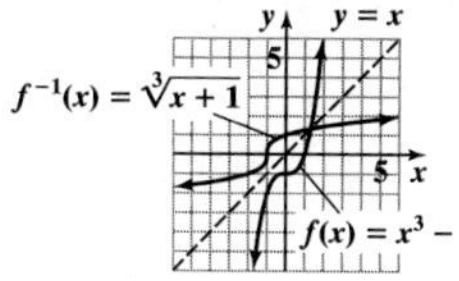

57. (a) $f^{-1}(x) = \sqrt{x-4}, x \geq 4$

$f(f^{-1}(x)) = f(\sqrt{x-4}) = (\sqrt{x-4})^2 + 4 = x$

$f^{-1}(f(x)) = f^{-1}(x^2+4) = \sqrt{(x^2+4)-4} = \sqrt{x^2} = x, x \geq 0$

(b) Domain of f = Range of f^{-1} = $\{x \mid x \geq 0\}$;
Range of f = Domain of f^{-1} = $\{x \mid x \geq 4\}$

(c)

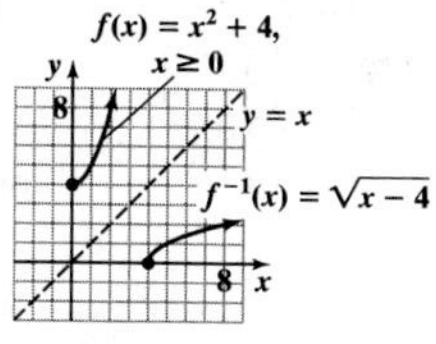

59. (a) $f^{-1}(x) = \dfrac{4}{x}$

$$f(f^{-1}(x)) = f\left(\frac{4}{x}\right) = \frac{4}{\left(\frac{4}{x}\right)} = x$$

$$f^{-1}(f(x)) = f^{-1}\left(\frac{4}{x}\right) = \frac{4}{\left(\frac{4}{x}\right)} = x$$

(b) Domain of f = Range of $f^{-1} = \{x \mid x \neq 0\}$; Range of f = Domain of $f^{-1} = \{x \mid x \neq 0\}$

(c)

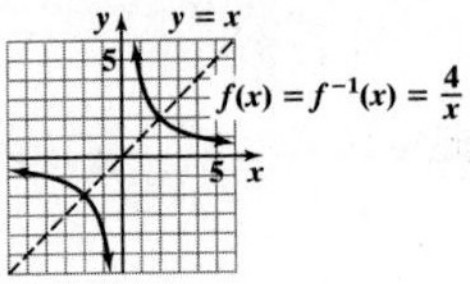

61. (a) $f^{-1}(x) = \dfrac{2x + 1}{x}$

$$f(f^{-1}(x)) = f\left(\frac{2x+1}{x}\right) = \frac{1}{\frac{2x+1}{x} - 2} = \frac{x}{(2x+1) - 2x} = x$$

$$f^{-1}(f(x)) = f^{-1}\left(\frac{1}{x-2}\right) = \frac{2\left(\frac{1}{x-2}\right) + 1}{\frac{1}{x-2}} = \frac{2 + (x-2)}{1} = x$$

(b) Domain of f = Range of $f^{-1} = \{x \mid x \neq 2\}$; Range of f = Domain of $f^{-1} = \{x \mid x \neq 0\}$

(c)

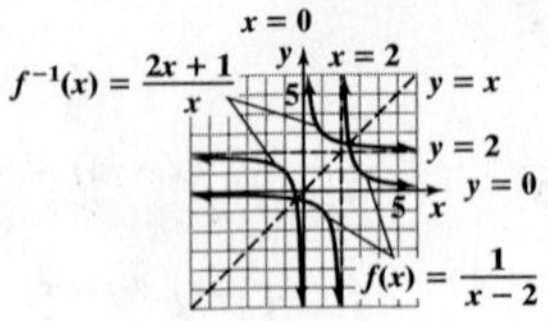

63. (a) $f^{-1}(x) = \dfrac{2 - 3x}{x}$

$$f(f^{-1}(x)) = f\left(\frac{2-3x}{x}\right) = \frac{2}{3 + \frac{2-3x}{x}} = \frac{2x}{3x + 2 - 3x} = \frac{2x}{2} = x$$

$$f^{-1}(f(x)) = f^{-1}\left(\frac{2}{3+x}\right) = \frac{2 - 3\left(\frac{2}{3+x}\right)}{\frac{2}{3+x}} = \frac{2(3+x) - 3\cdot 2}{2} = \frac{2x}{2} = x$$

(b) Domain of f = Range of $f^{-1} = \{x \mid x \neq -3\}$; Range of f = Domain of $f^{-1} = \{x \mid x \neq 0\}$

65. (a) $f^{-1}(x) = \dfrac{-2x}{x - 3}$

$$f(f^{-1}(x)) = f\left(\frac{-2x}{x-3}\right) = \frac{3\left(\frac{-2x}{x-3}\right)}{\frac{-2x}{x-3} + 2} = \frac{3(-2x)}{-2x + 2(x-3)} = \frac{-6x}{-6} = x$$

$$f^{-1}(f(x)) = f^{-1}\left(\frac{3x}{x+2}\right) = \frac{-2\left(\frac{3x}{x+2}\right)}{\frac{3x}{x+2} - 3} = \frac{-2(3x)}{3x - 3(x+2)} = \frac{-6x}{-6} = x$$

(b) Domain of f = Range of $f^{-1} = \{x \mid x \neq -2\}$; Range of f = Domain of $f^{-1} = \{x \mid x \neq 3\}$

67. (a) $f^{-1}(x) = \dfrac{x}{3x - 2}$

$$f(f^{-1}(x)) = f\left(\frac{x}{3x-2}\right) = \frac{2\left(\frac{x}{3x-2}\right)}{3\left(\frac{x}{3x-2}\right) - 1} = \frac{2x}{3x - (3x-2)} = \frac{2x}{2} = x$$

$$f^{-1}(f(x)) = f^{-1}\left(\frac{2x}{3x-1}\right) = \frac{\frac{2x}{3x-1}}{3\left(\frac{2x}{3x-1}\right) - 2} = \frac{2x}{6x - 2(3x-1)} = \frac{2x}{2} = x$$

(b) Domain of f = Range of $f^{-1} = \left\{x \,\middle|\, x \neq \frac{1}{3}\right\}$; Range of f = Domain of $f^{-1} = \left\{x \,\middle|\, x \neq \frac{2}{3}\right\}$

69. (a) $f^{-1}(x) = \dfrac{3x + 4}{2x - 3}$

$$f(f^{-1}(x)) = f\left(\frac{3x+4}{2x-3}\right) = \frac{3\left(\frac{3x+4}{2x-3}\right) + 4}{2\left(\frac{3x+4}{2x-3}\right) - 3} = \frac{3(3x+4) + 4(2x-3)}{2(3x+4) - 3(2x-3)} = \frac{17x}{17} = x$$

$$f^{-1}(f(x)) = f^{-1}\left(\frac{3x+4}{2x-3}\right) = \frac{3\left(\frac{3x+4}{2x-3}\right) + 4}{2\left(\frac{3x+4}{2x-3}\right) - 3} = \frac{3(3x+4) + 4(2x-3)}{2(3x+4) - 3(2x-3)} = \frac{17x}{17} = x$$

(b) Domain of f = Range of $f^{-1} = \left\{x \,\middle|\, x \neq \frac{3}{2}\right\}$; Range of f = Domain of $f^{-1} = \left\{x \,\middle|\, x \neq \frac{3}{2}\right\}$

71. **(a)** $f^{-1}(x) = \dfrac{-2x + 3}{x - 2}$

$$f(f^{-1}(x)) = f\left(\frac{-2x+3}{x-2}\right) = \frac{2\left(\frac{-2x+3}{x-2}\right) + 3}{\frac{-2x+3}{x-2} + 2} = \frac{2(-2x+3) + 3(x-2)}{-2x+3+2(x-2)} = \frac{-x}{-1} = x$$

$$f^{-1}(f(x)) = f^{-1}\left(\frac{2x+3}{x+2}\right) = \frac{-2\left(\frac{2x+3}{x+2}\right) + 3}{\frac{2x+3}{x+2} - 2} = \frac{-2(2x+3) + 3(x+2)}{2x+3-2(x+2)} = \frac{-x}{-1} = x$$

(b) Domain of f = Range of $f^{-1} = \{x \mid x \neq -2\}$; Range of f = Domain of $f^{-1} = \{x \mid x \neq 2\}$

73. **(a)** $f^{-1}(x) = \dfrac{2}{\sqrt{1 - 2x}}$

$$f(f^{-1}(x)) = f\left(\frac{2}{\sqrt{1-2x}}\right) = \frac{\frac{4}{1-2x} - 4}{2 \cdot \frac{4}{1-2x}} = \frac{4 - 4(1-2x)}{2 \cdot 4} = \frac{8x}{8} = x$$

$$f^{-1}(f(x)) = f^{-1}\left(\frac{x^2-4}{2x^2}\right) = \frac{2}{\sqrt{1 - 2\left(\frac{x^2-4}{2x^2}\right)}} = \frac{2}{\sqrt{\frac{4}{x^2}}} = \sqrt{x^2} = x, \text{ since } x > 0$$

(b) Domain of f = Range of $f^{-1} = \{x \mid x > 0\}$; Range of f = Domain of $f^{-1} = \left\{x \mid x < \frac{1}{2}\right\}$

75. **(a)** 0 **(b)** 2 **(c)** 0 **(d)** 1 **77.** 7 **79.** Domain of f^{-1}: $[-2, \infty)$; range of f^{-1}: $[5, \infty)$ **81.** Domain of g^{-1}: $[0, \infty)$; range of g^{-1}: $(-\infty, 0]$
83. Increasing on the interval $(f(0), f(5))$ **85.** $f^{-1}(x) = \frac{1}{m}(x - b), m \neq 0$ **87.** Quadrant I
89. Possible answer: $f(x) = |x|, x \geq 0$, is one-to-one; $f^{-1}(x) = x, x \geq 0$

91. **(a)** $r(d) = \dfrac{d + 90.39}{6.97}$

(b) $r(d(r)) = \dfrac{6.97r - 90.39 + 90.39}{6.97} = \dfrac{6.97r}{6.97} = r$

$d(r(d)) = 6.97\left(\dfrac{d + 90.39}{6.97}\right) - 90.39 = d + 90.39 - 90.39 = d$

(c) 56 miles per hour

93. **(a)** 77.6 kg

(b) $h(W) = \dfrac{W - 50}{2.3} + 60 = \dfrac{W + 88}{2.3}$

(c) $h(W(h)) = \dfrac{50 + 2.3(h - 60) + 88}{2.3} = \dfrac{2.3h}{2.3} = h$

$W(h(W)) = 50 + 2.3\left(\dfrac{W + 88}{2.3} - 60\right)$

$= 50 + W + 88 - 138 = W$

(d) 73 inches

95. **(a)** $\{g \mid 36{,}900 \leq g \leq 89{,}350\}$

(b) $\{T \mid 5081.25 \leq T \leq 18{,}193.75\}$

(c) $g(T) = \dfrac{T - 5081.25}{0.25} + 36{,}900$

Domain: $\{T \mid 5081.25 \leq T \leq 18{,}193.75\}$
Range: $\{g \mid 36{,}900 \leq g \leq 89{,}350\}$

97. **(a)** t represents time, so $t \geq 0$.

(b) $t(H) = \sqrt{\dfrac{H - 100}{-4.9}} = \sqrt{\dfrac{100 - H}{4.9}}$

(c) 2.02 seconds

99. $f^{-1}(x) = \dfrac{-dx + b}{cx - a}$; $f = f^{-1}$ if $a = -d$ **103.** No

107. $6xh + 3h^2 - 7h$ **108.**

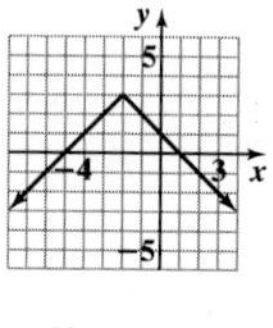

109. Zeros: $\dfrac{-5 - \sqrt{13}}{6}, \dfrac{-5 + \sqrt{13}}{6}$, x-intercepts: $\dfrac{-5 - \sqrt{13}}{6}, \dfrac{-5 + \sqrt{13}}{6}$

110. Domain: $\left\{x \,\middle|\, x \neq -\frac{3}{2}, x \neq 2\right\}$; Vertical asymptote: $x = -\frac{3}{2}$,
Horizontal asymptote: $y = 3$

6.3 Assess Your Understanding *(page 434)*

6. Exponential function; growth factor; initial value **7.** a **8.** T **9.** T **10.** $\left(-1, \frac{1}{a}\right)$; $(0, 1)$; $(1, a)$ **11.** 4 **12.** F **13.** b **14.** c
15. **(a)** 8.815 **(b)** 8.821 **(c)** 8.824 **(d)** 8.825 **17.** **(a)** 21.217 **(b)** 22.217 **(c)** 22.440 **(d)** 22.459 **19.** 1.265 **21.** 0.347 **23.** 3.320 **25.** 149.952
27. Neither **29.** Exponential; $H(x) = 4^x$ **31.** Exponential; $f(x) = 3(2^x)$ **33.** Linear; $H(x) = 2x + 4$ **35.** B **37.** D **39.** A **41.** E

43.

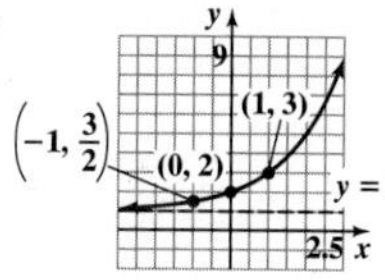

Domain: All real numbers
Range: $\{y \mid y > 1\}$ or $(1, \infty)$
Horizontal asymptote: $y = 1$

45.

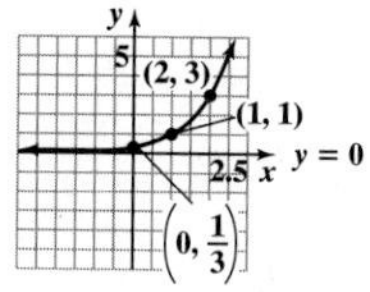

Domain: All real numbers
Range: $\{y \mid y > 0\}$ or $(0, \infty)$
Horizontal asymptote: $y = 0$

47.

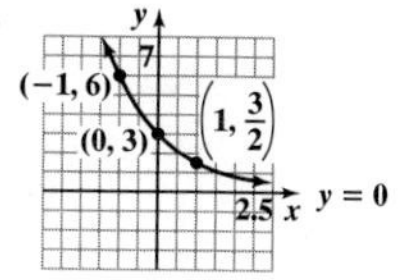

Domain: All real numbers
Range: $\{y \mid y > 0\}$ or $(0, \infty)$
Horizontal asymptote: $y = 0$

49.

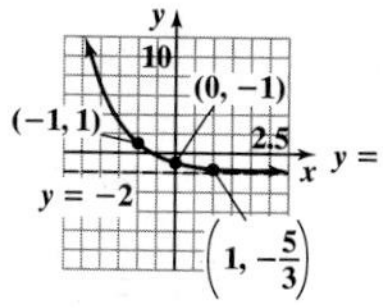

Domain: All real numbers
Range: $\{y \mid y > -2\}$ or $(-2, \infty)$
Horizontal asymptote: $y = -2$

51.

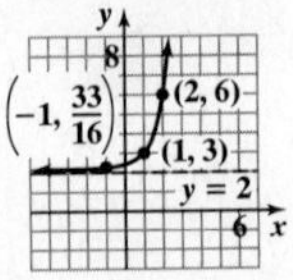

Domain: All real numbers
Range: $\{y \mid y > 2\}$ or $(2, \infty)$
Horizontal asymptote: $y = 2$

53.

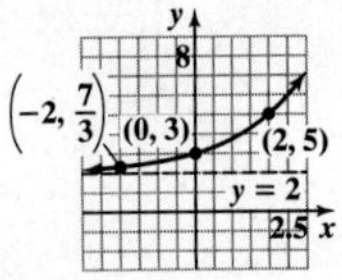

Domain: All real numbers
Range: $\{y \mid y > 2\}$ or $(2, \infty)$
Horizontal asymptote: $y = 2$

55.

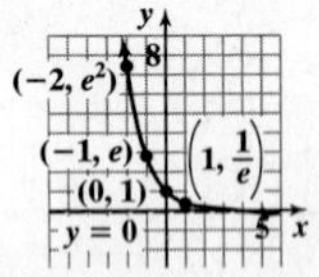

Domain: All real numbers
Range: $\{y \mid y > 0\}$ or $(0, \infty)$
Horizontal asymptote: $y = 0$

57.

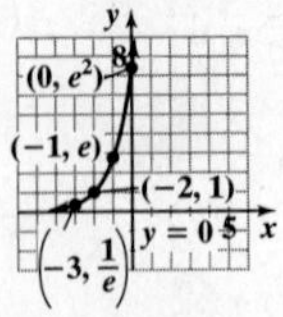

Domain: All real numbers
Range: $\{y \mid y > 0\}$ or $(0, \infty)$
Horizontal asymptote: $y = 0$

59.

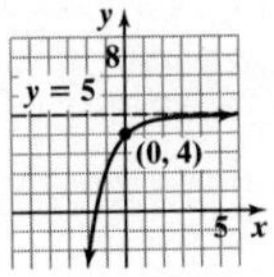

Domain: All real numbers
Range: $\{y \mid y < 5\}$ or $(-\infty, 5)$
Horizontal asymptote: $y = 5$

61.

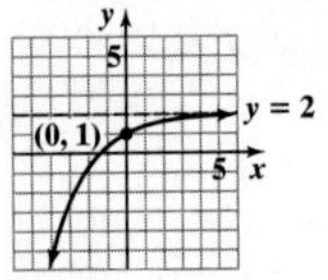

Domain: All real numbers
Range: $\{y \mid y < 2\}$ or $(-\infty, 2)$
Horizontal asymptote: $y = 2$

63. $\{3\}$ **65.** $\{-4\}$ **67.** $\{2\}$ **69.** $\left\{\frac{3}{2}\right\}$ **71.** $\{-\sqrt{2}, 0, \sqrt{2}\}$

73. $\{6\}$ **75.** $\{-1, 7\}$ **77.** $\{-4, 2\}$ **79.** $\{-4\}$ **81.** $\{1, 2\}$ **83.** $\frac{1}{49}$

85. $\frac{1}{4}$ **87.** 5 **89.** $f(x) = 3^x$ **91.** $f(x) = -6^x$ **93.** $f(x) = 3^x + 2$

95. (a) 16; (4, 16) **(b)** −4; $\left(-4, \frac{1}{16}\right)$ **97. (a)** $\frac{9}{4}$; $\left(-1, \frac{9}{4}\right)$ **(b)** 3; (3, 66)

99. (a) 60; (−6, 60) **(b)** −4; (−4, 12) **(c)** −2

101.

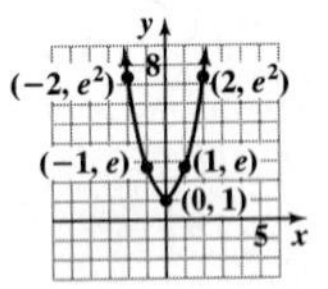

Domain: $(-\infty, \infty)$
Range: $[1, \infty)$
Intercept: (0, 1)

103.

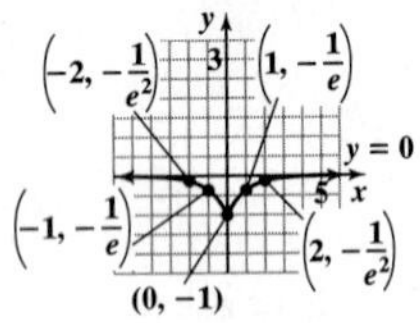

Domain: $(-\infty, \infty)$
Range: $[-1, 0)$
Intercept: $(0, -1)$

105. (a) 74% **(b)** 47% **(c)** Each pane allows only 97% of light to pass through. **107. (a)** \$16,231 **(b)** \$8626 **(c)** As each year passes, the sedan is worth 90% of its value the previous year. **109. (a)** 30% **(b)** 9% **(c)** Each year only 30% of the previous survivors survive again. **111.** 3.35 mg; 0.45 mg **113. (a)** 0.632 **(b)** 0.982 **(c)** 1 **(d)**

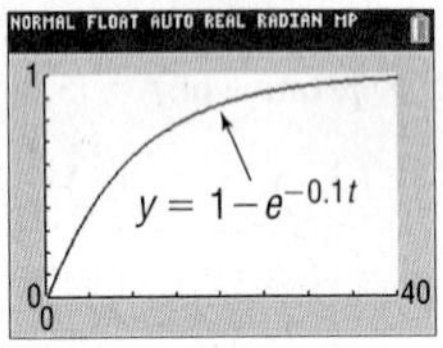

(e) About 7 min

115. (a) 0.0516 **(b)** 0.0888 **117. (a)** 70.95% **(b)** 72.62% **(c)** 100%
119. (a) 5.41 amp, 7.59 amp, 10.38 amp **(b)** 12 amp
(d) 3.34 amp, 5.31 amp, 9.44 amp
(e) 24 amp
(c), (f)

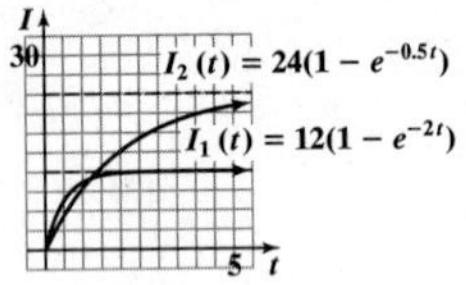

121. 36
123.

Final Denominator	Value of Expression	Compare Value to $e \approx 2.718281828$
1 + 1	2.5	$2.5 < e$
2 + 2	2.8	$2.8 > e$
3 + 3	2.7	$2.7 < e$
4 + 4	2.721649485	$2.721649485 > e$
5 + 5	2.717770035	$2.717770035 < e$
6 + 6	2.718348855	$2.718348855 > e$

125. $f(A + B) = a^{A+B} = a^A \cdot a^B = f(A) \cdot f(B)$ **127.** $f(\alpha x) = a^{\alpha x} = (a^x)^\alpha = [f(x)]^\alpha$

129. (a) $f(-x) = \frac{1}{2}(e^{-x} + e^{-(-x)}) = \frac{1}{2}(e^{-x} + e^x) = \frac{1}{2}(e^x + e^{-x}) = f(x)$

(b)

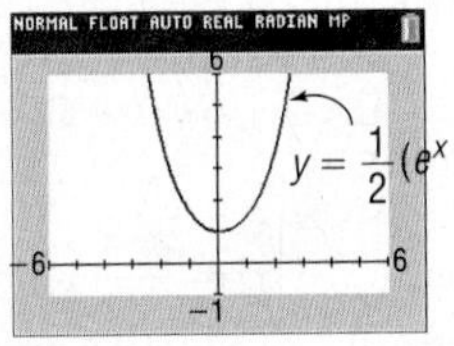

(c) $(\cosh x)^2 - (\sinh x)^2 = \left[\frac{1}{2}(e^x + e^{-x})\right]^2 - \left[\frac{1}{2}(e^x - e^{-x})\right]^2 = \frac{1}{4}[e^{2x} + 2 + e^{-2x} - e^{2x} + 2 - e^{-2x}] = \frac{1}{4}(4) = 1$

131. 59 minutes **135.** $a^{-x} = (a^{-1})^x = \left(\frac{1}{a}\right)^x$ **136.** $(-\infty, -5] \cup [-2, 2]$ **137.** $(2, \infty)$ **138.** $f(x) = -2x^2 + 12x - 13$

139. (a)

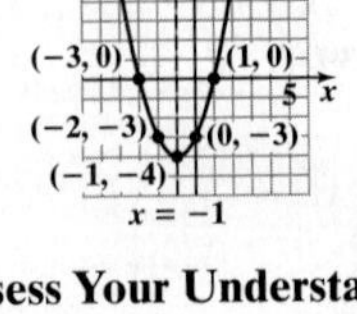

(b) Domain: $(-\infty, \infty)$; Range: $[-4, \infty)$
(c) Decreasing: $(-\infty, -1)$; Increasing: $(-1, \infty)$

6.4 Assess Your Understanding *(page 448)*

4. $\{x \mid x > 0\}$ or $(0, \infty)$ **5.** $\left(\frac{1}{a}, -1\right)$, $(1, 0)$, $(a, 1)$ **6.** 1 **7.** F **8.** T **9.** a **10.** c **11.** $2 = \log_3 9$ **13.** $2 = \log_a 1.6$ **15.** $x = \log_2 7.2$ **17.** $x = \ln 8$

19. $2^3 = 8$ **21.** $a^6 = 3$ **23.** $3^x = 2$ **25.** $e^x = 4$ **27.** 0 **29.** 2 **31.** −4 **33.** $\frac{1}{2}$ **35.** 4 **37.** $\frac{1}{2}$ **39.** $\{x \mid x > 3\}$; $(3, \infty)$

41. All real numbers except 0; $\{x \mid x \neq 0\}$; $(-\infty, 0) \cup (0, \infty)$ **43.** $\{x \mid x > 10\}$; $(10, \infty)$ **45.** $\{x \mid x > -1\}$; $(-1, \infty)$
47. $\{x \mid x < -1 \text{ or } x > 0\}$; $(-\infty, -1) \cup (0, \infty)$ **49.** $\{x \mid x \geq 1\}$; $[1, \infty)$ **51.** 0.511 **53.** 30.099 **55.** 2.303 **57.** −53.991 **59.** $\sqrt{2}$

61.

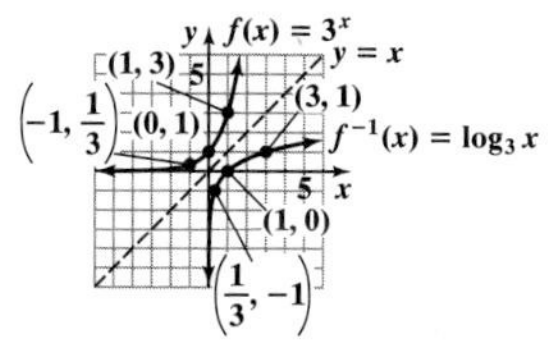

63.

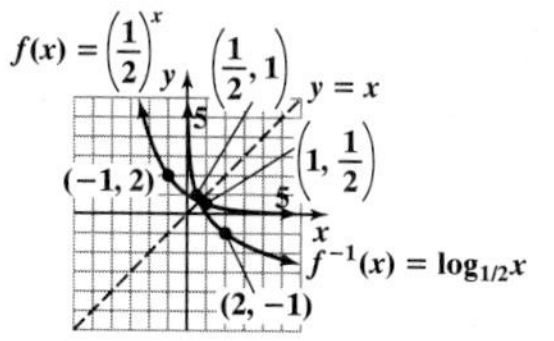

65. B **67.** D **69.** A **71.** E

73. (a) Domain: $(-4, \infty)$

(b)

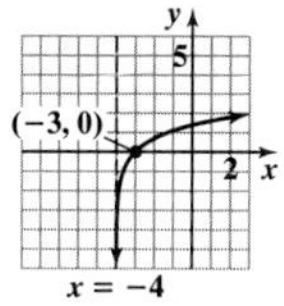

(c) Range: $(-\infty, \infty)$
Vertical asymptote: $x = -4$

(d) $f^{-1}(x) = e^x - 4$

(e) Domain of f^{-1}: $(-\infty, \infty)$
Range of f^{-1}: $(-4, \infty)$

(f)

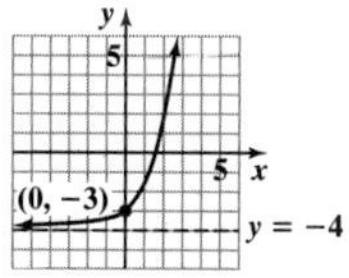

75. (a) Domain: $(0, \infty)$

(b)

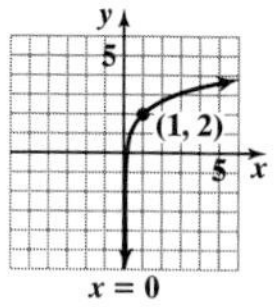

(c) Range: $(-\infty, \infty)$
Vertical asymptote: $x = 0$

(d) $f^{-1}(x) = e^{x-2}$

(e) Domain of f^{-1}: $(-\infty, \infty)$
Range of f^{-1}: $(0, \infty)$

(f)

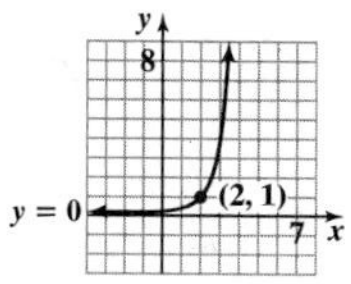

77. (a) Domain: $(0, \infty)$

(b)

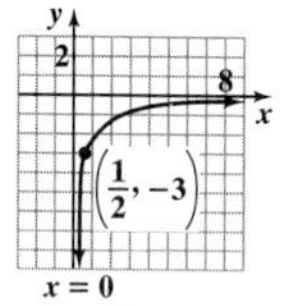

(c) Range: $(-\infty, \infty)$
Vertical asymptote: $x = 0$

(d) $f^{-1}(x) = \dfrac{1}{2}e^{x+3}$

(e) Domain of f^{-1}: $(-\infty, \infty)$
Range of f^{-1}: $(0, \infty)$

(f)

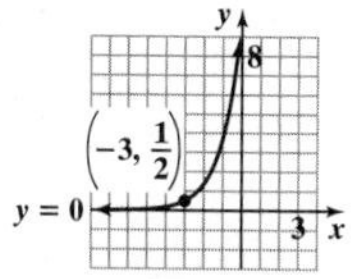

79. (a) Domain: $(4, \infty)$

(b)

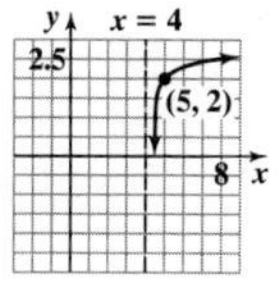

(c) Range: $(-\infty, \infty)$
Vertical asymptote: $x = 4$

(d) $f^{-1}(x) = 10^{x-2} + 4$

(e) Domain of f^{-1}: $(-\infty, \infty)$
Range of f^{-1}: $(4, \infty)$

(f)

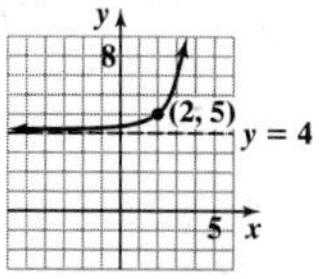

81. (a) Domain: $(0, \infty)$

(b)

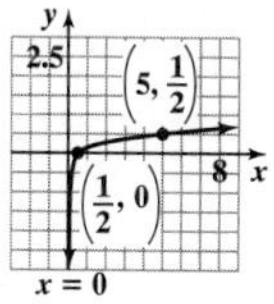

(c) Range: $(-\infty, \infty)$
Vertical asymptote: $x = 0$

(d) $f^{-1}(x) = \dfrac{1}{2} \cdot 10^{2x}$

(e) Domain of f^{-1}: $(-\infty, \infty)$
Range of f^{-1}: $(0, \infty)$

(f)

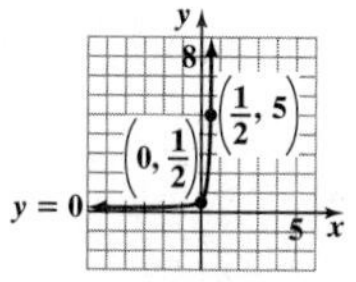

83. (a) Domain: $(-2, \infty)$

(b)

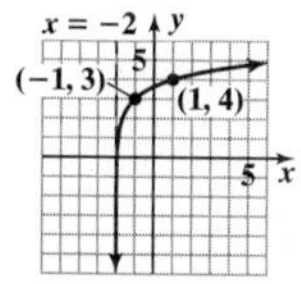

(c) Range: $(-\infty, \infty)$
Vertical asymptote: $x = -2$

(d) $f^{-1}(x) = 3^{x-3} - 2$

(e) Domain of f^{-1}: $(-\infty, \infty)$
Range of f^{-1}: $(-2, \infty)$

(f)

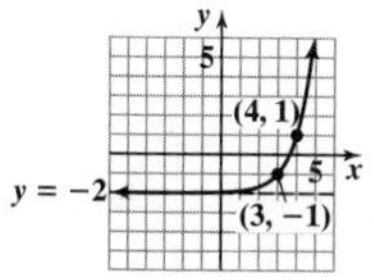

85. (a) Domain: $(-\infty, \infty)$

(b)

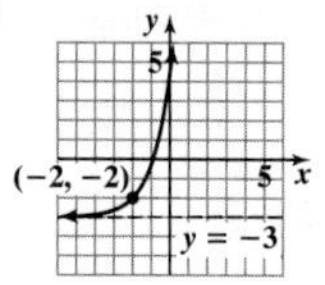

(c) Range: $(-3, \infty)$
Horizontal asymptote: $y = -3$

(d) $f^{-1}(x) = \ln(x + 3) - 2$

(e) Domain of f^{-1}: $(-3, \infty)$
Range of f^{-1}: $(-\infty, \infty)$

(f)

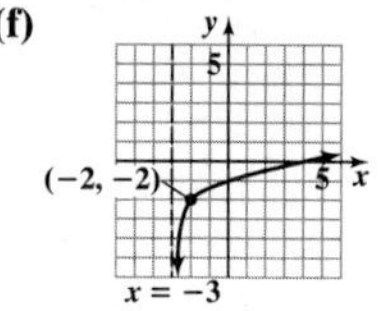

87. (a) Domain: $(-\infty, \infty)$

(b)

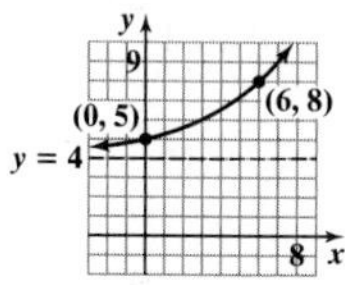

(c) Range: $(4, \infty)$
Horizontal asymptote: $y = 4$

(d) $f^{-1}(x) = 3\log_2(x - 4)$

(e) Domain of f^{-1}: $(4, \infty)$
Range of f^{-1}: $(-\infty, \infty)$

(f)

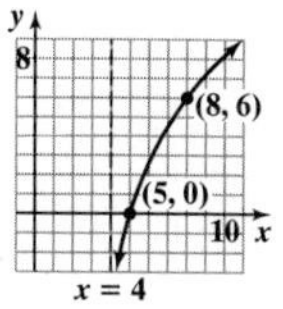

89. $\{9\}$ **91.** $\left\{\dfrac{7}{2}\right\}$ **93.** $\{2\}$ **95.** $\{5\}$ **97.** $\{3\}$

99. $\{2\}$ **101.** $\left\{\dfrac{\ln 10}{3}\right\}$ **103.** $\left\{\dfrac{\ln 8 - 5}{2}\right\}$

105. $\{-2\sqrt{2}, 2\sqrt{2}\}$ **107.** $\{-1\}$

109. $\left\{5 \ln \dfrac{7}{5}\right\}$ **111.** $\left\{2 - \log \dfrac{5}{2}\right\}$

113. (a) $\left\{x \,\middle|\, x > -\dfrac{1}{2}\right\}$; $\left(-\dfrac{1}{2}, \infty\right)$
(b) 2; (40, 2) **(c)** 121; (121, 3) **(d)** 4

115.

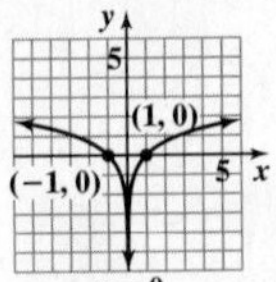

Domain: $\{x \mid x \neq 0\}$
Range: $(-\infty, \infty)$
Intercepts: $(-1, 0)$, $(1, 0)$

117.

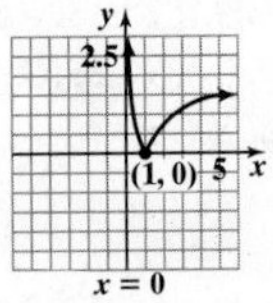

Domain: $\{x \mid x > 0\}$
Range: $\{y \mid y \geq 0\}$
Intercept: $(1, 0)$

119. **(a)** 1 **(b)** 2 **(c)** 3 **(d)** It increases. **(e)** 0.000316 **(f)** 3.981×10^{-8}
121. **(a)** 5.97 km **(b)** 0.90 km
123. **(a)** 6.93 min **(b)** 16.09 min
125. $h \approx 2.29$, so the time between injections is about 2 h, 17 min.

127. 0.2695 s 0.8959 s

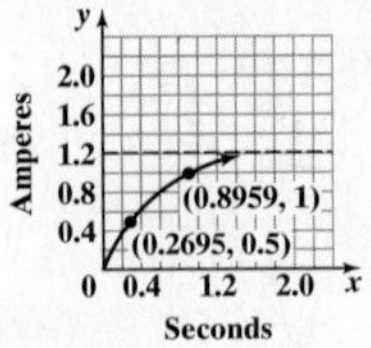

129. 50 decibels (dB) **131.** 90 dB **133.** 8.1 **135.** **(a)** $k \approx 11.216$ **(b)** 6.73 **(c)** 0.41% **(d)** 0.14%

137. Because $y = \log_1 x$ means $1^y = 1 = x$, which cannot be true for $x \neq 1$ **139.** Zeros: $-3, -\frac{1}{2}, \frac{1}{2}, 3$; x-intercepts: $-3, -\frac{1}{2}, \frac{1}{2}, 3$

140. 12 **141.** $f(1) = -5$; $f(2) = 17$ **142.** $3 + i$; $f(x) = x^4 - 7x^3 + 14x^2 + 2x - 20$; $a = 1$

6.5 Assess Your Understanding *(page 459)*

1. 0 **2.** M **3.** r **4.** $\log_a M$; $\log_a N$ **5.** $\log_a M$; $\log_a N$ **6.** $r \log_a M$ **7.** 7 **8.** F **9.** F **10.** F **11.** b **12.** b **13.** 71 **15.** −4 **17.** 7 **19.** 1 **21.** 1

23. 3 **25.** $\frac{5}{4}$ **27.** 4 **29.** $a + b$ **31.** $b - a$ **33.** $3a$ **35.** $\frac{1}{5}(a + b)$ **37.** $2 + \log_5 x$ **39.** $3 \log_2 z$ **41.** $1 + \ln x$ **43.** $\ln x - x$ **45.** $2 \log_a u + 3 \log_a v$

47. $2 \ln x + \frac{1}{2}\ln(1 - x)$ **49.** $3 \log_2 x - \log_2(x - 3)$ **51.** $\log x + \log(x + 2) - 2 \log(x + 3)$ **53.** $\frac{1}{3}\ln(x - 2) + \frac{1}{3}\ln(x + 1) - \frac{2}{3}\ln(x + 4)$

55. $\ln 5 + \ln x + \frac{1}{2}\ln(1 + 3x) - 3 \ln(x - 4)$ **57.** $\log_5 u^3 v^4$ **59.** $\log_3\left(\frac{1}{x^{5/2}}\right)$ **61.** $\log_4\left[\frac{x - 1}{(x + 1)^4}\right]$ **63.** $-2 \ln(x - 1)$ **65.** $\log_2[x(3x - 2)^4]$

67. $\log_a\left(\frac{25x^6}{\sqrt{2x + 3}}\right)$ **69.** $\log_2\left[\frac{(x + 1)^2}{(x + 3)(x - 1)}\right]$ **71.** 2.771 **73.** −3.880 **75.** 5.615 **77.** 0.874

79. $y = \frac{\log x}{\log 4}$

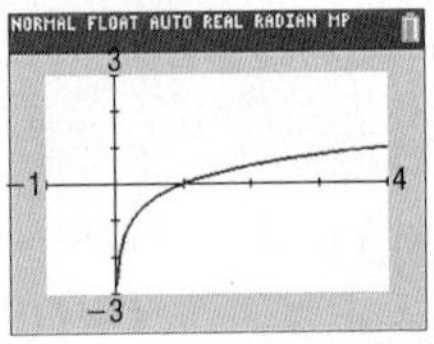

81. $y = \frac{\log(x + 2)}{\log 2}$

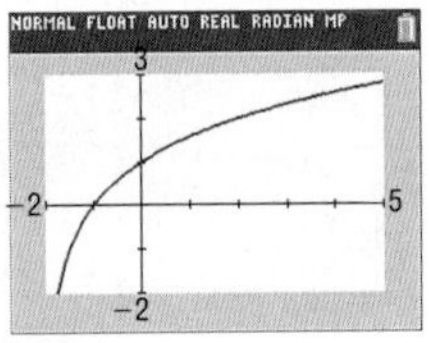

83. $y = \frac{\log(x + 1)}{\log(x - 1)}$

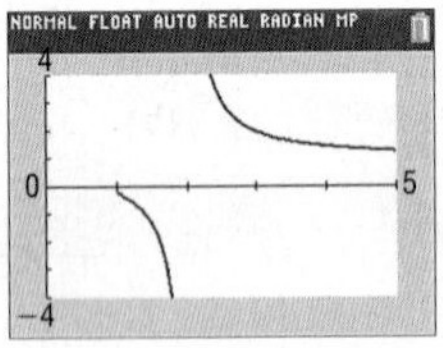

85. **(a)** $(f \circ g)(x) = x$; $\{x \mid x$ is any real number$\}$ or $(-\infty, \infty)$
(b) $(g \circ f)(x) = x$; $\{x \mid x > 0\}$ or $(0, \infty)$ **(c)** 5
(d) $(f \circ h)(x) = \ln x^2$; $\{x \mid x \neq 0\}$ or $(-\infty, 0) \cup (0, \infty)$ **(e)** 2

87. $y = Cx$ **89.** $y = Cx(x + 1)$ **91.** $y = Ce^{3x}$ **93.** $y = Ce^{-4x} + 3$

95. $y = \frac{\sqrt[3]{C}(2x + 1)^{1/6}}{(x + 4)^{1/9}}$ **97.** 3 **99.** 1

101. $\log_a(x + \sqrt{x^2 - 1}) + \log_a(x - \sqrt{x^2 - 1}) = \log_a[(x + \sqrt{x^2 - 1})(x - \sqrt{x^2 - 1})] = \log_a[x^2 - (x^2 - 1)] = \log_a 1 = 0$

103. $\ln(1 + e^{2x}) = \ln[e^{2x}(e^{-2x} + 1)] = \ln e^{2x} + \ln(e^{-2x} + 1) = 2x + \ln(1 + e^{-2x})$

105. $y = f(x) = \log_a x$; $a^y = x$ implies $a^y = \left(\frac{1}{a}\right)^{-y} = x$, so $-y = \log_{1/a} x = -f(x)$.

107. $f(x) = \log_a x$; $f\left(\frac{1}{x}\right) = \log_a \frac{1}{x} = \log_a 1 - \log_a x = -f(x)$

109. $\log_a \frac{M}{N} = \log_a(M \cdot N^{-1}) = \log_a M + \log_a N^{-1} = \log_a M - \log_a N$, since $a^{\log_a N^{-1}} = N^{-1}$ implies $a^{-\log_a N^{-1}} = N$; that is, $\log_a N = -\log_a N^{-1}$.

115. $\{-1.78, 1.29, 3.49\}$ **116.** A repeated real solution (double root) **117.** $-2, \frac{1}{5}, \frac{-5 - \sqrt{21}}{2}, \frac{-5 + \sqrt{21}}{2}$

118.

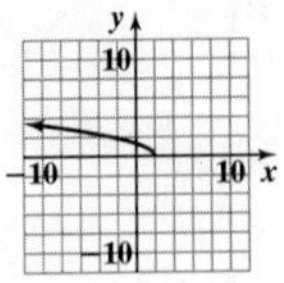

Domain: $\{x \mid x \leq 2\}$ or $(-\infty, 2]$
Range: $\{y \mid y \geq 0\}$ or $[0, \infty)$

6.6 Assess Your Understanding *(page 465)*

5. $\{16\}$ **7.** $\left\{\frac{16}{5}\right\}$ **9.** $\{6\}$ **11.** $\{16\}$ **13.** $\left\{\frac{1}{3}\right\}$ **15.** $\{3\}$ **17.** $\{5\}$ **19.** $\left\{\frac{21}{8}\right\}$ **21.** $\{-6\}$ **23.** $\{-2\}$ **25.** $\{-1 + \sqrt{1 + e^4}\} \approx \{6.456\}$

27. $\left\{\frac{-5 + 3\sqrt{5}}{2}\right\} \approx \{0.854\}$ **29.** $\{2\}$ **31.** $\left\{\frac{9}{2}\right\}$ **33.** $\{7\}$ **35.** $\{-2 + 4\sqrt{2}\}$ **37.** $\{-\sqrt{3}, \sqrt{3}\}$ **39.** $\left\{\frac{1}{3}, 729\right\}$ **41.** $\{8\}$

43. $\{\log_2 10\} = \left\{\dfrac{\ln 10}{\ln 2}\right\} \approx \{3.322\}$ **45.** $\{-\log_8 1.2\} = \left\{-\dfrac{\ln 1.2}{\ln 8}\right\} \approx \{-0.088\}$ **47.** $\left\{\dfrac{1}{3}\log_2\dfrac{8}{5}\right\} = \left\{\dfrac{\ln\frac{8}{5}}{3\ln 2}\right\} \approx \{0.226\}$

49. $\left\{\dfrac{\ln 3}{2\ln 3 + \ln 4}\right\} \approx \{0.307\}$ **51.** $\left\{\dfrac{\ln 7}{\ln 0.6 + \ln 7}\right\} \approx \{1.356\}$ **53.** $\{0\}$ **55.** $\left\{\dfrac{\ln \pi}{1 + \ln \pi}\right\} \approx \{0.534\}$ **57.** $\left\{\dfrac{\ln 3}{\ln 2}\right\} \approx \{1.585\}$

59. $\{0\}$ **61.** $\left\{\log_4(-2 + \sqrt{7})\right\} \approx \{-0.315\}$ **63.** $\{\log_5 4\} \approx \{0.861\}$ **65.** No real solution **67.** $\{\log_4 5\} \approx \{1.161\}$ **69.** $\{2.79\}$

71. $\{-0.57\}$ **73.** $\{-0.70\}$ **75.** $\{0.57\}$ **77.** $\{0.39, 1.00\}$ **79.** $\{1.32\}$ **81.** $\{1.31\}$ **83.** $\{1\}$ **85.** $\{16\}$ **87.** $\left\{-1, \dfrac{2}{3}\right\}$ **89.** $\{0\}$

91. $\{\ln(2 + \sqrt{5})\} \approx \{1.444\}$ **93.** $\left\{e^{\frac{\ln 5 \cdot \ln 3}{\ln 15}}\right\} \approx \{1.921\}$ **95. (a)** $\{5\}; (5, 3)$ **(b)** $\{5\}; (5, 4)$ **(c)** $\{1\}$; yes, at $(1, 2)$ **(d)** $\{5\}$ **(e)** $\left\{-\dfrac{1}{11}\right\}$

97. (a), (b)

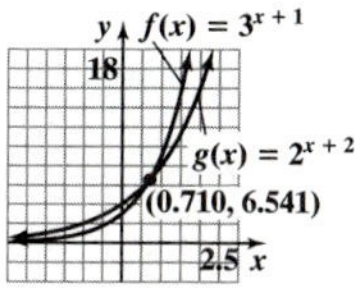

(c) $\{x \mid x > 0.710\}$ or $(0.710, \infty)$

99. (a), (b), (c)

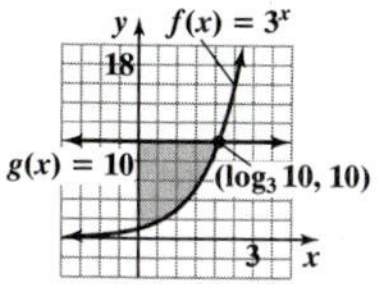

101. (a), (b), (c)

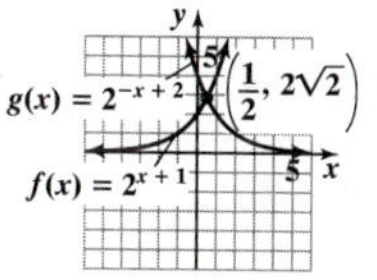

103. (a)

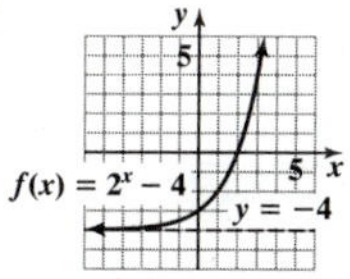

(b) 2 **(c)** $\{x \mid x < 2\}$ or $(-\infty, 2)$

105. (a) 2047 **(b)** 2059

107. (a) After 4.2 yr
(b) After 6.5 yr
(c) After 12.8 yr

110. $\left\{-3, \dfrac{1}{4}, 2\right\}$

111. one-to-one

112. $(f \circ g)(x) = \dfrac{x + 5}{-x + 11}$; $\{x \mid x \neq 3, x \neq 11\}$

113. $\{x \mid x \geq 1\}$, or $[1, \infty)$

6.7 Assess Your Understanding *(page 474)*

3. principal **4.** *I; Prt;* simple interest **5.** 4 **6.** effective rate of interest **7.** \$108.29 **9.** \$609.50 **11.** \$697.09 **13.** \$1246.08 **15.** \$88.72 **17.** \$860.72 **19.** \$554.09 **21.** \$59.71 **23.** 5.095% **25.** 5.127% **27.** $6\frac{1}{4}$% compounded annually **29.** 9% compounded monthly **31.** 25.992% **33.** 24.573% **35. (a)** About 8.69 yr **(b)** About 8.66 yr **37.** 6.823% **39.** 10.15 yr; 10.14 yr **41.** 15.27 yr or 15 yr, 3 mo **43.** \$104,335 **45.** \$12,910.62 **47.** About \$30.17 per share or \$3017 **49.** Not quite. Jim will have \$1057.60. The second bank gives a better deal, since Jim will have \$1060.62 after 1 yr. **51.** Will has \$11,632.73; Henry has \$10,947.89. **53. (a)** \$63,449 **(b)** \$44,267 **55.** About \$1019 billion; about \$232 billion **57.** \$940.90 **59.** 2.53% **61.** 34.31 yr **63. (a)** \$3686.45 **(b)** \$3678.79 **65.** \$6439.28

67. (a) 11.90 yr **(b)** 22.11 yr **(c)**

$$mP = P\left(1 + \frac{r}{n}\right)^{nt}$$
$$m = \left(1 + \frac{r}{n}\right)^{nt}$$
$$\ln m = \ln\left(1 + \frac{r}{n}\right)^{nt} = nt\ln\left(1 + \frac{r}{n}\right)$$
$$t = \frac{\ln m}{n\ln\left(1 + \frac{r}{n}\right)}$$

69. (a) 1.59% **(b)** In 2029 or after 21 yr **71.** 22.7 yr **76.** $R = 0$; yes **77.** $f^{-1}(x) = \dfrac{2x}{x - 1}$ **78.** $-2, 5$; $f(x) = (x + 2)^2(x - 5)(x^2 + 1)$ **79.** $\{6\}$

6.8 Assess Your Understanding *(page 486)*

1. (a) 500 insects **(b)** $0.02 = 2\%$ per day **(c)** About 611 insects **(d)** After about 23.5 days **(e)** After about 34.7 days **3. (a)** $-0.0244 = -2.44\%$ per year **(b)** About 391.7 g **(c)** After about 9.1 yr **(d)** 28.4 yr **5. (a)** $N(t) = N_0 e^{kt}$ **(b)** 5832 **(c)** 3.9 days **7. (a)** $N(t) = N_0 e^{kt}$ **(b)** 25,198 **9.** 9.797 g **11.** 9953 yr ago **13. (a)** 5:18 PM **(b)** About 14.3 min **(c)** The temperature of the pan approaches 70°F. **15.** 18.63°C; 25.1°C **17.** 1.7 ppm; 7.17 days, or 172 hr **19.** 0.26 M; 6.58 hr, or 395 min **21.** 26.6 days

23. (a) In 1984, 91.8% of households did not own a personal computer.
(b)

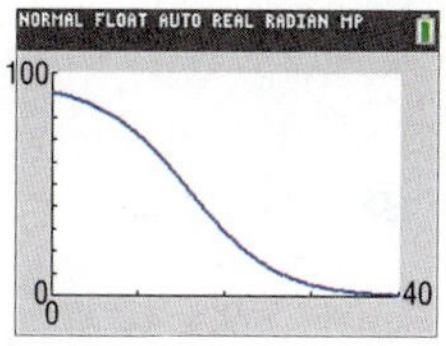

(c) 70.6% **(d)** During 2011

25. (a)

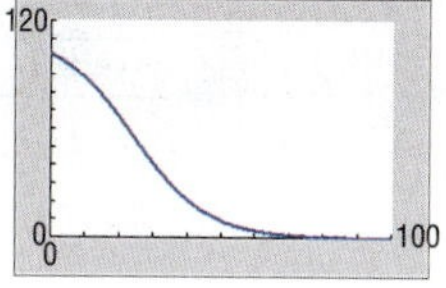

(b) 0.78, or 78%
(c) 50 people
(d) As n increases, the probability decreases.

27. (a) 9.23×10^{-3}, or about 0
(b) 0.81, or about 1
(c) 5.01, or about 5
(d) 57.91°, 43.99°, 30.07°

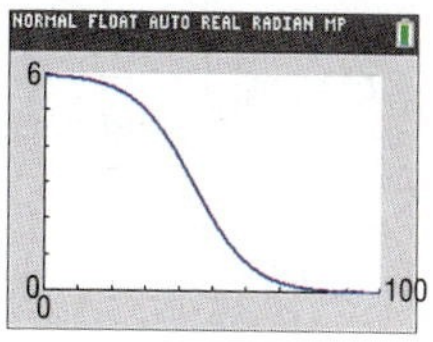

29. (a) $P(t) = 50(3)^{t/20}$ **(b)** 661 **(c)** In 48 days **(d)** $P(t) = 50e^{0.055t}$ **30.** $f(x) = -\dfrac{3}{2}x + 7$ **31.** Neither **32.** $2\ln x + \dfrac{1}{2}\ln y - \ln z$ **33.** $2\sqrt[3]{5}$

6.9 Assess Your Understanding *(page 493)*

1. (a) 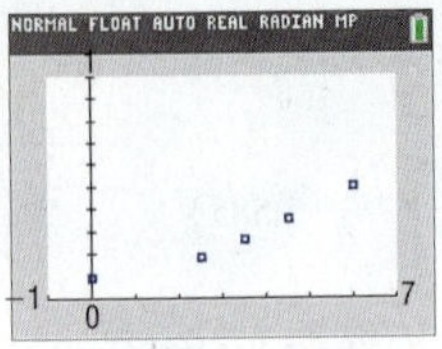

(d) 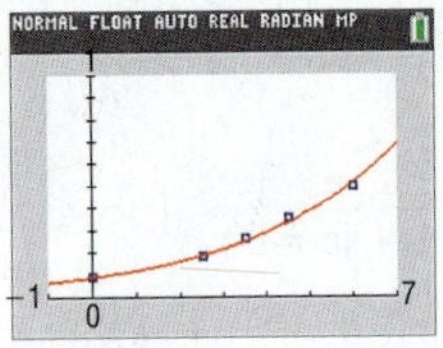

(b) $y = 0.0903(1.3384)^x$
(c) $N(t) = 0.0903e^{0.2915t}$
(e) 0.69
(f) After about 7.26 hr

3. (a) 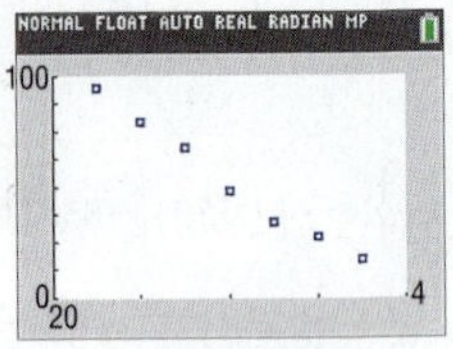

(d) 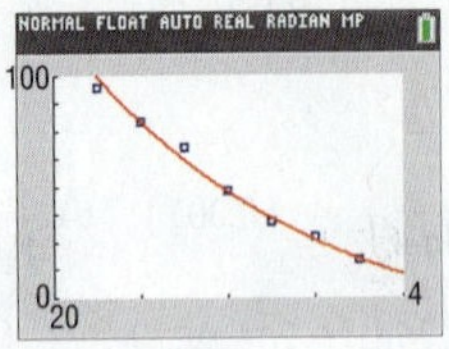

(b) $y = 118.7226(0.7013)^x$
(c) $A(t) = 118.7226e^{-0.3548t}$
(e) 28.7%
(f) $k = -0.3548 = -35.48\%$ is the exponential growth rate. It represents the rate at which the percentage of patients surviving advanced-stage breast cancer is decreasing.

5. (a) 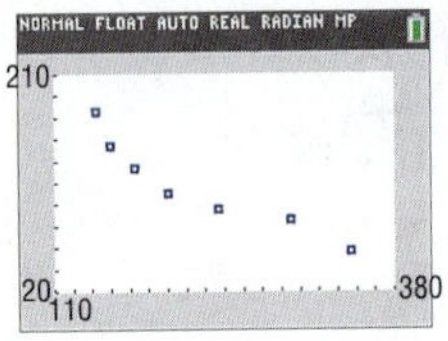

(c) 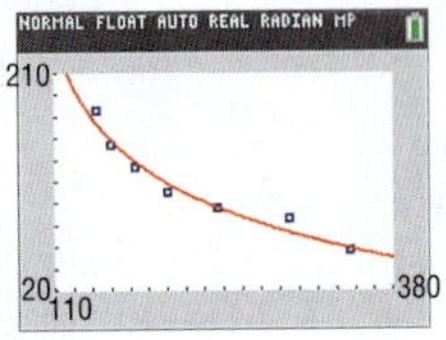

(b) $y = 330.0549 - 34.5008 \ln x$
(d) 185 billion pounds
(e) Under by 5 billion pounds

7. (a) 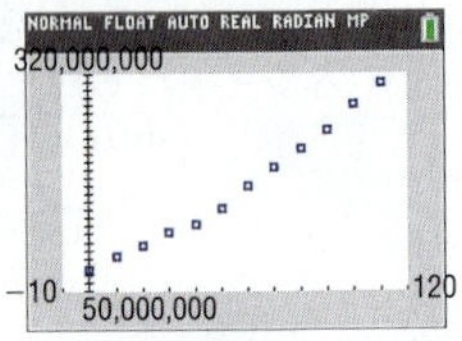

(c) 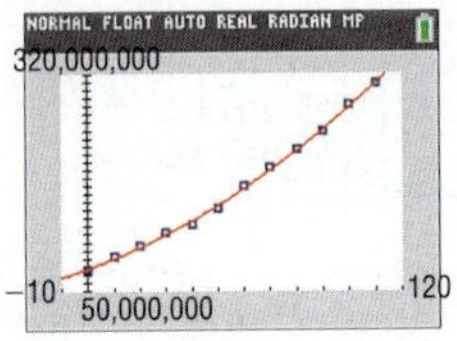

(b) $y = \dfrac{762{,}176{,}844.4}{1 + 8.7428e^{-0.0162x}}$
(d) 762,176,844
(e) Approximately 315,203,288 **(f)** 2023

9. (a) 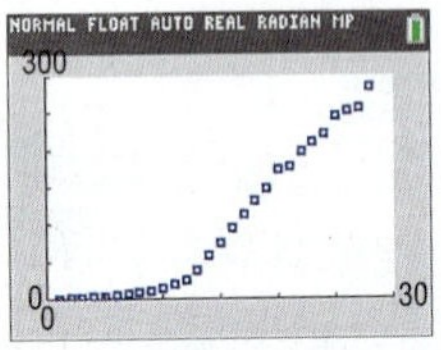

(c) 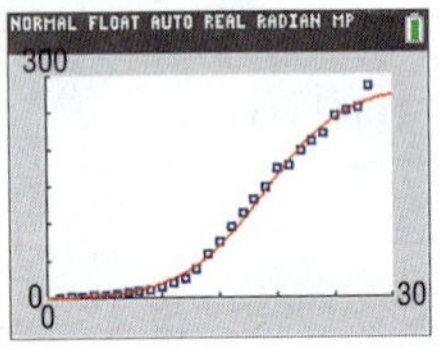

(b) $y = \dfrac{286.2055}{1 + 226.8644e^{-0.2873x}}$
(d) 286.2 thousand cell sites
(e) 281.3 thousand cell sites

11. (a) 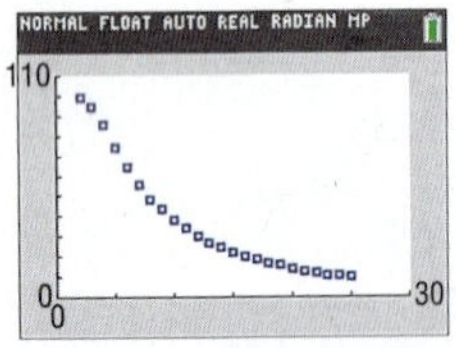

(d) 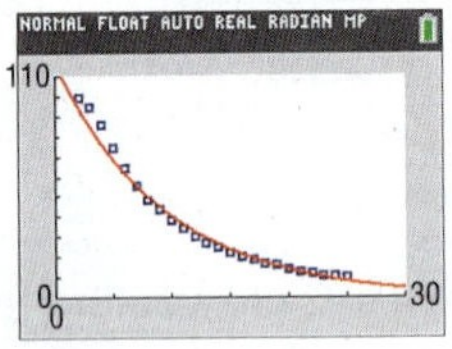

(b) Exponential
(c) $y = 115.5779(0.9012)^x$
(e) 5.1%

13. $f(x) = \frac{1}{3}(x + 3)(x + 1)^2(x - 2)$ **14.** $\frac{3\sqrt{2}}{2}$ **15.** $\sqrt{3}$ **16.**

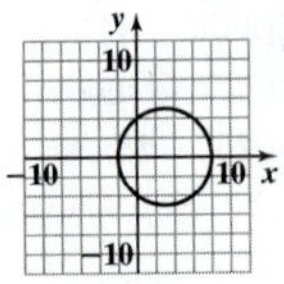

Review Exercises *(page 499)*

1. (a) -26 **(b)** -241 **(c)** 16 **(d)** -1 **2. (a)** $\sqrt{11}$ **(b)** 1 **(c)** $\sqrt{\sqrt{6} + 2}$ **(d)** 19 **3. (a)** e^4 **(b)** $3e^{-2} - 2$ **(c)** e^{e^4} **(d)** -17

4. $(f \circ g)(x) = 1 - 3x$, all real numbers; $(g \circ f)(x) = 7 - 3x$, all real numbers; $(f \circ f)(x) = x$, all real numbers; $(g \circ g)(x) = 9x + 4$, all real numbers

5. $(f \circ g)(x) = \sqrt{3 + 3x + 3x^2}$, all real numbers; $(g \circ f)(x) = 1 + \sqrt{3x} + 3x$, $\{x \mid x \ge 0\}$; $(f \circ f)(x) = \sqrt{3\sqrt{3x}}$, $\{x \mid x \ge 0\}$; $(g \circ g)(x) = 3 + 3x + 4x^2 + 2x^3 + x^4$, all real numbers

6. $(f \circ g)(x) = \dfrac{1+x}{1-x}$, $\{x \mid x \ne 0, x \ne 1\}$; $(g \circ f)(x) = \dfrac{x-1}{x+1}$, $\{x \mid x \ne -1, x \ne 1\}$; $(f \circ f)(x) = x$, $\{x \mid x \ne 1\}$; $(g \circ g)(x) = x$, $\{x \mid x \ne 0\}$

7. (a) one-to-one **(b)** $\{(2, 1), (5, 3), (8, 5), (10, 6)\}$ **8.**

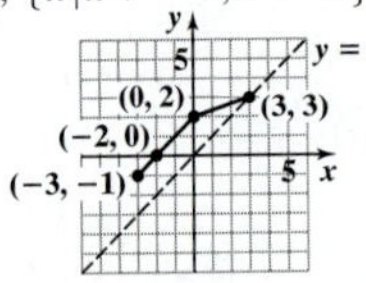

9. $f^{-1}(x) = \dfrac{2x + 3}{5x - 2}$

$$f(f^{-1}(x)) = \frac{2\left(\dfrac{2x + 3}{5x - 2}\right) + 3}{5\left(\dfrac{2x + 3}{5x - 2}\right) - 2} = x$$

$$f^{-1}(f(x)) = \frac{2\left(\dfrac{2x + 3}{5x - 2}\right) + 3}{5\left(\dfrac{2x + 3}{5x - 2}\right) - 2} = x$$

Domain of f = range of f^{-1} = all real numbers except $\frac{2}{5}$

Range of f = domain of f^{-1} = all real numbers except $\frac{2}{5}$

10. $f^{-1}(x) = \dfrac{x + 1}{x}$

$$f(f^{-1}(x)) = \frac{1}{\dfrac{x + 1}{x} - 1} = x$$

$$f^{-1}(f(x)) = \frac{\dfrac{1}{x - 1} + 1}{\dfrac{1}{x - 1}} = x$$

Domain of f = range of f^{-1} = all real numbers except 1

Range of f = domain of f^{-1} = all real numbers except 0

11. $f^{-1}(x) = x^2 + 2, x \geq 0$
$f(f^{-1}(x)) = \sqrt{x^2 + 2 - 2} = |x| = x, x \geq 0$
$f^{-1}(f(x)) = (\sqrt{x-2})^2 + 2 = x$
Domain of f = range of $f^{-1} = [2, \infty)$
Range of f = domain of $f^{-1} = [0, \infty)$

12. $f^{-1}(x) = (x-1)^3$;
$f(f^{-1}(x)) = ((x-1)^3)^{1/3} + 1 = x$
$f^{-1}(f(x)) = (x^{1/3} + 1 - 1)^3 = x$
Domain of f = range of $f^{-1} = (-\infty, \infty)$
Range of f = domain of $f^{-1} = (-\infty, \infty)$

13. (a) 81 **(b)** 2 **(c)** $\frac{1}{9}$ **(d)** -3
14. $\log_5 z = 2$ **15.** $5^{13} = u$ **16.** $\left\{x \middle| x > \frac{2}{3}\right\}; \left(\frac{2}{3}, \infty\right)$
17. $\{x|x < 1 \text{ or } x > 2\}; (-\infty, 1) \cup (2, \infty)$
18. -3 **19.** $\sqrt{2}$ **20.** 0.4
21. $\log_3 u + 2\log_3 v - \log_3 w$ **22.** $8\log_2 a + 2\log_2 b$

23. $2\log x + \frac{1}{2}\log(x^3 + 1)$ **24.** $2\ln(2x+3) - 2\ln(x-1) - 2\ln(x-2)$ **25.** $\frac{25}{4}\log_4 x$ **26.** $-2\ln(x+1)$ **27.** $\ln\left[\frac{16\sqrt{x^2+1}}{\sqrt{x(x-4)}}\right]$ **28.** 2.124

29.
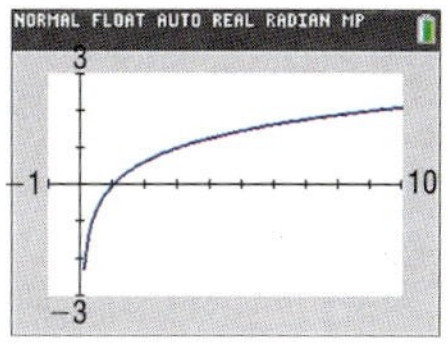

30. (a) Domain of f: $(-\infty, \infty)$
(b)
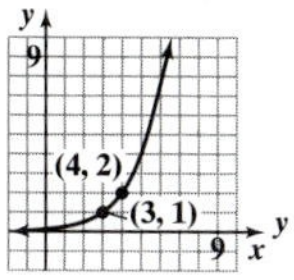

(c) Range of f: $(0, \infty)$
Horizontal asymptote: $y = 0$
(d) $f^{-1}(x) = 3 + \log_2 x$
(e) Domain of f^{-1}: $(0, \infty)$
Range of f^{-1}: $(-\infty, \infty)$
(f)
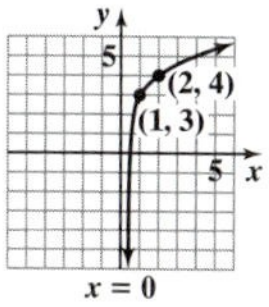

31. (a) Domain of f: $(-\infty, \infty)$
(b)
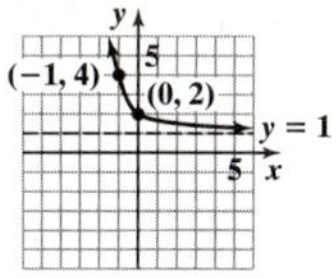

(c) Range of f: $(1, \infty)$
Horizontal asymptote: $y = 1$
(d) $f^{-1}(x) = -\log_3(x-1)$
(e) Domain of f^{-1}: $(1, \infty)$
Range of f^{-1}: $(-\infty, \infty)$
(f)
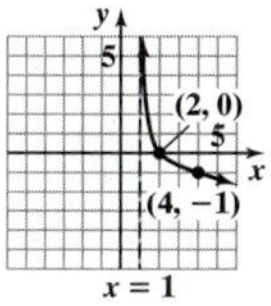

32. (a) Domain of f: $(-\infty, \infty)$
(b)
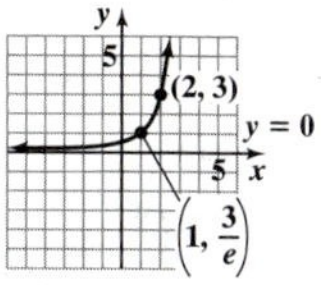

(c) Range of f: $(0, \infty)$
Horizontal asymptote: $y = 0$
(d) $f^{-1}(x) = 2 + \ln\left(\frac{x}{3}\right)$
(e) Domain of f^{-1}: $(0, \infty)$
Range of f^{-1}: $(-\infty, \infty)$
(f)
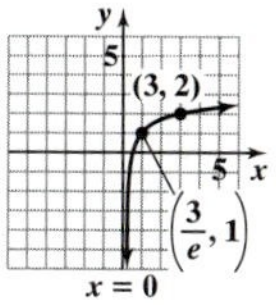

33. (a) Domain of f: $(-3, \infty)$
(b)
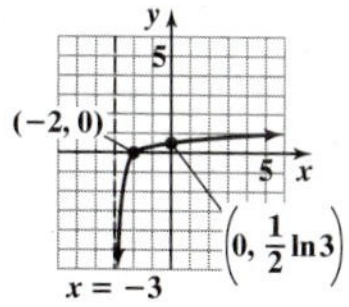

(c) Range of f: $(-\infty, \infty)$
Vertical asymptote: $x = -3$
(d) $f^{-1}(x) = e^{2x} - 3$
(e) Domain of f^{-1}: $(-\infty, \infty)$
Range of f^{-1}: $(-3, \infty)$
(f)
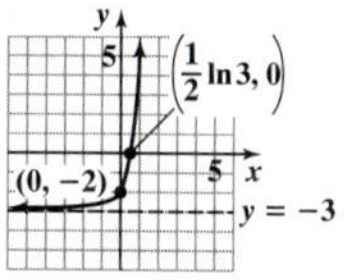

34. $\left\{-\frac{16}{9}\right\}$ **35.** $\left\{\frac{-1-\sqrt{3}}{2}, \frac{-1+\sqrt{3}}{2}\right\} \approx \{-1.366, 0.366\}$ **36.** $\left\{\frac{1}{4}\right\}$ **37.** $\left\{\frac{2\ln 3}{\ln 5 - \ln 3}\right\} \approx \{4.301\}$ **38.** $\{-2, 6\}$ **39.** $\{83\}$ **40.** $\left\{\frac{1}{2}, -3\right\}$

41. $\{1\}$ **42.** $\{-1\}$ **43.** $\{1 - \ln 5\} \approx \{-0.609\}$ **44.** $\left\{\log_3(-2+\sqrt{7})\right\} = \left\{\frac{\ln(-2+\sqrt{7})}{\ln 3}\right\} \approx \{-0.398\}$

45. (a), (e)
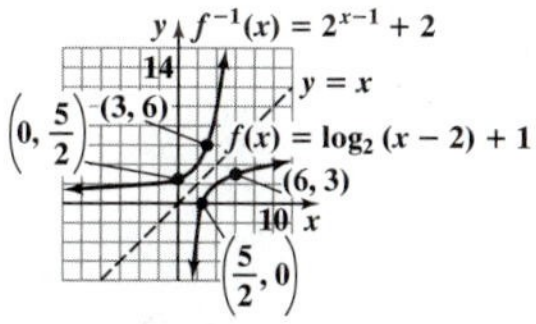

(b) 3; (6, 3)
(c) 10; (10, 4)
(d) $\left\{x \middle| x > \frac{5}{2}\right\}$ or $\left(\frac{5}{2}, \infty\right)$
(e) $f^{-1}(x) = 2^{x-1} + 2$

46. (a) 37.3 W **(b)** 6.9 dB **47. (a)** 11.77 **(b)** 9.56 in.
48. (a) 9.85 yr **(b)** 4.27 yr **49.** \$20,398.87; 4.04%; 17.5 yr
50. \$41,668.97 **51.** 24,765 yr ago **52.** 55.22 min, or 55 min, 13 sec
53. 7,967,521,519 **54.** 7.204 g; 0.519 g

55. (a) 0.3 **(b)** 0.8
(c)
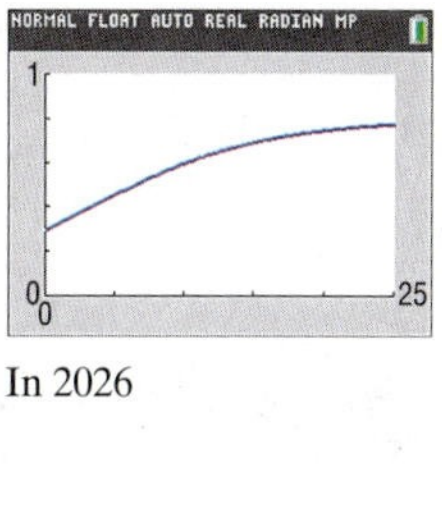

(d) In 2026

56. (a)
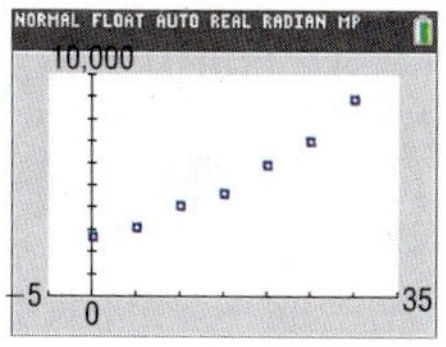

(b) $y = 2638.26(1.0407)^x$
(c) $A(t) = 2638.26e^{0.0399x}$
(d)
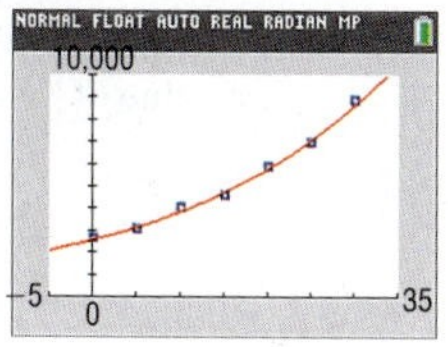

(e) 2021–22

57. (a)
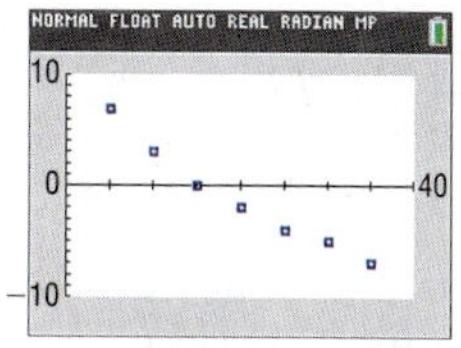

(b) $y = 18.921 - 7.096\ln x$
(c)
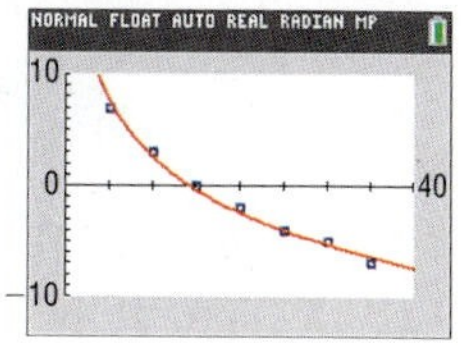

(d) Approximately −3°F

58. (a)
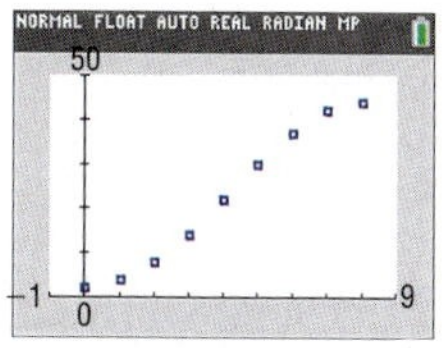

(b) $C = \frac{46.9292}{1 + 21.2733e^{-0.7306t}}$
(c)
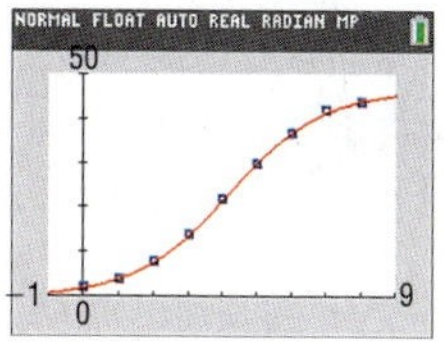

(d) About 47 people; 50 people
(e) 2.4 days; during the tenth hour of day 3
(f) 9.5 days

Chapter Test *(page 502)*

1. (a) $f \circ g = \dfrac{2x+7}{2x+3}$; domain: $\left\{x \middle| x \neq -\dfrac{3}{2}\right\}$ **(b)** $(g \circ f)(-2) = 5$ **(c)** $(f \circ g)(-2) = -3$

2. (a) The function is not one-to-one. **(b)** The function is one-to-one.

3. $f^{-1}(x) = \dfrac{2+5x}{3x}$; domain of $f = \left\{x \middle| x \neq \dfrac{5}{3}\right\}$, range of $f = \{y | y \neq 0\}$; domain of $f^{-1} = \{x | x \neq 0\}$; range of $f^{-1} = \left\{y \middle| y \neq \dfrac{5}{3}\right\}$

4. The point $(-5, 3)$ must be on the graph of f^{-1}. **5.** $x = 5$ **6.** $b = 4$ **7.** $x = 625$ **8.** $e^3 + 2 \approx 22.086$ **9.** $\log 20 \approx 1.301$

10. $\log_3 21 = \dfrac{\ln 21}{\ln 3} \approx 2.771$ **11.** $\ln 133 \approx 4.890$

12. (a) Domain of f: $\{x | -\infty < x < \infty\}$ or $(-\infty, \infty)$

(b)

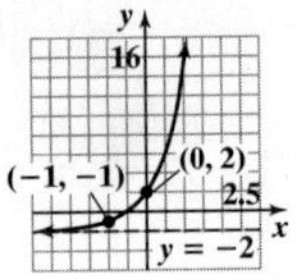

(c) Range of f: $\{y | y > -2\}$ or $(-2, \infty)$; Horizontal asymptote: $y = -2$

(d) $f^{-1}(x) = \log_4(x+2) - 1$

(e) Domain of f^{-1}: $\{x | x > -2\}$ or $(-2, \infty)$
Range of f^{-1}: $\{y | -\infty < y < \infty\}$ or $(-\infty, \infty)$

(f)

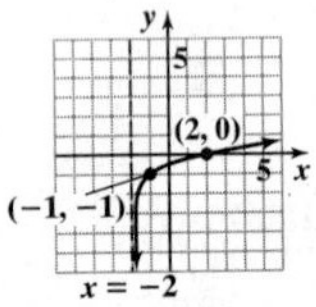

13. (a) Domain of f: $\{x | x > 2\}$ or $(2, \infty)$

(b)

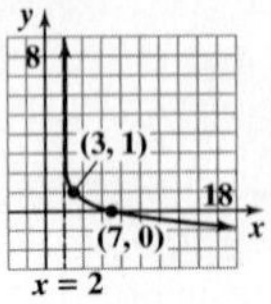

(c) Range of f: $\{y | -\infty < y < \infty\}$ or $(-\infty, \infty)$; vertical asymptote: $x = 2$

(d) $f^{-1}(x) = 5^{1-x} + 2$

(e) Domain of f^{-1}: $\{x | -\infty < x < \infty\}$ or $(-\infty, \infty)$
Range of f^{-1}: $\{y | y > 2\}$ or $(2, \infty)$

(f)

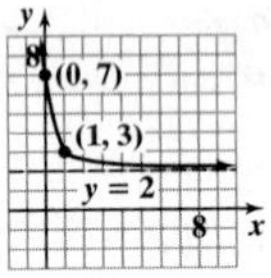

14. $\{1\}$ **15.** $\{91\}$ **16.** $\{-\ln 2\} \approx \{-0.693\}$ **17.** $\left\{\dfrac{1-\sqrt{13}}{2}, \dfrac{1+\sqrt{13}}{2}\right\} \approx \{-1.303, 2.303\}$ **18.** $\left\{\dfrac{3\ln 7}{1 - \ln 7}\right\} \approx \{-6.172\}$

19. $\{2\sqrt{6}\} \approx \{4.899\}$ **20.** $2 + 3\log_2 x - \log_2(x-6) - \log_2(x+3)$ **21.** About 250.39 days **22. (a)** \$1033.82 **(b)** \$963.42 **(c)** 11.9 yr

23. (a) About 83 dB **(b)** The pain threshold will be exceeded if 31,623 people shout at the same time.

Cumulative Review *(page 502)*

1. Yes; no **2. (a)** 10 **(b)** $2x^2 + 3x + 1$ **(c)** $2x^2 + 4xh + 2h^2 - 3x - 3h + 1$ **3.** $\left(\dfrac{1}{2}, \dfrac{\sqrt{3}}{2}\right)$ is on the graph. **4.** $\{-26\}$

5.

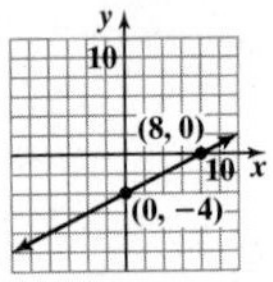

6. (a)

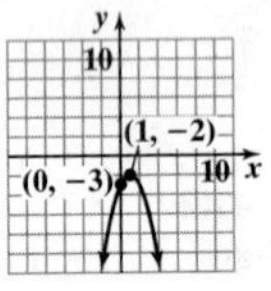

(b) $\{x | -\infty < x < \infty\}$ **7.** $f(x) = 2(x-4)^2 - 8 = 2x^2 - 16x + 24$

8.

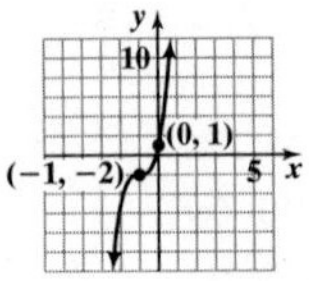

9. $f(g(x)) = \dfrac{4}{(x-3)^2} + 2$; domain: $\{x | x \neq 3\}$; 3

10. (a) Zeros: $-4, -\dfrac{1}{4}, 2$

(b) x-intercepts: $-4, -\dfrac{1}{4}, 2$; y-intercept: -8

(c) Local maximum value of 60.75 occurs at $x = -2.5$.
Local minimum value of -25 occurs at $x = 1$.

(d)

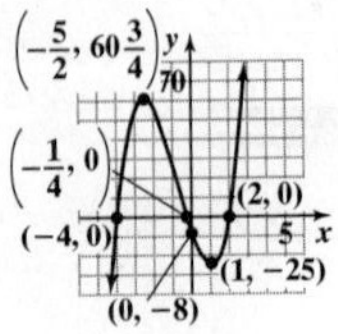

11. (a), (c)

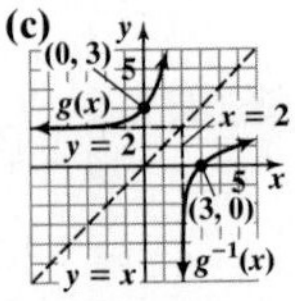

Domain g = range $g^{-1} = (-\infty, \infty)$
Range g = domain $g^{-1} = (2, \infty)$

(b) $g^{-1}(x) = \log_3(x-2)$

12. $\left\{-\dfrac{3}{2}\right\}$ **13.** $\{2\}$

14. (a) $\{-1\}$ **(b)** $\{x | x > -1\}$ or $(-1, \infty)$

(c) $\{25\}$

15. (a)

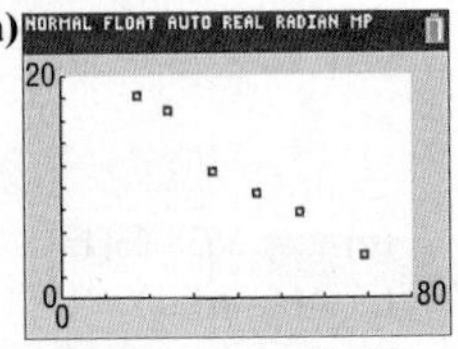

(b) Logarithmic; $y = 49.293 - 10.563 \ln x$

(c) Highest value of $|r|$

CHAPTER 8 Systems of Equations and Inequalities

8.1 Assess Your Understanding *(page 554)*

3. inconsistent **4.** consistent; independent **5.** $(3, -2)$ **6.** consistent; dependent **7.** b **8.** a **9.** $\begin{cases} 2(2) - (-1) = 5 \\ 5(2) + 2(-1) = 8 \end{cases}$

11. $\begin{cases} 3(2) - 4\left(\frac{1}{2}\right) = 4 \\ \frac{1}{2}(2) - 3\left(\frac{1}{2}\right) = -\frac{1}{2} \end{cases}$ **13.** $\begin{cases} 4 - 1 = 3 \\ \frac{1}{2}(4) + 1 = 3 \end{cases}$ **15.** $\begin{cases} 3(1) + 3(-1) + 2(2) = 4 \\ 1 - (-1) - 2 = 0 \\ 2(-1) - 3(2) = -8 \end{cases}$ **17.** $\begin{cases} 3(2) + 3(-2) + 2(2) = 4 \\ 2 - 3(-2) + 2 = 10 \\ 5(2) - 2(-2) - 3(2) = 8 \end{cases}$

19. $x = 6, y = 2; (6, 2)$

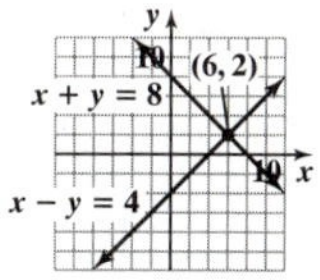

21. $x = 3, y = -6; (3, -6)$

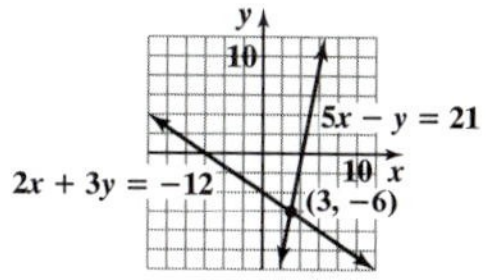

23. $x = 8, y = -4; (8, -4)$

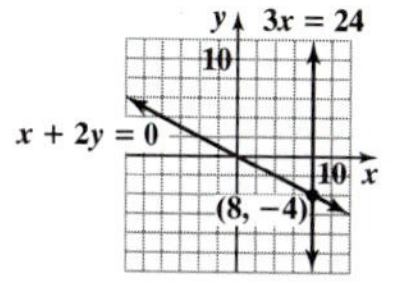

25. $x = \frac{1}{3}, y = -\frac{1}{6}; \left(\frac{1}{3}, -\frac{1}{6}\right)$

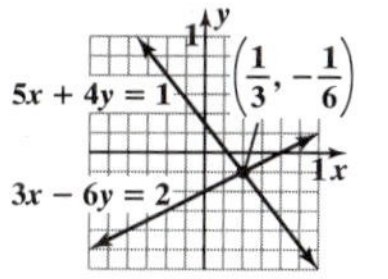

27. Inconsistent

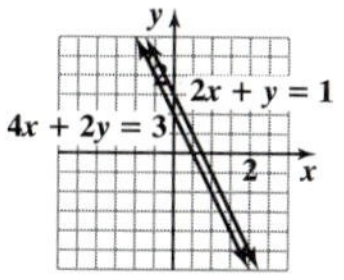

29. $x = \frac{3}{2}, y = 3; \left(\frac{3}{2}, 3\right)$

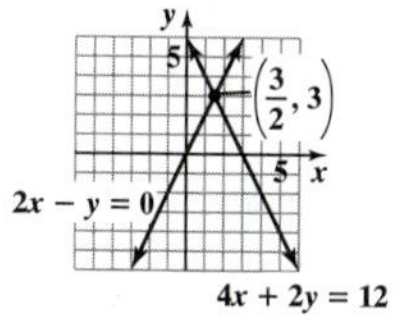

31. $\{(x, y) \mid x = 4 - 2y, y \text{ is any real number}\}$, or $\left\{(x, y) \,\middle|\, y = \frac{4 - x}{2}, x \text{ is any real number}\right\}$ **33.** $x = 1, y = 1; (1, 1)$ **35.** $x = \frac{3}{2}, y = 1; \left(\frac{3}{2}, 1\right)$

37. $x = 4, y = 3; (4, 3)$ **39.** $x = \frac{4}{3}, y = \frac{1}{5}; \left(\frac{4}{3}, \frac{1}{5}\right)$ **41.** $x = \frac{1}{5}, y = \frac{1}{3}; \left(\frac{1}{5}, \frac{1}{3}\right)$ **43.** $x = 8, y = 2, z = 0; (8, 2, 0)$

45. $x = 2, y = -1, z = 1; (2, -1, 1)$ **47.** Inconsistent **49.** $\{(x, y, z) \mid x = 5z - 2, y = 4z - 3; z \text{ is any real number}\}$ **51.** Inconsistent

53. $x = 1, y = 3, z = -2; (1, 3, -2)$ **55.** $x = -3, y = \frac{1}{2}, z = 1; \left(-3, \frac{1}{2}, 1\right)$ **57.** Length 30 ft; width 15 ft

59. There were 23 commercial launches and 58 noncommercial launches in 2013. **61.** 22.5 lb **63.** Smartphone: \$325; tablet: \$640

65. Average wind speed 25 mph; average airspeed 175 mph **67.** 80 \$25 sets and 120 \$45 sets **69.** \$9.96

71. Mix 50 mg of first compound with 75 mg of second. **73.** $a = \frac{4}{3}, b = -\frac{5}{3}, c = 1$ **75.** $Y = 9000, r = 0.06$ **77.** $I_1 = \frac{10}{71}, I_2 = \frac{65}{71}, I_3 = \frac{55}{71}$

79. 100 orchestra, 210 main, and 190 balcony seats **81.** 1.5 chicken, 1 corn, 2 milk

83. If x = price of hamburgers, y = price of fries, and z = price of colas, then $x = 2.75 - z$, $y = \frac{41}{60} + \frac{1}{3}z$, $\$0.60 \le z \le \$0.90$.

There is not sufficient information:

x	\$2.13	\$2.01	\$1.86
y	\$0.89	\$0.93	\$0.98
z	\$0.62	\$0.74	\$0.89

85. It will take Beth 30 hr, Bill 24 hr, and Edie 40 hr.

89.

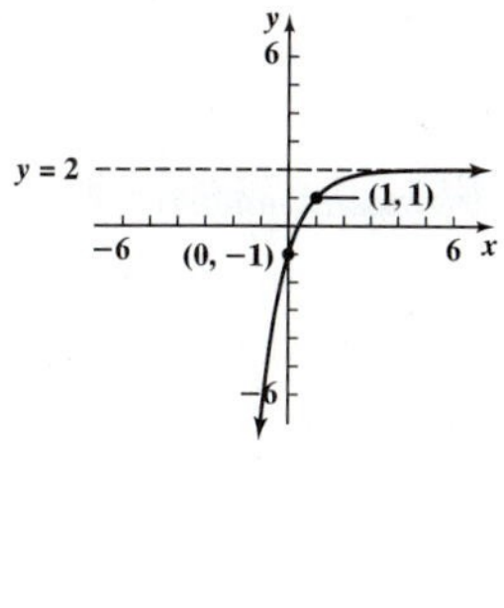

90. (a) $2(2x - 3)^3(x^3 + 5)(10x^3 - 9x^2 + 20)$ **(b)** $7(3x - 5)^{-1/2}(x + 3)^{-3/2}$

91. $\left[-\frac{2}{3}, 2\right]$ **92.** $f^{-1}(x) = \log_2(x - 1)$

APPENDIX Graphing Utilities

1 Exercises *(page A2)*

1. $(-1, 4)$; II **3.** $(3, 1)$; I **5.** $X\text{min} = -6$, $X\text{max} = 6$, $X\text{scl} = 2$, $Y\text{min} = -4$, $Y\text{max} = 4$, $Y\text{scl} = 2$
7. $X\text{min} = -6$, $X\text{max} = 6$, $X\text{scl} = 2$, $Y\text{min} = -1$, $Y\text{max} = 3$, $Y\text{scl} = 1$ **9.** $X\text{min} = 3$, $X\text{max} = 9$, $X\text{scl} = 1$, $Y\text{min} = 2$, $Y\text{max} = 10$, $Y\text{scl} = 2$
11. $X\text{min} = -11$, $X\text{max} = 5$, $X\text{scl} = 1$, $Y\text{min} = -3$, $Y\text{max} = 6$, $Y\text{scl} = 1$
13. $X\text{min} = -30$, $X\text{max} = 50$, $X\text{scl} = 10$, $Y\text{min} = -90$, $Y\text{max} = 50$, $Y\text{scl} = 10$
15. $X\text{min} = -10$, $X\text{max} = 110$, $X\text{scl} = 10$, $Y\text{min} = -10$, $Y\text{max} = 160$, $Y\text{scl} = 10$

2 Exercises *(page A4)*

1. (a) 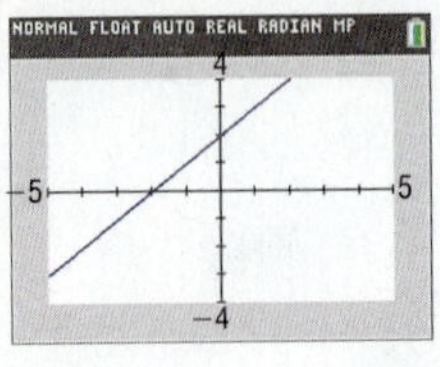

(b) 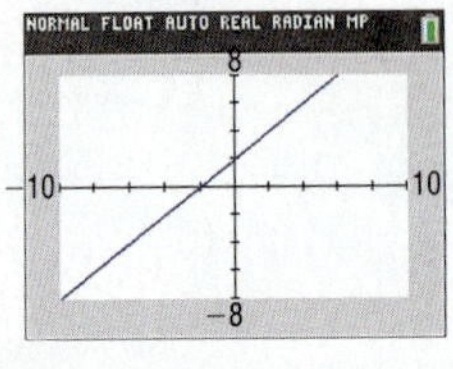

3. (a) 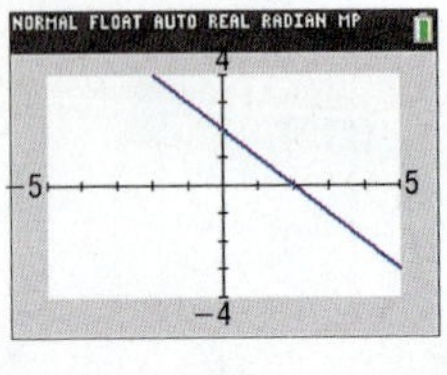

(b) 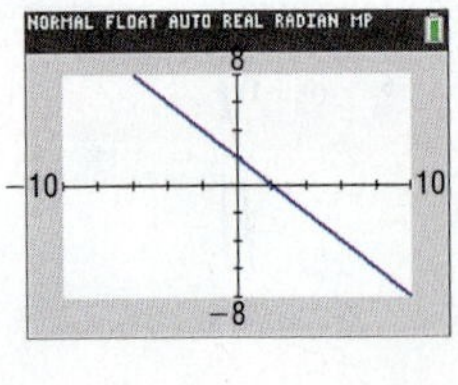

5. (a)
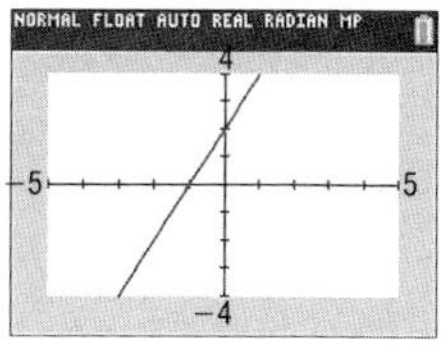

(b)
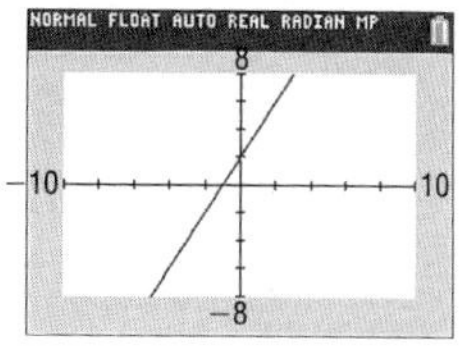

7. (a)
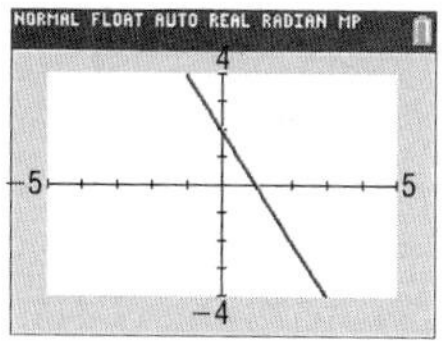

(b)
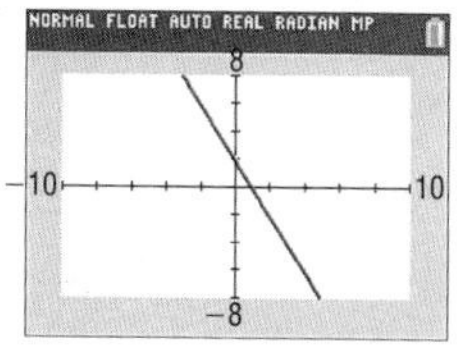

9. (a)
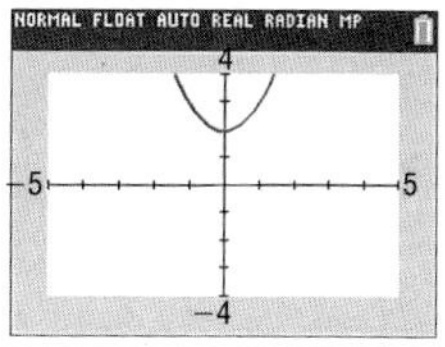

(b)
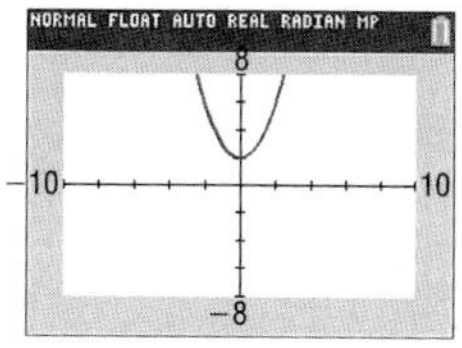

11. (a)
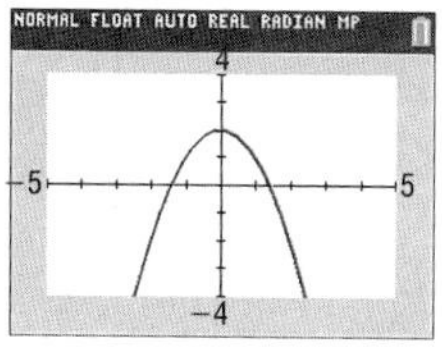

(b)
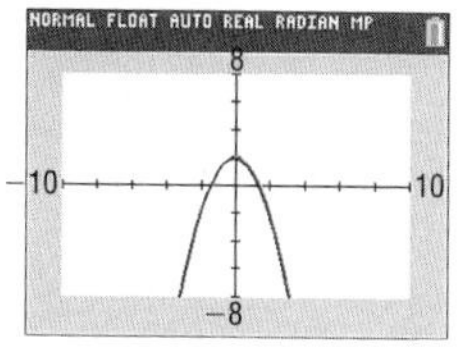

13. (a)
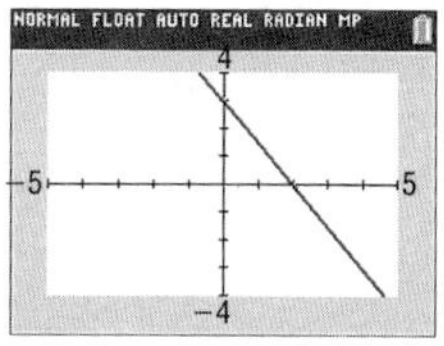

(b)
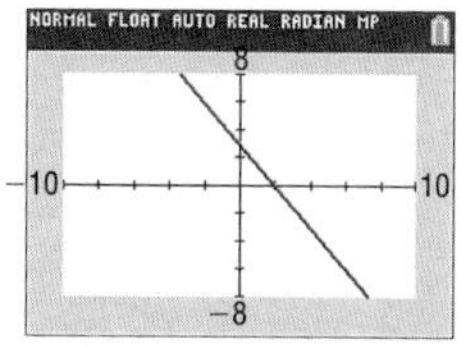

15. (a)
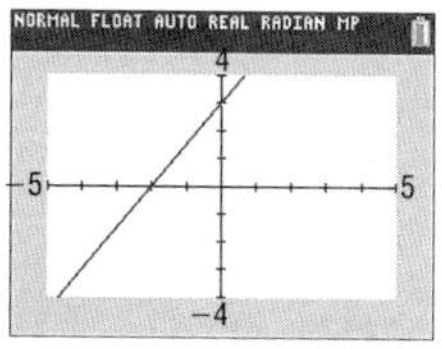

(b)
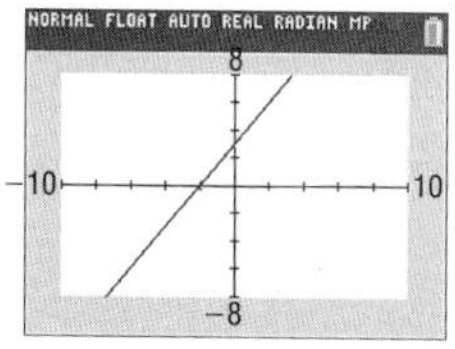

17.
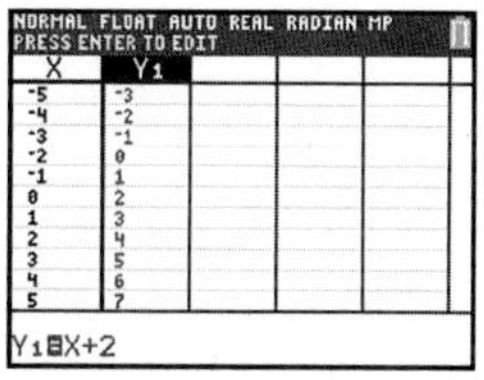

19.
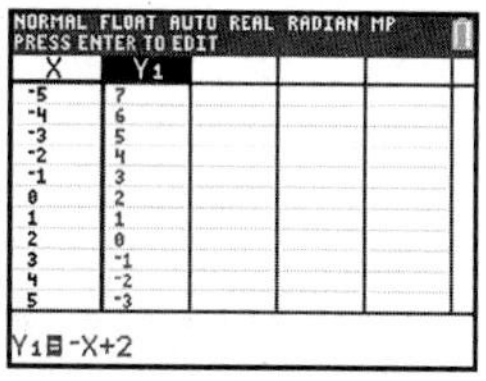

21.
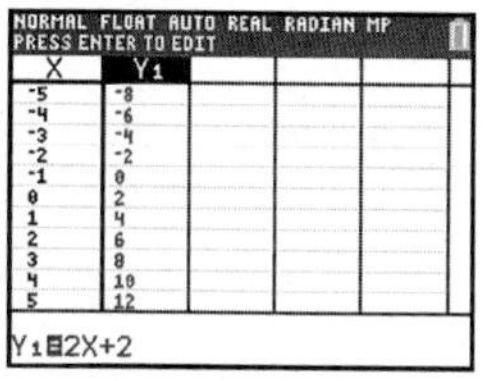

23.
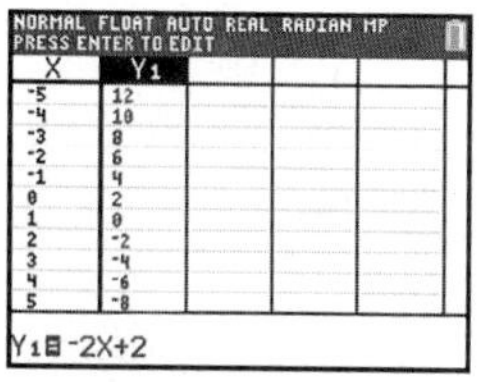

25.
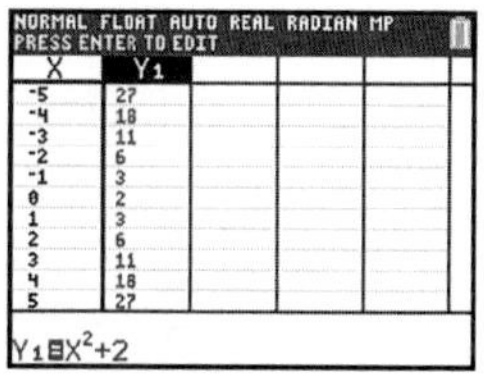

27.
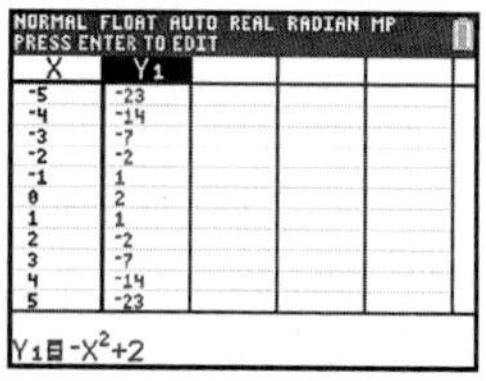

29.
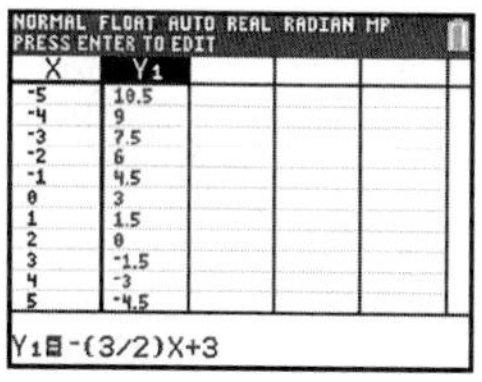

31.
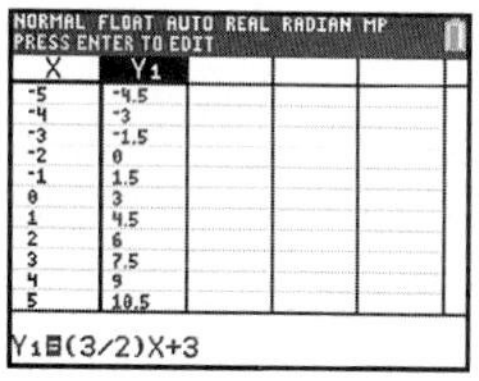

3 Exercises *(page A6)*

1. -3.41 **3.** -1.71 **5.** -0.28 **7.** 3.00 **9.** 4.50 **11.** 1.00, 23.00

5 Exercises *(page A8)*

1. Yes **3.** Yes **5.** No **7.** Yes **9.** Answers may vary. A possible answer is $Y\text{min} = 0$, $Y\text{max} = 10$, and $Y\text{scl} = 1$.

Credits

Photos

Front Matter Achieve Your Potential (Retain Your Knowledge), Mike Flippo/Shutterstock; Resources for Success, Sturti/E+/Getty Images

Chapter R Page 1, Andres Rodriguez/Fotolia; Page 31, Feraru Nicolae/Shutterstock.

Chapter 1 Pages 81 and 147, Rawpixel/Fotolia; Page 91, A.R. Monko/Shutterstock; Page 103(L), Stanca Sanda/Alamy; Page 103(R), Jeff Chiu/AP Images; Page 133, Nancy R. Cohen/Getty Images; Page 145, Hemera Technologies/Photos.com/Thinkstock.

Chapter 2 Pages 149 and 197, Maxx-Studio/Shutterstock; Page 160, Snapper/Fotolia; Page 166(L), Stockyimages/Fotolia; Page 166(R), Department of Energy; Page 180, Tetra Images/Alamy; Page 187, Tom Donoghue/Polaris/Newscom.

Chapter 3 Pages 198 and 271, Leigh Prather/Shutterstock/Asset Library; Page 212, NASA; Page 220, Exactostock/SuperStock; Page 259, Dreamstime LLC.

Chapter 4 Pages 273 and 320, Spaxiax/Shutterstock; Page 309, Geno EJ Sajko Photography/Shutterstock.

Chapter 5 Pages 321 and 400, John Foxx Collection/Imagestate/DK images; Page 367, Corbis.

Chapter 6 Pages 402 and 503, Rawpixel/Fotolia; Page 458, Hulton Archive/Handout/Getty Images; Page 467(L), Stockbyte/Getty Images; Page 467(R), Pascal Saez/Alamy; Page 472, Lacroix Serge/iStock/360/Getty Images; Page 487, Jupiterimages/Photos.com/360/Getty Images.

Chapter 7 Pages 506 and 542, Marcel Clemens/Shutterstock; Page 524, Thomas Barrat/Shutterstock.

Chapter 8 Pages 543 and 637, Wavebreakmedia/Shutterstock; Page 597, Library of Congress Prints and Photographs Division [LC-USZ62-46864].

Chapter 9 Pages 639 and 681, Denis Cristo/Shutterstock; Pages 663 and 676, Pearson Education.

Chapter 10 Pages 682 and 710, zentilia/Fotolia; Page 703, Georgios Kollidas/Getty Images.

Text

TI 84 Plus C screenshots courtesy of Texas Instruments. Chapter 3, page 201: Diamond price. Used with permission of Diamonds.com. Copyright © Martin Rapaport. All rights reserved. Chapter 6, page 495: Cell Phone Towers by CTIA-The Wireless Association. Copyright © 2013 by CTIA-The Wireless Association. Used by permission of CTIA-The Wireless Association®.

Subject Index

CONICS

Parabola

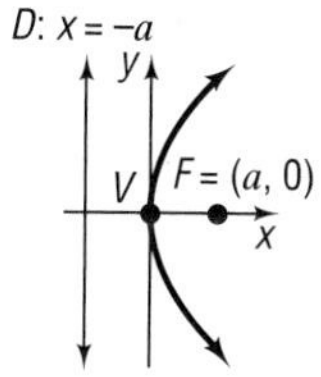

$y^2 = 4ax$

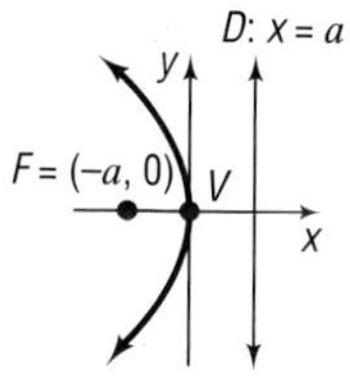

$y^2 = -4ax$

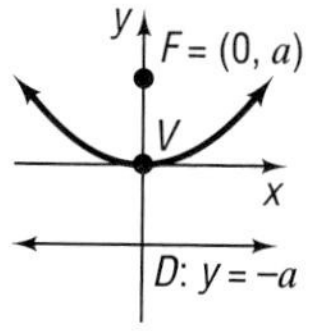

$x^2 = 4ay$

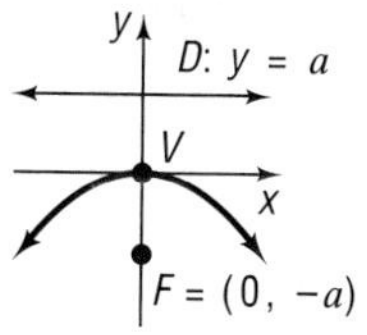

$x^2 = -4ay$

Ellipse

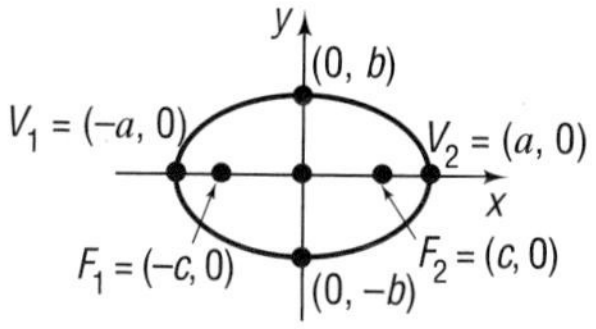

$$\frac{x^2}{a^2} + \frac{y^2}{b^2} = 1, \quad a > b, \quad c^2 = a^2 - b^2$$

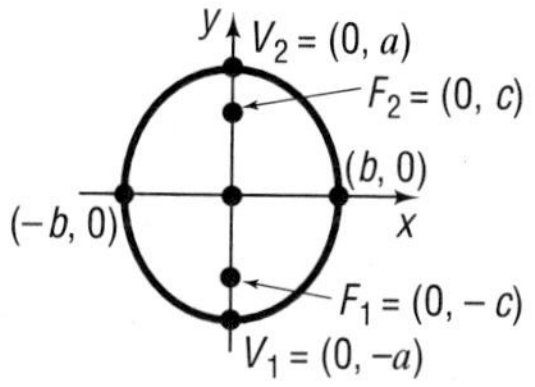

$$\frac{x^2}{b^2} + \frac{y^2}{a^2} = 1, \quad a > b, \quad c^2 = a^2 - b^2$$

Hyperbola

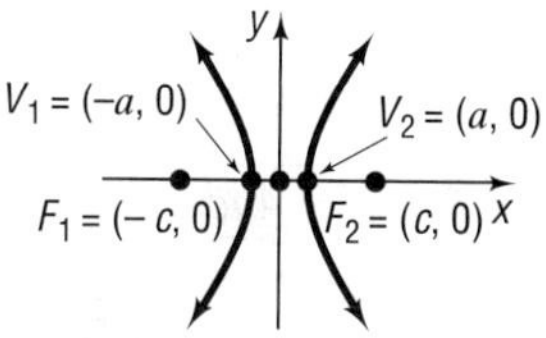

$$\frac{x^2}{a^2} - \frac{y^2}{b^2} = 1, \quad c^2 = a^2 + b^2$$

Asymptotes: $y = \frac{b}{a}x, \quad y = -\frac{b}{a}x$

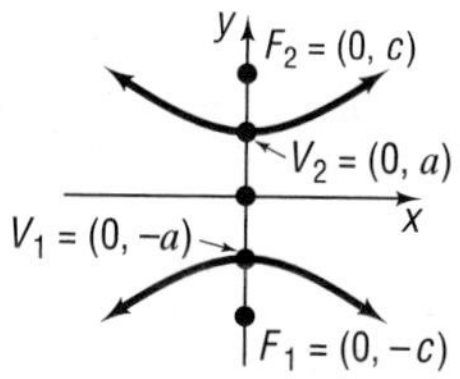

$$\frac{y^2}{a^2} - \frac{x^2}{b^2} = 1, \quad c^2 = a^2 + b^2$$

Asymptotes: $y = \frac{a}{b}x, \quad y = -\frac{a}{b}x$

PROPERTIES OF LOGARITHMS

$\log_a(MN) = \log_a M + \log_a N$

$\log_a\left(\frac{M}{N}\right) = \log_a M - \log_a N$

$\log_a M^r = r\log_a M$

$\log_a M = \frac{\log M}{\log a} = \frac{\ln M}{\ln a}$

$a^r = e^{r \ln a}$

PERMUTATIONS/COMBINATIONS

$0! = 1 \qquad 1! = 1$

$n! = n(n-1)\cdot \ldots \cdot(3)(2)(1)$

$P(n, r) = \frac{n!}{(n-r)!}$

$C(n, r) = \binom{n}{r} = \frac{n!}{(n-r)!r!}$

BINOMIAL THEOREM

$$(a + b)^n = a^n + \binom{n}{1}ba^{n-1} + \binom{n}{2}b^2a^{n-2} + \cdots + \binom{n}{n-1}b^{n-1}a + b^n$$

ARITHMETIC SEQUENCE

$$a_1 + (a_1 + d) + (a_1 + 2d) + \cdots + [a_1 + (n-1)d] = \frac{n}{2}[2a_1 + (n-1)d] = \frac{n}{2}[a_1 + a_n]$$

GEOMETRIC SEQUENCE

$$a_1 + a_1r + a_1r^2 + \cdots + a_1r^{n-1} = a_1 \cdot \frac{1 - r^n}{1 - r}$$

GEOMETRIC SERIES

If $|r| < 1, a_1 + a_1r + a_1r^2 + \cdots = \sum_{k=1}^{\infty} a_1r^{k-1} = \frac{a_1}{1 - r}$

FORMULAS/EQUATIONS

Distance Formula

If $P_1 = (x_1, y_1)$ and $P_2 = (x_2, y_2)$, the distance from P_1 to P_2 is

$$d(P_1, P_2) = \sqrt{(x_2 - x_1)^2 + (y_2 - y_1)^2}$$

Standard Equation of a Circle

The standard equation of a circle of radius r with center at (h, k) is

$$(x - h)^2 + (y - k)^2 = r^2$$

Slope Formula

The slope m of the line containing the points $P_1 = (x_1, y_1)$ and $P_2 = (x_2, y_2)$ is

$$m = \frac{y_2 - y_1}{x_2 - x_1} \qquad \text{if } x_1 \neq x_2$$

$$m \text{ is undefined} \qquad \text{if } x_1 = x_2$$

Point–Slope Equation of a Line

The equation of a line with slope m containing the point (x_1, y_1) is

$$y - y_1 = m(x - x_1)$$

Slope–Intercept Equation of a Line

The equation of a line with slope m and y-intercept b is

$$y = mx + b$$

Quadratic Formula

The solutions of the equation $ax^2 + bx + c = 0$, $a \neq 0$, are

$$x = \frac{-b \pm \sqrt{b^2 - 4ac}}{2a}$$

If $b^2 - 4ac > 0$, there are two unequal real solutions.
If $b^2 - 4ac = 0$, there is a repeated real solution.
If $b^2 - 4ac < 0$, there are two complex solutions that are not real.

GEOMETRY FORMULAS

Circle

r = Radius, A = Area, C = Circumference

$$A = \pi r^2 \qquad C = 2\pi r$$

Triangle

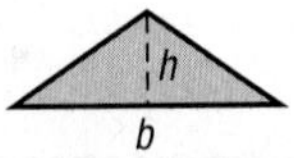

b = Base, h = Altitude (Height), A = area

$$A = \tfrac{1}{2} bh$$

Rectangle

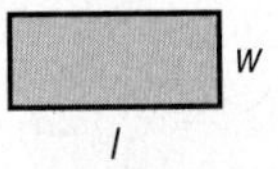

l = Length, w = Width, A = area, P = perimeter

$$A = lw \qquad P = 2l + 2w$$

Rectangular Box (closed)

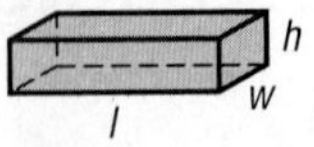

l = Length, w = Width, h = Height, V = Volume, S = Surface area

$$V = lwh \qquad S = 2lw + 2lh + 2wh$$

Sphere

r = Radius, V = Volume, S = Surface area

$$V = \tfrac{4}{3}\pi r^3 \qquad S = 4\pi r^2$$

Right Circular Cylinder (closed)

r = Radius, h = Height, V = Volume, S = Surface area

$$V = \pi r^2 h \qquad S = 2\pi r^2 + 2\pi rh$$

LIBRARY OF FUNCTIONS

Identity Function

$f(x) = x$

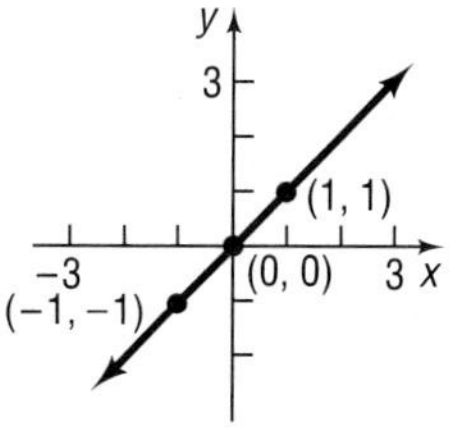

Square Function

$f(x) = x^2$

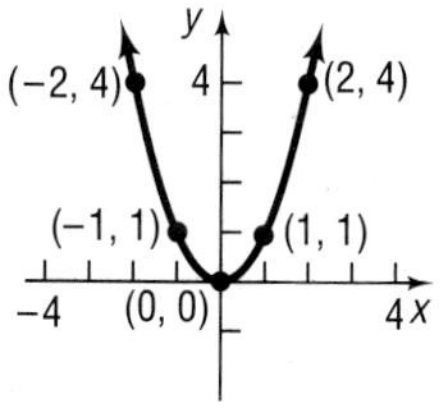

Cube Function

$f(x) = x^3$

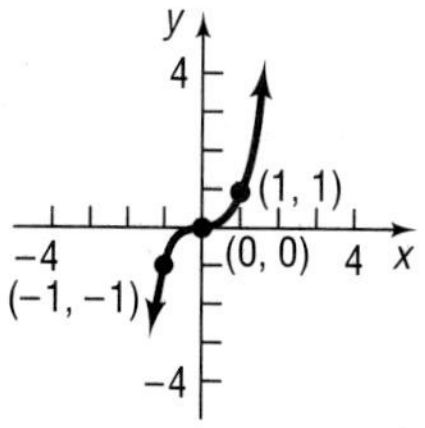

Square Root Function

$f(x) = \sqrt{x}$

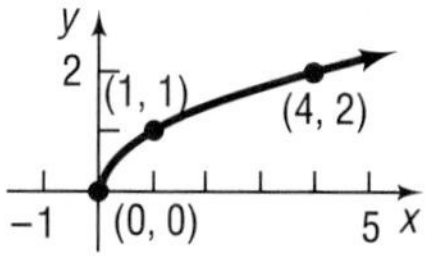

Reciprocal Function

$f(x) = \dfrac{1}{x}$

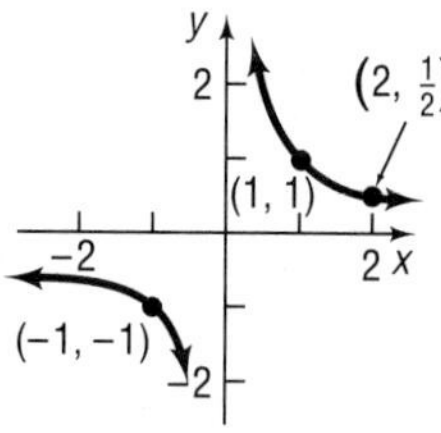

Cube Root Function

$f(x) = \sqrt[3]{x}$

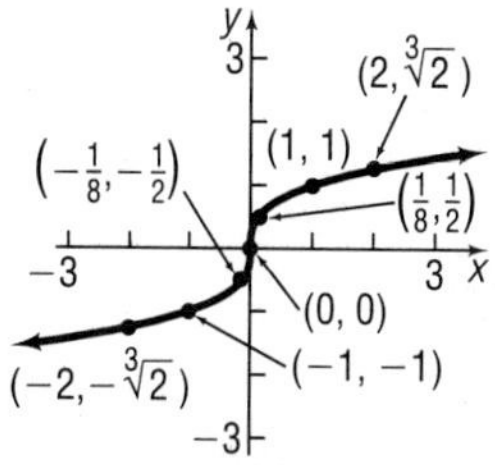

Absolute Value Function

$f(x) = |x|$

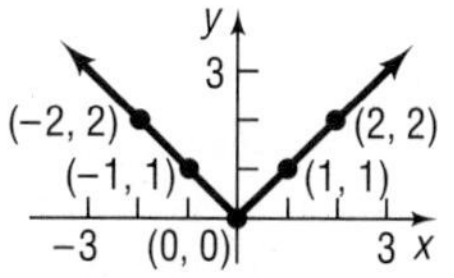

Exponential Function

$f(x) = e^x$

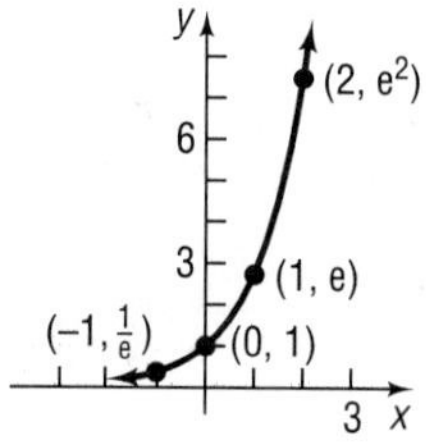

Natural Logarithm Function

$f(x) = \ln x$

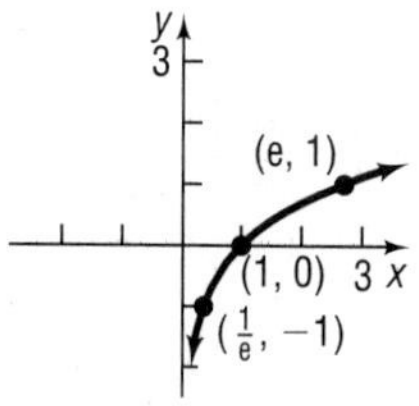

FUNCTIONS

Constant Function $f(x) = b$

Linear Function $f(x) = mx + b$, where m is the slope and b is the y-intercept

Quadratic Function $f(x) = ax^2 + bx + c, a \neq 0$

Polynomial Function $f(x) = a_n x^n + a_{n-1}x^{n-1} + \cdots + a_1 x + a_0$

Rational Function $R(x) = \dfrac{p(x)}{q(x)} = \dfrac{a_n x^n + a_{n-1}x^{n-1} + \cdots + a_1 x + a_0}{b_m x^m + a_{m-1}x^{m-1} + \cdots + b_1 x + b_0} \quad q(x) \neq 0$

Exponential Function $f(x) = a^x, a > 0, a \neq 1$

Logarithmic Function $f(x) = \log_a x, a > 0, a \neq 1$